International Handbook of Engineering Education Research

This comprehensive handbook offers a broad overview of contemporary research on engineering education and its practical application. Over the past two decades, the field of engineering education research (EER) has become a vibrant and impactful community with new journals, conferences, and doctoral and research programs established across the globe. The increased interest in this area has helped improve the education and training of the next generation of engineers, as well as support growth in the use of technology for teaching and learning, increased attention to broadening participation, diversity and inclusion in the field, and a wide international expansion of the field.

Drawing on the work of 100 expert contributors from over 20 countries, this volume covers both emergent and established areas of research within engineering education, giving voice to newcomers to the field as well as perspectives from established experts. Contents include:

- Sociocognitive and affective perspectives on engineering education.
- Technology and online learning in engineering education.
- Cultural and ethical issues, including diversity, equity, and inclusion in engineering education.
- Curriculum design, teaching practices, and teacher education at all levels.
- Research methods and assessment in engineering education.

This book offers an innovative and in-depth overview of engineering education scholarship and practice, which will be of use to researchers in engineering education, engineering educators and faculty, teacher educators in engineering education or STEM education, and other engineering and STEM-related professional organizations.

Aditya Johri is Professor of information sciences and technology (IST) in the College of Engineering and Computing (CEC) at George Mason University, USA.

International Handbook of Engineering Education Research

Edited by Aditya Johri

Routledge
Taylor & Francis Group
NEW YORK AND LONDON

Designed cover image: Aditya Johri
The cover image depicts the geographic spread and movement of authors. The black dots represent current location of authors. White dots and arrows represent movement of author across locations.

First published 2023
by Routledge
605 Third Avenue, New York, NY 10158

and by Routledge
4 Park Square, Milton Park, Abingdon, Oxon, OX14 4RN

Routledge is an imprint of the Taylor & Francis Group, an informa business

Library of Congress Cataloging-in-Publication Data
Names: Johri, Aditya, 1976- editor.
Title: International handbook of engineering education research / edited by Aditya Johri.
Description: New York, NY : Routledge, 2023. | Includes bibliographical references and index.
Identifiers: LCCN 2022060818 (print) | LCCN 2022060819 (ebook) | ISBN 9781032262765 (hardback) | ISBN 9781032262758 (paperback) | ISBN 9781003287483 (ebook)
Subjects: LCSH: Engineering—Study and teaching—Research.
Classification: LCC T65.3 .I496 2023 (print) | LCC T65.3 (ebook) | DDC 607.1—dc23/eng/20230309
LC record available at https://lccn.loc.gov/2022060818
LC ebook record available at https://lccn.loc.gov/2022060819

ISBN: 978-1-032-26276-5 (hbk)
ISBN: 978-1-032-26275-8 (pbk)
ISBN: 978-1-003-28748-3 (ebk)

DOI: 10.4324/9781003287483

Typeset in Bembo
by Apex CoVantage, LLC

Contents

About the Editor

Aditya Johri, PhD, Stanford University, is Professor of information sciences and technology (IST) in the College of Engineering and Computing (CEC) at George Mason University, USA. He studies how technology shapes learning across settings and the ethical implications of using technology in education. He publishes broadly in the fields of engineering and computing education, educational technology, and computer-supported collaborative work and learning. His research has been recognized with several best paper awards, and his co-edited volume, *The Cambridge Handbook of Engineering Education Research* (CHEER), received the 2015 Best Book Publication Award from Division I of AERA. His research is supported primarily by the US National Science Foundation (NSF), and he has personally supervised over $4M in external research funding, including an NSF Early CAREER Research Award. In 2022, he received the University Teaching Excellence Award and the Mentoring Excellence Award for undergraduate research advising at George Mason University. In 2021, he served as Fulbright-Nokia Distinguished Chair in ICT at Aalto University, Finland. While a faculty at Virginia Tech (2007–2013), he was recognized with the Engineering Dean's Faculty Fellow Award for Extraordinary Performance in Research (2013) and Outstanding New Assistant Professor Award (2010). More information is available at: http://mason.gmu.edu/~johri.

Advisory Board

Contributors

Abisola C. Kusimo is a PhD candidate in the Work, Technology, and Organizations group at Stanford University. Her work bridges the fields of sociology of work and economic development. Her cross-national research explores how educationally disadvantaged young adults understand their credentials, job search strategies, and relationships with a labor market intermediary in contexts of extreme and increasing job insecurity and precarity.

Adam R. Carberry is an associate professor at Arizona State University, where he serves as Program Chair for the Engineering Education Systems and Design program. He holds a BS in materials science engineering from Alfred University, MS in chemistry, and PhD in engineering education, both from Tufts University.

Aditya Johri is Professor of information sciences and technology in the College of Engineering and Computing, George Mason University. He is co-editor (with Barbara Olds) of the *Cambridge Handbook of Engineering Education Research* (CHEER).

Aileen Y. Huang-Saad is an associate professor of bioengineering at Northeastern University and serves as the director of Life Sciences and Engineering Programs at Northeastern's Roux Institute in Portland, Maine. She is Deputy Editor-in-Chief of *Springer's Biomedical Engineering Education*, and her current research areas are entrepreneurship, innovation, and transforming higher education.

Aive Pevkur is a senior lecturer in Tallinn University of Technology, Estonia, and a national and international expert in professional and public ethics and integrity. Her research interests are in practical philosophy and applied ethics, and she teaches courses on ethics and values in technology, engineering, and professional life.

Aleksandra A. Kazakova is a PhD candidate in philosophy of science and technology, UCAS (Beijing), and teaches engineering studies and philosophy for engineers in Bauman and Gubkin Universities (Moscow). She studied social theory in Lomonosov Moscow State University and had a research stay focused on sociology of engineering in KIT-ITAS, Germany.

Alexander A. Kist is Senior Fellow of the Higher Education Academy (SFHEA). His research interests include performance modelling, remote access laboratories, and engineering education. He has authored more than 140 scientific articles, and he is an elected executive member of the International Association of Online Engineering.

Alexandra Coso Strong is an assistant professor of engineering education at Florida International University, where she co-developed two different degree programs. Previously, she was a faculty member at Olin College of Engineering. Coso Strong holds graduate degrees in both aerospace and systems engineering, where she focused on human factors and education research.

Alison Gwynne-Evans is a senior lecturer at the University of Cape Town and teaches professional communication and ethics to engineering students. She aims to inspire students to be changemakers and leaders where they are.

Allison Godwin is an associate professor in the Robert Frederick Smith School of Chemical and Biomolecular Engineering at Cornell University. Her research focuses on how identity, among other affective factors, influences diverse students to choose engineering and persist in engineering. She also studies how different experiences within the practice and culture of engineering foster or hinder belongingness and identity development.

Aman Yadav is the Lappan-Phillips professor of computing education in the College of Education and College of Natural Science at Michigan State University. He also serves as the associate director of computing education for the CREATE for STEM Institute. His research and teaching focus on improving student experiences and outcomes in computer science and engineering at the K–16 level. His recently co-edited book, *Computational Thinking in Education: A Pedagogical Perspective*, tackles how to integrate computational thinking in relevant and meaningful ways in K–12 classrooms.

Amy Banzaert is the founding director of engineering studies and a lecturer in engineering at Wellesley College. She received her PhD in mechanical engineering from the Massachusetts Institute of Technology. Her work focuses on engineering education that is inclusive, project-based, interdisciplinary, and grounded in social justice.

Amy Sue Bix is Professor of history at Iowa State University. Her 2013 book *"Girls Coming to Tech!": A History of American Engineering Education for Women* (MIT Press) has won three major awards, including the 2015 Margaret Rossiter Prize from the History of Science Society.

Andrea Goncher is Instructional Assistant Professor in the Department of Engineering Education at the University of Florida. Her teaching interests include engineering education, human-centered design, and systems engineering.

Andrea Gregg is a director of online pedagogy and an assistant teaching professor in the Penn State Mechanical Engineering Department. She is a research practitioner, and much of her work has been in online higher education. Her interdisciplinary research is grounded in learner/learning experience design, especially as it relates to technology-mediated learning environments. Her PhD is in learning, design, and technology.

Andrés Felipe Valderrama Pineda is Associate Professor at Aalborg University in Denmark. He is part of the research group Design for Sustainability and Planning for Urban Sustainability. He researches and teaches design for sustainable transitions, sustainable transitions in mobility, and engineering education for sustainable transitions. He has formal training and is a permanent student in the fields of mechanical engineering, history of technology, engineering education, and sociotechnical systems design.

Andrew Katz is an assistant professor in the Department of Engineering Education at Virginia Tech. He studies decision-making in social-ecological systems using computational social science techniques along with engineering ethics education and methods for improving assessment in that domain.

Andrew Valentine is a teaching fellow (assistant professor) in the School of Computing and Information Systems at the University of Melbourne in Australia. His research focuses on engineering education, technology-based simulations in education, creativity in engineering, and identifying academic pathways to becoming an engineering education researcher.

Aparajith Ramnath is Associate Professor in the School of Arts and Sciences, Ahmedabad University. A historian of science, technology, and business, he is the author of *The Birth of an Indian Profession: Engineers, Industry, and the State, 1900–47* (Oxford University Press, 2017). Aparajith has been a Charles Wallace grant awardee, an international scholar of the Society for the History of Technology (SHOT), and winner of the Indian National Science Academy's Young Historian of Science Award (2018).

Ardeshir Raihanian Mashhadi is a sustainability scientist at Amazon Lab126. His research focuses on design for sustainability, circular economy, and understanding the role of empathy in design and design education.

Ashish Hingle is a PhD student in the College of Engineering and Computing at George Mason University. His research interests include engineering ethics, information systems, and student efficacy challenges in higher education.

Atasi Mohanty is an assistant professor of the Centre for Educational Technology at Indian Institute of Technology Kharagpur, India. Her research interests focus on educational psychology, curriculum and pedagogy, teacher education and professional development, human resource development, positive mental health and well-being, education for sustainable development, educational leadership, and human resource and sustainability.

Bruce Kloot is an associate professor in the Department of Mechanical Engineering at the University of Cape Town (UCT). He is active in UCT's Centre for Research in Engineering Education, a board member of the South African Society of Engineering Education, and the editor-in-chief of the *Southern Journal of Engineering Education*. His research in engineering education overlaps with higher education studies, and he is particularly interested in the relationship between engineering and society.

Buffy Smith, PhD, is the dean of Dougherty Family College and a professor of sociology at the University of St. Thomas. She earned her BA in sociology at Marquette University and her MS and PhD in sociology at the University of Wisconsin–Madison. Dr. Smith's primary research interests include examining racial and class disparities within the higher education system. She is the author of the book *Mentoring At-Risk Students through the Hidden Curriculum of Higher Education*.

Christine Winberg holds the South African National Research Foundation Chair in Work-Integrated Learning and leads the Professional Education Research Institute at the Cape Peninsula University of Technology in Cape Town, South Africa. Her research focus is professional and vocational education – with a particular focus on engineering education, the professional development of university teachers, and technical communication.

Cindy Rottmann is the associate director, research, at Troost ILead/ISTEP, University of Toronto. Her research examines engineering leadership in professional practice, engineering career paths, and the infusion of equity and social justice into engineering ethics education. She currently serves as the chair of the Engineering Leadership Division of ASEE. Dr. Rottmann holds bachelor's, master's, and doctoral degrees in educational leadership and policy studies.

Corrinne Shaw is an associate professor at the University of Cape Town in the Department of Mechanical Engineering and the Centre for Research in Engineering Education (CREE). She had a key role in the introduction of PhD and master's programs in engineering education in the Faculty of Engineering and the Built Environment. Her research draws on critical and systemic perspectives to explore learning and development.

Cory Brozina is an associate professor in the Rayen School of Engineering at Youngstown State University. He is also the associate chair of the school, as well as the director of the First-Year Engineering Program and the Industrial and Systems Engineering Program. His research focuses on student support and success and data analytics.

David Knight is an associate professor of engineering education and special assistant to the dean for strategic plan implementation at Virginia Tech. His research takes a systems-level perspective of how engineering education can become more effective, efficient, and inclusive and considers the intersection between policy and organizational contexts.

Denise R. Simmons, PhD, PE, is the associate dean of Workforce Development in the Wertheim College of Engineering at the University of Florida. Her research answers international calls for construction and civil engineering professionals to develop new competencies to navigate the changes of evolving workforce demographics, technology, and organizational structures.

Diana Bairaktarova is an associate professor in the Department of Engineering Education at Virginia Tech. Her research explores the (a) impacts of emerging technologies on learning environments and on student learning and abilities (mechanical reasoning, spatial and creative abilities) and the (b) effective teaching models of professional ethics and empathic design.

Diana A. Chen is an associate professor and a founding faculty member of Integrated Engineering at the University of San Diego. She aspires to shift engineering curricula and spaces to be more inclusive of diversity in all forms. Her scholarly interests include engineering education that contextualizes engineering sciences and design as sociotechnical, inclusive pedagogy, and sustainability and bio-inspired design in the built environment.

Diana A. Martin is a postdoctoral researcher in the Philosophy and Ethics Department at TU Eindhoven. Her research to date focuses on the incorporation of ethics and responsibility in engineering education. At TU Eindhoven, Diana is coordinating a Challenge-Based Learning course on Decisions Under Risk and Uncertainty, where students work on real-life sociotechnical problems in collaboration with external stakeholders from the university ecosystem.

Dominik May is an assistant professor in the Engineering Education Transformations Institute at the University of Georgia, GA, USA. He received his PhD from TU Dortmund University, Germany. His primary research focus lies on the development, introduction, practical use, and educational value of online laboratories and online experimentation. Dr. May is also President of the International Association of Online Engineering (IAOE) and serves as Editor-in-Chief for the *International Journal of Emerging Technologies in Learning* (iJET).

Emma Mercier is an associate professor of curriculum and instruction at the University of Illinois Urbana–Champaign. Her work focuses on computer-supported collaborative learning and using emerging technology to support the teaching of collaborative learning in classrooms.

Esther Matemba is a Tanzanian engineering education researcher and consultant working remotely from Perth with Lassonde Educational Innovation Studio (LEIS) at York University in Canada. Esther is a co-founder of the Engineering Education Research Network for Africa (EERN-Africa) and a governing board member of REEN (2021–2024).

Euan D. Lindsay is Professor of PBL and digitalization in engineering education at Aalborg University. His focus is the use of technology to flexibly support providing authentic learning experiences for student engineers. He is best known for his work as Foundation Professor of Engineering at Charles Sturt University.

Fiona Boyle is Head of Department STEM (Director of REEdI, Rethinking Engineering Education in Ireland) at Munster Technological University. Fiona works to drive agility and innovation in education through the use of immersive technologies and strong industry links to ensure the future skills needs of industry are met.

Gabriella Coloyan Fleming is a research associate in the Center for Equity in Engineering at the University of Texas at Austin.

Geoffrey L. Herman is the Stevens teaching associate professor with the Department of Computer Science. He earned his PhD in electrical and computer engineering from the University of Illinois at Urbana–Champaign. His research interests include creating systems for sustainable improvement in engineering and computer science education and studying conceptual change and development in engineering and computer science students.

Gordon D. Hoople is an associate professor and a founding faculty member of integrated engineering at the University of San Diego. His research in engineering education focuses on sociotechnical thinking, sustainability, and social justice. His recently co-authored book *Drones for Good: How to Bring Sociotechnical Thinking Into the Classroom* provides faculty with a hands-on approach for engaging students in challenging conversations at the intersection of technology and society.

Greses Pérez is the McDonnell Family Assistant Professor of engineering education in the Department of Civil and Environmental Engineering at Tufts University with secondary appointments in mechanical engineering and education. She received her PhD in science education with a focus on learning sciences and technology design from Stanford University. Her scholarship specializes in language, culture, and cognition for Black and Latinx/o/a students who experience cultural and linguistic mismatch between the practices of their communities and those in engineering.

Gunter Bombaerts is Assistant Professor in the philosophy and ethics group at Eindhoven University of Technology, the Netherlands. Gunter's interests focus on ethics and morality in system thinking, applied to energy systems, AI systems, and education as a system. He is Coordinator of a human and social sciences educational program at TU/e. He is also an enthusiastic teacher involved in several challenge-based learning courses that include philosophy and ethics.

Gustavo R. Alves obtained a PhD in computers and electrical engineering at the University of Porto (Portugal) in 1999. He is an associate professor at the Polytechnic Institute of Porto (Portugal). He has authored or co-authored 270+ publications, including book chapters and conference and journal papers with a referee process. His research interests include engineering education and remote laboratories. Dr. Alves currently serves as an associate editor for the *IEEE Journal of Latin-American Learning Technologies*.

Helen L. Chen, PhD, is a research scientist in the Designing Education Lab in the Department of Mechanical Engineering and the co-founder of the Integrative Learning Portfolio Lab at Stanford University. Her research focuses on engineering and entrepreneurship education, portfolio pedagogy, storytelling practices, and redesigning how learning is represented through digital records and alternative credentials.

Idalis Villanueva Alarcón, PhD, is an associate professor in the Engineering Education Department at the University of Florida. Her multiple roles as an engineer, engineering educator, engineering educational researcher, and professional development mentor for underrepresented populations have aided her in the design and integration of educational and physiological technologies to research "best practices" for student professional development and training. In addition, she has developed unique methodologies around hidden curriculum, academic emotions and physiology, and engineering makerspaces.

Inês Direito is a senior research fellow in the Centre for Engineering Education at University College London, United Kingdom. Her research interests lie in skills development, gender, diversity, and inclusion in engineering education.

Isabel Cardenas-Navia serves as Workcred's senior director of research to advance Workcred's research agenda and examine workforce credentialing issues and needs. Previously, Dr. Cardenas-Navia was the vice president of programs and previously the director of emerging workforce programs with the Business-Higher Education Forum (BHEF). She has a PhD in biomedical engineering and certificate in biomolecular and tissue engineering from Duke University and a BS in mechanical engineering from Yale University.

James Holly Jr. is an assistant professor of mechanical engineering and core faculty member within the Engineering Education Research program at the University of Michigan. He sees his teaching, research, and service as promoting pro-Blackness – affirming the humanity and epistemic authority of Black people – in engineering education.

James L. Huff is an associate professor of engineering education and an Honors College senior faculty fellow at Harding University, USA. He conducts research on identity and shame, examining how hidden phenomena drive individual behaviors that shape professional cultures. Dr. Huff earned a PhD in engineering education from Purdue University.

Jason Power is a lecturer in technology education at the University of Limerick. His research focuses on factors related to motivation within STEM learning environments. He is currently the PI of an EU-funded project that aims to support the use of evidence-based practice within university STEM learning environments and a co-PI of an EPA-funded project that aims to identify capacity and need for remanufacturing within industry in Ireland.

Jeffrey Buckley received his PhD from KTH Royal Institute of Technology and is a lecturer in the Department of Technology Education at the Technological University of the Shannon: Midlands Midwest. He is currently the associate editor for the *International Journal of Technology and Design Education.* His main research interest relates to the role of spatial ability in learning in technology and engineering education. He also conducts meta-research on the credibility of technology education research.

Jennifer R. Brown is an associate professor in the Chemical and Biological Engineering Department at Montana State University. Her research spans from technological fields to engineering education, including interests in rheology, transport dynamics in complex geometries, technical communication, and mentorship.

Jennifer M. Case is Professor and Head of the Department of Engineering Education at Virginia Tech. Prior to this, she was a professor in the Department of Chemical Engineering at the University of Cape Town, where she retains an honorary appointment. She completed postgraduate studies in the UK, Australia, and South Africa. Her research on student learning in engineering education, curriculum reform, and comparative higher education has been widely published, with over 60 peer-reviewed journal publications and two monographs.

Jillian Seniuk Cicek (she/her) is a settler living and working on Treaty 1 Territory and Homeland of the Métis Nation. She is invested in making space for Indigenous peoples and their voices and worldviews in engineering education and research. She's an assistant professor in the Centre for Engineering Professional Practice and Engineering Education, University of Manitoba, Canada.

Joel Alejandro Mejia is an associate professor with joint appointment in the Department of Biomedical and Chemical Engineering and the Department of Bicultural-Bilingual Studies at The University of Texas at San Antonio. His current work seeks to analyze and describe the assets, tensions, contradictions, and cultural collisions many Latino/a/x/é students experience in engineering through pláticas and testimonios.

Johanna Lönngren is an associate professor in science and engineering education and an educational developer at Umeå University, Sweden. She also coordinates the Emotions in Engineering Education Network. Her research focuses on education for sustainable development, wicked problems, and emotions in engineering education, applying discourse analytic and multimodal approaches.

Jon A. Leydens is a professor of engineering education research in the Department of Humanities, Arts, and Social Sciences at Colorado School of Mines, USA. Dr. Leydens's engineering education research interests are in three areas: sociotechnical thinking, communication, and social justice. He is the co-author of *Engineering Justice: Transforming Engineering Education and Practice*, which fills a gap in our understanding of how engineering and social justice complement each other in and outside the engineering curriculum.

Julie P. Martin is a faculty member in the Engineering Education Department at the Ohio State University. She is a fellow of the American Society for Engineering Education, a former program director at the US National Science Foundation, and the editor-in-chief of *Journal of Women and Minorities in Science and Engineering*.

Katherine L. Shirey is the founder of eduKatey, LLC, in Washington, DC. She holds a BA in physics, a BA in sculpture, and an MT in secondary science education from the University of Virginia, a PhD in education from the University of Maryland, and is a certified STEAM integration specialist.

Kelly J. Cross, Assistant Professor in the Biomedical Engineering Department at Georgia Tech and Emory University, is a data-informed, transformational, mission-focused, culturally responsive practitioner, researcher, and educational leader. Dr. Cross promotes inclusive excellence through collaboration, is an NSF CAREER awardee, has delivered multiple distinguished lectures, and has received a national mentoring award.

Kerrie A. Douglas is an associate professor of engineering education at Purdue University. Her research is focused on improving methods of evaluation and assessment in engineering learning contexts. In particular, she is focused on increasing the fairness of assessments for multicultural group of learners.

Knut Neumann is the director of the Department of Physics Education at the IPN, Leibniz Institute for Science and Mathematics Education, and a professor of physics education at the Christian-Albrechts-University in Kiel, Germany. His research is focused on assessing and developing students' science competencies as a multifaceted construct. His work on this topic has been published in numerous high-quality, peer-reviewed journals.

Lauri Malmi is a professor of computer science at Aalto University, Finland. His research focuses on computing education, especially pedagogical software to support programming education, as well as theoretical foundations in computing education research. He received the ACM SIGCSE Outstanding Contribution to Computer Science Education award in 2020.

Lizandra C. Godwin is an inclusive excellence postdoctoral scholar in the Electrical and Computer Engineering Department at the University of New Mexico. Her research topics include nanomaterial processing and interactions, novel manufacturing approaches, and equity in post-secondary engineering education.

Madeline Polmear is a Marie Sklodowska-Curie, EUTOPIA Science and Innovation Cofund fellow at the Vrije Universiteit Brussel, Belgium. Her research interests relate to engineering ethics education and the development of societal responsibility and professional competence through formal and informal learning. She has a PhD in civil engineering from the University of Colorado Boulder.

Margret A. Hjalmarson is a program director in the Division of Research on Learning in Formal and Informal Settings at the US National Science Foundation. She has expertise in engineering education and mathematics education, with a focus on teacher leadership, professional learning, and design-based research methods.

Maria Elena Oliveri, PhD, is a research associate professor and associate director of psychometric consulting at the Buros Center for Testing, University of Nebraska–Lincoln. Her expertise is on assessing workplace-relevant competencies, educational measurement, fairness, validity, and culturally responsive (antiracist) assessment. She has over 100 publications and presentations on these topics.

Maura Borrego is a professor of mechanical engineering and STEM education and director of the Center for Engineering Education at the University of Texas at Austin.

Michael Lachney is an assistant professor in Michigan State University's Educational Psychology and Educational Technology program. He has a background in science and technology studies and contributes to the field of computer science education. His research focuses on the cultural and racial politics of computing in school and community settings. His research has appeared in the journals *Learning, Media and Technology*, *Science as Culture*, and *Computer Science Education*, among others

Molly H. Goldstein is a teaching assistant professor and product design lab director in industrial and enterprise systems engineering at the University of Illinois, where she is a design educator and researcher. Her research focuses on student design behaviors and actions.

Monique S. Ross is an associate professor in the Engineering Education Department at the Ohio State University. Her research focuses on broadening participation in computing by exploring race, gender, and identity in the academy and industry. Dr. Ross earned a PhD in engineering education from Purdue University.

Nada Dabbagh is a professor and director of the Division of Learning Technologies in the College of Education and Human Development at George Mason University. Dr. Dabbagh's research is steeped in the learning sciences and focuses on the pedagogical ecology of technology-mediated learning environments, with the goal of understanding the social and cognitive consequences of learning systems design.

Nathalie Duval-Couetil is the director of the Certificate in Entrepreneurship and Innovation Program, the associate director of the Burton D. Morgan Center, and a professor in the Department of Technology Leadership and Innovation at Purdue University. Her research activities focus on entrepreneurship education, program evaluation, academic entrepreneurship, and gender and entrepreneurship.

Olga Pierrakos is Professor and Founding Chair of engineering at Wake Forest University, a program committed to educating the whole engineer. Social responsibility and a culture of inclusive excellence have enabled curricular innovation and a diverse student and faculty body. She was a program officer at the National Science Foundation and a founding faculty of engineering at James Madison University.

Pamela S. Lottero-Perdue is a professor of science and engineering education at Towson University. She holds a BS in mechanical engineering and an MEd and PhD in curriculum and instruction, specializing in science education, all from the University of Delaware.

Patric Wallin is Associate Professor in university pedagogy at the Norwegian University of Science and Technology, Norway. In his research, he uses critical pedagogy as an entry point to explore how to create spaces in higher education where it is possible to reimagine the university as a place for collaboration between students and academics with the common purpose to co-create knowledge and meaning by drawing on dialogic praxis and partnership.

Preethi Baligar is an assistant professor at KLE Technological University (KLE Tech), India, where she co-designed instruction for interdisciplinary, design-based, PBL and blended learning courses in first-year undergraduate engineering. She holds a Master's in Computer Science and a PhD in Engineering Education from KLE Tech.

Qin Zhu is Associate Professor in the Department of Engineering Education and Affiliate Faculty in the Department of Science, Technology, and Society and the Center for Human–Computer Interaction at Virginia Tech. His interests include engineering ethics, global engineering education, engineering cultures, and the ethics and policy of robotics and AI.

Rami Ghannam is a senior lecturer (associate professor) at the University of Glasgow and an IEEE Systems Society distinguished lecturer. He is a cross-disciplinary researcher with expertise in photonics, sensors, and technology-enhanced learning. In particular, he is interested in designing intelligent and interactive systems to improve the way people learn.

Robin Jephthah Rajarathinam is a doctoral student in the Department of Curriculum and Instruction at the University of Illinois. His research interests broadly focus on collaborative learning in engineering classroom contexts. He is specifically interested in using multimodal learning analytics to understand and develop meaningful collaborative learning experiences for students.

Robyn Paul is a PhD student at the University of Calgary. She has developed and implemented a mental wellness program to break down the hidden curriculum of engineering and create new narratives of what it means to be an engineer. Her other activities include bringing an ecofeminist approach to engineering, integrating regenerative sustainability into the curriculum, and advocacy to rewrite deeply ingrained and alienating engineering rituals to be more inclusive.

Roger G. Hadgraft is Professor and Director, Educational Innovation, at the University of Technology Sydney. He is a civil engineer with more than 30 years of experience in improving engineering education using problem- and project-based learning (PBL). He was instrumental in introducing a project-based curriculum in civil engineering at Monash University and in several disciplines at other universities. Roger has consulted on PBL to universities both nationally and internationally.

Roland Tormey is a senior scientist in learning sciences and the head of the Teaching Support Centre at EPFL in Switzerland. He researches diversity issues and emotions in teaching and learning. He has previously worked in teacher education and now focuses mostly on the teaching and learning of engineering.

Ron Ulseth, PhD, PE, is the founding director of both the project-based and work-based learning models at Iron Range Engineering. IRE has been named an emerging world leader in engineering education and has received the ABET Innovation award. He has taught eight or more engineering courses per year for the past 34 years.

Roy Pea is David Jacks professor of education and learning sciences at Stanford University, School of Education and Computer Science (Courtesy). His studies and publications in the learning sciences focus on advancing theories, research, tools, and social practices of technology-enhanced learning of complex domains, including his role as Co-Director and Co-PI of the NSF-funded LIFE Center (2004–2014). He is also Founder and Director of Stanford's PhD program in learning sciences and technology design.

Sarah Jayne Hitt is Lead for transferable skills at the New Model Institute for Technology and Engineering in Hereford, GB, and Visiting Professor at Edinburgh Napier University, and she specializes in integrating ethics, sustainability, and communication in engineering education. She is a senior fellow of the Higher Education Academy.

Sarah E. Zappe is the assistant dean of teaching and learning and director of the Leonhard Center for the Enhancement of Engineering Education at Penn State. Sarah is an educational psychologist with expertise in evaluation and faculty development. Her interests include entrepreneurship education, creativity, and the integration of art with engineering.

Shannon Chance, PhD, SFHEA, BArch, MArch, Registered Architect, LEED-AP, is Lecturer and Program Chair at TU Dublin and Visiting Professor at UCL. Shannon is Deputy Editor of the *European Journal of Engineering Education*. She has been Chair of the Research in Engineering Education Network and a Fulbright and Marie Curie fellow.

Stacy S. Klein-Gardner is an adjunct professor of biomedical engineering at Vanderbilt University. She holds a BSE in biomedical and electrical engineering from Duke University, an MS in biomedical engineering from Drexel University, and a PhD in biomedical engineering from Vanderbilt University. She is a fellow of ASEE.

Stephanie Cutler is an associate research professor and director of assessment and instructional support within the Leonhard Center for the Enhancement of Engineering Education at Penn State. Cutler holds degrees in mechanical engineering, industrial and systems engineering, and a PhD in engineering education from Virginia Tech.

Stephanie Masta is a member of the Sault Ste. Marie tribe of Chippewa Indians and an associate professor at Purdue University. Much of her research focuses on Indigenous peoples and their relationships to academic spaces. Stephanie's work is also invested in uncovering the intersections of colonialism and race within the academy.

Stephen Secules is Assistant Professor, Co-Graduate Program Director at the School of Universal Computing, Construction, and Engineering Education (SUCCEED) at Florida International University. Stephen Secules uses critical qualitative methods to look at everyday educational settings in engineering and shift them towards equity and inclusion. Stephen founded the Equity, Culture, and Social Justice in Education Division of the American Society for Engineering Education. He also serves on the editorial board of the *Journal of Engineering Education.*

Susan M. Zvacek (CollegeTeachingCoach.com) is an independent consultant, speaker, and online teacher focused on cultivating learning-centered teaching in higher education. Her disciplinary interests include online learning, engineering education, and instructional design. She has been an IEEE Distinguished Lecturer and Distinguished Speaker for the Association for Computing Machinery (ACM). Her publications are on topics such as course design, online learning, remote labs, and higher-order thinking, including co-authoring *Teaching and Learning at a Distance.*

Tom Goldfinch is Associate Dean (T&L) at the University of Sydney. His EER work follows a particular passion for recognition of Australian Indigenous perspectives and knowledge in engineering, contributing to many initiatives, including leadership of the landmark national project Embedding Indigenous Perspectives into Engineering Education (OLT ID13–2899).

Vanessa Svihla is an associate professor in organization, information, and learning sciences and in chemical and biological engineering at the University of New Mexico. As a learning scientist, their research, funded by the NSF CAREER award, focuses on how people use agency and learn to frame engineering design problems.

Victoria Sellers, PhD, is a postdoctoral research associate in the Department of Engineering Education at the University of Florida. She earned her BS degrees in geology and environmental health from East Carolina University, an MS in hydrogeology from Clemson University, and a PhD in engineering and science education, also from Clemson University. She has previously published research on hidden curriculum and community cultural wealth in engineering education.

Walter C. Lee, PhD, is an associate professor of engineering education and the director for reserach in the Center for the Enhancement of Engineering Diversity (CEED) at Virginia Tech. His research broadly focuses on issues related to diversity and equity in engineering education.

Xiaofeng Tang is Associate Professor and Associate Director for the Division of Engineering Education in the Institute of Education at Tsinghua University. He worked as Assistant Professor of Practice in the Department of Engineering Education at the Ohio State University. His research interests include engineering ethics, engineering cultures, and the development of engineering education in different nations.

Acknowledgments

I would like to acknowledge and thank, first and foremost, the contributors for their hard work. Without them, this volume would not exist. I am also grateful to the associate editors for shepherding the chapters through the review and revision process, and to the reviewers, for their insightful comments and recommendations.

Before deciding on the content of the handbook, I had conducted a community survey to collect ideas on what readers might like to see in this volume. I want to thank the 150 respondents. Overall, according to my rough calculations, around 300 people have participated and contributed to this handbook in some manner. I would like to express my gratitude to all – it does take a village!

The combined editorial expertise of the advisory board members Jennifer M. Case, Kristina Edström, Ahmad Ibrahim, Michael Loui, Sally Male, and Bill Williams was invaluable as they provided excellent advice on prospective contributors and helped make the volume more international in scope.

While working on this project, I spent a substantial time as Fulbright-Nokia Distinguished Chair at Aalto University, where I was hosted by Lauri Malmi. I want to express my thanks to Lauri for the many valuable discussions and useful advice related to this project. I also want to express my appreciation to Fulbright Finland and the Nokia Foundation for their support.

I would like to thank the US National Science Foundation for supporting this work directly through Award No. EEC-1941186, especially for providing funds to make the handbook open-access, and for further supporting my effort through awards DUE-1937950 and EEC-1939105. Any opinions, findings, and conclusions or recommendations expressed in this material are those of the author(s) and do not necessarily reflect the views of the National Science Foundation.

My current and past collaborators, especially my students, have been an invaluable source of knowledge and feedback, and I am thankful for their support.

Finally, my deepest gratitude is reserved for my mentors in EER: Wendy Newstetter, who was there when I started this journey at Georgia Tech 20 years ago, not knowing where I was headed; Susan Kemnitzer, for her unwavering support and mentorship while at NSF; Hayden Griffin, for taking on the role of a department chair in engineering education, a new field, at Virginia Tech and providing invaluable support to junior faculty like me; Jack Lohmann, for including me on various projects and for the opportunity to publish in the *Journal of Engineering Education (JEE)*; Barbara Olds, for providing countless pieces of advice and guidance and for co-authoring the *JEE* article and co-editing *The Cambridge Handbook of Engineering Education Research* (CHEER); Gary Downey, for many supportive conversations and for giving me the opportunity to edit a special issue of *Engineering Studies*, my first such endeavor; Michael Loui, for the advice, guidance, expertise, and many letters of support over the years; and Larry Shuman, for the words of encouragement, help with publications, letters, and for just being my champion.

All of you took a chance on me, and for that I am eternally grateful.

1

Introduction to the *International Handbook of Engineering Education Research (IHEER)*

Aditya Johri

1 Background and Purposes

The *International Handbook of Engineering Education Research (IHEER)* is a reference volume intended for those new to the field of engineering education research (EER) and current experts in the field. This short introduction to the handbook provides a background on its development and briefly describes its content.

Although interest in documenting and improving the education of engineers has existed for over a century, the institutionalization of EER as a discipline-based education research (DBER) field is a more recent phenomenon (Froyd & Lohmann, 2014; Loui & Borrego, 2019; National Research Council, 2012). Especially over the past couple of decades, the field has coalesced around the idea that a more research-focused approach to creating and implementing educational changes can lead to improved engineering education for all. Consequently, research-focused departments, doctoral programs, centers, journals, and conferences have proliferated not only in the United States, where significant funding is available for EER, but also across the world (Bernhard, 2018; Borrego & Bernhard, 2011).

This increasing and evolving footprint of EER has meant that many newcomers to the field, both students and faculty, are now looking for resources that can guide them as they conduct research or develop new educational resources. *IHEER* will serve this goal by providing readers with an overview of some of the major developments in the field over the past decade. In addition, emerging scholars and experts in the field will find the review of existing literature as well as the discussions of directions for future research useful. One highlight of this volume is the critical perspective authors take towards different topics, especially in Part 2, but also in other chapters interspersed throughout, to bring into focus new concerns and alternate viewpoints and methodologies.

Within the realm of publications in the field, *IHEER* appears almost a decade after the previous handbook, *The Cambridge Handbook of Engineering Education Research (CHEER)* (Johri & Olds, 2014), was published. *IHEER* was conceptualized as a volume that builds on *CHEER* by introducing new ideas, topics, and contributors rather than revising or revisiting content already covered. *IHEER* is also more international in scope, with authors and topics from many more non-US countries than *CHEER*. Education is a contextually applied science, and therefore, it is inconceivable that all research findings will be equally applicable across places and people. Yet a comparative approach can benefit all, and certain ideas and innovations should propagate universally. For this,

DOI: 10.4324/9781003287483-1

it is important that scholars from across the globe engage in a dialogue. This is already taking place through conferences and journals, and *IHEER* further cements some of those exchanges.

Given the large volume of publications in the field, *IHEER* only captures a slice of the research developments within EER. To determine the topics and contributors, a survey was distributed to community members and resulted in 150 responses. These responses were analyzed to create broad categories, and then, further subareas were identified. Finally, author teams were invited to submit chapters that underwent two rounds of reviews. The result is a handbook with 30 chapters authored by 100 scholars from 20 countries spread across five continents.

In addition to *IHEER* and *CHEER*, there are numerous sources that readers can consult to learn more about EER, including John Heywood's comprehensive book on the topic (Heywood, 2005) and special issues published in the *Journal of Engineering Education* and *European Journal of Engineering Education*, among other journals. Recently, several reviews of the fields have also been published (see Chapter 7). A more comprehensive overview and list of different publications, programs, scholars, and opportunities is available in sources listed under *Resources* at the end of this chapter, and additional readings and texts have been cited throughout this handbook.

2 Organization of the Handbook

In addition to the introduction and conclusion chapters, *IHEER* contains 30 chapters distributed equally across six sections. Any act of classification is a social exercise, and the organization of this handbook is no exception. The chapters have been organized according to some underlying logical similarities between them; in some instances, the alignment is stronger, but readers are cautioned against reading too much into the organization. Chapters across sections are meant to be in dialogue with each other and frequently refer to each other.

Part 1: Comparative Perspectives for Engineering Education Research

The first part of the handbook directly addresses the issue of taking an international perspective on engineering education and contains five chapters authored by teams of authors from across the globe. The first chapter is a compendium of viewpoints from authors from five countries. It provides a historical and global analysis of topics spanning engineering education during the colonial period, the entry of women into engineering education institutions, the teaching of ethics and technical education, and the consistent and persistent discussion of why and how we are training future engineers. In the second chapter, Tang et al. introduce and discuss comparative education that examines education across countries and uses the country as the unit for comparison. They provide a brief overview of the evolution of comparative education and outline its three main approaches for comparing education systems across countries: scientific, ameliorative, and interpretive. They then utilize a comparative approach to present exemplar EER studies that illustrate each one of the three main approaches of comparative education. Seniuk Cicek et al. tackle the contested notion of decolonization and its absence within engineering and related literature. They explore how decolonization is conceptualized within engineering education and identify four categories of decolonization. They further discuss drivers for and barriers to engaging in decolonial work within engineering education and make recommendations. In their chapter on engineering ethics education and research, Martin et al. make a plea for academics to engage explicitly with the value-laden nature of engineering and its contextual elements, especially global aspects. The chapter provides an overview of major actors within engineering ethics education research and surveys recent pedagogical and institutional practices to broaden engineering ethics education towards a global and culturally inclusive vision. In the final chapter of this part, Lindsay et al. address the disruption in engineering and engineering education that is taking place globally due to innovation in engineering processes, whereby engineering practice is ever evolving and requires new ways of tackling complex sociotechnical problems. They argue for and give exemplars from four different programs using the continuous improvement mindset to commence, continue, and sustain disruptive innovation in engineering education. They discuss potential barriers to innovation and how they can be overcome.

Part 2: Theoretical Orientations and Critical Approaches in Engineering Education Research

The second part of the handbook focuses on theoretical and critical orientations newly or further developed or deployed within the field. The part begins with a chapter outlining the role and uses of theory and theoretical frameworks. In the chapter, Goncher et al. discuss the role of theory in engineering education research (EER) and elaborate upon the utility of using theory and related underpinnings, such as paradigms and concepts. They outline and discuss three theories – social cognitive career theory, situated learning, and intersectionality – each corresponding with either the postpositivist, interpretivist, or critical paradigm, which are commonly used within EER to illustrate the use of theories. Lönngren et al. argue that affective constructs, especially emotions, need to find a more central place within EER. Through their chapter, they introduce the reader to the multidisciplinary field of emotion research and describe different disciplinary and theoretical perspectives on emotions, as well as methods and methodologies. They outline important and promising areas for future research and provide advice for researchers and doctoral students who plan to pursue engineering education emotions research. Huff and Ross advance an integrated conceptualization of engineering identity that considers the complexity of this theoretical construct, and locate engineering identity research within three foundational frameworks: (1) personal, (2) social, or (3) sociocultural. They further advocate strategies for EER to advance theory development on identity. Secules et al. review research from learning sciences and engineering education research to highlight considerations of power and culture as they intersect with knowledge, identity, agency, language practices, discourse, and sociomateriality. They elaborate on critical cultural analysis and demonstrate its utilities for examining, elucidating, and informing learning practices. Mejia and Martin take a critical view and argue that most US-based research related to diversity, equity, and inclusion (DEI) in engineering education is reductionist in its approach, and discuss recent scholarship that advocates for methodological activism and pluralistic approaches in engineering education research in order to truly address issues of DEI.

Part 3: Engineering Education Across Contexts and Participants

In Part 3, we take a look at engineering education beyond the traditional focus on undergraduate programs and curriculum. In recent years, there has been an increase in interest and research on the PK–12 level, and one of the actors here are teachers. The first chapter, by Carberry et al., examines teacher preparation, an important but often-overlooked aspect of preparing future engineers. In their chapter, the authors discuss the importance of and need for pre-college engineering teacher professional learning (PCE TPL). They present a case within a US context supporting the need and place for engineering teacher professional learning. The chapter also provides a foundation for future directions in pre-college engineering teacher professional learning. Fleming et al. review graduate engineering education, primarily within the US context. They present an overview of available data about graduate education and its contexts, and they discuss gaps in data. They discuss topics including students' graduate school experience, motivation to pursue graduate studies, skills development, and identity development, as well as career preparation. They also examine institutional practices that affect graduate enrollment and experience. Cutler and Strong review research on how faculty influence engineering education, who engineering faculty are, and how they can be better supported within their roles. They argue that a focus on faculty and incorporating faculty voices and narratives are essential if the EER community wants to further its efforts towards creating an educational system that is inclusive, student-centered, and equitable. In their chapter, Polmear et al. provide an overview of informal learning, discussing its definition, history, settings, and activities as relevant to engineering education. They further discuss the benefits and outcomes of informal learning, related to competency development and engagement of diverse learners. They end with implications and recommendations for engineering researchers and practitioners. In the chapter that follows, Chen et al. discuss innovations in engineering education and new approaches that acknowledge and recognize learning and competencies including non-degree credentials. The authors argue that non-degree credentials have the potential to broaden access to engineering and that engineering educators should care about the potential of non-degree credentials in their courses, degree programs, and institutions.

Part 4: Advancing Pedagogy and Curriculum in Engineering Education

The five chapters in Part 4 directly address pedagogical and curriculum issues within engineering with both practical and research-related discussions. In the first chapter, Chen et al. examine both the implementation of social justice into engineering curricula and the barriers preventing wider adoption. The chapter discusses three initiatives to infuse social justice into engineering at three different institutions to highlight entry points and barriers as well as the role institutional support has played in accelerating, enabling, and legitimizing the success of this integration. Alarcón et al. present a research-to-practice chapter for prospective and current engineering educators, scholars, and leaders interested in learning about how hidden curriculum (HC) transforms engineering education via social capital. They provide an overview of HC research, discuss its connection to social capital, and introduce an HC pathways model in engineering. Mercier et al. review the last decade of research on collaborative and cooperative learning along four key dimensions of these forms of pedagogy in classrooms: tasks, teams, tools, and teachers. And they end with propositions for future work in collaborative learning in engineering education practice and research. Zappe et al. review the research on integration for creativity, entrepreneurship, and leadership in engineering education. They identify gaps and advance recommendations for practitioners who teach in these areas. They end with a set of reflections on the major concerns and future directions relating to creativity, entrepreneurship, and leadership in engineering education. Hitt et al. review efforts that have been made to promote and develop the inclusion of the liberal arts within engineering education and vice versa. They discuss the historical background of these efforts, review relevant scholarship, and highlight innovative and creative approaches to integrating liberal arts and engineering.

Part 5: Engineering Education at the Intersection of Technology and Computing

Part 5 consists of five chapters at the intersection of engineering education, technology, and computing. Gregg and Dabagh review prior work on online learning and advance frameworks to help create productive learning environments. They also argue for a more strategic and directed effort by engineering educators to become leaders in online learning. Bairaktarova et al. discuss the state-of-the-art and applications of virtual and augmented realities (VR and AR), as well as wearable and haptic devices, in engineering education. They review the empirical research behind their use to examine how the integration of VR, AR, wearable, and haptic devices into the learning environments can enhance learning. May et al. take a closer look at engineering education research on online laboratories, with a focus on remote and virtual laboratories. Their chapter reviews prior work on the topic and discusses the future potential of online laboratories in advancing engineering learning. The final two chapters in this part focus on the intersection of engineering and computing education. The use of computing has become integral within engineering, as has the need to learn about computing. Yadav and Lachney discuss the historical development of computer science education and recent trends at primary and secondary levels, formally and informally. They show how recent developments in computer science education are moving away from perspectives that are overly celebratory or overly drab and towards techno-social realism. The chapter by Malmi and Johri discusses the history of computing education research (CER), a DBER field similar to EER, and focuses on two reviews of two specific subareas of research and practice at the undergraduate level – programming and tool development.

Part 6: Engineering Education Research Methods and Assessment

The final part of the handbook focuses on research methods and assessment. Holly Jr. et al. critique engineering education research methodology by articulating the ways in which anti-Black ideas are deeply embedded in the dominant approaches to knowledge production in the field. The chapter is grounded in the authors' experiences and perspectives as Black scholars who earned their doctorates in engineering education at the two oldest US engineering education programs. It builds on prior critiques of US EER and draws on Black intellectuals from various traditions. The second chapter, by Svihla et al., introduces and discusses the use of design cases in EER as a form of scholarship. Design cases report on an instructional problem, the process of designing a solution, and the learning design created to address the instructional problem. They illustrate the value of design cases by examining exemplars and also provide a template and practical advice for readers. The following two chapters concentrate on quantitative methods, where the first chapter provides an introductory background and the next chapter looks at advanced methods. The chapter by Hjalmarson et al. describes the basic considerations for conducting engineering education research studies using quantitative methodological approaches, with a focus on data quality and data appropriateness. The chapter by Katz et al. discusses opportunities to apply advanced statistical theory and computational techniques within EER. Intended for readers with some familiarity in quantitative research, the chapter discusses emerging quantitative methods and their considerations for quantitative research in four areas: (1) study design, (2) data collection and preparation, (3) data analysis, and (4) data equity. The part ends with a chapter on assessment by Douglas et al. The chapter advances a socioculturally informed, evidence-centered design (SCI-ECD) model for designing and validating assessment instruments of engineering competencies and provides practical exemplars to guide assessment developers and researchers.

3 Future Directions

IHEER has certain limitations, both in the range of topics that are covered and also in the treatment any given topic has received. Some of this is due to limitations of length, and other variations are due to different approaches taken by the authors. The concluding chapter, by Buckley et al., presents the collective reflection of its authors centered on a view towards the future of EER. They discuss the complexity that exists within and around EER and argue that EER can "always be more," and the community is faced with the question of what it wants EER to be and become.

The concluding chapter also discusses the limitations of the handbook in more detail. Here, based on my personal experience, I highlight some concerns. There is a need for a volume that focuses on implications, as the research is becoming increasingly insular, with little impact on students or other stakeholders. The criticism that EER is becoming increasingly devoid of any relevance is not entirely untrue. There is also the problem, as with many disciplinary fields, that there is limited experience of engineering as such within the field.

Finally, I hope this volume fits in with the overall EER oeuvre. Each publication venue in EER and each context of research has its unique flavor (Brozina et al., 2021; Valentine & Williams, 2021), and hopefully, this will continue to be the case, as this allows for a much broader range of work to be published. There is also a role for special issues and other review articles to bring specific topics into focus.

References

Bernhard, J. (2018). Engineering education research in Europe – coming of age. *European Journal of Engineering Education, 43*(2), 167–170.

Borrego, M., & Bernhard, J. (2011). The emergence of engineering education research as an internationally connected field of inquiry. *Journal of Engineering Education, 100*(1), 14–47.

Brozina, C., Katz, A., & Johri, A. (2021). Taking stock: Analysis of IJEE publications from 1996– 2020 to examine impact and coverage of topics. *International Journal of Engineering Education, 37*(5), 1146–1160.

Froyd, J. E., & Lohmann, J. R. (2014). Chronological and ontological development of engineering education as a field of scientific inquiry. In A. Johri & B. M. Olds (Eds.), *The Cambridge handbook of engineering education research* (pp. 3–26). Cambridge University Press.

Heywood, J. (2005). *Engineering education: Research and development in curriculum and instruction*. John Wiley and Sons.

Johri, A., & Olds, B. M. (Eds.). (2014). *The Cambridge handbook of engineering education research*. Cambridge University Press.

Loui, M. C., & Borrego, M. (2019). Engineering education research. In Fincher, S. A., & Robins, A. V. (Eds.). *The Cambridge handbook of computing education research* (pp. 292–321). Cambridge University Press.

National Research Council. (2012). *Discipline-based education research: Understanding and improving learning in undergraduate science and engineering*. National Academies Press.

Valentine, A., & Williams, B. (2021). Evolution of engineering education research in Portugal and Spain: A scientometric study. In *4th International Conference of the Portuguese Society for Engineering Education* (CISPEE) (pp. 1–8). IEEE.

Resources

Engineering education wiki created in 2011 by the American Society for Engineering Education's Student Division (ASEE SD), in collaboration with the Center for Engineering Learning & Teaching (CELT). The resource consists of links to programs, centers, researchers, societies, publication venues, etc. Primarily a resource for students and other interested individuals new to engineering education about the research happening in this field at institutions and centers all over the world.

http://engineeringeducationlist.pbworks.com/w/page/27578912/Engineering%20Education%20Community%20Resource

Research in Engineering Education Network (REEN). https://reen.co/

Part 1

Comparative Perspectives for Engineering Education Research

2

The Historically Global Forming of Engineering Education

Aparajith Ramnath, Amy Sue Bix, Christine Winberg, Aive Pevkur, and Eddie Conlon

1 Introduction

Engineering education research (EER) is a field where discussion of the historical underpinnings of the engineering profession are largely absent and few among us "engage deeply with the past in their research or teaching" (Wisnioski, 2015, p. 1). Even with increased attention to historically relevant topics recently (e.g., see the chapter on decolonization in this handbook), most discussions of social and institutional issues currently taking place do so in a historical vacuum. Concurrently, global comparative understanding is also largely missing. This is limiting for EER, as engineering is a profession where the global and the historic are intertwined more than in any other profession. The global movement of engineers across the world has been a historical reality. The genesis of the engineering profession in imperialist ambitions is a testament to this; consequently, engineering education across the world is still largely modeled after its inception within a few institutions (Jørgensen, 2007).

Inattention to history and the global root of engineering is problematic for EER for two reasons. One, many concerns that currently seem new or novel are actually manifestations of historical imbalances and deeply rooted biases and a capitalist approach to engineering work where the intentions of those with capital and power supersede all other concerns. A lack of understanding of the origins of problems results in reinventing the wheel when it comes to changing the profession and engineering education. We are often not cognizant of the path dependency of engineering. Second, a lack of focus on history and globality limits our understanding of how things have changed for the better. Without recognizing improvements, doomsday mentality related to engineering takes hold, and so does the thought that engineering education has not improved at all.

As the viewpoints in the chapter argue and show, a lot of work still needs to be done to make the education of engineers better, but a lot has changed as well, and there has been progress.

2 Colonialism, Global Connections, and the Gradual Diversification of Engineering Education in Colonial India

Aparajith Ramnath

2.1 Colonial Institutions

Engineering education in its modern form began in the Indian subcontinent in the 19th century, during the heyday of British colonialism. In 1847, the colonial authorities established the Thomason

DOI: 10.4324/9781003287483-3

College in the northern town of Roorkee in order to train subordinate engineers for the Ganges Canal Project. Irrigation works were hardly new to India: across the region, there were stepwells, tanks, reservoirs, canals, and anicuts (dams) built under earlier regimes, including, most recently, the Mughals. What was new about the Ganges Canal and the works that followed was their scale, and hence the attendant need for new techniques and the use of machinery. A corollary was the requirement for more formally trained technicians. Three more engineering colleges followed the one at Roorkee in the next two decades, one in each of the British presidencies: at Guindy (Madras), Sibpur (Bengal), and Poona (Bombay) (Kumar, 1995; Arnold, 2000, pp. 115–121; Bassett, 2016, p. 35). The primary function of these colleges was to supply the Indian Public Works Department (PWD) with junior engineers and subordinates at a time when the colonial state was wreathing the subcontinent in roads, railways, and canal systems. The engineering profession, in other words, was essentially geared towards building the infrastructure of the colonial state.

There was a clear racial hierarchy within the PWD. Its higher ranks were largely occupied by White Britons recruited in London, serving on higher salaries than their Indian-trained colleagues. Some of these British engineers were drawn from the Royal Engineer Corps and had been trained in the military engineering colleges at Woolwich, Addiscombe, and Chatham in England. But the majority of them were civilians, graduates of the Royal Indian Engineering College (RIEC) at Cooper's Hill near London, which had been established in 1871 at the expense of the Indian taxpayer. This was a pioneering institution: at this time, most civilian engineers working in Britain were trained not in colleges but by apprenticeship to their professional seniors. The RIEC was shut in 1906, by which time many British universities had begun offering degrees in engineering; their graduates were now eligible to be recruited in London for the Indian PWD (Ramnath, 2017, Chapter 3; Kumar, 1995; Prakash, 1999, Chapter 6).

The RIEC and the Indian engineering colleges had some common features. Firstly, they were shaped to a significant extent by British military engineers, who had undertaken the bulk of engineering work in India for the colonial state in earlier decades and now served as heads and instructors in these colleges (Arnold, 2000, pp. 116–119; Black, 2009; Ambirajan, 1995; Mital, 1986, p. 18 and Chapter 7). Secondly, these colleges focused overwhelmingly on civil engineering. Their curricula were oriented towards producing PWD engineers who, as officers of the colonial state, could function as gentlemen and generalists. The entrance exam for the RIEC included classical and modern languages; the program of study at the college covered subjects like mathematics, history, and geography alongside technical subjects (Kumar, 1995; Ramnath, 2017, Chapter 3). PWD engineers had to be able to design and execute many kinds of works: roads and bridges, government offices, dams, reservoirs, and canals. They also had to function as efficient bureaucrats, with duties ranging from personnel management to accounting and even, in some cases, the work of a magistrate (Ramnath, 2017, Chapter 3).

2.2 Indians Look beyond the Empire

What was conspicuous by its absence, then, was instruction in branches of engineering that were required for the development of industries. In the colonial engineering colleges, there were no degree programs in mechanical or electrical engineering before the interwar period (Kumar, 1995, p. 220). There were several government and private "industrial schools" scattered across the subcontinent, but these were largely geared towards providing vocational skills to students from artisanal backgrounds to prepare them for jobs in small-scale or cottage industries (McGowan, 2009, pp. 155–156). Barring exceptions such as the Victoria Jubilee Technical Institute (est. 1889), which catered to Bombay's textile industry, there were no avenues for students to seek training in advanced technological processes or the design and building of machinery (McGowan, 2009, p. 104; Kumar, 1995; Headrick, 1988, Chapter 9).

Indian nationalists did not fail to notice this lacuna. Over the decades that followed, they demanded that the government invest in technical education (Headrick, 1988, pp. 328–331). But their entreaties had little effect, and eventually, they tried to set up the necessary institutions with private support. These efforts dovetailed with the growth of the *swadeshi* movement. Peaking in the first decade of the 20th century, the movement was a drive for economic self-sufficiency, an exhortation to increase industrial production within the country and to consume locally produced goods (Metcalf & Metcalf, 2012, Chapter 5). Paradoxically, this impulse drove Indians to seek inspiration and expertise abroad. In their quest to set up local industries, they looked at Japan, Germany, and the United States (but notably not Britain) as models (Bassett, 2016, Chapters 1 and 2; Lubinski, 2014). In particular, as the historian Ross Bassett has shown, the Massachusetts Institute of Technology (MIT), established in the 1860s, captured the imagination of the Indian intelligentsia. In late-19th-century Poona, nationalist leader B. G. Tilak's newspaper, the *Mahratta*, introduced its readers to MIT, championing its approach as the kind of technical education Indians required (Bassett, 2016, Chapter 1). The fascination with MIT was to have important consequences for the Indian education system decades later.

In Bengal, the swadeshi movement led to the founding of a number of small industries financed and operated by Indians (Sarkar, 2010, Chapter 3). It also provided the impetus for the formation of the National Council of Education (NCE) and the Society for the Promotion of Technical Education (SPTE), which in turn established a number of institutions as an alternative to the government-sponsored education system. One of these was the Bengal Technical Institute (BTI), established near Calcutta in 1906 by the SPTE and taken over in 1910 by the NCE (Sarkar 2010, Chapters 3–4; Jadavpur University, n.d.). Hiralal Roy, a young professor who had taken an undergraduate degree in industrial chemistry at Harvard with financial support from the NCE, helped establish a chemical engineering degree program at the BTI in 1921 on the basis of an MIT program of study. A couple of years later, his college sent him to Germany, where he obtained a doctorate at the Technische Hochschule in Berlin. On his return, he had a long and successful teaching career, helping to institutionalize the field of chemical engineering in India through degree programs and a professional society. In 1947, he became the inaugural president of the Indian Institute of Chemical Engineers (Roy, 1989). During this time, the BTI developed into the College of Engineering and Technology, Bengal, and eventually into Jadavpur University in the 1950s (Jadavpur University, n.d.).

Another influential voice in modernizing engineering education in India was that of Sir M. Visvesvaraya (1861–1962). Hailing from the princely state of Mysore, he worked as a civil engineer for the Bombay PWD for a quarter of a century, served as Dewan (prime minister) of Mysore during the 1910s, and sat on many important committees during the interwar years. He undertook study tours abroad in every decade from the 1890s through the 1940s, visiting Japan and various European and North American countries, taking notes on their industries and their systems of education. During his time at the helm of the Mysore administration, a new engineering college was started in Bangalore, and advances were made in providing instruction in mechanical engineering (Visvesvaraya, 1951).

Visvesvaraya was an early advocate for the creation of advanced technological institutions in India. In 1921–1922, he headed a committee on technical education appointed by the Bombay government. The committee was split. Visvesvaraya and the other Indian members of the committee recommended – among many other measures – the creation of a technological institute along the lines of MIT, but they were outnumbered by their British colleagues, and the plan did not come to pass. However, Visvesvaraya had more success with another committee he chaired in 1930: this one, appointed by Bombay University, resulted in the founding of a Chemical Technology Institute as part of the university (Visvesvaraya, 1951, pp. 107–109; Bassett, 2016, p. 10).

The allure of MIT as a model was long-lived. During World War II, the British scientist A. V. Hill visited India and produced a report for the colonial government in which he insisted that

India needed colleges like MIT. This was taken up under Ardeshir Dalal, who became Member for Planning and Development in the colonial government in 1944. The following year, a committee chaired by business leader N. R. Sarkar recommended the creation of four technological institutes in India. While they did not explicitly invoke MIT, the members had in fact studied MIT closely (some had visited Cambridge, Massachusetts, while others were alumni of MIT). The Sarkar Committee's recommendations were to form the basis for the establishment of the Indian Institutes of Technology (IITs) beginning in the 1950s (Bassett, 2016, pp. 174–179 and 196–197).

2.3 Coda

In the post-Independence period, the IITs became the most prestigious Indian institutions offering an engineering education, although the colonial-era colleges continued to function alongside them. In subsequent decades, more public institutions were established, including seventeen Regional Engineering Colleges (RECs), each of them created jointly by the central and a state government (Department of Higher Education, n.d.).

A notable characteristic of the elite engineering colleges in post-colonial India was their outward orientation. Although the oldest IITs were established during the Cold War years with aid from countries as varied as the USSR, West Germany, the UK, and the United States, it was an American-inspired curriculum that eventually characterized all of them (Bassett, 2016, pp. 4 and 9–10). Perhaps for that reason, their students increasingly looked towards the United States as the place where they could put their skills to optimal use. In the latter part of the 20th century, a significant proportion of IIT graduates went on to the United States to pursue graduate degrees in engineering and technology, often staying behind to build their careers in America (Bassett, 2016, esp. Chapter 10 and Conclusion).

By the end of the 20th century, the IITs, the RECs, and the colonial-era colleges were supplemented by hundreds of new privately run engineering colleges across the country. As of 2021, there were more than 3,000 institutions in India where a student could earn an undergraduate or postgraduate degree in engineering or related fields (AICTE, 2021, p. 26, Table 2.3.4). Most engineering graduates do not aspire to a career in the PWD or, indeed, to work in infrastructure-related projects: it is the information technology, software, and semiconductor industries that hold the greatest allure. These industries are typically dominated by multinational enterprises, most of them either headquartered in or serving clients in the United States (see Bassett, 2016, Chapter 9 and Conclusion; Aneesh, 2006). If the last decades of the colonial period saw a diversification of engineering education catalyzed by global interactions, that process appears to have accelerated immeasurably in recent decades. A profession that once catered to the narrow needs of the colonial state and then grew to serve Indian industries is now widely identified with firms that provide services across the globe.

3 Engineering by Whom, for Whom?

Amy Sue Bix

The word "engineer" originated around the 1300s, referring to those who specialized in devising military constructions. In early modern Europe, Leonardo da Vinci was just one of many who made a living by designing new weapons and fortifications for wealthy patrons. Military engineers learned on the job through apprenticeship and practical experience, as did masters of Gothic cathedral design, mining, and manufacturing. The Industrial Revolution in Europe and the United States further situated machine knowledge inside all-male machine shops, canal building, and railroad work (Hunter, 1991; Reynolds, 1992).

Establishment of France's Ecole Polytechnique (1792) and America's West Point (1802) reified links to masculine worlds of hands-on construction work and military culture (Graber, 2008; Brown, 2000). Throughout most of the 1800s, engineering education in Europe and the United States remained almost universally the province of White men. Women had more generally faced obstacles for centuries in accessing higher education. Universities such as Oxford and Cambridge, Harvard, and William and Mary had been constructed by and for men. Classical study of Latin and Greek aimed to cultivate gentlemen, while training in law, medicine, and theology steered male students in directions barred for women.

In the United States, some of the first opportunities for women to pursue engineering training came at the public coeducational land grant colleges created by the Morrill Act in the late 1800s. Leaders assigned obviously gendered assumptions to different fields of study; land grant schools promised to advance their state's economic and social progress by training young men as engineers and "scientific farmers" while preparing young women to become teachers and wives for educated men (Geiger, 2015; Nienkamp, 2015).

But simply locating female students on campuses offering engineering opened up new possibilities. Not coincidentally, some of the first US women to earn engineering degrees graduated from various land grants: Elizabeth Bragg Cumming in civil engineering at California–Berkeley in 1876, and Elmina Wilson at Iowa State in 1892. Wilson remained to complete her master's in civil engineering in 1894 while also working as an assistant assigned to supervise the student drafting room. Iowa State then promoted Wilson to an instructor and, by 1902, to assistant professor of civil engineering who helped draft plans for a new campus water system. In 1893, Bertha Lamme completed a mechanical engineering degree at Ohio State, then designed motors and machines at Westinghouse (Bix, 2015).

During the early 1900s, a handful of women in the United States, one at a time, earned engineering degrees from land grant colleges and a few private schools. Nora Stanton Blatch earned a bachelor's degree with honors in civil engineering from Cornell in 1905. As the granddaughter of famous women's rights activist Elizabeth Cady Stanton, Blatch said she went into engineering precisely because so few women ever had. But in the late 1800s and early 1900s, the establishment of professional disciplinary organizations set new mechanisms for gatekeeping in engineering. Although Blatch worked as a draftsman and assistant engineer with New York City's water supply board, the American Society of Civil Engineers refused to accept her 1916 application for full membership.

Early 1900s American engineering was firmly set as masculine territory (Seely, 1999 and Seely, 2012). The creation of industrial research centers such as Bell Labs, Westinghouse, and General Electric positioned engineers within the male environment of corporate management (Reich, 1985). Civil engineering acquired masculinized glamour through books and radio programs that told stories of bold men who created railroads across deserts and pushed roads through jungle. Empire-building engineers assisted American military missions in the Philippines, linking technology to colonization and the supposedly natural dominance of "civilization" (Adas, 2006). Toy construction kits, such as Erector Sets, also celebrated the masculinity of engineering; ads literally opened with the line, "Hello, Boys!" Box art generally showed boys building projects, and marketing promised parents that hands-on play with Erector Sets would prepare young men to enter technical careers (Oldenziel, 1999).

Yet before 1940, at least 21 women had finished engineering degrees at Cornell alone. Isolation made their experience hard, and many women faced skepticism, ridicule, or opposition from parents, high school teachers, counselors, and both male and female classmates. Even engineering faculty and deans often openly doubted whether female students could succeed. A woman studying engineering seemed strange enough to get her photo on the front page of campus papers at Cornell, Iowa State, Minnesota, and elsewhere. News stories treated each one as an oddity (Bix, 2013).

World War II's manpower crisis in industry opened up some opportunities, at least temporarily, that would have been virtually unthinkable for American women outside the war emergency. Just as airplane factories and shipyards hired "Rosie the Riveter" to work the assembly lines, employers sought to hire female engineers, but too few had graduated in the years prior to Pearl Harbor. To fill the gap, Grumman Aircraft, Curtiss-Wright, General Electric, and a few other large companies created crash courses that prepared hundreds of women to enter war industries as engineering aides. In 1942, Curtiss-Wright recruited about 700 young women who had successfully completed college algebra to join the emergency program. About 100 of these "engineering Cadettes" became the first women to study at Rensselaer Polytechnic Institute. Others trained at Iowa State, Cornell, and other land grant schools that had previously graduated a handful of female engineers – but the Cadette program suddenly brought a hundred women to each engineering program all at once. Cadettes got a ten-month crash course in aeronautical engineering, including engineering mechanics, theory of flight, strength of materials, structural analysis, and more, often using the same textbooks as regular classes for full engineering majors. The women also spent four hours a week in college machine shops, learning to weld, solder, and operate machine tools. Teaching Curtiss-Wright women was a totally new experience for engineering faculty used to rooms full of men, but several acknowledged that the women were admirably serious about their studies and even superior in drafting accuracy. At Penn State, roughly one-third of Cadettes qualified for the dean's list, while some women at Minnesota volunteered to help with experimental flight simulation research. Cadettes made terrific public relations for the war effort. *Life* magazine ran photos of RPI women welding, drafting, and running wind tunnel tests. Cadettes justified stepping onto men's engineering territory as patriotic wartime service. After finishing training, Cadettes went to work at Curtiss-Wright, helping design parts, draft plans, check wiring diagrams, correct blueprints, and handle research computations (Bix, 2005). While Curtiss-Wright Cadettes did not spend enough time on campus to earn full engineering degrees, wartime publicity for these programs helped inspire more college women to declare engineering majors. By fall 1945, Purdue had 88 female engineering majors. Yet even the most academically successful women could not gain membership in the national honorary engineering fraternity. Tau Beta Pi issued women only a badge of merit and didn't admit women on equal terms until 1968.

The conservative postwar reset and the baby boom era reaffirmed traditional gender roles in American culture. While the number of women choosing to major in US engineering continued to rise in general during the postwar years, the flood of male engineering students entering on the GI Bill meant that women remained less than 1% of the nation's total engineering enrollment. In the 1950s, 10–15% of male college graduates got degrees in engineering; among women, less than 0.2% got degrees in engineering. A few Americans warned that was a mistake, that engineering classes in the Cold War Soviet Union were one-third female (Puaca, 2014).

Significantly, rather than just waiting passively for acceptance, female engineers organized to fight for change. In 1952, about 60 US women established the Society of Women Engineers, aiming to remind people that engineering was not actually all-male, that women engineers existed. SWE leaders believed that many girls just did not realize that women could and did go into engineering. To give them encouragement, SWE members gave talented high school students tours of their labs, spoke at high school career events, and personally wrote letters to encourage dozens of high school girls. These female engineers worried that even in elementary school, girls were indoctrinated with messages that engineering was for men. They knew that classmates, parents, and teachers would often discourage many interested girls, since the idea of a female engineer seemed strange. SWE members emphasized how many of them were married and had children, which they hoped would make girls and their parents think more positively about engineering as a happy, "normal" path. SWE also set up college chapters at Purdue, Drexel, and a growing number of other schools, offering to mentor current students. SWE leaders vividly remembered how intimidating it felt to be the only woman in an engineering class. As allies, SWE connected with a tiny but slowly widening pool of female faculty and staff in engineering programs.

A number of women thrived in engineering study in the 1950s and 1960s, lucky enough to find valuable encouragement from supportive family and friends. Some felt more accepted if they worked to integrate themselves more with their male classmates, to become "one of the boys." Others remained annoyed that many male classmates refused to take them seriously as intellectual equals and also slandered them as not properly feminine. Every time a woman switched majors or dropped out, critics pointed to that as supposed proof that women didn't belong.

For years to come, the campus climate for many female engineering majors remained chilly, and, in some programs, even toxic, discouraging some female students to the point of dropping out. Those making it to graduating frequently confronted problems in the job search, as hiring committees questioned women about when they would get married, doubted whether a female engineer could think about mechanical details "like a man," and offered salaries distinctly lower than men's.

In the late 1960s and 1970s, female undergraduates, graduate students, faculty, and staff at a number of schools began mobilizing to support each other and fight for better conditions. Women shared concerns about childcare problems, ways to recruit more female faculty, hiring biases, legal inequities, women's assertiveness training, and the strengths and difficulties of two-career marriages. They condemned the frequency of what would come to be called sexual harassment incidents, with X-rated messages posted in workplaces and improper propositions. They compared notes on times that male colleagues had excluded women from discussions or literally pushed them away from equipment. Female engineers who entered the field in the 1970s remembered improvising individual solutions to handle the sense of being an outsider and confronting a lack of respect in many quarters (Ettinger et al., 2019).

Especially since the 1970s, the National Science Foundation and the National Academy of Engineering, US engineering and science professional societies, universities, the Girl Scouts, major corporations, and dozens of other organizations have signed on to promote diversity in STEM fields. Such high-powered endorsements provide millions of dollars annually to support conferences, formal programs, and informal activities, all aimed at encouraging different populations to embrace the exciting potential of STEM. Working from the bottom up, local groups, teachers, counselors, parents, and children themselves have joined the mission for re-envisioning the future of engineering and science, careers long assumed to belong primarily to relatively well-off White men (Bix, 2019).

In the late 20th and 21st centuries, engineering institutions, organizations, and many individuals continued to challenge the social norms that long bounded engineering as a middle-class, White-male-dominated occupation. Yet after women's share of US engineering bachelor's degrees rose dramatically from 1% in 1970 to 21% in 2000, that growth has since stalled at 22% in 2018. Patterns in specific engineering fields vary widely, with relatively high female participation in environmental and biomedical engineering but persistently low figures in several other disciplines (Funk & Parker, 2018).

Women's involvement in computer engineering and science has shown particularly wide swings, reflecting a shifting history of education and jobs in those areas and a complex story of how gendering has been attached to computing itself (Misa, 2010). It is worth emphasizing US patterns represent only part of the story; female involvement in STEM varies widely in complex ways across the Americas, Europe, Asia, and the rest of the globe.

In the United States by 2018, women comprised only 14% of the engineering workforce, reflecting complex factors. Many researchers note ongoing retention problems, often attributed to a sense of workplace inequities, lack of support for family demands, plus discrimination and sexual harassment. Discouragement and obstacles facing Black and Latino/a Americans in STEM also persist, contributing to continued gaps in engineering participation by gender, race, and ethnicity, exacerbated by biases and economic inequalities (Funk & Parker, 2018). Over recent decades, researchers have analyzed such issues, often pointing to the "leaky pipeline" and "chilly climates" of STEM careers, "Silicon Valley bro culture," "implicit biases," "stereotype threat," and other educational, social, economic, and employment factors (Hill et al., 2010; Corbett & Hill, 2015). Yet despite ongoing issues, programs meant to draw underrepresented groups into STEM and support their success, such as Girls Who Code, the Society of Women Engineers, and numerous others, have clearly made a difference for many individuals. The challenges remain for the United States to continue investing in efforts to enrich engineering for the 21st century by promoting an intellectual, social, and economic climate welcoming talent from all backgrounds.

4 Educating Engineering Technicians in South African Universities of Technology

Christine Winberg

This viewpoint describes the educational provision for engineering technicians in South Africa, highlighting key developments over the post-apartheid period and laying out the current challenges for engineering education and research.

The Engineering Council of South Africa (ECSA) accredits three types of engineers, namely, professional engineers, engineering technologists, and engineering technicians. Engineering technicians have a range of "practical engineering skills – such as the ability to mill, turn, drill, solder and wire" (Lewis & Gospel, 2015). They manufacture components and sub-assemblies or build, calibrate, and operate machines across engineering fields. Tools, techniques, and technologies vary considerably, and engineering machines have become increasingly sophisticated, computer-integrated networked systems (Delahanty et al., 2020). Once qualified, technicians take up employment as computer technicians, operators in manufacturing plants, or undertake the repair and maintenance of electrical infrastructure. Some become building inspectors on construction sites, and some end up in engineering laboratories, doing benchwork, maintaining equipment, and assisting with experiments.

In developing countries, there is a shortage of all engineering professionals. South Africa has only 1 engineering professional per 2,600 capita (ECSA, 2021). This is much lower than comparable countries, such as Brazil (1:227) or India (1:157). The shortage of technicians can, in part, be attributed to the country's apartheid past, which denied Black South Africans access to the engineering professions. Table 2.1 shows the demographics of the engineering professions. In post-apartheid South Africa, there has been an effort to increase the number of Black engineering professionals, but there continues to be more White male professional engineers than any other population group.

A more detailed breakdown of the demographics shows that the situation is changing. In its 2019/2020 report, the Engineering Council of South Africa noted for the first time more Black professional engineers in the 20–29 age group than White engineers in the same age group. Gender representation has, however, been slower to change. There are very few women studying or

Table 2.1 Engineering Categories by Gender and "Race"

Categories	*Gender*		*"Race"*			*Totals*
	Male	*Female*	*African*	*White*	*Other*	
Engineer	18,584	1,188	2,251	16,056	1,465	19,772
Technologist	5,925	500	2,076	3,568	781	6,425
Technician	4,941	779	2,569	2,725	426	5,720

Source: ECSA, 2021, p. 65.

practicing engineering: only 28.5% of students enrolled in technician programs are female, and only 13.6% of registered engineering technicians are women. The attrition rate of women students in technical engineering studies has been attributed to their negative experiences on engineering programs in general, and gender discrimination specifically (Nel & Meyer, 2016).

The ideal ratio of engineers to technicians has been debated for years. The Engineering Council of South Africa (ECSA) recommends a 1:4 ratio, but actual figures are far off this benchmark. Engineering technicians consistently appear in the top ten of global scarce skills lists (WEF, 2020), surpassing professional engineers. Engineering technicians' skills are critical for developing economies (Banks & Chikasanda, 2015). South Africa's failing road, rail, and power infrastructure can partially be ascribed to the shortage of skilled technicians (Sutherland, 2020). Interruptions in power supply caused by the state-owned power utility's failure to maintain or replace its infrastructure have been exacerbated by the shortage of electrical engineering technicians (Tsikata & Sebitosi, 2010). There are acute shortages of technicians in renewable energy fields, which has kept South Africa dependent on coal power and delayed its migration to more sustainable energy production (Tsikata & Sebitosi, 2010). The need for technicians in South Africa can also be explained by the fact that the country is largely a user and adaptor of advanced technologies rather than a technology innovator (Blake, 2010). South Africa's economic regeneration is unlikely while there are critical shortages of technicians, and the number of registered technicians is on the decline (Sutherland, 2020). These are important considerations for engineering education.

Programs for engineering technicians are offered by South African universities of technology and some comprehensive universities; these diploma-level qualifications traditionally prepare students for direct entry into labor markets, supported by practice-oriented curricula, internships, and other forms of work-integrated learning. Technical engineering studies have become increasingly complex as technologies have evolved. Technical programs consequently demand regular review and updating of curricula (Banks & Chikasanda, 2015), while technicians' profiles have grown to include a range of more complex professional graduate attributes (Doorsamy & Bokoro, 2019). Technical engineering programs are designed for specific engineering disciplines and are accredited by the Engineering Council of South Africa, but many researchers and educators agree that there is a need to improve students' transitions into the world of work (Baldry, 2016). In the past, novice engineering technicians graduating from universities of technology secured employment within three to six months of graduation on average (van Broekhuizen, 2016). This trend has reversed, despite shortages in these fields, and many novice technicians currently struggle to find appropriate employment in the fields for which they are qualified. While there are many factors that affect the employment of graduates, the mismatch between the skills that graduates develop in their university studies and those that employers require from graduates in the 21st century has been highlighted as a contributing factor (Pauw et al., 2008). The skills mismatch has exacerbated South Africa's skills shortages and adversely affected the employment prospects of university of technology graduates more than it has their other higher education cohorts (Kraak, 2015). The skills mismatch contributing to the growing student unemployment among technical engineering graduates has been blamed on "academic drift." This is evident in how universities of technology have focused on converting diploma-level qualifications to degrees in an effort both to strengthen the scientific disciplinary knowledge base underpinning technical qualifications and to offer more prestigious qualifications instead of focusing on the employability of their graduates (Dell, 2016).

The inclusion of the basic and engineering sciences in technical engineering programs is not disputed, but how much science is needed, which concepts should be selected, and how these should be sequenced has caused considerable debate (Case, 2014). Shay (2013) proposes that the "logic of curriculum" in practice-oriented programs should necessarily be different from the logic of curriculum in theory-oriented programs (2013 p. 566). The issue of who should advise on an appropriate theory–practice mix in technical engineering programs is complicated. While practicing technicians are sometimes recruited onto curriculum advisory committees – and they do have a tacit understanding of how to train technicians – they do not have the educational knowledge to lead curricular development or propose appropriate pedagogies for technical education. It is traditional in many technical programs that the technicians themselves "neither set the entry and the performance standards of their occupation, nor control the educational process through which new recruits are trained" (Keefe & Potosky, 1997, p. 54). Very few technicians teach in technical engineering programs. This is because in order to be appointed as a lecturer in a South African university of technology, a minimum qualification of a master's degree is required. As not many engineering technicians have master's degrees, they are disqualified as faculty members, and students enrolled in technical engineering programs are therefore mainly taught by professional engineers and scientists. There are consequently very few technicians available in the university of technology sector to defend the worth of practical training in engineering technicians' qualifications. The neglect of practical training in technical programs was exacerbated during the pandemic, which closed access to internships in many engineering industries (Graham, 2022). The social distance between engineering technicians and engineering faculty is exacerbated in the South African context by the fault lines of race and class. Thus, there is the danger of reproducing the inequalities that are starkly evident in the Engineering Council's reports. In universities of technology, engineering technicians are usually found in undergraduate laboratories, and while there is evidence of the important roles they play in educating future technicians (Gqibani et al., 2018; Winberg, 2021), they are marginalized by departmental aspirations to offer professional engineering degrees rather than technician qualifications and the university's general shift towards valuing research over technical expertise.

The nature of practical knowledge in qualifications such as the engineering technician diploma has not been adequately theorized, and the specialist value of practical knowledge in technical engineering curricula is not fully understood (Wolff & Winberg, 2022). This makes students' call for the decolonization of technical engineering difficult (Fomunyam, 2017). Decoloniality has not been strongly foregrounded in engineering curriculum development, perhaps because, as Muller (2018) argues, engineers "translate knowledge into action to achieve practical goals" (2018 p. 2), and provided that these are recognized societal goals, a debate about by whom and how goals are determined is, to a certain extent, irrelevant. A first step towards decolonizing technician education and achieving "epistemological justice" might be to include practicing technicians, as well as student technicians and women students in particular, in decision-making processes (Winberg & Makua, 2019; Winberg & Winberg, 2017). A more robust conceptualization of practical knowledge that recognizes its complexity and value is needed.

The pressures on technical engineering programs come from many different directions and for different reasons: from professional bodies, from the field of practice, from the aspirations of universities of technology to offer more "academic" programs, and from students' need for epistemic justice, expanded life opportunities, and possibilities of meaningful work. Engineering educators and researchers have a role to play in developing principled responses to these multiple pressures – not least of which will be to discover intersections between sustainable technical practices, decoloniality and the common good, and more human-centered pedagogies. A variety of strategies will be required in order to address these challenges. Firstly, engineering educators need to involve practicing technicians in meaningful collaboration with engineering faculty in rethinking the education of technicians. Technicians could help engineering faculty to learn more about the world of technical engineering practice and begin to formulate a basis for appropriate curricula and pedagogies in the South African context. While technicians in practice are likely to have a tacit understanding of what is necessary in curriculum design and pedagogies of practice, the discourse of curriculum and pedagogy might be alien to many technicians, as well as to students enrolled on technical programs. Thus, engineering educators and researchers would need to make these concepts explicit in order for the team to develop a shared terminology for technical engineering education. Secondly, the practical help that engineering educators offer with regard to student engagement, lesson planning, and so on needs adaptation in technical engineering environments. Engineering faculty, technicians, and engineering educators need innovative pedagogies in these fields. Engineering education organizations might consider educational conference workshops and special interest groups for technical engineering. Keeping up-to-date with key developments in technical education could provide technicians in practice with the legitimacy and know-how needed to work with professional engineering colleagues and advise on educational matters. Finally, engineering educators will need the interpersonal skills to negotiate curricula with engineering faculty and others. Engineering educators might find themselves playing a mediating role between departmental colleagues, students, and practicing technicians.

Engineering educators and researchers have a particular role to play in supporting the transformative possibilities of technical higher education by understanding the curricula and pedagogical practices that might enable engaged learning towards competent, socially just, and sustainable practice in technical fields. Reclaiming practical knowledge is a necessary step towards addressing South Africa's failing infrastructure and resultant inequalities. In the face of increasing pressure on technical curricula, it is essential that the structures that are put in place and agents, such as policymakers, engineering educators, and researchers, recognize the value of practical knowledge. This is therefore an opportune time to reconceptualize the training of technicians, in particular the role that flows and transitions between technical engineering education and work practices could play in strengthening curricula, student learning, and innovative teaching.

5 Ethics Education for Technology Students in Estonia: The Influence of History

Aive Pevkur (https://orcid.org/0000-0003-1743-8327)

This viewpoint gives a short overview on the formation of ethics education for technology students in Estonia. The main aim of ethics education in universities is to develop proper values and attitudes in students considering their future profession and to prepare them for challenges they might face due to changes in technology and society. Ethical action itself depends on surrounding conditions, and especially on societal settings, and is influenced by values in a particular society, its ideology, and politics. Disruptions and changes in predominant ideologies frame understanding of right and wrong. Technology education might embed ethics education or not, but educational practices and life in general shape students' attitudes nonetheless. In this scenario, there are two relevant questions for ethics education. The first one is how much political change impacts the content of ethics education and subsequent professional practice. The second one concerns the impact of the past in shaping the current understanding of content and the need for ethics education in technology-related education. Estonia is an interesting example to shed a light on the teaching of ethics to students in technology-related curricula, as the country has faced drastic changes in history and political order.

Estonia is a small country close to Nordic countries and Russia, with a population of 1.33 million. The history of technology education reflects the historical processes the country has gone through, and the main changes in ethics education approaches are coherent with the processes in the society. Talking about ethics education in STEM and engineering curricula in small countries like Estonia, we cannot avoid talking about history. A ruling power, its ideology, and politics always influence the teaching of social sciences in the universities (TTÜ, 2018). Due to the size of Estonia, its history of technology ethics education can be studied through the history of one educational institution, the Tallinn University of Technology (TalTech). Established in 1918, Tallinn University of Technology is the only technical university in Estonia. TalTech, located in the capital city of Estonia, is a university for engineering, business, public administration, and maritime affairs. In the

University of Tartu, which is the biggest university in Estonia, there are also research fields related to technology, but the main focus is on science and less on engineering.

Before 1918, Estonia was a part of the Russian Empire, and many engineers got their technical education in universities outside of Estonia, mainly in St. Petersburg. The history of Estonian technology education can be traced back to the times of the First World War. While the country was still occupied by German forces in 1918, the Estonian Technical Society started to organize special technical courses in Estonia. In 1919, a private technical school called Tallinna Tehnikum was established to satisfy the needs of the young republic for engineers, architects, and technicians. In 1920, the school was reformed to give higher education in technical-related fields, and in 1938, the school was renamed Tallinn University of Technology (TTÜ, 2018). The Second World War was a period of enormous changes. Estonia lost its independence. The university lost assets, students, and teachers. Despite that, technology education continued and developed during the Soviet times (1944–1991). As STEM education was considered less ideological, the ruling power had a lower influence on the curricula and teaching; the same cannot be said about other subjects, especially the humanities, including philosophy and ethics.

There is a relative lack of sources describing engineering ethics education before 1991, when Estonia regained independence from the Soviet Union. The sources which could potentially describe engineering ethics education are, in practice, one of two types. The first ones are descriptive overviews of technology challenges and developments or aggregations of information about academic staff or students (Mägi, 1984). The second ones are about ideologically overloaded analyses of philosophical problems.

During the Soviet time, ethics education was considered an ideological task. Influencing students' attitudes was achieved through the teaching of Marxist–Leninist philosophy. In 1940, the institute of Marxist–Leninist philosophy was created in the university and existed in different forms until 1990 (TTÜ, 2018). In 1983, in the institute of dialectic and historical materialism, 12 people were employed; one of them had the highest scientific qualification (doctor of science), and eight were candidates of science, equivalent to the PhD (Laas, 1984). There is no evidence of courses in STEM ethics or engineering ethics during the Soviet times, but in all curricula, the courses of philosophy were compulsory. Scientific studies were about the Marxist–Leninist philosophy, for example, about the formation of ideologically knowledgeable people (Livshits, 1971). The content of philosophy courses was ideologized, and so was teaching of ethics-related issues (de George, 1967).

While in the Western Hemisphere, especially in the United States, discussions about the content of engineering ethics education have been ongoing for decades (Harris Jr. et al., 1996), the ethics education in Soviet universities was not a part of professional training, and the focus was on a moral education of a Soviet citizen (Zajda, 1988; Dunstan, 1981). That created a "double morality" (Khasanova et al., 2013), where officially people were committed to the high moral ideals preached by Communist Party officials, but in real life nobody took these high ideals seriously. People did not respect the state officials, because Estonia (and many other Soviet republics) was occupied and did not want to become a part of Soviet Union voluntarily. In this scenario, technical experts did not have to commit themselves publicly to the ideology – their expert knowledge did not depend on that. Unlike, say, journalists, who wore their ideology on their sleeves. Consequently, technical experts perceived ethics as inconsequential for their work, as it was shaped by the ruling ideology and ethics, like subjects in the humanities were for them insignificant, even alienated. In that situation, the professional ethical attitudes were shaped by values and beliefs of professors and technical elite, and that became more important than the content of official courses. While the society adopts double standards, a lot depends on persons, their charisma, and basic values, whether the professors encourage critical thinking or just repeat ideologically approved statements. Only in the 1980s did the teaching become less ideological and propagandistic. In lectures, people started to express their own thoughts (TTÜ, 2018).

In 1990, all ideological subjects were removed from the curricula (TTÜ, 2018), and transformation towards establishing a professional community of technical experts began. As early as in 1989, the engineering society adopted the first code of engineering ethics (Inseneriееtika . . ., 1989). The tradition of teaching philosophy remains, but the content was changed to basic knowledge in philosophy – metaphysics, theory of knowledge, philosophy of science. On January 23, 1995, lecturer of philosophy Aare Laanemäe published a short article in the university newspaper where he described the situation in the society as an ethical crisis and saw the reason for that in the heritage of the double morality in Soviet times. The article posted a question about the future of ethics education in the Tallinn University of Technology. He refers to a small survey he conducted among students, according to which up to 90% of students supported including ethics into the curricula. Laanemäe mentions that engineering ethics was already taught in some curricula (Laanemäe, 1995).

A contemporary general picture of teaching ethics-related courses in Estonian universities is described in the study by Saarniit and Pevkur (Saarniit & Pevkur, 2019). Insights into the developments of teaching ethics for technology students during the last 20 years were derived using data from the reference books of BA, MA, PhD, engineering, and applied studies in Tallinn University of Technology (Õppeteatmik, 2002, 2007) and data from the university's study information system (SIS -ÕIS[1]). The study conducted a review of ethics-related courses in curricula in years 2002, 2007, and 2022. In philosophy courses taught at the same time, there were metaphysics, theory of knowledge, and philosophy of science, and according to the course descriptions, ethics issues were not discussed during the course. Years 2002 and 2007 were similar regarding what was taught and to whom. As ethics courses were provided in three levels of education, BA, MA, and integrated engineering education (Eur-Ing[2], MSc, 3 + 2), PhD and applied programs were out of the scope. According to the study information system (SIS), philosophy was compulsory in almost all technology and engineering curricula (2002, $n = 38$; 2007, $n = 46$). For comparison, the only ethics course at that time, engineering ethics, was an elective course in six curricula in 2002 and in four curricula in 2007. The gene technology curricula had their own compulsory course of ethical, legal, and business aspects in gene technology.

In both observed years, engineering ethics was an elective course in all integrated engineering courses ($n = 3$). While in 2002, the engineering ethics course was in two BA and one MA curricula, in 2007 it remained only in one BA curricula as an elective course. In reference books (Õppeteatmik, 2002, 2007), there was a description of the engineering ethics course which did not change over five years, and there was only one teacher, Aare Laanemäe. In 2003, Laanemäe published a book, *Engineering Ethics*, in the University Publishing House (Laanemäe, 2003). Topics of science, technology, and ethics; humanistic and engineering professions and professional ethics; human, machine, and environment; computer ethics; engineering, AI, and ethics; codes of ethics; moral conflicts; ethical culture of European engineer (Eur-Ing) were covered. The main methodological approach in the book is descriptive, providing an overview of the main ideas. It is important to note that in all integrated engineering courses (Eur-Ing), the engineering ethics course was an elective course, which also covered culture, codes, and norms of the engineering profession.

Looking at the current situation, there have been changes. According to the study information system (SIS), a philosophy course is compulsory only in one technology-related BA curricula and an elective course in seven other curricula. In all integrated curricula in engineering faculty, the course of Philosophy of Culture and Built Environment Ethics is compulsory. The topics covered during the course are ecological crisis and environmental ethics; social aspects of designing the built environment; sustainability, risk, safety, and responsibility; engineering professional norms and requirements; and resolving dilemmas in professional life. Engineering ethics is an elective course in three BA curricula, and the content is designed to cover main engineering ethics topics but also to analyze problems in the specific field. For example, if students are from the IT faculty, questions of ethics challenges due to datafication, data protection, big data, AI, electronic wastes, privacy infringement, digital divide, and others are covered. The main task is to teach students to see their activity in a bigger societal and environmental context, to understand ethical challenges emerged due to technological achievements, and to realize their professional responsibility. Even though the number of people who select engineering ethics courses is small, the feedback on the courses has been good, and those who passed the course have admitted that it should be compulsory for all technology students.

In conclusion, in Europe, there is a strong tradition in technology-related fields to follow the law, which is seen as the minimum of proper behavior. The same approach is widespread in Estonia as well. Developing the best pedagogical approach for ethics education needs continuity and sustained development, which is a precondition for the implementation of right and proper values and attitudes into education for technology students. Understanding ethics as a necessary and trainable competence for technology students has not been rooted in higher education. Hopefully, the situation changes in the near future.

6 What Changes Do We Want in Engineering Education?

Edward Conlon

Although the theme of change has been a constant in engineering education, there are tensions among engineering educators about what change we need and how it is to be achieved. That much might be obvious, but it's worth restating as there is confusion about the purpose of change. The tensions primarily are between those who want change and those who are committed to the status quo and have a reverence for the engineering "core" and knowledge, which has a mathematical content, at the expense of everything else in the curriculum.

Some years ago, when debates about change in engineering education began to take off in my institution in Ireland, a colleague sent a shot across our bow when he circulated an email which said:

> A thorough grounding in Mathematics, Applied Mathematics (mechanics), Physics and Chemistry is an essential first step in any professional engineering programme.
>
> Students in their first year should be introduced to the various branches of the profession and the operation of engineering practice. This is often best achieved by case studies etc. delivered by practicing engineers.
>
> *A combination of lectures, laboratories and problem solving exercises has served the profession very well since the founding of the Ecole Nationale des Ponts et Chaussees some two hundred and fifty years ago. We see no conflict between this approach and the current views of educational theorists. (Italics added)*

This was meant as a warning to outsiders, such as those in teaching and learning centers (at whom this was directed) and other non-engineers (like myself), but what is more significant is the defense of a form of education established for an elite in French society. Ecole engineers were not just educated to design and build roads and bridges but were inculcated with highly elitist attitudes and to be members of a "corps d'elite" of "bureau chiefs and directors of industry" (Wickenden, 1929 in Bucciarelli et al., 2009, p. 106). But perhaps this is not surprising, as traditionalists in education have often defended access to knowledge as a mechanism of social stratification and selection. As Wheelehan has argued, they value less the intrinsic role of knowledge in society and are more concerned with "enculturation into traditional values and norms, based on a relationship of deference to traditional bodies of knowledge taught in traditional authoritarian ways that require submission to become the 'kind of person it is supposed to make you'" (Wheelahan, 2010, p. 7).

What is often at the heart of this is the defense of the foundational knowledge, seen to be at the core of engineering, which has made engineers what they are, and which is seen to be under threat from constructivist approaches committed to context-dependent learning and to real-world problem-solving entailing the use of teamwork and other skills emphasized by new approaches to engineering education.

What *really* is the purpose of these new approaches?

At the same time as this email was circulating, proponents of change in my institute were promoting an article about Olin College which stated:

> Throughout their time at Olin students will study the arts, humanities, and social sciences (AHS) and entrepreneurship in order to provide context for their engineering studies. Entrepreneurship is included in this list because students should not only appreciate the context in which they work but also be able to recognize and respond to human needs within this context.
> *(Somerville et al., 2005, p. 201)*

As I wrote at that time in an unpublished piece:

> This would seem to accept that the only way to meet human needs is through market-based approaches. One wonders if students in this programme are encouraged to be critical of markets and suggest that it is because of market failure that many human needs are not met. . . . What is problematic here is that while the context in which engineering takes place is seen to be changeable this change is conceived in very narrow terms.

New approaches are influenced by both constructivism and instrumentalism as both share a common focus on context: "both emphasise the contextual, situated and problem-orientated nature of knowledge creation and learning. Both sacrifice the complexity and depth of knowledge in curriculum in favour of 'authentic' learning in the workplace" (Wheelahan, 2010, p. 5). And while these new approaches may seem to represent a break with the traditionalists, Wheelahan (2010, p. 7) argues – convincingly, in my view – that all three approaches share an instrumental attitude to knowledge and do not see it as "a casually important objective in its own right because of the access it provides to the nature of the world and to society's conversation."

She defends the centrality of abstract theoretical knowledge to educational endeavors. It's the means that society "uses to think the 'not-yet-thought and unthinkable' and to imagine alternative futures . . . and make connections between objects and events that are not obviously related." It is the means society uses "to transcend the limits of individual experience to see beyond experiences to the nature of relations in the natural and social world" (p. 7). It provides us with the deeper understanding we need to make sense of the world. As Marx argued, there would be no need for science "if the outward appearances and essences of things coincided" (Marx, 1966, p. 817). He also argued that the point of understanding the world is to change it.

What is at stake here is the purpose of engineering education. What are we trying to do when we educate engineers, and what kinds of identifies are we trying to shape? There is confusion about this, even among those committed to change, tending to pull attempts to change engineering education "into opposing, even contradictory, directions" (Jamison et al.: 254). Jameson and others (2014), in a very useful piece, have told us that historically there are contesting models of engineering education in which dominant approaches (corresponding to the traditional and constructivist approaches earlier), which emphasize the technical and entrepreneurial dimensions of engineering education, have marginalized more integrated or social approaches focused on the public good and the role of engineers as change agents. The result is that students "have not been given the opportunity to understand the broader social and cultural aspects of the challenges facing engineering" (Jamison, 2013, p. 21).

This, for me, remains a key problem in debates about change in engineering education, and I found myself belonging to that historically marginal group which argues for social and cultural understanding to be added to the theoretical and practical components of engineering work. This requires, as Sterling (2004) has argued in relation to sustainability education, a focus on "the deeper levels of paradigm and purpose guiding policy and practice (which) . . . tend to be hidden from view and . . . most debate" (64) about change in education. To be clear, we need to know what we want to achieve when we say we want to change engineering education. Do we simply want to train out students to be more efficient and entrepreneurial and to insert themselves into the runaway world of globalized capitalism, or are we committed to providing students with the powerful and critical knowledge they need to address the substantial challenges facing engineering and the planet engineers share with others? The critical realist geographer John Huckle has described such knowledge in the following terms:

> To be critical it should reveal the structures and processes at work in the world that leads to injustice, a lack of democracy, and a failure to realize sustainable forms of development. It should reveal ideology that masks these structures and processes and should offer social alternatives or ways of realizing justice, democracy, and sustainability that can empower individuals and communities as they apply theory to practice.
>
> *(Huckle, 2017, p. 70)*

It can be suggested that some confusion about the purpose of engineering education is at the core of the process for accrediting engineering education in Ireland, with the result that while formally there is an aspiration for change, in practice, change is taking place at the margins and is driven by instrumental approaches.

Ireland is somewhat unique in having one professional body, Engineers Ireland (EI), representing all categories of engineers, with responsibility for accrediting all engineering programs. EI is one of the original signatories of the Washington Accord. Accrediting events, which occur every five years, involve a self-study by program teams, followed by an assessment event, during which accreditation assessors from industry and academia evaluate programs by reviewing evidence, touring facilities, and interviewing students, graduates, employers, and staff.[3]

Research focused on broadening the curriculum (Nicolaou et al., 2018; Martin, 2020; Murphy et al., 2019; Martin et al., 2021; Murphy et al., 2022) has confirmed the importance of accreditation to Irish engineering educators, with one study asserting that educators saw accreditation as the "most significant influence on program design" (Nicolaou et al., 2018). As one program leader put it to us in another project, "[c]hanging the program is usually done with an eye on what *Engineers Ireland* are looking for" (Murphy et al., 2022).

Key ingredient in the accreditation process are the learning outcomes (LOs) set for programs which range from foundational engineering knowledge related to engineering science and mathematics, problem-solving, and design to what some consider "softer" skills related to ethics, teamwork, and communication. In 2021, an outcome related to engineering management was added.

The ethics outcome requires that students have "an understanding and appreciation of the environmental, social and economic impacts of their judgements and promote the principles and practices of sustainable development" while "understanding of the importance of the engineer's role in society and the . . . importance of equality, diversity and inclusion." This provides a possible basis for the broadening of the curriculum in the manner argued for earlier. While there is evidence to suggest that an LO focused on ethics can increase the presence of ethics in the curriculum, we have doubts that the pressure exercised by accreditation bodies can translate into deeper curricular change (Martin et al., 2021).

Some research findings which arise from investigations of the accreditation process suggest that:

1 The integration of sustainability into engineering education has a predominant focus on technology and enhancing core disciplinary competencies. The social dimension of SD is marginalized (Nicolaou et al., 2018).
2 There is a strong disciplinary focus on core engineering competencies in program design and during accreditation. Murphy et al. (2019, p. 381) found "no evidence of systemic attention to a broadening agenda within the accreditation reports" and that change was mainly driven by "instrumental justification" for the inclusion of broadening content (378).
3 The focus on ethics in accreditation events is very narrow, with an emphasis on the code of ethics and less on the social impacts of engineering (Murphy et al., 2022). The most comprehensive study we have of ethics education, which includes a substantial interrogation of the accreditation process, claims that:

> [T]here appears to be more attention given to the procedural aspects related to how programmes prepare and display their evidence in support of outcome E (the ethics LO), than to ensuring that sufficient weight is given to ethics in the curriculum or exploring the broadness of its treatment.
>
> *(Martin, 2020, p. 246)*

It further finds that:

> While the lower weight given to outcome E in engineering programmes has been noticed by evaluators, it tends to be considered a common state of affairs, being that ethics does not need the same emphasis in the engineering curriculum compared to technical oriented outcomes.
>
> *(pg. 246–247)*

The study supports the view that the existence of an accreditation criterion dedicated to ethics does not necessarily lead to a curriculum which addresses the social and political dimensions of engineering practice in a broad manner, or that accreditation processes adequately interrogate the extent to which programs address this outcome.

So we face a contradiction of an aspiration to broaden engineering education being subverted by the culture of engineering which valorizes the technically proficient and the entrepreneurial. A key issue is that those educated and submerged in the dominant engineering culture are asked to evaluate progress in moving away from such a culture. A task they clearly struggle with as they don't seem to fully understand what is entailed in educating an ethical engineer (Martin, 2020).

In addition to acknowledging the struggles of those required to evaluate engineering programs, we need to understand there are powerful actors outside the academy who shape what happens in the engineering classroom and can often reinforce (or resist) pressures towards traditionalism or the instrumentalization of education (Slaughter, 2002). It is noteworthy that accreditation panels are composed of engineers and industry representations to the exclusion of the wider community impacted by what engineers do (Conlon, 2013).

In order to affect change, it can be suggested that we need better clarity about what we are trying to do when we say we want to change engineering education. And that we let those into the conversation who may have some insight about what it means to add social and cultural understanding to the theoretical and practical components of engineering work or who are substantially impacted by what engineers do.

Notes

1 https://ois2.ttu.ee/uusois/uus_ois2.tud_leht.
2 www.feani.org/feani/eur-ing-title/what-eur-ing-title
3 Full details of accreditation procedures can be found here: www.engineersireland.ie/Professionals/Membership/Become-a-member/Accredited-third-level-courses/Accreditation-criteria-procedure-and-training.

References

Adas, M. (2006). *Dominance by design: Technological imperatives and America's civilizing mission*. Harvard University Press.

All India Council for Technical Education (AICTE). (2021). *AICTE annual report & audited accounts 2020–21*. Retrieved November 6, 2022, from www.aicte-india.org/sites/default/files/AR/AICTE-Annual-Reports.pdf

Ambirajan, S. (1995). Science and technology education in South Asia. In R. MacLeod & D. Kumar (Eds.), *Technology and the Raj: Western technology and technical transfers to India, 1700–1947* (pp. 112–133). Sage.

Aneesh, A. (2006). *Virtual migration: The programming of globalization*. Duke University Press.

Arnold, D. (2000). *Science, technology and medicine in colonial India*. Cambridge University Press.

Baldry, K. (2016). Graduate unemployment in South Africa: Social inequality reproduced. *Journal of Education and Work, 29*(7), 788–812.

Banks, F., & Chikasanda, V. K. M. (2015). Technology education and developing countries. In J. P. Williams, A. Jones, & C. Buntting (Eds.), *The future of technology education: Contemporary issues in technology education*. Singapore: Springer, 217–238. https://doi.org/10.1007/978-981-287-170-1

Bassett, R. (2016). *The technological Indian*. Harvard University Press.

Bix, A. (2005). Engineering national defense: Technical education at Land-Grant Institutions during World War II. In A. Marcus (Ed.), *Engineering in a land-grant context: The past, present, and future of an idea* (pp. 105–133). Purdue University Press.

Bix, A. (2013). *Girls coming to tech! A history of American engineering education for women*. MIT Press.

Bix, A. (2015). Men in the food lab, women in the engine shop: Gendered stereotype breaking in land-grant technical programs In A. Marcus (Ed.), *Service as mandate: How Land-Grant Universities shaped the modern world* (pp. 19–57). University of Alabama Press.

Bix, A. (2019). Engineering girls: The evolution of advocacy for young women's Stem education. In S. Eckelmann, S. Fieldston, & P. M. Renfro (Eds.), *Growing up America: Youth and politics since 1945* (pp. 191–210). University of Georgia Press.

Black, J. (2009). The military influence on engineering education in Britain and India, 1848– 1906. *The Indian Economic and Social History Review, 46*(2), 211–239.

Blake, E. (2010). Software engineering in developing communities. In *Proceedings of the 2010 ICSE workshop on cooperative and human aspects of software engineering* (pp. 1–4). Association for Computing Machinery, Cape Town, South Africa. https://doi.org/10.1145/1833310.1833311

Brown, J. (2000). Design plans, working drawings, national styles: Engineering practice in Great Britain and the United States, 1775–1945. *Technology and Culture*, (41), 195–238.

Bucciarelli, L., Coyle, E., & Mc Grath, D. (2009). Engineering education in the US and EU. In S. H. Christensen, B. Delahousse & M. Meganck (Eds.), *Engineering in context*. Academia.

Case, J. (2014). Problematizing curriculum: Contemporary debates in engineering education. In M. Young & J. Muller (eds). *Knowledge, expertise and the professions*. Routledge.

Conlon, E. (2013). Broadening engineering education: Bringing the community in. *Science and Engineering Ethics*, 19, 1589–1594.

Corbett, C., & Hill, C. (2015). *Solving the equation: The variables for women's success in engineering and computing*. AAUW Report. www.aauw.org/app/uploads/2020/03/Solving-the-Equation-report-nsa.pdf

de George, R. T. (1967). The foundations of Marxist-Leninist ethics. In E. Laszlo (Ed.), *Philosophy in the Soviet Union: A survey of the mid-sixties* (pp. 49–60). Netherlands.

Delahanty, C. M., Genis, V., Herring, S., & Timby, T. A. (2020). Enhancing workforce readiness of engineering technicians. In *Proceedings of 2020 ASEE virtual annual conference*.

Dell, S. (2016, November 18). *Rebranding 'second class' universities of technology* (p. 437). University World News Global Edition. https://bit.ly/2flibHf

Department of Higher Education, Ministry of Education, Government of India. (n.d.). *National Institutes of Technology (NITs)*. Retrieved November 5, 2022, from www.education.gov.in/en/technical-education-5.

Doorsamy, W., & Bokoro, P. (2019, September). On the use of semantics for curriculum development of new Engineering Technology Programme. In *2019 IEEE AFRICON* (pp. 1–4). IEEE.

Dunstan, J. (1981). Soviet moral education in theory and practice. *Journal of Moral Education, 10*(3), 192–202.

Engineering Council of South Africa (ECSA). *Annual report 2020–2021*. www.ecsa.co.za/about/pdfs/ECSA%20Annual%20Report%202021.pdf

Ettinger, L., Conroy, N., & Barr II, W. (2019). What late-career and retired women engineers tell us: Gender challenges in historical context. *Engineering Studies, 11*, 217–242.

Fomunyam, K. G. (2017). Decolonising the engineering curriculum in a South African University of Technology. *International Journal of Applied Engineering Research, 12*(17), 6797–6805.

Funk, C., & Parker, K. (2018). *Women and men in STEM often at odds over workplace equity*. Pew Research Center Report. www.pewresearch.org/social-trends/2018/01/09/diversity-in-the-stem-workforce-varies-widely-across-jobs/

Geiger, R. (2015). Land-Grant colleges and the pre-modern era of American higher education, 1850–1890. In A. Marcus (Ed.), *Science as service: Establishing and reformulating American Land-Grant Universities, 1865–1930*. University of Alabama Press.

Gqibani, S., Clarke, N., & Nel, A. L. (2018). The order of skills development for technician and technologist training curricula. In *2018 IEEE global engineering education conference (EDUCON)* (pp. 1089–1095). IEEE.

Graber, F. (2008). Purity and theory: Theoretical tools at Ponts et Chaussees, circa 1800. *Technology and Culture*, (49), 860–883.

Graham, R. (2022). *Crisis and catalyst: The impact of COVID-19 on global practice in engineering education*. MIT. https://dspace.mit.edu/handle/1721.1/145955

Harris Jr., C. E., et al. (1996). Engineering ethics: What? Why? How? And when? *Journal of Engineering Education, 85*(2), 93–96.

Headrick, D. (1988). *The tentacles of progress: Technology transfer in the age of imperialism, 1850–1940*. Oxford University Press.

Hill, C., Corbett, C., & St. Rose, A. (2010). *Why so few? Women in science, technology, engineering, and mathematics*. AAUW report. www.aauw.org/app/uploads/2020/03/why-so-few-research.pdf

Huckle, J. (2017). Powerful geographical knowledge is critical knowledge underpinned by critical realism. *International Research in Geographical and Environmental Education, 28*(1), 70–84.

Hunter, R. (1991). Turnpike construction in Antebellum Virginia. In T. Reynolds (Ed.), *The engineer in America: a historical anthology from technology and culture* (pp. 43–68). University of Chicago Press.

Insenerieetika koodeks. (1989). www.insener.ee/wp-content/uploads/2021/03/Insenerieetika-koodeks.pdf

Jadavpur University. (n.d.). *History*. Retrieved October 24, 2022, from www.jaduniv.edu.in/about_us.php.

Jamison, A. (2013). *The making of green engineers*. Morgan and Claypool.

Jamison, A., Kolmas, A., & Holgaard, J. E. (2014). Hybrid learning: An integrative approach to engineering education. *Journal of Engineering Education, 103*, 253–273.

Jørgensen, U. (2007). Historical accounts of engineering education. In E. F. Crawley, J. Malmqvist, S. Östlund, & D. R. Brodeur (Eds.), *Rethinking engineering education* (pp. 216–240). Springer.

Keefe, J., & Potosky, D. (1997). Technical dissonance: Conflicting portraits of technicians. In S. R. Barley & J. Orr (eds). *Between craft and science: Technical workers in U.S. settings* (pp. 53–81). Cornell University Press.

Khasanova, G., et al. (2013). Ethics in training of engineers. In *2013 International conference on interactive collaborative learning (ICL)*.

Kraak, A. (2015). The value of graduate destination survey data in understanding graduate unemployment: A focus on the universities of technology. *South African Journal of Labour Relations, 39*(1), 93–113.

Kumar, A. (1995). Colonial requirements and engineering education: The public works department, 1847–1947. In R. MacLeod & D. Kumar (Eds.), *Technology and the Raj: Western technology and technical transfers to India, 1700–1947* (pp. 216–232). Sage.

Laanemäe, A. (1995, January). *Kas oht ikka varitseb? [eetikakursuse sisseviimise vajalikkusest TTÜs] // Tehnikaülikool 23*. https://digikogu.taltech.ee/et/Item/95e7c0cb-1330-4a66-b9f7-cdd530cc615b

Laanemäe, A. (2003). *Insenerieetika*. TTÜ Kirjastus.

Laas, J. (1984). TPI teadlaskonna osa Eesti NSV teadusorganisatsioonis. In *Kõrgema tehnilise hariduse ja tehnilise mõtte areng Eestis*, in Mägi, Vahur, Tallinna Polütehnilise Instituudi toimetised 590. https://digikogu.taltech.ee/et/Item/3ea16b13-fdbb-419d-a40b-8f40ebd066a3

Lewis, P. A., & Gospel, H. (2015). Technicians under the microscope: The training and skills of university laboratory and engineering workshop technicians. *Journal of Vocational Education & Training*, *67*(4), 421–441.

Livshits, Y. (1971). Obstsestvennii Porjadok I Litsnost v Svete Leninskih Idei (The Social System and Personality according to V.I. Lenin). In *Filosoofiaalaseid uurimusi*. Tallinna Polütehnilise

Lubinski, C. (2014). Liability of foreignness in historical context: German business in Preindependence India (1880–1940). *Enterprise & Society*, *15*(4), 722–758.

Mägi, V. (Ed.). (1984). Kõrgema tehnilise hariduse ja tehnilise mõtte areng Eestis. *Tallinna Polütehnilise Instituudi toimetised, 590*.

Martin, D. A. (2020). towards *a sociotechnical reconfiguration of engineering and an education for ethics: A critical realist investigation into the patterns of education and accreditation of ethics in engineering programmes in Ireland* [Doctoral thesis, Technological University Dublin]. https://doi.org/10.21427/7M6V-CC71

Martin, D. A., Conlon, E., & Bowe, B. (2021). A multi-level review of engineering ethics education: towards a socio-technical orientation of engineering education for ethics. *Science and Engineering Ethics*, *27*, 60.

Marx, K. (1966). *Capital, vol 3*. Lawrence and Wishart.

McGowan, A. (2009). *Crafting the nation in colonial India*. Springer.

Metcalf, B. D., & Metcalf, T. R. (2012). *A concise history of modern India* (3rd ed.). Cambridge University Press.

Misa, T. J. (Ed.). (2010). *Gender codes: Why women are leaving computing*. Wiley/IEEE Computer Society.

Mital, K. V. (1986). *History of the Thomason College of Engineering (1847–1949): On which is founded the University of Roorkee*. University of Roorkee.

Muller, M. (2018). Decolonising engineering in South Africa-Experience to date and some emerging challenges. *South African Journal of Science*, *114*(5–6), 1–6.

Murphy, M., Christenson, S., & Conlon, E. (2022). Social justice at an Irish practice-based university: In or out of place? In S. H. Christensen, A. Buch, E. Conlon, C. Didier, C. Mitcham, & M. Murphy (Eds.), *Engineering, social sciences, and the humanities: Have their conversations come of age?* Philosophy of Engineering and Technology Series, 42. Springer.

Murphy, M., O'Donnell, P., & Jameson, J. (2019). Business in engineering education: Issues, identities, hybrids, and limits. In S. H. Christensen, B. Delahousse, C. Didier, M. Meganck, & M. Murphy (Eds.), *The engineering-business nexus*, Philosophy of Engineering and Technology Series, 32. Springer.

Nel, H., & Meyer, J. (2016). Attraction, education and retention of technical women in South Africa. In *IEEE international conference on industrial engineering and engineering management (IEEM)* (pp. 174–178). IEEE.

Nicolaou, I., Conlon, E., & Bowe, B. (2018). Into the deep: The role of paradigms in understanding engineering education for sustainable development. *Irish Journal of Social, Economic and Environmental Sustainability*, *1*(2), 23–41.

Nienkamp, P. (2015). Engineering national character: Early Land-Grant College science and the quest for an American Identity. In A. Marcus (Ed.), *Science as service: Establishing and reformulating American Land-Grant Universities, 1865–1930*. University of Alabama Press.

Oldenziel, R. (1999). *Making technology masculine: Men, women, and modern machines in America, 1870–1945*. Amsterdam University Press.

Õppeteatmik. (2002). *Bakalaureuse-, inseneri- ja rakenduskõrgharidusõppe teatmik: 2002/2003*. Tallinn.

Õppeteatmik. (2007). *Bakalaureuse-, inseneri- ja rakenduskõrgharidusõppe teatmik: 2007/2008*. Tallinn.

Pauw, K., Oosthuizen, M., & Van Der Westhuizen, C. (2008). Graduate unemployment in the face of skills shortages: A labour market paradox. *South African Journal of Economics*, *76*(1), 45–57.

Prakash, G. (1999). *Another reason: Science and the imagination of modern India*. Princeton University Press.

Puaca, L. (2014). *Searching for scientific womanpower: Technocratic feminism and the politics of national security, 1940–1980*. University of North Carolina Press.

Ramnath, A. (2017). *The birth of an Indian profession: Engineers, industry, and the state, 1900–47*. Oxford University Press.

Reich, L. (1985). *The making of American industrial research: Science and business at GE and Bell, 1876–1926*. Cambridge University Press.

Reynolds, T. (1992). The education of engineers in America before the Morrill act of 1862. *History of Education Quarterly*, (32), 459–482.

Roy, H. (1989). *A chemical engineer recalls: Life and times of Hiralal Roy* (T. Roy, Ed.). Indian Institute of Chemical Engineers.

Saarniit, L., & Pevkur, A. (2019). Teaching ethics in academic curricula. The case of five disciplines in Estonian public universities. *Ethical Perspectives, 26*(1), 33–58.

Sarkar, S. (2010 [1973]). *The Swadeshi movement in Bengal 1903–1908*. Permanent Black.

Seely, B. (1999). The other re-engineering of engineering education, 1900–1965. *Journal of Engineering Education, 88*(3), 285–294.

Seely, B. (2012). Engineering education. In *Leadership in science and technology: A reference handbook*. SAGE.

Shay, S. (2013). Conceptualizing curriculum differentiation in higher education: A sociology of knowledge point of view. *British Journal of Sociology of Education, 34*(4), 1–20.

Slaughter, J. (2002). The political economy of curriculum making in American Universities. In S. Brint (Ed.), *The future of the city of the intellect*. Stanford University Press.

Somerville, M. et al. (2005, February). The Olin curriculum: Thinking toward the future. *IEEE Transaction on Education, 48*(1), 198–205.

Sterling, S. (2004). Higher education, sustainability and the role of systematic learning. In P. B. Corcoran & A. E. J. Wals (Eds.), *Higher education and the challenge of sustainability*. Kluwer.

Sutherland, E. (2020). The fourth industrial revolution – the case of South Africa. *Politikon, 47*(2), 233–252.

Tsikata, M., & A. B. Sebitosi. (2010). Struggling to wean a society away from a century-old legacy of coal based power: Challenges and possibilities for South African electric supply future. *Energy, 35*(3), 1281–1288.

TTÜ. (2018). *Tallinna Tehnikaülikool 1918–2018*. TTÜ kirjastus.

Van Broekhuizen, H. (2016, August 16). *Graduate unemployment and higher education institutions in South Africa*. Stellenbosch Economic Working Papers Series, WP08/2016. https://bit.ly/2kWy9W1

Visvesvaraya, M. (1951). *Memoirs of my working life*. M. Visvesvaraya.

Wheelahan, L. (2010). *Why knowledge matter in the curriculum: A social realist argument*. Routledge.

Wickenden, W. E. (1929). Research in the engineering colleges. *Science, 69*(1786), 312–318.

Winberg, C. (2021). The making of engineering technicians: Ontological formation in laboratory practice. *Engineering Studies, 13*(3), 226–248.

Winberg, C., & Makua, M. (2019). 'We teach them to be university students': The role of peer educators in the provision of ontological access to higher education. *Critical Studies in Teaching and Learning*, 7(2), 1–19.

Winberg, S., & Winberg, C. (2017). Using a social justice approach to decolonize an engineering curriculum. In *IEEE global engineering education conference (EDUCON)* (pp. 248–254). IEEE.

Wisnioski, M. (2015). What's the use? History and engineering education research. *Journal of Engineering Education, 104*(3), 244–251.

Wolff, K., & Winberg, C. (2022). Curricula under pressure: Reclaiming practical knowledge. *Teaching in Higher Education, 27*(8), 1058–1067.

World Economic Forum (WEF). (2020). *The future of jobs report*. www.weforum.org/reports/the-future-of-jobs-report-2020/

Zajda, J. (1988). The moral curriculum in the Soviet school. *Comparative Education, 24*(3), 389–404.

3

Gaps, Resources, and Potentials for Growth in Comparative Engineering Education Research

Xiaofeng Tang, Jennifer M. Case, and Andrés Felipe Valderrama Pineda

1 Introduction

The international mobility of the engineering workforce has attracted the attention of policymakers, business leaders, and educators for decades (Committee on Prospering in the Global Economy of the 21st Century, 2007; Bourn & Neal, 2008). Recent challenges on the global stage, such as the COVID-19 pandemic, tensions among major geopolitical powers, and pressing climate crises, underscore the role of engineers in fulfilling national and international needs. These global developments require engineering educators to stay informed of ideas and practices of educating engineers in different parts of the world, mindful of the political, social, economic, and cultural dynamics of different countries. Notably, national ambitions for strengthening the domestic engineering workforce are all reshaping the global landscape of engineering education.

In this chapter, we argue that researchers in engineering education are confronted with crucial needs and great opportunities to generate knowledge about global engineering education by engaging in comparative studies. We define "comparative" as studies that involve more than one country and use the country as the unit for analysis (Turner, 2019). A comparative approach can help researchers develop a more grounded understanding of the global transmission and adaptation of ideas about engineering teaching and learning. International comparisons also test and reveal the limitations of taken-for-granted assumptions of education that have been unproblematized in the Western context (Carnoy, 2019). Furthermore, comparative work can facilitate the growth of an international community of engineering education research (EER). We suggest that comparative studies of engineering education will be enriched by drawing conceptual and methodological frameworks developed in the field of comparative education.

The rest of the chapter is structured as follows. The remainder of this section presents the needs and values of comparative studies of engineering education. To assist researchers who are interested in beginning comparative work, the next two sections introduce the brief history of the field of comparative education and three main approaches developed in that field. We then survey recent comparative EER published in three languages: English, Chinese, and Spanish. Exemplar studies in EER illustrating the application of each one of the three approaches of comparative education are discussed. The following section lays out the strengths and limitations of the three main approaches of comparative education in studying engineering education. The subsequent section discusses

DOI: 10.4324/9781003287483-4

potential contributions to be achieved through engaging comparative perspectives in EER. We conclude this chapter by calling for researchers interested in the global scope of engineering education to proactively embrace comparative work.

Before proceeding with our analysis, we briefly present our positionalities. The three authors are all transnational in their academic experiences: one is of Chinese origin who recently accepted a position in a Chinese university after studying and teaching in the USA for 12 years, one taught in her native South Africa for decades before accepting an academic position in the USA, and the third one is a Colombian scholar with graduate degrees from European institutions and currently holding a position at a Danish university. Two elements of the authors' experiences are related to the analysis presented in this chapter. First, all three authors have experienced and observed universities in our home as well as other countries gravitate toward, and sometimes look up to, the US models of engineering education, including ways of administering programs and curricula and structures of accreditation informed by ABET. Second, our personal experiences also taught us the importance of considering local context in making sense of engineering education policy, contents, and methods, despite the appearance of "global isomorphism" (Klassen & Sá, 2020). Consequently, we were struck by the limited availability of high-quality comparative studies in the current engineering education literature, an endeavor we consider critical in understanding engineering education in various local contexts.

1.1 Moving Out of the National Silo

Several global developments have made comparative research of engineering education both more feasible and imperative. Firstly, the influx of international students in countries with renowned engineering education systems opens questions about engineering education in these students' countries of origin. This is worth noticing when a significant portion of international students in engineering are graduate students who have completed undergraduate training at home (Granovskiy, 2018). The literature on this topic tends to focus on accommodating and supporting international students in the host countries, whereas few studies examine the national systems of engineering education that produced them (Zhu & Cox, 2015). Secondly, as many engineering programs aspire to produce graduates for multinational corporations and/or cross-cultural teams, the teaching of foreign languages and cultures, along with topics like globalization, has become increasingly available in engineering programs. However, publications in the realm of "international engineering education" seem to focus primarily on pedagogical and organizational strategies that assist with students' cultural exchange (abundant examples can be found in the *Journal of International Engineering Education*); they tend to be less focused on investigating systems of engineering education in different national or cultural contexts.

Engineering educators tend to focus on the immediate concerns in their domestic contexts rather than the historical and spatial dimensions of engineering education as a globally connected enterprise, topics that tend to attract the attentions of historians. The history of cross-national influences on engineering education goes back to the colonial period when nations prioritized the production of engineers for colonial expansion and economic gain (Pedrosa & Kloot, 2018). While the training of engineers was directly governed by many local and national governments because of its importance for national and regional security and economic development, cross-national exchanges of ideas and models of engineering education have played a significant role throughout history, with French – and, to a lesser extent, German – models assuming earlier international leadership, which were then succeeded by British and American models in the 19th and 20th centuries, as newly independent nations sought to create their own systems of engineering education (Karvar, 1995; Reynolds & Seely, 1993).

Since the 1990s, significant changes have taken place in the global landscape of engineering education, caused in part by massive growth in engineering enrollments in emerging economies, most notably the economic powerhouses in the BRIC countries – Brazil, Russia, India, and China. In 2009, the total enrollment of engineering students in the BRIC countries was 75% more than the total of engineering students in the USA, Europe, South Korea, Japan, and Australia, countries that up to the 1990s had clearly dominated global engineering education (Loyalka et al., 2014). Meanwhile, international agreements for engineering education accreditation have driven global convergences in how engineering education is governed throughout the world (Case, 2017; Klassen & Sá, 2020).

The field of EER has grown against this backdrop of the worldwide increase in engineering enrollment and global convergences in the governance and practice of engineering education. Interests in improving the national quality of engineering education – a key mission that propelled the birth and growth of EER in many countries – are catalyzed by widely perceived urges to stay ahead in intense competitions of technological innovation, engineering workforce, and higher education in the global market (Lucena et al., 2008; Cao et al., 2021). However, when measured against the goal of devising globally competitive engineering education, most engineering education researchers do not seem to focus much on their "competitors" abroad; thus, published studies of engineering education are mostly conducted within a single national context, usually that of the main author's country of residence.

Besides the pragmatic concerns noted earlier, the scarcity of systematic comparison of engineering education in multiple national and cultural contexts, in our view, indicates missed opportunities to deepen reflections on engineering education in the researchers' own countries and to increase the impact of EER in broader intellectual communities. Here the metaphor "the fish can't see water" serves as a relevant methodological reminder for educational research, as the robustness of theoretical insights derived from education in a single national context often necessitates triangulation with similar educational phenomena in other nations characteristic of different political, economic, and cultural dynamics. For example, it has been pointed out that the famous US sociologist of higher education, Burton Clark, was able to present an influential and convincing interpretation of the key features of US higher education only after his earlier comparative research of other higher education systems (Välimaa, 2008). A similar case might be made with the famous *Wickenden Report* published by the American Society for Engineering Education (ASEE), formally titled "A Comparative Study of Engineering Education in the United States and Europe" (Wickenden, 1929), which made influential arguments about US engineering education by comparing it with European systems.

To scholars who are keen to reflect on the aims and approaches of EER, an important strength of the comparative approach is thus its ability to interrogate – through revealing and comparing different practices – the "blind spots" in one's native system. For example, Western educational researchers have come to think quite differently about the distinction between deep and surface approaches to learning thanks to studies of high-performing students in China who extensively utilized memorization for learning (Kember, 2016). This revealing function makes comparative research a potent vehicle for engineering education researchers seeking to question underexamined notions, such as "evidence-based" education (Riley, 2017). Similarly, notions of diversity, equity, and inclusion that are emphasized by many engineering education researchers in North America and Europe might be enriched by the insights of educators, students, and researchers in Africa, Asia, and South America through a comparative lens. Perhaps more importantly, studying engineering education through a comparative lens has the potential to generate profound understandings of the relationship between engineering education and broader societal parameters, understandings that are of interest not only to engineering educators but also to a broader range of scholarly communities concerned about education and society. After all, the education of engineers is intimately connected to society, especially to the engineering profession, and the structural and cultural shape of these connections will vary from country to country.

To illustrate these arguments further, we briefly unpack a rare example of a rich comparative study in the engineering education literature. The exemplar study, titled "Competencies Beyond Countries: The Re-Organization of Engineering Education in the United States, Europe, and Latin America," was published in the *Journal of Engineering Education* and authored by a team with cross-national and interdisciplinary expertise in history, cultural anthropology, and engineering education (Lucena et al., 2008). The study investigates visions of engineering competency in different countries and regions, as well as how these visions were or were not actualized in standards of engineering accreditation. Unlike many other cross-national/regional comparisons of engineering accreditation systems, Lucena et al. (2008) do not focus on listing student outcomes in various accreditation standards. Instead, the article centers on a more analytical question: Why is it that similar attempts to create a unified, outcome-based engineering accreditation system succeeded clearly in the USA and partially in the European Union yet failed in Latin America? The authors did not take for granted the success of the US accreditation system, nor did they assume the "correctness" of the US model simply because ABET achieved domestic and international success in promulgating outcome-based criteria for accreditation. Seeking a more grounded interpretation of the cross-national/regional differences, Lucena et al. (2008) inquire into the historical processes and institutional dynamics that produced success in certain cases and failures in others and, in this process, reveal structural features that account for differing receptions of a similar – and globally converging – idea for assessing engineering education, that is, outcome-based education.

Lucena et al. (2008) illustrate several strengths of comparative studies of engineering education. First, the study demonstrates that the meanings of engineers' global competency depend on national priorities, which are reflective of domestic economic needs, the status of industrial development, and the political inclinations of the engineering profession. Second, the study shows that efforts to reform engineering education are in important ways mobilized or constrained by the interplay between governments, businesses, professional organizations, and engineering educators. These lessons remind us that key concepts embraced in the US-based EER community, such as outcomes, evidence, and readiness, are the likely results of negotiations among stakeholders in specific discourses of engineering education. Hence, comparative perspectives should not only help engineering education researchers stay informed of foreign practices but also enhance their ability to generate more grounded theories of engineering teaching and learning by extending the scope of analysis to encompass such negotiations. Lucena et al. (2008) also showcase how comparative research of engineering education can contribute to a broader understanding of education and society. While the ABET EC2000 reform and the Bologna Process indicated the ambitions and readiness of the USA and the European Union to embrace a globalized engineering workforce, the unsuccessful bid for accreditation in Latin America reflected political fragmentations in the region, where competing national agendas slowed down regional integration. The comparison of engineering accreditation systems in this study thus provided a compelling case for the uneven globalization of higher education.

2 Comparative Education: A Brief History

The field of comparative education has developed vibrant intellectual communities and solid research infrastructure since the mid-20th century, whereas the origins of the field vary by different accounts. Many scholars consider the publication of French official Marc-Antoine Jullien's *Plan and Preliminary Views for a Work on Comparative Education* in 1817 as the beginning of comparative education (Manzon, 2011). In their book *Toward a Science of Comparative Education*, Noah and Eckstein (1969) sketch the history of the field in five stages. The first stage consisted of "travelers' tales" that incidentally reported the status of education in foreign countries. In the second stage, emissaries were sent by governments to intentionally study education abroad, seeking useful ideas and practices that might be borrowed to improve education at home. The activities in the third stage similarly consisted of sending visitors abroad, but the purposes for the visits were less concerned with borrowing educational ideas than building understanding and collaboration across countries. During the fourth stage, scholars of comparative education went beyond documenting foreign educational phenomena to investigating deeper political, social, and cultural dynamics that shape education in the target countries. During the fifth stage, the now-established field of comparative education, following the examples of more established social sciences, like economics and sociology, embarked on the process of reinventing itself as a science.

Noah and Eckstein's hope for a science of comparative education was echoed by several influential comparativists in the 1950s and 1960s, including Bereday, Holmes, and King. Bereday introduced a structured process of comparing education "through stages of description, juxtaposition, analysis, and interpretation" (Turner, 2019, p. 14). The enthusiasm for the science of comparative education during this period was partly fueled by North American scholars' training in quantitative techniques and their belief in positivist epistemologies, which hold that one can discover general laws of the relationship between education and society by comparing data from different countries. In the 1970s and 1980s, the strong influence of positivism on comparative education dimmed as more critical Marxist and neo-Marxist perspectives took over the spotlight. Carnoy's studies of education and income in different countries during this period laid the foundation for the argument that education serves as a cultural instrument for wealthy countries, many of whom were former or present colonizers, to keep the colonized and economically exploited countries at bay (Carnoy, 2019). Within countries that gained independence recently, Carnoy further points out, access to education serves to retain the concentration of power and resources among social and economic elites. Since the 1990s, the field of comparative education has entered a stage of "heterogeneity" that witnesses the co-existence of multiple standpoints and methodological traditions. The increasing impact of globalization has also driven some scholars to study education systems that transcend national borders, while others, inspired by postmodernism and post-structuralism, focus on comparing local educational practices instead of pursuing grander, national-level narratives (Manzon, 2011; Turner, 2019). As Turner (2019) puts it, the comparativists "have come to accept that there may be many ways of conducting research in comparative education" (p. 25).

3 Major Approaches in Comparative Education

Given the heterogeneity of objectives, standpoints, and methodological traditions in the field, any attempt to classify the major "approaches" of comparative education is likely to encounter justified opposition. This section presents one of such imperfect classifications as an entry point for interested engineering education researchers. Amid the succeeding, overlapping, and co-existence of multiple traditions of comparative education, three main approaches have endured since at least the 1950s, each characterized by somewhat-distinct purposes, epistemologies, and methods (Manzon, 2014). The first one might be termed a "scientific" approach. As Noah and Eckstein (1969) suggest, the scientific approach of comparative education attempts to produce generalized laws about education and society. Supported mainly by positivist epistemologies, the scientific approach emphasizes hypothesis-testing using cross-national, quantitative data, such as years of schooling, learning achievements, and income of graduates. The second approach might be called "ameliorative," which finds its predecessors among the educational emissaries in the 18th and 19th centuries. Inspired primarily by pragmatic epistemologies, the ameliorators seek to identify the best policy and pedagogical practices that can be borrowed to improve education in their home countries. Methods favored by the ameliorators range from content analysis of policy to qualitative and quantitative analysis of instructional strategies. A third, "interpretive," approach is often endorsed by the humanist traditions of educational research, which seek to understand the cultural and social contexts behind educational phenomena. The underlying epistemologies for the interpretive approach are influenced by cultural relativism. The interpretivists often use qualitative methods, like ethnography, for unpacking deeper cultural meanings of educational practice. In the remainder of this section, we illustrate each approach with a well-recognized study in the field of comparative education.

3.1 PISA Scores and National Economic Growth: An Illustrative Case for the Scientific Approach

The rise of large-scale international surveys and tests provided powerful support to advocates of the scientific approach of comparative education, who seek to test hypotheses with multinational data. Since the 1960s, the International Association for the Evaluation of Educational Achievement (IEA) has played a significant role in advancing scientific comparison of educational achievement across countries. Beginning in 1997, the OECD's Programme for International Student Assessment (PISA) has become another powerhouse for comparing educational performance through standardized tests.

The OECD report "The High Cost of Low Educational Performance: The Long-run Economic Impact of Improving PISA outcomes" exemplifies a scientific approach to predicting the relationship between national economic growth and the cognitive skills of its workforce (Hanushek & Woessmann, 2010). In this study, Hanushek and Woessmann sought to accurately portray the relationship between economic growth and educational achievement, replacing outdated metrics like average length of schooling with more precise measurement of the cognitive skills of national workforce, which was calculated from PISA scores. The resultant modeling produces (alluring) predictions of potential economic growth that can be gained from improving the education of the workforce: a boost of average PISA scores by 25 points among OECD countries can result in a net gain of 115 trillion dollars in accumulative growth in GDP.

3.2 Curriculum Standards from Top-Performing Countries: An Illustrative Case for the Ameliorative Approach

Schmidt et al.'s (2005) study of curriculum coherence exemplifies the focus of the ameliorative approach on discovering best practices of education in foreign countries. The study was motivated by US educators' concerns for the lack of coherence among national, state, and local curriculum standards, namely, that eclectic political processes in the USA resulted in curricula that looked like arbitrary "laundry lists" instead of well-organized systems of knowledge (Schmidt et al., 2005). To demonstrate more coherent ways of setting curricula, Schmidt et al. (2005) analyze curriculum standards from top-performing countries and regions in the Third International Mathematics and Science Study (TIMSS). The study finds that mathematics standards in the TIMSS top-performing countries and regions demonstrated clear and logical progression of knowledge, beginning with simpler, more foundational content in the lower grades and gradually developing into more complex content in the higher grades. This way, suggested the authors, learning in the earlier years builds the foundation for students to tackle more complex mathematics as they proceed in the curriculum. In comparison, the US mathematics standards showed less logic: most topics were covered from grade 1 to grade 8, leading to possible repetition and limited depth in student learning (Schmidt et al., 2005). The US science curriculum standards showed a similar lack of coherence when compared with those adopted by the top performers in TIMSS. While the latter introduced different scientific topics at different stages, Schmidt et al. (2005) notice an "absence of a clear pattern" (p. 551) in the US science standards.

Based on the cross-national comparison, Schmidt and colleagues made two suggestions for improving curriculum standards in the USA. First, the authors suggested enhancing the role of disciplinary specialists, such as university professors and mathematicians, in shaping curriculum standards. In a second – and bolder – suggestion, Schmidt et al. (2005) propose the relinquishment of state and local standards, arguing that "coherence and rigor might only be possible in the US if curriculum standards are national in scope" (p. 556).

3.3 Preschool in Three Cultures: An Illustrative Case for the Interpretive Approach

"The research methods used [should be] sufficiently searching to probe beyond the observable moves and counter-moves of pedagogy to the values and meanings which these [moves] embody" (Alexander, 2000, p. 266). This quote characterizes the gist of the interpretive approach of comparative education, which is also exemplified in *Preschool in Three Cultures: Japan, China, and the United States*, a seminal comparative study of early childhood education (Tobin et al., 1989). The authors termed their method "multivocal ethnography," with which daily operations of preschools in Japan, China, and the USA were videotaped by the research team, and the videos were subsequently viewed and commented, in turn, by insiders (administrators, teachers, parents, and students) affiliated with the featured preschools and by other parents, preschool teachers, and education researchers from each of the three countries. Through juxtaposing preschool practices in different countries and commentaries on these practices by multicultural audiences, the authors created a "dialogue" that revealed culturally specific ideas about education. For example, after watching scenes of a "difficult" child in a Japanese preschool who frequently challenged authorities and engaged in aggressive behaviors, many Chinese and American viewers expressed concerns over the teacher's failure to intervene. However, the supervisors and teachers at the Japanese preschool featured in the video approved of the teacher's non-provocative choices as pragmatic.

Beyond reporting views and practices of preschool in different cultures, Tobin et al. (1989) produce profound theoretical insights by probing the socioeconomic contexts lying behind the multicultural views and practices. The book argues that underneath differing approaches to educating young children in three countries was a similar motivation to prepare children as members of "low-fertility, educationally competitive, industrial societies" (p. 197). The authors pointed out that the evolution of economic relationships in urban China, Japan, and the USA made it difficult to pass down material wealth to the next generation, leading parents to invest instead in the cognitive and emotional development of children. The authors also noted distinctive ways in which members of different cultures reacted to the intense parental attention placed on children because of low fertility rates; in other words, in societies where birth rates are low, the average parental attention on a child is expected to rise accordingly (Riley, 2018). While the consequential promotion of children's individuality and ego was well accepted in the American culture, Chinese parents reacted ambivalently to these changes, being sensitive to the tension between children's individualistic development vis-à-vis collectivist cultural values in Chinese society.

4 A Survey of Comparative Engineering Education Research: 2010–2020

This section surveys the status of comparative engineering education research published in English, Chinese, and Spanish in the decade between 2010 and 2020. The time range was chosen to complement the *Cambridge Handbook of Engineering Education Research*, which was published in 2014. Although the *Cambridge Handbook* does not include a chapter explicitly dedicated to comparative research, several chapters in the handbook (e.g., "Global and International Issues in Engineering Education") take note of varying institutional structures and styles of engineering teaching and learning in different countries. Therefore, our survey seeks to present more recent developments in comparative studies of engineering education. The choice of surveying publications in three languages used by most of the world's population indicates our attempt to exercise the spirit of comparatists, namely, to read and compare scholarly works in different cultural communities (indicated by the language of publication) as a way to understand the respective priorities, educational contexts, and intellectual approaches characteristic of EER in different parts of the world.

For works published in English, we used the search string [(comparative OR cross-national) AND "engineering education"] in Web of Science, in addition to manually screening the titles of all the articles published in the *Journal of Engineering Education* and *European Journal of Engineering Education* during this period. For publications in the Chinese language, we used the search string [比较 (comparison) AND 工程(engineering) AND 教育(education)] in cnki.net. For works in Spanish, we reviewed publications in *Revista de Ingeniería* (Colombia), *Revista International de Educación en Ingeniería* (México), and *Revista Iberoamericana de Educación en Ingeniería* (RIEI). The search in Spanish literature was complemented with a wider search on the Internet and through the authors' contacts in the region.

Two inclusion criteria were applied to the initial findings: (1) publications should explicitly address engineering education, and (2) publications should focus on postsecondary education. According to these criteria, publications that treat STEM education as one entity or examine engineering education in the K–12 context were not included in this analysis. The search resulted in 54 publications in total: 25 in English, 25 in Chinese, and 4 in Spanish. A summary of resultant references and the main objects of comparison reported therein is presented in Table 3.1.

Table 3.1 Summary of Comparative Engineering Education Research Published in English, Chinese, and Spanish between 2010 and 2020

Language	*Reference*	*Object of Comparison*
English	Palma et al. (2011); Grenquist and Hadgraft (2013); Bradley (2013)	Accreditation
	Ku and Goh (2010); Gong et al. (2011); Ku et al. (2011); Khattak et al. (2012); Tang and Lord (2012); Lunev et al. (2013); Case et al. (2015); Gardelle et al. (2017)	Curriculum/facilities
	Cerda Suarez and Hernandez (2012); Cao (2015); Holmberg (2016); Santos et al. (2018); Polmear et al. (2019)	Pedagogy/instructor
	Lau (2013); Carr et al. (2015); Duffy et al. (2020)	Learning achievements

(*Continued*)

Table 3.1 (Continued)

Language	*Reference*	*Object of Comparison*
	Lahijanian et al. (2010); Barnard (2012); Kinnunen et al. (2016); Oda et al. (2018); Capretz (2019); Colomo-Palacios (2019)	Student perspectives
Chinese	Ma et al. (2010); Pu et al. (2010); Song et al. (2012); Cui (2013); Li, L., et al. (2013); Li, Y., et al. (2013); Fang (2014); Wang, et al. (2014); Liu and Zhu (2015); Zhang (2016); You et al. (2017); Du et al. (2019); Hu (2020); Wang, et al. (2020)	Accreditation
	Guo and Zhi (2010); Zhao and Lin (2011); Liu (2012); Shen (2013); Feng et al. (2014); Zhu et al. (2015)	Policy/vision
	Wu and Xu (2010); Wang (2016); Zhuang et al. (2020)	Curriculum/facilities
	Luo and Fan (2018)	Pedagogy/instructor
	Zhao and Chang (2020)	Student learning
Spanish	Hamid Betancur & Torres-Madronero (2015)	Accreditation
	Zartha Sossa (2013); ASIBEI (2019); Duque & Rangel Espejo (2021)	Curriculum/facilities

4.1 Comparative Engineering Education Research in English

The comparative papers published in English examined a wide range of topics in engineering education. Objects of comparison included accreditation, curriculum, educational facilities, instructors' perspectives, pedagogical choices, student learning achievement, as well as student perspectives on engineering learning.

The growth of international agreements on engineering accreditation – most notably the Washington Accord – has attracted scholarly interest in comparing national accreditation standards and systems (although this body of work in English was less prominent than in the Chinese literature shown next). Researchers also compared curricula of similar programs in different countries, although the samples were usually confined to a limited number of institutions in each country. A significant portion of the comparative studies examines engineering teachers and students, reflecting the focus of EER in North America and Europe. This body of work examines instructors' beliefs and pedagogical choices as well as the abilities, achievements, and perspectives of engineering students. Notably, explicit comparisons of student perspectives appeared only in English literature (and not in Chinese or Spanish literature). Overall, when compared with the extensive body of EER published in the same period, the number of comparative studies was minuscule (Williams et al., 2018). We also note that all the journal publications in English listed in Table 3.1 appeared in the *European Journal of Engineering Education*, while the rest of the English publications appeared in conference proceedings. We found no explicit cross-national comparison of engineering education published in the US-based *Journal of Engineering Education* during this period.

4.2 Comparative Engineering Education Research in Chinese

Driven by a wish to build a strong national system of engineering education, researchers in China have enthusiastically studied best practices from global "leaders," that is, nations and institutions that were considered homes of world-class engineering education. This enthusiasm is echoed in a plethora of publications that examine professional organizations, accreditation standards, engineering curricula, and instructional methods from advanced industrial countries like the USA, the UK, France, Germany, and Japan, as well as renowned engineering institutions like MIT, Cambridge, and École Polytechnique. To assess the status of comparative engineering education research, we limited the scope of our analysis to publications that had an explicit comparative intent, recognizing that there are many more publications in Chinese that report on foreign ideas and practices of engineering education without necessarily comparing them across countries.

Between 2010 and 2020, most comparative studies of engineering education published in Chinese focused on accreditation, as can be seen in Table 3.1 in the preceding section. This topical focus coincided with the emergence and development of engineering accreditation in China: pilot accreditation of engineering programs began in 2006, followed by China's application to join the Washington Accord in 2009 and its acceptance as a full signatory in 2016. The practical need of developing and operating a national accreditation system motivated scholarly investigations of other nations with established accreditation systems. However, the pragmatic urge to understand the standards and procedures of accreditation seemed to overshadow more in-depth analyses of structural factors that gave shape to accreditation systems in various industrial societies (as, for example, seen in the Lucena et al. (2008) study reviewed earlier). Besides accreditation, the comparative literature in Chinese examined policy initiatives and educational visions that drove reforms of engineering education in different countries, along with comparisons of curricula, learning facilities, pedagogical approaches, and student learning.

4.3 Comparative Engineering Education Research in Spanish

We only identified three journal publications in Spanish between 2010 and 2020 that met our inclusion criteria. Hamid Betancur and Torres-Madronero (2015) describe documents and procedures of engineering accreditation in Colombia, Costa Rica, Chile, Mexico, and the USA. Zartha et al. (2013) compare a set of quantitative indicators for 18 institutions of higher education from members of the Organization of American States. Duque and Rangel (2021) report the effort of a university to benchmark its engineering education against programs of similar institutions in other countries. Besides peer-reviewed research publications, the report of Ibero-American Society for Engineering Education (ASIBEI) represented a robust comparison of systems of engineering education in nine Iberoamerican countries: Argentina, Brazil, Colombia, Ecuador, Spain, Mexico, Paraguay, Portugal, and Uruguay (ASIBEI, 2019). This report details characteristics of the engineering curriculum, role of instructors, profiles of prospective engineering students, and graduate placement in member countries of ASIBEI. Common features of engineering competencies across the nine countries are also summarized in the report, making it a rare exemplar of comparative engineering education research that goes beyond superficial analyses of numerical indicators. The report nonetheless surprised readers with its late arrival, given that Lucena et al. (2008) mention the efforts by ASIBEI to create a regional profile 11 years before the appearance of the final report.

Overall, the volume of cross-national comparisons of engineering education published between 2010 and 2020, detailed in this analysis, has been small compared with the total amount of EER works published during this period. The thematic foci of the comparative studies, however, showed some breadth, ranging from macrolevel issues of governance and accreditation, midlevel topics like curricula, to microlevel issues, like instructional choices and student performance. Notably, issues of diversity, equity, and inclusion, which received extensive attention in the EER community during the same period, were not yet reflected in the body of comparative work.

Considering the methods used for comparison, we found that most published comparative studies of engineering education during this period had confined themselves to simply "benchmarking" or "describing" relevant policy, curriculum, and instructional methods, while very few studies were carried out following the major methodological approaches of comparative education. This status of methodological underdevelopment seems to hold EER scholars back from systematically assessing engineering education across countries and from investigating the underlying structural forces that influence international similarities and differences in engineering education. That said, across this survey, we did manage to identify a few studies of comparative engineering education that demonstrated the values and potential of systematically utilizing methods of comparative education. We now turn to illustrate each of these three comparative research approaches with a closer look at selected EER examples that we identified through our survey.

4.4 Comparative Engineering Education Research Using the Scientific Approach

The scientific approach of comparative education was utilized in a few studies that compared the achievement and attributes of engineering students across countries. Zhao and Chang (2020) compare the learning behaviors of engineering students in China and the USA using data collected with two standardized instruments: the Student Experience in the Research University International Consortium (SERU-I) and the University of California Undergraduate Experience Survey (UCUES).

Zhao and Chang intended to characterize engineering students' learning behaviors and to compare these characteristics with those of students in science, humanities, and social sciences. Two forms of comparison were presented in their paper: first, a comparison of engineering students (both Chinese and American) with college students in non-engineering majors; second, a comparison of Chinese and American engineering students' learning behaviors. Due to the fact that the proportion of engineering student respondents was not comparable between the Chinese and American samples, the authors declared that they did not statistically compare the results between Chinese and US students (Zhao & Chang, 2020).

The study compared seven dimensions of learning behaviors, including (1) allocation of study time, (2) reflective learning, (3) team-based cooperative learning, (4) learning through interaction with faculty outside classes, (5) challenge-based learning,[1] (6) class participation, and (7) task completion–based learning (Zhao & Chang, 2020). The authors found that:

- Engineering students in both countries spend more time studying than non-engineering majors. Engineering students in China spend more time taking classes than peers in other majors, whereas US engineering students spend more time studying outside classes in comparison with other majors.
- In both countries, engineering students engage less in learning through reflection, class participation, and task completion than students in humanities and social sciences majors in the same country.
- Engineering students in China participate similarly with other majors in team-based and challenge-based learning, whereas US engineering students are more likely to work in teams than other majors and less likely to engage in challenge-based learning than science majors.
- In both countries, students of all majors score low on learning through interaction with faculty outside classes.

Overall, Zhao and Chang (2020) find no definitive patterns of learning behaviors that characterize engineering students and distinguish them from students in other majors. The authors interpreted this finding from three dimensions. First, most educational objectives in engineering focus on the lower end of Bloom's taxonomy (remembering and understanding), while higher-level skills like analyzing, synthesizing, applying, evaluating, and innovating are underemphasized. Second, drawing from the second author's own educational experiences in Chinese and American universities, the authors suggested that undergraduate engineering curricula in China's research-intensive universities have a clear "theoretical orientation," which leaves limited curricular space for practical learning. This observation was referenced to contextualize the result that engineering students in China spend significantly more time studying inside than outside classes. Finally, Zhao and Chang (2020) suggest that engineering students' inactive class participation in both countries might result in part from the prevalence of traditional, lecture-based modes of engineering teaching.

4.5 Comparative Engineering Education Research Using the Ameliorative Approach

Polmear et al. (2019) explore engineering educators' perceptions of ethics teaching in the USA, non-US Anglo countries (such as Australia and Canada), and across a wide range of Western European countries. The study centers on three key questions, delivered via an online survey. First, a close-ended question inquired whether instructors think students receive sufficient ethics education in their programs, to which the majority responded "no." A second, close-ended question asked the respondents to check off from a list ethics-related topics that were taught in their programs. A third question invited open-ended responses on general issues about teaching ethics to engineering students.

The study was implicitly driven by an ameliorative approach, as the US-based research team sought best practices from international counterparts that could be incorporated into engineering education in the USA. The authors compared responses from US instructors with those from non-US Anglo countries and with those from Western European countries, respectively. The first comparison showed that engineering ethics education in the non-US Anglo countries had greater coverage of macroethics topics, such as environmental protection, sustainability, risk, and practices of the engineering profession, than the US curricula. The authors suggested that the difference might reflect a greater prominence of macroethics topics in the accreditation standards adopted by non-US Anglo countries.

The second comparison showed that the Western European countries taught environmental protection and sustainability more than the US programs did, while making fewer references to topics like professional codes of ethics and safety. To interpret this latter difference, the authors drew on other research to argue that Western European countries, most notably the Netherlands, differed from the "traditional American approach" that focuses on professional codes and instead paid more attention to the social and environmental contexts of engineering as a way to broaden the scope of engineering ethics education.

Notably, the analysis of open-ended responses pointed to a broadly existing consensus among instructors in different countries that engineering students did not get sufficient education in ethics. Polmear et al. (2019) also find that educators across different countries faced challenges in getting students to value ethics education. Accordingly, the authors suggested that ethics should be integrated into technical coursework.

4.6 Comparative Engineering Education Research Using the Interpretive Approach

Zhao and Lin (2011) propose "macro policy systems" as a lens to interpret models of engineering education in different nations. The authors contended that to understand education systems that were designed to produce innovative and practical engineers, one ought not to confine the scope of investigation to colleges and universities but should examine the broader "education system" – ranging from preschool to higher education – and how education is supported by the policy system characteristic of political structures and processes, as well as the relationship between the elite and the mass in each national context. Hence, Zhao and Lin (2011) investigate the relationship between models of engineering education and characteristics of the labor market and social welfare policies in 18 countries and regions across the globe. According to this analysis, Zhao and Lin divided the 18 countries and regions into five models of engineering education:

- The **continental/conservative** model aims to produce a high-quality and practice-oriented engineering workforce through close collaboration between educational and business institutions in countries with proactive social welfare policies. Zhao and Lin (2011) name Germany, France, and Italy as exemplars of this model.
- The **Anglo/liberalist** model, typified by the USA, the UK, Australia, and Canada, prioritizes the education of engineering generalists, highlighted by critical thinking, innovation, and understanding of sociotechnical systems, to meet the demands of market-driven economies.
- The **Nordic/social democratic** model is exemplified by Finland, Norway, Denmark, Sweden, and Switzerland, countries with high levels of human capital and student autonomy. The Nordic/social democratic model of engineering education, as described by Zhao and Lin (2011), emphasizes theoretical learning and academic research for the preparation of high-tech workers.
- The **East Asian** model, seen in Japan, Korea, and Taiwan, stresses the role of engineering in serving national and regional (economic) needs. The authors suggested that the co-existence of Western ideas (e.g., autonomy) and Confucian doctrines (e.g., loyalty) led to a fusion of intellectual freedom and social responsibility in the training of East Asian engineers (Zhao & Lin, 2011).
- Finally, Zhao and Lin (2011) discuss engineering education in **developing countries** (e.g., India and Brazil) that had shared characteristics of industrialization with China. The authors suggested

that engineering education systems in these developing countries were emulated from advanced industrial countries, but these systems had not yet matured into distinctive models of their own.

5 Strengths and Limitations of Three Approaches for Comparative Engineering Education Research

The strengths and limitations of the three approaches of comparative education in the context of EER are summarized in Table 3.2. As we hope to convey in the following paragraphs, there is no one best way to conduct comparative research on engineering education; rather, choices of approaches and methods should be aligned with the purposes of the research. The scientific approach of comparison has the potential to enhance EER in three ways. First, the collection and analysis of multinational data (via surveys and tests) can contribute to understanding international students as well as the distinctive characteristics of domestic students in comparison to international peers. Second, the need to collect and compare data across countries brings about opportunities to form international collaborations. Third, as Zhao and Chang (2020) illustrate, cross-national comparison of engineering and non-engineering students could potentially reveal distinct features of engineering teaching and learning, which lays the groundwork for future research (e.g., studies of the "boundaries" of engineering education across national contexts).

Meanwhile, researchers following the scientific approach in comparing engineering education are confronted by two main limitations. Firstly, comprehensive international data collection can be costly and time-consuming. What is more, unlike in the case of primary and secondary education, at present there is no standardized international test for engineering; thus, international comparisons following the scientific approach are often confined to self-report surveys, which limits the scope and depth of comparison. The second major limitation stems from well-known critiques of the positivist epistemology underpinning the scientific approach: in short, hypothesis-testing based on quantitative data necessitates certain processes of "abstraction" that leave out many meaningful details. For example, the name "engineering degree" means different professional credentials and varying lengths of study in different countries. Variations of this kind are not easily visible in quantitative analysis, which is often favored by the scientific approach.

An important strength of the ameliorative approach is its congruence with existing practices of EER. After all, the improvement of engineering education has been a key driving force for EER, and numerous reports on engineering education in different countries have been produced in the spirit of improving engineering teaching and learning. Given the emphasis placed on the continuous improvement of engineering education in different countries, the ameliorative approach is also likely to enhance the practical relevance of comparative engineering education research. In addition, studies following the ameliorative approach are often driven by specific and focused questions in educational practice, which provide clear guides for the selection of objects for comparison. Hence, this approach is more accessible for researchers who do not have extensive training in comparative education but are familiar with the practical aspects of engineering education.

However, the accessibility of the ameliorative approach is maintained at the cost of methodological consistency. Indeed, improvement can result from many sources, and there is not a clearly delineated framework for conducting comparative research for the amelioration of engineering education. Yet the lack of a consistent methodological framework is likely to confine the research findings to piecemeal recommendations. A second limitation of the ameliorative approach relates to researchers' relatively superficial consideration of the context for implementing the educational practices learned from abroad, when research in engineering education has pointed out that the context of implementation can influence the success of educational innovation as much as the actual focus of the innovation (Litzinger & Lattuca, 2014).

Table 3.2 Strengths and Limitations of Three Approaches for Comparative Engineering Education Research

	Strengths	*Limitations*
Scientific approach	• Understanding students • International collaboration • Inspiring future research	• Cost and feasibility • Positivist epistemology
Ameliorative approach	• Congruence with existing practice • Practical relevance • Focus and accessibility	• Inconsistent methodology • Context of implementation
Interpretive approach	• Grounded-ness • Global trends	• Feasibility and affordability • Knowing why vs. knowing how

A major strength of the interpretive approach of comparative research is its "grounded-ness," that is, this approach not only attempts to describe educational practices in different countries but also seeks to understand and interpret their meanings and implications. Therefore, the interpretive approach helps avoid imposing decontextualized concepts of engineering education to different countries. Grounded interpretations and comparisons of national engineering education systems also reveal overall trends of global engineering education development. This is the second major strength of the interpretive approach.

Of course, it is not an easy task to decode educational culture abroad, which often requires time-consuming and sophisticated training in local knowledge and languages. The feasibility and affordability of interpretive comparison hence form an important limitation. Another challenge for the interpretive approach might be characterized as the conflict between "knowing why" and "knowing how." While the interpretive approach provides culturally relevant interpretations of engineering education policies, structures, and ways of teaching and learning, it is not necessarily easy to derive actions for improving engineering education from these interpretations.

6 Potential Contributions of Comparative Engineering Education Research

This section draws together our findings to argue that comparative studies have the potential to make important contributions to the research, practice, and governance of engineering education in six ways:

1 ***Improving understanding of domestic and international stakeholders.*** Through creating knowledge about policies, engineering teaching and learning, instructors and students, as well as engineering professions and related organizations in different countries, comparative studies help researchers and educators better understand and communicate with stakeholders in engineering education, especially with international faculty and students, and transnational employers that conduct business in the global market.

2 ***Revealing diverse national systems of engineering education.*** Comparative analyses of the various ways in which local political wills, economic imperatives, and cultural norms interact with the processes and outcomes of engineering education help reveal the global diversity of engineering education.

3 ***Providing space and languages for international EER collaboration and exchange.*** The presentation of multinational and cross-national cases creates space, as well as common languages, for researchers in different countries and regions to exchange research strategies, questions,

and findings, thus facilitating a more engaged and inclusive global EER community. Also, the promotion of a comparative research agenda is likely to encourage engineering education researchers to formulate productive collaboration with broader communities of researchers (in comparative education and beyond) who have expertise in cross-national comparison.

4 ***Unpacking international and national governance of engineering education.*** Through global economic and technological exchanges, the impact of domestic engineering workforce preparation is often seen beyond national borders. Meanwhile, the governance of engineering education is deeply rooted in the political and economic dynamics within the national context. Thus, comparative studies help engineering educators better grasp the dynamics between main players (e.g., international organizations, governments, professional organizations, corporations, and educational institutions) that shape the governance of engineering education. Given the strong ties between engineering education and industrial and professional organizations, comparative engineering education research has unique opportunities to produce knowledge about education governance that is not easily available in other domains of (higher) education research.

5 ***Exploring common and unique challenges of broadening participation in engineering education in different contexts.*** Engineering educators in many parts of the world endeavor to recruit students from broader populations than those who have traditionally dominated the profession. Comparative research might shed light on similarities and differences that are characteristic of challenges to broadening participation in engineering in different parts of the world. Whereas the compound challenge of diversity, equity, and inclusion in the USA has motivated some EER scholars to incorporate the concept of intersectionality, in countries like South Africa, Colombia, and Brazil, the challenge also needs to encompass differences in regions and economic status. The entry of immigrants occupies center stage in debates about justice and inclusion in some European countries, and an emerging agenda of gender equity is being explored by engineering educators in China.

6 ***Creating new knowledge on the interactions between engineering (technology) and society around the globe.*** Comparativists often remind us that the systems and objectives of education are shaped by societal expectations. In a time with an expanding and fast-changing global order, comparative studies of evolving notions of engineering competency in different countries would help us appreciate societal expectations of engineering in different parts of the world, and such appreciation lays the groundwork for articulating and assessing the role of engineering in addressing global challenges.

Table 3.3 displays examples in which comparative studies of engineering education reviewed in this chapter contribute to one or several of the potential areas presented in the preceding list. The table shows that existing comparisons of engineering education systems, pedagogies, and learning styles help contribute to understanding of stakeholders, national systems, governance, and societal expectations of engineering education, as well as to intellectual exchanges between international EER scholars. As already noted earlier, the comparative works reviewed in this chapter have not explicitly addressed broadening participation in engineering education.

7 Conclusion

This chapter attempted to demonstrate the untapped potential of comparative studies in engineering education. While exemplar studies show the value of cross-national comparison, our review of recent EER literature in three major publishing languages found a limited number of comparative studies, and thus, there is much scope for enhancing and expanding this body of work. Considering the present availability of online tools for surveys and interviews, it is unlikely that difficulties in accessing multinational data are the major impediment to comparative research. Instead, we suspect

Table 3.3 Contributions of Comparative Engineering Education Research Illustrated by Studies Reviewed in This Chapter

Potential Contribution of Comparative Engineering Education Research	*Illustrations from Studies Reviewed in This Chapter*
A. Understanding stakeholders	Polmear et al. (2019) contribute to the understanding of engineering instructors in the USA and in European countries. Zhao and Chang (2020) contribute to the understanding of engineering students in China and the USA.
B. Diverse national systems	Lucena et al. (2008) show how international practices and standards influence engineering education in Europe and America. Zhao and Lin (2011) exhibit how the global landscape of engineering education is enriched by different "styles" of economic, welfare, and professional policies.
C International collaboration and exchange	Polmear et al. (2019) and Zhao and Chang (2020) use standardized survey instruments that enable the comparison of multinational data on engineering teaching and learning.
D. Education governance	Lucena et al. (2008) and Zhao and Lin (2011) reveal characteristics of engineering education governance across different nations, regions, and cultures.
E. Broadening participation	N/A.
F Engineering and society	Lucena et al. (2008) demonstrate how conceptions of engineering competency are contextualized in geopolitical dynamics and national economic structures.

that the limitation might be attributed to the self-identity of the EER community, which often focuses on issues bounded in a national context, due in part to the priorities of funding agencies and academic journals. However, we contend that an excessively inward-looking research agenda might hamper the capacity of researchers' national systems of engineering education when knowledge of alternative ways of organizing and delivering engineering teaching and learning is lacking.

To assist engineering education researchers to engage in more systematic comparative work, we have provided an overview of the field of comparative education and introduced three major approaches that are often used in that field. Each of the three approaches has been applied in EER and, in each case, has generated productive findings. However, if EER aims to be a truly global field of scholarly pursuit, a comparative perspective is going to play a much more significant role than it currently does, assisting in engineering educators' appreciation and reflection on the opportunities and limitations present in their locations.

Note

1 Challenge-based learning in Zhao and Chang (2020) refers to student learning through addressing academically challenging tasks. The related survey items examined students' willingness to select academically challenging courses and projects, sometimes at the expense of their grades.

References

Alexander, R. (2000). *Culture and pedagogy: International comparisons in primary education*. Blackwell.
ASIBEI. (2019). *Perfil de ingreso y egreso del ingeniero iberoamericano*. ASIBEI.

Barnard, S., Hassan, T., Bagilhole, B., & Dainty, A. (2012). 'They're not girly girls': An exploration of quantitative and qualitative data on engineering and gender in higher education. *European Journal of Engineering Education, 37*(2), 193–204. https://doi.org/10.1080/03043797.2012.661702.

Bourn, D., & Neal, I. (2008). *The global engineer: Incorporating global skills within UK higher education of engineers.* Engineers Against Poverty/Development Education Research Centre, Institute of Education, University of London.

Bradley, C. (2014). Information use skills in the engineering programme accreditation criteria of four countries. *European Journal of Engineering Education, 39*(1), 97–111. https://doi.org/10.1080/03043797.2013.833173

Cao, G. H. (2015). Comparison of China-US engineering ethics educations in Sino-Western philosophies of technology. *Science and Engineering Ethics, 21*(6), 1609–1635. https://doi.org/10.1007/s11948-014-9611-3

Cao, Y., Ma, X., Case, J. M., Jesiek, B. K., Knight, D. B., Oakes, W. C., . . . Zhao, H. (2021). Visions of engineers for the future: A comparison of American and Chinese policy discourses on engineering education innovation. In *Proceedings of 2021 ASEE virtual annual conference*. ASEE.

Capretz, L. F., Waychal, P., & Jia, J. (2019). Comparing the popularity of testing careers among Canadian, Chinese, and Indian students. In *2019 IEEE/ACM 41st international conference on software engineering: Companion proceedings (ICSE-Companion)* (pp. 258–259). IEEE.

Carnoy, M. (2019). *Transforming comparative education: Fifty years of theory building at Stanford.* Stanford University Press.

Carr, M., Fidalgo, C., de Almeida, M. E. B., Branco, J. R., Santos, V., Murphy, E., & Fhloinn, E. N. (2015). Mathematics diagnostic testing in engineering: An international comparison between Ireland and Portugal. *European Journal of Engineering Education, 40*(5), 546–556. https://doi.org/10.1080/03043797.2014.967182

Case, J. M. (2017). The historical evolution of engineering degrees: Competing stakeholders, contestation over ideas, and coherence across national borders. *European Journal of Engineering Education, 42*(6), 974–986.

Case, J. M., Fraser, D. M., Kumar, A., & Itika, A. (2016). The significance of context for curriculum development in engineering education: A case study across three African countries. *European Journal of Engineering Education, 41*(3), 279–292. https://doi.org/10.1080/03043797.2015.1056103

Cerda Suarez, L. M., & Hernandez, W. (2012). Leadership and performance in higher education: A comparative analysis in Portugal and Spain. *European Journal of Engineering Education, 37*(6), 592–599.

Colomo-Palacios, R., Ben Yahia, N., & Larrucea, X. (2019). Gender diversity among computing students: Reflections from Norway, Spain and Tunisia. [Teem'19: Seventh international conference on technological ecosystems for enhancing multiculturality]. In *7th International conference on technological ecosystems for enhancing multiculturality (TEEM)*, Leon, SPAIN.

Committee on Prospering in the Global Economy of the 21st Century. (2007). *Rising above the gathering storm: Energizing and employing America for a brighter economic future.* National Academies Press.

Cui, J. (2013). Research on demands for higher engineering curriculum reform by society – based on the international comparison of the engineering professional organizations. *Higher Education of Sciences, 4*, 64–70.

Du et al. (2019). An international comparative study on accreditation standards for engineering master's education. *Journal of World Education, 476*, 48–52.

Duffy, G., Sorby, S., & Bowe, B. (2020). An investigation of the role of spatial ability in representing and solving word problems among engineering students. *Journal of Engineering Education, 109*(3), 424–442. https://doi.org/10.1002/jee.20349.

Duque-Duque, D. A., & Rangel-Espejo, Y. C. (2021). *Estudio comparativo del programa de Ingeniería Industrial de la Universidad Católica de Colombia con Universidades a nivel Europa, Asia, América, Oceanía y África.* Universidad Católica de Colombia.

Fang, Z. (2014). A comparative study of accreditation of engineering education in signatories of Washington Accord. *Higher Education Development and Evaluation, 4*(30), 66–76 + 119.

Feng et al. (2014). A study of key features of excellent engineering talents: An international comparative perspective. *Science and Technology in Chinese Higher Education, 5*, 48–51.

Gardelle, L., Gil, E. C., Benguerna, M., Bolat, A., & Naran, B. (2017). The spread of European models of engineering education: The challenges faced in emerging countries. *European Journal of Engineering Education, 42*(2), 203–218. https://doi.org/10.1080/03043797.2016.1241983.

Gong, X., Cardella, M. E., & Lei, Q. (2011). Comparative study of first-year engineering honors programs between US and China. In *ASEE annual conference and exposition*, Vancouver, Canada.

Granovskiy, B. (2018). *Science, technology, engineering, and mathematics (STEM) education: An overview*. Congressional Research Service.

Grenquist, S., & Hadgraft, R. G. (2013). Are Australian and American engineering education programs the same? The similarities and differences between Australian and American engineering accreditation procedures. In *ASEE international forum*, Atlanta, GA.

Guo, S., & Zhi, X. (2010). A comparative study of higher engineering education in China and the US. *Journal of Northwestern Polytechnical University (Social Sciences), 30*(1), 78–81.

Hamid Betancur, N. E., & Torres-Madronero, M. C. (2015). Accreditation of engineering programs in the region: A comparative analysis. *Revista Educación En Ingeniería, 10*(19), 80–89.

Hanushek, E. A., & Woessmann, L. (2010). *The high cost of low educational performance: The long-run economic impact of improving PISA outcomes*. OECD Publishing.

Holmberg, M., & Bernhard, J. (2017). University teachers' perspectives on the role of the Laplace transform in engineering education. *European Journal of Engineering Education, 42*(4), 413–428. https://doi.org/10.1080/03043797.2016.1190957.

Hu, D. (2020). Empowering the future: International comparison and standard reconstruction of engineering education professional accreditation system reform. *Journal of Tianjian University (Social Sciences), 22*(3), 223–229.

Karvar, A. (1995). Model reception in the domain of engineering education: Mediation and negotiation. *History and Technology, an International Journal, 12*(2), 81–93.

Kember, D. (2016). Understanding and teaching the Chinese learner: Resolving the paradox of the Chinese learner. In R. B. King, & A. B. Bernardo (Eds.), *The psychology of Asian learners* (pp. 173–187). Springer.

Khattak, H., Ku, H., & Goh, S. (2012). Courses for teaching leadership capacity in professional engineering degrees in Australia and Europe. *European Journal of Engineering Education, 37*(3), 279–296.

Kinnunen, P., Butler, M., Morgan, M., Nylen, A., Peters, A.-K., Sinclair, J., Kalvala, S., & Pesonen, E. (2018). Understanding initial undergraduate expectations and identity in computing studies. *European Journal of Engineering Education, 43*(2), 201–218. https://doi.org/10.1080/03043797.2016.1146233.

Klassen, M., & Sá, C. (2020). Do global norms matter? The new logics of engineering accreditation in Canadian universities. *Higher Education, 79*(1), 159–174.

Ku, H., Ahfock, T., & Yusaf, T. (2011). Remote access laboratories in Australia and Europe. *European Journal of Engineering Education, 36*(3), 253–268. https://doi.org/10.1080/03043797.2011.578244.

Ku, H., & Goh, S. (2010). Final year engineering projects in Australia and Europe. *European Journal of Engineering Education, 35*(2), 161–173.

Lahijanian, A., Devi, S. R., & Mukhapadhyny, B. (2010). Comparative attitude of civil engineering students towards environmental education in India and Iran. *Journal of Food Agriculture & Environment, 8*(3–4), 1168–1173. <Go to ISI>://WOS:000286040100087.

Lau, K., Agogino, A. M., & Beckman, S. L. (2013). Global characterizations of learning styles among students and professionals. In *ASEE international forum*, Atlanta, GA.

Li, L., Yan, X., Yi, S., & Chen, B. (2013). Analysis and assessment of basic standard for graduates based on professional accreditation of engineering education: Learning from basic standard for graduates in Occident. *Journal of Architectural Education in Institutions of Higher Learning, 22*(4), 27–33

Li, Y., Du, X., Xue, S., & Gao, X. (2013). Discussion on the internationalization idea for civil engineering by comparing the Sino-US civil engineering education. *China Science and Technology Information, 5*, 116–117.

Litzinger, T. A., & Lattuca, L. R. (2014). Translating research to widespread practice in engineering education. In A. Johri & B. M. Olds (Eds.), *Cambridge handbook of engineering education research* (pp. 375–392). Cambridge University Press.

Liu, H. (2012). The reasons why higher engineering education in France, the United States, Germany and Russia is "excellence". *University Education Science, 132*, 46–50.

Liu, Q., & Zhu, J. (2015). A comparison of standards for engineering master's in Europe. *Xue Wei Yu Yan Jiu Sheng Jiao Yu, 10*, 71–77.

Loyalka, P., Carnoy, M., Froumin, I., Dossani, R., Tilak, J. B., & Yang, P. (2014). Factors affecting the quality of engineering education in the four largest emerging economies. *Higher Education, 68*(6), 977–1004.

Lucena, J., Downey, G., Jesiek, B., & Elber, S. (2008). Competencies beyond countries: The re-organization of engineering education in the United States, Europe, and Latin America. *Journal of Engineering Education, 97*(4), 433–447.

Lunev, A., Petrova, I., & Zaripova, V. (2013). Competency-based models of learning for engineers: A comparison. *European Journal of Engineering Education, 38*(5), 543–555.

Luo, X., & Fan, C. (2018, February). Review of research on engineering ethics education in China and other parts of the world – a co-word analysis based on CNKI and WOS databases. *Forum of Chinese Science and Technology*, 169–179.

Ma, Y., Yang, X., & Zheng, X. (2010). International comparative study of graduate level engineering education accreditation. *Research in Higher Education of Engineering, 4*, 60–63.

Manzon, M. (2011). *Comparative education: The construction of a field.* Comparative Education Research Centre, Springer.

Manzon, M. (2014). Comparing places. In M. Bray, B. Adamson, & M. Mason (Eds.), *Comparative education research: Approaches and methods* (pp. 97–137). Springer.

Noah, H. J., & Eckstein, M. A. (1969). towards *a science of comparative education.* The Macmillan Company.

Oda, S., Yamazaki, A. K., & Inoue, M. (2018). A comparative study on perceptions of cultural diversity in engineering students. In *10th International conference on education and new learning technologies (EDULEARN)*, Palma, SPAIN.

Palma, M., de los Rios, I., & Minan, E. (2011). Generic competences in engineering field: A comparative study between Latin America and European Union. In *3rd World conference on educational sciences (WCES)*, Bahcesehir Univ, Istanbul, Turkey.

Pedrosa, R. H. L., & Kloot, B. (2018). Engineering graduates in South Africa and Brazil: A common good perspective. In P. Ashwin & J. Case (Eds.), *Higher education pathways: South African undergraduate education and the public good* (pp. 274–287). African Minds.

Polmear, M., Bielefeldt, A. R., Knight, D., Canney, N., & Swan, C. (2019). Analysis of macroethics teaching practices and perceptions in engineering: A cultural comparison. *European Journal of Engineering Education, 44*(6), 866–881.

Pu, Y., Yang, L., & Zou, J. (2010). A comparative study of engineering education accreditation in China and Japan. *Heilongjiang Education, 3*, 39–42.

Reynolds, T. S., & Seely, B. E. (1993). Striving for balance: A hundred years of the American Society for Engineering Education. *Journal of Engineering Education, 82*(3), 136–151.

Riley, D. (2017). Rigor/Us: Building boundaries and disciplining diversity with standards of merit. *Engineering Studies, 9*(3), 249–265.

Riley, N. E. (2018). Good mothering in China: Effects of migration, low fertility, and birth constraints. In D. Poston Jr. (Ed.), *Low fertility regimes and demographic and societal change.* Springer. https://doi.org/10.1007/978-3-319-64061-7_7.

Santos, B., Virtudes, A., Goncalves, J., & Amin, S. (2018). Multicultural education in civil engineering: Planning and transportation approach. In *10th International conference on education and new learning technologies (EDULEARN)*, Palma, SPAIN.

Schmidt, W. H., Wang, H. C., & McKnight, C. C. (2005). Curriculum coherence: An examination of US mathematics and science content standards from an international perspective. *Journal of Curriculum Studies, 37*(5), 525–559.

Shen, X. (2013). A comparative study of engineering education reform in China and Germany. *Journal of Jilin Normal University (Humanities and Social Sciences), 41*(3), 109–111.

Song, S., Yang, S., Fu, G., Xu, K., & Niu, Y. (2012). Comparative study on safety engineering education and accreditation criteria between China and U.S. *China Safety Science Journal, 22*(12), 23–28.

Tang, Y., & Lord, S. M. (2012). Comparison of practical training experiences for electronics engineers in China and the USA: Case study of Southeast University and the University of San Diego. In *2012 ASEE international forum* (pp. 17–19).

Tobin, J. J., Wu, D. Y., & Davidson, D. H. (1989). *Preschool in three cultures.* Yale University Press.

Turner, D. A. (2019). Comparative and international education: Development of a field and its method and theory. In C. C. Wolhuter & A. W. Wiseman (Eds.), *Comparative and international education: Survey of an infinite field* (pp. 11–28). Emerald Publishing Limited. https://doi.org/10.1108/S1479-367920190000036015.

Välimaa, J. (2008). On comparative research in higher education. In A. Amaral, I. Bleiklie, & C. Musselin (Eds.), *From governance to identity* (pp. 141–155). Springer. https://doi.org/10.1007/978-1-4020-8994-7_11.

Wang, H., Li, C., Zhao, S., & Li, X. (2020). A comparative study of educational certification and professional certification for registered engineers in civil engineering between China and the UK. *Journal of North China University of Water Resources and Electric Power (Social Science Edition), 36*(2), 46–52

Wang, J., Zhang, F., Xia, S., Wang, B., Zhang, J., & Tang, X. (2014). On the "substantial equivalence" of international accreditation and domestic accreditation for chemical engineering degree. *Research in Higher Education of Engineering, 5*, 1–4.

Wang, Y. (2016). A comparison of general education for engineering students in China and US: Cases of Harbin Institute of Technology and University Wisconsin Madison. *Asia Pacific Education, 10*, 116.

Wickenden, W. E. (1929). *A comparative study of engineering education in the United States and in Europe*. Lancaster Press.

Williams, B., Wankat, P. C., & Neto, P. (2018). Not so global: A bibliometric look at engineering education research. *European Journal of Engineering Education, 43*(2), 190–200.

Wu, Z., & Xu, L. (2010). The research on the international comparison and the curriculum standards of solar energy engineering education. *Research in Higher Education of Engineering, 6*, 77–81.

You, Y., Wang, Y., & Xie, H. (2017). A comparative study on western countries' engineering education professional certification standards. *Journal of Higher Education Research, 40*(2), 73–79

Zartha Sossa, J. W., Orozco Mendoza, G. L., Arango Alzate, B., Vélez Salazar, F. M., Cortes, I., Agudelo, A., . . . Ríos Jaramillo, L. M. (2013). Análisis comparativo de programas de pregrado en ingeniería industrial en algunos países miembros de la oea. *Latin American & Caribbean Journal of Engineering Education*, 7(1).

Zhang, Y. (2016). A comparative study on the criterion system of program accreditation of higher engineering education between US and Japan. *Shanghai Journal of Educational Evaluation, 1*, 52–55

Zhao, L., & Chang, T. (2020). Analysis on characteristics of engineering undergraduates' learning behavior in China and the USA. *Journal of Higher Education, 41*(5), 97–109

Zhao, X., & Lin, J. (2011). An international comparative study on engineering talents training models. *Research in Higher Education of Engineering, 2*, 33–41.

Zhu, J., & Cox, M. F. (2015). Epistemological development profiles of Chinese engineering doctoral students in US institutions: An application of Perry's theory. *Journal of Engineering Education, 104*(3), 345–362.

Zhu, L., Lv, Z., & Li, W. (2015). The "computing" strategy of the great powers: Computing engineering and talent cultivation vision of Germany, the US and Russia. *Research in Higher Education of Engineering, 4*, 10–12.

Zhuang, W., Zhang, P., Liu, H., & Wan, W. (2020). Mechanical engineering laboratory construction for accreditation: A comparison based on Chinese and Irish Mechanical Engineering Laboratories. *Jiaoyu Guancha, 9*(29), 94–96, 100.

4

Decolonization in Engineering Education

Jillian Seniuk Cicek, Stephanie Masta, Tom Goldfinch, and Bruce Kloot

1 Introduction

Within the last several decades, a sociopolitical movement has emerged centered on decolonization. This movement serves as a response to both the embedded and the ongoing practices of coloniality present within the modern state. At the same time, the notion of decolonization remains contested, particularly within education. To address the relationship between decolonization and education, Shahjahan et al. (2022) conducted a comparative review of decolonization literature across disciplines and global higher education. The authors noted that the meaning and implementation of decolonization varied based on context (i.e., sociopolitical setting, institutional type, curriculum, student, and instructor positionality). They distinguished between "inward-facing" and "outward-facing" initiatives, which had geographic implications:

> *Outward*-facing strategies of decolonizing placed an emphasis on looking outside of the institution to decolonize, by centring local/Indigenous communities, building within communities, and reallocating institutional resources to local Indigenous communities to actualize communities' priorities and vision. Such an outward approach was prevalent in regions like Central and South America, such as Brazil, Colombia, Mexico, Nicaragua, and Ecuador. . . . This outward approach differed from other global regions, such as Oceania (e.g., New Zealand), that prioritized *inward*-facing strategies – targeting changes *within* higher education institutions through classroom based curricular and pedagogical shifts as well as bringing Indigenous/local knowledge systems and practices into higher education institutions rather than empowering existing work within Indigenous communities.
>
> *(Shahjahan et al., 2022, p. 84)*

Shahjahan et al. (2022) identified significant absence of literature on decolonization, particularly in the STEM disciplines. Our chapter addresses this absence by providing an entry point into the conversation on decolonization within engineering education. We begin by discussing the varied meanings of decolonization, followed by a review and then overview of the decolonization literature within engineering education. We then consider the drivers for, and barriers to, engaging in decolonial work within engineering education and research. We end with a discussion of limitations and recommendations, followed by concluding remarks.

DOI: 10.4324/9781003287483-5

As this chapter centers on decolonization, we note our positionalities as authors. The writing team is composed of four scholars whose work is directly, or tangentially, related to engineering education and decolonization. One author identifies as Indigenous, and three authors identify as non-Indigenous. We come from geographic regions in Australia, Canada, South Africa, and the United States. We acknowledge that our worldviews and perspectives form our meaning-making of decoloniality within our specific contexts. What we write in this chapter is not meant to be an extensive global history on decolonization efforts but rather an overview of the way decolonization is encountered in engineering education via academic spaces and publications. Together, we take responsibility for our words and perspectives in this chapter and are accountable for what we share. We emphasize that we are not historians or experts in this area but rather engineering educators and researchers working to learn about and practice decolonization in our respective spaces. We recognize that this work requires lifelong learning, listening, honesty, humility, and individual commitment rooted in place.

2 The Meanings of Decolonization

In this section we present a general definition of *decolonization*, which provides some context before we move into a specific discussion of how decolonization manifests in engineering education.

In the strict sense of the word, "decolonization" refers to the release of a country or region from the status of a colony to become politically independent. The word came into use after World War II (Betts, 2004) to refer to the rapid process unfolding in the Global South, particularly in Africa. This process continued for three decades (1945–1975), and in this relatively short time span, the "great political and diplomatic structure" (Darwin, 2000, p. 7) of the colonial empires was dismantled, along with most of its apparatus (such as garrisons and treaty ports). However, this trajectory of decolonization mainly applies to a type of colonialism known as *exploitation colonialism* (Veracini, 2013), so named because the Indigenous inhabitants of the colonial territory are exploited for labor and natural resources are plundered for raw materials. For this type of colonialism, the number of settlers from the colonial empires is small compared to the Indigenous populations.

In the case of *settler colonialism*, large numbers of settlers immigrate to the territory with the intention of building permanent, self-supporting settlements. In the process, the Indigenous populations are subjugated and either decimated or assimilated into colonial culture (Wolfe, 2006). Compared to the wave of decolonization that occurred in most of Africa, the process in settler-colonized nations occurred much earlier. For example, the settler population in the United States declared independence from Britain in 1776 and was the first colony in the Americas to break away from its mother country. Decolonization in Canada, if we focus on the political process, occurred over several decades. Settler autonomy from Britain widened until it was concluded by the Canada Act as recently as 1983. Australia followed a similar trajectory with a steadily growing settler population since the first British colony was established in 1788. Australia became a self-governed settler federation, independent from Britain in 1901, though it remained very close in terms of trade, allegiances, and governance processes for many decades following.

What is left out of this trajectory of decolonization is Indigenous inhabitants who have endured conflict, dispossession of land, and cultural oppression for hundreds of years. Wilmer (1993, p. 5) indicates that "indigenous peoples represent the unfinished business of decolonization" (cited in Tuhiwai Smith, 1999, p. 7). In other words, while the international community might have recognized the sovereignty of a country when the settler population declared political independence, Indigenous peoples were disenfranchised and continue to suffer the damaging effects of settler colonialism. Even though agreements, political assurances, and even "protection" laws and treaties (in the case of Canada, New Zealand, and the United States, for example) have been drawn up (Dow & Gardiner-Garden, 1998), Indigenous peoples continue to be denied sovereignty, autonomy, and rights to remain on their lands and practice their culture to this day.

There have been recent political and social movements to reconcile the historical wrongdoings that have perpetuated present and lasting trauma on Indigenous peoples in settler colonial nations (e.g., the Truth and Reconciliation Commission of Canada and the *94 Calls to Action* (2015), and the Native Title Land Rights Act in Australia (AIATSIS, n.d.)). While for some these "decolonization" movements can be understood as involving inclusion, reconciliation, and Indigenization, Tuck and Yang (2012) take a self-determining approach and declare that "decolonization in the settler colonial context must involve the repatriation of land . . . that is, all of the land, and not just symbolically" (p. 7). Paperson (2014) agrees: "decolonization is not just symbolic; its material core is repatriation of native life and land, which is incommensurable with settler re-inhabitation of native land" (p. 124). In this way, "decolonization" has an array of meanings and interpretations from inclusion of Indigenous values and knowledges into Western structures (e.g., Gaudry & Lorenz, 2018) to "land back" (e.g., Tuck & Yang, 2012) that may not necessarily be reconciled.

These trajectories of decolonization take as their point of departure particular forms of colonialism – exploitation and settler colonialism. However, there are two further sets of literature that intersect them. The first emerged from a critique of the relationship of dependence between some African nations and previous colonial powers. Various authors questioned the exploitation, undue political control, and underdevelopment faced after decolonization (Nkrumah, 1965; Rodney, 1973; wa Thiong'o, 1986). While these countries nominally gained independence, an unequal relationship in the context of economic trade and political influence was seen to be a direct result of colonialism. The focus of this trajectory of decolonization is economic and political justice in the context of international relations.

The second set of literature emerged from Latin America and focuses on colonial knowledge structures and the dominant influence of Eurocentric thought (Quijano, 2007). The manifestation of colonialism in this sense is referred to as "coloniality" (Maldonado-Torres, 2007), with the undoing or "untangling" (Mignolo, 2007) of knowledge (and social, cultural, etc.) production from these knowledge (social, cultural) structures called "decoloniality" (Maldonado-Torres, 2016).

Given these multiple intersecting trajectories, it is understandable that "decolonization" has rapidly become a buzzword (Fomunyam, 2017a; Omanga, 2020) or perhaps a "plastic" word in the sense of van der Laan (2001) in that it is "alleable and all-purpose, ubiquitous . . . yet resilient, and remarkably durable" (p. 349). Despite Tuck and Yang's (2012) insistence that "[d]ecolonization is not a metaphor," "decolonization" is increasingly being applied in a symbolic and metaphorical sense. Calls to "decolonize the curriculum" or "decolonize engineering and engineering education" unfortunately can result in decolonization losing its impetus as a social and political movement. In response to this situation, we urge colleagues who are interested in working in this area to understand their colonial history and what is at stake in their specific locale.

3 Decolonization in Engineering Education: Literature Review

How is "decolonizing" conceptualized in engineering education literature? A natural literature review was conducted[1] through which we identified four categories or themes (with several subcategories) that characterize how decolonization lives in this literature. These four categories are: (1) decolonizing as . . .; (2) Indigenous knowledges in engineering and technology; (3) decolonizing engineering education – pedagogy and curriculum; and (4) decolonizing engineering education – engagement, integration, and impact.

In the following sections, we discuss the literature vis-à-vis these four overarching categories and their subcategories. Though *engineering education* remains central to our scope, we include references to STEM education where relevant. Generally, we cite each reference in only one category, although most conceptualizations of decolonizing are multifaceted. Authors' professional global contexts are noted in the textual citations.

3.1 Decolonization As . . .

The first category we discuss is *decolonizing as . . .*, where authors write about decolonizing in six subcategories that we have identified as *equality, emancipation, liberation, localization, co-creation, equity; a response to Western hegemonic domination, neoliberalism, and the global crises in the natural and human world; a paradigm shift, epistemic transformation, epistemic justice, social justice; a methodology; land back*; and *as/for sustainability*. In the following sections, these categories are elaborated on in the various ways and contexts they are expressed.

3.1.1 Decolonization as . . . Equality, Emancipation, Liberation, Localization, Co-creation, Equity

Decolonization is conceptualized by some authors as *equality, emancipation, liberation, localization, co-creation*, and *equity*. This work largely comes out of South Africa, with articles from Kenya, Zimbabwe, Brazil, and the United States as well.

Fomunyam (2017a, 2017b, 2018; SA) argues that decolonizing in engineering education is a liberating process that requires both faculty and students as co-creators in knowledge formation. It is about localization, equality, and emancipation: "putting African and South African experiences at the centre of engineering education and engaging engineering students in ways that are emancipatory and empowering" (Fomunyam, 2018, p. 1533). Cooke et al. (2019; SA) compare colonial and decolonial curricula, where a

> [a] colonialised curriculum is, by its very nature, one that is shackled by a narrow, parochial approach, and the dominance of a few over many. A decolonised curriculum should, therefore, attempt to break those shackles and to espouse diversity, plurality, and a deep appreciation of "other."
>
> *(p. 39)*

They argue that internationalization can advance decolonization despite its tendency to perpetuate a colonized episteme.

Gupta (2020; SA) argues that decolonization "implies knowledge transformation from western thought to local (African and Global South in context of African continent) in order to enable the sustainable development" (p. 3247). Decolonization is intellectualized by Pido et al. (2019, p. 93; Kenya) as "bring[ing] African technologies into focus as foundational, precedent-setting coequals with other technologies . . . from the earliest stone tools to mechanized, electrified and electronic equipment." Mudondo (2020; Zimbabwe) emphasizes the value of African knowledge to technical and vocational education and training and its "ontological role" in socioeconomic prosperity.

Decolonial engineering is presented by Cruz (2021a; Brazil) as "committed to empowering people and constructing with them other possible sociotechnical orders (and ethical-political realities)" (p. 701). The author examines Brazilian grassroots engineering (GE) that emerged this century based on the work of Paulo Freire (2005). GE is linked to university technical teams, where knowledge is co-constructed with local communities, empowering communities' self-determination. GEs are considered decolonial "for their theoretical basis and intentions, methodological implementation in engineering interventions, and emancipatory outcomes" (701) and can be likened to participatory design and design for values. Cruz (2021a; Brazil) defines decolonial engineering education as a choice and a challenge:

> As Latin-American liberating and decolonial tradition says, other worlds (or socio-technical orders) are possible and have always been. However, bringing about such worlds requires bold steps and political choices. In engineering practice, teaching, and research, one such step or

> possible way is going decolonial. Taking or picking it, though, means going against the status quo, which makes inevitable fighting for being academically acknowledged, institutionally valued, and having one's research and interventions funded.
>
> *(pp. 701–702)*

In Cruz (2021b; Brazil), the author theorizes that technical design plays an important role in overcoming colonialism, conceptualized as a "triple mutually reinforcing and shaping imprisonment" of power, knowledge, and being (p. 1,847). Guanes (2021; US) argues for decolonization for equity, writing, "We need to decolonize engineering education to foster collectivity, care, and empathy; we need to become an equitable community where diverse ways of knowing and being are acknowledged and valued."

3.1.2 Decolonization as . . . a Response to Western Hegemonic Domination, Neoliberalism, and the Global Crises in the Natural and Human World

Decolonization is conceptualized by other authors as *a response to Western hegemonic domination, neoliberalism, and the global crises in the natural and human world*. Work in this area is written by authors from the United States, Canada, India, Singapore, Peru, Columbia, United Kingdom, Chile, and Mexico.

Daza (2012; US) presents a Spikvakian decolonizing perspective to challenge the neoliberal scientism that has infiltrated the NSF grants culture, arguing that neoliberal scientism is "colonizing research – narrowly defining what is and is not knowledge production and who are and are not researchers" (p. 773). Cole and O'Riley (2017; Canada, Peru) argue for a dismantling of Western STEM epistemology and knowledge imperialism, including the way in which knowledge is shared by re(storying) STEM with Indigenous knowledges:

> At this time of escalating global ecological crisis and social inequity, there is an urgent need in education for Indigenous knowledges and ways of thinking to reimagine and reshape the mainstream "progress narrative" of capitalist-rationalist modernity that has privileged mind over body, heart and spirit; human over more-than-human; and overlooked the knowledges, worldviews and self-determination of Indigenous peoples as well as their more-than-human relations. . . . This paper is a call to widen the Eurocentric knowledge base and framework of mainstream education to include as "equivalent" Indigenous and other Other(ed) worldviews and epistemologies. This (re)storying of STEM is based on the teachings of our Indigenous ancestors and current research with Indigenous communities in British Columbia and Peru as we work to regenerate more complex, culturally-inclusive possibilities for living together on a shared and finite planet.
>
> *(p. 24)*

Lunney Borden and Wiseman (2016; Canada) push back against STEM as a driver for the "economic imperative" (p. 140) and their observations that attempts to integrate STEM stories remain based in Western assumptions and philosophies. Drawing on their engagement with Indigenous peoples and communities, they reconceptualize STEM "not as a framework for teaching and learning but rather as an artifact that emerges from teaching and learning" and encourage "telling different stories in relation to what in English we refer to as *STEM* . . . and in languages indigenous to Canada as something more like learning or living or being in the world in particular places" (p. 140). Mutch et al. (2021; Canada) question the worlds we teach in engineering design "as a starting place for settler-descendent North American educators to begin a self-critique of current approaches that need continued work" (p. 1). Ochoa-Duarte and Peña-Reyes (2020; Columbia) offer *buen vivir*, based on Latin American ideas and centered on biocentrism, postcapitalism, decolonialism, and depatriarchalization, as an alternative to the Global North–centered UN Sustainable Development Goals (SDG) and to support the sociotechnical education of engineers. They present the duality of engineering as a "hegemonic system" or, through principles of ethics, integrality of knowledge, interculturality, autonomy, and sustainability, as "*buen vivir* to communities" (p. 3).

Raina (2017; India) writes about the first decades of decolonization and the Cold War and the establishment of the elite institutions of technology in India and what Krige (2006) calls the "Americanisation of Higher Education" (as cited in Raina, 2017, p. 49). Similarly, de Roock and Baildon (2019; Singapore) make a case for "decolonized or desettled STEM perspectives" in Singapore, drawing attention to the "neoliberal underpinnings of STEM discourse" in an online educational portal used for primary and secondary students. They argue that this pedagogy "leverages student interests, self-discovery, and diverse identities to recruit them into (neoliberal) figured worlds of learning, work, and citizenship that narrowly imagine a disciplined society and workforce appealing to global capital" (p. 285). Eichhorn (2020; UK) writes about the false narrative of Western engineering and technical superiority and the "developed versus undeveloped world" construct perpetuated in the Global North. He argues that "our exploitation in the Global North, which assumes historical superiority as a basic premise, will fail in tackling major issues. Change is needed through a decolonization of engineering projects, and western engineering curricula that are used to train future professionals" (p. 204).

Rodríguez-Seeger et al. (2021; Chile) intellectualize decolonization as a reaction to/against Western hegemony and a response to the global crises and their local impacts. They describe the Indigenous Peoples Program (PPI), an Indigenous-centered engineering program formed in 2019 at Universidad de Chile which aims to revalue Indigenous knowledges and cultures and address the "environmental disasters, material poverty, cultural loss and spiritual emptiness" left in the wake of the dominant, hegemonic culture. Castillo-González et al. (2021; Mexico) warn of the dangers and coloniality of Western technology:

> Digital platforms, machine-learning, and big data analysis are functional to an economic, political, and social system that is based on a capitalist, western, and patriarchal hegemony that exploits the resources of the South of the world, concentrates wealth, erodes the Commons on the Internet and racializes, sexualizes, and discriminates against various social groups through sophisticated algorithmic practices.
>
> *(p. 369)*

They argue that a new decolonial, global, ethical engineering education is required to "dismantle the violent and colonizing practices that underlie data capitalism" (Castillo-González et al., 2021, p. 370).

3.1.3 Decolonization as . . . a Paradigm Shift, Epistemic Transformation, Epistemic Justice, Social Justice

Others conceptualize decolonization as *a paradigm shift, epistemic transformation, epistemic justice,* and *social justice*. We see this work in South Africa and the United States.

In the South African context, Graham et al. (2017; SA) conceptualize decolonization as unearthing the power imbalances in the making of knowledge, which requires an epistemic shift: "Changing content, curricula and academic structures will not decolonize higher education, rather the recognition of the power imbalances inherent in the production and validation of knowledge is required. If that is achieved the former can follow" (p. 98). Fomunyam (2018; SA) argues:

> Decolonizing in engineering education is not integrating or the token inclusion of indigenous ceremony, but rather, about a paradigm shift from a culture of denial to the making of space

> for indigenous political philosophies and knowledge systems as they resurge, thereby shifting cultural perceptions and power relations in real ways.
>
> *(p. 1533)*

Walwyn and Combrinck (2021; SA) demonstrate that epistemic transformation – "chang[ing] those aspects of the knowledge system that cause psychological exclusion, oppression and unworthiness" (p. 600) – is central to decoloniality and the teacher is the catalyst for epistemic justice. They argue that "all aspects of the curriculum need to be reconsidered and reconceptualized for students who come from different ontologies" (p. 607).

Mejia and Pulido (2018; US) offer "a new paradigm by inverting the logic portrayed in many studies involving research that identifies Latinx as a monolithic group" (p. 3). They rewrite " 'what knowledge do Latinx students need?' to 'what do Latinx students offer to the construction of knowledge in engineering?' " (p. 3). They introduce *rasquachismo* – the resilience and ingenuity in daily life manifesting in one's own cultural production, identity, and survival. This is an assets-based paradigm valuing what Latinx students bring to engineering education to support "participation and inclusion for underrepresented students" and "expand the taxonomy of engineering education research" (Mejia & Pulido, 2018, p. 3). Riley (2018; US) offers Song's justification for multiculturalism, "freedom from domination," as one approach to a "culturally responsive engineering," which may look like "a decolonizing engineering, queer engineering, feminist engineering, crip, and so on" (p. 11). She offers this as an "invitation to explore deeply what might constitute epistemic justice for engineering and its disciplines of knowledge" (Riley, 2018, p. 11) and makes an argument for addressing colonialism (among other injustices) by understanding systemic generational injustice. Moore et al. (2021; US) decry a reviewer's centering of Western ways of knowing, making knowledge, and communicating, citing this as "a form of epistemic violence" and "cultural imperialism" (p. 5).

3.1.4 Decolonization as . . . a Methodology

Decolonization is also intellectualized as a methodology. This characterization is predominantly taken by authors publishing in the United States, with one publication from Mexico, and with many authors referencing decolonizing methodologies modelled by other authors. For example, references were made to González y González and Lincoln's (2006) decolonizing qualitative research, for example, Villa and Gonzalez (2011; Mexico) and Velarde et al. (2021; US). Carrigan et al. (2018; US) referenced Harrison's (1991) decolonizing anthropology. Several authors reference Tuhiwai Smith's (1999, 2012) decolonizing Indigenous methodologies, for example, Affolter et al. (2019; US), Baker et al. (2021; US), and Walther et al. (2015; US).

Narratives and storytelling are also put forward as decolonizing approaches. Pawley (2013; US), who cites Tuhiwai Smith (1999) and Denzin et al. (2008), discusses narratives as a decolonizing methodology "that have a history of use in indigenous communities and communities of color (as indeed in white communities)" (p. 10), and the responsibilities that come with using narratives in research. Foster's (2016; US) doctoral dissertation is a narrative, feminist, intersectional study of three Indigenous women who work and lead as engineers and technologists and whose lived realities lift nontraditional Native American voices in engineering and promote the power of story. Handley and Moje (2018; US) use narratives and counter-narratives of Latina youth on engineering problem definition to delaminate engineering identity, referencing McCarty et al. (2013) decolonizing activist ethnography.

3.1.5 Decolonization as . . . Land Back

Authors in South Africa and the United States bring forward decolonization as "land back." Zembylas et al. (2019–20; SA) write about decolonizing the university, arguing it is not just about decolonizing the knowledge produced but also about decolonizing the physical space – the land – referencing "Decolonization Is Not a Metaphor" by Tuck and Yang (2012) – and all physical artifacts on the land. They write:

> South Africa has made little progress in any aspect of redressing the legacy of dispossession. Despite billions of rands that have been spent there is very little to be shown in terms of the real transformation of the land ownership regime.
>
> *(p. 28)*

Jalali and Matheis (2017; US) cite a "spectrum of views in liberation" with Fanon's (2004) prioritization of "taking control of the land" (p. 10) at one end of the spectrum of decolonization. Valle et al. (2021; US) cite Paperson's (2014) view of decolonization as "the repatriation of land, the regeneration of relations, and the forwarding of Indigenous and Black and queer futures" (p. 9).

3.1.6 Decolonization as . . . as/for Sustainability

Finally, decolonization is conceptualized by some authors *as*, or *for*, *sustainability*. We see this in New Zealand, where Morgan (2008; NZ) explores the mātauranga Māori worldview for a holistic approach to sustainability assessment. In Canada, Mante et al. (2020; Canada) write about advancing engineering education towards sustainable development through the integration of Indigenous knowledges, perspectives, and design principles.

3.2 Indigenous Knowledges in Engineering and Technology

A second category in the literature was classified as *Indigenous knowledges and technology*. There are three subcategories here, including *place-based knowledges*, *decolonizing frameworks*, and *decolonized/ Indigenous technologies*.

3.2.1 Place-Based Knowledges

Several authors in the United States and one in Canada discuss *place-based knowledges* as a subcategory of *Indigenous knowledges in engineering and technology*. In the United States, Page-Reeves et al. (2019; US) offer wayfinding to understand Indigenous STEM professionals' success in historically White institutions, cultural reconnection, and accumulation of "experiential wisdom" (p. 178). Howard and Kern (2019; US) explore conceptualizing wayfinding as support for a multicultural approach to STEM education (p. 1,135). McGowan and Bell (2020; US) conceptualize engineering education as the development of sociotechnical literacy at the intersection of critical design theory and critical pedagogies of place:

> [A] two-way process of decolonizing the designed world by reading the inequities embodied in designed and constructed spaces in order to restory these spaces for more just and equitable redesign that looks at engineering problems holistically and with a lens on their differing impacts on individuals within communities.
>
> *(p. 995)*

In Canada, Marker (2019; Canada), critiques this notion of "success" as "disquieting decolonization" (p. 1,149), engendering Native Americans into colonial STEM. He argues rather for the centering of Indigenous place-based *knowledges* to advance STEM and transform and decolonize universities.

3.2.2 Decolonizing Frameworks

A second subcategory of *Indigenous knowledges in engineering and technology* is *decolonizing frameworks*. Here, we see several articles from authors in the United States, as well as articles from South Africa, Canada, and the United Kingdom.

In the United States, Foster and Jordan (2014; US) describe a Navajo framework for learning and evaluate Navajo and Western engineering education to understand the philosophies and epistemologies within both. The framework is offered to support the recruitment and retention of Navajo students to engineering and increase diversity, which they argue is needed to solve "modern-day grand challenges" (p. 3). Lakota teachings of the Four Directions prayer and the Medicine Wheel are shared in Catalano (2016; US) to enrich an engineering design paradigm. Eglash et al. (2020b; US) describe a workshop with high school students and offer a framework for translating Indigenous knowledge systems in a generative cycle to diversify STEM inputs and outputs and resurge Anishinaabe ways of being and knowing. This work is further discussed as transformative decolonization in Eglash et al. (2020a; US). Velez et al. (2022; US) present an Indigenous evaluation framework developed by an Indigenous consortium, grounded in Indigenous worldviews, to evaluate an environmental program. They quote Waapalaneexkweew (2018) to describe how "Indigenous evaluation is grounded in being 'caretakers of knowledge, community, or family . . . and relational interactions and responsibilities to all things in nature, the spirit world, and each other'" (p. 1), and is absent of the domination and power dynamics of Western evaluation methods.

In South Africa, Winberg and Winberg (2017, SA) argue that a systematic approach is required and suggest approaching curricula as an activity system to determine where change is needed. They recommend a social justice framework. In Canada, Seniuk Cicek et al. (2019b; Canada) discuss reinterpreting engineering graduate attributes with the Anishinaabe sacred hoop framework. In the United Kingdom, Skopec et al. (2021; UK) advance that decolonizing efforts must be both grass-roots and institutionally led and "fragility reactions" must be managed. They propose a framework for epistemic fragility – meritocracy, individualism, centrality, objectivity, and authority – to interpret faculty responses to decolonization.

3.2.3 Decolonized/Indigenous Engineering and Technologies

Decolonized/Indigenous engineering and technologies is the third subcategory in *Indigenous knowledges in engineering and technology*, where we find articles by authors in the United States and in Canada.

In the United States, Hess and Strobel (2013; US) conduct a scholarly literature search for examples of "ethno-engineering" and center Indigenous engineering on the context of social justice. They discuss sustainability as implicit in ethno-engineering: "Often-times, existing indigenous engineering strategies have proven to be environmentally sustainable . . . and although the explicit notion of sustainability may be absent, implicitly, sustainability concerns are commonly present in the broader holistic worldview of the indigenous people" (p. 59). Droz (2014; US) describes a bio-cultural design method that brings together Anishinaabe ways of being and knowing with ecological engineering "to create resilient designed systems," critical for "building allied strategies for bioregional resilience and sustainability" and "develop[ing] sustainable, resilient communities and nations" (p. 122). She explains the integrity of place to the Anishinaabek:

> [I]ndigenous resilience is an emergent property of the interconnected relationships within a place. . . . Knowledge of places is linked to knowledge of self and community, and the health of places is inextricable from the health of people and community. It is land-based knowledge – inextricable from community relationship-building – that fosters the resilience of the Anishinaabek.
>
> *(Droz, 2014, p. 122)*

Mejia and de Paula (2019; US) discuss the cultural and social practices and knowledges of the Mbyá-Guaraní of South America and how Indigenous engineering is enacted to broaden the conception of engineering beyond a Western-centric determination. They demonstrate how Indigenous knowledge is "participatory, experiential, process-oriented, and ultimately spiritual" (15, p. 36) and share Battiste's (2011) conceptualization of Indigenous knowledge, which is:

> [F]ar more than the binary opposite of Western knowledge . . . Indigenous knowledge benchmarks the limitations of Eurocentric theory – its methodology, evidence, and conclusions – reconceptualizes the resilience and self-reliance of Indigenous peoples, and underscores the importance of their own philosophies, heritages, and educational processes. Indigenous knowledge fills the ethical and knowledge gaps in Eurocentric education, research, and scholarship.
>
> *(as cited in Mejia & de Paula, 2019, p. 3)*

Mejia and de Paula (2019) argue that "the decolonization of the [engineering] curriculum is necessary to develop critically conscious engineers" (p. 5). In Canada, Fritz et al. (2011; Canada) highlight Indigenous design and technologies, and Kinsner (2015; Canada) references *Best Practices Using Indigenous Knowledge* (Boven & Morohashi, 2002).

3.3 Decolonizing Engineering Education – Pedagogy and Curriculum

Decolonizing engineering education – pedagogy and curriculum is the third category we characterized in the decolonizing literature. It has five subcategories, which are *pedagogical responses*, *decolonizing service learning*, *engineering/STEM outreach*, *guiding frameworks*, and *training*.

3.3.1 Pedagogical Responses

The first subcategory of *decolonizing engineering education – pedagogy and curriculum* is *pedagogical responses*, with a concentration of authors writing from Australia, South Africa, and Canada, and with one article from the United States.

In Australia, Goldfinch et al. (2010; AU) develop three tutorials designed to foster cultural awareness in engineering students using critical pedagogies and principles of cultural influence. Aboriginal peoples, knowledge, and culture are emphasized specifically via a guest lecturer. Authors consciously address students' biases and stereotypes and work to foster an attitude of equality for Indigenous knowledges and respect for Indigenous cultures. Motivated by a history of poor Indigenous consultation and engagement in engineering mining projects, Goldfinch et al. (2014; AU) facilitate undergraduate students to conduct research on past and present approaches to Aboriginal culture and heritage in engineering, considerations by government, and the significance for engineering education. The authors offer that focusing on legal obligations and industry case studies mitigates student concerns of topic legitimacy. Interaction with Aboriginal peoples (e.g., as key project stakeholders) also increases students' enthusiasm. A culture of respect and open-mindedness helps facilitate discussions between students and community members.

Baillie (2011; AU) reports on a large project supported by the Australian Teaching and Learning Council to develop curricula and pedagogies for "Engineering Education for Social and Environmental Justice" (p. 2). A multidisciplinary team critiques current engineering education practices in Western Australia and brings key ideas, texts, and epistemologies from their diverse disciplines to develop a knowledge base for a socially and environmentally just engineer. Learning about and respecting Indigenous peoples, cultures, and knowledges is considered critical, particularly as many engineers in Western Australia work in the mining industry with projects on Indigenous lands.

Local cultural awareness is linked to developing global cultural competence. Indigenous awareness and learning are identified as two of several knowledge hubs for curricular case studies for engineering preparation and practice. Kutay and Leigh (2017; AU) and Kutay et al. (2021; AU) provide knowledge on Indigenous engineering and technologies, explore the enhancement of engineering education through the incorporation of Indigenous knowledges, and provide guidance for developing these processes in respectful ways. Prpic and Bell (2018; Australia) collaboratively design a new course reframed as an on-country two-way learning exchange with the Gunditjmara community at the Budj Bim National Heritage Landscape at Lake Condah in southwest Victoria.

In South Africa, Langdon (2013; SA) uses "participatory learning and action (PLA) [decolonizing] techniques to address issues of community and identity" (p. 310) to support students in thinking in "unfamiliar, more reflexive way[s]" (p. 310). Motala (2017; SA) uses storytelling to introduce African knowledge as a decolonizing pedagogy. Steenkamp and Muyengwa (2017; SA) concur that practical engineering examples from South Africa will assist in decolonizing the curriculum. Baron (2018; SA) proposes using reflexivity and conversation theory to decolonize engineering curricula, where students are the "Nunataks" (p. 1), or reference points from which curricula emerge. Course curricula are thus built on "second order science" (p. 1), which diffuses the singular, dominant discourse and "realist view" (p. 1) of science traditionally found in engineering courses to a heterarchical-situated conversation tied to students' unique ways of knowing. Thus, instructors and students co-create a recursive, inclusive, and ethical curriculum, with ethics found "in both the method and choice of content" (p. 21).

Zembylas et al. (2019–20; SA) introduce a "hauntological decolonizing pedagogy," (p. 45) using GIS in an engineering course to explore "how the curriculum can be used to animate occluded injustices of the past by means of a micro-instance of activism in the form of a storytelling intervention" (p. 28) and "make visible the coloniality of apartheid" (p. 43). Gupta (2020; SA) offers that participation in seminars, workshops, and discussions and use of local case studies are ways to decolonize. A new course, *African Insights*, is being offered at the University of Johannesburg. There is an improved community engagement module at North-West University (forthcoming book chapter) and a curriculum design partnership in UCT's Department of Chemical Engineering that replaces a conventional chemical industry project with a local community project with a distinct decolonization focus (Agrawal, Heydenrych & Harding, 2022; SA).

In Canada, Friesen and Herrmann (2018; Canada) present plans to incorporate Indigenous knowledge, perspectives, and design principles in engineering curricula. Friesen et al. (2019; Canada) describe the work of an Indigenous engineer-in-residence, and Friesen (2019) reflects on Indigenizing the curriculum. Frey et al. (2018, 2019; Canada) describe the redesign of the first-year engineering program and addition of a cultural foundation course to introduce the importance of inclusivity and intercultural competencies in engineering and to enable contextualized inclusion of Indigenous examples in subsequent curriculum. Students are required to complete an Indigenous history course in the summer before entry. Eikenaar (2018; Canada) describes the first steps in Indigenizing the curriculum. Eikenaar et al. (2022; Canada) discuss a project integrated into a second-year communication course that "familiarizes students with the process of consulting with Indigenous communities by conducting research on community histories and priorities, while reflecting on the ethical, practical, legal, and intercultural demands of engineering communication" (p. 1). Boudreau and Anis (2019; Canada) reference Indigenous Elders as clients for capstone design projects. Perks and Dimitrova (2019; Canada) reference civil engineering capstone projects in partnership with Indigenous communities and governments that have been successful in preparing students for Indigenous engagement and increasing their awareness of small northern communities in Canada. The projects are hoped to also attract Indigenous youth into engineering. Seniuk Cicek et al. (2019; Canada) introduce a transdisciplinary design build course taught in partnership with Shoal Lake 40 First Nation.

Irish and Romkey (2021; Canada) introduce students to an Indigenous Ethics of Care Framework (Whyte & Cuomo, 2017) and Zoe Todd's (2017, 2018) Indigenous worldview, where rivers, fish, and ancient decayed beings in oil sands "are considered 'kin' and are upheld in kinship relations" (p. 4). They incorporate multiple competing worldviews by discussing the interests of Wet'suwet'en hereditary chiefs and of the land, plants, and animals to expose power relations and the pressures and problems in negotiations between corporate, government, Indigenous groups, and "the unspoken-for more-than-human actors" (p. 4). Johnson (2021; Canada) develops activities to support transformative learning so students recognize their responsibility as citizen engineers and the development of responsible public policy. Seniuk Cicek et al. (2022; Canada) introduce a new course to decolonize and Indigenize engineering that exclusively presents and supports Indigenous worldviews. In the United States, Dandridge et al. (2019; US) recognize decolonizing curricular reading lists.

3.3.2 Decolonizing Service Learning

Decolonizing service learning is the second subcategory in *decolonizing engineering education – pedagogy and curriculum*, with most articles written by authors in the United States, and one from an author in New Zealand.

Grommes (2004; US) offers an example of a partnership between two universities and a tribal college to educate students on sustainability and Indigenous cultures and to partner on a community project (the American Indian Housing Initiative) to develop sustainable and efficient technologies. Benning et al. (2014; US) describe how the Pine Ridge Indian Reservation and faculty and students from Oglala Lakota College and South Dakota School of Mines and Technology collaborated on the design, research, and testing of a sustainable, renewable energy food production system. Students learn that the Lakota did not have a word for *sustainability* as it is so ingrained in all aspects of Lakota life that it does not exist as a separate concept, that trust is built through developing relationships, that projects take more time to interact with multiple stakeholders. Wilson et al. (2015; US) use Nguzo Saba and Africentricity, borderlands, cultural community wealth, and critical race theories to conceptualize service learning as mentoring. They highlight *sitio y lengua* – decolonizing spaces and discourses – in "design[ing] a Mentor Service-Learning model [that] will provide a culturally responsive academic experience [to support Latine students'] academic growth in STEM while building and strengthening students' identities as members of their community" (p. 7). Garcia's (2018) article on decolonizing Hispanic-serving institutions is referenced (e.g., Salgado et al., 2021; US). Koh (2020; US) explains how service learning "reproduces colonial and globalist tendencies that ultimately undermine already-marginalized communities" (p. 2). Koh and Rossmann (2021; US) offer that service learning could be decolonized by deprioritizing student outcomes to prioritize community needs.

Hughes et al. (2017; NZ) discuss an inaugural design project in collaboration with the Koukourārata rūnanga that helps provide the community with design plans for future development and is hoped to inspire youth to engage in STEM. The course promotes "design thinking, concept generation and selection, collaborative work, sustainability, and engaging with Aotearoa/New Zealand's indigenous Māori communities in the spirit of respectful co-creation" (p. 10).

3.3.3 Engineering/STEM Outreach

Several authors write about *engineering/STEM outreach*, which is the third category in *decolonizing engineering education – pedagogy and curriculum*. These authors are writing from New Zealand, Canada, and the United States.

In New Zealand, Mitchell (2006; NZ) evaluates a tutoring program for Māori and Pasifika students demonstrating positive results for targeting students and student outcomes. Morgan (2006; NZ) describes Ngā hoa o te kupenga rorohiko, an innovative distance learning program to increase Māori student participation in engineering. Murray and Morgan (2009; NZ) discuss an award-winning participation strategy for Indigenous engineering students based on the Māori participation strategy of 5 "R"s – readiness, recruitment, retention, research, and role modelling – at the University of Auckland. It fosters equity, community–student relationships, student-led recruitment and mentoring, and a supportive Indigenous student and faculty cohort and is "justified by the university's Treaty of Waitangi obligations, equity considerations, and also by the view that our Māori and Pasifika students make very good engineers in practice" (p. 605). Leydens et al. (2017; US, NZ) evaluate this strategy using an engineering-for-social-justice (E4SJ) framework. The work "highlights the importance of acknowledging underrepresented students' culture and values not just outside the engineering curriculum, but within it" (p. 14). The strategy peaked in 2009 and 2010 but has declined since. The authors hope that this participation strategy will be resumed.

In Canada, Herrmann (2012, 2014; Canada) discusses the Indigenous access program at the University of Manitoba (2022), which has existed for 35 years. Kinsner et al. (2014; Canada) describe the Verna Kirkness space discovery camp for Indigenous high school students. Stacey et al. (2017; Canada) reflect on a decade of STEM outreach for Indigenous youth and the importance of community relationships, delivery, and leadership styles, and balancing Western and Indigenous worldviews. Wiseman and Herrmann (2019) challenge "colonial definitions of program success framed by the short-term thinking of participation, enrolment, retention, and graduation rates of Indigenous students" (p. 9) in discussing how to unlearn colonialism in access programs. Dornian et al. (2020; Canada) describe a pre-university coding workshop designed for Indigenous and women in engineering outreach. Huntinghawk et al. (2020; Canada) have built a cybersecurity mini curriculum as a community engagement initiative with First Nations communities in Manitoba. Gerrard and Boyer (2022) reflect on engaging Indigenous youth in a virtual land–based STEM leadership program.

In the United States, Fick et al. (2013; US) describe the vertical integration of civil and environmental design through a regenerative community project on the Pine Ridge Indian Reservation to increase enrolment of Oglala Lakota College students in engineering. Navickis-Brasch et al. (2014; US), and later, Pieri et al. (2020; US), describe collaborative projects with tribal colleges to culturally center students on their communities and connect them to engineering and STEM.

Jordan (2015; US) and Jordan et al. (2017, 2018, 2019) explore "how Navajo engineers experience, understand, and apply engineering design and practice in the context of their culture and community" (Jordan et al., 2019, p. 355). The intersection of Navajo worldviews and engineering design and practice informs culturally responsive engineering design curricula for K–12 Navajo students. The authors hold that culturally relevant pedagogy can braid engineering with culture for a stronger, more diverse, and relevant profession. A Tohono O'odham woman interviews three community members to explore how they perceive and connect to engineering to develop culturally relevant curricula for Tohono O'odham students (Anderson & Jordan, 2018; US). Preliminary findings point to engineering as a cultural strategy to survive and as a process for community development. Tohono O'odham values are mapped to engineering design traits.

3.3.4 Guides and Frameworks

The fourth subcategory in *decolonizing engineering education – pedagogy and curriculum* is *guides and frameworks*, with several each coming out of Australia and Canada.

In Australia, Leigh et al. (2014; AU) collate evidence of Aboriginal engineering and technologies to incorporate them into engineering curricula and prepare engineering students for respectful and culturally relevant interaction with Aboriginal peoples in engineering practice. They use a Venn diagram with three circles to conceptualize how Aboriginal, Western, and engineering ways of knowing overlap (see Figure 4.1).

Kennedy et al. (2016; AU) provide a model (Figure 4.2) for establishing inclusive learning spaces and including knowledge about relevant Aboriginal practices and principles in engineering content that incorporates the "artifact space" from Figure 4.1. An outcome is the recognition of Aboriginal engineering and the importance of relationships. As cited in Leigh et al., 2015, "[e]ngineering education that is inclusive of Indigenous perspectives cannot be achieved without sustained and productive relationships between Indigenous Communities and Engineering Schools" (p. 8).

Kennedy et al. (2016; AU) provide *A Beginner's Guide to Incorporating Aboriginal Perspectives into Engineering Curricula* as "a set of reference points for making decisions about how to build a genuine two-way relationship with the right Aboriginal people and Aboriginal communities that will contribute relevant Aboriginal perspectives to your Engineering discipline" (p. 9). The work is guided by five components: (1) Start with a new philosophy, (2) Explore Engineering through three perspectives, (3) Consider an Aboriginal worldview, (4) Engage with Aboriginal people and their communities, and (5) Tailor the learning environment (p. 10). They teach about the five "rights" of engaging with Indigenous communities: (1) right people, (2) right place, (3) right language, (4) right time, and (5) right way (pp. 19–23) (see Figure 4.2).

Ruta et al. (2021; Canada) offer a conceptual framework for developing an engineering education curriculum in partnership with Indigenous peoples and communities based on five themes categorized from 43 academic and gray literature texts identified in a rapid scoping review (see Figure 4.3).

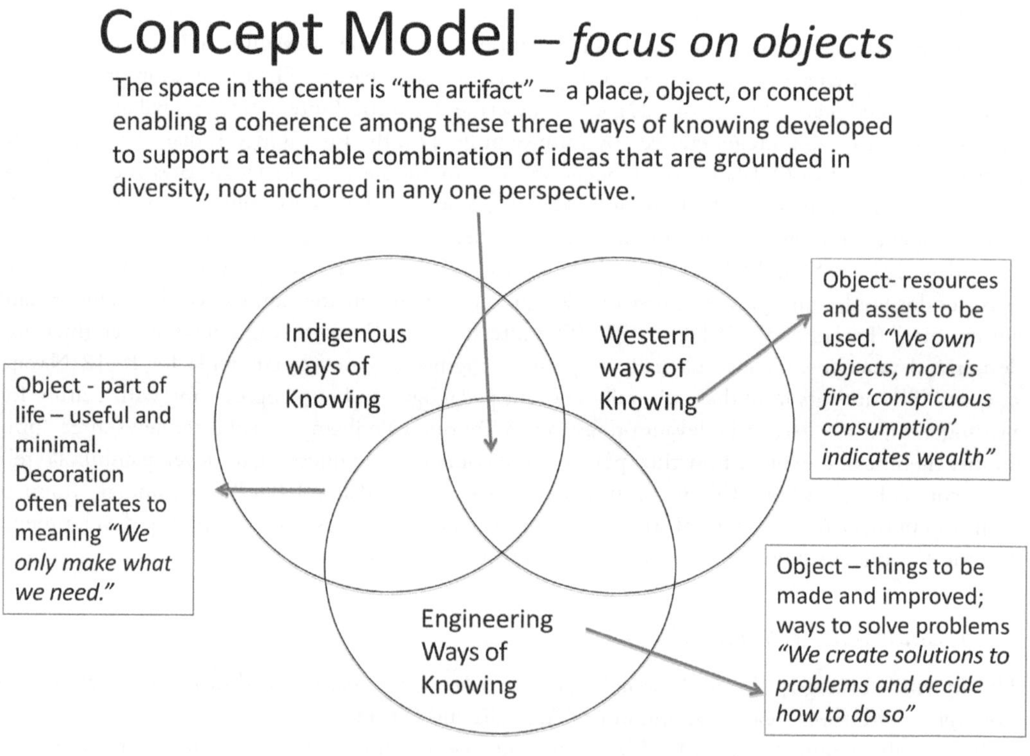

Figure 4.1 The artifact space – developing knowledge of Indigenous engineering.
Source: Leigh et al. (2014).

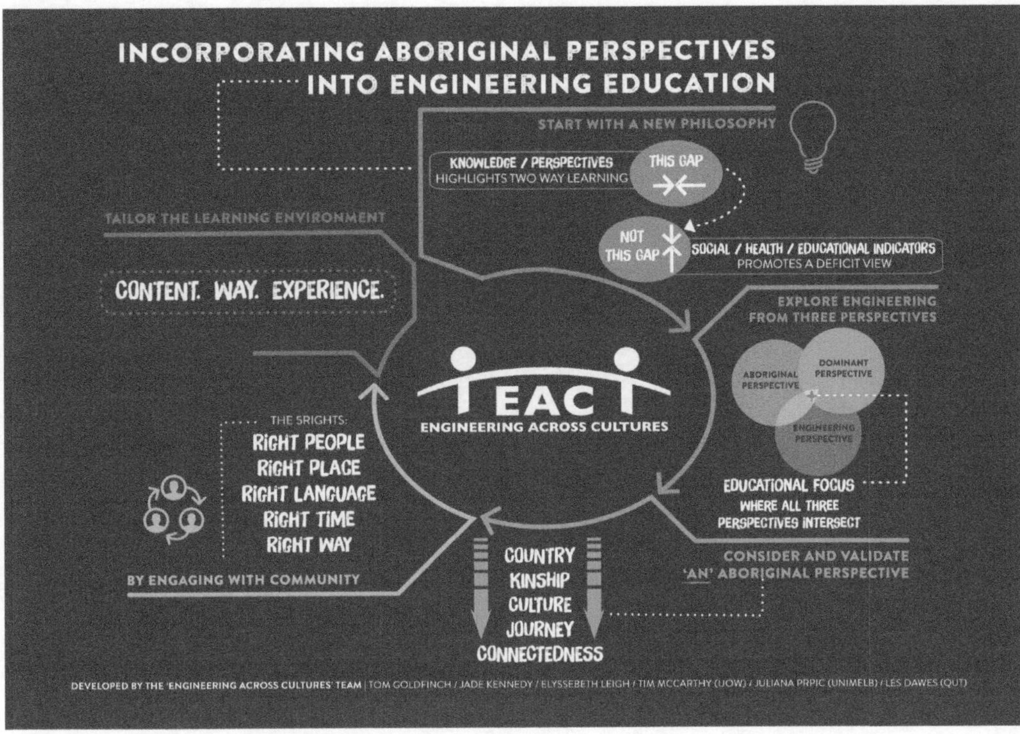

Figure 4.2 Model for embedding Indigenous perspectives in engineering education.

Source: Kennedy et al. (2016).

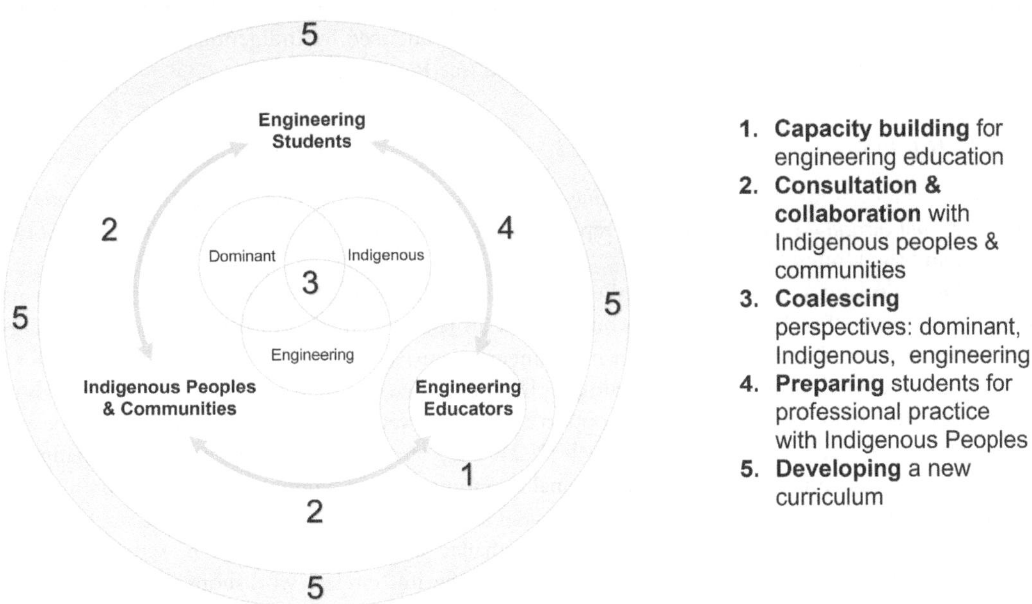

Figure 4.3 Conceptual diagram for developing a new curriculum that includes Indigenous peoples, knowledges, and perspectives in engineering education.

Source: Ruta et al. (2021).

Figure 4.4 Pictorial depiction of the guiding touchstones.
Source: Dubiel et al. (2021).

Dubiel et al. (2021; Canada), in consultation with Shoal Lake 40 First Nation community members, offer four touchstones to guide "how the engineering process can adapt to make space for Indigenous ways of knowing and being" (p. 6). The touchstones, *process*, *sustainability*, *means*, and *relationships*, are interwoven and omnipresent; they are influenced by Indigenous teachings and can be referred to at any point during the design process (see Figure 4.4).

3.3.5 Training

Finally, several authors write about *training* as a subcategory in *decolonizing engineering education – pedagogy and curriculum* with workshops reported on in Australia and New Zealand, Canada, South Africa, and the United States.

In New Zealand and Australia, Hughes et al. (2018; NZ, AU) offer a workshop to engineering educators called, "He Awa Whiria – a braided rive in te reo Māori", emphasizing weaving Indigenous and Western knowledges together in engineering curricula. Goddard et al. (2021; AU) offer a workshop to explore project-based learning with explicit focus on remote Indigenous homelands to assist educators in designing educational opportunities to develop students' cultural competence and risk management skills. Prpic and Goldfinch (2021; AU) offer a second workshop emphasizing the importance of relationships between Aboriginal communities and engineering educators.

In Canada, Seniuk Cicek et al. (2019a; Canada) and Seniuk Cicek et al. (2021a; Canada) reflect on a series of faculty workshops designed to enhance engineering education with Indigenous cultures, pedagogies, knowledges, and perspectives. Sharing circles, workshops, panels, and collaboratoriums centering on Indigenous peoples and initiatives have been offered at the Canadian Engineering Education Association (CEEA-ACEG) annual conference beginning in 2017 (Seniuk Cicek et al., 2020). In the United States and South Africa, Lord et al. (2019b; US, South Africa) and

Lord et al. (2019a; US) offer workshops to begin a dialogue on decolonizing engineering education by dismantling engineering as an "objective" activity.

3.4 Decolonizing Engineering Education – Impact on Students, Instructors, and Community Members

The fourth and final category of literature is organized under the heading *decolonizing engineering education – impact on students, instructors, and community members*. This category is divided into two subcategories: *engagement* and *integration*.

3.4.1 Engagement

The first subcategory of *decolonizing engineering education – impact on students, instructors, and community members* is *engagement*, with work coming from Australia. Godfrey and King (2010; AU) explore Indigenous student recruitment and retention in engineering. Their guiding principles include that success cannot only be quantitatively defined ("bums on seats") but that different types of initiatives are also needed for diverse approaches; communication consultation is required. They find no national strategy for recruitment and question if this would be valuable. Challenges include lack of connection, engagement, and relevance of engineering for Aboriginal communities and few qualified Aboriginal high school students. Opportunities include inclusive curricula, project-based learning, access programs, mentoring, and outreach. Authors recommend building relationships with the university Indigenous support unit, school mentoring groups, subject teachers and career advisers, local Indigenous communities, Indigenous engineers, and industry, and designing inclusive curriculum with hands-on opportunities for assisting local Indigenous communities with their priorities.

Goldfinch and Kennedy (2013; AU) present a range of strategies for growing genuine engagement between Aboriginal Community and engineering education identified through consultation with Illawarra Aboriginal communities. Misconceptions and mispractice, or four "don'ts" when engaging with Indigenous communities, include: (1) consultation and engagement for the wrong reasons, that is, meeting minimum regulatory/policy requirements, or "ticking boxes"; (2) unsustainable initiatives and a fly-in-fly-out approach to outreach, leading to abrupt discontinuities in engagement; (3) a deficit view of Indigenous cultures and communities; and (4) Indigenous culture and community as a singular entity (p. 5).

Five principles of country conceptualizing Aboriginality are also identified: (1) *Country* – refers to one's nature and natural surroundings. It includes lands and waters; trees and plant life; animals, birds, fish, and reptiles. (2) *Kinship* – reflects the system by which people are related to each other. It defines one's roles and responsibilities and obligations and commitments to the relationship. (3) *Culture* – is said by the Dharawal to be present in your everyday being. It is represented in art; song and dance; language, stories, and oral histories. (4) *Journey* – refers to the lived experiences that occur and have occurred "on country." This is presented through one's story and one's families' stories, one's history, past, present, and future. (5) *Connectedness* – speaks of the interrelationship of everything and distinguishes how nothing can be considered in isolation (pp. 5–6) (see Figure 4.2).

Goldfinch and Kennedy (2013; AU) offer how differing worldviews can create conflict for engineering education:

> Western cultures which have shaped engineering education in Australia emphasize facts and concepts in a decontextualised manner. In contrast, Indigenous worldviews tend to emphasize value, spirituality, and holistic understandings. . . . The clash of cultures can create significant

> challenges in education when these differing values and worldviews are not recognized and accommodated.
>
> *(p. 8)*

Engineering is seen as having a negative impact on country and in community, with little understanding of the profession in communities. Positive steps include supporting engagement through sustainable initiatives, moving away from a deficits view of Indigenous people, and recognizing the diversity of Indigenous Australia.

In Goldfinch and Hollis (2016; AU) and Hollis and Goldfinch (2017; AU), findings from an online survey of engineering graduates identify that encounters with Aboriginal culture heritage were common on engineering project sites. Participants had minimal formal education beyond grade school with engaging with Aboriginal communities and reported low confidence levels. Authors conclude that education and training are needed to meet the challenges of large infrastructure projects and the preservation of Aboriginal heritage.

3.4.2 Integration

The second subcategory of *decolonizing engineering education – impact on students, instructors, and community members* is *integration.* Much of this work comes out of Canada, with one article each from Australia and South Africa.

In Canada, Johnson (2016; Canada) explores intercultural competence and knowledge of social risk in mining engineering students through surveys and focus groups, identifying three sequential threshold concepts impeding students' development: recognizing different forms of knowledge, understanding values are tied to culture, and respecting varied conceptualizations of well-being and quality of life. In 2020, Johnson (2020; Canada) presents several theories towards a framework to support the development of intercultural competence curricula in engineering mining education.

Gibson (2016; Canada) theorizes a road built by engineers that is significant to both the Stl'atl'imx of the lower Lillooet River Valley and 38 years of grade 10 non-Indigenous students who traverse the road as part of their curriculum as "material culture or 'things' that both produce, and are products of, complex relationships with the human, non-human and natural world" (p. 26). Gibson frames how Stl'atl'imx decolonize their landscape and regain identity and belonging and offers a way of understanding the process of making and meaning-making of an engineering artifact.

Mante et al. (2019; Canada) introduce a study on the impact of integrating Indigenous knowledges and perspectives in engineering curricula on students' learning. Thomsen et al. (2021; Canada) and Kilada et al. (2021; Canada) share initial findings through student stories that highlight the impact of the Elder's teachings on students, the importance of Indigenous knowledges and perspectives in engineering education, and making space for student reflection. Knowledge exchange is recognized as dependent on relationships, which involves identifying each other's historical backgrounds, situational contexts, and values, and requires active listening, genuine curiosity, empathy, and time.

Seniuk Cicek et al. (2020, 2021b; Canada) share the Indigenous initiatives in accredited engineering programs across Canada and critically consider the work through Gaudry and Lorenz's (2018) spectrum of Indigenization. They find efforts in reconciliation and decolonization are predominantly in grassroots initiatives, with institutional initiatives focusing largely on inclusion. Forrest and Seniuk Cicek (2021; Canada) examine Indigenous design and the integration of Indigenous design with Western design. They conclude that "by incorporating Indigenous knowledges with western approaches, social, environmental, and economic sustainability would be more thoroughly assessed in the engineering design process" (p. 6). Zacharias et al. (2022; Canada) begin work to survey Indigenous land–based learning in engineering education. Preliminary findings demonstrate that land-based approaches are not widespread in engineering education in Canada. Diversified knowledge gathering is required to better understand the current landscape.

In Australia, Goldfinch, Jolley, et al. (2016; AU) and Goldfinch et al. (2017; AU) interview engineering educators to understand the conditions for implementing a curricular model for incorporating Aboriginal perspectives in engineering education. They find that engineering educators, though positive about doing this work, are motivated by how engineers can help Aboriginal peoples (a deficits' perspective) rather than recognizing mutual knowledge-sharing and learning. Challenges include lack of confidence in building relationships, competing priorities, lack of experience and contacts, and misunderstanding the time commitment needed. Benefits for engaging in the work are recognized, including new perspectives on sustainability, community engagement, engineering project management, and Aboriginal student recruitment and retention.

In South Africa, Mkansi et al. (2018; SA) survey students' understandings of how decolonization is applicable to engineering education at the University of the Witwatersrand. They find "strongly held views" and "not one common accepted meaning," however, decolonization is perceived as applicable, and curricular reform central to achieving it (p. 1,018). The authors share that curricula should be inclusive, epistemologically diverse, include African scientific knowledge, and decenter, but not remove Western science, and that decolonization requires both faculty and students. Further research is recommended on Indigenous scientific knowledge, African languages in engineering education, how students should/could engage in decolonization, and the impact of engineering education on local society.

4 Overview of the Literature

Overall, the literature we found on decolonization in engineering education is published by authors in Australia, Brazil, Canada, Chile, Columbia, India, Kenya, Mexico, New Zealand, Peru, Singapore, South Africa, the United Kingdom, United States, and Zimbabwe. The publications from Brazil, Chile, Columbia, India, Kenya, Peru, Singapore, and Zimbabwe are all single articles or the work of one author (as in the case of Brazil). There are two articles each from Mexico and the United Kingdom.

Predominantly, articles are published by authors out of Australia, Canada, South Africa, and the United States, with some notable trends. Authors from Australia are absent from dialogue in our first two categories of decolonization, *decolonization as . . .* and *Indigenous knowledges in engineering and technology*. However, authors publishing in Australia have strong representation in the third and fourth categories, *decolonizing engineering education – pedagogy and curriculum* and *decolonizing engineering education – impact on students, instructors, and community members*. In the latter, they are the only country represented in the subcategory *engagement*. Authors writing from the United States are represented in the first three categories; they have no articles in the fourth category, *decolonizing engineering education – impact on students, instructors, and community members*. Within these publications, they have particularly strong representation in the subcategories *a methodology* and *decolonizing service-learning*. Although one article from the United States is represented in the subcategory *pedagogical responses*, overall, this is an area where there is a notable absence of authors from the United States. Authors from Canada and South Africa have broad representation throughout the four categories of decolonization. Authors from Canada, like Australia, have the strongest representation in the latter two categories and are particularly active in the *integration* subcategory.

The literature we found centering decolonizing in the engineering education space to some extent reflects Shahjahan et al.'s (2022) observations of outward- and inward-facing strategies and philosophies of decolonization, however, not completely. Strategies that emphasized "looking outside of the institution to decolonize, by centering local/Indigenous communities, building within communities, and reallocating institutional resources to local Indigenous communities to actualize communities' priorities and vision" (p. 12), were, like in Shahjahan et al. (2022), largely concentrated in Latin America, Asia, Africa. It is in the spaces of *decolonization as . . . equality, emancipation, liberation, localization, co-creation, and equity* and *a response to Western hegemonic domination, neoliberalism, and the global crises in the natural and human world* where we largely find this outward-facing work and the single articles by authors writing from Brazil, Chile, Columbia, India, Kenya, Peru, Singapore, and Zimbabwe. Authors from these geographical locales are absent from the other characterizations (especially pedagogical translations) of decolonization in this chapter, where articles from Australia, New Zealand, and Canada are concentrated. So this is reflective of what Shahjahan et al. (2022) have observed in their review. However, there are also authors from South Africa who have representation across the decolonization literature in engineering education, as do authors working in the United States and Canada, which is not in keeping with Shahjahan et al. (2022).

Interestingly, the other two subsections of *decolonization as . . .*, which are *a paradigm shift, epistemic transformation, epistemic justice, social justice*, and *a methodology*, are arguably inward-facing. However, not as Shahjahan et al. (2022) describe as concentrated on curricular initiatives but rather in a reflexive and epistemic way, where the way one knows, how one accepts knowledge is created or shared, or who owns or controls knowledge is challenged, and a shift is required, or demanded. In this space, we hear authors from the United States, which is more in keeping with Shahjahan et al.'s (2022) inward-facing global divide, but we also hear authors from South Africa and Mexico, which deviates from Shahjahan et al. (2022).

Shahjahan et al. (2022) found that inward-facing strategies "targeting changes within higher education institutions through classroom based curricular and pedagogical shifts as well as bringing Indigenous/local knowledge systems and practices into higher education institutions" (p. 12) were more concentrated in Oceania, which, with the addition of a concentration from North America and representation of authors from South Africa, we also found in the decolonizing engineering education literature. In fact, our latter three categorizations of decolonization in the engineering education literature, which include *Indigenous knowledges in engineering and technology, decolonizing engineering education – pedagogy and curriculum*, and *decolonizing engineering education – impact on students, instructors, and community members*, generally can be categorized as inward-facing initiatives. Here, there are no authors from Latin America, Asia, or Africa, but there are authors from South Africa. So overall, both the definitions of the inward-facing binary that Shahjahan et al. (2022) offer and the global binary they see in the literature they reviewed in higher education do not exactly fit what we see in the decolonization literature in engineering education, though the binary itself offers us an interesting way to intellectualize the literature.

Largely, the conceptualization of *decolonization* in the literature reflects both sets of literature that we argue at the outset of this article intersects the trajectories of decolonization. The literature we found reflects both a movement for economic and political justice in the critique of the relationship of dependence between some African nations and previous colonial powers and the exploitation, undue political control, and underdevelopment faced after decolonization (Nkrumah, 1965; Rodney, 1973; wa Thiong'o, 1986). It also reflects a call for the undoing or "untangling" (Mignolo, 2007) of the dominant structures of knowledge, social, and cultural production and power referred to by Maldonado-Torres (2016) as "decoloniality." The literature reviewed here also demonstrates the array of meanings and interpretations of decolonization that we put forth in the opening of the chapter, from "inclusion" to "land back," accentuating the complexities of this work. Finally, the literature offers us an understanding of both the drivers of, and the barriers to, decolonization.

5 Drivers for Decolonization

In the literature we observed drivers for decolonization, found at both grassroots and organizational levels. We characterized these drivers in five broad categories: *educational mandates*, *policy mandates*, *industry mandates*, *social justice*, and *sustainability*.

5.1 Educational Mandates

One of the drivers for decolonization is education mandates, where educational programs must adapt to changing cohort demographics and/or expectations. Literature from Australia, Canada, New Zealand, and South Africa demonstrate curricular initiatives to decolonize engineering education. There is less evidence of this in the United States, apart from *decolonizing service learning*, which is strongly represented by authors from the United States. Student-driven educational imperatives, such as improving student motivation, retention, and local relevance, are particularly prominent in the South African context, where they are positioned as potential outcomes of a decolonized curriculum (Winberg & Winberg, 2017; SA; Ruric Vogel & Human-Vogel, 2018; SA; Lourens, 2017; SA). Kanyarusoke and Ngonda (2017; SA) argue for a complete approach in that

> Decolonizing Engineering Education means a sum total of teaching, learning, training and assessing that produces a group of motivated, knowledgeable and skilled professionals able to understand, own up and address the region's problems and challenges using new and/or established engineering methods.
>
> *(p. 401)*

Motivation, retention, and relevance are also key educational drivers, where decolonization is linked to broadening the appeal of engineering education to minoritized and Indigenous students. This is a feature prevalent in the *engineering/STEM outreach* literature and particularly prevalent in Australia and New Zealand. Low numbers of Aboriginal and Torres Strait Islander (AU) engineers and the increasing recognition of Indigenous lands in industry projects have prompted efforts to decolonize curriculum (Morgan, 2006; NZ; Hughes et al., 2017; NZ). Awareness of the need for educational programs that are inclusive of Aboriginal and Torres Strait Islander values and perspectives has grown over the past decade in Australia, sparked at a national level by the Engineers Without Borders Challenge (Engineers without Borders, 2010), which presented some of the earliest focus on Indigenous community groups (Cutler et al., 2010; AU). Awareness received a significant boost following a 2013 federally funded project (Goldfinch, Kennedy, et al., 2016; AU) and a 2015 national summit on Indigenous engineering (Prpic, 2015; AU), which formed a focal point for interested academics and precipitated the formation of communities of practice around curricular inclusion of Indigenous knowledges and perspectives in curriculum (Goldfinch & Kennedy, 2013; Leigh et al., 2014; Prpic & Bell, 2018; Campbell et al., 2020; AU).

A similar small but growing community of practice is found in the Canadian Engineering Education Association–Association Canadienne de l'éducation en génie (CEEA-ACÉG). Articles published in the annual conference proceedings referencing Indigenous peoples and knowledges focus on *curriculum, student learning, design, sustainability, outreach, access, and faculty development*. The 2022 conference opened with a keynote address on *Decolonizing Engineering Education* by Herrmann, Métis Professional Engineering, and Director of the Indigenous Access program (ENGAP) at the University of Manitoba (2022).

5.2 Policy Mandates

Another driver of decolonization are policy mandates, where change is necessitated by institutional or government directives. Much of the decolonization literature in Canada, Australia, and New

Zealand references policy mandates at institutional or government level as key drivers. Indeed, most institutions in Australia have policies in place which mandate action on decolonization to some extent (Hollis & Goldfinch, 2017; Universities of Australia & IHEAC, 2011; AU). Decolonization in Canada is largely connected to Indigenization prompted by the Truth and Reconciliation Commission (TRC) of Canada *Calls to Action* (2015). Elsewhere, a strong focus on biculturalism in New Zealand that has emerged in recent decades has seen decolonization-aligned policies and declarations from the federal government through to engineering industry (Black, 2018; NZ).

5.3 Industry Mandates

Industry mandates are another driver of decolonization, where decolonization of education is aligned with shifts occurring in industry. These imperatives are particularly apparent in the Australian and New Zealand contexts, where legislation and land rights agreements (Australia) or formal treaties (NZ) mandate land use agreements with Indigenous landowners and representative bodies in engineering projects that impact these territories (Goldfinch et al., 2014; AU). Such agreements have also precipitated employment targets within these projects and demands for greater incorporation of Indigenous knowledges, establishing significant demand for Indigenous graduates of engineering and allied fields and those who are familiar with these worldviews. Both outcomes are cited as key external drivers in research examining Indigenous perspectives in engineering (Hollis & Goldfinch, 2017; Wilkinson et al., 2021; AU).

5.4 Social Justice

While a social justice focus underpins many of the drivers mentioned previously, some efforts to decolonize engineering education exist with this as the primary driver. Connection to social justice and recognizing the minoritization of cultures, knowledges, technologies, and perspectives are prevalent in the literature. The United States particularly appears driven primarily by a social justice focus. The literature here is largely focused on decolonizing methodologies and Latine decolonization in engineering education; there was less literature on Native American–centered decolonization found in our narrative literature search, however, the American Indian Science and Engineering Society (AISES, 2022) has an annual conference and has been actively promoting Indigenous peoples in STEM for over 45 years.

It is worth noting that Indigenous inclusion is a relatively small component of the decolonization debate in South Africa, which was sparked by the "Rhodes Must Fall" student protests of 2015 (Kamanzi, 2015). This was a campaign demanding the removal of the Cecil John Rhodes statue from the main campus on the University of Cape Town, as students saw Rhodes as a symbol of institutional racism and exclusion. Indeed, the notion of "Indigenizing" engineering education was entirely absent from these protests. It is only more recently that a comparable discourse on Indigenous peoples has begun in South Africa, with the opening of the Khoi and San Centre at the University of Cape Town in 2020, launched to "foreground indigenous knowledges, rituals, languages and ways of knowing" (Swingler, 2021).

5.5 Sustainability

There is a strong connection to sustainability as a driver for decolonization. Although in this chapter only two articles are named in the same-named subcategory, many authors implicate sustainability as an incentive for learning about and advancing Indigenous and minoritized knowledges and approaches in engineering and engineering education. Indeed, the *decolonization as . . . a response to Western hegemonic domination, neoliberalism, and the global crises in the natural and human world* articles are arguably a more profound and razored reminder of the connection of sustainability to decolonization. These motivations are rooted in the current global crises, and the conceptualization of (and, at times, arrogance in) engineers' role in solving these critical problems.

6 Barriers to Decolonizing

In addition to the drivers of decolonization, there are several barriers to decolonization observed in the literature. These barriers include *false narrative of western engineering and roots in colonial knowledge, poverty and survival mentality, rematriation of land, STEM as the tool of the colonizer, "decolonization" as a watchword,* and *lack of resources and literature.*

6.1 False Narrative of Western Engineering and Roots in Colonial Knowledge

The false narrative of Western engineering and roots in colonial knowledge are exposed by Hoople et al. (2020; US), who argue that most engineering classes (e.g., thermodynamics) "have roots in the Industrial Revolution and are characterised by particularly ethnocentric (White), masculine, and colonial knowledge" (p. 2). Eichhorn (2020; UK) argues that the refusal to recognize the false narrative of Western engineering and technical superiority and learning to work with true global interdisciplinarity and diversity are not only a barrier to decolonization but also a threat to human existence.

6.2 Poverty and Survival Mentality

Poverty and survival mentality are barriers raised by Kant et al. (2015; US), who interview Native Americans to determine why they are among the most underrepresented groups in engineering in the United States. They learn that "the effects of poverty and the resulting survival mentality divert attention from what are perceived to be privileged pursuits such as engineering education" (p. 31). Colston et al. (2019; US) also name poverty as a barrier, citing financial, academic, belonging, and career information as challenges for Indigenous engineering students.

6.3 Rematriation of Land

Rematriation of land is cited as a barrier to true decolonization. Koh and Rossmann (2021; US) reference Tuck and Yang (2012) and Santiago-Ortiz (2019) in arguing that though critical consciousness is important, "without the literal rematriation of land critical pedagogies themselves are not decolonial" (Koh & Rossmann, 2021, p. 9). They also reference Gaudry and Lorenz (2018), academics from Canada who "point out that 'decolonial indigenization' would actually require an overhaul of the academy and reorientation of knowledge production" (Koh & Rossmann, 2021, p. 10). Taken in context, to decolonize engineering education, a complete overhaul of existing systems and rematriation of land is required.

6.4 STEM as the Tool of the Colonizer

de Roock and Baildon (2019; Singapore) argue that STEM is a tool of the colonizer and link it to today's global crises: "STEM education discourses are closely aligned with, and reinforce, neoliberal social relations . . . [and] Principles of solidarity and social justice are sidelined" (p. 388). They argue that efforts to diversify STEM conceal the true reasons for the perpetuated hegemony and unfairly put the onus on discriminated populations to gain self-efficacy or skills to meet these challenges. They look for a future of "radically democratic and decolonized alternatives" to the present neoliberal capitalist world perpetuated via STEM. Pido et al. (2019; Kenya) further the argument for this barrier by conceptualizing technology as a colonial mechanism. They offer a critical examination of the exclusion of African traditional and Indigenous knowledges and non-mechanized and non-digitized technology. Mudondo (2020; Zimbabwe) argues that the enslavement and colonization of technology and technical and vocational education and training in industrial engineering have hindered socioeconomic prosperity in Africa. Castillo-González et al. (2021; Mexico) position data capitalism as a new form of planetary colonialism that perpetuates "environmental deterioration and the vulnerability of historically excluded and subordinated populations" (p. 369).

6.5 *Decolonization as a "Watchword"*

Ansari et al. (2021; US, Canada) express concern about "the use of 'decolonizing' as a watchword without responsible definition." This is further defined as a "checkbox" or "normative frameworks, which then get applied in a template-like fashion" (p. 1). This is an argument that is also made in reference to the Truth and Reconciliation (2015) efforts in Canada.

6.6 *Lack of Resources and Literature*

Skopec et al. (2021; UK) discuss lack of resources or literature in addressing decolonization in engineering and privilege in higher education. Rogers and Valdez (2021; US) concur: "Though there have been efforts in fields related to systems engineering, these have mainly been outside of the US . . . curriculum changes are needed and . . . course material needs to be contextualized" (p. 8).

7 Recommendations

As one comes to terms with one's own colonial history, a genuine movement towards decolonization in engineering education and research might include substantial partnerships with nondominant, local, or Indigenous peoples, communities, and organizations; involvement of nondominant, local, or Indigenous leaders in making decisions about land use; and the centering of nondominant, local, or Indigenous worldviews and perspectives when developing engineering projects, curricula, research, and educational spaces and places. These types of activities can and should occur at both individual and systemic levels. We offer this advice: connect to local and Indigenous peoples and build authentic and lasting relationships and collaborations. Be cognizant of sociopolitical events and dynamics in your area. Efforts to decolonize engineering education and research are being taken forward in different ways, as demonstrated by the literature cited in this chapter; reach out to learn from local individuals doing the work.

It is also important to understand that decolonization is more than the inclusion of nondominant, local, or Indigenous peoples and perspectives – it requires decolonizing from within (wa Thiong'o, 1986). It requires a fundamental shift in thinking about whose knowledges and belief systems are accessed and how knowledge is acquired and shared. It challenges the notion of objectivity and the purposes and practices in engineering. It challenges the perpetuation of the status quo: decolonization is not merely including nondominant or Indigenous students and faculty in engineering education and the profession as it exists as the Western entity it is today. It requires centering non-Western methodologies, pedagogies, and practices. Deeper explorations and understandings of non-Western epistemic, ontological, and axiological worldviews are needed to drive genuine decolonization to create decolonial spaces and places. Engineering, in many places, is born out of male domination and privilege; it absolutely has a role to play and must decolonize itself. It is a fundamental yet for some, unintuitive challenge that engineering educators and researchers should champion.

In doing this work, particularly if undertaking inward-facing work to decolonize higher education institutions through pedagogical initiatives, integrating nondominant knowledge systems and practices into engineering education, or undertaking service learning with minoritized or Indigenous communities, we must be careful not to perpetuate colonizing behaviors and extractive partnerships. This is communicated specifically in the literature found in the *guides and frameworks*, the *decolonizing service-learning*, *engagement*, and *integration* subcategories. It is important to remember that the involvement of minoritized or Indigenized peoples is not necessarily decolonizing – these situations are not mutually exclusive.

Perhaps ideologically foreign to traditional engineering identity, the importance of building and *maintaining* relationships is repeatedly stressed in the decolonizing engineering education literature. Community outcomes must be prioritized in building decolonized relationships, which take time and resources to foster. As such, this work does not live easily in the academy, which traditionally drives a research output imperative, nor in the profession, which is largely driven by efficiency and economics.

And critically, the importance of *place* cannot be emphasized enough. Place is fundamental to decolonizing work and should guide all efforts in this area. This is especially important when looking at the decolonial frameworks in the *guides and frameworks* subcategory we share in this chapter. The information is shared to learn from what others are doing, but with an understanding that decolonizing work must be *localized* and done in awareness, acknowledgment, respect, and partnership with local peoples, cultures, social and political structures, stories, and histories.

The awareness and careful navigation of the decolonial space, and being mindful of not perpetuating colonial knowledges, attitudes, values, and behaviors, are both the challenges and gifts of doing this work. They should be one's guide. Know your heart when engaging in this work. Share your motivations honestly – with yourself and with those with whom you are working. Be open and willing to learn about and understand that there are other knowledges out there that we would all benefit to learn. But do this humbly and, as advised by Kirkness and Barnhardt (2001), with respect, in relationship, practicing reciprocity, and taking responsibility for what you learn, what you do, and what you say.

8 Limitations

The literature in this chapter largely reflects the decolonial work in engineering education found in more traditional forms of academic literature. We hear voices in this literature from both marginalized and dominant populations, although arguably voices from dominant populations are louder. As well, all the literature here was retrieved from publications within the structures of the "academy," which houses dominant, colonial structures, to which all these authors had access. Thereby, in this review, we do not hear the voices or learn from the perspectives of those who do not have this access.

Further to this, the gray literature is not well represented here. This area, centering decolonizing engineering education and, by extension, the academy, is arguably not well-served by the Western academic structure that supports a distinct type of scholarly dissemination (and numerous other anti-productive structures for this area, including funding, promotion, etc.) that is rooted in the Western colonial knowledge monoculture. We acknowledge the constraints of finding information on decolonization because we recognize that many individuals and communities doing this work do not live in the academic space. Other good, meaningful, and important work may well be underway to decolonize engineering education that is not represented here, existing in alternative places and spaces to the academy.

Finally, we also recognize the varied contexts from which scholars engaging in decolonial work draw. Literature reviews do not always accurately capture these nuances.

9 Conclusions

The purpose of this chapter was to explore how decolonization is conceptualized within the engineering education literature. As noted in the beginning, we bring to this chapter our own positionalities, which significantly inform our understanding and meaning-making in the decolonial space. Given this and the relative newness of the decolonial practices demonstrated within this space, our chapter should be viewed as the start of a conversation on decolonization in engineering education, not a final word. Acknowledging this, we offer some closing thoughts for engineering educators and researchers who want to move in a decolonial direction.

While there is movement to decolonize engineering education and the profession, the call for decolonization needs to grow louder. There has been an increase in pedagogy and scholarship in the literature in the last few decades; however, current efforts are at the beginning stages in some places, and the work to move towards a space of decolonization is challenging. Decolonization requires confronting and undoing colonizing practices while acknowledging one's own complicity. It requires reframing what knowledge is and who can create it, mostly by de-emphasizing Western worldviews and ways of knowing.

Decolonization changes societies' relationships with nationalism and capitalism and with living beings and the land. When colonization is the origin story for the dominating voices of nations, it is difficult to advance decolonization. However, should all engineering educators recognize the fundamental value of nondominant knowledges and perspectives for engineering and understand the urgency of our profession's role in advancing these diverse knowledges and values systems, they will take up the call for decolonization, and decolonization will be less niche and more mainstream in the very near future.

Acknowledgments

We are grateful to the Indigenous colleagues and allies who are leading the efforts to decolonize engineering education and research and who guide and support our work. We acknowledge and are grateful for all the leaders in decolonizing spaces, including the many voices represented in this chapter and the many more who are not. Tom would particularly like to acknowledge Mr. Jade Kennedy, a Yuin man and cultural knowledge holder, and communities in the Illawarra region of NSW for the transformative influence they have on his learning and on the engineering education landscape in Australia. We are grateful to Marie Speare, Engineering Librarian at the University of Manitoba, for her help with the literature review. We are thankful to the team of reviewers, including Randy Herrmann, Jennifer M. Case, and Stephen Secules, for their thoughtful and critical perspectives that were instrumental in shaping this chapter. We are thankful to Aditya Johri and the editorial team of this handbook for the opportunity to write this chapter and their guidance throughout the process. Finally, we are grateful for each other and for our learnings from the rich discussions and hard work we engaged in over the last 16 months.

Note

1 The literature review included searches of Scopus, Web of Science (including BIOSIS, Current Contents Connect, Data Citation Index, Derwent Innovations Index, KCI-Korean Journal Database, MEDLINE, Russian Science Citation Index, Web of Science Core Collection, Zoological Record), Compendex (Engineering Village), and ERIC databases. Title, abstracts, and keywords in Scopus and Web of Science were searched using the following search terms: (decolon* or indigenizing or indigenising) AND (engineer* or stem) AND ("post-secondary" or universit* or college or "higher education" or "higher ed" or education or curricul*). Similar keyword searches were performed in the Compendex along with relevant subject headings as well as combining the terms with the phrases "Indigenous knowledge(s)" or "Indigenous

worldview(s)" or "Indigenous materials." The ERIC search strategy included the subject heading "Indigenous Populations" and the words (decolon* or indigeni*) and "higher education." The databases were searched from database inception to May 2022.

We conducted manual search of Google Scholar and ASEE, CEEA-ACÉG, REES, SASEE, and SEFI conference proceedings (American Society for Engineering Education; Canadian Engineering Education Association-Association Canadienne de l'éducation en génie; Research in Engineering Education Symposium; South African Society for Engineering Education; European Society for Engineering Education). SEFI proceedings predetermined categories searched included "equity and diversity," "ethics in engineering education," "ethics and leadership," "internationalization," "sustainability," "the global engineer." AAEE conference proceedings (Australasian Association for Engineering Education) were found via Scopus, reference searches, and a search of the 2021 REES-AAEE online program. Inclusion for all publications was reference to decolonization in engineering education or STEM and was determined via titles and abstracts and, at times, full paper review. The authors' work and knowledge of publications supplemented the literature review.

References

Affolter, E. A., Litzler, E., Margherio, C., & Knaphus-Soran, E. (2019). Aligning your research methods with your social justice values. In *2019 Collaborative network for engineering and computing diversity conference*. https://peer.asee.org/31741

Agrawal, A., Heydenrych, H., & Harding, G. (2022). Decolonising an engineering curriculum through a community-driven design project. *Southern Journal of Engineering Education, 1*, 39–60. Retrieved from https://sjee.org.za/ojs/index.php/sjee/article/view/10

AIATSIS. (n.d.). *The Mabo case*. https://aiatsis.gov.au/explore/mabo-case

American Indian Science and Engineering Society (AISES). (2022). https://conference.aises.org/

Anderson, I., & Jordan, S. S. (2018). Engineering connections in a Native American community and culture. In *2018 ASEE annual conference & exposition*. https://peer.asee.org/30405

Ansari, A., Hoffmann, A. L., Gurses, S., & Sloane, M. (2021). Technology, equity and social justice roundtable. In *IEEE international symposium on technology and society (ISTAS)*. https://ieeexplore.ieee.org/document/9629208

Baillie, C. (2011). A multidisciplinary approach to curriculum development for engineering graduates who are socially and environmentally just. In *2011 ASEE annual conference & exposition*. https://peer.asee.org/17353

Baker, J. D., Keahiolalo, R. M., Keahi-Wood, K., Cogbill, J., Naeole, C., & Turner, H. (2021). Developing a family engagement plan for Native Hawaiian and Pacific Islander STEM students in higher education: A review and critique of the literature. *Asia Pacific Viewpoint, 62*(1), 86–99. https://doi.org/10.1111/apv.12288

Baron, P. (2018). Heterarchical reflexive conversational teaching and learning as a vehicle for ethical engineering curriculum design. *Constructivist Foundations, 13*(3), 309–319. http://constructivist.info/13/3/309

Battiste, M. (2011). *Reclaiming Indigenous voice and vision*. UBC Press.

Benning, J., Surovek, A., Dolan, D. Wilson, L., Thompson, A., & Pyatt, R. (2014). Cultural considerations in service learning with American Indian Reservation community stakeholders. *2014 IEEE Frontiers in Education Conference (FIE)*. https://doi.org/10.1109/FIE.2014.7044323

Betts, R. F. (2004). *Decolonization*. Routledge. ISBN. 9781134368372

Black, J. (2018, September 6). *Te Ao Rangahau*. Engineering New Zealand Te Ao Rangahau. www.engineeringnz.org/news-insights/te-ao-rangahau/

Boudreau, J., & Anis, H. (2019). Client-based cornerstone design projects. *2019 Canadian Engineering Education Association Conference (CEEA-ACÉG)*. https://doi.org/10.24908/pceea.vi0.13750

Boven, K., & Morohashi, J. (2002). *Best practices using indigenous knowledge*. Nuffic & UNESCO/MOST. http://web.uvic.ca/fnpp/bestpractices.pdf

Campbell, J. L., Michael, R. N., & Crough, J. (2020). Embedding Indigenous perspectives in a first-year engineering design challenge course. In *Australasian association for engineering education (AAEE 2020) annual conference*. https://search.informit.org/doi/10.3316/informit.725847413523972

Carrigan, C., Tanguay, S. K., Yen, J., Ivy, J. S., Margherio, C., Riskin, E. A., Grant, C. S., & Horner-Devine, M. C. (2018). Building and breaching boundaries: An intersectional coherent group approach to advancing women faculty in engineering. In *2018 ASEE annual conference & exposition*. https://peer.asee.org/30162

Castillo-González, M. C., Sánchez-Hernández, A. L., Lugo, N., & Pérez-Lezama, C. V. (2021). Decolonial ethics in training in computational engineering. A qualitative study of professors' and students' perspectives. In *2021 world engineering education forum/global engineering deans council (WEEF/GEDC)*, Madrid, Spain (pp. 369–374). https://doi.org/10.1109/WEEF/GEDC53299.2021.9657297

Catalano, G. D. (2016). Learning about design from the Lakota Nation. In *2016 ASEE annual conference & exposition*. https://peer.asee.org/25533

Cole, P., & O'Riley, P. (2017). Performing survivance: (Re)storying STEM education from an Indigenous perspective. *Critical Education, 8*(15), 24–40. http://ojs.library.ubc.ca/index.php/criticaled/article/view/186260

Colston, N. M., Turner, S. L., Mason Chagil, G., Jacobs, S. C., & Johnson, S. (2019). Exploring the career thinking of Native American engineering students. In *2019 ASEE annual conference & exposition*. https://peer.asee.org/32811

Cooke, L. A., Lungu, B., & Samuels, L. (2019). Virtual engagement: A nexus between internationalisation and decolonisation of the curriculum in the PEESA III Project? In *Balkan region conference on engineering and business education*. https://doi.org/10.2478/cplbu-2020-0005

Cruz, C. C. (2021a). Brazilian grassroots engineering: A decolonial approach to engineering education, *European Journal of Engineering Education, 46*(5), 690–706. https://doi.org/10.1080/03043797.2021.1878346

Cruz, C. C. (2021b). Decolonizing philosophy of technology: Learning from bottom-up and top-down approaches to decolonial technical design. *Philosophy & Technology, 34*, 1847–1881. https://doi.org/10.1007/s13347-021-00489-w

Cutler, S., Borrego, M., & Loden, D. (2010). Evaluation of the engineers without borders challenge at Western Australia Universities. *Engineers Australia*. https://search.informit.org/doi/10.3316/informit.307681549810543

Dandridge, T. M., Al Yagoub, H. A., Cleare, S., Major, J. C., Raye, S. L., Wright, C. E., & Pawley, A. L. (2019). Engaging in STEM education equity work through a course: Studying race, class, and gender theory in engineering education. In *2019 collaborative network for engineering and computing diversity conference*. https://peer.asee.org/31757

Darwin, J. (2000). Diplomacy and decolonization. *Journal of Imperial and Commonwealth History, 28*(3), 5–24. https://doi.org/10.1080/03086530008583096

Daza, S. L. (2012). Complicity as infiltration: The (im)possibilities of research with/in NSF engineering grants in the age of neoliberal scientism. *Qualitative Inquiry, 18*(9), 773–786. https://doi.org/10.1177/1077800412453021

Denzin, N. K., Lincoln, Y. S. & Tuhiwai Smith, L. (2008). *Handbook of critical and Indigenous methodologies*. SAGE. ISBN: 9781412918039.

de Roock, R. S., & Baildon, M. (2019). MySkillsFuture for students, STEM learning, and the design of neoliberal citizenship in Singapore. *Cognition and Instruction, 37*(3), 285–305. https://doi.org/10.1080/07370008.2019.1624545

Dornian, D., Moshirpour, M., & Behjat, L. (2020). Description of pre-university coding workshop: Recruiting for diversity. *2020 Canadian Engineering Education Association (CEEA-ACÉG) Conference*. https://doi.org/10.24908/pceea.vi0.14152

Dow, C., & Gardiner-Garden, J. (1998, April 6). *Indigenous affairs in Australia, New Zealand, Canada, United States of America, Norway and Sweden*. Parliament of Australia. www.aph.gov.au/About_Parliament/Parliamentary_Departments/Parliamentary_Library/Publications_Archive/Background_Papers/bp9798/98Bp15

Droz, P. (2014). Biocultural engineering design: An Anishinaabe analysis for building sustainable nations. *American Indian Culture and Research Journal, 38*(4), 105–126.

Dubiel, C., Seniuk Cicek, J., Greene, R., Bailey, S., & Delijani, F. (2021). An adapted engineering design process: Guiding touchstones. In *2021 Canadian engineering education association (CEEA-ACÉG) conference*. https://doi.org/10.24908/pceea.vi0.14971

Eglash, R., Bennett, A., Babbitt, W., Lachney, M., Reinhardt, M., & Hammond-Sowah, D. (2020a). Decolonizing posthumanism: Indigenous material agency in generative STEM. *British Journal of Educational Technology, 51*(4), 1334–1353. https://doi.org/10.1111/bjet.12963

Eglash, R., Lachney, M., Babbitt, W., Bennett, A., Reinhardt, M., & Davis, J. (2020b). Decolonizing education with Anishinaabe arcs: Generative STEM as a path to indigenous futurity. *Education Technology Research and Development, 68*, 1569–1593. https://doi.org/10.1007/s11423-019-09728-6

Eichhorn, S. (2020). How the west was won: A deconstruction of politicised colonial engineering. *The Political Quarterly*, *91*(1), 204–209. https://doi.org/10.1111/1467-923X.12773

Eikenaar, J. H. (2018). Indigenizing the applied science curriculum: First steps. In *2018 Canadian engineering education association (CEEA-ACÉG) conference*. https://doi.org/10.24908/pceea.v0i0.12955

Eikenaar, J. H., Forssman, N., Webb, G., Patterson, L., & Eisenstein, A. (2022). Preparing engineering students for their professional obligations for meaningful engagement with Indigenous communities in Canada. In *2019 Canadian engineering education association (CEEA-ACÉG) conference*.

Engineers without Borders. (2010). *EWB challenge 2010 – Kooma Nation at Murra*. www.youtube.com/watch?v=3xHv9dY9-_k.

Fanon, F. (2004). *The wretched of the earth* (R. Philcox, Trans). Grove Press.

Fick, D. R., Gribb, M. M., & Tinant, C. J. (2013). The impact of project-based service learning in a native American community on Student Performance in Civil Engineering Capstone Design. In *2013 IEEE frontiers in education conference (FIE)*. https://doi.org/10.1109/FIE.2013.6684826.

Fomunyam, K. G. (2017a). Decolonising teaching and learning in engineering education in a South African university. *International Journal of Applied Engineering Research*, *12*(23), 13349–13358.

Fomunyam, K. G. (2017b). Decolonising the engineering curriculum in a South African university of technology. *International Journal of Applied Engineering Research*, *12*(17), 6797–6805.

Fomunyam, K. G. (2018). Decolonising the mind in engineering education in South African higher education. *International Journal of Mechanical Engineering and Technology*, *9*(10), 1529–1541. www.iaeme.com/IJMET/issues.asp?JType=IJMET&VType=9&IType=10

Forrest, R., & Seniuk Cicek, J. (2021). Rethinking the engineering design process: Advantages of incorporating Indigenous knowledges, perspectives, and methodologies. In *2021 Canadian engineering education association (CEEA-ACÉG) conference*. https://ojs.library.queensu.ca/index.php/PCEEA/article/view/14911

Foster, C. H. (2016). *Hybrid spaces for traditional culture and engineering: A narrative exploration of Native American women as agents of change* [Doctoral dissertation, Arizona State University]. https://hdl.handle.net/2286/R.I.38505

Foster, C. H., & Jordan, S. S. (2014). A philosophy of learning engineering and a Native American philosophy of learning; An analysis for congruency. In *2014 ASEE annual conference & exposition*. https://peer.asee.org/19976

Freire, P. (2005). *Pedagogy of the oppressed* (M. B. Ramos, Trans). Continuum International Publishing.

Frey, J. B., Maw, S., Bens, S., Bugg, J., & Sparling, B. (2019). Design of a completely new first year engineering program at the University of Saskatchewan – Part II. In *2019 Canadian engineering education association (CEEA- ACÉG) conference*. https://doi.org/10.24908/pceea.vi0.13793

Frey, J. B., Maw, S., Bens, S., & Sparling, B. (2018). Design of a completely new first year engineering program at the University of Saskatchewan. In *2018 Canadian engineering education association (CEEA-ACÉG) conference*. https://doi.org/10.24908/pceea.v0i0.13091

Friesen, M. R. (2019). Indigenizing the engineering curriculum: Some of the scenery along the way. *Manitoba Consulting Engineer*, *10*(MMXIX), 17–22. www.acec-mb.ca/resources/manitoba-consulting-engineer-volume-10/

Friesen, M. R., & Herrmann, R. (2018). Indigenous knowledge, perspectives, and design principles in the engineering curriculum. In *2018 Canadian engineering education association (CEEA- ACÉG) conference*. https://doi.org/10.24908/pceea.v0i0.12964

Friesen, M. R., Ibrahim, N., McSorley, G., & Mattucci, S. 2019. Engineers-in-residence programs as a framework for industry engagement in undergraduate engineering education: Challenges and opportunities. In *2019 Canadian engineering education association (CEEA- ACÉG) conference*. https://doi.org/10.24908/pceea.vi0.13701

Fritz, D. W., Strong, D. S., & Bryant, J. T. (2011). Ancestral engineering: towards the integration of archaeology and engineering. In *Proceedings of the Canadian engineering education association (CEEA- ACÉG)*. https://doi.org/10.24908/pceea.v0i0.3772

Garcia, G. A. (2018). Decolonizing Hispanic-serving institutions: A framework for organizing. *Journal of Hispanic Higher Education*, *17*(2), 132–147. https://doi.org/10.1177/1538192717734289

Gaudry, A., & D. Lorenz. (2018). Indigenization as inclusion, reconciliation, and decolonization: Navigating the different visions for indigenizing the Canadian academy. *AlterNative: An International Journal of Indigenous Peoples*, *14*, 218–227.

Gerrard, D., & Boyer, E. (2022). Reflections on respect, relevance, reciprocity and responsibility: Engaging Indigenous youth in a virtual land-based STEM leadership program. In *2019 Canadian Engineering Education Association (CEEA- ACÉG) conference*.

Gibson, E. S. L. (2016). Remembering tomorrow: Wagon roads, identity and the decolonization of a First Nations landscape. *Public History Review, 23*, 25–42. ISSN 1833–4989

Goddard, G., Prpic, J. K., & Roberts, G. (2021). Aboriginal perspectives in engineering education practice and research: Barriers and enablers for building student understanding and cultural intelligence through remote project-based learning. In *Research in engineering education symposium and Australasian association for engineering education conference*.

Godfrey, E., & King, R. (2010). Scoping the opportunities to aid recruitment and retention of Indigenous students in engineering. In *Australasian association for engineering education annual conference*.

Goldfinch, T., & Hollis, X. (2016). Current engineering education and training practices on Aboriginal cultural heritage. In *Australasian association for engineering education annual conference*.

Goldfinch, T., Ilango, A., Roland, A., & Willis, J. (2014). Australian Indigenous culture and heritage in engineering project planning and the implications for engineering education. In *Australasian association for engineering education annual conference*.

Goldfinch, T., Jolley, L., Prpic, J. K., & Leigh, E. (2016). Australian engineering educators' perceptions of indigenous cultures and challenges of minority inclusion. In *44th SEFI conference*.

Goldfinch, T., & Kennedy, J. (2013). Understanding Indigenous consultation and engagement in engineering education. In *Australasian association for engineering education annual conference*.

Goldfinch, T., Kennedy, J., Leigh, E., Dawes, L., Prpic, J., & McCarthy, T. (2016). *Embedding Indigenous perspectives into engineering education*. Australian Government. https://Indigenousengineering.wordpress.com/

Goldfinch, T., Layton, C., & McCarthy, T. (2010). Encouraging cultural awareness in engineering students. In *20th Australasian association for engineering education annual conference*.

Goldfinch, T., Prpic, J. K. Jolly, L., Leigh, E., & Kennedy, J. (2017). Australian engineering educators' attitudes towards Aboriginal cultures and perspectives. *European Journal of Engineering Education, 42*(4), 429–444. https://doi.org/10.1080/03043797.2017.1328588

González y González, E. M., & Lincoln, Y. S. (2006). Decolonizing qualitative research: Non-traditional reporting forms in the academy. In N. K. Denzin & M. Gardina (Eds.), *Qualitative inquiry and the conservative challenge* (pp. 193–214). Left Coast Press.

Graham, B., Lortan, D., Maistry, S., & Walker, M. (2017). Using epistemic justice as a framework for underpinning articulation between technical and vocational education and training colleges and higher education engineering programmes. In *4th Biennial South African society for engineering education (SASEE) conference*.

Grommes, A. (2004). Learning from Native cultures: A cross cultural exchange. In *2004 ASEE annual conference & exposition*. https://peer.asee.org/13224

Guanes, G. (2021, June 30). *Decolonizing engineering education: A possible path* towards *a more collective, caring, and empathetic field*. SEFI. www.sefi.be/2021/06/30/decolonizing-engineering-education-a-possible-path-towards-a-more-collective-caring-and-empathetic-field/

Gupta, K. (2020). Some insights on social inclusion in engineering education. In *5th NA international conference on industrial engineering and operations management Detroit*. www.ieomsociety.org/detroit2020/papers/645.pdf

Handley, J., & Moje, E. B. (2018). "What the problem really was . . .": A preliminary exploration of youth problem definition in everyday contexts. In *2018 collaborative network for engineering and computing diversity conference (CoNECD)*. https://peer.asee.org/29504

Harrison, F. (1991). Ethnography as politics. In F. Harrison (Ed.), *Decolonizing anthropology: Moving further toward an anthropology for liberation*. American Anthropological Association.

Herrmann, R. (2012). The application of support mechanisms to increase the success of Aboriginal students. In *2012 Canadian engineering education association (CEEA-ACÉG) conference*. https://doi.org/10.24908/pceea.v0i0.4704

Herrmann, R. (2014). The engineering access program for Indigenous (First Nation/Metis/Inuit) students. In *Proceedings 2014 IEEE Canada international humanitarian technology conference (IHTC)*. https://ieeexplore.ieee.org/stamp/stamp.jsp?arnumber=7147545

Hess, J., & Strobel, J. (2013). Indigenous ways of doing: Synthesizing scholarly literature on ethno-engineering. *International Journal of Engineering, Social Justice, & Peace, 2*(2), 55–80. ISSN 1927–9434

Hollis, X., & Goldfinch, T. (2017). Experience, education, and training on Aboriginal cultural heritage in engineering. *2017 Australasian Journal of Engineering Education, 22*(1), 54–62. https://doi.org/10.1080/22054952.2017.1367362

Hoople, G. D., Mejia, J. A., Chen, D., & Lord, S. M. (2020). Reimagining energy year 2: Integrating CSPs into course development. In *2020 ASEE virtual annual conference*. https://doi.org/10.18260/1-2-35135

Howard, M. A., & Kern, A. L. (2019). Conceptions of wayfinding: Decolonizing science education in pursuit of Native American success. *Cultural Studies of Science Education, 14*, 1135–1148. https://doi-org.uml.idm.oclc.org/10.1007/s11422-018-9889-6

Hughes, M. W., Bello-Mendoza, R., Cunningham, M., Sharp, K., McMeeking, S., & Manning, R. (2017). Towards integration of the Māori world view and engineering: A case study on student design projects for the Koukourārata community, Aotearoa/New Zealand. In *2017 Australasian association for engineering education (AAEE) conference*.

Hughes, M. W., Prpic, J. K., Goldfinch, T., & Kennedy, J. (2018). He Awa Whiria: Weaving Indigenous and western perspectives and creating inclusion in Australasian engineering education. In *2018 Australasian association for engineering education (AAEE) conference*.

Huntinghawk, F., Richard, C., Plosker, S., & Srivastava, G. (2020). Expanding cybersecurity knowledge through an Indigenous lens: A first look. In *2020 IEEE Canadian conference on electrical & computer engineering (CCECE)*. https://doi.org/10.1109/CCECE47787.2020.9255753

Irish, R., & Romkey, L. (2021). Using actor network theory to explore sustainability issues in an engineering & society course. In *2021 Canadian engineering education association (CEEA-ACÉG) virtual conference*. https://doi.org/10.24908/pceea.vi0.14960

Jalali, Y., & Matheis, C. (2017). Liberation in education: What role do liberatory praxis and theory play in fostering critical thinking? In *2017 ASEE annual conference & exposition*. https://doi.org/10.18260/1-2-28629

Johnson, A. (2020). A theoretical basis for addressing culture in undergraduate mining education. *International Journal of Engineering, Social Justice, & Peace*, 7(1), 1–20. https://doi.org/10.24908/ijesjp.v7i1.13586

Johnson, E. A. (2021). From critical thinking to reflection to citizen engineers. In *2021 Canadian engineering education association (CEEA-ACÉG) virtual conference*. https://doi.org/10.24908/pceea.vi0.14820

Johnson, E. A. (2016). *Mitigating social risk in the extractive sector: Developing intercultural competence as a tool for negotiating western-Indigenous perspectives within the undergraduate mining engineering curriculum* [Doctoral thesis, Queen's University]. https://qspace.library.queensu.ca

Jordan, S. S. (2015). CAREER: Engineering design across Navajo culture, community, and society. In *ASEE annual conference & exposition*. https://doi.org/10.18260/p.23668

Jordan, S. S., Betoney, C. A., Pangan, T. J. D., Anderson, I., & Fernandez, J. A. (2018). Results from the implementation of culturally relevant engineering design curriculum for the Navajo Nation and future directions. In *2018 ASEE annual conference & exposition*. https://peer.asee.org/30081

Jordan, S. S., Foster, C. H., Anderson, I. K., Betoney, C. A., & Pangan, T. J. D. (2019). Learning from the experiences of Navajo engineers: Looking toward the development of a culturally responsive engineering curriculum. *Journal of Engineering Education, 108*, 355–376. https://doi.org/10.1002/jee.20287

Jordan, S. S., White, K., Anderson, I. K., Betoney, C. A., Pangan, T. J. D., & Foster, C. H. (2017). Culturally relevant engineering design curriculum for the Navajo Nation. In *2017 ASEE annual conference & exposition*. https://peer.asee.org/27902

Kamanzi, B. (2015, March 29). *'Rhodes must fall' – Decolonisation symbolism – What is happening at UCT, South Africa?* The Postcolonist. http://postcolonialist.com/civil-discourse/rhodes-must-fall-decolonisation-symbolism-happening-uct-south-africa/

Kant, J. M., Thunder, W. H. H. I., Burckhard, S. R., & Meyers, R. T. (2015). Why don't more American Indians become engineers in South Dakota? *International Journal of Engineering, Social Justice, & Peace*, 4(1–2), 17–34. https://doi.org/10.24908/ijesjp.v4i1-2.5992

Kanyarusoke, K. E., & Ngonda, T. 2017. Decolonising engineering education in sub-Saharan Africa: Some perspectives. In *4th Biennial South African society for engineering education conference (SASEE)*.

Kennedy, J., Goldfinch, T. Leigh, E., Kaya Prpic, J., Mccarthy, T., & Dawes, L. (2016). *A beginner's guide to incorporating Aboriginal perspectives into engineering curricula*. Australian Government Office for Learning and Teaching. https://indigenousengineering.files.wordpress.com/2016/07/eac-v4-small.pdf

Kilada, G., Thomsen, V., Seniuk Cicek, J., Mante, A. A., & Herrmann, R. (2021). The impact of Indigenous knowledges and perspectives in engineering education: One student's story. In *2021 Canadian engineering education association (CEEA-ACÉG) virtual conference*. https://doi.org/10.24908/pceea.vi0.14909

Kinsner, W. (2015). Expanding the body of knowledge concept for professional practitioners. In *2015 Canadian engineering education association (CEEA-ACÉG) conference*. https://doi.org/10.24908/pceea.v0i0.5828

Kinsner, W., Masood, N., Herrmann, R., Nasri, M., Van Dang, H., & Marcynuk, K. (2014). Pre-university outreach through a space camp and a Kirkness discovery camp. In *2014 Canadian engineering education association (CEEA-ACÉG) conference*. https://doi.org/10.24908/pceea.v0i0.5934

Kirkness, V. J., & R. Barnhardt (2001). First Nations and higher education: The four R's -respect, relevance, reciprocity, responsibility. In R. Hayoe & J. Pan. Hong Kong (Eds.), *Knowledge across cultures: A contribution to dialogue among civilizations*, Comparative Education Research Centre, The University of Hong Kong.

Koh, R. (2020). Engagement in practice: Community engaged capstone design experience. In *2020 ASEE virtual annual conference*. https://doi.org/10.18260/1-2-34534

Koh, R., & Rossmann, J. S. (2021). Strategic disruptions toward a more liberatory engineering education. In *2021 ASEE virtual annual conference*. https://peer.asee.org/37733

Krige, J. (2006). *American hegemony and the postwar reconstruction of science in Europe*. The MIT Press.

Kutay, C., & Leigh, E. (2017). Aboriginal engineering for an enduring civilisation. In *Australasian association for engineering education conference*. https://opus.lib.uts.edu.au/bitstream/10453/121485/1/Kutay%20and%20Leigh%20Camera%20Ready.pdf

Kutay, C., Leigh, E., & Herkness, S. (2021). First peoples engineering – Creating cultural spaces. In *Research in engineering education symposium (REES) & Australasian association for engineering education (AAEE)*.

Langdon, G. S. (2013). Sustainability, citizenship and ethics within the mainstream engineering curriculum? In *2nd Biennial conference of the South African society for engineering education (SASEE)*.

Leigh, E., Goldfinch, T., Dawes, L., Prpic, J., McCarthy, T., & Kennedy, J. (2015). Shifting the focus. In *2015 Australasian association for engineering education (AAEE)*.

Leigh, E., Goldfinch, T., Prpic, J. K., Dawes, L., Kennedy, J., & McCarthy, T. (2014). Shared values: Diverse perspectives – engaging engineering educators in integrating Indigenous engineering knowledge into current curricula. In *2014 annual conference of the Australasian association for engineering education (AAEE)*.

Leydens, J. A., & Morgan, T. K. K. B., & Lucena, J. C. (2017). Mechanisms by which Indigenous students achieved a sense of belonging and identity in engineering education. In *2017 ASEE annual conference & exposition*. https://doi.org/10.18260/1-2-28661

Lord, S. M., Mejia, J. A., Chen, D. A., & Hoople, G. D. (2019a). Starting a dialogue on decolonizing engineering education. In *2019 Frontiers in education (FIE)*. https://doi.org/10.1109/FIE43999.2019.9028477.

Lord, S. M., Mejia, J. A., Luckett, K., Wolmarans, N., Mochekoane, N., & Mukhar, V. (2019b). Decolonizing engineering education: Where do we start? In *Research in engineering education symposium (REES)*.

Lourens, A. (2017). Feedback from first-year industrial engineering students: Influences on student retention. In *4th Biennial South African society for engineering education (SASEE) conference*.

Lunney Borden, L., & Wiseman, D. (2016). Considerations from places where Indigenous and western ways of knowing, being, and doing circulate together: STEM as artifact of teaching and learning. *Canadian Journal of Science, Mathematics and Technology Education*, *16*(2), 140–152. https://doi.org/10.1080/14926156.2016.1166292

Maldonado-Torres, N. (2007). On the coloniality of being: Contributions to the development of a concept. *Cultural Studies*, *21*(2–3), 240–270.

Maldonado-Torres, N. (2016). *Outline of ten theses on coloniality and decoloniality*. Fondation Frantz Fanon. https://fondation-frantzfanon.com/wp-content/uploads/2018/10/maldonado-torres_outline_of_ten_theses-10.23.16.pdf

Mante, A., Friesen, M., & Atamanchuk, K. (2020). Advancing engineering education towards design for sustainable development at the University of Manitoba: Initial assessment. In *2020 Canadian engineering education association (CEEA-ACÉG) conference*. https://ojs.library.queensu.ca/index.php/PCEEA/article/view/14184

Mante, A. A., Seniuk Cicek, J., Friesen, M., & Kavanagh, L. (2019). Assessing the impact of integrating Indigenous knowledges and perspectives in engineering curricula on student's learning: The development of a classroom case study. In *2019 Canadian engineering education association (CEEA-ACÉG) conference*. https://doi.org/10.24908/pceea.vi0.13881

Marker, M. (2019). Indigenous STEM success stories as disquieting decolonization: Thoughts on new times and, old thoughts about place-ness. *Cultural Studies of Science Education, 14*, 1149–1154. https://doi.org/10.1007/s11422-018-9887-8

McCarthy, T., Wyman, L., & Nicholas, S. (2013). Activist ethnography with Indigenous youth: Lessons from humanizing research on language and education. In *Humanizing research: Decolonizing qualitative inquiry with youth and communities* (pp. 81–103). SAGE.

McGowan, V. C., & Bell, P. (2020). Engineering education as the development of critical sociotechnical literacy. *Science & Education, 29*, 981–1005. https://doi-org.uml.idm.oclc.org/10.1007/s11191-020-00151-5

Mejia, J. A., & de Paula, M. N. (2019). "Ingeniero como vos": An analysis of the Mbyá-Guaraní practices associated with engineering design. In *2019 ASEE annual conference & exposition*. https://doi.org/10.18260/1-2-31921

Mejia, J. A., & Pulido, A. L. (2018). (Fundamental) fregados pero no jodidos: A case study of latinx rasquachismo. In *2018 ASEE annual conference & exposition*. https://peer.asee.org/29650

Mignolo, W. D. (2007). Delinking: The rhetoric of modernity, the logic of coloniality and the grammar of decoloniality. *Cultural Studies, 21*(2–3), 449–514.

Mitchell, S. (2006). SPIES: An evaluation of tutoring programs for Māori and Pasifika students at the Faculty of Engineering, University of Auckland. In *Australasian association for engineering education (AAEE) annual conference*.

Mkansi, M., Emwanu, B., & Kuchwa-Dube, C. (2018). The concept of decolonisation within engineering education. In *International conference on industrial engineering and operations management Pretoria*.

Moore, K., Walton, R., & Jones, N. N. (2021). Redressing inequities within our margin of maneuverability: A narrative inquiry study. In *2021 ASEE virtual annual conference content access*. https://peer.asee.org/37648

Morgan, T. K. K. B. (2006). Ngā hoa o te kupenga rorohiko – an innovative distance learning programme to address the paucity of Māori student enrolments in engineering. In *Australasian association for engineering education (AAEE) annual conference*.

Morgan, T. K. K. B. (2008). *The value of a Hapū perspective to municipal water management practice: Mauri and potential contribution to sustainability decision making in Aotearoa New Zealand* [Thesis]. http://librarysearch.auckland.ac.nz/primo_library/libweb/action/display.do?fn=search&doc=uoa_voyager1833889&vid=UOA2_A.

Motala, S. (2017). In/between science and art: Posthumanist ruminations on geomatics education. In *4th Biennial South African society for engineering education (SASEE) Conference*.

Mudondo, T. (2020). Towards a new philosophy for repositioning and repurposing technical and vocational education and training for industrial engineering and operations management systems and development in Africa: A case of Zimbabwe. In *2nd African international conference on industrial engineering and operations management*.

Murray, M. R., & Morgan, T. K. K. B. (2009). An Indigenous approach to engineering effective learning opportunities. In *20th Australasian association for engineering education (AAEE) conference*.

Mutch, S., Borland, M., & Mercer, K. (2021). Engineering, patriarchy, and the pluriverse: What world of many worlds do we design? What worlds do we teach? In *2021 Canadian engineering education association (CEEA-ACÉG) conference*. https://doi.org/10.24908/pceea.vi0.14915

Navickis-Brasch, A. S., Kern, A. L., Fiedler, F., Cadwell, J. R., Laumatia, L., Haynie, K. C., & Meyer, C. (2014). Restoring water, culture, and relationships: Using a community-based participatory research methodology for engineering education. In *2014 ASEE annual conference & exposition*. https://10.18260/1-2-22980

Nkrumah, K. (1965). *Neo-colonialism: The last stage of imperialism*. International Publishers.

Ochoa-Duarte, A., & Peña-Reyes, J. I. (2020). Engineering education for buen vivir in the context of 4th Industrial Revolution. In *2020 IEEE world conference on engineering education (EDUNINE)*. https://doi.org/10.1109/EDUNINE48860.2020.9149566

Omanga, D. (2020). *Decolonization, decoloniality, and the future of African Studies: A conversation with Dr. Sabelo Ndlovu-Gatsheni*. Social Science Research Council.

Page-Reeves, J., Marin, A., Moffett, M. DeerInWater, K., & Medin, D. (2019). Wayfinding as a concept for understanding success among Native Americans in STEM: "*Learning how to map through life*". *Cultural Studies of Science Education, 14*, 177–197. https://doi.org/10.1007/s11422-017-9849-6

Paperson, L. 2014. A ghetto land pedagogy: An antidote for settler environmentalism. *Environmental Education Research, 20*(1), 115–130.

Pawley, A. L. (2013). Learning from small numbers" of underrepresented students' stories: Discussing a method to learn about institutional structure through narrative. In *2013 ASEE annual conference & exposition*. https://peer.asee.org/19030

Perks, A., & Dimitrova, R. (2019). How to win friends and influence industry collaborators. In *2019 Canadian engineering education association (CEEA-ACÉG) conference*. https://doi.org/10.24908/pceea.vi0.13867

Pido, D., Khamala, M., & Pido, O. (2019). Decolonizing 'technology' in Kenyan Higher Education and implications for development. In M. N. Amutabi (Ed.), *A new kind of development in Africa* (pp. 91–1030. CEDRED.

Pieri, R. V., Allard, A. J., Allery, T. A., Vallie, A., Bowen, B., Haefner, K., & Nelson, L., Luecke, D., & Parker, M. M. (2020). Pre-engineering collaboration as a tool to facilitate decolonization of Native American students. In *2020 ASEE virtual annual conference content access*. https://doi.org/10.18260/1-2-35071

Prpic, J. K. (2015). *National indigenous engineering summit*. Final report.

Prpic, J. K., & Bell, D. (2018). Designing an on-country engineering education experience in collaboration with the Gunditjmara community. In *29th Australasian association for engineering education (AAEE) annual conference*.

Prpic, J. K., & Goldfinch, T. (2021). Aboriginal perspectives in engineering education practice and research – understanding and appreciating relationships. In *Research in engineering education symposium (REES) & Australasian association for engineering education (AAEE) conference*.

Quijano, A. (2007). Coloniality and modernity/rationality. *Cultural Studies, 21*(2–3), 168–178.

Raina, D. 2017. Engineering science education and the Indian Institutes of Technology: Reframing the context of the 'cold war and science' (1950–1970). *Contemporary Education Dialogue, 14*(1), 49–70. https://doi.org/10.1177/0973184916678698

Riley, D. M. (2018). Refuse, refute, resist: Alt-right attacks on engineering and STEM education diversity scholarship. In *2018 collaborative network for engineering and computing diversity (CoNECD)*. https://peer.asee.org/29572

Rodney, W. (1973). *How Europe underdeveloped Africa*. Bogle-L'Ouverture.

Rodríguez-Seeger, C., Sáez-Hueichapan, D., Fuenzalida-Artigas, A., Ñancupil-Quirilao, I., Lienqueo, M. E., Contreras-Painemal, C., & Díaz-Alvarado, F. (2021). Decolonizing the training of engineers and scientists: The case of the Faculty of Physical Sciences and Mathematics at Universidad de Chile. *Scholarship of Teaching and Learning in the South, 5*(1), 87–106. https://doi.org/10.36615/sotls.v5i1.181

Rogers, C. C., & Valdez, R. S. (2021). Designing for diversity, equity, and inclusion in systems engineering education. In *2021 ASEE virtual annual conference*. https://doi.org/10.18260/1-2-36929

Ruric Vogel, F., & Human-Vogel, S. (2018). The relevance of identity style and professional identity to academic commitment and academic achievement in a higher education setting. *Higher Education Research & Development, 37*(3), 620–634. https://doi.org/10.1080/07294360.2018.1436526

Ruta, A., J. Seniuk Cicek, A. Mante, M. Speare, & R. Herrmann. (2021). Ten calls to action to integrate Indigenous knowledges and perspectives into the Biosystems Engineering Program at the University of Manitoba. *Canadian Biosystems Engineering/Le genie des biosystems au Canada, 63*(1), 9.1–9.17. https://doi.org/10.7451/CBE.2021.63.9.1

Salgado, H., Urquidi Cerros, Y. A., Kendall, M. R., & Strong, A. C. (2021). Faculty perceptions of, and approaches towards, engineering student motivation at Hispanic-serving institutions. In *2021 ASEE virtual annual conference*. https://peer.asee.org/37186

Santiago-Ortiz, A. (2019). From critical to decolonizing service-learning: Limits and possibilities of social justice-based approaches to community service-learning. *Michigan Journal of Community Service Learning, 24*(2), 43–54. https://doi.org/10.3998/mjcsloa.3239521.0024.204

Seniuk Cicek, J., Burgart, A. L., Rogalski, D., Gauthier, S., Mattucci, S., Bazylak, J., Mante, A., Robinson, M., Herrmann, R., & Staub-French, S. (2020). Indigenous initiatives in engineering education in Canada: Collective contributions. In *2020 Canadian engineering education association (CEEA-ACÉG) conference*. https://doi.org/10.24908/pceea.vi0.14162

Seniuk Cicek, J., Friesen, M., & Bailey, S. (2019). Transdisciplinary design build studio course in collaboration with Shoal Lake 40 First Nation: Development and assessment. In *Frontiers in education (FIE) conference*. https://doi.org/10.1109/FIE43999.2019.9028515

Seniuk Cicek, J., Herrmann, R., Forrest, R., & Monkman, K. (2022). Decolonizing and Indigenizing engineering: The design & implementation of a new course. In *2022 Canadian engineering education association (CEEA-ACÉG) conference*.

Seniuk Cicek, J., Mante, A. A., Friesen, M., & Herrmann, R. (2019a). Enhancing engineering education with Indigenous cultures, pedagogies, knowledges, and perspectives: A series of faculty workshops. In *2019 Canadian engineering education association (CEEA-ACÉG) conference*. https://doi.org/10.24908/pceea.vi0.13835

Seniuk Cicek, J., Mante, A., Geddert, C., & Fontaine, L. (2019b). Engineering education re-interpreted using the Indigenous sacred hoop framework. In *Research in engineering education symposium (REES)*. www.proceedings.com/49995.html

Seniuk Cicek, J., Mante, A., Herrmann, R., & Friesen, M. (2021a). Indigenous knowledges and perspectives in engineering education: Team reflections on a series of faculty workshops. In *Research in engineering education symposium (REES) & Australasian association for engineering education (AAEE)*.

Seniuk Cicek, J., Steele, A., Gauthier, S., Mante, A. A., Wolf, P., Robinson, M., & Mattucci, S. (2021b). Indigenizing engineering education in Canada: Critically considered. *Teaching in Higher Education*, *26*(7–8), 1038–1059. https://doi.org/10.1080/13562517.2021.1935847

Shahjahan, R. A., Estera, A. L., Surla, K. L., & Edwards, K. T. (2022). "Decolonizing" curriculum and pedagogy: A comparative review across disciplines and global higher education contexts. *Review of Educational Research*, *92*(1), 73–113. https://doi.org/10.3102/00346543211042423

Skopec, M., Fyfe, M., Issa, H., Ippolito, K., Anderson, M., & Harris, M. (2021). Decolonization in a higher education STEMM institution – is "epistemic fragility" a barrier? *London Review of Education*, *19*(1), 1–21. https://doi.org/10.14324/LRE.19.1.18

Stacey, L., Maksymyk, C., Scherer, M., & Wells, M. (2017). Stem outreach for Indigenous youth – Lessons learned in over a decade of programming. In *Canadian engineering education association (CEEA-ACÉG) conference*. https://doi.org/10.24908/pceea.v0i0.7383

Steenkamp, H., & Muyengwa, G. (2017). Industrial engineering as a career choice at the University of Johannesburg. In *4th Biennial South African society for engineering education (SASEE) conference*.

Swingler, H. (2021, October 6). Year-old San and Khoi Centre adds invaluable indigenous knowledge to archive. *University of Cape Town News*. www.news.uct.ac.za/article/-2021-10-06-year-old-san-and-khoi-centre-adds-invaluable-indigenous-knowledge-to-archive

Thomsen, V., Seniuk Cicek, J., Mante, A. A., Bailey, S., & Delijani, F. (2021). One settler student's story on incorporating Indigenous knowledges in engineering curricula. In *2021 Canadian engineering education association (CEEA-ACÉG) virtual conference*. https://doi.org/10.24908/pceea.vi0.14880

Todd, Z. (2017). Fish, kin and hope: Tending to water violations in amiskwaciwâskahikan and Treaty Six territory. *Afterall*, *43*, 103–107.

Todd, Z. (2018). Refracting the state through human-fish relations: Fishing, indigenous legal orders and colonialism in north/western Canada. *Decolonization: Indigeneity, Education & Society*, 7(1), 60–75.

Truth & Reconciliation Commission of Canada (2015). *Truth & reconciliation commission of Canada: Calls to action*. https://www2.gov.bc.ca/assets/gov/british-columbians-our-governments/indigenous-people/aboriginal-peoples-documents/calls_to_action_english2.pdf

Tuck, E., & K. W. Yang. (2012). Decolonization is not a metaphor. *Decolonization: Indigeneity, Education & Society*, *1*(1), 1–40.

Tuhiwai Smith, L. (1999). *Decolonizing methodologies: Research and Indigenous peoples*. Zed Books. 199 pp. ISBN 978-1-85649-624-7

Tuhiwai Smith, L. (2012). *Decolonizing methodologies: Research and Indigenous peoples* (2nd ed.). Zed Books. ISBN 978-1-84813-951-0

Universities Australia & Indigenous Higher Education Advisory Council (IHEAC) (Australia). (2011). *Guiding principles for developing Indigenous cultural competency in Australian universities*. Universities of Australia. http://hdl.voced.edu.au/10707/193461

University of Manitoba. (2022). *Engineering access program*. Price Faculty of Engineering. https://umanitoba.ca/engineering/engineering-access-program

Valle, J., & Bowen, C. L., & Riley, D. M. (2021). Liberatory potential of labor organizing in engineering education. In *2021 ASEE virtual annual conference*. https://peer.asee.org/37459

van der Laan, J. M. (2001). Plastic words: Words without meaning. *Bulletin of Science, Technology & Society*, *21*(5), 349–353.

Velarde, C., & Svihla, V., & Gutierrez, E. K. (2021). A tool for informing community-engaged projects. In *2021 ASEE virtual annual conference*. https://doi.org/10.18260/1-2-36625

Velez, C., Nuechterlein, B., Connors, S., RedShirt Tyon, G., Roane, T. M., & Mays, D. C. (2022). Application of the Indigenous evaluation framework to a university certificate program for building cultural awareness in science, technology, engineering, and mathematics. *Evaluation and Program Planning*, *92*. https://doi.org/10.1016/j.evalprogplan.2022.102066

Veracini, L. (2013). 'Settler colonialism': Career of a concept. *The Journal of Imperial and Commonwealth History*, *41*(2), 313–333.

Villa, C. G., & Gonzalez, E. (2011). The right to education for female engineering students in Mexico: Cultural considerations in their retention. In *2011 ASEE annual conference & exposition*. https://peer.asee.org/18952

wa Thiong'o, N. (1986). *Decolonising the mind: The politics of language in African literature*. East African Educational Publishers.

Waapalaneexkweew (N. R. Bowman-Farrell, Mohican/Lunaape). (2018). Looking backward but moving forward: Honoring the sacred and asserting the sovereign in Indigenous evaluation. *The American Journal of Evaluation*, *39*(4), 543–568. https://doi.org/10.1177/1098214018790412

Walther, J., Pawley, A. L., & Sochacka, N. W. (2015). Exploring ethical validation as a key consideration in interpretive research quality. In *2015 ASEE annual conference & exposition*. https://peer.asee.org/24063

Walwyn, D. R., & Combrinck, C. (2021). Epistemic justice during a global pandemic: Transforming curricula and pedagogical practices to improve student experiences of innovation studies. *Industry and Higher Education*, *35*(5), 598–608. https://doi.org/10.1177/0950422220987088

Whyte, K. P., & Cuomo, C. (2017). Ethics of caring in environmental ethics: Indigenous and feminist philosophies. In: S. M. Gardiner & A. Thompson (Eds.), *The Oxford handbook of environmental ethics* (pp. 234–247). Oxford University Press.

Wilson, J. A., Gaines, J. E., & Cooper, D. (2015). Bulls-eye mentoring: Developing a program intervention in the college of engineering. In *2015 ASEE annual conference & exposition*. https://peer.asee.org/23650

Wilkinson, C., Macfarlane, A. H., Hikuroa, D. C. H., McConchie, C., Payne, M., Holmes, H., Mohi, R., & Hughes, M. W. (2021). Landscape change as a platform for environmental and social healing. *Kōtuitui: New Zealand Journal of Social Sciences Online*, *17*(3), 352–377. https://doi.org/10.1080/1177083X.2021.2003826

Wilmer, F. (1993). *The Indigenous voice in world politics*. Sage.

Winberg, S., & Winberg, C. (2017). Using a social justice approach to decolonize an engineering curriculum. In *2017 IEEE global engineering education conference (EDUCON)*.

Wiseman, D., & Herrmann, R. (2019). Good, now keep going: Challenging the status quo in STEM pipeline and access programs. *Cultural Studies of Science Education*, *14*, 129–137. https://doi.org/10.1007/s11422-018-9863-3

Wolfe, P. (2006). Settler colonialism and the elimination of the native. *Journal of Genocide Research*, *8*(4), 387–409.

Zacharias, K., Seniuk Cicek, J., Wallace, N., & Mercer, K. (2022). Surveying land-based learning for engineering education: Preliminary steps. In *2022 Canadian engineering education association (CEEA-ACÉG) conference*.

Zembylas, M., Bozalek, V., & Motala, S. (2019–2020). A pedagogy of hauntology: Decolonizing the curriculum with GIS. *Capacious: Journal for Emerging Affect Inquiry*, *2*(1–2).

5

Developing a Global and Culturally Inclusive Vision of Engineering Ethics Education and Research

Diana A. Martin, Alison Gwynne-Evans, Aleksandra A. Kazakova, and Qin Zhu

1 Introduction

Engineering ethics education has traditionally emphasized the individual responsibility of engineers, as laid out in professional codes of ethics, while neglecting the collective nature of engineering practice and the global and societal responsibilities of the profession. The chapter expands the scope of established approaches to engineering ethics education by accounting for global and culturally inclusive practices. It aims to (1) identify regional authorship patterns as well as the major institutional and national actors in engineering ethics education research, (2) survey the methods used to conduct research on global and culturally inclusive practices in engineering ethics education, and (3) survey recent pedagogical and institutional practices to broaden engineering ethics education towards a global and culturally inclusive vision.

We start by making the case for a global vision of engineering ethics education rooted in an explicit commitment to value-ladenness, and we position the term "global" in this context. We then describe our methodological approach, which combines bibliometric analysis with a systematic literature review. In the following sections, we present a mapping of regional authorship patterns in engineering ethics education research (Section 4), followed by an overview of the methodological approaches employed in research describing global and culturally inclusive practices (Section 5) and the presentation of the learning goals, teaching practices, and institutional strategies reported in the literature (Section 6). We conclude with a list of recommendations for deepening the work and directions traced in this chapter.

2 Background and Rationale for a Global and Culturally Inclusive Engineering Ethics Education

The prevalent approach to engineering ethics education emphasizes the role and responsibilities of the individual engineer within the engineering sphere according to their professional duties as laid out in ethical codes (Bielefeldt et al., 2016; Colby & Sullivan, 2008; Herkert, 2000; Hess & Fore, 2018; Martin et al., 2021). This approach focuses more on the application of ethical tools and resources (such as ethics codes) in addressing the professional ethical concerns in engineering practice than on questioning the moral quality of engineering per se. It is distanced from macro-ethical concerns with ensuring social justice, care, or promoting the collective responsibilities of

DOI: 10.4324/9781003287483-6

engineering (Herkert, 2001). At the heart of our argument is the assumption that engineering practice is not a neutral endeavor but one which is intimately linked with the values of the communities in which engineering is practiced and of engineers themselves. As such, educational practices, institutional strategies, and educational research need to engage explicitly with the value-laden nature of engineering and its situatedness in local and global contexts.

2.1 The Value-Laden Nature of Engineering Practice

Engineering artifacts are more than the outcome of the application of technical and scientific knowledge. They embed social, cultural, and political values and can have socioeconomic and political effects (Winner, 1986, 1990; Bijker, 1995). Marzano (1993), now a retired chief design officer of Electrolux, stated that "design is a political act," such that "every time we design a product, we are making a statement about the direction the world will move in." Engineering practice can implicitly incorporate and propagate dominant stereotypes, beliefs, or biases, therefore excluding the needs or characteristics of different categories of users and so exacerbating inequality and injustice.

For example, Michelfelder et al. (2017) highlight how the prevalent male perspective in technological design leads to artifacts that perpetuate cultural stereotypes or that may be unsafe for specific categories of users. This is the case of artificial hearts that were shown to fit only 20% of women due to the smaller size of their chest cavity (Deng, 2015), or the first generation of voice recognition technologies which failed to recognize women's voices due to their higher pitch (Palmiter Bajorek, 2019). Looking at how algorithms discriminate against Afro-Americans, Umoja Noble (2018) notices a systematic algorithmic culture of oppression. When entering the term "Black girls" in online search engines, Umoja Noble found that the top entries were sexual. This bias goes beyond search engines to affect electoral politics and financial markets (Pasquale, 2015). Financial services nowadays rely on predictive algorithms to determine the suitability of a credit application, estimating the probability that an applicant will default. They do so by determining the percentage of individuals who defaulted based on similar characteristics to the applicant. Pasquale found that race is a key variable considered in predictive analysis, resulting in a low percentage of credit applications approved for Afro-Americans. Bias is not the only problem. There is also an uneven distribution of risks brought by technological developments (Okereke, 2010), as "the populations most at risk from anthropogenic climate change are low-income people in developing countries, who do not gain much from the activities that give rise to climate change" (Hansson, 2017, p. 162).

2.2 Characteristics of a Value-Laden Approach

Engineering in general and its underlying rationale have historically been understood in terms of the assumed universal and universalizing character of technological globalization – be it transfer of ideas, artifacts, or modes of action. However, sociohistorical studies have revealed the diversity of practices that are merged under the umbrella term "engineering" (Christensen et al., 2012). These practices are context-specific with respect to the evolution of national schools, occupational demarcation, or the sectoral and functional divisions within the profession (Meiksins & Smith, 1996).

The globalization of engineering manifests itself through the increased mobility of engineers, the international reach of engineering firms and projects, the outreach of professional societies, and the impact of global accords on national educational systems (Washington Accord, Seoul Accord, EUR-ACE, etc.). The Accreditation Board for Engineering and Technology (ABET) is considered a key global player in regulating technical education, in particular regarding ethics requirements (Akera, 2017). Its activity and global role were critically debated over the last decade (Bucciarelli, 2008; Hess & Fore, 2018; Matos et al., 2017; Slaton & Riley, 2015), highlighting the domination exercised by the United States in the engineering education landscape (Anwar & Richards, 2013;

Klassen, 2018). The conceptual, methodological, and pedagogical debates around engineering ethics education evolved predominantly in Anglophone countries (see further Section 3.1). More recently, we see incipient debates in countries outside of the direct reach of ABET (Gwynne-Evans et al., 2021) and in the countries looking to establish accreditation models rooted in national qualification processes.

The discussion on the globalization of engineering highlights the extent to which the image, content, and cultural meaning of the profession (e.g., the status of professional societies and their ethical codes), as well as its reproduction through educational systems (e.g., the philosophies of science and technology underlying the engineering curricula), are conditioned by the cultural-historical status of developed and developing countries (Gispen, 1990; Mitcham & Kazakova, 2020; Pfotenhauer & Jasanoff, 2017a).

This is further evident when contextualizing engineering in the "non-Western," post-colonial, or post-socialist societies. Along with the deficits of the local translation of technologies and the transfer of "best practices" around the world (Akubue, 2017; Pfotenhauer & Jasanoff, 2017b), the globalization of engineering raises reflection not only along the break lines of cultures (Jing & Doorn, 2020) but also – through contrast and feedback – within the dominant culture of teaching engineering ethics (Luegenbiehl, 2004; Luegenbiehl & Clancy, 2017). Despite a sophisticated level of self-critique, the established dominance of the West-centric approach to moral action in engineering persists.

The challenges to the dominant West-centric approach are rooted in the prevailing values and inner contradictions of the neoliberal technocratic culture of modernity. This culture has combined in a paradoxical way the cult of individual creativity and success (as exemplified by the figure of the outstanding inventor and entrepreneur) with a fatalistic belief in the "inevitable progress" of technology (through selection on the markets as well as accumulation of scientific knowledge) that permits and even endorses the anonymizing of individual action and responsibility. A fascination with innovation may be seen to focus moral reflection on the extraordinary cases (rather than on mundane practices), discrete decisions (rather than communication processes), professional roles (rather than multifaceted interactions), individual inputs and heroic responses (rather than collective routines and collaborative practices), short-term consequences (rather than path dependency), evaluation of risks (rather than precaution), microethics (rather than contextualizing a moral action at the macro level). All these alternatives are discussed both in the empirical findings in science and technology studies and in the conceptual research in the philosophy of engineering, fostered by communities such as the Forum on Philosophy, Engineering and Technology, the Society for Philosophy and Technology or the Engineering, Social Justice and Peace network. Nevertheless, the advancements in the empirical understanding of engineering ethics education are still insufficiently translated into teaching practices. This is partly due to methodological gaps and partly due to the cultural inertia of technical education (Mitcham & Englehardt, 2019).

Thus, engineering ethics education requires the engagement with values in an explicit way. A driving force of engineering ethics education were the political and commercial imperatives at work within the dominant culture. This focus has depoliticized the role of the individual engineer from that of social activist or changemaker (Forbes et al., 2021; Karwat et al., 2015; Riley, 2008) to a "prisoner of the capitalist machine" (Conlon, 2019) or a "technological barbarian" (Barry, 2012). This professional identity emphasizes the compliance and conformity to technical and safety requirements. It can thus contribute to the individual complacency with personal achievements rather than the focus on influencing political structures or power relations. The political is value-laden, whereby the focus on a particular set of values may displace other values and may require a shift in role from *expert* (authority and thus privileged) to *stakeholder* (and thus one of several parties with vested interests). When the focus is shifted to include the "how" and "why" of values, this opens an uncomfortable space where power is negotiated and subverted (Martin et al., 2021). The focus of ethics education in engineering is thus shifted from the vision of engineering – which is highly political – to the individual accountability for discreet actions.

2.3 The Status Quo of Engineering Ethics Education and Research

The domination of the Western perspective in engineering ethics education is manifest in the curricula (Mitcham, 2009; Kanemitsu, 2018; Clancy & Hohberger, 2019; Balakrishnan et al., 2021; Zhang & Zhu, 2021), in textbooks (Fu et al., 2018; Balakrishnan et al., 2021; Bielefeldt et al., 2021), in ethical frameworks (Bielefeldt, 2019; Zhu & Jesiek, 2020; Balakrishnan et al., 2021; Jing & Doorn, 2020; Zhu, 2021), through topics borrowed from Western codes of ethics (Luegenbiehl, 2018), and in research (Balakrishnan et al., 2021; Nasir et al., 2021). As Zhang and Zhu (2021) point out, we have been witnessing the adoption of the "American-style" of engineering ethics education and research as a "global paradigm."

This paradigm seems to take for granted the cultural homogeneity and the familiarity of students with the value structure of the Western society (Luegenbiehl, 2010; Zhu & Jesiek, 2020). This is potentially problematic due to differences concerning the nature of ethical judgments and actions across cultures (Wang & Thompson, 2013; Clancy & Hohberger, 2019; Zhu & Jesiek, 2020), as well as the neglect of non-Western cultural beliefs and values (Jing & Doorn, 2020) and global development issues (Bielefeldt et al., 2021). Given the increasing complexity of societal problems and the need for value-pluralist technological responses, a global education paradigm seeking homogeneity is a disempowering form of education. By accepting the status quo of the dominance of West-centric beliefs, values, and approaches, engineering ethics education faces the risk of perpetuating a form of cultural imperialism (Clancy & Hohberger, 2019). Changing the status quo of engineering education and research requires a critical and self-reflective stance that considers socioeconomic and cultural contexts, geographical differences, prevalent biases, and cognitive assumptions.

In the context of increasing attempts to decolonize the curriculum and to dismantle the status quo of the prevalent Western perspective in engineering education, the chapter has three major aims. First, it aims to identify prevailing authorship patterns and the major regional actors in engineering ethics education research. This allows us to better understand how global engineering ethics education research really is. Second, the chapter aims to map the methods used to conduct research on global and culturally inclusive practices in engineering ethics education. This allows us to capture any regional differences in the use of research methods. Thirdly, the chapter aims to survey recent attempts to broaden practices towards a global and culturally inclusive vision of engineering ethics education. This allows us to present teaching and institutional strategies for broadening engineering ethics education that can serve as inspiration to instructors and program leaders. The chapter thus provides an examination of the current state of global and culturally inclusive engineering ethics education, as documented through research, as well as highlighting the major players who are part of this debate. It identifies the research methods used to report on global and culturally inclusive educational approaches, as well as the learning goals, teaching practices, and institutional strategies supporting this vision.

3 Methodology

Before setting our research strategy, a methodological starting point is to clarify what is meant by the term *global* and our positionality as researchers.

3.1 Reflection on Terminology

Our contribution attempts to include two facets of the term "global": the *techno-economic* facet and the *sociocultural* facet.

First, global interconnectedness has led to ample collaboration opportunities between corporations in different countries. The global techno-economic order has also allowed multinational corporations to reduce their costs in materials and human resources and maximize their economic returns. Nevertheless, critics point out that a global economy may lead to a new form of colonialism by further exploiting historically marginalized and disadvantaged communities. Therefore, political theorists such as Amartya Sen (1999, 2009) and Martha Nussbaum (2006, 2011) have proposed a list of human capabilities that are fundamental to human well-being globally. Universalist human rights or values have further been integrated into tackling grand challenges encountered by humanity, such as climate change and ethical dilemmas in AI. A fundamental assumption here is that solving these problems of a global scale requires that nations find common ground while putting aside the "conflicts" (or differences) between them. However, building global consensus, not to mention actionable programs, has been difficult due to the concern that most global initiatives are led by Western countries, and the values they deem fundamental do not necessarily resonate with non-Western cultures or developing countries. When engineers and policymakers collaborate on global initiatives, it is not viable to ignore the cultural differences.

Concerns regarding the value of cultural diversity on the global market led to the second dimension of the term "global," pertaining to the sociocultural facet. It acknowledges that countries have their unique historical and cultural traditions that generate different responses to "global forms" (e.g., neoliberalism). However, cultural anthropologists challenge the approach portraying non-Western cultures as passive recipients. Instead, cultures can also (pro)actively select global forms and creatively integrate them with locally situated ethics and politics to serve their own needs. The term *global assemblages* describes the (pro)active agency of responding to the totalizing power of globalization (Collier, 2006). This implies that any globalizing effort will inevitably have to be localized. Therefore, a global project benefits from being sensitive and adaptive to diverse value systems prevalent in different cultures. This sensitivity in turn needs to be translated into the engineering education curricula.

Striving to reach a global standard may lead to the adoption of a singular viewpoint, that of the most powerful actor. In contrast, we find the goal of (co)developing an intercultural community of practice, whereby resolutions to global challenges are collectively conceived and developed. This distinguishes *ethical pluralism* from *ethical relativism*, by allowing multiple interpretations of a single idea which are "irreducibly different from one another but connected and coherent with one another (not simply compatible) by way of their shared point of origin and reference" (Ess, 2006, p. 218). Ethical pluralism emphasizes positive engagement between two (or more) traditions and allows one tradition to enhance and elaborate on the characteristics of the other (Ess, 2006). As such, a global vision of engineering ethics education incorporates processes of global learning and identity development facilitated through engineering education.

Global learning is a term coined by the United Nations University in the 1970s. Such learning was envisioned to be social and integrative, by enabling people to exchange and synthesize information across borders of difference (Landorf & et al., 2018). The term aimed to convey "both the sense of learning as a global process that must include all levels of society, and the sense of learning to think globally, in the recognition that the world is a finite, closely interconnected, global system" (Soedjatmoko & Newland, 1987, p. 221). As MacCleoud (2018) points out, global learning implies fostering a sense of agency and responsibility towards addressing global challenges, applying knowledge in culturally appropriate ways, and developing a global mindset and sense of global citizenship. As such, the term has a broad scope that encompasses philosophies, cognitive orientations, skills, actions, and/or ways of being (MacCleoud, 2018, p. 116).

Ethics within engineering can be understood as the combination of several elements pertaining to the concepts, knowledge, skills, attitudes, and values that are relevant to decision-making in a particular context (Gwynne-Evans et al., 2021). It influences students' understanding of their professional identity both in their individual capacity and as corporate and political role-players. Engineering practice, far from being neutral and objective, is profoundly affected by the engineers' understanding of their role and identity within a specific context and of the values associated with this role. At the core of identity development processes, we find practices requiring engagement, agency, and tenacity. In terms of identity, the global engineer has been defined by Giovannelli and Sandekian (2017) as "someone who practices engineering: (1) with forethought of its far-reaching consequences, both physical and social; (2) with an appreciation of international colleagues and/or in international offices; and (3) with cultural sensitivity, so that personal interactions are both pleasant and effective."

As such, there is a distinction between the globalization of ethics, as instantiated in the prevalence of the Western-centric stance in engineering education, and a global pursuit or vision of engineering ethics. The former has been criticized for equating the globalization of ethics with "a cultural turn that promotes passivity and distance from those it claims to protect" (Morrison & Sacchetto, 2018, p. 1119). The latter aims at inclusive and socially just practices and epistemologies reached in an active and participatory manner.

3.2 Reflections on Audience and Positionality

The positioning of "local" alongside "global" focuses attention on the audience as practitioners of ethics education research. In the same way that research is specific to a context, the writing up of research is necessarily situated within cultural and theoretical frames. In setting out the aims and research questions, we considered the duty to write not only for those looking to get an overview of the existing research on engineering ethics education but also for the engineering educators reflecting on their localized positions and contexts. By this, we intend to affirm aspirations and endeavors around what is possible in terms of research as to encourage the identification and incorporation of local contributions and culturally specific references alongside studies that are already recognized as influential. We envisioned going beyond the American-centric body of research (Figure 5.1) to present practices from other geographical and cultural contexts, as well as profiling pedagogical and institutional attempts to open up engineering education towards global learning, traditions, and challenges.

Our other intended audience are engineering ethics education researchers. Enhancing the cultural relevance and representativeness of engineering ethics education research towards a global approach requires identifying ways to broaden the theoretical framing, research questions, demography, methods, scope, authorship, and audience of future research studies. We invite our audience to engage with the question of "how can engineering ethics education research be more global and culturally inclusive?" and take it further than we have succeeded in this chapter. This self-critical endeavor can be complemented by reflecting on the role played by research funding mechanisms and the policy bodies determining priority research areas.

Research goes hand in hand with building the academic community. It becomes crucial to expand the community of engineering ethics education in an intentional way, in line with values that are chosen purposefully and in a participatory manner, to serve the global community as a whole.

3.3 Methods

The chapter pursues three research questions:

RQ1: What are the authorship patterns and major institutional and national actors publishing engineering ethics education research?

RQ2: What are the research methods used to report global and culturally inclusive practices in engineering ethics education research?

RQ3: What are the learning goals, teaching practices, and institutional strategies for global and culturally inclusive approaches, as reported in engineering ethics education research?

To address these questions, the study employed a bibliometric analysis (RQ1) and a systematic literature review (RQ2–3). We were guided by the methodological recommendations for systematic literature reviews developed by Borrego et al. (2014) and those for bibliometric research reported by Williams et al. (2018).

After setting the research questions, we discussed the criteria for selecting relevant studies. Afterwards, we selected a database for generating the research publications by region of the authors' affiliation for the bibliometric analysis and the review. The research publications were screened for relevance and categorized by research methodology, methods, research questions and aims, sample size, geographical setting, as well as the learning goals, teaching methods, thematic content, and institutional initiatives reported. Finally, we analyzed the research publications and reflected upon the factors that may enable the adoption of global and culturally inclusive approaches and the benefits of this broadening.

The literature review and bibliometric analysis shared the first stages up to the screening process. The analysis by categories was conducted for the literature review. In addition, for the bibliometric analysis, we also tracked the funding details and the number of citations of each publication.

3.3.1 Selection Criteria

Considering RQ1, our team agreed to include in the bibliometric analysis sources that explicitly addressed conceptual, methodological, pedagogical, or institutional issues in engineering ethics education. Among these documents, to address RQ2–3, a further narrowing down has been conducted to include in the literature review the sources that addressed global and cultural aspects from a conceptual, methodological, pedagogical, or institutional perspective. We included in the search the term "responsibility" and its cognates so as not to overlook educational practices that are explicitly guided by the principle of socially responsible engineering and that fall beyond the framework of traditional courses in engineering ethics emphasizing individual conduct.

We included conference proceedings, book chapters, and journal articles. The rationale for including all three publication types was to highlight emerging trends and activities in the regions with less-developed publishing infrastructures, thus ensuring a higher coverage of sources.

Considering the temporal range, we included for the bibliometric analysis publications from 2014 up to November 2021. We set this range to cover the research published since the launch of the *Cambridge Handbook of Engineering Education Research*, the predecessor of this volume, until the review process took place. For the literature review, we narrowed the analysis from eight to four years, covering research published between 2018 and November 2021. The rationale for the different temporal ranges is that the bibliometric analysis benefits from a bigger database of publications, while the literature review can be managed more thoroughly by our team of four researchers with a smaller database.

We set the focus on undergraduate engineering education and excluded publications that reported pedagogical practices or research at other study levels or disciplines.

3.3.2 Database Selection and Search Query

We opted for the Scopus database due to the compatibility of the metadata retrieved for the collected publications (including institutional affiliations of multiple authors as well as lists of references) with the software program VOSviewer used for the bibliometric analysis.

We undertook two Scopus searches of sources published from 2014 to 2017 and 2018 to November 2021 that mentioned "engineering" and "education" together with "ethics" or "responsibilit*", divided across six regional clusters. The title, abstract, and keywords were included in the search.

As such, the following search strings were created:

TITLE-ABS-KEY ((responsibilit* OR ethics) AND engineering AND education) AND PUBYEAR >2013 AND PUBYEAR <2018 and TITLE-ABS-KEY((responsibilit* OR ethics) AND engineering AND education) AND PUBYEAR >2017 AND PUBYEAR <2022.

Overall, the search resulted in 65 countries of affiliation listed. For the geographical clustering, we divided the list of countries into larger regions. For example, in the case of the Africa region, we used the following search string: TITLE-ABS-KEY ((responsibilit* OR ethics) AND engineering AND education) AND PUBYEAR >2017 AND PUBYEAR <2022 AND (LIMIT-TO (AFFILCOUNTRY "South Africa") OR (AFFILCOUNTRY "Nigeria") OR (AFFILCOUNTRY "Egypt") OR (AFFILCOUNTRY "Cameroon") OR (AFFILCOUNTRY "Ethiopia") OR (AFFILCOUNTRY "Ghana") OR (AFFILCOUNTRY "Namibia").

The search query resulted in a set of 2,166 academic documents published during the period 2014–2021. This corresponds to 1,006 academic publications between 2014 and 2017 and 1,160 academic publications between 2018 and November 2021.

The publication set was filtered by manually screening the abstracts to identify sources that explicitly addressed conceptual, methodological, pedagogical, or institutional issues in engineering ethics education. After considering the selection criteria and eliminating double entries and publications meeting the exclusion criteria, the list comprised 684 unique academic documents (300 for the period of 2014–2017 and 384 for the period of 2018–2021).

3.3.3 Bibliometric Analysis

For the bibliometric analysis, we focused on the publications on engineering ethics education from the 2014–November 2021 set of 684 documents mentioned previously. We used the VOSviewer program, which is based on the VOS (visualization of similarities) method that allows clustering of different entities by citations, co-citations, and co-authorships in a corpus of documents. The tool had been used for different purposes in the scientometric research (McAllister et al., 2021), such as in analyzing international collaborations in a specific area (He & Yu, 2019), as well as in the fields related to our study, such as mapping the terms and topics in computer and information ethics (Heersmink et al., 2011) or the sustainability debate in software engineering education (Jimenez et al., 2020). Recently, the VOSviewer program has been used to map the engineering ethics debates taking place in the largest engineering education conferences in combination with content analysis of the openly accessible curricula (Nasir et al., 2021). Our work resonates with other bibliometric research in engineering education (Williams et al., 2018; Valentine & Williams, 2021; Nasir et al., 2021), including modelling the scholarly output and impact. It maps the community of engineering ethics education research by revealing patterns in international collaborations and authorship, as well as major regional actors.

3.3.4 Literature Review

For the review process, we focused on the 2018–November 2021 subset of 384 academic documents. We further screened the full texts of these documents to identify publications about undergraduate engineering ethics education addressing global and culturally oriented conceptual frameworks, themes, pedagogies, learning goals, or institutional strategies. The screening identified

67 publications fitting the criteria. Considering the distribution by region and the institutional affiliation of the first author, there are 24 studies representing North America, 18 studies representing Asia, 18 studies representing Europe and the UK, 2 studies representing Central and South America, 2 studies representing Australia, 2 studies representing the Middle East, and 1 study representing Africa. We analyzed each document to identify the focus on global and cultural aspects as central to the study, the geographical scope of the study, the research methodology and methods used, as well as the learning goals, teaching practices, and institutional strategies mentioned in reference to global and cultural aspects of engineering ethics education.

3.4 Limitations

Although English has become a common language of published research, we recognize that language can be a limitation of the research undertaken in this study. Our team is comprised of native Romanian, English, Russian, and Chinese speakers. The search was conducted for academic documents written in English. Except for one publication which featured an English abstract but whose main text was written in Spanish, all publications were written in English. This may lead to the omission of important publications written in languages other than English that may reveal more pedagogical practices and research studies set in the regions analyzed. We are aware of a special issue on engineering ethics education published by the Japanese Society for Engineering Education in 2021, with articles in the Japanese language. There may be more omissions that we are unaware of due to our limited knowledge of foreign languages. We recommend the replication of the study in different regions and languages, conducted by research teams active in those regions and with a finer knowledge of the local engineering ethics education community and of the publication avenues that accept non-English work.

4 Authorship Patterns and Regional Actors in Engineering Ethics Education Research

In response to RQ1, the following section presents the results of the bibliometric analysis of the academic documents on engineering ethics education published between 2014 and 2021. It provides a picture of the established regional players and authorship patterns in engineering ethics education research.

4.1 Geographical Overview of Engineering Ethics Education Research

The analysis revealed the prevalence of publications authored with an institutional affiliation in the United States, comprising 51% of the list of sources included in the analysis. Based on the author affiliation and considering the scholarly output and citation count of publications in engineering ethics education, the list of influential institutions is topped by five US-based universities: University of Colorado Boulder, Purdue University, Colorado School of Mines, Tufts University, and the Virginia Polytechnic Institute and State University (Figure 5.1).

4.2 Co-Authorship National Networks

The United States is also the country with the most collaborations with researchers based outside its borders that resulted in engineering ethics education publications. Approximately 14% of all US-based publications were co-authored with institutions in other countries. This percentage exceeds the overall number of publications from any other country (Figure 5.2).

Institution	Scholarly Output ↓	Views Count	Field-Weighted Citation Impact (excl. self-citations)	Citation Count (excl. self-citations)
University of Colorado Boulder	41	791	1.89	259
Purdue University	40	936	1.88	314
Colorado School of Mines	27	613	1.82	198
Tufts University	25	508	1.58	132
Virginia Polytechnic Institute and State University	21	461	1.04	109

Figure 5.1 Top 5 institutions publishing in engineering ethics education (2014–2021).

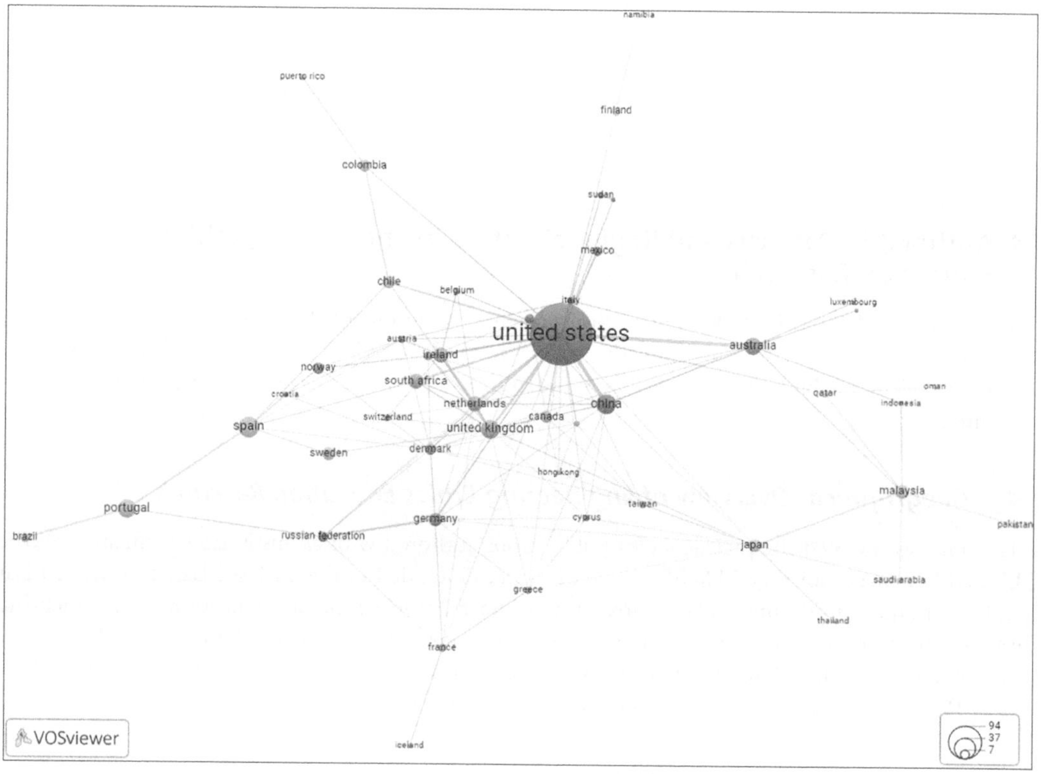

Figure 5.2 Co-authorship network in engineering ethics education (2014–2021).

Table 5.1 Countries by Link Strength of International Co-Authorship in Engineering Ethics Education

Ranking	*Country*	*Documents*	*Linked Countries*	*Total Link Strength*
1	United States	349	22	63.00
2	United Kingdom	30	13	17.00
3	Australia	26	9	16.00
4	Netherlands	21	11	15.00
5	China	34	7	14.00
6	Ireland	20	7	12.00
7	South Africa	19	4	8.00
8	Malaysia	16	6	8.00
9	Spain	38	8	7.00
10	Portugal	29	3	7.00
11	Chile	13	4	7.00
12	Germany	15	3	6.00
13	Japan	14	5	6.00
14	Denmark	11	12	5.00
15	Colombia	13	3	4.00
16	Canada	12	9	4.00
17	Sweden	16	4	3.00
18	Norway	12	5	2.00

Figure 5.2 presents the co-authorship network of 48 interconnected countries as the nodes weighted by the number of published documents. The closer the nodes are to each other, the stronger is the collaboration in their research. The fractional counting method was chosen, assigning fractionalized weight to a link for each of the multiple authors. The number of co-authored documents, number of countries of collaboration, and share of authors from the respective country in the collectives thus determine the total link strength of a node.

Table 5.1 lists the countries with ten or more published documents sorted by the total link strength of international co-authorship. Institutions from the United States serve as an important node of the network, reaching out to all the regional or language-based clusters.

The bibliometric analysis supports existing claims of the dominance of the American and Anglo-speaking research in the field of engineering ethics education. While the figures may obscure the international background of the authors themselves, they reveal the institutions and countries that dominate engineering ethics education research. Thus, they invite us to reflect on who is left behind in this field and what may be the enablers or barriers to conducting engineering ethics education research in order to widen participation and empower researchers from outside the most influential regions or institutions.

5 Overview of Research on Global and Culturally Inclusive Engineering Ethics Education

In response to RQ2, this section reviews the research methodology of the set of 67 academic documents published between 2018 and 2021 that address global and culturally oriented themes, pedagogies, institutional strategies, or learning goals.

When inquiring how the global and culturally inclusive engineering ethics education practices are being conveyed through research, we note both a geographical disparity and regional differences (Table 5.2).

Table 5.2 Methodological Overview of Academic Documents Addressing Global and Culturally Inclusive Engineering Ethics Education

Research Methodology	*Number of Publications*	*Geographical Focus*	*Methods*	*Number of Participants*
Mixed methods	6 (Alves et al., 2021; Bielefeldt et al., 2019, 2021; Hsu, 2020; Man & Sunyu, 2019; Poursharif et al., 2021)	*Global:* 2 (US Bielefeldt et al., 2021; Portugal, Alves et al., 2021); *US:* 1 (Bielefeldt et al., 2019); *UK:* 1 (Poursharif et al., 2021); *China:* 1 (Man & Sunyu, 2019); *Taiwan:* 1 (Hsu, 2020).	*Surveys:* 6 (Alves et al., 2021; Bielefeldt et al., 2019, 2021; Hsu, 2020; Man & Sunyu, 2019; Poursharif et al., 2021) *Textual analysis: 3* *open-ended survey questions: US, (Bielefeldt et al., 2019); UK, (Poursharif et al., 2021); *student assignments: Taiwan (Hsu, 2020) *Interviews:* 3 US (Bielefeldt et al., 2019, 2021); China, (Man & Sunyu, 2019) *Participant observation:*1 Portugal (Alves et al., 2021).	*<50:* 2 UK (Poursharif et al., 2021), Portugal (Alves et al., 2021); *51–100:* 1 Taiwan (Hsu, 2020); *101–500:* 1 China (Man & Sunyu, 2019); *501–1,000:* 0 *1,001–3,000:* 1 US (Bielefeldt et al., 2021); >3,000: 1 US (Bielefeldt et al., 2019).
Quantitative	12 (Balakrishnan et al., 2019, 2020; Clancy, 2020; Clancy & Hohberger, 2019; Lazarus, 2018; Luegenbiehl, 2018; Ngo & Chase, 2021; Polmear et al., 2019; Sattar et al., 2021; Snyder et al., 2020; Xiaofeng, 2021; Zhu & Jesiek, 2020)	*Single country focus: 7* 3 China: China (Clancy & Hohberger, 2019; Xiaofeng, 2021), US (Luegenbiehl, 2018); 2 US (Ngo & Chase, 2021; Snyder et al., 2020); 1 Australia (Lazarus, 2018); 1 Pakistan (Sattar et al., 2021); *Two-country focus:* 3 China and US: China (Clancy, 2020); Malaysia and Indonesia: Malaysia (Balakrishnan et al., 2020); Japan and Malaysia: Malaysia (Balakrishnan et al., 2019) *Multi-country focus:* 2 1 China, France, Germany, India, Japan, and Mexico: US (Zhu & Jesiek, 2020); 1 globe clusters: US (Polmear et al., 2019).	*Surveys*: 12	*<50:* 1 US (Ngo & Chase, 2021); *51–100:* 2 US (Zhu & Jesiek, 2020); China (Xiaofeng, 2021); *101–500:* 6 US (Luegenbiehl, 2018; Snyder et al., 2020); China (Clancy, 2020; Clancy & Hohberger, 2019); Malaysia (Balakrishnan et al., 2019, 2020); *501–1,000:* 0 *1,001–3,000:* 2 Australia (Lazarus, 2018); US (Polmear et al., 2019).

Qualitative	15 (Balakrishnan et al., 2021; Birzer & Hamilton, 2019; Boni et al., 2019; Fu et al., 2018; Gwynne-Evans et al., 2021; Kazakova, 2019; Ku et al., 2019; Laato & Sutinen, 2020; Martin et al., 2021b; Nasir et al., 2021; Roca et al., 2018; Terrón-López et al., 2020; Weiss et al., 2021; Zabaniotou, 2020; Zhang & Zhu, 2021)	*Single-country focus:* 9 Spain (Boni et al., 2019; Roca et al., 2018; Terrón-López et al., 2020; Ireland (Martin et al., 2021b); Austria (Weiss et al., 2021); Finland (Laato & Sutinen, 2020); China (Zhang & Zhu, 2021); Malaysia (Balakrishnan et al., 2021); South Africa (Gwynne-Evans et al., 2021); *Two-country focus:* 3 *2 US and China (Fu et al., 2018; Ku et al., 2019; US); *Australia and US: Australia (Birzer & Hamilton, 2019); *Multi-country focus:* 3 *India, Russia, and US: Russia (Kazakova, 2019); *Mediterranean region: Greece (Zabaniotou, 2020); *Global: Pakistan (Nasir et al., 2021)	*Interviews*: 5 US (Ku et al., 2019); Ireland (Martin et al., 2021b); Spain (Boni et al., 2019); China (Zhang & Zhu, 2021); Malaysia (Balakrishnan et al., 2021); *Textual analysis:* 7 Russia (Kazakova, 2019); Pakistan (Nasir et al., 2021); Spain (Roca et al., 2018); Finland (Laato & Sutinen, 2020); Austria (Weiss et al., 2021); US (Fu et al., 2018); South Africa *Action research:* 2 Greece (Zabaniotou, 2020); South Africa (Gwynne-Evans et al., 2021) *Case study*: 2 Spain (Terrón-López et al., 2020); Australia (Birzer & Hamilton, 2019) *Participant observation*: 1 Australia (Birzer & Hamilton, 2019); *Bibliometric analysis*: Pakistan (Nasir et al., 2021).	*<50:* 9 US (Fu et al., 2018); Ireland (Martin et al., 2021b); Finland (Laato & Sutinen, 2020); Spain (Boni et al., 2019); Australia (Birzer & Hamilton, 2019; China (Zhang & Zhu, 2021); Malaysia (Balakrishnan et al., 2021); Russia (Kazakova, 2019); South Africa (Gwynne-Evans et al., 2021); *51–100:* 1 Spain (Terrón-López et al., 2020) *101–500:* 2 US (Ku et al., 2019); Pakistan (Nasir et al., 2021) *Nonspecified:* 3 Austria (Weiss et al., 2021); Greece (Zabaniotou, 2020); Spain (Roca et al., 2018)
Other: descriptive	28 (Al Mamun, 2020; Børsen et al., 2021; Bramstedt, 2020; Dong, 2021; Friesel et al., 2021; Hansen, 2021; Hughes et al., 2020; Jordan et al., 2019, 2021; Kanemitsu, 2018; Leal et al., 2020; Lemaître, 2019; Li et al., 2019; Lord et al., 2018; MacCleoud, 2018; Malheiro et al., 2019; Masten et al., 2021;	*Single-country focus:* 18 US: US (Jordan et al., 2021; Lord et al., 2018; Ngo & Chase, 2021), China (Li et al., 2019); China (Dong, 2021; Sun et al., 2021; Wang & Yan, 2019); Spain (Muñoz López et al., 2021; Roca et al., 2018); Portugal (Leal et al., 2020; Malheiro et al., 2019); Austria (Bramstedt, 2020; Weiss et al., 2021); Sweden (Hansen, 2021); UK (Poursharif et al., 2021); Malaysia (Al Mamun, 2020); Japan (Kanemitsu, 2018)		

(Continued)

Table 5.2 (Continued)

Research Methodology	*Number of Publications*	*Geographical Focus*	*Methods*	*Number of Participants*
	Muñoz López et al., 2021; Ngo & Chase, 2021; Peters et al., 2020; Poursharif et al., 2021; Roca et al., 2018; Siegfried et al., 2020; Sun et al., 2021; Wang & Yan, 2019; Weiss et al., 2021; Zabaniotou, 2020);	*Two-country focus: 1* France, Iceland: Iceland (Siegfried et al., 2020); *Multi-country focus: 6* *France, Germany, the UK, Hungary, Vietnam, Morocco, Algeria, and Tunisia: France (Lemaître, 2019); *Denmark, Germany, Austria, the Netherlands, Spain: Denmark (Friesel et al., 2021); *US, France, Russia, Germany: Denmark (Børsen et al., 2021) *Mediterranean region: Greece (Zabaniotou, 2020); *Australasia, China, Europe, India, and North America: US (Hughes et al., 2020); *Countries in the Association of international educators: US (MacCleoud, 2018); *n.a.* 4 *conference reports (Jordan et al., 2019; Peters et al., 2020); *special issue editorial (Masten et al., 2021); *in position paper (Aponte et al., 2018)		

Other: philosophy	5 (Francis, 2021; Hansen, 2021; Jing & Doorn, 2020; Laato & Sutinen, 2020; Zhu, 2021;	1 China and the Netherlands (Jing & Doorn, 2020); US (Zhu, 2021); Finland (Laato & Sutinen, 2020); Denmark (Hansen, 2021); UK (Francis, 2021)
Other: position paper	5 (Hughes et al., 2020; Lantada, 2020; Lemaître, 2019; Torres Díaz et al., 2021)	Colombia (Aponte et al., 2018; Torres Díaz et al., 2021); UK (Hughes et al., 2020); Spain (Lantada, 2020); France (Lemaître, 2019)
Other: literature review	4 (Bielefeldt, 2019; Birzer & Hamilton, 2019; Shen & Li, 2021; Zhang & Zhu, 2021	China (Shen & Li, 2021; Zhang & Zhu, 2021); US (Bielefeldt, 2019); Australia (Birzer & Hamilton, 2019)
Other: forecast	1 (Gürdür Broo et al., 2021)	UK (Gürdür Broo et al., 2021)

Note: $n = 67$ documents.[1]

In terms of regional differences, the study demonstrated that the United States had the highest number of studies with more than 1,000 participants (Bielefeldt et al., 2019; Polmear et al., 2019; Bielefeldt et al., 2021). This points to the role of funding bodies such as the National Science Foundation and dedicated resources for conducting large-scale research. As large-scale studies are resource-intensive, this can further exacerbate the diminished visibility of perspectives on engineering ethics education rooted in other regional and cultural settings which do not benefit from a similar financial support.

The most common research approach describes examples of practice in a non-systematic manner (28 studies). In this case, there is either no methodology or, when mentioned, it referred to aspects related to the setup or organization of the learning experience or educational process. Only 33 of the 67 publications analyzed present a methodological research approach. These studies were conducted in a total of 15 countries (United States, China, Spain, Malaysia, Australia, Portugal, UK, Ireland, Finland, Greece, Austria, Russia, Pakistan, South Africa, Taiwan).

6 Pedagogy of Global and Culturally Inclusive Engineering Ethics Education

In response to RQ3, this section explores the learning goals, teaching practices, and institutional strategies for global and culturally inclusive approaches, as reported in the set of 67 academic documents.

6.1 Learning Goals towards a Global and Culturally Inclusive Engineering Ethics Education

The literature review identified 28 learning goals mentioned in connection to teaching engineering ethics in a global and culturally inclusive manner (Table 5.3). The most frequently mentioned goals refer to respect and sensitivity to cultures, diversity, and inclusion (20 mentions); awareness and action towards global social justice, equity, and public good (20 mentions); and a culturally sensitive understanding of engineering and the cultural situatedness of practice (19 mentions). These goals are pursued through a combination of theoretical content via ethical theories or problems of a socioeconomical nature affecting specific regions or manifest at the global level and experiential learning via real-life projects in collaboration with developmental NGOs or local communities or immersion in projects set in regions from the developing world.

The goals focused on character include the self-improving spirit or self-cultivation (Wang & Yan, 2019; Jing & Doorn, 2020; Sun et al., 2021; Zhu, 2021); developing the capacity for critical self-reflection, a critical spirit, critical consciousness, or re-evaluation of beliefs (Lord et al., 2018; Li et al., 2019; Terrón-López et al., 2020); the tolerance for ambiguity, uncertainty, or discomfort (Lord et al., 2018; Terrón-López et al., 2020); creativity (Boni et al., 2019; Terrón-López et al., 2020); patriotism (Clancy, 2020; Shen & Li, 2021); resilience (Boni et al., 2019; Jordan et al., 2021); and the craftsman spirit cultivation (Shen & Li, 2021; Xiaofeng, 2021). With one mention each, we find goals related to the development of an upright mind and well manners (Wang & Yan, 2019), a model worker spirit (Shen & Li, 2021), harmony (Jing & Doorn, 2020), humility (Aponte et al., 2018), helpfulness (Bielefeldt et al., 2019), generosity, fairness, and confidence (Boni et al., 2019), curiosity and openness (MacCleoud, 2018), resourcefulness and agility (Jordan et al., 2021), deference to excellence (Zhu, 2021), and the forging of lasting cross-cultural bonds (Aponte et al., 2018). The goals related to emotional development include empathy (Aponte et al., 2018; Bielefeldt et al., 2019; Birzer & Hamilton, 2019; Boni et al., 2019; MacCleoud, 2018; Snyder et al., 2020; Zhu, 2021), compassion (Aponte et al., 2018; Boni et al., 2019), and sympathy (Zhu, 2021).

The analysis points to the geographical boundness of several learning goals. For example, goals addressing the National Academy of Engineering (NAE) Grand Challenges are specific to studies

Table 5.3 Goals of Global and Culturally Inclusive Engineering Ethics Education

Goal	*Occurrence*	*Goal*	*Occurrence*
Respect and sensitivity to cultures, diversity, and inclusion	20	Collaboration and work in empowering ways with marginalized communities towards capacity building	9
Awareness and action towards global social justice, equity, and public good	20	Intercultural communication	8
Culturally sensitive understanding of engineering and the cultural situatedness of practice (i.e., how engineering is practiced in a specific geographical or cultural context)	19	Awareness and use of ethical theories situated in cultural context	8
Pursuit of user-centered design towards inclusive solutions that incorporate the perspectives and needs of those with different cultural and national backgrounds	18	Facilitation of the adoption of participatory methods	8
Character and attitude development	16	Pursuit of peace engineering	7
Understanding and reflection on the effects of technologic or scientific development on humankind at the global level or for a specific regional or cultural context	16	Emotional development	7
Addressing challenges of the UN Sustainable Development Goals (SDG)	16	Ability to navigate cultural differences and cross-cultural contexts of engineering practice	6
Development of the sense of responsibility and agency towards the global community	15	Immersion in another culture, language, economic, and social framework	5
Knowledge and understanding of multicultural traditions, values, and perspectives	13	Uptake of culturally appropriate application of science to global issues on a local level	5
Encourage the development of critical infrastructure and improvement of global conditions and quality of life (i.e., access to water, sanitation)	12	Upholding international professional standards and norms of engineering as global engineers	3
Awareness and critical reflection of underlying ideologies, dominant assumptions, cultural norms, and prevalent biases	13	Endorsing and contributing to the democratization and decolonization of knowledge and scientific practices	3
Multicultural teamwork	11	Addressing the National Academy of Engineering Grand Challenges	3
Global citizenship and global community building	9	Consideration of contentious topics, cultural and diplomatic conflicts, global disputes, or politically sensitive topics across cultures or regions	2
Climate action	9	Awareness and pursuit of universal human rights	2

Note: *n* = 67 research publications.

focused on the US context of engineering ethics education and are absent from any other geographical context. The studies based in the United States show a high emphasis on prompting awareness and action towards global social justice, equity, and public good (9/18 studies). Within the EU and the UK, no study mentioned goals with an explicit focus on universal human rights or the democratization and decolonization of knowledge and scientific practices. The highest occurrence in this region is for goals related to the UN SDGs and content related to sustainability (11/24 studies), the pursuit of user-centered design and inclusive solutions (9/24 studies), and awareness of the effects of a technology or scientific development at the global level or for a specific regional or cultural context (8/24 studies). For studies based in Asia, popular goals include character and attitude development, with several traits mentioned (7/18 studies) and fostering a culturally sensitive understanding of engineering and the cultural situatedness of practice (7/18 studies). No study mentioned goals targeting the UN SDG challenges, universal human rights, the democratization and decolonization of knowledge and scientific practices, emotional development, intercultural communication or engagement with contentious topics related to cultural, and diplomatic conflicts, global disputes, or politically sensitive issues across cultures or regions. In Central and South America, both studies mentioned as goals the awareness and action towards global social justice, equity, and public good.

There are variations in the frequency of goals related to global and culturally inclusive engineering ethics education. As such, there is an average of 9.5 goals mentioned in publications based in Australia, 9 goals in publications based in Central and South America, 5.2 goals in publications based in North America, 3.6 goals in publications based in the EU and UK studies, 3 goals in publications based in Asia, 2 goals for studies based in the Middle East, and 1 goal for the publication based in Africa.

6.2 Teaching Methods for a Global and Culturally Inclusive Engineering Ethics Education

A similar variety was recorded in the teaching methods used for conveying a global and cultural vision of engineering ethics education (Table 5.4). There are 28 teaching methods associated with this approach, leading to a broadened manner of teaching ethics when compared with the most popular teaching methods reported in the literature (Martin et al., 2021a). We note a comprehensive focus on experiential, theoretical, and reflective approaches.

The immersive element is an important mark of a global and culturally inclusive engineering ethics education. It is achieved through experiential approaches that bring students in contact with a diverse range of stakeholders, communities, and different cultures. The teaching methods recorded in our study include real-life projects, field trips (Birzer & Hamilton, 2019), international internships (Boni et al., 2019), humanitarian and service projects, operating start-ups in different regions (Birzer & Hamilton, 2019), and "study abroad" program components (MacCleoud, 2018). An emerging pattern of experiential approaches is the prevalence of real-life projects of an interdisciplinary manner conducted in collaboration with external stakeholders, such as developmental NGOs, local governments, minority groups, or groups representing communities from different countries (Roca et al., 2018; Birzer & Hamilton, 2019; Leal et al., 2020; Terrón-López et al., 2020; Poursharif et al., 2021). In some cases, this can take the form of humanitarian engineering, service learning, or social entrepreneurship (Birzer & Hamilton, 2019; Lemaître, 2019; Ngo & Chase, 2021).

Experiential methods tend to place a high emphasis on collaborative approaches for promoting a global and culturally inclusive engineering ethics education. Such collaborations aim to involve local stakeholders or communities from abroad as partners in the design process (Leal et al., 2020; Terrón-López et al., 2020). When the aims relate to servicing underprivileged communities, there

Table 5.4 Teaching Methods Associated with Global and Culturally Oriented Engineering Ethics Education

Teaching Method	*Number of Mentions*	*Teaching Method*	*Number of Mentions*
Real-life projects	15	Humanitarian engineering	2
Case studies and role-plays (culturally situated, global topics)	10	Research projects and research partnerships	2
Interdisciplinary projects	7	Systems thinking methods	2
Collaborative and participatory projects	7	Internships, including international internships	2
Reflection and observation	5	Guest lectures	1
Peer learning and mutual interviews	5	Study abroad and study away	1
Discussions, dialogue, debates, and critical questioning	5	Privilege walks	1
Global classrooms	4	Critical literacy	1
Fiction for specific purposes and scenario-building	3	Cultural service	1
International field experience	2	Service learning	1
Design competitions and summits	2	Forecast and future casting	1
Ecodesign	2	Social systems approaches	1
Field trips and on-the-scene learning	2	Social surveying	1
Humanist, religious, and cultural readings	2	Social entrepreneurship and operating start-ups	1

is an emphasis on participatory techniques and knowledge transfer and sharing (Roca et al., 2018; Li et al., 2019; Leal et al., 2020). The learning is accompanied by reflection and reassessment of students' beliefs, questioning of dominant worldviews, assumptions, and biases (Bielefeldt, 2019; Bielefeldt et al., 2021; Ngo & Chase, 2021). Dialogue and peer learning are an important component for facilitating reflection (Lord et al., 2018; Boni et al., 2019; Wang & Yan, 2019). Studies originating in the United States show the regional popularity of this teaching method.

Collaborative practices are also encountered in global classrooms, with a focus on multicultural teamwork and cross-cultural communication (Aponte et al., 2018; Ku et al., 2019). This teaching approach is linked with the emergence of cross-country university consortia or research projects. The European Project Semester is an example of a pan-European initiative employing this teaching approach (Malheiro et al., 2019). The creation of European universities of technology is a recent policy initiative that also aims to foster global classrooms across the continent (Gunn, 2020).

Theoretical approaches make use of novel and non-mainstream methods for engineering education, such as storytelling and future-casting (Gürdür Broo et al., 2021), scenario-building (Hansen, 2021; Weiss et al., 2021), critical literacy techniques (Lord et al., 2018), and humanist, cultural, or religious readings (Bielefeldt et al., 2021; Laato & Sutinen, 2020). An overarching goal of theoretical approaches is to enhance the awareness and respect for other cultures and traditions, as well as of envisioning culturally relevant strategies for better futures.

There are also approaches using theoretical frameworks, such as systems thinking (Lagun Mesquita & Missimer, 2020; Zabaniotou, 2020), social systems (Lagun Mesquita & Missimer, 2020), or social surveying (Zhang & Zhu, 2021).

6.3 *Institutional Strategies for Global and Culturally Inclusive Engineering Ethics Education*

The analysis identified 12 institutional strategies mentioned in the support of a global and culturally inclusive engineering ethics education. Institutional strategies include:

- Collaborations with non-governmental, outreach, or minority support organizations, such as Engineers without Borders, the Society of Women Engineers, the National Society of Black Engineers, the Hispanic Engineering and Science Society, the American Indian Science and Engineering Society, Out in STEM, Cives Mundis, Project Everest Ventures, Global Brigades, the WindAid Institute (Jordan et al., 2021; Bielefeldt et al., 2019; Lord et al., 2018; Poursharif et al., 2021; Zabaniotou, 2020; Terrón-López et al., 2020; Boni et al., 2019; Birzer & Hamilton, 2019; Li et al., 2019).
- Introducing new courses, such as interdisciplinary courses (MacCleoud, 2018); User-Centered Design, Circuits, Engineering and Social Justice, Engineering Peace (Lord et al., 2018); Engineering for People Design Challenge (Poursharif et al., 2021); Digital Theology, Theological Engineering, or Engineering and Theology (Laato & Sutinen, 2020); and Introduction to Energy in Global Development, Water, Climate Change, and Health (Li et al., 2019).
- Collaborations with governmental bodies, organizations, or city councils (Roca et al., 2018; Boni et al., 2019; Li et al., 2019; Leal et al., 2020).
- Developing new programs, such as Peace Engineering (Jordan et al., 2021; Aponte et al., 2018), "study abroad" programs (Bielefeldt et al., 2019), or the Meridies program (Boni et al., 2019).
- Establishing cross-university partnerships or consortia (MacCleoud, 2018; Lantada, 2020; Friesel et al., 2021).
- Creating dedicated course inserts (Bielefeldt et al., 2019; Li et al., 2019; Laato & Sutinen, 2020).
- Creating new departments and research centers, such as CERES (Center for Engineered Resilience and Ecological Sustainability), the WHY Laboratory Center for Water and the Environment (Jordan et al., 2021), the Institute for Global Initiatives (MacCleoud, 2018).
- Creating new learning spaces, such as an innovation space (Jordan et al., 2021; Aponte et al., 2018).
- Introducing dedicated staff roles, such as a director of global engagement or global learning coordinators (MacCleoud, 2018).
- Staff-oriented strategies focused on staff development (MacCleoud, 2018) or faculty engagement in community-based participatory research (Masten et al., 2021).
- Change of the institutional mission or vision to incorporate global and culturally oriented elements (MacCleoud, 2018).
- Introducing reward schemes, such as the Paul Simon Award for Campus Internationalization (MacCleoud, 2018).

7 Forward-Looking Conclusion

Although pervasive in both accreditation frameworks, research studies, and pedagogical practices conveyed through research, the dominant, non-culturally aware perspective has been challenged and is not impervious to change. Developing a global vision of engineering ethics education needs to address the necessary bias of that vision. In our contribution, we argued that the teaching and research of engineering ethics are not neutral endeavors, and any attempt at globalizing the engineering curriculum needs to locate itself both locally and globally.

Whereas research in engineering ethics education has historically emphasized its manner of operation as an objective and rule-based (or principle-based) system, decoupled from individuals and their context, the recent research surveyed for this study shows the emergence of non-mainstream approaches that engage in a more nuanced way with explicit or implicit values and principles for engineering design, decision-making, and action. These values and principles are integral to the specific traditions and priorities of the communities that an engineer is part of, be they professional, cultural, or political. Recognizing and negotiating these values require care as a researcher, given their influence on the research process and goals. For this, engineering ethics education research may consider how being value-laden affects the researcher's gaze and her observer stance.

The design of research studies conducted in the spirit of an ethics of care is intimately tied to the values and priorities of both the researcher and the research community in which it is practiced, in addition to those of other stakeholders. This requires that the positionality of research studies is explicitly acknowledged in order to profile the value of localized stances and the need to supplement "grand narratives" with "contextual narratives." As noted by Morrison and Sacchetto (2018), the legitimacy of research can be founded on its relations to the social context and its ability to narrow the gap between researchers and respondents. Care ethics may help the global community of researchers and educators to co-address at local and cross-regional levels the macro-issues of power and social justice, emphasize the beneficiaries of engineering processes and technology, and open practices to the co-creation of knowledge and technologies.

Furthermore, for developing a global vision for engineering ethics education research and globalizing the engineering curriculum, we suggest the following recommendations:

- Pedagogical content may consider explicitly engaging with cultural topics, localization of cases, problems, challenges, and solutions or challenges of a global nature, in the form of dedicated course or curricular inserts.
- As teaching ethics within engineering involves communication through the means of a discourse, it is important to connect the ethics discourse to both the industry discourse and the local contextual discourse. Connecting the ethics discourse to the industry requires a critical and analytical engagement with professional policy documents, internal ethics codes of local and multinational companies, and concepts of corporate social responsibility. Connecting the ethics discourse with the local context requires the identification of concepts of relevance to different cultures and communities, as well as their positioning in relation to traditional concepts purporting to ethics. Concepts such as "ubuntu," "maat" or "Confucianism" are potentially loaded terms but carry with them implicit meaning and significance that need to be excavated.
- Given the focus on participatory and multistakeholder interactions both in teaching and in institutional strategies for developing a global and culturally inclusive vision, we encourage further research exploring the impact of these initiatives on building (self-)awareness of the local strengths and cultural specificities within the partnering groups and communities. Multistakeholder collaborations in engineering education are encouraged to explicitly reflect on the cultural transfer between the involved parties and to foster a reciprocal transfer of expertise as an integral part of these educational initiatives.
- Research studies are encouraged to make their positionality and demographics explicit and engage with the challenges of global context (cultural, national, and/or political boundaries). For example, during the peer-review process, the *Journal of Engineering Education* encourages researchers to critically reflect on the extent to which participants are representative of the overall population in terms of demography (Pawley, 2017). Globalizing engineering education research is also associated with experimenting with different methods and methodological approaches. If ethics is seen to include knowledge, skills, values, attitudes, and concepts,

research into ethics in engineering is encouraged to practice innovation as regards method and argument, to better model the diversity of teaching ethics practices.
- Engineering (ethics) education working groups and local chapters need to be constituted to add value beyond their local or original sphere of influence and collaborate on capacity development, outreach, and research dissemination. An example in this sense is the activity of the SEFI ethics working group, REEN, and the organization strategy for the REES 2021 conference[2].
- Research-funding bodies are encouraged to support and be oriented at transnational or transdisciplinary approaches, scope, and consortia composition.
- Repositories of pedagogical resources should strive to internationalize their scope, to curate resources from different geographical and cultural contexts or address culturally situated themes. An example is the International Division of the Online Ethics Center for Engineering Science.

Finally, we prepared a list of 15 articles published between 2018 and November 2021 that we recommend to those new to the field or interested to adopt global or culturally inclusive practices in their teaching or research. These readings are highlighted with (*) in the Reference section.

Acknowledgments

The authors want to thank the reviewers of the chapter for their rich insights that helped us think in more depth and with more clarity at the message conveyed. The guidance provided by Sarah Jayne Hitt, Ashish Hingle, and Andrew Valentine has been extremely valuable for developing the manuscript in its current form. We also express our gratitude to Aditya Johri, who trusted us with this important work. The authors are also extending their appreciation to the participants of the workshop hosted by ESDIT–Intercultural Ethics conference, where a draft version of the chapter was discussed: Andreas Spahn, Monamie Bhadra Haines, Nick Travaglini, Gideon Haan. Aleksandra A. Kazakova expresses gratitude to the team of STS Center and Katerina Guba in European University at St. Petersburg for their fruitful comments and methodological advice on this study during her research stay. Qin Zhu acknowledges that this material is based upon work supported by the National Science Foundation under Grant No. 2124984.

Notes

1 In parentheses, we note the country of institutional affiliation of the first author.
2 The papers presented at REES 2021 fell outside the temporal range of analysis. Nevertheless, we want to acknowledge the important contribution of the conference organizers in promoting culturally diverse perspectives and regions.

References

Akera, A. (2017). Setting the standards for engineering education: A history. *Proceedings of the IEEE, 105*(9).

Akubue, A. I. (2017). International technology transfer. In S. O. Hansson (Ed.), *The ethics of technology: Methods and approaches.* Rowman & Littlefield.

Al Mamun, A., Muthalif, A. G. A., Ahmad, Z., Ismail, A. F., Ismail, K., Maleque, M. A. (2020). Islamisation of engineering education – a case at IIUM. *Universal Journal of Educational Research, 8*(2), 355–361. https://doi.org/10.13189/ujer.2020.080204.

Alves, C. A., Leão, C. P., Uebe-Mansur, A. F., & Kury, M. I. R. A. (2021). The knowledge and importance of Lean Education based on academics' perspectives: An exploratory study, *Production Planning & Control, 32*(6), 497 510. https://doi.org/10.1080/09537287.2020.1742371.

Anwar, A. A., & Richards, D. J. (2013). Is the USA set to dominate accreditation of engineering education and professional qualifications? *Proceedings of the Institution of Civil Engineers, 166*(CE1), 42–48.

Aponte, C., Agi, K., & Jordan, R. (2018). International profile of a global engineer. In *2017 7th world engineering education forum (WEEF)* (pp. 470–477). https://doi.org/10.1109/WEEF.2017.8467050.

Balakrishnan, B., Azman, M. N. A., & Indartono, S. (2020). Attitude towards engineering ethical issues: A comparative study between Malaysian and Indonesian engineering undergraduates. *International Journal of Higher Education*, *9*(2), 63–69.

(*) Balakrishnan, B., Tochinai, F., & Kanemitsu, H. (2019). Engineering ethics education: A comparative study of Japan and Malaysia. *Science and Engineering Ethics*, *25*(4), 1069–1083. https://doi.org/10.1007/s11948-018-0051-3.

Balakrishnan, B., Tochinai, F., Kanemitsu, H., & Altalbe, A. (2021). Engineering ethics education from the cultural and religious perspectives: A study among Malaysian undergraduates. *European Journal of Engineering Education*, *46*(5), 707–717. https://doi.org/10.1080/03043797.2021.1881449

Barry, J. (2012). *The politics of actually existing unsustainability: Human flourishing in a climate-changed, carbon-constrained world*. Oxford University Press.

Bielefeldt, A. R. (2019). Human rights as a lens for engineering ethics? In *Paper presented at 2019 ASEE annual conference & exposition*, Tampa, FL. https://doi.org/10.18260/1-2-32905

Bielefeldt, A. R., Canney, N. E., Swan, C., & Knight, D. (2016). Efficacy of macroethics education in engineering. In *2016 ASEE annual conference and exposition*.

Bielefeldt, A. R., Polmear, M., Knight, D., Canney, N., & Swan, C. (2021). Educating engineers to work ethically with global marginalized communities. *Environmental Engineering Science*, 320–330. https://doi.org/10.1089/ees.2020.0269

Bielefeldt, A. R., & Rulifson, G., & Canney, N. E. (2019). Social responsibility related to global experiences and interests of U.S. engineering students. In *Paper presented at 2019 ASEE annual conference & exposition*, Tampa, FL. https://doi.org/10.18260/1-2-33274

Bijker, W. E. (1995). *Of bicycles, bakelites, and bulbs: Toward a theory of sociotechnical change*. The MIT Press.

Birzer, C. H., & Hamilton, J. (2019). Humanitarian engineering education fieldwork and the risk of doing more harm than good. *Australasian Journal of Engineering Education*, *24*(2), 51–60. https://doi.org/10.1080/22054952.2019.1693123

Boni, A., Sastre, J. J., & Calabuig, C. (2019). Educating engineers for the public good through international internships: Evidence from a case study at Universitat Politècnica de València. *Science and Engineering Ethics*, *25*(6), 1799–1815. https://doi.org/10.1007/s11948-015-9728-z.

Borrego, M., Foster, M. J., & Froyd, J. E. (2014). Systematic literature reviews in engineering education and other developing interdisciplinary fields. *Journal of Engineering Education*, *103*(1), 45–76. https://doi.org/10.1002/jee.20038.

Børsen, T., Serreau, Y., Reifschneider, K., Baier, A., Pinkelman, R., Smetanina, T., & Zandvoort, H. (2021). Initiatives, experiences and best practices for teaching social and ecological responsibility in ethics education for science and engineering students. *European Journal of Engineering Education*, *46*(2), 186–209. https://doi.org/10.1080/03043797.2019.1701632.

Bramstedt, K. (2020). Partnering ethics and chemistry in secondary and university STEM education via an innovatively designed periodic table of chemical elements. *Acta bioethica*, *26*(1), 101–106. https://doi.org/10.4067/S1726-569X2020000100101.

Bucciarelli, L. (2008). Ethics and engineering education. *European Journal of Engineering Education*, *33*(2), 141–149.

Christensen, S. H., Mitcham, C., Li, B., & An, Y. (Eds.). (2012). *Engineering, development and philosophy: American, Chinese and European perspectives*. Springer.

(*) Clancy, R. F. (2020). The ethical education and perspectives of Chinese engineering students: A preliminary investigation and recommendations. *Science and Engineering Ethics*, *26*, 1935–1965. https://doi.org/10.1007/s11948-019-00108-0.

Clancy, R. F., & Hohberger, H. (2019). The moral foundations of Chinese engineering students: A preliminary investigation. In *Paper presented at 2019 ASEE annual conference & exposition*, Tampa, FL. https://doi.org/10.18260/1-2-33408.

Colby, A., & Sullivan, W. (2008). Ethics teaching in undergraduate engineering education. *Journal of Engineering Education*, *97*(3), 327–338.

Collier, S. J. (2006). Global assemblages. *Theory, Culture & Society, 23*(2–3), 399–401. https://doi.org/10.1177/026327640602300269

Conlon, E. (2019). Prisoners of the capitalist machine: Captivity and the corporate engineer. In S. Christensen, B. Delahousse, C. Didier, M. Meganck, & M. Murphy (Eds.), *The engineering-business nexus. philosophy of engineering and technology* (Vol. 32). Springer. https://doi.org/10.1007/978-3-319-99636-3_3.

Deng, M. (2015, September 4). *One size fits few: Artificial hearts leave many out (Op-Ed).* Live Science. https://www.livescience.com/52093-artificial-hearts-too-large-for-many-to-receive.html

Dong, M. (2021). Application of safety and stability design of engineering components in material mechanics teaching. In *2022 2nd International conference on artificial intelligence and education (ICAIE)* (pp. 684–687). https://doi.org/10.1109/ICAIE53562.2021.00150.

Ess, C. (2006). Ethical pluralism and global information ethics. *Ethics of Information Technology, 8*, 215–226.

Forbes, M. H., & Dalrymple, O. O., & Lord, S. M., & Baillie, C., & Hoople, G. D., & Mejia, J. A. (2021, July). The engineering exchange for social justice (ExSJ): Advancing justice through sociotechnical engineering and equitable partnership exchanges. In *Paper presented at 2021 ASEE virtual annual conference content access, virtual conference.* https://peer.asee.org/37861

Francis, N. (2021). Civil engineers' role in saving the world: Updating the moral basis of the profession. *Proceedings of the Institution of Civil Engineers – Civil Engineering, 174*(5), 3–9.

Friesel, A., Watty, R., Florensa, I., Cruz, J., Klamert-Schmid, J., Lackner, M., de Souza Daneluz, A. A., & Geraedts, H. (2021). A collaboration of five International Engineering Institutions on Innovation Engineering projects. In *2021 30th annual conference of the European association for education in electrical and information engineering (EAEEIE)* IEEE. 2018 28th EAEEIE Annual Conference. https://doi.org/10.1109/EAEEIE50507.2021.9530925.

Fu, J., Lei, Q., Grzybowski, D. M., & Cheng, D. (2018). What should be taught in engineering ethics education under globalization? Based on the comparative analysis of university textbooks in China and the United States. In *Paper presented at 2018 ASEE annual conference & exposition*, Salt Lake City, Utah. https://doi.org/10.18260/1-2-31236.

Giovannelli, L., & Sandekian, R. (2017). Global engineering: What do we mean by it and how are we preparing our students for it? In *American society for engineering education (ASEE) annual conference and exposition proceedings.*

Gispen, K. (1990). *New profession, old order: Engineers and German society, 1815–1914.* Cambridge University Press.

Gunn, A. (2020). The European universities initiative: A study of alliance formation in higher education. In A. Curaj, L. Deca, & R. Pricopie (Eds.), *European higher education area: Challenges for a new decade* (pp. 13–30). Springer.

Gürdür Broo, D., Boman, U., & Törngren, M. (2021). Cyber-physical systems research and education in 2030: Scenarios and strategies. *Journal of Industrial Information Integration, 21*, 100192. https://doi.org/10.1016/j.jii.2020.100192.

(*) Gwynne-Evans, A., Junaid, S., & Chetty, M. (2021). Repositioning ethics at the heart of engineering graduate attributes. *Australasian Journal of Engineering Education, 26*(1), 7–24. https://doi.org/10.1080/22054952.2021.1913882.

Hansen, K. S. (2021). Optimistic fiction as a tool for ethical reflection in STEM. *Journal of Academic Ethics*. 19, 425–439. https://doi.org/10.1007/s10805-021-09405-5.

Hansson, S. O. (2017). Technology and distributive justice. In S. O. Hansson (Ed.), *The ethics of technology: Methods and approaches* (pp. 51–66). Rowman & Littlefield.

He, X., & Yu, D. (2019). Bibliometric analysis of international cooperation in biomass energy research. *Current Science, 117*(11), 1785–1792. https://doi.org/10.18520/cs/v117/i11/1785-1792.

Heersmink, R., van den Hoven, J., van Eck, N. J., & van Berg, J. d. (2011). Bibliometric mapping of computer and information ethics. *Ethics and Information Technology, 13*(3), 241–249. https://doi.org/10.1007/s10676-011-9273-7.

(*) Herkert, J. R. (2000). Engineering ethics education in the USA: Content, pedagogy and curriculum. *European Journal of Engineering Education, 25*(4), 303–313.

Herkert, J. R. (2001). Future directions in engineering ethics research: Microethics, macroethics and the role of professional societies. *Science and Engineering Ethics*, 7, 403–414. https://doi.org/10.1007/s11948-001-0062-2

(*) Hess, J. L., & Fore, G. A. (2018). A systematic literature review of US engineering ethics interventions. *Science and Engineering Ethics, 24*(2), 551–583.

Hsu, Y.-C. (2020). A pilot study to incorporate collaboration and energy competency into an engineering ethics course. *Education Sciences, 10*(3), 72. https://doi.org/10.3390/educsci10030072.

Hughes, J., Plaut, E., Wang, F., von Briesen, E., Brown, C., Cross, G., Kumar, V., & Myers, P. (2020). Global and local agendas of computing ethics education. *In Proceedings of the 2020 ACM conference on innovation and technology in computer science education (ITiCSE '20)* (pp. 239–245). Association for Computing Machinery, New York, NY, USA. https://doi.org/10.1145/3341525.3387423.

Jimenez Ramos, L. M., Duarte Acosta, N., Berrio Lopez, J. P., & Pablo Velez Uribe, J. (2020). To train software engineers with principles of sustainable development: A bibliometric study. In *Paper presented at the 2020 Congreso Internacional De Innovacion y Tendencias En Ingenieria, CONIITI 2020 – conference proceedings*. https://doi.org/10.1109/CONIITI51147.2020.9240263

(*) Jing, S., & Doorn, N. (2020). Engineers' moral responsibility: A Confucian perspective. *Science and Engineering Ethics, 26*, 233–253. https://doi.org/10.1007/s11948-019-00093-4.

Jordan, R., Agi, K., Arora, S., Christodoulou, C. G., Schamiloglu, E., Koechner, D., Schuler, A., Howe, K., Bidram, A., Martinez-Ramon, M., & Lehr, J. (2021). Peace engineering in practice: A case study at the University of New Mexico. *Technological Forecasting and Social Change, 173*, 121113. https://doi.org/10.1016/j.techfore.2021.121113.

Jordan, R., & Nair, I., & Agi, K., & Koechner, D. M. (2019). How do we frame peace engineering education? A complex but vital question. In *2019 ASEE annual conference & exposition*, Tampa, FL. https://doi.org/10.18260/1-2-32893.

Kanemitsu, H. (2018). New trends in engineering ethics – A Japanese perspective. In A. Fritzsche & S. Oks (Eds.), *The future of engineering. Philosophy of engineering and technology* (p. 31). Springer. https://doi.org/10.1007/978-3-319-91029-1_17.

Karwat, D. M. A., Eagle, W. E., Wooldridge, M. S. & Princen, T. E. (2015). Activist engineering: Changing engineering practice by deploying praxis. *Science and Engineering Ethics, 21*, 227–239.

Kazakova, A. (2019). Responsibility in biomedical engineering education. *Információs Társadalom, XIX*(4), 50–60. https://doi.org/10.22503/inftars.XIX.2019.4.4.

Klassen, M. (2018). *The politics of accreditation: A comparison of the engineering profession in five anglospere countries*. University of Toronto. https://tspace.library.utoronto.ca/bitstream/1807/82894/1/Klassen_Mike_201803_MA_th esis.pdf

Ku, S. T., & Yang, X., & Wang, S. (2019). When eagle and dragon learn together: Engineering ethics education in the era of U.S.-China trade war. In *Paper presented at 2019 ASEE annual conference & exposition*, Tampa, FL. https://doi.org/10.18260/1-2-33557.

Laato, S., & E. Sutinen. (2020). Integrating theological aspects to engineering education. In *2020 IEEE frontiers in education conference (FIE)* (pp. 1–9). https://doi.org/10.1109/FIE44824.2020.9274089.

Lagun Mesquita, P., & Missimer, M. (2020). Supporting engineering students in analysing social sustainability of a product: Lessons learned. *Proceedings of the Design Society: DESIGN Conference, 1*, 1755–1764. https://doi.org/10.1017/dsd.2020.266.

Landorf, H., Doscher, S. P., & Simons-Lane, B. (2018). The origin and meaning of global learning. *Internationalisation of Higher Education, 1*, 1–18. https://goglobal.fiu.edu/_assets/docs/landorf_doschersimons-lane-the-origins-of-global-learning.pdf

Lantada, A. D. (2020). Engineering education 5.0: Continuously evolving engineering education. *International Journal of Engineering Education, 36*(6), 1814–1832.

Lazarus, D. (2018). Promoting self-responsibility: Learning from Australian maritime engineering student and alumni in developing employability competencies. *IOP Conference Series Earth and Environmental Science, 175*(1), 012224.

Leal, D., Alves, J. L., Fernandes, A., & Rangel, B. (2020). We won't waste you: A research project to introduce waste and social sustainability in design thinking. *2020 IEEE Global Engineering Education Conference (EDUCON)*, 1959–1963. https://doi.org/10.1109/EDUCON45650.2020.9125180.

Lemaître, D. (2019). Training engineers for innovation: Pedagogical initiatives for new challenges. *European Journal of Education, 54*, 566–576. https://doi.org/10.1111/ejed.12365

Li, H., & Xie, Y., & Yang, S., & Xu, R. (2019). A new approach in abolishing poverty: A case study and construction strategy for integrating inclusive innovation into engineering ethics education. *Paper presented at 2019 ASEE annual conference & exposition*, Tampa, FL. https://doi.org/10.18260/1-2-3197

(*) Lord, S. M., Mejia, J. A., Hoople, G., Chen, D., Dalrymple, O., Reddy, E., Przestrzelski, B., & A. Choi-Fitzpatrick. (2018). Creative curricula for changemaking engineers. In *2018 world engineering education forum – global engineering deans council (WEEF-GEDC)* (pp. 1–5). https://doi.org/10.1109/WEEF-GEDC.2018.8629612.

(*) Luegenbiehl, H. (2010). Ethical principles for engineers in a global environment. In I. van de Poel & D. Goldberg (Eds.), *Philosophy and engineering: An emerging agenda* (pp. 147–159). Springer.

Luegenbiehl, H. C. (2004). Ethical autonomy and engineering in a cross-cultural context. *Techné: Research in Philosophy and Technology, 8*(1), 57–78. https://doi.org/10.5840/techne20048110

Luegenbiehl, H. C. (2018). Chinese student perceptions of engineering ethics. *Boston Studies in the Philosophy and History of Science*, 237–245. https://doi.org/10.1007/978-3-319-62450-1_20

Luegenbiehl, H. C., & Rockwell, F. C. (2017). *Global engineering ethics*. Elsevier.

MacCleoud, H. (2018). Incorporating intercultural and global competencies into higher education STEM programming. In C. F. Maguire & R. D. Sheardy (Eds.), *Citizens first! Democracy, social responsibility and chemistry* (pp. 109–137). https://doi.org/10.1021/bk-2018-1297.ch008.

Malheiro, B., Guedes, P., Silva, M. F., & Ferreira, P. (2019). Fostering professional competencies in engineering undergraduates with EPS@ISEP. *Education Sciences, 9*(2), 119. https://doi.org/10.3390/educsci9020119.

Man, Z., & Sunyu, W. (2019). Current situation and problems of engineering ethics education at colleges and universities in China. In *2019 Annual SEFI conference proceedings*.

Martin, D. A., Bombaerts, G., & Johri, A. (2021). Ethics is a disempowered subject in the engineering curriculum. In *Proceedings of the SEFI 2021 Conference: Blended learning in engineering education: Challenging, enlightening – and lasting?* (blz. 357). European Society for Engineering Education (SEFI). www.sefi.be/wp-content/uploads/2021/12/SEFI49th-Proceedings-final.pdf.

(*) Martin, D. A., Conlon, E., & Bowe, B. (2021a). A multi-level review of engineering ethics education: towards a socio-technical orientation of engineering education *for* ethics. *Science and Engineering Ethics, 27*, 60. https://doi.org/10.1007/s11948-021-00333-6.

Martin, D. A., Conlon, E., & Bowe, B. (2021b). Using case studies in engineering ethics education: The case for immersive scenarios through stakeholder engagement and real life data. *Australasian Journal of Engineering Education, 26*(1), 47–63. https://doi.org/10.1080/22054952.2021.1914297

Marzano, S. (1993). Chocolate for breakfast. In *Design renaissance, international design congress*, Glasgow.

Masten, S. J., Harris, A., Kearns, J., Borrion, A., Peters, C. A., & Gadhamshetty, V. R. (2021). Global environmental engineering for and with historically marginalized communities. *Environmental Engineering Science, 38*(5), 285–287.

Matos, S., Riley, D. M., & Akera, A. (2017). WannABET? Historical and organizational perspectives on governance in engineering education. *Paper presented at 2017 ASEE annual conference & exposition*, Columbus, Ohio. https://doi.org/10.18260/1-2-29110.

McAllister, J. T., Lennertz, L., & Atencio Mojica, Z. (2021). Mapping a discipline: A guide to using VOSviewer for bibliometric and visual analysis. *Science & Technology Libraries*, 1–30. https://doi.org/10.1080/0194262x.2021.1991547.

Meiksins, P., & Smith, C. (Eds.). (1996). *Engineering labour: Technical workers in comparative perspective*. Verso.

Michelfelder, D. P., Wellner, G., & Wiltse, H. (2017). Designing differently: Toward a methodology for an ethics of feminist technology design. In S. O. Hansson (Ed.), *The ethics of technology: Methods and approaches* (pp. 193–218). Rowman & Littlefield.

(*) Mitcham, C. (2009). A historico-ethical perspective on engineering education: From use and convenience to policy engagement. *Engineering Studies, 1*(1), 35–53.

Mitcham, C., & Englehardt, E. E. (2019). Ethics across the curriculum: Prospects for broader (and deeper) teaching and learning in research and engineering ethics. *Science and Engineering Ethics, 25*(6), 1735–1762.

Mitcham, C., & Kazakova, A. A. (2020). Let us now think engineering: An interview with Carl Mitcham. *Philosophy of Science & Technology/Filosofiâ Nauki i Tehniki, 25*(2).

Morrison, C., & Sacchetto, D. (2018). Research ethics in an unethical world: The politics and morality of engaged research. *Work, Employment and Society, 32*(6), 1118–1129. https://doi.org/10.1177/0950017017726947

Muñoz López, N., Valero Martín, J. I., Biedermann, A., Santolaya Sáenz, J. L., & Fernández-Vazquez, A. (2021). Projecting more sustainable product and service designs. *Sustainability, 13*(21), 11872. https://doi.org/10.3390/su132111872

Nasir, O., Muntaha, S., Javed, R. T., & Qadir, J. (2021). Work in progress: Pedagogy of engineering ethics: A bibliometric and curricular analysis. In *2021 IEEE global engineering education conference (EDUCON)* (pp. 1553–1557). https://doi.org/10.1109/EDUCON46332.2021.9453923.

Ngo, T. T., & Chase, B. (2021). Students' attitude toward sustainability and humanitarian engineering education using project-based and international field learning pedagogies. *International Journal of Sustainability in Higher Education, 22*(2), 254–273. https://doi.org/10.1108/IJSHE-06-2020-0214.

Nussbaum, M. (2006). *Frontiers of justice: Disability, nationality, species membership*. Harvard University Press.

Nussbaum, M. (2011). *Creating capabilities*. Harvard University Press

Okereke, C. (2010). Climate justice and the international regime. *WIREs Climate Change, 1*, 462–474. https://doi.org/10.1002/wcc.52.

Palmiter Bajorek, J. (2019). Voice recognition still has significant race and gender biases. *Harvard Business Review*.

Pasquale, F. (2015). *The Black box society: The secret algorithms that control money and information*. Harvard University Press.

Pawley, A. L. (2017). Shifting the "default": The case for making diversity the expected condition for engineering education and making whiteness and maleness visible. *Journal of Engineering Education, 106*, 531–533. https://doi.org/10.1002/jee.20181.

Peters, A.-K., Bengtsson, S., Cajander, Å., Daniels, M., Grande, V., Lönngren, J., & Salminen-Karlsson, M. (2020). Care ethics to develop computing and engineering education for sustainability. In *2020 IEEE frontiers in education conference (FIE)*, Uppsala (pp. 1–4). https://doi.org/10.1109/FIE44824.2020.9274188.

Pfotenhauer, S., & Jasanoff, S. (2017a). Traveling imaginaries: The 'practice turn' in innovation policy and the global circulation of innovation models. In D. Tyfield, R. Lave, S. Randalls, & C. Thorpe (Eds.), *The Routledge handbook of the political economy of science* (pp. 416–428). Routledge.

Pfotenhauer, S., & Jasanoff, S. (2017b). Panacea or diagnosis? Imaginaries of innovation and the 'MIT model' in three political cultures. *Social Studies of Science*, *47*(6), 783–810. https://doi.org/10.1177/0306312717706110

(*) Polmear, M., Bielefeldt, A., Knight, D., Canney, N., & Swan, C. (2019). Analysis of macroethics teaching practices and perceptions in engineering: A cultural comparison. *European Journal of Engineering Education, 44*(6), 866–881.

Poursharif, T. P. Doss, R. Broadbent, & Knight, G. (2021). Developing global engineers through interdisciplinary PBL and design thinking. In *IEEE global engineering education conference (EDUCON)* (pp. 194–198). https://doi.org/10.1109/EDUCON46332.2021.9453869.

(*) Riley, D. (2008). *Engineering and social justice*. Synthesis Lectures on Engineers, Technology and Society. Morgan and Claypool Publishers.

Roca, E., Villares, M., & Oller, E. (2018). Collaborative workshop: Sustainable civil engineering proposals for real settings. A: International Conference on Civil Engineering Education. In *EUCEET 2018: 4th international conference on civil engineering education: Challenges for the third millennium. Barcelona: International centre for numerical methods in engineering (CIMNE)* (pp. 76–81).

Sattar, H., Syed, T. H., Naseem, A., Ahmad, Y., Mirza, M. Z., & Raza, M. (2021). Uncovering the indirect impact of work ethic on engineering students' productivity through positive and negative organizational behaviors and workaholism. *Sustainability, 13*(5), 2922. https://doi.org/10.3390/su13052922.

Sen, A. (1999). *Development as freedom*. Knopf.

Sen, A. (2009). *The idea of justice*. Allen Lane

(*) Shen, J., & Li, H. (2021). The "real images" of engineering ethics education: An analysis based on systematic literature review (SLR). In *2021 IEEE global engineering education conference (EDUCON)* (pp. 719–725). https://doi.org/10.1109/EDUCON46332.2021.9454004.

Siegfried, R., Haraldur, A., Liliane, E., & Thordur, V. F. (2020). Decision skills in engineering programs – a key for a VUCA era. In *2020 IFEES world engineering education forum – global engineering deans council (WEEF-GEDC)* (pp. 1–5). https://doi.org/10.1109/WEEF-GEDC49885.2020.9293669.

Slaton, A. E., & Riley, D. (2015). The wrong solution for STEM education. *Inside Higher Education*. www.insidehighered.com/views/2015/07/08/essay-criticizes-proposed-changes-engineering-accreditation-standards

Snyder, S. A., Bairaktarova, D., & Staley, T. W. (2020). Parallel disciplines: Exploring the relationship between global and ethics learning of undergraduate engineering students. In *Proceedings of the 48th annual SEFI Conference engaging engineering education* (pp. 495–510).

Soedjatmoko & Newland, K. (1987). The United Nations university: A new kind of university. *The Washington Quarterly*, *10*(3), 215–224.

Sun, J., Wang, C., & Hua, F. (2021). Considerations for intelligent manufacturing engineering talent development based on engineering ethics education. In *2021 IEEE world conference on engineering education (EDUNINE)* (pp. 1–6). https://doi.org/10.1109/EDUNINE51952.2021.9429129.

Terrón-López, M-J., Velasco-Quintana, P. J., Lavado-Anguera, S., & Espinosa-Elvira, M. d. C. (2020). Preparing sustainable engineers: A project-based learning experience in logistics with refugee camps. *Sustainability*, *12*(12), 4817. https://doi.org/10.3390/su12124817.

Torres Díaz, G., Leyva, D., Yepez, C., Pacheco Torres, P. (2021). Knowledge transfer and the challenges of engineering training in the face of globalization [Transferencia de conocimiento y los retos de la formación de ingenieros ante la globalización]. In *Encuentros: Revista de Ciencias Humanas, Teoría Social y Pensamiento Crítico* (pp. 97–106).

Umoja Noble, S. (2018). *Algorithms of oppression. How search engines reinforce racism*. New York University Press.

Valentine, A., & Williams, B. (2021). Evolution of engineering education research in Portugal and Spain: A scientometric study. In *2021 4th international conference of the Portuguese society for engineering education (CISPEE)* (pp. 1–8). https://doi.org/10.1109/CISPEE47794.2021.9507212.

Wang, G., & Thompson, R. G. (2013). Incorporating global components into ethics education. *Science and Engineering Ethics*, *19*(1), 287–298.

Wang, Q., & Yan, P. (2019). Development of ethics education in science and technology in technical universities in China. *Science and Engineering Ethics*, 1721–1733. https://doi.org/10.1007/s11948-019-00156-6

Weiss, A., Vrecar, R., Zamiechowska, J., & Purgathofer, P. (2021). Using the design of adversarial chatbots as a means to expose computer science students to the importance of ethics and responsible design of AI technologies. In C. Ardito et al. (Eds.), *Human-computer interaction – INTERACT 2021*. INTERACT 2021. Lecture Notes in Computer Science (Vol. 12934). Springer. https://doi.org/10.1007/978-3-030-85613-7_24

(*) Williams, B., Wankat, P. C., & Neto, P. (2018). Not so global: A bibliometric look at engineering education research. *European Journal of Engineering Education*, *43*(2), 190–200. https://doi.org/10.1080/03043797.2016.1153043

Winner, L. (1986). Do artifacts have politics? In L. Winner (Ed.), *The whale and the reactor: A search for limits in an age of high technology* (pp. 19–39). University of Chicago Press.

Winner, L. (1990). Engineering ethics and political imagination. In P. Durbin (Ed.), *Broad and narrow interpretations of philosophy of technology: Philosophy and technology*, 7, 53–64. Kluwer.

Xiaofeng, D. (2021). Research on the path of craftsman spirit cultivation education of science and Engineering College Students under the background of information technology integration. In *2021 International conference on internet, education and information technology (IEIT)* (pp. 52–55). https://doi.org/10.1109/IEIT53597.2021.00019.

Zabaniotou, A. (2020). New forms of social learning in Mediterranean higher engineering education: Change lab for gender equality transformation, methodology, design principles. *Sustainability*, *12*(16), 6618. https://doi.org/10.3390/su12166618

Zhang, H., & Zhu, Q. (2021). Instructor perceptions of engineering ethics education at Chinese engineering universities: A cross-cultural approach. In *Technology in society* (p. 65). Elsevier.

Zhu, Q. (2021). Self-cultivation of the Confucian engineer: What engineering ethics education can learn from Confucian moral theory. In P. H. Wong & T. X. Wang (Eds.), *Harmonious technology*. Routledge.

(*) Zhu, Q., & Jesiek, B. K. (2020). Practicing engineering ethics in global context: A comparative study of expert and novice approaches to cross-cultural ethical situations. *Science and Engineering Ethics*, *26*, 2097–2120. https://doi.org/10.1007/s11948-019-00154-8.

6

Disrupting Engineering Education

Euan D. Lindsay, Roger G. Hadgraft, Fiona Boyle, and Ron Ulseth

1 Introduction

In 2004, Richard Riley, former US Secretary for Education, famously said (Gunderson et al., 2004, p. 506):

> We are currently preparing students for jobs that don't yet exist, using technologies that haven't been invented, in order to solve problems, we don't even know are problems yet.

This statement is as true now as it was in 2004. Future-proofing engineering graduates is a goal of engineering schools across the globe and an ongoing requirement of industry and society. We need our graduates to be able to continuously think outside the box, upskill in emerging technologies, processes, and systems, all while meeting the day-to-day demands of their jobs. We strive to produce engineers of the future for workplaces of the future. So how do we continue to innovate engineering education to produce these future-proofed graduates? What process do we follow? How do we disrupt our current models?

Firstly, when we speak about disrupting engineering education, there are misinterpretations of what *disruption* is. This is a mistake commonly encountered, as the core concepts of disruption theory remain widely misunderstood (Christensen et al., 2018). However, we can start by understanding that while all disruptions are innovative, not all innovations are disruptive. Disruption theory essentially describes "the phenomenon by which an innovation transforms an existing market or sector by introducing simplicity, convenience, accessibility, and affordability, where complication and high cost are the status quo" (Christensen Institute, 2022).

Christensen identifies that there are two types of innovations seen in organizations (including education) – disruptive and sustaining innovations (Christensen et al., 2003). Disruptive innovations are "innovations that make products and services more accessible and affordable, thereby making them available to a larger population" (Christensen et al., 2018; Christensen et al., 2015), with sustaining innovation targeted to "improve products and services along dimensions of performance that mainstream customers care about and that markets have historically valued" (Christensen et al., 2018).

Essentially, disruptive innovation tends to create "good enough" products, while sustaining innovation creates "superior products." Arguably, sustaining innovations are more frequently seen in engineering education, rather than disrupting innovations, with the field using the term "disruption" in

DOI: 10.4324/9781003287483-7

a more colloquial sense as opposed to being aligned to Christensen's disruption innovation theory. This is a misconception commonly seen in the literature (Christensen et al., 2018).

From a strategic perspective, it is likely that a blend of both types of innovation is important in higher education, with a structured approach to identifying innovation opportunities being crucial. Therefore, to address this requirement, we have considered Doblin's 10 types of innovation as a framework to identify areas in which engineering education could innovate. This chapter presents a theoretical framework to categorize and present such innovations. Examples of innovation are identified not only from the affiliated organizations of the authors of this chapter (identified as emerging leaders in engineering education (Graham, 2018)) but also from the extensive literature in this space.

2 The Emerging Themes of Disruption

In 2018, Ruth Graham identified recent innovators in engineering education – *the global state of the art* (Graham, 2018). Three interesting trends identified were that there is a shift in innovation from the Global North to the Global South, there is a shift to programs that emphasize socially relevant, outward-facing curricula, and there is a new generation of leaders who can deliver programs at scale.

Graham identified the current leaders as Olin College (USA), MIT, Stanford, Aalborg University (DK), TU Delft (Netherlands), UCL (UK), Purdue, NUS (Singapore), Cambridge (UK), and Chalmers (Sweden). Emerging leaders identified were Singapore University of Technology and Design (SUTD), Iron Range Engineering (USA), Charles Sturt University (Australia), Tsinghua University (China), and Arizona State University (USA). Graham's report also identified "places to watch" (Box 2, page 12).

Project-based learning plays a key role in all these institutions, building students' practice skills as well as broadening engineering programs to include multidisciplinary problem-solving to engage students in engineering-like activity as early in the curriculum as possible. The second significant trend is the move to online content delivery and assurance. This forces careful thought about what is core and what is elective. What do students need to learn before they tackle a given project? Teaching students to seek the expertise they need is an important skill needed for their future workplaces, where lectures will not be available. Designing curricula around these two key innovations requires other changes to the normal university practices.

Key engineering education innovations use more than one kind of innovation. This includes the many forms of practice-based learning and flexible curricula, which are supported by modified learning environments and staffing structures, leading to gradual changes in perceptions and expectations by both students and academics.

Looking at our own innovations and those referenced by Graham (2018), we propose five themes that are emergent, unifying ideas, giving shape to the changes we see in contemporary engineering programs; they are also predictors of future changes that might better deliver the requirements for industry and for students:

1. Developing the *student engineer* identity from early in the program (a logical extension of practice-based curricula).
2. Understanding *perceptions and expectations* from students, employers, academics, and others.
3. Adopting a *flexible curriculum.*
4. Modifying the *learning environment.*
5. Starting with a *greenfield site* – entirely new programs.

2.1 The "Student Engineer" Identity

In general, engineering programs are evolving from content-centric models of what the students should *know* towards capability-centric models of what graduates should be able to *do*. Students are expected to be undertaking a rich engineering learning experience – the whole is greater than the sum of the parts. This disruption occurs where the focus is on ingraining the identity of the engineer. Identity can be powerful (Tonso, 2014), as the individual traverses the trajectory of "what I think I want to be" to "what I am." Authentic educational experiences that align that identity with the practice of the profession lead to more robust motivation, confidence, and passion, all of which lead to improved performance. Authentic problems involve the kinds of problems engineers face in the workplace in contrast with the less-valuable kinds they face in a traditional classroom (Jonassen, 2014).

Research in the area of engineering identity has increased in recent years (Morelock, 2017). Morelock identified several things that contribute to the establishment of the student engineer identity. These are exposure to experiences and the connection of these to a student's "aspects of self." The "aspects of self" can relate to gender, academic identity, race, or one's beliefs or values and, indeed, the experiences one encounters throughout one's life. It is not yet clear how students can connect these aspects of themselves to the engineering profession.

If the profession can be thought of in the *three domains* of technical, professional (transversal/social), and design, then educational experiences that either emulate or directly encounter the environments in which those domains are practiced will provide the opportunity of building their engineer identity. Reflective practice (Schön, 1983) is where the student engineer spends focused time processing the learning experiences, engaging their beliefs and values with the lived actions of engineering practice, across all three of the aforementioned domains.

Through structured and purposeful reflection activities, identity is developed (Johnson et al., 2015; Morelock, 2017). Furthermore, Duarte et al. (2016) found that students' own characteristics (e.g., motivation, self-efficacy) were more important when it comes to the development of self-learning skills (learner autonomy) compared to a particular teaching–learning transaction, and teaching strategies should account for this. Therefore, by applying a heutagogical teaching strategy, capability, capacity, and autonomy can be amplified in students, making them more prepared for the complex and ever-changing work environment (Blaschke, 2012).

2.2 Perceptions and Expectations

Disruption can also occur across the many ways that people perceive the profession and the educational processes. The stakeholders in engineering education, from industry to students, from faculty to accrediting bodies, have perceptions of what engineering education is and what it should be. The gaps between these expectations afford opportunities to innovate and disrupt. An example is bringing consulting engineers into the academic setting to work side by side with students while meeting the needs of the client (Mann et al., 2021).

Contemporary students, whose lives have been shaped by ubiquitous availability of information and entertainment, have neither the patience nor the desire to be "taught" in the traditional lecture and problem set model. Conversely, 13 years of school that have focused on face-to-face teacher-led education has left them oddly unprepared for new models of active learning. A transition semester is required to shift students from passive to active, self-directed learners. This is a critical transition for all learners in the 21st century, with lifelong learning identified as one of the top 10 critical skills for all workers (World Economic Forum, 2020).

2.3 The Flexible Curriculum

Few things have changed as much as how students engage with the content that underpins the process of becoming an engineer. Each discipline has its own specialized knowledge (fluid mechanics, thermodynamics, circuit analysis, etc.), and every engineering discipline also relies on a broad collection of common knowledge, such as systems engineering, project management, and people management.

For the last 500 years or so, we relied on textbooks and lectures as the dominant modes for knowledge transmission. Students practiced new skills in tutorials, laboratories, and design classes. In the last 30 years, knowledge has moved online. However, many academics and students still cling to the old methods, assuming there will be lectures, problem sets, and textbooks, when we have moved into a world of knowledge at our fingertips. Learning can now happen just in time (JIT) as well as just in case (JIC) (Killi & Morrison, 2015; Liberatore et al., 2017; Riskowski, 2015; Wilkie, 2013).

However, Killi and Morrison have shown that it is important to couple this JIT teaching and learning approach with the optimum time (just in need, JIN) when students' motivation is highest to learn the particular content (Killi & Morrison, 2015). This JIT/JIN learning is fundamentally transforming how to become an engineer. Students will progress more quickly from student to engineer, on a constant trajectory of learning on the job. Universities will need to disrupt their education programs to cater to this rapid acceleration into the workplace. The four- or five-year full-time, on-campus curriculum may be a thing of the past.

Further, drastic step changes in modes of learning, as enacted during the global COVID-19 pandemic, will continue to accelerate this change. Students have realized the benefits as well as the difficulties of remote learning. Universities are also adapting pathways into and out of engineering programs, enabling students to take a variety of routes both before and after university study. Recognition of prior learning is being adapted to cater for microcredentials and MOOCs (Rampelt & Suter, 2017; Shimson & Verstelle, 2017), as well as more formal, long-form programs of study.

In relation to MOOCs, since their emergence in 2011, they have continued to offer opportunities for learners to develop and enhance their knowledge (Rivas et al., 2020; Shimson & Verstelle, 2017). Presumed to have reached their pinnacle towards the end of the last decade, with academics believing the craze had passed, the pandemic has resulted in a resurgence of MOOCs. They have, for many, been part of an important educational response to the pandemic (Impey & Formanek, 2021; Van Melle & de Bie; Yang & Lee, 2021).

2.4 The Learning Environment

Despite parts of the curriculum moving online, most universities still see the face-to-face experience as an important part of university life. In fact, this is where students begin to learn the collaboration skills that will be essential to their future success (Trevelyan, 2014). Most Western universities have invested enormous sums of money into new learning spaces, learning commons, pervasive Wi-Fi, coffee carts, microwaves, and other features to make the campus experience attractive to students (Fraser, 2014).

Layered on top of these new learning spaces are matching social processes that are changing the way that students learn. Student societies are available in a bewildering variety, from social purposes to professionally focused ones, helping students develop their skills beyond the curriculum (e.g., the robotics club and the space society). Student societies that extend across multiple institutions are also prevalent, such as the Golden Key Honor Society, IEEE Eta Kappa Nu, and the Board of European Students of Technology (BEST). Start-ups are multiplying at many universities, with structured support through staffed centers (Miller & Dorning, 2018).

Students are also encouraged to learn collaboratively outside of class through formalized peer learning, such as PASS (peer-assisted study sessions), where trained student leaders help less-experienced students master difficult topics (Arendale, 2016). Students are assisted in other ways, through pastoral support, aided by learning analytics, which can detect non-engagement early, enabling a student to be given a helping hand long before they have failed a key assessment task. Students are also assisted by an increasing trend to bring industry adjuncts on campus in various roles – guest lectures, project and studio leaders and clients, engineers-in-residence, professors of practice, and so on. These industry colleagues boost student engagement and motivation and keep the academic program connected to essential topics, skills, and examples.

2.5 The Greenfield Environment

Many of the examples of disruption have taken place in a greenfield environment – those where new models could be established from scratch. This occurs reasonably regularly, although the pathways to that founding vary from philanthropy (Miller & Dorning, 2018) to government support (Magnanti, 2018) to industry patronage (Grose, 2016) to serving as an external sandbox for established institutions (TEDI, n.d.). Greenfield programs are much more able to innovate in the structure of their programs and the processes they change around it, but they do so at the cost of additional challenges to their brand and profit model (key issues in Doblin, to be explained shortly). While these schools are free of the need to subsidize other colleges, or a research agenda, they are also not able to amortize governance and administrative costs across the rest of the institution.

Working in a greenfield environment allows for some parts of the disruption trajectory to be accelerated. New programs and ideas can be developed without the "baggage" of existing habits and cultures. The price for this is that some of the "baggage" is actually "luggage" – useful practices that must, instead, be invented from scratch rather than just simply applied through existing, supportive work practices. This is the trade-off made by most start-up organizations, balancing the benefits of agility with those of scale. It is significant to note that the emerging leaders identified in Graham (2018) are almost all greenfield sites, introducing engineering as a new discipline to their institution or creating an entirely new institution.

A consequence of establishing a new program is that by their nature, they start small. Small programs share many advantages with new programs; there is a smaller footprint involved when the structure, process, and systems innovations are required. Many key disruptions initially emerge as small innovations in large institutions; it is then their trajectory through configuration innovations that see them move to scale. Some institutions keep their programs deliberately small, embracing the value of the learning environment for those students who can pay the fees required to make such programs viable. Other institutions see themselves as not yet large, investing upfront in their programs in the belief that the students and the revenues will flow once the advantages of their offering become well-known.

3 Matching Innovation to Disruption – Applying the Doblin Framework

There are many institutions doing innovative things in the engineering space. While there are themes emerging at the frontiers of engineering education, one theme that is missing in the literature is explicit reflection about the innovation process. There are many models of innovation present in the business literature. It is clear from inspection that many of the engineering education innovators are intuitively aligning with these models, but there is little evidence that there has been explicit thinking about the innovation process.

We saw an opportunity to address this absence, particularly in the hope of showing how innovation could be transferred between contexts. The "it won't work here" mindset can be very powerful in resisting change in the face of successful exemplars, and the opportunity to decouple an innovation from the specific operating environment by viewing it through the lens of a framework of innovation was too strong to resist.

We chose the Doblin 10 types of innovation framework as our lens (Doblin, 2021). The Doblin model is not the only model of innovation, but it resonated with us when we used it as a reflective

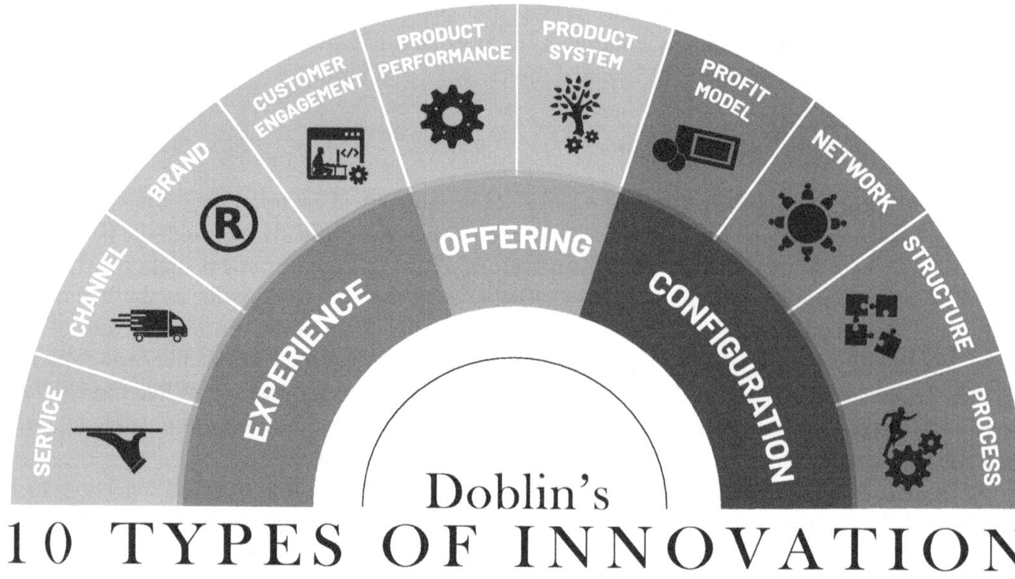

Figure 6.1 The Doblin innovation framework.

tool. We found it very useful in identifying explicitly many of the things that we were already doing implicitly and for providing a vocabulary to explain the different dimensions of innovation that we have observed in engineering education. It also provided useful insights into why our innovations had been successful when others had failed, and a pathway forward to the transferability of contexts that we seek.

3.1 The Doblin Framework

Doblin's 10 types of innovation model provides a framework to better conceptualize these transformations (Desjardin, 2020; Keeley et al., 2013). This framework (Figure 6.1) distinguishes three categories of innovation – experience, configuration, and offering – allowing for insights into the ways in which the different curriculum dimensions represent innovation in engineering education.

Experience innovations include transforming the services delivered to customers, the channels through which the products or services are delivered, expanding the brand, and improving customer engagement. *Configuration* innovations include reconsidering the profit model, exploiting networks of suppliers and collaborators, changing the structure of the organization, and finding superior processes for doing the work. *Offering* innovations includes addressing product performance and developing complementary products and services to build a product ecosystem.

The five themes discussed so far (the student engineer, expectations and perceptions, flexible curricula, the learning environment, and greenfield sites) can be understood within this framework. The student engineer, in a practice-based learning context, is a change of *experience*. It ticks the customer engagement box, which also requires or implies changes in service and channel and an opportunity for brand enhancement.

Flexible curricula are a natural partner for practice-based learning, providing students with the opportunity to engage in just-in-time learning, as will be the norm in the workplace. Flexible curricula change the student experience, altering the channel through which learning is delivered, with supporting services. Again, this is an opportunity for brand enhancement.

The learning environment must also complement the practice-based learning approach, supported by flexible curricula. Once again, these support the student experience, for example, learning commons provide additional learning services to support the learning channel and may feature in brand marketing.

Greenfield sites, and, indeed, brownfield sites, focus the attention on the configuration and offering aspects of innovation. These are the options that are easy to underestimate when innovation is considered. The profit model must be addressed at some point, networks should be developed with suppliers and collaborators (often neglected as we go it alone), the structure of the academic and administrative team needs consideration, and teaching and administration processes need training and enhancement. Product performance must be designed, and complementary products and services put in place, for example, engagement with industry and internships.

Finally, perceptions and expectations can be (must be) altered across many of the ten dimensions. Students and academics must commit to new ways of learning and new ways of interacting and assessing. These require new skills from both parties. There are opportunities for new branding, to differentiate the new program from its traditional competitors.

Now consider a project-based curriculum and how it requires innovation across all these ten dimensions.

3.2 An Example

Each of these ten types of innovation is expressed in educational environments. Impactful change requires innovation across many of the ten Doblin innovation types, as in a project-based learning (PBL) curriculum, which is designed to deliver the kinds of graduate capabilities being requested by engineering employers (ACED, 2021; Kolmos et al., 2004).

1 Altering the curriculum is a change of ***channel*** (*experience innovation*). Students now engage in learning through more design tasks and fewer lectures and tutorials. This may be accompanied by another channel innovation, such as greater reliance on self-directed learning, supported by educational technology.

2 Such curriculum changes are designed to alter the ***product performance***, through graduates with improved workplace capabilities (an *offering* innovation).

3 To achieve these changes, it is also necessary to change our teaching ***processes*** (*configuration*). Our academic staff must develop new skills and confidence to facilitate design-led learning.

4 This shift to a project-based curriculum creates new ***brand*** opportunities, through marketing a new kind of engineering education, attracting those students who are looking for a practice-focused education (*experience*).

5 We might also consider new ways of ***engaging with customers*** through outreach to schools, in-house design competitions for schools, and other means to engage high school students in the new educational model (*experience*). Similarly, we might engage industry through guest lectures, industry-sponsored projects, internships, and so on.

6 There are further opportunities suggested by the ten types of innovation. Students could be provided with better ***service*** to support their learning. Students often struggle with project-based learning at the start. *Service* could include training and handbooks to help them through this first stage of learning. The curriculum might also be structured so that senior students work with junior students, enabling knowledge transfer to the new students.

7 ***Networking*** is another *configuration* opportunity. For example, CDIO is a global network of educators committed to a certain style of project-based learning (CDIO, 2021; Crawley et al., 2014). Through the network, members can access syllabus guidelines and examples of good practice, such as projects, from around the world. Annual conferences support knowledge transfer between individuals and institutions.

8 The ***profit** model* must also be considered (*configuration*). Project-based learning is often seen as more expensive than traditional methods, yet some universities have operated successfully for almost 50 years using this curriculum model (Kolmos et al., 2004). Teaching ***processes*** must adapt to make sure that costs are contained. In fact, the greatest value is achieved when students become truly responsible for their own, and each other's, learning. We see this in autonomous teams, such as the Formula SAE competition, where students work together to complete the designated mission (SAE International, 2021).

9 Finally, we come to the ***product system*** (*offering*). Can the project-based curriculum become part of a larger ecosystem of products? The traditional approach would be to develop a follow-on set of postgraduate qualifications. Other approaches could include using the projects as central to engaging with industry in terms of undergraduate projects, postgraduate projects, industry projects, consulting projects. Thus, the engineering school becomes a knowledge factory, engaging with the community on a range of levels, where the project is the central concept and enabling process.

10 As the product system changes, the ***structure*** of the organization can adapt to truly bring teaching and research together in a seamless way (*configuration*). PBL is a research-oriented means of teaching. The undergraduates become part of the research team.

PBL is an example of a mature innovation that has evolved to touch all of Doblin's dimensions. Other innovations are more directly clustered within different parts of the Doblin framework.

3.3 Experience Innovations

Online learning is one innovation currently transforming engineering education. Universally available, online engineering knowledge means students now need ways of knowing what they need to know. We are moving from a "just in case" learning system towards a "just in time" system. Students do need fundamentals, but they don't need four years of fundamentals. Quickly they need to move into problem-solving mode through projects (Prince, 2004; Prince & Felder, 2006).

There are many successful examples of PBL, and all the most innovative engineering schools engage students in projects, usually from the first semester onwards (Bertel et al., 2021; Froyd et al., 2012). Students also need *real* projects, and they will get these through industry projects on campus and by students working in industry, for example, through internships and work placements.

There are emerging educational models where working is a key part of engineering education (Lindsay & Morgan, 2021; Ulseth et al., 2021; Boyle et al., 2022b; Morris et al., 2022; Fitzgerald et al., 2022). This is a return to cadetships from the 60s and 70s, when students worked full-time and studied part-time (very demanding, with high attrition), or via sandwich programs where they alternated between work and study. Other models are bringing industry on campus so that students are engaged in industry-like situations from week 1 (Mann et al. (2021), Ulseth et al. (2021)).

The REEdI project in Ireland (Boyle et al., 2022a) uses technology to change the student experience, using virtual and augmented reality (VR/AR) experiences into the curriculum to improve both authenticity and accessibility for the students. Similarly, COVID-19 emergency teaching shone a light on the challenge of delivering meaningful hands-on engineering experiences for students learning remotely with technology (Graham, 2022).

Other themes that emerged from Graham's work include the development of transversal skills (Ulseth et al., 2021), student-centered cultures, entrepreneurial mindsets (Svanström et al., 2012), sustainability (van Grunsven et al., 2021), equity and diversity (van Grunsven et al., 2021), ethical development (Lavi et al., 2021; Adams et al., 2021), and creativity and innovation (Loh et al., 2021). All these require changes to the student experience if they are to be implemented successfully.

All these are dimensions of disruption in engineering education. They have the power to create meaningful impact on student learning experiences and bring value to the organizations the graduates will serve. However, *branding* and messaging become essential. Beyond just implementing strategies in any of these areas, there must be effective *communication* to potential future students and their families, as well as to the companies the graduates will serve. Furthermore, we cannot underestimate the importance of bringing the wider academic community within our institutions along with us on the journey. Without acceptance by these stakeholders, the disruption will not be sustainable (Merton et al., 2004).

3.4 Configuration Innovations

Networking is probably the most overlooked path to innovation. Despite the significant overlap between the many programs offered by our engineering schools, academics often teach "their" subject in isolation. Consortia could work together to share resources, such as online tutorials, worked examples, projects topics, electives, reducing preparation time, while improving quality and coverage, but for the most part, we only go as far as adopting a textbook to replace bespoke lecture notes.

The CDIO consortium is a successful exception to this rule: an international community of project-based learning practitioners. CDIO has subsequently expanded to over a hundred institutions worldwide, easily the most successful example of networking in engineering education. There have been other consortia, for example, the coalitions funded by the NSF in the 1990s (Borrego, 2007; Coward et al., 2000), and the European Union has also funded many consortia (e.g., Erasmus and similar programs). However, except for CDIO, most fail after funding comes to an end. Are consortia destined to fail? There needs to be a return on investment for these innovations, for example, in the form of decreased costs (more efficient teaching practices), more students (economies of scale), better students (reduced cost of teaching), and so on.

A smaller-scale example is the EWB Challenge (Engineers Without Borders, 2021). Each year, the Challenge focuses on a community in a developing country, requiring first-year students to find engineering solutions to improve the lives of people in that country. This is usually the first chance for students to see the connection between engineering and social benefit. This is an interesting example of networking, given that it was initiated by a non-university organization, somehow overcoming the natural tendency of academics and students to steer away from "not invented here."

A related opportunity is academic development, enabling new pedagogical ideas to be introduced more quickly, such as alternative forms of assessment, group-based collaborative learning, professional skills, reflective practice, project-based learning, studios, and so on. Australia had considerable success through federal government funding of university consortia from 1990 to 2010 (Universities Australia, 2021). Sadly, these programs were discontinued, but they left behind many academics who had upskilled themselves in collaboration with colleagues from other universities. The Human Capital Initiative run by the Higher Education Authority in Ireland is another example of government funding university consortia (Higher Education Authority Ireland, 2022).

Business models are another key challenge for universities to consider. Universities tend to focus on teaching degrees. We recruit students into those degrees and graduate some of them (usually as few as 60% of them). However, at the heart of our degrees are subjects. These are our basic products. Could we make some of our subjects into more profitable products? Further, what are students paying for? They are no longer paying for lectures and tutorials. They may be paying for access to a community of learners and bespoke, block mode learning, on-campus challenges, global opportunities, and so on. Certainly, this is a very different model to delivering content to 500 students in a lecture theater. The teaching needs to be much more student-focused and career-directed. The question is, can this be translated into an offering that students (and governments) can and will pay for?

Other industries have already been through this transformation – gyms do not make their own exercise bikes or free weights but instead focus on the experience and support they can craft around them. In higher education, the textbook is an illustration of a mature disruption of this type; a small handful of publishers (Pearson, Wiley, Macmillan, Cengage, etc.) provides a standardized resource to us all. Many of these publishers are now looking to develop smart online tutorial systems and are competing with a host of new educational start-ups (EdX, Coursera, Khan Academy, Codecademy, etc.) as they do so.

3.5 Offering Innovations

When considered through the Doblin lens, there are few examples of innovations in the offering category. While there is a rich diversity of "product" that different institutions sell – a technical education, a credential, graduate employment, a learning community, a pedigree – most of this differentiation manifests in other forms of innovation. Engineering degrees are somewhat of a commodity: the four- or five-year accredited degree that makes an engineer. The innovations/ differentiations seem to be in how institutions deliver that product and in how students experience it.

Engineering degrees are not usually coupled with a complementary product or service. Double degrees may fall into this category, as do co-op work placements. Lifelong learning has been talked about for 30+ years but has not yet manifested as a complementary product or service. We have not moved from "education as a product" to "education as a lifelong service."

4 Disruption Isn't Meant to Be Easy . . .

Disruption happens as the "entrant" delivers a more functional model of learning to small, often neglected, audiences. The entrants are largely ignored by the "incumbents" as they move up market. Disruption occurs when the mainstream audiences begin demanding the disruptive model. One thing that makes it hard is that disruption can take a very long time to occur. By Christensen's definition, it might be argued that disruption isn't occurring yet in engineering education. There certainly are many examples of entrant disruptors and, as stated previously, many functional features of effective models. However, to date, there has not been widespread mainstream demand for or adoption of radical new teaching models (Sorby et al., 2021).

4.1 Continuous Improvement Mindset in Disruption

Implementing disruptive models of education is neither linear nor simple. Often, those looking to implement change want to find a model that works elsewhere and then copy and paste that model. This is a recipe for disaster, as it fails to acknowledge the differences in social context between the two locations. It also assumes that the model being copied is linear and static when in fact disruptive models are non-linear and dynamic. They are, instead, complex adaptive systems.

Complex adaptive systems require an instrumentation and control approach where there is continual monitoring of the system, resulting in appropriate actions from the input – probe, sense, and respond, as Snowden and Boone (2007) have said. Intentionally implementing feedback is a component in any engineering design cycle; thus, disruption, like engineering design, requires such a continuous improvement approach. The effectiveness of the learning can thus be continually improved during the implementation of the disruptive curriculum (Noor, 2013). Thus, those looking to implement should treat any innovation model as a complex adaptive system.

Ruth Graham identified the attributes of emerging and sustaining engineering education world leaders through Graham (2018) and CEEDA (2021). Of the many attributes identified, the focus on continuous improvement aligns with complex adaptive systems.

> Looking across the innovative approaches described in the Special Issue, two features stand out. The first is an emphasis on continuous change. An appreciation that change is not a single moment or stage, but an ongoing process, runs as a thread through the articles, and can be seen in the continuous improvement model adopted at Iron Range Engineering that led to the establishment of the Bell program (Ulseth et al., 2021).
>
> *(Graham, 2021)*

When considering nearly any model of learning, an essential component of the model is reflection (Dewey, 1933; Kolb, 1984; Schön, 1987). Reflection is a metacognitive act whereby the learner connects the learning to prior learning, assesses its value and contexts, and projects its future value. Reflective practice is essential for the ability of emerging engineering practitioners to nimbly meet the ever-changing needs they encounter while tackling the world's complex problems (Sheppard et al., 2008). Yet reflection is woefully missing as an intentionally developed skill in most engineering education (Ulseth & Johnson, 2017). At best, reflection is tacitly developed in traditional engineering models. The act of reflection is core to the engineering design process and the probe-sense-respond approach to complex problem-solving. Therefore, disruptors in engineering education need a level of reflection built not only into the curriculum but also into the continuous improvement processes of the program team.

4.2 The Journey of Innovation

Disruptive models need to use multiple Doblin dimensions as they move towards maturity. The kinds of innovation that are necessary to begin a disruption differ from those that are required to maintain it and to scale it; a deliberate trajectory is required. What may start as a different classroom experience may then require different channels to deliver that offering, networks within and beyond the institution to be scalable, and may eventually need to change the processes and even the structure of the institution.

What distinguishes the great from the good are those that provide value when innovating at the configuration level. It is not enough to just innovate to stop your disruption from being squashed; it is necessary that the innovation provides value on the configuration dimensions as well.

It is on these trajectories that the importance of leadership becomes clear. Successful and sustainable disruptions have leaders that work across all of Doblin's innovation types. In some places, this could be a single champion who is able to adapt roles as the project matures, but this is seldom the case. Most disruptions are instead successful with a team who can shepherd the disruption as it evolves through the different types of innovation and different leaders who are accountable for the kinds of work associated with each. Disruption is a journey, a vector that moves from the disruptive idea that manifests on a single dimension to the disruptive model that manifests across multiple dimensions. To make this journey successful, it is essential to approach it with the perspective that it will evolve over time; a continual improvement mindset is required.

4.3 Barriers to Disruption

Practicing disruptors encounter common roadblocks. One of the biggest is when the incumbents decide not to ignore but instead seek to eliminate the innovators, a common phenomenon carried out by entrenched and powerful curriculum or accreditation committees. Others include rigid physical infrastructure, technology limitations, enrolment sizes (too big or too small), government regulations, and industry demands. One of the most common barriers to disruption is in fact one of the key themes of disruption – expectations. While those driving the change may have already moved their expectations, other stakeholders can often provide inertia or resistance.

External accreditation is often presented as a roadblock to implementation; the perception that "it will never get accredited" undermines the implementation of change. This perception is in fact particularly pernicious because it is built upon a mistaken understanding of the goals of accrediting bodies. Accrediting bodies desire the better attributes of the newer models and are often dismayed by the lack of change made by those who support the incumbent model. This view is a carry-forward from a previous era of accreditation, where accreditation was more of a "protector of the status quo" (Froyd et al., 2012). Unfortunately, external accreditation is of such value to institutions that they may be unwilling to take risks in this space, even when those risks would be welcome.

Another perceived barrier to innovation is the issue of scale. Many engineering schools operate at very large scales, with thousands of students. There is a misconception that because these institutions operate at scale, any innovation must also commence at scale. Disruptions must be scalable, but they can start small, even within large institutions, such as the NEET program at MIT (Lavi et al., 2021).

A real challenge is that of the "pioneer cohort." When a disruptive model is implemented, the student engineers in the model are the pioneers that can make or break the success of the innovation. To their benefit, they usually develop a level of resilience that is unparalleled in a traditional program, as they must persist through the constantly changing environment around them – a powerful outcome in transversal skill development. However, they lack a ready-made identity, which is inherent when people before them have succeeded both in the model and after graduation.

This lack of identity often results in a built-in lack of confidence in the new program and the people delivering it. Learning to be an engineer is an inherently difficult quest in any model and requires change and resilience on the part of the students. Being part of a pioneer cohort provides the alternative of making the new model the scapegoat – they can question the validity of the program rather than engage with and overcome the challenges they encounter. Understanding the attributes of pioneer cohorts can lead new disruptors to be ready for and expect these challenges, oftentimes even mitigating the issues before they emerge.

4.4 The Role of Faculty in Disruption

Key stakeholders in the disruption process are the faculty who teach engineering programs. There is a well-entrenched traditional model of teaching engineering which is thoroughly embedded in the processes and structures of the institution. While some innovators will happily be early adopters of any change that comes along, the majority of faculty will be more circumspect in embracing change. Therefore, influence is required to ensure academics have a willingness to embrace and understand the template models of disruptive engineering programs available. They need to have a willingness to change and upskill, to be able to effectively drive the curriculum reforms required in their context. In addition, they need to understand the impact of disruptive technologies in the teaching and learning process and champion the inevitable cultural changes that are required.

We now need faculty to start thinking the way we want our engineering graduates of the future to think. They need a broad knowledge in not only engineering but also socioeconomic factors, they need an entrepreneurial attitude, they need to be disruptive thinkers focused on innovation and value creation for their university, and they need to be capable themselves of working in interdisciplinary teams of specialists, engineers, and stakeholders. If they don't think like this, how can they instill this mindset in our future engineers?

4.5 Helpful Change Models

At a more macro level, it is important for us to be aware of how higher education institutes are complex ecosystems and how an understanding of this complexity can help plan and drive innovation across the configuration, offering, and experience for a particular organizational context. Otherwise, we could end up perplexed as to why our organizations, our leaders, our departments, our curricula, our customers, or our teaching strategies remain stagnant and impervious to sustained innovation. Therefore, an understanding of organizational theories can be advantageous. Manning (2017) describes this complexity in higher education and how organizational theory was, and can be, further used to help higher education institutes to reinvent themselves and become more innovative.

Leadership style and change management have critical parts to play in the innovation process. There are many different change and leadership models referenced in the literature (Borrego & Henderson, 2014; Bush, 2015; Froyd, 2014; Kolmos et al., 2016; Vlachopoulos, 2021). Bush describes the linkages between organization theory with different types of leadership in higher education and how the connection (among other things) has an impact on innovation and change management. The author surmises that the four aspects of organizations that theorists study are structure, goals, culture, and context. One could argue that these aspects are closely aligned to Doblin's configuration subdomain, organizational design ("Make form follow function and align infrastructure with core qualities and business processes").

The link between culture, change, innovation, and leadership has been referenced in engineering education reform literature since the early 2000s (Merton et al., 2004) and is consistently identified as an important connection in higher education in general (Blanco-Portela et al., 2017; Borrego & Henderson, 2014; Bush, 2015; Froyd, 2014; Reinholz & Apkarian, 2018; Vlachopoulos, 2021).

5 But It Can Be Delightful . . .

5.1 A New Mindset for New Challenges

Disruption is a mindset as much as a process. For most of us, we are disrupting our business before someone else does. This may take several years, as we roll in new programs, across multiple disciplines. There will be transition issues, sometimes for hundreds of students, as old subjects are phased out and new ones replace them. There can be turmoil and unhappiness for both students and academics.

However, there is also delight when students see learning in a new light (Hadgraft et al., 2018):

> Open-ended scope, freedom, and creativity. I liked how I had freedom to learn using my own practical experiences, instead of a regimented assessment schedule.

We are training a new mindset that sees the world from a disruptive point of view. Increasingly complex engineering problems require new solutions; there are no old solutions to copy. Disruption must be a habit in every classroom if we are to graduate professionals with this mindset.

Our old programs have, instead, mostly relied on applying known theory to known problems, as if the problems in the world were unchanging. These curricula were inspired by the post–World War II explosion in science and engineering. Our current world faces many challenges, at increasingly large scales. The Sustainable Development Goals (SDGs) (United Nations, 2021) describe the global challenges that we will face over the next 50 years, and we need engineers who can engage with these problems. Engineering graduates need to demonstrate awareness and application of the SDGs and understand their importance for the engineer's role in society. It is important to build the SDGs into individual subjects in engineering degrees. For example, a key part of the REEdI program is that students are required to demonstrate awareness and application of the SDGs in their on-campus projects, workplace tasks, and industry projects (O'Sullivan et al., 2022).

5.2 Maintaining the Disruption

Ongoing disruption is hard to maintain. It's easy to run this year's project like last year's, go through the motions, take it easy. Industry partnerships help us break out of this laziness. Industry problems change regularly and usually have a level of difficulty that stretches students beyond what they think they can achieve. These problems create an edgy environment that can stress some students (and academics!). Our job is to project-manage our students, keep them on track, help them be less stressed, but encourage them to seek novel solutions to these problems. We should encourage excellence through innovation.

This changes how we design subjects, semesters, whole curricula. Our fondness for outcomes-based education has tended to constrain our subject design, handcuffing us with a set of learning outcomes that can be too specific, often focusing only on technical outcomes. Instead, we need meta-learning outcomes that define program-level concepts, not the textbook contents page. For example: engage with stakeholders to identify a problem; apply design and systems thinking to respond to the identified problem; apply technical skills to develop, model, and/or evaluate designs; demonstrate effective collaboration and communication skills; and conduct critical self-, peer-, and group review for continuous improvement (Hadgraft et al., 2018).

Self- and peer review are essential skills for us as much as for our students. We must put our practice under the microscope, seeking constantly to refine and improve our practice. Agile methodology is one approach that relies on constant review of progress (wrike.com, 2022). The semester is broken into a series of sprints, for example, weekly or fortnightly. Teams present their work at each scrum or stand-up meeting to report progress and to state what will be accomplished next. Any difficulties can be quickly resolved at those meetings. This keeps students on track and resolves difficulties that they might have. In this way, the academics project-manage the student teams to completion. This is helpful for those students who might feel a bit lost when confronted with a problem that they have not seen before.

However, we also need to be on the lookout for new possibilities, new ways of disrupting ourselves, before we become too comfortable with how the subject, major, or program works. Students, too, should be on the lookout for new tools and new ways of working together and new people to work with. Diversity brings new challenges as well as new ways of looking at problems. Problems are increasingly transdisciplinary. Without this disruptive mindset, we fall back into old ways of thinking, turning out last year's design, when a new solution is required.

Of course, disruption can be uncomfortable. We know that each problem we face will bring challenges we haven't seen before. However, the key shift is in our problem-solving capabilities. We are shifting from a focus on being able to solve specific technical problems (circuit analysis, thermodynamics) to a focus on multidisciplinary problems, requiring multi-stakeholder input. There are recognized problem-solving processes for these sorts of problems, such as Fleming (2021) or Fogler and LeBlanc (2007); it's just that traditional programs have not taught them. Equipped with these meta-problem-solving skills, new problems are not just less challenging; they become positively exciting!

6 And It Never Ends

Higher education is a well-established marketplace, with heavily entrenched traditional models of learning. There is a long history of innovation in higher education, some successful and some less so. The confluence of ever more powerful technology, changing expectations of industry and students, and external shocks, such as pandemics, means that engineering education is ripe for wide-ranging disruption.

Doblin's ten types of innovation provide a framework to consider the ways in which innovation and disruption take place in engineering education. While we may project to the world that we offer a wide range of different offerings to potential students, the Doblin lens shows us that what we offer are mostly different experiences and configurations of a substantially similar offering. Reflecting on the kinds of innovations we pursue assists us in identifying what can truly become disruptive.

Sustainable scalable disruption requires innovation in the configuration domains. Disruptions within existing institutions require innovation in the processes and/or the structures of those institutions, as they reinvent the way they go about their operations. Disruptions beyond an existing institution require innovation in the network dimension, sharing the ideas beyond a single organization.

Disruptions that begin as start-up organizations avoid the challenges of reinventing an existing culture within existing structures and processes, but they do so at the expense of a lack of existing infrastructure and with the significant challenge of having to innovate (often unsuccessfully) in the profit model dimension. A blank page is by no means essential for disruption – it merely changes the nature of the disruption trajectory.

What is essential is the mindset of continuous improvement. Disruption can only progress when there is a commitment to making the experience, configuration, and offering better and a willingness to adapt to the different needs and risks of innovating in each of these dimensions. For the early adopters among us, change is comfortable, and we are willing to try and fail and try again. The late adopters among us support the sustainable disruptions that have the full spectrum of innovations covered. Fortunately, the mindset for disrupting engineering education is the engineering design mindset – the mindset that we want to instill in our graduates. Continuously improving what we do will enable our graduates to continuously improve what they do and to better serve their profession, their communities, and their societies as they practice as professional engineers.

7 Lastly, We Leave You With This Thought . . .

Being disruptive is not about a single product – it is much more. We have shown through this chapter that a combination of innovations, inspired by Doblin's framework, will produce impactful and powerful results. We suggest that action research is the paradigm that best fits the insights from the Doblin model. Disruption changes over time, and so does the kind of innovation that you need.

Whether you are an academic, an action researcher, or tasked with a change management initiative at your organization, we encourage you to take this framework and utilize it as a diagnostic tool to analyze your current state or use it to formulate a proposed new endeavor. No doubt the opportunities uncovered will look different depending on the context of your institution. Let Doblin's framework give you insights, let it highlight the opportunities, and let it illuminate the research questions that are ready to be explored. The value of this new model is the novelty of applying the Doblin lens. Our "call to action" is to use a deliberate process for innovation. As we've shown, successful innovators have disrupted across many dimensions, though from all appearances they have done so intuitively. Now, the intuitive approach is no longer necessary. The Doblin framework is available for future disruptors to use as a guide, deliberately and explicitly.

References

ACED. (2021). *Engineering change – The future of engineering education in Australia*. Australian Council of Engineering Deans. Retrieved November 22, 2021, from www.aced.edu.au/downloads/2021%20Engineering%20Change%20-%20The%20future%20of%20engineering%20education%20in%20Australia.pdf

Adams, R. S., Brightman, A. O., DeBoer, J., Jamieson, L. H., Oakes, W. C., Riley, D. M., & Rudin, P. (2021). Cultures of engagement and innovation: Realizing Purdue's public mission of access and impact in engineering education. *Advances in Engineering Education, 9*(3).

Arendale, D. (2016). *Postsecondary peer cooperative learning groups annotated bibliography*. www.arendale.org/peer-learning-bib/

Bertel, L. B., Askehave, I., Brohus, H., Geil, O., Kolmos, A., Ovesen, N., & Stoustrup, J. (2021). Digital transformation at Aalborg University: Interdisciplinary problem-and project-based learning in a post-digital age. *Advances in Engineering Education, 9*(3).

Blanco-Portela, N., Benayas, J., Pertierra, L. R., & Lozano, R. (2017). towards the integration of sustainability in Higher Education Institutions: A review of drivers of and barriers to organisational change and their comparison against those found of companies. *Journal of Cleaner Production, 166*, 563–578.

Blaschke, L. M. (2012). Heutagogy and lifelong learning: A review of Heutagogical practice and self-determined learning. *International Review of Research in Open and Distance Learning, 13*, 56–71. https://doi.org/10.19173/irrodl.v13i1.1076

Borrego, M. (2007). Development of engineering education as a rigorous discipline: A study of the publication patterns of four coalitions. *Journal of Engineering Education, 96*, 5–18.

Borrego, M., & Henderson, C. (2014). Increasing the use of evidence-based teaching in STEM higher education: A comparison of eight change strategies. *Journal of Engineering Education, 103*, 220–252.

Boyle, F., Moolman, J., Stephens, R., & Walsh, J. (2022b). Using immersive technologies to rethink engineering education in Ireland. In M. E. Auer, A. Pester, & D. May (Eds.), *Learning with technologies and technologies in learning – experience, trends and challenges in higher education*. Springer.

Boyle, F., Walsh, J., Riordan, D., Geary, C., Kelly, P., & Broderick, E. (2022a). REEdI design thinking for developing engineering curricula. *Education Science, 12*. https://doi.org/10.3390/educsci12030206

Bush, T. (2015). Organization theory in education: How does it inform school leadership. *Journal of Organizational Theory in Education, 1*(1).

CDIO. (2021). *CDIO: Conceive, design, implement, operate*. Retrieved December 13, 2021, from www.cdio.org

CEEDA. (2021). *Collaborative engineering education in the digital age*. www.ceeda.org/

Christensen, C. M., Aaron, S., & Clark, W. (2003). Disruption in education. *Educause Review, 38*, 44–55.

Christensen, C. M., McDonald, R., Altman, E. J., & Palmer, J. E. (2018). Disruptive innovation: An intellectual history and directions for future research. *Journal of Management Studies, 55*(7), 1043–1078.

Christensen, C. M., Raynor, M. E., & McDonald, R. (2015, December)). What is disruptive innovation? *Harvard Business Review*

Christensen Institute. (2022). *Disruptive innovation*. Retrieved July 4, 2022, from www.christenseninstitute.org/key-concepts/disruptive-innovation-2/

Coward, H. R., Ailes, C. P., & Bardon, R. (2000). *Progress of the engineering education coalitions*. US National Science Foundation. Retrieved March 1, 2023, from https://www.nsf.gov/pubs/2000/nsf00116/nsf00116.txt

Crawley, E. F., Malmqvist, J., Östlund, S., Brodeur, D. R., & Edström, K. (2014). The CDIO approach. In *Rethinking engineering education* (pp. 11–45). Springer.

Desjardin, J. (2020). *10 Types of innovation: The art of discovering a breakthrough product*. Retrieved December 26, 2021, from http://VisualCapitalist.com

Dewey, J. (1933). *How we think. A restatement of the relation of reflective thinking to the educative process* (revised ed.). Heath.

Doblin. (2021). *Ten types of innovation*. Retrieved December 16, 2021, from https://doblin.com/ten-types

Duarte, M., Leite, C., & Mouraz, A. (2016). The effect of curricular activities on learner autonomy: The perspective of undergraduate mechanical engineering students. *European Journal of Engineering Education, 41*(1), 91–104. https://doi.org/10.1080/03043797.2015.1056101

Engineers without Borders. (2021). *EWB challenge*. Retrieved December 13, 2021, from www.ewbchallenge.org

Fitzgerald, N., O'Sullivan, K., Morris, I., Walsh, J., Boyle, F., O'Connell, E., & Corkery, G. (2022). An enhanced assessment methodology for significant work-based placement. In *14th Annual international conference on education and new learning technologies*, Mallorca, Spain.

Fleming, N. (2021). *Smashing the state of dumb-stuck*. MoshPit.

Fogler, H. S., & LeBlanc, S. E. (2007). *Strategies for creative problem-solving*. Prentice Hall. www.engin.umich.edu/scps/

Fraser, K. (Ed.). (2014). *The future of learning and teaching in next generation learning spaces*. Emerald.

Froyd, J. E. (2014). Change strategies and leadership frames for advancing engineering education. In *QScience proceedings: Engineering leaders conference*.

Froyd, J. E., Wankat, P. C., & Smith, K. A. (2012). Five major shifts in 100 years of engineering education. In *Proceedings of the IEEE, 100 (Special Centennial Issue)* (pp. 1344–1360). https://doi.org/10.1109/JPROC.2012.2190167

Graham, R. (2018). *The global state of the art in engineering education*. MIT. neet.mit.edu

Graham, R. (2021). Engineering education: Resilience and purposeful change. *Advances in Engineering Education*, *9*(3). https://advances.asee.org/engineering-education-resiliance-and-purposeful-change/?utm_source=rss&utm_medium=rss&utm_campaign=engineering-education-resiliance-and-purposeful-change

Graham, R. (2022). *Crisis and catalyst: The impact of COVID-19 on global practice in engineering education*. https://dspace.mit.edu/handle/1721.1/145955

Grose, T. (2016). Challenge accepted. *Prism*, *26*(4), 13.

Gunderson, S., Jones, R., & Scanland, K. (2004). *The jobs revolution: Changing how American works* (2nd ed.). Copywriters Inc.

Hadgraft, R. G., Francis, B., Lawson, J., Jarman, R., & Araci, J. T. (2018). Summer studios – lessons from a 'small bet' in student-led learning. In *SEFI 2018*, Copenhagen.

Higher Education Authority Ireland. (2022). *Human capital initiative*. Retrieved July 4, 2022, from https://hea.ie/skills-engagement/human-capital-initiative/

Impey, C., & Formanek, M. (2021). MOOCS and 100 Days of COVID: Enrollment surges in massive open online astronomy classes during the coronavirus pandemic. *Social Sciences & Humanities Open*, *4*(1). https://doi.org/10.1016/j.ssaho.2021.100177

Johnson, B., Ulseth, R., Smith, C., & Fox, D. (2015). The impacts of project based learning on self-directed learning and professional skill attainment: A comparison of project based learning to traditional engineering education. In *Proceedings – frontiers in education conference, FIE*.

Jonassen, D. (2014). Engineers as problem solvers. In Johri & B. Olds (Eds.), *Cambridge handbook of engineering education research* (pp. 103–118). Cambridge University Press. https://doi.org/10.1017/CBO9781139013451.009

Keeley, L., Walters, H., Pikkel, R., & Quinn, B. (2013). *Ten types of innovation: The discipline of building breakthroughs*. John Wiley & Sons.

Killi, S., & Morrison, A. (2015). Just-in-time teaching, just-in-need learning: Designing towards optimized pedagogical outcomes. *Universal Journal of Educational Research*, *3*, 742–750. https://doi.org/10.13189/ujer.2015.031013

Kolb, D. A. (1984). *Experiential learning: Experience as the source of learning and development*. Prentice-Hall.

Kolmos, A., Fink, F., & Krogh, L. (Eds.). (2004). *The Aalborg PBL model: Progress, diversity & challenges*. Aalborg University Press.

Kolmos, A., Hadgraft, R. G., & Holgaard, J. E. (2016). Response strategies for curriculum change in engineering. *International Journal of Technology and Design Education*, *26*(3), 391–411.

Lavi, R., Bathe, M., Hosoi, A., Mitra, A., & Crawley, E. (2021). The NEET ways of thinking: Implementing them at MIT and assessing their efficacy. *Advances in Engineering Education*, *9*(3).

Liberatore, M., Morrish, R., & Vestal, C. (2017). Effectiveness of just in time teaching on student achievement in an introductory thermodynamics course. *Advances in Engineering Education*, *6*(1), 1–15.

Lindsay, E., & Morgan, J. (2021). The CSU engineering model: Educating student engineers through PBL, WPL and an online, on demand curriculum. *European Journal of Engineering Education*, *46*(X). https://doi.org/10.1080/03043797.2021.1922360

Loh, A. P., Law, E., Putra, A. S., Koh, E., Zuea, T. K., & Tat, K. E. (2021). Innovation, design & entrepreneurship in engineering education. *Advances in Engineering Education*, *9*(3).

Magnanti, T. L. (2018). Building a new academic institution: The Singapore University of Technology and Design. In *Accelerated universities* (pp. 103–127). Brill Sense.

Mann, L., Chang, R., Chandrasekaran, S., Coddington, A., Daniel, S., Cook, E., & Smith, T. D. (2021). From problem-based learning to practice-based education: A framework for shaping future engineers. *European Journal of Engineering Education*, *46*(1), 27–47.

Manning, K. (2017). *Organizational theory in higher education* (2nd ed.). Taylor & Francis.

Merton, P., Froyd, J., Clark, C. M., & Richardson, J. (2004). Challenging the norm in engineering education: Understanding organizational culture and curricular change. In *American society for engineering education annual conference & exposition*, Salt Lake City, Utah.

Miller, R. K., & Dorning, A. M. (2018). Olin College of Engineering: Reinventing engineering education in the United States. In *Accelerated universities* (pp. 86–102). Brill Sense.

Morelock, J. R. (2017). A systematic literature review of engineering identity: Definitions, factors, and interventions affecting development, and means of measurement. *European Journal of Engineering Education*, *42*(6), 1240–1262. https://doi.org/10.1080/03043797.2017.1287664

Morris, I., O'Sullivan, K., Fitzgerald, N. P., Walsh, J., Boyle, F., O'Connell, E., & Corkery, G. (2022). Co-design and co-development of an overarching work placement framework with industry. In *16th Annual international technology, education and development conference*, Online.

Noor, A. K. (2013). Envisioning engineering education and practice in the coming intelligence convergence era – a complex adaptive systems approach. *Central European Journal of Engineering*, *3*(4), 606–619. https://doi.org/10.2478/s13531-013-0122-9

O'Sullivan, K., Morris, I., Fitzgerald, N., Walsh, J., Boyle, F., O'Connell, E., & Corkery, G. (2022). Implementation of the United Nations sustainable development goals in the project framework for rethinking engineering education in Ireland. In *14th Annual international conference on education and new learning technologies*, Mallorca, Spain.

Prince, M. (2004). Does active learning work? A review of the research. *Journal of Engineering Education*, *93*(3), 223–231. https://doi.org/10.1002/j.2168-9830.2004.tb00809.x

Prince, M. J., & Felder, R. M. (2006). Inductive teaching and learning methods: Definitions, comparisons, and research bases. *Journal of Engineering Education*, *95*(2), 123–138.

Rampelt, F., & Suter, R. (2017). Recognition of prior learning – outcome-oriented approaches to the recognition and assessment of MOOC-based digital learning scenarios. In *EDULEARN17 proceedings*

Reinholz, D. L., & Apkarian, N. (2018). Four frames for systemic change in STEM departments. *International Journal of STEM Education*, *5*(3). https://doi.org/10.1186/s40594-018-0103-x

Riskowski, J. L. (2015). Teaching undergraduate biomechanics with just-in-time teaching. *Sports Biomechanics*, *14*(2), 168–179. https://doi.org/10.1080/14763141.2015.1030686

Rivas, M. J., Baker., R. B., & Evans, B. J. (2020). Do MOOCs make you more marketable? An experimental analysis of the value of MOOCs relative to traditional credentials and experience. *American Educational and Research Association (AERA) Open*, *6*(4), 1–16. https://doi.org/10.1177/2332858420973577

SAE International. (2021). *Formula SAE*. Retrieved November 22, 2021, from www.fsaeonline.com

Schön, D. A. (1983). *The reflective practitioner: How professionals think in action*. Temple Smith.

Schön, D. A. (1987). *Educating the reflective practitioner: Toward a new design for teaching and learning in the professions*. Jossey-Bass.

Sheppard, S. D., Macatangay, K., Colby, A., & Sullivan, W. M. (2008). *Educating engineers: Designing for the future of the field*. Jossey-Bass.

Shimson, G., & Verstelle, M. (2017). *Online learning lab innovation report 16/17*. Centre for Innovation, Leiden University. Retrieved July 5, 2022, from www.academia.edu/33154395/Online_Learning_Lab_Innovation_Report_16_17

Snowden, D. J., & Boone, M. E. (2007). A leader's framework for decision making. *Harvard Business Review*, 69–76.

Sorby, S., Fortenbury, N., & Bertoline, G. (2021, September 13). *Stuck in 1955, engineering education needs a revolution*. Issues in Science and Technology. https://issues.org/engineering-education-change-sorby-fortenberry-bertoline/

Svanström, M., Palme, U., Wedel, M. K., Carlson, O., Nyström, T., & Edén, M. (2012). Embedding of ESD in engineering education: Experiences from Chalmers University of Technology. *International Journal of Sustainability in Higher Education*, *13*(3), 279–292.

TEDI. (n.d.). *Founding partners*. Retrieved July 5, 2022, from https://tedi-london.ac.uk/about/founding-partners/

Tonso, K. (2014). Engineering Identity. In A. Johri & B. Olds (Eds.), *Cambridge handbook of engineering education research* (pp. 267–282). Cambridge University Press. https://doi.org/10.1017/CBO9781139013451.019

Trevelyan, J. (2014). *The making of an expert engineer.* Taylor & Francis.

Ulseth, R., & Johnson, B. (2017). Self-directed learning development in PBL engineering students. *The International Journal of Engineering Education, 33*(3), 1018–1030.

Ulseth, R., Johnson, B., & Kennedy, C. (2021). Iron range engineering. *Advances in Engineering Education, 9*(3).

United Nations. (2021). *Sustainable development goals.* United Nations. Retrieved September 30, 2021, from http://sdgs.un.org

Universities Australia. (2021). *Learning & teaching repository.* Retrieved March 2, 2023, from https://ltr.edu.au/

van Grunsven, J. B., Marin, L. A. V. I. N. I. A., Stone, T. W., Roeser, S., & Doorn, N. (2021). How to teach engineering ethics? A retrospective and prospective sketch of TU Delft's approach to engineering ethics education. *Advances in Engineering Education, 9*(4).

Van Melle, L., & de Bie, T. *10 Years of MOOCs in education- from hype to resurgence.* Centre for Innovation, Leiden University. Retrieved July 5, 2022, from www.centre4innovation.org/stories/10-years-of-moocs-in-education-from-hype-to-resurgence/

Vlachopoulos, D. (2021). Organizational change management in higher education through the lens of executive coaches. *Education Sciences, 11*(269). https://doi.org/10.3390/educsci11060269

Wilkie, B. J. (2013). *An assessment of a just-in-time training intervention in a manufacturing organization* (Publication Number 31) [University of Southern Mississippi]. https://aquila.usm.edu/dissertations/31

World Economic Forum. (2020). *These are the top 10 job skills of tomorrow – and how long it takes to learn them.* Retrieved December 16, 2021, from www.weforum.org/agenda/2020/10/top-10-work-skills-of-tomorrow-how-long-it-takes-to-learn-them/

wrike.com. (2022). *What is agile methodology in project management?* Retrieved July 6, 2022, from www.wrike.com/project-management-guide/faq/what-is-agile-methodology-in-project-management/

Yang, Q., & Lee, Y.-C. (2021). The critical factors of student performance in MOOCs for sustainable education: A case of Chinese universities. *Sustainability, 13*. https://doi.org/10.3390/su13148089

Part 2

Theoretical Orientations and Critical Approaches in Engineering Education Research

7

The Role and Use of Theory in Engineering Education Research

Andrea Goncher, Ashish Hingle, Aditya Johri, and Jennifer M. Case

1 Introduction

There have been scholarly and practice-oriented publications related to engineering education for almost as long as there have been formal engineering programs of study, but in the last few decades, engineering education research (EER) has emerged as a distinct field of inquiry (Jesiek et al., 2009). EER has a substantial literature outlining how research in this field should be undertaken and what counts as quality research (Streveler & Smith, 2006; Johri & Olds, 2014; Klassen & Case, 2022). The field draws on a range of social science theories and methodologies to undertake investigations to inform the improvement of engineering education, and thus, it can be stated that EER is largely applied social research, that is, it is "research that attempts to solve a concrete problem or address a specific policy question and that has a direct, practical application" (Neuman, 2014, p. 1). This does not preclude advancing scientific understanding, but the use of knowledge to solve a problem inspires that goal (Stokes, 2011).

Since each field or discipline has many tacit practices through which knowledge is generated and shared, newcomers, especially those from different epistemologies, often face "conceptual difficulties" in their effort to learn a new domain. In work that examined these difficulties during the early formation of the EER field, Borrego (2007) identified five features that were problematic, in terms of conceptual boundary-crossing, for researchers who were trained primarily in engineering disciplines and were starting out in EER:

1 The open-ended nature of research questions ("why" or "how" questions as opposed to closed-form yes-or-no questions);
2 The need for a theoretical framework to guide educational research;
3 The use of qualitative or mixed research methods as a complement to quantitative approaches;
4 Difficulty in definition and measurement of constructs or variables for educational research; and
5 The need to work within interdisciplinary configurations to make up for the lack of individuals trained in both engineering and social science methods.

In the 15 years since the publication of that article, there have been more systematic review articles that provide a critical and comprehensive overview of development within the field. These analyses have examined the work bibliometrically, using citations or other relevant information, and through

DOI: 10.4324/9781003287483-9

qualitative analysis of papers (Brozina et al., 2021). They provide a window on development within EER in terms of methods, topics, and co-authorship (Malmi et al., 2018); Wankat et al., 2014; Williams et al., 2014; Xian & Madhavan, 2014). Overall, there is now substantially published literature in the field, with many discussions in relation to how to pose research questions, what methods to use, and so on. The field has made progress, in terms of codified knowledge, that has potentially made it easier for those trained in other disciplines to socialize into EER. However, except for Malmi et al. (2019), which refers to the use of theories, none of the other studies have looked specifically at theoretical frameworks. Yet theoretical development is often seen as paramount for the advancement of any field (Kemnitzer, 2008), and from early on, EER scholars also noted the need for both theoretical and methodological sophistication to inform its status as a field (Froyd & Lohmann, 2014).

In discussing what they consider as a "theory deficit" in higher education research, Hamann and Kosmützky (2021) give a description which could easily apply to EER: "A lot of research in the field is done in a demand-oriented mode for policy and practice, often based on normative questions" (p. 468). They caution that this theory deficit limits the capacity of the field for ongoing growth and development. To this end, they propose the notion of "theory work":

> With this notion we suggest understanding theories not as sterile and disembodied knowledge and, ultimately, uniform sets of propositions, but as tools that can be used for different purposes. Consequently, working with theories can be conceived of as a craft that can – and should – be made explicit, maintained, and trained.
>
> *(p. 469)*

Our main aim in this chapter is, therefore, to advance this orientation towards "theory work" for EER by building on previous scholarship on the role of theory in engineering education, by attending to the role it can play, and by looking at theories and theoretical perspectives that some recent scholars in the field have used to make this work explicit. We also discuss terminology and constructs that underpin theoretical work, such as paradigms, concepts, frameworks, and models (Hughes et al., 2019).

2 Theories and Theoretical Frameworks

One of the unintended consequences of a move towards greater engagement with theory at a relatively quick pace within EER has been a lack of clarity within the field on the role of theory and, increasingly, paucity of common ground around what constitutes a theory and when designing a research study or presenting findings (Johri, 2010). These are thorny matters and not limited to engineering education; for instance, our allied field, computing education, is grappling with similar questions (Nelson & Ko, 2018).

A *theory* is a system of interconnected abstractions or ideas that condenses and organizes knowledge in relation to a phenomenon in the world (Neuman, 2014). A theory or theoretical explanation uses a collection of carefully considered concepts, has logical consistency, and is embedded within a larger arrangement of similar explanations. In other words, any social theory is a compact way to describe and expand our understanding of the social world such that our understanding moves from a specific situation (e.g., students from the first-year engineering program at university X get better grades on a standardized assessment than students from the same program at university Y) towards general understanding and knowledge (e.g., students who are taught using open-ended problem-solving followed by multiple-choice exams tend to do better on standardized exams than those in a similar program who are taught primarily using lectures and multiple-choice exams) (Neuman, 2014).

Theory serves many useful functions in relation to research and writing. First, theory situates our findings or ideas in an existing scholarly conversation and shows clearly how our findings or arguments build upon or transcend prior work (Jawitz & Case, 2009; Case, 2008). Second, theory helps us provide succinct and coherent explanations for real-world behavior. *Causality*, the primary criterion of a theory in the natural sciences, is the ability to explain how observed phenomena are connected and is the ultimate goal of any theory (Fiske, 2004). The usefulness of theory is its interpretative function. Through it, we contextualize the world around us and our actions in the world. Third, theory shapes the design of a study and guides us in constructing knowledge right from the very beginning of a research endeavor. Underpinning a research design with theory helps avoid regurgitation and results in creative outcomes to help build and grow a body of knowledge. Fourth, theory assists us in the diffusion of research results by providing a guiding framework, a narrative structure, a discourse, or a schema to assist us in talking about our work and its usefulness. Coherence is the basic idea that arguments flow from each other, and contradiction is avoided. A theory tells a good story by revolving around an interesting problem with an equally or more exciting answer and is parsimonious in the telling of the story by being simple and effective.

From a traditionally positivist perspective, theory should be testable by other scholars, and it should be fertile in generating interest and scholarship (Fiske, 2004); (Jamieson & Lohmann, 2009; Watson, 2009). In this paradigm, theories help in addressing four central goals of empirical work: analysis, explanation, prediction, and prescription; theories that do this work can themselves be further classified as (1) theory for analyzing, (2) theory for explaining, (3) theory for predicting, (4) theory for explaining and predicting, and (5) theory for design and action (Gregor, 2006). We refer interested readers to Appendix 1 on p. 488 in Hamann and Kosmützky (2021) for further articulation of different conceptions of theory and theorizing, drawing on the work of Merton, Abend, and Krause.

Another useful way to understand theory is to look at what it is not. Sutton and Staw (1995) outline five common elements appearing in research papers that signify superficial uses of theory: references, data, diagrams, variables, and hypotheses. They argue that theory must be used, not simply referenced. They argue that authors need to go beyond providing a token laundry list of theoretical references and, instead, need to present an "explanation of why the theory or approach leads to a new or unanswered theoretical question" (p. 373). They argue that theories must be USED in theory-informed research to (1) explain why certain patterns were observed in the data, (2) specify how variables were generated and how they are connected, (3) spell out the underlying logic behind visual representations used in a paper, and (4) present logical arguments as support for why certain relationships should occur.

A *theoretical framework* refers to a selection of theories assembled to inform a research design. This is the most common understanding of the term, although some researchers make a further distinction between theoretical, conceptual, and analytical frameworks as follows (see Imenda, 2014; Neuman, 2014; Wimmer & Dominick, 2013; Magana, 2022 for further details):

- Theoretical frameworks must consist of elements of a theory or a set of theories and must talk about how that shapes the research and how the research might add to the framework, clarify elements of it, or make some contribution to it.
- Conceptual frameworks include sets of concepts that can be derived from theories but also from personal experience or other empirical work that is not necessarily theory-driven.
- Analytical frameworks are a combination of theoretical and conceptual frameworks and constructed primarily to guide empirical work or analysis but can also be used in the review of prior work.

In addition, like theories, although the most common use of frameworks is to guide the research study, they are also used to interpret and discuss the findings. This is especially the case when the findings are meant to extend or revise the original framework.

3 Paradigms

Theories are always embedded within broader paradigms (Wimmer & Dominick, 2013). A research paradigm offers a structure to undertake research by integrating the philosophical assumptions and providing models and techniques to assist with the methodology, including data collection and analysis (Kivunja & Kuyini, 2017). According to Neuman (2014), "[i]t organizes concepts, theoretical frameworks, and research methods" (pg. 60). A *paradigm* is thus an overarching term that encapsulates the researcher's worldview, how they see the phenomenon at hand, and how they justify going about research. Embedded within paradigms are both ontology and epistemology. *Ontology* refers to our views on the nature of the world, and *epistemology* refers to our views on how we come to know about the world (Baillie & Douglas, 2014). All these choices are often not made explicit by researchers, but attempting to do so will assist with the selection of concepts and theories we use and how we use them. Within educational research in particular, the role of paradigms has been contested for a long time (Rist, 1977), including in discipline-based research (Szyjka, 2012), and many articles provide guidance on the application of paradigms (Van Merriënboer & De Bruin, 2014).

The debate around the most appropriate paradigms in social sciences, their pros, cons, and usefulness, is ongoing and probably never-ending, but at this time within the academic community, most people can agree that there are three paradigms of research and scholarship under which most of us operate (Crotty, 1998). The first, the *postpositivist paradigm*, is the most common and widely used approach within the social sciences, especially in North America and Europe, and in this approach, the social sciences are seen as akin to the natural sciences, with similar assumptions about objectivity and facts and the research process (Jawitz & Case, 2009). Positivism is thus typically a starting point for EER researchers who were trained in science or engineering (Douglas et al., 2010; Kellam & Jennings, 2021). Here it should also be noted that increasingly researchers use the term *postpositivism*, which is a more refined take on positivism that acknowledges fallibility as a core part of the research process.

The second paradigm is *interpretivism*, and for interpretive researchers, social life is different from the natural world in fundamental ways, and they argue that, rather than borrowing scientific principles from the natural sciences, there is a need to develop unique ways of examining human social life by understanding the meanings humans attach to their experiences. The third paradigm is the *critical approach*, which shares many features with interpretivism but focuses on inequality and injustice in society and aims for change by putting knowledge into action. Why does the paradigm we function under matter? Primarily because encapsulating our worldview guides how we can and want to know the world and what we want to say about it; in other words, it shapes any form of empiricism we undertake, but more importantly, it shapes the ideas we have about the world.

4 Epistemological Tensions and Theoretical Pluralism

The social world is complex, and each theory or paradigm can only illuminate a particular "slice" of that reality. The argument has thus been made for the need for the deployment of multiple theories in higher education research (Ashwin, 2009) since there is no one theory that can capture all potentially important aspects of a phenomenon like engineering education. This is likely a challenge for those schooled in engineering, where there tends to be more agreement on the nature of scientific phenomena and how they might be conceptualized.

The tension between different ways of knowing, or epistemological tensions, are increasingly visible within EER. Kellam and Jennings (2021) point to the difficulties in having positivist-oriented reviewers giving comments asking for generalizability and statistical power in relation to their proposal, which was framed in the postmodern tradition. They survey recent research in EER and analyze "voice," showing what they identify as a tension between positivist and postmodern voices in the field. Seniuk et al. (2018) provide a collaborative autoethnography reflecting on how these epistemological tensions influence doctoral students in the field.

In this chapter, we want to move past this tension and advocate for theoretical pluralism in this field, similar to the calls for methodological pluralism already so well established (Beddoes, 2014; Case & Light, 2011). Not everyone needs to use every theory; in fact, this most probably would not make sense, given that we need scholars who can develop a deep understanding of particular theoretical orientations, but we need to advance mutual understanding of what people are doing with theory, what particular theories are good for, and where their limitations reside. The following section aims to explore this further.

5 The Use of Theories in Engineering Education Research

We conducted a study looking at *gray literature* to illustrate recent use of theory in EER (Hingle et al., 2022). *Gray literature* refers to information sources within the field that do not appear in traditional venues, such as conferences and journals, and that include outputs such as dissertations, research proposals, and technical reports, as well as, increasingly, online repositories and archives, such as arXiv (Adams et al., 2016). In the chapter, we specifically analyzed abstracts of funded US National Science Foundation (NSF) research and US-based PhD dissertations, focusing on the time period 2011–2021 (Hingle et al., 2022). These gray literature sources, that is, research not published in traditional venues, are essential to understanding research and topic trends within the domain as they are milestones for research and scholarship. Gray literature, especially dissertations and research projects, are also representative of the most recent and current scholarship in the area. Our chapter focused on mapping the use of *theoretical underpinnings*, including, "model," "conceptual framework," "theoretical framework," and "theory" (Passey, 2020), in dissertations and NSF-funded research. We used this approach as a starting point to understand any trends in how researchers undertaking EER intended to shape (Jesiek et al., 2011), define, and ground their study's focus, methods, and contextual considerations with the use of theory. In the rest of the section, we focus on a few theories identified through our analysis, and for readers interested in a larger list of theories that have been used across engineering and computing education research, we refer them to Appendix A.

The term "theoretical framework," which signals a mobilization of existing theories towards a particular study, informing its research design and facilitating a shared understanding of how research and scholarship are interpreted, was prevalent in the gray literature we explored. We identified particular theories used and found some broad families of theories that appeared prevalently:

- Social cognitive theory
- Motivation theory
- Identity theory
- Socialization theory
- Change theory

While these theories are typically put to work in conceptualizing learning in EER, they are theories that play on a much broader stage in the overall social sciences (for a review of theories focused specifically on learning practice, see Newstetter & Svinicki, 2014). One set of theories points to key theorizations of the individual, focusing on cognition, motivation, identity, and socialization.

It will be observed here that EER has tended to theorize at the individual level when it comes to the unit of analysis, as is common across many other fields of education research, given the largely psychological leaning of the field (Woolfolk et al., 2008). Even theories of socialization are predominantly looking at how the individual comes to be socialized. Identity theories also typically understand the formation of identity in the social context, but the focus for theorizing is on the individual.

We further identified a set of theories from the preceding list that aim to integrate theory and empirical research versus theories used largely to interpret and understand findings (Gregor, 2006). These "middle-range" theories (Merton, 1949; Lenz et al., 1995) utilize theory with analytical elements to guide empirical research. This analysis led to the following groupings of theories with similar orientations towards particular aspects of engineering education:

On Learning

- Cognitive load theory – a learning theory that describes the limitations on how many ideas a learner can hold in their head at one time.
- Communities of practice – on how novices learn to become experts within the context of joining a community.
- Self-efficacy – a learning theory that emphasizes the need for learners to self-regulate their learning.
- Situated learning – a learning theory that emphasizes the context in which learning occurs.
- Socio-constructivist learning theory – a variation of constructivist learning theory that recognizes the agency of the individual learner in making sense of things but also acknowledges the significance of the social context.

On Individual Choice (Including Careers)

- Social cognitive career theory – on the process of career choice.
- Expectancy-value theory – a theory of how individuals make decisions based on both the likelihood of success and the value of the task to the individual.

On Power Relations in Society

- Intersectionality theory – on how various markers of difference (race, class, gender, etc.) combine to place individuals in positions of marginality in the context of oppressive power relations.
- Social capital theory – a theory that understands social power relations in terms of capital that is not only economic, and is mobilized through habitus to obtain advantage in the social field.
- Community cultural wealth – a theory that contrasts itself to Bourdieu's theory of capital by emphasizing that those from seemingly low socioeconomic statuses draw on resources within the community that allow them to advance.

On Organizational Change

- Collective impact framework – an organizational change theory that emphasizes the need for collective and coordinated action.
- Diffusion of innovation – a theory that describes the spread of new ideas.
- Kotter's 8-step model of change – a model of change that outlines distinct steps for organizational change to be effective.

It is not surprising that a range of learning theories is present, since a core of the work continues to look at how to improve student learning outcomes in engineering education. The theoretical bases for studies of student choice of career also signal a longstanding interest for investigations in EER. It is interesting that this analysis did not yield any distinct theories of teaching or curriculum.

The prevalence of sociological theories in relation to power relations in society and, specifically speaking, to the marginalization of race, gender, and other such markers of diversity signals a significant area where the field of EER has been growing dramatically in its focus and theoretical grasp. Finally, the appearance of organizational change theories signals that funding priorities in this area in the United States are shifting the field productively.

It is interesting to compare this analysis with the overview of theoretical areas that was developed some years ago in the exercise to produce an EER taxonomy, led by Finelli and colleagues (Finelli et al., 2015). The taxonomy signals a significant focus on affective, cognitive, and developmental theories (of the individual). Critical theory and intersectionality are briefly featured in the taxonomy, but no other sociological theories of power relations appear. There is no mention of organizational change theories or epistemological theories (Baillie & Douglas, 2014). Given that the work on the taxonomy was done around 2012–2013, this probably reflects the changes in the use of theory within EER.

Having looked at the range of theories, we now dig deeper into a few selected theories, each aligned with one of the larger paradigms, to examine their use in-depth, including the nature and range of concepts used and some exemplar uses of the theory. Our aim here is to build on the point earlier about the need for epistemological pluralism in EER. There is no one "right" paradigm, and similarly, there is no "right" theory. What matters is the question that the researcher is asking and how they mobilize both theory and method to accomplish their aim.

5.1 Social Cognitive Career Theory – Theorizing Within a Postpositivist Paradigm

Our first example, from the postpositivist tradition, is *social cognitive career theory* (SCCT), which has been employed by researchers in both dissertations and research proposals. We characterize this theory under the postpositivist paradigm because of the foundational definition and characterization of the theory (Lent et al., 2002). According to the proponents of the theory, SCCT incorporates three central *variables,* that is, *factors that can change*, from general social cognitive theory: (1) self-efficacy, (2) outcome expectations, and (3) personal goals. These three variables are seen as basic "building blocks" of career development and represent key *mechanisms*, that is, *a system of interacting components*, by which people are able to exercise personal agency (p. 261). Lent et al. (2002) also state that the SCCT framework organizes career-related interest, choice, and performance into three interlocking *models*, that is, *system of related concepts*. In their paper, they present an overview of these models, focusing on the interplay among the central social cognitive variables in guiding career development and discussing how these *variables* operate in concert with other personal and contextual aspects of persons (p. 264). In the same chapter, on pages 266 and 269, they present figures that outline different hypothesized connections for each factor, and discuss mediator and moderator variables. Finally, even when discussing reviews of research on SCCT, the authors highlight the predictive power of the SCCT, the support for causal relations, and the quantifiable differences across social groups (p. 280; also see Sheu et al., 2010). The use of words such as *models*, *variables*, *prediction*, and *causal relations* all points to the positivist tradition, where the goal is to undertake social science research in a scientific manner that is modeled after the natural and physical sciences and where prediction and causal modeling is the primary goal.

Within engineering and computing education research, many scholars have used SCCT and primarily in the positivist tradition to examine gender differences in career choices, in student motivation for specific careers, and for predicting academic interests and goals of students (Carrico & Tendhar, 2012; Inda et al., 2013; Lent et al., 2008). We illustrate the use of SCCT through an example which applies the framework to a dissertation study. We selected this example to highlight how the framework was applied in the research design and analysis that allowed the researcher to utilize empirical results to extend the analytic generalizability of social cognitive career theory. Young (2017) used SCCT to guide the qualitative study on engineering student participants in a co-op development program (CDP).

Using theory to modify or extend an existing theoretical framework is important to guide further research rather than using it as cited work (Malmi et al., 2018). Young's use of social cognitive career theory (SCCT) allowed her to situate her findings and provide explanations of real-world behavior and practice. The use of theory in this study is a good example of how she used SCCT to situate the study's findings in existing scholarship and extend the analytic generalizability of the theory.

Young (2017) used the framework to predict and discuss results related to participants' career decision-making and how their choices were influenced by engagement in a co-op development program. The study's research questions were grounded in SCCT and operationalized how learning experiences shaped outcome expectations and career actions. The theoretical framework in this example played an important role in the concluding rationale and implications for future research, policy, and practice. The outcomes of this work informing student-centered university–industry partnerships aimed at supporting workforce development and, grounded by SCCT, is a good example of the conceptualization of theory to structure education and the workplace.

Another contribution from this example resulted in the creation of operationalized definitions of learning experiences embedded within SCCT and linked the identified student experiences that led to an early career within the engineering industry. Our earlier discussion on the potential ambiguity in defining the terminology, for example, *concept*, *construct*, and *framework*, associated with theory in EER can pose difficulties for researchers trying to utilize theory for interpretive function. Young's operationalized definitions can provide guidance for researchers who intend to understand and use SCCT in their research design and analysis.

5.2 Situated Learning – Theorizing Within an Interpretivist Paradigm

Compared to SCCT, *situated learning* has a broader realm as a theory, and its definition is less precise and states primarily that humans learn in practice through legitimate, recognized, peripheral participation in a community where they are guided towards developing expertise by those who are established practitioners (Lave & Wenger, 1991). The constructs or concept clusters associated with situated learning are "cognitive apprenticeship," "communities of practice," and "legitimate peripheral participation," among others (see Collins et al., 1989; Lave, 1991, 1993; Wenger, 1999). Within EER, aspects of situated learning are commonly used (Johri & Olds, 2011; Johri et al., 2014), and scholars have deployed it for research on engineering practice (Buch, 2015), transition to professional work (Lutz & Paretti, 2021; Reich et al., 2015), disciplinary participation (Peters, 2018), conceptual change (Bornasal et al., 2018), and engineering leadership, (Rottmann et al., 2018) among other purposes.

From within our dataset of dissertations, one exemplar is Radhakrishnan (2020). Radhakrishnan (2020) draws from the theoretical framework of situated learning and communities of practice (CoP) to discuss the outcomes of a three-phase professional development program for aspiring engineering teachers at an alternative school in western Kenya. This dissertation study uses theory in the broader realm to contribute to practice through the design and implementation of a professional development program, guided by learning as participation in a community of practice. Situated learning provided a structural framework for this dissertation work, focusing on the ideas that learning is a social process and constructed dynamically within particular social and physical environments. Situated learning and communities of practice provided the framework for both the design and evaluation of this research on professional development programs for engineering teachers in a specific context. Understanding the *context*, that is, Tumaini context, was necessary to understand the culture of the community. The results of Radhakrishnan's (2020) work are design outcomes that are based on a situated understanding of the theory in a "fragile context" (p. 47) and design principles that are transferable in comparable settings. Radhakrishnan's use of situated learning as a critical framework and communities of practice to frame the study design provides a good example of how frameworks can complement one another. It is important to explicate how theory is used in a study, especially for research intended to generate new knowledge. The theoretical framework should be the base for conducting research. Radhakrishnan leveraged the critical facets of situated learning and communities of practice to explain, inform, and provide support for the intended outcomes of the study. For example, CoP focuses on the formation of community, interconnections to practice, and construction of identify, all of which align with the significance of the study and Radhakrishnan's overarching goal of long-term success and sustainability. This is a good example of how a theoretical framework can be used to contextualize the study and provide links between theory and the experiences of real-world practice and behavior.

5.3 Intersectionality – Theorizing Within a Critical Paradigm

Critical theories are used to describe and critique existing social contexts or structures and aim for change. Intersectionality, a theoretical perspective that emerged from feminist studies and is now used across a range of critical theories, argues that lived experiences and social categories of gender vary, often dramatically, by race, ethnicity, class, sexual orientation, age, and other interconnected dimensions of difference and inequality (Bilge, 2010; Dhamoom, 2011; McCall, 2005). Coined by legal scholar Crenshaw (1989), the term *intersectionality* emerged when women of color formulated a critique of gender analyses based on experiences of White middle-class women that ignored the voices and stories of women of color and those without economic privilege. These scholars argued that gender never operates independently of other markers of identity or dimensions of difference (i.e., race, ethnicity, class, sexuality, age, nationality, etc.). Particular contexts may make one or another of these markers more or less salient for certain individuals or groups (Hulko, 2009). The intersectional approach guides the analysis towards examination of the mutual constitution of identities, social practices, and structures that produce and maintain hierarchies of difference (Bilge, 2010). Intersectionality examines the construction of social categories and its impact on discrimination or disadvantage. One context for examining intersectionality is engineering identity. Studies that employ intersectionality as a theoretical framework examine how categories such as race, gender, ability, sexual orientation, and other biological, social, or cultural categories intersect. In her ethnographic study of an engineering college, Tonso (2007) points out that "engineer" is itself a powerful social category that operates to privilege some forms of behavior and marginalize others in particular contexts.

Other types of critical theoretical frameworks employed in engineering education include critical agency, community cultural wealth, funds of knowledge, and Bourdieuian frameworks (Mejia et al., 2018). We selected another example dissertation study which applies a critical agency framework and highlights using EER to further existing theory and the field. Godwin's (2014) dissertation work builds on critical agency theory by validating and refining the framework to understand how critical engineering agency is developed within a community of practice. The research in this example built on previous work that identified relationships of CEA and used a qualitative phase to explain and interpret results, utilizing the interpretive power of theory (Malmi et al., 2019).

We selected another example dissertation study which applies a critical agency framework and highlights using EER to further existing theory and the EER field. Godwin's (2014) dissertation work builds on critical agency theory by validating and refining the framework to understand how critical engineering agency (CEA) is developed within a community of practice. The research in this example built on previous work that identified relationships of CEA and used a "mixed methods" approach to explain and interpret results, utilizing the interpretive power of theory (Malmi et al., 2019).

Godwin (2014) takes a pragmatist approach in her study to leverage mixing approaches and methods. The "mixed method" approach allows for the triangulation of results and the ability to explicate *new* theory. While triangulating results is useful in validating the results of most studies, it is particularly helpful in this work that aims to draw connections between individual frameworks, for example, agency, identity, role confidence, and social cognitive career theory. In this example, Godwin first highlights the issues women face in engineering and aims to address the gender gap through affective framing. We know that the intersectional approach guides an analysis of multiple categories and the way in which they shape each other. To understand the ways in which women choose engineering requires an understanding of multiple factors, including gender and identities. The critical engineering agency (CEA) framework utilizes multiple identities as well as student agency beliefs. Godwin provides rationale for why the CEA framework is appropriate to her study in that it allowed her to study multiple factors and go beyond individual women in engineering experiences. One way to use a theoretical framework is to apply the theory in a novel context (Malmi et al., 2019); Godwin also points out that CEA has been used in math and physics education but is a more novel application in EER.

Overall, this dissertation used a primary theoretical framework, that is, *critical engineering agency*, to serve the purpose of the study; "understand the framework of Critical Engineering Agency and highlight the ways this framework can empower women to choose engineering" (p. 25); inform the specific research questions, for example, "how does CEA, as an explanatory framework describe students' choice of engineering?" (p. 27); guide the research design and analysis; and develop an explanatory structural equation model for CEA so educators could understand student choices and beliefs related to engineering.

6 Recommendations and Conclusion

In this chapter, we have discussed the role of theory in engineering education research (EER) by elaborating on the utility of theory and working through a few key constructs, such as theory, theoretical framework, and paradigm. We have highlighted three different paradigmatic traditions of research in social sciences and how they have been deployed within EER, including a theory or theoretical framework within each paradigm. We have also provided a list in Appendix A of theories that have recently found use within engineering and computing education research.

Can a theory that has been posited within a specific paradigm only be used within that paradigm? Not necessarily. Researchers often use concepts or constructs across paradigms, but it is uncommon for highly positivist theories to be easily adopted into an interpretivist paradigm and especially in a critical paradigm. There have been attempts, though, to take theories from an interpretivist tradition and formalize them for use within a positivist tradition. For instance, many researchers have formalized situated learning, especially concepts such as "community of practice."[1] Finally, it would be antithetical for those working within the critical paradigm to formalize their theoretical work, although an interpretive stance might work for an empirical exploration. Similarly, for many interpretive scholars, a critical stance in their work is easier to deploy.

The use of theoretical frameworks provides a foundation for conducting research with the goal of advancing knowledge in the engineering education field. The successful use of framework(s) should help novice and experienced researchers organize, explain, describe, and interpret information. A recommended process, or step-by-step guide, is provided in the following text to facilitate readers in their implementation efforts. We also understand the context may vary depending on the project or study, so it is important to note that the following guidelines are meant as scaffolding to use a theoretical framework to ground and guide your research, and with experience, most researchers develop their own preferred underpinnings.

1. Examine your position and the discourse around your research philosophy. Consider how a theoretical framework will be used to define, frame, or ground your study given the paradigm(s) under which you will operate.
2. Identify possible suitable framework(s) and deliberate on its appropriateness, ease of application, and explanatory power. Although the literature review and theoretical framework should be two separate sections in your paper, we recommend that you read both the original sources and secondary sources of the theoretical framework(s). A study (Malmi et al., 2019) analyzing the use of theoretical frameworks found that very few published papers cited, or accurately cited, the original source. Secondary sources can also be very informative to your study, and talking with researchers that have applied, or tried to apply, a potential theoretical framework can also provide insight into how your approach can be successful in terms of aligning with your study's problem, purpose, and significance.
3. Consider the purpose(s) of your selected framework. Will you use it to develop your research questions? Will you use it to analyze your data or evaluate your design? Your theoretical framework(s) should provide guidance for how your work addresses the gap in the literature and should then facilitate the framing and formulation of your research questions.
4. Verify that your theoretical framework aligns with your study's problem, purpose, and significance. Now, consider how your theoretical framework(s) will guide you in selecting a method for data collection and analysis. Identifying a theoretical framework yet failing to explain how it is applied to research design can be a struggle for novice researchers, as well as more experienced researchers in the engineering education field. It is important to consider how it informs your data collection and analysis.
5. Use the appropriate theoretical framework to inform the discussion of your findings and recommendations based on the data analysis. You should also demonstrate how, and why, the theoretical framework you selected was more appropriate to your study (compared to alternative theoretical frameworks). Thinking about how you will apply the theoretical framework – especially in steps 4 and 5 – will be critical. Without a theoretical framework for your study, you may be presented with no perspectives or an endless number of perspectives on how to interpret your findings and how to focus on potential explanations. Using a theoretical framework as a lens will help you make sense of your data and provide conclusions, implications, and recommendations based in theory.

To conclude this chapter, we consider the developments that have been seen in the field in the decade or so since Johri (2010)'s editorial calling for more attention to theory in EER. There are regular special issues by journals that publish research produced within particular theoretical traditions or that compare and contrast different theories appropriate for specific problems in the field (e.g., most recently, a two-volume special issue in *Studies in Engineering Education* on "Theory and Methods in Engineering Education"; Vol. 1, Issue 2 (2020), and Vol. 2, Issue 1 (2021)). The field now has conference papers that use a range of theory across venues. The suggestion to have constructive review processes for journals that are directed towards feedback and revision to highlight theory use and development has been implemented and even institutionalized through programs such as peer support (Jensen et al., 2022).

The suggestion that researchers should be encouraged to share data and combine smaller studies they have undertaken into larger articles that are more theoretically grounded has not had much success (Johri et al., 2016). The field is still grappling with the issue of democratizing data, especially qualitative data that relies on interpretive analysis. The use of theories in empirical work as well as proposed empirical work within EER can be best described as "muddy waters," and by that, we mean that there is a lack of consistency in how to use theories, including the use of the same theory across studies. Dissertations and proposals are not finished products, and hence, it is not at all surprising to see how theories are used therein. It is for this reason that they also serve as an indicator of what those new to the field or those proposing new research are thinking about these issues.

A final aspiration outlined in the editorial was for the field to develop disciplinary theories unique to engineering education that might be taken up by other fields and to show a clear path from theoretical work to practice; the field has not had much success with either of these aspirations. How does one go about doing this, and how has this been done in the field in recent years, especially in research studies and not just published journal or conference articles, where much of the presentation is post hoc?

These observations about EER are not meant to be criticisms but a reflection on where the field stands and the directions it might take. As a relatively new and vibrant field, EER is changing in nature (as evident by many of the chapters in this volume), and therefore, the uptake of theory as well as contribution to theory are both becoming more common and are likely to keep growing in the future.

In the end, having made arguments for a practical and usable notion of theory, we acknowledge the inherent limitations of taking a highly instrumentalist approach to research. Instead, we want the community to recognize the role theory can play in expanding our limitations as scholars – "the contingency that has made us what we are, the possibilities of no longer seeing, doing or thinking what we are, do or think?" (Mahon, 1992, p. 122). Theory-making and theoretical thinking can also be a work that brings joy and beauty to our scholarship. It can also, as has been argued by many recently, be a force for change. As Ball (1995, p. 266) contends, theory can actually serve as a means for thinking differently and

> offers a language for challenge, and modes of thought, other than those articulated for us by dominant others. It provides a language of rigour and irony rather than contingency. The purpose of such theory is to de-familiarise present practices and categories, to make them seem less self-evident and necessary, and to open up spaces for the invention of new forms of experience.

This does not imply that theory is simply critical but, instead, that it is full of possibilities and freedom to work "'on and against' prevailing practices of ideological subjection" (Ball, 1995, p. 267).

Acknowledgments

This work is partly supported by US National Science Foundation Awards No. EEC-1941186, 1937950, 1939105. Any opinions, findings, and conclusions or recommendations expressed in this material are those of the authors and do not necessarily reflect the views of the funding agencies. We want to thank the associate editor and reviewers for their feedback.

Note

1 This trend hasn't always been welcomed by the originator of a theory. For instance, Jean Lave in many instances has had issues with formalization of situated learning as she believes that dilutes the whole idea of it being a *practice* theory (where the meaning people make of it is through their participation and not some external measure).

References

Adams, J., Hillier-Brown, F. C., Moore, H. J. et al. (2016). Searching and synthesising 'grey literature' and 'grey information' in public health: Critical reflections on three case studies. *System Review, 5*, 164. https://doi.org/10.1186/s13643-016-0337-y

Ashwin, P. (2009). *Analysing teaching – learning interactions in higher education: Accounting for structure and agency*. Continuum.

Baillie, C., & Douglas, E. P. (2014). Confusions and conventions: Qualitative research in engineering education. *Journal of Engineering Education, 103*(1), 1–7.

Ball, S. J. (1995). Intellectuals or technicians? The urgent role of theory in educational studies. *British Journal of Educational Studies, 43*(3), 255–271.

Beddoes, K. (2014). Methodology discourses as boundary work in the construction of engineering education. *Social Studies of Science, 44*(2), 293–312.

Bilge, S. (2010). Recent outlooks on intersectionality. *Diogenes, 225*, 58–72.

Bornasal, F., Brown, S., Perova-Mello, N., & Beddoes, K. (2018). Conceptual growth in engineering practice. *Journal of Engineering Education, 107*(2), 318–348.

Borrego, M. (2007). Conceptual difficulties experienced by trained engineers learning educational research methods. *Journal of Engineering Education, 96*(2), 91–102.

Brozina, C., Katz, A., & Johri, A. (2021). Taking stock: An analysis of IJEE publications from 1996–2020 to examine impact and coverage of topics. *International Journal of Engineering Education, 37*(5), 1146–1160.

Buch, A. (2015). Studying engineering practice. In *Engineering identities, epistemologies and values* (pp. 129–145). Springer.

Carrico, C., & Tendhar, C. (2012). The use of the social cognitive career theory to predict engineering students' motivation in the produced program. In *Proceedings of 2012 ASEE annual conference & exposition*. https://doi.org/10.18260/1-2-22111.

Case, J. (2008). *Education theories on learning: An informal guide for the engineering education scholar*. Higher Education Academy Engineering Subject Centre, Loughborough University.

Case, J. M., & Light, G. (2011). Emerging research methodologies in engineering education research. *Journal of Engineering Education, 100*(1), 186–210.

Collins, A. Brown, J. S., & Newman, S. E. (1989). Cognitive apprenticeship: Teaching the crafts of reading, writing, and mathematics. In *Knowing, learning, and instruction: Essays in honor of Robert Glaser* (pp. 453–494). Lawrence Erlbaum Associates, Inc.

Crenshaw, K. W. (1989). *Demarginalizing the intersection of race and sex: A Black feminist critique of antidiscrimination doctrine, feminist theory and antiracist politics* (pp. 139–167). University of Chicago Legal Forum.

Crotty, M. J. (1998). The foundations of social research: Meaning and perspective in the research process. Sage Publications.

Dhamoom, R. K. (2011). Considerations on mainstreaming intersectionality. *Political Research Quarterly, 64*, 230–243.

Douglas, E. P., Koro-Ljungberg, M., & Borrego, M. (2010). Challenges and promises of overcoming epistemological and methodological partiality: Advancing engineering education through acceptance of diverse ways of knowing. *European Journal of Engineering Education, 35*(3), 247–257.

Finelli, C., Borrego, M., & Rasoulifar, G. (2015, October). Development of a taxonomy of keywords for engineering education research. *Journal of Engineering Education, 104*(4), 365–387.

Fiske, S. T. (2004). Mind the gap: In praise of informal source of formal theory. *Personality and Social Psychology Review, 8*(2), 132–137.

Froyd, F., & Lohmann, J. R. (2014). Chronological and ontological development of engineering education as a field of scientific inquiry. In A. Johri & B. M. Olds (Eds.), *Cambridge handbook of engineering education research* (pp. 3–26). Cambridge University Press. https://doi.org/10.1017/CBO9781139013451.003

Godwin, A. F. (2014). *Understanding female engineering enrollment: Explaining choice with critical engineering agency* [Doctoral dissertation, Clemson University]. All Dissertations. 1787. https://tigerprints.clemson.edu/all_dissertations/1787

Gregor, S. (2006). The nature of theory in information systems. *MIS Quarterly*, 611–642.

Hamann, J., & Kosmützky, A. (2021). Does higher education research have a theory deficit? Explorations on theory work. *European Journal of Higher Education*, 468–488. https://doi.org/10.1080/21568235.2021.2003715

Hingle, A., Goncher, A., Johri, A., & Case, J. (2022). A mapping review of the use of frameworks in engineering education research grey literature. In *2022 IEEE frontiers in education conference (FIE)*, Uppsala, Sweden (pp. 1–6). https://doi.org/10.1109/FIE56618.2022.9962609

Hughes, S., Davis, T. E., & Imenda, S. N. (2019). Demystifying theoretical and conceptual frameworks: A guide for students and advisors of educational research. *Journal of Social Sciences*, *58*(1–3), 24–35.

Hulko, W. (2009). The time- and context-contingent nature of intersectionality and interlocking oppressions. *Affilia: Journal of Women and Social Work*, *24*(1), 44–55.

Imenda, S. (2014). Is there a conceptual difference between theoretical and conceptual frameworks? *Journal of Social Science*, *38*(2), 185–195.

Inda, M., Rodríguez, C., & Peña, J. V. (2013). Gender differences in applying social cognitive career theory in engineering students. *Journal of Vocational Behavior*, *83*(3), 346–355.

Jamieson, L., & Lohmann, J. (2009). Creating a culture for scholarly and systematic innovation in engineering education. *Phase 1 Report, ASEE.*

Jawitz, J., & Case, J. (2009). Communicating your findings in engineering education: The value of making your theoretical perspective explicit. *European Journal of Engineering Education*, *34*(2), 149–154.

Jensen, K., Ko, E., Lichtenstein, G., Watts, K., Bates, R., & Benson, L. (2022, August). Building a community of mentors in engineering education research through peer review training. In *2022 ASEE annual conference & exposition.*

Jesiek, B., Borrego, M., Beddoes, K., Hurtado, K., Rajendran, P., & Sangam, D. (2011). Mapping global trends in engineering education research, 2005–2008. *International Journal of Engineering Education*, *27*(1), 77–90.

Jesiek, B. K., Newswander, L. K., & Borrego, M. (2009). Engineering education research: Discipline, community, or field? *Journal of Engineering Education*, *98*(1), 39–52.

Johri, A. (2010). Creating theoretical insights in engineering education. *Journal of Engineering Education*, *99*(3), 183–184.

Johri, A., & Olds, B. M. (2011). Situated engineering learning: Bridging engineering education research and the learning sciences. *Journal of Engineering Education*, *100*(1), 151–185.

Johri, A., & Olds, B. M. (Eds.). (2014). *The Cambridge handbook of engineering education research.* Cambridge University Press.

Johri, A., Olds, B. M., & O'Connor, K. (2014). Situative frameworks for engineering learning research. In A. Johri & B. Olds (Eds.), *Cambridge handbook of engineering education research* (pp. 47–66). Cambridge University Press.

Johri, A., Yang, S., Vorvoreanu, M., & Madhavan, K. (2016). Perceptions and practices of data sharing in engineering education. *Advances in Engineering Education*, *5*(2), n2.

Klassen, M., & Case, J. M. (2022). Productive tensions? Analyzing the arguments made about the field of engineering education research. *Journal of Engineering Education*, *111*(1), 214–231. https://doi.org/10.1002/jee.20440

Kellam, N., & Jennings, M. (2021). Uncovering epistemologies and values of our qualitative engineering education research community: Listening for voices. *Studies in Engineering Education*, *2*(1).

Kemnitzer, S. (2008). *The need for theory-based research in engineering education.* Video interview recorded at ASEE 2008, available online at CASEE video channel on YouTube: www.youtube.com/watch?v=nfsh08jHEKs.

Kivunja, C., & Kuyini, A. B. (2017). Understanding and applying research paradigms in educational contexts. *International Journal of Higher Education*, *6*(5), 26–41.

Lave, J. (1991). Situated learning in communities of practice. In L. B. Resnick, J. M. Levine, & S. D. Teasley (Eds.), *Perspectives on socially shared cognition* (pp. 63–82). American Psychological Association.

Lave, J. (1993). The practice of learning. In S. Chaiklin, & J. Lave (Eds.), *Understanding practice: Perspectives on activity and context* (pp. 3–34). Cambridge University Press.

Lave, J., & Wenger, E. (1991). *Situated learning: Legitimate peripheral participation.* Cambridge University Press.

Lent, R. W., Brown, S. D., & Hackett, G. (2002). Social cognitive career theory. *Career Choice and Development*, *4*(1), 255–311.

Lent, R. W., Lopez Jr, A. M., Lopez, F. G., & Sheu, H. B. (2008). Social cognitive career theory and the prediction of interests and choice goals in the computing disciplines. *Journal of Vocational Behavior*, *73*(1), 52–62.

Lenz, E. R., Suppe, F., Gift, A. G., Pugh, L. C., & Milligan, R. A. (1995). Collaborative development of middle-range nursing theories: Toward a theory of unpleasant symptoms. *Advances in Nursing Science*, *17*(3), 1–13.

Lutz, B., & Paretti, M. C. (2021). Exploring the social and cultural dimensions of learning for recent engineering graduates during the school-to-work transition. *Engineering Studies*, *13*(2), 132–157.

Magana, A. J. (2022). The role of frameworks in engineering education research. *Journal of Engineering Education*, *111*, 9–13. https://doi.org/10.1002/jee.20443

Mahon, M. (1992). *Foucault's Nietzschean genealogy/truth, power, and the subject*. SUNY Press.

Malmi, L., Adawi, T., Curmi, R., de Graaff, E., Duffy, G., Kautz, C., Kinnunen, P., & Williams, B. (2018). How authors did it – a methodological analysis of recent engineering education research papers in the European Journal of Engineering Education. *European Journal of Engineering Education*, *43*(2), 171–189.

Malmi, L., Sheard, J., Kinnunen, P., Simon, S., & Sinclair, J. (2019, August 12–14). Computing education theories: What are they and how are they used? In *International computing education research conference (ICER '19)*, Toronto, ON, Canada. ACM, New York, NY, USA, 11 pages. https://doi.org/10.1145/3291279.3339409

McCall, L. (2005). The complexity of intersectionality. Signs, *30*(3), 1771–1800.

Mejia, J. A., Revelo, R. A., Villanueva, I., & Mejia, J. (2018). Critical theoretical frameworks in engineering education: An anti-deficit and liberative approach. *Education Sciences*, *8*(4), 158.

Merton, R. (1949, 2012). On sociological theories of the middle range. Chapter 39. In C. Calhoun (Ed.), *Classical sociological theory* (pp. 351–542). John Wiley & Sons.

Nelson, G. L., & Ko, A. J. (2018). On use of theory in computing education research. In *Proceedings of the 2018 ACM conference on international computing education research* (pp. 31–39).

Neuman, W. L. (2014). *Basics of social research: Quantitative and qualitative approaches*. Pearson Canada.

Newstetter, W. C., & Svinicki, M. D. (2014). Learning theories for engineering education practice. In A. Johri & B. M. Olds (Eds.), *Cambridge handbook of engineering education research* (pp. 29–46). Cambridge University Press.

Passey, D. (2020). Theories, theoretical and conceptual frameworks, models and constructs: Limiting research outcomes through misconceptions and misunderstandings. *Studies in Technology Enhanced Learning*, *1*(1), 95–114.

Peters, A. K. (2018). Students' experience of participation in a discipline – a longitudinal study of computer science and IT engineering students. *ACM Transactions on Computing Education (TOCE)*, *19*(1), 1–28.

Radhakrishnan, D. (2020). *Enabling untrained teachers to be engineering facilitators: A design-based research study of teacher professional development in fragile contexts (Version 1)*. Purdue University Graduate School. https://doi.org/10.25394/PGS.12746969.v1

Reich, A., Rooney, D., Gardner, A., Willey, K., Boud, D., & Fitzgerald, T. (2015). Engineers' professional learning: A practice-theory perspective. *European Journal of Engineering Education*, *40*(4), 366–379.

Rist, R. C. (1977). On the relations among educational research paradigms: From disdain to detente. *Anthropology & Education Quarterly*, *8*(2), 42–49.

Rottmann, C., Reeve, D., Klassen, M., Kovalchuk, S., Liu, Q., Olechowski, A., & Santia, M. (2018). Examining the engineering leadership literature: Community of practice style. In *Proceedings of 2018 ASEE annual conference & exposition*.

Seniuk Cicek, J., & Friesen, M. (2018). Epistemological tensions in engineering education research: How do we negotiate them? In *2018 IEEE frontiers in education conference (FIE)* (pp. 1–5). IEEE.

Sheu, H. B., Lent, R. W., Brown, S. D., Miller, M. J., Hennessy, K. D., & Duffy, R. D. (2010). Testing the choice model of social cognitive career theory across Holland themes: A meta-analytic path analysis. *Journal of Vocational Behavior*, *76*(2), 252–264.

Stokes, D. E. (2011). *Pasteur's quadrant: Basic science and technological innovation*. Brookings Institution Press.

Streveler, R., & Smith, K. (2006). Conducting rigorous research in engineering education. *Journal of Engineering Education*, *95*(2), 103–105.

Sutton, R. I., & Staw, B. M. (1995). What theory is not. *Administrative Science Quarterly*, *40*(3), 371–384.

Szyjka, S. (2012). Understanding research paradigms: Trends in science education research. *Problems of Education in the 21st Century*, *43*(1), 110–118.

Tonso, K. (2007). *On the outskirts of engineering: Learning identity, gender and power in engineering practice*. SENSE Publishers.

Van Merriënboer, J. J., & De Bruin, A. B. (2014). Research paradigms and perspectives on learning. In *Handbook of research on educational communications and technology* (pp. 21–29). Springer.

Wankat, P., Willaims, N., & Neto, P. (2014). Engineering education research in European Journal of Engineering Education and Journal of Engineering Education: Citation and reference discipline analysis. *European Journal of Engineering Education, 39*(1), 7–17.

Watson, K. (2009). Change in engineering education: Where does research fit? *Journal of Engineering Education, 98*(1), 3–4.

Wenger, E. (1999). *Communities of practice: Learning, meaning and identity*. Cambridge University Press.

Williams, B., Neto, P., & Wankat, P. (2014). Taking a snapshot: Four bibliometric indicators to track engineering education research evolution. *Chemical Engineering Education, 17*(1), 2–3.

Wimmer, R. D., & Dominick, J. R. (2013). *Mass media research*. Cengage learning.

Woolfolk, A. E., Hoy, A. W., Hughes, M., & Walkup, V. (2008). *Psychology in education*. Pearson Education.

Xian, H., &. Madhavan, K. (2014). Anatomy of scholarly communication in engineering education: A big-data bibliometric analysis. *Journal of Engineering Education, 103*(3), 486–514.

Young, G. (2017). *Preparing students for professional work environments through university- industry partnerships: A single case study of the co-op development program* [Doctoral dissertation, Virginia Tech University]. http://hdl.handle.net/10919/85414

Appendix A

A List of Theories from the Social Sciences and Allied Field That Have Found Application in Computing and Engineering Education Research

Theories identified in "Review and Use of Learning Theories within Computer Science Education Research: Primer for Researchers and Practitioners (2019),"

Theories identified in "A Mapping Review of the Use of Frameworks in Engineering Education Research Grey Literature (2022)" are marked ⋆.

Theories included in both lists are marked ⋆⋆.

Theories

Achievement goal theory
Affordance theory
Analytical behaviorism
Approaches to learning (deep vs. surface)
Associative learning
Change theory⋆
Chunking
Classical conditioning (Rescorla–Wagner)
Cognitive apprenticeship
Cognitive flexibility theory
Cognitive load theory⋆⋆
Cognitive theory of multimedia learning
Cognitivism
Collective impact framework⋆
Communities of practice⋆⋆
Community cultural wealth⋆
Conditions of learning
Connectionism
Connectivism
Constructionism
Constructivism
Constructivist theory⋆

Conversation theory
Diffusion of innovation*
Discovery learning
Dual coding theory
Elaboration theory
Emotional intelligence
Engagement taxonomy
Error quotient (EQ)
Expectancy theory
Expectancy–value theory*
Experiential learning
Expertise theory
Flow (Csikszentmihalyi)
Functional context theory
Gestalt theory
Grit theory
Hierarchy of needs (Maslow)
Identity formation
Information processing theory
Instrumental conditioning
Interest
Interference theory (Bergström)
Intersectionality theory*
Invitational education
Kotter's 8-step model of change*
Latent learning
Learning edge momentum
Learning styles
Learning trajectories
Mastery learning
Mental models
Metacognition
Mindsets
Model of hierarchical complexity
Model of programming errors
Motivation theories**
Operant conditioning
Problem-based learning (PBL)
Programming gene
Programming plans
Progression of early computational thinking (PECT)
Psychological behaviorism
Scaffolding
Schema theory
Self-determination theory
Self-efficacy theory**
Sensory theory
Situated cognition
Situated learning**

Social capital theory*
Social cognitive theory*
Social cognitive career theory*
Social development theory
Social identity theory*
Social learning theory
Socialization theory*
Socio-constructivist learning theory*
Spatial reasoning skills
Split attention effect
Stage theory of cognitive development
Subsumption theory
Threshold concepts
Threshold skills
Transformative learning
Variation theory of learning
Working memory
Zone of proximal development
Zone of proximal flow

8
Emotions in Engineering Education

Johanna Lönngren, Inês Direito, Roland Tormey, and James L. Huff

1 Introduction

Engineers and engineering students often identify their work as rational, beyond emotion, and engineering is often characterized as purely scientific, involving technical solutions to real-world problems (Cech, 2018; Lönngren, Adawi, et al., 2021; Polmear et al., 2018). However, engineering education and practice are embedded in contexts with complex social relationships, power structures, and conflicting value systems (Cech, 2018). Dealing with engineering problems in these contexts requires knowledge and competencies to collaboratively explore diverse perspectives on a problem and develop socially, ecologically, and economically sustainable technological solutions (Holmén, 2020; Lönngren et al., 2016; Van den Beemt et al., 2020). Engineering education researchers have, for example, demonstrated how emotions matter in engineering ethics and sustainability, social justice work, technological risk management, problem-solving, student development, retention, as well as diversity and inclusion (Hess et al., 2020; Kellam et al., 2018; Lönngren, Adawi, et al., 2021; Roeser, 2012a). These findings are also in line with research in science education (Davidson et al., 2020; Sinatra et al., 2014), sustainability education (Ojala, in print), and many other educational contexts (Pekrun & Linnenbrink-Garcia, 2014a; Zembylas & Schutz, 2016).

Dealing with engineering and sustainability problems requires, for example, critical emotional awareness (Ojala, in print), empathy (Bairaktarova & Plumlee, 2022; Hess & Fila, 2016), emotional intelligence (Lappalainen, 2015), emotional engagement (Gelles et al., 2020), and an ability to navigate conflicting emotion norms (Lönngren, Adawi, et al., 2021). In the engineering education literature, *empathy* has been described as necessary for (a) enabling engineers to design artifacts and processes that meet user's needs, (b) working with communities and clients whose background is different to their own, (c) communicating effectively with colleagues and clients, (d) building teams and resolving conflicts, and (e) responding appropriately to the ethical dilemmas that engineers face (Hess & Fila, 2016). A range of specific emotions (including joy, frustration, pride, shame, and guilt) has also been identified as important in the work and learning of engineers (e.g., Bates & Wilson, 2008; Huff et al., 2021; Warner, 2006).

Emotions are also important for engineering educators and other academic staff. For example, educators are expected to manage their own emotions (Adams & Turns, 2020; Decuir-Gunby et al., 2009; Lawless, 2018), understand and deal with students' emotions (Husman et al., 2015), engage students in critical emotional praxis (Chubbuck & Zembylas, 2008), cultivate a constructive

DOI: 10.4324/9781003287483-10

emotional climate (Bates & Wilson, 2008; Giannakos et al., 2014), and build positive, emotional, and caring relationships in the classroom (Nair & Bulleit, 2020; Quinlan, 2016; Tormey, 2021).

Clearly, to reform engineering education for the 21st century – and equip students with the knowledge, competencies, and confidence to contribute to solving future sustainability challenges – it is vital for engineering education research to engage with emotions in teaching and learning.

1.1 Purpose and Outline

This chapter introduces the complex and multidisciplinary field of research on emotions in engineering education (EEE), which draws on psychological, sociological, and philosophical perspectives and employs a wide range of research methods. Thus, we hope to support researchers new to EEE in navigating and contributing to this nascent field of research.

The chapter starts with a discussion of how emotions are defined in different disciplinary contexts and how emotions, components of emotion, and emotion-related phenomena can be measured. We then provide an overview of theoretical perspectives that are commonly applied in the multidisciplinary emotion research literature. Equipped with a broad understanding of what emotion research can entail, we turn our attention to engineering education and the nascent field of EEE. We summarize four dominant themes in the existing literature, which were identified in a recent scoping review (Lönngren, Bellocchi, et al., 2021): (1) academic emotions, (2) emotions and ethics, (3) emotional intelligence and other socio-emotional competencies, and (4) mental health. Based on existing research and our own experiences of conducting EEE research, we then provide advice for researchers and doctoral students who plan to pursue EEE research. Finally, we outline currently underdeveloped research areas, arguing that more EEE research is needed that employs sociological perspectives, mixed- and multi-methods approaches that do not (solely) rely on self-report measures, studies focused on cultural and gender differences in how emotions are experienced and expressed in engineering education, mental health, as well as engineering educators' and other staff members' emotions and emotion practices.

2 Challenges in Defining Emotion

For centuries, philosophers, physicians, psychologists, and more recently, neuroscientists have studied the relationship between emotion and reason (Lazarus, 1999). From Cartesian philosophical perspectives of the relationship between mind and body to current discussions of whether human emotion is a cognitive or noncognitive phenomenon (England, 2019), the study of human emotion is entangled with disciplinary, ideological, cultural, and political ideas about what it means to be human and how we should live our lives. Moreover, historians of emotion have shown that the study of emotion is influenced by constantly evolving sociocultural and disciplinary trends (Frevert, 2014). Even within the same research discipline, such as psychology, consensus is difficult to reach (Kleinginna & Kleinginna, 1981), and a wide range of definitions and conceptualizations is today used in emotion research (Bellocchi, 2019). Most emotion researchers, however, "generally agree that emotions are episodes with multiple components that are shaped by evolutionary and social contexts and can be expressed in a variety of ways" (Shuman & Scherer, 2014, p. 19).

This general agreement on defining emotion as *episodical* (relatively short-lived) and *componential* (consisting of multiple factors, processes, or components) is mirrored in Kleinginna and Kleinginna's (1981) attempt to provide a consensual yet theoretically flexible definition of emotion, which was based on an analysis and compilation of 92 different definitions available at that time:

> Emotion is a complex set of interactions among subjective and objective factors, mediated by neural/hormonal systems, which can (a) give rise to affective experiences such as feelings of

> arousal, pleasure/displeasure; (b) generate cognitive processes such as emotionally relevant perceptual effects, appraisals, labeling processes; (c) activate widespread physiological adjustments to the arousing conditions; and (d) lead to behavior that is often, but not always, expressive, goal-directed, and adaptive.
>
> *(p. 355)*

Today, Scherer's (2005) component model of emotion is widely cited in emotion research. In this model, Scherer described emotion in terms of five components: (1) motor expression, such as gestures and facial expressions; (2) neurophysiology, including arousal and biomarkers; (3) subjective feeling; (4) motivation; and (5) cognition. Further, *emotion* was defined as "an episode of interrelated, synchronized changes in the states of all or most of the five organismic subsystems in response to the evaluation of an external or internal stimulus event as relevant to major concerns of the organism" (Scherer, 2005, p. 697). Another component model was proposed by Thoits (1990), who described *emotions* as subjective experiences resulting from the interrelation of four components: (1) situational cues, (2) physiological changes, (3) expressive gestures, and (4) words referring to emotions.

Irrespective of which definition is employed, emotion is often distinguished from related phenomena, such as affect, mood, and feeling. While precise definitions of these terms vary between disciplines and individual studies, most researchers agree on a few basic tenets. For example, *affect* is often used as a broad construct encompassing emotions, feelings, moods, sentiments, as well as non-emotional constructs, such as motivation, interest, and attitudes (Shuman & Scherer, 2014). According to Zembylas (2021), *affect* "encompasses a variety of sensorial processes, experiences, and relations and refers generally to the body's capacities to act, to engage, to resist, and to connect; the term 'emotion' is often used to signal social and cultural constructs and conscious processes" (p. 770). *Feeling* is, at least in component models, typically defined as corresponding to the conscious and subjective experience component of emotion (Shuman & Scherer, 2014). Finally, *mood* is understood as more diffuse and longer-lasting than emotion. Also, while emotions are generally considered to be "about something" (e.g., being angry at something another person has done), moods may not have such an object, and they can, for example, be caused by hormonal changes alone (Fridja, 2008; Shuman & Scherer, 2014). We do not propose that engineering education researchers should agree on consensus definitions for each of these concepts – which would unnecessarily limit the scope of research questions that can be addressed. However, as for any other topics researched in engineering education, it is important to clearly define emotions and any related terms that are used in a study. For a detailed discussion of challenges in defining emotions in education research, we refer readers to Shuman and Scherer's (2014) overview in the International Handbook of Emotions in Education (Pekrun & Linnenbrink-Garcia, 2014a).

In an ongoing systematic review of EEE research, Lönngren et al. (Lönngren, Bellocchi, et al., 2021; Lönngren et al., forthcoming) found that many engineering education researchers do not define what they mean by *emotion* and sometimes even use the term interchangeably with related – but distinct – concepts, such as *affect* and *feeling*. This lack of conceptual clarity in the field is problematic, especially given the interdisciplinary nature of engineering education research, where researchers draw on a wide range of disciplines (e.g., education, sociology, psychology, philosophy, management, etc.). Different disciplines tend to employ different types of emotion theories, which influences what emotion phenomena can be studied and how results can be interpreted and transferred between empirical contexts. Thus, in an interdisciplinary field such as engineering education, researchers can be expected to draw on a range of conceptualizations of emotions and other affective phenomena. These conceptualizations need to be made explicit to allow readers to adequately interpret and judge research findings (Turner, 2009).

3 Theorizing Emotions

Most educational emotion research has so far been informed by psychological, sociological, and critical theories (Chubbuck & Zembylas, 2008). Existing EEE research has almost exclusively relied on psychological theories, while sociological and critical perspectives have not yet been widely used (Lönngren et al., forthcoming). In this section, we describe both types of theories, aiming to support readers in making informed decisions regarding which types of theories to use in which types of EEE studies.

3.1 Psychological Theories – Linking Emotion and Cognition

From psychological perspectives, emotions are conceptualized as complex, intrapersonal phenomena that result in physiological, neurological, and cognitive changes in individuals. These perspectives are commonly used to explore (a) the function of emotion in mediating a person's response to their environment and (b) the ways in which emotion and cognition interact in this process.

In education research, emotions are often understood as mediating how students and educators respond to events related to teaching and learning (Shuman & Scherer, 2014). For example, a student who performs less well than expected on an exam may experience physiological changes (e.g., adrenaline, heart rate, blood flow) which are part of an experience of anxiety or anger and which may lead to reactions to the situation (e.g., argue with the instructor or decide not to invest energy in the course; Zeidner, 2014). Emotion researchers in this tradition highlight the role of *appraisal*, that is, the processes through which individuals evaluate whether a phenomenon or situation is in line with their own values and goals (Moors et al., 2013).

Appraisal is theorized to occur through two processes. First, *primary appraisal* is described as rapid, automatic, and unconscious. It does not generate emotions *per se* but locates the person on a valence dimension (ranging from displeasure to pleasure) and an activation dimension (ranging from low- to high-energy responses) (Murphy & Zajonc, 1993; Russell, 2003; Zajonc, 1980). Thus, primary appraisal regulates readiness (Fridja, 1986, 2007) to respond to a given situation. *Secondary appraisal* involves cognitive evaluation of a situation, resulting in more nuanced placement on the valence and energy dimensions – allowing the person to experience distinct emotions, such as boredom, fear, anger, or awe (Oatley et al., 2006).

3.1.1 Emotion-Related Phenomena Based on Psychological Theories

In education research (including EEE), cognitive appraisal theory underlies, for example, the widely used framework of *academic emotions* (Pekrun & Linnenbrink-Garcia, 2012, 2014b), describing four groups of emotions. First, *achievement emotions* are associated with appraisal of one's own academic performance, for example anxiety, pride, or shame regarding exam results. Second, *epistemic emotions* are linked to appraisal of cognitive processes involved in the development of new knowledge. For example, students who encounter facts or ideas that are not readily integrated into their existing mental models may experience curiosity, anxiety, or frustration as they try to make sense of the new information. Third, *topic emotions* involve appraisal of the topic or subject matter that is studied. Examples include climate anxiety (anxiety and distress about the implication of climate change) and love of mathematics. Finally, *social emotions* are linked to appraisal of social relationships in educational settings. They may include appraisal related to others' achievement (e.g., awe, envy, admiration), psychological safety (e.g., trust, confidence, anxiety), or affection (e.g., love, joy, loneliness).

Indeed, some of these emotions are not simply related to academic settings but can be related to other types of performance, including aspects of performing the roles of an engineer. Engineers may experience pride and shame related to achievements, envy, and anger related to working in social teams, and curiosity and frustration related to problem-solving (Davis, 2017). Thus, these appraisals are relevant both to the *engineering* and the *education* dimensions of the domain in question.

In Western cultures and philosophy, the predominant traditions of thought have generally conceptualized emotional appraisals as a source of bias and a negative force in human judgment: "more primitive, less intelligent, more bestial, less dependable, and more dangerous than reason" (Solomon, 2008, p. 3). By the 1990s, however, this has begun to change as emotion was increasingly seen as important and potentially valuable both in contributing to reason and judgement (e.g., Damasio, 1994) and to social life (e.g., Hoffman, 2000).

One way of conceptualizing how emotion links to reason and judgment was the idea of *emotional intelligence*, which sought to articulate a model of how emotion and cognition were linked and how these links could be regulated. This idea became particularly influential at this time and widely popularized through the work of Goleman (1995, 2013, 1998), especially in management and leadership disciplines. The original emotional intelligence model, as articulated by Mayer et al. (2000), conceptualized emotional intelligence as involving the ability to (a) perceive and express emotions, (b) understand emotions and emotional change processes, (c) use emotions to facilitate particular types of cognition (e.g., using positive emotions to facilitate creative thinking), and (d) regulate emotions in oneself and others. In this model, *emotional intelligence* is defined as "the subset of social intelligence that involves the ability to monitor one's own and others' feelings and emotions, to discriminate among them and to use this information to guide one's thinking and actions" (Salovey & Mayer, 1990, p. 189). While Mayer and Salovey (1993) conceived emotional intelligence as an innate set of *cognitive* abilities (i.e., not really an emotional phenomenon), other researchers (e.g., Bar-On, Goleman, and Petrides) developed models which saw emotional intelligence as being closer to a personality trait (Corcoran & Tormey, 2012). Petrides et al. (2004) define *trait emotional intelligence* as "a constellation of emotion-related dispositions and self-perceived abilities" (p. 575), which could also be described as *trait emotional self-efficacy* – it concerns peoples' beliefs about their own emotions (Petrides & Mavroveli, 2018).

Another approach to linking emotion to reason and judgment is found in research on moral reasoning, where emotion is described as providing an initial ethical appraisal of a situation that can contribute to ethical or moral action. In psychology, this perspective is associated with the social intuitionist perspective of Haidt (2001, 2003), and in philosophy, it is associated with the work of Nussbaum (2001, 2004) and Roeser (2012a). In engineering, for example, Roeser (2012a) argues that emotions improve judgment since

> we need moral emotions in order to be aware of moral aspects of risky technologies. . . . Purely rational reflection would not be able to provide us with the imaginary power that we need to envisage future scenarios and to take part in other people's perspectives and to evaluate their destinies.
>
> *(p. 106)*

3.2 Sociological and Critical Theories – Linking Emotion and Social Contexts

Sociological and critical theories conceptualize emotions not as uniquely biological or psychological but as primarily social phenomena. These theories can be used to study (1) "the social nature of emotions" (Bericat, 2016, p. 495), treating emotions as social constructions, and (2) "the emotional nature of social reality" (*ibid.*), treating emotions as contributing to the construction of social reality. Studying the *social nature of emotions*, sociologists and social psychologists have, for example, used discourse analytic approaches to explore how emotions emerge in interaction in and across diverse cultural, temporal, spatial, and relational contexts. They have also explored how linguistic descriptions and bodily expressions of emotions are used in interaction to negotiate social realities and relationships (Edwards, 1999; Pepin, 2008; Wetherell, 2013). From this perspective, emotions are understood as complex "intersections of language, desire, power, bodies, social structures, subjectivity, materiality and trauma" (Zembylas, 2016, p. 546). Studying the *emotional nature of social reality*, feminist and critical scholars have explored how social structures – such as cultural ideologies, beliefs, and social norms – constrain and construct interpretations, expressions, and arousal of emotion (Stets & Turner, 2008; J. H. Turner & Stets, 2005). They have explored, for example, how emotions are constructed as separated from reason and rationality (Ritzer, 2011) and how conceptualizations of emotion may reproduce – and resist – power structures and social inequalities (Ahmed, 2014; Boler, 1999; Zembylas, 2007b).

3.2.2 Emotion-Related Phenomena Based on Sociological and Critical Theories

In sociological research on emotions in education, Bourdieu's work has been highly influential. Most importantly, his work on habitus has been leveraged to challenge pervading dualism between concepts such as mind/body, objective/subjective, and emotion/cognition (Bourdieu, 1990; Cottingham, 2016; Zembylas, 2007a). Another influential idea based on the work of Bourdieu is the notion of *emotional capital*, which Cottingham (2016) defined as "one's trans-situational, emotion-based knowledge, emotion management skills, and feeling capacities, which are both socially emergent and critical to the maintenance of power" (p. 454). While emotional capital is similar to emotional intelligence in that it involves identifying emotion as a resource, it locates this resource not within individuals but in macro-social structures, unequally distributed and linked to social power. In education research, this notion "offers a tool for thinking about ways in which emotion practices are regulated within an educational context" (Zembylas, 2007a), for example, in discourses about the importance of fostering emotional intelligence and regulation in individual students.

Research on emotional capital has also been linked to Hochschild's (1979, 1983) work on feeling rules and emotional labor. *Feeling rules* are social norms regarding who is expected to feel which emotions, how to feel them, and in which situations (Hochschild, 1979). *Emotional labor* is the effort professionals perform when they express emotions that are socially expected but not aligned with their inner feelings, or when they try to correct inner feelings to align with social norms and expectations (Hochschild, 1983). Emotional labor has been shown to be pervasive in educational settings since teachers often suppress negative emotions to "convey support, encouragement, and a safe place for their students" and "sustain an outward appearance that produces a particular state of mind in their students" (Fraser & Brandt, 2013, p. 146). Educational researchers have, for example, explored teachers' emotional labor in higher education (Lawless, 2018), science education (Zembylas, 2004), social justice education (Rivera Maulucci, 2013), and many other contexts and disciplines.

Emotions in education have also been explored from feminist and critical perspectives. Many educational researchers in this field have drawn on Boler's (1999) work on emotions and power in education, exploring "how emotions are an invisible presence in education, and how emotions are disciplined to maintain social control" (p. 22), thus upholding gendered and racialized power structures. Ahmed's (2014) work on *emotional politics* and *affective economies* has also been highly influential in research on emotions in education. Ahmed theorized emotions as cultural practices that bind communities together while simultaneously positioning others on the outside; emotions are "produced, circulated and capitalised on to achieve political purposes such as unity or conflict" (Zembylas, 2007a, p. 458). Ahmed's theories have, for example, been used to explore how "emotions are strategically and politically used to frame [educational] policies" (Lindgren & Rönnberg,

2018, p. 57). Other researchers have explored the political effects of specific emotions in education, for example, how disgust can contribute to racial discrimination (Matias & Zembylas, 2014) and how shame can contribute to constructing affective connections in intercultural education (Zembylas, 2008).

In conclusion, education research has demonstrated the usefulness of sociological and critical perspectives in exploring the role of emotions in teaching and learning. Unfortunately, we found very few studies employing these perspectives in the EEE literature (Lönngren et al., forthcoming). Cech and Sherick's (2015) work provides an example of how engineering education researchers can study cultural and structural dimensions of EEE. Through the concept of the *ideology of depoliticization*, they explored engineering students' disengagement (that is, lack of activating emotions) with ethics and social justice issues and the societal consequences such disengagement can have. Another example can be found in Adams and Turns's (2020) case study of educational innovation, which included analyses of, for example, discourses of innovation, distributed structures of course coordination, emotions triggered when innovators break social norms, and innovators' emotional labor. Finally, a study by Lönngren et al. (2021) employed positioning theory to explore engineering students' reflections on how to deal with a sustainability problem. They showed how engineering students negotiated and related to conflicting discourses of engineering as (1) purely rational (that is, unemotional) and (2) requiring empathy (that is, involving emotionality).

4 Methods and Methodologies for Emotion Research

Research on emotions can use a broad range of methods and methodological approaches (Lindblom-Ylänne, 2019; Zembylas & Schutz, 2016), employing different "*measurement paradigms*" (Shuman & Scherer, 2014, p. 17, italics in original). Irrespective of which combination of methods is chosen, however, researchers need to ensure proper alignment between research methods, definitions, and theories of emotions. Different theoretical perspectives point researchers' attention to different aspects of emotions, which can be investigated through different types of methods and methodologies and will result in findings that are applicable to, and relevant for, different types of challenges in engineering education (Shuman & Scherer, 2014). To guide new EEE researchers in purposefully selecting and combining methods, we provide an overview of methods used for studying different aspects of emotions and emotion-related phenomena and examples of how these types of methods have been used in existing EEE research.

4.1 Types of Emotion Measures

So far, most emotion research in education has employed *self-report measures*, that is, research subjects' descriptions of their own emotions. Self-report measures can be collected with quantitative (e.g., surveys), qualitative (e.g., interviews), and multi- or mixed-methods studies, and they are generally relatively easy to collect and analyze. Self-report measures are particularly useful for exploring the subjective feeling, motivation, and cognition components in Scherer's (2005) model, but they can also be adapted for analyzing motor expression and conscious, neurophysiological processes. However, self-report measures are only useful if (1) subjects are cognitively aware of what they feel and what they want to achieve; (2) subjects' interpretations and reporting of their own emotions is not unduly influenced by, for example, a desire to please the researchers; and (3) subjects and researchers share similar linguistic and cultural ways of talking about emotions (Pekrun & Bühner, 2014; Shuman & Scherer, 2014).

To analyze the motor expression component of emotion, *observational methods* are particularly useful, focusing on participants' emotional behavior, including verbal expressions (i.e., speech), as well as non-verbal expressions (e.g., facial, vocal, or bodily displays). These methods can be used in experimental settings, but they have been more often used in non-experimental settings and field research. In fact, the specific situation in which emotional behavior is displayed can provide information about "potentially emotion-eliciting events and the context in which they occur" (Reisenzein et al., 2014, p. 584). Therefore, situational descriptions are often used to inform analysis and interpretation of emotional behavior (J. C. Turner & Trucano, 2014). An important limitation in using observational methods is that they primarily provide information about *emotional behavior* – rather than the actual emotions participants experience (Reisenzein et al., 2014).

Finally, *physiological methods* can be used to measure the ways in which emotions influence research participants' physical bodies (e.g., heart rate, blood pressure, sweating, or cortisol levels). They are well-suited to attend to the short-term neurophysiologic component in Scherer's (2005) model. They may be particularly useful for exploring nonconscious emotional processes in teaching and learning (Immordino-Yang & Christodoulou, 2014). However, interpreting physiological data is challenging, since "it is not yet fully understood how specific psychophysiological changes relate to particular emotions" (Immordino-Yang & Christodoulou, 2014, p. 616). For example, a high level of physiological arousal, measured as increased heart rate and blood pressure, could signal anxiety but also excitement. In addition, physiological measures can only provide data about phenomena that are easily measurable. Solely relying on these types of measures may also result in low reliability and replicability as important situational factors may be missed. Physiological measures should therefore always be combined with self-report and observational measures to allow researchers to triangulate and make sense of physiological data in relation to specific educational situations (Fulmer & Frijters, 2009; Immordino-Yang & Christodoulou, 2014; Villanueva et al., 2018).

In the EEE literature, most published research has relied on self-report measures. Of these, quantitative approaches have been used most often, followed by qualitative, multi-methods, and mixed-methods approaches. The literature also includes a substantial number of non-empirical, conceptual studies and a few studies employing physiological measures (Lönngren, Bellocchi, et al., 2021).

4.2 Quantitative Methods

The EEE literature (as well as the broader literature on emotions in education) is dominated by studies employing quantitative methods, typically relying on self-report measures. This dominance mirrors a strong focus in the EEE literature on emotional intelligence (Section 5.3), a theme that is often explored through psychometric instruments developed for experimental research in psychology. The most-used instruments are summarized in Table 8.1 (for emotional intelligence) and Table 8.2 (for other emotion-related phenomena). Arguably, the convenient and seemingly objective approach these instruments offer may appeal to engineering education researchers who have a background in science or engineering disciplines. Indeed, these instruments have been shown to be useful for exploring specific emotional phenomena, such as emotional intelligence or exam anxiety. However, used in isolation, these instruments do not provide enough information to develop a complete understanding of complex emotional phenomena and how these phenomena may play out in different cultural and situational contexts (Pekrun & Bühner, 2014).

4.3 Physiological Methods

Physiological methods have not yet been widely used in EEE research. Notable exceptions, however, are found in the work by Villanueva et al. (2015, 2018), who used measures of electrodermal activity, cortisol levels, and serum amyloid A (SAA) proteins to study students' engagement and emotions in different educational settings, including exams. Villanueva et al. (2016, 2019) have also provided experimental protocols for conducting these types of studies. Additional examples of this type of

Table 8.1 Examples of Commonly Used Quantitative Instruments for Researching Emotional Intelligence

Instrument	*Purpose/Aim*	*Source*
Mayer-Salovey-Caruso Emotional Intelligence Test (MSCEIT, Mayer & Caruso, 2002)	Tests emotional intelligence as a set of four abilities through a set of questions which are scored as having correct answers.	Operated by a commercial publisher; use requires payment.
Schutte Emotional Intelligence Scale/ Assessing Emotions Scale (Schutte et al., 1998)	A 33-item self-report questionnaire aiming to test emotional intelligence abilities by asking respondents how good they are at particular tasks.	Published in Schutte et al. (1998) and freely available to use.
Bar-On Emotional Quotient Inventory (EQ-I, Bar-On, 1997)	Tests EI as a mixed set of skills, competencies, and abilities tested by participants ranking their agreement with a set of 133 statements.	Operated by a commercial publisher; use requires payment.
Trait Emotional Intelligence Questionnaire (TEIQue, Petrides et al., 2007)	Tests emotional intelligence as a facet of personality through participants ranking agreement with a set of statements. Available in long format (153 items) and short format (30 items).	Available through the London Psychometrics Laboratory at University College London; donations requested, but not required.

Table 8.2 Examples of Commonly Used Quantitative Instruments for Researching Emotional Phenomena Other Than Emotional Intelligence

Instrument	*Purpose/Aim*	*Source*
Achievement Emotions Questionnaire (AEQ, Pekrun et al., 2005)	Assesses college students' emotions related to attending class (80 items), studying (75 items), and being tested (77 items) through a self-report questionnaire in which participants rate their agreement with short sentences.	Published in the technical manual; freely available.
Positive and Negative Affect Schedule (PANAS, Watson et al., 1988)	Assesses mood with two scales measuring positive and negative affect through a 20-item questionnaire in which respondents indicate how often they have felt specific emotions over the past week. A 60-item version is also available.	Published in the technical manual; freely available.

research are found in science education. For example, pulse rate and blood oxygen saturation measures have been used to analyze teachers' expression of emotions in relation to physiological changes (Tobin et al., 2016) and the role of emotions when sensitive and controversial topics are discussed in science education (Calderón, 2016).

There may be several reasons for the relative lack of EEE research employing physiological measures, in addition to the general challenges associated with these measures described earlier. First, physiological measures are difficult to collect in authentic classroom settings (Immordino-Yang & Christodoulou, 2014; Sinatra et al., 2014), not least due to ethical and data protection issues. Second, in interdisciplinary collaborations, scholars trained in engineering may be more likely to take on the role of designing the technological aspects of measurement systems rather than analyzing the role of emotions in education. Finally, much EEE research to date has focused on emotional intelligence, which is typically measured through quantitative instruments.

4.4 Qualitative Methods

Even though quantitative methods have been most frequently used in EEE research so far, the EEE literature also includes studies employing a range of qualitative methods, including self-report measures (e.g., interviews), artifact analysis (e.g., student writing or course descriptions), and observation (Lönngren, Bellocchi, et al., 2021). Interviews have, for example, been used to explore emotions in terms of engineering students' lived experience, identity development, or positioning (e.g., Huff et al., 2021; Kellam et al., 2018; Lönngren et al., 2020). Artifact analysis of course documents and written reflections have been used to inform a study on engineering students' experiences of perplexity in a design innovation course (e.g., Ge & Leifer, 2020). Finally, observations have been used to study how engineering students and educators express emotions in social interaction, for example, during a lecture or student group work (e.g., Lönngren, Adawi, et al., 2021; Tanu et al., 2017; Wells, 2005).

The field of EEE research may benefit from using qualitative methods more often and more intentionally. Qualitative methods may be particularly useful for expanding EEE research on emotions in social interaction, for example, exploring how emotions and emotion norms are co-constructed in specific educational contexts. Qualitative observational methods can be used to study co-construction of emotions in real time, as it unfolds during observed interactional episodes (Hufnagel & Kelly, 2018; for example, Lönngren, Adawi, et al., 2021), as well as over time through ethnographic observations (e.g., Chubbuck & Zembylas, 2008; Zembylas, 2004). Artifact analysis can provide insights about how co-construction occurs asynchronously over time. For example, researchers could analyze how emotions play out as students produce texts (e.g., Zembylas et al., 2008) or physical products in design projects. Finally, interviews can elicit real-time emotional co-production (e.g., between researcher and participant) and relate back to previously experienced emotions, thus allowing researchers to explore longitudinal social construction of emotions across different contexts (i.e., typically also including/engaging people who are not actively involved in the research; for example, Chubbuck & Zembylas, 2008).

4.5 Multi- and Mixed Methods

EEE researchers can choose from a wide range of methods to explore emotions, components of emotions, and emotion-related phenomena. However, studying emotions by employing a single type of data may lead to incomplete understandings of emotional phenomena. In fact, educational researchers have stressed the need for multi- and mixed-methods approaches in emotion research to be able to do justice to the inherent complexity of emotions, the multitude of ways in which emotions can be defined and theorized, and the diverse roles emotions play in teaching and learning (Lindblom-Ylänne, 2019; Schutz et al., 2016). In addition, mixed- and multi-method research studies can allow researchers to combine benefits and mitigate drawbacks associated with individual methods (Fulmer & Frijters, 2009). It is also important, however, to be aware of the challenges associated with synthesizing data collected through different approaches: mixed- and multi-methods studies need to be carefully designed regarding each individual method, combination of data and findings across methods, and alignment of theoretical perspectives associated with each method (Choudhary & Jesiek, 2016). The nascent body of EEE research includes a few studies employing mixed-methods (e.g., Hess et al., 2020; Tafur Arciniegas, 2015) and multi-methods (e.g., Leicht et al., 2009; Villanueva et al., 2018) approaches. As the field matures, we expect that many more studies will benefit from intentional and purposeful combination of different research methods.

5 Prominent Themes in the EEE Literature

Having discussed theoretical and methodological perspectives that frame the scope of possibilities in conducting emotion research, we now turn our attention to extant research on EEE. Specifically, we describe four prominent themes in the EEE literature, based on a recent scoping review of the literature (Lönngren, Bellocchi, et al., 2021).

5.1 Academic Emotions

While the framework of academic emotions is seldom applied explicitly, the EEE literature includes studies on all four types of academic emotions (achievement, epistemic, topic, and social emotions).

Achievement emotions. In the wider educational literature on emotions, there is a strong focus on achievement emotions. This literature is dominated by research on achievement in terms of educational outcomes, most often test anxiety. But achievement emotions have also been related to educational processes, such as students' perceived ability to focus during a lecture or to engage with a practical task (Pekrun & Perry, 2014). Surprisingly, the focus on achievement emotions appears to be less pronounced in the EEE literature, but there are a few studies on the influence of emotions and emotion-related competencies on achievement (Anand et al., 2016; Rizwan et al., 2019; Skipper & Brandenburg, 2013; Villavicencio, 2011), students' emotional experiences of examination (Villanueva et al., 2019), students' coping strategies for dealing with achievement emotions (Bélanger et al., 2007; Deveci, 2016), and ways in which educators can help students reduce test anxiety (Bellinger et al., 2015).

Epistemic emotions. Epistemic emotions can be triggered when learners' views of knowledge and learning are challenged. In engineering education, students' views are likely to be challenged when they are confronted with uncertainty and ambiguity, or when they are required to take responsibility for their own learning (Ge & Leifer, 2020; Lönngren et al., 2019). These situations are more likely to occur when educators use pedagogical approaches that emphasize active learning, teamwork, interdisciplinary interaction, and open-ended problem-solving (Owens et al., 2020). It is therefore not surprising that much of the existing EEE research related to epistemic emotions has focused on design tasks (Adams & Turns, 2020; Ge & Leifer, 2020; Villanueva et al., 2018), teamwork (Leicht et al., 2009; Sunderland et al., 2014), and problem- or project-based learning (Chance & Williams, 2020; Deveci & Nunn, 2016).

Social emotions. Pedagogical approaches that challenge students epistemologically may also pose social challenges. For example, engineering students have been shown to experience social anxiety related to their role and status in teamwork and anxiety related to speaking in front of peers and teachers (Mohd Radzuan & Kaur, 2011; Vitasari et al., 2011). Emotions have also been shown to be important in student–teacher relationships (Tormey, 2021). Another strand of EEE research has focused on emotions related to diversity, equity, and inclusion issues, such as students' sense of belonging (Rohde et al., 2018), experiences and expressions of emotion in underrepresented groups (Decuir-Gunby et al., 2009), and a range of emotions related to students' social identities (Martin et al., 2019). However, these studies have only begun to scratch the surface of the complex array of ways in which social emotions emerge in, and impact on, in engineering education. There is a clear need for more research in this area, for example, on emotions in student–teacher relationships (Tormey, 2021), the emotional experiences of social inequalities (Rodriguez, 2017), and emotions in social justice education (Chubbuck & Zembylas, 2008; Zembylas, 2012).

Topic emotions. Topic emotions have so far mostly been researched in relation to broad disciplinary topics, focusing on students' emotions related to mathematics (Jaltare & Moghe, 2020; Jamil et al., 2011), programming (Giannakos et al., 2014), and writing (Quinto & MacAyan, 2020). More research is needed to understand emotions related to specific topics in engineering education, such as emotions triggered in teaching and learning about sustainability, social justice, inequality, or norm criticism (Kalonaitytė, 2014; Lönngren et al., 2019; Ojala, 2013; Zembylas & Chubbuck, 2009). Research should also explore a range of topic emotions, beyond the current dominance of research on anxiety. Research focused on ethics is relatively strong in EEE, which will be discussed in the next section.

5.2 Emotions and Ethics

Historically, emotions have been regarded as problematic in moral decision-making since they were thought to introduce "biases that threaten objectivity and rationality" (Roeser, 2012a, p. 107). Three distinct but related challenges to this view have been offered in the EEE literature, focusing on (1) care ethics, (2) emotional empathy, and (3) other moral emotions.

Care ethics. This perspective has its roots in the work of Gilligan (1982), who argued that an understanding of moral judgment as rationalistic and individualistic reflects masculine biases. She showed that boys typically make moral judgments by applying values and rules, while girls typically focus on social relationships and consider moral problems in terms of how their social network should respond. Gilligan's (1982) work helped launch a feminist *ethics of care* (Fisher & Tronto, 1990; Noddings, 1988, 2012), exploring moral judgments in terms of peoples' vulnerability and the situated relationships in which people interact. EEE research on care ethics is not yet well developed (Van der Poel & Royakkers, 2011), but a few studies have explored it in relation to, for example, engineering design (Pantazidou & Nair, 1999) and social justice in engineering practice (Riley et al., 2009). While emotion was not the central focus of this work, care ethics was originally developed as a counterpoint to overly cognitive accounts in mainstream moral psychology and philosophy (Gilligan, 1982), and it is therefore clearly relevant for EEE research.

Emotional empathy. Defined as "an emotional state triggered by another's emotional state or situation, in which one feels what the other feels or would normally be expected to feel in his [*sic.*] situation" (Hoffman, 2008, p. 440), *emotional empathy* is crucial for understanding other peoples' vulnerability. In engineering education research, Walther, Miller et al. (2017), and Walther et al. (2020) developed a model of empathy as a learnable skill. Hess and Fila (2016) and Hess et al. (2020), on the other hand, defined *empathy* not as a skill but as cognitive or affective positioning of oneself with respect to others. They also explicitly linked empathy to engineering students' ethical development. Finally, Lönngren et al. (2021, 2020) identified empathy as an emotion norm in engineering education, allowing students to position themselves as empathetic and, thus, emotional human beings. Moreover, the role of empathy on engineering design and design thinking has been explored by Bairaktarova et al. (2016) and Bairaktarova and Plumlee (2022). One of the pedagogical techniques used to strengthen engineering students' understanding of users' needs, the empathy map technique, focuses on categorizing users' emotions and feelings as a guiding premise of empathic design (Bairaktarova et al., 2016).

Other moral emotions. In recent years, the focus in EEE research has been broadened from empathy to a wider range of moral emotions. For example, Huff et al. (2018, 2021) explored the moral emotion *shame* through interpretative phenomenological analysis, and Gelles et al. (2020) identified *anger* as an important emotion in fueling advocacy against unjust academic structures. In the wider literature, moral emotions have been explored in relation to individuals' ethical judgment and behavior in several ways. First, moral emotions have been theorized as sources of *moral insight*, which, in turn, can be processed cognitively and contribute to intentional risk assessment in engineering: "Emotions such as fear, sympathy, and compassion help to grasp morally salient features of risky technologies, such as fairness, justice, equity, and autonomy that get overlooked in conventional, technocratic approaches to risk" (Roeser, 2012b, p. 820). Sunderland (2014; Sunderland et al., 2013, 2014) used this approach to explore emotions in engineering ethics education. Second, moral emotions have been conceptualized as sources of *moral intuitions*, which may be less amenable to cognitive processing but still influence ethical judgment and behavior (Greene & Haidt, 2002; Haidt, 2003). There is today general agreement on the importance of emotions in engineering ethics education, but more research is needed to better understand how different types of emotional information, experiences, and processes influence ethics learning in engineering education.

5.3 Emotional Intelligence and Other Socio-emotional Competencies

The *emotional intelligence* concept was first developed by Salovey et al. (2008), who highlighted the synergistic relationship between emotion and reason:

> Humans are not, in any practical sense, predominantly rational beings, nor are they predominantly emotional beings. They are both. Thus people's abilities to adapt and cope in life depend on the integrated functioning of their emotional and rational capacities.
>
> *(p. 535)*

Salovey et al. (2008) described emotional intelligence in terms of four components: (1) ability to perceive, appraise, and express one's own emotions, as well as perceive and appraise others' emotions; (2) ability to use emotions to facilitate cognitive activities, such as problem-solving; (3) ability to understand and analyze emotions, including the ability to label emotions with appropriate words and recognize relationships between emotions; and (4) ability to manage emotions, in oneself and others.

Reviewing the EEE literature, Lönngren et al. (Lönngren, Bellocchi, et al., 2021; Lönngren et al., forthcoming) identified emotional intelligence as one of the most frequently researched topics. EEE studies in this area have focused on (1) emotional intelligence and other socio-emotional competencies of engineering students (e.g., Bhave et al., 2020; Botello Ojeda & Fragoso Luzuriaga, 2015; Carballeira et al., 2019; Luisa Casado et al., 2016; Tekerek & Tekerek, 2017) and (2) emotional intelligence in association with other variables, such as self-regulation (Saibani et al., 2015), coping strategies (Bélanger et al., 2007), teamwork (Deveci, 2015; Lee et al., 2018), academic performance (Anand et al., 2016; Rizwan et al., 2019; Skipper & Brandenburg, 2013), leadership (Lappalainen, 2015), entrepreneurship (Anesukanjanakul et al., 2019), and employability (Xu, 2013).

While emotional intelligence receives a lot of attention in EEE, the concept has sometimes been used uncritically, and some authors have referred to popular work (e.g., Goleman, 1995) rather than scientific publications (Lönngren et al., forthcoming). Uncritical use of the concept is problematic for several reasons. First, it risks perpetuating a dualistic understanding of emotion as different and separated from reason, since "to argue that emotion needs to be included in education . . . through emotional intelligence skills . . . is to assume that emotion is not already part of reason" (Zembylas, 2016, pp. 542–543). This risk is particularly problematic in engineering education, where educators struggle to teach topics such as ethics, sustainability, and human-centered design. Discussions about these topics are difficult to reconcile with the prevailing rationality discourse in engineering (Lönngren, Adawi, et al., 2021; Roeser, 2012a), and it would therefore be unfortunate if emotional intelligence research contributed to strengthening these discourses. Second, calls for emotional intelligence are often based on problematic assumptions that students and educators should engage in individual, emotional self-control and self-improvement – they are expected to conform with dominant emotion norms, irrespective of whether those norms are beneficial for teaching, learning, and responsible engineering practice. Increasing homogeneity in emotion practices can also lead to reduced diversity, inclusion, and creativity in engineering classrooms (Boler, 1999; Webb, 2010; Zembylas, 2007a, pp. 456–458). Finally, emotional intelligence has often been theorized as an antidote to undesirable, "untamed," and even dangerous emotion. Such emotions have often been associated with women and underrepresented groups, who risk being stereotyped as overly emotional and – consequently – less able to assume positions of political or financial power. Thus, uncritical use of emotional intelligence risks perpetuating existing power hierarchies and inequalities (Boler, 1999).

In addition to emotional intelligence, EEE researchers have explored a range of other socio-emotional skills, including motivation and emotional regulation (e.g., Cheng, 2017), transversal competencies (e.g., Luisa Casado et al., 2016), professional skills and generic skills (e.g., Pertegal-Felices et al., 2010), and self-efficacy (e.g., Lappalainen, 2015). There is also an emerging body of research on grit in engineering education (Direito et al., 2021; Duckworth et al., 2007). Unfortunately, like emotional intelligence, these concepts have often been used uncritically and inconsistently (Direito et al., 2021; Zembylas, 2016). It would therefore be useful if future EEE research on emotional intelligence could adopt critical and sociological perspectives to develop a more nuanced understanding of the social and educational consequences that may result from teaching emotional intelligence in engineering education.

5.4 Mental Health

The World Health Organization (2022) defines *mental health* as "a state of mental well-being that enables people to cope with the stresses of life, realize their abilities, learn well and work well, and contribute to their community" (para. 1). In line with this definition, we understand mental health as a multidimensional construct that involves (1) competency in completing tasks, (2) interpersonal connection with others, and (3) intrapersonal peace with oneself. A large body of psychological research highlights how emotions and emotion-related phenomena (e.g., emotion regulation, anxiety, depression) are relevant to understanding mental health (Berking & Wupperman, 2012; Cisler et al., 2010; Joormann & Stanton, 2016). It is therefore not surprising that mental health is a relatively common research focus in the EEE literature (Lönngren, Bellocchi, et al., 2021), where mental health is often explored in relation to students' academic performance and performance-related emotions, particularly anxiety. Some studies focus on overall academic performance (e.g., Jamil et al., 2011; Villavicencio, 2011); others explore performance in specific tasks, such as, programming (Jaltare & Moghe, 2020), working in high-voltage laboratories (Güneş & Özsoy-Güneş, 2016), exams (Bellinger et al., 2015; Ramming & Mosier, 2018), delivering presentations (Mohd Radzuan & Kaur, 2011), or technical writing (Quinto & MacAyan, 2020).

Most of the published EEE studies on mental health aimed to understand how students could better regulate their emotions to meet the demands of existing tasks in the existing curriculum. The curriculum itself, on the other hand, has often been perceived as, fixed and existing tasks have seldom been scrutinized. Much of this research has also focused on what is healthy and desirable for the engineering profession (i.e., competent engineers) and less on what is healthy for individuals within engineering education and practice. Hypotheses or research questions in this line of research have typically conceptualize emotion as a phenomenon that, if not well regulated, would inhibit learning as marked by academic performance or engagement. In other words, most of the existing EEE research mental health has employed deficit framings of mental health, for example, associating certain emotional states with mental health disorders. Positive and neutral emotional states associated with mental health, such as enjoyment or relaxation, have received less attention.

There is also an emerging body of EEE research on mental health in relation to *care* (Section 5.2) rather than students' performance and productivity (e.g., Berdanier et al., 2020; Danowitz & Beddoes, 2020; Stefl, 2020). This strand of research is often focused on social emotions (Section 5.1), as well as issues of *identity* and *belonging*. For example, Wilson and Wilson (2020) analyzed how students from underrepresented groups experienced emotions that exacerbated their sense of isolation in engineering education. Further, Huff et al. (2021) and Sharbine et al. (2021) explored how a specific emotion, *shame*, can threaten students' sense of belonging and well-being. Employing a more conceptual approach, Tormey (2021) proposed a model for student–teacher affective relationships in higher education in which emotions were related to affection, attachment, and psychological safety. Finally, Jensen and Cross (2021) highlighted the complex relationships between engineering identity, cultures of stress, and mental health, amplifying the need to investigate mental health in engineering education for its own sake, rather than in the interest of increased performance in academia and engineering practice.

6 Practical Advice for EEE Research

Based on our review of the existing EEE literature (Lönngren, Bellocchi, et al., 2021; Lönngren et al., forthcoming) and our own EEE research (e.g., Direito et al., 2021; Huff et al., 2021; Lönngren, Adawi, et al., 2021; Tormey, 2021), we offer four points of advice for researchers and doctoral students who plan to pursue EEE research.

1 *Get familiar with the existing EEE literature, as well as emotion literature from the wider field of educational research.* Only if we are aware of previous work can we contribute to the collective endeavor of developing new knowledge – which then can be used to achieve real change in engineering education practice. It is also helpful to read emotion literature from other disciplines, such as psychology, philosophy, or sociology. Such reading helps develop a broader understanding of the wide range of emotion-related phenomena that can be researched and the large array of theoretical and methodological approaches that are available.
2 *Define what you mean by "emotion."* As discussed in Sections 2 and 3, different disciplines define emotion in different ways. Therefore, just like any other theoretical concepts used across disciplines, emotions and emotion-related phenomena need to be properly conceptualized, enabling others to understand the assumptions on which the research is based. Moreover, it is important to make *informed* and *intentional* choices of how to conceptualize emotions. Due to the contested nature of "emotion" in different disciplines, definitions and theories of emotion are not neutral and may position the research within long-standing academic debates. For example, defining emotions as purely neurobiological phenomena – ignoring the ways in which emotions are shaped by language, culture, and power relations – also means *defining out* of emotions many of the features that are most interesting to social scientists and many educational researchers. Unfortunately, clear conceptualizations of emotion are today often missing not only in the EEE literature (Lönngren et al., forthcoming) but also in other emotion research (Pepin, 2008). One reason may be that emotions are ubiquitous in all human practices, which may lead us to assume that everyone thinks of emotions in the same way.
3 *Ensure appropriate alignment between theoretical perspectives, data collection methods, analytic approaches, and research questions.* Such alignment is a prerequisite for high-quality research in engineering education (Huff et al., 2020; Sochacka et al., 2018; Walther, Sochacka, et al., 2017). Alignment is also important in emotion research, since different theoretical perspectives "suggest different structures and measurement of emotions" (Shuman & Scherer, 2014, p. 25). While we found relatively high degrees of alignment in the EEE literature (Lönngren et al., forthcoming), we suspect that much of this alignment is due to the availability and frequent use of published measurement instruments (Tables 1 and 2) for which standard procedures for data collection and analysis have been established. In addition, around one-third of published EEE studies (Lönngren et al., forthcoming) did not conceptualize emotions or emotion-related phenomena in any way. For that part of the literature, it was not possible to assess alignment between theories, methods, and research purposes, which means that misalignment may be much more common than the levels found in our literature review. As the field matures, we hope for more intentional conceptualization and a greater variety of theoretical perspectives and research approaches. These developments, we believe, will greatly increase the need for EEE researchers to pay careful attention to theoretical and methodological alignment.

4 *Commit to emotion as an intentional, primary research focus.* In previous EEE studies, emotions have often been studied as incidental phenomena, that is, emotions emerged as salient phenomena in research that originally intended to explore something else. For example, many of the existing studies have focused on improving learning of professional competencies, where emotions are identified as barriers or facilitators for such learning. Intentional research design is crucial for achieving alignment between theories, methods, and research questions. It also enables researchers to explore emotion-related phenomena in more depth, which in turn creates opportunities for leveraging EEE research for educational change. Finally, intentional design is important for safeguarding ethical conduct. Participants may share sensitive personal data, such as information about their mental health, and we need to ensure that these data are handled appropriately.

7 Widening the Scope of EEE Research

In Section 5, we have described four dominant themes in the EEE literature. Here, we outline six thematic, theoretical, and methodological opportunities for widening the scope of EEE research.

First, EEE research needs to attend to a diversity of emotional phenomena, beyond emotional intelligence. The strong dominance of research on emotional intelligence today risks perpetuating a narrow understanding of the role of emotions in engineering and engineering education. Some examples of under-researched emotional phenomena in research from sociological and critical perspectives are *feeling rules* and *emotional labor* (Hochschild, 1979, 1983) and emotional capital (Cottingham, 2016; Zembylas, 2007a). In research from psychological perspectives, *social emotions* (Pekrun & Linnenbrink-Garcia, 2012; Tormey, 2021) deserve more attention. The concept of social emotions may also provide an opportunity to bridge individual (psychological) and sociological perspectives in EEE research.

Second, EEE research should explore the role of emotions in different educational settings and with diverse participants. Today, mirroring the wider field of engineering education research, most of the existing EEE research has focused on higher education. While a few studies have started to explore other educational levels, including secondary education (McEneaney & Nieswandt, 2017; Sánchez-Martín et al., 2017), primary education (Campbell & Jane, 2012; McMahon, 2012), preschool (Ismail et al., 2017), pre-kindergarten (Lippard et al., 2017), informal education (Ofori-Boadu et al., 2019), as well as transitions between educational levels (Budny et al., 2010; Du-Plessis & Steyn, 2006), studies in these contexts are still scarce. Most EEE research has also concentrated on students, with only a few studies focusing on engineering and technology educators' emotions (Jha & Singh, 2012; McMahon, 2012; Rodriguez, 2017) or emotions in professional engineering contexts (Guntzburger et al., 2018; Lappalainen, 2015). More research is clearly needed on educators' and other staff members' emotions as well as emotion practices in diverse educational settings.

Third, cross-disciplinary research is needed to understand differences and similarities in the role of emotions in teaching and learning across engineering disciplines. In EEE, emotions have been studied in a variety of engineering disciplines, such as architectural (Saibani et al., 2012) and construction engineering (Owusu-Manu et al., 2019), biomedical (Hess et al., 2020) and chemical engineering (Botello Ojeda & Fragoso Luzuriaga, 2015), computer engineering (Bélanger et al., 2007; Pertegal-Felices et al., 2010), industrial engineering (Lee et al., 2018), manufacturing engineering (Brubaker et al., 2019), and marine engineering (Chung et al., 2019). These studies have shown that emotions are important in many engineering disciplines. However, the existing studies were not conducted in such a way to allow direct comparison across contexts, and we are therefore not yet able to draw conclusions regarding similarities and differences in how emotions influence teaching and learning across engineering disciplines. This is unfortunate since research has shown that the specific ways in which emotions influence teaching and learning can be discipline-specific

(Goetz et al., 2006). Hess et al.'s (2020) study on empathy in biomedical engineering provides a clear example of discipline-specific emotions – the study focused on students' emotions as they are confronted with having to test and euthanize research animals as part of their laboratory coursework (and future profession). A better understanding of disciplinary variation of the role of EEE would also contribute to developing a better understanding of how EEE research can be transferred and translated between educational contexts.

Fourth, EEE research needs to pay more attention to cultural contexts in which the research is conducted. Emotions may be experienced and expressed differently across cultural contexts, which creates challenges for interpreting cross-cultural differences and may result in misrepresentation of emotions in non-Western contexts (DeCuir-Gunby & Williams-Johnson, 2014). EEE research should therefore be conducted within and across diverse cultural contexts and employ theories and methodologies that are culturally relevant for the respective contexts (*ibid.*). While the existing EEE literature has been produced by authors in many different countries, research conducted by authors affiliated with institutions in North America, Europe, and Asia is still strongly over-represented (Lönngren, Bellocchi, et al., 2021). This distribution of authors reflects similar, and equally problematic, trends in engineering education research at large (Williams et al., 2014).

Fifth, we see a need for more EEE research drawing on sociological perspectives (Lönngren, Adawi, et al., 2021). Such research would, for example, allow the field to explore the role of emotions in creating, maintaining, or challenging power hierarchies in engineering education:

> Paying attention to the politics of emotions in [education] means analyzing and challenging the cultural and historical emotion norms with respect to what emotions are, how they are expressed, who gets to express them and under what circumstances. . . . [T]here is always something "political" in which teachers and students are caught up as they relate emotionally to one another across classroom spaces, because power relations are unavoidable; there are always emotion norms caught up in subject-matter epistemologies and pedagogies, emotion discourses and emotional expressions in the classroom.
>
> *(Zembylas, 2016, p. 545)*

Sixth, there is a need for more research employing methods that do not (solely) rely on self-report measures. For example, EEE researchers could explore research methods employing measurement of physiological markers (Villanueva et al., 2018) and observation of emotions in social interaction (Lönngren, Adawi, et al., 2021). Due to the complexity of emotions, there is also a need for more multi- and mixed-methods EEE research (Schutz et al., 2016).

8 Concluding Thoughts

Despite the well-documented importance of emotions for engineering education and practice, most engineering education today still prioritizes cognitive aspects of learning. Hoping to contribute to a better understanding of emotional aspects and their importance for engineering education, we have introduced widely used disciplinary and theoretical perspectives in emotion research, as well as a range of methods and methodologies that can be used to explore emotions, components of emotions, and emotion-related phenomena. We have also outlined the emerging field of EEE research, describing four dominant themes in the literature: (1) academic emotions, (2) emotions and ethics, (3) emotional intelligence and other socio-emotional competencies, and (4) mental health (Lönngren, Bellocchi, et al., 2021; Lönngren et al., forthcoming). Finally, we have provided practical advice for EEE research and outlined areas for future research. There clearly is a large and rapidly growing interest in EEE, and we see many opportunities for further research and practical applications as we strive to reform engineering education for the 21st century. We invite other researchers and doctoral students to join the emerging conversation and to connect with other EEE researchers in the Emotions in Engineering Education Network (EEEN, n.d.).

References

Adams, R., & Turns, J. (2020). The work of educational innovation: Exploring a personalized interdisciplinary design playbook assignment. *International Journal of Engineering Education, 36*(2), 541–555.

Ahmed, S. (2014). *The cultural politics of emotion* (2nd ed.). Edinburgh University Press.

Anand, A. F. J., Ramgopal, C. N., & Swaminathan, V. D. (2016). Emotional intelligence and academic performance of graduate engineering students. *Research Journal of Pharmaceutical, Biological and Chemical Sciences*, 7(3), 2540–2551.

Anesukanjanakul, J., Cholsuk, D., & Rattamanee, K. (2019). Predictors of entrepreneurial intentions: Mediating role of self-efficacy. *Journal of Computational and Theoretical Nanoscience, 16*(11), 4798–4806. https://doi.org/10.1037/0021-9010.90.6.1265

Bairaktarova, D., Bernstein, W., Reid, T., & Ramani, K. (2016). Beyond surface knowledge: An exploration of how empathic design techniques enhances engineers understanding of users' needs. *International Journal of Engineering Education, 32*(1), 111–122.

Bairaktarova, D., & Plumlee, D. (2022). Creating space for empathy: Perspectives on challenges of teaching design thinking to future engineers. *International Journal of Engineering Education, 38*(2), 512–524.

Bar-On, R. (1997). *Bar-on emotional quotient inventory (EQ-i) A test of emotional intelligence*. Multi-Health Systems.

Bates, R., & Wilson, D. (2008, June 22–25). Confronted by students: Dealing with anger, frustration and more pleasant emotions in office hours and in class. In *ASEE annual conference and exposition*, Pittsburgh. https://doi.org/10.18260/1-2--4427

Bélanger, F., Lewis, T., Kasper, G. M., Smith, W. J., & Harrington, K. V. (2007). Are computing students different? An analysis of coping strategies and emotional intelligence. *IEEE Transactions on Education, 50*(3), 188–196. https://doi.org/10.1109/TE.2007.900029

Bellinger, D. B., DeCaro, M. S., & Ralston, P. A. (2015). Mindfulness, anxiety, and high-stakes mathematics performance in the laboratory and classroom. *Consciousness and Cognition, 37*, 123–132. https://doi.org/10.1016/j.concog.2015.09.001

Bellocchi, A. (2019). Emotions and teacher education. In G. Noblit (Ed.), *The Oxford research encyclopedia of education*. Oxford University Press.

Berdanier, C. G., Whitehair, C., Kirn, A., & Satterfield, D. (2020). Analysis of social media forums to elicit narratives of graduate engineering student attrition. *Journal of Engineering Education, 109*(1), 125–147. https://doi.org/10.1002/jee.20299

Bericat, E. (2016). The sociology of emotions: Four decades of progress. *Current Sociology, 64*(3), 491–513. https://doi.org/10.1177/0011392115588355

Berking, M., & Wupperman, P. (2012). Emotion regulation and mental health: Recent findings, current challenges, and future directions. *Current Opinion in Psychiatry, 25*(2), 128–134. https://doi.org/10.1097/YCO.0b013e3283503669

Bhave, S. Y., Mardhekar, V., Mane, S., & Itkarkar, S. (2020). Study on socioemotional aspects of engineering girl students. *Journal of Indian Association for Child & Adolescent Mental Health, 16*(2), 90–105. https://doi.org/10.1177/0973134220200207

Boler, M. (1999). *Feeling power: Emotions and education*. Routledge.

Botello Ojeda, A. G., & Fragoso Luzuriaga, R. (2015). Emotional intelligence in chemical engineering students: A case study at the Benemerita Autonomous University of Puebla. *Dilemas Contemporaneos-Education Politica y Valores, 2*(2).

Bourdieu, P. (1990). *The logic of practice*. Polity Press.

Brubaker, E. R., Maturi, V. R., Karanian, B. A., Sheppard, S., & Beach, D. (2019, June 20–23). Integrating mind, hand, and heart: How students are transformed by hands-on designing and making. In *ASEE annual conference and exposition*, Louisville. https://doi.org/10.18260/1-2--32988

Budny, D., Allen, A., & Tartt, J. (2010, June 20–23). Dealing with failure and making the transition between high school and college. In *ASEE annual conference and exposition*, Louisville.

Calderón, O. (2016). Oximetry: A reflective tool for the detection of physiological expression of emotions in a science education classroom. *Cultural Studies of Science Education, 11*(3), 653–667. https://doi.org/10.1007/s11422-016-9731-y

Campbell, C., & Jane, B. (2012). Motivating children to learn: The role of technology education. *International Journal of Technology and Design Education, 22*(1), 1–11. https://doi.org/10.1007/s10798-010-9134-4

Carballeira, A. M., Marrero Carreira, B., & Abrante Rodríguez, D. (2019). Inteligencia emocional y ajuste psicológico en estudiantes: Nivel académico y rama de estudios. *Universitas Psychologica, 18*(4), 1–14. http://dx.doi.org/10.11144/Javeriana.upsy18-4.ieap

Cech, E. A. (2018). The (mis)framing of social justice: Why ideologies of depoliticization and meritocracy hinder engineers' ability to think about social injustices. In J. C. Lucena (Ed.), *Engineering education for social justice: Critical explorations and opportunities* (pp. 67–84). Springer.

Cech, E. A., & Sherick, H. M. (2015). Depoliticization and the structure of engineering education. In S. H. Christensen, C. Didier, A. Jamison, M. Meganck, C. Mitcham, & B. Newberry (Eds.), *International perspectives on engineering education* (pp. 203–216). Springer.

Chance, S. M., & Williams, B. (2020, April 27–30). Here you have to be mixing: Collaborative learning on an engineering program in Ireland as experienced by a group of young Middle Eastern women. In *IEEE global engineering education conference, EDUCON* (pp. 1787–1794), Porto. https://doi.org/10.1109/EDUCON45650.2020.9125207

Cheng, K. C. (2017). *Future time perspective, socio-emotional regulation, and diurnal cortisol patterns in post-secondary engineering students* (1948887543) [Doctoral thesis, Arizona State University].

Choudhary, N., & Jesiek, B. K. (2016). State of mixed methods research in engineering education: In-depth examination of JEE articles, 2010–2015. 1–9. https://doi.org/10.1109/FIE.2016.7757554

Chubbuck, S. M., & Zembylas, M. (2008). The emotional ambivalence of socially just teaching: A case study of a novice urban schoolteacher. *American Educational Research Journal, 45*(2), 274–318. https://doi.org/10.3102/0002831207311586

Chung, C.-C., Cheng, Y.-M., Shih, R.-C., & Lou, S.-J. (2019). Research on the learning effect of the positive emotions of "ship fuel-saving project" APP for engineering students. *Sustainability (Switzerland), 11*(4). https://doi.org/10.3390/su11041136

Cisler, J. M., Olatunji, B. O., Feldner, M., & Forsyth, J. P. (2010). Emotion regulation and the anxiety disorders: An integrative review. *Journal of Psychopathology and Behavioral Assessment, 32*(1), 68–82. https://doi.org/10.1007%2Fs10862-009-9161-1

Corcoran, R. C., & Tormey, R. (2012). *Developing emotionally competent teachers: Emotional intelligence and pre-service teacher education.* Peter Lang AG International Academic Publishers.

Cottingham, M. D. (2016). Theorizing emotional capital. *Theoretical Sociology, 45*, 451–470. https://doi.org/10.1007/s11186-016-9278-7

Damasio, A. (1994). *Descartes' error: Emotion, rationality and the human brain.* Putnam.

Danowitz, A., & Beddoes, K. (2020, October 22–24). A snapshot of mental health and wellness of engineering students across the western United States. In *Frontiers in education*, Uppsala (online).

Davidson, S. G., Jaber, L. Z., & Southerland, S. A. (2020). Emotions in the doing of science: Exploring epistemic affect in elementary teachers' science research experiences. *Science Education, 104*(6), 1008–1040. https://doi.org/10.1002/sce.21596

Davis, M. (2017). In praise of emotion in engineering. In D. Michelfelder, B. Newberry, & Q. Zhu (Eds.), *Philosophy and engineering* (Vol. 26, pp. 181–194). Springer.

Decuir-Gunby, J. T., Long-Mitchell, L. A., & Grant, C. (2009). The emotionality of women professors of color in engineering: A critical race theory and critical race feminism perspective. In *Advances in teacher emotion research: The impact on teachers' lives* (pp. 323–342). Springer.

DeCuir-Gunby, J. T., & Williams-Johnson, M. R. (2014). The influence of culture on emotions: Implications for education. In R. Pekrun & L. Linnenbrink-Garcia (Eds.), *International handbook of emotions in education* (pp. 539–557). Routledge.

Deveci, T. (2015). Freshman students' emotional intelligence and team-work satisfaction levels. A comparative study: Gender and nationality. *Yuksekogretim Dergisi, 5*(1), 35–43. https://doi.org/10.2399/yod.15.007

Deveci, T. (2016). Boredom among freshman engineering students in a communication course. *Journal of Teaching English for Specific and Academic Purposes, 4*(2), 261–276.

Deveci, T., & Nunn, R. (2016). Development in freshman engineering students' emotional intelligence in project-based courses. *Asian ESP Journal, 12*(2), 54–92.

Direito, I., Chance, S., & Malik, M. (2021). The study of grit in engineering education research: A systematic literature review. *European Journal of Engineering Education, 46*(2), 161–185. https://doi.org/10.1080/03043797.2019.1688256

Duckworth, A. L., Peterson, C., Matthews, M. D., & Kelly, D. R. (2007). Grit: Perseverance and passion for long-term goals. *Journal of Personality and Social Psychology, 92*(6), 1087–1101. https://doi.org/10.1037/0022-3514.92.6.1087

Du-Plessis, I., & Steyn, T. (2006, June 28–July 1). Academic survival: Experiences of a group of at-risk engineering students in the School of Engineering, University of Pretoria. In *SEFI annual conference*, Uppsala.

Edwards, D. (1999). Emotion discourse. *Culture & Psychology, 55*(3), 271–291. https://doi.org/10.1177/1354067X9953001

EEEN. (n.d.). *Emotions in engineering education network.* www.umu.se/institutionen-for-naturvetenskapernas-och-matematikens-didaktik/forskning/eeen/

England, R. (2019). The cognitive/noncognitive debate in emotion theory: A corrective from Spinoza. *Emotion Review, 11*(2), 102–112.

Fisher, B., & Tronto, J. C. (1990). Toward a feminist theory of caring. In E. Abel & M. Nelson (Eds.), *Circles of care* (pp. 36–54). SUNY Press.

Fraser, J., & Brandt, C. B. (2013). Trading zones in environmental education: The emotional life of the environmental educator. In M. E. Krasny & J. Dillon (Eds.), *Creating transdisciplinary dialogue* (pp. 133–158). Peter Lang AG International Academic Publishers.

Frevert, U. (2014). Defining emotions: Concepts and debates over three centuries. In *Emotional lexicons: Continuity and change in the vocabulary of feeling 1700–2000.* Oxford University Press.

Fridja, N. H. (1986). *The emotions.* Cambridge University Press.

Fridja, N. H. (2007). *The laws of emotion.* Erlbaum.

Fridja, N. H. (2008). The psychologists' point of view. In M. Lewis, J. M. Haviland-Jones, & L. F. Barrett (Eds.), *Handbook of emotions* (3rd ed.). The Guilford Press.

Fulmer, S. M., & Frijters, J. C. (2009). A review of self-report and alternative approaches in the measurement of student motivation. *Educational Psychology Review, 21*, 219–246. https://doi.org/10.1007/s10648-009-9107-x

Ge, X., & Leifer, L. (2020). When tough times make tough designers: How perplexing experiences shape engineers' knowledge and identity. *International Journal of Engineering Education, 36*(2), 650–663.

Gelles, L. A., Youmans, K. L., & Villanueva, I. (2020, September 20–24). Sparking action: How emotions fuel or inhibit advocacy around hidden curriculum in engineering. In *SEFI annual conference proceedings* (pp. 1566–1575), Twente (online).

Giannakos, M. N., Jaccheri, L., & Leftheriotis, I. (2014). Happy girls engaging with technology: Assessing emotions and engagement related to programming activities. In *LCT 2014, part I, 8523 LNCS.* Springer.

Gilligan, C. (1982). *In a different voice; psychological theory and women's development.* Harvard University Press.

Goetz, T., Frenzel, A. C., Pekrun, R., & Hall, N. C. (2006). The domain specificity of academic emotional experiences. *The Journal of Experimental Education, 75*(1), 5–29. https://doi.org/10.3200/JEXE.75.1.5-29

Goleman, D. (1995). *Emotional intelligence.* Bantam Books.

Goleman, D. (1998). *Working with emotional intelligence.* Bantam Books.

Goleman, D. (2013). *Focus: The hidden driver of excellence.* Harper.

Greene, J., & Haidt, J. (2002). How (and where) does moral judgment work? *Trends in Cognitive Sciences, 612*(517–523). https://doi.org/10.1016/S1364-6613(02)02011-9

Güneş, A., & Özsoy-Güneş, Z. (2016). The investigation of danger perceptions and anxiety levels of electric-electronic engineering students in high voltage laboratory environment. *Turkish Online Journal of Educational Technology*, 989–996.

Guntzburger, Y., Johnson, K. J., Martineau, J. T., & Pauchant, T. C. (2018). Professional ethnocentrism and ethical risk management efficacy: How engineer's emotional openness mediates this complex relationship. *Safety Science, 109*, 27–35. https://doi.org/10.1016/j.ssci.2018.05.004

Haidt, J. (2001). The emotional dog and its rational tail: A social intuitionist approach to moral judgement. *Psychological Review, 108*, 814–834. https://doi.org/10.1037/0033-295X.108.4.814

Haidt, J. (2003). The moral emotions. In R. J. Davidson, K. R. Scherer, & H. H. Goldsmith (Eds.), *Handbook of affective sciences* (pp. 852–870). Oxford University Press.

Hess, J. L., & Fila, N. D. (2016, June 26–28). The development and growth of empathy among engineering students. In *ASEE annual conference & exposition*, New Orleans. https://doi.org/10.18260/p.26120

Hess, J. L., Miller, S., Higbee, S., Fore, G. A., & Wallace, J. (2020). *Empathy and ethical becoming in biomedical engineering education: A mixed methods study of an animal tissue harvesting laboratory*. https://doi.org/10.1080/22054952.2020.1796045

Hochschild, A. (1979). Emotion work, feeling rules, and social structure. *American Journal of Sociology, 85*, 551–575.

Hochschild, A. (1983). *The managed heart: Commercialization of human feeling*. University of California Press.

Hoffman, M. L. (2000). *Empathy and moral development: Implications for caring and justice*. Cambridge University Press.

Hoffman, M. L. (2008). Empathy and prosocial behaviour. In M. Lewis, J. M. Haviland-Jones, & L. Feldman Barrett (Eds.), *Handbook of emotion* (3rd ed., pp. 440–455). The Guilford Press.

Holmén, J. (2020). *Navigating sustainability transformations: Backcasting, transdisciplinarity and social learning* [Doctoral thesis, Chalmers University of Technology].

Huff, J. L., Okai, B., Shanachilubwa, K., Sochacka, N. W., & Walther, J. (2021). Unpacking professional shame: Patterns of White male engineering students living in and out of threats to their identities. *Journal of Engineering Education, 110*(2), 414–436. https://doi.org/10.1002/jee.20381

Huff, J. L., Shanachilubwa, K., & Secules, S. (2018, June 24–27). Shame amid academic success: An interpretative phenomenological analysis case study of a student's experience with emotions in engineering. In *ASEE annual conference and exposition*, Salt Lake City. https://doi.org/10.18260/1-2—30959

Huff, J. L., Walther, J., Sochacka, N. W., Sharbine, M. B., & Kamanda, H. (2020). Coupling methodological commitments to make sense of socio-psychological experience. *Studies in Engineering Education, 1*(2), 1–13. https://doi.org/10.21061/see.29

Hufnagel, E., & Kelly, G. J. (2018). Examining emotional expressions in discourse: Methodological considerations. *Cultural Studies of Science Education, 13*, 905–924. https://doi.org/10.1007/s11422-017-9806-4

Husman, J., Cheng, K. C., Puruhito, K., & Fishman, E. J. (2015, June 14–17). Understanding engineering students stress and emotions during an introductory engineering course. In *ASEE annual conference and exposition*, Seattle. https://doi.org/10.18260/p.24958

Immordino-Yang, M. H., & Christodoulou, J. A. (2014). Neuroscientific contributions to understanding and measuring emotions in educational contexts. In R. Pekrun & L. Linnenbrink-Garcia (Eds.), *International handbook of emotions in education* (pp. 607–624). Routledge.

Ismail, I. M., Hashim, S., Anis, S. K., Ismail, A., & Ismail, M. E. (2017, January). Implementation of a development in cognitive, psychomotor and socio emotional elements through games to achieve national preschool curriculum standards. In *Proceedings of IEEE ICEED* (pp. 143–147). https://doi.org/10.1109/ICEED.2017.8251182

Jaltare, V., & Moghe, K. (2020). Effect of expressive writing on math anxiety of engineering students. *IOP Conference Series: Materials Science and Engineering, 804*, 1. http://dx.doi.org/10.1088/1757-899X/804/1/012057

Jamil, S. J., Saad, S., Jaafar, W. N. W., & Abdullah, N. (2011, December 7–8). An of investigation of mathematics anxiety among diploma engineering students. In *Third international congress on engineering education (ICEED)*, Kuala Lumpur.

Jensen, K. J., & Cross, K. J. (2021). Engineering stress culture: Relationships among mental health, engineering identity, and sense of inclusion. *Journal of Engineering Education, 110*(2), 371–392. https://doi.org/10.1002/jee.20391

Jha, A., & Singh, I. (2012). Teacher effectiveness in relation to emotional intelligence among medical and engineering faculty members. *Europe's Journal of Psychology, 8*(4), 667–685. https://doi.org/10.5964/ejop.v8i4.483

Joormann, J., & Stanton, C. H. (2016). Examining emotion regulation in depression: A review and future directions. *Behaviour Research and Theraphy, 86*, 35–49. https://doi.org/10.1016/j.brat.2016.07.007

Kalonaityté, V. (2014). *Normkritisk pedagogik – För den högre utbildningen*. Studentlitteratur.

Kellam, N., Gerow, K., Wilson, G., Walther, J., & Cruz, J. (2018). Exploring emotional trajectories of engineering students: A narrative research approach. *International Journal of Engineering Education, 34*(6), 1726–1740.

Kleinginna, P. R., & Kleinginna, A. M. (1981). A categorized list of emotion definitions, with suggestions for a consensual definition. *Motivation and Emotion*, *5*(4), 345–379. https://doi.org/10.1007/BF00992553

Lappalainen, P. (2015). Predictors of effective leadership in industry – should engineering education focus on traditional intelligence, personality, or emotional intelligence? *European Journal of Engineering Education*, *40*(2), 222–233. https://doi.org/10.1080/03043797.2014.944102

Lawless, M. (2018). Documenting a labor of love: Emotional labor as academic labor. *Review of Communication*, *18*(2), 85–97. https://doi.org/10.1080/15358593.2018.1438644

Lazarus, R. (1999). The cognition – emotion debate: A bit of history. In *Handbook of cognition and emotion* (pp. 1–19). Wiley.

Lee, C. T., Bristow, M., & Wong, J. C. (2018). Emotional intelligence and teamwork skills among undergraduate engineering and nursing students: A pilot study. *Journal of Research in Interprofessional Practice & Education*, *8*(1), 1–16. https://doi.org/10.22230/jripe.2018v8n1a260

Leicht, R. M., Lewis, A., Riley, D. R., Messner, J. I., & Darnell, B. (2009). Assessing traits for success in individual and team performance in an engineering course. *Construction Research Congress: Building a Sustainable Future*, 1358–1367. https://doi.org/10.1061/41020(339)138

Lindblom-Ylänne, S. (2019). Research on academic emotions requires a rich variety of research methods. *Studies in Higher Education*, *44*(10), 1803–1803. https://doi.org/10.1080/03075079.2019.1665333

Lindgren, J., & Rönnberg, L. (2018). The emotional politics of quality assurance reform: Shifting affective atmospheres in Swedish higher education policy. *Higher Education Policy*, *31*(1), 55–73. https://doi.org/10.1057/s41307-017-0045-9

Lippard, C. N., Lamm, M. H., & Riley, K. L. (2017). Engineering thinking in prekindergarten children: A systematic literature review. *Journal of Engineering Education*, *106*(3), 454–474. https://doi.org/10.1002/jee.20174

Lönngren, J., Adawi, T., & Berge, M. (2020, October 21). "I don't want to be influenced by emotions" – Engineering students' emotional positioning in discussions about wicked sustainability problems. In *Frontiers in education*, Uppsala, Sweden. https://dblp.org/rec/conf/fie/LonngrenAB20

Lönngren, J., Adawi, T., & Berge, M. (2021). Using positioning theory to study the role of emotions in engineering problem solving: Methodological issues and recommendations for future research. *Studies in Engineering Education*, *2*(1), 53–79. https://doi.org/10.21061/see.50

Lönngren, J., Adawi, T., & Svanström, M. (2019). Scaffolding strategies in a rubric-based intervention to promote engineering students' ability to address wicked problems. *European Journal of Engineering Education*, *44*(1–2), 196–221. https://doi.org/10.1080/03043797.2017.1404010

Lönngren, J., Bellocchi, A., Berge, M., Bøgelund, P., Direito, I., Rahman, N. F. A., Huff, J., Mohd-Yusof, K., Murzi, H., & Tormey, R. (forthcoming). *Emotions in engineering education: A systematic review of the literature.*

Lönngren, J., Bellocchi, A., Bøgelund, P., Direito, I., Huff, J., Mohd-Yusof, K., Murzi, H., & Tormey, R. (2021, December 5–8). Emotions in engineering education: Preliminary results from a scoping review. In *Research in engineering education symposium*, Perth (online).

Lönngren, J., Svanström, M., Ingerman, Å., & Holmberg, J. (2016). Dealing with the multidimensionality of sustainability through the use of multiple perspectives – A theoretical framework. *European Journal of Engineering Education*, *41*(3), 342–352. https://doi.org/10.1080/03043797.2015.1079811

Luisa Casado, M., Lopez-Fernandez, D., & Lapuerta, V. (2016). Socio-emotional competencies in engineering education. *International Journal of Engineering Education*, *32*(4), 1660–1678.

Martin, C. C., LeDoux, J. M., & Newstetter, W. C. (2019, June 16–19). Is optimal distinctiveness theory useful for increasing belonging in educational settings? In *ASEE annual conference and exposition*, Tampa.

Matias, C. E., & Zembylas, M. (2014). 'When saying you care is not really caring': Emotions of disgust, whiteness ideology, and teacher education. *Critical Studies in Education*, *55*(3), 319–337. https://doi.org/10.1080/17508487.2014.922489

Mayer, J. D., & Caruso, D. R. (2002). *Mayer-Salovey-Caruso emotional intelligence test (MSCEIT) user's manual.* Multi-Health Systems.

Mayer, J. D., & Salovey, P. (1993). The intelligence of emotional intelligence. *Intelligence*, *17*(4), 433–442. https://doi.org/10.1016/0160-2896(93)90010-3

Mayer, J. D., Salovey, P., & Caruso, D. (2000). Models of emotional intelligence. In R. J. Sternberg (Ed.), *Handbook of intelligence* (pp. 396–420). Cambridge University Press. https://doi.org/10.1017/CBO9780511807947.019

McEneaney, E., & Nieswandt, M. (2017, June 25–28). The critical role of group affect in engineering design tasks in high school biology (fundamental). In *ASEE annual conference and exposition*, Columbus.

McMahon, A. P. (2012, June 10–13). Mental models elementary teachers hold of engineering design processes: A comparison of two communities of practice. In *ASEE annual conference and exposition*, San Antonia.

Mohd Radzuan, N. R., & Kaur, S. (2011). Technical oral presentations in English: Qualitative analysis of Malaysian engineering undergraduates' sources of anxiety. *Procedia – Social and Behavioral Sciences*, *29*, 1436–1445. https://doi.org/10.1016/j.sbspro.2011.11.383

Moors, A., Ellsworth, P. C., Scherer, K. R., & Frijda, N. H. (2013). Appraisal theories of emotion: State of the art and future development. *Emotion Review*, *5*(2), 119–124. http://dx.doi.org/10.1177/1754073912468165

Murphy, S. T., & Zajonc, R. B. (1993). Affect, cognition, and awareness: Affective priming with optimal and suboptimal stimulus exposures. *Journal of Personality and Social Psychology*, *64*(5), 723–739. https://doi.org/10.1037/0022-3514.64.5.723

Nair, I., & Bulleit, W. M. (2020). Pragmatism and care in engineering ethics. *Science and Engineering Ethics*, *26*(1), 65–87. https://doi.org/10.1007/s11948-018-0080-y

Noddings, N. (1988). An ethic of caring and its implications for instructional arrangements. *American Journal of Education*, *96*(2), 215–230.

Noddings, N. (2012). The caring relation in teaching. *Oxford Review of Education*, *38*(6), 771–781. https://doi.org/10.1080703054985.2012.745047

Nussbaum, M. (2001). *Upheavals of thought: The intelligence of emotions*. Cambridge University Press. https://doi.org/10.1017/CBO9780511840715

Nussbaum, M. (2004). Emotions as judgments of value and importance. In R. C. Solomon (Ed.), *Thinking about emotion: Contemporary philosophers on emotion* (pp. 183–199). Oxford University Press.

Oatley, K., Keltner, D., & Jenkins, J. M. (2006). *Understanding emotions* (2nd ed.). Blackwell.

Ofori-Boadu, A. N., Deng, D., Stevens, C. M., Gore, K., & Borders-Taylor, I. (2019, June 16–19). Learning experiences and self-efficacy of minority middle-school girls during a bio-char modified cement paste research program at an HBCU. In *ASEE annual conference and exposition*, Tampa.

Ojala, M. (2013). Emotional awareness: On the importance of including emotional aspects in education for sustainable development (ESD). *Journal of Education for Sustainable Development*, 7(2), 167–182. https://doi.org/10.1177/0973408214526488

Ojala, M. (in print). Climate-change education and critical emotional awareness (CEA): Implications for teacher education. *Educational Philosophy and Theory*. https://doi.org/10.1080/00131857.2022.2081150

Owens, D. C., Sadler, T. D., Barlow, A. T., & Smith-Walters, C. (2020). Student motivation from and resistance to active learning rooted in essential science practices. *Research in Science Education*, *50*, 253–277. https://doi.org/10.1007/s11165-017-9688-1

Owusu-Manu, D., Edwards, D. J., Kukah, A. S., Pärn, E. A., El-Gohary, H., & Aigbavboa, C. (2019). An assessment of the level of emotional intelligence attributes of undergraduate built environment students in developing countries. *Industry and Higher Education*, *33*(2), 108–115. https://doi.org/10.1177/0950422218798551

Pantazidou, M., & Nair, I. (1999). Ethic of care: Guiding principles for engineering teaching & practice. *Journal of Engineering Education*, *88*(2), 205–212. https://doi.org/10.1002/j.2168-9830.1999.tb00436.x

Pekrun, R., & Bühner, M. (2014). Self-report measures of academic emotions. In R. Pekrun & L. Linnenbrink-Garcia (Eds.), *International handbook of emotions in education* (pp. 561–579). Routledge.

Pekrun, R., Goetz, T., & Perry, R. P. (2005). *Achievement emotions questionnaire (AEQ) – User's manual*. University of Munich.

Pekrun, R., & Linnenbrink-Garcia, L. (2012). Academic emotions and student engagement. In S. Christenson, A. Reschly, & C. Wylie (Eds.), *Handbook of research on student engagement*. Springer. https://doi.org/10.1007/978-1-4614-2018-7

Pekrun, R., & Linnenbrink-Garcia, L. (2014a). *International handbook of emotions in education*. Routledge. https://doi.org/10.4324/9780203148211

Pekrun, R., & Linnenbrink-Garcia, L. (2014b). Introduction. In R. Pekrun & L. Linnenbrink-Garcia (Eds.), *International handbook of emotions in education* (pp. 1–10). Routledge.

Pekrun, R., & Perry, R. P. (2014). Control-value theory of achievement emotions. In *International handbook of emotions in education* (pp. 120–141). Routledge.

Pepin, N. (2008). Studies on emotions in social interactions. Introduction. *Bulletin VALS-ASLA, 88*, 1–18. http://doc.rero.ch/record/11876/files/bulletin_vals_asla_2008_088.pdf

Pertegal-Felices, M. L., Castejón-Costa, J. L., & Jimeno-Morenilla, A. (2010). Personal and emotional skill profiles in the professional development of the computer engineer. *International Journal of Engineering Education, 26*(1), 218–226.

Petrides, K. V., Furnham, A., & Frederickson, N. (2004). Emotional intelligence. *The Psychologist, 17*(10), 574–577.

Petrides, K. V., & Mavroveli, S. (2018). Theory and applications of trait emotional intelligence. *Psychology: The Journal of the Hellenic Psychological Society, 23*(1), 24–36. https://doi.org/10.12681/psy_hps.23016

Petrides, K. V., Pita, R., & Kokkinaki, F. (2007). The location of trait emotional intelligence in personality factor space. *British Journal of Psychology, 98*(2), 273–289. https://doi.org/10.1348/000712606x120618

Polmear, M., Bielefeldt, A., Knight, D., Swan, C., & Canney, N. (2018, June 24–27). Faculty perceptions of challenges to educating engineering and computing students about ethics and societal impacts. In *ASEE annual conference & exposition*, Salt Lake City.

Quinlan, K. M. (2016). How emotion matters in four key relationships in teaching and learning in higher education. *College Teaching, 64*(3), 101–111. https://doi.org/10.1080/87567555.2015.1088818

Quinto, E. J. M., & MacAyan, J. V. (2020). Interactive influence of anxiety and goal orientation in predicting language task performance of Filipino engineering students. *Asian ESP Journal, 16*(21), 335–359.

Ramming, C. H., & Mosier, R. (2018, June 24–27). Time limited exams: Student perceptions and comparison of their grades versus time in engineering mechanics: Statics. In *ASEE annual conference & exposition*, Salt Lake City.

Reisenzein, R., Junge, M., Studtmann, M., & Huber, O. (2014). Observational approaches to the measurement of emotions. In R. Pekrun & L. Linnenbrink-Garcia (Eds.), *International handbook of emotions in education* (pp. 580–606). Routledge.

Riley, D., Pawley, A. L., Tucker, J., & Catalano, G. D. (2009). Feminisms in engineering education: Transformative possibilities. *NWSA Journal, 21*(2), 21–40.

Ritzer, G. (2011). *The McDonaldization of society 6* (Vol. 6). Pine Forge Press.

Rivera Maulucci, M. S. (2013). Emotions and positional identity in becoming a social justice science teacher: Nicole's story. *Journal of Research in Science Teaching, 50*(4), 453–478. https://doi.org/10.1002/tea.21081

Rizwan, A., Alsulami, H., Elnahas, N., Bashir, M., Bawareth, F., Kamrani, R., & Noorelahi, R. (2019). Impact of emotional intelligence on the academic performance and employability of female engineering students in Saudi Arabia. *International Journal of Engineering Education, 35*(1, A), 119–125.

Rodriguez, A. J. (2017). How do teachers prepare for and respond to students' evoked emotions when addressing real social inequalities through engineering activities? *Theory Into Practice, 56*(4), 263–270. https://doi.org/10.1080/00405841.2017.1350497

Roeser, S. (2012a). Emotional engineers: Toward morally responsible design. *Science and Engineering Ethics, 18*, 103–115. https://doi.org/10.1007%2Fs11948-010-9236-0

Roeser, S. (2012b). Moral emotions as guide to acceptable risk. In S. Roeser, R. Hillerbrand, P. Sandin, & M. Peterson (Eds.), *Handbook of risk theory*. Springer.

Rohde, J. A., Benson, L., Potvin, G., Kirn, A., & Godwin, A. (2018, June 24–27). You either have it or you don't: First year engineering students' experiences of belongingness. In *ASEE annual conference and exposition*, Salt Lake City. https://doi.org/10.18260/1-2-31320

Russell, J. A. (2003). Core affect and the psychological construction of emotion. *Psychological Review, 110*(1), 145–172. https://doi.org/10.1037/0033-295X.110.1.145

Saibani, N., Muhamad, N., Abd Wahab, D., & Sahari, J. (2012). Level of emotional intelligence (EQ) scores among engineering students during course enrollment and course completion. In S. R. S. Abdullah, S. A. Osman, & N. Mohamad (Eds.), *Universiti Kebangsaan Malaysia teaching and learning congress 2011, Vol II* (Vol. 60, pp. 479–483). Universiti Kebangsaan Malaysia. https://doi.org/10.1016/j.sbspro.2012.09.410

Saibani, N., Sabtu, M. I., Harun, Z., Wan Mahmood, W. M. F., Muhamad, N. I., Wahab, D. A., & Sahari, J. (2015). Comparison of emotional intelligence scores among engineering students of an academic programme. *Journal of Engineering Science and Technology, 10*, 41–51.

Salovey, P., Detweiler-Bedell, B. T., Detweiler-Bedell, J. B., & Mayer, J. D. (2008). Emotional intelligence. In M. Lewis, J. M. Haviland-Jones, & L. Feldman Barrett (Eds.), *Handbook of emotions* (3rd ed., pp. 533–547). The Guilford Press.

Salovey, P., & Mayer, J. D. (1990). Emotional intelligence. *Imagination, Cognition and Personality, 9*(3), 185–211. https://doi.org/10.2190/DUGG-P24E-52WK-6CDG

Sánchez-Martín, J., Álvarez-Gragera, G. J., Dávila-Acedo, M. A., & Mellado, V. (2017). What Do K-12 students feel when dealing with technology and engineering issues? Gardner's multiple intelligence theory implications in technology lessons for motivating engineering vocations at Spanish secondary school. *European Journal of Engineering Education, 42*(6), 1330–1343. https://doi.org/10.1080/03043797.2017.1292216

Scherer, K. R. (2005). What are emotions? And how can they be measured? *Social Science Information, 44*, 695–729.

Schutte, N. S., Malouff, J. M., Hall, L. E., Haggerty, D. J., Cooper, J. T., Golden, C. J., & Dornheim, L. (1998). Development and validation of a measure of emotional intelligence. *Personality and Individual Differences, 25*(2), 167–177. https://doi.org/10.1016/S0191-8869(98)00001-4

Schutz, P. A., Decuir-Gunby, J. T., & Williams-Johnson, M. R. (2016). Using multiple and mixed methods to investigate emotions in educational contexts. In M. Zembylas & P. A. Schutz (Eds.), *Methodological advances in research on emotion and education* (pp. 217–229). Springer.

Sharbine, M. B., Huff, J. L., Sochacka, N. W., & Walther, J. (2021, December 5–8). Professional shame as a socio-psychological mechanism for marginalization in engineering education. In *Research in engineering education symposium*, Perth (online).

Shuman, V., & Scherer, K. R. (2014). Concepts and structures of emotions. In R. Pekrun & L. Linnenbrink-Garcia (Eds.), *International handbook of emotions in education* (pp. 13–35). Routledge. https://doi.org/10.4324/9780203148211.ch2

Sinatra, G. M., Broughton, S. H., & Lombardi, D. (2014). Emotions in science education. In R. Pekrun & L. Linnenbrink-Garcia (Eds.), *International handbook of emotions in education* (pp. 415–436). Routledge. https://doi.org/10.4324/9780203148211

Skipper, C. O., & Brandenburg, S. (2013). Emotional intelligence and academic performance of engineering students. *Engineering Project Organization Journal, 3*(1), 13–21. https://doi.org/10.1080/21573727.2012.738669

Sochacka, N. W., Walther, J., & Pawley, A. L. (2018). Ethical validation: Reframing research ethics in engineering education research to improve research quality. *Journal of Engineering Education, 107*(3), 362–379. https://doi.org/10.1002/jee.20222

Solomon, R. C. (2008). The philosophy of emotion. In M. Lewis, J. M. Haviland-Jones, & L. Feldman Barrett (Eds.), *Handbook of emotions* (3rd ed., pp. 3–16). The Guilford Press.

Stefl, S. K. (2020). *Women engineering faculty well-being: An interpretative phenomenological analysis* [Doctoral thesis, Clemson University].

Stets, J. E., & Turner, J. H. (2008). The sociology of emotion. In M. Lewis, J. M. Haviland-Jones, & L. F. Barrett (Eds.), *Handbook of emotion* (3rd ed.). The Guilford Press.

Sunderland, M. E. (2014). Taking emotion seriously: Meeting students where they are. *Science and Engineering Ethics, 20*, 183–195. https://doi.org/10.1007/s11948-012-9427-y

Sunderland, M. E., Ahn, J., Carson, C., & Kastenberg, W. E. (2013, June 23–26). Making ethics explicit: Relocating ethics to the core of engineering education. In *ASEE annual conference & exposition*, Atlanta.

Sunderland, M. E., Taebi, B., Carson, C., & Kastenberg, W. (2014). Teaching global perspectives: Engineering ethics across international and academic borders. *Journal of Responsible Innovation, 1*(2), 228–239. https://doi.org/10.1080/23299460.2014.922337

Tafur Arciniegas, M. (2015). *Understanding how adults approach technological challenges: A sequential mixed methods research* [Doctoral thesis, Purdue University].

Tanu, E. D., Quan, G. M., Gupta, A., & Turpen, C. A. (2017, June 25–28). The role of empathy in supporting teaching moves of engineering design peer educators. In *ASEE annual conference & exposition*, Columbus. http://dx.doi.org/10.18260/1-2--29004

Tekerek, M., & Tekerek, B. (2017). Emotional intelligence in engineering education. *Turkish Journal of Education, 6*(2), 88–95. https://doi.org/10.19128/turje.306499

Thoits, P. (1990). Emotional deviance: Research agenda. In T. D. Kemper (Ed.), *Research agendas in the sociology of emotions* (pp. 180–203). State University of New York Press.

Tobin, K., King, D., Henderson, S., Bellocchi, A., & Ritchie, S. M. (2016). Expression of emotions and physiological changes during teaching. *Cultural Studies of Science Education, 11*(3), 669–692. https://doi.org/10.1007/s11422-016-9778-9

Tormey, R. (2021). Rethinking student-teacher relationships in higher education: A multidimensional approach. *Higher Education, 82*, 993–1011. https://doi.org/10.1007/s10734-021-00711-w

Turner, J. C., & Trucano, M. (2014). Measuring situated emotion. In R. Pekrun & L. Linnenbrink-Garcia (Eds.), *International handbook of emotions in education* (pp. 643–658). Routledge. https://doi.org/10.4324/9780203148211

Turner, J. H. (2009). The sociology of emotions: Basic theoretical arguments. *Emotion Review, 1*(4), 340–354. https://doi.org/10.1177/1754073909338305

Turner, J. H., & Stets, J. E. (2005). Conceptualizing emotions sociologically. In *The sociology of emotions* (pp. 1–25). Cambridge University Press.

Van den Beemt, A., MacLeod, A., Van der Veen, J., Van de Ven, A., van Baalen, S., Klaassen, R., & Boon, M. (2020). Interdisciplinary engineering education: A review of vision, teaching, and support. *Journal of Engineering Education, 109*, 508–555. https://doi.org/10.1002/jee.20347

Van der Poel, I., & Royakkers, L. (2011). *Ethics, technology and engineering, an introduction*. Wiley.

Villanueva, I., Campbell, B. D., Raikes, A. C., Jones, S. H., & Putney, L. G. (2018). A multimodal exploration of engineering students' emotions and electrodermal activity in design activities. *Journal of Engineering Education, 107*(3), 414–441. https://doi.org/10.1002/jee.20225

Villanueva, I., Husman, J., Christensen, D., Youmans, K., Khan, M. T., Vicioso, P., Lampkins, S., & Graham, M. C. (2019). A cross-disciplinary and multi-modal experimental design for studying near-real-time authentic examination experiences. *Journal of Visualized Experiments, 151*. https://doi.org/10.3791/60037

Villanueva, I., Raikes, A., Ruben, N., Schaefer, S., & Gunther, J. (2015, October 22–25). The use of physiological tools to identify changes in affective responses for graduate students recently admitted into a scientific discipline. In *Frontiers in education conference (FIE), conference proceedings*, Madrid. https://doi.org/10.1109/FIE.2014.7044316

Villanueva, I., Valladares, M., & Goodridge, W. (2016). Use of galvanic skin responses, salivary biomarkers, and self-reports to assess undergraduate student performance during a laboratory exam activity. *Journal of Visualized Experiments, 2016*(108). https://doi.org/10.3791/53255

Villavicencio, F. T. (2011). Critical thinking, negative academic emotions, and achievement: A mediational analysis. *The Asia-Pacific Education Researcher, 20*(1), 118–126.

Vitasari, P., Wahab, M. N. A., Herawan, T., & Sinnadurai, S. K. (2011). Representation of social anxiety among engineering students. *Procedia – Social and Behavioral Sciences, 30*, 620–624. https://doi.org/10.1016/j.sbspro.2011.10.120

Walther, J., Brewer, M. A., Sochacka, N. W., & Miller, S. E. (2020). Empathy and engineering formation. *Journal of Engineering Education, 109*(1), 11–33. https://doi.org/10.1002/jee.20301

Walther, J., Miller, S. E., & Sochacka, N. W. (2017). A model of empathy in engineering as a core skill, practice orientation, and professional way of being. *Journal of Engineering Education, 106*(1), 123–148. https://doi.org/10.1002/jee.20159

Walther, J., Sochacka, N. W., & Benson, L. (2017). Qualitative research quality: A collaborative inquiry across multiple methodological perspectives. *Journal of Engineering Education, 106*(3), 398–430. https://doi.org/10.1002/jee.20170

Warner, S. A. (2006). Keeping joy in technology education. *Technology Teacher, 65*(7), 6–11.

Watson, D., Clark, L. A., & Tellegen, A. (1988). Development and validation of brief measures of positive and negative affect: The PANAS scales. *Journal of Personality and Social Psychology, 54*(6), 1063.

Webb, D. (2010). Paulo Freire and 'the need for a kind of education in hope.' *Cambridge Journal of Education, 40*(4), 327–339. https://doi.org/10.1080/0305764X.2010.526591

Wells, T. L. (2005). *Emotion and culture in a collaborative learning environment for engineers* [Doctoral thesis, The University of Texas].

Wetherell, M. (2013). Affect and discourse – What's the problem? From affect as excess to affective/discursive practice. *Subjectivity, 6*(4), 349–368. https://doi.org/10.1057/sub.2013.13

WHO. (2022). *Mental health: Strengthening our response*. www.who.int/news-room/fact-sheets/detail/mental-health-strengthening-our-response

Williams, B., Neto, P., & Wankat, P. C. (2014). Taking a snapshot: Four bibliometric indicators to track engineering education research evolution. *International Journal of Engineering Pedagogy*, *4*(4), 16–22. https://doi.org/10.3991/ijep.v4i4.3843

Wilson, K. F., & Wilson, K. (2020). How do students in different minority groups experience first-year engineering? *Australasian Journal of Engineering Education*, *25*(1), 66–78. https://doi.org/10.1080/22054952.2020.1796044

Xu, Y. (2013). The characteristic of college students' employability and its relationship with emotional intelligence. *Journal of Chemical and Pharmaceutical Research*, *5*(12), 70–76.

Zajonc, R. B. (1980). Feeling and thinking: Preferences need no inferences. *American Psychologist*, *35*(2), 151–175. https://doi.org/10.1037/0003-066X.35.2.151

Zeidner, M. (2014). Anxiety in education. In R. Pekrun & L. Linnenbrink-Garcia (Eds.), *International handbook of emotions in education* (pp. 265–288). Routledge.

Zembylas, M. (2004). Emotion metaphors and emotional labor in science teaching. *Science Education*, *88*(3), 301–324. https://psycnet.apa.org/doi/10.1002/sce.10116

Zembylas, M. (2007a). Emotional capital and education: Theoretical insights from Bourdieu. *British Journal of Educational Studies*, *55*(4), 443–463. http://dx.doi.org/10.1111/j.1467-8527.2007.00390.x

Zembylas, M. (2007b). The power and politics of emotions in teaching. In P. A. Schutz & R. Pekrun (Eds.), *Emotion in education* (pp. 293–309). Academic Press. https://doi.org/10.1016/B978-0-12-372545-5.X5000-X

Zembylas, M. (2008). The politics of shame in intercultural education. *Education, Citizenship and Social Justice*, *3*(3), 263–280. https://doi.org/10.1177/1746197908095135

Zembylas, M. (2012). Critical emotional praxis: Rethinking teaching and learning about trauma and reconciliation in schools. In P. Trifonas & B. Wright (Eds.), *Critical peace education*. Springer. https://doi.org/10.1007/978-90-481-3945-3_7

Zembylas, M. (2016). Making sense of the complex entanglement between emotion and pedagogy: Contributions of the affective turn. *Cultural Studies of Science Education*, *11*, 539–550. https://doi.org/10.1007/s11422-014-9623-y

Zembylas, M. (2021). Dewey's account of habit through the lens of affect theory. *Educational Theory*, *71*, 767–786. https://doi.org/10.1111/edth.12505

Zembylas, M., & Chubbuck, S. (2009). Emotions and social inequalities: Mobilizing emotions for social justice education. In P. A. Schutz & M. Zembylas (Eds.), *Advances in teacher emotion research: The impact on teachers' lives* (pp. 343–363). Springer. https://doi.org/10.1007/978-1-4419-0564-2_17

Zembylas, M., & Schutz, P. A. (2016). *Methodological advances in research on emotion and education*. Springer.

Zembylas, M., Theodorou, M., & Pavlakis, A. (2008). The role of emotions in the experience of online learning: Challenges and opportunities. *Educational Media International*, *45*(2), 107–117. https://doi.org/10.1080/09523980802107237

9

Advancing an Integrative Perspective of Identity in Engineering Education

James L. Huff and Monique S. Ross

1 Introduction: Relevance of Identity in Engineering Education

Identity, in the context of engineering education and practice, is a well-studied phenomenon marked by an extensive body of literature. As put by Tonso (2014), "engineering education and engineering identity are two sides of the same coin that mutually shape one another" (p. 278). Identity is a useful conceptual tool to understand chronic problems of engineering education and practice, such as attrition and inequity. Yet the field of identity research in engineering education stands at an important precipice for evolution into how we can leverage identity research to enact systemic change in engineering programs. Identity research will evolve not by simply growing the number of studies on identity in engineering education but by developing an integrated perspective of what it means to become an engineer, by expanding the scope of why we should study identity, and by framing identity studies to cohesively dialogue with the broader literature on professional identity.

On the face of it, identity is a concept that seems readily understood. Identity, as put by Vignoles et al. (2012), "involves peoples' explicit or implicit responses to the question, 'Who are you?'" (p. 2). This definition, albeit concise, illustrates the complexity of processes that occur on an intrapsychic level within individuals who are considering this question. Such a conceptualization recognizes that individuals are engaging this question in ways that enter their consciousness and in ways beneath the surface of their awareness. This conceptualization also highlights that, while identity certainly does involve individuals, broader societal contexts structure or elicit these questions from such individuals. And in relation to the singular–plural ambiguity of the second-person pronoun in the question "Who are you?" identity resides in the agentic and interior worlds of individuals, in the structural and social worlds of groups that comprise such individuals, and in the dynamic dialectic that occurs between agentic individuals and structural contexts (Vignoles et al., 2012). While such a pluralistic conception only amplifies the complexity of the identity construct, it also enables opportunities for expanding the horizon of identity inquiries in engineering education research.

The complexity of identity is useful to understand chronic issues that vex engineering educators and engineering education researchers. How might more students persist in finishing their engineering degrees and entering the workplace? Who do dominant practices of engineering programs or workplaces include and exclude? Researchers may understand these questions and more through the concept of identity. By interrogating the collective and sociocultural narratives of identity that permeate professional spaces, we gain insight into how we might reauthor dominant narratives to

DOI: 10.4324/9781003287483-11

be equitable for individuals of diverse social backgrounds. Investigating how students make sense of their learning and professional practice as a phenomenon of understanding who they are as individuals allows us to elucidate how students can positively construct an identity of who they are as engineers.

However, engineering education researchers need no defense on the relevance and timeliness of investigating identity as a phenomenon. Indeed, we are aware of two previous handbook chapters (Godwin et al., 2020; Tonso, 2014) that characterized the state of the field of engineering identity research and two systematic literature reviews developed for the same purpose (Morelock, 2017; Rodriguez et al., 2018), endeavors that were only possible with a corpus of literature on engineering identity. Each of these reviews addressed diverse theoretical frames of identity, varying from examining the individual to the collective culture that constrains or propels individuals. Authors of these reviews recommended that future research include more theoretical and operational precision in framing identity (Morelock, 2017; Rodriguez et al., 2018), more quantitative studies (Godwin et al., 2020; Morelock, 2017; Rodriguez et al., 2018), and more connection between identity research and interventions that secure a positive and inclusive engineering identity (Godwin et al., 2020; Tonso, 2014).

Studies of identity in engineering education grew substantially in the four years immediately preceding Morelock's (2017) systematic literature review (2010–2014), based on peer-reviewed sources gathered in April 2014. Since 2014, the number of publications related to identity in engineering education has continued to increase dramatically. For example, we gathered all articles, which included *identity* in the title or as an author-supplied keyword, published through 2021 in the *Australasian Journal of Engineering Education, European Journal of Engineering Education*, and the *Journal of Engineering Education* (n = 54). This international sample journals connected to engineering education professional societies (i.e., AAEE, SEFI, ASEE) published approximately 75% of these articles centered on engineering identity (n = 40) between 2015 and 2021 and 50% (n = 27) between 2019 and 2021. Such a publication trend demonstrates how engineering education researchers have increased their collective examination of identity as a way to understand relevant phenomena within engineering education research and practice.

In the sections that follow, we explore the state of engineering education research by examining three key perspectives of understanding identity as (1) personal, (2) social, and (3) sociocultural. Beyond simply describing these frameworks, we evaluate their treatments of systemic problems in engineering education (e.g., persistence, equity, engagement). We then illustrate the benefits and trade-offs of each identity perspective by considering how each framing might influence the study design and knowledge claims of a common example – the case of studying marginalized students. Finally, we conclude with recommendations on how the already-robust field of identity research in engineering education and practice can evolve to address chronic and systemic issues of engineering education.

2 Theoretical Perspectives of Identity

We organize our review of literature on engineering identity by discussing three perspectives with which investigators might examine the construct: as a personal phenomenon experienced by an individual, as a social phenomenon in which an individual adopts a role within a context, and as a sociocultural phenomenon in which the context is at the center of the investigation. These three broad perspectives are not meant to describe the totality of identity conceptualizations. However, they do align with not only dominant frameworks in the field of engineering education research (Villanueva & Nadelson, 2017) but also with categorizations in the field of identity research more generally (Vignoles et al., 2012). For each perspective, we discuss key theoretical conceptualizations and demonstrate how these perspectives manifest in the engineering education literature (refer to

Table 9.1 Perspectives of Identity in Engineering Education Research (EER)

Theoretical Perspective	*Central Question in EER*	*Nature of Identity*	*Influential Frameworks in EER*
Personal	Who am *I* as an engineer?	Individuals personally construct or experience identity.	Erikson, 1959; also Arnett, 2004; Marcia, 1966; McAdams, 2012; Oyserman & James, 2012.
Social	Who am I *as an engineer*?	Social contexts primarily constrain or otherwise define identities that individuals adopt (or not).	Tajfel & Turner, 1986; Hogg & Terry, 2000; Stets & Burke, 2000; also Gee, 2000; Carlone & Johnson, 2007.
Sociocultural	Who are *we (or they) as engineers*?	Sociocultural contexts construct identity for groups of individuals.	Holland et al., 2001; also Faulkner, 2000; Gee, 2000; Stevens et al., 2008.

Table 9.1). Additionally, we explore the ways that each perspective is useful to investigate identity, depending on the aims of the particular studies.

2.1 Personal Identity

In his theory of psychosocial development, Erik Erikson (1959) framed identity as a *personal* phenomenon. Specifically, he understood identity to be the central focus of an individual in the stage of adolescence, and that through this formative period, an individual would define themself in a consistent, unitary manner. Although other perspectives on identity later emerged to frame the construct as something that was more defined by social and contextual factors (Huff, 2019; Tonso, 2014), Eriksonian perspectives on identity that center on the individual have persisted in psychological literature.

Building on Erikson's work, Marcia (1966) developed the concept of identity statuses as a way to recognize how individuals engage with their identity through both commitment and exploration. Further, Kroger et al. (2010) named three significant dimensions of identity with which individuals interact through commitment and exploration: vocation, ideology, and sexuality. Arnett (2004) extended identity explorations to be the core feature of emerging adulthood, that is, the period following adolescence in which individuals are seeking to explore identities and achieve autonomy in relation to their social environment. Arnett further highlighted that emerging adults are particularly attentive to exploring identity in relation to love and work. Erikson's conceptualization of identity also influenced subsequent theorization of possible selves (Oyserman & James, 2012), that is, the identities of the future that an individual envisions for themself. Such a concept readily links to one's motivation to become a certain identity, including through career pathways.

However, while personal identity theory recognizes that individuals do shape their understandings of themselves through contexts such as the workplace, there is little research in psychology that examines how individuals integrate personal identity across multiple domains. One useful construct to considering how an individual's core personal identity might interact with their professional identity is contextual identity integration, which "involves the fit of the multiple identity domains that individuals consider important to who they are, or are forced to deal with due to social-structural factors" (Syed & McLean, 2016, p. 111). Syed and McLean (2016) develop their general understanding of identity integration as a phenomenon connected to McAdam's (2012) framing of narrative identity. Indeed, the perspective of narrative identity is quite distinct from the previously discussed work of Marcia (1966) as it defines *identity* to be "an internalized and evolving story of the self, providing a person's life with some semblance of unity, purpose, and meaning" (McAdams, 2012, p. 100). From such a perspective, an individual's roles in multiple contexts might cohesively merge through the narrative that they develop to unify how they enact identity in contextual domains.

Thus, from a personal identity perspective, one's identity in relation to their profession is a dimension of a holistic, personal identity. When approaching engineering identity research from a personal identity orientation, an investigator seeks to understand how individuals engage the question "Who am *I* as an engineer?" The study's focus is on the individual, and the context of participation in engineering education and practice is peripheral to understanding something more holistic than professional practice alone. Furthermore, an individual's goals in exploring their identity in relation to their profession are more related to understanding their overall sense of identity rather than simply developing as a proficient professional.

Adopting a personal identity perspective in engineering education research is useful when we want to understand identity as a developmental phenomenon that occurs within individuals. For example, in their study on how first-year engineering students identified with their field, Meyers et al. (2012) built on Arnett's (2004) conceptualization of emerging adulthood to survey the extent that students were exploring or committing to the identity of being an engineer. Although the goal of their study was to conceptualize an Eriksonian stage model of identity development, their results did not support a clear pathway. However, they did demonstrate a positive association between persistence in the plans to work as an engineer and the extent to which students identified with this profession. Anchored in the concepts of identity commitment and exploration (Marcia, 1966; Luyckx et al., 2008), Li et al. (2021) demonstrated the validity of applying an adapted form of the Utrecht-Management of Identity Commitments Scale (U-MICS; Crocetti et al., 2008) to assess identity formation of engineering students. Both of these works envisioned identity as a developmental phenomenon on the terms of student, and both administered surveys that exclusively assessed identity in the particular context of being and becoming an engineer.

Personal identity has been functional for engineering education research that is also interested in answering research questions related to motivation (Bennett & Male, 2017; Godwin & Kirn, 2020; Kajfez et al., 2016). For example, in their study that assessed the motivation and teacher-identity formation of engineering graduate teaching assistants, Kafjez et al. (2016) generated a context-specific developmental model called the longitudinal model of motivation and identity (LMMI). Anchored in the foundations of the possible selves theory (Oyserman & James, 2012) and self-determination theory (Deci & Ryan, 2002), this model demonstrated how future versions of a faculty identity were synergistic with motivational constructs of graduate teaching assistant performance (e.g., autonomy, competence, relatedness).

Personal identity is particularly conducive to examine identity as a holistic phenomenon in an individual. In their interpretative phenomenological analysis (IPA) study of identity in engineering students who transitioned from school to workplaces, Huff et al. (2019) found their participants to demonstrate a holistic sense of commitment to identity through their pursuit to practice engineering but a delayed sense of exploration in relation to non-professional domains of their identities. Connecting their findings to theoretical perspectives of Arnett (2004), Marcia (1966), and McAdams et al. (1996), Huff et al. (2019) questioned the overall benefits of rushing to commitment toward engineering. Such rigid identity commitment might come at the expense of overall identity development, which occurs amid social relationships that intensive professional education settings generally undervalue. Huff and colleagues (2021) build on this work to further examine the experience of *professional shame* in engineering education, an emotional state that involves a feeling that the whole, or global, self is inadequate when failing to meet perceived identity-relevant standards in a professional domain.

Studies adopting a personal identity approach have yielded many insights into the power of identity development as a personal and agentic journey. By understanding the complexity of identity in the human stakeholders of engineering education and practice, we may better understand how to care for these stakeholders' holistic needs rather than only improving their professional competencies.

2.2 Social Identity

Social identity provides a framework of how one understands who they are in relation to others in a group context. Tajfel and Turner (1986) understood social identity as a way to characterize how individuals evaluate and ultimately classify themselves as belonging to certain groups or not. Through social comparison, an individual self-categorizes as either being similar enough to be part of the in-group or different enough to be part of the out-group (Stets & Burke, 2000). This social comparison can either result in an affirmation of their status in the in-group or accentuate the differences between the self and the in-group (Hogg & Terry, 2000; Stets & Burke, 2000).

Claims to a social identity mean being like others in the group, often resulting in uniformity of perception and action among group members. Such uniformity can yield social stereotyping of in-group and out-group members, which, in turn, fosters homogeneity (Haslam et al., 1999; Stets & Burke, 2000). This homogeneity then reinforces the stereotypes and reproduces and validates the norms and values of the group. The negotiation of the individual with group engagement or membership is a function of self-categorization, often described as belongingness, which aids an individual in their self-categorization. If a person determines that they do indeed belong to or fit with the group, they will have achieved identity salience or the activation of an identity (Stets & Burke, 2000).

Upon activation of a social identity, the person must assume the group's norms and values and also function to maintain the group's identity standard. The individual becomes categorized into a certain predefined and socially constructed role, which forms the parameters of how others might recognize them and how they might perform that role. This activated identity provides a definition of who a person perceives they are in terms of the defining characteristics of the social identity they have assumed (Stets & Burke, 2000). Social identities are then not only descriptive and prescriptive but also evaluative in nature – meaning, they regulate behaviors and norms in that social category. Specifically, social identities lead members of that category to not only adopt the behaviors and norms but also maintain them across the group (Terry et al., 1999). Thus, a social identity perspective helps us understand how group identities guide choices and behaviors and provide strength in the collective. However, this theoretical perspective also provides context for the oftentimes-destructive nature of maintaining norms and values in a categorization.

While not explicitly linked to the body of research stemming from the social identity conceptualization of Tajfel and Turner (1986), we might also classify Gee's (2000) perspective of identity as having similar aims of the social identity framework. Gee (2000) conceptualized identity as anchored in four dimensions: nature, institutions, discourse, affinity. Although his work stratifies the social and sociocultural perspectives of identity research, Gee (2012) provided a means of connecting the four dimensions to inclusive and exclusive practices that could foster beliefs around belonging to the in-group. He gave a concrete means of understanding how language, for instance, can act as an invitation to be part of the in-group or how nature or our natural identities can serve as a barrier to the in-group. Investigators who adopt Gee's perspective of identity, particularly in science education (Carlone & Johnson, 2007) and engineering education, are concerned with how an individual navigates the institutional, affinity, or discursive contexts of engineering programs and workplaces (Allie et al., 2009; Blair et al., 2017; Capobianco et al., 2012; Faber et al., 2020; McNair et al., 2011), in particular, whether their appearance, engagement, or lexicon afford them membership to engineering in-groups.

The social identity perspective accounts for the dominant trend of engineering identity research, seeking to answer the question, "Who am I *as an engineer*?" Engineering education researchers who have conducted studies from a social identity perspective have made some of the boldest advancements in developing identity theory within engineering education contexts. For example, by adapting Carlone and Johnson's (2007) model of science identity, which included three components of performance, recognition, and competence, Godwin et al. (2016) administered a large-scale survey study to demonstrate that engineering identity was well-established based on three interrelated components: *performance/competence, interest, recognition*. Their work gave the engineering education research community a means of measuring engineering identity with strong validation evidence for items in a survey instrument to measure the extent that individuals identified as engineers.

A social identity perspective is useful to characterize how individuals navigate systems of engineering education and practice, emphasizing not only the individuals themselves but also the ways that a student or professional interacts with the group identities within engineering cultures. In their grounded theory investigation of civil engineering students, McCall et al. (2020, 2021) sensitized themselves to social identity perspectives to generate contextually robust claims of how civil engineering students negotiated their identities amid their professional socialization. Their findings yielded a model for understanding the specific identity navigations of students based on the educational processes that occur within engineering programs. By making the negotiation of identity visible, their model is meant to "challenge existing sociocultural norms that constrain conceptions of what engineering is and who can enter the profession" (McCall et al., 2021, p. 394).

The social identity perspective also anchors engineering education research that examines how individuals navigate participation in multiple groups, particularly systemically marginalized identities in engineering degree programs and workplaces. In their IPA study of Black women engineers, Ross et al. (2021) examined the lived experiences of identity in participants who identified as Black, as women, and as engineers. Rather than framing the participants' membership in historically marginalized social categories of race and gender, Ross et al. (2021) demonstrated how the intersectional power of their participants' identifications – as Black, women, and engineers – forged identities of resilience amid the hardships they encountered in engineering workplaces.

Social identity provides a means of understanding a person's ability to navigate the complex relationship between self and acceptance to the engineering in-group. This perspective provides a means to understand identity development and salience with careful consideration for the impact and influence of norms, values, acceptance, and rejection within the engineering community.

2.3 Sociocultural Identity

As previously reviewed by Tonso (2014), key theoretical movements within the fields of anthropology and cultural studies informed the foundations of engineering identity research from a sociocultural perspective. One key framework that remains influential in engineering education research is Holland et al.'s (2001) model of cultural productions of identity, which emphasizes the ways that individuals are both identified (by others) and identify (themselves) with certain culturally produced identities. Holland et al.'s (2001) conceptualization of identity informs the research focuses of multiple investigations in engineering education research (Bahnson et al., 2021; Gonsalves et al., 2019; Stevens et al., 2008), and their framework influenced Stevens et al.'s (2008) well-regarded concept of *Becoming an Engineer*, in which they defined identity as "a double-sided process of positioning ourselves and being positioned by others" (p. 357).

Specifically, in their person-centered ethnography, Stevens et al. (2008) captured how engineering identities changed over time with sensitivity to how institutional programs shaped the identifications of their members. Differences in this conceptualization from the personal and social identity perspectives are subtle but important. Personal identity focuses on individuals and their internalized conceptualizations of who they are. Social identity focuses on a person in relation to others and the processes that accompany in these interactions. Sociocultural identity demonstrates a strong focus on the structural environment of programs operating on the identity processes of individuals (Stevens et al., 2008). For example, using a sociocultural identity perspective, Stevens et al. (2008) challenged programs to afford inclusive pathways to becoming an engineer rather than the once-dominant and restrictive metaphor of the pipeline. This framing of engineering identity from a sociocultural perspective has served as the basis of other investigations that demonstrate sociocultural claims and criticisms of engineering programs (Danielak et al., 2014; Sochacka et al., 2016).

From a sociocultural perspective of identity, we might learn how dominant forces in engineering cultures structurally constrain the full participation of individuals with minoritized identities. Here, the boundary between examining the culture and the identities that are privileged or marginalized within the culture becomes blurry. Prior research on sociocultural identity in engineering education tends to examine how institutional structures shape a collective engineering identity, interrogating the question, "Who are *we (or they)* as engineers?"

Engineering education research that examines identity from a sociocultural position often aims to elucidate systemic explanations of how powerful social forces that dominate the collective identity of engineers tend to marginalize students or professionals (Du, 2006; Faulkner, 2000, 2007; Pawley, 2009; Secules et al., 2018; Sochacka et al., 2021; Tonso, 2006). For example, Faulkner's (2000, 2007) ethnographic research on gendered patterns of engineering workplaces frames engineering identity with a clear demarcation of the technical and social, noting how discourse of engineering identity (Gee, 2000, 2012) tends to privilege technical features while excluding social elements that are, in fact, deeply intertwined with engineering work (Trevelyan, 2007). Tonso's (2006) ethnographic investigation into engineering teams highlights the idiosyncratic and gendered identities of engineering students within a specific campus culture as a way of illuminating who belonged to the culture of engineering at that campus. And Downey, Lucena, and their colleagues completed several studies from an anthropological perspective that demonstrated the identities of engineering based on how various nations identified explicit and implicit competencies (Downey & Lucena, 2004; Lucena et al., 2008).

While a sociocultural approach might elucidate identity development of an individual, it does so by foregrounding an understanding of the cultural setting in which identity processes occur. Such approaches prioritize an understanding of the context and structures as a mode of understanding those externals factors on identity development.

3 Identity Framing at Work: Students from Marginalized Backgrounds

To illustrate the benefits and trade-offs of each of the previously described ways of framing identity, we examine the possible choices that an investigator could make in designing a study to examine the identity experiences of students who hold at least one identification in a marginalized social category in engineering education. We demonstrate this example based on the proliferation of such studies in the engineering identity literature (Rodriguez et al., 2018). In this discussion, we consider the nature of the social reality under investigation as a way to examine *theoretical validation*, a core investigative choice that anchors subsequent decisions about research methods (Walther et al., 2013, 2017). Our goal is not to discuss methodological choices of the investigation but rather to recognize how any of the aforementioned frameworks might anchor an identity study, with each framework establishing utilitarian benefits and trade-offs. Because we are examining the context of marginalization, we begin our discussion by first considering the dominant treatment to such studies, that is, a social identity perspective. Then, we move towards the two other perspectives to understand the same context with different ways framing the social reality.

Adopting a ***social identity*** perspective to understand the experiences of students from multiple marginalized backgrounds represents the ostensible choice for such an investigation. Indeed, such a perspective, anchored in understanding the person in relation to a specific context, undergirds the assumptions already conceptualized when identifying the choice to study marginalized individuals in engineering education. First, such phrasing recognizes individuals based on their connection to a broader social identity of groups. At the outset, social identity informs our assumptions, not only in relation to engineering itself, but also in relation to the intersections of identities of gender, race, ethnicity, sexual orientation, first-generation status, or any other number of marginalized identifications in engineering education. Second, by framing identifications as *marginalized*, we are designating certain social identifications as members of the *out-group* in engineering education, a concept that is central to social identity theory (Tajfel & Turner, 1986).

The choice to frame our phenomenon through a social identity perspective carries some advantages that could advance important knowledge claims in engineering education research and practice. For example, if the underlying goals of our investigation are to bolster retention of students within engineering degree programs, then a social identity perspective can help us examine the relationship between engineering programs, as they currently stand, and students' choices to persist with or depart such programs. Furthermore, if the goal of our study is to align the identifications of students with an aspirational vision of what it means to be an engineer (e.g., for students to pursue engineering practice), then a social identity perspective perhaps best supports such an investigation of identity.

However, if we define *engineering identity* too narrowly, adopting a social identity perspective can reify additional barriers to participation for those who are seen as and see themselves as deviating from the norms, values, and perceptions of who can perform in engineering. Prior work suggests that majority-shaped perceptions of what it means to be an engineer demonstrate gendered and classist assumptions (Berge et al., 2019). Further, the absence of equity for African American, Indigenous, Latiné/x/a/o, and Hispanic students in engineering education leaves a racialized landscape (Harper et al., 2009; Holly et al., 2022). Extant research on identity has taken care to explore ontological beliefs about what or who an engineer is, and such inquiry gave us a view into norms, values, and perceptions of the discipline (Hatmaker, 2013; Jorgenson, 2002). Likewise, deconstructing the identity framework further into subconstructs (e.g., interest, performance/competence, and recognition) gave us a means of parsing out the factors that contribute to or detract from the development of an engineering identity (Capobianco et al., 2012; Godwin et al., 2016; Godwin & Kirn, 2020; Patrick et al., 2018). But how has choosing to adopt an engineering identity pushed students to make a binary choice between being an engineer and upholding other marginalized identities in engineering education and practice? Ross et al. (2021) described how Black women that were in the engineering industry for ten or more years merged their social identities (Black and woman) with their engineering identity, but what about those who did not remain in the engineering industry? Could it be that they were not willing or able to merge their identities and instead grappled with the incongruence between who they were and who they wanted to be? When adopting a social identity perspective to frame investigations, investigators might inadvertently perpetuate narratives of marginalization by upholding the engineering group identity as fixed and unchangeable while pushing students to conform to an identity that causes them to sacrifice central elements of who they are.

By adopting a ***sociocultural identity*** perspective, investigators may focus on the socially constructed realities that constrain marginalized individuals who are attempting to navigate systems of engineering education rather than on the identities of individuals themselves. The roles of the individuals in the study are to inform and demonstrate a broader sociocultural environment that shapes or constrains identity, in this case, how systems of engineering education serve to marginalize or privilege individuals based on social identifications.

Studying identity phenomena from a sociocultural identity perspective is useful when the investigator aims to demonstrate claims related to how sociocultural norms and messages of engineering education programs (e.g., Faulkner, 2007; Pawley, 2009; Secules et al., 2018; Sochacka et al., 2021). Moreover, in the case of studying the experiences of students who come from marginalized backgrounds, such a perspective can anchor the investigation to understand the social systems of education that are doing the marginalizing. If we desire to promote education systems that are equitable and just for the students that comprise them, then a sociocultural perspective can inform us of how our messages and norm-setting behaviors may be problematic to individuals who do not align with the dominant narratives of engineering education and practice. Adopting sociocultural perspectives of identity comes with compelling affordances that can lead to improving systems of education.

However, such a perspective also comes with trade-offs related to its intentional de-emphasis of individuals within sociocultural systems. For engineering educators who work with students, there is an immediate need to know how to support the choices, success, and agency of individual students who are navigating systems that are marginalizing to them. Sociocultural perspectives of identity illuminate problematic features of dominant narratives that pervade engineering education and remove the invisibility of such narratives in ways that undermine their dominance (McLean & Syed, 2015). Yet such a perspective can leave the individual needs of the marginalized student unmet by the education system that they must navigate. Thus, when adopting a sociocultural perspective of identity, we encourage investigators to connect their findings to engineering education research that does highlight strategies which support individual and agentic needs in navigating engineering education.

Finally, by adopting a ***personal identity*** perspective on the identity experiences of students from marginalized social backgrounds, investigators can emphasize the lived experiences of such students on their own terms rather than limiting the relevance of their identities to how they participate in engineering education and practice. Such a choice in framing identity allows for the investigator to regard the individual students as a whole person who is developing in ways that are more comprehensive and more complex than their development of competencies as engineers. Such a framework may also allow investigators to examine features of developing a positive identity through identity commitment (Li et al., 2021) or sustaining motivation to become engineers (Godwin & Kirn, 2020; Kajfez et al., 2016), but we encourage investigators to consider how such constructs operate within the holistic development of the individual rather than their development in a professional domain. Thus, career choice and persistence or engagement in a major are manifestations of overall human development rather than the end outcomes themselves. Upholding a personal identity perspective enables the investigator to honor that individuals carry interior worlds that hold relevance beyond their professional identities. Such framing is conducive to producing insights of how students can exercise personal agency in making career decisions or the ways in which an individual can navigate choices that facilitate emotional resilience and well-being (Huff et al., 2021). By understanding the personal identities of engineering students, we come to regard them as holistic individuals with human needs. Such understanding is necessary if we want to understand how engineering education programs can foster holistic care and compassion for the individuals therein.

The trade-offs of adopting a personal identity perspective lie on the reverse side of the benefits of maintaining a social or sociocultural perspective on identity. The emphasis on whole individuals, particularly among those marginalized in engineering education and practice, may come at the cost of not recognizing or challenging the systems that embed them. While identity reflects processes that are well-understood to occur within individuals, focusing narrowly on intrapsychic experiences means that we may risk focusing on the failures or shortcoming of the individuals rather than challenging the failures of the systems. Furthermore, adopting a personal identity perspective may lead to outcomes that meet the needs of whole individuals but at the expense of practical outcomes of education, such as supporting student success and retention within a major. Therefore, we encourage researchers who adopt a personal identity perspective to develop their knowledge claims in dialogue with other approaches to studying engineering identity that emphasize social or sociocultural perspectives. We further amplify the need for investigators to frame personal identity with an anti-deficit approach that explores the assets of these students and their approaches to persistence and success in engineering (Harper, 2010) on the terms of their lived experience and not on the terms of engineering professional standards.

In summary, the framing by which investigators understand identity significantly limits or advances how we produce relevant theoretical understanding in our studies, regardless of our methodological choices. Which perspective is the most suitable for advancing research on identity in engineering education and practice? We contend that each of these perspectives carries affordances and tradeoffs. As such, researchers should align their identity perspectives choice with their goals. We do not recommend that investigators attempt to take on all perspectives in a singular investigation but rather demonstrate self-awareness as to how one's study is in dialogue with other research in engineering education that frames identity from diverse theoretical perspectives.

4 Recommendations for Identity Research in Engineering Education

The body of identity literature in engineering education research has advanced from a topical interest of few scholars to a wide and diverse field of research. In the following subsections, we offer three recommendations on how engineering identity research might evolve to a transformative force to develop well-being of engineering individuals and justice and advance equity within engineering cultures.

4.1 Cultivate an Integrated Perspective on Identity in Engineering: Moving Beyond a Definition and Defense

As we demonstrated in the previous section, identity is a multifaceted construct that we cannot fully understand by any singular perspective. We have reviewed three perspectives that we believe to represent the existing body of engineering education research, but more perspectives (e.g., neuroscientific, postmodern) serve to advance knowledge claims beyond the scope of what we have reviewed here. We encourage investigators, when designing inquiries, to locate the core theoretical premises that undergird how they understand the identity phenomenon. While engineering education researchers have demonstrated a keen awareness of how they theoretically frame identity, such framing often appears in a way that defends the rationale for choosing a particular framework to investigate research questions. Although this approach is helpful in guiding the reader towards understanding the benefits of a theoretical framework, such argumentation often minimizes and overshadows the limitations of the perspective.

Rather than defending frameworks as comparatively superior to others, we can include clear, reflective exposition of how our theoretical perspectives allow us to focus on certain aspects of social realities while acknowledging and deliberately stating that such a focus moves some features of identity to the periphery. The defense of the theoretical positioning is critical to understanding the approach, but we should clearly state what we sacrifice by this decision.

We do call for an integrated perspective of identity in engineering, but this is a feat that one investigator or one article cannot accomplish. Rather, as we discuss our knowledge claims based on empirical findings for a particular study, we can advance toward an integrative perspective of identity in engineering education by positioning our work in dialogue with related work that adopts different perspectives of identity.

4.2 Expand the Motivations for Studying Identity: Leaning Into Anti-Deficit Framing

The body of research on identity in engineering education primarily focuses on possible explanations for why students may engage, persist in, or be retained by the profession. The approach to such studies ranges from examining the systemic inequities within engineering cultures to advancing psychological perspectives of why students may or may not identify as – and thus continue their professional formation as – engineering students or engineers. Such an impetus for identity research certainly serves to advance the profession of engineering, but it presumes that students becoming engineers is an inherently positive outcome.

Conducting identity research to foster perseverance in students or to create more inclusive spaces for students are worthwhile goals of engineering education research, but we must now push further to expand into other branches of identity theory that do not presume that engineering identity is a decisively positive achievement. Leveraging the previously discussed perspectives, coupled with an anti-deficit framing, we might grow the body of knowledge around those that do not identify as an engineer or reject the identity due to incongruence with their other identities. Such an investigative orientation might result in questions like: How do students and professionals distance themselves from what it means to be an engineer? How might we promote healthy decision-making processes for leaving engineering education and practice rather than assuming that all forms of leaving are negative? When students leave engineering education or practice, how might their identities as engineers transform subsequent domains of education or practice? How might we promote the well-being of all students and professionals in engineering, regardless of the extent to which they commit to the field?

Furthermore, we might strengthen the field of identity research by examining the assets of students from systemically marginalized groups rather than only on their marginalization. Such a consideration builds on asset-focused counternarratives of identity (e.g., Rincón & Rodriguez, 2021; Ross et al., 2021; Verdín, 2020). What roles can marginalized students assume within engineering? In absence of ascribing to the dominant narratives of engineering identity, might we instead explore what they wish to accomplish and contribute to elevate the discipline without restricting the boundaries of how they assume such an identity?

The dominant focus of prior identity research has helped us understand engagement, retention, and persistence of engineering students. However, what if this commitment to advancing the engineering profession has also acted as an impediment to engagement for those who find themselves oppressed by the system? Expanding the body of work in the realm of sociocultural identity could yield a more in-depth understanding of how engineering cultures forge or dismantle identities within marginalized people. In what ways does oppression manifest in engineering education? Likewise, inquiry into personal and social identity experiences can advance understanding of the discipline from a nondominant perspective, yielding the nuance needed to actively make engineering education more equitable based on the needs of individual and the social identifications they represent.

By opening the scope of identity research to address a range of questions that do not presume the goal of advancing engineering identities as a positive outcome, we do not compromise the work that does seek to promote such identity development. Rather, we open the scope of ways that engineering education can facilitate positive spaces for overall human development, beyond but also including individuals' developmental trajectories of who they become as engineers.

4.3 Build a Cohesive Field of Scholarship on Identity in Engineering

Finally, we recommend advancing the scope of identity research by committing to dialogue with other research that makes claims about engineering identity. Beyond ensuring that we cite other engineering education research, this commitment to dialogue means that we allow other perspectives in engineering identity to inform our theoretical framing and that our findings dialogue with the claims of others when we make sense of the findings. In other words, we engage other engineering identity not only in the *background literature* section but also in the *theoretical framework* and *discussion* sections that commonly limit the treatment of literature to core frameworks of identity outside of engineering education research. In short, we should ensure that we support our methodological choices with the work of other identity scholars in engineering education when framing our work and when making sense of our findings. How does our work confirm, conflict, and extend the body of knowledge on identity in engineering education?

Committing to such a dialogue holds critical implications for how we report the findings of empirical studies on identity. Authors should clearly state their understandings of identity and their knowledge claims related to identity. In the extant literature on identity in engineering education, many authors describe what they learned about identity-relevant constructs, but not always on how the results shaped their theoretical understandings of identity itself. Have their findings confirmed or extended their presuppositions of identity? Or do they stand in tension with prior understandings of identity? Such trend in engineering education research may result from the dominance of qualitative studies on identity in engineering education (Godwin et al., 2020; Morelock, 2017; Rodriguez et al., 2018) that intentionally do not generate broad-sweeping claims through statistical inferencing. As qualitative researchers ourselves, we contend that while qualitative research should uphold an idiographic focus of generating knowledge claims, it should still advance concerns of generalization – not through making inferential claims of how broadly applicable the findings are, but by enriching contextual insights that position the findings in dialogue with other insights of engineering education research (Kirn et al., 2019; Smith et al., 2022) and by extending our understanding of engineering identity. Regardless of our chosen methodologies, the substance of our discussions should not stop at summarizing descriptive findings but should also include novel claims about the identity frameworks that informed our investigations.

5 Concluding Statement

In engineering education research, we stand at a critical opportunity to expand the scope of how we understand identity in the context of the engineering profession. Such a task is not possible by an isolated investigator but rather by a body of complex yet cohesive work. In quick fashion, the field of engineering education research has converged onto the construct of identity to primarily identify explanations that support retaining students on their pathways to becoming engineers. We instead wish to amplify the perspective of Hanson (2014):

> When we think of students as a human form of capital, the view potentially restricts our intellectual terrain. We run the risk of limiting ourselves to questions about what students know or how they perform prescribed tasks. We lose sight of the notion that schools allow people to forge new selves.
>
> *(p. 10)*

When we examine identity as a phenomenon of retention alone, we limit our understanding of how we can affect holistically positive change for engineering programs and students. By cultivating an integrative perspective on identity research, expanding our motivations to study identity, and committing to deeper dialogue within the field engineering identity research, we advance a transformative, equitable, and caring system of engineering education.

Acknowledgments

We are grateful to the anonymous reviewers and to Jillian Seniuk Cicek for their constructive dialogue with us, which sharpened both our thinking on identity and the overall chapter. We are grateful to anonymous reviewers from the 2019 Research in Engineering Education Symposium, whose feedback shaped early portions of the ideas contained in the section on personal identity. Finally, we express our gratitude to the editorial team of this handbook, especially to Aditya Johri, for the invitation and overall vision for this chapter.

This work was supported through funding by the National Science Foundation (Awards Nos. 1329225, 1748384, 1752897, 1845884, 2045392). Any opinions, findings, and conclusions or recommendations expressed in this material are those of the author(s) and do not necessarily reflect the views of the National Science Foundation.

References

Allie, S., Armien, M. N., Burgoyne, N., Case, J. M., Collier-Reed, B. I., Craig, T. S., . . . Wolmarans, N. (2009). Learning as acquiring a discursive identity through participation in a community: Improving student learning in engineering education. *European Journal of Engineering Education, 34*(4), 359–367. https://doi.org/10.1080/03043790902989457

Arnett, J. J. (2004). *Emerging adulthood: The winding road from the late teens through the early twenties*. Oxford University Press.

Bahnson, M., Perkins, H., Tsugawa, M., Satterfield, D., Parker, M., Cass, C., & Kirn, A. (2021). Inequity in graduate engineering identity: Disciplinary differences and opportunity structures. *Journal of Engineering Education, 110*(4), 949–976. https://doi.org/10.1002/jee.20427

Bennett, D., & Male, S. A. (2017). An Australian study of possible selves perceived by undergraduate engineering students. *European Journal of Engineering Education, 42*(6), 603–617. https://doi.org/10.1080/03043797.2016.1208149

Berge, M., Silfver, E., & Danielsson, A. (2019). In search of the new engineer: Gender, age, and social class in information about engineering education. *European Journal of Engineering Education, 44*(5), 650–665. https://doi.org/10.1080/03043797.2018.1523133

Blair, E. E., Miller, R. B., Ong, M., & Zastavker, Y. V. (2017). Undergraduate STEM instructors' teacher identities and discourses on student gender expression and equity. *Journal of Engineering Education, 106*(1), 14–43. https://doi.org/10.1002/jee.20157

Capobianco, B. M., French, B. F., & Diefes-Dux, H. A. (2012). Engineering identity development among pre-adolescent learners. *Journal of Engineering Education, 101*(4), 698–716. https://doi.org/10.1002/j.2168-9830.2012.tb01125.x

Carlone, H. B., & Johnson, A. (2007). Understanding the science experiences of successful women of color: Science identity as an analytic lens. *Journal of Research in Science Teaching: The Official Journal of the National Association for Research in Science Teaching, 44*(8), 1187–1218. https://doi.org/10.1002/tea.20237

Crocetti, E., Rubini, M., & Meeus, W. (2008). Capturing the dynamics of identity formation in various ethnic groups: Development and validation of a three-dimensional model. *Journal of Adolescence, 31*(2), 207–222. https://doi.org/10.1016/j.adolescence.2007.09.002

Danielak, B. A., Gupta, A., & Elby, A. (2014). Marginalized identities of sense-makers: Reframing engineering student retention. *Journal of Engineering Education, 103*(1), 8–44. https://doi.org/10.1002/jee.20035

Deci, E. L., & Ryan, R. M. (2002). *Handbook of self-determination research*. The University of Rochester Press.

Downey, G. L., & Lucena, J. C. (2004). Knowledge and professional identity in engineering: Code-switching and the metrics of progress. *History and Technology, 20*(4), 393–420. https://doi.org/10.1080/0734151042000304358

Du, X. Y. (2006). Gendered practices of constructing an engineering identity in a problem-based learning environment. *European Journal of Engineering Education, 31*(1), 35–42. https://doi.org/10.1080/03043790500430185
Erikson, E. H. (1959). *Identity and the life cycle. Psychological issues 1.* International Universities Press.
Faber, C. J., Kajfez, R. L., McAlister, A. M., Ehlert, K. M., Lee, D. M., Kennedy, M. S., & Benson, L. C. (2020). Undergraduate engineering students' perceptions of research and researchers. *Journal of Engineering Education, 109*(4), 780–800. https://doi.org/10.1002/jee.20359
Faulkner, W. (2000). Dualisms, hierarchies, and gender in engineering. *Social Studies of Science, 30*(5), 759–792. https://doi.org/10.1177/030631200030005005
Faulkner, W. (2007). "Nuts and bolts and people": Gender-troubled engineering identities. *Social Studies of Science, 37*(3), 331–356. https://doi.org/10.1177/0306312706072175
Gee, J. P. (2000). Chapter 3: Identity as an analytic lens for research in education. *Review of Research in Education, 25*(1), 99–125. https://doi.org/10.3102/0091732X025001099
Gee, J. P. (2012). *Social linguistics and literacies: Ideology in discourses* (4th ed.). Routledge.
Godwin, A., Cribbs, J., & Kayumova, S. (2020). Perspectives of identity as an analytic framework in STEM education. In C. C. Johnson, M. J. Mohr-Schroeder, T. J. Moore, & L. D. English (Eds.), *Handbook of research on STEM education* (pp. 267–277). Routledge.
Godwin, A., & Kirn, A. (2020). Identity-based motivation: Connections between first-year students' engineering role identities and future-time perspectives. *Journal of Engineering Education, 109*(3), 362–383. https://doi.org/10.1002/jee.20324
Godwin, A., Potvin, G., Hazari, Z., & Lock, R. (2016). Identity, critical agency, and engineering: An affective model for predicting engineering as a career choice. *Journal of Engineering Education, 105*(2), 312–340. https://doi.org/10.1002/jee.20118
Gonsalves, A. J., Silfver, E., Danielsson, A., & Berge, M. (2019). "It's not my dream, actually": Students' identity work across figured worlds of construction engineering in Sweden. *International Journal of STEM Education, 6*(13), 1–17. https://doi.org/10.1186/s40594-019-0165-4
Hanson, C. (2014). Changing how we think about the goals of higher education. In C. Hanson (Ed.), *In search of self: Exploring student identity development* (pp. 7–14). Jossey-Bass.
Harper, S. R. (2010). An anti-deficit achievement framework for research on students of color in STEM. *New Directions for Institutional Research, 2010*(148), 63–74. https://doi.org/10.1002/ir.362
Harper, S. R., Patton, L. D., & Wooden, O. S. (2009). Access and equity for African American students in higher education: A critical race historical analysis of policy efforts. *The Journal of Higher Education, 80*(4), 389–414. https://doi.org/10.1080/00221546.2009.11779022
Haslam, S. A., Oakes, P. J., Reynolds, K. J., & Turner, J. C. (1999). Social identity salience and the emergence of stereotype consensus. *Personality and Social Psychology Bulletin, 25*(7), 809–818. https://doi.org/10.1177/0146167299025007004
Hatmaker, D. M. (2013). Engineering identity: Gender and professional identity negotiation among women engineers. *Gender, Work & Organization, 20*(4), 382–396. https://doi.org/10.1111/j.1468-0432.2012.00589.x
Hogg, M. A., & Terry, D. J. (2000). Social identity and self-categorization in organizational contexts. *The Academy of Management Review, 25*(1), 121–140. https://doi.org/10.2307/259266
Holland, D., Lachicote, W., Skinner, D., & Cain, C. (2001). *Identity and agency in cultural worlds.* Harvard University Press.
Holly, J., & Quigley, L. T. (2022). Reckoning with the harm of anti-Blackness in engineering education: A reparatory justice research approach. *Journal of Women and Minorities in Science and Engineering, 28*(2), 95–110.
Huff, J. L. (2019, July 10–12). *Identity in practice: Examining personal identities of engineering graduates in the transition to the workplace* [Paper presentation]. Research in Engineering Education Symposium, Cape Town, South Africa.
Huff, J. L., Okai, B., Shanachilubwa, K., Sochacka, N. W., & Walther, J. (2021). Unpacking professional shame: Patterns of White male engineering students living in and out of threats to their identities. *Journal of Engineering Education, 110*(2), 414–436. https://doi.org/10.1002/jee.20381
Huff, J. L., Smith, J. A., Jesiek, B. K., Zoltowski, C. B., & Oakes, W. C. (2019). Identity in engineering adulthood: An interpretative phenomenological analysis of early-career engineers in the United States as they transition to the workplace. *Emerging Adulthood*, 7(6), 451–467. https://doi.org/10.1177/2167696818780444

Jorgenson, J. (2002). Engineering selves: Negotiating gender and identity in technical work. *Management Communication Quarterly*, *15*(3), 350–380. https://doi.org/10.1177/0893318902153002

Kajfez, R. L., Matusovich, H. M., & Lee, W. C. (2016). Designing developmental experiences for graduate teaching assistants using a holistic model for motivation and identity. *The International Journal of Engineering Education*, *32*(3), 1208–1221.

Kirn, A., Huff, J. L., Godwin, A., Ross, M., & Cass, C. (2019). Exploring tensions of using interpretative phenomenological analysis in a domain with conflicting cultural practices. *Qualitative Research in Psychology*, *16*(2), 305–324. https://doi.org/10.1080/14780887.2018.1563270

Kroger, J., Martinussen, M., & Marcia, J. E. (2010). Identity status change during adolescence and young adulthood: A meta-analysis. *Journal of Adolescence*, *33*(5), 683–698. https://doi.org/10.1016/j.adolescence.2009.11.002

Li, N., Fan, W., Wiesner, M., Arbona, C., & Hein, S. (2021). Adapting the Utrecht-management of identity commitments scale to assess engineering identity formation. *Journal of Engineering Education*, *110*(4), 885–901. https://doi.org/10.1002/jee.20417

Lucena, J., Downey, G., Jesiek, B., & Elber, S. (2008). Competencies beyond countries: The re-organization of engineering education in the United States, Europe, and Latin America. *Journal of Engineering Education*, *97*(4), 433–447. https://doi.org/10.1002/j.2168-9830.2008.tb00991.x

Luyckx, K., Schwartz, S. J., Berzonsky, M. D., Soenens, B., Vansteenkiste, M., Smits, I., & Goossens, L. (2008). Capturing ruminative exploration: Extending the four-dimensional model of identity formation in late adolescence. *Journal of Research in Personality*, *42*(1), 58–82. https://doi.org/10.1016/j.jrp.2007.04.004

Marcia, J. E. (1966). Development and validation of ego identity status. *Journal of Personality and Social Psychology*, *3*(5), 551–558.

McAdams, D. P. (2012). Narrative identity. In S. J. Schwartz, K. Luyckx, & V. Vignoles (Eds.), *Handbook of identity theory and research (Volume 1: Structures and processes)* (pp. 99–116). Springer.

McAdams, D. P., Hoffman, B. J., Day, R., & Mansfield, E. D. (1996). Themes of agency and communion in significant autobiographical scenes. *Journal of Personality*, *64*(2), 339–377.

McCall, C., McNair, L. D., & Simmons, D. R. (2021). Advancing from outsider to insider: A grounded theory of professional identity negotiation in undergraduate engineering. *Journal of Engineering Education*, *110*(2), 393–413. https://doi.org/https://doi.org/10.1002/jee.20383

McCall, C., Shew, A., Simmons, D. R., Paretti, M. C., & McNair, L. D. (2020). Exploring student disability and professional identity: Navigating sociocultural expectations in US undergraduate civil engineering programs. *Australasian Journal of Engineering Education*, *25*(1), 79–89. https://doi.org/10.1080/22054952.2020.1720434

McLean, K. C., & Syed, M. (2015). Personal, master, and alternative narratives: An integrative framework for understanding identity development in context. *Human Development*, *58*(6), 318–349. https://doi.org/10.1159/000445817

McNair, L. D., Newswander, C., Boden, D., & Borrego, M. (2011). Student and faculty interdisciplinary identities in self-managed teams. *Journal of Engineering Education*, *100*(2), 374–396. https://doi.org/10.1002/j.2168-9830.2011.tb00018.x

Meyers, K. L., Ohland, M. W., Pawley, A. L., Silliman, S. E., & Smith, K. A. (2012). Factors relating to engineering identity. *Global Journal of Engineering Education*, *14*(1), 119–131.

Morelock, J. R. (2017). A systematic literature review of engineering identity: Definitions, factors, and interventions affecting development, and means of measurement. *European Journal of Engineering Education*, *42*(6), 1240–1262. https://doi.org/10.1080/03043797.2017.1287664

Oyserman, D., & James, L. (2012). Introduction: Toward an integrative view of identity. In S. J. Schwartz, K. Luyckx, & V. Vignoles (Eds.), *Handbook of identity theory and research (Volume 1: Structures and processes)* (pp. 117–145). Springer.

Patrick, A. D., Borrego, M., & Prybutok, A. N. (2018). Predicting persistence in engineering through an engineering identity scale. *International Journal of Engineering Education*, *34*(2-A), 351–363.

Pawley, A. L. (2009). Universalized narratives: Patterns in how faculty members define "engineering". *Journal of Engineering Education*, *98*(4), 309–319. https://doi.org/10.1002/j.2168-9830.2009.tb01029.x

Rincón, B. E., & Rodriguez, S. (2021). Latinx students charting their own STEM pathways: How community cultural wealth informs their STEM identities. *Journal of Hispanic Higher Education*, *20*(2), 149–163. https://doi.org/10.1177/1538192720968276

Rodriguez, S. L., Lu, C., & Bartlett, M. (2018). Engineering identity development: A review of the higher education literature. *International Journal of Education in Mathematics, Science and Technology*, *6*(3), 254–265. https://doi.org/10.18404/ijemst.428182

Ross, M. S., Huff, J. L., & Godwin, A. (2021). Resilient engineering identity development critical to prolonged engagement of Black women in engineering. *Journal of Engineering Education*, *110*(1), 92–113. https://doi.org/10.1002/jee.20374

Secules, S., Gupta, A., Elby, A., & Turpen, C. (2018). Zooming out from the struggling individual student: An account of the cultural construction of engineering ability in an undergraduate programming class. *Journal of Engineering Education*, *107*(1), 56–86. https://doi.org/10.1002/jee.20191

Smith, J. A., Flowers, P., & Larkin, M. (2022). *Interpretative phenomenological analysis: Theory, method, and research (second edition)*. Sage.

Sochacka, N. W., Guyotte, K. W., & Walther, J. (2016). Learning together: A collaborative autoethnographic exploration of STEAM (STEM+ the Arts) education. *Journal of Engineering Education*, *105*(1), 15–42. https://doi.org/10.1002/jee.20112

Sochacka, N. W., Walther, J., Rich, J. R., & Brewer, M. A. (2021). A narrative analysis of stories told about engineering in the public discourse: Implications for equity and inclusion in engineering. *Studies in Engineering Education*, *2*(2), 54–77. http://doi.org/10.21061/see.55

Stets, J. E., & Burke, P. J. (2000). Identity theory and social identity theory. *Social Psychology Quarterly*, 224–237. https://doi.org/10.2307/2695870

Stevens, R., O'Connor, K., Garrison, L., Jocuns, A., & Amos, D. M. (2008). Becoming an engineer: Toward a three dimensional view of engineering learning. *Journal of Engineering Education*, *97*(3), 355–368. https://doi.org/10.1002/j.2168-9830.2008.tb00984.x

Syed, M., & McLean, K. C. (2016). Understanding identity integration: Theoretical, methodological, and applied issues. *Journal of Adolescence*, *47*, 109–118. https://doi.org/10.1016/j.adolescence.2015.09.005

Tajfel, H., & Turner, J. C. (1986). The social identity theory of intergroup behaviour. In S. Worchel & W. G. Austin (Eds.), *The psychology of intergroup behaviour* (pp. 7–24). Nelson Hall.

Terry, D. J., Hogg, M. A., & White, K. M. (1999). The theory of planned behaviour: Self-identity, social identity and group norms. *British Journal of Social Psychology*, *38*(3), 225–244. https://doi.org/10.1348/014466699164149

Tonso, K. L. (2006). Teams that work: Campus culture, engineer identity, and social interactions. *Journal of Engineering Education*, *95*(1), 25–37. https://doi.org/10.1002/j.2168-9830.2006.tb00875.x

Tonso, K. L. (2014). Engineering identity. In A. Johri & B. M. Olds (Eds.), *Cambridge handbook of engineering education research* (pp. 267–282). Cambridge.

Trevelyan, J. (2007). Technical coordination in engineering practice. *Journal of Engineering Education*, *96*(3), 191–205. https://doi.org/10.1002/j.2168-9830.2007.tb00929.x

Verdín, D. (2020). "I learned to divide at 25" A counter-narrative of how one Latina's agency and resilience led her toward an engineering pathway. In E. M. Gonzalez, F. Fernandez, & M. Wilson (Eds.), *An asset-based approach to advancing Latina students in STEM* (pp. 147–164). Routledge.

Vignoles, V., Schwartz, S. J., & Luyckx, K. (2012). Introduction: Toward an integrative view of identity. In S. J. Schwartz, K. Luyckx, & V. Vignoles (Eds.), *Handbook of identity theory and research (Volume 1: Structures and processes)* (pp. 1–27). Springer.

Villanueva, I., & Nadelson, L. (2017). Are we preparing our students to become engineers of the future or the past? *The International Journal of Engineering Education*, *33*(2), 639–652.

Walther, J., Sochacka, N. W., Benson, L. C., Bumbaco, A. E., Kellam, N., Pawley, A. L., & Phillips, C. M. (2017). Qualitative research quality: A collaborative inquiry across multiple methodological perspectives. *Journal of Engineering Education*, *106*(3), 398–430. https://doi.org/10.1002/jee.20170

Walther, J., Sochacka, N. W., & Kellam, N. N. (2013). Quality in interpretive engineering education research: Reflections on an example study. *Journal of Engineering Education*, *102*(4), 626–659. https://doi.org/10.1002/jee.20029

10

Critical and Cultural Analysis of Engineering Learning

Stephen Secules, Greses Pérez, Roy Pea, and Aditya Johri

1 Introduction

Understanding how people learn engineering and how to prepare better engineers has been a central concern since the formalization of engineering education (Froyd et al., 2012). Yet from the 1960s to the 1980s, "engineering education journals and conferences remained focused on the mechanics of classroom instruction" (Felder et al., 2005, p. 9). In the 1980s and 1990s, engineering education began to be guided by theories of cognition and conducted via social science research methodologies. A more recent theoretical and empirical focus on learning has been influenced by disciplines beyond engineering education, notably the interdisciplinary learning sciences field. In prior reviews drawing on this interdisciplinary perspective, Johri and Olds (2011) and Johri et al. (2014) outlined and discussed analytical aspects of situated learning in engineering. A situated approach conceives of cognition and knowledge "as distributed among people and their environments, including objects, artifacts, tools, books, and the communities of which they are a part" (Greeno et al., 1996, p. 17), where people draw on their mind, the natural and designed physical environment, other people, and social norms to accomplish goals (Gutiérrez, 2008; Lahlou, 2018; Nasir et al., 2020; Pea, 1993b).

This chapter adds to the focus on engineering learning and expands on these prior reviews of situated learning by highlighting critical and cultural aspects and approaches to the analysis of engineering learning. Our orienting constructs for the discussion are power and culture, two aspects of engineering learning which some studies background. With this orientation, we discuss a number of additional dimensions of critical cultural analysis: knowledge, identity, agency, language practices, discourse, and sociomateriality. We note how an analysis of power and culture extends our analytical attention towards central aspects of learning that otherwise tend to be overlooked.

2 Orienting Constructs

2.1 Critical Perspectives and Power

We contrast critical perspectives on learning with colloquial accounts and perspectives on learning from individualistic disciplines such as psychology that do not centrally attend to power. Stemming from critical theory (Blake & Masschelein, 2003), critical perspectives are approaches to examining systems of power and oppression. Originally, critical theorists dealt with class politics and the

DOI: 10.4324/9781003287483-12

reproduction of the power relations of social class (Willis, 1977). Founding critical theorists are sometimes critiqued for this singular focus on social class. Other related critical theories speak to the axes of oppression that were left out by that earlier exclusive focus on class concerns. Critical race theory deals with the power relations in race and White supremacy (Ladson-Billings & Tate, 1995). Feminism focuses on gendered systems of power and patriarchy; Black and Latinx feminism deals with the subversion of the power of women of color (Ohito, 2020). Further critical theories such as queer theory and crip theory address power in systems of heteronormativity and ableism, respectively (Pawley & Slaton, 2015). Additionally, intersectionality helps extend the single critical analyses to consider the ways multiple systems of power and oppression intersect in social systems, settings, and experiences (Crenshaw, 1989, 2016).

In a broad sociological sense, power is what dominant groups hold that allows them to pursue their interests in spite of others' opposition or resistance (Giroux, 1983). Power is also the opposite of oppression, for example, the system of race affords power to White people and, to a lesser extent, those with proximity to them (e.g., some Asian people in United States STEM contexts), as it oppresses groups with less proximity to White people, that is, Black, Latinx, and Indigenous people in United States contexts. While critical theorists sometimes provide circular definitions (i.e., power is what powerful people hold over less-powerful people), there are useful discussions about different forms of power (authoritarian, distributed), how it is wielded (passive, active), and what happens in reaction to power wielding (self-conforming, rebellion) (Foucault, 1982). Power can be a useful construct for local instances of learning, for analyzing who is rendered powerful through interactions, discourse, environmental affordances, or epistemological values. When we position some learning sciences or education research as critical, we mean that it contends with such power relations. These local power analyses are often situated within larger contexts of power relations and can therefore help us explain broader phenomena of inequities in everyday contexts. Critical theorists argue that fundamental change towards a more equitable society can occur only by an analysis that begins by acknowledging the existence of systems of oppression. Relative to other research perspectives, critical theorists move towards reflective, action-oriented, and participatory approaches to their research.

In critical perspectives, learning environments are cultural and political sites of contestation and struggle among differentially empowered cultural and economic groups (Giroux, 1983). Critical theorists argue for a re-examination of the current state of affairs regarding how we teach, what we teach, and how our status quo is perpetuated. Giroux (1983) argues that in the intersection of theory and practice, "various groups come together and raise the fundamental question of how they may enlighten each other, and how through such an exchange (of theoretical positions) a mode of practice might emerge in which all groups may benefit" (p. 240). If engineering learning is conceived in terms of access to valued participation within the complex systems of relations that characterize contemporary society, then research on engineering learning must centrally involve attention to the organizing of processes through which people move along trajectories into their futures. These processes include the conditions in which people become recognized, or not, as valued participants in social worlds (de Royston et al., 2020; Gee, 1999; Penuel & O'Connor, 2010; Taylor, 1992). Critical approaches elucidate these processes and possibilities.

2.2 Culture

Although culture is often a disputed construct (McDermott & Varenne, 2006) as it is pervasive and complex, it is useful and worthy of our effort to operationalize it. The word *culture* encompasses a range of meanings that intersect learning settings. Nasir et al. (2022) define *culture* as

> the constellations of practices communities have historically developed and dynamically shaped in order to accomplish the purposes they value, including the tools they use, the social networks

> with which they are connected, the ways they organize joint activity, and their ways of conceptualizing and engaging with the world.

Clancey (1997) summarized the interconnections of culture and learning by highlighting their pervasiveness, stating:

> The idea that knowledge is a possession of an individual is as limited as the idea that culture is going to the opera. Culture is pervasive; we are participating in a culture and shaping it by everything we do. Knowledge is pervasive in all our capabilities to participate in our society; it is not merely beliefs and theories describing what we do.
>
> *(p. 271)*

Thus, *culture* includes our knowledge, practices, and development towards societal participation. In this way, culture is at once intersubjective – agreed upon and built across individuals – and subjective, or individually interpreted. This duality is highlighted in Holland et al.'s (1998) concept of *figured worlds*, which emphasizes how overlapping cultural worlds are figured (i.e., created and given meaning) by its many participants. In engineering education, several relevant "worlds" or different overlapping cultures include the disciplinary culture of engineering, divergent cultures of specific sub-disciplines of engineering, university cultures, classroom cultures, and national cultures. Culture can also be usefully defined in opposition to *structure* (indicating the environment, policies, systems – the tangible or hierarchical framework that people move through) and *agency* (indicating choice, improvisation, free will – the ways that individuals can choose to engage in culture and structure).

McDermott and Varenne's (2006) framework for culture suggests three stages of cultural analysis for examining "educational problems," which in this case could include attrition, poor grades, school failure, etc. Consistent with this chapter's focus on power, we could see McDermott and Varenne's three stages as increasingly incorporating dimensions and aspects of power. First, the frame of culture as a deficit (or individual trait) fairly simply and uncritically roots a problem in an individual's innate traits (or, problematically, "lack of culture"). In engineering education, we often label this stage as a "deficit framework." Although explicitly, deficit views are widely known to be disfavored in engineering education circles, the larger tradition of locating engineering problems within individual traits is more prominent than such collective agreement might suggest. Individual psychological constructs such as grit, self-efficacy, identity, motivation, or other cognitive assessments highlight individual rather than relational or systemic aspects. While these psychological constructs can be useful in a cultural or a critical analysis, if uncoupled from a broader reckoning with the ways in which a particular context helps create students as not gritty, lacking in self-efficacy, or unmotivated, they root successes and failures in individual traits, re-inscribing existing power dynamics, and are unable to provide critical feedback for improving institutional systems and cultural interactions.

Second, the interpretive frame of culture as difference (or socialized difference) complicates the individual trait analysis with a power analysis for how broad differences in socialization (including privilege and oppression) shape the experience and abilities of different sets of students. Here we might locate some simple versions of "assets" frameworks which argue for the valuable and different perspectives of some particular demographic, often one or more minoritized groups. This increased interrogation of power ends up creating more valuable scholarship that can, in many cases, help demonstrate the valued perspectives of marginalized students. However, purely socialized difference perspectives tend to underemphasize the power dynamics that help create privilege, oppression, and the undervaluing of particular students' perspectives in a classroom. Socialized difference lenses are a step forward from the individual trait views and thus can help provide a "teachable moment" to those in the community who may still cling to deficit views. But for the most part, they leave oppressive systems uncritiqued and unchanged.

Finally, consider the frame of culture as the construction or production of an educational fact. Cultural production focuses specifically on the structures that enable a novel cultural production shift away from an oppressive norm (Carlone & Johnson, 2012). Cultural construction examines the processes of culture that reproduce categories of educational problems and locates students within them (McDermott & Varenne, 2006). Each of these latter two frameworks have the advantage of integrating power into analyses of educational culture for deepening our understanding of the enactment of oppressive systems. These forms of cultural analysis can yield specific critiques and suggest actionable reforms. An example of a critical analysis of the construction of an educational and cultural fact will be provided in the first example study.

3 Critical Cultural Analysis of Engineering Learning

In subsequent sections, we review critical cultural approaches to analyzing learning in engineering. We introduce specific constructs to help clarify those critical approaches, including knowledge, identity, agency, language, discourse, and sociomateriality. We then illustrate critical cultural analysis regarding several of these constructs through two focal empirical examples.

3.1 Knowledge

A critical cultural analysis can consider the dimension of valued knowledge or what forms of knowing are considered most valuable (Warren et al., 2020). The official curriculum inside a classroom helps shape what students are intended to know, whereas null curriculum (what is not taught or not said) and the hidden curriculum (what unspoken strategies or lessons help one navigate the classroom) are also ways that classroom curricula can convey power (see Chapter 18, this volume; Villanueva, 2018). A particular curriculum will align with a particular expectation for the normative student, what prior knowledge the student brings to the classroom, what terminology requires defining, and what cultural references or idioms should be familiar (or not used) for effective participation. The *funds of knowledge* framework establishes an assets-based framework that contends with the power associated with typical classrooms. Funds of knowledge research originally focused on resisting the dominant curriculum devaluing Mexican American students (Moll et al., 2009; Wilson-Lopez et al., 2016). Although some perspectives on funds of knowledge reduced it to a simplistic, uncritical analysis of students' differing unique cultural perspectives, revisiting the historical context of the framework's development suggests critical focus on how teachers conceive of their students and how they can reorient curriculum (Secules & Mejia, 2021).

More broadly, epistemology, or "how you know what you know," factors into many aspects of critical cultural analyses of education. Epistemologies are important because disciplines such as engineering and education are inherently concerned with producing knowledge through ways of knowing that are collectively sanctioned. The learning sciences field has discussed epistemology in terms of how students reason about physics (Hammer, 1996), engage in argumentation (Osborne & Patterson, 2011), and relate humans to the natural world (Medin & Bang, 2014). Engineering has begun to discuss the epistemologies and philosophies of engineers (Bucciarelli, 2003; Montfort et al., 2014). There are divergences among scholars concerning whether to discuss epistemology as relatively stable/static, as dynamic resources (Elby, 2009), or as rooted in evolving cultural understandings (Bang & Medin, 2010). Epistemic practices (e.g., discipline-specific knowledge domain practices) are also constructed through social interactions among people through their concerted activity toward shared sensemaking and "professional vision" (Goodwin, 1994). The disciplinary-specific uses of language, including signs and symbols, are characteristic of epistemic cultures (Goodwin, 2013; Knorr-Cetina, 1999; Stevens, Johri, & O'Connor, 2014). Through discourse processes, members of a group frame opportunities to define what counts as knowledge and how to adjudicate knowledge claims. Such discourse requires communicative knowledge about how to participate in a cultural group and includes not only the functional aspects of relevant semantics but also relevant background knowledge. Further, scholars have argued that to not include a diversity of voices (in what is created through engineering, by whom, and for whom) is to deny the field of the wealth of knowledge and experiences which different voices bring to the field (Pérez et al., 2021).

3.2 Identity and Agency

Engineering identity is often discussed in terms of survey statements, like "I identify as an engineer," or some combination of similar constructs. This unitary view of identity should be complicated by additional interactional factors, such as Stevens et al.'s (2008) three-dimensional view, including engineering identity, accountable disciplinary knowledge, and navigation. Interactions and culture are at play in how identities are constructed and made visible. Further, a unitary view of engineering identity leaves out the broader context of how power intersects with that choice. An effective disciplinary identity can be constructed by a classroom environment (Carlone et al., 2011; Esmonde, 2009; Gresalfi et al., 2009), and a disciplinary identity can be in alignment or in conflict with a personal identity, such as racial identity (Berhane et al., 2020; Carlone & Johnson, 2007; Martin, 2007; Nasir, 2002). For Black students in STEM disciplines, there is additional identity work embodying emotional and mental labor associated with identifying with a discipline normatively associated with White people. Any identity is personal and performative/semiotic (Gee, 2000). For example, we signal and perform our gender as we enter classroom spaces and co-create gendered classroom cultures for others (Secules, Gupta, Elby & Turpen, 2018). A simplistic analysis of identity ignoring power dimensions is likely to miss the ways in which specific identities are not easily or equally accessible to all students. For more on engineering identity, see Chapter 9 (this volume).

Agency could be seen as one important operationalization of power, that is, the range of actions that are possible for a specific person in a particular context. Classroom activity structures are comprised of complex and unpredictable sets of social interactions, and thus the effective agency of each classroom participant is defined in relation to other participants. Both enacted and perceived agencies are important for student experience (Ko et al., 2014). We can think about dominant power and how marginalized individuals can create resistance (e.g., as individuals narrating their experience: Secules, Gupta, Elby & Tanu, 2018), but we also must consider the degrees of agency people hold within a system. Professors and instructors often see their agency as constrained by ABET accreditation or county standards, by what their students will or can do, by supervisors or standardized tests or evaluators, yet their agency is, in most cases, significantly greater than the agency experienced by their students. Professors also have agency to reproduce or resist a dominant cultural norm for their students (Carlone & Johnson, 2012).

3.3 Example Study on Knowledge, Culture, Identity, and Agency

Secules, Gupta, Elby, and Turpen (2018) described an introductory programming course for electrical engineers. Doing embedded observations inside the course over multiple semesters, Secules (researcher) noticed a pattern where one or a small subset of students became disengaged and discouraged during activities. In a post-semester interview, a participant who tended to be slower than his classmates in lab said he thought he "just didn't have the brain for programming." While this phenomenon, which Secules et al. broadly named being "not cut out for engineering," could have been interpreted purely psychologically, as self-efficacy or beliefs about one's abilities, Secules et al. were inspired by the cultural construction framework to investigate the cultural and structural root causes of the phenomenon of feeling "not cut out for engineering." These roots included seating arrangements, gender and status interpretations, interactions between students, and student interaction with instructors.

The knowledge valued in the classroom, the epistemological dimensions, was a specific interest to co-authors Gupta and Elby and structured many of the initial protocol designs. There was express attention to whether programming/coding was viewed as authentic engineering and how success or failure of a code was both a tangible and a knowledge-building act. Secules also attuned towards the knowledge individuals brought into the classroom and how it was valued. The students who tended to come to believe they were "not cut out for engineering" in this course were most likely to enter without a programming background but having other foundational electrical engineering knowledge (e.g., electrical systems in buildings). They may have believed they were one of the only ones without that valued knowledge since others (typically more privileged in terms of knowledge and identity) were prominently performing their fluidity with coding vocabulary in lecture and their speed in completing lab tasks. A major feature of knowledge presentations came in PowerPoint lecture texts, which tended to feature esoteric language without scaffolding using more shared vocabulary. The interactions and texts in class communicated a particular epistemological underpinning to the course, that a formal knowledge of coding was valued over an intuitive or layperson's understandings of electrical and computational concepts.

Many identities were at play in the programming course. Secules et al. mapped out seating patterns that included gender, which was prominently noticed and even "spotlighted" (McLoughlin, 2005) by instructors and students, and relative status in terms of programming experience (see Figure 10.1 and 10.2). The gendered knowledge performances of men in the course helped concretize the effective identities of students as capable of programming or "not cut out for" it. Students were able to identify both gender and status patterns; however, these correlations were sometimes inaccurate and speak to the limited knowledge we have when identifying each other. For instance, some women students were not looked upon with programming expertise even when they were typically finishing lab quicker and more efficiently than their peers (a marker of status for men in the lab). Some men were perceived as having programming expertise because they expressed confidence and bravado in lab, but privately they expressed insecurity that they lacked as much programming experience as they claimed.

Secules et al. looked at these inequities that interrelated with knowledge and identity domains as an opportunity to consider the typical unit of education research analysis. When looking at the individual, one might diagnose problems in terms of a learner's individual deficits (i.e., individual trait lens) or as the unfortunate socialization of broader systems in society that shape the individual (socialization lens). While these two lenses have some purchase, they narrow the understanding of the complex interplay happening within classroom spaces, and they are unactionable, as it generally does not work to "fix" the student or society. Instead, Secules et al. suggested looking at culture as the unit of analysis. In this case, *culture* includes the patterns of interactions that tended to take place, the normative assumptions and interpretations, the roles and course structures that were taken for granted as ordinary in that setting, and a larger set of engineering and educational cultural influences beyond the classroom. In particular, many of the laboratory challenges seemed to be rooted in a culture of competition and meritocracy as students (particularly men) attempted to perform competence for one another in terms of coding speed and success. This culture had negative consequences on students with less programming background (typically all the women and some of the men) who could not keep up with the norms of the class competition on individual assignments and who experienced a double bind on paired assignments. While apprenticed learning is theorized as having great potential for enabling a novice to become more familiar with a community of practice (e.g., coding), in cases where the value of competition and performance outweighs learning, a novice learner was either consistently sidelined (in expert–novice pairing) or overtly stressed (in novice–novice pairs).

```
Slide 27
/* example #02 electric field calculation
* Written by Phil
* Version 1.3 Last updated Sept 3, 2014 */
#include <stdio.h>
#define CHARGE -1.602e-19     /* new feature */
#define EPSILON_0 8.854e-12
#define PI 3.141592654
/*Pre-processor directive "define" generates symbolic constants. Constants
are replaced in code by numeric values before compilation.*/
```

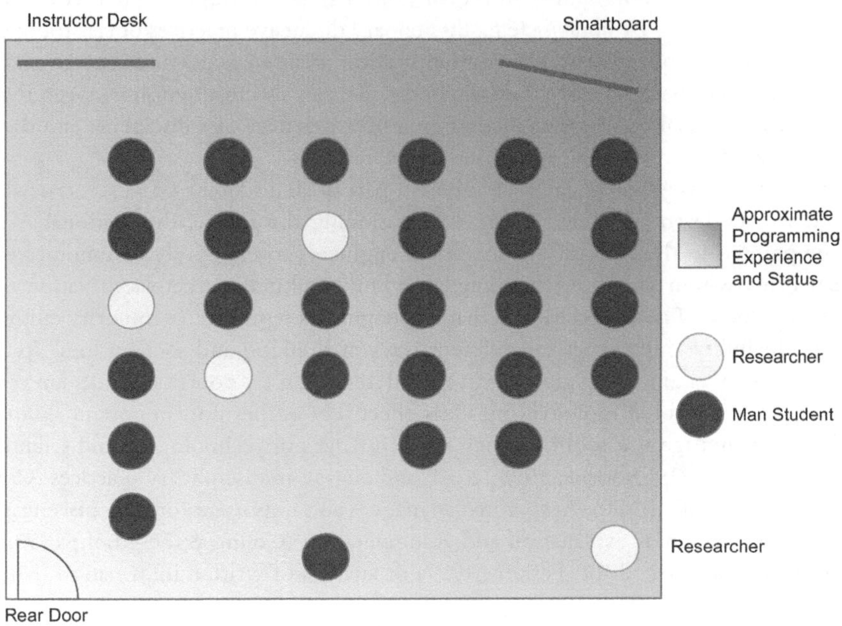

Figure 10.1 and 10.2 Artifacts from programming course for electrical engineers. Top: Example PowerPoint slide. Bottom: Typical seating chart, gender, and approximate status.

Source: Figure adapted from Secules, Gupta, Elby, and Turpen (2018).

Finally, this example foregrounds the limited but significant agency (individual power) of each actor in a classroom setting. Some actors had more agency than others – the instructor had significant agency to set course structures and to invoke a value system for the class. He also had some agency towards the cultural norms, although he was outnumbered by the students and the normative interactions and interpretations they made. Dominant/privileged students had the agency to set many aspects of the course culture – some dominant male students made a game of trying to distract class time using advanced questions. When asked, the professor would comment, "That's way beyond the scope of the course," but would proceed to give an extended detailed answer, derailing the course from more basic content while amusing the dominant students. The result was multiple negative consequences for nondominant students who had significantly limited agency to protest or disrupt that norm. In labs, nondominant students also had significantly limited agency, finding themselves either slowest in terms of individual labs or in a double bind in terms of their group lab pairings. While marginalized students have limited agency, a parallel study by these authors (Secules, Gupta, Elby, & Tanu, 2018) emphasized that marginalized students can have other forms of agency, including the power to narrate and reinterpret the meaning of events.

3.4 Language Practices and Discourse

Situated learning scholars have found that an understanding of language and discourse is crucial for the empirical study of engineering learning. By *language* we mean the dynamic and complex communicative practices or repertoire of groups of speakers (e.g., the African American vernacular English of the Texana communities in North Carolina) or the ways people convey meaning in meaningful social contexts. By *discourse* (e.g., disciplinary discursive practices or classroom discourse) we mean the verbal or nonverbal acts of communication (written texts, interactions) and the contexts they create when speakers enact their discourse. We can also distinguish between the ("capital D") Discourse, the official or sanctioned communicative practices of a discipline, and the ("lowercase d") discourse of classrooms and interpersonal interactions.

Communication through language is an inherent part of all disciplines, and it is crucial for engineering and science learning (Kelly, 2014). In envisioning the future, the National Academy of Engineering pointed to the necessity of successful engineers to effectively communicate with the public, particularly within a growing multilingual and multicultural marketplace (National Academy of Engineering, 2005). The expectation is that US engineers represent "a minority culture" which will require "flexibility and respect for different ways of thinking and social values" (p. 152). For instance, in a study of engineering practices in six different industry contexts, Anderson et al. (2010) found that engineers pointed to conveying ideas effectively as the most important skill to have in their profession. Language is a tool by which we construct our technological and scientific reality (Halliday & Martin, 1993; Osborne, 1998, 2019) and engage in disciplinary practices (Cornelius & Herrenkohl, 2004; Lemke, 1990). At its core, language is also important for students' engagement in disciplinary practices such as explanation and argumentation (Collins & Ferguson, 1993; Ohlsson, 1996; Osborne & Patterson, 2011; Perkins, 1997) or analysis of written information (Millar, 2006; Norris & Phillips, 2003).

Due to the field's history, engineering education has primarily focused on dominant languages and disciplinary nomenclature (e.g., English, German; academic varieties) as language practices (Dong et al., 2004; Mabogunje & Leifer, 1997; Song & Agogino, 2004). Although engineers work in societies where language diversity is the norm (more than 300 languages are spoken in the United States, US Census, 2020), they are often trained only in dominant languages (e.g., English). When engineers learn about their technical expertise in languages other than English, it is typically considered a foreign language competency, even if they are heritage speakers – those who have learned the language informally in their communities (Crawley et al., 2014). In discussing language training in engineering, the focus is typically on developing ways of speaking for the purpose of working in international settings where languages other than English are spoken instead of "taking advantage of people's capabilities and languages" (National Academy of Engineering, 2002) for doing engineering.

Even though scholars agree on the significance of language for learning in context, there is disagreement on what language is and the role power plays in notions of language. For instance, sociolinguists and anthropologists have challenged a portrayal of bilinguals as two monolinguals inside one mind (Flores & Rosa, 2015; Heller, 2007; Pennycook, 2006). Cook (2016) argues for deconstructing the idea of separate languages by focusing on language practices instead, or "the overall system of a mind or a community that uses more than one language" (p. 7). These scholars understand languages and their boundaries (the imaginary separation that determines the differences between English and Spanish) as social constructions. These boundaries between how some speakers talk versus others are rooted in historical and political processes, leading to narrow definitions of languages as discrete and named entities, associated with structures of power around nations and states (García, 2009; Rosa, 2016; Valdés et al., 2015), as well as disciplines such as engineering and science. These power dynamics position nondominant speakers (e.g., Spanish speakers in the United States) and their talk differently in a society reigned by normative ideologies (Blommaert & Rampton, 2011). These social norms of standard language in the classroom (e.g., the idea of a singular academic way of speaking in STEM or of a dominant standard language such as English as the lingua franca) may fail students from underrepresented groups because they create a divide between the realities of their communities and school (Brown, 2021; Calabrese-Barton et al., 2013; Fang, 2005; Lemke, 1990).

In contrast with a view of language defined by clear boundaries, sociocultural theorists emphasize a spectrum of shared and contrasting language practices across people, groups, and contexts (García & Wei, 2014, 2015; Valdés et al., 2015). Bilingual science education scholars highlight the ambiguous nature of language boundaries, dictated by dominant groups, and the ways these divisions often unjustifiably recognize only certain language practices in engineering and science as legitimate forms of participation (Hamman, 2018; Poza, 2017). Brown (2004) demonstrated differential appropriations of disciplinary discourses that resulted in the diminution of students' opportunities to learn, indicating that interactions are dominated by power disparities between the profession's discourses and those of ethnic minorities. Power structures in engineering position the discursive practices of dominant groups as the legitimate forms of participation in the discipline, denying traditionally underrepresented groups access to opportunities for equitably engaging in engineering practices. Science education scholars have suggested the benefits of effectively teaching students to navigate between their day-to-day experiences and their realities in the classroom through crossing cultural borders between the practices of their locale and disciplinary practices (Aikenhead & Jegede, 1999). Other researchers have expanded this notion by suggesting the appropriation of day-to-day discursive practices into the teaching and learning of science and engineering to develop a deeper appreciation of the disciplinary ways of speaking and knowing (Brown, 2004; Gee, 2006; Lee & Fradd, 1996). In their study of innovation in Indigenous learning, Rosado-May et al. (2020) argue for co-constructing forms of knowledge across cultures by "respecting and using Indigenous forms of learning and knowledge systems, in conjunction with Western approaches" (p. 92).

Both societal-level Discourse and classroom discourse can be representations of the "culture of power" of dominant society (Delpit, 1988; Lee & Luykx, 2014). According to Delpit (1988), Black and Brown students and those from low socioeconomic backgrounds must be explicitly taught about the hidden and visible rules behind the practices of power. Although Delpit talks about Discourse as it applies to broader societal narratives, the knowledge and practices of fields such as engineering are also a form of Discourse. Dominant Discourses associated with engineering include a valuing of technical knowledge over the sociocultural and political factors which influence the work of engineers (Cech, 2014; Pérez et al., 2021; Trevelyan, 2010).

Disciplinary practices emerge through processes in which participants define how ideas are conveyed and what kinds of contributions are valued and legitimized within the community (Cunningham & Kelly, 2017). Scholars have referred to "the several languages of design" to acknowledge the multiplicity of ways in which engineers tailor their speech for different purposes with audiences (Dym, 1999; Dym et al., 2005). According to the authors, effective engineering requires uses of multiple modalities (e.g., inscriptions, mathematical models, statements, etc.) to communicate ideas in the forms of human cognition and cultural practices known as engineering. The authors argue for expanding the language of engineering beyond mathematics to incorporate multiple modes of communicating, including sketches and graphical representations. However, these perspectives have nonetheless reinforced normative views of language by focusing on the usage of particular words to indicate levels of design creativity (Dym et al., 2005). This stance perpetuates a limited idea of valued language in engineering instead of more expansive language practices generated in the context of communities (Creese & Blackledge, 2015; García, 2017; García & Wei, 2015).

3.5 Sociomateriality

Physical and material aspects of engineering and engineering learning processes are important but sometimes overlooked in studies focused on cognition or communication. A dual focus on social and material aspects (or, sociomateriality) can help alert us to changes brought about by shifting material affordances of learning practices and consequent changes in social affordances, and thus in the overall engineering practice (Guerrettaz et al., 2021; Johri, 2011; Johri, 2012; Pea, 1993a). The rise of online learning exemplifies new sociomaterial configurations, including AR/VR and simulations (Hopwood et al., 2016), that have changed how engineers learn (see Chapter 23, this volume). Another example is the rise of Makerspaces, which borrow some features from labs or after-school clubs but are unique in other sociomaterial aspects and have spawned new language and cultural practices (Kumpulainen & Kajamaa, 2020). As sociomaterial affordances vary, so do the opportunities for learning and associated outcomes of learning in these new configurations (Johri, 2011; Johri, 2022). Sociomateriality also foregrounds how the products of engineers and engineering – the artifacts, objects, and infrastructures they create – shape our world (Styhre et al., 2012). For instance, climate change and sustainability are a core concern of engineering from a natural world perspective but are also important in their collective impacts on lives and livelihood, migration, transportation, energy, and food supply chains; engineers need to understand and respond to these sociomaterial problems (Baumann & Lindkvist, 2021). Since engineering is a highly sociotechnical practice, materials play an essential role in coordinating engineers' distributed practice (Trevelyan, 2010).

The conveying of language and concepts through media and representations is another form of sociomateriality (Björkman & Harris, 2018; Borning et al., 2020; Johri, 2020; Pea, 1992; Styre, 2017). For example, inscriptions – "graphical representations recorded in and available through some medium such as paper and computer monitors" (Roth & McGinn, 1998, p. 35) – connect language and other forms of representations to the physical world (Latour, 1992; Roth, 2013). Inscriptions are a way to articulate and bring together the practices of engineering and those of communities (Roth, 2013; Johri et al., 2013). In particular, the existence or production of "pictorial inscription, and the gestures it affords, contribute to the emergence of culture and engineering language" (Roth & Lawless, 2002). This view of inscriptions centers on interaction analysis of language practices and associated gesture uses as crucial for understanding their development in engineering, where "use is experienced in concrete relations rather than existing in ephemeral 'mental frameworks' and 'mental constructions'" (Roth, 2013; Wittgenstein, 1997). Within engineering and science, expertise and competencies develop through the use and conversational exchange of inscriptions with others in multi-turn media-rich learning conversations (Pea, 1993a). Therefore, it is insufficient for novices to simply observe inscriptions or inscription-related activities, such as books or lectures, since they need to "actively participate in engineering-related talk where specific inscriptions are used or discussed (Johri et al., 2013, p. 14)."

The situated aspects of sociomaterial *contexts* where learning takes place are also highlighted through a critical cultural lens. Learning contexts and environments can send important signals to learners about what is accepted behavior, sanctioned discourse, valued knowledge and language practices, and recognized participation in a specific context. Formal learning contexts and informal learning contexts come with different sets of cultural rules, norms, and logic, which make visible different legitimate performances (McDermott, 1993). Harrison and Dourish (1996) suggested the term "space" for environments defined by their material and geometric properties. They suggested "place" for sociomaterial contexts defined by the ways that human activities occur within that environment. "Space" describes geometrical arrangements that structure, constrain, and enable certain forms of movement and interaction that are significant for learning and engineering practices. For instance, a lectern and stadium-style seating may invoke a hierarchical power relationship of a presenter onstage; a circle of chairs or roundtable may invoke a democratic relationship. "Place" focuses on features that are not inherent to the material structure but acquire recognizable and persistent social meaning during the course of interaction. The material constraints of spaces in engineering are not neutral in their material and geometric properties but imbue the power relations of economic, political, and social forces (Costanza-Chock, 2018, 2020; Winner, 2017). Places are socially constructed and therefore can be socially changed (Calabrese Barton et al., 2021; Calabrese Barton & Tan, 2019; Leander et al., 2010). Perhaps the circle of chairs includes a leader or dominant participant who controls the discourse and makes their center position the focus of attention. In this configuration, the space of the circle of chairs with its democratic affordances has been socially constructed as a place which endows more power to select participants.

3.6 Example Study on Language, Discourse, and Sociomateriality

Pérez researched a translingual engineering design summer program for elementary and middle school students in Northern California where students learned about engineering design in three lessons facilitated in English, Spanish, and both languages, respectively (Joehnk, 2021). Through video analysis and semi-structured interviews, the authors documented factors beyond flexible language norms in the classroom or translanguaging that influenced students' uses of their language resources (e.g., speaking in Spanish or Spanglish to explain their ideas). Translanguaging as a theory and pedagogy proposes to free the bilingual child from the social constructions of language boundaries (García, & Wei, 2015), allowing learners freedom to engage in language practices of their communities (e.g., explaining their prototypes drawing on the home language practices) in the classroom when engaging in disciplinary talk. In a post-interview, students mixing language practices and transgressing boundaries (e.g., speaking Spanish in the English-only lesson) to explain their ideas in engineering stated they engaged in language mixing as a playful, "deficient," or "accidental" act.

While this phenomenon could be interpreted from a cognitivist perspective of learning and language (where people "have" language in their minds), the authors instead were inspired by sociocultural theories and situated learning that position people and communities as "doing" and being through language (practices). The goal of this theoretical position was to investigate engineering talk and learning in social interaction within meaningful contexts and to alternatively imagine learning environments that are linguistically inclusive for multicompetent learners (those who live between languages and cultures, Pérez, 2022). In imagining the alternative reality of the translingual informal learning contexts ("alternative" because of the absence of opportunities for nondominant language practices in engineering classrooms), the authors considered students' engagement in the creation of the learning activity itself – designing efficient modes of transportation or alternative energies. They also pondered the learners' prior experiences with language in and out of school or the engineering language practices in the classroom and those in their communities. Finally, the authors considered the kinds of messages students have received about legitimized ways of engineering talk or the Discourses of the discipline about valued engineering discursive practices.

In considering context in and out of the study setting, the authors focused on power disparities and potential incongruities between the disciplinary Discourses (e.g., conceptions of what it means to do and be an engineer) and those of underrepresented communities. They highlighted students' dexterity in explaining their prototypes by connecting scientific ideas with the sociomateriality. For example, Sharky (a Mexican American elementary student who grew up in a Spanish-speaking household) drew parallels between his household fan and the design of the windmill blades in his team's prototype. Speaking a nondominant English variety known by students as Spanglish, Sharky made this connection to explain air flow across the blades while moving his fingers through the different blades. The authors also pointed to the awareness about audience and ways to best convey the message as shown in students' single lexical switches within a sentence to explain the same idea (air flow in blades) in both languages, depending on the language background of those listening to the student's explanation.

Even though students demonstrated understanding of language as a tool and medium for learning and conveying ideas, broader social norms in engineering learning influenced their perceptions and use of nondominant languages in class and whether they exploited opportunities to draw on their available repertoire. Even though the program allowed students to use the ways of talking in their communities, many students did not engage in nondominant language practices, and those who did (including Sharky) saw it as something accidental or deficient. During the interviews, students pointed to disciplinary and social expectations about language use as contributing to their perceptions of engaging with language mixing.

4 Conclusion

With our positioning of knowledge, identity, agency, language practices, discourses, and materiality as topics for critical cultural analysis, one might ask, "What is not critical cultural analysis?" Indeed, even these topics are not comprehensive, and we agree that any topic is open to a critical and cultural analysis. To take a critical perspective on something is to think about how power is affecting it, and each facet of engineering learning that we investigate could uncover new and important aspects of power. Something ordinary, like a lectern or a circle of chairs or the insistence on a formal address for instructors, can provide subtle but significant connotations of power. Critical analysis enables the scrutiny of power dynamics across many everyday and seemingly innocuous aspects of learning. Engineering education research and practice always involves language, but a critical cultural analysis can probe the power differentials between dominant and nondominant ways of speaking in engineering. A cultural perspective problematizes norms, "makes the familiar strange," and considers practices and ways of being that are overlapping between the individual, familial, ethnic, disciplinary, and institutional. We contend that an analysis of engineering education not examining power will likely perpetuate a status quo inequity, while an analysis of power without attention to the cultural, interactional, and situated nuances of engineering learning will likely miss the subtleties of interactions through which those inequities become perpetuated.

Although we have noted the consequential possibilities that a critical cultural analysis of engineering learning can afford, we also note its relatively small uptake to date. Much of the scholarship cited derives from the learning sciences or science/math education communities. While there is potential for inspiration and collaboration across disciplines, the engineering-specific issues outlined here will require an engineering-specific analysis. Contrary to a vision of research insight and impact that is predicated on increasing scale, we suggest an opportunity for the community to have greater understanding of and impact on the everyday and ubiquitous instances of engineering learning. While the studies we have outlined in the preceding text have merely scratched the surface of prospective applicability of critical cultural analysis, we suggest a vision for collective impact by critical inquiry into the consequences of engineering learning practices in local settings and looking in partnership to make learning more equitable and inclusive.

We suggest the following for readers hoping to engage a critical cultural analysis of engineering learning. For engineering education researchers seeking to further a critical cultural analysis in their work, we hope the topics suggested forge a starting point for further reading. In addition to the theories and scholarship discussed, we suggest an engagement with the methodological approaches associated with the learning sciences, including analyzing classroom video (Derry et al., 2010), discourse and interaction analysis, ethnography, design-based research, and participatory and action research. Those who are looking for inspiration for further studies may look at some of the example topics outlined and see what adaptations of context, theory, and method will help extend our understanding of the topic. For the intellectual community of authors, reviewers, mentors, and funders who may have the opportunity to produce or support critical cultural analysis scholarship, we ask for an open mind regarding this often-local and nuanced work that promises outsized impact on our collective understanding. Finally, for educational practitioners and other stakeholders of engineering learning, we hope the topics help attune readers to mechanisms for the reproduction of the inequities of the engineering discipline and sites of resistance and change.

Acknowledgments

Dr. Secules's work was partially supported by US National Science Foundation Award no. 1939105. He would also like to thank the co-authors and study participants involved in the example studies they shared, including Becca (participant) and the many students in the engineering class, co-authors Ayush Gupta, Andy Elby, and Chandra Turpen, and the US National Science Foundation and PI Wes Lawson for grant no. 1245745 for supporting that work. Dr. Johri's effort was partially supported by US National Science Foundation Awards no. 1941186, 1937950, and 1939105. Any opinions, findings, and conclusions or recommendations expressed in this material are those of the authors and do not necessarily reflect the views of the funding agencies. We would also like to thank the reviewers and editors for their feedback, which helped improve this chapter.

References

Aikenhead, G. S., & Jegede, O. J. (1999). Cross-cultural science education: A cognitive explanation of a cultural phenomenon. *Journal of Research in Science Teaching, 36*(3), 269–287. https://doi.org/10.1002/(SICI)1098-2736(199903)36:3<269::AID-TEA3>3.0.CO;2-T

Anderson, K. J. B., Courter, S. S., McGlamery, T., Nathans-Kelly, T. M., & Nicometo, C. G. (2010). Understanding engineering work and identity: A cross-case analysis of engineers within six firms. *Engineering Studies, 2*(3), 153–174. https://doi.org/10.1080/19378629.2010.519772

Bang, M., & Medin, D. (2010). Cultural processes in science education: Supporting the navigation of multiple epistemologies. *Science Education, 94*(6), 1008–1026.

Baumann, H., & Lindkvist, M. (2021). A sociomaterial conceptualization of flows in industrial ecology. *Journal of Industrial Ecology,* 1–12. https://doi.org/10.1111/jiec.13212

Berhane, B., Secules, S., & Onuma, F. (2020). Learning while Black: Identity formation and experience for five Black men who transferred into engineering undergraduate Programs. *Journal of Women and Minorities in Science and Engineering, 26*(2), 93–124. https://doi.org/10.1615/JWomenMinorScienEng.2020024994

Björkman, L., & Harris, A. (2018). Engineering cities: Mediating materialities, infrastructural imaginaries and shifting regimes of urban expertise. *International Journal of Urban and Regional Research, 42*(2), 244–262.

Blake, N., & Masschelein, J. (2003). Critical theory and critical pedagogy. In *The Blackwell guide to the philosophy of education* (pp. 38–56). Blackwell Publishing Ltd.

Blommaert, J., & Rampton, B. (2011). Language and superdiversity. *Diversities, 13*(2), 1–21.

Borning, A., Friedman, B., & Logler, N. (2020). The 'invisible' materiality of information technology. *Communications of the ACM, 63*(6), 57–64.

Brown, B. A. (2004). Discursive identity: Assimilation into the culture of science and its implications for minority students. *Journal of Research in Science Teaching, 41*(8), 810–834. https://doi.org/10.1002/tea.20228

Brown, B. A. (2021). *Science in the city: Culturally relevant STEM education*. Harvard Education Press.

Bucciarelli, L. L. (2003). *Engineering philosophy*. DUP Satellite; an imprint of Delft University Press.

Calabrese Barton, A., Kang, H., Tan, E., O'Neill, T. B., Bautista-Guerra, J., & Brecklin, C. (2013). Crafting a future in science: Tracing middle school girls' identity work over time and space. *American Educational Research Journal, 50*(1), 37–75.

Calabrese Barton, A., Schenkel, K., & Tan, E. (2021). Collaboratively engineering for justice in sixth grade STEM. *Journal of Research in Science Teaching, 1*(31), 1–31. https://doi.org/10.1002/tea.21691

Calabrese Barton, A., & Tan, E. (2019). Designing for rightful presence in STEM: The role of making present practices. *Journal of the Learning Sciences, 28*(4–5), 616–658. https://doi.org/10.1080/10508406.2019.1591411

Carlone, H. B., Haun-Frank, J., & Webb, A. (2011). Assessing equity beyond knowledge- and skills-based outcomes: A comparative ethnography of two fourth-grade reform-based science classrooms. *Journal of Research in Science Teaching, 48*(5), 459–485. https://doi.org/10.1002/tea.20413

Carlone, H. B., & Johnson, A. (2007). Understanding the science experiences of successful women of color: Science identity as an analytic lens. *Journal of Research in Science Teaching, 44*(8), 1187–1218. https://doi.org/10.1002/tea

Carlone, H. B., & Johnson, A. (2012). Unpacking "culture" in cultural studies of science education: Cultural difference versus cultural production. *Ethnography and Education*, 7(2), 151–173. https://doi.org/10.1080/17457823.2012.693691

Cech, E. A. (2014). Culture of disengagement in engineering education? *Science, Technology, & Human Values, 39*(1), 42–72. https://doi.org/10.1177/0162243913504305

Clancey, W. J. (1997). The conceptual nature of knowledge, situations, and activity. In P. Feltovich, K. Ford, & A. Hoffman (Eds.), *Human and machine expertise in context* (pp. 247–291). AAAI Press.

Collins, A., & Ferguson, W. (1993). Epistemic forms and epistemic games: Structures and strategies to guide inquiry. *Educational Psychologist, 28*(1), 25–42. https://doi.org/10.1207/s15326985ep2801_3

Cook, V. (2016). Premises of multi-competence. In V. Cook & L. Wei (Eds.), *The Cambridge handbook of Linguistic multi-competence* (pp. 1–25). Cambridge University Press. https://doi.org/10.1017/CBO9781107425965.001

Cornelius, L. L., & Herrenkohl, L. R. (2004). Power in the classroom: How the classroom environment shapes students' relationships with each other and with concepts. *Cognition and Instruction, 22*(4), 467–498. https://doi.org/10.1207/s1532690Xci2204_4

Costanza-Chock, S. (2018). *Design justice:* towards *an intersectional feminist framework for design theory and practice* (SSRN Scholarly Paper ID 3189696). Social Science Research Network. https://papers.ssrn.com/abstract=3189696

Costanza-Chock, S. (2020). *Design justice: Community-led practices to build the worlds we need*. The MIT Press. https://openresearchlibrary.org/content/41b0401f-8601-4b5c-b434-6b4a32a44c8c

Crawley, E. F., Malmqvist, J., Östlund, S., Brodeur, D. R., & Edström, K. (2014). Rethinking Engineering Education. In E. F. Crawley, J. Malmqvist, S. Östlund, D. R. Brodeur, & K. Edström (Eds.), *Rethinking engineering education* (pp. 11–45). Springer International Publishing. https://doi.org/10.1007/978-3-319-05561-9_2

Creese, A., & Blackledge, A. (2015). Translanguaging and identity in educational settings. *Annual Review of Applied Linguistics, 35*, 20–35. https://doi.org/10.1017/S0267190514000233

Crenshaw, K. (1989). Demarginalizing the intersection of race and sex: A Black feminist critique of antidiscrimination doctrine, feminist theory and antiracist policies. *The University of Chicago Legal Forum, 1989*(1), 139–167. https://chicagounbound.uchicago.edu/uclf/vol1989/iss1/8

Crenshaw, K. (2016). Mapping the margins: Intersectionality, identity politics, and violence against women of color. *Stanford Law Review, 43*(6), 1241–1299.

Cunningham, C. M., & Kelly, G. J. (2017). Epistemic practices of engineering for education. *Science Education, 101*(3), 486–505. https://doi.org/10.1002/sce.21271

de Royston, M. M., Barron, B., Bell, P., Pea, R., Stevens, R., & Goldman, S. (2020). Learning pathways: How learning is culturally organized. In *Handbook of the cultural foundations of learning* (pp. 195–211). Routledge.

Delpit, L. (1988). The silenced dialogue: Power and pedagogy in educating other people's children. *Harvard Educational Review, 58*(3), 280–299. https://doi.org/10.17763/haer.58.3.c43481778r528qw4

Derry, S. J., Pea, R. D., Barron, B., Engle, R. A., Erickson, F., Goldman, R., . . . Sherin, B. L. (2010). Conducting video research in the learning sciences: Guidance on selection, analysis, technology, and ethics. *The Journal of the Learning Sciences, 19*(1), 3–53.

Dong, A., Hill, A. W., & Agogino, A. M. (2004). A document analysis method for characterizing design team performance. *Journal of Mechanical Design, 126*(3), 378–385. https://doi.org/10.1115/1.1711818

Dym, C. L. (1999). Learning engineering: Design, languages, and experiences*. *Journal of Engineering Education, 88*(2), 145–148. https://doi.org/10.1002/j.2168-9830.1999.tb00425.x

Dym, C. L., Agogino, A. M., Eris, O., Frey, D. D., & Leifer, L. J. (2005). Engineering design thinking, teaching, and learning. *Journal of Engineering Education, 94*(1), 103–120. https://doi.org/10.1002/j.2168-9830.2005.tb00832.x

Elby, A. (2009). Defining personal epistemology: A response to Hofer & Pintrich (1997) and Sandoval (2005). *Journal of the Learning Sciences, 18*(1), 138–149. https://doi.org/10.1080/10508400802581684

Esmonde, I. (2009). Ideas and identities: Supporting equity in cooperative mathematics learning. *Review of Educational Research, 79*(2). https://doi.org/10.3102/0034654309332562

Fang, Z. (2005). Scientific literacy: A systemic functional linguistics perspective. *Science Education, 89*(2), 335–347. https://doi.org/10.1002/sce.20050

Felder, R. M., Shepard, S. D., & Smith, K. A. (2005). A new journal for a field in transition. *Journal of Engineering Education, 94*(1), 7–10.

Flores, N., & Rosa, J. (2015). Undoing appropriateness: Raciolinguistic ideologies and language diversity in education. *Harvard Educational Review, 85*(2), 149–171.

Foucault, M. (1982). The subject and power. *Critical Inquiry, 8*(4), 777–795. https://doi.org/10.1086/448181

Froyd, J. E., Wankat, P. C., & Smith, K. A. (2012). Five major shifts in 100 years of engineering education. *Proceedings of the IEEE, 100*(Special Centennial Issue), 1344–1360.

García, O. (2009). *Bilingual education in the 21st century: A global perspective*. John Wiley & Sons.

García, O. (2017). Translanguaging in schools: Subiendo y Bajando, Bajando y Subiendo as afterword. *Journal of Language, Identity & Education, 16*(4), 256–263. https://doi.org/10.1080/15348458.2017.1329657

García, O., & Wei, L. (2014). *Translanguaging*. Palgrave Macmillan UK. https://doi.org/10.1057/9781137385765

García, O., & Wei, L. (2015). Translanguaging, bilingualism, and bilingual education. *The Handbook of Bilingual and Multilingual Education, 223*, 240.

Gee, J. P. (1999). *An introduction to discourse analysis: Theory and method*. London:

Gee, J. P. (2000). Identity as an analytic lens for research in education. *Review of Research in Education, 25*(1), 99–125.

Gee, J. P. (2006). Oral discourse in a world of literacy. *Research in the Teaching of English, 41*(2), 153–159.

Giroux, H. A. (1983). *Theory and resistance in education: A pedagogy for the opposition*. Bergin & Garvey.

Goodwin, C. (1994). Professional vision. *American Anthropologist*, 606–633.

Goodwin, C. (2013). The co-operative, transformative organization of human action and knowledge. *Journal of Pragmatics*. 46(1), 8–23.

Greeno, J., Collins, A., & Resnick, L. (1996). Cognition and learning. In R. Calfee & D. Berliner (Eds.), *Handbook of educational psychology* (pp. 15–46). Macmillan.

Gresalfi, M., Martin, T., Hand, V., & Greeno, J. (2009). Constructing competence: An analysis of student participation in the activity systems of mathematics classrooms. *Educational Studies in Mathematics, 70*, 49–70. https://doi.org/10.1007/s10649-008-9141-5

Guerrettaz, A. M., Engman, M. M., & Matsumoto, Y. (2021). Empirically defining language learning and teaching materials in use through sociomaterial perspectives. *The Modern Language Journal, 105*(S1), 3–20.

Gutiérrez, K. D. (2008). Developing a sociocritical literacy in the third space. *Reading Research Quarterly, 43*(2), 148–164.

Halliday, M. A. K., & Martin, J. R. (1993). *Writing science: Literacy and discursive power*. Routledge.

Hamman, L. (2018). Translanguaging and positioning in two-way dual language classrooms: A case for criticality. *Language and Education, 32*(1), 21–42. https://doi.org/10.1080/09500782.2017.1384006

Hammer, D. (1996). Misconceptions or P-prims: How may alternative perspectives of cognitive structure influence instructional perceptions and intentions. *Journal of the Learning Sciences, 5*(2), 97–127. https://doi.org/10.1207/s15327809jls0502_1

Harrison, S., & Dourish, P. (1996). Re-place-ing space: The roles of place and space in collaborative systems. In *Proceedings of the 1996 ACM conference on computer supported cooperative work* (pp. 67–76).

Heller, M. (2007). *Bilingualism: A social approach*. Palgrave Macmillan.

Holland, D. C., Lachicotte, W., Skinner, D., & Cain, C. (1998). *Identity and agency in cultural worlds*. Harvard University Press.

Hopwood, N., Rooney, D., Boud, D., & Kelly, M. (2016). Simulation in higher education: A sociomaterial view. *Educational Philosophy and Theory, 48*(2), 165–178.

Joehnk, G. A. (2021). *Beyond representation: Uncovering the role of language and cognition for multicompetent students in engineering and science* [Doctoral dissertation, Stanford University].

Johri, A. (2011). The socio-materiality of learning practices and implications for the field of learning technology. *Research in Learning Technology, 19*(3), 207–217.

Johri, A. (2012). Learning to demo: The sociomateriality of newcomer participation in engineering research practices. *Engineering Studies, 4*(3), 249–269.

Johri, A. (2020). Artificial intelligence and engineering education. *Journal of Engineering Education, 109*(3), 358–361.

Johri, A. (2022). Augmented sociomateriality: Implications of artificial intelligence for the field of learning technology. *Research in Learning Technology*, 30. https://doi.org/10.25304/rlt.v30.2642

Johri, A., & Olds, B. (2011). Situated engineering learning: Bridging engineering education research and the learning sciences. *Journal of Engineering Education, 100*(1), 151–185.

Johri, A., Olds, B. M., & O'Connor, K. (2014). Situative frameworks for engineering learning research. In A. Johri & B. Olds (Eds.), *Cambridge handbook of engineering education research* (pp. 47–66). Cambridge University Press.

Johri, A., Roth, W-M., Olds, B. M. (2013). The role of representations in engineering practices: Taking a turn towards inscriptions. *Journal of Engineering Education, 102*(1), 2–19.

Kelly, G. J. (2014). Discourse practices in science learning and teaching. In N. G. Lederman & S. K. Abell (Eds.), *Handbook of research on science education* (Vol. 2, pp. 321–336). Lawrence Erlbaum Associates.

Knorr-Cetina, K. D. (1999). *Epistemic cultures: How the sciences make knowledge*. Harvard University Press.

Ko, L. T., Kachchaf, R. R., Hodari, A. K., & Ong, M. (2014). Agency of women of color in physics and astronomy: Strategies for persistence and success. *Journal of Women and Minorities in Science and Engineering, 20*(2), 171–195. https://doi.org/10.1615/JWomenMinorScienEng.201400819

Kumpulainen, K., & Kajamaa, A. (2020). Sociomaterial movements of students' engagement in a school's makerspace. *British Journal of Educational Technology, 51*(4), 1292–1307.

Ladson-Billings, G., & Tate, W. (1995). Toward a critical race theory of education. *Teachers College Record, 97*(1), 47–68.

Lahlou, S. (2018). *Installation theory: The societal construction and regulation of behaviour*. Cambridge University Press.

Latour, B. (1992). Where are the missing masses? The sociology of a few mundane artifacts. In W. E. Bijker & J. Law (Eds.), *Shaping technology/building society: Studies in sociotechnical change* (pp. 225–258). MIT Press.

Leander, K. M., Phillips, N. C., & Taylor, K. H. (2010). The changing social spaces of learning: Mapping new mobilities. *Review of Research in Education, 34*(1), 329–394.

Lee, O., & Fradd, S. (1996). Literacy skills in science learning among linguistically diverse students. *Science Education, 80*(6), 651–671.

Lee, O., & Luykx, A. (2014). Science education and student diversity- race, ethnicity, language, culture, and socioeconomic status. In N. G. Lederman & S. K. Abell (Eds.), *Handbook of research on science education* (Vol. 1, pp. 171–197). Lawrence Erlbaum Associates.

Lemke, J. L. (1990). *Talking science: Language, learning, and values*. Ablex Pub. Corp.

Mabogunje, A., & Leifer, L. J. (1997). Noun phrases as surrogates for measuring early phases of the mechanical design process. In *International design engineering technical conferences and computers and information in engineering conference* (Vol. 80456, p. V003T30A016). American Society of Mechanical Engineers.

Martin, D. B. (2007). Mathematics learning and participation in the African American context: The co-construction of identity in two intersecting realms of experience. In N. S. Nasir & P. Cobb (Eds.), *Improving access to mathematics* (pp. 146–158). Teachers College Press.

McDermott, R. (1993). The Acquisition of a child by a learning disability. In *Understanding practice: Perspective on activity and context* (pp. 269–305). https://doi.org/10.1080/1355800021016105

McDermott, R., & Varenne, H. (2006). Reconstructing culture in educational research. In L. A. Hammond & G. Spindler (Eds.), *Innovations in educational ethnography: Theory, methods, and results*. L. Erlbaum Associates. https://doi.org/10.4324/9780203837740

McLoughlin, L. A. (2005). Spotlighting: Emergent gender bias in undergraduate engineering education. *Journal of Engineering Education, 94*(4), 373–381. https://doi.org/10.1002/j.2168-9830.2005.tb00865.x

Medin, D. L., & Bang, M. (2014). *Who's asking? Native science, western science, and science education*. MIT Press.

Millar, R. (2006). Twenty first century science: Insights from the design and implementation of a scientific literacy approach in school science. *International Journal of Science Education, 28*(13), 1499–1521. https://doi.org/10.1080/09500690600718344

Moll, L. C., Amanti, C., Neff, D., & Gonzalez, N. (2009). Funds of knowledge for teaching: Using a qualitative approach to connect homes and classrooms. *Theory into Practice, 31*(2), 132–141.

Montfort, D., Brown, S., & Shinew, D. (2014). The personal epistemologies of civil engineering faculty. *Journal of Engineering Education, 103*(3), 388–416. https://doi.org/10.1002/jee.20050

Nasir, N. I. S. (2002). Identity, goals, and learning: Mathematics in cultural practice. *Mathematical Thinking and Learning, 4*(2–3), 213–247.

Nasir, N. S., Lee, C. D., Pea, R. D., & de Royston, M. M. (Eds.). (2020). *Handbook of the cultural foundations of learning*. Routledge.

Nasir, N. S., Rosebery, A. S., Warren, B., & Lee, C. D. (2022, in press). Learning as a cultural process: Achieving equity through diversity. In R. K. Sawyer (Ed.), *The Cambridge handbook of the learning sciences* (3rd ed.). Cambridge University Press.

National Academy of Engineering. (2002). *Diversity in engineering: Managing the workforce of the future*. The National Academies Press. www.nap.edu/catalog/10377/diversity-in-engineering-managing-the-workforce-of-the-future

National Academy of Engineering. (2005). *Educating the engineer of 2020: Adapting engineering education to the new century* (p. 11338). National Academies Press. https://doi.org/10.17226/11338

Norris, S. P., & Phillips, L. M. (2003). How literacy in its fundamental sense is central to scientific literacy. *Science Education, 87*(2), 224–240. https://doi.org/10.1002/sce.10066

Ohito, E. O. (2020). Some of us die: A Black feminist researcher's survival method for creatively refusing death and decay in the neoliberal academy. *International Journal of Qualitative Studies in Education, 0*(0), 1–19. https://doi.org/10.1080/09518398.2020.1771463

Ohlsson, S. (1996). Learning to do and learning to understand: A lesson and a challenge for cognitive modeling. In P. Reiman & H. Spade (Eds.), *Learning in humans and machines* towards *an interdisciplinary learning science* (pp. 37–62). Pergamon. https://hal.archives-ouvertes.fr/hal-00699807

Osborne, J. (1998). Science education without a laboratory. In J. Wellington (Ed.), *Practical work in school science: Which way now* (Vol. 139, pp. 156–173). http://ebookcentral.proquest.com Created from stanford-ebooks on 2018-11-24 10:25:21. http://link.springer.com/10.1007/BF02972560

Osborne, J. (2019). *Verbal conversation*. Stanford University.

Osborne, J., & Patterson, A. (2011). Scientific argument and explanation: A necessary distinction? Scientific argument and explanation. *Science Education, 95*(4), 627–638. https://doi.org/10.1002/sce.20438

Pawley, A. L., & Slaton, A. E. (2015). The power and politics of STEM research design: Saving the "small n." In *Proceedings of the American society for engineering education annual conference*. https://doi.org/10.18260/p.24901

Pea, R. D. (1992). Augmenting the discourse of learning with computer-based learning environments. In E. de Corte, M. Linn, & L. Verschaffel (Eds.), *Computer-based learning environments and problem-solving* (pp. 313–343) (NATO Series, subseries F: Computer and System Sciences). Springer-Verlag GmbH.

Pea, R. D. (1993a). Learning scientific concepts through material and social activities: Conversational analysis meets conceptual change. *Educational Psychologist, 28*(3), 265–277.

Pea, R. D. (1993b). Practices of distributed intelligence and designs for education. In G. Salamon (Ed.), *Distributed cognitions: Psychological and educational considerations* (pp. 47–87). Cambridge University Press.

Pennycook, A. (2006). *Global Englishes and transcultural flows*. Routledge.

Penuel, W. R., & O'Connor, K. E. V. I. N. (2010). Learning research as a human science: Old wine in new bottles. *National Society for the Study of Education, 109*(1), 268–283.

Pérez, G. (2022). Multicompetent learners. In *Proceedings of the 16th international conference of the learning sciences – ICLS 2022*. International Society of the Learning Sciences, Hiroshima, Japan.

Pérez, G., Sheppard, S., Muller, C. B., & Nittala, S. (2021, July 26). *Contextual social awareness in design: Engineering education as a catalyst for change*. American Society of Engineering Education, Virtual meeting.

Perkins, D. N. (1997). Epistemic games. *International Journal of Educational Research, 27*(1), 49–61. https://doi.org/10.1016/S0883-0355(97)88443-1

Poza, L. (2017). Translanguaging: Definitions, implications, and further needs in burgeoning Inquiry. *Berkeley Review of Education, 6*(2), 101–128.

Rosa, J. (2016). Racializing language, regimenting Latinas/os: Chronotope, social tense, and American raciolinguistic futures. *Language & Communication, 46*, 106–117.

Rosado-May, F. J., Urrieta Jr, L., Dayton, A., & Rogoff, B. (2020). Innovation as a key feature of indigenous ways of learning: Individuals and communities generating knowledge 1. In *Handbook of the cultural foundations of learning* (pp. 79–96). Routledge.

Roth, W.-M. (2013). The social nature of representational engineering knowledge. In A. Johri & B. M. Olds (Eds.), *Cambridge handbook of engineering education research* (pp. 67–82). Cambridge University Press. https://doi.org/10.1017/CBO9781139013451.007

Roth, W.-M., & Lawless, D. (2002). Science, culture, and the emergence of language. *Science Education, 86*(3), 368–385. https://doi.org/10.1002/sce.10008

Roth, W. M., & McGinn, M. K. (1998). Inscriptions: Toward a theory of representing as social practice. *Review of Educational Research, 68*(1), 35–59.

Secules, S., Gupta, A., Elby, A., & Tanu, E. (2018). Supporting the narrative agency of a marginalized engineering student. *Journal of Engineering Education, 107*(2), 186–218.

Secules, S., Gupta, A., Elby, A., & Turpen, C. (2018). Zooming out from the struggling individual student: An account of the cultural construction of engineering ability in an undergraduate programming class. *Journal of Engineering Education, 107*(1), 56–86. https://doi.org/10.18260/p.26239

Secules, S., & Mejia, J. A. (2021). Contextualizing the past to guide the future: Situating three critical frameworks for educational culture. *Studies in Engineering Education, 1*, 156–171. http://doi.org/10.21061/see.51

Song, S., & Agogino, A. M. (2004). Insights on designers' sketching activities in new product design teams. In *Volume 3a: 16th international conference on design theory and methodology, 2004* (pp. 351–360). https://doi.org/10.1115/DETC2004-57474

Stevens, R., O'Connor, K., Garrison, L., Jocuns, A., & Amo, D. M. (2008). Becoming an engineer: Toward a three dimensional view of engineering learning. *Journal of Engineering Education, 97*(3), 355–368.

Styhre, A., Wikmalm, L., Ollila, S., & Roth, J. (2012). Sociomaterial practices in engineering work: The backtalk of materials and the tinkering of resources. *Journal of Engineering, Design and Technology, 10*(2), 151–167.

Styre, A. (2017). Thinking about materiality: The value of a construction management and engineering view. *Construction management and economics, 35*(1–2), 35–44.

Taylor, C. (1992). *Multiculturalism and "the politics of recognition"*. Princeton University Press.

Trevelyan, J. (2010). Reconstructing engineering from practice. *Engineering Studies, 2*(3), 175–195. https://doi.org/10.1080/19378629.2010.520135

US Census. (2020). *American community survey: DP02 selected social characteristics in the United States*. US Census Bureau.

Valdés, G., Poza, L., & Brooks, M. D. (2015). Language acquisition in bilingual education. In W. E. Wright, S. Boun, & O. García (Eds.), *The handbook of bilingual and multilingual education* (pp. 56–74). John Wiley & Sons, Ltd.

Villanueva, I. (2018, June). What does hidden curriculum in engineering look like and how can it be explored? In *Proceedings of the American society of engineering education annual conference and exposition, minorities in engineering division*.

Warren, B., Vossoughi, S., Rosebery, A. S., Bang, M., & Taylor, E. V. (2020). Multiple ways of knowing: Re-imagining disciplinary learning. In N. Nasir, C. Lee, R. Pea, & M. M. de Royston (Eds.), *Handbook of the cultural foundations of learning* (pp. 277–294). Routledge.

Willis, P. (1977). *Learning to Labor: How working class kids get working class jobs*. Columbia University Press.

Wilson-Lopez, A., Mejia, J. A., Hasbun, I. M., & Kasun, G. S. (2016). Latina/o adolescents' funds of knowledge related to engineering. *Journal of Engineering Education, 105*(2), 278–311. https://doi.org/10.1002/jee.20117

Winner, L. (2017). Do artifacts have politics? In *Computer ethics* (pp. 177–192). Routledge.

Wittgenstein, L. (1997). *Philosophische Untersuchungen/Philosophical investigations* (2nd ed.). Oxford: Blackwell (First published in 1953).

11

Critical Perspectives on Diversity, Equity, and Inclusion Research in Engineering Education

Joel Alejandro Mejia and Julie P. Martin

1 Introduction

For many decades now, science, technology, engineering, and mathematics (STEM) fields have called to broaden the participation of women and minoritized populations – or populations that have less power or representation because of historically established social constructs (e.g., race, gender, national origin) that resulted in systemic disadvantages. Engineering is just one STEM field that has sought to increase the number of minoritized populations (Su, 2010) with unimpressive results. Although more minoritized students are earning engineering degrees, the percentage of minoritized engineering graduates employed in engineering occupations in the United States continues to be stagnant (National Science Board, 2018). Centering diversity, equity, and inclusion (DEI) in engineering education research is fundamental for the transformative changes in organizational culture needed to increase the percentage of minoritized engineering graduates employed in engineering occupations in the United States, as well as the participation of minoritized engineers in many other areas. This trend of participation is also present in other countries, where diversity-based discourse has gained momentum. According to Pineda and Mishra (2022), different arguments have been used worldwide to increase diversity in higher education, including economic benefit, social justice–oriented actions, and equity. Although the framing of DEI efforts may not be the same as in the United States, interest in increasing participation and representation of those who have been historically minoritized seems to be driven primarily by the Global North (Pineda & Mishra, 2022)

Recently, scholars – particularly in the United States – have asserted that DEI initiatives need to occur through anti-racist approaches (Coley et al., 2021; Cross, 2020; J. Holly Jr., 2020; Long III, 2020), but DEI "success" is still usually measured by how many underrepresented individuals are able to enroll or complete programs (Hurtado et al., 2012). Countries where colonization, as well as *de jure* and *de facto* segregation, played an important role in shaping the social fabric seem to connect with this anti-racist approach to DEI and decolonization efforts, such as those in South Africa (Pineda & Mishra, 2022; Williams et al., 2021) and countries in the Caribbean (Sappleton & Adams, 2022). Nonetheless, many institutions may opt for DEI discourse to maintain – rather than disrupt – the established historical norms and prevent radical change. For instance, DEI work has not been able to address high poverty rates, systemic bias, or unfair remedial placement that impacts minoritized students in engineering education. Most DEI initiatives continue to focus on increasing the numbers of minoritized students and diversifying the student body (Hurtado et al., 2012) even

DOI: 10.4324/9781003287483-13

though engineering education scholars have argued for increased and different approaches to DEI work (Long III & Mejia, 2016). And even with this focus, DEI efforts have failed to increase the number of engineering faculty of color present in many US campuses (Cross, 2020).

As we explore in this chapter, one fundamental problem is the multiple meanings of DEI. In the United States, the phrase "diversity, equity, and inclusion" can broadly refer to research or practices that focus on individual constructs without always remembering to question privilege, personal interests, or systemic hurdles as well. In other parts of the world, according to Pineda and Mishra (2022), the semantics of diversity vary across regions, and no universalized consensus exists on DEI, but most of its application is directed toward STEM and medicine in the United States and Canada, while in the UK, Ireland, Oceania, and the rest of Europe, DEI efforts are focused on teaching practices (Pineda & Mishra, 2022). Similarly, the meaning of diversity seems to focus more on race and ethnicity in the United States, Canada, the UK, and Ireland, while gender and cultural diversity seem to be more prominently aligned to DEI efforts in Oceania and the rest of Europe. On the contrary, the Global South embraces different meanings for DEI that range from quality evaluation and disability to multiculturalism, race, and gender (Pineda & Mishra, 2022). Although DEI emerged in the United States to inform strategic plans to diversify the student body, calls to situate the internationalization of DEI work are becoming more common but certainly need to be amplified (Özturgut, 2017; Pineda & Mishra, 2022). In an effort to expand DEI work that addresses more than broadening the participation of minoritized populations, engineering education researchers are examining what DEI-based research means and entails for the field.

We discuss how the terms "diversity," "equity," and "inclusion" have been utilized in engineering educational research and systems. The term "diversity" possesses multiple meanings in engineering education research, meanings dependent on the end goals of the groups using the term as well as the premises behind the term; the same can be said about "equity" and "inclusion." These semantic divides make it difficult for us to discuss what DEI work means, because scholars are using the same words but defining them – and measuring their outcomes – quite differently. The meaning of a term reflects material, institutional, and political agendas. We assert here that motivations for and framing of DEI research in engineering education research need to move beyond economic benefits.

Throughout this chapter, we interrogate the meaning of "diversity, equity, and inclusion" within the context of engineering education research, primarily from a US perspective. We follow a US perspective because of our training and situatedness, but also to expand on the argument that DEI efforts are primarily applied to STEM research contexts in the United States and Canada as opposed to other regions where the focus is on teaching (Pineda & Mishra, 2022). We also describe how the framing of DEI work reflects the sociopolitical agendas of those of us engaged in this work. DEI efforts are often framed in neoliberal terms, especially in STEM, and perhaps even more so in engineering. We believe we need to transform the world of engineering education research at the same time we (re)interpret what DEI means. As de Sousa Santos (2015) argues, transformation requires a collective (re)interpretation of the world. DEI work and its (re)interpretation is a collective task to revolutionize how DEI in engineering education research is both theorized and put into practice. Moreover, we argue that DEI should be used not only to rebrand institutions of higher education in the Global North but to relinquish the power that these institutions yield over the global discourse. We end with a call to action for researchers engaged in DEI work and those who participate in DEI-based research, specifically inviting others to center their work on methodological activism and engage in pluralistic research approaches.

2 Authors' Positionalities

We start by reflecting on our positionalities and how they influenced our work on this chapter. Using Secules and colleagues' (2021) framework, we each examined our positionality with respect to the research topic, epistemology, ontology, methodology, researcher-as-instrument, and communication, and we included our reflection here. In doing this, we discovered that the salient lived experiences at play in our process and the final product were quite different for each of us; Alex primarily reflected on his ethnic identity and its relationship to his educational experiences, and Julie primarily reflected on how her identity as an academic was related to the experience of co-authoring with Alex for the first time.

Alex: I self-identify as a gay Latino – more specifically Mexican American – who currently resides in the United States Southwest but spent his formative years (3–15 years old) in a small town in Mexico. My experience is informed not just by my educational background but also by my lived experience: unwillingly immigrating to the United States; experiencing deficit ideologies in school in the United States for not speaking English; being denied the opportunity to participate in STEM spaces; observing neoliberal realities imposed on my community (and its resulting inequities) after the North American Free Trade Agreement (NAFTA) was signed between Canada, Mexico, and the United States; and eventually becoming an *atravesado* in engineering like many other Latinos/as/xs because we are often seen as invalid inhabitants of engineering – a transgressor who does not belong in a White-dominated space (Mejia et al., 2022). The research topics and questions I choose to work on are informed by this lived experience because my identity and my close proximity to issues of diversity, equity, and inclusion (DEI) are strongly tied to who I am as a researcher and individual. I do recognize, however, that my training in engineering education involved immersing myself in the colonial project of education, and I am "a product of colonial educational systems" (Secules et al., 2021, p. 28). Nonetheless, I continue – to the best of my ability – to reject remnants of coloniality in my own epistemology. The lens I use in this chapter is without a doubt influenced by my place-based training in the United States, and I may sometimes consciously or unconsciously base my framing of DEI on ideologies prescribed primarily by the Global North (Pineda & Mishra, 2022). In this work, although mostly informed by US-centric framings, I invite readers to think about diversity, equity, and inclusion not as a *rebranding act* for institutions of higher education and engineering education programs but to examine their own approaches to DEI issues.

Julie: I will come out and say the thing I think needs to be said. I do not have a PhD in engineering education. Alex does. I was already a faculty member when the first PhD programs were being formed in the United States. With graduate training in materials science and engineering, my engineering education research work originated with one of the National Science Foundation (NSF)–funded Rigorous Research in Engineering Education cohorts. Essentially, this was a five-day crash course in what Alex has spent years learning in graduate school. I spent the first five, six, or maybe even seven years as a faculty member (in one of the PhD programs, nonetheless) apologizing for becoming a qualitative researcher. I really did. Look at my early publications (on second thought, please don't!) – I downplayed small sample sizes and lamented "lack" of generalizability. It was not until I was tenured that I felt comfortable moving from the interpretive paradigm (itself a major stretch from my formal training) to one that was critical. By that time, I was the program director for engineering education in NSF's engineering directorate, the editor-in-chief of *Journal of Women and Minorities in Science and Engineering*, a journal that focuses on DEI scholarship, and had long held several DEI-related positions in national professional organizations. So to say I came to the critical paradigm late seems like an understatement. And here I was, co-authoring a chapter on critical DEI scholarship with Alex Mejia, who, as far as I knew, was a critical DEI researcher from Day 1. We are different generations of engineering education scholars. Could I keep up? Would my ideas pale in comparison? I was nervous. My admiration for him as a person and his work convinced me it would be okay. To write 10,000+ words with someone you have never collaborated with before is – well, it *could* have been really difficult, but actually it was not. I think that was because of the mutual respect we had for each other's ideas and each other's places on our individual and collective scholarly journeys. Initially, I think we had two different visions for what the chapter could be. Ultimately, I think we merged them nicely. Starting with the outline and continuing into the final draft, we worked hard to give each other the space to say what we each thought needed to be said. We trusted in each other's experiences, in each other's positionalities in the world of DEI scholarship. That spirit has shaped our chapter. I hope it also helped bridge the "generation" gap in engineering education DEI scholars.

Readers will note that we used our experiences as Americans at institutions in the United States to write about the state of DEI engineering education scholarship from a US-centric perspective. This focus is very much a product of our academic history and expertise, rather than a reflection of the value we place on worldwide engineering education scholarship.

3 Views on Diversity

Across engineering education literature in the United States, interpretations of the term "diversity" are inconsistent. Most researchers associate diversity with access to engineering (Chubin et al., 2005; May & Chubin, 2003; McGee & Bentley, 2017; Samuelson & Litzler, 2016). Engineering education researchers have also framed diversity in terms of (a) numbers (Fletcher et al., 2021; Revelo et al., 2017), (b) culture (Atadero et al., 2018; Ikram et al., 2016; Godfrey & Parker, 2010), (c) economic benefits (Commission on the Advancement of Women and Minorities in Science, Engineering, and Technology Development, 2001; Vandenberghe, 2021), and (d) hierarchy ("Diversity," 2008). Researchers have adopted these four understandings of diversity in engineering education research, but rarely all at the same time in the same study or program.

The first framing of *diversity* involves counting the number of individuals included in a sample while sometimes considering specific demographic and nondemographic criteria such as race, gender, age, or ethnicity as attributes that encompass diversity in the sample. Conceptualizing diversity solely in numeric terms can be dangerous, particularly if diversity is defined at the collective level and not the individual level or used as a justification to disregard systemic barriers in the name of statistical significance (Revelo et al., 2017). As indicated by Qin et al. (2014), "[d]iversity is concerned with differences (e.g., personal attributes) between people; however, there are numerous attributes that differentiate people" (p. 136). Some engineering education research studies have strategically agglomerated data under the premise that statistical significance is necessary to demonstrate differences (Revelo et al., 2017; Revelo & Stepin, 2018). The justification for this agglomeration of data creates the impression that minoritized groups may be representative of monolithic populations (Revelo et al., 2017). As researchers in engineering education continue to engage in DEI-based research, it can be important to recognize that increasing numbers in a sample do not necessarily mean that differences among individuals have been taken into consideration. And it is important to consider that demographic characteristics represented in numbers may not create the changes desired as we researchers move toward an internationalization of DEI work. A contextual understanding of power, privilege, and systemic barriers is needed, including paying attention to socioeconomic status, first-generation status, and immigration status, and the implications of these factors in accessing and participating in higher education globally (Özturgut, 2017).

The second framing of diversity involves describing individuals in terms of cultural markers. Engineering education researchers have engaged with social theory to understand diversity in terms of culture. However, engineering education researchers have often oversimplified the construct of culture, assuming culture to be static and regularly descriptive of a system of behavioral norms (Secules & Mejia, 2021). The oversimplification of culture as a static construct has led to the uncritical analysis of how deficit ideologies have permeated the engineering education literature (Mein & Esquinca, 2017; Mejia et al., 2018). That is, it is assumed that individuals (primarily individuals from low-income, historically minoritized groups) carry deficits that need to be fixed. These deficits, according to proponents of deficit thinking, are the result of the inadequacy of the home to prepare individuals and their subsequent failure in school (Valencia, 1997, 2010). Because researchers have chosen to describe minoritized populations in engineering education research in deficit-oriented ways, essentialization of their cultures has occurred in DEI-based research. Although conceptualizations of diversity in terms of culture in engineering education research has increased over the years, we believe that it should be done carefully, given that pragmatic interpretations of culture can lead to oppressive marginalization. For example, a narrow framing of culture can lead to the perception that sociocultural processes occur in isolation, thus ignoring the material and lived realities of individuals and (re)producing deficit ideologies that hurt marginalized populations.

When engineering education researchers seek to describe diversity through differences in culture, understanding culture becomes the vehicle by which researchers define, analyze, and evaluate a population. Qualitative research approaches in engineering education have shaped how culture is studied in the field. Godfrey and Parker (2010), for example, described the cultures of engineering in terms of ways of knowing, doing, and being. They analyzed how faculty and students recognized one another as members of the world of engineering by the ways in which they behaved, believed, or acted. They warned readers about essentializing culture and normalizing certain attitudes, behaviors, and belief systems of engineering and engineers, which can lead to the lack of recognition and the invalidation of other forms of knowing, doing, and being, thus negatively impacting minoritized students. For example, they illustrated that engineering students and faculty often argue that engineering solutions are gender- and race-free (Godfrey & Parker, 2010), thus denying the validity and existence of ways of knowing and being that deviate from the normative White, Eurocentric worldviews.

The third framing researchers use to approach diversity is through the economic motive argument. That is, they promote the notion that increasing diverse ideas, cultures, or races will result in economic benefits. These arguments are primarily seen in studies that seek to examine the practice of engineering in a globalized world and the importance of diversity to promote innovation, efficiency, and improved quality (Chubin et al., 2005; LaFave et al., 2015; Vandenberghe, 2021). Both the concept of diversity as well as the concept of economic benefit have been central to engineering (Chen et al., 2019). In a field where profits are highly valued (Chen et al., 2019) and meritocracy and objectivity are the norm (Cech, 2013; Slaton, 2015; Riley, 2008), diversity is lauded as a beacon for innovation. Roughly, the economics-driven diversity claim goes something like this: diverse engineers bring diverse ideas and diverse approaches to the table, which encourages more innovative results and higher profits, or more productivity, which leads to more prestige, funding, and publications.

We strongly assert that engineering education researchers need to consider their ethical reasons for doing DEI work when diversity is approached through an economic motive argument. Chen et al. (2019) argue that when the economic argument is used for DEI initiatives, it negatively impacts women and minoritized populations. For example, women and minoritized individuals are seen as people that "take care" of others and are therefore better suited to undertake tasks that involve affective labor. However, affective labor is typically not beneficial to the institution unless it produces capital (Weeks, 2007), and it negatively affects women and minoritized individuals by devaluating diversity efforts if capital gains are not achievable. Thus, efforts to diversify the institution become viable and feasible if, and only if, they will lead to an increased revenue. When we, researchers, choose to center economic motive arguments for diversity, we risk detrimental impacts on historically marginalized populations. The economic motive argument for diversity is not exclusive to the United States context; this is an argument that has been widely used across the globe (Dlouhy & Froidevaux, 2022; Reader, 2006; Vandenberghe, 2021). In the United States, however, this narrow conceptualization of diversity in capitalist terms has created a system where institutions of higher education seek to improve the numbers of minoritized individuals on their campuses to have access to streams of funding (such as the case of Hispanic-serving institutions in the United States) (Garcia, 2018, 2020) or promote projects in the name of development while ignoring the systemic barriers imposed by the institutions themselves.

Although measuring diversity remains a challenge, measuring economic benefits seems more straightforward. It may be that this perceived "ease" of tracking outcomes is one of the reasons that diversity is often framed and approached in terms of economic impact. In engineering education research, this economic framing creates an even more conspicuous disparity between the privileged (i.e., the researcher) and the oppressed (i.e., marginalized populations). Understandably, tensions and hostility emerge when power dynamics are ignored, particularly when the researcher benefits the most, instead of the community being studied. Researchers recognize that diversity can be a tool to achieve economic success because diversity in the workplace or in particular projects signifies progress, access to global markets, and as a result, more profit (Qin et al., 2014).

The fourth way in which researchers frame diversity is in terms of stratified hierarchies. Hierarchies allow for social functioning, including adoption of values, beliefs, or behaviors. That is, diversity defined by stratified hierarchies acknowledges the historical ways in which groups have existed, particularly through the lenses of power, privilege, and wealth. As indicated by Phillips et al. (2018), "[h]ierarchy is a social dimension that emerges only within group contexts and cannot be perceived within a single individual" (p. 8). Studies have shown that visual cues for perceptions of hierarchy help people understand whether or not they belong, determine the roles people play in a group, and contribute to the identification of in-group dominance (Phillips et al., 2018). Several studies in engineering education have reported on the experiences of historically marginalized populations in engineering through an analysis of hierarchies (Faulkner, 2000, 2007). For example, Tonso (2006) explored how the establishment of hierarchies among engineering students impacted the dynamics of teamwork. Interactions among students are impacted and lead to the development of identities – as identifiers – that dictate who belongs in engineering and who does not. These established hierarchies and identities ultimately serve to marginalize certain individuals by the creation of in- and out-groups (Faulkner, 2000). Recognizing that stratified hierarchies impact diversity in different ways is important because it acknowledges the historical and sociocultural practices (i.e., the socialization processes) that impact diversity.

We situate our work and our understanding of diversity inside these framings and approaches to diversity as it has been described in engineering education research. We argue that looking at diversity through one of these conceptualizations instead of analyzing diversity through a critical understanding of all these frames will not result in transformational changes. As we design DEI initiatives and research, we advocate for approaching diversity holistically, and by that we mean that we, researchers, deem that a consideration of diversity that encompasses all the framings presented here is necessary. Framing diversity as a complex set of variables provides us as researchers with an opportunity to understand big systems and context while we study how individuals situate themselves in space and time and relative to one another. For example, research in engineering education has focused on – and erroneously embraced – the idea that Latinos/as/xs are a monolithic group, thus highlighting the lack of understanding within group differences (Revelo et al., 2017). If we reject DEI research typically grounded on framings of diversity in terms of numbers, we can move the field past these simplistic notions about diversity.

4 Views on Equity

Framings for the term "equity" also vary and tend to be context-specific. The meaning of the term "equity" reflects material, institutional, and political agendas. Brand (2015) asserts that "equity is not a static concept but one that different social groups actively construct to make claims on the state to support their own interests" (p. 249). Thus, it is important for researchers and others who engage in equity work to consider their positionality (Secules et al., 2021). Positionality involves one's personal values, beliefs, views, experiences, education, and perspectives that dictate how one understands the world around them and their position in it. We are aware that asking researchers to engage in deep reflection on their own positionality is in direct conflict with those in higher education who still maintain that researchers can design studies, collect data, and interpret results without bias. As the researcher engages with the world, their definitions, priorities, and actions will take form according to what is conceptualized as equity and, to a certain degree, their own interests.

It is important to bear in mind that equity can be bounded by space and time and deliberately defined according to what the researcher holds most important. The researcher determines the population that will be studied, the criteria for such selection, and eventually, how to analyze and interpret the data collected; researchers choose to deeply engage with historicity, criticality, positionality, or situatedness, or they choose not to, thus furthering inequities (Secules et al., 2021; Secules & Mejia, 2021). Failing to critically think about identity, community membership, and power dynamics prevents the researcher from fully embracing the complex nature of equity research (González et al., 2011; Secules & Mejia, 2021). To achieve equity, researchers in engineering education should critically think about and define how the practices, policies, and systems affect the experiences, outcomes, and access of those who participate in the research, and work toward dismantling those structures.

In engineering education research, researchers tend to frame equity in two broad ways: (a) a movement toward methodologies that engage researchers in providing equal access throughout the research process and (b) a movement toward research resulting in the weakening or disruption of imbalanced power structures to provide equal access to engineering. These two fronts are interrelated. The first is scholarship-oriented, while the second one is action-oriented. Research that incorporates pluralistic, cultural, and decolonial methodological practices contributes to the challenging of deficit ideologies embedded in past and present research activities.

First, we believe researchers in engineering education should continue to frame equity in dynamic and shifting ways as sociopolitical and sociohistorical contexts emerge and change. Researchers who are committed to the idea of providing equal access throughout the research process and also who acknowledge historical, political, social, and material privilege and oppression of the groups involved in the research process (whether these are in-groups or out-groups) end up doing research that impacts the researchers as well the participants. As Artiles (2019) has rightly claimed, "[e]quity research cannot disregard history" (p. 326). Equity-mindedness requires a recognition of the historical context, who participates in the research and to what end, the impact of that history on the research process, the interpretation of results and the implications for equal access, and the resulting actions taken to address injustices done in the past. For example, the insistence on the "colorblindness" in engineering education research that has persisted throughout the years has created a racialized version of minoritized groups while ignoring the White supremacist nature of the discipline (Mejia et al., 2020; Pawley et al., 2018).

Second, researchers in engineering education who seek to dismantle systemic structures that prevent equal access to engineering for all individuals tend to adopt more critical action-research practices. Researchers in engineering education have called for critical equity research that weakens or disrupts imbalanced power structures (Cross, 2020; J. Holly Jr., 2020; Long III, 2020; Mejia et al., 2018; Pawley, 2017; Slaton & Pawley, 2018) to reach a stage of liberation where the minds are liberated and the playing field is leveled for all (Freire, 2003). The movement to dismantle power structures has led to the conceptualizations of equity in engineering education to be described in terms of providing access to engineering spaces (Capraro et al., 2013; Ferreira, 2002; Villanueva Alarcón et al., 2021), achieving gender parity (Bilimoria & Liang, 2012; Franzway et al., 2009; Sharp et al., 2012; Slaton, 2015), countering deficit notions (Harper, 2010; Mejia et al., 2018; Secules & Mejia, 2021), and transforming approaches to engineering education research that move toward asset-based framings (Castaneda & Mejia, 2018; Martin et al., 2022b; Mejia et al., 2018; Mejia et al., 2019;

Wilson-Lopez et al., 2016). Researchers focused on equity in terms of access ask how easy or how difficult it may be for individuals to enter certain spaces, procure resources, secure opportunities, or achieve equal opportunity. Researchers focused on equity in terms of gender parity may seem to be limiting their research to one population, but in fact, they draw from critical perspectives to argue for the dismantling of structural and institutional barriers that negatively impact multiple populations (Slaton, 2015). Gender equity work in engineering education has served the purpose of pushing the field to critically analyze the programs, policies, and practices that impact not only women but also Black, Indigenous, and people of color (BIPOC) (DeAro et al., 2019; Torres, 2012). Finally, researchers seeking to counter deficit notions in both methodologies and research approaches refer to the importance of (re)framing how communities are portrayed while acknowledging them as holders and creators of knowledge (Delgado Bernal, 2002; Delgado Bernal & Villalpando, 2002). Although these trends are increasing in the field, we note that only a small amount of research in engineering education addressed equity in terms of curriculum, and we continue to see curriculum that fails to provide culturally responsive engineering education (Gelles et al., 2020; Gelles et al., 2021; Hoople et al., 2020), pointing to opportunities for engineering education research to contribute to changing the trend.

5 Views on Inclusion

Inclusion refers to the fair, equitable, and healthy participation of all individuals within an organization or community while increasing the presence of marginalized populations. Inclusion also involves the presence of spaces where individuals feel valued and acknowledged (Özturgut, 2017). Historically, inclusion has been defined according to the needs of the majority, and it has benefited people in positions of power the most while disregarding the experiences of minoritized individuals (Qvortrup & Qvortrup, 2018). More recently, inclusion efforts in engineering education research have examined this history, asking who participates in the research, for what purpose, for what ends, and for whose ultimate benefit. Researchers in engineering education have adapted their conceptualizations of inclusion as historical and sociopolitical contexts of engineering that are more deeply understood by the research community.

While efforts to increase students' opportunities to learn engineering and participate in engineering inclusively have grown, the promise of "engineering for all" remains elusive. For example, in the United States context, educational policies such as the 1988 reauthorization of the Elementary and Secondary Education Act (ESEA) – and ultimately No Child Left Behind (NCLB) – are credited as the fuel that ignited the movement toward national testing and standardization that disproportionally affected minoritized students by creating a one-size-fits-all approach to education (Penfield & Lee, 2010). The one-size-fits-all approach has negatively impacted engineering education research as well. We can see how these one-size-fits-all policies in engineering and technology assessments – particularly in K–12 education settings – that use standardized assessments on the national level in the United States, such as the National Assessment of Educational Progress (NAEP) Technology and Engineering Literacy (TEL) Assessment (National Academies of Sciences et al., 2017, 2018), are developed for and privilege native speakers of English. While such assessments may yield valid scores for students from the dominant culture who are native English speakers, there is little doubt that assessments yield valid scores for minoritized students, especially students from emergent bilingual populations (Penfield & Lee, 2010) and even students with disabilities or neurodivergent. Similar standardization projects have increased around the world, bringing with them a myriad of challenges for inclusion, most importantly the ways in which standardization has shaped educational policies (Al'Abri, 2011; Roldán Vera & Robles Valle, 2020; Ross & Gibson, 2007). As individuals and organizations that hold power in the decision-making process throughout research activities – not only as external participants of the communities being studied but also as members of those groups – researchers would be wise to ask the critical question: "Who benefits and who loses as a result of my conceptualization of inclusion?" Ignoring the history and sociopolitical contexts that surround inclusion initiatives keeps the door open for numerous pitfalls and contradictory results between the aims of inclusion and the intended beneficiary.

Diversity in engineering education research may seek to broaden the participation of minoritized individuals in terms of numbers, but inclusion demands accountability for those diversity efforts to transform equitable access to engineering. While many engineering education researchers have sought to engage in inclusive research, some of the research seems to focus primarily on educational practices rather than engaging methodological dimensions of research inclusion, such as levels of inclusion, arenas of inclusion, and degrees of inclusion (Qvortrup & Qvortrup, 2018). Inclusion demands different strategies aimed at achieving social justice while validating individual diversity (i.e., lived experiences) (Puritty et al., 2017) as well as promoting opportunities that enhance human capabilities (Leydens & Lucena, 2017). As noted by Puritty and colleagues (2017), *inclusion* refers to how individuals "are treated and how they feel" (p. 1101) and requires that researchers acknowledge that structural biases exist and influence how individuals experience their lived reality in those spaces.

Inclusion provides the space for all individuals to feel respected, engaged, motivated, acknowledged for their contributions, and valued. As researchers, focusing on inclusion means intentionally making an effort to remove barriers that may otherwise be present while critically engaging in reflexive practices that contribute to the elimination of research bias. These multiple frames of reference are necessary to consider if DEI work is to be implemented in engineering education research.

6 Calls to Action

Leading scholars have recently begun calling for researchers in the field to reject using deficit-oriented approaches to research and instead to shift motivation for DEI research to social justice concerns. In doing so, our research will be stronger when it recognizes the complex aspects of diversity – not just the numerical or perceived "culturally salient" or "measurable" characteristics – and explicitly addressing equity and inclusion as inherently valuable, regardless of economic outcomes. By working in interrelated (a) critical, (b) methodologically robust, and (c) pluralistic ways, engineering education researchers can advance the movement toward socially just research that results in more transformational consequences for historically marginalized people.

6.1 The Call to Be Critical

Patrick et al.'s (2022) state-of-the-art review identifies six elements of critical research methods from 22 exemplar studies in STEM higher education. These elements are:

1. Establishing a critical reflection of author positionality.
2. Acknowledging and describing the historical or social context for oppression.
3. Engaging participants.
4. Maintaining consistency of critical approaches across the theory, methods, and discussion of published manuscripts.
5. Providing a deep explanation of critical race theory tenets emphasized in the study.
6. Providing implications sections in published work that seek to disrupt power structures.

Here we highlight the first two elements identified in the review. We refer readers to their review for more details and examples. In accordance with these authors' suggestion, we encourage researchers to look beyond engineering education to other STEM and/or science education studies and higher education fields for other exemplars (Calabrese Barton & Tan, 2009; Harper, 2010; Marx,

2016; McGee & Bentley, 2017; Mejia et al., 2018; Peterson, 2021; Secules et al., 2021; Secules & Mejia, 2021; Tan, 2011; Villanueva Alarcón et al., 2021).

Scholars have challenged engineering education researchers to do a better job of discovering, naming, and exploring the historical, social, and institutional contexts of their work in many ways, such as identifying anti-Black racism (Holly Jr., 2020), heteronormativity (Hughes, 2017; Yang et al., 2021), and ableism (Peterson, 2021) as origins of the oppression that has led to the need for DEI work in the field (Martin et al., 2022b). This contextualization has appeared in multiple sections of published manuscripts, including the introduction or background of a manuscript (e.g., Cech & Waidzunas, 2021; Martin & Garza, 2020; Yang et al., 2021; Patrick et al., 2022) incorporated in the author positionality statement (e.g., Secules et al., 2021; Sellers et al., 2022), or in more than one section (e.g., McGee et al., 2021).

Recent work in the field has also urged engineering education researchers to engage in reflexive positionality work and pushed for positionality statements to become the norm in published research. A meaningful positionality statement goes beyond simply stating the authors' social identities; it includes a discussion of how those identities influenced all aspects of the research design and implementation. While positionality statements are becoming more common in qualitive engineering education research, they are much less common in quantitative studies. Scholars have asserted that positionality is important, regardless of methodology, because it affects multiple areas of the research (Hampton et al., 2021), including the research questions asked, epistemology, ontology, methodology, researcher-as-instrument, and communication (Secules et al., 2021). Secules and colleagues (2021) recently articulated three ways in which positionality has been used in engineering education: (1) acknowledging practice, (2) establishing transparency of self-attributes, and (3) contextualizing methodology. At least one journal in the field, *Journal of Women and Minorities in Science and Engineering*, now requires a positionality statement as a condition of publication for all manuscripts (Martin et al., 2022a), and others, such as the *Journal of Engineering Education*, encourage it.

Researchers have shown how positionality statements offer an opportunity for engineering education researchers to describe how their DEI-focused work leads to liberatory praxis (Mejia et al., 2018), which is a laudable ultimate goal of DEI research. The liberation process is characterized by achieving *concientização* through dialogue and thus gaining the ability to hold the most critical possible view of reality (Freire, 2003). Liberation is reached when the researcher engages in scholarship that is guided not just by theory but also by critical reflexivity in combination with action (i.e., praxis; Mejia et al., 2018; c.f. Freire, 2003). Liberatory praxis is characterized by the state in which the researcher has reached critical consciousness in all aspects of the research process (i.e., theory, reflection, and action) and enacts critical consciousness in all aspects of their work, including DEI research. This is especially important to highlight because not all scholars possess the same amount of privilege. Scholars acknowledge that publishing positionality statements has the potential to carry a personal and professional risk, and researchers should retain control of which identities they choose to publish (Hampton et al., 2021; Martin et al., 2022a; Secules et al., 2021).

Positionality statements can – and perhaps should – take different forms in order to be effective and liberatory. Recently published work in the field includes positionality statements written in the third person (e.g., Burt et al., 2018) and first person (e.g., Hughes, 2017; Mondisa, 2020). Positionality statements for co-authored works can be effectively written in a collective sense (e.g., Cech et al., 2021; McGee & Bentley, 2017) or include statements by individual authors (e.g., Cross et al., 2021; Martin & Garza, 2020) or a combination of the two (Sellers et al., 2022). For example, here we highlight a few examples from recent literature.

Patrick et al. (2022) name the positionality statement by López et al. (2018) as an exemplar because of the way the authors connect their identities and commitment to critical race theory (CRT) to research decisions such as their commitment to preserving the experiences of Black students in their sample (despite being a small proportion of their sample). Leyva's (2021) positionality statement stands out because of the adherence to Milner's framework for positionality in educational research that is based on CRT, which includes researching the self, researching the self in relation to others, engaged reflection and representation, and shifting from self to system (Milner, 2016). The author's statement discusses how the author addressed each element of the framework.

Similarly, Sellers et al. (2022) used a three-tiered model for reflexivity developed by Sochacka et al. (2009) in which researchers examined ontological and epistemological assumptions, the influence of their personal values, and the influence of their prior experiences. This article's positionality statements also include an acknowledgment of how recent US events (i.e., the murder of George Floyd by police and the #BlackLivesMatter movement) in combination with their positions of relative power as White women influenced their methodological choices. McGee (2021) provides excellent contextualization of the historical context of her work under a US president who was antiscience, and writes a unique positionality statement that establishes her motivation for the work in the context of a particular event, the 2013 death of Ella Kissi-Debra, a 9-year-old Black girl who died of respiratory failure after suffering from asthma. She goes on to establish how that example embodies the struggle of Black and Brown people in the United States who are "situated in disproportionately hazardous environments (mentally and physically) and are offered the solution of being resilient against omnipresent forces of anti-Blackness, racism (individual, economic, structural, environmental), and greed," and yet who are "systematically excluded from conducting the research that could incorporate equity into STEM" (p. 11).

6.2 The Call to Be Methodologically Robust

Engineering education scholars have urged researchers to be methodologically robust by thoughtfully choosing and honoring theories and research methods (Mejia et al., 2018; Secules & Mejia, 2021) and for the political purpose of empowering marginalized populations through liberatory praxis (Mejia et al., 2018). That is, they have called for researchers to embrace methodological activism because it results in "equitably consequential" (Calabrese Barton et al., 2017) research. *Methodological activism* refers to the intentional use of research design choices as a form of activism to contribute to a political purpose (Martin et al., 2022b; Ong, 2005; Sellers et al., 2022; Zuberi & Bonilla-Silva, 2008). We echo the calls to action articulated by Cross (2020), Holly Jr. (2020), Secules and Mejia (2021), Martin and colleagues (2022b), Mejia and colleagues (2018), and many others for engineering education researchers to carefully consider their choices of research questions, theory, and methods/methodologies that empower marginalized populations and have potential to enact social change. Adhering to methodological activism means that we all work toward more participatory and liberatory research (Mejia et al., 2022).

Research questions as methodological activism. A central aspect of methodological activism is the way in which research design decisions are made: how individuals and communities are framed, why the research is done with a particular population, and for what purpose. Harper (2010) was one of the first researchers to call for framing research questions in STEM education in ways that avoided deficit framing of marginalized individuals and groups. His anti-deficit examples (also called asset-based examples) continue to be relevant in engineering education. In the last few years, multiple examples of asset-based questions have been published, particularly when studying the experiences of people of color. Here are a few examples that serve as models:

- What are the motivational factors that influence Black men in graduate engineering programs at predominantly White institutions to persist? (Burt et al., 2020)
- What are the institutional and noninstitutional factors that lead to persistence in engineering among women of color faculty? (McGee et al., 2021)

- What forms of resilience, particularly coping strategies, played a role in Black women's experiences of managing within-group tensions? (Leyva, 2021)
- Hypothesis: Minoritized students' funds of knowledge will support their interest in engineering, foster self-efficacy beliefs, and contribute to their choice to pursue an engineering major. (Verdín et al., 2021)

In framing questions that reject deficit notions of individual participants, their families, or their communities, Martin et al. (2022b) called for more researchers to ask questions that critique the system within which participants and the study are situated. The following examples from recently published research illustrate this idea:

- How do mathematical contexts in P–12 education at a historically White university function as White patriarchal spaces to shape Black women's within-group tensions? (Leyva, 2021)
- What ecological and sociological barriers within a college of engineering promote non-normative student role strain in Black male students pursuing graduate engineering degrees? (Burt et al., 2018)
- How can academic mentoring, as a construct, be race-reimaged? (Villanueva et al., 2019)

Theory as methodological activism. Engineering education scholars have urged other researchers to reconsider how the voices of marginalized populations are portrayed in their research. We echo their voices in calling for intentional and critical reflection, as noted by scholars such as Secules et al. (2021), as well as a holistic integration of reflection, theory, and action to achieve liberatory actions that serve historically marginalized groups, as indicated by Mejia et al. (2018). Drawing from Freire's (2003) critical pedagogy, Mejia et al. (2018) urge engineering education researchers to use their methodology not only to describe events or experiences but also to "ask questions of power, privilege, and oppression" and identify ways to "fight alongside" those who are research participants (p. 10). Choosing theories that align with methodological activism includes using and honoring theoretical paradigms that were designed by or for Black, Indigenous, and People of Color (BIPOC) or marginalized communities and ones that overtly articulate causes of oppression. Currently, this often means drawing on theories from other disciplines and applying them to engineering education.

What follows are just a few salient examples of recent work that draws on theories from other fields and applies them to engineering education. Yang's (2021) work with LGBTQ engineering students draws on queer theory and resistance theories, including queer resistance and transformational resistance. Hughes's (2017) conceptual framework uses theories of internal processes around sexual orientation disclosure and sexual orientation covering. Hancock et al. (2021) apply dis/ability critical race theory (DisCrit) to early childhood education via fieldwork by preservice teachers. Recent studies with people of color include Thomas and colleagues' (2016) as well as Cross and colleagues' (2021) use of womanist theory (also called Black feminist thought or womanism). Both are examples of research that centers on the history and experiences of women of color (and especially Black women). McGee's (2021) use of structural racism unambiguously names the sources of oppression perpetrated on the Black population in the United States and names the Trump presidential administration and its antiscience policies as the major current source of oppression again the Black STEM doctoral students she studies. Mejia et al. (2022) call on engineering education researchers to validate and honor the voices of historically marginalized populations, particularly Latinos/as/xs, by rejecting any type of epistemological injustice created by systemic oppression through Borderlands Theory.

Methodological activism also involves rejecting deficit notions of marginalized populations. More engineering education researchers than ever are rejecting deficit-based theoretical stances (implicitly or explicitly). They are employing theories that explicate educational assets held by marginalized groups, such as people of color in Western countries. Among these, Yosso's (2005) community cultural wealth is growing in use in the US-based literature. Denton, Borrego, and Bocklage's systematic literature review of community culture wealth (CCW) in STEM education (2020) and Denton and Borrego's (2021) scoping review of Moll and colleagues' (1992) funds of knowledge (FoK) in STEM education contain numerous useful examples from engineering education and some direction for future research. While most CCW and FoK work in the field to date has been qualitative (e.g., Mejia, 2014; Mejia et al., 2014; Wilson-Lopez et al., 2016), recently some scholars have undertaken quantitative work (see, for example, Hiramori et al., 2021 for CCW; and Verdín et al., 2019; Verdín et al., 2021 for FoK).

Other recent work has developed new theories to advance DEI research in engineering education. For example, Burt et al. (2020) developed the theory of Black men's graduate engineering motivation and use their article to challenge engineering colleges to "make true commitments to broadening participation, which means centering students – in this case, Black men – as contributors to STEM" (p. 2). Leyva (2021) uses Critical Race Theory (CRT) and intersectionality to propose a framework for mathematics education as White patriarchal space using ideological, institutional, and relational dimensions in a study. He argues for "structural disruption[s]" that allow Black women to engage in within-group peer collaboration where they are brought together in affinity spaces in which they are not isolated (Leyva, 2021, p. 142).

In another example, Rodriguez et al. (2020) have combined existing theories to develop a new conceptual framework for the US context. Their conceptual framework for computing identity development for Latina undergraduate students encourages educators to draw on the wealth of knowledge from Latinas' backgrounds and experiences to form computing identity, which include (among other dimensions) community cultural wealth and funds of knowledge possessed by Latinas, as well as intersectionality and multiple forms of oppressions experienced by Latinas. Mejia et al. (2022) draw from Chicana feminist theory to argue for engineering education research that values the *conocimiento* and *testimonio* of Latinos/as/xs by legitimizing their voices and engaging in liberatory praxis. Lee and Matusovich (2016) built on Tinto's (1987) theory of student departure to develop a new conceptual model of cocurricular support (MCCS) that incorporates the breadth of support required for retaining undergraduate engineering students and advocates/facilitates for collaboration between student support practitioners and educational researchers. All these examples of methodological activism are unique; they each responded to their specific context, geography, participant goals, researcher abilities, project time frame, funding constraints, and community liberatory aims.

Methods/methodology as methodological activism. Patrick, Martin, and Borrego's review identifies authentic, equitable involvement from participants with marginalized identities as an important methodological component for critical DEI work (Patrick et al., 2022). We reiterate several examples from their review and bring in a few more recent examples that have been published since that time. We highlight several recent studies that provide ideas for how researchers can look beyond interviewing to other, perhaps multiple, sources of data. Sharing power with participants as co-authors is one way to practice equity while doing DEI research in engineering education. Patrick et al. (2022) identified two articles that engaged participants as co-authors. One article they point to is Secules et al. (2018) because the publication describes Secules's prolonged engagement with Tanu over a period of a few years via interviews. Then once an initial manuscript draft was written, Secules asked Tanu to become a co-author to contribute reflections that became part of the published work. Patrick et al. (2022) also point to Martin and Garza (2020) where the participant was a co-author but was engaged in the research much earlier in the process, co-constructing the manuscript from inception. That article centers Garza's (participant and coresearcher) voice using autoethnography that includes first-person accounts of her salient life events. Holly Jr.'s (2021) critique of Martin and Garza's collaboration points to potential for such efforts to further complicate issues of power, reward, and risk in shared authorship with participants. In another example, Holly Jr. (2020) conducted a critical autoethnography investigating his experiences teaching engineering to Black boys in an after-school program. Data included in the study came from interviews, journal entries, videos, lesson plans, and student artifacts, among other sources.

Neither the deep impact that participants have as co-authors on the quality of research nor the impact the act of authorship has on participants should be surprising. Smith and Lucena's (2016a, 2016b) ethnographic research provided examples of how prolonged engagement with participants was designed to not only recognize marginalized participants as experts in their own experiences but also to directly benefit them as a result of their participation in the research. These researchers acquainted their participants with the funds of knowledge framework, then facilitated individual and group activities whereby participants' experiences were validated and where participants could adapt the framework to their individual educational journeys. Through a different approach, Mejia et al. (2022) drew from the authors' *testimonios* to individually and collectively reflect on the sociohistorical, political, and situational events that became the backbone of their lived and material realities both inside and outside the bounds of engineering. The authors describe how through *testimonios* it is possible to acknowledge the voices of marginalized communities, build *conocimiento*, and develop critical consciousness as both participants and researchers.

Rethinking how participants can share their experiences beyond the traditional interview that tends to privilege the researcher with preset interview questions and direction is another way to enact methodological activism. Recent "mixed methods" studies provide additional illustrations of data collection approaches that go beyond traditional interviewing. Yang et al. (2021) used a "mixed methods" approach to examine LGBTQ+ engineering students' resistance and community, choosing focus groups over interviews for the qualitative data collection because of the potential for focus groups to foster connection between participants, a choice consistent with the research topic. Villanueva et al. (2019) used an engaging electrodermal activity in combination with interviews in qualitative–dominant design and "mixed methods" design. In this unique study, researchers collected electrodermal data while interviewing "womxn" graduate students and faculty about their reactions to the content. Leyva's recent publication (2021) focuses on counterstories of the resilience and resistance of Black and Latina women who are STEM students and/or aspiring engineers experiencing the White patriarchal spaces of mathematics education. This research similarly utilizes prolonged participant engagement and multiple sources of data in a cross-case methodology. Data collection included a short autobiographical document, questionnaire, focus group, and individual interviews.

6.3 The Call to Be Pluralistic

Lastly, we point to recent scholarship that asserts the need for DEI research to be pluralistic. A pluralistic approach to research honors and recognizes members of marginalized communities that have been historically oppressed. Pluralistic DEI research acknowledges that DEI work is not "new." Rather, it is a continuation of long struggles that emerged from liberatory movements formulated, driven, and accomplished by BIPOC people. As indicated by Flick (2018), researchers need to develop a "new sensitivity" that takes into consideration the "diversification of life worlds" (p. 4) since knowledge is situated in local experience, practices, and culture and not in deductive methodologies. Current research in engineering education frames DEI work as a measurable data point and not as an aspiration, which contradicts the essence of pluralistic research approaches.

We cannot ignore the fact that remnants of colonialism and patriarchy are still not only present in our society but also permeate through the work we perform, how we perform it, and for whom we perform it. These remnants are observed through militarism in engineering, which was a fundamental piece of the history of engineering and grounded on colonial settler perspectives. These perspectives have created the idea that engineers are problem-solvers (Pawley, 2009) and can decide what is best for others in paternalistic ways. Often, these actions are done on what is perceived to be "best for others" through a process in which the engineer/researcher is the solution-provider to a problem framed by the engineer/researcher themselves. Engineering education researchers still hold a great deal of power over the populations with whom they work, and the methodological decisions researchers make are important (Secules et al., 2021). However, pluralism requires rejecting perspectives rooted in colonialism and patriarchy, and taking actions that are based on mutual understanding, co-construction of knowledge, and collaboration with the groups participating in DEI research.

There is no single articulation or consensus to fight colonialism, patriarchy, or neoliberalism in engineering education research – all of which are opposite to a pluralistic view of local knowledge and lived experiences – because as researchers, we all continue to be motivated by individualistic goals (e.g., obtaining funding is critical for us to be able to do our work, fund our students, and earn tenure). On one hand, we have invoked DEI as a pathway to achieve social change in engineering. On the other hand, we are part of a system where its elitist nature has obscured the purpose of DEI research through performative work (i.e., actions that do not benefit oppressed groups and in turn uplift the visibility of the oppressor). De Sousa Santos (2018) argues that partial interventions can produce effects opposite to those that were initially intended. Thus, partial DEI interventions and approaches that focus on individual gain and not on a pluralistic view should be analyzed carefully and reconsidered.

None of this comes without risks. Researchers working on DEI initiatives who are willing to decenter themselves ultimately elevate the voices of the whole (Mejia et al., 2018). It is important for engineering education researchers to acknowledge that BIPOC scholars have been working endlessly to bring DEI awareness to a community of primarily White researchers whose work has simultaneously impeded the progress of DEI research by not agreeing to listen and relinquish their own neoliberal framings of DEI. Often, we see BIPOC engineering education researchers being gaslighted and their DEI efforts going unrecognized. We can counter this gaslighting by decolonizing ourselves, our methods, and our practices and problematizing the fact that epistemological injustice (de Sousa Santos, 2015) continues to exist in engineering education research. Nonpluralistic approaches to DEI research become epistemological when we assume that the only valid knowledge that exists is the one that is ratified by its own supremacy. We believe that engineering education research does not need a revolutionary approach to DEI work; it needs to revolutionize how we think about DEI.

7 Closing Thoughts

As the field of engineering education engages more in DEI initiatives, we urge scholars to consider for whom and for what purposes the DEI initiatives and research are done. The concepts presented in this chapter seek to encourage engineering education researchers to critically reflect on the ways in which they can shift the motivations for DEI research to those of social justice concerns and move towards enacting methodological activism and pluralism in engineering education research. We encourage the field to acknowledge that our closeness to Whiteness – and the historical influence of Whiteness in engineering education – has become a container for the theories, methodologies, and actions that are allowable in DEI initiatives.

DEI work in engineering education requires critical scholarship that seeks to question why things are the way they are to better understand such issues and how to address them. We challenge ourselves and our colleagues to ask: "How does my positionality influence my experiences and the ways in which I interpret DEI work?" Most importantly, we challenge ourselves and each other to think about the impact of the DEI work that is being done on those that it is intended to support. Reflecting on those who have left engineering (i.e., Latinos/as/xs, BIPOC people), DEI research has not been enough to support them or create a pathway for them to stay in the field. They will probably never return. Thus, what is the real purpose of DEI work, and for whom is it done? We hope that the questions and calls presented in this chapter ignite an interest in challenging dominant paradigms in DEI research. We also hope this chapter creates more space to welcome counternarratives that embrace the complexity of DEI and more questions that have yet to be asked about this important research and work. Ultimately, as engineering education researchers, we need to think about the goals of DEI research: Are we working to diversify engineering education students, or are we working to homogenize "diverse" students?

References

Al'Abri, K. (2011). The impact of globalization on education policy of developing countries: Oman as an example. *Literacy Information and Computer Education Journal*, *2*(4), 491–502.

Artiles, A. J. (2019). Fourteenth annual Brown lecture in education research: Reenvisioning equity research: Disability identification disparities as a case in point. *Educational researcher*, *48*(6), 325–335.

Atadero, R. A., Paguyo, C. H., Rambo-Hernandez, K. E., & Henderson, H. L. (2018). Building inclusive engineering identities: Implications for changing engineering culture. *European Journal of Engineering Education*, *43*(3), 378–398.

Bilimoria, D., & Liang, X. (2012). *Gender equity in science and engineering: Advancing change in higher education*. Routledge.

Brand, A. L. (2015). The politics of defining and building equity in the twenty-first century. *Journal of Planning Education and Research*, *35*(3), 249–264.

Burt, B. A., Roberson, J. J., Johnson, J. T., & Bonanno, A. (2020). Black men in engineering graduate programs: A theoretical model of the motivation to persist. *Teachers College Record*, *122*(11), n11.

Burt, B. A., Williams, K. L., & Smith, W. A. (2018). Into the storm: Ecological and sociological impediments to Black males' persistence in engineering graduate programs. *American Educational Research Journal*, *55*(5), 965–1006. https://doi.org/10.3102/0002831218763587

Calabrese Barton, A., & Tan, E. (2009). Funds of knowledge and discourses and hybrid space. *Journal of Research in Science Teaching: The Official Journal of the National Association for Research in Science Teaching*, *46*(1), 50–73.

Calabrese Barton, A., Tan, E., & Greenberg, D. (2017). The makerspace movement: Sites of possibilities for equitable opportunities to engage underrepresented youth in STEM. *Teachers College Record*, *119*(6), 1–44.

Capraro, M. M., Capraro, R. M., & Lewis, C. W. (2013). *Improving urban schools: Equity and access in K-12 STEM education for all students*. Information Age Publishing Inc.

Castaneda, D. I., & Mejia, J. A. (2018). Culturally relevant pedagogy: An approach to foster critical consciousness in civil engineering. *Journal of Professional Issues in Engineering Education and Practice*, *144*(2), 02518002. https://doi.org/10.1061/(ASCE)EI.1943-5541.0000361

Cech, E. (2013). The (mis)framing of social justice: Why ideologies of depoliticization and meritocracy hinder engineers' ability to think about social injustices. In J. Lucena (Ed.), *Engineering education for social justice* (pp. 67–84). Springer.

Cech, E., Montgomery, G., Settles, I., Elliott, K., Cheruvelil, K., & Brassel, S. (2021). The social is professional: The effects of team climate on professional outcomes for LGBTQ persons in environmental science. *Journal of Women and Minorities in Science and Engineering*.

Cech, E., & Waidzunas, T. (2021). Systemic inequalities for LGBTQ professionals in STEM. *Science Advances*, 7(3), eabe0933.

Chen, D. A., Mejia, J. A., & Breslin, S. (2019). Navigating equity work in engineering: Contradicting messages encountered by minority faculty. *Digital Creativity*, *30*(4), 329–344.

Chubin, D. E., May, G. S., & Babco, E. L. (2005). Diversifying the engineering workforce. *Journal of Engineering Education*, *94*(1), 73–86.

Coley, B. C., Simmons, D. R., & Lord, S. M. (2021). Dissolving the margins: LEANING INto an antiracist review process. *Journal of Engineering Education*, *110*(1), 8–14. https://doi.org/https://doi.org/10.1002/jee.20375

Commission on the Advancement of Women and Minorities in Science, Engineering, and Technology Development. (2001). Land of plenty: Diversity as America's competitive edge in science, engineering, and technology. *Leadership and Management in Engineering*, *1*(4), 27–30.

Cross, K. J. (2020). Racism is the manifestation of White supremacy and antiracism is the answer. *Journal of Engineering Education, 109*(4), 625–628. https://doi.org/https://doi.org/10.1002/jee.20362

Cross, K. J., Mendenhall, R., Clancy, K. B., Imoukhuede, P., & Amos, J. (2021). The pieces of me: The double bind of race and gender in engineering. *Journal of Women and Minorities in Science and Engineering, 27*(3).

de Sousa Santos, B. (2015). *Epistemologies of the South: Justice against epistemicide*. Routledge.

de Sousa Santos, B. (2018). *The end of the cognitive empire: The coming of age of epistemologies of the south*. Duke University Press.

DeAro, J., Bird, S., & Ryan, S. M. (2019). NSF ADVANCE and gender equity: Past, present and future of systemic institutional transformation strategies. *Equality, Diversity and Inclusion: An International Journal, 38*, 131–139.

Delgado Bernal, D. (2002). Critical race theory, Latino critical theory, and critical raced-gendered epistemologies: Recognizing students of color as holders and creators of knowledge. *Qualitative Inquiry, 8*(1), 105–126.

Delgado Bernal, D., & Villalpando, O. (2002). An apartheid of knowledge in academia: The struggle over the "legitimate" knowledge of faculty of color. *Equity & Excellence in Education, 35*(2), 169–180.

Denton, M., & Borrego, M. (2021). Funds of knowledge in Stem education: A scoping review. *Studies in Engineering Education, 1*(2).

Denton, M., Borrego, M., & Boklage, A. (2020). Community cultural wealth in science, technology, engineering, and mathematics education: A systematic review. *Journal of Engineering Education, 109*(3), 556–580. https://doi.org/10.1002/jee.20322

Diversity. (2008). In J. W. A. Darity (Ed.), *International encyclopedia of the social sciences* (2nd ed., Vol. 2, pp. 419–420). Macmillan Reference USA.

Dlouhy, K., & Froidevaux, A. (2022). Evolution of professionals' careers upon graduation in STEM and occupational turnover over time: Patterns, diversity characteristics, career success, and self-employment. *Journal of Organizational Behavior*, 1–18. https://doi.org/10.1002/job.2615

Faulkner, W. (2000). Dualisms, hierarchies and gender in engineering. *Social Studies of Science, 30*(5), 759–792.

Faulkner, W. (2007). Nuts and bolts and people' gender-troubled engineering identities. *Social Studies of Science, 37*(3), 331–356.

Ferreira, M. (2002). Ameliorating equity in science, mathematics, and engineering: A case study of an after-school science program. *Equity & Excellence in Education, 35*(1), 43–49.

Fletcher, T. L., Jefferson, J. P., Boyd, B. N., & Cross, K. J. (2021). Missed opportunity for diversity in engineering: Black women and undergraduate engineering degree attainment. *Journal of College Student Retention: Research, Theory & Practice, 0*(0). https://doi.org/10.1177/1521025120986918

Flick, U. (2018). *An introduction to qualitative research: Theory, method and applications* (6th ed.). Sage.

Franzway, S., Sharp, R., Mills, J. E., & Gill, J. (2009). Engineering ignorance: The problem of gender equity in engineering. *Frontiers: A Journal of Women Studies, 30*(1), 89–106.

Freire, P. (2003). *Pedagogy of the oppressed*. Bloomsbury.

Garcia, G. A. (2018). Decolonizing hispanic-serving institutions: A framework for organizing. *Journal of Hispanic Higher Education, 17*(2), 132–147.

Garcia, G. A. (2020). *Hispanic serving institutions (HSIs) in practice: Defining "servingness" at HSIs*. IAP.

Gelles, L. A., Lord, S. M., Hoople, G. D., Chen, D. A., & Mejia, J. A. (2020). Compassionate flexibility and self-discipline: Student adaptation to emergency remote teaching in an integrated engineering energy course during COVID-19. *Education Sciences, 10*(11), 304. www.mdpi.com/2227-7102/10/11/304

Gelles, L. A., Mejia, J. A., Lord, S. M., Hoople, G. D., & Chen, D. A. (2021). Is it all about efficiency? Exploring students' conceptualizations of sustainability in an introductory energy course. *Sustainability, 13*(13), 7188.

Godfrey, E., & Parker, L. (2010). Mapping the cultural landscape in engineering education. *Journal of Engineering Education, 99*(1), 5–22.

González, N., Wyman, L., & O'Connor, B. H. (Eds.). (2011). *The past, present, and future of "funds of knowledge"*. Wiley-Blackwell.

Hampton, C., Reeping, D., & Ozkan, D. S. (2021). Positionality statements in engineering education research: A look at the hand that guides the methodological tools. *Studies in Engineering Education, 1*(2), 126–141. https://doi.org/10.21061/see.13

Hancock, C. L., Morgan, C. W., & Holly Jr, J. (2021). Counteracting dysconscious racism and ableism through fieldwork: Applying DisCrit classroom ecology in early childhood personnel preparation. *Topics in Early Childhood Special Education*, *41*(1), 45–56.

Harper, S. R. (2010). An anti-deficit achievement framework for research on students of color in STEM. *New Directions for Institutional Research*, *148*(148), 63–74.

Hiramori, D., Knaphus-Soran, E., & Litzler, E. (2021). Critically quantitative: Measuring community cultural wealth on surveys. In *2021 ASEE virtual annual conference content access*.

Holly Jr., J. (2020). Disentangling engineering education research's anti-Blackness. *Journal of Engineering Education*, *109*(4), 629–635. https://doi.org/https://doi.org/10.1002/jee.20364

Holly Jr, J. (2021). Criticality is crucial: Fidelity in what we say and what we do. *Studies in Engineering Education*, *2*(2), 46–53. https://doi.org/10.21061/see.78

Hoople, G. D., Chen, D. A., Lord, S. M., Gelles, L. A., Bilow, F., & Mejia, J. A. (2020). An integrated approach to energy education in engineering. *Sustainability*, *12*(21), 9145.

Hughes, B. E. (2017). "Managing by not managing": How gay engineering students manage sexual orientation identity. *Journal of College Student Development*, *58*(3), 385–401. https://doi.org/10.1353/csd.2017.0029

Hurtado, S., Alvarez, C. L., Guillermo-Wann, C., Cuellar, M., & Arellano, L. (2012). A model for diverse learning environments. In M. B. Paulsen (Ed.), *Higher education: Handbook of theory and research* (Vol. 27, pp. 41–122). Springer.

Ikram, A., Su, Q., Fiaz, M., & Shabbir, M. W. (2016). Cultural diversity and challenges for female entrepreneurs: Empirical study of an emerging economy. *International Journal of Economics and Management Engineering*, *10*(10), 3385–3388.

LaFave, J. M., Kang, H.-S., & Kaiser, J. D. (2015). Cultivating intercultural competencies for civil engineering students in the era of globalization: Case study. *Journal of Professional Issues in Engineering Education and Practice*, *141*(3), 05014008.

Lee, W. C., & Matusovich, H. M. (2016). A model of co-curricular support for undergraduate engineering students. *Journal of Engineering Education*, *105*(3), 406–430.

Leydens, J. A., & Lucena, J. C. (2017). *Engineering justice: Transforming engineering education and practice*. John Wiley & Sons.

Leyva, L. A. (2021). Black women's counter-stories of resilience and within-group tensions in the white, patriarchal space of mathematics education. *Journal for Research in Mathematics Education*, *52*(2), 117–151.

Long III, L. L. (2020). Toward an antiracist engineering classroom for 2020 and beyond: A starter kit. *Journal of Engineering Education*, *109*(4), 636–639. https://doi.org/https://doi.org/10.1002/jee.20363

Long III, L. L., & Mejia, J. A. (2016). Conversations about diversity: Institutional barriers for underrepresented engineering students. *Journal of Engineering Education*, *105*(2), 211–218.

López, N., Erwin, C., Binder, M., & Chavez, M. J. (2018). Making the invisible visible: Advancing quantitative methods in higher education using critical race theory and intersectionality. *Race Ethnicity and Education*, *21*(2), 180–207.

Martin, J. P., Desing, R., & Borrego, M. (2022a). Positionality statements are just the tip of the iceberg: Moving towards a reflexive process. *Journal of Women and Minorities in Science and Engineering*, *28*(4), v–vii. https://doi.org/10.1615/JWomenMinorScienEng.2022044277

Martin, J. P., & Garza, C. (2020). Centering the marginalized student's voice through autoethnography: Implications for engineering education research. *Studies in Engineering Education*, *1*(1), 1–19.

Martin, J. P., Stefl, S. S., & Slaton, A. E. (2022b). Learning in public and a path towards methodological activism: A conversation on equity research. *Journal of Women and Minorities in Science and Engineering*, *28*(1), 75–87.

Marx, S. (2016). *Qualitative research in STEM: Studies of equity, access, and innovation*. Routledge.

May, G. S., & Chubin, D. E. (2003). A retrospective on undergraduate engineering success for underrepresented minority students. *Journal of Engineering Education*, *92*(1), 27–39.

McGee, E. (2021). Fear, fuel, and fire!: Black STEM doctoral students' career decisions during the Trump presidency. *International Journal of Qualitative Studies in Education*, 1–22.

McGee, E. O., & Bentley, L. (2017). The troubled success of Black women in STEM. *Cognition and Instruction*, *35*(4), 265–289. https://doi.org/10.1080/07370008.2017.1355211

McGee, E. O., Main, J. B., Miles, M. L., & Cox, M. F. (2021). An intersectional approach to investigating persistence among women of color tenure-track engineering faculty. *Journal of Women and Minorities in Science and Engineering, 27*(1).

Mein, E., & Esquinca, A. (2017). The role of bilingualism in shaping engineering literacies and identities. *Theory into Practice, 56*(4), 282–290.

Mejia, J., Revelo, R., Villanueva, I., & Mejia, J. (2018). Critical theoretical frameworks in engineering education: An anti-deficit and liberative approach. *Education Sciences, 8*(158), 1–13. https://doi.org/10.3390/educsci8040158

Mejia, J., Ruiz, D., Popov, V., Esquinca, A., & Gadbois, D. (2019). Asset-based practices in engineering design (APRENDE): Development of a funds-of-knowledge approach for the formation of engineers. In *ASEE annual conference & exposition*, Tampa, FL.

Mejia, J. A. (2014). *A sociocultural analysis of Latino high school students' funds of knowledge and implications for culturally responsive engineering education*. Utah State University.

Mejia, J. A., Revelo, R. A., & Pawley, A. L. (2020). Thinking about racism in engineering education in new ways [Commentary]. *IEEE Technology and Society Magazine, 39*(4), 18–27.

Mejia, J. A., Villanueva Alarcón, I., Mejia, J., & Revelo, R. A. (2022). Legitimized tongues: Breaking the traditions of silence in mainstream engineering education and research. *Journal of Women and Minorities in Science and Engineering, 28*(1).

Mejia, J. A., Wilson, A. A., Hailey, C. E., Hasbun, I. M., & Householder, D. L. (2014). Funds of knowledge in Hispanic students' communities and households that enhance engineering design thinking. In *American society for engineering education annual conference*, Indianapolis.

Milner, H. R. (2016). Race, culture, and researcher positionality: Working through dangers seen, unseen, and unforeseen. *Educational Researcher, 36*(7), 388–400. https://doi.org/10.3102/0013189x07309471

Moll, L. C., Amanti, C., Neff, D., & Gonzalez, N. (1992). Funds of knowledge for teaching: Using a qualitative approach to connect homes and classrooms. *Theory into Practice, 31*(2), 132–141.

Mondisa, J.-L. (2020). The role of social capital in African American STEM mentoring relationships. *Journal of Women and Minorities in Science and Engineering, 26*(2).

National Academies of Sciences, Engineering, & Medicine. (2017). *Evaluation of the achievement levels for mathematics and reading on the national assessment of educational progress*. National Academies Press.

National Academies of Sciences, Engineering, & Medicine. (2018). *English learners in STEM subjects: Transforming classrooms, schools, and lives*. National Academies Press. http://search.ebscohost.com/login.aspx?direct=true&scope=site&db=nlebk&AN=2037444

National Science Board. (2018). *Science and engineering indicators 2018*. National Science Foundation. www.nsf.gov/statistics/indicators/

Ong, M. (2005). *Achieving multi-sitedness and social justice: "Methodological activism" and other dilemmas in physics education research*. American Educational Research Association, Montreal, QC.

Özturgut, O. (2017). Internationalization for diversity, equity, and inclusion. *Journal of Higher Education Theory & Practice, 17*(6).

Patrick, A., Martin, J., & Borrego, M. (2022). Critical research methods in STEM higher education: A state-of-the-art review. *Journal of Women and Minorities in Science and Engineering, 28*(3), 1–26.

Pawley, A. L. (2009). Universalized narratives: Patterns in how faculty members define "engineering". *Journal of Engineering Education, 98*(4), 309–319.

Pawley, A. L. (2017). Shifting the "default": The case for making diversity the expected condition for engineering education and making whiteness and maleness visible. *Journal of Engineering Education, 106*(4), 531–533.

Pawley, A. L., Mejia, J. A., & Revelo, R. A. (2018). Translating theory on color-blind racism to an engineering education context: Illustrations from the field of engineering educationresearch with underrepresented populations. In *ASEE annual conference and exposition*, Salt Lake City.

Penfield, R. D., & Lee, O. (2010). Test-based accountability: Potential benefits and pitfalls of science assessment with student diversity. *Journal of Research in Science Teaching: The Official Journal of the National Association for Research in Science Teaching, 47*(1), 6–24.

Peterson, R. J. (2021). We need to address ableism in science. *Molecular Biology of the Cell, 32*(7), 507–510. https://doi.org/10.1091/mbc.E20-09-0616

Phillips, L. T., Slepian, M. L., & Hughes, B. L. (2018). Perceiving groups: The people perception of diversity and hierarchy. *Journal of Personality and Social Psychology*, *114*(5), 1–20. https://doi.org/http://dx.doi.org/10.1037/pspi0000120

Pineda, P., & Mishra, S. (2022). The semantics of diversity in higher education: Differences between the Global North and Global South. *Higher Education*, 1–22. https://doi.org/https://doi.org/10.1007/s10734-022-00870-4

Puritty, C., Strickland, L. R., Alia, E., Blonder, B., Klein, E., Kohl, M. T., McGee, E., Quintana, M., Ridley, R. E., & Tellman, B. (2017). Without inclusion, diversity initiatives may not be enough. *Science*, *357*(6356), 1101–1102.

Qin, J., Muenjohn, N., & Chhetri, P. (2014). A review of diversity conceptualizations: Variety, trends, and a framework. *Human Resource Development Review*, *13*(2), 133–157.

Qvortrup, A., & Qvortrup, L. (2018). Inclusion: Dimensions of inclusion in education. *International Journal of Inclusive Education*, *22*(7), 803–817.

Reader, J. (2006). *Globalization, engineering, and creativity*. Morgan and Claypool.

Revelo, R. A., Mejia, J. A., & Villanueva, I. (2017). Who are we? Beyond monolithic perspectives of Latinxs in engineering. In *American society for engineering education annual conference and exposition*, Columbus, OH.

Revelo, R. A., & Stepin, N. (2018). Within-group differences of engineering identity for Latinx engineering students. In *2018 IEEE frontiers in education conference (FIE)*, San Jose, CA.

Riley, D. (2008). *Engineering and social justice: Synthesis lectures on engineers, technology, and society*. Morgan and Claypool.

Rodriguez, S. L., Lu, C., & Ramirez, D. (2020). Creating a conceptual framework for computing identity development for Latina undergraduate students. In *An asset-based approach to advancing Latina students in STEM* (pp. 25–39). Routledge.

Roldán Vera, E., & Robles Valle, A. (2020). The education system of Mexico: Inequality, standardization, and compensation. *The Education Systems of the Americas*, 1–26.

Ross, E. W., & Gibson, R. J. (2007). *Neoliberalism and education reform*. Hampton Press.

Samuelson, C. C., & Litzler, E. (2016). Community cultural wealth: An assets-based approach to persistence of engineering students of color. *Journal of Engineering Education*, *105*(1), 93–117.

Sappleton, S. J., & Adams, D. (2022). On decolonizing US education: Lessons from the Caribbean and South Africa. *Professional Educator*, *45*(1), 42–57.

Secules, S., Gupta, A., Elby, A., & Tanu, E. (2018). Supporting the narrative agency of a marginalized engineering student. *Journal of Engineering Education*, *107*(2), 186–218.

Secules, S., McCall, C., Mejia, J. A., Beebe, C., Masters, A. S., L. Sánchez-Peña, M., & Svyantek, M. (2021). Positionality practices and dimensions of impact on equity research: A collaborative inquiry and call to the community. *Journal of Engineering Education*, *110*(1), 19–43.

Secules, S., & Mejia, J. A. (2021). Contextualizing the past to guide the future: Situating three critical theoretical frameworks for educational culture. *Studies in Engineering Education*, *1*(2).

Sellers, V., Martin, J. P., & Seraphin, M. (2022). A narrative inquiry approach to community cultural wealth of Black men in engineering. *Journal of Women and Minorities in Science and Engineering*, *28*(4), in press.

Sharp, R., Franzway, S., Mills, J., & Gill, J. (2012). Flawed policy, failed politics? Challenging the sexual politics of managing diversity in engineering organizations. *Gender, Work & Organization*, *19*(6), 555–572.

Slaton, A. E. (2015). Meritocracy, technocracy, democracy: Understandings of racial and gender equity in American engineering education. In A. E. Slaton (Ed.), *International perspectives on engineering education* (pp. 171–189). Springer. https://doi.org/https://doi.org/10.1007/978-3-319-02904-7

Slaton, A. E., & Pawley, A. L. (2018). The power and politics of engineering education research design: Saving the 'Small N'. *Engineering Studies*, *10*(2–3), 133–157.

Smith, J. M., & Lucena, J. C. (2016a). "How do I show them I'm more than a person who can lift heavy things?" the funds of knowledge of low income, first generation engineering students. *Journal of Women and Minorities in Science and Engineering*, *22*(3).

Smith, J. M., & Lucena, J. C. (2016b). Invisible innovators: How low-income, first-generation students use their funds of knowledge to belong in engineering. *Engineering Studies*, *8*(1), 1–26. https://doi.org/10.1080/19378629.2016.1155593

Sochacka, N., Walther, J., Jolly, L., & Kavanagh, L. (2009). Confronting the methodological challenges of engineering practice research: A three-tiered model of reflexivity. In *Proceedings of the 2009 research in engineering education symposium (REES)*.

Su, L. K. (2010). Quantification of diversity in engineering higher education in the United States. *Journal of Women and Minorities in Science and Engineering, 16*(2), 161–175.

Tan, M. (2011). Mathematics and science teachers' beliefs and practices regarding the teaching of language in content learning. *Language Teaching Research, 15*(3), 325–342.

Thomas, L. D., Watt, D. L., Cross, K. J., Magruder, J. A., Easley, C. R., Monereau, Y.-A. J., Phillips, M. R., & Benjamin, A. M. (2016). As purple is to lavender: Exploring womanism as a theoretical framework in engineering education. In *2016 ASEE annual conference & exposition*.

Tinto, V. (1987). *Leaving college: Rethinking the causes and cures of student attrition*. University of Chicago Press.

Tonso, K. L. (2006). Teams that work: Campus culture, engineer identity, and social interactions. *Journal of Engineering Education, 95*(1), 25–37.

Torres, L. E. (2012). Lost in the numbers: Gender equity discourse and women of color in science, technology, engineering and mathematics (STEM). *International Journal of Science in Society, 3*(4).

Valencia, R. R. (1997). Conceptualizing the notion of deficit thinking. *The Evolution of Deficit Thinking: Educational Thought and Practice, 19*(1), 1–12.

Valencia, R. R. (2010). *Dismantling contemporary deficit thinking: Educational thought and practice*. Routledge.

Vandenberghe, J. (2021). How professional engineers can contribute to attraction and retention of minority groups into the engineering profession through equity, diversity, inclusion, and decolonization efforts. *The Canadian Journal of Chemical Engineering, 99*(10), 2116–2123.

Verdín, D., Smith, J. M., & Lucena, J. (2019). Recognizing engineering students' funds of knowledge: Creating and validating survey measures. In *ASEE annual conference & exposition*, Tampa, FL.

Verdín, D., Smith, J. M., & Lucena, J. (2021). Funds of knowledge as pre-college experiences that promote minoritized students' interest, self-efficacy beliefs, and choice of majoring in engineering. *Journal of Pre-College Engineering Education Research (J-PEER), 11*(1), 192–213. https://doi.org/10.7771/2157-9288.1281

Villanueva, I., Di Stefano, M., Gelles, L., Osoria, P. V., & Benson, S. (2019). A race re-imaged, intersectional approach to academic mentoring: Exploring the perspectives and responses of womxn in science and engineering research. *Contemporary Educational Psychology, 59*. https://doi.org/10.1016/j.cedpsych.2019.101786

Villanueva Alarcón, I., Downey, R. J., Nadelson, L., Choi, Y. H., Bouwma-Gearhart, J., & Tanoue, C. (2021). Understanding equity of access in engineering education making spaces. *Social Sciences, 10*(10), 384.

Weeks, K. (2007). Life within and against work: Affective labor, feminist critique, and post-Fordist politics. *Ephemera: Theory and Politics in Organization*, 7(1), 233–249.

Williams, B. C., Squire, D. D., & Tuitt, F. A. (2021). *Plantation politics and campus rebellions: Power, diversity, and the emancipatory struggle in higher education*. SUNY Press.

Wilson-Lopez, A., Mejia, J. A., Hasbún, I. M., & Kasun, G. S. (2016). Latina/o adolescents' funds of knowledge related to engineering. *Journal of Engineering Education, 105*(2), 278–311.

Yang, J. A., Sherard, M. K., Julien, C., & Borrego, M. (2021). Resistance and community-building in LGBTQ+ engineering students. *Journal of Women and Minorities in Science and Engineering, 27*(4).

Yosso, T. J. (2005). Whose culture has capital? A critical race theory discussion of community cultural wealth. *Race Ethnicity and Education, 8*(1), 69–91.

Zuberi, T., & Bonilla-Silva, E. (2008). *White logic, white methods: Racism and methodology*. Rowman & Littlefield Publishers.

Part 3

Engineering Education Across Contexts and Participants

12

Professional Learning for Pre-College Engineering Teachers

Adam R. Carberry, Stacy S. Klein-Gardner, Pamela S. Lottero-Perdue, and Katherine L. Shirey

1 Introduction

There is an ever-growing need for trained science, technology, engineering, and mathematics (STEM) professionals all over the world (Fayer et al., 2017). Pre-college education, including pre-school (beginning at around age 3) to secondary school (ending at around age 18), has the potential to excite, prepare, and educate students about further education and careers in STEM while promoting STEM literacy for all students (NASEM, 2020). Engineering has been the component of STEM that has been least prominent in pre-college education. Meeting the needs of the world for a greater number and more diverse engineering workforce will require a complementary workforce of qualified educators who have engineering-specific content knowledge and specialized, pedagogical content knowledge to teach engineering to pre-college students (Ball et al., 2008; Magnusson et al., 1999; Shulman, 1986; Sun & Strobel, 2014).

This chapter aims to better understand pre-college engineering teacher professional learning (PCE TPL) intended to meet the call for qualified pre-college engineering educators. We have scoped this chapter purposefully to provide a focused discourse around teacher professional learning for engineering education within the United States (US). It is our hope that our presentation of PCE TPL in the US will provide a starting point or case for a broader global discussion of how we prepare pre-college teachers to teach engineering as well as other related science, technology, and mathematics subjects. The decision to position the chapter in this way leverages our place in and knowledge of engineering education within the US. It also provides a focus that allows for greater depth of discussion in lieu of a broad discussion of global PCE TPL. We have included references to pre-college (referred to in the US as K–12) STEM education when it focuses on engineering education and want to recognize resources available for science and mathematics TPL (e.g., Loucks-Horsley et al., 2010). It is beyond the scope of this chapter to address the myriad meanings of "STEM" education (Martín-Páez et al., 2019) or to dissect what has been done in other related fields (e.g., computer science). Additionally, our discussion of PCE TPL certification and accreditation is highly relevant to those teaching or aiming to teach within public schools but may not be as applicable for private schools that typically do not require such credentials.

DOI: 10.4324/9781003287483-15

In what follows, we explore PCE TPL by asking and answering the following key questions in five sections of the chapter, respectively:

1 What is the case to be made for the need for PCE TPL?
2 How has PCE TPL changed over time?
3 What is the current landscape for PCE TPL?
4 What major levers have influenced PCE TPL?
5 What does the future landscape for PCE TPL look like?

Throughout the chapter we embed vignettes from pre-college teachers, educators, and administrators across the US. We acquired these quotes through direct solicitation of their thoughts and experiences to be inclusive of their voices in this discussion. Our aim is for these vignettes to remind readers that while we discuss PCE TPL in broad ways throughout the chapter, PCE TPL impacts the experiences and perspectives of individual teachers and administrators who, in turn, directly impact student learning.

2 Making the Case for PCE TPL

> I was excited for the transformative process of changing how I taught science and, therefore, engineering. My district embedded engineering concepts and lessons in a very authentic way for students. I believe that the quality of the curriculum and the professional development provided made the buy-in for teachers abundant. To me, teaching engineering to all students makes sense and is necessary for our society. I also wanted to attract more girls to engineering at an early age, so I developed an all-girls engineering after-school club at my school.
>
> – *Alison Roe, Elementary School Educator, Maryland*

The vignette from Alison Roe exemplifies some of the arguments for PCE TPL. When it is done well with high-quality curricula, teachers become partners in the process of learning and teaching engineering. This partnership helps motivate teachers to reach all students through engineering, including those who are underrepresented in many engineering fields (Fry et al., 2021). This focus on students echoes the National Academies of Sciences, Engineering, and Medicine's (2020) Consensus Study Report "Building Capacity for Teaching Engineering in K–12," which identified four goals for K–12 (ages 5–18) engineering education: (1) develop engineering literacy, (2) improve mathematics and science achievement through the integration of concepts and practices across the STEM fields, (3) improve college and career readiness, and (4) prepare a small percentage of students for matriculation in postsecondary engineering programs. Achieving these goals would mean empowering pre-college students' academic success, providing innovative leadership, and helping humanity overcome our most challenging problems. This will require drawing upon knowledge and research of teachers and students within engineering education (e.g., Hynes et al., 2017; Sneider & Ravel, 2021) to help forge new ground in pre-college engineering teacher preparation.

It is crucial that pre-college educators be well-prepared to teach engineering, as collective teacher efficacy significantly influences student success (Tschannen-Moran & Barr, 2004). This begins with the recruitment and preparation of pre-college teachers from a variety of backgrounds to teach engineering to today's youth. These backgrounds include those who are preparing to become teachers, that is, pre-service teachers (PSTs), and teachers who are currently teaching, that is, in-service teachers (ISTs). Most PSTs and ISTs in the US have enrolled in or graduated from programs that focus on early childhood (ages birth to 10), elementary (ages 5 to 11), or secondary education (ages 12 to 18). Some PSTs and ISTs may have earned a degree with a focus in one or two teaching content areas, particularly those in programs focused on secondary education. It is important to note that not all who enter into the pre-college teaching profession do so through traditional four-year education programs. Some, including the authors of this chapter, earn degrees in other fields (e.g., engineering or a science discipline) before considering a pathway into pre-college teaching. Those looking to make a transition to teaching within public schools can earn state certification later, as part of a master's program, or work in private schools, where no such certification is required.

The most recent report of the National Survey of Science and Mathematics Education (NSSME+; the plus is used for additional focus on computer science) provided multiple insights regarding the status of science and mathematics education in the US from a representative sample of science, computer science, and mathematics ISTs (Banilower et al., 2018). The study did not include teachers outside of these categories (e.g., non–computer science career and technical education [CTE] teachers). Engineering was positioned as a subcategory of science within the report due to the inclusion of engineering within the *Next Generation Science Standards ("NGSS")* (NGSS Lead States, 2013) and *A Framework for K–12 Science Education* (NRC, 2012). The NGSS standards have been adopted by 20 of 50 states in the US, and the *Framework* was used by 24 additional states to develop their own standards (National Science Teachers Association [NSTA], n.d.). These collective standards task elementary, middle school, and high school science teachers with teaching or integrating engineering into their science curriculum.

The NSSME+ report also examined science, math, and computer science teachers' preparedness to teach engineering. The report indicated that "few teachers at any grade level have had coursework in engineering" (Banilower et al., 2018, p. 13), specifically 3% of elementary teachers, 10% of middle school teachers, and 13% of high school teachers. Few ISTs felt fairly (14%) or very well (3%) prepared to teach engineering. The majority (51%) felt inadequately prepared to teach engineering. The feeling of preparation varied depending on what was being asked of teachers and/or the specific engineering concept. Preparation to teach engineering content for middle and high school science teachers was disaggregated into helping students develop possible solutions, defining engineering problems, and optimizing design solutions. This further refinement of preparedness revealed middle school science teachers to feel slightly more prepared than their high school counterparts, but the vast majority of middle and high school teachers still felt somewhat or inadequately prepared to teach engineering.

3 PCE TPL Changes Over Time

> I remember my supervisor of science advising me that "STEM education is the wave that we must catch as soon as possible." It was clear that to address STEM education at the elementary level, teachers needed professional learning experiences (PLEs) on the *T* and the *E* of STEM. We developed a partnership with a local university and Engineering is Elementary (EiE) to integrate engineering into our elementary science curriculum. PLEs began with a subset of pilot teachers who then became lead teachers who facilitated subsequent PLEs on the science-engineering integrated units. Watching teachers transfer their new knowledge and pedagogy into other content areas [and] after-school and summer school programs has been the biggest reward. I've gotten to see how teachers, administrators, and supervisors have genuinely embraced STEM education as a benefit to their students. Perhaps the most satisfying effect of engineering teacher training has been the transformed teachers and their instruction in our elementary science classrooms over the past ten years.
>
> – *Amy Ryan, Elementary Science Curriculum Specialist, Maryland*

Amy Ryan's supervisor referred to the "wave of STEM education" in 2008, which was when her school system decided to start integrating engineering into the elementary science curriculum. She witnessed change in teaching engineering as a part of science education over time, beginning when such integration was novel and unfamiliar to teachers to the present day, when some elementary educators have been teaching engineering for well over a decade and lead TPL for other teachers in the system. This is but one example of the growth of PCE TPL during the last two decades. What we know anecdotally is that the *number of* TPL opportunities for pre-college engineering teachers has grown tremendously all over the world. There is no single clearinghouse for such opportunities, but this growth has been influenced by support from professional organizations, emerging pre-college engineering learning standards, inclusion of engineering in science content standards, and national-level programs and resources providing curricula and enacting corresponding professional learning experiences.

US-based ***professional organizations*** have made changes to their structures or focus to include supporting pre-college engineering teachers and teacher educators. The American Society for Engineering Education (ASEE) formed the Pre-College Engineering Education Division in 2003, marking a widening in the society's focus from primarily university-level engineering education to including pre-college audiences and research. In 2014, the ASEE board of directors established an association-wide committee for pre-college engineering education, now called the ASEE Commission on P–12 Engineering Education (ASEE, n.d. a). The International Technology and Engineering Education Association (ITEEA) chose to rename itself in 2010 to include engineering (formerly International Technology Educators Association) to be more inclusive to both pre-college technology and engineering education. ITEEA today offers a variety of engineering and STEM TPL opportunities, including summer institutes around their Engineering by Design (EbD) courses and the Authorized Training Institute designed to certify TPL coaches for their EbD courses, through their ITEEA STEM Center for Teaching and Learning (ITEEA, n.d.). The NSTA began offering TPL opportunities in both STEM and engineering following the publication of NGSS in 2013. One notable change was the addition of their annual STEM Forum and Expo established one year prior to the introduction of NGSS. These changes are examples of how professional organizations are signaling the increasing presence and need for engineering within pre-college education.

Another sign demonstrating growth for PCE TPL is the emergence of national-level academic engineering ***standards***. State-adopted academic standards, derived under the influence of national and expert recommendations, are a primary driver of educational curricula and the day-to-day work priorities of pre-college schools and teachers. Although the word "engineering" did not appear in the *National Science Education Standards* (NSES) (NRC, 1996), engineering design ideas have now been explicitly stated in *A Framework for K–12 Science Education* (NRC, 2012) and NGSS that arose from that framework (NGSS Lead States, 2013). This has promoted the growth of engineering instruction as it is integrated with science content. Such integrative or integrated-STEM education aims to teach individual subjects, that is, science and engineering, while capitalizing on the synergy that comes from using the STEM disciplines together (Moore et al., 2014). The NSSME+ Report (Banilower et al., 2018) notes that nearly all sampled elementary teachers teach a self-contained class where they are responsible for teaching science with engineering. The NGSS suggests the integration of engineering into traditionally offered, graduation-requirement science courses like biology, chemistry, physics, and Earth science (Blank et al., 2004). Thus, a huge proportion of US pre-college elementary teachers and secondary science teachers whose school systems align with NGSS are asked by these standards to teach engineering in the context of science.[1]

Standards in the engineering and technology education community have also been updated. This includes ITEEA's modified Standards for Technology and Engineering Literacy (STEL) (Daugherty et al., 2021) and the recently released A Framework for P–12 Engineering Education Learning introduced and promoted to shape what pre-college learners should know with respect to engineering education (Advancing Excellence in P-12 Engineering Education [AE^3] & ASEE, 2020).

Professional organizations and standards provide infrastructure, support, and purpose for engineering education. So do individual ***programs and resources*** funded by industry, universities, and/or funding agencies. Many such programs now exist, including Project Lead the Way (PLTW) (started in 1997) (Blias & Adelson, 1998; PLTW, n.d.), Engineering Is Elementary (EiE) (started in 2003) (EiE, n.d.; Cunningham, 2008), TeachEngineering (TE) (started in 2003) (TE, n.d.; Zarske et al., 2005), Engineer Your World (started in 2008) (EYW, n.d.; Farmer et al., 2012), and Engineering for US All (e4usa) (started in 2018) (e4usa, n.d.; Reid & Greisinger, 2021). Each of these programs involves professional learning experiences for PSTs, ISTs, or both in addition to providing curricular resources for teacher use (Daugherty & Custer, 2012). The professional learning experiences offered by these programs prepare teachers to use the program's curricular materials, often by modeling the use of the curricular materials and preparing teachers to implement the curriculum in their classrooms.

4 The Current Landscape for PCE TPL

The previously noted changes over time have helped create the current landscape for PCE TPL. This section separates and summarizes PCE TPL for PSTs and ISTs.

4.1 PCE TPL for Pre-Service Teachers

> Many pre-service teachers come to our program with a mindset that they don't know what STEM is and they don't see themselves as an engineering or STEM person. The main thing we try to do is to get teachers to unpack their own previous experiences with what might be, or count as, engineering or science. For example, we have them take pictures from their everyday life, which they then use to inspire science questions and engineering problems. This helps them recognize that science and engineering are all around them. We want them to leave their pre-service experiences thinking, "Yes, I can do STEM" and "Yes, we are all involved in STEM every day." This shift is designed to help guide them into STEM spaces, ultimately to improve the learning of their students.
>
> – Dr. Christopher Wright, Pre-Service Teacher Educator, Pennsylvania

Dr. Christopher Wright's vignette opens our discussion of PCE TPL for pre-service teachers by highlighting how teacher education can show future teachers that STEM and engineering are connected and integral to real life, including both the teachers' lives and the lives of their students. This serves as a starting point for teacher educators who prioritize growing pre-service teachers' engineering practices and habits of mind, as well as preparing them to teach engineering to K–12 students.

There are a variety of college-level programs or pathways that prepare future PSTs to teach engineering content. Individuals interested in teaching engineering can choose the mechanism by which they may enter the profession. There is no current requirement in the US that pre-college teachers have an engineering degree or have taken any engineering-specific courses. Prospective teachers in the US generally become authorized to teach through one of four methods:

1. Earn an education degree if currently holding no degrees;
2. Seek alternative certification for teaching if holding a non-education degree;
3. Obtain advanced degrees (e.g., master's or doctorate) if holding an education degree and interested in adding an endorsement or new license; or
4. Earn certification reciprocity if already certified and interested in additional state certifications.

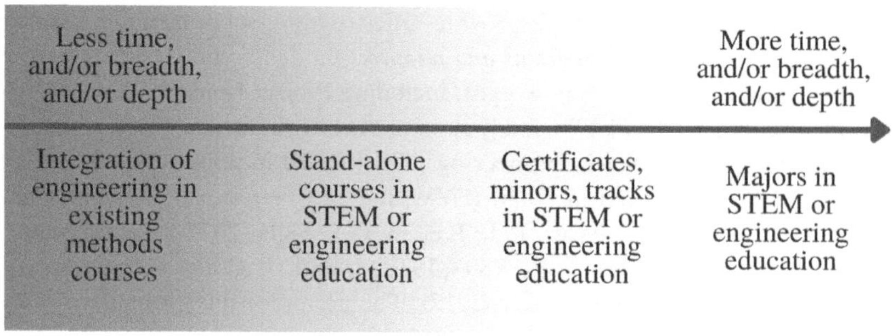

Figure 12.1 Current means of including PCE TPL in pre-service education programs.

These options vary state-by-state, and shades of nuance exist even from program-to-program within a state.

Pre-service education programs found within education units use a variety of approaches (Figure 12.1). Some programs involve more time on engineering professional learning, which provides a higher chance for greater depth to the engineering content and professional learning. Others aim for wider breadth of engineering content, instruction, and connections to other STEM subjects. Worth noting is that it is impossible to assume that every educator across the country has the same opportunities to learn engineering due to extended time required, expertise needed, and geographic access. We cite some examples in the US for each of these programs across all levels of pre-college education to provide a sense of what is currently offered, but note that what is presented is not an exhaustive list.

The most common approach taken at the early childhood and elementary levels to prepare PSTs is the ***integration of engineering*** units of instruction within required science teaching methods courses (Purzer & Quintana-Cifuentes, 2019). This tactic adds engineering explicitly or implicitly within the existing structure of these programs without adding a course. For example, Towson University has been integrating engineering into elementary science methods and content courses and early childhood courses for over a decade (Lottero-Perdue, 2017). Courses employing such tactics are difficult to track because course names typically do not include the term *engineering*. Research on such learning experiences has demonstrated improved engineering teaching efficacy beliefs, deepened content knowledge, refined views of engineering, improved implementation, and increased intentions to teach engineering (Nesmith & Cooper, 2021; Radloff et al., 2019; Yesilyurt et al., 2021). It is also recommended that integration of engineering into such programs occur throughout the curriculum rather than within a single course (Capobianco et al., 2021).

Other teacher preparation programs have expanded their curricular offerings for education majors to include single or multiple courses dedicated to engineering education. For example, Tufts University offers two courses around the development of knowledge and reasoning in engineering and practice of teaching engineering and design that provide students in the teacher education program opportunities to see examples and design lessons that teach specific engineering or design concepts to pre-college students. These offerings expand beyond traditional engineering design-based courses to include maker education approaches (e.g., Harlow et al., 2018). Such courses have been shown to improve self-efficacy towards pedagogical content knowledge, engagement, and disciplinary knowledge (Perkins Coppola, 2019).

Additional programs provide ***minors, concentrations, certificates, and even majors in engineering education or STEM education***. Millersville University in Pennsylvania offers a minor in integrative STEM education methods for early childhood education teachers (Brusic et al., 2019; Millersville University, n.d.). This program is designed to teach teachers how to integrate science, technology, engineering, and mathematics in pre-K through grade 4 in developmentally appropriate ways using hands-on materials. North Carolina State University (NCSU) has a similar concentration in STEM education for elementary majors aimed to prepare future teachers to provide strong STEM-focused instruction (NCSU, n.d. b), while Purdue University offers a graduate certificate in K–12 integrated STEM education that focuses significantly on engineering (Purdue, n.d. b).

A review of "science, technology, engineering, and mathematics (STEM) educational methods" programs and majors within the National Center for Education Statistics (NCES) revealed 23 programs or majors in the US (US Department of Education, n.d.). This database provides some information about the number of STEM education programs but does not include minors or tracks. There is no option to search for "engineering education," yet there are 86 additional programs in the US that result from a search of "technology teacher education/industrial arts teacher education." Several of these programs provide degrees in integrative STEM education or technology and engineering education that offer initial certification. Some examples include bachelor's degrees in technology and engineering from the College of New Jersey (TCNJ) (n.d.), engineering technology teacher education from Purdue University (n.d. a), and technology, engineering, and design education from NCSU (n.d. c). The NCSU program is also an example of a program that offers an equivalent minor for other majors and an advanced master's degree (NCSU, n.d. a), while the Purdue University degree is an example of a program that provides a pre-engineering teaching certificate for PLTW as part of the degree requirements. Most programs at any level offer initial certification routes (often in CTE) as well as graduate courses for career advancement through professional learning.

PST preparation programs are not isolated to only offerings within education units. Many engineering units have joined these efforts by offering engineering education or interdisciplinary degrees. A prime example is Ohio Northern University (n.d.), which offers an engineering education major for prospective students interested in teaching engineering at the high school level. TCNJ (n.d.) provides another unique program offering a dual-major option in integrative STEM (iSTEM) for engineering or K–6 education majors. This approach pairs iSTEM majors with one of six majors, including elementary education or special education. An alternative to the dual major is offering an interdisciplinary program like the Engineering Plus program at University of Colorado Boulder (n.d.), which allows engineering students to learn about their chosen discipline while including a secondary area of emphasis (e.g., teacher preparation). Such offerings align with STEM major recruitment programs like the UTeach program out of the University of Texas at Austin (n.d.), which offers pathways for STEM undergraduates (and degree holders) to obtain teacher certification.

Influencing all PCE TPL programs are the pedagogical approaches that teacher educators take when teaching how to teach engineering. A ubiquitous approach involves helping teachers understand engineering practices and habits of mind by engaging in engineering design challenges themselves as learners (Hanson et al., 2021; Kim et al., 2019). Another approach is engaging in field experiences teaching engineering design challenges to pre-college students in classrooms or afterschool programs, typically with lesson planning and instructional support from teacher educators and/or mentor teachers (e.g., Capobianco et al., 2021). These experiences may be supported by various readings and discussions from teacher practitioner journals, research articles, standards documents, and/or methods texts that support engineering education.

4.2 PCE TPL for In-Service Teachers

> Initially, I was a good bit terrified, and honestly, also amused, when I was asked to join e4usa. "How am I, a musician, going to do this in a way that is credible in terms of

> students' experiences and educational needs and not find myself so outside my own sense of self?" Turns out that my sense of what engineering is was way off. It's not simply an advanced application of math and physics. Engineering is all about helping people solve problems. The use of things like math and physics is similarly just circumstance. Now, I very much see the process used to solve problems through engineering as parallel to the creative process. I find I am able to help my students not just see the world through a different lens but also see their role in the world differently. That they can make a difference. They can change it. That their potential is truly limitless.
>
> – *Richard Maxwell, High School Educator, Arizona*

PSTs are not the only ones with options to receive PCE TPL. Like Richard Maxwell, ISTs from all kinds of disciplinary backgrounds also need pedagogical guidance on how to incorporate engineering into the classroom with limited resources (e.g., time, money, materials, and knowledge) (Porter et al., 2019). A variety of certificates, advanced degrees, and workshops is available for ISTs with varying amounts of dedicated time, yielding varying degrees of breadth (amount of content covered) and depth (amount of detail or complexity explored) (Figure 12.2).

IST offerings include many of the same approaches as noted for PSTs but may also include topics such as developing teacher and student engineering identities (Kouo et al., 2020); exploring stereotype threat, implicit bias, and imposter syndrome as well as how those topics impact both teachers and their students' choices to participate in engineering (Dalal et al., 2021; Kouo et al., 2020); and assessing students' engineering design work, particularly through a rubric-based method that encourages both formative and summative assessment (Groves et al., 2014). Schools want ISTs to learn to engineer as their students do – where they create and solve real problems, draw from their background knowledge, and work with one another in teams (Lieberman, 1995). We again cite some examples in the US but note that what is presented is not an exhaustive list.

One set of IST options involves ***learning while teaching***. The Tufts University Teacher Engineering Education Program (TEEP) (n.d.) certificate is a four-course, 16-credit sequence designed to build knowledge in engineering and expertise in teaching engineering. The certificate is offered online by the Tufts' Center for Engineering Education and Outreach. Another example is Virginia Tech's (n.d.) graduate certificate, master's, and doctoral programs in integrative STEM education. The department offering these programs allows ISTs to teach, take online classes, and earn a master's degree in lieu of a pre-service licensure option. Johns Hopkins University (n.d.) is one of several institutions that offer graduate coursework enabling Pre-K-to-6 teachers to earn Maryland's STEM instructional leadership endorsement – a credential earned after teachers have already earned initial certification.

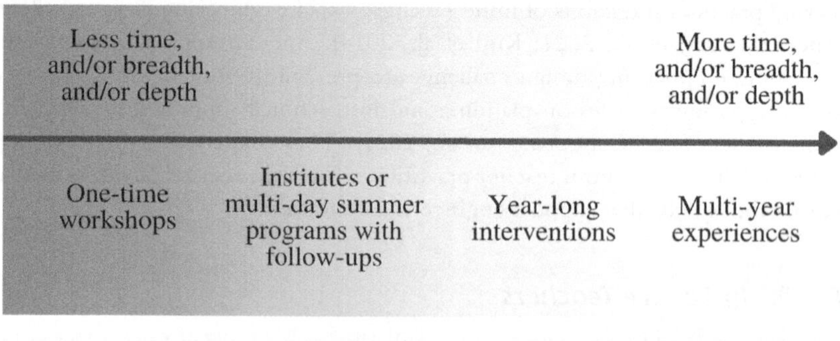

Figure 12.2 Current means of including PCE TPL in in-service education programs.

Other options have been made available to teachers to combat the limited availability, required time, and needed funding of higher education programs. This has led to the more common option of enrolling in ***in-service professional learning experiences***. These experiences contribute to the ongoing education of teachers already in the workforce. Such opportunities are designed to help bridge potential disconnects between current practices and innovations in teaching and learning. For example, the Science Learning through Engineering Design (SLED) Summer Institute for teachers (Lehman & Capobianco, 2012). Many licensed teachers use professional learning opportunities to fulfill professional learning hours to support their license renewal processes. The 2018 NSSME+ indicates that 25%, 34%, and 23% of elementary, middle school, and high school science teachers, respectively, reported that their last three years included at least one professional learning experience with a heavy emphasis on engineering (Banilower et al., 2018). The report expressed surprise that "fewer than a third of K–12 science teachers have attended professional development that focused heavily on deepening their understanding of how engineering is done" (p. 55).

TPL in engineering education for ISTs includes interventions ranging from in-school professional development to out-of-school research experiences for teachers (e.g., NSF Research Experience for Teachers program, RET, n.d.). One in-service strategy is to provide teachers with an "educative" curriculum they can use – and perhaps that the school or school system has decided to adopt – to teach engineering concepts. Examples include EiE, PLTW, and e4usa, which include robust curricula that span multiple units and/or grade levels and may even provide curricula for whole courses of study (e.g., a one-credit high school class). The curricula for these programs are used as the basis for a paired professional learning experience designed specifically to help teachers learn about engineering and how to teach engineering at the same time they learn to implement the particular curriculum (Cunningham, 2008; Kouo et al., 2020; Nathan et al., 2011; Singer et al., 2016).

Not all curricular professional learning experiences are as large in scope or directly connected to a particular curriculum. For example, TeachEngineering is an extensive repository of vetted engineering lessons curated by the University of Colorado at Boulder. Lessons are fully accessible and free to be used by any teacher, regardless of attendance in a TeachEngineering professional learning experience, which is designed to assist teachers in the use of TeachEngineering resources. Additional avenues for in-service teachers include district supported professional learning via in-school specialists/support staff, hiring external consultants to train staff members, or independently seeking professional learning opportunities to satisfy personal goals and licensure renewal requirements. Many independent opportunities can be found through resources provided or offered by professional organizations (e.g., NSTA annual STEM Forum and Expo for teachers or the scholarships and databases provided by the National Education Association, NEA). Engineering-specific organizations like the STEM Ecosystems by the Teaching Institute for Excellence in STEM (TIES) (n.d.), private providers such as the Knowles Teacher Initiative (n.d.), and public providers such as museums (e.g., the Exploratorium) also offer for-free and for-fee TPL programs in engineering integration.

Research exploring the effectiveness of professional learning experiences for ISTs has highlighted the importance of structure (Guzey et al., 2014), namely, providing teachers with hands-on, real-world experiences that allow for a walk-through of curricular materials within a safe and nurturing environment for practice (Hynes & Dos Santos, 2007). Opportunities should be provided that promote collaboration, interaction, and collegiality with peers and outside experts (Guskey, 2003; Hynes et al., 2007; Mesutoglu & Baran, 2021). Effective professional learning experiences can empower teachers, increase self-efficacy toward engineering, improve content knowledge, reduce perceived barriers toward teaching engineering, and increase recognition of engineering (Hammack et al., 2020; Havice et al., 2018; Singer et al., 2016).

Access to PCE TPL for ISTs since the start of the COVID-19 pandemic has changed dramatically. These learning experiences had largely been in-person until March 2020, when providers (and all educators) were suddenly thrust into designing creative ways to offer professional learning experiences through alternative delivery mechanisms. For example, e4usa offered fully online professional learning opportunities that featured a mix of synchronous and asynchronous sessions and used a kit of supplies that was shipped in advance. The ASEE Pre-College Engineering Education Conference for P–12 Educators, which has been previously held as a one-day, in-person workshop for teachers since 2004, was offered online for the first time in 2021. These online offerings are not perfect, but they were better than having nothing at all as they increased teachers' ability to access professional learning from anywhere Brown et al., 2021; Lockee, 2021).

Finally, the content of professional learning for ISTs is similar to that for PSTs in its frequent inclusion of engineering design challenges in which ISTs learn design challenges as students (often before teaching those same design challenges). ISTs are not in need of field experiences but may try out what they learned from professional learning experiences on the students they teach in their classrooms or after-school clubs. An interesting approach within the aforementioned TEEP program are opportunities for teachers to learn to notice salient aspects of classroom interactions within others and their own videos of engineering instruction (Watkins et al., 2021).

The sum of these efforts presented throughout this section and the previous for PSTs and ISTs highlights the variation of offerings available to future and current teachers. Important to note is that there is no one-size-fits-all approach to PCE TPL (Hammack et al., 2020).

5 Levers That Influence PCE TPL

> My perspective about the learning process and my teaching pedagogies took a dramatic shift in the summer of 2009. A colleague told me about this summer program at a local university called Research Experiences for Teachers (RET), and she thought I would enjoy the challenge. The outcome from this program was that my lessons became more student-centered, and the embedded engineering was more student-engaging. A direct result of these changes is, the number of students taking physics at the school has more than doubled, and an introduction to engineering design class has been implemented at the school.
>
> – *Mike Kiser, High School Educator, Tennessee*

Mike Kiser's engagement in an engineering-focused RET experience was transformative. The rigorous professional learning experience changed how he taught, that is, more student-centered instruction. It also led to the inclusion of engineering challenges in his physics classes and the addition of an engineering design class at this school. RET programs like the one in which Mike Kiser enrolled depend upon funding, in this case, from the National Science Foundation. Funding is one of three levers that we have identified that have influenced and continue to impact PCE TPL. This section explores how specific standards, licensure, and funding have and continue to help shape PCE TPL.

5.1 Standards for PCE TPL

Any and all teachers learning to teach pre-college engineering can face hurdles in their learning. They may (productively) struggle with the role of engineering in the core sciences (Chandler et al., 2011; Shirey, 2015), value too highly student engagement over learning engineering skills, content, and mindsets (Dare et al., 2014), or view instruction of engineering as tinkering or trial-and-error (Roehrig et al., 2012). The abundance of professional learning opportunities and the need to address early adoption issues have led organizations and researchers to orchestrate mechanisms to ensure the quality of learning experiences for teachers. One such mechanism is the Standards for Preparation and Professional Development for Teachers of Engineering (Reimers et al., 2015), which begins by defining three types of engineering literacy: engineering design, engineering careers, and engineering and society. The five subsequent standards aim to ensure that teachers develop sufficient engineering literacy themselves to teach their students.

- Standard A requires TPL to augment teachers' content knowledge by providing opportunities for them to experience engineering design practices so that they may reflect on the content of engineering from a learner perspective and then think about classroom implementation.
- Standard B requires TPL to help teachers develop a robust repertoire of teaching practices to support growing engineering pedagogical content knowledge.
- Standard C suggests that TPL should offer opportunities for teachers to seek and reflect upon the relationships between engineering and engineering instruction, including content and pedagogy, in the context of other classroom content, student learning practices, disciplinary practices, and teaching experiences.
- Standard D focuses on curriculum and assessment. TPL should help teachers learn about the merits and benchmarks of doing engineering while practicing and reflecting on their own assessment strategies for engineering skills, processes, and products.
- Standard E requires TPL to align to research, including having a sustained duration, a content focus, local coherence with objectives, local collaboration with school colleagues, and opportunities for active engagement.

An associated matrix for identifying the level to which a particular professional learning opportunity teaches these standards enables TPL providers to consider a wide range of potential topics for inclusion in their learning experiences. According to the matrix, no one TPL opportunity should endeavor to meet every substandard of every standard. ASEE has recently established a process using these standards to provide ASEE endorsement for high-quality professional learning programs for ISTs through the Engineering Teacher Professional Development Endorsement (ASEE, n.d. b).

An example program using these standards to create their TPL is e4usa. e4usa intends to demystify and democratize engineering for all, including teachers and students. TPL began in 2019, with each teacher attending a one-week, in-person professional learning workshop. A fully remote, online professional learning experience for new teachers was offered a year later, which included a mix of asynchronous assignments and synchronous sessions spanning two weeks. This new experience proved effective and continues to be the delivery method, especially as the program has grown and in-person experiences have become cost-preventative (Kouo et al., 2021). The e4usa first-year teacher program received the American Society for Engineering Education (ASEE) endorsement in fall 2021. Table 12.1 summarizes how the program met each of the standards.

5.2 Licensure and Credentialing

One way to examine the status of PCE TPL is to examine a key end point in order to teach within public schools in the US, that is, required licensure or credentialing tests. There has been some growth in engineering items integrated into some tests. To date, engineering is not prevalent within existing exams to become an elementary or middle school teacher, nor is there a stand-alone engineering test for elementary teachers.

Table 12.1 e4usa Example: Aligning a Professional Learning Experience to the Standards for Preparation and Professional Development for Teachers of Engineering

Standard	*Examples of How the Standard Is Reflected by teachers*
Standard A Engineering Content and Practices	• Attend to two content/practice-based curricular threads, (e.g., *engineering design* and *engineering and society*) throughout the experience. • Engage in teams. • Participate in multiple engineering design challenges using a variety of tools, including sketching and computer-aided design (CAD). • Analyze and debate about ethics case studies.
Standard B Pedagogical Content Knowledge for Teaching Engineering	• Learn to teach engineering design challenges that rely on locally sourced problems that need to be solved. • Receive mentorship from program coaches (veteran teachers) and local university or industry liaisons.
Standard C Engineering as a Context for Teaching and Learning	• Discuss and reflect upon connections between 21st-century skills and engineering design challenges. • Engage in design challenges that connect engineering with science, technology, and mathematics. • Focus on grand challenges, such as improving water quality.
Standard D Curriculum and Assessment	• Review the curriculum and other associated materials. • Recognize embedded flexibility in the curriculum and modify based on teacher and student needs. • Analyze the curriculum and relevant standards to ensure inclusivity of all learners at an appropriate level of support and challenge.
Standard E Alignment to Research, Standards, and Educational Practices	• Participate over the course of an entire school year. • Engage in a community of practice with other e4usa educators.

Engineering largely appears in licensure related to CTE but varies from state to state. For example, California has no initial teaching credential available in engineering education or STEM but does offer an industrial and technology education initial licensure. North Carolina currently has no teaching license in STEM or engineering education available. Both states align to NGSS. This disparity from state to state is further complicated by the variety of popular licensure tests (e.g., PRAXIS, Pearson, and edTPA) and the requirement of most US states and territories predicating new teacher licensures on passing a disciplinary PRAXIS test. Examples include Test No. 5442 for middle school science teachers (ETS, 2020b) and Test No. 5485 for secondary physical science teachers (ETS, 2021), which includes 14% focus on nature and impact of science and engineering. Test No. 5051 for technology education teachers (ETS, 2017) includes a 20% focus on technological design and problem-solving. There are no such tests to date that are specifically focused on engineering education, and no engineering questions within the required elementary PRAXIS bundle (ETS, 2020a). We suspect that licensure and other credentials would arise if more teaching positions became specifically focused on engineering education across all of pre-college education.

5.3 Funding

> National Science Foundation (NSF) grant efforts play a critical role in providing the funding to conceptualize, develop, implement, and assess professional development opportunities for pre-college engineering teachers. Our challenge is in the institutionalization of these promising programs and practices through long-term institutionalized funding and the continued support at the school district level of teachers engaged in these grant efforts.
>
> – Claire Duggan, Center for STEM Education Director, Massachusetts

In the beginning of this section on levers that influence PCE TPL, we shared a vignette from Mike Kiser, a teacher who directly benefited from a grant-funded RET program. Claire Duggan's vignette not only describes the multiple ways in which NSF grant funding supports PCE TPL efforts but also reminds readers that funding for PCE TPL must be sustainable through ongoing funding provided by institutions and the government.

Federal funding by the US for pre-college engineering (or, more broadly, STEM) education at all levels is estimated to be between $2.8 and $3.4 billion annually, with less than half of that funding going to pre-college engineering education efforts (Granovsky, 2018). The Department of Education, the National Science Foundation (NSF), and the Department of Health and Human Services are the primary agencies responsible for STEM education efforts among 13–15 total federal agencies managing 105–254 funded STEM education programs (Granovsky, 2018).

Federally funded initiatives like Race to the Top (RTTT), NSF GK–12, and the Carl D. Perkins CTE Act have supplied financial support to grow STEM-related programs across the US. Many of these funding sources were designed to sunset (e.g., RTTT and NSF GK–12) but they inspired change toward long-term investment in STEM. Additional sources have included university support funneled through centers (e.g., Tufts' Center for Engineering Education Outreach or NCSU's the Engineering Place), federal or state grants, private philanthropic funding, or operating funds within school systems to support the curriculum acquisition and/or development of professional learning experiences. Sustained initiatives have also emerged, such as the NSF RET program, which began making awards in 2002 and currently provides $5.8 million per year in new funding to support authentic summer research experiences for K–14 educators to foster long-term collaborations between universities, community colleges, school districts, and industry partners (NSF, n.d.).

There have been and continue to be ebbs and flows in terms of funding to support pre-college engineering education in the US. The retirement of some programs, emergence of new programs, and identification of additional sources of funding stress the importance of committing to and investing in the sustainability of pre-college engineering education efforts.

6 The Future Landscape for PCE TPL

> I believe the challenge for engineering educators is to both inspire students to the engineering profession while at the same time preparing them for the rigor of post-secondary studies. The multitude of engineering professional pathways is both incredible and overwhelming for the pre-college student. It is our challenge to provide students a sampling of experiences that allow them to see that at its core engineering is an awesome blend of creative and technical talents. Professional learning should help engineering teachers learn to (1) build and utilize local connections with community, higher education, and business establishments in pursuit of authentic engineering projects, connections,

> and experiences; (2) develop students' leadership, collaboration, and project management skills; (3) implement assessment that emphasizes process over product; and (4) build an inclusive classroom culture that encourages creativity, curiosity, and confidence in every student.
>
> – Jim Muscarella, High School Educator, Pennsylvania

Jim Muscarella's vignette represents a vision of the possible outcomes of PCE TPL. He envisions secondary engineering teachers as being able to help students understand what engineering entails, assist in student growth of their creative and technical expertise, situate engineering learning in local contexts, implement process-oriented assessment, and support an inclusive environment. This section builds on Jim's sentiments by focusing on the future of PCE TPL. We're aware that predicting the future landscape for PCE TPL is an impossible task. Instead, this section presents emerging and innovative ways to address PCE TPL paired with our aspirations for what high-quality PCE TPL for PSTs and ISTs might look like based on existing offerings. We consider who learns and who teaches engineering, how we might broaden the purposes for engineering education within PCE TPL, emerging approaches and tools in PCE TPL, and what levers we might hope for to support PCE TPL and engineering education.

6.1 Who Learns and Who Teaches Engineering?

Engineering education has become increasingly more and more open to younger and younger engineering learners over time. Attention was first placed within colleges of engineering and then in high schools. We have now come to understand that even preschoolers can engineer (Moore et al., 2014). Engineering education has also become increasingly open to all students. The idea of who can engineer has shifted from the exclusive Sputnik-era mindset to an inclusive "for us all" perspective. Engineering education now aims to create opportunities for all to become engineers, if they so choose, and for all future citizens to learn about the design of technologies and who designs them.

What this suggests is that teachers of all levels must be prepared to teach engineering to students. What is perhaps more novel is the idea that teachers of all backgrounds, ranging from different grade-level expertise to school setting and former life experiences, have unique talents and strengths to offer as teachers of engineering (Carberry et al., 2021; Dalal et al., 2020). We urge those who offer PCE TPL to embrace the assets that all sorts of individuals bring to engineering education learning experiences (Superfine, 2021). The newness here is in seeing diverse backgrounds not as deficient, that is, without experience doing or teaching engineering, but rather as having assets to offer that can enhance the engineering education that students experience. To extend an example shared earlier, the inquisitive music teacher's deep knowledge of his craft and associated technologies creates a fertile space for identifying and solving problems (Dalal et al., 2022).

Similarly, we must not merely acknowledge that all students can learn to engineer if given the right learning experiences. This is reminiscent of tabula rasa thinking, that is, filling young brains with good engineering experiences. PCE TPL must help teachers learn to expect that many of engineering's key practices and dispositions are embodied and inherent within the students we teach, even the youngest learners. We urge such asset/strengths-based framing to become a regular feature within PCE TPL, as has been suggested for pre-college engineering education (Martin & Wendell, 2021). We need to value all engineering practices, including the use of empathy, lived experiences, or other relevant educational experiences (Lottero-Perdue & Settlage, 2021).

6.2 Broadening and Making Explicit the Purpose(s) for Engineering

It is essential for engineering educators who lead PCE TPL to acknowledge that engineering education has multiple purposes, including both technical skills and encouraging student literacy, inclusion, and empowerment in engineering. The myriad purposes should be taught explicitly, with an aim towards including more than one purpose for teaching engineering in a given PCE TPL experience. Purposes for teaching engineering include (1) practicing the design process and habits of mind, (2) emulating or learning about engineers, (3) reinforcing scientific or mathematical understandings, (4) providing the integrative "glue" that connects STEM or STEAM, (5) serving as a tool for social justice, and (6) empowering change in one's environment (Bybee, 2013; Cunningham, 2009; Cunningham, 2007; National Research Council, 2009; National Research Council, 2011; Tan et al., 2019). The purpose for any one unit or design challenge may be combinations of these purposes. Embedded in many of these purposes is inclusion of culturally relevant pedagogical practice and service learning within engineering. Both have been included in engineering education to date but perhaps not emphasized as much as the future warrants (Calabrese Barton et al., 2021; Lima & Oakes, 2006). These purposes, which can be combined with others, importantly suggest that engineering has the power to empower. We encourage the inclusion of these purposes within PCE TPL wherever and whenever possible. Successful integration of these purposes will make teachers more likely aware of how to teach engineering for social justice and service, enabling them to create engineering learning experiences that empower their students and serve others.

6.3 Emerging Approaches and Tools in PCE TPL

We mentioned earlier that the COVID-19 pandemic pushed PCE TPL for ISTs online with varying degrees of success. Some online aspects of TPL were found to be effective and have continued. We suspect that future PCE TPL for ISTs may continue to leverage at a minimum hybrid (online plus face-to-face) approaches in order to reach more ISTs. University courses may be similarly positioned to offer online or hybrid options for PSTs' learning. We encourage future research studying these different modalities and how they may be beneficial or detrimental to teacher learning and access.

An emerging approach from mathematics and science education that could serve as a model for future PCE TPL is the use of mixed-reality simulations. This approach allows teachers to practice teaching engineering discussions with groups of avatars with a "simulation specialist" or "interactor" who is a highly trained human-in-the-loop (Dieker et al., 2013). Such simulations have been used to study and allow IST elementary teachers and PST middle school teachers to practice cross-team post-testing engineering argumentation discussions (Lottero-Perdue et al., 2020). They have also allowed PST middle school teachers to practice managing idea fixation while supporting students as they consider how to improve their designs (Lottero-Perdue & Mikeska, 2022).

6.4 Levers to Create and Use

Levers have influenced the current landscape of PCE TPL, so it's safe to assume that new levers will need to be created and used if we want to see continued growth of such offerings for both PSTs and ISTs. This starts by first garnering inspiration from the experiences of other related disciplines who, at some point in their histories, were faced with the same fight engineering currently engages in today. For example, science did not become a formal focus of the curriculum until the late 19th century (DeBoer, 2019). Science was secondary to mathematics and literacy until changing societal contexts and subsequent concerns stressed the need for K–12 science education (NASEM, 2019). Success in including science subjects into K–12 education required a great deal of professional learning to support teachers in teaching these new subjects. A more recent example are the efforts made by computer science, which began after and has already surpassed engineering in its influence on the pre-college curricula in the US. Advocacy from entities such as the College Board and NSF have greatly benefitted computer science and will be essential for engineering to find its place in pre-college education.

We previously discussed standards, licensure and credentialing, and funding as levers that have helped shape the current landscape. Updates and changes to standards, new licensure and credentialing options, and expanded funding will still be major factors in achieving our envisioned future landscape of PCE TPL. Changes that we expect to see in these areas include more state departments including standards for engineering education, science and elementary licensure that includes greater attention to engineering, and funding with an eye towards ethical scaling and sustainability. Each will require greater emphasis and resources towards PCE TPL.

Emergent levers should aim to take advantage of new technologies and social media. Providing online, hybrid, virtual reality, augmented reality, etc., PCE TPL will allow such opportunities to reach a far greater percentage of teachers around the globe. There is energy in these spaces because of the COVID-19 pandemic. This energy, paired with the energy that already exists around improving overall STEM education, suggests that pre-college engineering education is potentially at the precipice of becoming commonplace. It will be critical that advocates for pre-college engineering education take advantage to ensure engineering is not somehow left out of the evolving pre-college education curriculum.

It's also important that these efforts not forget those who have provided support for such efforts to date. A better conception of PCE TPL programs, variations across the teacher preparation landscape, and pre-college engineering education research will assist colleges of education to understand current trends in teacher preparation. This would also help scholars evaluate the success of the varying levels of course or program time, depth, and breadth on teacher and student outcomes. Organizations like ASEE and ITEEA need to continue to provide advocacy for and guidance on how to better prepare pre-college teachers to teach engineering. Building a repository of PCE TPL certified programs through endorsement programs, like the ASEE ETPDE, will help improve the overall quality of such experiences across the US and potentially abroad.

7 Conclusion

The goal for pre-college engineering education to support engineering literacy efforts has been slowly growing over the past two decades in the US. Signs of change and current state efforts in pre-college engineering education for teachers highlight the growth that has occurred. These efforts provide a foundation for amplifying guidance and encouragement for future offerings to the engineering education community at large. We advocate for continued growth of professional learning opportunities that help more teachers to educate their students to gain engineering literacy and to consider engineering as a path forward.

Including engineering within compulsory pre-college education still faces an uphill battle as schools seek solutions for making space in the curriculum and identifying qualified teachers prepared to teach such content. Multiple entry points for professional learning must be provided to accommodate PSTs, ISTs, and others (e.g., currently practicing engineers or other college majors looking to make a career change) interested in or tasked with teaching engineering at the pre-college level. The past and present demonstrate an unclear trajectory for what could or should come next. We suggest a need to continue to grow measures of quality in engineering TPL and work towards remaining intentionally separated from CTE. Accomplishing these next steps will be the responsibility of the community to devise sustainable and scalable professional learning solutions if we are to truly reach the full potential for pre-college education to impact the engineering profession and overall engineering literacy.

Acknowledgments

The authors would like to thank the educators and administrators who shared their voices in vignettes throughout the chapter. Special acknowledgment is owed to every educator and mentor who provides intentional modeling and support that contributes to bringing along new engineers, engineering educators, researchers, and the whole of the engineering education ecosystem.

Note

1 There is an ongoing debate in the US between teaching integrated/integrative engineering as part of a science or STEM course versus teaching engineering as a stand-alone course. We appreciate this debate and recognize the impact such decisions have on already-crowded curricula and fidelity of content taught but do not purport to take a stance on this debate in this chapter.

References

Advancing Excellence in P-12 Engineering Education (AE[3]) & American Society for Engineering Education (ASEE) (2020). *Framework for P-12 engineering learning: A defined and cohesive educational foundation for P-12 engineering*. https://doi.org/10.18260/1-100-1153-1

American Society for Engineering Education (ASEE). (n.d. a). *ASEE PreK-12 engineering education – ASEE commission on P-12 engineering education*. https://sites.asee.org/p-12committee/

American Society for Engineering Education (ASEE). (n.d. b). *Engineering teacher professional development endorsement (ETPDE)*. https://etpde.asee.org/

Ball, D. L., Thames, M. H., & Phelps, G. (2008). Content knowledge for teaching what makes it special? *Journal of Teacher Education*, *59*(5), 389–407. https://doi.org/10.1177/0022487108324554

Banilower, E. R., Smith, P. S., Malzahn, K. A., Plumley, C. L., Gordon, E. M., & Hayes, M. L. (2018). *Report of the 2018 NSSME+*. Horizon Research Inc. www.horizon-research.com/report-of-the-2018-nssme

Blais, R. R., & Adelson, G. I. (1998). Project lead the way models a program for changing technology education. *Tech Directions*, *58*(4), 40.

Blank, R. K., Langesen, D., Laird, E., Toye, C., & Bandeira de Mello, V. (2004). Meeting NCLB goals for highly qualified teachers: Estimates by state from survey data. *Education Policy Analysis Archives*, *12*, 70. https://doi.org/10.14507/epaa.v12n70.2004

Brown, C., Correll, P., & Stormer, K. J. (2021). The "new" normal: Re-imagining professional development amidst the COVID-19 pandemic. *Middle School Journal*, *52*(5), 5–13. https://doi.org/10.1080/00940771.2021.1978787

Brusic, S., Marcum-Dietrich, N., Shettel, J., & White, J. (2019). Integrative STEM for PK-4 preservice teachers. In *International society for technology in education conference*, Philadelphia, PA.

Bybee, R. W. (2013). *The case for STEM education: Challenges and opportunities*. NSTA Press.

Calabrese Barton, A., Greenberg, D., Kim, W. J., Brien, S., Roby, R., Balzer, M., Turner, C., & Archer, L. (2021). Disruptive moments as opportunities towards justice-oriented pedagogical practice in informal science learning. *Science Education*, *105*(6), 1229–1251. https://doi.org/10.1002/sce.21682

Capobianco, B. M., Radloff, J., & Clingerman, J. (2021). Facilitating preservice elementary science teachers' shift from learner to teacher of engineering design-based science teaching. *International Journal of Science and Mathematics Education*, *20*, 747–767. https://doi-org.proxy-tu.researchport.umd.edu/10.1007/s10763-021-10193-y

Carberry, A. R., Dalal, M., Nagda, M., & McCarthy, B. (2021). *Expanding the STEM teacher pool: A history teacher's experience teaching a high school engineering course*. National Association of Research in Science Teaching (NARST) Virtual Conference.

Chandler, J., Fontenot, A. D., & Tate, D. (2011). Problems associated with a lack of cohesive policy in K-12 pre-college engineering. *Journal of Pre-College Engineering Education Research*, *1*(1), *5*. https://doi.org/10.7771/2157-9288.1029

Cunningham, C. (2007). Elementary teacher professional development in engineering: Lessons learned from Engineering is Elementary. In *Workshop of the NAE/NRC committee on engineering education*, Washington, DC.

Cunningham, C. (2008). Elementary teacher professional development in engineering: Lessons learned from Engineering is Elementary. In *American society for engineering education annual conference & exposition*. Pittsburgh, PA. https://doi.org/10.18260/1-2-4414.

Cunningham, C. M. (2009). Engineering is elementary. *The Bridge, 30*(3), 11–17.

Dalal, M., Carberry, A. R., & Maxwell, R. (2022). Broadening the pool of pre-college engineering teachers: The path experienced by a music teacher. *IEEE Transactions on Education*. 65(3), 344–355. https://doi.org/10.1109/TE.2022.3141984

Dalal, M., Carberry, A. R., Warmington, D., & Maxwell, R. (2020). A case study exploring transfer of pedagogical philosophy from music to engineering. In *ASEE/IEEE frontiers in education virtual conference (FIE)*. https://doi.org/10.1109/FIE44824.2020.9274128

Dalal, M., Klein-Gardner, S. S., Kouo, J., Reid, K., Beauchamp, C., O'Neal, B., Lopez-Roshwalb, J., & Pines, D. (2021, January). Stereotypes and implicit biases in engineering education: Will students need to whistle Vivaldi? In *The collaborative network for engineering and computing diversity conference*, Arlington, VA. https://peer.asee.org/36122

Dare, E. A., Ellis, J. A., & Roehrig, G. H. (2014). Driven by beliefs: Understanding challenges physical science teachers face when integrating engineering and physics. *Journal of Pre-College Engineering Education Research, 4*(2), *5*. https://doi.org/10.7771/2157-9288.1098

Daugherty, J. L., & Custer, R. L. (2012). Secondary level engineering professional development: Content, pedagogy, and challenges. *International Journal of Technology and Design Education, 22*(1), 51–64. https://doi.org/10.1007/s10798-010-9136-2

Daugherty, M., Carter, V., & Sumner, A. (2021). Standards for technological and engineering literacy and STEM education. *Technology and Engineering Teacher, 80*(5).

DeBoer, G. (2019). *A history of ideas in science education*. Teachers College Press.

Dieker, L. A., Rodriguez, J. A., Lignugaris, B., Hynes, M. C., & Hughes, C. E. (2013). The potential of simulated environments in teacher education: Current and future possibilities. *Teacher Education and Special Education, 37*(1), 21–33. https://doi.org/10.1177/0888406413512683

Educational Testing Service (ETS). (2017). *The Praxis(R) study companion: Technology education* (5051). www.ets.praxis

Educational Testing Service (ETS). (2020a). *The Praxis(R) study companion: Elementary education three subject bundle – mathematics, social studies, and science (5901, subtests 5903, 5904, 5905)*. www.ets.praxis

Educational Testing Service (ETS). (2020b). *The Praxis(R) study companion: Middle school science* (5442). www.ets.praxis

Educational Testing Service (ETS). (2021). *The Praxis(R) study companion: Physical science* (5485). www.ets.praxis

Engineering for Us All. (n.d.). *Homepage*. https://e4usa.org/

Engineering Is Elementary. (n.d.). *Homepage*. www.eie.org/

Engineering Your World. (n.d.). *Homepage*. https://engineeryourworld.org/

Farmer, C., Allen, D. T., Berland, L. K., Crawford, R. H., & Guerra, L. (2012, June). Engineer your world: An innovative approach to developing a high school engineering design course. In *American society for engineering education annual conference & exposition*, San Antonio, TX. https://doi.org/10.18260/1-2-21291

Fayer, S., Lacey, A., & Watson, A. (2017). STEM occupations: Past, present, and future. *Spotlight on Statistics*, 1–35.

Fry, R., Kennedy, B., & Funk, C. (2021). *STEM jobs see uneven progress in increasing gender, racial and ethnic diversity*. Pew Research Center, 1–28.

Granovsky, B. (2018). *Science, technology, engineering, and mathematics (STEM) education: An overview*. Congressional Research Service. https://crsreports.congress.gov/product/pdf/R/R45223

Groves, J. F., Abts, L. R., & Goldberg, G. A. (2014). Using an engineering design process portfolio scoring rubric to structure online high school engineering education. In *American society for engineering education annual conference & exposition*, Indianapolis, IN. https://doi.org/10.18260/1-2-23254

Guskey, T. (2003). What makes professional development effective? *Phi Delta Kappan, 84*, 748–750.

Guzey, S. S., Tank, K., Wang, H. H., Roehrig, G., & Moore, T. (2014). A high quality professional development for teachers of grades 3–6 for implementing engineering into classrooms. *School Science and Mathematics, 114*(3), 139–149. https://doi.org/10.1111/ssm.12061

Hammack, R., Gannon, P., Foreman, C., & Meyer, E. (2020). Impacts of professional development focused on teaching engineering applications of mathematics and science. *School Science and Mathematics, 120*(7), 413–424. https://doi.org/10.1111/ssm.12430

Hanson, J. R., Hardman, S., Luke, S., & Lucas, B. (2021). Developing pre-service primary teachers' understanding of engineering through engineering habits of mind and engagement with engineers. *International Journal of Technology and Design Education, 32*, 1469–1494. https://doi.org/10.1007/s10798-021-09662-w

Harlow, D. B., Hansen, A. K., McBeath, J. K., & Leak, A. E. (2018). Teacher education for maker education: Helping teachers develop appropriate PCK for engaging children in educative making. In S. Uzzo, S. Graves, E. Shay, M. Harford, & R. Thompson (Eds.), *Pedagogical content knowledge in STEM: Advances in STEM education* (pp. 265–280). Springer. https://doi.org/10.1007/978-3-319-97475-0_14

Havice, W., Havice, P., Waugaman, C., & Walker, K. (2018). Evaluating the effectiveness of integrative STEM education: Teacher and administrator professional development. *Journal of Technology Education, 29*(2), 73–90.

Hynes, M., & Dos Santos, A. (2007). Effective teacher professional development: Middle-school engineering content. *International Journal of Engineering Education, 23*(1), 24.

Hynes, M. M., Mathis, C., Purzer, S., Rynearson, A., & Siverling, E. (2017). Systematic review of research in P-12 engineering education from 2000–2015. *International Journal of Engineering Education, 33*(1), 453–462.

International Technology and Engineering Education Association (ITEEA). (n.d.). *Professional learning – STEM center for teaching and learning.* www.iteea.org/STEMCenter/EbD-PL.aspx

Johns Hopkins University (n.d.). *Graduate certificate – STEM instructional leader (Pre-K-6).* https://education.jhu.edu/academics/gc-stem-instructional-leader/

Kim, E., Oliver, J. S., & Kim, Y. A. (2019). Engineering design and the development of knowledge for teaching among preservice science teachers. *School Science and Mathematics, 119*(1), 24–34. https://doi.org/10.1111/ssm.12313.

Knowles Teacher Initiative. (n.d.). *Homepage.* Retrieved December 31, 2021, from https://knowlesteachers.org/

Kouo, J. L., Dalal, M., Berhane, B. T., Ladeji-Osias, J. K., Reid, K., Beauchamp, C., Carberry, A. R., & Klein-Gardner, S. S. (2020). WIP: Initial investigation of effective teacher professional development among experienced and non-experienced engineering teachers. In *American society for engineering education annual virtual conference & exposition.* https://doi.org/10.18260/1-2-35547

Kouo, J. L., Klein-Gardner, S. S., Dalal, M., & Eagle, E. (2021). When a pandemic requires a pivot in the modality of teacher professional development (Work in Progress). In *American society for engineering education (ASEE) virtual annual conference & exposition.* https://peer.asee.org/38054

Lehman, J. D., & Capobianco, B. M. (2012). Creating shared instructional products for integrating engineering education in the Science Learning through Engineering Design (SLED) partnership. In *Association for science teacher education annual conference*, Clearwater, FL.

Lieberman, A. (1995). Practices that support teacher development: Transforming conceptions of professional learning. *Phi Delta Kappan, 76*(8), 591–596.

Lima, M., & Oakes, W. C. (2006). *Service-learning: Engineering in your community.* Great Lakes Press.

Lockee, B. B. (2021). Shifting digital, shifting context: (re)considering teacher professional development for online and blended learning in the COVID-19 era. *Education Technology Research and Development, 69*, 17–20. https://doi.org/10.1007/s11423-020-09836-8

Lottero-Perdue, P. S. (2017). *Pre-service elementary teachers learning to teach science-integrated engineering design PBL.* In J. Saye & T. Brush (Eds.), *Developing and supporting PBL practice: Research in K-12 and teacher education settings* (pp. 105–131). Purdue University Press.

Lottero-Perdue, P. S., & Mikeska, J. N. (2022). Simulated engineering teaching experiences: Preservice teachers learning to facilitate discussions to help students become informed designers. In *American society for engineering education conference & exposition*, Minneapolis, MN.

Lottero-Perdue, P. S., Mikeska, J., & Orlandi, E. (2020). Development and teacher perceptions of an avatar-based performance task for elementary teachers to practice post-testing argumentation discussions in engineering design. In *American society for engineering education annual virtual conference & exposition.* https://doi.org/10.18260/1-2-34444

Lottero-Perdue, P. S., & Settlage, J. (2021). Equitizing engineering education by valuing children's assets: Including empathy and an ethic of care when considering trade-offs after design failures. *Journal of Pre-College Engineering Education Research, 11*(1), *4*. https://doi.org/10.7771/2157-9288.1280

Loucks-Horsley, S., Stiles, K. E., Mundry, S., Love, N., & Hewson, P. W. (2010). *Designing professional development for teachers of science and mathematics.* Corwin Press.

Magnusson, S. J., Borko, H., & Krajcik, J. S. (1999). Nature, sources, and development of pedagogical content knowledge for science teaching. In J. Gess-Newsome & N. Lederman (Eds.), *Science teacher knowledge* (pp. 95–132). Kluwer Academic Publishers.

Martin, L., & Wendell, K. B. (2021). Reflections on asset-based pre-college engineering education to promote equity: An introduction to the special issue. *Journal of Pre-College Engineering Education Research, 11*(1), 3. https://doi.org/10.7771/2157-9288.1325

Martín-Páez, T., Aguilera, D., Perales-Palacios, F. J., & Vílchez-González, J. M. (2019). What are we talking about when we talk about STEM education? A review of literature. *Science Education, 103*(4), 799–822. https://doi.org/10.1002/sce.21522

Mesutoglu, C., & Baran, E. (2021). Integration of engineering into K-12 education: A systematic review of teacher professional development programs. *Research in Science & Technological Education, 39*(3), 328–346. ttps://doi.org/10.1080/02635143.2020.1740669

Millersville University. (n.d.). *Integrative STEM education methods.* www.millersville.edu/programs/integrative-stem-education-methods.php

Moore, T. J., Stohlmann, M. S., Wang, H. H., Tank, K. M., Glancy, A. W., & Roehrig, G. H. (2014). Implementation and integration of engineering in K-12 STEM education. In S. Purzer, J. Strobel, & M. Cardella (Eds.), *Engineering in pre-college settings: Synthesizing research, policy, and practices* (pp. 35–60). Purdue University Press.

Nathan, M. J., Atwood, A. K., Prevost, A., Phelps, L. A., & Tran, N. A. (2011). How professional development in Project Lead the Way changes high school STEM teachers' beliefs about engineering education. *Journal of Pre-College Engineering Education Research, 1*(1), 1–15. https://doi.org/10.7771/2157-9288.1027

National Academies of Sciences, Engineering, and Medicine (NASEM). (2019). *Science and engineering for grades 6–12: Investigation and design at the center.* National Academies Press.

National Academies of Sciences, Engineering, and Medicine (NASEM). (2020). Building capacity for teaching engineering in K-12 education. National Academies Press. https://doi.org/10.17226/25612.

National Research Council (NRC). (1996). *National science education standards.* National Academies Press.

National Research Council. (2009). *Engineering in K-12 education: Understanding the status and improving the prospects.* National Academies Press.

National Research Council. (2011). *Successful K-12 STEM education: Identifying effective approaches in science, technology, engineering, and mathematics.* National Academies Press.

National Research Council. (2012). *A framework for K-12 science education: Practices, crosscutting concepts, and core ideas.* National Academies Press.

National Science Foundation (NSF). (n.d.). *Research experiences for teachers in engineering and computer science.* https://beta.nsf.gov/funding/opportunities/research-experiences-teachers-engineering-and-computer-science-0

National Science Teaching Association (NSTA). (n.d.). *About the next generation science standards.* https://ngss.nsta.org/about.aspx

Nesmith, S. M., & Cooper, S. (2021). Connecting engineering design and inquiry cycles: Impact on elementary preservice teachers' engineering efficacy and perspectives toward teaching engineering. *School Science and Mathematics, 121*(5), 251–262. https://doi.org/10.1111/ssm.12469

NGSS Lead States (2013). *Next generation science standards: For states, by states.* National Academies Press. www.nextgenscience.org/next-generation-science-standards.

North Carolina State University (NCSU). (n.d. a). *Concentration: Engineering and technology education.* https://ced.ncsu.edu/graduate/programs/masters/engineering-and-technology-education/

North Carolina State University (NCSU). (n.d. b). *Elementary education.* https://ced.ncsu.edu/undergraduate/programs/elementary-education/

North Carolina State University (NCSU). (n.d. c). *Technology, engineering, and design education.* https://ced.ncsu.edu/undergraduate/programs/technology-engineering-and-design-education/

Ohio Northern University. (n.d.). *Engineering education.* www.onu.edu/academics/engineering-education

Perkins Coppola, M. (2019). Preparing preservice elementary teachers to teach engineering: Impact on self-efficacy and outcome expectancy. *School Science and Mathematics, 119*(3), 161–170. ttps://doi.org/10.1111/ssm.12327

Porter, T., West, M. E., Kajfez, R. L., Malone, K. L., & Irving, K. E. (2019). The effect of teacher Professional development on implementing engineering in elementary schools. *Journal of Pre-College Engineering Education Research, 9*(2), *5*. https://doi.org/10.7771/2157-9288.1246

Project Lead The Way. (n.d.). *Homepage*. www.pltw.org/

Purdue University. (n.d. a). *Engineering-technology teacher education degree*. https://polytechnic.purdue.edu/degrees/engineering-technology-teacher-education

Purdue University. (n.d. b). *K-12 integrated STEM education graduate certificate*. www.education.purdue.edu/graduate-students/prospective-students/graduate-programs/science-education/k-12-integrated-stem-education-graduate-certificate/

Purzer, S., & Quintana-Cifuentes, J. P. (2019). Integrating engineering in K-12 science education: Spelling out the pedagogical, epistemological, and methodological arguments. *Disciplinary and Interdisciplinary Science Education Research*, *1*(1), 1–12. https://doi.org/10.1186/s43031-019-0010-0

Radloff, J., Guzey, S., Eichinger, D., & Capobianco, B. (2019). Introducing engineering design to an introductory preservice elementary biology content course using a life science compost modeling task. *Journal of College Science Teaching*, *49*, 45–52.

Reid, K. J., & Greisinger, T. M. (2021, October). The evolution of the precollege e4usa (engineering for us all) engineering curriculum. In *IEEE frontiers in education conference*, Lincoln, NE. https://doi.org/10.1109/FIE49875.2021.9637424

Reimers, J. E., Farmer, C. L., & Klein-Gardner, S. S. (2015). An introduction to the standards for preparation and professional development for teachers of engineering. *Journal of Pre-College Engineering Education Research*, *5*(1), 40–60. https://doi.org/10.7771/2157-9288.1107

Roehrig, G. H., Moore, T. J., Wang, H. H., & Park, M. S. (2012). Is adding the E enough? Investigating the impact of K-12 engineering standards on the implementation of STEM integration. *School Science and Mathematics*, *112*(1), 31–44. https://doi.org/10.111/j.1949-8594.2011.00112.x

Shirey, K. L. (2015, June). The engineering education epistemology of a science teacher (RTP, Strand 1). In *American society for engineering education annual conference & exposition*, Seattle, WA. https://doi.org/10.18260/p.24867

Shulman, L. (1986). Those who understand: Knowledge growth in teaching. *Educational Researcher*, *15*(2), 4–14.

Singer, J. E., Ross, J. M., & Jackson-Lee, Y. (2016). Professional development for the integration of engineering in high school STEM classrooms. *Journal of Pre-College Engineering Education Research*, *6*(1), 30–44. https://doi.org/10.7771/2157-9288.1130

Sneider, C. I., & Ravel, M. K. (2021). Insights from two decades of P-12 engineering education research. *Journal of Pre-College Engineering Education Research*, *11*(2), 63–98. https://doi.org/10.7771/2157-9288.1277

Sun, Y., & Strobel, J. (2014). From knowing-about to knowing-to: Development of engineering pedagogical content knowledge by elementary teachers through perceived learning and implementing difficulties. *American Journal of Engineering Education*, *5*(1), 41–60. https://doi.org/10.19030/ajee.v5i1.8610

Superfine, A. C. (2021). An asset-based perspective on prospective teacher education. *Journal of Mathematics Teacher Education*, *24*, 331–333. https://doi.org/10.1007/s10857-021-09503-6

Tan, E., Calabrese Barton, A., & Benavides, A. (2019). Engineering for sustainable communities: Epistemic tools in support of equitable and consequential middle school engineering. *Science Education*, *103*(4), 1011–1046.

TeachEngineering (TE). (n.d.). *Homepage*. www.teachengineering.org/

Teaching Institute for Excellence in STEM (TIES). (n.d.). *Homepage*. www.tiesteach.org/

The College of New Jersey (TCNJ). (n.d.). *Integrative STEM education*. https://istem.tcnj.edu/about/academic-programs/bachelor-of-science-bs-in-technology-education/

Tschannen-Moran, M., & Barr, M. (2004). Fostering student learning: The relationship of collective teacher efficacy and student achievement. *Leadership and Policy in Schools*, *3*(3), 189–209. https://doi.org/10.1080/15700760490503706

Tufts University. (n.d.). *Engineering education – teacher engineering education program (TEEP)*. https://asegrad.tufts.edu/academics/explore-graduate-programs/engineering-education

University of Colorado Boulder. (n.d.). *Engineering plus*. www.colorado.edu/eplus/

University of Texas at Austin. (n.d.). *UTeach*. https://uteach.utexas.edu/

US Department of Education. (n.d.). *Institute of Educational Sciences, National Center for Education Statistics*. Retrieved October 19, 2022, from https://nces.ed.gov

Virginia Polytechnic Institute and State University. (n.d.). *Integrative STEM education*. https://liberalarts.vt.edu/departments-and-schools/school-of-education/academic-programs/integrative-stem-education.html

Watkins, J., Portsmore, M. D., & Swanson, R. D. (2021). Shifts in elementary teachers' pedagogical reasoning: Studying teacher learning in an online graduate program in engineering education. *Journal of Engineering Education, 110*(1), 252–271. https://doi.org/https://doi.org/10.1002/jee.20369

Yesilyurt, E., Deniz, H., & Kaya, E. (2021). Exploring sources of engineering teaching self-efficacy for preservice elementary teachers. *International Journal of STEM Education, 8*(1), 1–15. https://doi.org/10.1186/s40594-021-00299-8

Zarske, M., Reitsma, R., Cyr, M., Shaw, N., Mooney, M., Sullivan, J., & Klenk, P. (2005, June). The TeachEngineering digital library: Engineering comes alive for K 12 youth. In *American society for engineering education annual conference & exposition*, Portland, OR. https://doi.org/10.18260/1-2-15513

13

Engineering Graduate Education in the United States

Gabriella Coloyan Fleming, Maura Borrego, and David Knight

1 Introduction

Within engineering education research, graduate education within the US context is unique in a number of ways. There is less research on graduate education than most other education levels. We believe this is because graduate student experiences are highly influenced by their local context, including the graduate program and adviser/supervisor. Further, and perhaps because of the importance of local context, there is no culture of benchmarking data or transparency in completion rates and time to degree. Other aspects, such as how students are classified as master's versus doctoral students when their ultimate goal is a PhD, further complicate the graduate education data landscape.

Compared with other disciplines, engineering graduate education is also unique. Engineering tends to have a larger proportion of international students, lab and group-based collaborative research, and more collaborative relationships between students and adviser (i.e., more co-authorship). Thus, graduate education research from beyond engineering varies in its relevance to engineering. Since there is a lot of ground to cover, we scope this chapter to focus on studies of engineering graduate students (or including engineering in their coverage of STEM) in the United States conducted by those identifying as engineering education researchers.

Internationally, some of the biggest contextual graduate education differences are in student funding and required coursework (e.g., Australia and the United States (Deters et al., 2021)). For example, it is much more common in the United States for master's and doctoral students to complete approximately two years of coursework before focusing on research (if a thesis or dissertation is part of their program requirements); in many other countries, graduate education is focused almost exclusively on conducting research and is much more individually directed. These differences and varying needs in the global labor market lead to engineering PhD earners working in different sectors of the labor force in different countries (Mason et al., 2022). The graduate funding situation and dynamics in the United States are also quite complex, since students may be funded through a combination of mechanisms controlled by different stakeholders, and interactions with different types of stakeholders can lead to very different student experiences. For example, the federally funded National Science Foundation (NSF) fellowship allows students to have flexibility in their projects and allows them to publish their results, whereas Department of Defense or industry-funded projects can have higher expectations for deliverables, less freedom in what topics the student researches, and some restrictions on publishing due to intellectual property concerns. In other parts of the world, it

DOI: 10.4324/9781003287483-16

is more common for engineering graduate students to be fully funded once admitted to a graduate program. Given these important contextual differences, we made the choice to scope this chapter to review the literature on engineering graduate education within US contexts. We encourage readers to look to our colleagues' work from around the globe for a broader view of this topic for features and considerations that do not necessarily translate across national contexts.

We have conducted a narrative review, which allows us to identify what has been previously published but can be limited by our experiences and perspectives as researchers in this area (Grant & Booth, 2009). We draw primarily from peer-reviewed journal articles and books in engineering education, as these tend to have more in-depth investigation into issues on graduate education. However, we do cite several conference papers, which are more commonly used for evaluation of specific programs for graduate students. The chapter is laid out in the following sections: "Why Engineering Students Pursue Graduate Education," "Recruitment of Graduate Students," "Engineering Students' Skills Development During Graduate Education," "Engineering Students' Identity Development During Graduate Education," "Supporting Engineering Graduate Students' Skill and Identity Development," "Why Engineering Students Leave Graduate Education," "Career Trajectories of Engineering Graduate Students," "Challenges of Reforming Graduate Education in Engineering," and "Opportunities for Future Research in Graduate Education in Engineering."

2 Why Engineering Students Pursue Graduate Education

We begin with what motivates engineering students to undertake graduate education. The most widely known preparation is undergraduate research (Dukhan & Jenkins, 2007), but it is difficult to understand whether the experience simply increases interest of students who have already self-selected into research. In a longitudinal study that controlled for self-selection bias, Eagan et al. (2013) found that STEM students who participated in undergraduate research experiences were significantly more likely to be interested in pursuing a STEM graduate degree. In addition to learning hands-on lab, writing, and presentation skills, students in formalized programs also learn about the *process* of doing research. For example, Issen et al. (2007) reported that students found it rewarding to overcome research obstacles and said the experience was pivotal in their decisions to attend graduate school. Additionally, the social community during and after the program, in which students made long-lasting friendships, was just as important as the research community during the short summer experience (Issen et al., 2007). In a summer research program specifically for minoritized students, May (1997) found that 89% of participants attended or planned to attend graduate school, a statistically significant finding compared to students who did not participate. Additionally, 60% of student researchers said that program participation helped them find funding for graduate school, a key concern for prospective students.

Undergraduate research experiences can affect students' interest in graduate school without them realizing it. Zydney et al. (2002) surveyed alumni of the University of Delaware, asking them about their experiences as undergraduates without disclosing that the purpose of the survey was to assess long-term effects of undergraduate research experiences. Over half of those surveyed who did research said it was "very" or "extremely" important (4 or 5 out of a 5-point Likert scale), with scores increasing according to the number of semesters of undergraduate research completed. Around 87% of alumni who went on to earn PhDs participated in undergraduate research, compared to only 8% of students who did not complete any research (Zydney et al., 2002).

However, not all students who participate in research decide to attend graduate school. Willis et al. (2013) found that some mechanical engineering students were less likely to be interested in graduate school at the end of their research experience than they were at the beginning, as they found research "tedious" and "slow," favoring working in industry where they could see more immediately applied results from their work (Willis et al., 2013). Though some may find it discouraging that students decide not to further pursue research, this decision-making is a natural part of students' discovery process. As discussed in Section 6, "Why Engineering Students Leave Graduate Education," many advisers hope that students can figure out early if graduate school is right for them, and having an undergraduate research experience is one mechanism for helping students make that decision.

There are also several non-research factors that contribute to students' decisions to go to graduate school. In a study of engineers 5 to 25 years after earning their PhDs, London et al. (2014) found that motivations to attend graduate school included passion for research or a particular technical area, the opportunity to pursue an academic career, or influence from a mentor. McGee et al. found that Black engineering PhD students were motivated to pursue advanced degrees due to passion for their fields, desires to be role models for other Black students interested in STEM, and aspirations to benefit society (McGee et al., 2016). Another important consideration is cost. As Kennedy et al. (2016) found in a study on students' knowledge of financial resources to attend graduate school, the second-most commonly cited factor when deciding between graduate school offers was financial incentives (personal fit within the institution was the most important factor). They also found that a lack of knowledge about financial resources was a deterrent for students considering graduate school, as many undergraduate students did not know about funding mechanisms, such as assistantships, and perceived that there are not as many scholarships available for graduate school as there are for an undergraduate degree. Students who did receive funding, however, found this to be an incentive to attend graduate school, though it seemed limited to PhD students and not available to master's students (Kennedy et al., 2016).

Engineering students' skills and personal factors also contribute to their decision to pursue graduate education. Ro et al. (2017) identified three factors: mathematics proficiency, self-assessment of teamwork and leadership skills, and co-curricular experiences. Students with higher math proficiency and self-assessed leadership skills were more likely to attend graduate school, whereas students with higher self-assessed teamwork skills were less likely to attend graduate school. Additionally, students who spent more time on non-engineering community volunteer work were more likely to attend a non-engineering graduate program (Ro et al., 2017). Two-thirds of study participants in non-engineering master's programs were in business or management, possibly as preparation for careers in engineering management. Borrego et al. (2018) identified personal factors that influenced students' intentions to go to graduate school. The most influential positive factor they identified was self-efficacy, which includes students' belief in their abilities to learn new skills, conduct independent research, and complete their graduate degrees. Other positively correlated factors were outcome expectations (e.g., perceived time to degree completion, impact of having a graduate degree on future job opportunities) and support (e.g., positive interactions with graduate students as an undergraduate, faculty adviser encouragement to attend graduate school, positive mentoring experiences). The more barriers a student encountered (e.g., lack of information about graduate school, perceived inability to pay for graduate school, worries about the competitiveness of application, anticipated low level of future support), the less likely they were to attend graduate school (Borrego et al., 2021).

While many engineering graduate students matriculate directly after earning their undergraduate degrees ("direct-pathway students"), a small percentage begin graduate school after several years in the workforce. These "returners" are an understudied student population, and their work experience gives them unique perspectives and preparation for graduate school. In a quantitative study on returners, Mosyjowski et al. (2017) found that prior to graduate school enrollment, returners reported less confidence in their ability to complete their degree than direct-pathway students, though confidence levels in the two groups were similar after beginning their degrees. Additionally, returners reported higher costs related to finances, work–life balance, and navigating a new environment (Mosyjowski et al., 2017).

3 Recruitment of Graduate Students

Engineering graduate students, especially at the doctoral level, tend to be well-funded. For example, nearly three-quarters of engineering doctoral students in the United States were funded via research assistantships or other grants and fellowships, and less than 10% funded their programs via personal means (Nettles et al., 2006). There is a complex ecosystem of potential funding sources for students, and so many are funded through research grants, making the US graduate recruiting environment fiercely competitive. This competitive arena is particularly present within highly ranked institutions, where a program's prestige is tied to their consistent ability to attract talented graduate students (Evans, 1993; Posselt, 2014; Rutter et al., 2016; Wall Bortz et al., 2020). There are several driving forces, largely stemming from the fact that there have been enormous financial investments to support university-based research engines, and graduate education and the research enterprise are strongly coupled (National Research Council, 1995, 2014). For example, the *US News and World Report*'s rankings of the best engineering schools operationalizes graduate student enrollments within the "faculty resources" category (25% of the ranking score), which considers more doctoral degrees awarded and higher graduate student-to-adviser ratios to be characteristics of higher-ranked institutions. Graduate student selectivity (10% of the ranking score) is operationalized by programs' acceptance rates and entering graduate students' average quantitative GRE (standardized exam for graduate school admissions) score (Morse et al., 2021).

Despite the highly competitive nature of graduate recruitment processes within engineering and their enormous financial implications, there has been surprisingly little systems-level research from the perspective of institutions, with single-institution studies being the predominant research design. One exception is a study (Wall Bortz et al., 2020) of STEM graduate recruitment across institutions which found that programs adopt the same kinds of strategies, even when those strategies do not align with program leaders' stated values or graduate students' priorities. Financial resources comprise the main mechanism used by programs to recruit graduate students and include fellowship offers, multi-year funding packages, research assistantship guarantees, or "top-off" stipends that can act as signing bonuses. However, as the prior section in this chapter highlighted, other factors, such as academic considerations, research interests, adviser fit, location, and program supports, may be even more important for students' decision processes (Le & Tam, 2008). As such, there is an opportunity to enhance and demonstrate such considerations throughout recruitment. Adviser involvement is critical for the graduate recruitment process within STEM fields, a factor of which many faculty members may not be aware (Baron, 1987; Bersola et al., 2014; Evans, 1993).

Exploring graduate recruitment processes is particularly important for a field that has been relatively stagnant with respect to diversifying the student body. The engineering graduate student population is even less diverse than the engineering undergraduate population, which has downstream implications for efforts to broaden participation in engineering. Main et al. (2020) demonstrated correlations between the number of women of color faculty members in a program and the number of women of color who complete bachelor's degrees in engineering. Ong et al. (2011) showed that mentors play an important role in women of color STEM students' decisions to attend graduate school. Mondisa (2018) found that same-race mentors bring some unique relationship- and identity-building approaches to mentoring of African American STEM undergraduates which support their educational and career persistence. Thus, there are multiple positive-feedback loops between diversifying graduate education in engineering so that more faculty members and other mentors can support the next generation and diversify the engineering field more broadly. In a systems-level approach, Fleming et al. (2023) illustrated the institutional pathways (and therefore limited institutional mobility) from bachelor's degrees at highly ranked, not highly ranked, non-US, and minority-serving institutions to PhDs at these same types of institutions for Black/African American and Hispanic/Latino PhD earners.

Despite these many reasons to diversify graduate education in engineering, programs have struggled to meet this goal. Engineering graduate recruiting is competitive, and doubly so for recruiting racially minoritized and women students within engineering. Posselt (2014) asserted that "attracting academically accomplished students from underrepresented backgrounds has become a way that programs evaluate themselves against one another, such that diversity itself is associated with prestige" (p. 501). To understand this more specific competitive arena, Wall Bortz et al. (2021) compared programs' recruitment and yield strategies with decision-making factors of minoritized students. Offering "diversity" fellowships was the most commonly mentioned strategy for attracting racially minoritized students and women in some engineering contexts, yet program leaders also cited a small pool of minoritized graduate students as the limiting factor (as opposed to a lack of available financial resources). Recent national-scale research focused on educational pathways of Black and Hispanic engineering doctoral recipients problematizes this "lack of supply" perception and instead argues that programs can broaden their recruitment efforts by considering bachelor's degree recipients from a range of institutional types and rankings (Fleming et al., 2023; Wall Bortz et al., 2021). Coupled with this idea, many researchers have shown that relying on GRE scores and grade point averages as admissions criteria systematically excludes racially minoritized and women students rather than considering their diverse experiences as valuable assets. Thus, broadening participation in graduate education within engineering requires changes to the "gatekeeping" system that currently characterizes admissions processes.

Wall Bortz et al. (2021) also noted a variety of other non-monetary recruitment strategies that aligned with minoritized students' priorities as well as with prior literature. Such strategies include personal contact with faculty and program personnel (Sowell et al., 2015), leveraging professional faculty networks, including those at historically Black colleges and universities (HBCUs) and other minority-serving institutions (MSIs) (Sowell et al., 2015), and conveying a positive and supportive campus culture, including connecting with faculty and peers (Griffin & Muñiz, 2011). Relative to White and Asian students, Bersola et al. (2014) found that racially minoritized admitted students place more importance on factors such as the diversity of faculty, students, and the community; quality of the campus infrastructure; urbanicity; and life considerations, such as childcare and housing. Student-facing strategies, such as hosting open house events, recruiting in intentional locations, forming and supporting cohorts of minoritized students, and offering a range of professional development activities, were all raised by minoritized students in the Wall Bortz et al. (2020) study. The literature has shown that cohort-based strategies are particularly important for recruiting and supporting minoritized students (Bostwick & Weinberg, 2018), which is a core feature of bridge programs that have been extremely successful in supporting the enrollment of minoritized students into STEM graduate programs (Gámez et al., 2021). Given all these different considerations, unless the overall financial package differs significantly between institutions, money is not likely to sway a minoritized student from an initial preference (Bersola et al., 2014; Freeman, 1984; Jackson & Chapman, 1984; Wall Bortz et al., 2021). Many of these student-facing strategies would also support graduate student retention, which is another pressing issue (Nicole & DeBoer, 2020; Sowell et al., 2015). Therefore, investing time and resources to support recruitment and yield likely will have multiplicative effects on a graduate program's culture of support of racially minoritized and women students.

4 Engineering Students' Skills Development During Graduate Education

We turn now to what and how students learn once they matriculate into an engineering graduate program. Although graduate students in some programs learn from coursework, most of the engineering education research on learning during graduate study focuses on the research group/

laboratory environment. A research group comprises students conducting research under the supervision of one or more advisers, often sharing equipment and methods. These groups can vary significantly in size, membership (i.e., undergraduate researchers, postdocs, and technicians), and climate (Crede & Borrego, 2012). Through their interaction with research group members, graduate students learn important skills, such as presenting research, receiving and responding to feedback, and solving and troubleshooting problems (Burt, 2017).

A small body of research within engineering education has focused on specific skills that develop during engineering graduate students' programs. In a study of engineering PhD earners working in academia and industry, London et al. (2014) found that having a PhD provided additional knowledge, skills, and attributes, including the ability to conduct scientific work and a deeper understanding of fundamental concepts. Science and engineering graduate students and postdocs who mentor undergraduate researchers develop specific skills related to mentoring, including understanding students' needs, building positive working relationships with mentees, developing interpersonal skills, and specific character traits such as patience, flexibility, and humility (Ahn & Cox, 2016). Using a survey-based approach to understand students' perceptions of different kinds of skill development, Grote et al. (2021) focused on four different sets of skills: (1) research skills, (2) communication skills, (3) peer training and mentoring skills, and (4) teamwork and project management skills. Each of these different skill sets could position a student for a variety of different kinds of careers, but the results found a correlation between students' predominant graduate study funding mechanisms and their perceptions that they had opportunities to develop certain skill sets (Grote et al., 2021). Receiving a fellowship is often viewed as prestigious because of the autonomy and, sometimes, pay rate associated with such awards, but these results suggest that receiving a fellowship could come at the cost of having fewer opportunities to develop a range of career-relevant skills. This finding is similar to research by Kinoshita et al. (2020), which used national-scale data to show that women and racially minoritized students funded via fellowships were more likely than other students to report no job offers when they completed the Survey of Earned Doctorates.

Also within the engineering education literature, writing as a skill has been highlighted as essential for graduate students' academic and career success, even though many engineering graduate students select the field because it may seemingly emphasize other kinds of skills, such as math or statistics skills. Research conducted by Berdanier (2019) highlighted distinct rhetorical moves within engineering graduate students' National Science Foundation Graduate Research Fellowship Program proposal materials to help students visualize different argumentation patterns that can be applied within their writing (Berdanier, 2019). Her subsequent research demonstrated linkages between prospective and current engineering graduate students' attitudes about writing and the actual rhetorical patterns that appeared in their writing (Berdanier, 2021). Writing skills are crucial for PhD students, as they need to write dissertations in order to obtain their degrees. A dissertation writing workshop for racially minoritized PhD students enabled them to understand the utility of their dissertations in relation to their career paths, adjust their perceptions (particularly around perfectionism) about writing their dissertations, and improve how to plan writing (Miller et al., 2020).

There is also research into what STEM graduate students learn from interdisciplinary training programs. This analysis used a curriculum design framework to understand the intended outcomes, evidence, and learning experiences. Among interdisciplinary graduate traineeship programs, including at least one engineering discipline, 73% listed various technical skills and knowledge, including grounding in multiple disciplines, 54% sought to cultivate in students a broad perspective of their interdisciplinary domain and ability to integrate multiple disciplines, 49% had the goal of creating an interdisciplinary environment for students, 42% focused on teamwork skills needed to collaborate across disciplines, and 24% addressed interdisciplinary written and oral communication skills. Specialized coursework and team-based research were the most common approaches for cultivating these skills (Borrego & Cutler, 2010; Borrego & Newswander, 2010).

There is also a recent and increasing focus on professional development for graduate students. In the United States, instructors generally do not receive such training, and students who are interested in an academic career find that the professional development available to them as a graduate student is insufficient to prepare them for the teaching aspects of those positions. Coso Strong and Sekayi (2018) found that many advisers and departments are not supportive of teaching or other non-research activities; students needed to actively seek out professional development on how to teach, such as through teaching certificate programs. Focused on developing the professoriate, preparing future faculty (PFF) initiatives are sponsored by institutions for their own and/or external students and by professional societies. Funding has been available for institutions to host PFF programs, particularly aimed at increasing faculty gender and racial/ethnic diversity. These initiatives fall under a wide variety of formats, including workshops (Tormey et al., 2020), formal mentoring programs, formal courses (in which students earn credit), short courses and seminars, structured teaching practicum, reading and writing assignments, formal networking experiences, and research mentoring practicum. Though formal PFF programs were established in 1993, there are few, if any, reported studies on the efficacy of these programs in developing engineering faculty members; rather, reports on PFF programs have focused on best practices and program content (Diggs et al., 2017). Some PFF programs are targeted specifically for students from minoritized populations (Diggs & Mondisa, 2022).

5 Engineering Students' Identity Development During Graduate Education

Identity has emerged as a lens for studying graduate student retention and interest in various roles within and beyond graduate study, two areas that we focus on specifically in subsequent sections. To avoid overlap with the identity chapter in this volume, we focus on studies of graduate students' teacher, researcher, engineer, and scholar identities. This role identity approach can be heavily influenced by the roles undertaken by faculty members and, therefore, is often tied to future faculty programs and research questions.

Svyantek et al. (2015) examined the influence of electronic portfolios in graduate student role identity development, finding a mismatch in teacher identity between where students were at the time of the study and where they would like to be in the future. Participating students were much more confident in the trajectory of their researcher identities. One of the co-authors, Kajfez and colleagues (2016), extended this work to study teacher identity and motivation in graduate teaching assistants. Their longitudinal model of motivation and identity includes future faculty identity and recommends that graduate teaching assistants interact with faculty members who may serve as role models.

Kirn, Perkins, and collaborators (Bahnson et al., 2021) culminated their many qualitative and quantitative studies of graduate student identity with a survey instrument measuring engineer, researcher, and scientist identities, each with their own recognition, interest, and performance/competence subscales. For a national US sample, they reported significant differences by engineering discipline, gender, and race/ethnicity within engineer identity, but not for researcher or scientist identities. Following a similar approach of adapting performance/competence, interest, and recognition identity constructs to the graduate level, Choe and Borrego (2020) related engineer and researcher identities to interest in academic, industry, and government careers among doctoral and thesis master's students. Their results suggest a positive relationship between engineer identity and industry career interest and a correspondingly negative relationship of engineer identity with academic and government career interest. Gelles and Villanueva (2020) found that engineering doctoral identity evolves during graduate school, and engineering PhD students think of people who are earning (fellow students) or have earned (faculty) PhDs as "insiders," and non-PhD earners as "outsiders," with defending their dissertations being a rite of passage to prove themselves to other "insiders." The study's participants described only engineering PhDs as innovative and creative problem-solvers, devaluing the skills and abilities of non-PhD engineers.

In general, there are many more quantitative studies of where graduate students actually end up in the workforce than there are qualitative studies exploring their decision processes. One notable exception is Burt's (2019) work, which integrates many aspects of peer networks, adviser relationships, and social identities into his theoretical model of engineering professorial intentions, which seeks to explain why doctoral students decide whether to pursue academic careers.

International students play important roles within the engineering graduate education ecosystem in many countries. For example, South Africa's doctoral programs had an enrollment of 40% international students in 2016 (Herman & Meki Kombe, 2019). In Canada, 28% of graduate students are international (Universities Canada, 2014). Further, over the past decade, graduate schools in several nations have increased their number of international students. For instance, the number of international graduate students in South Korea doubled from 2009 to 2017 (Ministry of Education of the Republic of Korea, 2018). In addition, the Institute of International Education (2016) reported an influx of international graduate students to US graduate programs over the past two decades, with over half of science and engineering PhDs earned by temporary visa holders (National Science Foundation, 2017). These students face many difficulties related to acculturation (the process of adapting to the new societal norms and behaviors of a host culture) upon arriving in their graduate programs, including facing acculturation stress, gaining cultural competency, and mastering another language (Burdett & Crossman, 2012; Newberry et al., 2011; Wang, 2008; Watkins & Green, 2003). Women international students face additional stressors in US graduate programs, such as feeling excluded in their classes and research groups, needing to work harder to overcome stereotypes, and experiencing tokenization. They work to overcome these barriers by speaking up in study groups and creating social networks where they can provide and receive support from others with similar experiences (Dutta, 2015). In addition to societal integration, international engineering graduate students also go through a process of integration into a new profession, which has its own culture. While they may have been engineers (by education or employment) in their home country, the professional norms of being in the United States can be very different (Newberry et al., 2011). For example, in Japan, engineering identity is strongly linked to one's employer, whereas in France it is tied to where one earned their degree. German engineering identity is tied to a collective social responsibility, which developed as a reaction to technology's role during World War II (Newberry et al., 2011). These international engineering PhD earners go on to contribute to the US engineering workforce, with almost 70% of temporary visa holders planning to stay in the United States after graduation (Sanderson et al., 2000). Models of idea generation show the potential impact that these graduates have: for each 10% increase in international students, there is a resulting 5% increase in the number of patent applications, 7% university patent grants, and 5% non-university grants (Chellaraj et al., 2008).

6 Supporting Engineering Graduate Students' Skill and Identity Development

Unlike in undergraduate education, in which students can have many people providing support, engineering graduate students' primary support comes from their academic program and their research adviser. There are different processes for student–adviser matching, which can sometimes depend on disciplinary norms (Artiles et al., 2023). In many engineering disciplines, students begin their graduate studies without an adviser, and the student–adviser matching process takes place without assistance from the graduate program. In other disciplines (e.g., civil engineering), a program may assign a temporary adviser, and students and advisers find matches without the assistance of the program. Finally, some graduate programs have a formal matching process (most commonly in chemical engineering), in which the graduate program facilitates creating student–adviser pairs (Artiles & Matusovich, 2022a, 2022b; Artiles et al., 2023). During the matching process, students consider future career prospects as well as a potential adviser's funding, research area, personality, and average graduation time; advisers consider students' academic credentials and perceptions of students' research abilities (Joy et al., 2015). New graduate students learn about prospective advisers through information systems in the forms of research seminars, one-on-one interviews, rotation programs (particularly in biomedical engineering), undergraduate research experiences, and independent study courses (Artiles et al., 2023). However, simply accessing these information systems does not equally provide students with the same benefits in navigating the adviser selection process, as students with prior research experience better understand how to use this information to develop criteria for choosing an adviser (Artiles & Matusovich, 2022a).

Advisers often serve to enculturate their advisees to the norms of academia (Boyle & Boice, 1998) and socialize them to the professoriate by demonstrating the different duties a faculty position entails, such as supervising students, managing a research lab, serving on committees, and obtaining external funding (Saddler & Creamer, 2009). Students also learn about many aspects of conducting research from their advisers, including uncertainty, time commitment, publishing, and competition (Saddler, 2009). Advisers are students' primary example for developing a faculty prototype, or their idea of what it means to be a professor, which can influence their interests in academic careers (Burt, 2019) and how they themselves are as advisers, if and when they become faculty (Lee, 2008). There are different approaches to advising, which are not mutually exclusive: functional (acting as a project manager), enculturation (encouraging students to join their discipline's community), critical thinking (encouraging the student to "question and analyze their work"), emancipation (encouraging the student to "question and develop themselves"), and developing a quality relationship (inspiring and caring for the student) (Lee, 2008, p. 267). Students feel supported by advisers who are approachable, foster good working relationships, and frequently communicate with them (De Welde & Laursen, 2008). However, as De Welde and Laursen (2008) found, not all students consider their advisers "mentors": only half of participants in their study on STEM PhD students said they viewed their adviser as a mentor, although an additional 21% said they still viewed their non-mentor adviser as a good adviser. Moreover, they found that advisers do not always provide their students with career advice, with 36% of participants reporting they received no career advice and 20% receiving some advice but less than they would have liked.

In the United States, women and racially minoritized graduate students can have very different experiences than their peers from majority groups (e.g., White and Asian men) (McGee, 2021a). Advising can be a racialized experience: in a study of Black men engineering PhD students, Burt et al. (2016) found that microaggressions from advisers caused the students to question their ability to engage in and feel less comfortable in engineering. In a study of Black engineering and computing PhD students, racialized experiences caused significant stress and strain, leading to academic performance anxiety, impostor syndrome, and poor physical and mental health (McGee, Griffith, et al., 2019). Identity as a scientist, engineer, or researcher is impacted by a student's relationships with their peers and advisers, and these impacts vary for engineering PhD students from different gender and racial/ethnic groups (Perkins et al., 2020). Asian women's poor relationships with their advisers partially cause low science interest, and women of color report stronger benefits from positive relationships with their advisers than their peers (Perkins et al., 2020). Women and racially minoritized graduate students face discrimination, often in the forms of racism or sexism, from peers, advisers, and other faculty (Corneille et al., 2019; De Welde & Laursen, 2011; Fabert et al., 2011; McGee, Griffith, et al., 2019), which reinforces their self-perceptions as insiders or outsiders in their programs (Bahnson, Satterfield, et al., 2022). While such discrimination is rooted in a larger historical and societal context, students are often not aware of the systemic nature of these exclusionary practices (Bahnson, Satterfield, et al., 2022). Engineering education researchers interested in conducting research on race- or sex-based discrimination should be aware of the Discrimination in Engineering Graduate Education (DEGrE) scale, a validated survey including sections on interactions with a student's adviser and other faculty, sexism, and lab culture (Bahnson, Hope, et al., 2022).

7 Why Engineering Students Leave Graduate Education

When talking about engineering graduate education, in addition to discussing the reasons that students pursue a graduate degree and their experiences during degree completion, it is also important to recognize that not all students who begin a degree complete one. There are several reasons that doctoral students decide to leave prior to completing their PhD. It is difficult to quantify retention rates, as some students begin with the intention of completing a PhD but leave after completing a master's degree (colloquially, "mastering out") and are therefore counted together with terminal master's students. There are not many studies of attrition in engineering graduate students. In a study that included graduate students from engineering and other fields, Gardner (2009) found that common reasons students provide for departing prior to PhD completion are personal problems (family and physical/mental health issues), departmental issues (poor advising, lack of financial support, department policies and politics), and graduate school being a poor fit for them personally. "Poor fit" also encompasses students who began their PhDs with a specific goal, but that goal changed over time and the students no longer needed a PhD for their new career goals. Students do not necessarily feel that the time spent figuring out what they wanted is a waste of time but rather a natural part of the maturation process (Gardner, 2009; Zerbe & Berdanier, 2019). Advisers, on the other hand, do not always view such changes of mind so positively and perceive students changing their minds after beginning a PhD as a waste of adviser time and resources. Advisers also cite very different reasons for student departure, including students lacking ability, drive, or motivation; students who should not have begun graduate school in the first place (i.e., the students should have figured out prior to starting their PhDs that it was not right for them); and personal problems. It is notable that there is only a small overlap (i.e., personal issues) between the reasons provided by faculty compared to reasons provided by students to explain departures.

These themes are not stand-alone reasons for attrition but are rather interconnected. Berdanier et al. (2020) conducted a study which used the social media website Reddit to gather reasons for engineering PhD student attrition, and six interconnected themes emerged: adviser role and relationship, student support network, quality of life and work, cost (both time and money), perception by others should they depart, and lacking or changing goals (Berdanier et al., 2020). Though the primary sources of students' issues were problems with their advisers, students were more likely to depart if they were experiencing more than one of these themes. Additionally, a good relationship with one's adviser when experiencing other detrimental factors can complicate a student's decision to depart without a PhD, as students feel guilty about letting their adviser down. While it might seem that students' reasons for early degree departures are the accumulation of events, a single critical event, such as an incident with an adviser, change in funding, or medical event, can also precipitate such decisions (Zerbe et al., 2022). These critical events can take place inside or outside the university setting and appear in either a routine or unexpected context.

Attrition rates between students from different demographic groups are far from equal. US women engineering PhD students have an estimated attrition rate of 35%, compared with just 24% for men (Council of Graduate Schools, 2008). Two studies sponsored by the Council of Graduate Schools (CGS) investigated this further. Sowell et al. found that attrition rates are even higher for racially minoritized students, with 36% of racially minoritized STEM PhD students withdrawing from their programs prior to completion. Black/African American students had a lower completion rate than Hispanic/Latino students (Sowell et al., 2015). The other CGS study found that White

STEM PhD students had a ten-year PhD completion rate of 55%, while Black/African American and Hispanic/Latino students had ten-year completion rates of 47% and 51%, respectively (Sowell et al., 2008). The rates of students considering early degree departures are even higher. Bahnson and Berdanier (2022) surveyed engineering PhD students and research-based master's students and found that 70% had considered leaving their programs in the previous month alone, with women considering leaving at higher rates than men, and US PhD students considering leaving at higher rates than international PhD students. As previously discussed, student–adviser relationships play an important role in students' decisions to complete their PhD. Interviews with advisers and racially minoritized graduate students revealed that students placed higher value on engaging in tasks associated with a personal sense of identity, while advisers placed higher value on tasks that provided mastery and utility, such as preparing presentations and writing grants (Artiles & Matusovich, 2020). Artiles and Matusovich posit that this mismatch between what advisers and racially minoritized students value could possibly lead to a communication discrepancy and contribute to higher attrition rates for racially minoritized students. Similarly, Gardner found that women and students of color also have higher rates of attrition because of being less integrated with their peers and program faculty (Gardner, 2009).

The other side of the attrition coin is student retention. Crede and Borrego (2014) conducted a survey of engineering graduate students across multiple institutions across the United States to determine the student demographic differences in relation to completing their PhD. Factors that were related to student intentions to complete their degree included their perception of their adviser valuing their work, project ownership, and climate. They uncovered differences between students from different regions in the world; notably, students from the Middle East and India reported the highest rates of feeling their work is valued by their adviser and ownership of their projects. Students from the United States had the most positive view of their group climate, while students from East Asia had the lowest climate scores. Additionally, students in more competitive groups (for example, with competition for resources, such as adviser time, funding, or equipment) were less likely to complete their degree (Borrego et al., 2018).

8 Career Trajectories of Engineering Graduate Students

Many of the professional development resources for graduate students have focused on preparation for academic careers. Yet master's- and doctoral-level engineering graduates end up in a variety of employment sectors, including industry, entrepreneurial/start-ups, government, academia, nonprofit, and postdoctoral positions across all those sectors (Fiegener, 2010; NSF & NCSES, 2017; Turk-Bicakci et al., 2014). A survey of engineering master's and doctoral students found that most students simultaneously consider multiple careers within and outside of academia (Choe & Borrego, 2020). US engineering doctoral recipients enter industry positions at rates (38%) nearly as high as academia (45%, including faculty and postdoctoral research positions) (Fiegener, 2010), with a majority (59%) of new US engineering doctorates beginning employment in private, for-profit industry (NSF & NCSES, 2017), and an additional 8% entering into government roles. There is much less career trajectory data for master's students, even though there are over four times as many engineering master's recipients as doctoral recipients (NSF & NCSES, 2015). One survey study from Wendler et al. (2012) found that nearly 25% of 1,500 engineering master's graduates entered into government careers. Considering all of STEM, 81% of master's recipients will enter industry or government jobs (NSF & NCSES, 2017).

There are also noteworthy differences by race and gender as well as expected salaries across sectors. In a study of US engineers in 2010, more than half of male engineering doctorate holders worked in the for-profit industry sector (Turk-Bicakci et al., 2014). Black, Hispanic, and White women were more likely to work in government than other demographic groups, and there are more Black women PhDs in government than any other career sector (Turk-Bicakci et al., 2014). The gender wage gap for graduate-level engineers working in government is less than in industry (Buffington et al., 2016), and government and industry careers have been linked with higher salaries as compared with academic careers (Yang & Webber, 2015). There is also evidence of differences by gender and race in attaining a job offer at the time a PhD student completes their program. Kinoshita et al. (2020) showed that women engineering doctorate holders, particularly those who were married, and racially minoritized engineering doctorate holders were more likely than their peers to have no job offers upon completing NSF's Survey of Earned Doctorates at the end of their programs. Thus, despite claims that there are "supply" or "pipeline" challenges for diversifying engineering, there is still evidence of systematic differences in job offers by gender and race, even among individuals in the United States with the highest levels of education.

The *Graduate STEM Education for the 21st Century* report by the US National Academies of Sciences, Engineering, and Medicine (2018) lamented that many academic institutions do not have a student-centered STEM graduate education system with respect to helping students prepare for a variety of career pathways. According to the report, in an ideal STEM graduate education system:

- Students would be encouraged and given time & resources to explore diverse career options, perhaps through courses, seminars, internships, and other kinds of real-life experiences. . . .
- Graduate programs would develop course offerings and other tools to enable student career exploration and to expose students to career options. . . .
- Institutions would help students identify advisors and mentors who can best support their academic and career development. Faculty advisors would not stigmatize those who favor nonacademic careers.

(National Academies of Sciences, Engineering, and Medicine, 2018, pp. 3–4)

Despite evidence that decreasing percentages of graduates seek entry into academic careers (Fiegener, 2010; NSF & NCSES, 2015), and those who do face increased competition for a limited number of tenure track positions (Larson et al., 2014), the majority of research and resources focus on academic career paths (Main & Wang, 2019), reinforcing a culture that privileges academic positions over nonacademic career pathways (St. Clair et al., 2017; Thiry et al., 2007). Many engineering PhD students begin their graduate studies interested in academic careers (Choe & Borrego, 2020) but lose interest as time goes on because of perceived norms and pressures of working in academia, including stressful environment, pressure to find funding, and work–life balance (McGee, Naphan-Kingery, et al., 2019). As Burt's (2020) article describing the journey of one graduate student's developing interests in the professoriate argues, however, much more research is needed to understand how and why graduate students from marginalized backgrounds choose to pursue a career in academia. Overall, the messages graduate students receive about suitable PhD employment and their understanding of their own preparation leaves much room for improvement. Borrego et al. (2021) conducted interviews with STEM graduate students and highlighted a power differential between advisers and students that makes it difficult for students to express interest in nonacademic career plans. These interviews also described how students find it challenging to articulate the skills and preparation needed to work in industry, even when it was their intended career. Engineering graduates also make career choices based on nonacademic lived experiences. In studies on STEM PhD students of color, McGee and collaborators found that students' career paths were influenced by President Donald Trump's anti-science policies and the COVID-19 pandemic, with some students expressing interest in work on social and racial justice (McGee, 2021b; McGee et al., 2021). For students who are interested in tenure-track faculty careers, the prestige of their PhD institution plays a significant role in where they end up: only roughly 15% of engineering faculty work at a more prestigious institution than where they earned their PhDs (Wapman et al., 2022), and a non-reportable, small number of Black/African American or Hispanic/Latino faculty who earned their PhD outside the US News and World Report's top 25 engineering PhD programs works at a top 25 program (Fleming et al., 2023).

A few prior studies address skills and preparation for PhD engineers in the nonacademic workforce. Cox (2019) asked 40 engineering PhDs in the United States who had entered the workforce across an array of sectors to reflect on the most important skills necessary for career success. They identified communication as most important, followed by teamwork, problem-solving, and deep technical knowledge. Engineering PhDs in industry pointed "particularly to the ability to transition within and across roles in an organization and moving between technical and non-technical roles" (p. 31) as well as confidence-building, scientific/research skills, expertise, and flexibility to design their careers as advantages unique to PhD holders working in industry. Participants from this study recommended students be exposed to interdisciplinary and collaborative research experiences, have ample opportunities to practice communication and presentation skills, and also be exposed to industry PhDs to ask them about life in industry careers (Cox, 2019). In a similar study surveying 100 engineering PhDs working in industry, Watson and Lyons (2011) found that the most important skills needed by entry-level PhD engineers at the respondents' organizations were learning and working independently, working in teams, written and oral communication, and solving problems, while the least important skills were marketing products/processes, managing others, identifying customer needs, and writing peer-reviewed papers. Even though writing specifically peer-reviewed papers was not ranked as highly needed, analysis of job postings showed that written communications skills were highly sought after by companies (Watson & Lyons, 2011). Related studies have also pointed to the benefits of added exposure to teaching, research, professional skills, and industry expectations as part of the doctoral process in STEM (Cox et al., 2011; London et al., 2014). In a national-scale quantitative study, Main et al. (2021b) explored the role of post-PhD early career management training on individuals' subsequent career paths as leaders in industry. Findings specific to women PhD holders in STEM showed that expanding such professional development opportunities for women can result in boosting opportunities for women to hold leadership roles. Amelink and Artiles (2021) surveyed US racially minoritized engineering PhD students and found that internships and related interactions helped students figure out their career goals and understand available nonacademic career options by learning new ways to utilize their skill sets.

In sum, prior literature has shown that most graduate-level engineers spend at least some of their careers outside of academia, yet we know very little about how to prepare them for these careers. Cox's (2019) landmark study identifies important skills needed in industry and makes suggestions for interventions. There are pockets of experiences that have been shown to support preparation for a variety of career paths. For example, Borrego et al. (2021) and Denton et al. (2020) showed that internship experiences can be promising for industry and government career preparation. Informal mentoring interactions around career decisions also can make it easier for students to express and discuss nonacademic career aspirations with program faculty (Denton et al., 2020). The rest of the details supporting nonacademic career paths remain to be filled in by future research, particularly on ways that engineering graduate programs can strategically support and position both master's and PhD students for nonacademic career pathways.

Despite the growth in nonacademic pathways for engineering PhDs, a considerable proportion of new PhDs begins in postdoctoral positions. There are many reasons that PhD earners choose to accept a postdoctoral position, including that other jobs were not available, postdoc training is expected for one's field, and desiring additional training in the same or a different field (Main et al., 2021a; Stephan & Ma, 2005). Postdoctoral work is a common and beneficial preparation for a tenure-track academic position. Wang and Main (2021) found that STEM PhD earners who completed a postdoc were 13% more likely to obtain a tenure-track position than those who did not.

They also found that early career average salaries are similar for people who have and have not completed a postdoc (Main et al., 2021a). Denton et al. (2022) explored engineering and physical sciences postdocs in academic, industry, and government sectors, finding that for US PhDs eligible for government postdoctoral positions, the likelihood of attaining a tenure-track position was comparable to that of academic postdocs, but the potential long-term salary was higher, particularly if they ended up in industry or government permanent positions. When hiring a postdoc, advisers expect postdocs to know about the scientific process, have certain levels of mastery in field-specific methodologies and techniques, and be strong written and oral communicators (Bahnson, Berdanier, et al., 2022). During their appointment, advisers expect postdocs to master new technical skills, contribute to publications and grant proposals, and learn how to navigate the academic ecosystem in preparation for achieving a faculty position and, eventually, tenure (Bahnson, Berdanier, et al., 2022).

People from certain demographic groups are more likely to complete a postdoc. In an analysis of 19 years of PhD earners in science and engineering using the Survey of Earned Doctorates, Stephan and Ma (2005) found that women in engineering were more likely to engage in a postdoc than men, though the reason is unclear. Additionally, they found that people with a temporary visa were more likely to engage in a postdoc than US citizens and permanent residents. This finding is not surprising, given that many jobs in engineering industry and at national laboratories are restricted to US citizens and permanent residents and a postdoc is an option for people to obtain a visa to continue working in the United States. In another study, Main et al. (2021a) found that PhD earners from higher-ranked universities and from programs with higher percentages of graduates who go on to postdoctoral employment are more likely to become a postdoc.

9 Challenges of Reforming Graduate Education in Engineering

Given all that we know from prior research about how to improve graduate education, why is it so difficult to change? Institutions with large graduate engineering enrollments have been around for many years, which makes it extremely challenging to make changes. Particularly in institutional contexts that have strong shared governance between administrators and faculty members, organizations are intentionally designed to withstand sudden changes and shifts caused by external pressures. As the US National Academies (2018) wondered in their report *Graduate STEM Education for the 21st Century*

> dramatic innovations in research technologies, changes in the nature of work, shifts in demographics, and growth in occupations needing STEM expertise all raise questions about how well the current STEM graduate education system can adapt to these changes to continue meeting the nation's needs.
>
> *(p. 1)*

What we do not know, as a National Academies Working Group (2017) articulated in its analysis on graduate student mentoring, is how to effectively change graduate education to develop integrated networks across organizational layers – institutions, departments, programs, and individual advisers. Graduate education is complicated to change because there is a need to understand how to change the entire system.

Graduate education tends to be controlled at the individual discipline or department level as opposed to at higher levels of the organization, such as the college or university (De Valero, 2001). As has been shown by many researchers, graduate student socialization tends to happen at this discipline or departmental level (Gardner, 2007; Golde, 1998) as all processes tied to students' time in programs tend to occur here, including admissions, funding, and degree requirements, all of which are influenced by disciplinary norms and practices (Golde, 2005). Because disciplines drive faculty members' behaviors and attitudes much more than institutions (Bowen & Schuster, 1986), it is quite difficult for any one university to challenge the norms of a particular discipline (Abbott, 2001). The net result is that the same discipline at two different institutions tends to be more similar in their processes than two different disciplines at the same institution. This decentralization is even more pronounced for engineering, as a large proportion of students are funded via research assistantships, which tend to be managed and controlled at the individual adviser level. Relative to life and physical sciences, engineering graduate education is less coupled to the undergraduate education enterprise from a funding perspective (i.e., in the form of teaching assistantships) (Knight et al., 2018), so colleges of engineering have even fewer internal resource mechanisms to incentivize or demand changes in graduate education. Thus, US graduate education in engineering is a highly decentralized process, which makes integrated reform strategies extremely challenging.

There are also challenges to collecting, sharing, and monitoring data about graduate education. The graduate education community does not have the same culture of data reporting and benchmarking as seen at the undergraduate level. Characteristics of enrolled students and graduate degree recipients are reasonably straightforward to obtain, since many of the same databases report on all degree levels. Examples in the United States include Integrated Postsecondary Education Data System (National Center for Education Statistics), American Society for Engineering Education (ASEE), and National Center for Science and Engineering Statistics (National Science Foundation). Publicly reported completion rates, however, are notoriously difficult to find. This may be in part because completion rates are so low. The Council of Graduate Schools (2012) estimates that master's completion rates are less than 70%, and the National Academies reports completion rates of only 60% for doctoral degrees (Ostriker et al., 2015) – both values are over a decade old despite calls for more recent data and concerted efforts for more transparent data reporting (National Academies of Sciences, Engineering, and Medicine, 2018). Further, retention rates across graduate programs can be difficult to calculate, given the variation in how students without master's degrees who are admitted to PhD programs are classified in information systems, particularly when they "master out" by completing a master's degree but are not retained in their PhD program. Greater transparency in graduate program retention rates would likely lead to better-informed decisions by prospective students, deeper discussions in the field of who leaves a graduate program and why, and redoubled efforts to improve student retention at the graduate level.

10 Opportunities for Future Research in Graduate Education in Engineering

Each of the areas we reviewed in the prior sections has enormous potential for additional research. The engineering education research community focused on US graduate education within the engineering disciplines is growing but is still quite small relative to the entire community. Particularly because of the enormous investment in graduate education and its importance for the overall research enterprise, the body of literature focused on graduate education in engineering is surprisingly small. As expectations for employment will continue emphasizing higher levels of education, understanding graduate education in engineering will become even more important in the future.

We want to explicitly point to a few topics that received little or no attention in prior work; these areas represent gaps in the literature base and are opportunities for future research in this area. Much of the prior work on engineering graduate students has focused on doctoral-level processes, experiences, and pathways. However, enrollments in master's programs are far higher than enrollments in PhD programs, yet we know very little about this stage of education. Understanding recruitment processes and why students may choose to engage in an engineering master's degree program, for example, represent areas of research that have not been undertaken. As colleges and universities build out professionally oriented master's degree programs to better serve industry needs, engineering education researchers can learn a lot about these programs that bring together a unique mixture of students with a wide range of backgrounds and goals (Stewart & Chen, 2009). While in this chapter we have relied extensively on engineering education research conducted by individuals who identify as engineering education researchers, there is also important work being done by researchers in their role as practitioners of graduate education. In these cases, the emphasis is on delivering high-quality programs rather than publishing about them. Program details and evaluation evidence are more often published as conference papers, as in the case of dissertation institute, a week-long dissertation-writing workshop for racially minoritized students (Cruz et al., 2018; Cruz et al., 2019; Hasbún et al., 2016); the NSF Integrative Graduate Education and Research Traineeship (IGERT) program (Borrego & Cutler, 2010; Haapala et al., 2021; Wang et al., 2020); and the NSF Research Trainee (NRT) Program (Denton et al., 2020; Duval-Couetil & Yi, 2021). There is an opportunity for future research investigating the long-term effects of such programs and synthesis of multiple graduate education innovations through systematic review or similar means, which is of particular interest, given the substantial amount of funding that is allocated to such programs. Another area for future research is considering what methods are used to investigate issues in graduate education (particularly, how large-scale quantitative data is difficult to obtain). An example of this is early degree departures, since institutions report this differently: as Berdanier et al. (2020) point out, only some departures are captured because many are characterized as master's degree conferrals.

We also note that engineering graduate education is characterized by very high percentages of international students: non-US students comprise over half of the enrolled graduate student population (National Science Foundation, 2017). Much of US engineering education research has prioritized domestic students' experiences and trajectories, and so there is an enormous opportunity for research to understand processes, experiences, and trajectories of international students, which often comprise the majority of enrollments (Silva et al., 2016). Research that has taken a closer look at these students tends to aggregate international students into one group, which does not consider the different cultures and prior systems of education experienced by these students. Disaggregating research approaches that consider international students' home countries or regions more specifically can enhance understanding of graduate education. Moreover, in considering international differences, we see an opportunity for comparative graduate education research that explores differences across national and continental systems of graduate education (e.g., McQueen, 1994), which connects to a theme advanced in a different chapter. In Europe, for example, the Bologna Process created the European Higher Education Area (EHEA), which encourages the 49 member countries to offer (if not require) pedagogical training for instructors to cultivate an "inclusive and innovative approach to learning and teaching" (European Commission). Different education systems stand to benefit from learning from one another, and the engineering education research community can play a role in advancing such work.

As we note in a prior section, much of the prior research on graduate education within engineering has focused on academic career trajectories, which represents a misalignment with the predominant career paths of graduate degree holders in engineering. Although developing the professoriate, particularly for individuals with marginalized identities, has critically important feedback loops for the future of education and the field, we see a critical gap in the engineering education research literature on understanding nonacademic career paths of engineering graduate students. There are substantial opportunities for understanding how such individuals succeed once they enter a range of work sectors, how they uniquely contribute to the workforce, and how programs can best support such pathways. Disaggregating each of these ideas to consider a range of social identities represents important future work.

Finally, the engineering education research community can continue addressing important systemic issues within graduate education. As we explained in prior sections, it is challenging to disentangle different elements of the complex system of graduate education. When thinking about advising processes, for example, there needs to be a consideration of admissions and funding processes as well as considerations of both adviser autonomy and student agency. Systems-level research that interrogates this complexity across a wide range of engineering disciplinary and institutional contexts can help programs identify focused areas in need of reform as well as strategies for becoming more efficient with limited resources while also maintaining an eye on inclusivity. Finally, as is the case for nearly all aspects of engineering education research, US graduate education research within engineering must continue addressing systemic issues of diversity, equity, and inclusion. This lens must be applied to all aspects of the system, from recruitment processes to admissions, funding, and advising processes; to program-level experiences and supports processes; to preparation for a wide range of career trajectories; and to experiences of graduate degree holders within those subsequent career destinations. The engineering education research community can play an important role in building out new, sorely needed datasets and subsequent understanding around graduate education in engineering.

References

Abbott, A. (2001). *Chaos of disciplines*. University of Chicago Press.

Ahn, B., & Cox, M. F. (2016). Knowledge, skills, and attributes of graduate student and postdoctoral mentors in undergraduate research settings. *Journal of Engineering Education*, *105*(4), 605–629.

Amelink, C. T., & Artiles, M. S. (2021). Minority student experiences in engineering graduate programs: Socialization and impact on career trajectories. In *2021 ASEE virtual annual conference content access*, Virtual.

American Society for Engineering Education. *Profiles of engineering and engineering technology*. https://shinyapps.asee.org/apps/Profiles/

Artiles, M., & Matusovich, H. (2020). Examining doctoral degree attrition rates: Using expectancy-value theory to compare student values and faculty supports. *International Journal of Engineering Education*, *36*(3), 1071–1081.

Artiles, M., & Matusovich, H. (2022a). Achieving autonomy, competence, and relatedness in the advisor selection process in a chemical engineering doctoral program. *International Journal of Engineering Education*, *38*(5).

Artiles, M., & Matusovich, H. (2022b). Doctoral advisor selection in chemical engineering: A comparative study of two doctoral programs. *Studies in Engineering Education*, *2*(2).

Artiles, M. S., Knight, D. B., & Matusovich, H. M. (2023). Doctoral advisor selection processes in science, math, and engineering programs in the United States. *International Journal of STEM Education*, *10*(1), 1–16.

Bahnson, M., & Berdanier, C. G. P. (2022). Current trends in attrition considerations of engineering master's and Ph.D. students at research-intensive universities in the United States. *International Journal of Engineering Education*.

Bahnson, M., Berdanier, C. G. P., & Ross, M. (2022). Postdoctoral supervisors' expectations of the knowledge, skills, and attributes required for and developed during postdoctoral training. *International Journal of Engineering Education*, *38*(5(A)), 1274–1290.

Bahnson, M., Hope, E. C., Satterfield, D., Wyer, M., & Kirn, A. (2022). Development and initial validation of the Discrimination in Engineering Graduate Education (DEGrE) Scale. *Journal of Diversity in Higher Education*. https://doi.org/10.1037/dhe0000429

Bahnson, M., Perkins, H., Tsugawa, M., Satterfield, D., Parker, M., Cass, C., & Kirn, A. (2021). Inequity in graduate engineering identity: Disciplinary differences and opportunity structures. *Journal of Engineering Education*, *110*(4), 949–976. https://doi.org/10.1002/jee.20427

Bahnson, M., Satterfield, D., Wyer, M., & Kirn, A. (2022). Interacting with ruling relations: Engineering graduate student experiences of discrimination. *Studies in Engineering Education*, *3*(1). https://doi.org/10.1037/t85787-000

Baron, P. B. (1987). Graduate student recruitment. *Communicator*, *8*, 12. https://eric.ed.gov/?id=ED289392

Berdanier, C. G. (2019). Genre maps as a method to visualize engineering writing and argumentation patterns. *Journal of Engineering Education*, *108*(3), 377–393.

Berdanier, C. G. (2021). Linking current and prospective engineering graduate students' writing attitudes with rhetorical writing patterns. *Journal of Engineering Education, 110*(1), 207–229.

Berdanier, C. G., Whitehair, C., Kirn, A., & Satterfield, D. (2020). Analysis of social media forums to elicit narratives of graduate engineering student attrition. *Journal of Engineering Education, 109*(1), 125–147.

Bersola, S. H., Stolzenberg, E. B., Love, J., & Fosnacht, K. (2014). Understanding admitted doctoral students' institutional choices: Student experiences versus faculty and staff perceptions. *American Journal of Education, 120*(4), 515–543.

Borrego, M., Choe, N. H., Nguyen, K., & Knight, D. B. (2021). STEM doctoral student agency regarding funding. *Studies in Higher Education, 46*(4), 737–749. https://doi.org/10.1080/03075079.2019.1650737

Borrego, M., & Cutler, S. (2010). Constructive alignment of interdisciplinary graduate curriculum in engineering and science: An analysis of successful IGERT proposals. *Journal of Engineering Education, 99*(4), 355–369.

Borrego, M., Knight, D. B., Gibbs Jr, K., & Crede, E. (2018). Pursuing graduate study: Factors underlying undergraduate engineering students' decisions. *Journal of Engineering Education, 107*(1), 140–163.

Borrego, M., & Newswander, L. K. (2010). Definitions of interdisciplinary research: Toward graduate-level interdisciplinary learning outcomes. *The Review of Higher Education, 34*(1), 61–84.

Bostwick, V., & Weinberg, B. (2018). *Nevertheless she persisted?* Gender Peer Effects in Doctoral STEM Programs.

Bowen, H. R., & Schuster, J. H. (1986). *American professors: A national resource imperiled.* ERIC.

Boyle, P., & Boice, B. (1998). Best practices for enculturation: Collegiality, mentoring, and structure. *New Directions for Higher Education, 101*, 87–94. https://doi.org/10.1002/he.10108

Buffington, C., Cerf, B., Jones, C., & Weinberg, B. A. (2016). STEM training and early career outcomes of female and male graduate students: Evidence from UMETRICS data linked to the 2010 census. *American Economic Review, 106*(5), 333–338.

Burdett, J., & Crossman, J. (2012). Engaging international students: An analysis of the Australian Universities Quality Agency (AUQA) reports. *Quality Assurance in Education, 20*(3), 207–222. https://doi.org/10.1108/09684881211240286

Burt, B. A. (2017). *Learning competencies through engineering research group experiences.* Studies in Graduate and Postdoctoral Education.

Burt, B. A. (2019). Toward a theory of engineering professorial intentions: The role of research group experiences. *American Educational Research Journal, 56*(2), 289–332.

Burt, B. A. (2020). Broadening participation in the engineering professoriate: Influences on Allen's journey in developing professorial intentions. *Journal of Engineering Education, 109*(4), 821–842.

Burt, B. A., McKen, A. S., Burkhart, J. A., Hormell, J., & Knight, A. J. (2016). Racial microaggressions within the advisor-advisee relationship: Implications for engineering research, policy, and practice. In *2016 ASEE annual conference & exposition.*

Chellaraj, G., Maskus, K. E., & Mattoo, A. (2008). The contribution of international graduate students to US innovation. *Review of International Economics, 16*(3), 444–462.

Choe, N. H., & Borrego, M. (2020). Master's and doctoral engineering students' interest in industry, academia, and government careers. *Journal of Engineering Education, 109*(2), 325–346.

Corneille, M., Lee, A., Allen, S., Cannady, J., & Guess, A. (2019). Barriers to the advancement of women of color faculty in STEM: The need for promoting equity using an intersectional framework. *Equality, Diversity and Inclusion: An International Journal, 38*(3), 328–348.

Coso Strong, A., & Sekayi, D. (2018). Exercising professional autonomy: Doctoral students' preparation for academic careers. *Studies in Graduate and Postdoctoral Education, 9*(2), 243–258.

Council of Graduate Schools. (2008). *Ph.D. completion and attrition: Analysis of baseline data.*

Council of Graduate Schools. (2012). *The role and status of the masters degree in STEM.* https://cgsnet.org/role-and-status-masters-degree-stem

Cox, M. (2019). *Demystifying the engineering PhD.* Academic Press.

Cox, M. F., Adams, S. G., & Omer, I. S. (2011). Creating comprehensive educational experiences for the 21st century Ph. D. In *Higher education and human capital* (pp. 1–17). Brill Sense.

Crede, E., & Borrego, M. (2012). Learning in graduate engineering research groups of various sizes. *Journal of Engineering Education, 101*(3), 565–589.

Crede, E., & Borrego, M. (2014). Understanding retention in US graduate programs by student nationality. *Studies in Higher Education, 39*(9), 1599–1616.

Cruz, J. M., Artiles, M. S., Lee-Tomas, G., Matusovich, H. M., & Adams, S. G. (2018). The dissertation institute: Evaluation of a doctoral student writing workshop. In *2018 IEEE frontiers in education conference (FIE).*

Cruz, J. M., Artiles, M. S., Matusovich, H. M., Lee-Thomas, G., & Adams, S. G. (2019). Revising the dissertation institute: Contextual factors relevant to transferability. In *2019 ASEE annual conference & exposition.*

De Valero, Y. F. (2001). Departmental factors affecting time-to-degree and completion rates of doctoral students at one land-grant research institution. *The Journal of Higher Education, 72*(3), 341–367.

De Welde, K., & Laursen, S. L. (2008). The "ideal type" advisor: How advisors help STEM graduate students find their 'scientific feet'. *The Open Education Journal, 1*(1).

De Welde, K., & Laursen, S. L. (2011). The glass obstacle course: Informal and formal barriers for women Ph. D. students in STEM fields. *International Journal of Gender, Science and Technology, 3*(3), 571–595.

Denton, M., Borrego, M. J., Chang, C.-N., Boklage, A., & Arroyave, R. (2020). Nonacademic career pathways for engineering doctoral students: An evaluation of an NSF research traineeship program. In *2020 ASEE virtual annual conference content access.*

Denton, M., Borrego, M., & Knight, D. B. (2022). US postdoctoral careers in life sciences, physical sciences and engineering: Government, industry, and academia. *PloS One, 17*(2), e0263185. https://doi.org/10.1371/journal.pone.0263185

Deters, J. R., Holloman, T. K., Pearson, A., & Knight, D. B. (2021). *Understanding Australian and United States engineering education research (EER) contexts through the eyes of early-career EER researchers.* Australasian Association of Engineering Education.

Diggs, A. B., & Mondisa, J.-L. (2022). STEM future faculty development programs for minoritized populations: Understanding characteristics and opportunities. *The Journal of Faculty Development, 36*(1), 15–22.

Diggs, A. B., Mondisa, J.-L., & Robert, D. (2017). Toward a systematic review of the preparing Future Faculty Program initiatives. In *2017 ASEE annual conference & exposition*

Dukhan, N., & Jenkins, M. (2007). Undergraduate research as a motivation for attending graduate school. In *2007 Annual conference & exposition.*

Dutta, D. (2015). Sustaining the pipeline: Experiences of international female engineers in US graduate programs. *Journal of Engineering Education, 104*(3), 326–344.

Duval-Couetil, N., & Yi, S. (2021). Education and evaluation for the NRT: Accounting for numerous requirements, multiple disciplines, and small cohorts. In *2021 ASEE virtual annual conference content access.*

Eagan Jr, M. K., Hurtado, S., Chang, M. J., Garcia, G. A., Herrera, F. A., & Garibay, J. C. (2013). Making a difference in science education: The impact of undergraduate research programs. *American Educational Research Journal, 50*(4), 683–713.

European Commission. *The Bologna process and the European higher education area.* Retrieved December 13, 2021, from https://ec.europa.eu/education/policies/higher-education/bologna-process-and-european-higher-education-area_en

Evans, L. M. (1993). *Graduate student recruitment.* The University of Arizona. https://repository.arizona.edu/handle/10150/186185#?

Fabert, N., Cabay, M., Rivers, M. B., Smith, M. L., & Bernstein, B. L. (2011). Exaggerating the typical and stereotyping the differences: Isolation experienced by women in STEM doctoral programs. In *2011 ASEE annual conference & exposition.*

Fiegener, M. K. (2010). *Numbers of doctorates awarded continue to grow in 2009; Indicators of employment outcomes mixed.* InfoBrief. NSF 11–305. National Science Foundation.

Fleming, G. C., Patrick, A., Grote, D., Denton, M., Knight, D., Lee, W., Borrego, M., & Murzi, H. (2023). The fallacy of "There are no candidates": Institutional pathways of Black/African American and Hispanic/Latino doctorate earners. *Journal of Engineering Education, 112*(1), 170–194.

Freeman, H. (1984). The impact of no-need scholarships on the matriculation decision of academically talented students. In *Proceedings of the American association of higher education.*

Gámez, R., Packard, B. W.-L., & Chavous, T. M. (2021). Graduate bridge programs as nepantla for minoritized students in STEM: Navigating challenges with non-bridge peers and faculty. *Journal of Diversity in Higher Education.*

Gardner, S. K. (2007). "I heard it through the grapevine": Doctoral student socialization in chemistry and history. *Higher Education, 54*(5), 723–740.

Gardner, S. K. (2009). Student and faculty attributions of attrition in high and low-completing doctoral programs in the United States. *Higher Education, 58*(1), 97–112.

Gelles, L. A., & Villanueva, I. (2020). Co-constructing engineering doctoral identities through career prospects. In *2020 IEEE frontiers in education conference (FIE)* (pp. 1–5). https://doi.org/10.1109/fie44824.2020.9274164

Golde, C. M. (1998). Beginning graduate school: Explaining first-year doctoral attrition. *New Directions for Higher Education, 101*, 55–64.

Golde, C. M. (2005). The role of the department and discipline in doctoral student attrition: Lessons from four departments. *The Journal of Higher Education, 76*(6), 669–700.

Grant, M. J., & Booth, A. (2009). A typology of reviews: An analysis of 14 review types and associated methodologies. *Health Information & Libraries Journal, 26*(2), 91–108.

Griffin, K. A., & Muñiz, M. M. (2011). The strategies and struggles of graduate diversity officers in the recruitment of doctoral students of color. *Equity & Excellence in Education, 44*(1), 57–76.

Grote, D., Patrick, A., Lyles, C., Knight, D., Borrego, M., & Alsharif, A. (2021). STEM doctoral students' skill development: Does funding mechanism matter? *International Journal of STEM Education, 8*(1), 1–19.

Haapala, K. R., Hutchins, M. J., Rivera, J. L., Kumar, V., Clarke, A. R., Eatmon, T., Harris, R. A., Durfee, M. H., Mihelcic, J. R., & Shonnard, D. R. (2021). Education, research, and training aspects of the Sustainable Futures NSF IGERT project. In *2007 North midwest section meeting*.

Hasbún, I. M., Matusovich, H. M., & Adams, S. G. (2016). The Dissertation Institute: Motivating doctoral engineering students toward degree completion. In *2016 IEEE frontiers in education conference (FIE)*.

Herman, C., & Meki Kombe, C. L. (2019). The role of social networks in the transitional experiences of international African doctoral students at one university in South Africa. *Higher Education Research & Development, 38*(3), 508–521.

Institute of International Education. (2016). *Open doors 2016 report on international educational exchange*. Retrieved October 28, 2022, from www.iie.org/~/media/Files/Corporate/Open-Doors/Open-Doors-2016-Presentation.pdf?la=en

Issen, K. A., Moosbrugger, J. C., Howard, A. J., Ingraham, M. D., Reardon, B. A., & Sabini, L. M. (2007). Transforming student perspectives through summer undergraduate research. In *ASME international mechanical engineering congress and exposition*.

Jackson, R., & Chapman, R. (1984). *The influence of no-need aid and other factors on college choices of high ability students*. College Board Annual Forum.

Joy, S., Liang, X. F., Bilimoria, D., & Perry, S. (2015). Doctoral advisor-advisee pairing in STEM fields: Selection criteria and impact of faculty, student and departmental factors. *International Journal of Doctoral Studies, 10*, 343.

Kajfez, R. L., Matusovich, H. M., & Lee, W. C. (2016). Designing developmental experiences for graduate teaching assistants using a holistic model for motivation and identity. *The International Journal of Engineering Education, 32*(3), 1208–1221.

Kennedy, M. S., Lanier, S. K., Ehlert, K. M., High, K. A., Pegues, K. K., & Sharp, J. L. (2016). Understanding the role of knowledge related to financial resources on decisions to attend graduate school. In *2016 IEEE frontiers in education conference (FIE)*.

Kinoshita, T. J., Knight, D. B., Borrego, M., & Wall Bortz, W. E. (2020). Illuminating systematic differences in no job offers for STEM doctoral recipients. *PloS One, 15*(4), e0231567. https://doi.org/10.1371/journal.pone.0231567

Knight, D., Kinoshita, T., Choe, N., & Borrego, M. (2018). Doctoral student funding portfolios across and within engineering, life sciences and physical sciences. *Studies in Graduate and Postdoctoral Education, 9*(1), 75–90.

Larson, R. C., Ghaffarzadegan, N., & Xue, Y. (2014). Too many PhD graduates or too few academic job openings: The basic reproductive number R0 in academia. *Systems Research and Behavioral Science, 31*(6), 745–750.

Le, K. N., & Tam, V. W. (2008). How to retain postgraduate students in engineering programmes: A practical perspective. *European Journal of Engineering Education, 33*(5–6), 511–524.

Lee, A. (2008). How are doctoral students supervised? Concepts of doctoral research supervision. *Studies in Higher Education, 33*(3), 267–281.

London, J., Cox, M. F., Ahn, B., Branch, S., Zephirin, T., Torres-Ayala, A., & Zhu, J. (2014). Motivations for pursuing an engineering PhD and perceptions of its added value: A US-based study. *International Journal of Doctoral Studies*, *9*, 205. https://doi.org/10.28945/2050

Main, J. B., Tan, L., Cox, M. F., McGee, E. O., & Katz, A. (2020). The correlation between undergraduate student diversity and the representation of women of color faculty in engineering. *Journal of Engineering Education*, *109*(4), 843–864. https://doi.org/10.1002/jee.20361

Main, J. B., & Wang, Y. (2019). Board 97: Is postdoctoral training linked to faculty careers and higher salaries among engineering Ph. Ds? In *2019 ASEE annual conference & exposition*.

Main, J. B., Wang, Y., & Tan, L. (2021a). The career outlook of engineering PhDs: Influence of postdoctoral research positions on early career salaries and the attainment of tenure-track faculty positions. *Journal of Engineering Education*, *110*(4), 977–1002. https://doi.org/10.1002/jee.20416

Main, J. B., Wang, Y., & Tan, L. (2021b). Preparing industry leaders: The role of doctoral education and early career management training in the leadership trajectories of women STEM PhDs. *Research in Higher Education*, 1–25.

Mason, S., Lévesque, M., Meki-Kombe, C., Abel, S., Balaban, C., Chiappa, R., Grund, M., Joubert, B., Kuchumova, G., Mantai, L., Main, J. B., Motshoane, P., Qi, J., Steyn, R., & Zheng, G. (2022). Global labour market developments. In M. Nerad, D. Bogle, U. Kohl, C. O'Carroll, C. Peters, & B. Scholz (Eds.), towards *a global core value system in doctoral education*. UCL Press.

May, G. S. (1997). An evaluation of the research experiences for undergraduates program at the Georgia Institute of Technology. In *Proceedings frontiers in education 1997 27th annual conference. Teaching and learning in an era of change*.

McGee, E. (2021a). *Black, Brown, bruised: How racialized STEM education stifles innovation*. Harvard Education Press. https://doi.org/10.1177/15210251211018178

McGee, E. (2021b). Fear, fuel, and fire! Black STEM doctoral students' career decisions during the Trump presidency. *International Journal of Qualitative Studies in Education*, 1–22. https://doi.org/10.1080/09518398.2021.1930246

McGee, E., Fang, Y., Ni, Y., & Monroe-White, T. (2021). How an antiscience president and the COVID-19 pandemic altered the career trajectories of STEM PhD students of color. *Aera Open*, *7*, 23328584211039217. https://doi.org/10.1177/23328584211039217

McGee, E. O., Griffith, D. M., & Houston, S. L. (2019). "I know I have to work twice as hard and hope that makes me good enough": Exploring the stress and strain of Black doctoral students in engineering and computing. *Teachers College Record*, *121*(4), 1–38. https://doi.org/10.1177/016146811912100407

McGee, E. O., Naphan-Kingery, D., Mustafaa, F. N., Houston, S., Botchway, P., & Lynch, J. (2019). Turned off from an academic career: Engineering and computing doctoral students and the reasons for their dissuasion. *International Journal of Doctoral Studies*, *14*(1).

McGee, E. O., White, D. T., Jenkins, A. T., Houston, S., Bentley, L. C., Smith, W. J., & Robinson, W. H. (2016). Black engineering students' motivation for PhD attainment: Passion plus purpose. *Journal for Multicultural Education*, *10*(2), 167–193.

McQueen, D. H. (1994). How long does it take to earn a doctorate? An international comparison. *European Journal of Engineering Education*, *19*(2), 215–225.

Miller, D. A., Artiles, M. S., & Matusovich, H. M. (2020). The role of writing clusters in minority students' beliefs about writing. *Papers & Publications: Interdisciplinary Journal of Undergraduate Research*, *8*(1), 11.

Ministry of Education of the Republic of Korea. (2018). *Overview statistics*. Retrieved October 28, 2022, from http://english.moe.go.kr/main.do?s=english

Mondisa, J.-L. (2018). Examining the mentoring approaches of African-American mentors. *Journal of African American Studies*, *22*(4), 293–308.

Morse, R., Vega-Rodriguez, J., Castonguay, A., Brooks, E., & Hines, K. (2021). *Methodology: 2022 best engineering schools rankings*. www.usnews.com/education/best-graduate-schools/articles/engineering-schools-methodology

Mosyjowski, E. A., Daly, S. R., Peters, D. L., Skerlos, S. J., & Baker, A. B. (2017). Engineering PhD returners and direct-pathway students: Comparing expectancy, value, and cost. *Journal of Engineering Education*, *106*(4), 639–676.

National Academies of Sciences, E., & Medicine. (2018). *Graduate STEM education for the 21st century.* https://nap.nationalacademies.org/catalog/25038/graduate-stem-education-for-the-21st-century

National Center for Education Statistics. (2022). *Integrated postsecondary education data system.* https://nces.ed.gov/ipeds/

National Center for Science and Engineering Statistics, & National Science Foundation. (2017). *National survey of college graduates (NSCG).* https://www.nsf.gov/statistics/2020/nsf20300/

National Research Council. (1995). *Research doctorate programs in the United States: Continuity and change.* National Academies Press.

National Research Council. (2014). *Furthering America's research enterprise.* The National Academies Press. https://doi.org/10.17226/18804

National Science Foundation. *National center for science and engineering statistics.* ncses.nsf.gov

National Science Foundation. (2017). *2015 doctorate recipients from U.S. universities.* National Science Foundation and National Center for Science and Engineering Statistics (NCSES).

National Science Foundation, & National Center for Science and Engineering Statistics. (2015). *Science and engineering degrees: 1966–2012. Detailed statistical tables NSF 15–326.* www.nsf.gov/statistics/2015/nsf15326/

National Science Foundation, & National Center for Science and Engineering Statistics. (2017). *Survey of doctorate recipients, survey year 2017.* https://ncsesdata.nsf.gov/doctoratework/2017/

Nettles, M. T., Millett, C. M., & Millett, C. M. (2006). *Three magic letters: Getting to Ph. D.* JHU Press.

Newberry, B., Austin, K., Lawson, W., Gorsuch, G., & Darwin, T. (2011). Acclimating international graduate students to professional engineering ethics. *Science and Engineering Ethics, 17*(1), 171–194.

Nicole, F., & DeBoer, J. (2020). A systematized literature review of the factors that predict the retention of racially minoritized students in STEM graduate degree programs. In *2020 ASEE virtual annual conference content access.*

Ong, M., Wright, C., Espinosa, L., & Orfield, G. (2011). Inside the double bind: A synthesis of empirical research on undergraduate and graduate women of color in science, technology, engineering, and mathematics. *Harvard Educational Review, 81*(2), 172–209.

Ostriker, J. P., Kuh, C. V., & Voytuk, J. A. (2015). *A data-based assessment of research-doctorate programs in the United States.* National Academies Press.

Perkins, H. L., Bahnson, M., Tsugawa-Nieves, M. A., Satterfield, D. J., Parker, M., Cass, C., & Kirn, A. (2020). An intersectional approach to exploring engineering graduate students' identities and academic relationships. *International Journal of Gender, Science and Technology, 11*(3), 440–465.

Posselt, J. R. (2014). Toward inclusive excellence in graduate education: Constructing merit and diversity in PhD admissions. *American Journal of Education, 120*(4), 481–514. https://doi.org/10.1086/676910

Ro, H. K., Lattuca, L. R., & Alcott, B. (2017). Who goes to graduate school? Engineers' math proficiency, college experience, and self-assessment of skills. *Journal of Engineering Education, 106*(1), 98–122.

Rutter, R., Roper, S., & Lettice, F. (2016). Social media interaction, the university brand and recruitment performance. *Journal of Business Research, 69*(8), 3096–3104.

Saddler, T., & Creamer, E. (2009). Socialization to the professoriate through research collaboration: Examining what engineering doctoral students aspiring to faculty careers learn from faculty mentors. In *2009 Annual conference & exposition.*

Saddler, T. N. (2009). Exploring what engineering doctoral students, aspiring to faculty careers learn about research from faculty mentors. In *2009 39th IEEE frontiers in education conference.*

Sanderson, A. R., Dugoni, B. L., Hoffer, T. B., & Myers, S. L. (2000). *Doctorate recipients from United States universities.* Summary Report, 1999.

Silva, P., Woodman, K., Taji, A., Travelyan, J., Samani, S., Sharda, H., Narayanaswamy, R., Lucey, A., Sahama, T., & Yarlagadda, P. K. (2016). Support services for higher degree research students: A survey of three Australian universities. *European Journal of Engineering Education, 41*(5), 469–481.

Sowell, R., Allum, J., & Okahana, H. (2015). *Doctoral initiative on minority attrition and completion.* Council of Graduate Schools, 1. https://cgsnet.org/ckfinder/userfiles/files/DIMAC_2015_final_report_PR.pdf

Sowell, R., Zhang, T., Bell, N., & Redd, K. (2008). *PhD completion and attrition: Analysis of baseline data.* Council of Graduate Schools.

St. Clair, R., Hutto, T., MacBeth, C., Newstetter, W., McCarty, N. A., & Melkers, J. (2017). The "new normal": Adapting doctoral trainee career preparation for broad career paths in science. *PloS one, 12*(5), e0177035.

Stephan, P., & Ma, J. (2005). The increased frequency and duration of the postdoctorate career stage. *American Economic Review*, *95*(2), 71–75.

Stewart, R. A., & Chen, L. (2009). Developing a framework for work integrated research higher degree studies in an Australian engineering context. *European Journal of Engineering Education*, *34*(2), 155–169.

Svyantek, M. V., Kajfez, R. L., & McNair, L. D. (2015). Teaching vs. research: An approach to understanding graduate students' roles through ePortfolio reflection. *International Journal of ePortfolio*, *5*(2), 135–154.

Thiry, H., Laursen, S. L., & Liston, C. (2007). Valuing teaching in the academy: Why are underrepresented graduate students overrepresented in teaching and outreach? *Journal of Women and Minorities in Science and Engineering*, *13*(4).

Tormey, R., Hardebolle, C., & Isaac, S. (2020). The teaching toolkit: Design of a one-day pedagogical workshop for engineering graduate teaching assistants. *European Journal of Engineering Education*, *45*(3), 378–392.

Turk-Bicakci, L., Berger, A., & Haxton, C. (2014). *The nonacademic careers of STEM PhD holders*. STEM Am Inst Res.

Universities Canada. (2014). *Internationalization at Canadian universities: Quick facts*. Retrieved October 28, 2022, from www.univcan.ca/universities/facts-and-stats/internationalization-at-canadian-universities-quick-facts/

Wall Bortz, W. E., Knight, D. B., Lyles, C. H., Kinoshita, T., Choe, N. H., Denton, M., & Borrego, M. (2020). A competitive system: Graduate student recruitment in STEM and why money may not be the answer. *The Journal of Higher Education*, *91*(6), 927–952. https://doi.org/10.1080/00221546.2019.1706017

Wall Bortz, W. E., Lyles, C. H., Knight, D., Kinoshita, T., Borrego, M., & Denton, M. (2021). Responsible recruitment: Broadening participation in STEM through graduate student recruitment. In *Annual conference for the American education research association (AERA)*.

Wang, C., Clarkson, C. M., Andler, J., Handwerker, C. A., Korey, M., Frost, K. D., & Reeves, M. S. (2020). Lessons learned from the NSF IGERT program: Cultivating student motivation in the interdisciplinary and international contexts. In *2020 ASEE virtual annual conference content access*.

Wang, J. (2008). Toward understanding international students: A study conducted at Middle Tennessee State University. *The Southeastern Librarian*, *56*(3), 4.

Wang, Y., & Main, J. B. (2021). Postdoctoral research training and the attainment of faculty careers in social science and STEM fields in the United States. *Studies in Graduate and Postdoctoral Education*. https://doi.org/10.1108/sgpe-04-2020-0025

Wapman, K. H., Zhang, S., Clauset, A., & Larremore, D. B. (2022). Quantifying hierarchy and dynamics in US faculty hiring and retention. *Nature*, 1–8.

Watkins, S., & Green, R. (2003). Speaking and writing proficiency of international graduate students in elective, mentoring environments. *Journal of Engineering Education*, *92*(2), 147–154.

Watson, J., & Lyons, J. (2011). Aligning academic preparation of engineering Ph. D. programs with the needs of industry. *The International Journal of Engineering Education*, *27*(6), 1394–1411.

Wendler, C., Bridgeman, B., Markle, R., Cline, F., Bell, N., McAllister, P., & Kent, J. (2012). Pathways through graduate school and into careers. *Educational Testing Service*, 1–47.

Willis, D. A., Krueger, P. S., & Kendrick, A. (2013). The influence of a research experiences for undergraduates program on student perceptions and desire to attend graduate school. *Journal of STEM Education: Innovations and Research*, *14*(2).

Yang, L., & Webber, K. L. (2015). A decade beyond the doctorate: The influence of a US postdoctoral appointment on faculty career, productivity, and salary. *Higher Education*, *70*(4), 667–687.

Zerbe, E., & Berdanier, C. G. (2019). "If I knew what else I should do, I would have left by now:" Two engineering PhD students' experiences with Master's-level departure. In *IEEE frontiers in education conference*.

Zerbe, E., Sallai, G. M., Shanachilubwa, K., & Berdanier, C. G. (2022). Engineering graduate students' critical events as catalysts of attrition. *Journal of Engineering Education*. https://doi.org/10.1002/jee.20481

Zydney, A. L., Bennett, J. S., Shahid, A., & Bauer, K. W. (2002). Impact of undergraduate research experience in engineering. *Journal of Engineering Education*, *91*(2), 151–157.

14

The Overlooked Impact of Faculty on Engineering Education

Stephanie Cutler and Alexandra Coso Strong

1 Introduction

Over the last two decades, the engineering education community has conducted research and pursued efforts to create systemic and lasting change to engineering students' educational experiences. These efforts are central to the conversation sparked within the other chapters of this handbook, including, but not limited to: *How can we help students learn x? How do we promote sense of belonging and engineering identity development among students? How do we acknowledge and overcome our history as an exclusionary discipline? How do we prepare students to be leaders and change agents upon graduation?* The answers to these questions and the resulting changes to the system impact and are impacted by many stakeholders.

Nevertheless, these questions, as written, focus almost exclusively on students, with the "we" acting as an implicit identifier for engineering faculty. Engineering faculty play a critical role in the student experience inside and outside the classroom and, as such, are central to efforts of educational change (Dika & Martin, 2018; Herman & Loewenstein, 2017; Simmons & Lord, 2019). However, there is little research into who these engineering faculty members are, what impacts their sense of belonging and identity development within academia, or how they learn to make decisions that will positively impact the student experience. Overall, we argue that engineering faculty are largely overlooked within the engineering education literature.

Current research on engineering faculty predominantly focuses on faculty actions and perceptions as engineering educators. While vital research is being conducted to study engineering students, this research does not always lead to changes in the engineering classroom or within engineering colleges more broadly (Henderson et al., 2012; Herman & Loewenstein, 2017). A recent focus within engineering education has been the adoption of evidence-based instructional practices, along with ways to revolutionize engineering departments to make large-scale cultural changes to one part of the academic system (Borrego et al., 2010; Finelli & Froyd, 2019; Lord, Berger, et al., 2017). However, even with these new initiatives, faculty, who are ultimately responsible for implementing new practices (or not), are examined most often as actors within the system rather than as people who are engaged in the larger academic context outside of a particular initiative. To better support faculty teaching and development and to promote positive change in engineering education, additional research about engineering faculty is needed.

DOI: 10.4324/9781003287483-17

The purpose of this chapter is to critically examine the limited research on engineering faculty and their roles in students' engineering educational experiences. Henderson and colleagues (2011) highlighted a need for interdisciplinary research across three distinct research communities (i.e., disciplinary-based STEM education, faculty development, and higher education researchers). Specifically, they called for research to focus not only on student learning (a focus of engineering education and other discipline-based STEM research communities) but also on pedagogical skills, professional development cultural norms, organizational structures, and policies impacting faculty and higher education practices more broadly. This chapter leverages these findings by not only incorporating research from the engineering education community but also introducing faculty-focused research from other disciplines.

In particular, we explore existing research about the ways in which faculty influence engineering education (e.g., what they do in the classroom, who they are as educators, how they approach research advising and mentorship, how they approach administrative positions), who faculty are (e.g., diversity of roles and responsibilities, academic and social identities), and how we can better support them within their roles (e.g., as educators and researchers). Our use of the term "faculty" aligns with the broad categorization of professors within higher education in the United States (US). These individuals may hold a PhD, the highest-level terminal degree in their field, or a master's degree in some institutional contexts. We do not limit our definition to tenured/tenure-track faculty. As we will discuss, the specifics of job responsibilities, appointment types, and other factors impact the day-to-day work and overall career pathway for individuals holding one of these roles. We chose to structure the chapter as a survey of the existing research and key research needs. To that point, additional stakeholders, such as educational developers, graduate students, postdocs, and administrators, will be touched upon but are not within the scope of this chapter. Lastly, due to the variation in higher education contexts around the world, this chapter has a primarily US-centric lens. The framing of the chapter, though, can provide a structure for other regional and national examinations of research on engineering faculty. Where possible, international research was included; nevertheless, we did not delve into these differences and primarily focused on the US higher education system.

1.1 Positionality

As two faculty and engineering education scholars who wear many hats in the workplace (Lee et al., 2017), we view this chapter as an opportunity to deeply explore the experiences of critical stakeholders within the educational change process and to advance the engineering education community's understanding of existing research on faculty. Like many within this stakeholder group, both of us hold faculty titles but play different roles at our institutions. We have held faculty and/or staff positions at multiple institutions, and our research and educational change work is strongly influenced by our previous educational experiences.

Dr. Stephanie Cutler is an associate research professor within the Leonhard Center for the Enhancement of Engineering Education at Penn State and previously worked as a research specialist within the Rothwell Center for Teaching and Learning Excellence at Embry-Riddle Aeronautical University–Worldwide. Both positions focus on faculty development, primarily with respect to teaching. Cutler holds degrees in mechanical engineering (BS) and industrial and systems engineering (MS) and a PhD in engineering education from Virginia Tech. She is a White woman whose research has many interests across engineering education, including projects exploring the peer review process and how it impacts the field of engineering education, faculty identity, and other action research collaborations with engineering faculty exploring innovations in their classrooms. Her core responsibilities center on faculty and graduate student development as well as assessment and evaluation.

Dr. Alexandra Coso Strong is an assistant professor of engineering education at Florida International University (FIU) but previously was an engineering faculty member at Olin College (an engineering-focused baccalaureate institution). She is the founding faculty of two degree programs at FIU and served as the inaugural graduate program director for the doctoral program. Coso Strong holds three degrees in engineering and began conducting engineering education research while pursuing her master's in systems engineering. She is a White-presenting Latina scholar whose research examines educational change from a systems perspective. Her projects explore faculty and graduate student experiences across institutional contexts and disciplines. She also conducts faculty development programming, but not as a core component of her position.

2 Faculty Influences on Engineering Education

To achieve sustained adoption of educational change projects, Henderson and colleagues (2011) emphasize the need to acknowledge and understand the complexities of the academic context. A faculty member's knowledge of instructional practices, beliefs, and classroom actions influences how students learn, their feelings of belonging within engineering, and their persistence (Aragón et al., 2017; Finelli et al., 2014; Simmons & Lord, 2019). Still, the influence of faculty on engineering education extends beyond their classroom actions to their participation in educational change projects, perceptions on curriculum design, and mentoring and advising of students. This section examines the multitude of ways faculty can and do impact the student experience, by exploring existing research and research needs on four topics: (1) educational change efforts, (2) instructional beliefs and practices, (3) educational structures and culture, and (4) mentoring and advising.

2.1 Impact on Engineering Education Through Change Efforts

Researchers from various fields have investigated how to move educational practices forward using the principles of educational change and change theory. One area that has received specific attention in engineering is shifting instructional practices from more lecture-based approaches to more student-centered, evidence-based approaches (e.g., active learning). Traditional instructional practices within engineering have been characterized by the instructor lecturing (usually at a board or, more recently, PowerPoint slides) while students take notes. Active learning strategies (Felder & Brent, 2009; Prince, 2004), on the other hand, encourage students to "actively" engage with the course content and each other. For example, a student may pair with another student to solve problems, creating opportunities for deeper interaction with the material. These strategies have gained traction as more effective instructional practices for student learning (Prince, 2004; Theobald et al., 2020). However, this change is rooted in the behavior and actions of faculty, and many instructors (not just in engineering) have been resistant to implementing such strategies.

In the 2010s, work exploring academic change through the implementation of evidence-based instructional strategies became pervasive within engineering education (Besterfield-Sacre et al., 2014; Borrego et al., 2010; Nguyen et al., 2017; Pembridge & Jordan, 2016). Much of this work was funded by the National Science Foundation (NSF) to explore strategies for lasting instructional change and factors that could foster wider adoption of evidence-based instructional practices. This work leveraged multiple frameworks and theories of change, with Roger's diffusion of innovation (DOI) (2003) and Henderson's four categories of change strategies (2011) being among the most common. Roger's DOI framework, for example, was developed to explore the diffusion of technological innovations. The framework includes elements exploring the characteristics of the innovation and types of adopters, which can be extrapolated to different types of innovations and adopters (e.g., new, evidence-based instructional strategies as innovations and faculty as adopters).

However, this framework has been critiqued within change research for pushing innovations from the outside into organizations and for its reliance on seeking adopter buy-in as compared with empowering adopter ownership of the innovation (Dearing & Singhal, 2020; Singhal & Svenkerud, 2018). Henderson's categories of change strategies are more focused on change in academic settings, showcasing various types of change approaches and different stakeholder groups for the different levels of change. One category within this framework describes the need for developing a shared understanding among stakeholders. Development of a shared understanding has been central in considering how to enable buy-in from stakeholders at multiple levels and is a focal point of recent NSF calls for proposals (Lord, Berger, et al., 2017). While each of these frameworks and others have enabled change approaches within engineering, as is discussed throughout this handbook, revisions to or new change frameworks may be necessary to support more inclusive, equitable, and culturally responsive teaching environments.

Beyond change frameworks, researchers have explored the supports and barriers for engineering faculty to pursue instructional change (e.g., Finelli et al., 2014; Shadle et al., 2017). Shadle and colleagues (2017), for instance, defined 18 faculty-identified barriers to change, including time constraints as a commonly cited barrier, as well as large class sizes, feelings of loss of autonomy within the classroom, feelings that new strategies would not be rewarded within the promotion process, and student resistance. Additionally, 15 drivers for faculty change were also identified, including promoting engagement, faculty wanting students to succeed, encouraging professional development, enhanced teaching satisfaction, and more. To further enable these drivers and overcome these barriers, the Royal Academy of Engineering developed a framework for teaching rewards for faculty within the UK (Graham, 2018). This and similar work capture what Henderson and colleagues (2011) define as the complexities of the academic work environment that can affect faculty's agency and willingness to engage in educational change.

Within the academic work environment, faculty do not work alone and are impacted by multiple levels of the university structure. "Improving engineering education will require not only a commitment on the part of faculty, but also significant support, resources, and recognition on the part of the university" (Ambrose & Norman, 2006, p. 31). Recently, the NSF has recognized these different structural impacts and the culture that is embedded at each administrative level of the university. This recognition led to the creation of the RED (Revolutionizing Engineering Departments) Program (Lord, Berger, et al., 2017; Lord, Camacho, et al., 2017) to encourage department-level cultural change for engineering and engineering education. One key insight from the initial RED projects is the importance of communities of practice (Cross et al., 2021; Tomkin et al., 2019) and the value of creating a sense of community to foster innovation. Additional research suggests that for change to happen, the university must support the faculty through policies and structures that involve (1) institutional leadership, (2) finance and academic departmental influence and configurations, (3) faculty training and development programs, (4) physical facilities, and (5) incentives to learn, develop, and maintain new practices (Piskadlo, 2016).

In practice and in the literature, faculty are key constituents in engineering education change efforts. From existing change frameworks, we have a foundation of considerations for facilitating educational change in engineering. Insights from these change frameworks and large-scale change efforts should be leveraged to further propagate approaches to change. From a research perspective, work to adapt these frameworks to academic settings or bring in new frameworks is needed to further support change in engineering education. Lastly, the current research has primarily focused on innovation and change in classroom practices, with new approaches and structures aimed at the departmental level. Still, to move forward with more inclusive and equitable higher education institutions, additional research is needed to understand the structures and policies that can empower faculty towards change.

2.2 Impact Within the Classroom Through Beliefs and Practices

Even with the existing research on change, educational change within engineering continues at the glacial pace typically associated with the academy, and faculty continue to resist change in their classrooms. Much of this resistance may be a result of faculty's epistemological beliefs and the (mis) alignment of those beliefs with the evidence base for new instructional practices. Faculty are people with unique experiences and pathways that affect their overall work in academia, including their teaching. From these experiences, each faculty member develops complex beliefs that they carry into their classrooms, and some have formed inflexible beliefs about how to be an effective teacher, based on their success as a student in similar classrooms and, in some cases, years of teaching with sufficiently high student evaluations. The specific construct of teacher beliefs has been extensively explored within education generally, especially among K–12 teachers (e.g., Fives & Buehl, 2012; Gow & Kember, 1993). This section will further explore the existing research on teaching and learning beliefs, to highlight what we know about the effect of these beliefs on classroom environments and discuss existing work within engineering education.

Starting from the student perspective, research has demonstrated that beliefs about learning more generally have been shown to impact the student experience, as can be illustrated through research on fixed and growth mindsets (Dweck, 2016). Fixed and growth mindsets, or implicit theories, refer to beliefs that someone's intelligence, characteristics, and abilities are innate (in the case of fixed mindset, one is born with certain abilities that cannot be changed or can be changed very little) or are malleable (in the case of growth mindset, one can use directed practice and exert effort to grow and change). Encouraging a growth mindset in students has been shown to combat stereotype threat (Steele, 2010) and help with the overall success of students who are members of groups systemically minoritized within STEM and engineering (e.g., Aguilar et al., 2014; Canning et al., 2019; Rattan et al., 2015).

These student beliefs are compounded with faculty beliefs, which influence the instructional practices of faculty and the student experience in the classroom. Aragón et al. (2018) explored STEM faculty mindset (as broadly defined as beliefs about general intelligence) in the context of the adoption of active learning, using their EPIC implementation model (exposure, persuasion, identification, commitment, and implementation). STEM faculty with a growth mindset were more likely to implement active learning. STEM faculty with a fixed mindset were less likely to be convinced of the value of active learning and were less likely to implement these strategies. Canning and colleagues (2019) found that in courses where faculty held a growth mindset with respect to their students, students outperformed their peers from courses taught by instructors with a fixed mindset with respect to their students. Additionally, in courses with instructors who held a growth mindset, the achievement gap between students from minoritized populations and their counterparts who are members of majority groups was cut in half.

Within EER, there has been some exploration of faculty beliefs with respect to teaching and learning. Borrego and colleagues (2013), for instance, explored faculty epistemological beliefs around how students learn (or learn best) in different engineering sciences courses. Many faculty participants believed problem-solving helped students learn, and they recognized the shortcomings of lecture. Still, these participants struggled in how to make changes to their classes to better support learning and to align their beliefs with their practices. More recently, Ross and colleagues (2017) explored how a professional development workshop could impact faculty beliefs and shift perspectives from teacher-centered teaching to student-centered teaching. While they saw significant changes in faculty beliefs (towards more student-centered teaching), they did not see a shift in faculty practices in the classroom. Future research is needed to further explore the relationship between the espoused beliefs of faculty (i.e., what they perceive and say are their beliefs) and their enacted beliefs (i.e., instructional practices).

There is also a need to explore other beliefs held by faculty that can impact classroom practices as well as other elements of faculty life. For example, Aragón et al. (2017) applied the EPIC implementation model to explore how STEM faculty ideologies (colorblind or multicultural) impacted the adoption of inclusive teaching practices. After completing professional development around inclusive teaching, faculty with a multicultural ideology were more likely to implement inclusive practices over their colleagues with a colorblind ideology. Within engineering education, more researchers are beginning to explore a broader range of faculty beliefs, including perspectives of diversity and inclusion (Cross & Cutler, 2017; Grifski et al., 2021).

Overall, when thinking about faculty as teachers, we need to keep in mind the different beliefs, attitudes, and perspectives that are informing the practices they implement and the reasoning behind those choices. Better understanding of faculty beliefs and perspectives can better inform faculty professional development opportunities and the change approaches and frameworks used by engineering education scholars. With more informed approaches, we can continue to develop and enable faculty to further improve the student experience in the classroom and beyond.

2.3 Impact on Educational Structures and Culture

Institutional structures are like a nesting doll of contexts and influences that impact both students and faculty. Faculty commonly work within a department within a college within a higher education institution, each of which is impacted by accrediting agencies (e.g., ABET), state legislature, and the cultural and political environment of the country. Each of these elements (and more) contributes to the structures that comprise the overall education of students (e.g., the program curriculum) as well as the overall experiences of faculty (e.g., factors impacting promotion and tenure).

One of the areas that is underexplored within engineering is the influence faculty have on many of these structural elements (e.g., the overall curricular design, policies, and procedures within an institution). NSF's RED Program (Lord, Berger, et al., 2017; Lord, Camacho, et al., 2017) is one example of a program seeking to impact change at the broader departmental level, as a way to enhance the student experience inside and outside the classroom. Still, the literature is limited in regards to studies of faculty and administrators, with the few studies exploring specific perceptions of certain initiatives and change strategies (Besterfield-Sacre et al., 2014; Cech et al., 2016). Few examples of research exist exploring spaces outside of the classroom that impact the culture of a program and opportunities for change (e.g., Berger et al., 2021; Briody et al., 2019). Briody and colleagues (2019), for instance, explored faculty-student interactions from the perspectives of faculty, students, and staff within an engineering department at a research-intensive public US university. Their qualitative study identified social distance between faculty and students (i.e., due to unequal status within the hierarchical culture of the department) as a barrier to these critical interactions (Briody et al., 2019). Additional work is needed, however, to deeply understand the impact of faculty and administrators on educational structures and the reverse, the impact of these structures on faculty actions and instructional practices.

The impact of administration is also largely understudied in engineering education in comparison to other disciplinary-based education fields. Within medical education and criminal justice, for example, research explores doctoral admissions processes and the role of graduate program directors (e.g., Kim et al., 2015; Puscas, 2016). Within STEM, Gomez and colleagues (2019) examined the strategies used and the challenges experienced by program directors to support their students and programs. While the study includes some engineering program directors, they are only a subset of the participant pool. Within engineering education, Schimpf and colleagues' (2012) study of parental leave policies, on the other hand, does provide insight into the role of department heads and university policy administrators on policies and procedures. With its focus on faculty experiences as opposed to policies regarding students, this article serves as an example of how we may choose to explore the impact of faculty serving in administrative roles on faculty or students' experiences.

The academic workplace is a complex mechanism in which faculty are navigating and affecting. Additional research is needed both on how the broader institutional structures impact faculty work and how faculty can impact these systems and structures. Within these systems, the administrators (typically former or current faculty members) are typically understudied but could be critical sources of insights for advancing educational change as they are in positions of power and could lead or support structure or policy changes. Overall, future research efforts to enable educational change at the department, college, and institutional level should consider institutional structures and policies as well as faculty stakeholders in various administrative roles.

2.4 Impact Through Mentoring and Advising

Beyond the classroom, one of the key education roles for faculty is as a mentor. Extensive research exists around mentoring in academia and beyond, including resource pages (e.g., National Academy of Sciences et al., 1997; Wisconsin Center for Education Research, 2021) and research (National Academy of Sciences, 2019) to help guide faculty as mentors. However, many of these resources and studies focus on the mentoring of graduate students, as this is often a primary responsibility for many faculty members at research-extensive institutions and closely tied to faculty research success. Contrasting this vast literature, few studies explore faculty perceptions, attitudes, or beliefs about their role as a mentor, with far fewer studies examining actual faculty mentoring practices. When thinking about future research around mentoring, researchers should consider different elements of mentoring, such as different groups of mentees, like undergraduate researchers (National Academy of Sciences et al., 1997) or postdoctoral researchers (e.g., Mena, 2015); mentoring outside of the research context, such as mentoring in how to be a mentor (Tise et al., 2018) or how to teach (Calkins & Kelley, 2005); or the different mentoring roles of faculty, including as an undergraduate academic adviser (Allen et al., 2013; National Academy of Sciences et al., 1997).

2.5 Faculty Influences on Engineering Education Summary

This section outlined different ways faculty impact the educational experiences of students, from teaching to defining programmatic structures. In addition, it introduced research on broader educational change efforts and the role of faculty within those efforts. Overall, the role of faculty members within an educational system is critical to change and innovation efforts within engineering education. However, as will be discussed in the next section, to deeply understand the beliefs and actions of faculty, we need to also understand them as people within the context of their positions and roles as educators.

3 Who Faculty Are in Their Roles and as People

Most commonly, discussions of the engineering faculty population tend to perpetuate a default of White male-identifying, tenure-line faculty at research-extensive (e.g., R-1), predominantly White institutions (Pawley, 2017). This default limits our field's understanding of (1) the roles faculty hold and contexts they inhabit, (2) their role identities, (3) their pathways into and through faculty roles, and (4) their personal, social, and cultural identities. This section will explore existing research and research needs within engineering education, and education more broadly, in these four areas.

3.1 Roles and Responsibilities in Context

Faculty roles have long been described in terms of three categories of responsibilities: research, teaching, and service (Adams, 2002). These responsibilities have been approached separately (e.g., research on faculty teaching) or in explorations of the "whole of the academic role" (Sutherland, 2017). Yet within the research over the last 30 years, the core responsibilities have not substantially evolved and how we communicate about faculty roles has been slow to change (Barber, 1987). Still, in parts of the engineering education literature, we have begun to look at the whole of a faculty position, incorporating considerations of administration and leadership roles alongside the most commonly discussed responsibilities (Cutler et al., 2020; Edalgo et al., 2021; Lee et al., 2018). This expansion of our definition of faculty roles enables discussions of job crafting (Berg et al., 2007; Bodnar et al., 2021), or how faculty modify their roles to achieve professional goals, and what faculty need in terms of support to achieve their professional goals (Cutler et al., 2020; Harlow et al., 2020). This subsection explores how our current definitions of faculty roles are insufficient in describing the diversity of roles faculty hold, how we need to expand our use of theory to explore faculty responsibilities and work, and lastly, how context plays a critical role in our research on faculty.

There has been a rise over the last several years in the hiring of professional track faculty (Farrell, 2018; Fitzmorris et al., 2020). These faculty hold roles that emphasize one category of responsibilities over others. For example, instructional faculty teach as their primary responsibility. Depending on their contract, they may also have service, research, and/or leadership responsibilities, but they are predominantly evaluated on their teaching (Coso Strong et al., 2022; Fitzmorris et al., 2020). These professional track faculty also tend to hold positions not on the tenure line, which has raised concerns about how they are treated by their tenure-line colleagues, the extent to which their voice is heard in their department, how they will be promoted, and the extent to which they are able to impact educational change at their institutions (Coso Strong et al., 2019; Fitzmorris et al., 2020; O'Meara, Templeton, et al., 2018). Within engineering specifically, Fitzmorris and colleagues (2020) presented narratives from participants who had experienced disrespect and exclusion within departmental governance, the day-to-day culture, and overarching university policies. The researchers conducted interviews to explore the experiences of 13 instructional faculty in electrical engineering across multiple research-extensive institutions. In a quote that resonated with other studies (e.g., Haviland et al., 2017), one of the participants explained, "I would argue that NTT [non-tenure-track] faculty are almost treated as second-class citizens, that we weren't good enough or smart enough or whatever" (Fitzmorris et al., 2020, p. 7). While our understanding of engineering instructional faculty is growing (e.g., Bracho Perez et al., 2021; Urquidi Cerros et al., 2021), research is quite limited on research faculty, professional faculty who work in centers (e.g., research centers, centers for teaching and learning), or faculty who hold non-tenure-track positions but are not captured by these categories.

As an example, over the past decade or so, there has been a rise in the hiring of contingent (adjunct) faculty. These faculty are typically hired on a course-by-course basis each semester and work part-time for university. Though many of these faculty are teaching foundational engineering courses, they are not included within traditional institutional structures, limiting their voices and ability to make change within the system. Additional research is needed about adjunct faculty in engineering and the ways they engage with their universities, courses, and students.

As we continue to explore these evolving position types, it is critical to recognize how assumptions about faculty roles impact the appropriateness of certain theories and methodologies. Fitzmorris and colleagues' (2020) work builds on work of education scholars (e.g., Crick et al., 2020; Kezar, 2013) who determined that new theoretical models were needed when examining the experiences of professional track faculty to overcome potential biases and mitigate comparisons using the lens of tenure-line roles. Kezar (2013), for instance, extended job satisfaction and performance frameworks to capture the experiences of over 100 professional track faulty across three institutions. Within engineering education, theoretical development work focused on faculty is limited and highly needed to capture the diversity of roles.

Faculty responsibilities are not only a function of a particular role but also their institutional context. Commonly, in studies about faculty, as with many studies about students (Pawley, 2017), we are provided limited information about the context or the context is not used as a lens through which to explore and understand the results. This lack of integration of context can influence a reader's interpretation, the transferability of the work, and the overall implications of the findings. In addition, while there may be studies that only focused on, for example, faculty from baccalaureate colleges, that focus is not always central to the focus of the paper.

To date, the work in engineering education on institutional contexts outside of predominantly White institutions is limited yet important to considerations of supporting faculty in diverse positions and roles. For instance, in Buswell's (2021) work, participants were doctoral graduates who chose to pursue a faculty position at a non-R1 institution. These faculty described a sense of resistance within their graduate programs as they pursued positions at non-R1s and even feelings of failure as they accepted those (Buswell, 2021). Given that studies have shown that graduate students develop conceptualizations of what it means to be a faculty member most often from their doctoral adviser (e.g., Bieber & Worley, 2006), these new faculty could have benefited from additional support as they made sense of differing tenure and promotion expectations, a teaching-focused work environment, and a different balance in responsibilities (Buswell, 2021). These differences across engineering faculty experiences can also arise, for example, for those in two-year institutions as well as minority-serving institutions (MSIs) in the United States. Recent studies by Kendall and colleagues (2021) illustrate engineering faculty members' experiences pursuing educational change at MSIs, isolating the needs and assets at their institutions. Yet engineering education remains behind higher education literature and even disciplinary education literature on examinations of the faculty experience within diverse institutional contexts in the United States and abroad (e.g., Hubbard & Stage, 2009; Nuñez et al., 2010).

These discussions of the diversity of faculty roles and positions are critical, especially as they begin to compound with the multiple identities of faculty, impacting faculty persistence within academia, as well as the extent to which they can positively impact students. As we begin to advance our understanding of faculty roles and contexts, we need to be mindful of the tendency to compare the experiences across roles and contexts and to recognize the value in studying a particular type of professional track faculty or institutional context on its own.

3.2 Role Identities

Taylor (1999) states that "traditional understandings of academics' sense of professionalism are neither fixed, nor closed . . . [but are] . . . social constructions – partial, patchy and incomplete" (p. 116). Much of the literature on academic identity aligns with this representation of a professional, academic identity, one that is a moving target with socially constructed boundaries (e.g., Archer, 2008; Hunter, 2020). This subsection borrows from Billot (2010), who explained that the "academic self" develops as an individual reflects on what comprises "the academic" along with "their past experiences and their understanding of the current situation." Literature has already demonstrated, as previously noted, that graduate students commonly glean their understanding of the faculty identity from their experiences in their doctoral programs (e.g., Bieber & Worley, 2006; Reybold, 2003). Thus, within engineering, where engineering faculty can be situated in diverse roles and contexts, there is a need to better articulate what it means to hold an engineering faculty identity and how we can communicate that identity to future faculty.

Vellamo and colleagues (2020) define engineering faculty identity using two dimensions – engineering profession (teaching component) and academic profession (research component). The components are complementary (i.e., developing as a researcher in the academic profession can support how one develops students as engineers). Yet these dimensions can also be in tension. In an exploration of designing engineering courses for social justice, for example, two engineering faculty (both engineering education scholars) explored their experiences through autoethnographic accounts. These critical examinations of their faculty identity development emphasize the "profound and lasting influence on what constitutes engineering content – and what does not" (Leydens & Lucena, 2017, p. 224) even as these scholars sought to incorporate social justice into their courses. Pawley (2009) noticed a similar pattern within narratives of what faculty perceived as "engineering" (i.e., applied science and math, solving problems, making things) and the nature of engineering work. These perceptions reflect the critical nature of understanding not only what it means to be an academic broadly but also how that identity intersects with one's identity as an engineer. Given the potential impact of faculty engineering identity on student instruction, there has been significant work on the engineering identity as it relates to students and some with professional engineers (e.g., Godwin & Kirn, 2020; Morelock, 2017). Yet research on engineering faculty identity specifically (Vellamo et al., 2020) and how those identities may shape engagement in educational change efforts is limited.

While these engineering narratives capture part of what Billot describes as past experiences (how faculty learned and developed beliefs about engineering), another critical component of the faculty identity lies within an individual's interactions with the context. Vellamo and colleagues (2020) explored the impact of organizational changes on engineering faculty identity in a Finnish university setting, and their results emphasize the importance of context in defining and making sense of one's identity. Gardner and Willey (2018, 2019) illustrated how an individual's interaction with their disciplinary context, through professional societies and conferences, impacts identity development. In their studies of engineering faculty transitioning into EER in Australian institutional contexts, the findings point to the critical nature of peer reviews and conference participation on the faculty's identity trajectory (Gardner & Willey, 2019). These findings align closely with discussions of the socially constructed nature of a faculty identity, as the faculty were affected by their interactions with other engineering education researchers and the feedback they received from those individuals.

Outside of engineering, Archer (2008) explored early-career faculty members' experiences both "becoming" and "unbecoming" academics in English universities. In her qualitative exploration, she articulates faculty experiences of "becoming" an academic as "not smooth, straightforward, or automatic, but can also involve conflict and instances of inauthenticity, marginalisation, and exclusion" (p. 387). Many of her interviewees discussed their identities in relation to the institution's requirements for particular quantitative outcomes (i.e., total number of publications, funded grants), which contributed to their feelings of inauthenticity. In an exploration of higher education in the United Kingdom and beyond, Lamont and Nordberg (2014) explain how faculty identities are impacted by both structure (as defined by the culture and contexts we inhabit) and agency, through a process of negotiation. Thus, within engineering, we may be able to advance approaches for overcoming our historically marginalizing disciplinary culture by critically examining our contexts and structures through the lens of engineering faculty identity development.

Overall, many of these studies of identity development explore spaces of transition (e.g., the early-career faculty, institutional changes). Outside of engineering, researchers have explored academic and faculty identity in instances where role conflict can occur, such as during large organizational structural and/or cultural changes (e.g., Quinney et al., 2017; Vellamo et al., 2020) or significant economical and societal challenges (e.g., limited government funding – Robinson, 2010). Additional research about engineering faculty identity may be beneficial for examining other moments of transition (e.g., post-tenure, transition to full professor, transition to administrative positions), in roles that have a teaching or research focus (i.e., professional track faculty, adjuncts), and in the context of how this identity impacts or is impacted by the social and cultural identities we hold.

3.3 Pathways Into, Through, and Out of a Faculty Career

Studies of faculty pathways are dominated by a focus on critical changes (i.e., starting a faculty career, going up for tenure) and early-career faculty. Explorations of the entry point into a faculty career are commonly focused on experiences of graduate students moving into faculty roles (e.g., Austin et al., 2009) or particular approaches for supporting new faculty (e.g., Brent & Felder, 2000). More recently, studies have focused on the hiring process as a means of supporting diversity among engineering faculty (e.g., Boyle et al., 2020; Simmons & Lord, 2019).

This work on entry points also serves as a reminder that much of the research on faculty pathways in engineering education implicitly (and sometimes explicitly) focuses on tenure-line, research-focused faculty at predominantly White, research-extensive institutions in the United States. More recently, there has been an increase in studies focused on transition points for interdisciplinary engineering faculty (e.g., Coso Strong et al., 2021; McCave et al., 2020). As the pathway becomes less linear, though, for instance, a faculty member begins their career in industry and then comes back to academia (Banik, 2016), the studies within engineering education are mostly limited to conference proceedings presenting single-person or small-group "lessons learned" and "best practices" (e.g., Birmingham, 2007; Gregg et al., 2005). Yet within other education and social science fields, studies of these less linear pathways and nontraditional entry points (i.e., not straight out of graduate school or a postdoc) are more common (e.g., Perry et al., 2019). Overall, additional work is needed to explore the pathways of engineering faculty who hold different roles and who arrived at their faculty career through different entry pathways.

The act of moving through the faculty career in engineering is also understudied. One exception are groups who have explicitly explored transitions from engineering to engineering education at different career stages (e.g., Gardner & Willey, 2018; Siddiqui et al., 2016). Another is the work of Pawley and Hoegh (2010), who sought to theorize the ways women traverse career pathways in STEM faculty positions. Aspects of how engineering faculty develop within their roles, which is discussed later in the chapter, have also been explored to some extent (e.g., Bird and Kellam (2013) explored the teaching journey of faculty). Outside of engineering, though, studies of faculty explore even more of the movement through, such as promotion and evaluation practices (e.g., Dolan et al., 2018; Glass et al., 2011), joint appointments (e.g., Hart & Mars, 2009), as well as transitions and pathways of specific faculty populations (e.g., Reybold & Alamia, 2008).

A critical component of moving through the faculty career occurs after promotion, when there is a shift and faculty, theoretically, have more freedom with respect to advancing their career (Canale et al., 2013). However, this transition from early-career to mid-career can be challenging. The fairly clear guidance for moving from assistant to associate professor titles is replaced with less-clear guidelines for moving from associate to full professor, and even less-clear guidance for those promoted on professional faculty tracks (Canale et al., 2013). Faculty's professional goals and motivations tend to change, while many also find themselves needing to balance more personal responsibilities (Rockinson-Szapkiw, 2019). Continued mentoring can help support faculty during this transition as faculty again seek to understand the expectations of their new position as a promoted, and in some cases, tenured, faculty member (Canale et al., 2013; Rispoli, 2019). Typically, during their mid-career, faculty make decisions to pursue academic leadership (Rojewski, 2018), change their research focus, or expand their service opportunities within the university and as part of their professional societies. Some universities help provide leadership programs for faculty (e.g., Davidson et al., 2001; Hornsby et al., 2012), but little research outlines how faculty make decisions with respect to the differing opportunities. Exploring faculty pathways beyond the early career is far less documented, especially in engineering, and additional research into the experiences, motivations, and career decisions of mid-career (and late-career) faculty is needed.

Lastly, studies of leaving a faculty career pathway are very limited within the engineering education literature, in contrast to higher education research (e.g., Park's (2015) study of turnover and O'Meara and colleagues' (2014) exploration of the decisions of those who depart from a faculty career). Our community needs to better understand not only the late-career decisions of engineering faculty but also the experiences and perceptions of those who choose to leave the academic career pathway earlier in their career. This understanding will allow us to pursue the continued development of more-inclusive and supportive cultures within departments and institutions (Rockquemore, 2016), which stand to have a positive impact on our students as well.

For many faculty, their academic journey lasts for their entire career, over 30 years in some cases. Exploring this journey from the decision to move into a faculty career through promotion and on to later career transitions should be the subject of future research. Additionally, engineering education would benefit from explorations of faculty decisions to change careers within academia or to leave academia to ensure that institutions are working to retain and support their faculty.

3.4 Faculty as People

One of the core purposes for crafting this chapter is to hold space within EER for discussions and explorations of faculty as people. The impact of engineering faculty on the student experience, learning, and identity development is well documented (e.g., Simmons & Lord, 2019). Faculty social identities, however, are often left out of discussions, for example, of faculty behavior in the classroom and willingness to pursue education innovation. Faculty are people too. Outside of their role identity, they bring many different social identities (e.g., race/ethnicity, gender identification, sexual orientation, religion, citizenship status, parental status, marital status) and cultural identities into their roles. As with students, within the historically exclusionary culture of engineering, faculty experience marginalization for their identities from other faculty, staff, leadership, and students.

Simmons and Lord (2019), among others, remind us of the low representation of women, Black, Latinx people among engineering faculty in the United States. Some researchers have sought to amplify the voices of faculty who are members of marginalized groups in an effort to pursue change within the academic culture. For instance, in higher education, studies explore inconsistent service loads of women and faculty of color, as compared with faculty from dominant groups, and the tensions this raises for promotion, work–life balance, and overall job satisfaction (e.g., Baez, 2000; C. Graham & McGarry, 2019; O'Meara, Jaeger, et al., 2018). Graham and McGarry's (2019) study, for example, examines the experiences of mid-career faculty with intersectional identities and articulates not only disproportionate service loads but also negative differential treatment.

> The collective message is that even for mid-career women faculty across African, Latinx, Asian, and Native American Diasporas, who survive the tenure and promotion process, and are perceived as thriving, there can be a tremendous toll on their physical and mental health.
>
> *(Graham & McGarry, 2019, p. 76)*

In the context of the COVID-19 pandemic and reoccurring acts of racial injustice in the United States, the broader education community has engaged in critical discussions about the experiences of faculty of color, as well as the need to focus on anti-racism inside and outside the classroom (e.g., Ash, 2020; Cross, 2020; Holly Jr., 2020). From the faculty perspective, much of the work on the experiences of faculty of color as they navigate the academic context has occurred outside of engineering (e.g., Cole et al., 2017; Hubbard & Stage, 2009; Murakami-Ramalho et al., 2010). Within engineering, the focus has been predominantly programmatic. Support structures through the NSF sponsored programs (e.g., ADVANCE program, the Launching Academics on the Tenure Track: An Intentional Community in Engineering (LATTICE) program) have created community and provided resources for faculty from systemically marginalized groups. However, as Simmons and Lord (2019) articulate, these programs are on soft money, and faculty would benefit if these and similar initiatives were institutionalized. To create an academic culture that is supportive and anti-racist, the engineering education community would benefit from leveraging existing education research and critically reflecting on how the structures and culture need to change.

The COVID-19 pandemic has also made salient the long-standing challenges faculty who are parents face, especially women or those that take on the main caregiving responsibilities (e.g., Fulweiler et al., 2021; Krukowski et al., 2021). During the pandemic, multiple studies documented the differences in productivity levels among faculty who had parental responsibilities and those who didn't (e.g., Ellinas et al., 2021; Krukowski et al., 2021). Parenting in the academy has been explored to some extent at different career stages (e.g., Rockinson-Szapkiw, 2019), with some work within engineering focused on parental leave policies (or lack thereof) and motherhood (Schimpf et al., 2012, 2013). Given the already-low representation of individuals who are members of systemically marginalized groups, there is a need to amplify the voices of these scholars and to center their stories in the broader context of the engineering faculty experience.

As we draw attention to the faculty experience, it is important to highlight the challenges and consequences of the professional pressures associated with the position. Burnout has been highlighted as a rising result of pandemic life for everyone, including those in academia (Flaherty, 2020). However, the pandemic is not the only cause of faculty burnout. Sabagh et al. (2018) highlight that job demands, such as workload and task characteristics, combined with lack of resources, such as social support and rewards, contribute to faculty burnout, reducing performance and retention of faculty. Moving forward, faculty burnout in engineering should be explored to find mechanisms for creating a thriving and healthy academic community. Research in this area will need to consider the origins of burnout, starting from graduate education, where burnout has been explored as a factor in graduate student persistence (e.g., Berta & Pembridge, 2019; Cornér et al., 2017). In addition, for graduate students, faculty become the model for their potential future self within an academic career. Graduate students observing the stress and burnout of the faculty may chose a different career pathway. Faculty must take care of themselves as a key first step to taking care of, and supporting, their undergraduate and graduate students. Institutions and faculty support programs should pay specific attention to faculty mental health and overall wellness. Further exploring the causes and potential mitigation strategies for burnout can help to better support faculty members' continued success.

The people who work as faculty bring their social identities with them to the workplace, just like everyone else. When working to support faculty, it is important to explore and understand the impacts of these identities on faculty as teachers and researchers as well as members of their professional fields. Each faculty member brings expertise and life experiences that are assets we should be recognizing and supporting, as these assets can serve as critical resources for, as an example, creating inclusive and equitable learning environments.

3.5 Who Faculty Are in Their Roles and as People Summary

As noted at the beginning of this chapter, the term *faculty* refers to a diverse group of people who fill many and a variety of roles and responsibilities within an institution. Further research is needed to examine these different roles and responsibilities and their impact on the faculty experience. Alternatively, our field may benefit from more assets-based explorations of what faculty bring to their roles (e.g., different social and cultural identities) as well as how they transform their roles to better support their students, the institution, and their research field. Lastly, as engineering continues to evolve both as a research field and as an educational system, longitudinal examinations about how the faculty work responsibilities and context change are an important area to inform decisions about the future of higher education.

4 Support for and Development of Faculty

Now that we've discussed faculty as people and the impact that they can have in the classroom, we move to a discussion of the work that has been done to help faculty in navigating their responsibilities. There are not always easy solutions to the individual and collective challenges faced. We cannot simply wait until next semester and redesign a course to better support faculty. This section focuses on the existing work and the research needs surrounding faculty professional development – one avenue for further supporting and empowering current and future faculty.

It is important to note at the start of this discussion that educational development (a.k.a. professional and organizational development, faculty development) is an established and growing field of scholarship (Little, 2020). Both authors of this chapter have been active within the Professional and Organizational Development Network (POD Network) and the American Society for Engineering Education (ASEE), exploring bridges between the fields (Cutler et al., 2017; Strong et al., 2016). The members of the educational development field work directly with faculty to make changes to educational systems and should be brought in for their expertise in translating educational research to faculty practice. These professionals (sometimes faculty themselves) are another under-researched group within academia, particularly within engineering education. Future work exploring the impact of, identity of, and approaches used by these professionals is needed within engineering education.

In recent years, scholars within engineering education have sought to create more space for people working in faculty professional development to share practices and research (e.g., Chua et al., 2016; Cutler et al., 2017). The networks established within these professional societies created the impetus for the ASEE Faculty Development Division as well as pushed forward the POD STEM Special Interest Group (SIG). The ASEE Faculty Development Division has helped bring together professionals working to aid faculty in their professional development, starting as an informal group in 2016 with the session "Faculty Developers on Faculty Development: Join the Conversation," then forming a Constituency Committee in 2018, and becoming a division in 2019. This division has highlighted faculty professional development networks within engineering education through the publication of over 100 papers at ASEE conferences (peer.asee.org). The experiences and knowledge of these groups of professionals can inform our community's attempts to support not only educational development efforts but also broader holistic professional development efforts moving forward.

4.1 Educator Development

The claim is regularly made that faculty are not trained as teachers before beginning their first academic appointment, and therefore, faculty development must prioritize the teaching responsibilities of faculty (Ambrose & Norman, 2006; Brent & Felder, 2000). Engineering education, as a field, began with engineering faculty who wanted to improve their teaching and the overall education of future engineers (Borrego, 2007; Jamieson & Lohmann, 2009). Though the field of engineering education has been working to improve engineering education and faculty development as a field has primarily focused on improving faculty as educators, progress to change faculty instructional practices is still slow. As outlined by Ambrose and Norman, the minimum expectations for engineering faculty as educators is that they (1) "understand their students as intellectual-social-emotional beings," (2) "understand the basic principles of learning," and (3) "understand the components of effective course design" (2006).

The first of these has gained special attention at the start of the COVID-19 pandemic. Even before, though, there have been multiple calls for more empathetic teaching (e.g., Arghode et al.,

2013; MacCarty, 2021; Strachan, 2020), where the challenges students are facing can be acknowledged and accounted for in their education. Additionally, as a result of long-standing racial injustice and calls for change within the existing educational structure (e.g., Cross, 2020; Holly & Masta, 2021; L. L. Long, 2020), faculty are being called on to grow their awareness of the diversity in their classroom and to create inclusive learning environments with a higher urgency than in the past. Educator development initiatives must embrace these new challenges for engineering faculty and create effective supports and resources to enable more inclusive classroom environments that are responsive to the individual needs of their students.

Ambrose also calls for engineering faculty to "understand the basic principles of learning." Within engineering education and faculty development, the focus on the basic principles of learning has been central in efforts to support active learning and principles of engagement. Allowing faculty to develop a more nuanced and advanced understanding of learning by broadening the basic principles regularly highlighted within the field of engineering education and educational development can help engineering faculty improve their teaching with a more research-informed approach. For example, Nelson and Brennan identified 14 threshold concepts for engineering educators (such as having a learner-centered focus, constructive alignment of assessments with learning outcomes, and inquiry into student learning) as well as the LENS faculty development model to not only highlight these threshold concepts but also aid engineering educators in moving from novice to expert in each concept (2021a, 2021b).

Finally, the third minimum expectation for engineering faculty is related to effective course design. When exploring early-career resources for faculty, guidelines exist for how to design a course with minimal effort by using resources from previous versions of the course (e.g., Fink et al., 2005). However, when we publish work around new courses in engineering education, it is rare for course design principles to be discussed (highlighted as a challenge with entrepreneurial support program reporting within Zappe et al., 2021). Explicit discussion of different course elements varies by project and dissemination. Additional discussion and support of research-based principles of course design could better support engineering educators at all levels in adopting and adapting different approaches and course techniques.

An important audience to mention here is future faculty in the form of graduate students. The development of graduate teaching assistants as teachers, for instance, has been explored within the professional development space (Gardner & Jones, 2011; Reeves et al., 2016) as well as within engineering education specifically (Fong et al., 2019; Harper et al., 2015). Where general graduate studies tend to focus on students developing into researchers (Gardner, 2008), engineering education scholars have contributed to research that can support the professional development of graduate students as teachers as well (e.g., Coso Strong & Sekayi, 2018; Torres Ayala, 2012). Teaching graduate students the basic principles of learning and course design places them a step ahead in their educator development.

Overall, extensive work has explored strategies, practices, and research around improving engineering education and how faculty impact that work. However, as members of a field that is dedicated to improving engineering education, we must remember that we do not represent all engineering faculty. Just as our students are not the same – and not the same as us – not all engineering faculty are the same, and educational developers must remember that it is important to understand faculty as intellectual-social-emotional beings, to use the basic principles of learning that we are trying to pass along, while using effective training design.

4.2 Researcher Development

Research is regularly noted as a tenure-line faculty member's primary responsibility, both with respect to how they spend their time and what they must focus on for promotion and tenure. In the area of researcher development (i.e., supports and resources for improving the knowledge, skills, and attitudes (KSAs) related to conducting and leading research), one primary example is at the core of the literature: graduate school. There exists research examining how to help undergraduate students take an interest in graduate school through undergraduate research (Follmer et al., 2016; Willis et al., 2013), the experiences of graduate students as they navigate the journey to independent researcher (e.g., Brown, 2016; S. Gardner, 2008; Gonzalez et al., 2019; Lovitts, 2005), and how graduate students decide on a faculty career (Amelink & Edwards, 2020). Throughout the graduate school experience, factors such as socialization (Amelink & Edwards, 2020; Jairam & Kahl Jr., 2012) and departmental culture (de Valero, 2001; Golde, 2005) can impact not only graduate students' decision to leave graduate school but also their decision to enter the professoriate. For additional discussion of research of graduate engineering education, see IHEER chapter on graduate engineering education.

Limited research, however, explores the transition from graduate student researcher to faculty researcher (Gelso, 2006; Kahn & Scott, 1997). The KSAs needed by a graduate student are not the same as those needed by faculty. Faculty require the research skills developed as a graduate student, along with a new perspective on the larger picture of running a research program. For example, graduate students tend to focus on one or two projects and are responsible for conducting all elements (i.e., data collection and analysis, documentation) of those projects. Faculty members, on the other hand, become supervisors of the graduate and undergraduate students who are conducting the research. All the while, they must ensure funding is available for those students, all projects are all moving forward, the results are being published, and the lab is stocked with the equipment needed to conduct the research (which is linked to managing budgets and resources) – essentially the transition from graduate student researcher to faculty researcher becomes equivalent to moving from worker to manager (Amundsen & McAlpine, 2011). Nevertheless, few professional development resources are typically provided to support the development of these new skills needed for running a successful research program.

One often-cited professional development support for new faculty is mentoring (Edalgo et al., 2021; Z. Long et al., 2018; Mondisa & Adams, 2020). As noted earlier, socialization and professional networks can impact on graduate student success. Continuing the socialization through mentoring for faculty members can aid in their overall success, as well as in finding new collaborations for research, creating informal opportunities to learn from colleagues, and improving the broader range of skills needed as a faculty researcher.

Lastly, a faculty member's research focus may shift over their career. Mid-career, and even late-career, faculty could potentially benefit from opportunities to explore new methods and topics within their field or new opportunities for collaboration. This exploration is typically supported through a sabbatical or faculty internship (Carraher et al., 2014) or, more recently, through funded programs such as the NSF's Mid-Career Advancement grant. However, figuring out how to navigate these and when to position these within one's career is not often discussed. Additional exploration to understand researcher development needs would enable effective professional development for faculty across their entire career.

4.3 Design of and Structures for Faculty Professional Development

When discussing faculty development, the focus is largely on educator development – the classroom or education-related responsibilities of faculty, as previously described at length. However, as already noted, faculty have multiple, often competing, responsibilities that can impact their overall success, including in the classroom, and the decisions they make regarding their development. Recently, there have been calls for more holistic development support for faculty (Cutler et al., 2020; C. Lee et al., 2018; Whittaker & Montgomery, 2014), including within STEM and engineering education (Cutler et al., 2020; Lee et al., 2018). Yet to achieve this more holistic design of professional development support, spaces and resources are needed within and across universities.

5 Call for Research on Engineering Faculty and Their Impact on Engineering Education

Throughout this chapter, we sought to raise awareness of the often-overlooked engineering faculty that play a key role in engineering education. Faculty can be the enactors of educational change practices, though some barriers to enacting changes have yet to be fully explored within engineering education. As a field, engineering education should recognize the myriad of factors (e.g., pedagogical beliefs, faculty identity development, the varying positions/roles and institutional contexts) that impact faculty as people and how faculty experiences and backgrounds influence their educational decisions and their interactions with students. This foundation will allow us to expand and refine professional development opportunities for faculty that capture their diverse responsibilities, experiences, and goals. We have an opportunity to build on existing interdisciplinary research on faculty, to identify appropriate theories and methodologies for use in this research, to redefine our community's understanding of engineering faculty, and ultimately, to enhance our ability to impact positive and sustainable change within engineering education. Given the complex nature of faculty careers and experiences and the gaps that exist in the current EER, we identified five key areas for future research:

- **Impact of faculty on instruction and education systems.** How do we encourage faculty to change their teaching practices and/or beliefs to adopt evidence-based strategies? How do/can faculty impact broader educational structures beyond their classroom?
- **Understanding of faculty roles, identities, and pathways.** What is the impact of differing faculty roles and responsibilities on classroom practices? What does it mean to be an engineering faculty member, and how does that vary for faculty with diverse social identities? What are the reasons faculty leave institutions or academia? How do faculty make decisions throughout their career?
- **Educational development in engineering.** How can educational/faculty developers aid in change processes? What can engineering education researchers learn from educational development to help with dissemination and educational change?
- **International contexts for engineering faculty.** What could we learn in the United States by bringing in an international lens on the faculty experience?
- **Systematic literature reviews.** With the limited space within a single chapter, what areas of focus in faculty research would benefit from a systematic literature review?

Acknowledgments

The authors would like to thank Yu Xia, the reviewers, and our network of amazing colleagues who have supported our explorations into the faculty experience.

References

Adams, K. A. (2002). *What colleges and universities want in new faculty*. Association of American Colleges and Universities.

Aguilar, L., Walton, G., & Wieman, C. (2014). Psychological insights for improved physics teaching. *Physics Today*, *67*(5), 43–49. https://doi.org/10.1063/PT.3.2383

Allen, E., Castillo, F., & Schiorring, E. (2013). The reflective engineering advisor: A paradigm for learning-centered student advising. In *2013 ASEE annual conference & exposition proceedings* (pp. 23.1228.1–23.1228.18). https://doi.org/10.18260/1-2-22613

Ambrose, S., & Norman, M. (2006). Preparing engineering faculty as educators. *Bridge-Washington National Academy of Engineering, 36*(2), 25.

Amelink, C. T., & Edwards, C. D. (2020). Exploring the socialization experiences of underrepresented engineering graduate students. *Journal of Women and Minorities in Science and Engineering, 26*(4), 357–379. https://doi.org/10.1615/JWomenMinorScienEng.2020032606

Amundsen, C., & McAlpine, L. (2011). New academics as supervisors: A steep learning curve with challenges, tensions and pleasures. In L. McAlpine & C. Amundsen (Eds.), *Doctoral education: Research-based strategies for doctoral students, supervisors and administrators* (pp. 37–55). Springer Netherlands. https://doi.org/10.1007/978-94-007-0507-4_3

Aragón, O., Eddy, S., & Graham, M. (2018). Faculty beliefs about intelligence are related to the adoption of active-learning practices. *CBE – Life Sciences Education, 17*(3), ar47.

Aragón, O. R., Dovidio, J. F., & Graham, M. J. (2017). Colorblind and multicultural ideologies are associated with faculty adoption of inclusive teaching practices. *Journal of Diversity in Higher Education, 10*(3), 201–215. https://doi.org/10.1037/dhe0000026

Archer, L. (2008). Younger academics' constructions of 'authenticity', 'success' and professional identity. *Studies in Higher Education, 33*(4), 385–403. https://doi.org/10.1080/03075070802211729

Arghode, V., Yalvac, B., & Liew, J. (2013). Teacher empathy and ScienceEducation: A collective case study. *EURASIA Journal of Mathematics, Science and Technology Education, 9*(2). https://doi.org/10.12973/eurasia.2013.921a

Ash, A. N. (2020). Anti-racism in higher education: A model for change. *Anti-Racism in Higher Education, 4*(3), 35.

Austin, A. E., Campa, H., Pfund, C., Gillian-Daniel, D. L., Mathieu, R., & Stoddart, J. (2009). Preparing STEM doctoral students for future faculty careers. *New Directions for Teaching and Learning, 2009*(117), 83–95. https://doi.org/10.1002/tl.346

Baez, B. (2000). Race-related service and faculty of color: Conceptualizing critical agency in academe. *Higher Education, 39*, 363–391.

Banik, G. (2016). Strategies and techniques for new tenure-track faculty to become successful in academia. In *2016 ASEE annual conference & exposition proceedings* (p. 25886).
https://doi.org/10.18260/p.25886

Barber, S. P. (1987). Faculty development needs as a function of status in the academic guild. *To Improve the Academy, 9.*

Berg, J., Dutton, J., & Wrzesniewski, A. (2007). *What is job crafting and why does it matter?* University of Michigan, Ross School of Business. https://positiveorgs.bus.umich.edu/wp-content/uploads/What-is-Job-Crafting-and-Why-Does-it-Matter1.pdf

Berger, E. J., Wu, C., Briody, E. K., Wirtz, E., & Rodríguez-Mejía, F. (2021). Faculty subcultures in engineering and their implications for organizational change. *Journal of Engineering Education, 110*(1), 230–251. https://doi.org/10.1002/jee.20370

Berta, M., & Pembridge, J. (2019). Exploring burnout among graduate teaching assistants. *2019* In *ASEE annual conference & exposition proceedings* (p. 32802). https://doi.org/10.18260/1-2-32802

Besterfield-Sacre, M., Cox, M. F., Borrego, M., Beddoes, K., & Zhu, J. (2014). Changing engineering education: Views of U.S. faculty, chairs, and deans. *Journal of Engineering Education, 103*(2), 193–219. https://doi.org/10.1002/jee.20043

Bieber, J. P., & Worley, L. K. (2006). Conceptualizing the academic life: Graduate students' perspectives. *The Journal of Higher Education*, 77(6), 1009–1035.

Billot, J. (2010). The imagined and the real: Identifying the tensions for academic identity. *Higher Education Research & Development, 29*(6), 709–721. https://doi.org/10.1080/07294360.2010.487201

Bird, S., & Kellam, N. N. (2013). Teaching journeys of engineering faculty: Stories of transition. *2013 ASEE Annual Conference & Exposition, 9.*

Birmingham, S. (2007). Leaving tenure behind: Lessons learned. In *2007 Annual conference & exposition proceedings* (pp. 12.1014.1–12.1014.9). https://doi.org/10.18260/1-2-2520

Bodnar, C. A., McCave, E. J., Smith-Orr, C., Coso Strong, A., Faber, C. J., & Lee, W. (2021). Defining academic engineering education roles within the United States. In *Proceedings of the research in engineering education symposium & Australasian association for engineering education conference* (p. 9).

Borrego, M. (2007). Development of engineering education as a rigorous discipline: A study of the publication patterns of four coalitions. *Journal of Engineering Education, 96*(1), 5–18. https://doi.org/10.1002/j.2168-9830.2007.tb00911.x

Borrego, M., Froyd, J. E., & Hall, T. S. (2010). Diffusion of engineering education innovations: A survey of awareness and adoption rates in U.S. engineering departments. *Journal of Engineering Education, 99*(3), 185–207. https://doi.org/10.1002/j.2168-9830.2010.tb01056.x

Borrego, M., Froyd, J. E., Henderson, C., Cutler, S., & Prince, M. (2013). Influence of engineering instructors' teaching and learning beliefs on pedagogies in engineering science courses. *International Journal of Engineering Education, 29*(6), 1456–1471.

Boyle, S., Phillips, C., Pearson, Y., DesRoches, R., Mattingly, S., Nordberg, A., Li, W., & Rifai, H. (2020). An exploratory study of intentionality toward diversity in STEM faculty hiring. In *2020 ASEE virtual annual conference content access proceedings* (p. 34124). https://doi.org/10.18260/1-2-34124

Bracho Perez, V., Salgado, H., Coso Strong, A., & Kendall, M. R. (2021). Engineering instructional faculty perceptions of students' background at Hispanic Serving Institutions. In *Proceedings of the ASEE/IEEE frontiers in education conference.*

Brent, R., & Felder, R. (2000). Helping new faculty get off to a good start. In *2000 ASEE annual conference & exposition* (p. 8). https://doi.org/10.18260/1-2-8424

Briody, E. K., Wirtz, E., Goldenstein, A., & Berger, E. J. (2019). Breaking the tyranny of office hours: Overcoming professor avoidance. *European Journal of Engineering Education, 44*(5), 666–687. https://doi.org/10.1080/03043797.2019.1592116

Brown, B. (2016). *A case study analysis of minority students' negotiation of STEM, racial/ethnic, and graduate student identities.* Purdue University.

Buswell, N. T. (2021). The purpose of a PhD in engineering: Where does teaching fit in? *Studies in Engineering Education, 1*(1), 83. https://doi.org/10.21061/see.8

Calkins, S., & Kelley, M. R. (2005). Mentoring and the faculty-TA relationship: Faculty perceptions and practices. *Mentoring and Tutoring, 13*(2), 259–280. https://doi.org/10.1080/13611260500105915

Canale, A. M., Herdklotz, C., & Wild, L. (2013). *Mid-career faculty support: The middle years of the academic profession* (p. 16) [Faculty Career Development Services, The Wallace Center]. Rochester Institute of Technology. www.rit.edu/provost/sites/rit.edu.provost/files/images/FCDS_Mid-CareerRpt.pdf

Canning, E. A., Muenks, K., Green, D. J., & Murphy, M. C. (2019). STEM faculty who believe ability is fixed have larger racial achievement gaps and inspire less student motivation in their classes. *Science Advances, 5*(2), eaau4734. https://doi.org/10.1126/sciadv.aau4734

Carraher, S. M., Crocitto, M., & Sullivan, S. (2014). A kaleidoscope career perspective on faculty sabbaticals. *Career Development International, 19*(3), 295–313. https://doi.org/10.1108/CDI-04-2013-0051

Cech, E., Waidzunas, T., & Farrell, S. (2016). Engineering deans' support for LGBTQ inclusion. In *2016 ASEE annual conference & exposition proceedings* (p. 26633). https://doi.org/10.18260/p.26633

Chua, M., Cutler, S., Hahn, L., Harris, E., Kappers, W., & Pulford, S. (2016). Faculty developers on faculty development: Join the conversation. In *ASEE annual conference & exposition*, New Orleans, LA.

Cole, E. R., McGowan, B. L., & Zerquera, D. D. (2017). First-year faculty of color: Narratives about entering the academy. *Equity & Excellence in Education, 50*(1), 1–12. https://doi.org/10.1080/10665684.2016.1262300

Cornér, S., Löfström, E., & Pyhältö, K. (2017). The relationships between doctoral students' perceptions of supervision and burnout. *International Journal of Doctoral Studies, 12*, 091–106. https://doi.org/10.28945/3754

Coso Strong, A., Kendall, M., Basalo, I., & Henderson, G. (2019). Impact of Faculty Development workshops on instructional faculty at Hispanic-serving institutions. In *ASEE annual conference & exposition.*

Coso Strong, A., Kendall, M., & Henderson, G. (2022). Voices of engineering faculty at the margins: Supporting professional agency through faculty development. In S. M. Linder, C. M. Lee, & K. A. High (Eds.), *Handbook of STEM faculty development.* Information Age Publishing.

Coso Strong, A., & Sekayi, D. (2018). Exercising professional autonomy: Doctoral students' preparation for academic careers. *Studies in Graduate and Postdoctoral Education, 9*(2), 243–258. https://doi.org/10.1108/SGPE-D-18-00005

Coso Strong, A., Smith-Orr, C., Bodnar, C., Lee, W., McCave, E., & Faber, C. (2021). Early career faculty transitions: Negotiating legitimacy and seeking support in engineering education. *Studies in Engineering Education*, *1*(2), 71. https://doi.org/10.21061/see.52

Crick, K. A., Larson, L. M., & Seipel, M. T. (2020). Non-tenure track faculty satisfaction: A self-determination model. *Journal of Career Assessment*, *28*(3), 425–445. https://doi.org/10.1177/1069072719870681

Cross, K. (2020). Racism is the manifestation of White supremacy and antiracism is the answer. *Journal of Engineering Education*, *109*(4), 625–628. https://doi.org/10.1002/jee.20362

Cross, K., & Cutler, S. (2017). Engineering faculty perceptions of diversity in the classroom. In *2017 ASEE annual conference & exposition proceedings* (p. 28253). https://doi.org/10.18260/1-2-28253

Cross, K., Mamaril, N., Johnson-Glauch, N., & Herman, G. (2021). Building cultures of collaboration that promote instructional change. *Studies in Engineering Education*, *2*(2), 1. https://doi.org/10.21061/see.48

Cutler, S., Harris, E., Kappers, W., Pulford, S., & Yasuhara, K. (2017). Faculty developers on faculty development: Continuing the conversation, a birds-of-a-feather session. In *ASEE annual conference & exposition*, Columbus, Ohio.

Cutler, S., Kottmeyer, A., Heinen, R., Xia, Y., Zappe, S., & Litzinger, T. (2020). A holistic assessment of the responsibilities and areas of support of engineering faculty. *International Journal for Academic Development*, 1–14. https://doi.org/10.1080/1360144X.2020.1855592

Davidson, V. J., Vance, J. M., & Niemeier, D. A. (2001). Women in engineering leadership institute: Academic leadership development plans. In *ICWES12: 12th international conference of women engineers and scientists*, Ottawa, Canada., Retrieved April 12, 2005.

de Valero, Y. F. (2001). Departmental factors affecting time-to-degree and completion rates of doctoral students at one land-grant research institution. *The Journal of Higher Education*, *72*(3), 341. https://doi.org/10.2307/2649335

Dearing, J. W., & Singhal, A. (2020). New directions for diffusion of innovations research: Dissemination, implementation, and positive deviance. *Human Behavior and Emerging Technologies*, *2*(4), 307–313. https://doi.org/10.1002/hbe2.216

Dika, S. L., & Martin, J. P. (2018). Bridge to persistence: Interactions with educators as social capital for Latina/o engineering majors. *Journal of Hispanic Higher Education*, *17*(3), 202–215. https://doi.org/10.1177/1538192717720264

Dolan, E. L., Elliott, S. L., Henderson, C., Curran-Everett, D., St. John, K., & Ortiz, P. A. (2018). Evaluating discipline-based education research for promotion and tenure. *Innovative Higher Education*, *43*(1), 31–39. https://doi.org/10.1007/s10755-017-9406-y

Dweck, C. (2016). *Mindset: The new psychology for success*. Ballantine Books.

Edalgo, S., High, K. A., Lichtenstein, G., Lee, C. M., & Main, J. B. (2021). Exploring how faculty mentoring influences faculty productivity. In *2021 ASEE annual conference & exposition proceedings*, Virtual Meeting.

Ellinas, E. H., Ark, T. K., Kaljo, K., Quinn, K. G., Krier, C. R., & Farkas, A. H. (2021). Winners and losers in academic productivity during the COVID-19 pandemic: Is the gender gap widening for faculty? *Journal of Women's Health*, jwh.2021.0321. https://doi.org/10.1089/jwh.2021.0321

Farrell, S. (2018, September). Whom are we serving, and how can we improve? Letter from the President. *ASEE Today*. www.asee-prism.org/asee-today-sep-4/

Felder, R. M., & Brent, R. (2009). Active learning: An introduction. *ASQ Higher Education Brief*, *2*(4), 1–5.

Finelli, C. J., Daly, S. R., & Richardson, K. M. (2014). Bridging the research-to-practice gap: Designing an institutional change plan using local evidence: Designing an institutional change plan using local evidence. *Journal of Engineering Education*, *103*(2), 331–361. https://doi.org/10.1002/jee.20042

Finelli, C. J., & Froyd, J. E. (2019). Improving student learning in undergraduate engineering education by improving teaching and assessment. *Advances in Engineering Education*, 1–30.

Fink, L. D., Ambrose, S., & Wheeler, D. (2005). Becoming a professional engineering educator: A new role for a new era. *Journal of Engineering Education*, *94*(1), 185–194. https://doi.org/10.1002/j.2168-9830.2005.tb00837.x

Fitzmorris, C. W., Shehab, R., & Trytten, D. (2020). As necessary as the cleaning crew: Experiences of respect and inclusion among full-time non-tenure-track electrical engineering faculty at research-intensive institutions. *IEEE Transactions on Education*, 1–10. https://doi.org/10.1109/TE.2020.2978643

Fives, H., & Buehl, M. M. (2012). Spring cleaning for the "messy" construct of teachers' beliefs: What are they? Which have been examined? What can they tell us? In K. R. Harris, S. Graham, T. Urdan, S. Graham, J. M. Royer, & M. Zeidner (Eds.), *APA educational psychology handbook, Vol 2: Individual differences and cultural and contextual factors*. (pp. 471–499). American Psychological Association. https://doi.org/10.1037/13274-019

Flaherty, C. (2020). Burning out. *Inside Higher Ed*. www.insidehighered.com/news/2020/09/14/faculty-members-struggle-burnout

Follmer, D. J., Zappe, S., Gomez, E., & Kumar, M. (2016). Examining student outcomes from a research experiences for undergraduates (REU) program: Year two results. In *2016 ASEE annual conference & exposition proceedings* (p. 26793). https://doi.org/10.18260/p.26793

Fong, C. J., Gilmore, J., Pinder-Grover, T., & Hatcher, M. (2019). Examining the impact of four teaching development programmes for engineering teaching assistants. *Journal of Further and Higher Education*, *43*(3), 363–380. https://doi.org/10.1080/0309877X.2017.1361517

Fulweiler, R. W., Davies, S. W., Biddle, J. F., Burgin, A. J., Cooperdock, E. H. G., Hanley, T. C., Kenkel, C. D., Marcarelli, A. M., Matassa, C. M., Mayo, T. L., Santiago-Vàzquez, L. Z., Traylor-Knowles, N., & Ziegler, M. (2021). Rebuild the academy: Supporting academic mothers during COVID-19 and beyond. *PLOS Biology*, *19*(3), e3001100. https://doi.org/10.1371/journal.pbio.3001100

Gardner, A., & Willey, K. (2018). Academic identity reconstruction: The transition of engineering academics to engineering education researchers. *Studies in Higher Education*, *43*(2), 234–250. https://doi.org/10.1080/03075079.2016.1162779

Gardner, A., & Willey, K. (2019). The role of peer review in identity development for engineering education researchers. *European Journal of Engineering Education*, *44*(3), 347–359. https://doi.org/10.1080/03043797.2018.1500526

Gardner, G., & Jones, M. G. (2011). Pedagogical preparation of the science graduate teaching assistant: Challenges and implications. *Science Educator*, *20*(2), 31–41.

Gardner, S. (2008). "What's too much and what's too little?": The process of becoming an independent researcher in doctoral education. *The Journal of Higher Education*, *79*(3), 326–350. https://doi.org/10.1353/jhe.0.0007

Gelso, C. J. (2006). On the making of a scientist-practitioner: A theory of research training in professional psychology. *Training and Education in Professional Psychology*, *S*(1), 3–16. https://doi.org/10.1037/1931-3918.S.1.3

Glass, C. R., Doberneck, D. M., & Schweitzer, J. H. (2011). Unpacking faculty engagement: The types of activities faculty members report as publicly engaged scholarship during promotion and tenure. *Journal of Higher Education Outreach and Engagement*, *15*(1), 7–29.

Godwin, A., & Kirn, A. (2020). Identity-based motivation: Connections between first-year students' engineering role identities and future-time perspectives. *Journal of Engineering Education*, *109*(3). https://doi.org/10.1002/jee.20324

Golde, C. M. (2005). The role of the department and discipline in doctoral student attrition: Lessons from four departments. *The Journal of Higher Education*, *76*(6), 669–700. https://doi.org/10.1353/jhe.2005.0039

Gonzalez, L. M., Wester, K. L., & Borders, L. D. (2019). Supports and barriers to new faculty researcher development. *Studies in Graduate and Postdoctoral Education*, *10*(1), 21–34. https://doi.org/10.1108/SGPE-D-18-00020

Gow, L., & Kember, D. (1993). Conceptions of teaching and their relationship to student learning. *British Journal of Educational Psychology*, *63*(1), 20–23. https://doi.org/10.1111/j.2044-8279.1993.tb01039.x

Graham, C., & McGarry, J. (2019). Career advancement experiences of mid-career women faculty and those across African, Latinx, Asian, and Native American Diasporas. In A. G. Welch, J. Bolin, & D. Reardon (Eds.), *Mid-career faculty: Trends, barriers, and possibilities* (pp. 61–85). BRILL. https://doi.org/10.1163/9789004408180

Graham, R. (2018). *The career framework for university teaching: Background and overview*. Royal Academy of Engineering.

Gregg, M., Knott, T., & Lo, J. (2005). Transitioning from adjunct to tenure track. In *2005 Annual conference proceedings* (pp. 10.1355.1–10.1355.7). https://doi.org/10.18260/1-2-14737

Grifski, J., Dringenberg, E., & Delpech, D. M. (2021). Thinking as argument: A theoretical framework for studying how faculty arrive at their deeply-held beliefs about inequity in engineering. In *2021 ASEE virtual annual conference content access proceedings*.

Harlow, A., Lo, S. M., Saichaie, K., & Sato, B. K. (2020). Characterizing the University of California's tenure-track teaching position from the faculty and administrator perspectives. *PLOS ONE*, *15*(1), e0227633. https://doi.org/10.1371/journal.pone.0227633

Harper, K., Zierden, H., Wegman, K., Kajfez, R., & Kecskemety, K. (2015). Teaching assistant professional development through design: Why they participate and how they benefit. In *2015 ASEE annual conference and exposition proceedings* (pp. 26.1469.1–26.1469.14). https://doi.org/10.18260/p.24806

Hart, J., & Mars, M. M. (2009). Joint appointments and the professoriate: Two houses but no home? *Innovative Higher Education*, *34*(1), 19–32. https://doi.org/10.1007/s10755-008-9092-x

Haviland, D., Alleman, N. F., & Allen, C. C. (2017). 'Separate but not quite equal': Collegiality experiences of full-time non-tenure-track faculty members. *The Journal of Higher Education*, *88*(4), 505–528. https://doi.org/10.1080/00221546.2016.1272321

Henderson, C., Beach, A., & Finkelstein, N. (2011). Facilitating change in undergraduate STEM instructional practices: An analytic review of the literature. *Journal of Research in Science Teaching*, *48*(8), 952–984. https://doi.org/10.1002/tea.20439

Henderson, C., Dancy, M., & Niewiadomska-Bugaj, M. (2012). Use of research-based instructional strategies in introductory physics: Where do faculty leave the innovation-decision process? *Physical Review Special Topics – Physics Education Research*, *8*(2), 020104. https://doi.org/10.1103/PhysRevSTPER.8.020104

Herman, G. L., & Loewenstein, J. (2017). Evidence-based change practices: Evidence-based change practices. *Journal of Engineering Education*, *106*(1), 4–13. https://doi.org/10.1002/jee.20152

Holly, J., & Masta, S. (2021). Making whiteness visible: The promise of critical race theory in engineering education. *Journal of Engineering Education*, *110*(4), 798–802. https://doi.org/10.1002/jee.20432

Holly Jr., J. (2020). Disentangling engineering education research's anti-Blackness. *Journal of Engineering Education*, *109*(4), 629–635. https://doi.org/10.1002/jee.20364

Hornsby, E. E., Morrow-Jones, H. A., & Ballam, D. A. (2012). Leadership development for faculty women at The Ohio State University: The president and provost's leadership institute. *Advances in Developing Human Resources*, *14*(1), 96–112. https://doi.org/10.1177/1523422311428758

Hubbard, S. M., & Stage, F. K. (2009). Attitudes, perceptions, and preferences of faculty at hispanic serving and predominantly black institutions. *The Journal of Higher Education*, *80*(3), 270–289. https://doi.org/10.1080/00221546.2009.11779013

Hunter, A. (2020). Snapshots of selfhood: Curating academic identity through visual autoethnography. *International Journal for Academic Development*, *25*(4), 310–323. https://doi.org/10.1080/1360144X.2020.1755865

Jairam, D., & H. Kahl Jr., D. (2012). Navigating the doctoral experience: The role of social support in successful degree completion. *International Journal of Doctoral Studies*, 7, 311–329. https://doi.org/10.28945/1700

Jamieson, J., & Lohmann, J. (2009). *Creating a culture for scholarly and systematic innovation in engineering education*. American Society for Engineering Education. https://onlinelibrary.wiley.com/doi/10.1002/j.2168-9830.2009.tb01019.x

Kahn, J., & Scott, N. (1997). Predictors of research productivity and science-related career goals among counseling psychology doctoral students. *The Counseling Psychologist*, *25*(1), 38–67.

Kendall, M. R., Coso Strong, A., Henderson, G., & Basalo, I. (2021). Perceptions of engineering faculty on educational innovation at Hispanic-serving institutions. *Journal of Women and Minorities in Science and Engineering*, *27*(6), 21–57. https://doi.org/10.1615/JWomenMinorScienEng.2021034722

Kezar, A. (2013). Departmental cultures and non-tenure-track faculty: Willingness, capacity, and opportunity to perform at four-year institutions. *The Journal of Higher Education*, *84*(2), 153–188. https://doi.org/10.1353/jhe.2013.0011

Kim, B., Stallings, R. P., Merlo, A. V., & Lin, A. W.-C. (2015). Mentoring in criminology and criminal justice doctoral education: Doctoral program coordinators' perspectives. *Journal of Criminal Justice Education*, *26*(4), 390–407. https://doi.org/10.1080/10511253.2015.1049630

Krukowski, R. A., Jagsi, R., & Cardel, M. I. (2021). Academic productivity differences by gender and child age in science, technology, engineering, mathematics, and medicine faculty during the COVID-19 pandemic. *Journal of Women's Health*, *30*(3), 341–347. https://doi.org/10.1089/jwh.2020.8710

Lamont, C., & Nordberg, D. (2014). Becoming or unbecoming: Contested academic identities. *British Academy of Management*, 15.

Lee, C., High, K., Linder, S., & Stefl, S. (2018). Holistic STEM faculty development. In *National association for research in science teaching (NARST) conference*. https://stemfacdev.org/publications/

Lee, D., Stefl, S., Linder, S., Lee, C., Jamil, F., & High, K. (2017). Board # 86: How many hats do you wear: Building research capacity for STEM faculty development workshop. In *2017 ASEE annual conference & exposition proceedings* (p. 27942). https://doi.org/10.18260/1-2-27942

Leydens, J. A., & Lucena, J. C. (2017). *Engineering justice: Transforming engineering education and practice*. Wiley. https://books.google.com/books?id=uKZFDwAAQBAJ

Little, D. (2020). Reflections on the state of the scholarship of educational development. *To Improve the Academy, 33*(1). https://doi.org/10.3998/tia.17063888.0033.104

Long, L. L. (2020). Toward an antiracist engineering classroom for 2020 and beyond: A starter kit. *Journal of Engineering Education, 109*(4), 636–639. https://doi.org/10.1002/jee.20363

Long, Z., Buzzanell, P., Kokini, K., Wilson, R., Batra, J., & Anderson, L. (2018). Mentoring women and minority faculty in engineering: A multidimensional mentoring network approach. *Journal of Women and Minorities in Science and Engineering, 24*(2), 121–145.

Lord, S., Berger, E., Kellam, N., Ingram, E., Riley, D., Rover, D., Salzman, N., & Sweeney, J. (2017). Talking about a revolution: Overview of NSF RED projects. In *2017 ASEE annual conference & exposition proceedings* (p. 28903). https://doi.org/10.18260/1-2-28903

Lord, S., Camacho, M., Kellam, N., & Williams, J. (2017). Institutional mentoring to incite a revolution through NSF's RED program. In *2017 ASEE annual conference & exposition proceedings* (p. 28538). https://doi.org/10.18260/1-2-28538

Lovitts, B. E. (2005). Being a good course-taker is not enough: A theoretical perspective on the transition to independent research. *Studies in Higher Education, 30*(2), 137–154. https://doi.org/10.1080/03075070500043093

MacCarty, T. (2021). Supporting online students: Why teaching with empathy matters. *How to Teach Online*. www.snhu.edu/about-us/newsroom/education/supporting-online-students-and-teaching-with-empathy

McCave, D. E. J., Bodnar, C. A., Smith-Orr, C. S., Coso Strong, A., Lee, W. C., & Faber, C. J. (2020). I graduated, now what? An overview of the academic Engineering Education Research job field and search process. In *2020 ASEE virtual annual conference* (p. 18). https://doi.org/10.18260/1-2-34741

Mena, I. (2015). The science and engineering postdoc experience through the lens of liminality. In *2015 ASEE annual conference and exposition proceedings* (pp. 26.1575.1–26.1575.17). https://doi.org/10.18260/p.24912

Mondisa, J.-L., & Adams, R. S. (2020). A learning partnerships perspective of how mentors help protégés develop self-authorship. *Journal of Diversity in Higher Education*. https://doi.org/10.1037/dhe0000291

Morelock, J. R. (2017). A systematic literature review of engineering identity: Definitions, factors, and interventions affecting development, and means of measurement. *European Journal of Engineering Education, 42*(6), 1240–1262. https://doi.org/10.1080/03043797.2017.1287664

Murakami-Ramalho, E., Nuñez, A., & Cuero, K. K. (2010). Latin@ advocacy in the hyphen: Faculty identity and commitment in a Hispanic-serving institution. *International Journal of Qualitative Studies in Education, 23*(6), 699–717. https://doi.org/10.1080/09518391003641924

National Academy of Sciences. (2019). *The science of effective mentorship in STEMM: Online guide*. www.nap.edu/resource/25568/interactive/tools-and-resources.html#section3

National Academy of Sciences, National Academy of Engineering, & Institute of Medicine. (1997). *Adviser, teacher, role model, friend: On being a mentor to students in science and engineering* (p. 5789). National Academies Press. https://doi.org/10.17226/5789

Nelson, N., & Brennan, R. (2021a). LENS: A model for engineering faculty development. In *Research in engineering education symposium & Australasian associate for engineering education conference*, Perth, WA.

Nelson, N., & Brennan, R. (2021b). Threshold concepts in the engineering educator's journey: A systematic review. In *Research in engineering education symposium & Australasian associate for engineering education conference*, Perth, WA.

Nguyen, K., DeMonbrun, R., Borrego, M., Prince, M., Husman, J., Finelli, C., Shekhar, P., Henderson, C., & Waters, C. (2017). The variation of nontraditional teaching methods across 17 undergraduate engineering classrooms. In *2017 ASEE annual conference & exposition proceedings* (p. 29021). https://doi.org/10.18260/1-2-29021

Nuñez, A.-M., Ramalho, E. M., & Cuero, K. K. (2010). Pedagogy for equity: Teaching in a Hispanic-serving institution. *Innovative Higher Education, 35*(3), 177–190. https://doi.org/10.1007/s10755-010-9139-7

O'Meara, K., Jaeger, A., Misra, J., Lennartz, C., & Kuvaeva, A. (2018). Undoing disparities in faculty workloads: A randomized trial experiment. *PLOS One, 13*(12), e0207316.

O'Meara, K., Lounder, A., & Campbell, C. M. (2014). To heaven or hell: Sensemaking about why faculty leave. *The Journal of Higher Education, 85*(5), 30.

O'Meara, K., Templeton, L., & Nyunt, G. (2018). Earning professional legitimacy: Challenges faced by women, underrepresented minority, and non-tenure-track faculty. *Teachers College Record*, 39.

Park, J. (2015). *Determinants of turnover intent in higher education: The case of international and U.S. faculty* [PhD thesis, Virginia Commonwealth University].

Pawley, A. (2009). Universalized narratives: Patterns in how faculty members define "engineering". *Journal of Engineering Education, 98*(4), 309–319. https://doi.org/10.1002/j.2168-9830.2009.tb01029.x

Pawley, A. (2017). Shifting the "default": The case for making diversity the expected condition for engineering education and making whiteness and maleness visible: Shifting the "default". *Journal of Engineering Education, 106*(4), 531–533. https://doi.org/10.1002/jee.20181

Pawley, A., & Hoegh, J. (2010). Modeling the career pathways of women engineering faculty through oral histories and participatory research methods. In *ASEE annual conference and exposition* (p. 12).

Pembridge, J., & Jordan, K. (2016). Balancing the influence of driving and restricting factors to use active learning. In *123rd Annual ASEE conference & exposition*, New Orleans, LA.

Perry, A. L., Dean, S. R., & Hilton, A. A. (2019). New faculty transitions and obstacles: An auto-ethnographic exploration. *The Journal of the Professoriate, 10*(2), 43–72.

Piskadlo, K. S. (2016). *A case study on the influence of organizational structures and policies on faculty implementation of learner-centered teaching* [PhD thesis, University of Massachusetts, Boston].

Prince, M. (2004). Does active learning work? A review of the research. *Journal of Engineering Education, 93*(3), 223–231. https://doi.org/10.1002/j.2168-9830.2004.tb00809.x

Puscas, L. (2016). *Viewpoint from a program director* they can't all walk on water. *Journal of Graduate Medical Education, 8*(3), 314–316. https://doi.org/10.4300/JGME-D-16-00237.1

Quinney, A., Lamont, C., Biggins, D., & Holley, D. (2017). Optimising disruptive approaches: Extending academic roles and identities in higher education. *Journal of Learning Development in Higher Education, 12*, 14.

Rattan, A., Savani, K., Chugh, D., & Dweck, C. S. (2015). Leveraging mindsets to promote academic achievement: Policy recommendations. *Perspectives on Psychological Science, 10*(6), 721–726. https://doi.org/10.1177/1745691615599383

Reeves, T. D., Marbach-Ad, G., Miller, K. R., Ridgway, J., Gardner, G. E., Schussler, E. E., & Wischusen, E. W. (2016). A conceptual framework for graduate teaching assistant professional development evaluation and research. *CBE – Life Sciences Education, 15*(2), es2. https://doi.org/10.1187/cbe.15-10-0225

Reybold, L. E. (2003). Pathways to the professorate: The development of faculty identity in education. *Innovative Higher Education*, 18.

Reybold, L. E., & Alamia, J. J. (2008). Academic transitions in education: A developmental perspective of women faculty experiences. *Journal of Career Development, 35*(2), 107–128. https://doi.org/10.1177/0894845308325644

Rispoli, M. (2019). Keeping the momentum: Mid-career faculty mentorship. In A. G. Welch, J. Bolin, & D. Reardon (Eds.), *Mid-career faculty: Trends, barriers, and possibilities* (pp. 178–196). BRILL. https://doi.org/10.1163/9789004408180

Robinson, G. (2010). *Identity work and sensemaking by faculty approaching tenure*. University of Pittsburgh.

Rockinson-Szapkiw, A. (2019). The academic mother at mid-career. In A. G. Welch, J. Bolin, & D. Reardon (Eds.), *Mid-career faculty: Trends, barriers, and possibilities* (pp. 41–60). BRILL. https://doi.org/10.1163/9789004408180

Rockquemore, K. A. (2016). How to retain a diverse faculty (essay). *Inside Higher Ed*, 7.

Rogers, E. (2003). *Diffusion of innovation* (5th ed.). Free Press.

Rojewski, J. M. (2018). *Perception of academic leadership held by mid-Career STEM and social science faculty* [PhD thesis, Michigan State University].

Ross, L., Judson, E., Krause, S., Ankeny, C., Culbertson, R., & Hjelmstad, K. (2017). Relationships between engineering faculty beliefs and classroom practices. In *2017 ASEE annual conference & exposition proceedings* (p. 28793). https://doi.org/10.18260/1-2-28793

Sabagh, Z., Hall, N. C., & Saroyan, A. (2018). Antecedents, correlates and consequences of faculty burnout. *Educational Research, 60*(2), 131–156. https://doi.org/10.1080/00131881.2018.1461573

Schimpf, C., Mercado Santiago, M., & Pawley, A. (2012). Access and definition: Exploring how STEM faculty, department heads, and university policy administrators navigate the implementation of a parental leave policy. In *2012 ASEE annual conference & exposition proceedings* (pp. 25.124.1–25.124.22). https://doi.org/10.18260/1-2-20884

Schimpf, C., Santiago, M. M., Hoegh, J., & Pawley, A. (2013). STEM faculty and parental leave: Understanding an institution's policy within a national policy context through structuration theory. *Science and Technology, 24.*

Shadle, S. E., Marker, A., & Earl, B. (2017). Faculty drivers and barriers: Laying the groundwork for undergraduate STEM education reform in academic departments. *International Journal of STEM Education, 4*(1), 8. https://doi.org/10.1186/s40594-017-0062-7

Siddiqui, J. A., Allendoerfer, C., Adams, R. S., & Williams, B. (2016). Integration of scholarship: Interconnections among three studies on becoming an engineering education researcher. *International Journal of Engineering Education, 32*(6), 2352–2377. https://doi.org/0949-149X/91

Simmons, D. R., & Lord, S. M. (2019). Removing invisible barriers and changing mindsets to Improve and diversify pathways in engineering. *Advances in Engineering Education*, 1–22.

Singhal, A., & Svenkerud, P. J. (2018). Diffusion of evidence-based interventions or practice-based positive deviations. *Journal of Development Communication, 29*(2), 54–64.

Steele, C. (2010). *Whistling Vivaldi: How stereotypes affect Us and what we can do.* W.W. Norton & Company.

Strachan, S. L. (2020). The case for the caring instructor. *College Teaching, 68*(2), 53–56. https://doi.org/10.1080/87567555.2019.1711011

Strong, A., Chua, M., & Cutler, S. (2016). Talking "faculty development" with engineering educators, then talking "engineering education" with faculty developers: A collaborative reflection on working across communities. In *2016 ASEE annual conference & exposition proceedings* (p. 26010). https://doi.org/10.18260/p.26010

Sutherland, K. (2017). Constructions of success in academia: An early career perspective. *Studies in Higher Education, 42*(4), 743–759.

Taylor, P. G. (1999). *Making sense of academic life: Academics, universities, and change.* The Society for Research into Higher Education and Open University.

Theobald, E. J., Hill, M. J., Tran, E., Agrawal, S., Arroyo, E. N., Behling, S., Chambwe, N., Cintrón, D. L., Cooper, J. D., Dunster, G., Grummer, J. A., Hennessey, K., Hsiao, J., Iranon, N., Jones, L., Jordt, H., Keller, M., Lacey, M. E., Littlefield, C. E., . . . Freeman, S. (2020). Active learning narrows achievement gaps for underrepresented students in undergraduate science, technology, engineering, and math. *Proceedings of the National Academy of Sciences, 117*(12), 6476–6483. https://doi.org/10.1073/pnas.1916903117

Tise, J., Hochstedt, K., Zappe, S., Gomez, E., & Kumar, M. (2018). Lessons learned from a chemical engineering REU: The importance of training graduate students who are supervising REU students. In *2018 ASEE annual conference & exposition proceedings* (p. 30760). https://doi.org/10.18260/1-2-30760

Tomkin, J. H., Beilstein, S. O., Morphew, J. W., & Herman, G. L. (2019). Evidence that communities of practice are associated with active learning in large STEM lectures. *International Journal of STEM Education, 6*(1), 1. https://doi.org/10.1186/s40594-018-0154-z

Torres Ayala, A. T. (2012). *Future engineering professors' conceptions of learning and teaching engineering* [PhD thesis, University of South Florida].

Urquidi Cerros, Y., Kayyali, M., Kendall, M. R., & Coso Strong, A. (2021). Motivational factors influencing engineering faculty's pursuit of instructional faculty positions at Hispanic-serving institutions. In *Proceedings of the ASEE/IEEE frontiers in education conference.*

Vellamo, T., Pekkola, E., Siekkinen, T., & Cai, Y. (2020). Technical identity in a merger process – between a rock and a hard place. In L. Geschwind, A. Broström, & K. Larsen (Eds.), *Technical Universities: Past, present and future* (Vol. 56, pp. 165–190). Springer International Publishing. https://doi.org/10.1007/978-3-030-50555-4

Whittaker, J. A., & Montgomery, B. L. (2014). Cultivating institutional transformation and sustainable STEM diversity in higher education through integrative faculty development. *Innovative Higher Education, 39*(4), 263–275. https://doi.org/10.1007/s10755-013-9277-9

Willis, D. A., Krueger, P. S., & Kendrick, A. (2013). The influence of a research experiences for undergraduates program on student perceptions and desire to attend graduate school. *Journal of STEM Education*, *14*(2), 8.

Wisconsin Center for Education Research. (2021). *Center for the improvement of mentored experiences in research (CIMER)* [University Research Project]. https://cimerproject.org/

Zappe, S. E., Cutler, S. L., & Gase, L. (2021). A systematic review of the impacts of entrepreneurial support programs in science, technology, engineering, and math fields. *Entrepreneurship Education and Pedagogy*, 251512742110404. https://doi.org/10.1177/25151274211040422

15

Informal Learning as Opportunity for Competency Development and Broadened Engagement in Engineering

Madeline Polmear, Shannon Chance, Roger G. Hadgraft, and Corrinne Shaw

1 Introduction

Engineering for the 21st century requires graduates to have a broad range of competencies and experiences that enable them to work in a technology-dependent and globalized workforce within a complex sociotechnical landscape (National Academy of Engineering, 2004; SEFI, 2016). Fueled by both economic (National Academies of Sciences, Engineering, and Medicine, 2018) and equity (Baber, 2015) arguments, the engineering profession also needs a diverse and inclusive workforce that represents the communities it serves and that protects the welfare of all (Chubin et al., 2005; National Academy of Engineering, 2002). However, it is increasingly recognized that the engineering curriculum alone cannot fulfill these aims (National Academies of Sciences, Engineering, and Medicine, 2018).

Throughout the world, the engineering curriculum is densely packed with content that is focused on engineering science and has evolved little over the past few decades (Froyd & Lohmann, 2014). Learning outside of the classroom provides opportunities for students to develop key competencies that are needed for employability (Fisher et al., 2017; Mtawa et al., 2019; Polmear et al., 2021a; Tan et al., 2021) and skills that help them persist through the degree program (Kuh et al., 2008). Learning outside of formal class time also supports the engagement of students who have been traditionally underrepresented in engineering (Espinosa, 2011). Given the role of informal learning in supporting competency development and the engagement of diverse students, it is crucial for engineering programs to integrate these learning opportunities and to recognize their value.

Learning occurs in myriad ways and settings that contribute to students' personal, academic, and professional development (Shulman, 2005). Formal learning provides structured and intentional opportunities in classrooms and lecture halls through which students acquire knowledge and learn what it means to be an engineer (Ainsworth et al., 2010). However, only a small portion of the educational experience occurs within the formal curriculum and class time (Bell et al., 2009; Lee & Matusovich, 2016). Learning is not confined to these formal settings but rather continues informally and sometimes implicitly throughout the daily lives and activities of students (Ainsworth et al., 2010; Denson et al., 2015; Kotys-Schwartz et al., 2011).

DOI: 10.4324/9781003287483-18

Informal learning occurs outside of the instructional and institutional setting in which individuals participate, often without being part of the academic program or any requirement for graduation (Trinder et al., 2008). The definition of *informal learning* is intentionally broad to capture the range of activities, organizations, and settings through which students learn informally. Across these contexts, informal learning is defined along the dimensions of being (1) non-didactive, (2) collaborative, (3) within a meaningful context, (4) driven by the learner's interest, and (5) without external assessment (Callanan et al., 2011).

One aim of this chapter is to introduce informal learning by synthesizing the various definitions and disciplinary perspectives that have contributed to the piecemeal development of literature on informal learning. A second aim is to demonstrate the benefits and outcomes of informal learning that transcend the various settings, contexts, and activities in which it takes place. Pulling these pieces together, the chapter shows that the potential of informal learning is still being realized in education and research. More engagement from administrators, educators, and institutions is needed to provide opportunities for informal learning and recognition for the students who are participating and gaining engineering skills and experiences. More research is needed to understand the various ways in which students engage in informal learning, the different experiences of students from diverse backgrounds, and ways to assess the outcomes of informal learning.

The chapter begins with an overview of informal learning, including established definitions, a history of informal learning in engineering, a description of activities and settings relevant to engineering education in the university context, and a profile of student participation in informal learning. The chapter focuses on two key benefits related to informal learning: (1) competency development within the university experience and for engineering practice and (2) broadened engagement for learners with considerations for diversity, equity, and inclusion (DEI) in an international context.

To realize these benefits, the chapter identifies implications and provides suggestions for engineering education and research. The chapter is intended to serve as a guide for educators, administrators, researchers, and graduate students interested in studying or implementing informal learning. Informal learning is an opportunity for engineering programs to broaden the skill set and holistic experiences of their students. Yet to fully leverage this potential, there is a need for more research on outcomes accrued by students who engaged in the range of informal learning activities and settings. Also needed are better mechanisms to recognize the outcomes that students gain.

2 Overview of Informal Learning

2.1 Definitions

Since informal learning occurs throughout an individual's life, there are varying definitions and disciplinary perspectives based on context. Formalizing the vocabulary surrounding informal learning in engineering education is needed to help the engineering education community readily identify studies on the topic and work across disciplinary and national boundaries (Kotys-Schwartz et al., 2011).

In addition to being lifelong, informal learning is conceptualized as life-wide since it occurs in activities and settings outside of the classroom (Meyers et al., 2013). It is characterized as being iterative and open-ended, often collaborative, and embedded in local context (Bell et al., 2009; Falk & Dierking, 2016; Griffin, 1998). Informal learning is immersive and spontaneous (Ainsworth et al., 2010). It can be structured or unstructured – particularly regarding the objectives, availability of support, and length of time spent – but it is almost always motivated and guided by the interests of the learner. Informal learning is generally not formally or externally assessed (Rogoff et al., 2016). Informal learning stands in contrast to formal learning, which occurs in classrooms, lecture halls, and laboratories, where every student is expected to achieve specific learning outcomes and performance indicators.

In educational research, there is a longer history of examining informal learning in the context of K–12 and science education. One foundational reference in this space is "Learning Science in Informal Environments: People, Places, and Pursuits" published by the National Research Council of the National Academies (Bell et al., 2009). This work provides a helpful starting place to understanding informal learning and identifies six "stands of learning" that typify informal environments. Learning in informal settings is characterized by (1) excitement, interest, and motivation to learn about science phenomena; (2) remembering, understanding, using, and generating science concepts, explanations, arguments, models, and facts; (3) observing, exploring, questioning, predicting, manipulating, testing, and making sense of the natural and physical world; (4) reflecting on science as a way of knowing and on one's own process of learning about phenomena; (5) participating in activities and learning practices with others, using science language and tools; and (6) thinking about oneself as a science learner and developing an identity as someone who knows about, uses, and sometimes contributes to science and technology (adapted from Bell et al., 2009, p. 4).

Although this chapter focuses on informal learning for university students, K–12 education provides insight into the historical development of informal learning, a larger body of scholarship, and a pathway into university engineering education.

2.2 History of Informal Learning in Engineering Rooted in K–12 Education

Informal learning has always existed in cultures and communities, and discussions of informal learning in educational theory emerged in the early 20th century through the work of Dewey and Vygotsky (Callanan et al., 2011). Informal education was explicitly distinguished from formal education in the work of Scribner and Cole in the 1970s, which ushered in a greater focus on informal learning inside and outside schools (Callanan et al., 2011). In the decades since, programming and research related to informal learning have focused on pre-university (i.e., K–12, also known as primary and secondary school; Callanan et al., 2011; Ehsan et al., 2018; Jagušt et al., 2018).

Informal learning in K–12 has long served as an opportunity to interest younger students in going to university and pursuing a STEM degree. A meta-analysis of research on out-of-class STEM programs on K–12 students indicated a positive effect on interest in STEM (Young et al., 2016). Evidence shows the importance of reaching students younger than secondary/high school and having multiple informal learning opportunities to expand pathways into STEM (Demetry & Sontgerath, 2020). Outreach programs offered through universities also benefit the university by providing a recruiting mechanism. In addition to the value of informal learning in students' STEM interest, participation in out-of-class activities is considered valuable for university admission in the USA (Clinedinst & Koranteng, 2017).

As university engineering education has evolved, from its roots in the Industrial Revolution and a practical focus then to a more theoretical approach (Seely, 1999), informal learning has served as an important entry point into engineering education, an opportunity to complement the curriculum, and support workforce development (Kovalchuk et al., 2017). Over time, the activities and settings in which students informally learn about engineering have also grown, as detailed in the following section.

2.3 Types of Informal Learning: Activities and Settings

Informal learning can happen in a range of settings and through a variety of activities. *Settings* describe the environment in which people learn informally, and *activities* describe the actions they are doing as part of their learning. Given this breadth, different typologies have been developed to categorize informal learning. Science education once again provides a starting point where scholars (Bell et al., 2009) have identified three main types of settings where people learn science: programmed settings, designed environments, and everyday environments. These three categories are also recognized in the literature on engineering education (Denson et al., 2015; Kotys-Schwartz et al., 2011).

The first type, programmed settings, tends to "emulate or complement formal school settings" (Denson et al., 2015, p. 11). Programmed settings frequently involve facilitators, a fixed group of participants for multiple sessions, and a plan with goals and learning objectives (Kotys-Schwartz et al., 2011). Programmed settings may intentionally complement a formal curriculum to extend benefits and reinforce the learning. At a general level, they include programs that are connected to schools and community organizations (Bell et al., 2009).

The second categorical environment for informal learning, designed settings, involves places that are deliberately curated to facilitate learning. Designed settings include science and technology museums, civic and community learning centers, and world fairs. With the rise of community maker spaces and maker fairs, this type of environment, designed to facilitate informal learning of engineering, is becoming more prevalent (Martin & Betser, 2020; Wilczynski, 2015).

The third category, everyday settings, is the most fluid and accessible. Individuals encounter science content throughout their lives and learn about the natural world through everyday encounters, cultural practices, Internet, and media (Bell et al., 2009). The distinguishing feature of this setting is that it often does not have the explicit goal of teaching or learning. Although everyday learning can be unexpected and opportunistic, it can also take shape through more deliberate activities, such as pursuing a science-related hobby.

Another prevalent scheme for categorizing informal learning settings in engineering education is organized into co-curricular and extracurricular (Simmons et al., 2018; Simmons et al., 2017). Co-curricular activities often extend the formal curriculum and may be explicitly tied to formal academic learning. Co-curricular activities can include co-ops, internships, service projects, and some activity in clubs and organizations. They are connected to (and reinforce or mirror) the formal academic curriculum, and they may even accrue credit towards graduation, but they are not a required component of the student's selected degree program. Usually, they are separated from academic coursework, ungraded, and occurring outside of class hours.

Extracurricular activities are less explicitly tied to the curriculum than co-curricular activities, even when they are provided by the academic institution. *Extracurricular activities* are consistently defined as engagement outside academics – and, more specifically, outside required coursework. Thus, extracurricular activities could include sports, jobs, community service, student governance, politics, arts, religion, hobbies, clubs, and other personal development or personal interest organizations.

In engineering, programmed settings for informal learning frequently include the types of co-curricular and extracurricular activities described previously (Fisher et al., 2017). This categorization mirrors the K–12 approach, which Kotys-Schwartz et al. (2011) recommend adapting in engineering education, and distinguishes (1) the "associated model," which is closely tied to and aligned with weekly objectives of the formal curriculum (curricular); (2) the "coordinated model," which relates to the general curriculum but is not tied to weekly outcomes (co-curricular); and (3) the "integrated model," which runs completely separate from the curriculum (extracurricular).

Within these typologies and broader categories, there is a wide range of activities and settings relevant to engineering education. Examples are displayed in Table 15.1. These informal learning opportunities are highlighted because they are commonly associated with universities and engineering programs (Kotys-Schwartz et al., 2011; Simmons et al., 2018). Due to their proximity to engineering education and practice, these activities and settings also provide opportunities for researchers

Table 15.1 Examples of Informal Learning Opportunities in Engineering Education

Type	*Definition within Informal Learning*	*Example Reference(s)*
Disciplinary professional society	Student chapters affiliated with engineering professional organizations that provide access to design competitions, networking events, and career resources.	Evans et al. (2001)
Design competition teams	Student teams that design and build a vehicle or device or develop a solution to an engineering challenge and compete against teams at other universities. The competitions are typically organized by professional societies, government agencies, and nonprofit organizations.	Wolfinbarger et al. (2021)
Service learning and community engagement	Community-based projects at a scale ranging from local to global that are not situated in a course.	Swan et al. (2013), Litchfield et al. (2016)
Research	Undergraduate research experience outside of class time or course credit.	Carter et al. (2016)
Identity-based organization (can be related to engineering or not)	Activity or society associated with a particular personal or group identity (e.g., race, ethnicity, national origin, religion, gender, sexual orientation).	Revelo Alonso (2015), Ross and McGrade (2016)
Living learning community	Program in which students live together in a campus residence hall and participate in curricular and co-curricular activities.	Maltby et al. (2016)
Study abroad	Educational program or opportunity (usually as a collaboration between universities) outside the country where the student is completing a degree.	Parkinson (2007), Berger and Bailey (2013), Klahr and Ratti (2000)
Internship, work placement	The learning environment is an authentic workplace setting that usually forms part of the curricular activities with associated learning objectives. The student is expected to observe, participate in, and complete tasks usually with supervision and/or mentorship.	Winberg et al. (2011)
Sports	Sports and athletic activities within or outside the higher education institution.	Muñoz-Bullón et al. (2017), Miller and Hoffman (2009)

and practitioners to study and leverage their potential for competency development and broadened engagement.

2.4 Informal Learning in the Workplace

The transition of engineering graduates into the workplace requires the recontextualization of the knowledge, skills, competencies, and practices that students acquire. Graduates are often described as underprepared for the demands and expectations of the workplace, particularly in recent research on the employability of graduates (for example, Trevelyan, 2019). Although many attempts have been made to close the gap between university undergraduate engineering curricula and practice, these efforts have had limited success (Trevelyan, 2019).

Work placement programs are a strategy for helping prepare engineering students to transition into the workplace after graduation. While such programs are offered by many universities around the world, Eraut (2004) makes the point that "it is usually the work that is structured and not the learning" (p. 247). Despite the complexities of the workplace aligning with outcomes of the formal curriculum, there are many studies (e.g., Jackson, 2013) reporting the value of work-integrated learning for development of the competencies and skills appropriate for the employability.

The "messy, complex, everyday complexities of work" (Dean et al., 2012, p. 11) provide valuable opportunities for informal learning. Such learning, according to Eraut (2004), can include *deliberative learning*, which is planned and intentional; *reactive learning*, which, "although . . . is intentional, occurs in the middle of the action, when there is little time to think" (Eraut, p. 250); and *implicit learning*, which occurs independently of conscious efforts to learn. Ngonda et al. (2022) identify factors that could facilitate or constrain student learning in an authentic work environment, which include the student's organizational environment, the type and scope of work allocated to the student, the availability of industry mentors, and self-efficacy and agency.

Although what is learned and how it is learned may be less predictive than in the formal curriculum, the workplace has the potential to provide unparalleled opportunities for the development of knowledge, competencies, and skills appropriate for engineering practice.

2.5 Student Participation in Informal Learning

During the higher education experience, formal learning comprises a small portion of students' time. In the USA, the National Survey of Student Engagement (NSSE) collects annual data at hundreds of universities regarding how first-year and senior (final-year) students participate in activities and spend their time (NSSE, 2020). Engineering students spend on average 19 hours per week preparing for class, with 42% of seniors reporting spending more than 20 hours per week, which is higher than students in other majors (NSSE, 2011). Outside of academic activities, engineering students spend an average of 6 hours per week on co-curricular activities (NSSE, 2011). The NSSE provides some data on the specific activities in which students are participating. Table 15.2 displays data from 281,136 first-year and senior (final-year) students from 491 universities in the USA in 2019. Large-scale quantitative research on engineering undergraduate student engagement in informal learning is limited, but additional data on engineering student participation in informal learning activities can be found in work by Wilson and colleagues (2014) and Simmons and colleagues (Simmons et al., 2018; Simmons, Ye, et al., 2018).

Table 15.2 is provided to show engineering student participation within out-of-class activities that are commonly offered at higher education institutions in the USA. A similar survey was conducted in Europe through the European Student Engagement Project (STEP), which examined how students engage inside and outside higher education curricula, with a focus on the development of transversal skills (European STEP, 2019). The survey included sports, peer mentoring, law clinics,

Table 15.2 Engineering Student Participation in Co-Curricular Activities

Activity	*Student Participation (%)*	
	First-Year	*Senior (Final-Year)*
Internship/co-op	-	48
Study abroad	-	14
Service learning	53	60
Research	5	22
Living learning community	13	22

artistic/cultural activities, student unions, and student associations as activities within the higher education institution that are considered part of student engagement.

South Africa similarly conducts a national survey, the South African Survey of Student Engagement (SASSE), adapted from the NSSE (SASSE, 2021). Survey data is collected from first-year and senior (final-year) students to measure engagement based on four themes, namely, academic challenge, learning with peers, experiences with staff, and the campus environment. In addition, students report their participation in 15 co-curricular activities that are seen to have high impact for learning (Kuh, 2008). In this survey, engineering is included in a category with science and technology, and this overall group reports the highest rates of participation in service learning, peer learning support, and working with students (e.g., group work) of all student categories.

2.6 Cultural and National Context

Opportunities for informal learning are not uniform across institutional contexts. Whereas higher education institutions in the USA leave room for student choice within the formal curriculum (providing several open electives in a student's graduation requirements, as well as general education requirements that allow for free choice within specified themes such as history, humanities, or languages), the curricula in other parts of the world may not leave much, or indeed any, free choice of modules for students.

In Europe, for instance, the Bologna process standardized the first (ordinary bachelor's) engineering degree into a three-year program, which is technically focused and affords the student little to no room for self-selection in the curriculum. Europe did, however, implement a wide-scale Erasmus program that facilitates credit transfer across European universities, thus enabling some students to study outside their home countries. Efforts are underway to align university curricula across the continent, through university alliances and a European universities initiative, which "will enable students to obtain a degree by combining studies in several EU countries and contribute to the international competitiveness of European universities" (O'Malley, 2021, para. 10). These programs provide a somewhat-higher level of flexibility for students, at the formal level.

In the USA, residential campuses (where most students live on or near campus) typically provide a range of informal, extracurricular, and co-curricular activities. This is also true of residential campuses in South Africa. Likewise, in the UK, "three quarters of students are classified as 'movers' or students who study away from their parental/guardian home, with the average student choosing to travel 91 miles for their university education" (Chipperfield, 2019, para. 1). To continue attracting students to make this commitment of time and money, Chipperfield (2019) explains, universities in the UK and elsewhere "are looking at their offer more holistically – ensuring a structured academic curriculum alongside an informal education program with a focus on developing skills, social events and a large range of sports activities." Residence life activities seek to build a sense of community and learning outside the formal classroom and have been an important part of universities' move toward informal learning in the UK, USA, and beyond.

In places where students do not reside on or near their campuses but rather continue living with their families or commuting from elsewhere, they may engage in paid employment or in community activities more than in university-sanctioned clubs and societies. Moreover, in non-Western parts of the world, informal learning may take forms not defined in this chapter. The format and type of learning may fall outside Anglicized or American perspectives and definitions. Even in Western places, opportunities to engage in "living learning communities" and professional "sororities and fraternities" may be limited. Yet individuals may still learn and develop engineering-related competencies by participating in other activities. In such scenarios, work placements, competitions, and service projects may be integrated into the required curriculum rather than offered as elective modules or optional activities.

In South Africa, elective modules and work placements are classified as part of the formal engineering curricula (ECSA, 2022), with a range of optional opportunities provided by both the university and external organizations. Available options include service-learning, competitions, and mentoring programs. Work placements are also integrated in some universities in Australia (Blicblau et al., 2016), the UK (Tennant et al., 2018), Ireland (GTI Futures Ltd., 2022), and elsewhere.

Across these activities and settings, project-based learning (PBL) can be employed. In engineering education today, much attention is paid to studying the process and outcomes of PBL. A systematic review conducted in Denmark by Chen et al. (2021) analyzed 108 empirical research papers on PBL implementation that were published between 2000 and 2019. Chen et al. found that this active learning format was being used and reported in the literature at four different levels. The authors called one of these the "project level," explaining that it is conducted outside the required curriculum. This level could also be labelled "co-curricular" or "extracurricular," as these characteristics distinguish it from PBL provided at the other levels (individual course, set of courses, or program curriculum). This example illustrates that scholars outside the USA may be using terms besides "informal learning" to describe similar or related concepts.

Although the systematic review by Chen et al. (2021) did not mention "informal learning" (Chen, 2022), it stated that "during the professional socialization process [for engineers], students could have opportunities to interact with peers, including in-team collaboration, after-class communication, and other formal or informal interactions" (p. 18). This highlights that the term *informal learning* may not be as prevalent outside the USA, and that scholars in other places may be studying associated issues but using other keywords (like PBL). Apparently, PBL research may be more common outside the USA than inside it, considering that the systematic review of PBL implementation conducted by Chen et al. (2021) identified 27 relevant publications from inside the USA, with 81 originating elsewhere. Researchers elsewhere may be more concerned with the group-based and hands-on aspects of learning than with considering if the activities accrue credit or not.

It is also worth considering that, because important informal learning may be happening completely outside the academic environment, educators may not notice the many ways they might harness the power of, or connect to, these nonacademic environments in ways that support students' informal learning of engineering. It is crucial, therefore, to broaden our understanding of what counts as learning, since these opportunities outside of the classroom can be formative in students' engineering knowledge and socialization.

3 Benefits and Outcomes of Informal Learning

Research on education has historically focused on formal learning. However, empirical and theoretical work in the past few decades has illuminated the important interplay between formal and informal learning in achieving desired outcomes. Informal learning via out-of-class activities supports a range of outcomes, which Kuh (1993) organized into five factors, based on a qualitative study of 149 seniors from 12 USA institutions. These five factors involve (1) **practical competence**, which includes self-management and contribution to society; (2) **personal competence**, which includes self-awareness, autonomy, confidence, social competence, and purpose; (3) **cognitive complexity**, which describes reflective thought and knowledge application; (4) **knowledge and academic skills**, which relate to the acquisition and valuation of skills; and (5) **altruism and estheticism**, which entail awareness of others and the ability to collaborate.

Notably, such benefits are not uniform or universal. In fact, Wilson and colleagues (2014) pointed out that evidence is mixed regarding the impact of co-curricular participation on academic outcomes. It is thus important to consider the type of informal learning, the targeted outcome, and the disciplinary and demographic characteristics of the students. In engineering education, learning outside of the classroom has emerged as an opportunity to develop nontechnical and professional competencies that are desired by employers but afforded limited space in the curriculum.

3.1 Persistence

Learning occurs within and beyond class time as students are involved inside and outside the classroom. Astin's (1984) theory of student involvement posits that the more energy a student puts into learning, the higher the learning gains will be – and that gains are in direct proportion to the quality of the effort expended. One implication of this theory is that participation in extracurricular activities contributes to the decision to persist through university (Astin, 1984). A systematic literature review on persistence of transfer students found student integration, including learning communities and campus involvement, was a key factor (Smith & Van Aken, 2020). Although the effects of informal learning in engineering often center on competency development and workforce preparation, as detailed in the following, it is important to note the benefits are also being realized during the undergraduate experience in helping students progress toward degree attainment.

3.2 Development of Nontechnical and Professional Competencies

Engineering practice is constantly evolving in response to technological, environmental, and societal changes. To keep pace with these changes and to anticipate future needs, engineers are expected to demonstrate a broad range of competencies. The past few decades have ushered in growing recognition that engineers need to develop professional, interpersonal, and intrapersonal skills to work effectively on international and interdisciplinary teams, account for the societal context of engineering solutions, design for a range of stakeholders, communicate with various audiences, and make ethical decisions (ABET, 2018; International Engineering Alliance, 2013; National Academy of Engineering, 2004).

However, the broadening of engineering competence has been accompanied by the tightening of curricular space. Engineering programs around the USA are reducing their credit hours to make the engineering degree more manageable in four years (Williamson & Fridley, 2017), and as we have noted, in some parts of the world, the engineering bachelor's is condensed into just three years. This has created a squeeze for programs to offer the requisite courses and to achieve and document the student outcomes mandated by accrediting agencies. With a finite number of hours in the curriculum, informal learning has emerged as a more flexible opportunity for developing the competencies and experiences that engineering students need to be workforce-ready. Informal learning can provide forms of engagement and cultivate competencies that are otherwise limited in the curriculum (Garrett et al., 2021).

The following subsections focus on three competencies (leadership, ethics, and communication) that are highly demanded in the engineering workforce but have been found to receive insufficient attention in the engineering curriculum (Bodmer et al., 2002; Grant & Dickson, 2006).

3.2.1 Leadership

Within engineering, leadership is recognized as a crucial competency for individual advancement, organizational innovation, and societal problem-solving (Klassen et al., 2020). Recent scholarship has highlighted informal learning as a way for students to develop and practice leadership skills. Activities outside of the classroom provide different forms of engagement that can support leadership development through experience (Knight & Novoselich, 2017). For example, participation in design competition teams can support students' identity development as engineering leaders, by providing them experience with shared decision-making, peer coaching, and task management on complex projects (Wolfinbarger et al., 2021).

A survey of engineering faculty members and administrators revealed consensus among respondents that leadership is most effectively developed in extracurricular activities (Novoselich & Knight, 2014). Civil engineering students similarly reported attaining leadership through out-of-class activities, especially female and first-generation students when compared to male and continuing generation students, respectively (Polmear et al., 2021a). With a focus on situated and self-driven learning, context, and application, informal settings provide opportunities for students to engage in the process and development of leadership skills while supplementing formal leadership instruction in the classroom.

3.2.2 Ethics

Engineers are expected to protect public welfare and demonstrate professional responsibility (National Society of Professional Engineers, 2019). The undergraduate experience is instrumental to professional ethical development, as future engineers are equipped with the skills and values of the profession. On a global scale, accreditation has served as a powerful lever for the integration of ethics in the formal engineering curriculum (ABET, 2018; International Engineering Alliance, 2013).

Despite significant growth in ethics education and research over the past few decades, research with engineering faculty members (Polmear et al., 2019), industry employers, and alumni (McGinn, 2003) has indicated that instruction in ethics is insufficient. Structural factors, such as limited curricular space, the assumed dichotomy of social and technical realms (i.e., sociotechnical dualism), and cultural norms, which include the marginalization of nontechnical skills and the educators who teach them, have challenged the integration of ethics in the curriculum (Martin & Polmear, 2022; Newberry, 2004; Polmear et al., 2018).

Most instruction and research have focused on the formal curriculum (Hess & Fore, 2018), but the quantity and quality of engineering students' participation in co-curricular activities also contribute to their ethical development (Finelli et al., 2012), and ethics can be an outcome of out-of-class engagement (Polmear et al., 2021b). Undergraduate engineering students reported exposure to ethical decision-making through co-curricular activities (Burt et al., 2011), and project-based, informal learning in Engineers Without Borders helped students develop ethical responsibility (Lee et al., 2017). Engineering educators have also reported that students in their program learn about ethics via co-curricular activities (Bielefeldt et al., 2020).

3.2.3 Communication

Communication is another learning outcome expected by accreditation agencies and a professional skill desired by employers. Engineering students have attributed communication skill development to co-curricular activities (Kovalchuk et al., 2017). Engineering students who participated in undergraduate research grants and projects reported higher communication skills compared to students with similar backgrounds and experiences who did not participate in these types of research activities (Carter et al., 2016). The social context of informal learning facilitates communication through opportunities to work with others and articulate ideas.

3.3 Broaden Engagement of Diverse Learners

Calls to increase the number of engineering graduates and diversify the engineering profession are often accompanied by recognition of the need to attract and retain students from demographic groups that have been traditionally underrepresented in engineering (Chubin et al., 2005; National Academy of Sciences, 2002). Participation in informal learning is one strategy to support engagement, persistence, and competence for all engineering students and can be particularly impactful for underrepresented students (Simmons, Van Mullekom, et al., 2018; Polmear et al., 2021a). However, the outcomes and benefits are not evenly distributed due to challenges with access and equity in informal learning (Bell et al., 2009).

There can be structural and cultural barriers to participation in informal learning. "Learning begets learning" (Noy et al., 2016, p. 56); thus, inequities in access to informal learning starting in pre-K to fifth-grade informal learning (Bell et al., 2009) can continue to be compounded through higher education. Lack of time as well as schedule and cost were the reasons most selected by undergraduate engineering students for not participating in out-of-class activities (Simmons, Ye, et al., 2018).

The broad view of participation described in the overview section does not, therefore, capture the variations and nuances across the population of engineering students. Prior research has shown that demographics influence students' involvement in co-curricular activities, and that there are differences across the type of activity and level of involvement. Among engineering students, women have reported higher engagement in out-of-class activities relative to men (Millunchick et al., 2021) and greater participation in living learning communities, fraternities/sororities related to their engineering field of study, service, international experiences, identity-based organizations, and engineering outreach support (Simmons, Van Mullekom, et al., 2018).

Research has shown that female students may find engineering more appealing when it has clear or explicit social or environmental relevance (Du & Kolmos, 2009; Kolmos et al., 2013), and the interest in social and environmental relevance is likely to influence a student's motivations to engage with informal learning as well. For example, engagement by women in Engineers Without Borders is twice as high as in engineering education as a whole (Amadei & Sandekian, 2010).

Students from low-income families are less involved in activities outside of class than their peers (Simmons, Ye, et al., 2018), and first-generation college students are less likely to be involved than continuing generation students (Simmons & Chau, 2021). However, recent work by Millunchick and colleagues (2021) indicates that demographic factors may be only part of the story. Participation in co-curricular activities can be predicted by a combination of utilizing proactive behaviors (including general socializing behavior and feedback-seeking behavior) and of knowledge of higher education systems (e.g., having university preparatory experience, family ties to the university, or relatives who studied at university and understand how to identify, name, and navigate university systems, opportunities, and support structures). In the study by Millunchick and colleagues, participation in design competition teams and professional societies was best predicted by proactive behavior – while participation in research was best predicted by a combination of demographics and knowledge of higher education.

Another consideration related to access and equity is the climate within the informal learning setting, which can support or impede inclusion. For example, design competition teams are one of the most common co-curricular activities among engineering students in the USA (Wilson et al., 2014). However, design competition teams can represent homogenous environments dominated by White males in which cultural and structural factors contribute to systematic exclusion (Walden et al., 2015). Women have felt discouraged from participating due to perceiving gendered stereotypes and disregard for their contributions (Foor et al., 2013).

The culture is also exclusionary for students who must work or commute and thus cannot fulfill the "pervasive ethos of commitment" within these teams (Foor et al., 2013, p. 18). On the other hand, informal learning in out-of-class settings can contribute to the academic success and persistence of groups traditionally underrepresented in engineering. For example, participation in the Society of Hispanic Professional Engineers provided a sense of community, family-like connection, and mentorship that supported Latino/a/x students' engineering identity development (Revelo Alonso, 2015). Inconclusive research findings regarding the benefits of informal learning in different environments and for different groups suggest the need for further research to understand these contextual variations and increase access and inclusion for all students.

4 Considerations for Practice

This section provides considerations and recommendations for practitioners to facilitate informal learning in ways that realize its value related to competency development and broadened engagement.

4.1 Designing Informal Learning Opportunities

Regardless of the specific setting or activity, there are guiding recommendations for informal learning (Bell et al., 2009). The environment should be developed for specific learning objectives, be interactive, provide different ways to engage, encourage learners to draw on their past knowledge and experience, and stimulate lifelong learning. More purposeful and effective use of informal learning environments can support the shift in engineering education away from traditional delivery-focused transmission-of-content models toward constructivist approaches, which focus on the student, rather than content, and leverage how students learn through social interactions and past experiences (Hein, 1991).

These more innovative constructivist approaches use active, collaborative, and increasingly informal learning to help improve the analytical, problem-solving, technical, and collaboration skills needed to solve contemporary engineering challenges (Chang et al., 2009). Moreover, engaging with engineering outside formal curricula allows students to "'experience engineering' in an authentic environment" (Kotys-Schwartz et al., 2011, p. 1) and to develop crucial competencies.

4.2 Designing Informal Learning Spaces

Built environments can also support various types of learning and serve as "third teachers" for immersive and experiential learning. Campos Calvo-Sotelo (2010) asserts that universities must be designed to support the social and psychological development of students as well as their intellectual growth. The author's principles for planning an educational campus include helping people bond with the place and with each other – thus fostering a community of learning by stimulating personal contact – with spaces and buildings serving multiple functions, to bring disparate factors together.

Such places also should promote the psychological well-being of community members by providing spatial, emotional, and intellectual harmony (Campos Calvo-Sotelo, 2010). Campus spaces can expose people to nature and art and provide lessons in sustainability, having to do with geography, climate, and biodiversity (Chance, 2010, 2012; Fox, 2007). Moreover, campus designs can help

Table 15.3 Learning Modes vs. Categories of Spaces Matrix

	Presentation	*Seminar*	*Brainstorm*	*Study*	*Simulation*	*Contemplative*	*Social*
Amphitheater	✓✓	X	✓	-	XX	-	✓
Library	XX	XX	X	✓✓	X	✓	-
Learning space	-	-	✓✓	✓✓	✓✓	-	✓✓
Simulation space	✓	✓	✓	✓✓	✓✓	✓✓	-
Green spaces	XX	XX	✓✓	✓✓	XX	✓✓	✓✓
Reflection space	XX	XX	X	XX	XX	✓✓	X
Café and cafeteria	XX	XX	✓✓	✓	XX	✓✓	✓✓
Atrium	-	-	✓✓	✓	-	✓✓	✓✓
Circulation	XX	XX	✓	X	X	✓✓	✓✓
Iconic place	X	X	✓✓	✓	X	✓✓	✓✓

Legend: *XX*, not recommended; *X*, unsuitable; –, suitable; ✓, recommended; ✓✓, highly recommended.
Source: Content adapted from Carreira and Heitor (2014).

tie members to larger social, cultural, and political contexts, and they can encourage increasingly innovative modes of learning and teaching.

In a follow-up to the work by Campos Calvo-Sotelo (2010), Carreira and Heitor (2014) investigated how the design of spaces on university campuses affects learning. They focused specifically on how spaces influence social interactions related to acquiring, transmitting, generating, and sharing knowledge. Carreira and Heitor developed a matrix for evaluating the learning supported by various types of spaces (see Table 15.3). Informal learning that involves a mix of contemplation and social interaction, these researchers found, can be best supported in the purposeful design of green spaces, cafés and cafeterias, atriums, circulation spaces, and iconic places on campus.

From the field of engineering education research, Chang et al. (2009) reported a preliminary study regarding the use and benefits of campus spaces at the University of Melbourne. The spaces under investigation were designed to support the learning of information by enabling student-centered and small-group learning. One of the ways that planners achieved this was by providing casual-feeling bar- or café-style seating. Communal spaces were arranged for small groups rather than individual learners and designed with attention to light, color, density of activity, provision of electric sockets, pervasive Wi-Fi, and the like. These communal spaces were placed close to the students' formal classrooms, and they were made available for students' use around the clock via swipe-card access.

In similar fashion, Chance and Cole (2014, 2019) provide a case study of techniques used in one K–12 school district in the USA to build, teach, and operate with environmental sustainability at the core. Readers interested in learning more about designing learning spaces are further directed to Fraser (2014) and Strange and Banning (2001). Fraser provides a reference book on the next generation of learning spaces. Strange and Banning consider the role of design and space (physical environments) as well as humans and organizational environments to foster success, promote safety and inclusion, and build a community of learners.

4.3 Recognizing Student Participation in Informal Learning

As described in the section "Benefits and Outcomes of Informal Learning," engineering students develop a range of competencies via out-of-class learning that contribute to their undergraduate experience and workforce preparation. However, since this learning is outside the curriculum, it is often not recognized or assessed. Although students can include activities, organizations, and jobs on their résumés and curriculum vitae, specific skills they acquire are typically not reflected in formal documents or academic transcripts. As a result, it can be difficult for employers to evaluate applicants' competencies. One approach that has emerged to address this challenge is a co-curricular or experiential transcript, an electronic documentation of student participation in learning outside the curriculum (Parks & Taylor, 2016).

Another system that has gained traction is microcredentialing, the validation of skills gained through learning activities that are linked to workforce demands (European Commission, 2022). Microcredentials can be earned by students and practitioners alike; they provide evidence of learning and/or achieving specified outcomes by way of short courses or modules that are transparently assessed (Ruddy & Ponte, 2019). Typically, a certification or digital "badge" is conferred on those who successfully complete the course. The certificates that one can earn via LinkedIn Learning and subsequently post to one's LinkedIn profile illustrate the popularity of relatively new microcredentialing programs (Du, 2021). Readers are referred to Chapter 16 within this handbook on non-degree credentials for a detailed account of how such credentials can shape education, employment, and equity within engineering.

Practitioners and programs should consider how to recognize competencies gained through informal learning to leverage the value of these learning opportunities and support students in their employability.

5 Considerations for Research

This section provides considerations and recommendations for researchers to examine informal learning in ways that realize its value related to competency development and broadened engagement. The section begins with theoretical perspectives that have been historically employed to understand students' outcomes and experiences and framework that may aid future investigations. The section also offers areas for future research. Research on assessment in informal learning has been highlighted as scarce (Kotys-Schwartz et al., 2011) and is important to understand students' competency development. Mental health is also highlighted since it is a growing area of focus in engineering education generally, and more work is needed to understand the effect of informal learning on mental health.

5.1 Theoretical Perspectives

Various theoretical perspectives have been developed and applied to understand students' experiences and outcomes in university education, both inside (formal) and outside (informal) the classroom. Research on students in university often follows development or impact approaches (Kuh, 1993).

Development approaches describe discrete, and somewhat-linear, developmental stages during which changes in the cognitive, affective, and behavioral domains are examined. They are rooted in psychology and emphasize intrapersonal, rather than environmental, influences. Kuh (1993) cites Baxter Magolda's (1992) study of co-curricular influences on cognitive development as an example of this approach.

On the other hand, impact approaches emphasize the interaction between the individual and the environment to explain outcomes associated with the university experience. Foundational research using this approach includes Astin's (1984, 1993) input-environment-output (I-E-O) model, with a focus on student involvement. In addition, Tinto (1987) applied the impact approach with a focus on social and academic integration to develop a framework for understanding students' decisions to depart from university. Weidman (1989) extended the I-E-O model to conceptualize undergraduate student socialization.

These models, and more recent work that has employed and revised them (e.g., Terenzini & Reason, 2005), share four basic elements: student characteristics, institutional characteristics, student interactions with faculty members and peers, and interactions with the academic environment (Pascarella, 1985). Within the impact approach, informal learning is conceptualized as part of the academic environment or organizational context and peer environment. Recent scholarship in engineering education has employed this approach to understand the impact of informal learning. Millunchick and colleagues (2021) used Weidman's (1989) model of socialization to examine undergraduate engineering students' participation in co-curricular activities based on their pre-university preparation and knowledge of how higher education systems work, proactive behavior, and demographics. As another example, Lee and Matusovich (2016) employed Tinto's model of departure to develop a conceptual model of co-curricular support for undergraduate engineering students. Knight and Novoselich oriented their study of curricular and co-curricular influences on undergraduate engineering students' leadership in Terenzini and Reason's (2005) I-E-O model.

Engagement is another theoretical lens for examining informal learning; it draws in part on Astin's theory of involvement (1984) and Tinto's theory of integration (1987). Engagement broadly links activities and experiences within higher education to their outcomes. Engagement research, and the work from which it builds, demonstrates that the impact of higher education is dependent on students' efforts and involvement in formal/curricular and informal/co-curricular opportunities (Pascarella & Terenzini, 2005).

Engagement serves as the conceptual framework for the National Survey of Student Engagement (NSSE) (National Survey of Student Engagement, 2013) used in the USA and replicated elsewhere. The resulting student survey, the College Student Report, collects data from first-year (freshman) and final-year (senior) students in the USA and Canada on how they spend their time and what they gain during their undergraduate experience. Since 2000, six million students from 1,650 institutions have participated in the survey, providing a profile of student engagement inside and outside the classroom.

Applications of this theory in the context of engineering education research have demonstrated the contribution of faculty members to student engagement via experiences in-class and out-of-class (Chen et al., 2008), the link between co-curricular participation and academic engagement (Wilson et al., 2014), the role of communities outside the classroom in shaping student engagement and outcomes (Allendoerfer et al., 2012), the development of ethics as an outcome of out-of-class engagement (Polmear et al., 2021b), and the attainment of outcomes across a range of out-of-class activities and student demographic groups (Simmons, Van Mullekom, et al., 2018). Research on informal learning through the lens of engagement indicates the importance of students' efforts and involvement associated with their outcomes, while also accounting for the different types of activities and characteristics of students, thus providing more granularity.

Another theoretical perspective relevant to informal learning in engineering education is "situative learning." The foundation of this theory is that all learning happens in a particular place, at a given time, and thus, knowledge develops through a social context (Johri & Olds, 2011). The situative perspective is applied to informal learning because it emphasizes active participation, community membership, and student self-direction (Newstetter & Svinicki, 2013). Johri and colleagues (2016) detailed three analytical features of situative learning and their implications for informal learning: (1) **the social and material context** accounts for the tools and representations in a setting and their contribution to learning; (2) **activities and interactions** describe the teamwork and modality of informal learning that shape students' engagement; and (3) **participation and identity** examine the role of community and participation in identity formation. Situative learning has been used to frame community and service-learning, design competitions, and internships (Newstetter & Svinicki, 2013).

This brief overview of theoretical perspectives relevant to informal learning captures the various ways in which the higher education experience has been conceptualized in terms of how students navigate the university experience and what they gain as a result. Broadening the lens to outside the classroom puts into focus the myriad ways students learn and develop during the undergraduate experience and the outcomes they attain.

5.2 Areas for Future Research

Given the piecemeal development of literature in informal learning (Bell et al., 2009), the number of activities and settings that fall within informal learning, and the range of outcomes that can be developed, there are many directions for future research. To continue realizing the potential of informal learning for competency development and broadened engagement, the following sections highlight assessment and mental health.

5.2.1 Assessment

Despite the growing visibility and popularity of informal engineering settings, "little research has been conducted to actually define what constitutes appropriate content for informal learning models or to assess the degree to which these informal experiences impact students" (Kotys-Schwartz et al., 2011, p. 1). There is evidence that informal learning activities provide skills and experiences that support engineering students' competency development and their transitions into the workforce (Kovalchuk et al., 2017). However, research in this space often relies either upon students' self-reported skills and gains or upon internal evaluations of program objectives.

Based on an extensive review of informal learning in engineering, Kotys-Schwartz and colleagues (2011) concluded there is a dearth of validated tools for assessing gains and benefits. Furthermore, informal learning can be tacit and unintentional, and it can occur in settings out of the reach of traditional assessment. Competency development in engineering happens in a complex social system, which contributes to "accidental competency," in which this wider context, including informal learning outside of the classroom, affects students' professional formation (Walther et al., 2011).

Assessment thus remains a challenge, and there is a need for instruments and methods to capture the short- and long-term impacts of informal learning across a range of settings (Noy et al., 2016). Given both the existing evidence on the benefits of informal learning and the broad range of activities that fall within informal learning, assessment is a promising future direction of research. Scholars should consider the interpersonal, intrapersonal, and professional competencies that students gain by participating, while accounting for the context in which these competencies are being developed, and for which groups of students.

Assessment and evaluation are critical components in broadening the participation of students who have been historically underrepresented and marginalized in engineering (Holloman et al., 2021). Such assessment can determine the effectiveness of informal learning programs and interventions and their impact on aims, such as recruiting and retaining diverse students or improving the experience of diverse undergraduates. A literature review on assessment in engineering education related to broadening participation indicated a focus on K–12 programs while identifying a need to plan outcomes, collect data, and implement change (Holloman et al., 2021).

Another question in assessing informal learning is what constitutes informal learning and what activities and settings are valued within academia and the workforce. It is important to consider how the value of informal learning opportunities is being weighted, both formally through evaluation and informally through messaging students receive. The latter connects to the hidden curriculum: the tacit lessons and attitudes that students learn related to what they should value and how they should behave (Hafferty, 1998; Villanueva et al., 2018); the hidden curriculum is further discussed in Chapter 18 of this handbook. Future research could explore the messages that students receive related to if and how they should participate in informal learning, as this could help illuminate the role of informal learning in engineering education and identify what is enabling or impeding students' engagement.

5.2.2 Mental Health

Mental health is a growing area of attention and scholarship in engineering education. For example, Jensen and Cross (2021) found high self-reported levels of stress, anxiety, and depression among engineering students in the USA and noted the relationship between inclusion, engineering identity, and mental health. A review of literature on engineering graduate students' mental health found the important role that social support, faculty member interaction, and belonging play in mental health; the authors called for additional research in this area (Bork & Mondisa, 2022).

Participation in co-curricular activities has been associated with subjective well-being, which describes individuals' general satisfaction and emotional state (Hossan et al., 2021). Extracurricular engagement has also been found to support well-being and belonging (Winstone et al., 2022). Although the studies by Hossan et al. and Winstone et al. examined to what extent students participated in such activities, future work could untangle the multiple informal learning environments to understand their potential effect on mental health while also accounting for access and impact for students from diverse backgrounds.

6 Conclusions and Future Directions

Engineering education scholars are constantly examining how to better prepare graduates for workforce demands and societal needs. Challenges related to sustainability, technological development, and employability (Hadgraft & Kolmos, 2020) and calls for DEI have highlighted the importance of developing competencies to address these challenges and to provide pathways for diverse learners, which have resulted in needing to look outside the formal curriculum to educate future engineers. Informal learning offers non-didactive, collaborative, and contextual opportunities for students to learn – opportunities that are driven by students' interest (Callanan et al., 2011). Most research on informal learning has focused on science and K–12 education, but the interplay between undergraduate engineering education and informal learning is critical.

Research in higher education demonstrates the need to examine the impact of the undergraduate experience holistically by accounting for learning in both formal and informal settings. By better understanding the diversity of informal learning environments, the ways they are experienced by students, and the outcomes students gain, educators can design better, more effective ties between formal and informal learning. The heterogeneity of informal learning activities and settings available today is a strength that can be built upon, but students do not have equal access to these opportunities. There is a need to broaden participation in informal learning among those who have already joined engineering – and to use informal learning to help attract more, and increasingly diverse, participants to join engineering.

There is also a clear need to build awareness and understanding of the many structural and cultural barriers that prevent access and to recognize the compounding effect that existing inequities cause. Students with the highest levels of social and economic capital are also those who can afford the time away from family obligations and paid employment, which will let them spend time and money in clubs and other optional learning activities.

Looking at issues of equity and inclusion and recognizing that much of the research on informal learning comes from the USA and other English-speaking countries, there is a need to expand our definitions and understandings of informal learning. Engineering education can benefit from research that asks students in a diversity of national locations, cultural contexts, and demographic groups about out-of-class activities they engage in, where they apply, build, gain, or develop their engineering skills, knowledge, and values. It is likely educators and researchers are not identifying or recognizing the full range of settings.

Informal learning represents an opportunity to prepare current and future generations of engineers for 21st-century challenges by cultivating the requisite competencies and engaging students with a range of backgrounds and experiences. Research and practice can leverage the benefits of informal learning, helping design effective and inclusive learning activities, examining outcomes across settings and learners, and extending our understanding of what counts as learning.

References

ABET. (2018). *Criteria for accrediting engineering programs, 2019–2020 | ABET*. www.abet.org/accreditation/accreditation-criteria/criteria-for-accrediting-engineering-programs-2019-2020/

Ainsworth, H. L., Eaton, S. E., & Clydesdale, J. (2010). *Formal, non-formal and informal learning in the sciences*. Onate Press. https://central.bac-lac.gc.ca/.item?id=sciences&op=pdf&app=Library

Allendoerfer, C., Wilson, D., Bates, R., Crawford, J., Jones, D., Floyd-Smith, T., Plett, M., Scott, E., & Veilleux, N. (2012). Strategic pathways for success: The influence of outside community on academic engagement. *Journal of Engineering Education, 101*(3), 512–538. https://doi.org/10.1002/j.2168-9830.2012.tb00060.x

Amadei, B., & Sandekian, R. (2010). Model of integrating humanitarian development into engineering education. *Journal of Professional Issues in Engineering Education and Practice, 136*(2), 84–92. https://doi.org/10.1061/(ASCE)EI.1943-5541.0000009

Astin, A. W. (1984). Student involvement: A developmental theory for higher education. *Journal of College Student Personnel, 25*(4), 297–308.

Astin, A. W. (1993). *Assessment for excellence: The philosophy and practice of assessment and evaluation in higher education*. ORIX PRESS.

Baber, L. D. (2015). Considering the interest-convergence dilemma in STEM education. *The Review of Higher Education, 38*(2), 251–270.

Baxter Magolda, M. B. (1992). Cocurricular influences on college students' intellectual development. *Journal of College Student Development, 33*(3), 203–213.

Bell, P., Lewenstein, B., Shouse, A. W., & Feder, M. A. (2009). *Learning science in informal environments: People, places, and pursuits* (Vol. 140). National Academies Press.

Berger, E., & Bailey, R. (2013). Designing short-term study abroad engineering experiences to achieve global competencies. In *2013 ASEE international forum proceedings* (pp. 21.19.1–21.19.24). https://doi.org/10.18260/1-2-17224

Bielefeldt, A. R., Lewis, J., Polmear, M., Knight, D., Canney, N., & Swan, C. (2020). Educating civil engineering students about ethics and societal impacts via cocurricular activities. *Journal of Civil Engineering Education, 146*(4), 04020007. https://doi.org/10.1061/(ASCE)EI.2643-9115.0000021

Blicblau, A. S., Nelson, T. L., & Dini, K. (2016). The role of work placement in engineering students' academic performance. *Asia-Pacific Journal of Cooperative Education, 17*(1), 31–43.

Bodmer, C., Leu, A., Mira, L., & Rütter, H. (2002). Successful practices in international engineering education. In *SPINE final report, Benchmarking Study*, Zurich.

Bork, S. J., & Mondisa, J.-L. (2022). Engineering graduate students' mental health: A scoping literature review. *Journal of Engineering Education, 111*(3), 665–702. https://doi.org/10.1002/jee.20465

Burt, B., Carpenter, D., Finelli, C., Harding, T., Sutkus, J., Holsapple, M., Bielby, R., & Ra, E. (2011). Outcomes of engaging engineering undergraduates in co-curricular experiences. In *2011 ASEE annual conference & exposition proceedings* (pp. 22.1132.1–22.1132.11). https://doi.org/10.18260/1-2-18498

Callanan, M., Cervantes, C., & Loomis, M. (2011). Informal learning. *WIREs Cognitive Science, 2*(6), 646–655. https://doi.org/10.1002/wcs.143

Campos Calvo-Sotelo, P. (2010). The concept of "Educational Campus" and its application in Spanish universities. OECD publications. https://doi.org/10.1787/5kmbjxzp5ghc-en

Carreira, M. B., & Heitor, T. (2014). Learning spaces in the university context. In *The 7th knowledge cities world summit* (p. 254).

Carter, D. F., Ro, H. K., Alcott, B., & Lattuca, L. R. (2016). Co-curricular connections: The role of undergraduate research experiences in promoting engineering students' communication, teamwork, and leadership skills. *Research in Higher Education, 57*(3), 363–393. https://doi.org/10.1007/s11162-015-9386-7

Chance, S. (2012). Planning for environmental sustainability: Learning from LEED and the USGBC. *Planning for Higher Education, 41*(1).

Chance, S., & Cole, J. T. (2014). Designing buildings to enhance performance and learning. In T. C. Chan, K. E. Lane, & M. D. Richardson (Eds.). *Marketing the green school: Form, function, and the future*. Advances in Educational Marketing, Administration, and Leadership (AEMAL) Book Series. IGI Global.

Chance, S., & Cole, J. T. (2019). Designing buildings to enhance performance and learning. In *Architecture and design: Breakthroughs in research and practice*. "Critical Explorations" Research anthology series. IGI Global.

Chance, S. M. (2010). *University leadership in energy and environmental design: How postsecondary institutions use the LEED® green building rating system* [PhD dissertation, The College of William and Mary, Williamsburg, VA].

Chang, R. L., Stern, L., Sondergaard, H., & Hadgraft, R. G. (2009, January). Places for learning engineering: A preliminary report on informal learning spaces. In *Proceedings of the Research in engineering education symposium, Palm Cove, QLD* (Vol. 12, p. 2009). Retrieved December.

Chen, H., Kusimo, A., & Cardenas-Navia, I. (2023). Chapter 16: Enabling a skilled and diverse engineering workforce with non-degree credentials. In A. Johri (Ed.), *International handbook of engineering education research*. Routledge.

Chen, H. L., Lattuca, L. R., & Hamilton, E. R. (2008). Conceptualizing engagement: Contributions of faculty to student engagement in engineering. *Journal of Engineering Education, 97*(3), 339–353. https://doi.org/10.1002/j.2168-9830.2008.tb00983.x

Chen, J. (2022). *Professional identity development for engineering students in a PBL context* [PhD dissertation, Aalborg University, Denmark].

Chen, J., Kolmos, A., & Du, X. (2021). Forms of implementation and challenges of PBL in engineering education: A review of literature. *European Journal of Engineering Education, 46*(1), 90–115.

Chipperfield, C. (2019). *What are Residence Life programmes and why do they matter?* Retrieved October 17, 2022, from https://wonkhe.com/blogs-sus/what-are-residence-life-programmes-and-why-do-they-matter/

Chubin, D. E., May, G. S., & Babco, E. L. (2005). Diversifying the engineering workforce. *Journal of Engineering Education, 94*(1), 73–86. https://doi.org/10.1002/j.2168-9830.2005.tb00830.x

Clinedinst, M., & Koranteng, A. M. (2017). *State of college admission*. National Association for College Admission Counseling.

Dean, B. Sykes, C., & Turbill, J. *'So, what did you do?' A performative, practice-based approach to examining informal learning in WIL 2012* (pp. 1–13). Retrieved July 26, 2022, from https://ro.uow.edu.au/commpapers/2780

Demetry, C., & Sontgerath, S. (2020). Longitudinal evaluation of a university-based outreach program for middle school girls yields evidence of positive engineering recruitment outcomes. *Journal of Women and Minorities in Science and Engineering, 26*(1), 30.

Denson, C. D., Hailey, C., Stallworth, C. A., & Householder, D. L. (2015). Benefits of informal learning environments: A focused examination of STEM-based program environments. *Journal of STEM Education, 16*(1), 11–15.

Du, X. (2021). *Embedding LinkedIn learning MOOCs to enhance students' educational experience and employability*. https://doi.org/10.34190/EEL.20.082

Du, X., & Kolmos, A. (2009). Increasing the diversity of engineering education – a gender analysis in a PBL context. *European Journal of Engineering Education, 34*(5), 425–437.

ECSA (Engineering Council of South Africa). (2022, May 18). *Policy on accreditation of engineering programmes E-01-POL REVISION 1*. www.ecsa.co.za/ECSADocuments/Shared%20Documents/E-01-POL%20%20Policy%20on%20Accreditation%20of%20Engineering_Programmes.pdf.

Ehsan, H., Ohland, C., Dandridge, T., & Cardella, M. (2018). Computing for the critters: Exploring computational thinking of children in an informal learning setting. In *2018 IEEE frontiers in education conference (FIE)* (pp. 1–5). https://doi.org/10.1109/FIE.2018.8659268

Eraut, M. (2004). Informal learning in the workplace. *Studies in Continuing Education, 26*(2), 247–273. https://doi.org/10.1080/158037042000225245

Espinosa, L. (2011). Pipelines and pathways: Women of color in undergraduate STEM majors and the college experiences that contribute to persistence. *Harvard Educational Review, 81*(2), 209–241. https://doi.org/10.17763/haer.81.2.92315ww157656k3u

European Commission. (2022). *A European approach to micro-credentials | European Education Area.* https://education.ec.europa.eu/node/1558

European STudent Engagement Project. (2019). *How do European higher education institutions recognise student engagement?* https://uni-foundation.eu/uploads/2019_European_STEP_Preliminary_Report.pdf

Evans, M. D., Evans, D. M., & Sherman, L. D. (2001). Seven keys to a successful ASCE student chapter or club: Guide for student leaders and faculty advisors. *Journal of Professional Issues in Engineering Education and Practice, 127*(2), 65–74. https://doi.org/10.1061/(ASCE)1052-3928(2001)127:2(65)

Falk, J. H., & Dierking, L. D. (2016). *The museum experience revisited.* Routledge.

Finelli, C., Holsapples, M., Ra, E., Bielby, R., Burt, B., Carpenter, D., Harding, T., & Sutkus, J. (2012). An assessment of engineering students' curricular and co-curricular experiences and their ethical development. *Journal of Engineering Education, 101*(3), 469–494.

Fisher, D. R., Bagiati, A., & Sarma, S. (2017). Developing professional skills in undergraduate engineering students through cocurricular involvement. *Journal of Student Affairs Research and Practice, 54*(3), 286–302. https://doi.org/10.1080/19496591.2017.1289097

Foor, C. E., Walden, S. E., Shehab, R. L., & Trytten, D. A. (2013). "We weren't intentionally excluding them . . . just old habits": Women, (lack of) interest and an engineering student competition team. *2013 IEEE Frontiers in Education Conference (FIE)*, 349–355. https://doi.org/10.1109/FIE.2013.6684846

Foor, C. E., Walden, S., Trytten, D., & Shehab, R. (2013). "You choose between TEAM A, good grades, and a girlfriend – You get to choose two!" – How a culture of exclusion is constructed and maintained in an engineering design competition team. *2013 ASEE Annual Conference & Exposition Proceedings*, 23.18.1–23.18.22. https://doi.org/10.18260/1-2-19027

Fox, W. (2007). Human relationships, nature, and the built environment: Problems that any general ethics must be able to address. In J. Pretty, A. S. Ball, T. Benton, J. S. Guivant, D. R. Lee, D. Orr, M. J. Pfeffer, & H. Ward (Eds.), *The SAGE handbook of environment and society* (pp. 107–123). SAGE.

Fraser, K. (2014). The future of learning and teaching in next generation learning spaces. In *The future of learning and teaching in next generation learning spaces* (Vol. 12, pp. xv–xxiv). Emerald Group Publishing Limited.

Froyd, J. E., & Lohmann, J. R. (2014). Chronological and ontological development of engineering education as a field of scientific inquiry. In A. Johri & B. M. Olds (Eds.), *Cambridge handbook of engineering education research* (1st ed., pp. 3–26). Cambridge University Press. https://doi.org/10.1017/CBO9781139013451.003

Garrett, S. D., Martin, J. P., & Adams, S. G. (2021). Developing nontechnical professional skills in African American engineering majors through co-curricular activities. *IEEE Transactions on Education*, 8.

Griffin, J. (1998). Learning science through practical experiences in museums. *International Journal of Science Education, 20*(6), 655–663.

Grant, C. D., & Dickson, B. R. (2006). Personal skills in chemical engineering graduates: The development of skills within degree programmes to meet the needs of employers. *Education for Chemical Engineers, 1*(1), 23–29. https://doi.org/10.1205/ece.05004

GTI Futures Ltd. (2022). *How to get a work experience placement in engineering.* Retrieved October 17, 2022, from https://gradireland.com/careers-advice/engineering/how-get-work-experience-placement-engineering

Hadgraft, R. G., & Kolmos, A. (2020). Emerging learning environments in engineering education. *Australasian Journal of Engineering Education, 25*(1), 3–16. https://doi.org/10.1080/22054952.2020.1713522

Hafferty, F. W. (1998). Beyond curriculum reform: Confronting medicine's hidden curriculum. *Academic Medicine: Journal of the Association of American Medical Colleges, 73*(4), 403–407. https://doi.org/10.1097/00001888-199804000-00013

Hein, G. E. (1991). *Constructivist learning theory.* Exploratorium. www.exploratorium.edu/education/ifi/constructivist-learning

Hess, J. L., & Fore, G. (2018). A systematic literature review of us engineering ethics interventions. *Science and Engineering Ethics, 24*(2), 551–583. https://doi.org/10.1007/s11948-017-9910-6

Holloman, T. K., Lee, W. C., London, J. S., Ash, C. D. H., & Watford, B. A. (2021). The assessment cycle: Insights from a systematic literature review on broadening participation in engineering and computer science. *Journal of Engineering Education, 110*(4), 1027–1048.

Hossan, Md. R., Khan, M. H. A., & Hasan, Md. T. (2021). Co-curricular activities and subjective well-being among university students. *SN Social Sciences*, *1*(8), 207. https://doi.org/10.1007/s43545-021-00214-6

International Engineering Alliance. (2013). *Graduate attributes and professional competencies*. www.ieagreements.org/assets/Uploads/Documents/Policy/Graduate-Attributes-and-Professional-Competencies.pdf

Jackson, D. (2013). The contribution of work-integrated learning to undergraduate employability skill outcomes. *Asia-Pacific Journal of Cooperative Education*, *14*(2), 99–115.

Jagušt, T., Botički, I., & So, H.-J. (2018). A review of research on bridging the gap between formal and informal learning with technology in primary school contexts. *Journal of Computer Assisted Learning*, *34*(4), 417–428. https://doi.org/10.1111/jcal.12252

Jensen, K. J., & Cross, K. J. (2021). Engineering stress culture: Relationships among mental health, engineering identity, and sense of inclusion. *Journal of Engineering Education*, *110*(2), 371–392. https://doi.org/10.1002/jee.20391

Johri, A., Bland, L., & Kusano, S. (2016). Informal learning in engineering. In *2016 ASEE annual conference & exposition proceedings* (p. 25687). https://doi.org/10.18260/p.25687

Johri, A., & Olds, B. M. (2011). Situated engineering learning: Bridging engineering education research and the learning sciences. *Journal of Engineering Education*, *100*(1), 151–185. https://doi.org/10.1002/j.2168-9830.2011.tb00007.x

Klahr, S. C., & Ratti, U. (2000). Increasing engineering student participation in study abroad: A study of U.S. and European programs. *Journal of Studies in International Education*, *4*(1), 79–102. https://doi.org/10.1177/102831530000400106

Klassen, M., Reeve, D., Evans, G. J., Rottmann, C., Sheridan, P. K., & Simpson, A. (2020). Engineering: Moving leadership from the periphery to the core of an intensely technical curriculum. *New Directions for Student Leadership*, *2020*(165), 113–124. https://doi.org/10.1002/yd.20373

Knight, D. B., & Novoselich, B. J. (2017). Curricular and co-curricular influences on undergraduate engineering student leadership: Influences on engineering student leadership. *Journal of Engineering Education*, *106*(1), 44–70. https://doi.org/10.1002/jee.20153

Kolmos, A., Mejlgaard, N., Haase, S., & Holgaard, J. E. (2013). Motivational factors, gender and engineering education. *European Journal of Engineering Education*, *38*(3), 340–358.

Kotys-Schwartz, D., Besterfield-Sacre, M., & Shuman, L. (2011). Informal learning in engineering education: Where we are – Where we need to go. *2011 Frontiers in Education Conference (FIE)*, T4J-1-T4J-7. https://doi.org/10.1109/FIE.2011.6142836

Kovalchuk, S., Ghali, M., Klassen, M., Reeve, D., & Sacks, R. (2017). Transitioning from university to employment in engineering: The role of curricular and co-curricular activities. In *2017 ASEE annual conference & exposition proceedings* (p. 29043). https://doi.org/10.18260/1-2-29043

Kuh, G. D. (1993). In their own words: What students learn outside the classroom. *American Educational Research Journal*, *30*(2), 277–304. https://doi.org/10.2307/1163236

Kuh, G. D. (2008). *High-impact educational practices: What they are, who has access to them, and why they matter*. Association of American Colleges and Universities.

Kuh, G. D., Cruce, T. M., Shoup, R., Kinzie, J., & Gonyea, R. M. (2008). Unmasking the effects of student engagement on first-year college grades and persistence. *The Journal of Higher Education*, *79*(5), 540–563. https://doi.org/10.1080/00221546.2008.11772116

Lee, T., Berg, D., & Buchanan, E. (2017). Board # 85: Exploring, documenting, and improving humanitarian service learning through engineers without borders USA. In *2017 ASEE annual conference & exposition proceedings* (p. 27940). https://doi.org/10.18260/1-2-27940

Lee, W. C., & Matusovich, H. M. (2016). A model of co-curricular support for undergraduate engineering students: Model of co-curricular support. *Journal of Engineering Education*, *105*(3), 406–430. https://doi.org/10.1002/jee.20123

Litchfield, K., Javernick-Will, A., & Maul, A. (2016). Technical and professional skills of engineers involved and not involved in engineering service. *Journal of Engineering Education*, *105*(1), 70–92. https://doi.org/10.1002/jee.20109

Maltby, J. L., Brooks, C., Horton, M., & Morgan, H. (2016). Long term benefits for women in a science, technology, engineering, and mathematics living-learning community. *Learning Communities Research and Practice*, *4*(1), 19.

Martin, D., & Polmear, M. (2022). Chapter 8: The two cultures of engineering education: Looking back and moving. In S. H. Christensen, A. Buch, E. Conlon, C. Didier, C. Mitcham, & M. Murphy (Eds.), *Engineering, social science, and the humanities: Has their conversation come of age?* Springer Nature.

Martin, L., & Betser, S. (2020). Learning through making: The development of engineering discourse in an out-of-school maker club. *Journal of Engineering Education, 109*(2), 194–212. https://doi.org/10.1002/jee.20311

McGinn, R. E. (2003). "Mind the gaps": An empirical approach to engineering ethics, 1997–2001. *Science and Engineering Ethics, 9*(4), 517–542. https://doi.org/10.1007/s11948-003-0048-3

Meyers, E. M., Erickson, I., & Small, R. V. (2013). Digital literacy and informal learning environments: An introduction. *Learning, Media and Technology, 38*(4), 355–367. https://doi.org/10.1080/17439884.2013.783597

Miller, K. E., & Hoffman, J. H. (2009). Mental well-being and sport-related identities in college students. *Sociology of Sport Journal, 26*(2), 335–356.

Millunchick, J. M., Brennan-Wydra, E., Henderson, T., Johnson, A., & Finelli, C. J. (2021). The role of college knowledge and proactive behavior on participation in cocurricular activities. *Journal of Engineering Education, 110*(1), 114–142. https://doi.org/10.1002/jee.20380

Mtawa, N., Fongwa, S., & Wilson-Strydom, M. (2019). Enhancing graduate employability attributes and capabilities formation: A service-learning approach. *Teaching in Higher Education, 26*, 1–17. https://doi.org/10.1080/13562517.2019.1672150

Muñoz-Bullón, F. (2017). The influence of sports participation on academic performance among students in higher education. *Sport Management Review*, 15.

National Academies of Sciences, Engineering, and Medicine. (2018). *Measuring the 21st century science and engineering workforce population: Evolving needs*. The National Academies Press. https://doi.org/10.17226/24968

National Academy of Engineering. (2004). *The engineer of 2020: Visions of engineering in the new century*. The National Academies Press.

National Academy of Sciences. (2002). *Diversity in engineering: Managing the workforce of the future*. National Academy Press. www.nap.edu/read/10377/chapter/1

National Society of Professional Engineers. (2019). *Code of ethics for engineers*. https://www.nspe.org/sites/default/files/resources/pdfs/Ethics/CodeofEthics/NSPECodeofEthicsforEngineers.pdf

National Survey of Student Engagement. (2011). *Fostering student engagement campuswide – Annual results 2011*. Indiana University Center for Postsecondary Research. https://files.eric.ed.gov/fulltext/ED527089.pdf

National Survey of Student Engagement. (2013). *NSSE's conceptual framework*. Indiana University Center for Postsecondary Research. https://nsse.indiana.edu/nsse/about-nsse/conceptual-framework/index.html

National Survey of Student Engagement. (2020). *Engagement insights: Survey findings on the quality of undergraduate education- annual results 2019*. Indiana University Center for Postsecondary Research. https://scholarworks.iu.edu/dspace/bitstream/handle/2022/25321/NSSE_2019_Annual_Results.pdf?sequence=1&isAllowed=y

Newberry, B. (2004). The dilemma of ethics in engineering education. *Science and Engineering Ethics, 10*(2), 343–351. https://doi.org/10.1007/s11948-004-0030-8

Newstetter, W. C., & Svinicki, M. D. (2013). Learning theories for engineering education practice. In A. Johri & B. M. Olds (Eds.), *Cambridge handbook of engineering education research* (pp. 29–46). Cambridge University Press. https://doi.org/10.1017/CBO9781139013451.005

Ngonda, T., Shaw, C., & Kloot, B. (2022). Perceived influence of mechanical engineering students' work placement experiences on their occupational competency and self-efficacy. *International Journal of Mechanical Engineering Education, 50*(1), 197–216.

Novoselich, B., & Knight, D. (2014). Developing engineers who lead: Are student, faculty, and administrator perspectives aligned? In *2014 ASEE annual conference & exposition proceedings* (pp. 24.401.1–24.401.17). https://doi.org/10.18260/1-2-20292

Noy, M. V., James, H., & Bedley, C. (2016). *Reconceptualizing learning: A review of the literature on informal learning* (p. 84). Rutgers, the State University of New Jersey.

O'Malley, B. (2021). *Challenges for 'European Universities' revealed by survey – University World News*. Retrieved October 17, 2022, from https://eutopia-university.eu/english-version/news/on-the-media/university-world-news-challenges-for-%E2%80%98european-universities-revealed-by-survey#:~:text=The%20European%20Universities%20consist%20of,according%20to%20the%20European%20Commission.

Parkinson, A. (2007). Engineering study abroad programs: Formats, challenges, best practices. In *2007 Annual conference & exposition proceedings* (pp. 12.653.1–12.653.17). https://doi.org/10.18260/1-2-1706

Parks, R., & Taylor, A. (2016). Innovative credentialing: Employers weigh in on co-curricular transcripts. *College and University, 91*(2), 63.

Pascarella, E. T. (1985). Students' affective development within the college environment. *The Journal of Higher Education, 56*(6), 640–663. https://doi.org/10.2307/1981072

Pascarella, E. T., & Terenzini, P. T. (2005). *How college affects students: A third decade of research* (Vol. 2). Jossey-Bass.

Polmear, M., Bielefeldt, A. R., Knight, D., Canney, N., & Swan, C. (2019). Analysis of macroethics teaching practices and perceptions in engineering: A cultural comparison. *European Journal of Engineering Education, 44*(6), 866–881. https://doi.org/10.1080/03043797.2019.1593323

Polmear, M., Bielefeldt, A. R., Knight, D., Swan, C., & Canney, N. (2018). Faculty perceptions of challenges to educating engineering and computing students about ethics and societal impacts. In *2018 ASEE annual conference & exposition proceedings* (p. 30510). https://doi.org/10.18260/1-2-30510

Polmear, M., Chau, A. D., & Simmons, D. R. (2021a). Intersection of diversity, out-of-class engagement, and engineer of 2020 outcomes for civil engineering students. *Journal of Management in Engineering, 37*(4), 04021019. https://doi.org/10.1061/(ASCE)ME.1943-5479.0000901

Polmear, M., Chau, A. D., & Simmons, D. R. (2021b). Ethics as an outcome of out-of-class engagement across diverse groups of engineering students. *Australasian Journal of Engineering Education, 26*(1), 64–76. https://doi.org/10.1080/22054952.2020.1836752

Revelo Alonso, R. A. (2015). *Engineering familia: The role of a professional organization in the development of engineering identities of Latina/o undergraduates.* University of Illinois. www.ideals.illinois.edu/handle/2142/78420

Rogoff, B., Callanan, M., Gutiérrez, K. D., & Erickson, F. (2016). The organization of informal learning. *Review of Research in Education, 40*(1), 356–401. https://doi.org/10.3102/0091732X16680994

Ross, M., & McGrade, S. (2016). An exploration into the impacts of the national society of black engineers (NSBE) on student persistence. In *2016 ASEE annual conference & exposition proceedings* (p. 27280). https://doi.org/10.18260/p.27280

Ruddy, C., & Ponte, F. (2019). Preparing students for university studies and beyond: A micro-credential trial that delivers academic integrity awareness. *Journal of the Australian Library and Information Association, 68*(1), 56–67. https://doi.org/10.1080/24750158.2018.1562520

SASEE (South African Survey of Student Engagement) (2021). *South African survey of student engagement: University of Cape Town, Institutional Report (2021).* University of the Free State, Centre for Teaching and Learning.

Seely, B. E. (1999). The other re-engineering of engineering education, 1900–1965. *Journal of Engineering Education, 88*(3), 285–294. https://doi.org/10.1002/j.2168-9830.1999.tb00449.x

SEFI. (2016). *SEFI position paper on developing graduate engineering skills.* SEFI.

Shulman, L. S. (2005). Signature pedagogies in the professions. *Daedalus, 134*(3), 52–59.

Simmons, D. R., & Chau, A. D. (2021). Factors that predict participation in out-of-class activities for STEM students. *Journal of STEM Education: Innovations and Research, 22*(1). https://jstem.org/jstem/index.php/JSTEM/article/view/2444

Simmons, D. R., Creamer, E. G., & Yu, R. (2017). Involvement in out-of-class activities: A mixed research synthesis examining outcomes with a focus on engineering students. *Journal of STEM Education: Innovations and Research, 18*(2). https://jstem.org/jstem/index.php/JSTEM/article/view/2238

Simmons, D. R., Van Mullekom, J., & Ohland, M. W. (2018). The popularity and intensity of engineering undergraduate out-of-class activities. *Journal of Engineering Education, 107*(4), 611–635. https://doi.org/10.1002/jee.20235

Simmons, D. R., Ye, Y., Ohland, M. W., & Garahan, K. (2018). Understanding students' incentives for and barriers to out-of-class participation: Profile of civil engineering student engagement. *Journal of Professional Issues in Engineering Education and Practice, 144*(2), 04017015. https://doi.org/10.1061/(ASCE)EI.1943-5541.0000353

Smith, N. L., & Van Aken, E. M. (2020). Systematic literature review of persistence of engineering transfer students. *Journal of Engineering Education, 109*(4), 865–883. https://doi.org/10.1002/jee.20357

Strange, C. C., & Banning, J. H. (2001). *Educating by design: Creating campus learning environments that work.* The Jossey-Bass Higher and Adult Education Series. Jossey-Bass.

Swan, C., Paterson, K., & Bielefeldt, A. R. (2013). Community engagement in engineering education as a way to increase inclusiveness. In A. Johri & B. M. Olds (Eds.), *Cambridge handbook of engineering education research* (pp. 357–372). Cambridge University Press. https://doi.org/10.1017/CBO9781139013451.023

Tan, L. M., Laswad, F., & Chua, F. (2021). Bridging the employability skills gap: Going beyond classroom walls. *Pacific Accounting Review*, *34*(2), 225–248. https://doi.org/10.1108/PAR-04-2021-0050

Tennant, S., Murray, M., Gilmour, B., & Brown, L. (2018). Industrial work placement in higher education: A study of civil engineering student engagement. *Industry and Higher Education*, *32*(2), 108–118. https://doi.org/10.1177/0950422218756384

Terenzini, P. T., & Reason, R. D. (2005). Parsing the first year of college: A conceptual framework for studying college impacts. In *Annual meeting of the association for the study of higher education*, Philadelphia, PA.

Tinto, V. (1987). *Leaving college: Rethinking the causes and cures of student attrition*. University of Chicago Press, 5801 S.

Trevelyan, J. (2019). Transitioning to engineering practice. *European Journal of Engineering Education*, *44*(6), 821–837. https://doi.org/10.1080/03043797.2019.1681631

Trinder, K., Guiller, J., Margaryan, A., Littlejohn, A., & Nicol, D. (2008). *Learning from digital natives: Bridging formal and informal learning*. The Higher Education Academy.

Villanueva, I., Carothers, T., Di Stefano, M., & Khan, M. T. H. (2018). "There is never a break": The hidden curriculum of professionalization for engineering faculty. *Education Sciences*, *8*(4), 157. https://doi.org/10.3390/educsci8040157

Villanueva Alarcón, I., Sellers, V., Paul, R. M., & Smith, B. (2023). Chapter 18: Transforming engineering education through social capital in response to hidden curriculum. In A. Johri (Ed.), *International handbook of engineering education research*. Routledge.

Walden, S., Foor, C., Pan, R., Shehab, R., & Trytten, D. (2015). Leadership, management, and diversity: Missed opportunities within student design competition teams. In *2015 ASEE annual conference and exposition proceedings* (pp. 26.1059.1–26.1059.23). https://doi.org/10.18260/p.24396

Walther, J., Kellam, N., Sochacka, N., & Radcliffe, D. (2011). Engineering competence? An interpretive investigation of engineering students' professional formation. *Journal of Engineering Education*, *100*(4), 703–740. https://doi.org/10.1002/j.2168-9830.2011.tb00033.x

Weidman, J. (1989). *Undergraduate socialization: A conceptual approach* (Vol. 5, pp. 289–322).

Wilczynski, V. (2015). Academic maker spaces and engineering design. In *2015 ASEE annual conference and exposition proceedings* (pp. 26.138.1–26.138.19). https://doi.org/10.18260/p.23477

Williamson, D., & Fridley, K. (2017). A survey of credit hour requirements in BS civil, electrical, and mechanical engineering ABET accredited programs. In *2017 ASEE annual conference & exposition proceedings* (p. 27522). https://doi.org/10.18260/1-2-27522

Wilson, D., Jones, D., Kim, M. J., Allendoerfer, C., Bates, R., Crawford, J., Floyd-Smith, T., Plett, M., & Veilleux, N. (2014). The link between cocurricular activities and academic engagement in engineering education. *Journal of Engineering Education*, *103*(4), 625–651. https://doi.org/10.1002/jee.20057

Winberg, C., Engel-Hills, P., Garraway, J., & Jacobs, C. (2011). *Work-integrated learning: Good practice guide – HE Monitor No. 12*. Council on Higher Education (CHE).

Winstone, N., Balloo, K., Gravett, K., Jacobs, D., & Keen, H. (2022). Who stands to benefit? Wellbeing, belonging and challenges to equity in engagement in extra-curricular activities at university. *Active Learning in Higher Education*, *23*(2), 81–96. https://doi.org/10.1177/1469787420908209

Wolfinbarger, K. G., Shehab, R. L., Trytten, D. A., & Walden, S. E. (2021). The influence of engineering competition team participation on students' leadership identity development. *Journal of Engineering Education*, jee.20418. https://doi.org/10.1002/jee.20418

Young, J. R., Ortiz, N., & Young, J. L. (2016). STEMulating interest: A meta-analysis of the effects of out-of-school time on student STEM interest. *International Journal of Education in Mathematics, Science and Technology*, *5*(1), 62. https://doi.org/10.18404/ijemst.61149

16

Enabling a Skilled and Diverse Engineering Workforce with Non-Degree Credentials

Helen L. Chen, Abisola C. Kusimo, and Isabel Cardenas-Navia

1 Credentials in Engineering

Over the last decade, there has been a proliferation of different kinds of non-degree credentials beyond the traditional degree, including certificates, certifications, and licenses. These credentials differ in who awards them, what learning or skills they recognize, time to completion, accreditation criteria, and their longevity or revocation process (Corporation for a Skilled Workforce, George Washington Institute of Public Policy, & Workcred, 2020; Workcred, 2021). Online learning has not only increased access to training and professional skill development but also widened the constellation of educational providers beyond academic institutions (see Gregg & Dabbagh, Chapter 22, this volume). Increase in informal learning opportunities in engineering have further made the issue of credentialing important within the field (see Polmear et al. Chapter 15, this volume).

The 2020 Strada-Gallup Survey of a nationally representative sample of approximately 14,000 students documents the growing prevalence of non-degree credentials in the US workforce; 40% of working adults have completed a non-degree credential or education program, compared with 46% who have completed a college degree. Overall, 20% of adults report a non-degree credential as their highest level of education, which is more than double the number of adults that report an associate degree as their highest level of education. Approximately one-third of the survey respondents have a professional license or certification, and one-fifth have earned an educational certificate. Demographic analysis showed that non-degree credentials are being awarded to individuals from all different educational backgrounds (Strada Center for Education Consumer Insights, 2021).

In the United States, the science, technology, engineering, and mathematics (STEM) workforce includes over 36 million people who hold jobs that require STEM knowledge and expertise and represent nearly a quarter of the total US workforce. STEM education in a variety of forms provides workers with the necessary knowledge, skills, and abilities needed to contribute to the STEM workforce (National Science Board, 2021).

Community colleges are often sought out as a pathway into a four-year institution as well as an entry point into the science and engineering (S&E) workforce by offering associate's degrees and, increasingly, non-degree credentials, such as certificates which may be offered independently or embedded within the degree program. As students make progress towards their degree – even if they do not complete it – they will be better positioned to demonstrate and reflect the industry-relevant skills that they acquired through the certification. In 2019, community colleges awarded 258,000

DOI: 10.4324/9781003287483-19

certificates in science and engineering technologies, highlighting how non-degree credentials are being positioned as postsecondary education programs which can support individuals to earn a first credential (National Science Board, 2021).

A growing number of non-degree credentials are also being used to supplement the training and education of individuals who already hold a four-year degree (Columbus, 2019). The number of non-degree credentials awarded in engineering is likely to expand since National Science Board numbers do not account for the additional certificates, certifications, badges, and licenses issued annually by professional societies, certification bodies, state licensure boards, and companies. In the engineering workforce, non-degree credentials play distinct roles by providing value that can be distinguished from an engineering degree.

1.1 Relevance to the Current Engineering Educator Ecosystem

Engineering was first established as an educational program in the United States in 1802. It was modeled after the first engineering schools in France, for example, the "establishment of the US Military Academy at West Point in 1802 . . . was to train an elite of engineers; it was modeled on the French Ecole Polytechnique, founded a few years earlier in 1797" (Collins, 2019, p. 217). From its earliest formalizations as a field of both learning and work, engineering has been elusive to most people. Women, people of color, lower-income, and first-generation degree seekers were often excluded, which is still true for many undergraduate engineering programs today: 24% of enrolled undergraduate engineering students identified as female, 45% as non-White, and nearly 10% as part-time students (ASEE, 2021). Similarly, for those in engineering occupations, Pew Research Center (2021) found that 15% of those in engineering jobs in the United States identified as female and 29% as non-White.

Non-degree credentials have the potential to change this dynamic by offering more inclusive pathways to the engineering workforce relative to degrees which may lead to greater representation and accessibility for a wider group of individuals. According to the US Census Bureau's 2019 American Community Survey of 50 million employed college graduates ages 25 to 64, of the 37% who reported holding a bachelor's degree in science or engineering, only 14% worked in a STEM occupation (Cheeseman Day & Martinez, 2021). While engineering continues to have one of the highest persistence rates at the bachelor's level and computer and information sciences at the associate degree level, improvement in the overall persistence and retention rates for the 2.3 million people entered college for the first time in fall 2020 are still below pre-pandemic levels and uneven across institution sectors (National Student Clearinghouse Research Center, 2022).

Engineering educators are uniquely positioned to support diverse learners across their academic and professional trajectories at a moment when the US Bureau of Labor Statistics (2022a) has estimated that the demand for engineering graduates is expected to grow 6% from 2020 to 2030 and a projection of 146,000 new jobs will be added. However, the retirement patterns and aging of the science and engineering workforce are well documented (National Science Board, 2014, 2016) and will need to be addressed, particularly in the context of other risk factors, including limitations in the ability to recruit and retain undergraduate students (e.g., Seymour & Hewitt, 1997; Seymour & Hunter, 2019), women (Fouad et al., 2020; Frehill, 2012), and other talent from abroad (National Academy of Sciences, 2003).

1.2 Global Implications of Non-Degree Credentials

The United States is the most credentialed society in the world (Collins, 2019), and therefore, understanding how non-degree credentials operate in the United States and facilitate mobilization into engineering programs and the workforce has important implications for non-US contexts (Gaston & Van Noy, 2022).

Diaz et al. (2022) examined the shift in the Americas from degrees to non-degree credentials (or alternative credentials) and identified the strengths of non-degree credentials. They note that the wage and employment benefits seen in non-degree credentials in the United States have also been demonstrated in Latin America. Further, they argue that certain occupations are more open to non-degree credentials than others, including electro-technology engineers, engineers (a variety of specialties), and architects, planners, and designers. These occupations are likely to value skills over credentials, which can be signaled by non-degree credentials as well as degrees.

For example, in regions of the Global South,[1] where many educational institutions are run by the government and industry's confidence in graduates' practical technical skills is low, non-degree credentials can have a large impact in bridging the gap between education and employment. Learners can earn non-degree credentials to signal specific skills, earned technical competencies, as well as professional development and employability skills.

In this context, non-degree credentials may better facilitate the placement of engineering graduates into technical work by complementing their current training and providing employers with more quality checkpoints beyond the standard unitary degree or certificate. Nevertheless, future work will need to explore how engineering employers in different countries interpret non- degree credentials, assign value to novel non-degree credentials, how non-degree credentials impact hiring and promotional decisions, as well as how traditional higher education institutions cooperate with non-degree-credentialing organizations. A growing number of scholars are exploring the relationships among the global engineering education curriculum and experiences, credentials, and hiring practices (Diaz et al., 2022; Kusimo & Sheppard, 2019; Matemba & Lloyd, 2017; Trevelyan, 2014).

2 Overview of Non-Degree Credentials

Non-degree credentials in engineering are often stacked, embedded, or aligned with degree programs. They may be offered by educational programs at colleges and universities to supplement and enhance their educational offerings for both degree-seeking and non-degree-seeking learners. The professional engineer (PE) license is perhaps the most widely used model, where a four-year bachelor's degree is a prerequisite along with the requirement to pass an assessment and professional practice.

Non-degree credentials which are "stacked" this way are designed to meet the needs of students, employers, and communities by (1) providing greater flexibility for students; (2) addressing the evolving and emerging needs of employers and potentially strengthening partnerships between academia and industry; (3) increasing access to postsecondary education and persistence to a degree or credential, particularly for underserved populations; and (4) providing another means for colleges and universities to respond to technology advancements that may require additional education and training (US Department of Education, Office of Career, Technical, and Adult Education, 2021).

Table 16.1 provides an overview of the different types of credentials often applied in engineering, the types of organizations that issue them, and other distinguishing characteristics. The credentials listed in Table 16.1 – degrees, licenses, certificates, and certifications – are well-established, and there is typically consensus on their main distinguishing characteristics. For example, certifications are third-party, independent competency assessments related to an occupation. They are typically time-limited and renewable, indicating that their holders are current in their mastery of competencies represented by the certification.

Certificates are much more variable as they can be awarded for multiple reasons, including participation, completion of a series of courses, or mastery of competencies. Many higher education institutions award certificates (in addition to degrees), which also vary widely in the number of courses needed for completion.

Two other non-degree credentials – badges and microcredentials – are still being defined in different ways by different credential issuers and therefore were not included in Table 16.1. For the purposes of this chapter, badges are defined as either (1) digital representations of an already-awarded credential, that is, a badge is issued for a certificate earned for an online training course; or (2) original digital credential recognizing skills, achievement, or learning.

Similarly, for the purposes of this chapter, a *microcredential* is defined as a non-degree credential recognizing a smaller set of skills, achievement, or learning than other non-degree credentials issued by an organization. Therefore, a microcredential for a certification body would represent a narrower

Table 16.1 Characteristics of Different Types of Credentials for Engineering

	Degrees	*License*	*Certificate**	*Certification*
Awarded by	Education institutions (colleges and universities)	Government agencies	Education and training providers, employers, labor unions, and industry associations	Industry certification bodies
Awarded for	Course of study	Meeting requirements of an occupation	An exam at the end of a training or education course or a onetime assessment	Third-party, independent competency assessment
Indicates	Education, successfully passed courses	Legal permission	Education/ knowledge/skills	Skill mastery/ competencies
Time to complete	Variable, generally 2 years or more	Variable	Variable, generally less than 2 years	Variable
Time and renewal requirements	No time limit, no renewal requirement	Time-limited, renewal generally required	Often no time limit, no renewal requirement	Time-limited, includes recertification
Revocation process	Cannot be revoked	Can be revoked for incompetence or unethical behavior	Cannot be revoked	Can be revoked for incompetence or unethical behavior
Examples	Bachelor of Science in Engineering, Associate's Degree in Engineering Technology	Professional Engineer License	Amazon Web Services Certified DevOps Engineer Professional; IEEE Continuing Education Course Certificates	Certified Systems Engineering Professionals; Project Management Professionals
Standards for accreditation	National, regional, or programmatic (ABET)	State law defines scope of practice	ANSI/ASTM E2659–18, a globally recognized American National Standard	ANSI/IS/IEC 17024:2012, an international and national standard

* There are many types of certificates. Some examples include certificates of participation, certificates of achievement, certificates of completion for apprenticeships, and assessment-based certificates.
Source: Adapted from Workcred (2021).

set of competencies than a certification, whereas a microcredential for a university would represent mastery of fewer learning outcomes than a certificate.

These different characteristics can support learners and workers in different ways. Table 16.1 also reflects the expansion of the credentialing landscape within academia and industry, presumably in response to evolving learner needs.

2.1 Issuers of Non-Degree Credentials

Non-degree credentials in engineering are issued by a wide variety of organizations (see Table 16.2): higher education institutions issue badges, certificates, and microcredentials; companies issue badges, certificates, certifications, and microcredentials; professional associations issue badges, certificates, certifications, and microcredentials; state agencies issue licenses; and other nonprofit and for-profit providers issue badges, certificates, certifications, and microcredentials.

As reflected in Table 16.2, non-degree credentials in engineering are issued by a wide variety of organizations. Notably, higher education institutions (colleges and universities, 18%; community colleges, 12%) issue fewer than half of all non-degree credentials awarded (Strada Center for Education Consumer Insights, 2021).

Approximately half of degree holders have also been awarded a non-degree credential. Nearly half of working-age adults with a college degree have combined a non-degree credential with a bachelor's (47%) or associate's (51%) degree (Strada Center for Education Consumer Insights, 2021). Individuals benefit from holding both non-degree credential(s) and degree(s). Looking more closely at the STEM workforce, for those workers in S&E-related occupations, certifications and licenses are held by over half of those with a bachelor's degree (69%) and those without (53%) (National Science Board, 2021). In comparison to people with only a college degree, those with college degrees and non-degree credentials reported stronger agreement that their combined education helped them achieve their goals, was worth the cost, and made them more attractive job candidates. Consistent with human capital theory (Becker, 1993), individuals who hold a certification and/or a license are more likely to be employed, earn more, and believe they have a good job or career compared with someone who does not – at every educational level (Strada Education Network, Gallup, Inc., and Lumina Foundation, 2020).

2.2 Important Functions of Non-Degree Credentials in Engineering

More specifically, non-degree credentials in engineering provide at least five potential functions to their holders (detailed in the section that follows), with many providing more than one: signaling specialized skills within a degree program, signaling technical competency, signaling interdisciplinary and out-of-degree specialization, signaling professional development, and enabling non-degree

Table 16.2 Type of Non-Degree Credentials Most Commonly Awarded by Type of Institution

Credential/Institution	*Badges*	*Microcredentials*	*Certificates*	*Certifications*	*Licenses*
Higher education	+	+	+		
Companies	+	+	+	+	
Professional associations	+	+	+	+	
State agencies					+
Nonprofit and for-profit providers	+	s	+	+	

pathways to engineering careers. These functions are realized independently of a degree and represent knowledge, skills, and abilities which are relevant to specific engineering occupations.

2.2.1 Signaling Specialized Skills Within a Degree Program

Non-degree credentials can signal that their holder has developed specialized skills within a degree program. In a report investigating how microcredentials could improve engineering education in New Zealand, non-degree credentials in engineering education can be thought of as "at least, a complement to more substantial qualifications, and may further facilitate a shift from a one-badge (e.g., a bachelor's degree) to multiple 'badges' of competency" (Mischewski, 2017, p. 14). This aligns with "signaling theories," which suggest that job seekers who are awarded a credential upon completion of education or additional training may possess skills or characteristics that are valuable to potential employers (Bills, 2003).

As students make progress towards a degree, they can acquire traceable and stackable marketable professional and technical knowledge and skills along a career pathway that is aligned with their academic interests. For example, the University System of Georgia has developed the nexus degree, which represents a new degree program and a targeted academic credential that can be earned in conjunction with a bachelor's degree or independently. The nexus degree is aimed at individuals who wish to specialize or transition into a high-demand career sector, such as cybersecurity, mechatronics, or blockchain.

By using this approach, the nexus degree addresses a recommendation made by the American Society of Engineering Education (ASEE) Corporate Member Council (ASEE, 2020) to strengthen the relationship between academic and industry partners through curriculum focusing on content on emerging technologies, experiential learning, such as internships, co-ops, apprenticeships, and credentialing via certificates and badging. This recommendation was a response to the ASEE survey of 350 recent engineering graduates which found that only 5% of respondents felt very prepared and only 18% felt somewhat prepared in security knowledge related to data, cyber, etc. The practice of embedding non-degree credentials into bachelor's degree programs offers learners the ability to gain and signal more technical knowledge and skills along with a strong academic background in engineering (Workcred, 2020).

Taken in the context of the "unbundling" of higher education (Cliff et al., 2022; Gehrke & Kezar, 2015; Higgs, 2019), non-degree credentials could also compel institutions to utilize digital technologies that help with specialization, collaboration, avoiding duplication of efforts, and finding economies of scale. In practice, non-degree credentials support the reality of all engineering learners wishing to differentiate themselves from other learners both within and outside of their program, leading them to a closer "skills fit" for future work roles and tasks that most align with their interests. Desires for differentiation among engineering learners is happening for a multitude of reasons, for example, to specialize in areas for a particular work role or passion project, to feel a greater sense of ownership over a rigid degree program, and to highlight personal individuality and pursue interests that create a more well-rounded educational experience.

This specialization is commonly seen through certificate programs which are offered alongside baccalaureate engineering programs or as postgraduate offerings. For example, at the postgraduate level, Johns Hopkins University offers postgraduate short-term certificate programs, including for cybersecurity and systems engineering. These programs are intended for individuals who already have an undergraduate degree in engineering or another related technical field. Certificate programs within engineering degrees can also offer an individual the opportunity to develop skills and knowledge related to the practice of engineering but which are not strictly technical. For example, the University of Colorado–Boulder offers undergraduate certificates in engineering leadership and engineering management in their College of Engineering and Applied Science. When combined with a degree, these types of non-degree credentials demonstrate that an individual has all the technical skills typically associated with their degree program as well as an understanding of the broader context needed to practice as an engineer.

Badges which can be embedded in an engineering degree program can also allow an individual to showcase their individual pathway within that degree program. Since badges are often represented digitally, they allow issuers to include information about the skills represented by that badge (such as the description, criteria, and evidence) and store them in the badge metadata. This means that a university could issue different badges to learners enrolled in the same senior engineering project course based on which skills were earned through their specific projects. Faculty can also use a series of badges embedded in courses to show proficiency of skills that are specific to the course or even to a specific learner. Open badges can be used to provide context for how the competencies in the badge were assessed, list the specific competencies represented by the badge, or provide evidence of learning beyond technical competencies, such as leadership or teamwork (Hickey & Schenke, 2019; Hickey et al., 2020; Hickey & Buchem, 2021). While much of this research is not specific to engineering education, it is likely to apply to engineering courses or projects.

2.2.2 Signaling Technical Competency

Non-degree credentials can signal specific technical competencies. This means that individuals at all levels of education can signal new technical abilities they have developed. For example, universities are working to develop new non-degree credentials to support regional workforce needs: MIT is leading the AIM Photonics Academy to develop a certificate program in advanced manufacturing and integrated photonics to support students at regional colleges who seek to enter a career in photonics.

Often, these types of non-degree credentials are narrowly focused. For example, the American Society of Civil Engineering offers certificates in multiple areas, including structural earthquake engineering for buildings, geographic information systems for asset management, and port engineering. When used for this purpose, certificates typically signal completion of a curriculum or training program and are more typically offered by higher education institutions and professional engineering associations. Certifications can also be used to signal specific technical competencies and are typically agnostic of where those skills were acquired. Certification bodies, which can include professional engineering associations, commonly offer these non-degree credentials.

In rapidly changing engineering fields, competencies captured by non-degree credentials are much more likely to be up-to-date than degree programs for multiple reasons. First, many certification bodies offering technical credentials revise their assessments at regular intervals, and holders must pass updated assessments to maintain their certifications. Second, a two- or four-course certificate program can be easier to update regularly than an entire degree program, potentially resulting in more current and timely curricula. Engineering-related jobs in these fields may require both degree and non-degree credentials as evidence that professionals are staying current on advancements in their fields and maintaining relevant skills and abilities.

2.2.3 Signaling Interdisciplinary/Out-of-Degree Specialization

Non-degree credentials can signal interdisciplinary/out-of-degree specialization. These credentials can complement engineers at all educational levels because they do not require specific technical prerequisites. Many different types of credentials can play this role, including certificates, certifications, and badges, and they can focus on broadly applicable skills, such as big data analytics or logistics.

For example, project management is a highly sought-after skill set which can be used to support planning, coordination, and change management. The Project Management Institute is an organization which offers certifications for individuals seeking to provide evidence of or improve their project management abilities. While they offer their own online education/training programs to prepare for their certifications, they also partner with more than 800 organizations that they approve to provide education/training aligned with their certifications, including the California Institute of Technology and the University of Texas at Austin's Center for Professional Education.

Alternatively, these non-degree credentials may focus on regulatory frameworks and/or societal or policy implications relevant to engineering professionals. For example, George Washington University offers a postgraduate certificate in environmental systems management that encompasses policy, technical, regulatory, and social considerations relevant to professionals working in environmental management.

2.2.4 Signaling Professional Development

Non-degree credentials can signal professional development after earning an engineering degree. According to the US Bureau of Labor Statistics (2022b), the median age of employees working in architectural, engineering, and related services was 43.3 years, and nearly 67% were 35 years or older. These individuals are likely to have earned their engineering degree more than a decade ago. While some may have returned to earn another degree, the majority of these individuals will use non-degree education and training programs to signal their continued professional development.

Professional societies for engineers as well as other providers have developed a substantial number of non-degree credentials to support the need for continued professional development. For example, the Institute of Electrical and Electronics Engineers (IEEE) offers a credentialing program consisting of certificates and badges for engineers seeking professional development. Additionally, some credentials are specifically targeted for engineers at different points in their career pathway, such as the certificates offered by the IEEE Electronics Packaging Society, one of which is aimed at early-career engineers and two of which are more advanced.

Higher education institutions also offer these opportunities, both through their traditional graduate programs or through continuing education. For example, the Ohio State University offers boot camps and certificates through its College of Engineering Professional and Distance Education, which include programs in sustainable engineering, corrosion, and data analytics.

While other non-degree credentials can be used, certifications are the main credential that can signal continued professional development. Like the professional engineer (PE) license, evidence of continued competence is necessary in order to remain valid. While some certifications call for more frequent recertification, the majority will require evidence of continuing education or re-taking the assessment every three years. This stipulation means that certifications and PE licenses are the only non-degree credentials in engineering which signal lifelong professional development.

2.2.5 Enabling Non-Degree Pathways to Engineering Careers

Non-degree credentials create a pathway into engineering outside of traditional degree programs. While the majority of the 21.7 million people who work in S&E occupations and S&E-related occupations hold a bachelor's degree or higher, one-third of these workers who use S&E skills in their jobs do not have a bachelor's degree (National Science Board, 2021). Some of these individuals may have entered these roles based on their work experience, but others are likely to have earned non-degree credentials and used the competencies gained to enter engineering occupations.

Badges, certificates, and certifications are the most likely to support this type of occupational entry. For example, some postgraduate certificate programs in engineering do not necessarily require a baccalaureate degree in engineering, unlike full master's degree programs. As a result, these certificate programs can support individuals with non-engineering degrees to earn non-degree credentials in engineering and more easily enter an engineering occupation. This approach may serve to expand the diversity of engineering professionals rather than limiting it to those who already have a bachelor's degree in engineering (more on this point in Section 3).

One prominent example of this approach is the Massachusetts Institute of Technology MicroMasters Program in Principles of Manufacturing. While an engineering background is recommended, there are no educational prerequisites for enrollment in the program. Additionally, individuals can earn individual certificates as they progress through the curriculum, which can benefit learners who are not able to complete the entire program. This approach of incremental credentialing exemplifies why non-degree credentials provide a more comprehensive portrait of knowledge and skills attainment than traditional measures of educational attainment (i.e., degrees) (Columbus, 2019).

Non-degree credentials offer a strategy for individuals to be recognized for the skills they have developed, regardless of whether those skills are products of a degree program or not. For example, the Education Design Lab, a nonprofit organization that addresses equity disparities in the education-to-workforce trajectory, has developed a series of digital badges which learners can earn to showcase skills like initiative, resilience, time management, and persistence that have been developed informally through their lived experiences. As employers increasingly consider and implement skills-based hiring practices, non-degree credentials can provide an entry to individuals who have not earned an engineering degree but have the skills relevant to work in an engineering occupation.

This quality of allowing individuals to represent their skills more fully is a significant advantage of non-degree credentials and is likely to increasingly benefit people of color and first-generation learners who have been underrepresented as engineering degree-holders. These skills include both the technical skills necessary to perform in an engineering occupation, as well as the 21st-century skills employers are demanding (Pistrui et al., 2020).

3 Benefits of Non-Degree Credentials

Within the field of engineering, non-degree credentials represent a timely opportunity to strengthen strategic relationships between industry and academia, rapidly respond to the evolving needs of the engineering workforce, and broaden access to engineering concepts, tools, and skills. Trends in demographic representation within engineering have been stubbornly stagnant for decades (ASEE, 2021), while the demand for high-quality engineering talent around the world has continued to increase in most developed countries, where accurate data on the engineering workforce is available (Royal Academy of Engineering, 2020).

High-quality engineers are often in demand by employers who find their critical thinking and problem-solving skills invaluable. However, the COVID-19 pandemic has shifted the global science and engineering enterprise and highlighted important capacity-building areas, including investments in research and development, creating a more inclusive STEM workforce, and improvements in primary and secondary STEM education as well as recruiting, training, and retaining talent (National Science Board, 2022). Researchers argue that:

> [T]o prosper in the Industry 4.0 ecosystem, individuals and organizations will be required to develop 21st century skill sets. The talent pipeline is failing to provide sufficient quantities of workers and calls for stepping up Industry 4.0 reskilling have become ever more urgent.
>
> *(Pistrui et al., 2020)*

In what follows, five key benefits of non-degree credentials in engineering are highlighted for their significance and contributions to the engineering workforce.

3.1 Promoting Lifelong and Life-Wide Learning of Engineering Skills for All

In addition to encouraging lifelong learning to enhance the quality of current engineers and trainees, non-degree credentials would allow non-engineering majors to be recognized for their acquired engineering knowledge, skills, and abilities by quantifying their skills through a credential. Whether through nontraditional, informal, or experiential means, non-degree credentials offer a way to reward and acknowledge these individuals. The potential for this new technology and method of tracking acquired skills has a global reach that extends beyond implications in the United States.

A report from Mastercard Foundation (2020) on the state of secondary education in sub-Saharan Africa stated: "By mapping and benchmarking skills acquired, national qualifications frameworks have the potential to enable youth to move between informal training and formal education" (p. 8). The systems currently being developed to support non-degree credentials may offer insight into how prior learning can also be recognized and validated in regions of the world where acquiring jobs and higher wages is not only vital to survival (Afeti & Adubra, 2012) but is also a means to recognize both formal and informal learning experiences. In fact, previous scholars have noted that the relationship between employment and higher educational attainment may be stronger and thus may matter even more in less-developed countries (Berntson et al., 2006, p. 226).

In a joint report by the Massachusetts Institute of Technology and Harvard University's Center for International Development, Ellis et al. (2016) elaborated on how years of formal education do not directly translate into increased economic prosperity for most emerging economies (as they tend to in advanced economies such as the United States), but rather, pushing for more specific and specialized schooling has a greater net positive impact on a country's GDP (Hausmann et al., 2013). Leveraging non-degree credentials to build portfolios, chart specialized career paths, and recognize prior learning experiences may be a disruptive innovation for learners all over the world hoping to enter technical fields and command a higher wage.[2]

Promotions of "engineering skills for all" has led to initiatives around diversity, equity, and inclusion in engineering gaining renewed traction in higher education institutions and the workforce. At the heart of the desire to increase representation within the field lie two important categorical spheres for engineering: (1) supporting learners in pursuit of engineering degrees and (2) supporting the engineering workforce that encompasses a wide array of educational pathways, credentials, and work roles.

3.2 Supporting Engineering Learners and Degree Seekers

Two important functions identified for non-degree credentials (Section 2) that benefit engineering learners and degree seekers are signaling specialized skills within and outside of a degree program, which may lead to more equitable outcomes for engineering learners. One lever for improving representation within engineering education is through increasing successful retention efforts.

Scholars in engineering education have shown that one reason women and people of color are more motivated to join engineering is when it shows potential for social impact in their communities (Bielefeldt, 2018; Brubaker et al., 2017), which often clusters them into engineering majors with clearer prosocial motives, such as civil and environmental engineering, biomedical and bioengineering, and chemical engineering. One potential mechanism for why underrepresented engineering learners have high attrition rates within engineering is that they do not see the direct relevance of their skills to the type of careers or communities they care about (Lyon & Green, 2020; Rulifson & Bielefeldt, 2017). Non-degree credentials allow learners to acquire specialized skills in fields that complement personal learning goals and interests, both those within and outside their degree programs. Examples of specialized skills may include a LEED certification, a Six Sigma certification, or a project management certificate.

Financial constraints imposed by engineering programs are another barrier to representation that non-degree credentials may help alleviate by broadening the pathways into engineering education for a wider variety of learners, for example, those with diverse backgrounds, multiple academic interests, pursuing school part-time, and participating in distance/online learning. In particular, for part-time learners and those that do not go on to complete their degree program, non-degree credentials can be earned in place of individual course credit. Thus, even if a learner does not earn a degree, they still earn digital credentials from a university reflecting a part of their skill sets. Additionally, learners who choose to major in other academic fields (e.g., due to financial or time constraints) with hopes of still joining the engineering workforce would be able to formally demonstrate their specific engineering competencies and skills acquired throughout their academic journey. Examples of engineering competencies typically covered outside of the formal engineering education program include cybersecurity analysis and user experience design.

3.3 Diversifying the Engineering Workforce

The three remaining functions identified for non-degree credentials (Section 2) which may benefit the engineering workforce include (1) signaling technical competency; (2) documenting ongoing professional development, skilling, and reskilling over one's engineering career; and (3) enabling non-degree pathways into engineering careers. In this section, we expand upon how each of these functions can lead to use cases with more equitable outcomes for the engineering workforce.

First, non-degree credentials enable staying up-to-date in technical competencies and emerging fields in a fast-changing world, where policy, environmental considerations, and technological breakthroughs force engineers either into lifelong continuous learning or obsolescence. With engineering being such an expensive major, it's likely that even with a degree, the quality of training across academic institutions is not standardized: access to state-of-the-art facilities, multimedia learning materials, and industry-experienced staff is often reserved for only a few elite programs around the world. Non-degree credentials may enable more individuals, both degree and non-degree holders in engineering, to get more practical and up-to-date technical competencies that are valuable in industry by providing "just-in-time" skills and training throughout one's career. Individuals historically underserved and underrepresented in engineering can leverage non-degree credentials to help signal their technical abilities, especially in emerging fields their academic institutions may not have had the resources to support (e.g., AI/machine learning, robotics, and renewable energy).

Second, pursuing professional development opportunities represents ways of showing sustained commitment to their field and an adaptability to the changing ways of work, particularly to certain professional societies and fields. Non-degree credentials that represent achievements and commitment to professional development can help with disparities of perception about which individuals are committed members of their field. This function is especially valuable for employers that seem to universally agree that "employability" and "mobility" skills are both highly important and yet lacking in engineering. This deficit perspective is more likely to be attributed to job seekers coming from low-income and historically underrepresented backgrounds in the world of work (Pager et al., 2009). For example, a study on employers' perspective of the employability skills of Malaysian engineering graduates found dissatisfaction in their "employability skills" of communication, language (often a command of English and other languages), interpersonal skills, and teamwork (Saad & Majid, 2014). These employability skills align with the competencies identified by Passow and Passow (2017) as critical to prepare students for professional practice and the engineering workplace.

Third, and finally, non-degree credentials assist STARs (skilled through alternative routes, that is, individuals that pursued non-degree educational pathways) in entering the engineering workforce

(Blair et al., 2020). This is an exciting opportunity, as it aids in democratizing engineering skills, such that the hurdles limiting individuals from entry into higher education do not also limit their ability to enter the engineering workforce in technical or supporting roles. The explosion in non-degree credential program offerings, both stand-alone organizations and those embedded within higher education institutions, has meant many more avenues for those seeking to enter engineering and adjacent fields through multiple educational pathways – many of whom are more likely to be female, people of color, and older individuals (Lyon & Green, 2020). For example, in the United States, there are a higher proportion of women pursuing non-degree credentials via computer science coding boot camps than in four-year undergraduate degree programs (36% vs. 17.9%, respectively) (Eggleston, 2017). In their longitudinal qualitative study, Lyon and Green (2020) show that nearly half of the women studied had interest and exposure to programming before or during college; however, either due to the logistical challenges of switching majors, fear of stereotyping, or not considering how to translate personal hobbies into a career, they did not formally pursue the computer science degree. Nevertheless, all the women in the study were seeking full-time roles in the software development field, despite not having computer science degrees. Thus, non-degree credentials have an important role to play in helping those in part-time higher education programs, those with informal learning experiences, and career switchers in search of better economic opportunities – all of whom stand to diversify and add richness of experiences and perspectives to the ever-evolving needs in the engineering workforce.

3.4 Dynamically Responding to Evolving Needs in the Engineering Workforce

The future of work is dynamic, and so are the necessary skills for individuals to thrive. Non-degree credentials introduce agility into a rigid educational system that, for decades, has insisted on teaching core engineering principles (i.e., calculus-based math, physics, and chemistry), often at the expense of immediate practical relevance to graduates of the program. Non-degree credentials are innately flexible and equip students with more relevant skills and competencies not covered in traditional engineering education. For example, digital badges earned during a degree program can provide students with the flexibility to signal mastery of atypical engineering competencies to more easily enter career paths that utilize those skills. Additionally, certifications embedded within degree programs enable students who do not complete their degree but do earn the certification to still receive a credential that can lead to immediate employment.

Meaningful collaborations between industry and academia offer win-win scenarios for both learners and employers. Learners can exercise agency over their specialization and create a curriculum that is more agile and responsive to trends and yet does not compromise the integrity of their degree. Learners can more easily "fine-tune" their courses and cultivate a rich and unique set of skills tailored towards specific needs in the workforce, making them more competitive in the labor market and a greater asset as a future member of professional engineering societies.

3.5 Strengthening Relationships With Industry and Professional Engineering Societies

Universities have leveraged non-degree credentials as ways to build industry relevance into the engineering curriculum. This may involve strengthening connections with university alumni by inviting alumni to teach skills that are important to industry and are not currently addressed in the formal curriculum. It can also involve embedding a non-degree credential into the engineering curriculum. For example, Ohio University's bachelor of science in engineering technology and management (ETM) degree prepares engineering graduates for their careers by incorporating technical, business, and leadership skills into the ETM curriculum. Seniors are required to complete the Association of Technology, Management, and Applied Engineering's (ATMAE) Certified Manufacturing Specialist (CMS) examination for graduation. This certification covers 16 areas of manufacturing, including technical and managerial areas, which include topics such as Lean Six Sigma, automation, robotics, tooling, CAM, project management, leadership, and supervision, all of which are covered in ETM coursework (Workcred, 2020).

4 Challenges and Opportunities for Non-Degree Credentials

While there are significant rewards at stake for introducing non-degree credentials into engineering education, there is still much work to be done to ensure that learners, educators, and employers can all maximize those benefits. Some of the benefits of non-degree credentials in engineering education have already begun to be realized through pilot initiatives and incremental adoption and scaling. However, challenges related to determining the quality of non-degree credentials, limitations in what kinds of outcomes data exist, as well as processes for systematic collection of this information, exist. Meaningful interpretation of what non-degree credentials represent and communicate about an individual's knowledge, skills, and abilities is another important consideration. These challenges also represent opportunities for thoughtful consideration in how non-degree engineering credentials are designed, implemented, and evaluated, as well as a starting point to inform a research agenda around the value of the various kinds of non-degree credentials for whom and under what circumstances.

4.1 Assessing Quality

For many engineering degree programs, accreditation often serves to ensure that a college or university program is meeting the quality standards of the profession. Currently, no singular approach exists for non-degree credentials. In fact, different non-degree credentials use a range of approaches offered by various organizations and rely on a spectrum of standards and benchmarks. While there are some quality frameworks which are applicable to all non-degree credentials, none are widely adopted and used across all non-degree credentials.

The most widely recognized quality frameworks for non-degree credentials are ISO/IEC 17024: 2012, Conformity assessment – General requirements for bodies operating certification of persons; the National Commission for Certifying Agencies' (NCCA) Standards for Accreditation of Certification Programs; the Education Quality Outcomes Standards Board (EQOS) framework; the National Skills Coalition quality framework; and the Rutgers Education and Employment Research Center quality framework. These frameworks consider elements of both process and outcomes when considering the quality of a non-degree credential. For example, they might consider how the competencies of a credential are developed (e.g., are subject matter experts involved in their development?) or employment outcomes of individuals who hold the credentials (Van Noy, 2020).

Concerns about quality are increasingly important, as new engineering non-degree credentials are developed. A surge of new actors, including private companies such as Amazon, boot camp providers like General Assembly, professional societies like IEEE, and private universities like MIT, has begun creating and offering engineering-related non-degree credentials without an established governing framework for assessing their quality. This means that learners must take on the risk of deciding which programs to invest their resources in, with the hopes of receiving the intended positive outcomes.

Quality control in engineering curriculum is particularly critical in a field where changes in environmental conditions, advances in technology, and new legislation each affect the work. For example, what constitutes a high-quality environmental engineer differs substantially between the passage of the 1948 US Clean Water Act and the more recent 2015 Paris Agreement.[3] Non-degree credentials are able to supplement the necessary learnings of highly qualified environmental engineers in this dynamic context by offering learners access to additional skills within LEED certifications, renewable energy technology certificates, and nanodegrees in reservoir geomechanics.

Nevertheless, the quality of our engineering workforce is closely tied to the quality of their training. Quality control in engineering education is largely ruled by accreditation and licensing, which can play a large role in who gets hired and who is seen as qualified on teams to lead large projects. Specific standards for appraising graduate attributes and professional competencies and accrediting engineering curriculum and qualifications are often guided by the Washington, Sydney, and Dublin Accords under the International Engineering Alliance and organizations such as ABET. A comparable accreditation body or other acceptable methods for ensuring outcomes, training, and continuous improvement for non-degree credentials have yet to be established,[4] and this leads to further complications of appropriate valuation by both the earner of the non-degree credential and the employer that must interpret its meaning for the organization. In the absence of clear data to offer insight into selecting a non-degree credential, faculty may consider reaching out to credentialing organizations (or other faculty within their own institution) to better understand the competencies and value of non-degree credentials they are considering, as well as other important questions. While non-degree credentials have many characteristics which are likely to support equity and diversity, there is currently little data on the demographics of the engineering professionals who hold non-degree credentials.

4.2 Limitations of Evidence

One challenge in applying existing quality frameworks is that there is currently little or no outcomes data for the majority of non-degree credentials (Gardner, 2022). As noted previously, the highly fragmented non-degree credential market means that most non-degree credential providers do not share their data with anyone and likely lack the infrastructure to do so. Some non-degree credential issuers may voluntarily offer self-reported data on their websites or marketing materials, but there is almost no third-party, validated data on employment outcomes or wage outcomes for any single non-degree credential.

While there is an effort by nonprofit organizations such as the National Student Clearinghouse (2021) to determine this data for certifications, future work could explore employer demands for non-degree credentials and the impact on salary and hiring into the engineering workforce for both degree and non-degree holders, particularly related to understanding the return on investment of non-degree credentials on wages and employment benefits.

Non-degree credentials have the potential to disrupt the traditional educational system and how we understand employer hiring decisions. On the other hand:

> [I]f the ladder of educational opportunity rises high at the doors of some [individuals] and scarcely rises at the doors of others, while at the same time formal education is made a prerequisite to occupational and social advance, then education may become the means, not of eliminating race and class distinctions, but of deepening and solidifying them.
>
> *(US President's Commission on Higher Education, 1947)*

This quotation highlights just how education may also be a vehicle for reproducing inequities in society. Without proper attention to the limitations of the scant evidence on non-degree credentials, we run the risk of misinterpreting potential equity implications of non-degree credentials.

Equity implications for non-degree credentials, like their degree counterparts, present both opportunities to increase access within engineering and to reproduce systems of inequity currently present in the field (Keep & Mayhew, 2010; Ball, 2013). Even with some jobs in the United States loosening degree requirements, there are still many structural barriers that prevent individuals from entering certain fields within the engineering workforce. For example, even though individual cybersecurity hiring managers expressed ambivalence about college degrees as necessary prerequisites for cybersecurity analyst roles, hiring guidelines in certain work settings (i.e., higher education institutions) still have human resources policies on the books requiring a minimum bachelor's degree qualification (Gallagher, 2021; McKinsey & Company, 2020). As another example, even with a college degree and a non-degree credential from completing a coding boot camp, because they did not have "traditional" backgrounds or experience in computer science, participants in the boot camp described their job search as "brutal," having to submit hundreds of applications for software engineering roles (Lyon & Green, 2020, p. 112). Formal education and training alone are not the magic bullets to solve equity issues in the labor market (Keep & Mayhew, 2014). However, we believe that both the potential economic and personal benefits (e.g., increased autonomy, self-efficacy, and mastery) associated with furthering one's education still make the pursuit of non-degree credentials an important one.

4.3 Drawing Meaningful Interpretations of What Non-Degree Credentials Represent and Communicate

Emerging non-degree credentials are changing the employment landscape for both the learner and the employer, who must evaluate and interpret these credentials particularly in relation to traditional degrees. Understanding the role that non-degree credentials play in upskilling and reskilling the engineering workforce will support individuals, employers, and policymakers to make educated investments when selecting a credential. This research is likely to be highly specific for different engineering occupations; non-degree credentials which support engineers working on autonomous vehicles are unlikely to be the same ones that support engineers working in water treatment facilities. As a result, partnerships with professional or industry associations should take care to eliminate or reduce any conflicts of interest if those organizations offer their own non-degree credentials. This research should also consider how different demographic groups access, use, and benefit from non-degree credentials to support their professional development.

While non-degree credentials stand to increase access to the engineering profession by providing more offerings and pathways to gaining relevant knowledge, skills, and abilities, a critical challenge lies in how both the learner and employer come to understand and make meaning of the value of each non-degree credential. With a wide variety of non-degree credential providers, credential types, and signaling functions (e.g., competency, exposure, current skills, or achievement), evaluation schemas also grow in complexity. For example, how does one reasonably compare the credentials of an undergraduate student who earned a minor in cybersecurity to a coding boot camp graduate who earned a cybersecurity certificate? Non-degree credentials provide a formalized way of recognizing the many learning experiences that individuals encounter; however, they do not clarify how employers will make meaning of novel experiences or how they ought to be compared to the traditional ways of how education and competencies are recorded and recognized.

Designers of non-degree credentials must take into consideration both the learner and employer and understand the different needs that both will have in how the non-degree credentials are earned, communicated, and interpreted.[5] A qualitative study from Laryea et al. (2021) on learners who earned non-degree credentials from MOOCs found that learners often had ambiguous interpretations of their awarded non-degree credentials. While some learners posted on LinkedIn, others opted to discuss their new credentials in interviews, and still others were less sure about how to translate the value of their non-degree credential for a promotion or hiring and did not think it worth mentioning at all.

Future research is needed to understand the signaling role of non-degree credentials intentionally integrated into a degree program, as a complement to a degree, or completely independent of a degree. To better understand the engineering workforce, the role of non-degree credentials as a pathway into engineering jobs should be examined. There is currently very little data known about individuals with only non-degree credentials who work in engineering roles, as workers without a college degree are often excluded from most national datasets on engineering and engineering occupations. Therefore, research resulting in data on this workforce would provide a baseline for understanding the size and demographics of the engineering workforce with non-degree credentials. Furthermore, research examining which non-degree credentials are accessible and lead to good engineering jobs, if any, would provide guidance for the millions of individuals who do not have a four-year degree (a majority of people) but are interested in engineering occupations.

All stakeholders in the engineering education-to-workforce continuum will benefit from addressing these questions: employers will better understand how to use non-degree credentials in their hiring, development, and promotion processes, particularly if non-degree credentials are shown to be inclusive; individuals will better understand how and which non-degree credentials will support their goals and resource constraints; and faculty will better understand how to integrate non-degree credentials into engineering degree programs, advise students seeking non-degree credentials, and develop quality non-degree credentials. Similarly, colleges and universities will have better metrics to develop and support new or existing non-degree credential programs; policymakers will have data on which to make informed decisions; and credentialing organizations will be able to use data to improve the value of their credentials or eliminate those with poor outcomes. The constellation of stakeholders who can both benefit from and contribute to the design, implementation, and use of non-degree credentials will continue to grow, and developing this body of research will be critical to ensure their value for the success of current and future learners. Collectively, understanding these contextual factors that contribute to and influence how non-degree credentials are incorporated into a learners' academic and professional trajectory may inform the development of unique credential-seeking personas and highlight insights for how best to support them over time.

5 Conclusion

Non-degree credentials have been, and will continue to be, earned by engineers and those in the engineering workforce. Given their versatility and benefits, as well as the tremendous technological change continuing to occur in the workplace, the demand for these credentials is likely to expand.

This means that non-degree credentials in engineering education will not only continue to transform the way that individuals seek to gain education and skills; their widespread use will also require a corresponding evolution in how learning is recorded, recognized, and communicated by institutions as well as the learners themselves. Currently, engineering education is largely captured through degrees, experiences, and non-degree credentials represented on an academic transcript, résumé, or online professional networking profile. In the future, these traditional records may no longer be sufficient as learners seek out new platforms and processes that allow them to effectively communicate their knowledge, skills, and abilities in formats that can be easily shared and are comprehensive, verifiable, and trusted.

The growing number of organizations which issue non-degree credentials, as well as a corresponding increase in non-degree credentials issued, in addition to degrees, will result in too many engineering credentials for individuals and employers to easily understand and distinguish. The breadth of providers will make it difficult to judge the quality of any single credential, particularly for non-degree credentials for which no accreditation standards exist.

This creates a foundational need to establish an infrastructure to collect and review outcomes data for credentials. Funding and oversight must be provided by credible agencies, organization, policymakers, and leaders. With this underpinning framework in place, peer-reviewed research and robust evaluation processes will be incentivized, resulting in more transparency on the quality of non-degree credentials in engineering, allowing individuals greater agency and ownership in how they can be used to support and advance their professional and personal goals across a lifelong and life-wide learning career.

Acknowledgments

The authors would like to thank the four anonymous reviewers that offered thoughtful feedback, questions, and considerations for this chapter. We would also like to thank Janet Forte from Workcred for her careful edits and review of this work.

Notes

1 "The phrase 'Global South' refers broadly to the regions of Latin America, Asia, Africa, and Oceania. It is one of a family of terms, including 'Third World' and 'Periphery,' that denote regions outside Europe and North America, mostly (though not all) low-income and often politically or culturally marginalized" (Dados & Connell, 2012: 12).
2 Human capital factors are not the sole drivers of employment opportunities and pathways for individuals. Social and cultural capital have long been studied by scholars as ultimately determining who gets hired for jobs (Rivera, 2012; Brown & Hesketh, 2004; Brown et al., 2011). For more on other considerations, please see Section 4, "Challenges and Opportunities of Non-Degree Credentials."
3 The Paris Agreement is an international treaty of countries agreeing to adopt certain practices and policies to limit the effects of climate change. Or the recently proposed Recovering America's Wildlife Act, which could stand to be one of the most significant conservation laws in decades.
4 QA Commons is an organization that is trying to tackle this issue of assessing quality of non-degree credentials. For more information, visit their website at https://theqacommons.org/.
5 Credential Engine is an organization that is working towards the standardization of non-degree credential evaluation. For more information, visit their website at https://credentialengine.org.

References

Afeti, G., & Adubra, A. L. (2013). *Lifelong technical and vocational skills development for sustainable socioeconomic growth in Africa.* Association for the Development of Education in Africa Triennale on Education and Training in Africa, Ouagadougou, Burkina Faso, February 12–17, 2012.

American Society for Engineering Education. (2020). *ASEE Corporate Member Council survey for skills gaps in recent engineering graduates.* https://workforcesummit.asee.org/wp-content/uploads/2021/10/CMC-ebook.pdf

American Society for Engineering Education. (2021). *Profiles of engineering and engineering technology.* https://ira.asee.org/wp-content/uploads/2021/11/Total-by-the-Number-2020.pdf

Ball, S. J. (2013). *Foucault and education: Disciplines and knowledge.* Routledge.

Becker, G. S. (1993). Nobel lecture: The economic way of looking at behavior. *Journal of Political Economy, 101*(3), 385–409.

Berntson, E., Sverke, M., & Marklund, S. (2006). Predicting perceived employability: Human capital or labour market opportunities? *Economic and Industrial Democracy, 27*(2), 223–244.

Bills, D. B. (2003). Credentials, signals, and screens: Explaining the relationship between schooling and job assignment. *Review of Educational Research, 73*(4), 441–469. https://doi.org/10.3102/00346543073004441

Bielefeldt, A. R. (2018). Professional social responsibility in engineering. In I. Muenstermann (Ed.), *Social responsibility.* IntechOpen. https://doi.org/10.5772/intechopen.73785

Blair, P. Q., Castagnino, T. G., Groshen, E. L., Debroy, P., Auguste, B., Ahmed, S., Garcia Diaz, F., & Bonavida, C. (2020, March). *Searching for STARs: Work experience as a job market signal for workers without bachelor's degrees.* NBER Working Paper No. 26844. National Bureau of Economic Research. https://doi.org/10.3386/w26844

Brown, P. & Hesketh, A. (2004). *The mismanagement of talent. employability and jobs in the knowledge economy.* Oxford University Press.

Brown, P., Lauder, H. & Ashton, D. (2011). *The global auction: The broken promises of education, jobs and incomes.* Oxford University Press.

Brubaker, E. R., Schar, M., & Sheppard, S. (2017, June). Impact-driven engineering students: Contributing behavioral correlates. In *Paper presented at 2017 ASEE annual conference & exposition*, Columbus, Ohio. https://doi.org/10.18260/1-2-28473

Cheeseman Day, J., & Martinez, A. (2021, June 2). From college to jobs: Pathways in science, technology, engineering, and math (STEM). STEM majors earned more than other STEM Workers, *America Counts: Stories Behind the Numbers*, United States Census Bureau. www.census.gov/library/stories/2021/06/does-majoring-in-stem-lead-to-stem-job-after-graduation.html

Cliff, A., Walji, S., Jancic Mogliacci, R., Morris, N., & Ivancheva, M. (2022). Unbundling and higher education curriculum: A cultural-historical activity theory view of process. *Teaching in Higher Education, 27*(2), 217–232. https://doi.org/10.1080/13562517.2019.1711050

Collins, R. (2019). *The credential society: An historical sociology of education and stratification.* Columbia University Press.

Columbus, R. (2019). *Nondegree credentials, work-based learning, and the American working class.* www.aei.org/research-products/report/nondegree-credentials-work-based-learning-and-the-american-working-class/

Corporation for a Skilled Workforce, George Washington Institute of Public Policy, and Workcred. (2020). *Understanding certifications.* https://workcred.org/Documents/Understanding-Certifications-Report-Dec-2020.pdf

Dados, N., & Connell, R. (2012). The global south, contexts. *American Sociological Association, 11*(1), 12–13.

Diaz, M., Lim, J. R., Cardenas-Navia, I., & Elzey, K. (2022). *A world of transformation: Moving from degrees to skills-based alternative credentials* (Technical Note N° IDB-TN-02347). Inter-American Development Bank. http://dx.doi.org/10.18235/0004299

Eggleston, L. (2017). *2017 Coding bootcamps outcomes & demographics report.* www.coursereport.com/reports/coding-bootcamp-job-placement-2017#overview

Ellis, L. E., Nunn, S. G., & Avella, J. T. (2016). Digital badges and micro-credentials: Historical overview, motivational aspects, issues, and challenges. In D. Ifenthaler, N. Bellin-Mularski, & D. K. Mah (Eds.), *Foundation of digital badges and micro-credentials: Demonstrating and recognizing knowledge and competencies* (pp. 3–21). Springer International Publishing.

Fouad, N. A., Kozlowski, M. B., Singh, R., Linneman, N. G., Schams, S. S., & Weber, K. N. (2020). Exploring the odds: Gender differences in departing the engineering profession. *Journal of Career Assessment, 28*(3), 446–461. https://doi.org/10.1177/1069072719876892

Frehill, L. M. (2012). Gender and career outcomes of U.S. engineers. *International Journal of Gender, Science and Technology, 4*(2), 148–166. https://genderandset.open.ac.uk/index.php/genderandset/article/view/199.

Gallagher, S. (2018, December). *Educational credentials come of age: A survey on the use and value of educational credentials in hiring.* https://cps.northeastern.edu/wp-content/uploads/2021/03/Educational_Credentials_Come_of_Age_2018.pdf.

Gardner, L. (2022, September 16). The mystery of short-term credentials. *The Chronicle of Higher Education.* www.chronicle.com/article/the-mystery-of-short-term-credentials.

Gaston, P. L., & Van Noy, M. (2022). *Credentials: Understand the problems, identify the opportunities, create the solutions.* Stylus Publishing.

Gehrke, S., & A. Kezar. (2015). Unbundling the faculty role in higher education: Utilizing historical, theoretical, and empirical frameworks to inform future research. In M. B. Paulsen (Ed.), *Higher education: Handbook of theory and research* (Vol. 30, pp. 93–150). Springer International Publishing. https://doi.org/10.1007/978-3-319-12835-1_3

Gregg, A., & Dabbagh, N. (2023). Chapter 22: Engineering education and online learning. In A. Johri (Ed.). *International handbook of engineering education research.* Routledge.

Hausmann, R., Hidalgo, C. A., Bustos, S., Coscia, M., Simoes, A., & Yildrid, M. A. (2013). *The atlas of economic complexity: Mapping paths to prosperity.* Massachusetts Institute of Technology and Center for International Development, Harvard University. MIT Press.

Hickey, D. T., & Buchem, E. (2021). Advancing sustainable educational ecosystems with open digital credentials and badges. In S. Jagannathan (Ed.), *Educational technologies for sustainable development: How upskilling data, analysis, and digital innovations foster lifelong learning*. Routledge.

Hickey, D. T., & Schenke, K. (2019). Open digital badges and reward structures. In K. A. Renninger & S. E. Hidi (Eds.), *The Cambridge handbook on motivation and learning* (pp. 209–237). Cambridge University Press.

Hickey, D. T., Uttamchandani, S. L., & Chartrand, G. T. (2020). Competencies in context: New approaches to capturing, recognizing, and endorsing learning. In M. J. Bishop, E. Boling, J. Elen, & V. Svihla (Eds.), *Handbook of research in educational communications and technology* (pp. 547–592). Springer.

Higgs, J. (2019). Edupunks and universities: Employability engagement possibilities. In J. Higgs, W. Letts, & G. Crisp (Eds.), *Education for employability* (Vol. 2, pp. 189–202). Brill.

Keep, E., & Mayhew, K. (2010). Moving beyond skills as a social and economic panacea. *Work, Employment and Society, 24*(3), 565–577.

Keep, E., & Mayhew, K. (2014). Inequality–'wicked problems', labour market outcomes and the search for silver bullets. *Oxford Review of Education, 40*(6), 764–781.

Kusimo, A. C., & Sheppard, S. (2019, August). Manufacturing vulnerability: How resource scarcity hinders team coordination during manufacturing in Nigerian factories. In *Proceedings of the ASME 2019 international design engineering technical conferences and computers and information in engineering conference*. Volume 7: 31st International Conference on Design Theory and Methodology. Anaheim, CA, USA. American Society of Mechanical Engineering. https://doi.org/10.1115/DETC2019-97820

Laryea, K., Paepcke, A., Mirzaei, K., & Stevens, M. L. (2021). Ambiguous credentials: How learners use and make sense of Massively Open Online Courses. *The Journal of Higher Education, 92*(4), 596–622. https://doi.org/10.1080/00221546.2020.1851571

Lyon, L. A., & Green, E. (2020). Women in coding boot camps: An alternative pathway to computing jobs, *Computer Science Education, 30*(1), 102–123. https://doi.org/10.1080/08993408.2019.1682379

Mastercard Foundation. (2020, July). *Secondary education in Africa: Preparing youth for the future of work*. https://doi.org/10.15868/socialsector.35972

Matemba, E., & Lloyd, N. (2017). Internationalisation of professional engineers: A review of globalisation of engineering education and accreditation-challenges from an African perspective. *The International Journal of Engineering Education, 33*(6), 2083–2097

McKinsey & Company. (2020). *Beyond hiring: How companies are reskilling to address talent gaps*. www.mckinsey.com/business-functions/people-and-organizational-performance/our-insights/beyond-hiring-how-companies-are-reskilling-to-address-talent-gaps.

Mischewski, B. (2017). *Micro-credentials: A model for engineering education*. Tertiary Education Commission, Wellington, New Zealand. www.voced.edu.au/content/ngv%3A79731

National Academy of Sciences. (2003). *Envisioning a 21st century science and engineering workforce for the United States: Tasks for university, industry, and government*. The National Academies Press. https://doi.org/10.17226/10647.

National Science Board. (2014). *Science and engineering indicators 2014* (NSB 14–01). National Science Foundation. www.nsf.gov/statistics/seind14/content/overview/overview.pdf

National Science Board. (2016). *Science and engineering indicators 2016* (NSB-2016–1). National Science Foundation. www.nsf.gov/statistics/2016/nsb20161/#/report

National Science Board. (2021). *The STEM labor force of today: Scientists, engineers and skilled technical workers. Science and Engineering Indicators 2022* (NSB-2021–2). National Science Foundation. https://ncses.nsf.gov/pubs/nsb20212

National Science Board. (2022). *Science and engineering indicators 2022: The state of U.S. science and engineering* (NSB-2022–1). National Science Foundation. https://ncses.nsf.gov/pubs/nsb20221

National Student Clearinghouse. (2021). *Industry credentials: Pilot program studies matching up industry credential attainment, enrollment and degree data to identify new learner pathways*. www.studentclearinghouse.info/onestop/wp-content/uploads/IndustryCredentialsWhitepaper.pdf

National Student Clearinghouse Research Center. (2022, June). *Persistence and retention: Fall 2020 beginning post-secondary student cohort*. https://nscresearchcenter.org/wp-content/uploads/PersistenceRetention2022.pdf

Pager, D., Bonikowski, B., & Western, B. (2009). Discrimination in a low-wage labor market: A field experiment. *American Sociological Review, 74*(5), 777–799.

Passow, H. J., & Passow, C. H. (2017). What competencies should undergraduate engineering programs emphasize? A systematic review. *Journal of Engineering Education, 106*(3) 475–526. https://doi.org/10.1002/jee.20171

Pew Research Center. (2021, April). *STEM jobs see uneven progress in increasing gender, racial and ethnic diversity.* www.pewresearch.org/science/2021/04/01/stem-jobs-see-uneven-progress-in-increasing-gender-racial-and-ethnic-diversity/

Pistrui, D., & Kleinke, D. K., & Das, S., & Bonnstetter, R., & Gehrig, E. T. (2020). The industry 4.0 talent pipeline: A generational overview of the professional competencies, motivational factors, and behavioral styles of the workforce. In *Paper presented at 2020 ASEE virtual annual conference content access, virtual online.* https://doi.org/10.18260/1-2-35341

Polmear, M., Chance, S., Hadgraft, R. G., & Shaw, C. (2023). Chapter 15: Informal learning as opportunity for competency development and broadened engagement in engineering. In A. Johri (Ed.), *International handbook of engineering education research.* Routledge.

Rivera, L. A. (2012). Hiring as cultural matching: The case of elite professional service firms. *American Sociological Review*, 77(6), 999–1022.

Royal Academy of Engineering. (2020). *Global engineering capability review.* https://engineeringx.raeng.org.uk/programmes/engineering-skills-where-they-are-most-needed/global-engineering-capability-review

Rulifson, G., & Bielefeldt, A. (2017). Motivations to leave engineering: Through a lens of social responsibility. *Engineering Studies, 9*(3), 222–248. https://doi.org/10.1080/19378629.2017.1397159

Saad, M. S. M., & Majid, I. A. (2014). Employers' perceptions of important employability skills required from Malaysian engineering and information and communication technology (ICT) graduates. *Global Journal of Engineering Education, 16*(3), 110–115.

Seymour, E., & Hewitt, N. M. (1997). *Talking about leaving: Why undergraduates leave the sciences.* Westview Press.

Seymour, E., & Hunter, A. (2019). *Talking about leaving revisited: Persistence, relocation, and loss in undergraduate STEM education.* Springer.

Strada Center for Education Consumer Insights. (2021). *Examining the value of non-degree credentials.* https://stradaeducation.org/report/pv-release-july-28-2021/

Strada Education Network, Gallup, Inc., & Lumina Foundation (2020). *Certified value: When do adults without degrees benefit from earning certificates and certifications?* https://go.stradaeducation.org/certified-value

Trevelyan, J. (2014). towards a theoretical framework for engineering practice. In B. Williams, J. Figueiredo, & J. Trevelyan (Eds.), *Engineering practice in a global context: Understanding the technical and the social* (pp. 33–60). CRC Press, Taylor & Francis Group.

U.S. Bureau of Labor Statistics. (2022a, April). Architecture and engineering occupations [Table]. *Occupational Outlook Handbook.* www.bls.gov/ooh/architecture-and-engineering/home.htm

U.S. Bureau of Labor Statistics. (2022b, January). Table 11b. Employed persons by detailed occupation and age [Table]. *Labor Force Statistics from the Current Population Survey.* www.bls.gov/cps/cpsaat11b.htm

U.S. Department of Education, Office of Career, Technical, and Adult Education. (2021). *Introduction to stackable credentials* (OCTAE Program Memorandum 21–1). https://s3.amazonaws.com/PCRN/file/introduction-to-stackable-credentials.pdf

U.S. President's Commission on Higher Education. (1947). *Higher education for American democracy (Volume I: Establishing the goals).* U.S. Government Printing Office.

Van Noy, M. (2020). *Identifying high quality industry certifications.* Rutgers Education and Employment Research Center, Issue Brief. https://smlr.rutgers.edu/faculty-research-engagement/education-employment-research-center-eerc/eerc-projects/identifying-high

Workcred. (2020, November). *Embedding certifications into bachelor's degrees: Certification-degree pathways project.* https://workcred.org/Documents/Embedding-Certifications-Into-Bachelors-Degrees-Certification-Degree-Pathways-Project-Framework-Report-November-2020.pdf

Workcred. (2021). *How do credentials differ?* https://workcred.org/Interactive-Graphics/How-Do-Credentials-Differ-Certification.aspx

Part 4

Advancing Pedagogy and Curriculum in Engineering Education

17

Institutionalizing Social Justice in Engineering Curricula

Diana A. Chen, Gordon D. Hoople, Jon A. Leydens, and Cindy Rottmann

1 Introduction

The reasons social justice is part of engineering and belongs in engineering education have been covered sufficiently elsewhere (Baillie & Catalano, 2009; Leydens et al., 2022; Leydens & Lucena, 2018; Riley, 2008); therefore, here we only briefly note a few reasons, organized around key questions. First, who solves engineering problems? Downey and colleagues observe that "engineering problems do not solve themselves; they are always solved by people. Once people are introduced to the problem-solving situation, it takes on human as well as technical dimensions" (Downey et al., 2006, p. 109). Also, for whom are engineering problems solved? Engineering education helps prepare students for a profession that is diverse yet generally client-centered and public-facing, involving problem-solving *by people for people* (Stevens et al., 2014; Trevelyan, 2014). Stakeholders of engineering problem-solving vary from users to community members, citizens to any others affected by engineering designs and services. Clearly, as a sociotechnical profession, engineering is already embedded in social structures, which often feature questions of equal opportunity and fairness in resource distribution. In other words, social justice is not something that is added to engineering or engineering education; it is already *inherent* in both. Teaching social justice in the context of an engineering education is a matter of making what has been rendered invisible visible (Leydens et al., 2022). In that sense, *not* teaching social justice is also a choice, one that also comes with consequences for the next generation of engineers. In 1989, Ursula Franklin argued, "The viability of technology, like democracy, depends in the end on the practice of justice and on the enforcement of limits to power" (Franklin, 1999, p. 5). Since then, the scholarship around engineering and social justice has seen substantial development; however, engineering curricula have been comparably slower to adapt.

To begin to fill that curricular gap, this chapter examines how social justice has emerged in several diverse curricular contexts over the past decade. We explore three case studies of social justice work at the authors' home institutions, two in the United States and one in Canada. These case studies were chosen as they span a range of intervention sizes. The first focuses on a module – a conceptually bounded set of lessons within a single course – integrated into multiple courses at the University of Toronto. The second case examines a semester-long course at Colorado School of Mines. The last case focuses on the development of a new degree program at the University of San Diego. These case studies do not encompass the full breadth of engineering curricular interventions related to social

DOI: 10.4324/9781003287483-21

justice but serve as a starting point to explore some unique approaches to integrating social justice in the engineering curriculum. For each case study, we discuss entry points and barriers associated with integrating social justice into the engineering curriculum and the role institutional support has played in accelerating, enabling, and legitimizing the success of this integration. This content is followed by a discussion of what compels engineering educators across contexts to address social (in) justice, what prevents us from engaging in this work, and how we can support engineering education practitioners in this work.

Before continuing, it is important to examine the authors' mutual understanding of social justice. *Social justice* is a complex term – one that eludes an easy or consistent definition. Terminology that overlaps with or is synonymous with "social justice" varies from context to context. For example, engineering societies' codes of ethics often mention "public welfare" but do not include concerns about human rights or ending oppression. In so doing, these codes appear to assume that the economic, environmental, and values-based status quo is socially just, or they elect not to engage issues of social justice. Meanwhile, the term "ethics" in Denmark has a social justice connotation that doesn't exist in the United States (Green-Pedersen & van Kersbergen, 2002), while "community engagement" in the United States has been perverted in many circles to often have elements of White saviorism (Hickmon, 2015; Mitchell et al., 2012).

In *Engineering and Social Justice*, Riley explores the difficulties of identifying a single definition. We follow her lead and, rather than choosing a single definition, build upon a synthesis of key elements that run across multiple social justice definitions: "the struggle to end different kinds of oppression, to create economic equality, to uphold human rights or dignity, and to restore right relationships among all people and the environment" (Riley, 2008, p. 4). It is important to begin conversations on this topic by discussing how those participating in the discussion conceptualize social justice.

While discussions about and challenges to integrating more social considerations into the engineering curriculum date back to the 1890s (Gianniny, 2004; Leydens et al., 2022), there is reason for cautious optimism that we may be on the verge of more systemic change. A global pandemic and racial reckoning after the murder of George Floyd encouraged many engineering educators in the United States to (re)examine issues of social injustice, including everything from global vaccine distribution to revealing previously (in)visible contributions from people of color to engineering innovations. Also, external pressure, such as the public's and regulators' concerns with the increasing power of big technology firms and increasing concerns about the local and global effects of climate change, has encouraged many engineering educators across national contexts to directly reckon with the ways in which engineers are complicit in perpetuating systemic injustice. Such catalysts raise the question of how to engage social justice work within the context of engineering education.

The three case studies in this chapter provide diverse engagement approaches, but they are limited to two US-based schools and one Canadian school, revealing a North American bias. We reached out to social justice–minded engineering educators in other national contexts, but those invitations were either not accepted or were rendered complex due to differences in cultural and institutional contexts around the use of the term "social justice." Hence, we limited ourselves to a globally narrow scope to take advantage of authors' experientially derived depth of contextual understanding. We accept that terminology differs from nation to nation, and even locality to locality, but we believe the institutional diversity addressed in our chapter exists across national contexts, providing engineering educators outside of North America with organizationally resonant examples on which to build or extend their own practices. To understand the three approaches used in our respective institutional contexts, we first discuss relevant conceptual frameworks and paradigms for social justice in the context of engineering curricula.

2 Framing Socially Just Change in University Contexts

Social justice as a term has at least two distinct paradigmatic roots – liberal and critical. In this chapter, we use the terms "liberal" and "critical" in ways that reflect their philosophical roots. Within Western academic norms, liberal social justice can be traced back to John Rawls (1971), and critical social justice can be traced back to Karl Marx (1906/2010). Rawls advocates for procedural fairness on the basis of equality (providing equal opportunity for all), while Marx advocates for distributive fairness on the basis of equity (eliminating structural barriers that unfairly advantage some). Within engineering, liberal notions of social justice have been translated into conversations about representation, retention, diversity, and inclusion, while critical notions of social justice touch on power relations and structural inequity.

While these paradigmatic roots may restrict our ability to talk across social justice frameworks, both are needed if we intend to institutionalize social justice in both the formal and hidden (Snyder, 1970) engineering curriculum. This process of institutionalization also requires us to touch on other curricular and institutional elements. We have selected three such elements to help us reflect on socially just change at our own institutions: (1) objectives, (2) theories of change, and (3) institutional context.

First, we explore a key question for engineering educators: What are the **objectives or purposes** of social justice research and action? We frame this discussion using Leydens and Lucena's notion of engineering for social justice (E4SJ):

> [E]ngineering practices that strive to enhance human capabilities (ends) through an equitable distribution of opportunities and resources while reducing imposed risks and harms (means) among agentic citizens of a specific community or communities.
>
> *(Leydens & Lucena, 2018, p. 15)*

To understand socially just objectives, we look within the E4SJ framework to Nussbaum's human capabilities model (2001, 2007, 2011), which assesses the degree to which a community development project met its intended goals and identifies the intended outcomes of social justice work in accessible, development terms. By accentuating human capabilities that serve "as a benchmark for a minimally decent life" (Nussbaum, 2007, p. 22), the model addresses *why* it is important to attain an equitable distribution of opportunities and resources and to reduce imposed risks and harms. While other social justice objectives exist, we follow Leydens, Lucena, and other scholars who have connected the capabilities model to technological design (Frediani & Boano, 2012; Murphy & Gardoni, 2012; Nichols & Dong, 2012; Oosterlaken, 2012; van den Hoven, 2012), because it provides resonance, credibility, and focus to engineering educators who may be new to social justice education.

Second, when it comes to **theories of change**, we ask *how* key actors have conceptualized and experienced change over the course of their program development process. How much change is enough to constitute socially just change? Must it be system-wide, as Graham proposes (2012a, 2012b), or do we make the road by walking, as Horton et al. (1990) and Begay-Campbell (2010) suggest? Change can be conceptualized in multiple ways, drawing on the work of theorists from engineering, education, and management (Bamford & Forrester, 2003; Fosfuri & Rønde, 2009; Godfrey, 2014; Graham, 2012a, 2012b; Horton et al., 1990). Following Bamford and Forrester (2003), we ask whether change is planned, emergent, or imposed. Our answers to this question may differ not only by initiative type but also by our respective social, professional, and institutional locations. For instance, some of us occupy senior administrative roles authorizing us to drive institutional change, while others are positioned within implementation or advisory roles. Similarly, some of us are viewed as professional insiders permitted a level of decision-making authority, while others are

Table 17.1 Leading Socially Just Change in Engineering Education

	Authors	*Key Concepts*	*Case Study Inquiry*
Objectives	Leydens and Lucena (2018); Nussbaum (2007, 2011)	E4SJ; human capabilities framework	What are our goals as social justice engineering educators?
Theories of change	Graham (2012a, 2012b); Bamford and Forrester (2003); Godfrey (2014); Fosfuri and Rønde (2009); Horton et al. (1990); Begay-Campbell (2010)	Systemic, planned change vs. emergent, politicized change	How have key actors conceptualized change over the course of our program/course/module development?
Institutional context	Cech (2013); Riley (2008); Faulkner (2000, 2007)	Engineering ideologies, mindsets, and culture reifying depoliticization, meritocracy, and dualistic thinking	(How) have normative engineering mindsets, cultures, and ideologies constrained the success of our social justice initiatives?

characterized as disciplinary outsiders imposing social justice on a field that is not our own. Finally, compounding organizational and disciplinary positions, each of us is differentially privileged by demographic norms in both engineering and higher education. In short, competencies and theories of change aside, young, racially underrepresented, non-binary, and/or non-tenure-track instructors trained in the social sciences may need to approach socially just change differently than their older, White, male, tenured, technically trained, full professor colleagues.

Finally, given that our social justice initiatives take place in three distinct **institutional contexts**, we draw attention to the cultural waters within which we swim, including examinations of mindsets, cultures, and ideologies in engineering. For instance, for normative engineering mindsets, we draw on Riley's (2008) analysis of engineering humor as a value-laden discourse; for engineering cultures, we draw on Faulkner's (2000, 2007, 2009a, 2009b) examination of dualistic thinking in engineers' professional practice; and for ideologies, we draw on Cech's (2013, 2014) analysis of student disengagement in engineering education, paying specific attention to the operation of meritocracy, technical/social dualism, and de-politicization in engineering norms. These three conceptual anchors help us understand why social justice–minded engineering educators continue to struggle across institutional contexts after decades of equity, diversity, and inclusion efforts in the field. Table 17.1 identifies the key conceptual elements we use to frame our analysis of three social justice change initiatives in engineering education.

3 Case Studies: Engineering and Social Justice Curricula

How have some institutions made strides towards sustainable institutionalization of social justice while others risk remaining fledgling or even dissolving if key personnel leave the program? What foundations need to be in place for engineering and social justice programs to be vibrant and transformative over the long term?

In this section, we describe three different programs that have integrated social justice into the engineering curriculum and explore their entry points, examine their barriers and obstacles along the way and the support structures that facilitated success. We will focus on the authors' home contexts and institutions – *The Faculty of Applied Science and Engineering* at the University of Toronto; *Humanities, Arts, and Social Sciences* at Colorado School of Mines; and *Integrated Engineering* at University of San Diego. In particular, we explore how each program institutionalized social justice in the curriculum.

3.1 Ethics and Equity Project, University of Toronto

3.1.1 Institutional Context

The University of Toronto (UofT) is a large, public, R1 (research intensive) institution in Canada with a student population of approximately 64,000. It houses multiple professional schools, including the Faculty of Applied Science and Engineering (FASE). Engineering at UofT is highly competitive and academically driven, with a combined undergraduate and graduate student population of approximately 7,000 and an entering class of 2020 with a secondary school grade point average of 94.5%; 36.7% of the undergraduate population and 19.7% of tenured or tenure-stream faculty members are women (FASE, 2021). A growing number of equity, diversity, and inclusion (EDI) initiatives have been established over the past decade, including three elective courses at both undergraduate and graduate levels, ongoing outreach efforts to underrepresented groups of high school students, a standing committee of faculty council, an engineering EDI action group, and two senior administrative positions, one dedicated to diversity, inclusion, and professionalism and the other dedicated to access and inclusive pedagogy. FASE also houses several student clubs with an EDI focus. While some of these positions are institutionalized, most are led by a growing network of students, staff, and faculty members who do social justice work off the sides of our respective desks. An internal survey we conducted in 2018 highlighted 44 distinct, EDI-informed initiatives taking place in engineering, with an additional 55 programs, offices, courses, or EDI touchpoints located in the broader UofT community. We share only one of these 99 initiatives in the current case study.

3.1.2 Initiative Type – Research-Informed Module

The Engineering Ethics and Equity Case Study project we describe in this chapter is a research-informed module featuring curricular integration of case studies based on ethical and equity-based dilemmas faced by Canadian engineers. We deliberately used the term "equity" rather than "social justice" to facilitate Canadian Engineering Accreditation Board (CEAB) processes, but our understanding of equity reflects critical notions of social justice outlined earlier in this chapter. In particular, we define "equity" as "a process of naming and addressing historic and current power imbalances that systemically disadvantage marginalized groups" (Rottmann et al., 2021). The project spanned five years, involving four overlapping phases: (1) case study generation based on critical incident interviews with 22 Canadian engineers, (2) workshop development to integrate equity concepts into ethical case study learning, (3) assessment of the completed module in four undergraduate classes, and (4) curricular integration paired with professional development for engineering professors committed to delivering the module in their courses.

Each of the anonymized case studies features an engineer or engineering student who narrates a deeply contextualized first-person account of an ethical dilemma with implicit or explicit equity consequences. The narratives include background information on the individuals and organizations in question, a detailed description of the ethical dilemmas individuals struggled to navigate, a meditative reflection activity examining possible ways forward, and a consideration of the EDI consequences for each action. The narratives end with the actor leaving this momentary reflective state to ask the reader, "What should I do?"

The cases feature a specific (anonymized) individual and may thus be considered microethical (Herkert, 2005), but the workshop we use to scaffold student learning pushes students to connect microethical situations with the broader sociopolitical context. We scaffold this micro to macro bridging in two ways. First, we introduce students to four equity concepts, all of which foreground structural inequity, and ask them to select one of these concepts to apply to their micro-level dilemma. Second, we ask them to map out the actors identified in their case in institutional and societal contexts, with power relations drawn between them. Together, these two strategies help students realize that while ethical dilemmas feel personal, they are never caused by an individual actor, nor can they be solved by an individual actor. The three-hour workshops involve the following elements: a land acknowledgment, a mindful listening exercise on discrimination, an introductory lecture on ethical theories and equity concepts, a small group activity helping students connect one of the equity concepts to their own experiences in engineering, introduction to two ethical codes, and a guided small-group discussion of systemic features of one ethical case. In some courses, the learning outcomes from the module are assessed through quizzes, final examinations, and assignments. In other courses, the module is considered to be an unassessed supplement to existing course content.

In January 2018, we formalized case study instruction through an elective undergraduate course called "The Art of Ethical and Equitable Decision-Making in Engineering." While the course functions as an intensification of the module experience, we continue to integrate case study workshops into mandatory design and professional practice courses. Finally, we are currently under contract with the University of Toronto Press to publish the full set of case studies and a teaching guide.

3.1.3 Entry Points, Catalysts, and Drivers

The primary driver for this project was an accreditation visit in 2014. Equity (paired with ethics) is one of 12 graduate attributes required for engineering program accreditation in Canada (CEAB, 2021). While all engineering programs at UofT passed the accreditation process that year, the CEAB evaluation team indicated that we had room to grow when it came to addressing equity. As a result, senior administrators became increasingly motivated to support curricular innovation in this area. A secondary catalyst for this project involved an institutional inducement program called the Engineering Instructional Innovation Program (FASE-EIIP). Each year, faculty and staff compete for a small grant established by the dean to enhance innovation in engineering education. The year we applied for an EIIP grant, the FASE strategic plan included an objective to "enhance ethics and equity." We tailored our application to meet this institutional objective and succeeded in winning a grant. Four years later, we won a follow-up grant to publish the case studies as a textbook.

3.1.4 Institutional Supports and Barriers

Our project was supported by several intersecting factors: an explicit learning outcome in the national engineering accreditation system, increasingly politicized engineering students who were comfortable calling out discrimination, growing societal awareness of institutional racism through the Black Lives Matter movement, a supportive dean, growing institutional recognition that engineering educators are responsible for EDI, and a critical mass of faculty, staff, and students willing to do something about it.

Barriers have included ongoing microaggressions within and beyond the university, the meritocratic assumption held by some faculty, staff, and students that improving demographic representation would result in the erosion of excellence in our program, the related assumption that equity and excellence are incompatible, discomfort with EDI material among technically trained professors, concerns about a densely packed curriculum, related concerns about EDI replacing core technical material in this densely packed curriculum, limited institutional power of a voluntary EDI action group, and a longstanding process of professional socialization that privileges technical over social aspects of engineering, as well as the separation of the technical and social dimensions of engineering problems (technical/social dualism and de-politicization).

To draw out our point, we illustrate an example of differentiated uptake. Over the years, we have had greater success integrating the optional case study workshop into communications, leadership, and design courses than into "core" technical courses. The introduction of a new AI minor at the University of Toronto, paired with concerns raised at Faculty Council that students would not have access to field-specific ethics education, provided us with an opportunity to turn the tide by integrating our workshop into the required AI foundations course. While we did not experience any overt resistance to our workshop, we were not invited back the following semester or the following year. This experience contrasts with every other invitation we have had in communications, leadership, and design courses, in which workshops have continued to be integrated, either by the project team or by the course instructor. We deliberately included AI-specific examples of ethical responsibility, such as the use of "model cards" (M. Mitchell et al., 2019), and consulted with a small group of AI educators and students in the workshop design process but fell short of securing long-term curricular integration in this course. When we asked what we could have done differently, we were told that our workshop squeezed out technical aspects of the newly designed curriculum. Rather than forcing our way into technical courses, we have decided to reach all first-year students through their design, communications, and leadership courses.

3.1.5 Success/Impact

As we have navigated barriers and leaned on supports over the past few years, our ethics and equity project has grown. It has been integrated over 30 times in nine different courses, with especially high levels of integration in humanities/social science courses, design, and professional practice curricula. One of the greatest advantages of this resource-based initiative is that we have been able to extend our reach to engineering schools and employers beyond the University of Toronto. We have shared project findings through journal articles and conference papers in the United States and Canada (Rottmann et al., 2015, 2020, 2021; Rottmann & Reeve, 2020), were invited to present our project to an audience of European engineering ethics educators (*Centring Equity in Engineering Ethics Case Study Instruction*, 2021), and have shared our findings with industry partners affiliated with the University of Toronto. More recently, we have been invited to collaborate with colleagues at Penn State's College of Engineering to develop an engineering ethics, equity, and leadership course, and we have been invited to serve on two National Science Foundation (NSF) research advisory boards for projects linking ethics with equity. What we lack in institutional durability, we make up for in curricular agility and outreach. Since 2015, our module has been taken by 11,500 students and engineering graduates, demonstrating that even a micro-level innovation can have a substantial impact on engineering student development.

3.2 Engineering and Social Justice Course, Colorado School of Mines

3.2.1 Institutional Context

Colorado School of Mines (Mines) is a small, public R1 (doctoral university, very high research activity) institution in the United States. Its approximately 7,000 students, roughly 5,400 undergraduates and 1,600 graduate students, are primarily focused on engineering and applied sciences (*Mines by the Numbers*, 2021). Mines holds the highest admission standards in Colorado, as more than 13,000 applicants vie each year for about 1,500 spots in the entering class. Among undergraduates (BS degree) graduating in 2019–2020, 95% were either working in industry or going onto graduate school. Mines' tagline is "Earth, Energy, and the Environment," as many graduates go on to work in fields related to energy, particularly the oil and gas industry, and the environment. Approximately one-third of students are women, and over 28% of all students are from other underrepresented groups (*Mines by the Numbers*, 2021). The institution has undergraduate and graduate programs in humanitarian engineering (HE), including minors in engineering for community development and leadership in social responsibility (*Humanitarian Engineering*, 2021). Along with community and sustainability, one of the three pillars of the HE program is social justice. Growing interest in the HE program emerged partly due to student demand in the last 15 years. Among others, the HE program has been assisted by courses such as engineering and sustainable community development as well as engineering and social justice. Although initially developed in another department, these courses now reside in the Engineering, Design, and Society Department.

3.2.2 Initiative Type – New Course Development

This case study focuses on the development of the course engineering and social justice (ESJ), including its precursors, influences, and consequences. Before the use of the term *social justice* in the HE program, and even before the design of ESJ, two early precursors shaped those developments: the emergence of Engineers without Borders and student interest in leveraging engineering knowledge to help underserved populations.

In the early 2000s, Dr. Bernard Amadei gave a few talks to Mines students, discussing the emergence of Engineers without Borders (EWB) in the United States. A professor of civil engineering at the University of Colorado–Boulder, Dr. Amadei was in the process of growing and establishing EWB USA while working towards connecting with other universities that could start up their own EWB chapters (Amadei, 2021). His talks centered on not just leveraging engineering to attain a high salary but to do good for those who are underserved globally. The ability to make a difference using engineering skills and knowledge resonated with and inspired some Mines students and faculty. As different versions of EWB emerged at Mines, some students wanted to follow engineering for the greater good (or several variations on that theme) as a career track, as an alternative to or detour from the typical industry career track. However, such alternatives or detours were, at that time, largely uncharted territory. Around 2008, Mines professor Juan C. Lucena organized Mines' first workshop connecting students and faculty with the nonprofit sector, particularly those nonprofit professionals whose organizations leveraged engineering knowledge and skills to achieve their core missions. Between the emergence of EWB and student interest in diverse career tracks, there was momentum, but this momentum was not yet explicitly connected with social justice.

3.2.3 Entry Points, Catalysts, and Drivers

Our entry point into social justice involved our work with Mines students focused on EWB projects, which led to research and practice on engineering and sustainable community development (Lucena et al., 2010; Schneider et al., 2008, 2009). In that research, we had realized that of three components of sustainable community development practice – profit, planet, and people – the *people* component was often neglected or misunderstood. During this period, two parallel conversations emerged: one occurred with European scholars in the 2008 conference "Educating Engineers for Sustainable Development," in which sustainable development had been explored in more depth than in US contexts. However, the third component – people – was largely invisible for most of those scholars, and our paper "Where Is Community?" rendered the human component more visible (Schneider et al., 2008). At the same time, the second conversation occurred in the Engineering, Social Justice, and Peace network, wherein the *people* dimension was visible and well emphasized but less consideration was given to issues of community development. Our work connected these two conversations.

The turning point towards social justice came at a 2008 National Academy of Engineering (NAE) workshop (Hollander et al., 2010). Our sustainable community development work resulted in an invitation to the NAE workshop, where we encountered a group of scholars focused on exploring the nexus between engineering and social justice. That group, including scholars Caroline Baillie, George Catalano, Dean Nieusma, and Donna Riley, had in 2004 initiated a conference, and later would launch a journal, exploring that nexus (see Nieusma, 2013). In 2008, Riley's landmark book *Engineering and Social Justice* had just been published. After the conference, three Mines faculty members – Juan C. Lucena, Jen Schneider, and Jon A. Leydens – read Dr. Riley's book and realized that the mindsets in engineering that can keep social justice either at arm's distance or foster its development (Riley, 2008) were quite present – yet largely invisible – within our own institutional context. Also, social justice frameworks facilitated rich understandings of the disproportionate impacts of technological solutions on people and communities. At that point, we decided to create the ESJ course.

3.2.4 Institutional Barriers and Supports

For the ESJ course to earn permanent course status, an affirmative response would be required by our Undergraduate Council, which included a representative from each (almost exclusively engineering and applied science) department. Since Mines has had a conservative history in the oil and gas and mining industries, securing such a response to ESJ required establishing legitimacy and relevance. Multiple encounters with Mines faculty indicated that many were likely to be skeptical of such a social science and humanities course. For instance, one faculty member said that the word "engineering" had no place in a social science course title.

Such skepticism also emerged in the 2008 NAE Workshop. As noted in the summary of the 2008 NAE Workshop, participants engaged in "an intense discussion of the meaning of 'social justice'" (Hollander et al., 2010, p. 28). Several participants were adamantly opposed to exploring the nexus between engineering and social justice, while others indicated that exploration to be an ethical responsibility. Whereas some participants pointed to an incommensurability between engineering as a technical profession and social justice as a political and/or religious concept, other participants both disagreed with that technical/social dualism and accentuated intersections between technical knowledge and social/ethical dimensions that emerge when applying that engineering knowledge to real-world problems. Such problems relate to ecology, transportation, and other instances wherein disproportionate benefits and impacts can occur (Hollander et al., 2010). The closing session of the workshop involved one prominent participant pounding on the podium, insisting that engineering and social justice be kept separate. Similarly, some Mines faculty had told us that a course on ESJ did not make sense because the two exist in completely different disciplinary areas and fields of practice. Certainly, in the 2010 context, an ESJ course was perceived as an identity threat to some engineering faculty. Furthermore, Mines donors have expressed discomfort with social justice in the title of the HE program and in other campus initiatives.

3.2.5 Success/Impact

Riding strong student demand for courses like engineering and sustainable community development and for alternative career tracks, Lucena, Schneider, and Leydens wrote a proposal to the National Science Foundation (NSF) to investigate the (in)commensurability between engineering and social justice historically, philosophically, conceptually, and ethically as well as in the engineering curriculum (Lucena et al., 2009). Although that NSF funding provided time to research and create the ESJ course, it also provided *legitimacy* among our university colleagues who might otherwise have been more skeptical of ESJ. After faculty secured that NSF funding and developed the course, our Undergraduate Council approved ESJ.

What followed in the wake of the ESJ course was the result of significant work by multiple faculty, largely sparked and guided by student interest. Immediately, the ESJ course was well received, and many students expressed interest in a minor involving ESJ and related courses. Soon, we developed the humanitarian engineering minor, which included ESJ and spawned interest in an HE program that emphasized "educating engineers to serve communities by collaboratively identifying problems and providing solutions that are just, socially responsible, and sustainable" (*Humanitarian Engineering*, 2021). Also, due to student demand, the HE minor grew into two specialized undergraduate minors and two design-oriented engineering focus areas emphasizing community, sustainability, and social justice along with complementary concepts such as human-centered problem definition and design. Later, and also driven largely by student interest, an HE graduate program emerged, with engineering for social justice comprising a significant part of the graduate program introductory course. Some faculty involved with the initial 2009 NSF grant later went on to write an additional grant (Johnson et al., 2016), which investigated student perceptions on the integration of social justice in a technical course, called Introduction to Feedback Control Systems (Leydens et al., 2021). In 2015–2016, three Mines faculty won the Exemplar in Engineering Ethics Education Award from the National Academy of Engineering (NAE) for an initiative on "Enacting macroethics: making social justice visible in engineering education" (Leydens et al., 2016). In 2018, Leydens and Lucena published a book focused on integrating social justice in three components of the engineering curriculum: the engineering sciences, engineering design, and courses in the humanities and social sciences for engineers (Leydens & Lucena, 2018). Student-led, grassroots calls for social justice have emerged in spaces, such as the Socially Responsible Scientists and Engineers student organization, which has brought more visibility to social justice via events focused on issues such as homelessness (framed as *housing insecurity*) and environmental justice. In 2020, shortly after the murder of George Floyd, several Mines students wrote and signed a letter to the Mines administration, advocating for social justice to be more explicitly integrated in the engineering curriculum. A movement that began with a single course in 2010 has grown considerably, and multiple indicators – including sustained student interest, two minors, a graduate program, and strong research foundations – suggest that growth will continue.

3.3 Integrated Engineering Program, University of San Diego

3.3.1 Institutional Context

The University of San Diego (USD) is a four-year private, mid-sized, primarily residential liberal arts university located in San Diego, California, United States of America. The student population is majority undergraduate (approximately 5,500 undergraduate students), and USD is classified as an R2 institution (doctoral university, high research activity) and recognized as a "community engagement institution." USD brands itself as a contemporary Catholic university that embraces the Catholic social traditions and views peace and justice as inseparable from education, scholarship, and service. The university's strategic plan, *Because the World Needs Changemakers*, has catalyzed the integration of social justice topics into university-wide curricula. The Shiley-Marcos School of Engineering has an approximate enrollment of 600 students and awards primarily joint BS/BA undergraduate degrees. Students in all engineering majors complete a 4.5-year program that requires graduates to have a robust education in both engineering and the liberal arts.

3.3.2 Initiative Type – Program Development

The Integrated Engineering Department at the University of San Diego is a relatively new program, created in 2017, that aims to graduate students who are equipped to "tackle the world's most complicated problems with a sociotechnical mindset to design a sustainable future" (*Integrated Engineering*, n.d.). The Integrated Engineering program takes an interdisciplinary approach to engineering while infusing social justice and sustainability throughout the entire curriculum. The program's objectives are to develop graduates who (1) apply an interdisciplinary set of technical, leadership, and other professional skills to address important challenges facing society; (2) practice engineering with a holistic understanding of how engineers engage with stakeholders and impact society; and (3) have a critical awareness of their personal attitudes, behaviors, and values and the ways in which these align with their professional aspirations (*Educational Objectives and Student Outcomes*, n.d.). Integrated Engineering students, who earn a dual BS/BA degree in engineering, share a common set of required major courses and have an option to choose a concentration in sustainability, embedded software, biomedical engineering, or engineering and the law (Chen & Hoople, 2017). Students also have the option to co-design an "individual plan of study" concentration with a faculty adviser to facilitate their career goals.

In creating this program, faculty built upon the liberal arts mission of the university and sought to develop a more holistic approach to engineering coursework. Social justice is a core value within this program and has informed the design of much of the newly created curriculum. For example, two engineering courses were designed to satisfy the university core curriculum requirement for "diversity, inclusion, and social justice" (DISJ), which requires students to reflect on their own privilege and personal experience; critique historical and contemporary social, political, and economic factors that affect DISJ; and critically examine the intersections of power relationships that lead to systemic inequities. *User-Centered Design* introduces all first- and second-year engineering students to the concepts of social justice, privilege, and oppression in the context of engineering design (Chen et al., 2020; Chen, Mejia et al., 2019; Mejia et al., 2018; Momo et al., 2020). In the senior year, Integrated Engineering students take a deeper dive with *Engineering and Social Justice*, a writing-intensive course that critically analyzes engineering practice (Chen, Mejia et al., 2019; Mejia et al., 2018; Mejia & Revelo, 2021; Momo et al., 2020).

In addition to the courses previously described that explicitly address issues of social justice, faculty have consciously infused social concepts throughout the required courses in the major to help students see that engineering is a sociotechnical endeavor. These courses are important for providing a scaffold as well as continuity throughout the program for students to grapple with concepts of social justice. Program courses emphasize that engineering problems cannot be reduced to a set of abstract, technical concepts; rather, to be successful, engineers must recognize the ways in which social and technical elements of problems are inextricably linked (Chen, Peters et al., 2019; Hoople & Choi-Fitzpatrick, 2020; Leydens & Lucena, 2018; Mejia et al., 2018; Momo et al., 2020). This has proven to be a very successful entry point for engaging students in conversations about social justice. By contextualizing abstract technical problems, students can begin to link engineering to the world around them, even in codified courses like statics (Chen & Przestrzelski, 2019; Chen & Wodin-Schwartz, 2019; Momo et al., 2020). By then focusing on how engineering manifests in and impacts society, students begin to see how engineering is sociotechnical, rather than just technical, and how some engineering solutions have led to injustice (Chen et al., 2020; Mejia et al., 2018; Momo et al., 2020).

In addition to curricular efforts, the Integrated Engineering Department has also focused on building sustainable program pathways for faculty and students. An important criteria while hiring faculty in this department has been a demonstrated commitment to issues of social justice, diversity, ethics, and/or inclusion. Creating a culture of faculty engagement around these issues has supported the creation of a robust, innovative program that places issues of sustainability and social justice at the center of the student experience. This synergy between faculty led to an NSF grant award to learn from and embed culture-based pedagogies (culturally sustaining, responsive, and relevant pedagogies) into our engineering courses to create a more inclusive learning environment (Hoople et al., 2018). Students have also created a self-sustaining co-curricular branch through a student club whose mission is to build community within the major and help with program recruitment and retention.

3.3.3 *Entry Points, Catalysts, and Drivers*

The Integrated Engineering Department was created and developed alongside the award of a National Science Foundation (NSF) Revolutionizing Engineering and Computer Science Departments (RED) grant, with the goal of developing change-making engineers who practice engineering with consideration for social justice, peace, humanitarian advancement, and sustainable practices (Roberts et al., 2015). The principal investigators of the RED grant consisted of the school leadership at the time of submission (the dean, associate dean, and all three department chairs). In addition to the important financial support provided by this grant, the prestige and support of the National Science Foundation for this effort was an important entry point, catalyst, and driver for change.

Another catalyst for the creation of the program was the development of a more flexible pathway for students to graduate with an engineering degree. The length of the programs can be a surprise and deterrent to students, especially when there is very little room for elective courses. The creation of the Integrated Engineering program coinciding with new core curriculum requirements, beginning in the 2017–2018 academic year, allowed us to design courses from scratch that embedded both levels of the DISJ flag within our major, enabling students to have more flexibility in elective courses.

3.3.4 *Institutional Barriers and Supports*

To secure institutional support, the new program was strategically designed by the leadership team to align with the university's Catholic mission. The university leadership and the board of trustees were enthusiastic supporters of the proposal, which allowed the dean to hire three new faculty members – the authors of this case study (Chen and Hoople) and a professor of praxis who is an expert in engineering and social justice (Baillie & Pawley, 2012) – to design the new curriculum from the ground up.

While the program was well aligned with the university mission at large, there was a small but vocal contingent of engineering faculty who were strongly opposed to the Integrated Engineering program. As courses with sociotechnical and social justice objectives were introduced into the curriculum, the department as a whole, as well as individual Integrated Engineering faculty, publicly faced comments from other engineering faculty that the new degree would not be worthwhile due to its lack of technical rigor. Sadly, this sentiment is not unique to our efforts for integrating social justice into engineering (Riley, 2017).

Disparaging comments from colleagues have trickled down to students, influencing their attitudes and receptiveness to sociotechnical/social justice course content. For instance, some cohorts of undeclared engineering students were advised by some faculty to avoid declaring Integrated Engineering due to its lack of technical rigor. In one extreme case, a faculty member even falsely advised a recruiter that the educational quality of the Integrated Engineering program would not meet their entry criteria. In a school meeting early in the program's development, several faculty from other engineering programs suggested that, if our graduates could not acquire jobs, we should refund their tuition. For program faculty, the difficulty of navigating the politics of tenure and academia has also played a role in the evolution of course content (Chen, Mejia et al., 2019).

3.3.5 *Successes/Impacts*

While still a young program, Integrated Engineering has been quite successful. Not only are the large majority of graduates (from four cohorts) gainfully employed or currently pursuing graduate degrees (*Integrated Engineering Career Outlooks*, n.d.), but the faculty have also secured multiple NSF grants aligned with their social justice mission (Hoople et al., 2018; Mejia, 2021; Mejia & Popov,

2018). Additionally, the program underwent its initial accreditation through ABET with resounding success. We have also begun to develop a reputation as leaders in this work through presentations at national conferences and workshops.

Throughout these early years, it has become clear that the shared vision of the faculty has been instrumental to the success of the department. The close-knit relationships between department faculty allowed the program and faculty to thrive throughout the 2020 global pandemic (Chen, Gelles et al., 2021; Gelles et al., 2020). Looking ahead, the department is now on the precipice of enrollments spilling over into a second section for its introductory courses – a major milestone for our small program. While the path was not always easy, the future of the department looks bright.

4 Cross-Case Analysis

Reading these case studies makes one thing abundantly clear: there is no one-size-fits-all approach to integrating social justice into engineering curricula, for at least two related reasons. First, no single initiative fits every institutional context or student population, and second, change agents with greater access to institutional power and authority may succeed where their colleagues may fail. That said, a high-level comparison reveals important similarities in the pathways to institutionalization of social justice curriculum across different contexts. We discuss these similarities and differences now by comparing the three conceptual anchors introduced earlier in this paper: objectives, theories of change, and institutional contexts.

4.1 What Are Our Objectives as Social Justice Engineering Educators?

The primary objective in all three case studies was to help students develop a critical awareness of social issues within an engineering context. In the UofT case, the objective was to formalize social justice and equity by pairing them with a longstanding aspect of engineering education – professional ethics. The eventual institutionalization of this small-scale pedagogical innovation was facilitated by senior administrative drive to meet national accreditation requirements.

In the Mines case, the initial discussion was driven by students' inquiry, asking, "What is engineering for?" This inquiry led to the objective to augment students' understanding of how human capabilities (Nussbaum, 2007) might feature more prominently in engineering. The result of this is the ESJ course that addresses engineering's impact on people and communities and examines what human capabilities engineers might focus on to make the greatest difference. At Mines, students opt into the engineering and social justice course as an elective, meaning, that students share similar values and choose to be there; many of those students are in the humanitarian engineering minor, but not all, and among students, the course has a reputation as "life-changing" and/or "perspective-shifting."

In the USD case, the mission of the program was driven by faculty values, resulting in an objective to promote sociotechnical thinking (as opposed to techno-centric thinking) in engineers such that graduates would be motivated to work on humanity's most urgent challenges. In the process, students develop a holistic view on engineering – what it means to be an engineer and what engineering practice looks like. To do so, students develop a critical awareness of the impact of engineering on society and reflect on their alignment of personal and professional aspirations. Similar to the UofT case, students in all engineering majors are required to take a lower-division course that discusses social justice in an engineering context. This provides context for the Integrated Engineering major, and similar to the Mines case, students then demonstrate their buy-in when they choose to declare it as their engineering major.

4.2 How Have Key Actors Conceptualized Change Over the Course of the Initiative?

In all three cases, the combination of strategic leveraging of institutional elements and trailblazing by those on the ground was important for facilitating change. The three cases reveal that change can never be unilaterally planned in a complex context with actors that have differential power or decision-making authority. All our enacted theories of change depended on activist energies of the students, staff, and/or faculty who had personal drives to make change happen in ways that addressed the strategic priorities (e.g., accreditation, university initiatives) of the administration.

At UofT, the administration's call for pedagogical innovation proposals, combined with a national accreditation requirement of universally accessible education in engineering ethics and equity, provided a pathway for the modular initiative to be generated and widely implemented. These centralized, top-down supports paved the way for a productive partnership between a staff researcher trained in social justice education (Rottmann) and a senior faculty member with professional credibility and institutional authority (Reeve). To date, this small-scale intervention has touched 11,500 students and professionals directly, demonstrating the power of small, incremental approaches to enact socially just change. While this curricular change effort has yet to reach the scale of USD's program, it illustrates Begay-Campbell's (2010) theory of change, which asserts that we can lay the road while walking, learning, and growing with each step.

At Mines, the NSF grant award to develop the ESJ course was leveraged as a source of legitimacy with the undergraduate council. Once the course was approved, the trailblazing happened inside the classroom by forging new connections at the nexus of engineering and social justice. The course itself was groundbreaking. Many students found the course to be transformative, showing them a side of engineering they did not know existed, particularly regarding the complexities associated with how engineering shapes (and is shaped by) societal forces. The ESJ course showed students a new lens for engineering that they did not know was possible, in some cases rekindling their interest in and passion for engineering.

In USD's case, the university and school of engineering administrators were the visionaries that paved the way for the Integrated Engineering program to come to life. They were able to strategically leverage multiple levels of institutional change by earning a RED grant while taking advantage of several coinciding events. The founding faculty accepted this opportunity to trailblaze an improved education experience for engineering students based on their own experiences with racism and sociotechnical dualism. These faculty were taught when they were students that both social and technical realms were important and tied to each other, yet never taught how to connect the two spheres. Much of the Integrated Engineering curriculum (at multiple layers – within projects, courses, and the program as a whole) has arisen through faculty working together to deconstruct canonical engineering curricula and reconstruct fundamentals through a sociotechnical lens on behalf of students to help them see and continue to draw their own connections.

4.3 How Have Normative Engineering Mindsets, Cultures, and Ideologies within Institutional Contexts Constrained the Success of Our Social Justice Initiatives?

Through discussion, the four authors found a common theme in all our cases: normative engineering mindsets, cultures, and ideologies have shaped what it means to be an engineer (Cech, 2013; Faulkner, 2000; Riley, 2008). For instance, many engineering educators hold the ideology of technical/social dualism, which made it difficult for the authors in all three institutional contexts to integrate social justice work into the curriculum without many of our colleagues feeling like something central to the profession (i.e., its technical components or nature) was being lost. Similarly, the mindset equating math and science with rigor, and further equating rigor with excellence, made it difficult for some of us to frame social justice engineering as a significant, relevant, or even cognitively challenging aspect of engineering education. As may be expected by any social justice initiative that goes against the grain of professional norms, our initiatives were considered threatening or irrelevant to many faculty members and students.

The inertia of these deeply entrenched ideologies continues to be a key barrier in the broadening of engineering education to have a greater focus on social transformation. Additionally, new initiatives in engineering that focus on employability skills and fast-paced innovation, such as entrepreneurship, present competing demands for attention for educators. Compounding these economic and ideological complications is the challenge to demonstrate the success or failure of social justice interventions on the curriculum, particularly for engineering faculty who are accustomed to examining quantitative data with clear-cut answers. Nevertheless, all four authors have investigated the impact of our respective interventions (Chen & Przestrzelski, 2019; Gelles & Lord, 2020; Hoople et al., 2020; Hoople & Choi-Fitzpatrick, 2020; Leydens et al., 2021; Rottmann & Reeve, 2020) and found them to meet key learning objectives. Please consult the studies cited earlier for additional details.

For engineering educators who are social justice–curious and for experienced social justice educators who are new to engineering, it can be personally and structurally challenging to start with an ideological overhaul of the profession. Instead, we encourage them to start small, join a social justice committee, buy a coffee for a colleague doing this work, ask questions, and scan the three case studies for actions that seem manageable in their respective institutional contexts. Fortunately, with the development of engineering education as a field of study and practice, a new generation of academics has begun to integrate social justice into traditional engineering approaches from the start of their careers, slowly supplementing purely technical norms with sociotechnical practices. While the mythology of a purely technical profession has not dissipated or lost much power over the years, it is at least beginning to diversify through a wider range of curricular initiatives.

5 Discussion and Recommendations

The context-specific nature of these case studies means that it is not possible to generate a list of plug-and-play steps, ready to implement change across any institution. Every institution will have different opportunities, barriers, and entry points. Our goal in this chapter is to illustrate the breadth of how and where social justice can appear in engineering curricula to inspire the reader to consider how and what might fit into their own institutional context.

Social justice work is challenging and cannot easily be summarized in a few simple steps; the work required is dependent on the context of the institution in which it is attempted and the identities of the people attempting it (Chen, Mejia et al., 2019). Our case studies provide descriptive narratives of an empirical rather than theoretical nature in order to highlight contextual cues on what may be accomplished at each step along the way. We hope readers are able to identify with some aspects of the three initiatives in ways that resonate with their local contexts.

For faculty who are interested in doing this work, we pose the following questions to identify catalysts, institutional entry points, barriers, and supports:

1. Integrating social justice into the curriculum requires a mindset and pedagogical shift for many faculty. How does your own positionality and position in the academy affect how your efforts might be received at your institution? Do you have privilege that you can leverage for others, or are there allies that you can identify to work with?
2. When social justice initiatives are institutionalized and integrated into the engineering curriculum, rather than relegated to electives or left exclusively to humanities/social science faculty to

teach, the relevance and importance of social justice to the engineering discipline is elevated. Yet engineering faculty typically receive little pedagogical training, let alone support in developing a critical consciousness and building an expertise in social justice topics (Chen, Lord et al., 2021). Who can you partner with across disciplines, or what centers or resources do you have on campus to help you do this work?

3 Becoming a transformative change agent in any university context will be buffeted by better understanding the mindsets (Riley, 2008) and ideologies (Cech, 2013, 2014; Riley, 2008) that permeate – but are often invisible in – many engineering and engineering education contexts. Hence, faculty will need to spend some time learning how to leverage these mindsets and ideologies across the engineering curriculum (e.g., see Leydens & Lucena, 2018). Only then can we pass these lessons on to students directly and indirectly through our pedagogy and curricula. How can you help students develop their own critical awareness by encouraging them to identify the mindsets and/or ideologies in creative places (e.g., problem statements, course texts, syllabi, university communications, and other locations within the university context and beyond)?
4 To ensure long-term viability of social justice within the curriculum, programs should be aligned with institutional, departmental, and other elements of campus culture. What are the different terms used to describe concepts related to social justice at your institution or local/national context? What centers, initiatives, or funding sources on (or off) campus use these terms, and how might you capitalize on these within your context?

Moreover, we encourage those who are new to this work, or those who are at institutions without any existing support, to start small. Any social justice initiatives, even if minor, will begin the process of identifying allies and barriers. Beginning with small efforts can be a strategy to get buy-in from colleagues and eventually lead to more inspiration for larger changes. We encourage faculty to start with a module in a course, then possibly integrate these topics throughout a course, before creating stand-alone courses and reforming entire programs. In some cases, it may be possible to reach every student without reforming entire programs. We encourage our graduate readers to engage with their academic circles to learn what exists and begin gathering case studies they could use in their future work. Engineering academics can now engage with like-minded folks in various networks, such as at the *Engineering, Social Justice, and Peace* conference, and in special interest groups at conferences that welcome this work, such as *Equity, Culture, and Social Justice* within the American Society for Engineering Education (ASEE), the *Ethics* and *Gender and Diversity* groups within the European Society for Engineering Education (SEFI), and the Equity, Diversity, and Inclusion special interest group within the Canadian Engineering Education Association (CEEA). Our charge to the reader is not that you need to overhaul your program, but that each individual needs to start somewhere. As you do this work, continue to seek out allies, identify entry points, and begin to mobilize power.

6 Limitations

By choosing to focus on three case studies, we have privileged analytic depth over breadth in ways that cannot illustrate the full range of curricular initiatives in social justice engineering education. Similarly, given the North American context of the three case studies, we have privileged a relatively narrow set of national drivers for social justice education.

What did we miss by being North American–centric, and how can educators from other contexts make use of our findings? Many nations are signatories of the Washington Accord, making global engineering accreditation somewhat similar to ABET across national contexts. Similarities in professional accreditation aside, we likely missed a wide range of national policy drivers for social justice work in this chapter. Fortunately, our deliberate diversification of institutional contexts and levels of analysis – featuring small social justice–minded engineering schools with large-scale programmatic reform and large research-intensive schools with smaller social justice initiatives – allows engineering educators across contexts to tailor case study strategies to their own contexts, adapting and extending the types of curricular innovation that are most likely to work for them in their respective institutional and programmatic contexts.

Another limitation of choosing to explore this topic through case studies is that we cannot show the myriad ways in which engineering educators have examined the intersection of social justice and engineering, be it through the development of theoretical frameworks, qualitative interviews, or quantitative surveys (Canney & Bielefeldt, 2016; Kabo & Baillie, 2009; for example, Riley, 2008). We encourage readers to explore the many foundational social justice studies cited in this chapter.

7 Conclusion

Engineering culture has a long history of resistance to curricular change, particularly related to topics that are not viewed as sufficiently "rigorous" or technical (Riley, 2017). Understanding the normative ideologies and mindsets that we are working to overcome provides a framework through which we can develop strategies to begin this work. As seen in our case studies, there is no one-size-fits-all approach to integrating social justice concepts into the curriculum. Efforts to include social justice should be based on a theory of change appropriate for the local context and should be accessible to key actors. While this work is hard, engineering faculty need not, and in fact cannot, do it alone. Engineering faculty must engage with colleagues in various academic networks, building interdisciplinary partnerships with humanities and social science faculty. External collaborators and funding can also help build credibility within any institution. The last decade saw an explosion in scholarship around engineering and social justice. Since theory often precedes practice, we are hopeful that the next decade will see a proliferation of locally institutionalized social justice practices in engineering education.

Acknowledgments

Authors Chen and Hoople would like to thank Drs. Caroline Baillie, Mark Chapman, Susan Lord, Alex Mejia, and our dean, Chell Roberts, for their collaboration in creating the Integrated Engineering Department. Partial support for this work was provided by the National Science Foundation Award No. 1519453. Author Leydens would like to thank Juan C. Lucena for his input into the description of the Mines case in this chapter, and the National Science Foundation for its support through grants SES-0930213 and EEC-1441806. Author Rottmann would like to thank Professors Doug Reeve and Emily Moore for their contribution to this project, the UofT Faculty of Applied Science and Engineering for awarding us two grants in support of the project – (1) *EIIP18–03–Scaling up Engineering Ethics and Equity: Lesson plans for course integration* and (2) *EIIP13–04 – Engineering Case studies for Ethical and Equitable Decision-making*, the Troost ILead Community of Practice, the 22 engineers who participated in critical incident interviews, and the 12 faculty members who have integrated our case studies into their courses. Any opinions, findings, and conclusions or recommendations expressed in this material are those of the author(s) and do not necessarily reflect the views of the funding agencies.

References

Amadei, B. (2021). *Bernard Amadei*. www.colorado.edu/ceae/bernard-amadei

Baillie, C., & Catalano, G. (2009). Engineering and society: Working towards social justice, part I: Engineering and society. *Synthesis Lectures on Engineers, Technology, and Society*, *4*(1), 1–114.

Baillie, C., & Pawley, A. (2012). *Engineering and social justice: In the university and beyond*. Purdue University Press.

Bamford, D. R., & Forrester, P. L. (2003). Managing planned and emergent change within an operations management environment. *International Journal of Operations & Production Management*, *23*(5), 546–564. https://doi.org/10.1108/01443570310471857

Begay-Campbell, S. (2010). Walking in beauty on an ever-changing path: A leadership perspective from a native American woman engineer. *Leadership and Management in Engineering*, *10*(4), 150–152. https://doi.org/10.1061/(asce)lm.1943-5630.0000086

Canney, N. E., & Bielefeldt, A. R. (2016). Validity and reliability evidence of the engineering professional responsibility assessment tool. *Journal of Engineering Education*, *105*(3), 452–477.

CEAB. (2021). *Accreditation criteria and procedures*. Engineers Canada. https://engineerscanada.ca/sites/default/files/2021-11/2021%20Accreditation%20Criteria%20Book%20Word.pdf

Cech, E. A. (2013). The (mis)framing of social justice: Why ideologies of depoliticization and meritocracy hinder engineers' ability to think about social injustices. In J. C. Lucena (Ed.), *Engineering education for social justice: Critical explorations and opportunities* (pp. 67–84). Springer.

Cech, E. A. (2014). Culture of disengagement in engineering education? *Science, Technology & Human Values*, *39*(1), 42–72.

Centring equity in engineering ethics case study instruction. (2021, April 12). www.sefi.be/2021/04/12/centring-equity-in-engineering-ethics-case-study-instruction/?utm_source=newsletter&utm_medium=email&utm_campaign=SEFI+Ethics+Special+Interest+Group+Newsletter

Chen, D. A., Chapman, M. A., & Mejia, J. A. (2020). Balancing complex social and technical aspects of design: Exposing engineering students to homelessness issues. *Sustainability: Science Practice and Policy*, *12*(15), 5917.

Chen, D. A., Gelles, L. A., Lord, S. M., Hoople, G. D., Mejia, J. A., & Chapman, M. A. (2021). Lessons learned: How our agile department survived the COVID-19 pivot American society for engineering education. In *ASEE annual conference (Virtual)*. www.asee.org/public/conferences/223/papers/32647/view

Chen, D. A., & Hoople, G. D. (2017). Contextualizing a new general engineering curriculum in the liberal arts. In *2017 ASEE annual conference & exposition*. https://peer.asee.org/contextualizing-a-new-general-engineering-curriculum-in-the-liberal-arts

Chen, D. A., Lord, S. M., Mejia, J. A., & Hoople, G. D. (2021). A critical reflection on the challenges of implementing culturally sustaining pedagogy. In *2021 IEEE Frontiers in Education Conference (FIE)* (pp. 1–5).

Chen, D. A., Mejia, J. A., & Breslin, S. (2019). Navigating equity work in engineering: Contradicting messages encountered by minority faculty. *Digital Creativity*, *30*(4), 329–344.

Chen, D. A., Peters, M. R., Hoople, G. D., Mejia, J. A., & Lord, S. M. (2019). Vocation in the engineering curriculum: Challenging students to recognize their values. *Peer.asee.org*. https://peer.asee.org/vocation-in-the-engineering-curriculum-challenging-students-to-recognize-their-values.pdf

Chen, D. A., & Przestrzelski, B. (2019). Let the composites speak: Using statics to critically evaluate gerrymandering. In *2019 IEEE frontiers in education conference (FIE)* (pp. 1–4).

Chen, D. A., & Wodin-Schwartz, S. (2019). Contextualizing statics: Our process and examples. In *2019 ASEE annual conference & exposition*. https://peer.asee.org/contextualizing-statics-our-process-and-examples

Downey, G. L., Lucena, J. C., Moskal, B. M., Parkhurst, R., Bigley, T., Hays, C., Jesiek, B. K., Kelly, L., Miller, J., Ruff, S., Lehr, J. L., & Nichols-Belo, A. (2006). The globally competent engineer: Working effectively with people who define problems differently. *Journal of Engineering Education*, *95*(2), 107–122.

Educational objectives and student outcomes. (n.d.). Retrieved October 27, 2021, from www.sandiego.edu/engineering/programs/integrated-engineering/objectives-outcomes.php

FASE. (2021). *By the Numbers 2021*. University of Toronto. www.engineering.utoronto.ca/files/2021/09/By_The_Numbers_2021_Web-1.pdf

Faulkner, W. (2000). Dualisms, hierarchies and gender in engineering. *Social Studies of Science*, *30*(5), 759–792.

Faulkner, W. (2007). 'Nuts and bolts and people': Gender-troubled engineering identities. *Social Studies of Science*, *37*(3), 331–356.

Faulkner, W. (2009a). Doing gender in engineering workplace cultures. I. Observations from the field. *Engineering Studies*, *1*(1), 3–18.

Faulkner, W. (2009b). Doing gender in engineering workplace cultures. II. Gender in/authenticity and the in/visibility paradox. *Engineering Studies*, *1*(3), 169–189.

Fosfuri, A., & Rønde, T. (2009). Leveraging resistance to change and the skunk works model of innovation. *Journal of Economic Behavior & Organization, 72*(1), 274–289. https://doi.org/10.1016/j.jebo.2009.05.008

Franklin, U. M. (1999). *The real world of technology*. House of Anansi.

Frediani, A. A., & Boano, C. (2012). Processes for just products: The capability space of participatory design. In I. Oosterlaken & J. van den Hoven (Eds.), *The capability approach, technology and design* (pp. 203–222). Springer.

Gelles, L. A., & Lord, S. M. (2020). The final straw: Incorporating accessibility and sustainability considerations into material selection decisions. In *Proceedings of the ASEE annual conference*. https://peer.asee.org/the-final-straw-incorporating-accessibility-and-sustainability-considerations-into-material-selection-decisions

Gelles, L. A., Lord, S. M., Hoople, G. D., Chen, D. A., & Mejia, J. A. (2020). Compassionate flexibility and self-discipline: Student adaptation to emergency remote teaching in an integrated engineering energy course during COVID-19. *Education Sciences, 10*(11), 304.

Gianniny, O. A., Jr. (2004). A century of ASEE and liberal education (or how did we get here from there, and where does it all lead?). In D. F. Ollis, K. A. Neeley, & H. C. Luegenbiehl (Eds.), *Liberal education in twenty-first century engineering: Responses to ABET/EC 2000 criteria* (pp. 320–346). Peter Lang.

Godfrey, E. (2014). Understanding disciplinary cultures: The first step to cultural change. In A. Johri & B. M. Olds (Eds.), *Cambridge handbook of engineering education research* (pp. 437–455). Cambridge University Press Cambridge.

Graham, R. (2012a). *Achieving excellence in engineering education: The ingredients of successful change*. The Royal Academy of Engineering & Massachusetts Institute of Technology.

Graham, R. (2012b). The one less traveled by: The road to lasting systemic change in engineering education. *Journal of Engineering Education, 101*(4), 596–600. https://doi.org/10.1002/j.2168-9830.2012.tb01120.x

Green-Pedersen, C., & van Kersbergen, K. (2002). The politics of the 'third way': The transformation of social democracy in Denmark and the Netherlands. *Party Politics, 8*(5), 507–524.

Herkert, J. R. (2005). Ways of thinking about and teaching ethical problem solving: Microethics and macroethics in engineering. *Science and Engineering Ethics, 11*(3), 373–385.

Hickmon, G. (2015). Double consciousness and the future of service-learning. *Michigan Journal of Community Service Learning, 22*(1), 86–88.

Hollander, R., National Academy of Engineering, & Advisory Group for the Center for Engineering, Ethics, and Society. (2010). *Engineering, social justice, and sustainable community development: Summary of a workshop*. National Academies Press.

Hoople, G. D., Chen, D. A., Lord, S. M., Gelles, L. A., & Bilow, F. (2020). An integrated approach to energy education in engineering. *Sustainability: Science Practice and Policy*. www.mdpi.com/879028

Hoople, G. D., Chen, D. A., Lord, S. M., & Mejia, J. A. (2018). *Reimagining energy: Exploring inclusive practices for teaching energy concepts to undergraduate engineering majors* (1836504). National Science Foundation. www.nsf.gov/awardsearch/showAward?AWD_ID=1836504&HistoricalAwards=false

Hoople, G. D., & Choi-Fitzpatrick, A. (2020). *Drones for good: How to bring sociotechnical thinking into the classroom*. Morgan & Claypool Publishers.

Horton, M., Friere, P., Bell, B., Gaventa, J., & Peters, J. (1990). *We make the road by walking: Conversations on education and social change* (Vol. 32, p. 256). Temple University Press.

Humanitarian engineering. (2021). https://humanitarian.mines.edu/

Integrated engineering. (n.d.). Retrieved December 1, 2021, from www.sandiego.edu/engineering/undergraduate/integrated-engineering/

Integrated engineering career outlooks. (n.d.). Retrieved December 1, 2021, from www.sandiego.edu/engineering/undergraduate/integrated-engineering/career-outlooks.php

Johnson, K. E., Moskal, B. M., & Leydens, J. A. (2016). *Research initiation grant: Social justice in engineering with a focus on control systems* (1441806). National Science Foundation. www.nsf.gov/awardsearch/showAward?AWD_ID=1441806&HistoricalAwards=false

Kabo, J., & Baillie, C. (2009). Seeing through the lens of social justice: A threshold for engineering. *European Journal of Engineering Education, 34*(4), 317–325.

Leydens, J. A., Johnson, K. E., & Moskal, B. M. (2021). Engineering student perceptions of social justice in a feedback control systems course. *Journal of Engineering Education, 110*(3), 718–749. https://doi.org/10.1002/jee.20412

Leydens, J. A., & Lucena, J. C. (2018). *Engineering justice: Transforming engineering education and practice*. John Wiley & Sons.

Leydens, J. A., Lucena, J. C., & Johnson, K. (2016). *Enacting macroethics: Making social justice visible in engineering education*. Online Ethics Center for Engineering and Science. https://onlineethics.org/cases/nae-exemplars-engineering-ethics-education/enacting-macroethics-making-social-justice-visible

Leydens, J. A., Lucena, J. C., & Riley, D. M. (2022). Engineering education and social justice. In G. Noblit (Ed.), *Oxford research encyclopedia of education*. Oxford University Press. https://doi.org/10.1093/acrefore/9780190264093.013.1772

Lucena, J. C., Leydens, J. A., & Schneider, J. (2009). *Engineering and social justice: Research and education of (in) commensurable fields of practice* (No. 0930213 -). National Science Foundation. www.nsf.gov/awardsearch/showAward?AWD_ID=0930213&HistoricalAwards=false

Lucena, J. C., Schneider, J., & Leydens, J. A. (2010). *Engineering and sustainable community development* (C. Baillie, Ed.). Morgan & Claypool Publishers.

Marx, K., & Engels, F. (1906/2010). Bourgeois and Proletarians. In R. Altschuler (Ed.), *Seminal sociological writings: From Auguste Comte to Max Weber* (pp. 11–22.). Gordian Knot Books.

Mejia, J. A. (2021). *CAREER: Breaking the tradition of silence through conocimiento and consciousness raising among Latinx engineers* (No. 2151404). National Science Foundation.

Mejia, J. A., Chen, D. A., Dalrymple, O., & Lord, S. M. (2018). Revealing the invisible: Conversations about – Isms and power relations in engineering courses. In *2018 ASEE annual conference proceedings*. ASEE. https://peer.asee.org/revealing-the-invisible-conversations-about-isms-and-power-relations-in-engineering-courses

Mejia, J. A., & Popov, V. (2018). *Research: Asset-based practices in engineering design (APRENDE): Formation of engineers through a funds of knowledge approach* (No. 1826354). National Science Foundation.

Mejia, J. A., & Revelo, R. A. (2021). Critical literacies in practice: Deconstructing engineering through an engineering social justice course. In A. Wilson-Lopez, E. Tucker-Raymond, J. A. Mejia, & A. Esquinca (Eds.), *The literacies of design: Studies of equity and imagination with engineering and making*. Purdue University Press.

Mines by the numbers. (2021). www.mines.edu/about/by-the-numbers/

Mitchell, M., Wu, S., Zaldivar, A., Barnes, P., Vasserman, L., Hutchinson, B., Spitzer, E., Raji, I. D., & Gebru, T. (2019). Model cards for model reporting. In *Proceedings of the conference on fairness, accountability, and transparency* (pp. 220–229).

Mitchell, T. D., Donahue, D. M., & Young-Law, C. (2012). Service learning as a pedagogy of whiteness. *Equity & Excellence in Education: University of Massachusetts School of Education Journal, 45*(4), 612–629.

Momo, B., Hoople, G. D., Chen, D. A., Mejia, J. A., & Lord, S. M. (2020). Broadening the engineering canon: How culturally responsive pedagogies can help educate the engineers of the future. *Murmurations Emerg. Equity in Education, 2*, 6–21.

Murphy, C., & Gardoni, P. (2012). Design, risk and capabilities. In I. Oosterlaken & J. van den Hoven (Eds.), *The capability approach, technology and design* (pp. 173–188). Springer.

Nichols, C., & Dong, A. (2012). Re-conceptualizing design through the capability approach. In I. Oosterlaken & J. van den Hoven (Eds.), *The capability approach, technology and design* (pp. 189–201). Springer.

Nieusma, D. (2013). Engineering, social justice, and peace: Strategies for educational and professional reform. In J. C. Lucena (Ed.), *Engineering education for social justice: Critical explorations and opportunities* (pp. 19–40). Springer.

Nussbaum, M. C. (2001). *Women and human development: The capabilities approach*. Cambridge University Press.

Nussbaum, M. C. (2007). Human rights and human capabilities. *Harvard Human Rights Journal, 20*, 21.

Nussbaum, M. C. (2011). *Creating capabilities: The human development approach*. Belknap Press of Harvard University Press.

Oosterlaken, I. (2012). The capability approach, technology and design: Taking stock and looking ahead. In I. Oosterlaken & J. van den Hoven (Eds.), *The capability approach, technology and design* (pp. 3–26). Springer.

Rawls, J. (1971). *A theory of justice*. Harvard University Press.

Riley, D. (2008). Engineering and social justice. *Synthesis Lectures on Engineers, Technology, and Society, 3*(1), 1–152.

Riley, D. (2017). Rigor/us: Building boundaries and disciplining diversity with standards of merit. *Engineering Studies, 9*(3), 249–265.

Roberts, C., Huang, M., Lord, S., Olson, R., Camacho, M., & Perry, L. (2015). *IUSE/PFE RED: Developing changemaking engineers* (1519453). National Science Foundation. www.nsf.gov/awardsearch/showAward?AWD_ID=1519453

Rottmann, C., Moore, E., Chan, A., & Radebe, D. (2021). *Who can I turn to? Engineers navigating ethical dilemmas at work*. Paper presented at the IEEE ETHICS-2021. www.researchgate.net/publication/355008487_Who_can_I_turn_to_Engineers_navigating_ethical_dilemmas_at_work_Paper_presented_at_the_IEEE_ETHICS-2021_Waterloo_ON

Rottmann, C., & Reeve, D. (2020). Equity as rebar: Bridging the micro/macro divide in engineering ethics education. *Canadian Journal of Science Mathematics and Technology Education*, *20*(1), 146–165.

Rottmann, C., Reeve, D., Sacks, R., & Klassen, M. (2015). Engineering ethics education: More than a CEAB requirement. In *Proceedings 2015 Canadian engineering education association (CEEA15) conference*. Canadian Engineering Education Association Conference. https://ojs.library.queensu.ca/index.php/PCEEA/article/view/5809/pdf

Rottmann, C., Reeve, D., Sacks, R., & Klassen, M. (2020). Where's my code? Engineers navigating ethical issues on an uneven terrain. In *2018 ASEE annual conference & exposition proceedings*. 2018 ASEE Annual Conference & Exposition. https://doi.org/10.18260/1-2-31242

Schneider, J., Leydens, J. A., & Lucena, J. C. (2008). Where is "community"? Engineering education and sustainable community development. *European Journal of Engineering Education*, *33*(3), 307–319. https://doi.org/10.1080/03043790802088640

Schneider, J., Lucena, J. C., & Leydens, J. A. (2009). Engineering to help: The value of critique in engineering service. *IEEE Technology and Society Magazine*, *28*(4), 42–48.

Snyder, B. R. (1970). *The hidden curriculum* (p. 203). Knopf.

Stevens, R., Johri, A., & O'Connor, K. (2014). Professional engineering work. In *Cambridge handbook of engineering education research* (pp. 119–137). Cambridge University Press.

Trevelyan, P. J. (2014). *The making of an expert engineer: How to have a wonderful career creating a better world and spending lots of money belonging to other people*. CRC Press.

van den Hoven, J. (2012). Human capabilities and technology. In I. Oosterlaken & J. van den Hoven (Eds.), *The capability approach, technology and design* (pp. 27–36). Springer.

18

Transforming Engineering Education Through Social Capital in Response to Hidden Curriculum

Idalis Villanueva Alarcón, Victoria Sellers, Robyn Paul, and Buffy Smith

1 Introduction

We present this research-to-practice chapter for prospective and current engineering educators and scholars interested in learning about how individuals navigate hidden curriculum in engineering education by utilizing social capital. *Hidden curriculum* (HC) represents the unacknowledged and oftentimes "hidden" lessons or messages in a working or learning environment that hinder marginalized groups from navigating their environments successfully. HC propagates structurally (i.e., manifestation of systems of racial or other forms of bias across institutions and society; National Museum of African American History and Culture, 2021) through social networks and interactions where norms, values, and beliefs of a context are transferred to the learner. What an individual learns about their surrounding environment, in turn, influences how they respond to, react to, and act upon HC they experience. While not all HC is negative (Villanueva, Gelles, Di Stefano, et al., 2018), a failure to address the potential negative outcomes can lead to unintended consequences (e.g., attrition) in engineering.

1.1 What Is HC, Where Did It Come From, and How Have Researchers Described It?

Hidden curriculum (HC), a term originally coined by Philip Jackson in his book *Life in Classrooms* (1968), consisted of the behaviors that children learned in schools, such as learning manners, making an effort, and being punctual. These behaviors provide "a distinctive flavor of classroom life . . . which each student must master" to make their way "satisfactorily through the school" (p. 33–34). As Jackson (1968) stated, HC is formed through an "apparent shaping power of forces that had little or nothing to do with standard explanations of what goes on inside schools" (p. xv). Around the same time, Robert Dreeben (1968) examined the norms of school culture and concluded that students were taught to bury much of their personal identity in schools and accept their categorical treatments. While researchers did not fully understand the tenets of HC at that time, the work of these early scholars provided the foundation for understanding how systems and structures in schools derive from societal norms to train their students (Higham, 1959).

Over time, other theorists, such as Henry Giroux and Anthony Penna (1979), Michael Apple (1980), and Eric Margolis (2001), advanced the definition of HC to include what happens outside

DOI: 10.4324/9781003287483-22

the formal curriculum of school and higher education. These norms, values, and belief systems percolate not only in the formal curriculum but also through life in ways that guide and inform the student daily. The constant reproduction of HC, in turn, creates ways for people to hold on to power and reinforce their control over others (e.g., the ruling class over the working class). For a deeper explanation of the evolution of HC theories, refer to Kentli (2009).

Previous researchers have studied HC in undergraduate and graduate education (e.g., Smith, 2013; Lyles et al., 2022). Also, HC has garnered interest in disciplines, such as nursing, science, informational technology, medicine, and engineering (e.g., Allan et al., 2011; Abramovich & Bower, 2004; Bejerano & Bartosh, 2015; Hansson, 2018; Hafferty & Franks, 1994; Sellers & Villanueva Alarcón, 2021a; Villanueva et al., 2020), all of which provide a service to society. Educators and other professionals in these fields view HC as a mechanism to help create consciousness in the treatment of others (e.g., medical doctors with their patients) and their service-to-society roles through curriculum interventions.

We want to clarify the grammar and language we use to describe HC. We are aware of the discussion of using hidden "curriculum" versus "curricula"; however, we use "curriculum" throughout this chapter to encompass the multiple invisible norms, values, and beliefs that exist across educational and professional systems and structures. This is in alignment with HC scholarship, where researchers use both "hidden curriculum" and "*the* hidden curriculum" to describe the concept. Furthermore, although it is called *hidden* curriculum, individuals may be well aware of norms, values, and beliefs, yet individuals may not acknowledge or examine them. Norms, values, and beliefs become "hidden" when individuals uncritically accept or address them, which contributes to the unconscious and normalized part of educational systems and professionalization processes.

We note that HC is not always negative. Individuals process HC in many ways that are uniquely contextual and situational, which is contingent on the way that structures and systems communicate HC to individuals. For example, instructors in medicine use HC to teach students how to identify biases and potential inequity in their patient treatment (Hafferty & Franks, 1994). Nursing instructors use HC to debunk negative connotations and beliefs about the discipline and promote practices of care and empathy (Allan et al., 2011). Instructors can use HC as a counternarrative for positive educational and workforce change if deployed appropriately.

We, as researchers and educators, came together because of our shared interest and passion for deconstructing HC in higher education institutions and normative disciplines like engineering. We use our common goals and enthusiasm to provide an overview of HC, present findings from HC research, describe a practical example of a program designed to deconstruct HC, and give readers recommended actions. We framed this chapter as a research-to-practice because HC should not just stay in the "hows" and "whys" but rather should be used to enact action and justice (Martin, 1976). We are all passionate about this work, and we are confident that acknowledging and addressing HC in engineering will lead to transformative educational change.

1.2 Hidden Curriculum in Engineering

In engineering, the exploration of HC is still in its infancy. Tonso (2001) first discussed HC-related topics in US engineering education when she introduced the concept of "gender curriculum"; she used this term to describe implicit messaging present among women engineering undergraduate students as they described the disparities they experienced in their classroom activities (e.g., design projects). Erickson (2007) formally introduced the term "hidden curriculum" when she published a dissertation on the experience of women doctoral students in engineering and how they tied their sense of belonging to the implicit messages they received from others in their research environments.

To our understanding, the earliest known international research around HC in engineering was conducted by Tormey and colleagues (2015). These researchers compared the formal and hidden curriculum of ethics in engineering education to explore how "students learn implicitly through the social and organizational nature of their studies" (Tormey et al., 2015, p. 2). These authors showed that in engineering courses in Switzerland, students showed a bias towards higher levels of moral reasoning in their judgment of ethical dilemmas. Bejerano and Bartosh (2015) conducted a study with engineering students in New Zealand and found four gendered themes included in syllabi that reflect salient engineering values: women as incompetent, women as helpers, autonomy and separation, and masculine thinking. Rottmann and Reeve (2020) used an HC lens to address issues when case studies derailed in Canadian engineering courses designed to address ethics equity; they found that adding more critical analysis into case study learning and respectful dialogues instead of rational argumentation was important in avoiding pitfalls to moral relativism. Pehlivanli-Kadayifci (2019) explored HC among Turkish engineering faculty and found that jokes and other institutional structures ignore the presence of women and mock their contributions in engineering, posing several disadvantages to their representation. Thus, international HC research includes engineering ethics, as well as gender issues.

In the United States, Villanueva Alarcón[1] developed a structural framework and pathways model that allows researchers to investigate HC issues in engineering (NSF Award Nos. 1653140 and 2123016). Structural frameworks, in sociology, describe how groups or institutions have moving parts that are integrated in cohesive ways and are a function of common norms, customs, traditions, and cultures to promote solidarity and stability of a system (Parsons, 1977; Turner, 1985; Urry, 2012). Analogously, Villanueva (2017) suggested that individuals process HC by four factors that situate how they receive and respond to HC. Villanueva, Carothers, et al., 2018; Villanueva et al., 2020) developed and validated an instrument to identify the four factors: HC awareness (factor 1), emotions (factor 2), self-efficacy (factor 3), and self-/advocacy (factor 4). Villanueva and colleagues disseminated the validated UPHEME (Upending Previously Hidden Engineering Messages for Empowerment) survey to 58 colleges of engineering in the United States and Puerto Rico to 984 engineering faculty members and students between 2018 and 2019; an additional 120 individuals participated in follow-up research activities between 2020 and 2021. While analysis is still underway, Sellers and Villanueva Alarcón (2021a) performed a sub-analysis among 333 Black, Indigenous, and people of color of all intersecting identities (BIPOCx) in engineering who responded to UPHEME. They found that individuals cope with HC by changing their environment, negotiating their identities, or avoiding HC altogether. They also found that individuals with intersectional, marginalized racial, and gender identities avoided HC more than those in majority groups, yet majority groups in engineering (e.g., White) traded their personal identities the most in exchange for an engineering identity.

Other researchers have propagated the research from Villanueva (2017) and colleagues (Gelles et al., 2019; Gelles et al., 2020; Villanueva, Campbell, et al., 2018; Villanueva, Carothers et al., 2018; Villanueva, Gelles et al., 2018; Villanueva, Gelles, Di Stefano et al., 2018; Villanueva et al., 2019, 2020; Villanueva Alarcón & Sellers, 2022) in the United States and among international engineering education research and practice circles (Paul, Adeyinka, et al., 2021; Paul, Behjat, et al., 2021; Polmear et al., 2019, 2022; Rea et al., 2021; Villanueva, Campbell et al., 2018; Villanueva, Carothers et al., 2018; Villanueva, Gelles et al., 2018; Villanueva, Gelles, Di Stefano et al., 2018; Villanueva et al., 2019, 2020; Villanueva Alarcón & Sellers, 2022). In the United States, Polmear and others (2022) used Villanueva's HC model to uncover unexamined assumptions about leadership in engineering. They found that students conceptualize leadership in three ways, including whether individuals can develop leadership, how they practice it, and how they define it through their traits and behaviors. In Canada, Paul, Adeyinka, et al. (2021) and Paul, Behjat, et al. (2021) have used Villanueva et al.'s HC model (2020) to conceptualize ways to model situational HC. Paul, Behjat, et al. (2021) have also proposed an individual-based model to study Canadian engineering education programs, specifically those designed to tackle a given HC and encourage a sense of belonging and mindfulness in their students. Paul, Adeyinka, and others' (2021) and Paul et al. (2020) curriculum design, framed as a case study later in the chapter, recognizes that other researchers should study HC in engineering in ways that uncover impact on "specific demographics, rather than only having data about the population-level changes" (Paul, Behjat et al., 2021, p. 2).

These studies point to the pervasiveness of HC in engineering education internationally and in the United States. Also, these studies indicate how structures, embedded in fields like engineering, can cross boundaries, culture, language, systems of education, employment, and gatekeepers. It also alludes to the potential dangers that individuals can experience from HC, if not designed and delivered for the benefit of all. Since HC, according to Villanueva (2017) and others (Villanueva, Gelles, Di Stefano et al., 2018; Villanueva et al., 2019, 2020), is conceptualized as a structural framework that affects individuals differently, let us explore in more detail the factors that influence an individual's experience. Let us also explore how individuals cope with HC in various ways in engineering.

1.3 A Pathways Model to Explore Hidden Curriculum in Engineering

Villanueva and colleagues (NSF Award Numbers. 1653140 and 2123016) conceptualized HC as a structural framework that included several interconnected pathways via a validated instrument UPHEME. The instrument contains a four-factor model (Villanueva, Campbell et al., 2018; Villanueva et al., 2020) composed of HC awareness (factor 1), emotions (factor 2), self-efficacy (factor 3), and self-/advocacy (factor 4). Villanueva and colleagues found these factors to be main contributors to how individuals received, reacted to, and responded to HC (Villanueva et al., 2020). Villanueva and others note that there may be other, unexplored factors as well. A brief summary of these factors is provided in what follows, although the readers are encouraged to read Villanueva et al. (2020) for more details.

Factor 1: hidden curriculum awareness. Awareness is an important subcomponent of consciousness that helps individuals recognize and discern what and how information is being communicated. "Regardless of the level of awareness a person may have about an [HC] issue, these can't be brought up to full consciousness unless they are internalized first" (Villanueva et al., 2020, p. 1551).

Factor 2: emotions connected to hidden curriculum. In the context of HC, emotions signal to a person how external expressions, glances, gestures, and other behaviors connect to motivational outcomes (e.g., sense of belonging). These emotions assist individuals to focus on what factors are important when making decisions, learning, or socializing (Pekrun & Linnenbrink-Garcia, 2014). Individuals manifest emotions as: (a) valence (positive or negative emotions) or (b) activation level (focused or unfocused energy). When emotions are positively activated (e.g., enjoyment), individuals may experience an increase in "reflective processes, whereas negatively activated emotions (e.g., anger) may result in low levels of cognitive processing" (Villanueva et al., 2020, p. 1551). Readers can further explore the connection between emotions and engineering education in a chapter in this handbook, "Emotions in Engineering Education," by Lönngren and others (2023).

Factor 3: self-efficacy connected to hidden curriculum. An individual's emotions cannot lead to a decision or action unless they believe that they have an ability to cope with challenging scenarios, which is coping self-efficacy (Bandura, 1986). In the context of HC, individuals with higher self-efficacy can take actions like changing their environment, whereas lower self-efficacy leads to individuals avoiding HC (Sellers & Villanueva Alarcón, 2021a). At the same time, an individual's higher self-efficacy may not necessarily relate to greater awareness of HC unless someone has helped a person to "see the HC" around them (Villanueva, Gelles, Di Stefano, et al., 2018).

Factor 4: self- and other forms of advocacy in hidden curriculum. An individual may choose to self-advocate to cope with HC based on their level of self-efficacy. Those with higher self-efficacy are more likely to self-advocate by changing the environment around them; individuals with moderate self-efficacy will change themselves or their mindsets, and those with low self-efficacy will take no or minimal action (Sellers & Villanueva Alarcón, 2021a). Of course, individuals' levels of self-efficacy may be a response to the systems of power around them (Sellers & Villanueva Alarcón, 2021a).

To date, researchers have only presented the pathway model of HC on an individual level, with an ultimate goal of self-/advocacy, which serves as an "indication of a person's willingness to take action and speak up about a matter to improve their quality of life" and for others (Villanueva et al., 2020, p. 1553). Because HC is a structural framework, advocacy is not just an individual action but is also collective action, systemic action, or structural action. Villanueva's research group is currently exploring this thread. In this chapter, we build upon prior work from Villanueva, Campbell, et al. (2018), Villanueva, Carothers, et al. (2018), Villanueva, Gelles, Di Stefano, et al. (2018), Villanueva, Gelles, Youmans, et al. (2018), Villanueva et al. (2019, 2020) to present how social stakeholders in engineering (e.g., students, faculty members), either individually or collectively, acquired and acted upon HC.

1.4 The Relationship between Social Capital and Hidden Curriculum

Social capital allows individuals to unveil HC. Social capital includes information-sharing that holds groups of individuals and their sense of belonging together in relationships, networks, and competencies (Pooley et al., 2005). For this and other research, we use Pierre Bourdieu's (1986) conceptual framework of social capital. Bourdieu defines *social capital* as "the aggregate of the actual or potential resources which are linked to possession of a durable network of more or less institutionalized relationships of mutual acquaintance or recognition" (p. 248).

Bourdieu acknowledges the valuable social resources and benefits individuals and families acquire through social relationships and networks. Social capital provides access to institutional cultural capital, which is the knowledge that individuals use to decode, interpret, and navigate the culture of a given environment (Smith, 2013). Individuals acquire institutional cultural capital by building relationships with people (social capital) who have insider knowledge about how institutions function and what knowledge is valued. Thus, individuals need institutional cultural capital to decode, interpret, and understand HC (Smith, 2013).

It is also important to understand how social capital relates to one's sense of belonging. *Sense of belonging* refers to a feeling of connectedness to others (Rosenberg & McCullough, 1981). A sense of belonging takes on heightened significance in environments that individuals experience as different, unfamiliar, or foreign, as well as in contexts where individuals feel marginalized, unsupported, or unwelcomed (Freeman et al., 2007). Previous researchers have linked a strong sense of belonging to persistence (Soria & Stebleton, 2013). Sense of belonging has been found to both conceptually and statistically act as an indicator for social capital (Ahn & Davis, 2020). Both concepts are rooted in developing relationships through social networks and participation in social activities (Ahn & Davis, 2020).

Individuals develop relationships with weak or strong ties. Individuals use weak ties to cross a societal divide, such as class or race, allowing for information exchange, ideas, and innovation (Claridge, 2018). Individuals form strong ties within a group or community and are a source of social support (Claridge, 2018). Thus, engineering stakeholders (e.g., students and faculty members) build their sense of belonging by developing social capital via strong and weak ties. Social capital leads to institutional cultural capital, which helps individuals unveil HC. In the next section, we detail how engineering stakeholders use social capital to navigate HC, where HC is a barrier between social capital and institutional cultural capital in engineering.

2 How Engineering Stakeholders Use Social Capital to Navigate Hidden Curriculum in Engineering: A Study of the United States Context

Villanueva and others (2020) administered the UPHEME instrument to engineering students and faculty members across 58 colleges of engineering across the United States. Participants answered questions related to the four-factor model conceived by the research team. The instrument had both qualitative and quantitative questions. Out of 984 respondents, approximately a third (333) responded to all items of the UPHEME survey. Individuals identified as Black, Indigenous, people of color of all intersecting identities (BIPOCx), although some identified as White or of Eurocentric roots. We will only discuss the qualitative responses in this chapter, and we utilized an *expo facto* and inductive coding of participants' responses for this chapter.

In response to qualitative questions of UPHEME, individuals described an example of an HC message they experienced in engineering and strategies they used to navigate it. We found four prevalent HC messages in engineering: (1) engineering is difficult, (2) engineering is inflexible, (3) people feel underrepresented or undervalued in engineering, and (4) people feel supported in engineering. We also found that individuals utilized three categories of strategies to navigate HC messages, such as changing the environment, negotiating themselves, or avoiding the issue. We recommend that readers review Sellers and Villanueva Alarcón (2021a) for more details about the strategies.

From the 333 respondents, we noted that a sixth of the respondents (43 undergraduate and graduate students) explicitly acquired social capital to navigate HC; these participants also related social capital to a greater sense of belonging in engineering.

We categorized participants' responses of how they changed their engineering education environments using their social capital through relationships and associated resources with others. From the analysis, we identified three descriptions of individuals who utilized social capital to navigate HC: (a) seekers, collectors of social capital; (b) bridgers, sharers of social capital; and (c) agents, brokers of social capital. We present a detailed description of the findings and archetype character traits in the following sections.

2.1 Seekers: The Collectors of Social Capital

Seekers of social capital (n = 24, 56% of total participants) looked for support from others to navigate structural or situational HC in engineering. Of all seekers, the majority were self-reported men (n = 15, 63% of seekers), were White, or had an ethnicity that is Eurocentric (n = 19, 79% of seekers). Approximately 67% of participants self-reported as continuing-generation (parent or guardian completed some college) students (n = 16, 67% of seekers) and had a traditional, uninterrupted K–16 educational pathway (n = 18, 75% of seekers).

Most seekers (n = 18, 75% of seekers) experienced HC messaging that engineering is difficult but felt they could not express to others that they were struggling amid this difficulty. Others indicated that engineering was inflexible to them (n = 6, 25% of seekers), but felt that they could not communicate to others the nature of such an HC.

In terms of the strategies used, seekers indicated that they quickly became friends with others who knew how to build technical skills to successfully navigate their education, which increased their sense of belonging in engineering. For example, a Latino mechanical engineering graduate student stated that he had a "poor work ethic" and was "not performing well" in engineering classes when he started in engineering; this message that engineering is difficult prompted the student to see himself as having a deficiency that he needed to fix. The student opted to make "friends in the engineering department" to boost his skills, demonstrate his work ethic, and feel welcomed by his peers. Thus, this individual developed social capital in the form of weak ties with others who held institutional cultural capital needed to help him navigate and persist in the engineering program.

Seekers also used strategies to pursue help from other social stakeholders they perceived to have more power than them. These social stakeholders were advisers, professors, or mentors. For example, a White man and civil engineering undergraduate student similarly described difficulty with his engineering homework, but he strategized by "working with other students on the homework, as well as even going to [the] TA and professor [*sic*] office hours for help." This participant sought social capital through weak ties to elevate his technical skills and connect to the holders of knowledge and power over his performance. By choosing people who held more power than his peers, this individual understood that both skill-building and acquiring institutional cultural capital are important for his success and persistence in engineering.

Seekers become aware of HC in their surrounding systems and structures and use social capital as one way to navigate it. Seekers build social capital to identify the deficits they need to address and collect information to succeed and belong in engineering. Seekers quickly realize that institutional cultural capital is powerful and that the more social capital they develop to access institutional cultural capital, the more power they will have. You can visualize this in Figure 18.1, where seekers accept more social capital from others than they return. In socialized settings, seekers pair up with holders of knowledge (e.g., peers from majority groups, administrators, advisers, mentors) to learn about the intricacies of an unknown environment or setting and then use that knowledge for personal advantage. Seekers can hold different roles in engineering (e.g., students), and their goal is to understand the environment and the institutional capital around them.

While seekers' ultimate outcome is to persist and succeed in engineering, they do not often question or even recognize that HC guides their actions. By not questioning HC and improving their abilities, skills, and competencies, seekers may inadvertently perpetuate an ongoing cycle of meritocratic values, beliefs, or ideals in engineering. While seekers may indeed achieve professional success, their embodiment of the norms of engineering may result in severe consequences to their mental and emotional well-being that they may not address.

2.2 Bridgers: The Sharers of Social Capital

Whereas seekers build social connections to improve their skills or abilities to navigate HC in engineering, **bridgers** of social capital (n = 14, 33% of total participants) looked for current support from like-minded individuals to cope with HC directed at them and their personal identities. Most bridgers felt marginalized in engineering or that others undervalued them (n = 12, 85% of bridgers). A few bridgers experienced HC that engineering is difficult (n = 1, 7% of bridgers) or HC that engineering was inflexible (n = 1, 7% of bridgers). Most bridgers self-reported as women (n = 11, 79% of bridgers) and were from marginalized ethnic/racial groups (n = 8, 57% of bridgers). Most of these bridgers were continuing-generation students (n = 11, 79% of bridgers), and most pursued a traditional, uninterrupted K–16 educational pathway (n = 13, 93% of bridgers).

A strategy shared by bridgers, in addition to seekers, was that they found support with peers. However, a key difference was that bridgers found kindred peers who experienced similar situational HC to create stronger relationship ties (i.e., stronger social capital). Bridgers sought like-minded individuals to support each other as a community academically and interpersonally. For example, a Latino and chemical engineering undergraduate student described difficulty adapting to his "college setting and to navigate the language barrier" because he is an international student. He navigated his sense of "onlyness" by finding "faculty members that are not only international but also Hispanic." Thus, this individual surrounded himself with others who could understand his experiences and help him build social capital in engineering and develop strong ties. The social capital acquired by this participant also strengthened his sense of belonging with faculty members in his engineering program to survive this stage of his engineering education.

A White woman and environmental engineering undergraduate student discussed how she was marginalized because there was a "large population of males" in her classes. However, she coped with this HC by surrounding herself with "peers who don't believe in the stereotypes perpetuated by HC" so they "can all work together to better ourselves and our grades." This participant described bonding with peers where gender-based HC was not a decisive factor, and was able to work with her peers, as opposed to competing against them, to improve her grades. Thus, this participant and her peers collectively built stronger social capital through strengthened relationship ties to increase their access to institutional cultural capital and override gender-biased HC.

While bridgers may not necessarily coalesce to change their environments structurally, their focus is survival and equipping others to do the same. Bridgers understand that addressing their mental well-being is important, and they find support by being a member of a community. Bridgers tend to share social capital with others as much as they receive it in the form of social support, as depicted in Figure 18.1.

2.3 Agents: The Brokers of Social Capital

Agents of social capital (n = 5, 12% of total participants) aimed to change engineering education for future generations, particularly for members of marginalized groups. Most agents self-identified as women (n = 4, 80% of agents), were from majority ethnic/racial groups in engineering (n = 4, 80% of agents), were continuing-generation (n = 5, 100% of agents), and had traditional and uninterrupted K–16 educational pathways (n = 5, 100% of agents).

Agents acted as representatives or spokespersons of a marginalized group. A Black man and electrical engineering undergraduate student navigated HC of underrepresentation of Black people in engineering curriculum by relating "construct concepts of an African American famous engineer" even though his professor "kept stating that what I was talking about . . . was not a direct representation of what we were talking about [in class]." The participant "tried to introduce it more" to his teacher, and the participant "was able to teach the class" and his instructor concepts from the famous African American engineer, of which they were previously unaware. The participant navigated the HC so that he could become a source of social capital to others and they could be aware of African American contributions to engineering that were not previously in the formal course curriculum.

Among the strategies used by agents, they also want to be role models for others. They provided a voice for others and were not afraid to communicate concerns to other important social stakeholders (e.g., instructors). For example, a White woman and material engineering undergraduate student described how men in her program did not believe women should be engineers, but she noted that she wanted to continue in engineering because "if [she] stick[s] with it and help[s] make a difference in the field, more young women will be able to enter the field without fear." The participant noted that she intentionally persists in engineering so that she can make it better for future classes of women. The participant described being an agent of future social capital for other women, which inadvertently increased her own self-efficacy to become an engineer, with the altruistic goal to provide future women engineers with the hope that they can also persist.

Agents of HC enact strategies and practices that challenge their surrounding systems and structures. They take acquired institutional cultural capital and use it to raise awareness for others in the future, which bridgers do not consider. Individuals can experience agents' actions vicariously, such as seeing agents speak up for themselves and others and raising awareness of issues that others do not see (Sellers & Villanueva Alarcón, 2021a). Agents' sense of self, their personal and professional identities, and their sense of belonging with others in engineering (and not the profession) are strong. As such, agents are willing to suffer the consequences of their advocacy to disrupt the status quo, even

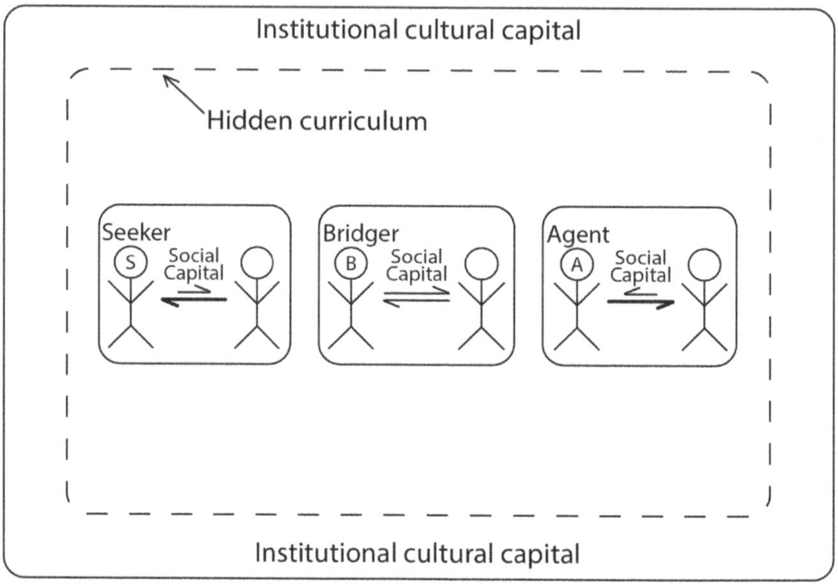

Figure 18.1 Conceptualization of social capital (SC) between seekers, bridgers, and agents, with arrows that indicate the amount and direction of social capital exchanged. Seekers receive more social capital than they lend to others, bridgers exchange social capital equitably with others, and agents lend more social capital than they receive. Individuals aim to build social capital to uncover HC, strengthen their social ties and networks to gain institutional capital, cultural capital, and/or a sense of belonging.

if it affects their own belonging to a non-supportive community or profession (strong social capital is independent of strength of relationship ties). Thus, agents tend to share more social capital with others than they receive, as depicted in Figure 18.1.

Therefore, we find that in traditional engineering programs in the United States, individuals build social capital to uncover HC, strengthen their social ties and networks to gain institutional cultural capital and/or a sense of belonging. In Figure 18.1, we conceptualize the role that seekers, bridgers, and agents must acquire social capital and navigate HC. While seekers navigate HC to persist and understand their learning and working environments, they internalize and perpetuate the status quo of norms and values in engineering. By forming relationships that share and give social capital and improve access to institutional cultural capital, bridgers and agents are less likely to recreate cultures of exclusion and marginalization in engineering.

With these findings, a question the reader may be wondering is: "How can engineering education intentionally equip individuals to navigate HC they may experience?" Let us explore a case study at the University of Calgary where engineering educators and researchers have created an intentional, formal curriculum to debunk the myth of rational-emotional dualism in engineering (i.e., engineering is rational and emotionless; Lönngren et al., 2021, 2023; Kellam et al., 2018) and promote a sense of belonging in students.

3 University of Calgary: An Engineering Program That Intentionally Aims to Tackle Hidden Curriculum Being Directed to Its Students

In engineering, using HC for positive educational change by bringing awareness to hidden messaging is just starting (e.g., Paul et al., 2021a, 2021b). In this section, we present a case study of a novel engineering program at the Schulich School of Engineering at the University of Calgary. The Engineering Attributes (EA) program aims to develop mental wellness as well as promote a sense of belonging, lifelong learning, and effective learning strategies through weekly modules across all first-year engineering courses (Paul et al., 2021a, 2021b). Leaders developed the program in 2019 to train first-year undergraduate engineering students to navigate their degree while debunking the myth of rational-emotional dualism (Lönngren et al., 2021; Kellam et al., 2018). The leaders of this program were aware of the pervasive HC of an engineering culture based on pure rationality without emotions (e.g., Huff et al., 2016; Husman et al., 2015; Lönngren et al., 2021; Kellam et al., 2018; Secules et al., 2021; Villanueva, Carothers et al., 2018; Villanueva et al., 2020). They wanted students to understand that engineers have emotions, and emotions are inextricably linked to cognition. Thus, there is a benefit of bringing emotions into the forefront of awareness instead of hiding them (e.g., Husman et al., 2015; Lönngren et al., 2021; Villanueva, Carothers et al., 2018; Villanueva et al., 2020). To do so, leaders created a curriculum based on theories of mindfulness, mental wellness, and learning strategies (Paul et al., 2021a, 2021b). Leaders present the modules in weekly short (10–15 minute) presentations across the first-year engineering curriculum. There are typically 12–15 modules; leaders deliver these modules across four or five of the core first-year engineering courses during the academic year. Leaders intentionally designed the modules to be regular, short tidbits of information so that they expose students to activities and strategies they need to mitigate and bring awareness to negative, pervasive HC messaging in engineering.

Within the EA program, the goal of the weekly modules is to remind students that they are human and to dismantle some of the dualisms and HC that exist in engineering (Paul et al., 2021a, 2021b). Colloquially, the program leaders call the program the *Engineers Have Feelings* project, as they encourage students, through reflective activities, to bring their emotional awareness to their engineering academics. Leaders remind students that they are more than just logical problem solvers, encourage students to bring emotional awareness to their critical thinking (Kellam et al., 2018), and to build community with their peers to support social capital building (Pooley et al., 2005). For each module, first-year engineering undergraduate students submit qualitative reflection responses on the weekly topic (see Table 18.1 for overview of topics). Leaders grade these responses for completion, and the responses count for a small percentage of the final grade for each first-year course, at the discretion of the instructor. The program leaders inform the students that they will only be checking a small set of responses each week. The program's emphasis is to support students to understand the importance of reflection for themselves, more than focusing on the grading of the reflections. Program leaders provide brief summaries of the previous week's reflection responses to encourage thoughtful responses and to help students know they are not alone in their feelings. In open-ended student feedback about the program, students often speak positively of the reflections' importance. For example, one student indicated:

> The most helpful part [of the EA program] has been the [reflection] quizzes because they allow me to reflect on myself, my habits, and my thinking. They help me realize that I have been in these situations and through the quizzes, I start to think about strategies to deal with these situations.

We summarize an outline of the main objectives of the curriculum and course sequence in Table 18.1, where we also provide examples of how the curriculum design parallels the four factors of HC pathways. Lastly, we provide specific tips on attending to each of the archetypes with examples and more details on the module content and examples of reflection questions.

Within HC pathways, there were several character traits that the EA curriculum addressed. For example, the four factors found by Villanueva et al. (2020) are integrated into the curriculum regularly. The first factor, HC awareness, is an important step to recognize what systems or individuals

Table 18.1 Overview of the Engineering Attributes Curriculum, University of Calgary, Schulich School of Engineering, 2020–2021

Term 1 - Fall 2020		*Term 2 - Winter 2021*	
Week	*Module*	*Week*	*Module*
1	Mental Health Continuum	2	Mental Health Continuum
1	Motivation	2	Emotions and Hidden Curriculum
2	Teamwork and Diversity	3	Teamwork and Diversity
4	Time Management	5	Finance
6	Academic Burnout	8	Exam Anxiety
7	Exam Anxiety	8	Academic Burnout
8	Metacognition	8	Errorful Learning
11	Resiliency	10	Resiliency
12	Final Portfolio	11	Motivation
		13	Final Portfolio

Source: Adapted from Paul, Adeyinka et al. (2021).

communicate and discerning unconscious misrepresentations in learning environments (Villanueva et al., 2020). Within the EA program, leaders equip students to navigate the engineering norms and become aware of unconscious beliefs that they might hold. For example, some students believe that engineering is difficult and that engineering programs fail a certain percentage of students. In the *Academic Burnout* module, instructors raise awareness of the challenge of an engineering education by helping students self-assess their emotional patterns; they help students understand that it is abnormal to feel continuously exhausted, have high anxiety, or lack motivation towards school. The instructors use these self-assessment talking points to provide students with tips for recovery, ways to seek help, and how to access support via campus resources.

The second factor, emotions, is important because emotions are required to process the HC that individuals are aware of (Villanueva et al., 2020). Program leaders designed the EA program around the mental wellness wheel (see Figure 18.2), which they adapted from Hettler (1976). Regularly during the program, instructors conduct check-ins on students' emotional well-being. Program leaders also introduce students to a full module on *Emotions and Hidden Curriculum*, where instructors talk about how the brain is interconnected with emotional processing and working memory and how integral emotions are to learning (e.g., Tyng et al., 2017), as well as the emotions that they experience (Gelles et al., 2020).

The third factor, self-efficacy, allows individuals to believe they can improve their environment (Sellers & Villanueva Alarcón, 2021a; Villanueva et al., 2020). One of the most popular EA modules, *Resiliency*, gives students the tools to change how they think. In this module, instructors summarize cognitive distortions and give students tools to manage these intrusive thoughts using the *catch it, challenge it, change it* framework, which program leaders modified from cognitive behavior therapy (CBT) techniques (Stallard, 2019). In the context of HC, these tools are valuable to help students be aware of HC messaging they receive, use self-efficacy tools to change their own internal dialogue, and challenge HC they receive.

Finally, students use their gained confidence (self-/advocacy) to take action or speak up for themselves (Villanueva et al., 2020). An early module in the EA program on *Teamwork and Diversity* discusses the idea of brave spaces where the program leaders supply tools and strategies to students to have difficult conversations. Students are encouraged to advocate for themselves while owning their intentions, being respectful because everyone defines *respect* differently, and engaging in "controversy with civility," where program leaders cordially frame challenges and conflict

(Arao & Clemens, 2013). These conversations provide students with the tools needed for self-/advocacy.

In consultation with Villanueva and colleagues (2020), Villanueva Alarcón & Sellers (2022) and Smith (2013), leaders re-evaluated the EA program to see how the curriculum (Paul, Adeyinka, et al., 2021; Paul, Behjat, et al., 2021) aligns with this description of seekers, bridgers, and agents. While these curricular elements are not prescriptive, we hope that these strategies will ignite similar ideas in other engineering classrooms, departments, and programs for transformative educational change.

3.1 Supporting Seekers: Acknowledge Emotions, Normalize Help-Seeking, and Provide Resources

As described in a previous section, seekers aim to build social capital to belong and succeed in engineering. Typically, they are looking for support to better navigate through HC in engineering. Aligned with this, one of the primary aims of the EA curriculum is to help students understand that they are not alone and to provide them with tools to navigate their educational degree.

The resiliency module described previously uses the *catch it, challenge it, change it* framework with reflection questions to give students an opportunity to reflect on their feelings and internal dialogue, consider how they can challenge this internal dialogue, and then create a plan to change it. This process helps students become aware of and acknowledge internalized thoughts based on norms in engineering, challenge these thoughts, and make a plan to bring them to conscious awareness so they can navigate HC of engineering. One student emphasized the importance of awareness in their reflection when they indicated:

> I feel that a way to change those feelings, is to talk about your feelings, be it through writing it down or speaking it. I feel that if you take a moment to write or prepare something, you can see things in a different light.

Another student acknowledged that they experienced HC messaging that made them feel like they did not belong in engineering: "Most of the time, I feel like I don't belong here, and that I made a mistake of choosing engineering." By asking them how they could reframe their internal dialogue and navigate HC, they emphasized how it was important to remember that everyone struggles:

> I could reframe my internal narrative by telling myself that everyone feels confused all the time and that it is not just me. I'd tell myself that the course is just really hard and that I am not stupid and that I belong here.

Finally, they concluded by reminding themselves, "I got accepted into engineering because I am capable and that I have what it takes. During these times, I am most likely stressed so taking a break would help me." This student confronted HC messaging that made them feel they did not belong, and they identified a strategy around reflection and reframing of their thoughts. In the future, they plan to take a break when they start having these thoughts.

Many students applied the tools given through the modules in their reflections. For example, the *power of yet* is discussed, where students add the word "yet" to their sentence. One student who felt like they were struggling more than everyone else concluded with, "I need to count my accomplishments rather than downgrade myself on my flaws. I need to know that I am not understanding chem [chemistry] YET, but I will understand it eventually." The *power of yet* is a helpful tool for seekers as it helps them navigate HC messaging that they are internalizing. Another student applied to join the

Solar Car team, but she wrote, "I couldn't help feeling that I had faked the interview, or that I didn't deserve to be on the team. I felt that, as a female, I didn't belong (all the other new recruits were male)." To overcome these feelings, she noted that it has been really helpful to find mentors to look up to: "I think talking to other women about this specific situation has helped a lot. . . . I usually think about how others have done the same thing before me and succeeded, so I should be able to as well." She sought out social capital from mentors who are women to help her gain knowledge and succeed in her role.

Also, the EA program emphasizes the importance of seeking support from peers, professors, and advisers if students are struggling. As seekers looked to gain social capital for their own personal advantage, they commonly reflected on seeking support from peers. A student doubted their coding abilities, and they planned "on asking the people around [them] if they understand something better than [them], and if they do, could they please explain it to [them] too" to help overcome their doubts. Another student who was nervous about looking "dumb" for asking questions, indicated that:

> Over time, I start to ask a few questions during class time. It is my opportunity for me this year with online learning, so I feel much better to ask questions. One thing that surprises me is that my friend is also supportive, some of them even give me answers [in the chat] before professors notice them.

It is evident these students seek more social capital to support them through their academics, and because of the EA reflections, they have committed to ask for help and talk to their peers more often.

Overall, the EA program provided a curriculum structure allowing seekers to gain social capital by first acknowledging their emotions, seeking help more often, and understanding the tools and resources that are available to them. This provides seekers with a foundation and helps them build social capital so they can begin to transition to becoming bridgers and agents.

3.2 Encouraging Bridgers: Supporting Group Dialogue That Allows for Sharing Personal Identities, Individual Sense of Belonging, and Community Building

Bridgers look to build social capital with like-minded individuals to navigate HC messages in engineering that target their identity. They aim to share social capital within their communities to help themselves persist and support others to understand HC messages directed to them. Within the EA program, the modules are delivered with three components: content, in-class interaction, and personal reflections. This structure creates many opportunities throughout the curriculum to allow for bridgers to connect with like-minded peers, as well as reflect on these conversations.

The teamwork and diversity module introduces the diversity wheel (see Figure 18.2) and discusses the multiple dimensions of identities to support bridgers in finding ways to relate to their peers. In the EA program, after introducing the wheel, students are asked to consider elements of their identity based on the prompts and then answer two questions: *What are the elements of your identity that are most important to you? What are the elements of your identity that others notice first?* After spending a couple minutes on this activity on their own, students go into groups to discuss these questions with their peers. This activity has been impactful in helping bridgers understand more ways that they can find kindred peers and support each other. Most people will first notice external elements of our identity to find community (e.g., perceived age, skin color, gender, etc.), when there are also many other elements of our identity that can be used to connect with and support each other. For example, one student identified themselves as creative and noted, "[W]hen working in groups, this can come as an advantage because we can all bounce around creative ideas and put our

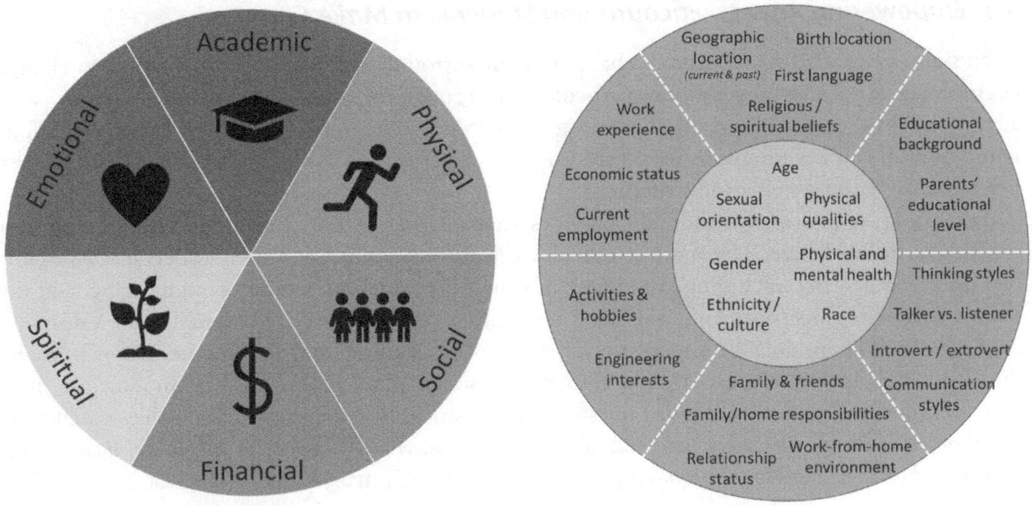

Figure 18.2 Wellness wheel and diversity wheel from EA modules.
Source: Left, adapted from Hettler (1976); right, with permission from Dr. Kim Johnson.

minds together to come up with truly useful solutions." Although HC often sends messaging that creativity is not important in engineering, this student found a way to use their creativity and the creativity of their group members to thrive in engineering problem-solving.

One student talked about how their identity would allow them to share social capital and support other group members. They wrote:

> [M]y own struggles with mental health can help me be more patient and supportive if a group member is having similar struggles. Likewise, my introversion can help me be more cognizant of shy or underspoken group members and lead me to take extra care in ensuring that their voices are heard.

This bridger understands the importance of supporting peers who have similar identities as they may have experienced similar HC in engineering, such as not being heard because they are a quieter group member. Through social capital sharing, this student aims to build a community where peers can support each other. Another student talked about resolving conflict due to diversity and stated that although "[d]ifferences should be appreciated and respected," it is important to remember that the "similarities in a group should always be the driving force because that creates a sense of unity and collectiveness." This student hopes to find commonalities that allow them to share social capital across their peer group to support each other and have greater success as a community. Although bridgers often seek kindred peers with similar identities, during the module, students are reminded that exclusion can also occur when only looking to find peers who are the same as themselves. It is important for bridgers to understand this nuance and build community across identities.

The EA program provides an opportunity for bridgers to reflect on how they can build community with their peers through engaging in vulnerable spaces. Bridgers reflect and emphasize the importance of social capital sharing to support each other when navigating engineering, particularly HC as it relates to their identities. Leveraging this and connecting it directly within the curriculum content and delivery pedagogies is powerful as it supports bridgers in thriving themselves and supporting their community of peers to thrive.

3.3 Empowering Agents: Encouraging Students to Make Change

Agents aim to change social capital, changing their surrounding systems and structures to change engineering for future generations, particularly marginalized groups. Agents raise awareness by advocating for themselves and others on issues that are often hidden. Through intentional curriculum design in the EA program, we create learning opportunities for agents to have the tools they need to change the systems and structures.

For example, one of the modules provides an overview of the concept of HC to empower students to make change. The module provides a definition of HC, and then leaders discuss examples from engineering, including the hierarchical and individualistic nature of engineering, and the assumption of engineering being a difficult and elite program (examples inspired from Villanueva et al., 2020). The module aims to empower students by talking about the importance of bringing these unconscious internal experiences into the conscious mind. Following this module are three reflection questions, where the third question pushes students to consider actions they could take: *Are there any topics you feel have been missing from your engineering education, or do you feel there is lesser value placed on some elements of engineering over others?* Although many students did not communicate significant advocacy in their answers, a few students were agents with a desire to change their engineering education experience. One student noted that the mental wellness modules have been helpful in changing engineering education but that it needs to be taken further. "I think our professors should take mental health more into consideration when conducting their classes (more interaction with peers, extended due dates, having profs be easily accessible to ask questions, etc.)." This student is advocating for change on behalf of themself and their peers to improve and change HC that exists in engineering classrooms.

The EA program also empowers agents to make change by hiring 4–6 students each summer to develop their own modules. The EA program inspires and encourages students who are hired to the summer research program to make changes to engineering culture and HC. During the summer research program, leaders give students autonomy to research a topic they are passionate about and create their own module. The students' choices of topics are often inspired by HC that have influenced them personally. For example, one student chose to develop a module on finances and scholarships. She navigated many financial barriers herself by applying for scholarships and saving money to attend university. To develop the module, the student researched the connection between mental health and finances, pulled together resources lists, and laid out tips based on her experience. Thus, the summer research program provides an opportunity to engage students who are agents in peer advocacy, where they can make more systems-level changes to engineering education curriculum.

Each of these three curricular elements described provides powerful examples of how these strategies connect seekers, bridgers, and agents to content and resources relevant to them. By connecting the curriculum directly to seekers, bridgers, and agents, we can ensure we are supporting as many students as possible to thrive in utilizing social capital to navigate the HC. The EA program aims to normalize help-seeking and provide resources, to foster community and belonging through vulnerable conversations, and to empower students to make change. These values are continuously repeated throughout the year in the weekly modules in the program. The program teaches students that a successful engineer is more than just being able to solve difficult problems and manage high workloads. As one student indicated, "I think the mental wellness program makes us more human and mentally healthy which in turn would make us work more efficiently." The EA modules bring awareness to HC in engineering and help students navigate it. As the program improves and continues to iterate, it will be important to engage faculty members, in addition to students, in the weekly modules so that they begin to deconstruct their own biases and HC that they unintentionally promote in their classrooms.

4 Lessons Learned about Hidden Curriculum in Engineering Through Social Capital

Engineering departments and colleges can create a more inclusive learning environment that can foster the growth and development of seekers to transition into bridgers and agents. In order to produce more bridgers and agents and collaborative learning communities, a competition-neutral pedagogical approach must be created (Secules, 2019). Within the context of the learning environment, faculty members and students must become aware of HC but also know that they have the power and tools to change it.

Since we understand that educational environments are part of a larger ecosystem, we have broken down some of our recommended strategies for department/college, researchers, faculty members, and students. However, we understand that these recommendations are only presented as starting points and encourage readers to adapt it to the context of their realities. Also, these recommendations are not meant to be prescriptive but rather as starting points for reflection and discussions. These recommendations are based on previous or ongoing research by Sellers and Villanueva Alarcón (2021b), as well as work by Paul, Adeyinka, et al. (2021), Paul, Behjat, et al. (2021), and Smith (2013).

4.1 Departments/Colleges

We recommend the following to university departments and colleges for how they can become aware of and address HC in those contexts:
[COMP: Please align below bulleted list also make sure they are aligned and differentiated with the sub-bullets as per given in MS.]

- ***Departments/colleges should take a critical look at their engineering cultures and how each member contributes to this culture.***
    - They can plan a department retreat to unpack the unwritten norms, values, and expectations of how they define success for their students in the department and in the profession. If some aspects of HC are revealed to be less affirming and inclusive, they should have a discussion on how to make those changes.
    - If there is agreement that the expectations, norms, and values are not affirming and inclusive, then the department should seek programming that will help all faculty members understand, navigate, and transform HC. For example, offer lunch and learn workshops on an HC topic led by an expert on the topic and/or an administrator or faculty member who has been trained on the topic. The lunch-and-learn workshops can be offered throughout the school year.
- ***Departments/colleges should consult with students on changes to the engineering culture.***
    - Hire a neutral, third-party evaluator (preferably outside the discipline) to conduct interviews and/or focus groups with students on prevalent norms, values, and expectations and things they would like to see changed in the department.
    - From the evaluator's recommendations, establish a plan that includes small suggestions for engineering cultural change directed to faculty and staff members and students.
    - Evaluate plans and revisit suggestions at least once a year.
    - Department can invite campus partners to come in monthly to talk about HC topics. For example, invite the academic counseling and health services to talk about managing stress and the importance of taking care of one's mental health.

4.2 Researchers

We recommend the following to researchers who want to extend HC work in engineering education:

- ***Researchers should consider how to unearth the status quo in engineering.***
    - Consider, "What norms, values, and beliefs may be present in engineering but are unexamined because they are ubiquitous?"
    - Also, ponder, "Who benefits or is harmed by these unexamined norms, values, and beliefs?"
- ***Researchers can utilize research designs to examine HC contextually in departments, classes, workplaces, and other engineering spaces.***
    - Researchers can ask themselves, "How can I design an inquiry that will allow me to see HC authentically in context?"
    - For example, critical ethnography to explore how HC occurs in first-year classes (similar to the University of Calgary example).
    - Use results from critical ethnography or other research designs to inform support for faculty to design and implement curricular change.
- ***Researchers can explore HC longitudinally to determine if stakeholders' experiences with HC change.***
    - Examine the impacts of a departmental cultural change plan to see if stakeholders report changes in their experiences with HC.
    - Use iterative results to continue but adjust aspects of initiative as necessary.

4.3 Faculty Members

We recommend the following to faculty members and researchers who want to become aware of and address HC in their interactions with students:

- ***Design class structure, syllabus, lesson plans, and assignments/projects to foster a strong sense of belonging in students.***
    - Create an assignment that rewards collaboration and not competition to earn full points on that assignment.
    - Create fun class activities that discuss the culture of engineering programs and why diversity is important. For instance, play a Kahoot! phone app game with HC-related topics.
- ***Faculty members can foster a strong sense of belonging in their classroom if they show their own vulnerability so that students can see that they are human and have emotions.***
    - Faculty members could spend five minutes during set class times for check-ins, where professors share their joys and challenges of the day and ask other students if they want to volunteer and share their joys and challenges of the day. Alternatively, faculty members can share with students if they used tutoring and other support services as a student; showing them how you sought support as a student may encourage them to persist in engineering and seek out necessary support.
    - If it has been a hard week for most students (multiple exams for the same engineering concentration), faculty members could consider giving the entire class an extension on the homework assignment for that week. Showing an empathetic stance may engender trust in students and foster a deeper sense of belonging.

- Provide extra credit to students for attending tutoring and other academic and social support services.
- Invite former students to come and speak to the class about the benefits of seeking out support and thriving in the department.

- ***Design instruction to encourage student feedback and evaluation.***
  - Empower students that they can offer feedback on the structure of the class and that they are co-creators of their learning experience. One way to do this is to leave a class topic open in the syllabus and poll students on what they want to learn that semester in class.
- ***Professors can reward academic excellence and collaboration.***
  - Professors could assign students into small groups and let these groups identify what are the underlying assumptions of the assignments and what is unclear about the assignments. Then, share those comments and feedback with the entire class, and the professor can provide additional clarity. This is one way to make HC visible for all students and, all students would receive the same information at one time since not all students feel comfortable going to office hours.

4.4 Students

We recommend the following to students who encounter HC in engineering contexts:

- ***Everyone experiences HC in engineering, but some may experience it more than others. Some may experience HC directed at their personal identities (race/ethnicity, gender, sexual orientation, etc.). Be aware that norms, values, and beliefs are difficult to identify.***
  - Ask, "What groups are not represented in my class/group/club/program?" and contemplate how you can include them.
  - Ask, "Why does this need to be this way in engineering?" or question, "Is the way things have always been done the best way to learn/complete them?"
  - Do not be afraid to ask professors questions about the rationale for some assignments, projects, or activities connected to the profession. It is important to understand what norms, values, and beliefs dominate in the engineering profession prior to graduation.
- ***Students may not immediately have the proper resources, support, or strategy to navigate HC, but it is important to communicate what students can do for themselves and others who may experience similar issues. Be aware that using social capital is a powerful strategy to navigate and overcome potentially negative HC.***
  - Ask, "What resources are available to students to mitigate this issue?" or "How can students leverage other people and their resources/connections to help me with my issue?" such as with peer or upper-class students, teaching assistants, advisers, career counselors, and affinity groups, among others.
- ***Engineering has a strong history of maintaining exclusive norms and culture. Students (and faculty) can change this culture, but it is not without its roadblocks. As such, be patient with the progress made and keep bridging or brokering social capital. Students and others can raise awareness of HC and help transform engineering towards more positive educational change.***

5 Final Thoughts

In this chapter, we summarized a four-factor model of HC where individuals must become aware of HC and utilize their emotions and self-efficacy to advocate for themselves. We also presented research findings about how individuals utilize social capital to unveil HC and advocate for themselves, specifically by collecting social capital, sharing it, and lending it. This chapter then connected research-to-practice by describing an engineering program that is attempting to create a new narrative to the emotional-rational dualism in engineering and support individuals to use social capital to advocate around HC. Finally, we presented recommendations and prompts, based on our current research, to assist administrators, faculty members, and students to become aware of and advocate for change around HC in engineering. We acknowledge, because of our research, that transforming the institutional culture of engineering education is a process that should begin with an awareness of HC. Yet awareness of HC is not enough. The four-factor model of HC, as we have shared in this chapter, is a structural framework by which individuals can use to derive strategies and for survival, navigation, or change in engineering. But understanding is not alone; it takes intentional action to create counterspaces and counternarratives to positively transform the landscape of engineering education and practice.

Acknowledgment

This material is based upon work supported by the National Science Foundation (NSF) No. EEC-1653140 and 2123016, given to the first author. Any opinions, findings, and conclusions or recommendations expressed in this material do not necessarily reflect those of the NSF. We want to give special thanks to the institutional liaisons, Dr. Hector Cruzado, Dr. Sindia Rivera-Jimenez, Dr. Heather Shipley, Dr. Kimberly Cook-Chennault, and Dr. Paul Barr, who assisted us with collecting participant data in the first stage of sampling. We want to thank Drs. Laura Gelles, Kate Youmans, Darcie Christensen, Marialuisa Di Stefano, Tarique Khan, Jamaal Downey, Cijy Sunny, and Taya Carothers for their early or ongoing contributions to this work. We also want to thank the participants for sharing their experiences with us and the readers of this work. Finally, we would like to thank Dr. Kim Johnston, the associate dean at the Schulich School of Engineering, University of Calgary, who is the brains behind the Engineering Attributes program and has been the biggest leader and advocate for bringing this HC-inspired mental wellness program into first-year engineering classes. We are so appreciative of this diverse group of authors who came together and the beautiful support and joy we provided each other through the writing process. A special thank-you to baby Riley, who was born to the third author almost exactly on the day the manuscript was initiated and started walking as the manuscript was completed. We look forward to seeing this manuscript walk on its own and the journeys it will take.

Note

1 As an act of resistance and counternarrative to the HC of scientific naming, I, Dr. Villanueva Alarcón, decided to include my father's (Villanueva) and mother's (Alarcón) surnames in subsequent authorships related to HC and other areas of research in 2020 to reflect her cultural identity. Please note that while some authorships have only her father's surname, she is the same author in the cited works of this chapter that include her mother's surname as well.

References

Abramovich, S., & Brouwer, P. (2004). Developing technology-mediated entries into hidden mathematics curriculum as a vehicle for "good learning" by elementary pre-teachers. *Journal of Computers in Mathematics and Science Teaching, 23*(3), 299.

Ahn, M. Y., & Davis, H. H. (2020). Sense of belonging as an indicator of social capital. *International Journal of Sociology and Social Policy, 40*(7/8), 627–642.

Allan, H. T., Smith, P., & O'Driscoll, M. (2011). Experiences of supernumerary status and the hidden curriculum in nursing: A new twist in the theory-practice gap? *Journal of Clinical Nursing, 20*(5–6), 847–855. https://doi.org/10.1111/j.1365-2702.2010.03570.x

Apple, M. W. (1980). Analyzing determinations: Understanding and evaluating the production of social outcomes in schools. *Curriculum Inquiry, 10*(1), 55–76. https://doi.org/10.1080/03626784.1980.11075203

Arao, B., & Clemens, K. (2013). From safe spaces to brave spaces. In *The art of effective facilitation: Reflections from social justice educators* (pp. 135–150). Stylus Publishing.

Bandura, A. (1986). The explanatory and predictive scope of self-efficacy theory. *Journal of Social and Clinical Psychology, 4*(3), 359–373. https://guilfordjournals.com/doi/abs/10.1521/jscp.1986.4.3.359

Bejerano, A. R., & Bartosh, T. M. (2015). Learning masculinity: Unmasking the hidden curriculum in science, technology, engineering, and mathematical courses. *Journal of Women and Minorities in Science and Engineering, 21*(2), 107–124. https://doi.org/10.1615/JWomenMinorScienEng.2015011359

Bourdieu, P. (1986). *The forms of capital.* In J. Richardson (Ed.), *Handbook of theory and research for the sociology of education* (pp. 241–258). Greenwood.

Claridge, T. (2018). Functions of social capital – bonding, bridging, linking. *Social Capital Research.* www.socialcapitalresearch.com/wp-content/uploads/2018/11/Functions-of-Social-Capital.pdf

Dreeben, R. (1968). Social relations in a secondary school. *Social Forces, 47*(2), 235–236. https://doi.org/10.1093/sf/47.2.235-a

Erickson, S. K. (2007). *Engineering the hidden curriculum: How women doctoral students in engineering navigate belonging.* Arizona State University.

Freeman, T. M., Anderman, L. H., & Jensen, J. M. (2007). Sense of belonging in college freshmen at the classroom and campus levels. *The Journal of Experimental Education, 75*(3), 203–220. https://doi.org/10.3200/JEXE.75.3.203-220

Gelles, L. A., Youmans, K. L., & Villanueva, I. (2019, April 14–17). Hidden curriculum advocacy and resources for graduate students in engineering. In *CONECD conference.*

Gelles, L. A., Youmans, K. L., & Villanueva, I. (2020). Sparking action: How emotions fuel or inhibit advocacy around hidden curriculum in engineering. In *European society of engineering education (SEFI), 47th annual conference: Varietas Delectat . . . complexity is the new normality, proceedings* (pp. 1566–1575). https://par.nsf.gov/servlets/purl/10162378

Giroux, H. A., & Penna, A. N. (1979). Social education in the classroom: The dynamics of the hidden curriculum. *Theory and Research in Social Education,* 7(1), 21–42. https://doi.org/10.1080/00933104.1979.10506048

Hafferty, F. W., & Franks, R. (1994). The hidden curriculum, ethics teaching, and the structure of medical education. *Academic Medicine, 69*(11), 861–871. https://doi.org/10.1097/00001888-199411000-00001

Hansson, L. (2018). Science education, indoctrination, and the hidden curriculum. In *History, philosophy and science teaching* (pp. 283–306). Springer.

Hettler, B. (1976). *The six dimensions of wellness.* National Wellness Institute. www.nationalwellness.org/page/Six_Dimensions

Higham, J. (1959). The cult of the 'American consensus': Homogenizing our history. *Commentary, 27,* 93–100.

Huff, J. L., Sullins, J., Sochacka, N. W., Youngblood, K. M., Wood, K. M., Miller, S. E., & Walther, J. (2016). Exploring shame in engineering education. In *Proceedings of the IEEE frontiers in education conference* (pp. 1–4).

Husman, J., Cheng, K., Puruhito, K., & Fishman, E. (2015, June.). Understanding engineering students' stress and emotions during an introductory engineering course. In *2015 ASEE annual conference and exposition proceedings.* https://doi.org/10.18260/p.24958

Jackson, P. W. (1968). *Life in classrooms.* Teachers College Press.

Kellam, N., Gerow, K., Wilson, G., Walther, J., & Cruz, J. (2018). Exploring emotional trajectories of engineering students: A narrative research approach. *International Journal of Engineering Education, 34*(6), 1726–1740.

Kentli, F. D. (2009). Comparison of hidden curriculum theories. *European Journal of Engineering Education, 1*(2), 83–88.

Lönngren, J., Adawi, T., & Berge, M. (2021). Using positioning theory to study the role of emotions in engineering problem solving: Methodological issues and recommendations for future research. *Studies in Engineering Education, 2*(1), 53–79.

Lönngren, J., Direito, I., Tormey, R., & Huff, J. (2023). Emotions in engineering education. In A. Johri (Ed.), *International handbook of engineering education* (in-press). Taylor and Francis, pp. 156–192.
Lyles, C., Huggins, N., & Robbins, C. K. (2022). Unveiling the hidden curriculum within graduate education. In D. J. Nguyen & C. W. Yao (Ed.), *A handbook for supporting today's graduate students* (pp. 86–100). Sterling.
Margolis, E. (2001). *The hidden curriculum in higher education*. Psychology Press.
Martin, J. R. (1976). What should we do with a hidden curriculum when we find one? *Curriculum Inquiry*, *6*(2), 135–151.
National Museum of African American History and Culture. (n.d.). *Being antiracist*. Retrieved December 8, 2021, from https://nmaahc.si.edu/learn/talking-about-race/topics/being-antiracist
Parsons, T. (1977). *Social systems and the evolution of action theory* (1st ed.). Free Press.
Paul, R., Dedemus, D., Boyce, M., & Johnston, K. (2020). The "engineers have feelings" project: Integrating mental wellness and lifelong learning skills in first-year undergraduate engineering courses. *Proceedings of the Canadian engineering education association (CEEA)*.
Paul, R., Adeyinka, O., Boyce, M., Eldib, G., Gaulin, K., Johnston, K., . . . Tormon, R. (2021). Impact of integrating mental wellness and personal learning reflections into first-year undergraduate engineering courses. *Proceedings of the Canadian engineering education association (CEEA)*.
Paul, R., Behjat, L., & Brennan, R. (2021). Using individual-based modeling to better understand the hidden curriculum of engineering. *Proceedings of the Canadian engineering education association (CEEA)*.
Pehlivanli-Kadayifci, E. (2019). Exploring the hidden curriculum of gender in engineering education: A case of an engineering faculty in Turkey. *International Journal of Engineering Education*, *35*, 1194–1205.
Pekrun, R., & Linnenbrink-Garcia, L. (Eds.). (2014). *International handbook of emotions in education*. Routledge. https://doi.org/10.4324/9780203148211
Polmear, M., Bielefeldt, A., Knight, D., Swan, C., & Canney, N. (2019). Hidden curriculum perspective on the importance of ethics and societal impacts in engineering education. In *2019 ASEE annual conference & exposition proceedings*. https://doi.org/10.18260/1-2-32887
Polmear, M., Clegorne, N., & Simmons, D. R. (2022). Uncovering the hidden curriculum of leadership education in civil engineering. *International Journal of Engineering Education*, *38*(1), 1–13.
Pooley, J. A., Cohen, L., & Pike, L. T. (2005). Can sense of community inform social capital? *The Social Science Journal*, *42*(1), 71–79. https://doi.org/10.1016/j.soscij.2004.11.006
Rea, S. C., & Shiekh, K., & Zhu, Q., & Nieusma, D. (2021, July). *The hidden curriculum and the professional formation of responsible engineers: A review of relevant literature in ASEE conference proceedings*. Paper presented at 2021 ASEE virtual annual conference content access, virtual conference. https://peer.asee.org/37866
Rosenberg, M., & McCullough, B. C. (1981). Mattering: Inferred significance and mental health among adolescents. *Research in Community & Mental Health*, *2*, 163–182.
Rottmann, C., & Reeve, D. (2020). Equity as rebar: Bridging the micro/macro divide in engineering ethics education. *Canadian Journal of Science, Mathematics and Technology Education*, *20*(1), 146–165. https://doi.org/10.1007/s42330-019-00073-7
Secules, S. (2019). Making the familiar strange: An ethnographic scholarship of integration contextualizing engineering educational culture as masculine and competitive. *Engineering Studies*, *11*(3), 196–216. https://doi.org/10.1080/19378629.2019.1663200
Secules, S., Sochacka, N. W., Huff, J. L., & Walther, J. (2021). The social construction of professional shame for undergraduate engineering students. *Journal of Engineering Education*, *110*(4), 861–884. https://doi.org/10.1002/jee.20419
Sellers, V. B., & Villanueva Alarcón, I. (2021a). What strategies do diverse women in engineering use to cope with situational hidden curriculum? In *2021 ASEE virtual annual conference*. https://peer.asee.org/38051
Sellers, V. B., & Villanueva Alarcón, I. (2021b). *Hidden curriculum pathways in engineering: From incident to action* [Unpublished manuscript]. Department of Engineering Education, University of Florida.
Smith, B. (2013). *Mentoring at-risk students through the hidden curriculum of higher education*. Lexington Books.
Soria, K. M., & Stebleton, M. J. (2013). Social capital, academic engagement, and sense of belonging among working-class college students. *College Student Affairs Journal*, *31*(2), 139.
Stallard, P. (2019). *Think good, feel good: A cognitive behavioural therapy workbook for children and young people*. John Wiley & Sons.

Tonso, K. L. (2001). "Plotting something dastardly": Hiding a gender curriculum in engineering. In *The hidden curriculum in higher education* (pp. 165–184). Routledge.

Tormey, R., Leduc, I., Isaac, S., Hardebolle, C., & Cardia, I. V. (2015). The formal and hidden curricula of ethics in engineering education. In *Proceedings of the 43rd annual SEFI conference*. www.sefi.be/wp-content/uploads/2017/09/56039-R.-TORMEY.pdf

Turner, J. H. (1985). *Herbert Spencer: A renewed appreciation* (Vol. 1). SAGE Publications, Incorporated.

Tyng, C. M., Amin, H. U., Saad, M. N., & Malik, A. S. (2017). The influences of emotion on learning and memory. *Frontiers in Psychology*, *8*, 1454.

Urry, J. (2012). *Sociology beyond societies: Mobilities for the twenty-first century*. Routledge.

Villanueva, I. (2017). *CAREER: Advocating for engineering through hidden curricula: A multi-institutional mixed method approach*. https://nsf.gov/awardsearch/showAward?AWD_ID=1653140&HistoricalAwards=false

Villanueva, I., Campbell, B. D., Raikes, A. C., Jones, S. H., & Putney, L. G. (2018). A multimodal exploration of engineering students' emotions and electrodermal activity in design activities. *Journal of Engineering Education*, *107*(3), 414–441. https://doi.org/10.1002/jee.20225

Villanueva, I., Carothers, T., Di Stefano, M., & Khan, M. (2018). "There is never a break": The hidden curriculum of professionalization for engineering faculty. *Education Sciences*, *8*(4), 157. https://doi.org/10.3390/educsci8040157

Villanueva, I., Di Stefano, M., Gelles, L., Youmans, K., & Hunt, A. (2020). Development and assessment of a vignette survey instrument to identify responses due to hidden curriculum among engineering students and faculty. *International Journal of Engineering Education*, *36*(5), 1–21.

Villanueva, I., Gelles, L. A., Di Stefano, M., Smith, B., Tull, R. G., Lord, S. M., Benson, L., Hunt, A. T., Riley, D. M., & Ryan, G. W. (2018, June 24–27). What does hidden curriculum in engineering look like and how can it be explored? In *Proceedings of the American society of engineering education annual conference & exposition, minorities in engineering division*, Paper #21884 (pp. 1–16).

Villanueva, I., Gelles, L. A., Youmans, K., & Di Stefano, M. (2018). Hidden curriculum awareness: A comparison of engineering faculty, graduate students, and undergraduates. *World Engineering Education Forum*, 1–6.

Villanueva, I., Revelo, R. A., & Mejia, J. A. (2019). Professional development of Latinx engineers on hidden curriculum: An exploratory study. In *IEEE frontiers in education conference* (pp. 1–5). https://ieeexplore.ieee.org/abstract/document/9028632/?casa_token=IlEutod0MsYAAAAA:56av6jlwQiST1cOOXdRMfi6yjeyFAjDszVw6TwcC1ruk5FSeOSdB7wa_8Ga_3_3DMNhoRj5h

Villanueva Alarcón, I., & Sellers, V. (2022). Faculty development in the third space: Influence of hidden curriculum amid engineering educators. In S. Linder, C. Lee, & K. High (Eds.), *Handbook of STEM faculty development*. Information Age Publishing, Inc.

19

Collaborative Learning in Engineering Education

Emma Mercier, Molly H. Goldstein, Preethi Baligar, and Robin Jephthah Rajarathinam

1 Introduction

The pedagogical practice of group work has a long history in higher education, having been the subject of formal research in the field for over a century (Johnson et al., 1998). Despite this, tension persists between different educational approaches in engineering, specifically between traditional lectures with accompanying homework, problem sets, and exams, and active learning approaches such as collaborative and cooperative learning (Bubou et al., 2017; Wieman, 2014). *Collaborative learning* is an active learning approach that more closely resembles the practices of engineering professionals in industry settings than traditional educational techniques (Cady & Reid, 2018; Prince, 2004). Because working effectively on a team is crucial to success in the engineering workplace (Johri et al., 2015; Jonassen et al., 2006), engineering accreditation agencies worldwide specify it as an explicit learning outcome (e.g., ABET Accreditation Board for Engineering and Technology, Criteria for Accrediting Engineering Programs, 2018).

Research in the learning sciences has demonstrated that collaborative, problem-based, and design-focused experiences produce deep learning, increase persistence in STEM fields, and aid the development of the collaborative skills that are essential for industry (e.g., Barron & Darling – Hammond, 2008; Nokes – Malach et al., 2019). As Sheppard et al. suggest in *Educating Engineers* (Sheppard & Macatangay, 2008), these best practices for developing effective learning opportunities should be adopted across engineering programs in order to prepare budding engineers to collaboratively address multidisciplinary and complex problems in their professional careers.

This chapter brings together research on collaborative and cooperative learning in education, the learning sciences, and engineering education. Our goal is to review current understandings of the purposes of collaborative learning, how collaboration leads to learning, and the conditions necessary to have successful collaborative learning experiences in engineering courses. The chapter is written both as a guide for how to think about implementing collaborative learning and to prepare the ground for future research in the field.

While the concept of active learning encompasses a large range of activities, we have limited our focus to learning activities that take place in groups where there is some sustained interaction between group members – from at least one class period or lesson to semester-long team projects. We also acknowledge a significant, if possibly temporary, increase in the use of online learning and group work during recent years. However, the scope of this chapter does not allow us to include

DOI: 10.4324/9781003287483-23

a detailed examination of the emerging literature on sustaining collaboration solely in an online context.

We use the 4Ts framework developed by the first author (Mercier & Higgins, 2015) to outline the crucial elements of collaborative learning in classrooms: teams, tasks, tools, and teachers. These four categories are useful in mapping the current state of the field and identifying areas in which further research is necessary. While laying out these categories, however, this framework also acknowledges their fundamentally overlapping nature (e.g., team interaction is influenced by the type of task being used, and so on).

We reviewed papers published between 2010 and 2022 in the *European Journal of Engineering Education*, the *International Journal of Engineering Education*, the *Journal of Engineering Education*, and *Studies in Engineering Education*. We arrived at this short list by seeking to represent the "top" journals in standard engineering education research using a combination of impact factor, h-index, and indexing databases while representing a variety of countries. We identified papers using the following keywords: *collaborative learning*, *cooperative learning*, *group work*, *capstone project*, *group-based*, and *design project*. We then categorized these papers using the 4Ts framework to provide insights into the current state of the field, explore implications for teaching, and highlight directions for future research.

2 Overview of Collaborative Learning

2.1 Collaborative and Cooperative Learning

The terms "collaborative learning" and "cooperative learning" are often used interchangeably (Hmelo-Silver et al., 2013), with early studies in engineering education preferring the latter term (Froyd et al., 2012; Smith et al., 1981a, 1981b). In this chapter, we will draw a distinction between the terms in order to clarify different aspects of these processes, recognizing that elements of both types of learning are necessary in engineering education and that the differences between them are neither entirely clear nor universally agreed upon in the field.

Smith et al. (2005) use the requirement in cooperative learning of structured, individual accountability to disambiguate from cooperative and collaborative learning. Alternatively, Dillenbourg et al. (1995) define the key feature of cooperative learning as individuals having responsibility for different elements of the task, while collaborative learning is defined as a high level of mutual engagement and coordinated effort to solve the problem or create joint understanding. Drawing on this literature, we define collaborative learning as requiring the co-construction of knowledge by groups. Cooperative learning, on the other hand, is defined as groups interdependently working to achieve a common goal, which results in activities that lend themselves to students taking different roles during a task, and possibly breaking the task down into subparts and completing them under a rubric of individual accountability (Davidson & Major, 2014; Smith et al., 2005). While this chapter is focused on collaborative learning, it should be noted that there is significant overlap in these two forms of pedagogy, and that each has a time and place in which it is the more effective option, and as such, we will discuss both throughout the chapter.

2.2 Why Do Collaborative and Cooperative Learning Work?

Drawing from different epistemic foundations, a range of theoretical perspectives has emerged to illuminate how collaboration leads to learning (Hmelo-Silver et al., 2013). Among these, the two key theoretical schools here are sociocultural theory and cognitive and sociocognitive theories. Despite both being fundamentally grounded in a constructivist approach to learning, which precedes from the idea of students as active constructors of their understandings rather than mere receptors of knowledge, each provides different explanations for the phenomena associated with successful learning experiences. Based in the Vygotskian view of learning, sociocultural theory focuses on socially mediated cognition, that is, the idea that learning occurs through socially constructed artifacts, including language. In this context, the effectiveness of collaborative learning comes from the opportunity to learn how to be an engineer by talking about engineering concepts, engaging in the processes of the engineering community, and using the tools, artifacts, and language of the engineering industry (e.g., authentic problem-solving tasks, software used by engineers). Cognitive theories posit that participating in a group initiates the types of cognition that result in learning, such as retrieval, rehearsal, and experiencing cognitive conflict when a peer has a differing idea or opinion (Webb, 2013). Students need to have opportunities to discuss ideas with their peers in order to engage with different perspectives, to rehearse and verbalize their own thinking, and to ask questions.

From a cooperative learning perspective, the success of group learning is grounded in social interdependence theory. First defined by Deutsch (1962), social interdependence theory states that social interdependence exists when goals are shared by individuals and each individual's outcomes are based on the actions of others. Social interdependence can be both competitive (some win and some lose) or cooperative (all win or all lose). Thus, cooperative learning occurs when groups have a common goal and collective success or failure relies on the actions of all group members.

2.3 The Benefits of Collaborative Learning

Research across the field of engineering education and the learning sciences emphasizes the importance of collaborative learning for a range of educational outcomes, including learning and problem-solving as well as persistence in STEM fields and career preparation (e.g., Barron & Darling-hammond, 2008; Hmelo-Silver & Chinn, 2016). Engineering education research specifically highlights the value of using collaborative tasks for learning and increased student engagement as well as for the development of collaborative skills for future careers (Borrego et al., 2013; Johri et al., 2015).

2.3.1 Increased Persistence in STEM Fields

Equitable participation in STEM fields and the experiences of minoritized students in STEM-related programs and professions have been matters of great concern for several decades (Adelman et al., 1998; Felder, 1994; NRC, 1994; Tao, 2016). Researchers have examined several issues related to engaging K–12 students in STEM courses as well as the nature of the experience of minoritized students in early undergraduate courses. These studies suggest that class sizes, dense theoretical content, and a focus on memorization or drill-and-practice assignments can be alienating both from peers and from the goals of the subject area. Consequently, students often feel that the content they are studying bears no connection to their original reasons for entering the field (Ballen et al., 2017; Pattison et al., 2020).

Collaborative problem-based activities have been identified as mechanisms for addressing many of the difficulties that arise in STEM courses (e.g., Fullilove & Treisman, 1990; Margolis & Fisher, 2002; Pattison et al., 2020). Kalaian et al. (2018) performed a meta-analysis of small-group learning pedagogies in engineering and technology education programs. They report an effect size of .45 in favor of group learning, which, for the complex field of education research, where multiple variables are likely to influence the effect being studied, is a substantial result (Kraft, 2020).

At the same time, as one might expect with such an effect size, results from individual research studies are mixed. Van Dusen and Nissen (2020) report on a study examining the intersectional nature of race/racism and gender/sexism using the LASSO database of 13,857 students in first-semester physics courses. The authors concluded that collaborative learning experiences improved equity, with all students learning more in collaborative than in lecture-style instruction; however, it did little to foster equality, as all groups improved equally. Another recent study (Mollet et al., 2021) reports that, although participating in peer learning activities was important for sustaining GPAs among minoritized STEM students, collaborative learning in groups where these students were in the minority were not seen to be helpful. This may indicate a need to pay more attention to status issues and address them directly in the creation of collaborative activities. Similarly, Stump et al. (2011) analyzed the results of two surveys of engineering students asked to report on their own collaboration, self-efficacy, knowledge-building behaviors, and course grade. Although these behaviors were significantly predictive of course grade, female students were more likely than male students to report using collaborative learning, and those who used collaborative learning were more likely to get a B than an A or a C. These results all point towards the potential benefits of collaborative learning, while also highlighting the complex nature of pedagogical innovation and recognizing gaps in our understanding of how best to create learning experiences that bring traditionally minoritized students into engineering and STEM more broadly.

2.3.2 Successful Learning Outcomes

Cooperative learning, collaborative learning, and group problem-solving activities are some of the most heavily researched topics in the field of education (Barron & Darling-Hammond, 2008), with most studies finding that these forms of learning yield positive effects. A series of meta-analyses and review articles provides a bird's-eye view of the literature, showing an overall positive effect of group activities on learning outcomes. For example, Kyndt et al. (2013) report on 65 studies conducted after 1995 that compared real-life classroom-based cooperative activities with traditional, lecture-based learning experiences. They reported an overall effect size on achievement of .54, another unusually high effect size for education research. Nokes-Malach et al. (2015) summarize findings from laboratory studies that show groups outperform individuals on tasks including memory, categorization, and problem-solving. However, they also note that the classroom-level data is more complex; most classroom-based studies agree with the lab studies' findings, but some show that not all groups have the same experiences. These meta-analyses suggest that collaborative learning experiences have the potential to create more productive learning outcomes but also indicate that success may depend on a large research team implementing the intervention, which may be difficult for individual instructors to replicate in a complex classroom environment. In the sections that follow, we will look closely at aspects of implementing collaborative learning in classrooms that determine whether and how it yields successful learning outcomes.

2.3.3 Collaborative Skills and the Need to Prepare for Careers

Industry professionals have increasingly recognized the importance of collaborative skills in the engineering workplace, where more and more work is done by teams and in groups. Engineers are more than ever required to work collaboratively to tackle global problems, leading to a call for more authentic learning experiences in engineering programs (e.g., (Agrawal & Harrington – Hurd, 2016; Liu & Wayno, 2008; Vergara et al., 2009). Researchers have begun to dedicate considerably more effort to both assessing and documenting what counts as collaborative skills (e.g., OECD, 2017) and how to teach them (Borge et al., 2018).

The PISA framework (OECD, 2017) for assessing collaborative problem-solving skills is based on four features of problem-solving that have been identified and analyzed in the literature: exploring and understanding, representing and formulating, planning and executing, and monitoring and reflecting. Researchers have directly connected these four features to three core collaboration skills: establishing and maintaining shared understanding, taking appropriate action to solve the problem, establishing and maintaining team organization to create a 12-part matrix of collaborative problem-solving skills. Similarly, Hesse et al. (2015) proposed a framework for teaching collaborative problem-solving skills that builds on the ATC21S™ criteria, which mandates that the skills should be measurable both in large-scale assessments and in classroom settings and that they must be teachable (OECD, 1999). Under this framework, two skill sets are highlighted: social process skills and cognitive process skills. Social process skills are divided into the categories of participation, perspective taking, and social regulation. Cognitive skills are divided into task regulation, learning, and knowledge building. Indicators for subelements of each of these skills are identified, with a description of what these would look like when present at low, medium, and high levels. This list of skills provides an accessible set of behaviors from which to teach students how to collaborate and provide insights to instructors on how to assess collaboration skills.

However, research indicates that merely engaging in collaborations may not sufficiently develop students' collaboration skills; it is important to provide opportunities for students to explicitly learn collaboration skills, to reflect on their own experiences, and to learn to diagnose and address difficulties within their group dynamics (Barron et al., 2009; Borge et al., 2018; Mercier et al., 2009). In Mercier's study, 19 stable groups who worked together weekly during the 7th, 8th, 9th, and 10th weeks of the semester showed low-level collaboration in week 7 (the first week working as a group), which remained low across the four weeks (Mercier et al., 2015). While this longitudinal analysis of interactions is rare, it resonates with other research that finds that groups who struggle in their interactions may not be able to repair and/or improve their processes without intervention (e.g., Barron, 2003; Kozlowski & Ilgen, 2006; Mercier et al., 2009).

Over time, researchers have examined different ways to support students during collaboration in order to improve their interactions. For example, Fischer et al. (2013) argue that external scripts can be used to scaffold the development of internal scripts, which allow learners to become better collaborators in future activities. Studies also indicate that students engage in higher-quality interactions when provided with guidance for how to engage in collaboration, whether in the form of scripts (e.g., Kester & Paas, 2005; Weinberger & Fischer, 2001), models (e.g., Rummel et al., 2009), or prompts and activities requiring them to reflect on the interactions (e.g., Cortez et al., 2009). A key feature of the model proposed by Fischer et al. (2013) is the appropriate fading of external scripts as students become more accustomed to collaborative activities and develop internal scripts to guide their actions. As yet, however, few studies have observed students for long enough to track the development of these internal scripts. Borge and White (2016) provide some insight into how students respond over time, reporting on the use of strategies and tools to improve collaboration over an 11-week science unit. Their findings suggest that students can learn to use strategies provided to them, and the quality of their interactions improved over time as they used these strategies more effectively with practice.

In sum, the literature identifies the necessity of teaching students to collaborate but shows that merely placing them in groups, while common, is somewhat ineffective. Techniques can be used to support collaboration, although more research is needed to indicate which are most useful and which are most useful in engineering courses.

2.4 Engineering Education Standards

In the United States, accreditation criteria have explicitly recognized collaboration as a crucial component of an engineering education (ABET, 2018). This was formalized with the change in criterion 3 (student outcomes) from "an ability to function on multidisciplinary teams" to "an ability to function effectively on a team whose members together provide leadership, create a collaborative and inclusive environment, establish goals, plan tasks, and meet objectives," which was implemented during the 2019–2020 cycle (ABET, 2019–2020, 2019).

Similarly, the European Network for Accreditation of Engineering Education specifies that bachelor's graduates should have the "ability to effectively communicate information, ideas, problems and solutions with engineering community and society at large" and should also be able to "function effectively in a national and international context, as an individual and as a member of a team and to cooperate effectively with engineers and non-engineers" (ENAEE, 2008). The International Engineering Alliance, through its Washington Accord, specifies that graduates should "function effectively as an individual, and as a member or leader in diverse teams and in multi-disciplinary settings" (IEA, 2021).

Clearly, international accreditation agencies recognize the role of interdisciplinary teamwork in professional engineering contexts and promote educational contexts in which students learn to effectively frame, address, and solve problems collaboratively. However, despite more than a decade of this heightened focus on teamwork and collaboration, industry experts still see a great deal of room for improvement. For example, a survey conducted by the Association of American Colleges and Universities reported that 63% of the 400 employers surveyed felt graduates were not prepared to work in teams effectively and were ill-prepared to use technologies to solve a problem (Hart Research Associates, 2015). There is still much to be done in effectively designing integrating experiences that will allow students to develop collaborative learning skills as part of their undergraduate training.

2.5 Addressing Diversity, Equity, and Inclusion in Collaborative Learning

Despite the many benefits of collaborative learning in engineering courses, research has illuminated negative aspects of teamwork related to gender (e.g., Beigpourian & Ohland, 2019; Laeser et al., 2003; Tonso, 1996), race/ethnicity (e.g., Blosser, 2020; Chen et al., 2015; Cross & Paretti, 2020), and sexual orientation (Cech & Rothwell, 2019; Cech & Waidzunas, 2021), where students can be further marginalized in engineering teamwork (Meadows et al., 2015). Indeed, marginalized students are apt to "experience group environments differently not because they lack sufficient skills or resources, but because cultural and social norms create barriers not typically experienced by students from dominant groups" (Meadows et al., 2015, p. 12). Oftentimes, minoritized students are burdened with "stereotype management" (McGee & Martin, 2011) which might contribute to their discomfort in course situations (Meadows et al., 2015), with a negative impact of their sense of belonging. Moreover, because many (quantitative) studies aggregate minoritized students into one group, failing to account for the intersectionality of multiple groups, we still have much to learn regarding the best ways to create an inclusive environment in collaborative learning. While this chapter will explicitly discuss DEI considerations in Section 3.1.2, "Group Composition," we recognize the need for additional research in collaborative learning.

2.6 Issues in Using Collaborative Learning

Despite decades of research indicating the value of collaboration in learning contexts and clear evidence that collaborative learning enhances classroom experiences, the actual implementation of collaborative learning has several drawbacks (Nokes-Malach et al., 2015). Teaching both short-problem-based collaborative activities and longer-term projects can present a number of challenges for instructors, many of which are understudied (e.g., Mercier et al., 2009; Takai & Esterman, 2019). These challenges are often associated with inequities in participation and poor social interaction skills but can be addressed through task design, classroom interventions, and spending time to develop social skills and group norms during the course. These issues can make teachers and instructors reluctant to use collaborative and cooperative learning pedagogies (Gillies & Boyle, 2010;

Shehab & Mercier, 2019). In Section 3.4, we discuss the role of the teacher in implementing collaborative learning in more detail, although note that it is an area that needs significant future research.

One of the most frequent complaints from students who are assigned collaborative learning projects is that one or more of their group mates did not complete their share of the work and/or did not participate in group activities (Tenenberg, 2019). Known as free riders, these students appear to sit back and allow others to complete the work. The free rider is left with a less-successful educational experience, and the rest of the team has to work harder to make up for the free rider's indifference, often becoming frustrated with and resentful of the task. While this can be particularly problematic in a multi-week project, it can also be harmful in shorter collaborative activities.

Another problem that arises in collaborative projects, sometimes in relation to a free rider issue, is the domination of the group by one or more students (Salomon & Globerson, 1989). In this situation, a single student or subgroup of students takes over and either does not engage in collaborative knowledge-building discussions yet makes decisions and assigns work to the other group members or completes all the work alone. This can negatively impact the educational experience of less-involved students, may result in all students missing the opportunity to learn more about collaborating, and may yield a poorly designed final product or less elaboration of ideas within a group if one or more member is rushing to complete the project.

Another related issue is that of social loafing, wherein some or all the students in a group reduce the amount of effort that they would typically put into a class assignment. Borrego et al. (2013), who deliver a systematic review of the literature on student teams in computer science and engineering, note that social loafing was the most frequently cited negative team behavior. Social loafing is often viewed as an issue of motivation. It can be addressed through assessment practices, although some argue that increasing the inherent value of the task for students is a more effective way of treating the problem (Karau & Williams, 1993, 1995).

Groupthink, in which agreement and cohesion between group members become more important than fully exploring a topic, is another well-documented issue in the research on teamwork in education (e.g., Bénabou, 2013; Janis, 1991). There is a significant body of literature on this topic, which is particularly prevalent within the organizational psychology literature. In previous studies, we have documented a series of issues that may arise in student teams when groupthink dominates the early stages of a design project (e.g., Mercier et al., 2009) and suggest that, in designing the phases of a project, strategies such as presenting multiple ideas be included.

These negative effects are associated with a variety of causes: tasks that are not truly group-worthy (see Section 3.2), students who do not understand the purpose of participation, students who do not know how to participate or to recruit other members' participation, or students who are struggling with the content. Different goals for the activity can also lead to these issues, as when students who are strongly focused on getting a high grade are required to work with students who are less invested in the course.

Research grounded in status theory points to the importance of attending to an individual's status within the group and how perceived status can impact the way group members interact (e.g., Cohen & Lotan, 1997). When a group collectively identifies one member as low status with respect to the task, they tend to exclude that member or disregard their ideas (Berger et al., 1980). This can be particularly relevant for interdisciplinary teams, where preconceptions about the value of the different disciplines may impact the way contributions are valued, work is assigned, and/or group assessments are conducted (Booker et al., 2009). Scholarship on Complex Instruction (e.g., Azizan et al., 2018; Cohen, 1994) indicates that it is possible to configure classroom instruction and tasks so as to identify the value group members bring to a project and thereby promote more equitable participation.

3 The Four Ts of Collaborative Learning in Classrooms

In our prior work, we identified four core aspects of collaborative learning in classrooms and recommended that more attention be paid to each of the four aspects so that this pedagogy could be implemented more successfully (Mercier & Higgins, 2015). As before, these four aspects are teams, tasks, tools, and teachers. Although the amount of research dedicated to each aspect has varied, all four overlap with one another to some degree. In the sections that follow, we take each of these four aspects and review relevant studies and the overall research stream. We have also included assessment as a subsection of teaching, since, although it is part of the teacher's role (e.g., Kaendler et al., 2015), drawing it out as a specific section allows us to highlight the importance of assessment to the collaborative learning experience.

Learning also takes place in a specific context, and attention to that context, both in terms of the norms that are established by the teacher and class as well as to the classroom design, available tools, etc., is essential to thinking about how to implement collaborative learning activities.

As noted earlier, we reviewed publications in four engineering education journals in order to generate an understanding of where the field is positioned in relation to these four core areas of collaborative learning. The distribution of papers is shown in Table 19.1. Of the 285 papers we collected from these journals since 2010 (since 2019 for *SEE*, the first year it was published), the majority were primarily related directly to teams (150), followed by tasks, assessment, teachers, and tools. Around 161 papers received one code, 85 two codes, and 11 received three codes. One paper (Marbouti et al., 2019) was classified as discussing teams, teachers, tasks, and assessment; 27 articles, all from *Studies in Engineering Education*, did not receive any classification. The majority of these were review papers, position papers, or papers discussing methodological issues in the field.

3.1 Teams

On a basic level, the literature agrees on certain prerequisites for functional collaborative teams. While they have been written about using a variety of terminology, there is a broad consensus that team members need to communicate effectively with each other, be responsive to other team members, and be willing to cooperate and compromise to accomplish tasks (Hmelo – Silver et al., 2013). Commitment to supporting teammates has also been identified as an essential element in successful groups (e.g., Bratman, 1992), which is particularly relevant in collaborative learning contexts, as making sure that all team members understand the assignment is essential if all students are to have equitable learning experiences.

Beyond these fundamental requirements, a significant cross-section of research in collaborative learning has focused on the functions of teams and the types of interactions that are most often associated with successful outcomes; it is generally accepted that the quality of students'

Table 19.1 Topics Covered by Papers about Collaboration in Engineering Education Journals

Journal	*Article Topics*				
	Teams	*Tools*	*Tasks*	*Teachers*	*Assessment*
Journal of Engineering Education	22	5	4	12	7
International Journal of Engineering Education	79	14	36	17	34
European Journal of Engineering Education	46	14	43	17	15
Studies in Engineering Education	3	0	0	0	0
Total	150	33	83	46	56

interactions is pivotal to successful learning experiences (Barron & Darling – Hammond, 2008; Hesse et al., 2015). High-quality interactions include co-creating a joint understanding of the problem space (Roschelle, 1992), creating joint visual representations (Mercier & Higgins, 2015), building on each other's ideas (Barron, 2003), and responding to questions with elaborated explanations (Webb, 2013).

Successful teams attend to two dimensions of collaborative interaction: the social or interactional space and the problem or cognitive space (Barron, 2003). The social or interactional space includes participation, perspective taking, and social regulation, while the cognitive space includes task regulation and learning, as well as knowledge building (Hesse et al., 2015). In the interactional space, it is important that students engage with peers, persist with the task, participate in perspective taking, that is, responding to teammates, listening, and adapting based on their perspectives, and maintaining social regulation. Social regulation is described in multiple ways across the literature and encompasses self-directed behaviors, such as negotiation, self-regulation, monitoring one's own understanding in relation to the group, as well as socially shared regulation of the group. Socially shared regulation refers to the meta-cognitive, deliberate processes required for groups to regulate their participation and behavior (Hadwin et al., 2019).

Research in engineering education also emphasizes the importance of teams; indeed, most of the papers we reviewed focused on this topic. Many of these papers highlight elements of team composition (e.g., Griffin et al., 2004; Mikic & Rudnitsky, 2016), the development of teamwork skills (e.g., El-Sakran et al., 2013; Hadley, 2014; Maturana et al., 2014), and the impact of diversity of group members or ideas (Fila & Purzer, 2014; Lau et al., 2012; Vanhanen & Lehtinen, 2014). We also found a large subset of papers examining interdisciplinary teams, many of which indicate that this is a substantially more complex form of collaboration that requires explicit attention (Goldberg & Malassigné, 2017; Gulbulak et al., 2020; Hoople et al., 2019; McNair et al., 2011; Shooter & Mcneill, 2002).

While significant work has gone into identifying the behaviors associated with successful groups, we also need to understand how to foster these behaviors (Mercier, 2017). Classroom evidence, alongside general experience, shows that many group activities are not productive, and that collaborative learning is rarely successful without significant support (e.g., Nokes – Malach et al., 2015). Preparing students to work effectively in groups is essential, not only for their experience in the course, but also for the development of collaborative skills that they will take into future courses and the workforce (Barron et al., 2009; Smith et al., 2005). Thus, in the next section, we present the different mechanisms for fostering behaviors that are typically demonstrated in successful groups.

3.1.1 Scripting Collaboration

The literature offers several mechanisms for fostering productive team behaviors: scaffolding and scripting for promoting social interactions, cognition, and metacognition (Ertmer & Glazewski, 2019; Reiser, 2018; van de Pol et al., 2010); engendering interdependence (Johnson & Johnson, 1984; Smith et al., 2005); and the ICAP framework: interactive, constructive, active, and passive (Chi & Wylie, 2014), which refers to the degree of student-centeredness. Each of these mechanisms will be briefly described in the subsequent part of this section.

Rooted in Vygotsky's zone of proximal development (Vygotsky, 1978), *scaffolding* describes temporary support designed by the instructors to improve the effectiveness and efficiency of learning and interaction. While scaffolds are elementally defined by contingency, fading, and transfer of responsibility (Wood et al., 1976), research also indicates that supportive elements like technology, tools, artifacts, resources, and even environments may be considered scaffolds when they aid student learning and interaction (Nadir, 2021; Puntambekar & Hubscher, 2005). Scaffolding can be provided at different levels: cognitive, linguistic, affective, social, metacognitive, and strategic planning

(Baxter & Williams, 2010; Belland, 2017; Belland et al., 2013), all of which are necessary during collaboration.

In engineering education, scaffolding (Wood et al., 1976) is especially important because it requires that students collaboratively solve complex and ill-structured problems, often in a project-based learning setting (Moallem et al., 2019; Tucker et al., 2020; Frank et al., 2018). Frank et al. (2018) demonstrated the effect of scaffolding on complex problem-solving and provide scaffolds for collaborative projects to be implemented during different phases, like problem definition, information evaluation, mathematical model development, and communication and argumentation. Consistent with this, Ge and Land (2004) promote the use of question prompts as scaffolding techniques to facilitate peer interactions during the four processes of ill-structured problem-solving: representation of the problem, development of solutions, justification of solutions, and monitoring and evaluation. The prompts are designed to direct teams' attention to the crucial points of the problem and solution space while encouraging different perspectives from the team members.

Scripting collaboration is another way to support students working in groups. Scripts can be either epistemic or social (Weinberger et al., 2005), guiding students to engage in the task and collaborate with their teammates. Epistemic scripts are often embedded in the task itself and can be seen as scaffolds. They may come in the form of a table or other representation that students fill out as they work through the task, which can guide the team to approach the task in a specific sequence or ensure that particular elements are not missed. Epistemic scripts may also guide students through the processes necessary for a particular task (e.g., look for evidence, look for ideas that contradict your claims, etc. or, in the case of introductory engineering courses, create a free-body diagram before you attempt to generate and solve an equation).

Social scripts aim to prompt the type of interactions that are most likely to foster knowledge building and help structure the way students work together. A classic example of a social script assigns a single paper to a pair of students. One student takes on the role of explainer, and the other of listener. The listener must try to understand what the explainer is telling them and add any missed details after the explainer finishes their part of the task (e.g., O'Donnell, 1999). More common classroom examples include "think-pair-share," where students are given time to think about a problem, then share or discuss with a peer, and finally share out to the class. This type of script can be adapted for other contexts and can be used effectively in problem-solving tasks in which students are instructed to read the problem alone, discuss it with their peers, generate solution options alone, share them with peers and determine a plan together, work on calculations alone, and then compare and complete the task. While this may feel unnecessary in higher education, often students need help understanding the different stages of a task or thinking about what they want to discuss with the group. This type of script also prevents students from simply charging ahead and trying to solve the problem without fully discussing it with their peers or grasping it themselves.

Scaffolding or scripting for social interactions creates interdependencies (Buchs et al., 2017; Marra et al., 2016), which in turn create pathways for productive dialogues between team members. The theory of social interdependence helps us understand the different types of interdependencies that instructors can draw on in designing collaborative tasks. Positive interdependence influences the degree of interaction between team members, which in turn impacts team effectiveness (Johnson & Johnson, 1984). A systematic review (Borrego et al., 2013) concluded that interdependence is one of five key elements of successful teamwork in engineering education. Interdependence mediates interactions between team members, fostering reciprocal relationships between them by creating a concrete agenda for collaboration. This agenda comes from different interdependencies categorized based on their focus: means or ends. When focused on means, goal and reward interdependencies are applicable. When focused on ends, resources and role interdependencies are applicable (Johnson & Johnson, 2008a). Yet all types of interdependencies are not theoretically equivalent (Ortiz et al., 1996), as particular combinations of interdependencies are more effective than others (Johnson &

Johnson, 2008b). We now present a few concrete examples of how interdependence has been introduced in engineering courses.

Common ways of structuring interdependence include dividing tasks, sharing limited resources, and assigning specific roles to individuals based on disciplinary and course-related objectives (Johnson & Johnson, 2008b). Pinho-Lopes and Macedo (2016) compare course implementation using collaborative and cooperative learning approaches for project-based learning in a geotechnics course in a civil engineering program. In the version of the course that follows cooperative learning pedagogy, role and task interdependence and individual and group accountability are applied through a set of instructional decisions based on the jigsaw system. Cámara-Zapata and Morales (2020) describe how the roles of presenter, evaluator, and observer were assigned and rotated between the team members in an introductory physics course on mechanics and thermodynamics. In line with these, Baligar et al. (in press) detail how the interdependencies of task, reward, goals, and resources can be applied to design instruction for engineering design problems at first-year undergraduate engineering education. The interventions and task aids designed for the four phases – information gathering, problem definition, concept generation, and detailed design – elicited better performance without aggravating interpersonal processes and individual well-being and learning. This is specifically important as increased interdependence often leads to conflicts and arguments (Opdecam & Everaert, 2018). In another recent study, Beddoes (2020) introduces a conceptual framework called interdisciplinary teamwork artifacts and practices (ITAPs), which proposes a set of artifacts and practices designed to engender interdependence, avoid conflict, and promote trust and shared teamwork and taskwork mental models for interdisciplinary teamwork in engineering education. The artifacts are aligned to orient, operate, level, propose, align, and structure taskwork knowledge, including task procedures and strategies and functions and forms, as well as teamwork knowledge, including roles, responsibilities, and team interactions.

Lastly, another practical framework that can influence the design and study of interactions in a collaborative learning environment is ICAP (interactive, constructive, active, and passive), which puts students' engagement with content and peers along a continuum from passive to interactive (Chi & Wylie, 2014). At the level of "interaction," this framework focuses on productive dialogues between peers who respond to each other constructively. It identifies the indicators of "interaction" as those that build on the views and perspectives of peers, seek elaboration, identify inconsistencies, generate arguments on the correctness of a view, and elicit justification for a view. This framework can also be used as a yardstick to assess the effectiveness of the scaffold/script in eliciting productive dialogues between peers in a collaborative learning setting.

In this way, instructional design using scaffolding/scripting has the potential to reduce dysfunctional team processes (Mathieu & Rapp, 2009) that lead to coordination and task allocation delays, specifically for students new to team-based problem-solving experiences like problem-based learning and project-based learning. Overall, positive interdependence, coupled with individual accountability, is known to mitigate social loafing and free riding, which, as discussed earlier, often negatively impacts team morale (Johnson & Johnson, 1984).

3.1.2 Group Composition

Significant attention has been given to group composition, with a focus on understanding how the different elements an individual brings to a collaborative group influence the group's experience. Individual features may range from personality traits (Boudreau & Anis, 2020), gender (Dasgupta & Stout, 2014), prior relationships or friendships (Maldonado et al., 2009), GPA, or prior performance in the course or related courses. There is some evidence that the gender composition of a team can be impactful; Dasgupta and Stout (2014) identify female-majority and sex-parity groups as initially important for female students despite having little or no effect on the anxiety of senior female students in the longer term. Meadows and Sekaquaptewa (2011) report on participation in presentations among mixed-gender groups, noting that male students talked more, and about more technical content than female students. In addition, male students rated their own leadership and performance higher if there were fewer males in their group.

Research also suggests that minority status in the classroom, which often correlates with negative peer and instructor perceptions of ability stemming from racist and sexist stereotypes, can have a negative impact on minoritized students' participation in teams, which detracts both from their learning experiences and the overall quality of group discourse. The influence is bidirectional, with students who fall into these categories being less willing to participate and students who fall into majority categories being less willing to listen to or build upon the ideas of minoritized students (see Esmonde, 2009 for a review). One common strategy is to ensure minoritized students are not isolated on teams (e.g., a female student in a majority male course is not in a team with only male students). In addition, both course-level interventions and addressing issues related to perceptions of peers' status directly have the potential to counteract these issues but require the instructor to examine their own biases and explicitly address them (e.g., Wagner, 2014; Webb et al., 2019). With this in mind, it is also important to note evidence suggesting that the experience of working on a diverse team improves attitudes towards minoritized students and those who do not speak English as fluently as native speakers (e.g., Cabrera et al., 1998; Godwin et al., 2017). Thus, working to mitigate the effects of minority status appears to be positive for all students, and likely to have long-term impacts on their future collaborations.

However, evidence on the specific impacts of individual-level variables is mixed, with contradictory results across studies. Drawing on the literature on team effectiveness, Borrego et al. (2013) note that the approach of trying to isolate the effect of a single variable on the fluid and emergent nature of collaborative groups is problematic. It is unlikely that a single variable will be predictive of behavior across situations, and behaviors will vary during a multi-week project as students deal with emergent issues. Instead, they argue for more attention on the alignment of processes and goals, which suggests that, for team formation, efforts to ensure teams are aligned in their goals for the course or project, as well as in their preferred processes (which may include simple things such as availability for meetings at times of the day and preferences for project topics), are paramount. Their study also highlights the importance of providing students with guidance in the development of their processes and in creating opportunities to restore broken group processes as necessary.

However, regardless of the general inconclusiveness of research on group formation, students need to be grouped somehow, and particularly in interdisciplinary or online courses that draw students from multiple time zones, there may be requirements for group composition that include ability to meet at certain times. In a review presented by Parker et al. (2019), the division of students into capstone teams was guided by six different approaches. The first of these approaches is the simple random assignment of students to teams, which is seen as the least time-consuming and a strategy to prevent groups being formed based on friendship or prior experiences together. However, Parker et al. (2019) report that it is the least frequently used and is associated with bad team experiences, with little to justify group membership, which can be particularly problematic over a long-term project. Two forms of student-formed teams were identified: no guidance and conditional self-formed teams. The first allows students to select their own teammates, and some may come to capstone course with complete or partially formed teams. Fully formed teams can join the course with a lot of motivation to work together on their chosen project, while partially formed teams who have members added may struggle with ownership and domination of those in the original group. It is also likely that some groups may be formed out of those who didn't come to the course with a group, which may be problematic for reasons that include poor prior performance by these students (and so they were not invited to join groups) or lack of motivation to participate in the course. Conditional self-formed teams are provided with a list of criteria by the instructor for team membership, and team members are selected to meet these criteria. This can result in teams with the necessary balance of skills or prerequisite knowledge for the project, while eliminating issues associated with friendship-based groupings. They are also likely to give students an experience more similar to those they will experience once employed, where group membership is often driven by project needs, not member preferences. Instructor-led team formation options fall into three categories: student skills, student preferences, and client preferences. For student skills, the instructor conducts an audit of skills (e.g., a mini-project, collection of résumés, asking students to apply for certain roles, etc.). While there is benefit in having students identify their own skills, this is a labor-intensive activity for instructors and may lead to teams where students work to their existing strengths rather than learn a new skill set. Instructor grouping by student preferences may focus on the project or topic preferences of students, whereby the instructor groups students based on a set of choices or rankings made by students. They may also use areas where students identify a need for development and place them in roles to develop those skills. Additionally, student preferences may be used to exclude certain groupings, where students identify people they do not want to be in a group with. Finally, instructor-led team formation can be based on client preferences, where project sponsors may meet with members of a class beforehand and then select those they wish to work with, potentially privileging students who perform well in interview situations. All instructor-led approaches have the potential to be time-consuming and may result in preferences not being met (due to an imbalance of preference for project types, team members, etc.)

Computer-supported team assignment tools can be used to account for a number of factors simultaneously and group students based on a more quantitative rubric which may be identified prior to the course by instructors (e.g., CATME, Layton et al., 2010). Of concern is students' willingness or eagerness to accept instructor-assigned teams, leading to a weaker sense of ownership. These concerns can be alleviated by engaging teams in specific team-building activities in the initial stages of projects (e.g., Hastings et al., 2018; Mercier et al., 2009). More recent work also points to the potential of having students participate in selection of the assignment criteria but allowing the tools' algorithm to create teams, leaving students more aware, and more content with, the group formation process (Hastings et al., 2022).

3.2 Tasks

Tasks are an integral component of collaborative learning. The tasks should scaffold and direct the groups' focus towards the prescribed learning objectives. Problem-based learning approaches provide an effective way of bringing real-world scenarios into the classroom by presenting the problem first and then finding the solution (Barrows & Tamblyn, 1980; Kolmos & Graaf, 2014). In engineering classrooms, collaborative tasks are often implemented using problem-based learning activities (Di Pietro et al., 2019; Perez – Poch et al., 2019; Zhang et al., 2018). Problem-based learning reduces students' cognitive load by allowing them to draw on their expertise via verbal interactions (Hmelo – Silver et al., 2013). These tasks differ based on the context in which they are being implemented (Boxtel et al., 2000).

There are two types of tasks instructors use to support learning, well-structured and ill-structured tasks (Jonassen, 2004). Well-structured tasks facilitate the students' interactions (Roth & Roychoudhry, 1992). They have convergent solutions and set well-defined parameters for solving the task. Students can be directed to exchange information through different task features like reflection (Turns et al., 2014). Assigning tasks that use methods like the "jigsaw" or that "script" the process provides students with a set of rules or instructions on how to interact with their peers and collaborate (Kagan, 1989; Dillenbourg, 2002). These tasks are often paired with cooperative learning, where students have set roles and their interactions are structured. However, care must be taken to avoid overstructuring the group interaction process, which could be detrimental to the intrinsic motivation of the students, thereby disturbing natural interaction processes (Dillenbourg, 2002).

The second category is ill-structured tasks. These tasks pose situations lacking necessary information and having multiple solutions and solution paths. Ill-structured tasks are more suited to collaborative learning (Cohen, 1994). The problem-based learning approach lends itself well to ill-structured tasks as it requires posing the problem first. Ill-structured tasks require more than the domain knowledge and justification skills required for solving problems (Shin et al., 2003). They require additional skills relating to regulation of cognition, which involves planning, monitoring, and re-evaluating goals (Chin & Chia, 2006). Studies have shown that leveraging tools to aid in the development of these skills has helped students effectively collaborate (Fidalgo – Blanco et al., 2018; Pazos et al., 2019).

Engineering tasks specifically designed for collaborative work that take full advantage of our understanding of how to support learning and interactions are much needed (Shehab & Mercier, 2017). In creating collaborative tasks, it is important to consider the difficulty of the task presented to the students so that it is neither too easy, leading to a single student completing it, nor too difficult for the group. The intermittent development zone (IDZ), a corollary to Vygotsky's zone of proximal development (ZPD), describes the space between what a learner can and cannot learn without the assistance of peers (Vygotsky, 1978; Fernández et al., 2015). Tasks should scaffold peers to interact with each other within the IDZ so as to facilitate collaboration. Tasks should also provide multiple layers of meaning, allowing students to gain multiple perspectives and to think about the problem and communicate their ideas (DeLiema et al., 2015). In addition, reflecting on elements of scripting interactions described in Section 3.1.1, there is a clear need to consider how best to support interaction through the task features.

An additional element of collaborative tasks is reflection, which involves students revisiting features of their experiences and attaching meaning to them that can guide future actions (Turns et al., 2014). Studies have shown that reflection is a necessary part of effective collaborative activities and design projects (Burkholder & Wieman, 2021). Students tend to improve their reflection specificity over time, and their reflections are a good predictor of academic performance (Anwar & Menekse, 2020). Interdisciplinary tasks and projects involving interdisciplinary teams allow students to gain different perspectives (Hoople et al., 2019). Continuous self- and peer assessment can be incorporated into tasks as a way to improve student behavior and the reliability of feedback from peers (Foong & Liew, 2020). Studies in engineering education journals have explored the role of peer assessment (Alba-Flores & Rios, 2019; Carberry et al., 2016) and the impact of peer assessment on teams (Foong & Liew, 2020; Mandala et al., 2018). Findings indicate that peer review increased students' awareness of collaborative behaviors, that the evaluations indicated improved quality of collaboration over time, and that both providing verbal feedback and spending more time providing feedback positively impacted students' learning experiences.

Although reflective processes and their benefits have been the subject of considerable research, other task features and how they foster student interaction and encourage collaboration and learning are less understood. More micro-level understanding of tasks and their influence on student interaction can help us understand how to leverage these tasks to facilitate collaboration in a more meaningful way among students.

In the engineering education journals we reviewed, many papers focused on both the tasks and teams dimensions. A small but important subset of studies explored how task elements impact the manner in which teams interact, in particular focusing on the role of agency in decision-making or task framing (Burkholder & Wieman, 2021; Svihla et al., 2021). Several papers report on issues that emerged in groups, some of which were attributed to limitations in the implementation of collaboration pedagogy. Successful groups were attributed to the design of implementations that focus on team spirit, assignment, and the rotation of roles within groups, as well as support for effective time management and interdependence strategies (e.g., Berge & Weilenmann, 2014; Missingham & Matthews, 2014; Pinho-Lopes et al., 2011).

3.3 Tools

The tools provided to students that support their collaborative learning are essential to the success of this pedagogical approach, although this was the least-discussed category in the engineering education journals. Johri et al. (2013) divide tools into two types: representational tools and relational tools. Representational tools are designed to support the creation of (joint) representations during collaborative activities, while relational tools support interaction between group members. This distinction allows us to consider how both needs will be met when we assign students to groups and either provide or guide students toward, tools that will allow them to represent their solution progress and communicate with each other.

The need for tools obviously differs by context and activity type. Classroom-based collaborations to be completed in a single session rarely need additional support for communication but can benefit from tools to support representations. While research points towards the value of high-tech tools (e.g., Berthoud & Gliddon, 2018; Mayer et al., 2013; Mellingsæter & Bungum, 2015), it is important to note that tools do not have to be technologically advanced to be helpful. Certain low-cost tools can have a profound impact on groups, such as A3 (12 × 17") sized whiteboards that can be provided to groups as a surface on which to visualize joint representations of problem-solving activities, rather than having each student write in their own personal device or notebook (Essick et al., 2016). In contrast, a semester-long project where students must meet outside of class time may require a suite of tools to help students communicate and manage their interactions, as well as tools to represent their solution processes (e.g., Colomo-Palacios et al., 2020; Kirschman & Greenstein, 2002; Oladirana et al., 2011).

Tools may also be categorized based on whether they were created explicitly for a particular course or topic (e.g., Caballé et al., 2014; Serrano – Cámara et al., 2016) or were already available to the general public but adaptable for use by student teams (e.g., Pazos et al., 2019). The literature points towards value in both approaches, with tools that are designed specifically for a context providing rich learning opportunities for students, but often requiring additional effort to implement or learn to use. Tools that are available more generally often have the benefit of being familiar but may not meet all the requirements of a team and may require trade-offs between instructor control and monitoring (such as communication through a learning management system) or familiarity and distraction (such as using a social media site to communicate) (Tlhoaele et al., 2016).

Recent literature examines new ways of using technology to augment collaborative learning in engineering. Examples of emerging technology that may be suitable for this purpose include tools that are designed to support teachers or instructors during collaboration activities (e.g., Lawrence & Mercier, 2019); collaborative simulation-based platforms to learn engineering concepts, such as online electronics breadboards (Andrews et al., 2017; Horwitz et al., 2017); augmented reality (AR) approaches to software concepts (Schiffeler et al., 2019); and AR-supported circuits and electronics (Villanueva et al., 2020). In addition, tools that support teaching large courses more generally, such as those designed to automatically create teams based on particular criteria (e.g., years of CAD experience, familiarity with a programming language, etc.) can make collaborative pedagogies easier to implement.

Imbricating tools within the rest of the framework is also essential. It is useful to consider whether students are prepared to use the tools they are given by the instructor to support their interactions and co-construction of knowledge, whether they are aware of the tools they can use, and whether the team members agree on how tools will be used over the course of a long-term project. It is also necessary to consider how tools can function to make tasks more accessible, and how teachers can use tools to support their teaching or gain insights into the processes of student groups, as well as to help with in-class intervention or grading.

3.4 Teachers

Teachers are a crucial element of collaborative learning, particularly in undergraduate engineering settings, as they are responsible for designing collaborative activities and overseeing their success (Lawrence & Mercier, 2019). However, relatively little research has been done on the role of teachers in collaborative learning (Webb, 2009; Kaendler et al., 2015). In practice, many educators who believe that they are employing collaborative learning strategies are actually missing crucial elements that would enable students to work together more effectively (Johnson et al., 1991). Meijer et al. (2020) have synthesized two principles for designing collaborative learning exercises from the larger collaborative learning literature: individual accountability with positive interdependence (based on Slavin, 1980; Johnson, 1981; Strijbos, 2011) and adherence to eight collaborative components: (1) interaction, (2) learning objectives and outcomes, (3) assessment, (4) task characteristics, (5) structuring, (6) guidance, (7) group constellation, and (8) facilities (de Hei et al., 2016). Arguably, teachers have the greatest influence over all these factors; the previous sections on teams, tasks, and tools all suggest the important role teachers play in developing, structuring, and executing these areas. In planning a class, teachers have a certain amount of latitude over how to reach course learning objectives and outcomes. In addition, they are often the ones in charge of designing course assessments. Teachers are responsible for aligning learning outcomes and assessment in ways that are facilitated with collaborative learning activities, a crucial component in successful collaborative learning. Additionally, instructors make decisions regarding group composition, roles within the group, and sometimes the arrangement of the room (Johnson et al., 1991). Thus, the effort and time the instructor invests in designing instruction for collaborative tasks will be influenced by the students' previous experience of engaging in collaborative tasks. Based on the students' proficiency in process aspects of collaborative work execution, it would make sense to design tasks with micro-level interdependency (resources, tasks, and roles) or macro-level (goals and rewards).

Outside of task development, tool selection, and team formation, teachers have a particularly instrumental role in supporting students' processes during collaborative learning. This support can involve monitoring students' interactions (e.g., Hmelo-Silver, 2004), guiding student activities, and providing feedback on their immediate or longer-term collaborative efforts (e.g., Mercier et al., 2009; Shehab & Mercier, 2019).

Across the field of engineering education, research on the role of teachers and instructors has also been limited, with less attention paid to this essential element as compared to other aspects of collaboration. Of particular relevance is the work of Gómez Puente et al. (2015), who report on professional development among instructors planning to implement design-based learning. During professional development sessions, the instructors redesigned their own teaching material and then implemented these new strategies in the second year of the study. Results indicated improvement in projects across a number of areas, suggesting that even a relatively minimal (seven-hour) intervention may allow teachers to more effectively implement these types of learning experiences.

3.4.1 Assessment of Collaborative Learning

An essential role teachers play in collaborative learning is deciding when and how to assess it. Assessment in engineering education is a complex topic (Pellegrino et al., 2001), which operates with the goal of educating and improving student performance rather than simply auditing student efforts (Wiggins, 1998). Educational assessment can be used to assist learning, determine individual achievement, and evaluate programs (Pellegrino et al., 2014) not only in general educational but also in collaborative learning environments. However, assessment has not been a key focus in collaborative learning research over the last two decades. Of 14 meta-studies on collaborative learning design, assessment was mentioned in two (de Hei et al., 2016).

We posit that many of the assessment issues raised by Pellegrino et al. (2014) are also issues in collaborative learning. In particular, we find that the four main considerations or tensions in assessment of collaborative learning are (1) formative vs. summative feedback, (2) individual vs. group (or a combination of both), (3) grading based on effectiveness of the group vs. overall "correctness," and (4) assessment of product vs. process. Irons and Elkington (2021), in their comprehensive detailing of the theoretical foundations and principles of formative feedback and assessment, provide practical suggestions on how feedback can be structured to maximize its reach and comprehension to students while suggesting that instructors should focus not only on the outputs of groupwork but also on the process. Concentrating on the social aspects of groupwork, Thistlethwaite et al. (2016), in the context of interprofessional education, devised an individual teamwork observation and feedback tool which assesses students' individual contribution to teamwork at two levels: basic version, which includes 11 observable behaviors, and an advanced version, which lists 10 observable behaviors. These versions are based on students' maturity in teamwork experiences. Thus, it becomes imperative that the instructors are aware of the quality indicators of effective and efficient collaboration.

Assessment of group processes often comes in the form of peer assessment, where peers are given the opportunity to evaluate their team members' contributions, generating scores which are sometimes used to provide an individual grade to each student. From a cooperative learning perspective, this can fulfill the goal of individual accountability, although it has the potential to be used as a punishment for team members who did not participate actively. Rather than solely relying on the summative evaluation of peers, opportunities for commentary on group processes during the course (e.g., Mercier et al., 2009) may be more effective for identifying and intervening when issues arise, thus supporting ongoing learning of collaborative skills.

An alternative approach, developed and described by Borge and colleagues (e.g., Borge & Goggins, 2014; Borge et al., 2018, 2020), engages students in assessments of their discussion board posts using a framework of effective collaborative communication behaviors. Teams self-assess their contributions to the group and have the opportunity to discuss how to improve their communication and collaboration, setting goals for how they would like to collaborate in their next activity. Thus, students are involved in self- and peer assessment, albeit in service of supporting their own development rather than assigning grades. While this system is currently limited to text-based communication, emerging use of AI and data analytics to assess collaboration may provide access to students' processes more readily in the future (e.g., Bachour et al., 2009; Paquette et al., 2018).

4 Conclusion

In this chapter, we have reviewed the literature on collaborative learning, highlighting four important areas that must be attended in implementing collaborative learning in classrooms (tasks, tools, teams, and teachers). Lab-based research shows that collaborative learning activities can be highly effective, and there is no doubt that engineering students need to develop the skills necessary to engage in collaborative problem-solving before joining the workforce. Under the right circumstances, they can do this through well-structured classroom experiences. Our review of the literature, however, reveals several challenges, including the complexities of studying and implementing collaborative learning successfully in classrooms, the reliance on student attitudes towards their collaborative experiences as metrics of success, and the lack of research into the development of collaborative skills over time and into the facilitation of student reflections and metacognition around their most effective group processes.

Studying and implementing collaborative learning are complex endeavors that take place in a complex environment where tasks, tools, and teachers all impact how students function in teams and achieve the goals of the task. In addition, while research has provided us with a good description of what groups look like when they are successful, we are still working to understand how best to get students to engage in the types of interactions that lead to successful outcomes. Research on the design of tasks that successfully support collaboration is still in its infancy (e.g., Tucker et al., 2020), while work on the role of teachers in effectively intervening is also relatively limited and has yet to fully become integrated into the literature of engineering education (Shehab & Mercier, 2019). While emerging literature (e.g., Gómez Puente et al., 2015) indicates that there is potential in short-term interventions for faculty to improve the quality of their teaching materials and courses when using collaborative learning, future research in this area of the field is essential (both in engineering and in education more generally).

A further challenge for the field is the extent to which studies rely on students' self-reported data, despite evidence that students are not necessarily the best judges of what counts as a successful learning experience. Students often prefer lectures, which may result in less learning, overactive learning activities, and projects which require more effort on their part but result in better learning outcomes (Deslauriers et al., 2019). For example, Mostafapour and Hurst (2020) report that students preferred taking a divide-and-conquer approach to their group tasks, which is often more efficient but is unlikely to provide the opportunities to engage in discussion of the material or to co-construct a solution, leaving them without the key skills these tasks were designed to develop. Research that relies on students' preferences may thus lead us in the wrong direction in terms of understanding what is best for learning outcomes. It is also important, given the discrepancy between students' attitudes towards learning experiences and the desired learning outcomes, that faculty implementing active learning engage students in discussion of learning theory and explain why the experiences that seem harder to students are designed to give them better learning outcomes and better prepare them for their future careers.

Finally, although programs are moving away from requiring a single collaborative project-based experience as a capstone or cornerstone project, there is still a limited range of opportunities across the curriculum for students to develop collaborative skills. While capstone projects can be highly effective, if it is students' first and only opportunity to work in groups, they are unlikely to know how to do so effectively. Additionally, having opportunities to reflect on and try different collaborative strategies will allow students to be better prepared for the workforce (e.g., Borge et al., 2018), which is not an option with a single project experience. As the field progresses, understanding more about how to foster collaborative learning skills early in an engineering program, and the manner in which scaffolding can be strategically removed as students become more adept at working in groups, is going to be important.

4.1 A Comment on Methods and Scope of the Research Reviewed

In this chapter, we brought together research from the fields of education and the learning sciences with research in engineering education. While there is significant overlap, the goals and methodologies of the two areas differ and make different types of contributions which should be acknowledged as we move forwards to consider the future work necessary in this field.

Historically, research in education and the learning sciences that describe collaborative learning focused on understanding the differences between individual and group performances, why groups differed in outcomes on the same tasks, and the types of interactions that are associated with successful learning. These studies are primarily grounded in constructivist and social constructivist theories of learning and draw on quantitative methods, such as performance comparisons between different configurations of learners or video analysis, and case study approaches to understand the nuances of interaction differences between groups. More recently, in the field of learning sciences, the adoption of design-based research (Anderson & Shattuck, 2012; Bell, 2004) and especially design-based implementation research (Coburn & Penuel, 2016; Fishman & Penuel, 2021; Penuel et al., 2011) has focused on larger-scale studies that partner with educators to align research goals with the needs of practitioners. For the most part, this moves the work away from the more detailed analysis of interaction and more towards understanding the complex nature of classroom implementation. Cooperative learning, seen less in the field of learning sciences, had a more applied focus from the beginning. Grounded in social interdependence theory, researchers in cooperative learning quickly focused on designing classroom interventions and supporting teachers in effectively using cooperative learning in the classroom (Johnson & Johnson, 2009).

In contrast, engineering education research emerged from the needs of engineering practice, with research grounded in experimentation within classroom contexts and responding to the learning concerns raised by those embedded in the teaching. The resulting research tends to rely on classroom assessments, intervention reports, and survey data, although recent research tends to use a more comprehensive range of approaches. The research in this field is often more readily applicable in a new context, although less focused on the reported mechanisms behind the successes, making it harder to abstract and generalize across findings.

There does appear to be a move closer towards each other, with the maturation of the field of engineering education aligned with adoptions of research–practitioner partnerships and design-based implementation research in the learning sciences, which allow for approaches that seek to understand how questions of practice can be used to address theoretical questions about learning and collaboration, rather than starting with theoretical questions. However, there is still much to be explored in terms of what elements of intervention are truly necessary for success in different contexts.

4.2 Future Research

While there has been significant research conducted in the areas of collaborative and cooperative learning in engineering education, this review has highlighted key areas where it will be important to focus on future research.

4.2.1 Teams

There is a large amount of work that illustrates what groups look like when they are successful, particularly over short periods of time (Barron et al., 2009; Mercier et al., 2009). One important area where more research is needed, however, is exploring groups over long periods of time in order to understand how collaboration develops within teams of students and how best to establish and support teams as they negotiate learning over multiple sessions, weeks, or even semesters.

Furthermore, understanding the nature of the experiences of minoritized students is of paramount importance. The research that currently exists points towards a complex relationship between group learning activities and minoritized status. In particular, research that creates distinctions between different minoritized groups, rather than aggregating them as is common, is important to understand more about how different students experience group work. In addition, while research points towards the value for future collaborators of engaging with a diverse team, it is important to understand and mitigate the additional burden placed on minoritized students in managing stereotype-related expectations and acting as a representative for their entire group.

4.2.2 Tasks

The design of collaborative tasks that are effective for groups is one area where there is a need for a large amount of future research. In particular, it is important that we develop an understanding of how to embed scaffolds within tasks for students to manage their own problem-solving processes, and how to fade those scaffolds over time to allow students to develop effective collaborative problem-solving skills when they face these problems in a work situation.

4.2.3 Tools

The potential for technology to support collaborative learning is just emerging, particularly as we learn more about how to use multi-modal data analytics to understand more about the nature of interaction and problem-solving processes (Paquette et al., 2018). This type of research holds the potential to gain more insight into the moment-to-moment development of shared understandings and the unfolding of collaborative interactions over time. It is also possible it will provide new ways to consider the quality of collaborative interactions and problem-solving. Once we have sufficient understanding of the elements of collaboration we can assess through indirect measures, such as screen activity, tone of voice, etc., we will be able to provide more targeted support for students and instructors. Initial versions of this type of work show promise (e.g., Paquette et al., 2018; Viswanathan & Vanlehn, 2018).

4.2.4 Teachers

Research on the role of teachers in setting up cooperative learning has always been important; however, it is only since the late 2000s that attention to the role of teachers during group activity has started to gain attention within the various fields attending to collaborative learning (Kaendler et al., 2015; Webb et al., 2019). There is much to be explored in this area –understanding the role of the teacher, how they should intervene in groups, and how to prepare people to use this form of pedagogy effectively.

Acknowledgments

We thank the anonymous reviewers of the chapter who gave us very useful guidance. We also acknowledge colleagues who collaborated with us on various parts of this work, including Monique S. Ross.

References

Accreditation Board for Engineering and Technology. (2018). *Criteria for accrediting engineering programs, 2018–2019*. Accreditation Board for Engineering and Technology (ABET) Inc.

Adelman, C., National Inst. for Science Education, M. W. I., & National Center for Improving Science Education, W. D. C. (1998). *Women and men of the engineering path: A model for analyses of undergraduate careers*. U.S. Department of Education.

Agrawal, A. K., & Harrington-Hurd, S. (2016). Preparing next generation graduates for a global engineering workforce: Insights from tomorrow's engineers. *Journal of Engineering Education Transformations* [Special Issue]. https://doi.org/10.16920/jeet/2016/v0i0/85428

Alba – Flores, R., & Rios, F. (2019). Incorporating peer review techniques to enhance students' communication skills and team performance in engineering capstone projects. *International Journal of Engineering Education, 35*(6B), 1969–1982.

Anderson, T., & Shattuck, J. (2012). Design-based research a decade of progress in education research? *Educational Researcher, 41*(1), 16–25. https://doi.org/10.3102/0013189X11428813

Andrews, J. J., Chamberlain, C. O. R. D., Koon, A., & von Davier, A. A. (2017, 18–22 June). Designing simulations for evaluating collaborative problem solving in electronics [conference session]. In *Making a difference: Prioritizing equity and access in CSCL*.

Anwar, S., & Menekse, M. (2020). Unique contributions of individual reflections and teamwork on engineering students' academic performance and achievement goals. *International Journal of Engineering Education, 36*(3), 1018–1033.

Azizan, M. T., Mellon, N., Ramli, R. M., & Yusup, S. (2018). Improving teamwork skills and enhancing deep learning via development of board game using cooperative learning method in reaction engineering course. *Education for Chemical Engineers, 22*, 1–13. https://doi.org/10.1016/j.ece.2017.10.002

Bachour, K., Kaplan, F., & Dillenbourg, P. (2009). An interactive table for supporting participation balance in face – to – face collaboration. *IEEE Transactions on Learning Technologies, 3*(3), 203–213. https://doi:10.1109/TLT.2010.18

Baligar, P., Joshi, G., Shettar, A., Kandakatla, R., & Patil, P. (in press). Effect of interdependence on team effectiveness during design problem-solving in first-year engineering. *Journal of Engineering Design*. Post publication: http://dx.doi.org/10.1080/09544828.2022.2139142

Ballen, C. J., Wieman, C., Salehi, S., Searle, J. B., & Zamudio, K. R. (2017). Enhancing diversity in undergraduate science: Self-efficacy drives performance gains with active learning. *CBE Life Sciences Education, 16*(4). https://doi.org/10.1187/CBE.16-12-0344/SUPPL_FILE/RAW_DATA_AS_AN_EXCEL_FILE.XLSX

Barron, B. (2003). When smart groups fail. *Journal of the Learning Sciences, 12*(3), 307–359. https://doi.org/10.1207/S15327809JLS1203_1

Barron, B., & Darling-Hammond, L. (2008). *Teaching for meaningful learning: A review of research on inquiry – based and cooperative learning*. George Lucas Educational Foundation.

Barron, B., Martin, C. K., Mercier, E., Pea, R., Steinbock, D., Walter, S., Herrenkohl, L., Mertl, V., Tyson, K., & Miyake, N. (2009). Repertoires of collaborative practice. *Computer Supported Collaborative Learning Practices, CSCL 2009 Community Events Proceedings – 9th International Conference, 2*, 25–27. https://doi.org/10.3115/1599503.1599513

Barrows, H. S., & Tamblyn, R. M. (1980). *Problem – based learning: An approach to medical education* (Vol. 1). Springer Publishing Company.

Baxter, J. A., & Williams, S. (2010). Social and analytic scaffolding in middle school mathematics: Managing the dilemma of telling. *Journal of Mathematics Teacher Education, 13*(1), 7–26.

Beddoes, K. (2020). Interdisciplinary teamwork artefacts and practices: A typology for promoting successful teamwork in engineering education. *Australasian Journal of Engineering Education, 25*(2), 133–141. https://doi.org/10.1080/22054952.2020.1836753

Beigpourian, B., & Ohland, M. W. (2019, June). *A systematized review: Gender and race in teamwork in undergraduate engineering classrooms*. Paper presented at 2019 ASEE annual conference & exposition, Tampa, Florida. https://doi.org/10.18260/1-2—32011

Bell, P. (2004). On the theoretical breadth of design – based research in education. *Educational Psychologist, 39*(4), 243–253. https://doi.org/10.1207/s15326985ep3904_6

Belland, B. R. (2017). Computer – based scaffolding strategy. In *Instructional scaffolding in STEM education* (pp. 107–126). Springer.

Belland, B. R., Kim, C., & Hannafin, M. J. (2013). A framework for designing scaffolds that improve motivation and cognition. *Educational Psychologist, 48*(4), 243–270.

Bénabou, R. (2013). Groupthink: Collective delusions in organizations and markets. *Review of Economic Studies, 80*(2), 429–462. https://doi.org/10.1093/restud/rds030

Berge, M., & Weilenmann, A. (2014). Learning about friction: Group dynamics in engineering students' work with free body diagrams. *European Journal of Engineering Education, 39*(6), 601–616. https://doi.org/10.1080/03043797.2014.895708

Berger, J., Rosenholtz, S. J., & Zelditch Jr, M. (1980). Status organizing processes. *Annual Review of Sociology, 6*, 479–508.

Berthoud, L., & Gliddon, J. (2018). Using wikis to investigate communication, collaboration and engagement in Capstone engineering design projects. *European Journal of Engineering Education, 43*(2), 247–263. https://doi.org/10.1080/03043797.2017.1332574

Blosser, E. (2020). An examination of Black women's experiences in undergraduate engineering on a primarily white campus: Considering institutional strategies for change. *Journal of Engineering Education, 109*(1), 52–71. https://doi.org/10.1002/JEE.20304

Booker, A., Goldman, S., & Mercier, E. M. (2009). Interdisciplinarity in learning technology design courses. In C. DiGiano, S. Goldman, & M. Chorost (Eds.), *Educating learning technology designers: Guiding and inspiring creators of innovative educational tools* (pp. 122–144). Routledge.

Borge, M., & Goggins, S. (2014). Towards the facilitation of an online community of learners: assessing the quality of interactions in yammer. In *Proceedings of the 11th ICLS*. Boulder: International Society of the Learning Sciences (ISLS). (pp. 753–760).

Borge, M., Ong, Y. S., & Rosé, C. P. (2018). Learning to monitor and regulate collective thinking processes. *International Journal of Computer-Supported Collaborative Learning, 13*, 1–32. https://doi.org/10.1007/s11412-018-9270-5

Borge, M., Soto, J. A., Aldemir, T., & Mena, J. A. (2020). Building multicultural competence by fostering collaborative skills. *Teaching of Psychology*. https://doi.org/10.1177/0098628320977421

Borge, M., & White, B. (2016). Toward the development of socio-metacognitive expertise: An Approach to developing collaborative competence. *Cognition and Instruction, 34*(4), 323–360. https://doi.org/10.1080/07370008.2016.1215722

Borrego, M., Karlin, J., McNair, L. D., & Beddoes, K. (2013). Team effectiveness theory from industrial and organizational psychology applied to engineering student project teams: A research review. *Journal of Engineering Education, 102*(4), 472–512. https://doi.org/10.1002/jee.20023

Boudreau, J., & Anis, H. (2020). Effect of personality traits in team dynamics and project outcomes in engineering design. *International Journal of Engineering Education, 36*(1B), 420–435.

Bratman, M. E. (1992). Shared cooperative activity. *The Philosophical Review, 101*(2), 327. https://doi.org/10.2307/2185537

Bubou, G. M., Offor, I. T., & Bappa, A. S. (2017). Why research – informed teaching in engineering education? A review of the evidence. *European Journal of Engineering Education, 42*(3), 323–335. https://doi.org/10.1080/03043797.2016.1158793

Buchs, C., Filippou, D., Pulfrey, C., & Volpé, Y. (2017). Challenges for cooperative learning implementation: Reports from elementary school teachers. *Journal of Education for Teaching, 43*(3), 296–306.

Burkholder, E., & Wieman, C. (2021). Impact of decision – making in capstone design courses on students' ability to solve authentic problems. *International Journal of Engineering Education, 37*(3), 650–662.

Caballé, S., Gañán, D., Gómez, M., Conesa, J., Rius, À., & Santanach, F. (2014). Experiences at the Open University of Catalonia with the virtualization of live collaborative learning in support for engineering education. *International Journal of Engineering Education, 30*(3), 663–674. http://www.uoc.edu

Cabrera, A. F., Nora, A., Bernal, E. M., Terenzini, P. T., & Pascarella, E. T. (1998, November 5–8). Collaborative learning: Preferences, gains in cognitive & affective outcomes, and openness to diversity among college students [Conference session]. In *Annual meeting of the association for the study of higher education.*

Cady, E., & Reid, P. P. (2018). *Understanding the educational and career pathways of engineers*. The National Academies Press. https://doi.org/10.17226/25284

Carberry, A. R., Brunhaver, S. R., Csavina, K. R., & McKenna, A. F. (2016). Comparison of written versus verbal peer feedback for design projects. *International Journal of Engineering Education, 32*(3), 1458–1471.

Cámara – Zapata, J. M., & Morales, D. (2020). Cooperative learning, student characteristics, and persistence: An experimental study in an engineering physics course. *European Journal of Engineering Education, 45*(4), 565–577. https://doi.org/10.1080/03043797.2019.1569593

Cech, E. A., & Rothwell, W. R. (2019). LGBT workplace inequality in the federal workforce: Intersectional processes, organizational contexts, and turnover considerations. *ILR Review, 73*(1), 25–60. https://doi.org/10.1177/0019793919843508

Cech, E. A., & Waidzunas, T. J. (2021). Systemic inequalities for LGBTQ professionals in STEM. *Science Advances*, 7(3), 933–948. https://doi.org/10.1126/SCIADV.ABE0933/SUPPL_FILE/ABE0933_SM.PDF

Chen, P., Hernandez, A., & Dong, J. (2015). Impact of collaborative project-based learning on self-efficacy of urban minority students in engineering. *Journal of Urban Learning, Teaching, and Research, 11*, 26–39.

Chi, M. T., & Wylie, R. (2014). The ICAP framework: Linking cognitive engagement to active learning outcomes. *Educational Psychologist, 49*(4), 219–243.

Chin, C., & Chia, L. G. (2006). Problem – based learning: Using ill – structured problems in biology project work. *Science Education, 90*(1), 44–67. https://doi.org/10.1002/sce.20097

Coburn, C. E., & Penuel, W. R. (2016). Research – practice partnerships in education: Outcomes, dynamics, and open questions. *Educational Researcher, 45*(1), 48–54.

Cohen, A. D. (1994). *Assessing language ability in the classroom*. Heinle & Heinle.

Cohen, E., & Lotan, R. (1997). Complex instruction : Equity cooperative learning classrooms. *Theory into Practice*, *38*(2), 80–86. http://www.tandfonline.com/doi/pdf/10.1080/00405849909543836

Colomo-Palacios, R., Samuelsen, T., Casado-Lumbreras, C., & Larrucea, X. (2020). Students' selection of teamwork tools in software engineering education: Lessons learned. *International Journal of Engineering Education*, *36*(1B), 309–316.

Cortez, C., Nussbaum, M., Woywood, G., & Aravena, R. (2009). Learning to collaborate by collaborating: A face-to-face collaborative activity for measuring and learning basics about teamwork. *Journal of Computer Assisted Learning*, *25*(2), 126–142. https://doi.org/10.1111/j.1365–2729.2008.00298.x

Cross, K. J., & Paretti, M. C. (2020). African American males' experiences on multiracial student teams in engineering. *Journal of Women and Minorities in Science and Engineering*, *26*(4), 381–411. https://doi.org/10.1615/JWOMENMINORSCIENENG.2020033004

Dasgupta, N., & Stout, J. G. (2014). Girls and women in science, technology, engineering, and mathematics: STEMing the tide and broadening participation in STEM careers. *Policy Insights from the Behavioral and Brain Sciences*, *1*(1), 21–29. https://doi.org/10.1177/2372732214549471

Davidson, N., & Major, C. H. (2014). Boundary crossings: Cooperative learning, collaborative learning, and problem-based learning. *Journal on Excellence in College Teaching*, *25*(3 & 4).

de Hei, M., Strijbos, J. W., Sjoer, E., & Admiraal, W. (2016). Thematic review of approaches to design group learning activities in higher education: The development of a comprehensive framework. *Educational Research Review*, *18*, 33–45. https://doi.org/10.1016/J.EDUREV.2016.01.001

DeLiema, D., Lee, V., Danish, J. A., Enyedy, N., & Brown, N. J. (2015). A microlatitudinal/microlongitudinal analysis of speech, gesture, and representation use in a student's repeated scientific explanations of phase change. *Knowledge and Interaction*, 149–175. https://doi.org/10.4324/9781315757360–15

Deslauriers, L., McCarty, L. S., Miller, K., Callaghan, K., & Kestin, G. (2019). Measuring actual learning versus feeling of learning in response to being actively engaged in the classroom. *Proceedings of the National Academy of Sciences of the United States of America*, *116*(39), 19251–19257. https://doi.org/10.1073/pnas.1821936116

Deutsch, M. (1962). Cooperation and trust: Some theoretical notes. In M. R. Jones (Ed.), *Nebraska symposium on motivation, 1962* (pp. 275–320). University of Nebraska Press.

Di Pietro, L., Botte, E., Granati, R., Moroni, S., Tomasi, M., Vozzi, G., & De Maria, C. (2019). Teaching design standards and regulations on medical devices through a collaborative project-based learning approach. *International Journal of Engineering Education*, *35*(6A), 1803–1815.

Dillenbourg, P. (2002). Over-scripting CSCL: The risks of blending collaborative learning with instructional design. In P. A. Kirschner (Ed.), *Three worlds of CSCL: Can we support CSCL* (pp. 61–91). Open Universiteit Nederland.

Dillenbourg, P., Baker, M., Blaye, A., & O'Malley, C. (1995). The evolution of research on collaborative learning. In P. Reimann & H. Spada (Eds.), *Learning in humans and machine:* towards *an interdisciplinary learning science* (pp. 189–211). Elsevier.

El-Sakran, T., Prescott, D., & Mesanovic, M. (2013). Contextualizing teamwork in a professional communication course for engineering students. *International Journal of Engineering Education*, *29*(2), 439–449.

Ertmer, P. A., & Glazewski, K. D. (2019). Scaffolding in PBL environments. *The Wiley Handbook of Problem-Based Learning*, 321–342. https://doi.org/10.1002/9781119173243.CH14

Esmonde, I. (2009). Ideas and identities: Supporting equity in cooperative mathematics learning. *Review of Educational Research*, *79*(2), 1008–1043. https://doi.org/10.3102/0034654309332562

Essick, R., West, M., Silva, M., Herman, G. L., & Mercier, E. (2016, June 26–August 28). Scaling-up collaborative learning for large introductory courses using active learning spaces, TA training, and computerized team management. In *ASEE annual conference and exposition, conference proceedings*. https://doi.org/10.18260/p.27342

European Network for Accreditation of Engineering Education. (2008). *EUR – ACE framework standards and guidelines*. www.enaee.eu/eur-ace-system/standards-and-guidelines/#standards-and-guidelines-for-accreditation-of-engineering-programmes

Felder, R. M., & Brent, R. (1994). *Cooperative learning in technical courses: Procedures, pitfalls, and payoffs*. National Science Foundation.

Fernández, M., Wegerif, R., Mercer, N., & Rojas-Drummond, S. (2015). Re-conceptualizing "scaffolding" and the zone of proximal development in the context of symmetrical collaborative learning. *Journal of Classroom Interaction, 50*(1), 54–72.

Fidalgo-Blanco, Á., Sein-Echaluce, M. L., & García-Peñalvo, F. J. (2018). Enhancing the main characteristics of active methodologies: A case with micro flip teaching and teamwork. *International Journal of Engineering Education, 35*(1), 397–408.

Fila, N. D., & Purzer, Ş. (2014). The relationship between team gender diversity, idea variety, and potential for design innovation. *International Journal of Engineering Education, 30*(6), 1405–1418.

Fischer, F., Kollar, I., Stegmann, K., & Wecker, C. (2013). Toward a script theory of guidance in computer – supported collaborative learning. *Educational Psychologist, 48*(1), 56–66. https://doi.org/10.1080/00461520.2012.748005

Fishman, B., & Penuel, W. (2021). Design – based implementation research. In F. Fischer, C. E. Hmelo – Silver, S. R. Goldman, & P. Reimann (Eds.), *International handbook of the learning sciences* (pp. 393–400). Routledge.

Foong, C. C., & Liew, P. Y. (2020). Investigating the reliability and usefulness of self– and peer assessments of a capstone design project. *International Journal of Engineering Education, 36*(6), 1850–1861.

Frank, B., Simper, N., & Kaupp, J. (2018). Formative feedback and scaffolding for developing complex problem solving and modelling outcomes. *European Journal of Engineering Education, 43*(4), 552–568. https://doi.org/10.1080/03043797.2017.1299692

Froyd, J. E., Wankat, P. C., & Smith, K. A. (2012). Five major shifts in 100 years of engineering education. *Proceedings of the IEEE, 100* [Special Centennial Issue], 1344–1360.

Fullilove, R. E., & Treisman, P. U. (1990). Mathematics achievement among African American undergraduates at the University of California, Berkeley: An evaluation of the mathematics workshop program. *The Journal of Negro Education, 59*(3), 463. https://doi.org/10.2307/2295577

Ge, X., & Land, S. M. (2004). A conceptual framework for scaffolding III-structured problem-solving processes using question prompts and peer interactions. *Educational Technology Research and Development, 52*(2), 5–22. https://doi.org/10.1007/BF02504836

Gillies, R. M., & Boyle, M. (2010). Teachers' reflections on cooperative learning: Issues of implementation. *Teaching and Teacher Education, 26*(4), 933–940. https://doi.org/10.1016/J.TATE.2009.10.034

Godwin, A., Kirn, A., & Rohde, J. A. (2017). Awareness without action: Student attitudes toward team diversity after engineering teaming experiences. *International Journal of Engineering Education, 33*(6), 1878–1891.

Goldberg, J. R., & Malassigné, P. (2017). Lessons learned from a 10–year collaboration between biomedical engineering and industrial design students in capstone design projects. *International Journal of Engineering Education, 33*(5), 1513–1520.

Gómez Puente, S. M., van Eijck, M., & Jochems, W. (2015). Professional development for design-based learning in engineering education: a case study. *European Journal of Engineering Education, 40*(1), 14–31. https://doi.org/10.1080/03043797.2014.903228

Griffin, P. M., Griffin, S. O., & Llewellyn, D. C. (2004). The impact of group size and project duration on capstone design. *Journal of Engineering Education, 93*(3), 185–193. https://doi.org/10.1002/j.2168–9830.2004.tb00805.x

Gulbulak, U., Ertas, A., & Cordell, M. (2020). The impact of transdisciplinarity on solving complex engineering problems in an ethnically diverse classroom. *International Journal of Engineering Education, 36*(6), 1976–1987.

Hadley, K. R. (2014). Teaching teamwork skills through alignment of features within a commercial board game. *International Journal of Engineering Education, 30*(6), 1376–1394.

Hadwin, A., Järvelä, S., & Miller, M. (2018). Self-regulation, co-regulation, and shared regulation in collaborative learning environments. In D. H. Schunk & J. A. Greene (Eds.), *Handbook of self-regulation of learning and performance* (pp. 83–106). Routledge/Taylor & Francis Group. https://doi.org/10.4324/9781315697048-6

Hart Research Associates. (2015). *Falling short? College learning and career success.* Association of American Colleges and Universities.

Hastings, E. M., Jahanbakhsh, F., Karahalios, K., Marinov, D., & Bailey, B. P. (2018). Structure or nurture? The effects of team-building activities and team composition on team outcomes. *Proceedings of the ACM on Human-Computer Interaction, 2*(CSCW). https://doi.org/10.1145/3274337

Hastings, E. M., Krishna Kumaran, S. R., Karahalios, K., & Bailey, B. P. (2022). A learner-centered technique for collectively configuring inputs for an algorithmic team formation tool. In *SIGCSE*

2022 – Proceedings of the 53rd ACM Technical Symposium on Computer Science Education, 1, 969–975. https://doi.org/10.1145/3478431.3499331

Hesse, F., Care, E., Buder, J., Sassenberg, K., & Griffin, P. (2015). A framework for teachable collaborative problem solving skills. In P. Griffin & E. Care (Eds.), *Assessment and teaching of 21st century skills* (pp. 37–56). Springer. https://doi.org/10.1007/978-94-017-9395-7_2

Hmelo-Silver, C. E. (2004). Problem-based learning: What and how do students learn? *Educational Psychology Review*, *16*(3), 235–266. https://doi.org/10.1023/B:EDPR.0000034022.16470.f3

Hmelo-Silver, C. E., & Chinn, C. A. (2016). Collaborative learning. In L. Corno & E. M. Anderman (Eds.), *Handbook of educational psychology* (pp. 349–363). Routledge/Taylor & Francis Group. https://psycnet.apa.org/record/2016-02743-026

Hmelo-Silver, C. E., Chinn, C. A., Chan, C., & O'Donnell, A. M. (Eds.). (2013). *The international handbook of collaborative learning*. Routledge.

Hoople, G., Choi-Fitzpatrick, A., & Reddy, E. (2019). Drones for good: Interdisciplinary project – based learning between engineering and peace studies. *International Journal of Engineering Education*, *35*(5), 1378–1391.

Horwitz, P., von Davier, A., Chamberlain, J., Koon, A., Andrews, J., & McIntyre, C. (2017). Teaching teamwork: Electronics instruction in a collaborative environment. *Community College Journal of Research and Practice*, *41*(6), 341–343.

International Engineering Alliance. (2021). *Graduate attributes and professional competencies*. www.ieagreements.org/assets/Uploads/IEA-Graduate-Attributes-and-Professional-Competencies-2021.1-Sept-2021.pdf

Irons, A., & Elkington, S. (2021). Enhancing learning through formative assessment and feedback. Routledge.

Janis, I. L. (1991). Groupthink. In E. Griffin (Ed.), *A first look at communication theory* (pp. 235–246). McGrawHill.

Johnson, D. W. (1981). Student – student interaction: The neglected variable in education. *Educational Researcher*, *10*(1), 5–10.

Johnson, D. W., & Johnson, R. T. (1984). Structuring groups for cooperative learning. *Organizational Behavior Teaching Review*, *9*(4), 8–17.

Johnson, D. W., & Johnson, R. T. (2008a). Social interdependence theory and cooperative learning: The teacher's role. In R. M. Gillies, A. F. Ashman, & J. Terwel (Eds.), *The teacher's role in implementing cooperative learning in the classroom* (pp. 9–37). Springer US. https://doi.org/10.1007/978-0-387-70892-8_1

Johnson, D. W., & Johnson, R. T. (2009). An educational psychology success story: Social interdependence theory and cooperative learning. *Educational Researcher*, *38*(5), 365–379. https://doi.org/10.3102/0013189X09339057

Johnson, D. W., Johnson, R. T., & Smith, K. A. (1991). *Cooperative learning: Increasing college faculty instructional productivity*. ASHE – ERIC Reports on Higher Education.

Johnson, D. W., Johnson, R. T., & Smith, K. A. (1998). Cooperative learning returns to college What evidence is there that it works? Change. *The Magazine of Higher Learning*, *30*(4), 26–35. https://doi.org/10.1080/00091389809602629

Johnson, R. T., & Johnson, D. W. (2008b). Active learning: Cooperation in the classroom. *The Annual Report of Educational Psychology in Japan*, *47*, 29–30. https://doi.org/10.5926/arepj1962.47.0_29

Johri, A., Olds, B., & O'Connor, K. (2015). Situative frameworks for engineering learning research. In A. Johri & B. Olds (Eds.), *Cambridge handbook of engineering education research* (pp. 47–66). Cambridge University Press. doi:10.1017/CBO9781139013451.006

Johri, A., Williams, C., & Pembridge, J. (2013). Creative collaboration: A case study of the role of computers in supporting representational and relational interaction in student engineering design teams. *International Journal of Engineering Education*, *29*(1), 33–44.

Jonassen, D. H. (2004). *Learning to solve problems: An instructional design guide* (Vol. 6). John Wiley & Sons.

Jonassen, D. H., Strobel, J., & Lee, C. B. (2006). Everyday problem solving in engineering: Lessons for engineering educators. *Journal of Engineering Education*, *95*(2), 139–151. https://doi.org/10.1002/j.2168-9830.2006.tb00885.x

Kaendler, C., Wiedmann, M., Rummel, N., & Spada, H. (2015). Teacher competencies for the implementation of collaborative learning in the classroom: A framework and research review. *Educational Psychology Review*, *27*(3), 505–536. https://doi.org/10.1007/s10648-014-9288-9

Kagan, S. (1989). The structural approach to cooperative learning. *Educational Leadership*, *47*(4), 12–15.

Kalaian, S. A., Kasim, R. M., & Nims, J. K. (2018). Effectiveness of small-group learning pedagogies in engineering and technology education: A meta-analysis. *Journal of Technology Education, 29*(2), 20–35. https://doi.org/10.21061/jte.v29i2.a.2

Karau, S. J., & Williams, K. D. (1993). Social loafing: A meta-analytic review and theoretical integration. *Journal of Personality and Social Psychology, 65*(4), 681–706. https://doi.org/10.1037/0022-3514.65.4.681

Karau, S. J., & Williams, K. D. (1995). Social loafing: Research findings, implications, and future directions. *Current Directions in Psychological Science, 4*(5), 134–140. https://doi.org/10.1111/1467-8721.EP10772570

Kester, L., & Paas, F. (2005). Instructional interventions to enhance collaboration in powerful learning environments. *Computers in Human Behavior, 21*(4), 689–696. https://doi.org/10.1016/j.chb.2004.11.008

Kirschman, J. S., & Greenstein, J. S. (2002). The use of groupware for collaboration in distributed student engineering design teams. *Journal of Engineering Education, 91*(4), 403–407. https://doi.org/10.1002/j.2168–9830.2002.tb00724.x

Kolmos, A., & de Graaff, E. (2014). Problem-based and project-based learning in engineering education: Merging models. In A. Johri & B. M. Olds (Eds.), *Cambridge handbook of engineering education research* (pp. 141–161). Cambridge University Press. https://doi.org/10.1017/CBO9781139013451.012

Kozlowski, S. W. J., & Ilgen, D. R. (2006). Enhancing the effectiveness of work groups and teams. *Psychological Science in the Public Interest, Supplement*, 7(3), 77–124. https://doi.org/10.1111/j.1529–1006.2006.00030.x

Kraft, M. A. (2020). Interpreting effect sizes of education interventions. *Educational Researcher, 49*(4), 241–253. https://doi.org/10.3102/0013189X20912798

Kyndt, E., Raes, E., Lismont, B., Timmers, F., Cascallar, E., & Dochy, F. (2013). A meta – analysis of the effects of face – to – face cooperative learning. Do recent studies falsify or verify earlier findings? *Educational Research Review, 10*, 133–149. https://doi.org/10.1016/J.EDUREV.2013.02.002

Laeser, M., Moskal, B. M., Knecht, R., & Lasich, D. (2003). Engineering design: Examining the impact of gender and the team's gender composition. *Journal of Engineering Education, 92*(1), 49–56. https://doi.org/10.1002/J.2168-9830.2003.TB00737.X

Lau, K., Beckman, S. L., & Agogino, A. M. (2012). Diversity in design teams: An investigation of learning styles and their impact on team performance and innovation. *International Journal of Engineering Education, 28*(2), 293–301.

Lawrence, L., & Mercier, E. (2019). A review of the evolving definition of orchestration: Implications for research and design. *Computer-Supported Collaborative Learning Conference, CSCL, 2*, 829–830. https://doi.org/10.22318/CSCL2019.829

Layton, R. A., Loughry, M. L., Ohland, M. W., & Ricco, G. D. (2010). Design and validation of a web-based system for assigning members to teams using instructor – specified criteria. *Advances in Engineering Education, 2*(1), 1–28. www.catme.org.

Liu, C., & Wayno, F. (2008, September). National science foundation workshop on the future power engineering workforce. In *workshop report*. https://www.cewd.org/documents/BestPractices/Engineering_Workforce_Report.pdf

Maldonado, H., Klemmer, S., & Pea, R. (2009). When is collaborating with friends a good idea? Insights from design education. In C. O'Malley, D. Suthers, P. Reimann, & A. Dimitracopoulou (Eds.), *Computer supported collaborative learning practices: CSCL2009 conference proceedings* (pp. 227–231). International Society of the Learning Sciences.

Mandala, M., Schunn, C., Dow, S., Goldberg, M., Pearlman, J., Clark, W., & Mena, I. (2018). Impact of collaborative team peer review on the quality of feedback in engineering design projects. *International Journal of Engineering Education, 34*(4), 1299–1313.

Marbouti, F., Mendoza-Garcia, J., Diefes-Dux, H. A., & Cardella, M. E. (2019). Written feedback provided by first-year engineering students, undergraduate teaching assistants, and educators on design project work. *European Journal of Engineering Education, 44*(1–2), 179–195. https://doi.org/10.1080/03043797.2017.1340931

Margolis, J., & Fisher, A. (2002). *Unlocking the clubhouse: Women in computing*. MIT Press.

Marra, R. M., Steege, L., Tsai, C. L., & Tang, N. E. (2016). Beyond "group work": An integrated approach to support collaboration in engineering education. *International Journal of STEM Education, 3*(1). https://doi.org/10.1186/s40594-016-0050-3

Mathieu, J. E., & Rapp, T. L. (2009). Laying the foundation for successful team performance trajectories: The roles of team charters and performance strategies. *Journal of Applied Psychology, 94*(1), 90–103. https://doi.org/10.1037/A0013257

Maturana, J., Tampier, G., Serandour, G., & Luco, R. (2014). Developing teamwork skills in first and second year engineering students. *International Journal of Engineering Education, 30*(5), 1225–1233.

Mayer, I., Warmelink, H., & Bekebrede, G. (2013). Learning in a game – based virtual environment: A comparative evaluation in higher education. *European Journal of Engineering Education, 38*(1), 85–106. https://doi.org/10.1080/03043797.2012.742872

McGee, E. O., & Martin, D. B. (2011). "You Would Not Believe What I Have to Go Through to Prove My Intellectual Value!" Stereotype management among academically successful black mathematics and engineering students. *American Education Research Journal, 48*(6), 1347–1389. https://doi.org/10.3102/0002831211423972

McNair, L. D., Newswander, C., Boden, D., & Borrego, M. (2011). Student and faculty interdisciplinary identities in self-managed teams. *Journal of Engineering Education, 100*(2), 374–396. https://doi.org/10.1002/j.2168–9830.2011.tb00018.x

Meadows, L. A., & Sekaquaptewa, D. (2011). The effect of skewed gender composition on student participation in undergraduate engineering project teams. In *ASEE annual conference and exposition, conference proceedings*. https://doi.org/10.18260/1–2–18957

Meadows, L. A., Sekaquaptewa, D., Paretti, M. C., Pawley, A. L., Jordan, S. S., Chachra, D., & Minerick, A. (2015). Interactive panel: Improving the experiences of marginalized students on engineering design teams. In *ASEE Annual Conference and Exposition*, 26–1007. https://doi.org/10.18260/P.24344

Meijer, H., Hoekstra, R., Brouwer, J., & Strijbos, J. W. (2020). Unfolding collaborative learning assessment literacy: A reflection on current assessment methods in higher education. *Assessment and Evaluation in Higher Education, 45*(8), 1222–1240. https://doi.org/10.1080/02602938.2020.1729696

Mellingsæter, M. S., & Bungum, B. (2015). Students' use of the interactive whiteboard during physics group work. *European Journal of Engineering Education, 40*(2), 115–127. https://doi.org/10.1080/03043797.2014.928669

Mercier, E. M. (2017). The influence of achievement goals on collaborative interactions and knowledge convergence. *Learning and Instruction, 50*, 31–43. https://doi.org/10.1016/J.LEARNINSTRUC.2016.11.006

Mercier, E., Goldman, S., & Booker, A. (2009). Focusing on process: Evidence and ideas to promote learning through the collaborative design process. In C. DiGano, S. Goldman, & M. Chorost (Eds.), *Educating learning technology designers: Guiding and inspiring creators of innovative educational tools* (pp. 36–61). Routledge. https://doi.org/10.4324/9780203357217

Mercier, E., & Higgins, S. E. (2015). The four Ts of the collaborative classroom. *CEUR Workshop Proceedings, 1411*, 15–19.

Mercier, E., Shehab, S., Sun, J., & Capell, N. (2015, July). The development of collaborative practices in introductory engineering courses. *Computer-Supported Collaborative Learning Conference, CSCL, 2*, 657–658.

Mikic, B., & Rudnitsky, A. (2016). An emergent approach to design thinking and collaborative team composition in an introduction to engineering course. *International Journal of Engineering Education, 32*(3), 1514–1521.

Missingham, D., & Matthews, R. (2014). A democratic and student-centred approach to facilitating teamwork learning among first – year engineering students: A learning and teaching case study. *European Journal of Engineering Education, 39*(4), 412–423. https://doi.org/10.1080/03043797.2014.881321

Moallem, M., Hung, W., & Dabbagh, N. (Eds.). (2019). *The Wiley handbook of problem-based learning*. John Wiley & Sons.

Mollet, A., Holmes, J., & Linley, J. L. (2021). An analysis of URM STEM students' GPAs relative to good teaching practices. *Journal of Women and Minorities in Science and Engineering, 28*(5), 1–24. https://doi.org/10.1615/jwomenminorscieneng.2021033546

Mostafapour, M., & Hurst, A. (2020). An exploratory study of teamwork processes and perceived team effectiveness in engineering capstone design teams. *International Journal of Engineering Education, 36*(1B), 436–449.

Nadir, H. M. (2021). *Investigating a teacher's scaffolding for design problem solving in project-based learning* (PhD Thesis). Indiana University.

Nokes-Malach, T. J., Richey, J. E., & Gadgil, S. (2015). When is it better to learn together? Insights from research on collaborative learning. *Educational Psychology Review, 27*(4), 645–656. https://doi.org/10.1007/s10648-015-9312-8

Nokes-Malach, T. J., Zepeda, C. D., Richey, J. E., & Gadgil, S. (2019). Collaborative learning: The benefits and costs. In J. Dunlosky & K. A. Rawson (Eds.), *The Cambridge handbook of cognition and education* (pp. 500–527). Cambridge University Press. https://doi.org/10.1017/9781108235631.021

NRC. (1994). Women scientists and engineers employed in industry. In *Women scientists and engineers employed in industry: Why so few?* The National Academies Press. https://doi.org/10.17226/2264

O'Donnell, A. M. (1999). Structuring dyadic interaction through scripted cooperation. In A. M. O'Donnell & A. King (Eds.), *Cognitive perspectives on peer learning* (pp. 179–196). Lawrence Erlbaum Associates Publishers.

Oladirana, M. T., Uziaka, J., Eisenbergb, M., & Schefferc, C. (2011). Global engineering teams – a programme promoting teamwork in engineering design and manufacturing. *European Journal of Engineering Education, 36*(2), 173–186. https://doi.org/10.1080/03043797.2011.573534

Opdecam, E., & Everaert, P. (2018). Seven disagreements about cooperative learning. *Accounting Education, 27*(3), 223–233. https://doi.org/10.1080/09639284.2018.1477056

Ortiz, A., Johnson, D., & Johnson, R. T. (1996). The effect of positive goal and resource interdependence on individual performance. *Taylor & Francis, 136*(2), 243–249. https://doi.org/10.1080/00224545.1996.9713998

The Organisation for Economic Co-operation and Development (OECD). (1999). *Programme for international student assessment (PISA).* OECD Library. https://www.oecd-ilibrary.org/

The Organisation for Economic Co-operation and Development (OECD). (2017). *Programme for international student assessment (PISA).* OECD Library. https://www.oecd-ilibrary.org/

Paquette, L., Bosch, N., Mercier, E., Jung, J., Shehab, S., & Tong, Y. (2018). Matching Data-Driven Models of Group Interactions to Video Analysis of Collaborative Problem Solving on Tablet Computers. *Rethinking Learning in the Digital Age: Making the Learning Sciences Count - Proceedings of the 13th International Conference of the Learning Sciences, ICLS 2018*, London, UK, June 23–27, 2018. https://repository.isls.org/handle/1/719

Parker, R., Sangelkar, S., Swenson, M., & Ford, J. D. (2019). Launching for success: A review of team formation for capstone design. *International Journal of Engineering Education, 35*(6B), 1926–1936.

Pattison, S., Gontan, I., Ramos-Montañez, S., Shagott, T., Francisco, M., & Dierking, L. (2020). The identity-frame model: A framework to describe situated identity negotiation for adolescent youth participating in an informal engineering education program. *Journal of the Learning Sciences, 29*(4–5), 550–597. https://doi.org/10.1080/10508406.2020.1770762

Pazos, P., Ringleb, S. I., Kidd, J., & Jones, R. (2019). Scaffolding project-based learning in an engineering and education partnership using open-access technology. *International Journal of Engineering Education, 35*(5), 1306–1315.

Pellegrino, A., Weiss, M. P., Regan, K., & Mann, L. (2014). Learning to collaborate: Exploring collective and individual outcomes of special and general educators. *Journal of Social Studies Research, 1*(1), 96–104.

Pellegrino, J. W. E., Chudowsky, N. E., & Glaser, R. E. (2001). *Knowing what students know: The science and design of educational assessment.* National Academies Press. https://doi.org/10.17226/10019

Penuel, W. R., Fishman, B. J., Haugan Cheng, B., & Sabelli, N. (2011). Organizing research and development at the intersection of learning, implementation, and design. *Educational Researcher, 40*(7), 331–337. https://doi.org/10.3102/0013189X11421826

Perez-Poch, A., Sánchez-Carracedo, F., Salán, N., & López, D. (2019). Cooperative learning and embedded active learning methodologies for improving students' motivation and academic results. *International Journal of Engineering Education, 35*(6A), 1851–1858.

Pinho-Lopes, M., & Macedo, J. (2016). Project-based learning in Geotechnics: Cooperative versus collaborative teamwork. *European Journal of Engineering Education, 41*(1), 70–90. https://doi.org/10.1080/03043797.2015.1056099

Pinho – Lopes, M., Macedo, J., & Bonito, F. (2011). Cooperative learning in a soil mechanics course at undergraduate level. *European Journal of Engineering Education, 36*(2), 119–135. https://doi.org/10.1080/03043797.2011.565115

Prince, M. (2004). Does active learning work? A review of the research. *Journal of Engineering Education, 93*(3), 223–231. https://doi.org/10.1002/J.2168–9830.2004.TB00809.X

Puntambekar, S., & Hubscher, R. (2005). Tools for scaffolding students in a complex learning environment: What have we gained and what have we missed? *Educational Psychologist, 40*(1), 1–12.

Reiser, B. J. (2018). Scaffolding complex learning: The mechanisms of structuring and problematizing student work. In *The journal of the learning sciences* (pp. 273–304). Psychology Press.

Roschelle, J. (1992). Learning by collaborating: Convergent conceptual change. *Journal of the Learning Sciences, 2*(3), 235–276. https://doi.org/10.1207/s15327809jls0203_1

Roth, W-M., & Roychoudhury, A. (1992). The social construction of scientific concepts or the concept map as device and tool thinking in high conscription for social school science. *Science Education, 76*(5), 531–557. https://doi.org/10.1002/sce.3730760507

Rummel, N., Spada, H., & Hauser, S. (2009). Learning to collaborate while being scripted or by observing a model. *International Journal of Computer – Supported Collaborative Learning, 4*(1), 69–92. https://doi.org/10.1007/s11412-008-9054-4

Salomon, G., & Globerson, T. (1989). When teams do not function the way they ought to. *International Journal of Educational Research, 13*(1), 89–99.

Schiffeler, N., Stehling, V., Hees, F., & Isenhardt, I. (2019, June). Effects of collaborative augmented reality on communication and interaction in learning contexts-results of a qualitative pre-study. In *2019 ASEE annual conference & exposition*.

Serrano – Cámara, L. M., Paredes – Velasco, M., Velázquez-Iturbide, J. Á., Alcover, C. M., & Castellanos, M. E. (2016). MoCAS: A mobile collaborative tool for learning scope of identifiers in programming courses. *International Journal of Engineering Education, 32*(2), 969–981.

Shehab, S., & Mercier, E. (2017). Designing engineering tasks for collaborative problem solving. *Computer-Supported Collaborative Learning Conference, CSCL, 2*, 235–266.

Shehab, S., & Mercier, E. (2019). Implementing collaborative learning in classrooms: Science teachers' perspectives. *International Journal of Innovation in Education, 5*(4), 265. https://doi.org/10.1504/ijiie.2019.102597

Sheppard, S. D., & Macatangay, K. A. W. M. (2008). *Educating engineers: Designing for the future of the field*. Carnegie Foundation for the Advancement of Teaching.

Shin, N., Jonassen, D. H., & McGee, S. (2003). Predictors of well-structured and ill-structured problem solving in an astronomy simulation. *Journal of Research in Science Teaching, 40*(1), 6–33. https://doi.org/10.1002/tea.10058

Shooter, S., & Mcneill, M. (2002). Interdisciplinary collaborative learning in mechatronics at Bucknell university. *Journal of Engineering Education, 91*(3), 339–344. https://doi.org/10.1002/j.2168-9830.2002.tb00712.x

Slavin, R. E. (1980). Cooperative learning. *Review of Educational Research, 50*(2), 315–342.

Smith, K. A., Johnson, D. W., & Johnson, R. T. (1981a). Structuring learning goals to meet the goals of engineering education. *Engineering Education, 72*(3), 221–226.

Smith, K. A., Johnson, D. W., & Johnson, R. T. (1981b). The use of cooperative learning groups in engineering education. *Frontiers Education Conference, IEEE*, 26–32.

Smith, K. A., Sheppard, S. D., Johnson, D. W., & Johnson, R. T. (2005). Pedagogies of engagement: Classroom-based practices. *Journal of Engineering Education, 94*(1), 87–101. https://doi.org/10.1002/J.2168-9830.2005.TB00831.X

Strijbos, J. W. (2011). Assessment of (computer-supported) collaborative learning. *IEEE Transactions on Learning Technologies, 4*(1), 59–73. https://doi.org/10.1109/TLT.2010.37

Stump, G. S., Hilpert, J. C., Husman, J., Chung, W. T., & Kim, W. (2011). Collaborative learning in engineering students: Gender and achievement. *Journal of Engineering Education, 100*(3), 475–497. https://doi.org/10.1002/j.2168-9830.2011.tb00023.x

Svihla, V., Peele – Eady, T., & Gallup, A. (2021). Exploring agency in capstone design problem framing. *Studies in Engineering Education, 2*(2), 96. https://doi.org/10.21061/SEE.69

Takai, S., & Esterman, M. (2019). A review of team effectiveness models and possible instruments for measuring design-team inputs, processes, and outputs. *International Journal of Engineering Education, 35*(6), 1684–1697.

Tao, Y. (2016). Where do they do engineering? Gender differences in employment at the position level among engineering doctorate recipients. *Journal of Women and Minorities in Science and Engineering, 22*(1), 69–89.

Tenenberg, J. (2019). Factors affecting free riding on teams: Implications for engineering education. *International Journal of Engineering Education, 35*(6A), 1703–1724.

Thistlethwaite, J., Dallest, K., Moran, M., Dunston, R., Roberts, C., Eley, D., Bogossian, F., Forman, D., Bainbridge, L., Drynan, D., & Fyfe, S. (2016). Introducing the individual Teamwork Observation and Feedback Tool (iTOFT): Development and description of a new interprofessional teamwork measure. *Journal of Interprofessional Care, 30*(4), 526–528. https://doi.org/10.3109/13561820.2016.1169262

Tlhoaele, M., Suhre, C., & Hofman, A. (2016). Using technology-enhanced, cooperative, group-project learning for student comprehension and academic performance. *European Journal of Engineering Education*, *41*(3), 263–278. https://doi.org/10.1080/03043797.2015.1056102

Tonso, K. L. (1996). Student learning and gender. *Journal of Engineering Education*, *85*(2), 143–150. https://doi.org/10.1002/J.2168-9830.1996.TB00223.X

Tucker, T., Shehab, S., & Mercier, E. (2020, June). The impact of scaffolding prompts on the collaborative problem solving of ill-structured tasks by undergraduate engineering student groups. In *ASEE annual conference and exposition, conference proceedings* (Vol. 2020, p. 1375).

Turns, J. A., Sattler, B., Yasuhara, K., Borgford-Parnell, J. L., & Atman, C. J. (2014). Integrating reflection into engineering education. In *ASEE annual conference and exposition, conference proceedings*. https://doi.org/10.18260/1-2-2066

Van Boxtel, C., Van der Linden, J., & Kanselaar, G. (2000). Collaborative learning tasks and the elaboration of conceptual knowledge. *Learning and Instruction*, *10*(4), 311–330. https://doi.org/10.1016/S0959-4752(00)00002-5

van de Pol, J., Volman, M., & Beishuizen, J. (2010). Scaffolding in teacher-student interaction: A decade of research. *Educational Psychology Review*, *22*(3), 271–296. https://doi.org/10.1007/S10648-010-9127-6

Van Dusen, B., & Nissen, J. (2020). Equity in college physics student learning: A critical quantitative intersectionality investigation. *Journal of Research in Science Teaching*, *57*(1), 33–57. https://doi.org/10.1002/tea.21584

Vanhanen, J., & Lehtinen, T. O. A. (2014). Software engineering problems encountered by capstone project teams. *International Journal of Engineering Education*, *30*(6), 1461–1475.

Vergara, C. E., Urban – Lurain, M., Dresen, C., Coxen, T., MacFarlane, T., Frazier, K., Briedis, D., Buch, N., Esfahanian, A. H., Paquette, L., Sticklen, J., Laprad, J., & Wolff, T. F. (2009). Aligning computing education with engineering workforce computational needs: New curricular directions to improve computational thinking in engineering graduates. In *Proceedings – frontiers in education conference, FIE* (pp. 17–22). https://doi.org/10.1109/FIE.2009.5350463

Villanueva, A., Zhu, Z., Liu, Z., Peppler, K., Redick, T., & Ramani, K. (2020, April). Meta – AR – app: An authoring platform for collaborative augmented reality in STEM classrooms. In *Proceedings of the 2020 CHI conference on human factors in computing systems* (pp. 1–14).

Viswanathan, S. A., & Vanlehn, K. (2018). Using the tablet gestures and speech of Pairs of students to classify their collaboration. *IEEE Transactions on Learning Technologies*, *11*(2), 230–242. https://doi.org/10.1109/TLT.2017.2704099

Vygotsky, L. S. (1978). *Mind and society: The development of higher mental processes*. Harvard University Press.

Wagner, E. A. (2014). Using a kinesthetic learning strategy to engage nursing student thinking, enhance retention, and improve critical thinking. *Journal of Nursing Education*, *53*(6), 348–351. https://doi.org/10.3928/01484834-20140512-02

Webb, N. M. (2009). The teacher's role in promoting collaborative dialogue in the classroom. *The British Journal of Educational Psychology*, *79*(Pt 1), 1–28. https://doi.org/10.1348/000709908X380772

Webb, N. M. (2013). Collaboration in the classroom. *International Guide to Student Achievement*, 215–217.

Webb, S., Massey, D., Goggans, M., & Flajole, K. (2019). Thirty-five years of the gradual release of responsibility: Scaffolding toward complex and responsive teaching. *The Reading Teacher*, *73*(1), 75–83. https://doi.org/10.1002/TRTR.1799

Wieman, C. E. (2014). Large-scale comparison of science teaching methods sends clear message. *Proceedings of the National Academy of Sciences of the United States of America*, *111*(23), 8319–8320. https://doi.org/10.1073/pnas.1407304111

Weinberger, A., Ertl, B., Fischer, F., & Mandl, H. (2005). Epistemic and social scripts in computer – supported collaborative learning. *Instructional Science*, *33*(1), 1–30. https://doi.org/10.1007/s11251-004-2322-4

Weinberger, A., & Fischer, F. (2001). *Scripts and scaffolds in problem-based computer supported collaborative environments: Fostering participation and transfer*. Institute for Empirical Pedagogy and Pedagogical Psychology, November. http://epub.ub.uni – muenchen.de/252/

Wiggins, G. (1998). *Educative assessment: Designing assessments to inform and improve student performance*. Jossey – Bass Publishers.

Wood, D., Bruner, J. S., & Ross, G. (1976). *The role of tutoring in problem solving*. Child Psychology & Psychiatry & Allied Disciplines.

Zhang, J., Xie, H., & Li, H. (2018). Project based learning with implementation planning for student engagement in BIM classes. *International Journal of Engineering Education*, *35*(1), 310–322.

20

Teaching for Creativity, Entrepreneurship, and Leadership in Engineering

Sarah E. Zappe, Aileen Y. Huang-Saad, Nathalie Duval-Couetil, and Denise R. Simmons

1 Introduction

Society is changing at a rapid pace. Although technological advances in engineering have been prolific, complex problems confronting our world, such as climate change, disease, social justice, and population growth, are significant, with solutions requiring leadership, creativity, and ingenuity. Concurrently, the engineering workplace is transforming. Today, graduates are unlikely to remain in the same company for their entire careers and instead will navigate between engineering, managerial, and leadership roles in start-ups, small companies, or large organizations throughout their careers. This requires that future engineers be technically proficient but also equipped with the professional skills needed to tackle complex and ambiguous problems in a variety of contexts.

Universities, government agencies, professional societies, and nonprofits have recognized this shift and have emphasized the importance of new skill sets and characteristics for contemporary engineers. For example, the National Academy of Engineering in the United States, in their pivotal publication *The Engineer of 2020* (National Academy of Engineering, 2005), emphasized the need for graduates to have practical ingenuity, creativity, communication skills, business acumen, leadership, ethical standards, and other qualities. In the United Kingdom, the Royal Academy of Engineering stated a key role for engineers is a "change agent providing creativity, innovation, and leadership to meet new challenges" (Royal Academy of Engineering, 2013, p. 4). In response, engineering schools have been actively working to reform instructional practice to meet these educational and professional needs by emphasizing the development of creativity, entrepreneurship, and leadership.

The purpose of this chapter is to provide an overview of how the domains of creativity, entrepreneurship, and leadership apply to engineering and why they are critical to the formation of contemporary engineers. While the need to increase a broad range of professional skills is widely recognized, we focus on these domains specifically because of their role in the innovation process, their independent trajectories into engineering education, and the overlapping characteristics that make teaching them a challenge.

DOI: 10.4324/9781003287483-24

2 The Relevance of Creativity, Entrepreneurship, and Leadership to Engineering Education

As the world shifts to a global, knowledge-based economy, the role of science and technology, and how we educate those that work to support these fields, has national and international importance. Numerous domestic and international studies have indicated contemporary economic success has and will result from technological advances (National Academy of Sciences et al., 2007, 2010; Wilson & Purushothaman, 2006). As a result, governments around the world are calling for engineering education reform (Li et al., 2003; National Academy of Sciences et al., 2007, 2010; Royal Academy of Engineering, 2007). In 2005, the US National Academies of Science convened a committee of presidents of major universities, Nobel laureates, and CEOs of Fortune 100 corporations to determine the top 10 actions policymakers could take to support the US science and technology community and their efforts to remain competitive in the 21st century (National Academy of Sciences et al., 2007). The committee's recommendations were shaped to address the need to provide infrastructure and resources to support science, math, and engineering education.

Three years later, the same committee published a follow-on report to update their findings, given growing economic instability (National Academy of Sciences et al., 2010). The committee reported not much had changed since their original report, and some support structures had worsened, thus putting Americans at risk of not being able to compete in a future job market that does not acknowledge geographical boundaries. In the follow-on report, the authors not only recognized the need to increase US human capital in STEM fields but also identified the importance of moving innovations to market at a faster speed than competitors, through entrepreneurship and innovation. While policymakers paved the path for national investment in innovation and human capital, the National Academy of Engineering (NAE) convened its own committee to develop a vision for the role of future engineers and inform undergraduate engineering education (National Academy of Engineers, 2005). This forward-thinking approach was intended to keep engineering education at pace with rapid societal and technological transformations. The NAE's review projected technological, societal, global, and geopolitical challenges for new engineers. It also identified desired attributes for engineers graduating in 2020 that went beyond basic analytical skills to those that would help engineers meet the requirements of working in a global economy. Today, engineers are expected to be "T-shaped," meaning, they are technologically knowledgeable but also able to collaborate with experts in other disciplines and apply knowledge from other fields to solve engineering problems (Barile et al., 2015).

The *Engineer 2020* initiative concluded innovation is critical for the United States to maintain economic leadership, and engineers are key to its processes and outcomes. Thus, engineering education must adapt to educate future innovators. Creativity, entrepreneurship, and leadership are critical domains of the innovation process (Duval-Couetil, 2013). Creativity spurs innovation, and innovative ideas are scaled through entrepreneurship and intrapreneurship, all of which is guided by leadership. Given the significant role each domain plays in the innovation process, creativity, entrepreneurship, and leadership have garnered varying levels of interest in engineering education and research over the last few decades (Figure 20.1). As a result, the implementation and assessment of each domain has evolved in often-separate threads of literature.

3 Integrating Creativity into Engineering Education

3.1 Creativity Research

While creativity is considered the first phase of the innovation process, exploration of creativity research and educational approaches in engineering education has been minimal. The study of

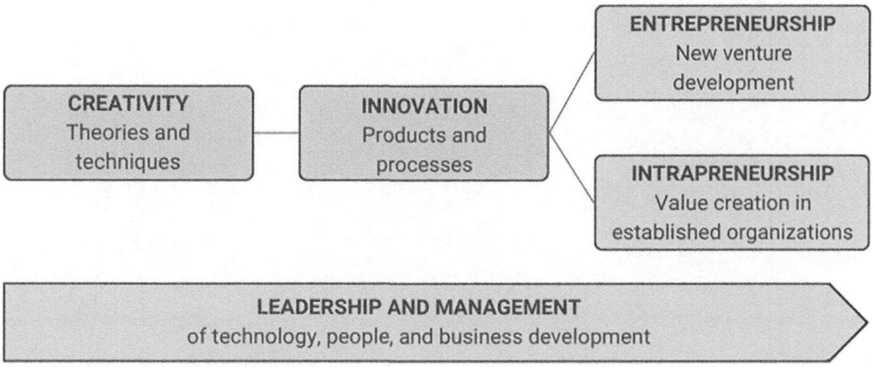

Figure. 20.1 Creativity, entrepreneurship, and leadership are intertwined in the innovation process. *Source:* Adapted from Duval-Couetil et al. (2013).

creativity has a deep history in disciplines such as philosophy and psychology (Runco & Albert, 2010). The earliest research in creativity in the 1800s focused on defining creativity, identifying creative people and their characteristics, and questioning whether creativity can be increased (Becker, 1995). These issues are still examined in recent research (e.g., Abdulla Alabbassi et al., 2021; Glăveanu & Beghetto, 2021; Kryshtanovych et al., 2021). Later work in the 1950s/1960s focused on the relationship between creativity and intelligence and whether these are distinct constructs (Runco & Albert, 2010).

Creativity research in psychology over the past century has grown exponentially, making it challenging for researchers to understand how the field has evolved. Using computational methods to synthesize over 38,000 articles relating to creativity, Mejia et al. (2021) found there are 12 clusters of creativity research in the past 100 years, which relate to organizations/teams, social psychology, industries/cities, idea generation, neuroscience, identity and multiculturalism, and others. The authors also examined recent trends in creativity research. Among many other areas, some of the trends that most relate to engineering include transformational leadership, drivers of innovative work behavior, creativity management, problem-based learning, design thinking, and STEAM (science, technology, engineering, arts, and mathematics). In another systematic overview study, Mi et al. (2020) used text mining of publication databases to determine what words clustered together, providing insight into areas of creativity research in education. The cluster of engineering education emerged as a topic of increasing interest in recent creativity research, and the authors identified this research cluster as likely to experience future growth.

Despite the conclusions of Mi et al. (2020), creativity research in engineering education appears to have been limited. Zappe et al. (2013) conducted a limited systematic review of articles published in the major engineering education journals between 2006 and 2011. During this period, only 16 articles with the words "creative" or "creativity" in the title were published in five major engineering education journals. Similarly, Figure 20.2 displays the number of American Society for Engineering Education (ASEE) annual conference papers published from 1996 to 2021 with "creative" or "creativity" in the title. While there does seem to be a weak trend with increasing numbers of creativity-related papers since 1996, creativity research does not seem to be a primary area of emphasis in engineering education broadly, as indicated by the relatively low number of papers each year. In contrast, creativity in engineering design research, as evident by recent publications in the field, has been much more prevalent, focusing on identifying where and how creativity fits into the design process (e.g., Howard et al., 2008), describing how creative ideas evolve during the design process (e.g., Starkey et al., 2016), and assessing creativity in the design process (e.g., Charyton &

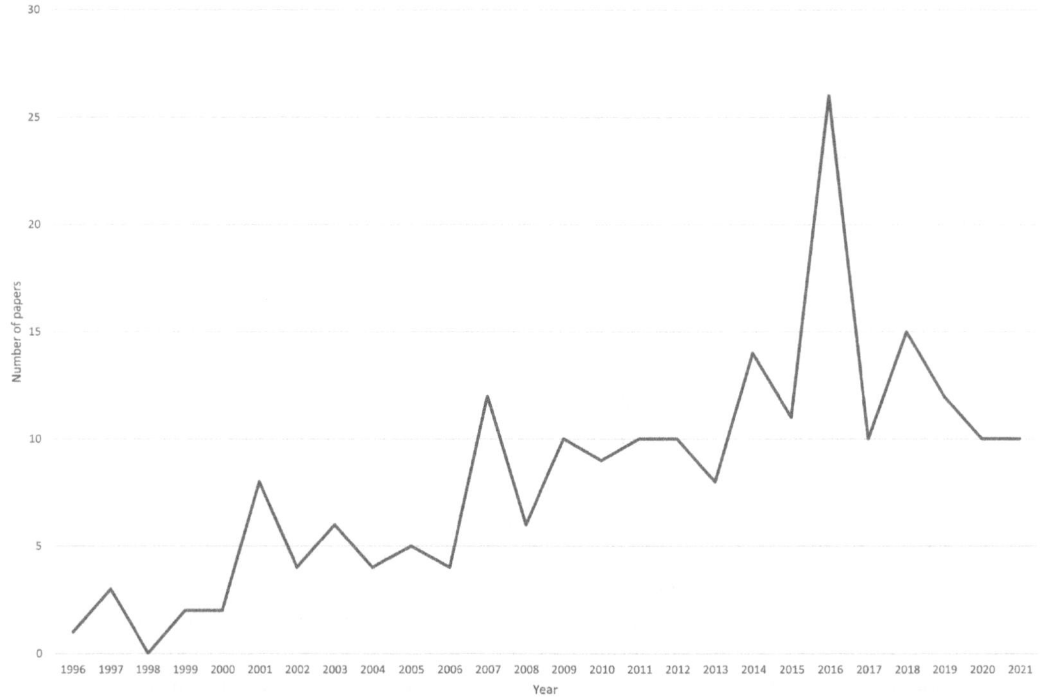

Figure. 20.2 ASEE conference papers with "creative" or "creativity" in the title by year.
Source: ASEE's Papers on Engineering Education Repository (PEER).

Merrill, 2009; Denson et al., 2015; Miller et al., 2021). Overall, these studies suggest a disconnect in engineering education, with creativity being explored more deeply in relation to engineering design yet ignored in relation to other engineering contexts. An area of future research, beyond the scope of this chapter, would be to further explore how creativity is being defined and researched specifically in engineering design.

3.2 What Is Creativity, and Why Is It Needed in Engineering?

Plucker et al. (2004) reviewed definitions of *creativity* to produce one of the most often-used definitions: "Creativity is the interaction among aptitude, process, and environment by which an individual or group produces a perceptible product that is both novel and useful as defined within a social context" (p. 90). While the emphasis on novelty and usefulness is common in conceptualizing creativity, within engineering, the focus of creativity is often on problem-solving. This problem-solving perspective is evident in creative process models, such as that developed by Mumford et al. (1991), with "problem" defined broadly as a task to be accomplished. Mumford et al. (2012) defined three critical assumptions that underlie their creative process model: (1) creative problem-solving is based on knowledge and information, (2) existing knowledge must be recombined and reorganized to generate novel ideas leading to problem solutions, and (3) ideas are evaluated and then used to plan and develop a creative product.

Despite the lack of emphasis on creativity in the engineering education literature, the ability to creatively problem-solve is needed in many engineering tasks, such as product development, prototyping, and design (Litzinger et al., 2015). Cropley (2017) provides a cogent argument as to why creativity is important:

> Engineering is a forward-looking, optimistic pursuit that seeks to develop new technological solutions to the stream of new and challenging problems that we must face as the world continues to develop. It follows that engineers themselves must have, as a core competency, the ability to find and develop these novel solutions.
>
> *(p. 213)*

Cropley continues, "Creativity needs to be nurtured in engineering education because without it, engineers are not fully equipped for their role as technological problem solvers" (p. 213).

3.3 Creativity in the Engineering Curriculum

Despite the importance of creativity to engineering, most universities do not emphasize creativity in the curriculum. In a study of instructors' and students' perceptions of creativity (Kazerounian & Foley, 2007), engineering students were found to be more likely than non-engineering students to feel their courses lacked the necessary characteristics needed to encourage creativity. According to the authors, "there was never an instance in the surveys when engineering students felt a creativity criteria was present in their education" (p. 765). In a cross-sectional study exploring engineering students' creative self-concepts (Zappe et al., 2015), engineering students in their senior year felt instructors had lower expectations for creative behaviors as compared to first-year students. Additionally, the study found the curriculum may influence women's perceptions of whether they are creative individuals, as suggested by the finding that female seniors had much lower scores on a scale of creative identity as compared to first-year females. Zappe and Tise (2019) continued this line of research, conducting a longitudinal study of students' creative self-concepts and perceptions. The results showed students' conceptions of themselves as creative individuals were stable across their academic career. However, senior students had lower ratings on scales that measured their perceived expectations for displaying creative behavior, the need to be creative in their role as engineering students, and the perceived value of being creative in personal and professional settings. The authors concluded that "students perceive engineering to be a discipline without strong expectations for creative behaviours" (p. 16). Atwood and Pretz (2016) found measures of creativity did not significantly predict student achievement in engineering courses, thus resulting in the conclusion that "creative performance is not strongly encouraged or rewarded in the curriculum" (p. 550).

3.4 Pedagogical Approaches for Integrating Creativity into Engineering

Building on the interactionist definition of *creativity* developed by Plucker et al. (2004), Figure 20.3 presents a model of how creative outcomes can be achieved in the engineering classroom. Instructional approaches to creativity need to be less structured, requiring students to be creative. Less-structured instructional approaches, such as problem-based learning, case-based learning, and experiential learning, are more likely to promote creative outcomes. Litzinger et al. (2015) suggested instructors can focus on individual steps of the creative process, such as conceptual combination, idea generation, or idea evaluation, as integrating the entire creative process into technical courses may be difficult. Törnkvist (1998) argued that the engineering curriculum needs to use more open forms of learning (such as problem-based learning), incorporate qualitative narratives, and require all instructors to be knowledgeable of learning theories. The classroom environment needs to be supportive, inclusive, and welcoming for students to have psychological safety and to feel free to engage

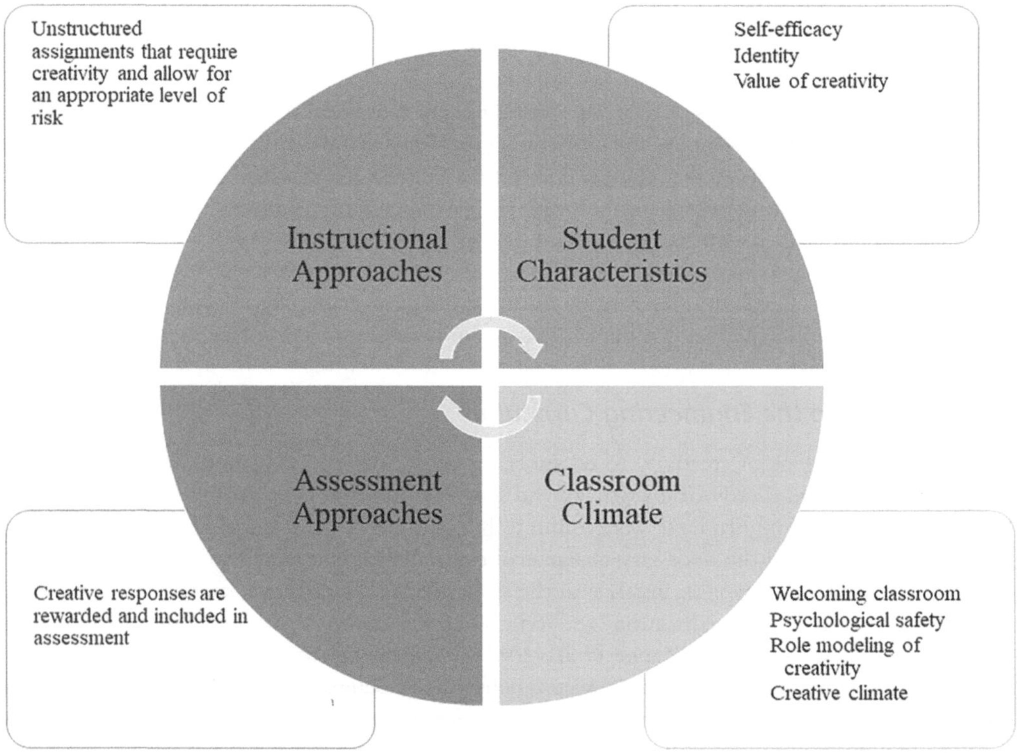

Figure 20.3 Person–environment model for creative outcomes in the classroom.

in creative tasks (Törnkvist, 1998; You, 2021). As Sternberg (2007) noted, students need to have the opportunity to engage in creativity, positive encouragement, and rewards for the demonstration of creative behavior.

All these classroom characteristics interact with student characteristics, as some students may have greater self-efficacy to engage in creative behavior (Tierney & Farmer, 2002), have stronger identities of being creative individuals (Jaussi et al., 2007), and value creativity more highly (Zappe & Tise, 2019; Waller & Strong, 2017). These characteristics and other factors, such as sense of belonging, gender, race, ethnicity, and international student status, are likely to impact creative outcomes; however, research on these in the engineering education context is extremely limited. Regardless of the student characteristics, however, instructional approaches not likely to produce creative outcomes include traditional lecture classes, where students do not engage with the class material, assignments that require algorithmic problem-solving, or projects that do not reinforce and reward creative behaviors (Beghetto, 2010). If conformity is what is assessed and rewarded, creative outcomes are not likely. As Beghetto states, "when it comes to assessing creativity, what you assess is typically what you get" (p. 453).

In the applied SOTL (scholarship of teaching and learning) engineering education literature, which includes some papers from the ASEE Annual Conferences, there are numerous examples on how instructors have integrated unstructured instructional approaches with the intention of promoting creativity in their students. While these papers were not helpful to examine for this chapter to investigate broad themes relating to research in creativity as a construct, a systematic literature review of the SOTL literature may provide further understanding on how creativity is being conceptualized by engineering instructors. While it is beyond the scope of this chapter to review these papers, future research might be worthwhile to determine (1) how engineering instructors have attempted to teach creativity, (2) how instructors have assessed the impact of their approaches, and (3) what impact the approaches have had on students' creativity. Additionally, investigation into co-curricular opportunities for students to be creative, such as Research Experiences for Undergraduates (REU) programs, may be another avenue for investigation (e.g., Huffstickler et al., 2017).

4 Integrating Entrepreneurship Into Engineering Education

4.1 Entrepreneurship Education

Interest in entrepreneurship education has grown significantly over the last several decades as governments seek to be more competitive in the global economy. Whereas big corporations dominated economic growth in the mid-20th century, new and small businesses are the driving force of economic growth and job creation in the 21st-century global economy (Neumark et al., 2011). In the late 1990s and early 2000s, small businesses created 60–80% of the net new jobs in the US economy, and most were the result of start-up companies in their first two years of operation (U.S. Small Business Administration, 2004). These innovative firms commercialized the radical innovative technologies that transform the way in which people do things and interact today.

The economic power of start-ups catalyzed interest in entrepreneurship education (Kuratko, 2009; Vesper & Gartner, 1997). Entrepreneurship education originated in US business schools (Katz, 2003; Vesper & Gartner, 1997), with the most documented instance of the first formal entrepreneurship course being attributed to Harvard Business School. The intent of Harvard's course was to support returning World War II veterans who were faced with fewer job prospects due to a collapsed weapons industry (Vesper & Gartner, 1997). Following the oil crises of the 1970s, entrepreneurship education became more common (Fayolle et al., 2016). By 2013, there were over 3,000 entrepreneurship courses offered around the world, and over 600 universities had created centers or institutes (Morris et al., 2013).

More recently, interest in teaching entrepreneurship has expanded beyond business schools to other academic disciplines (Wilson, 2008), including engineering (Gilmartin et al., 2014) and the arts (Essig & Guevara, 2016). In the United States, this was accelerated through the financial support of public and private foundations interested in the wide-scale adoption of entrepreneurship on university campuses (Blessing et al., 2008; Nnakwe et al., 2018; Torrance, 2013; Weilerstein & Shartrand, 2008; Huang-Saad et al., 2020). The Lemelson Foundation supported the creation of the National Collegiate Inventors and Innovators Alliance (NCIIA; Weilerstein & Shartrand, 2008), now referred to as VentureWell, which executes the foundation's vision for developing the next generation of innovators and inventors through programs, courses, and activities that promote involvement in technology commercialization and entrepreneurship (Weilerstein & Shartrand, 2008). Between 2003 and 2006, through the Kauffman Campus Initiative, the Kauffman Foundation concentrated their investments on interdisciplinary, entrepreneurship initiatives at 14 colleges and universities (Torrance, 2013). In 2005, the Kern Family Foundation focused attention on entrepreneurship in undergraduate engineering education with the establishment of the Kern Entrepreneurial Engineering Network (KEEN). KEEN focuses on cultivating an entrepreneurial mindset in students and broader outcomes beyond creating start-ups (Blessing et al., 2008; Wheadon & Duval-Couetil, 2017), This differs from Lemelson's focus on supporting a new generation of innovators through increasing access to strong technical and business skills (Weilerstein & Shartrand, 2008). In 2011 and 2012, the National Science Foundation made two significant investments in entrepreneurship, first, through the Stanford University Epicenter, which disseminated entrepreneurship education across engineering schools through practice and research (Epicenter, 2016), and second, through the

I-Corps program, which has become the largest entrepreneurship training program for academic researchers across the United States (Epstein et al., 2020).

International recognition of the role innovation and entrepreneurship play in economic development has also spurred global interest in entrepreneurship education. In 2000, the European Commission called for a greater focus on entrepreneurship to increase competitiveness (Wilson, 2008), and it has since advocated for the growth and formalization of entrepreneurship education in Europe (European Commission, 2015). In 2001, China's Ministry of Education piloted an entrepreneurship education program at nine universities (Li et al., 2003, Li & Li, 2015) and, in 2012, called specifically for innovation and enterprise education to become one of the most important directions for higher education. In the early 2000s, Chile's Economic Development Agency and National Council for Innovation and Competitiveness launched a comprehensive vision for promoting entrepreneurship in the country (Espinoza et al., 2019).

4.2 Why Entrepreneurship in Engineering Is Important

The growth of entrepreneurial training within engineering schools grew significantly in the early 2000s (Byers et al., 2005; Gilmartin et al., 2014; Shartrand et al., 2010), given recognition that the professional role of engineers was expanding (Rover, 2005; Yurtseven, 2002). As key players in the knowledge economy, it was recognized that engineers needed to be technically savvy but also able to address contemporary societal challenges by being entrepreneurial-minded (Brunhaver et al., 2018; Huang-Saad et al., 2018). There was evidence that recent STEM graduates were twice as likely to start a company than university faculty members (Åstebro et al., 2012) and able to generate significant economic impact (Roberts et al., 2015; Eesley & Miller, 2018). From an education perspective, this suggested that engineers benefitted not only from having deep technical knowledge focused on product development but also from entrepreneurship and professional skills that enable them to solve complex, interdisciplinary problems to play a role in the business of engineering (Royal Academy, 2007). These include the ability to identify opportunities, understand market trends and customer needs, advocate for ideas, lead teams, and have basic business literacy.

This broadening of competencies was reinforced by the Accreditation Board of Engineering and Technology (ABET) when it implemented Engineering Criteria (EC) 2000. Given concerns about the quality of engineering graduates, expressed by representatives of major US-based multinational companies, ABET shifted its criteria from educational inputs to outputs (Lucena et al., 2008). This outcomes-based evaluation process now accredits engineering programs at universities around the world and encompasses competencies viewed as pertinent to entrepreneurship and innovation (Duval-Couetil et al., 2013).

Modifications of accreditation standards have resulted in the growth of educational programs that deliver entrepreneurial knowledge, skills, and experiences to engineering students in the form of courses, minors, majors, and certificates, as well as co- and extracurricular activities. Some entrepreneurship initiatives are embedded within the engineering curriculum and target engineering students exclusively, while others are offered through campus-wide initiatives open to students in many majors. Course and program outcomes can vary widely, with some focused on raising awareness of entrepreneurship, developing an entrepreneurial mindset, or focusing more specifically on creating innovative products, technologies, business models, and ventures (Duval-Couetil et al., 2016).

There is some evidence these entrepreneurship education initiatives targeting engineering students are successful. A study at three US institutions found two-thirds of students felt entrepreneurship education could broaden their career prospects (Duval-Couetil et al., 2012). Those who took an entrepreneurship course had higher levels of self-efficacy and were more likely to get hands-on experience with market analysis, technology commercialization, and business communication and complete internships with start-up companies. Those who took one entrepreneurship course perceived significant increases in their knowledge of entrepreneurial terms and concepts, but those who took two or more courses increased their perceived ability to confidently perform entrepreneurial tasks. Interviews conducted with engineering alumni who had taken entrepreneurship courses and were in the professional world for 2–5 years reported entrepreneurship education helped them find employment, develop the ability to communicate, work with people from other disciplines, see the "big picture," and develop an entrepreneurial mindset (Duval-Couetil & Wheadon, 2013). A study of STEM students at two European universities also showed that participating in entrepreneurship programs increases student entrepreneurial intentions and attitudes (Souitaris et al., 2007). The most influential benefit was determined to be "inspiration."

4.3 What Is Entrepreneurship Education?

Approaches to teaching entrepreneurship have evolved, from informing students about entrepreneurial practices to emphasizing learning through doing. Scholars have differentiated methods for teaching and learning by categorizing them as learning "about," "for," or "through" entrepreneurship (Gibb, 2002; Hytti & O'Gorman, 2004; Lackéus, 2015; Rasmussen & Sørheim, 2006). Learning "about" entrepreneurship refers to learning the theory and concepts related to entrepreneurship. Learning "for" entrepreneurship involves learning the skills and knowledge to be entrepreneurial. Learning "through" entrepreneurship reflects the immersion of students in the entrepreneurial process. Some scholars consider entrepreneurship to be inherently experiential and, therefore, propose that it be used as a method, rather than content or a process (Neck et al., 2011). This pushes students to move beyond understanding and knowing, towards a more cognitive or "mindset" view of entrepreneurship education that is of interest to educators outside of business schools (Zappe et al., 2013a).

Equipping students with an entrepreneurial mindset is of particular interest to the engineering education community (McKenna et al., 2018), which is interested in the mindset pertinence and value to professional roles both within and outside of start-ups. This interest has led to several engineering education journal special issues and symposia dedicated to the scholarly exploration of entrepreneurship education from an engineering education viewpoint (*Advances in Engineering Education*, 2018, 7:1; *Entrepreneurship Education and Pedagogy*, 2020, 3:1). A broader mindset approach can encompass engineering education priorities, including design (Brunhaver et al., 2018) and experiential learning (Huang-Saad, 2009). Scholarship focused on understanding the relevance of entrepreneurship education to engineering (McKenna et al., 2018), as well as the Kern Family Foundation's significant financial investment in supporting educational institutions committed to integrating the KEEN entrepreneurial mindset framework into engineering curriculum continue to drive interest in this approach (London et al., 2018). Beyond KEEN's focus on developing entrepreneurial mindset through curiosity, connections, and creating value ("The 3Cs"), engineering scholars are exploring other constructs and approaches, for example, through ideation, open-mindedness, interest, altruism, empathy, help-seeking (Brunhaver et al., 2018).

4.4 Pedagogical Approaches for Entrepreneurship

Pedagogical approaches to entrepreneurship vary according to whether one is learning "about," "for," or "through" entrepreneurship (Falkäng & Alberti, 2000; Neck et al., 2011). Learning "about" entrepreneurship is passive and uses traditional didactic approaches, such as lectures and readings to reinforce specific knowledge (Kakouris & Liargovas, 2021). Teaching "for" and "through" entrepreneurship is more active and leverages experiential learning, where students practice entrepreneurial skills and develop competencies necessary to develop their own businesses. When students learn "through" entrepreneurship, they are immersed in the process and begin to develop entrepreneurial

	Instructional Activities	*Pedagogies*	*Entrepreneurial Competencies*
Competency ◄► Knowledge	Lectures Readings Role models/guest speakers Company visits	Didactic Learning through parables Field trips	Knowledge • Mental models • Declarative knowledge • Self-insight
	Case studies Simulations Group projects Co-curricular clubs and organizations	Problem-based learning Project-based learning Game-based learning Role-playing Collaborative learning Cooperative learning Social learning	Skills • Marketing skills • Resource skills • Opportunity skills • Interpersonal skills • Learning skills • Strategic skills • Critical thinking
	Business plan creation Working with entrepreneurs Consulting services Pitch competitions Mentorship by entrepreneurs Incubators Customer discovery Working with communities	Experiential learning Active learning Learning by doing Action learning Value creation Service learning	Through/Action • Entrepreneurial passion • Self-efficacy • Entrepreneurial identity • Proactiveness • Uncertainty/ambiguity • Tolerance • Innovativeness • Perseverance

Figure 20.4 Instructional examples of entrepreneurship education with respect to pedagogy and entrepreneurial competencies.

Source: Entrepreneurial KSAs adapted from Lackéus (2015).

attitudes, such as self-efficacy, proactiveness, and innovativeness (Lackéus, 2015). It should be noted that educational approaches to innovation and entrepreneurship education have been enhanced and informed by engineering-centric activities, including approaches to design (Woodcock et al., 2019; Huang-Saad, 2009; Schuelke-Leech, 2021) and the maker movement (Forest et al., 2014; Browder et al., 2017).

Figure 20.4 delineates common instructional activities in entrepreneurship education and the associated pedagogy for common entrepreneurial competencies in the context of knowledge, skills, and attitudes showing the range of knowledge-building capabilities associated with various learning activities. Beyond this, Gibb and Price's Compendium of Pedagogies for Teaching Entrepreneurship (2014) is a valuable resource for instructors seeking examples of how to operationalize the instructional activities and pedagogical practices outlined in Figure 20.4. Engineering education's history of design, hands-on practice, and learning sciences research (Froyd et al., 2012) has also influenced entrepreneurship and pedagogy directed at non-engineering audiences.

4.5 How Entrepreneurship Is Assessed

The real value of entrepreneurship education lies in the belief it will prepare graduates to thrive in a professional world, where bringing new ideas to market is highly rewarded either through the founding of new companies or generating value in established ones. As previously stated, there is some evidence that entrepreneurship education has a positive impact on engineering graduates through either entrepreneurial or intrapreneurial activities. However, it should be noted that several of the unique features of entrepreneurship education discussed previously also make assessing professional and educational outcomes particularly complex. These include the heterogeneity of program and curricular models (Duval-Couetil et al., 2016), teaching practices (Nabi et al., 2017), target audiences (Morris et al., 2013), and instructor backgrounds (Zappe et al., 2013b). This complexity is exacerbated with the expansion of entrepreneurship education to new disciplines, such as engineering, which are focused more on the entrepreneurial knowledge, skills, and mindsets that are transferable to a broad set of professional activities, rather than just the founding of start-ups (Sá & Holt, 2019). This leads to a lack of clear and measurable objectives, construct confusion, disconnects between theory and practice, different communities of practice (Huang-Saad et al., 2018; Zappe, 2018), all of which are difficult to manage when outcomes-based assessment is expected for accreditation (Froyd et al., 2012).

Regardless of discipline, there are two major challenges associated with evaluating the impact of entrepreneurship education: (1) the selection of evaluation criteria and (2) their effective measurement given the effect of time (Fayolle et al., 2006). Fundamentally, entrepreneurship is a professional outcome that can manifest at any time during one's career. While there are examples of students becoming entrepreneurs in college or immediately after graduation, research shows it is most common in one's early 40s (Azoulay et al., 2020), after acquiring domain expertise and professional experience. At a methodological level, establishing a causal relationship between an education intervention and a professional outcome is limited, given the abundance of confounding factors influencing career choices over time and challenges associated with conducting rigorous longitudinal research (Bauer, 2004; Yi & Duval-Couetil, 2021). Reviews of entrepreneurship education research have highlighted methodological limitations, including inferentially weak research designs, few validated assessment instruments, and weak statistical power, thereby offering minimal evidence of long-term impact (Bae et al., 2014; Martin et al., 2013; Nabi et al., 2017; Rideout & Gray, 2013). Measuring an engineering graduate's value generation within an existing organization (intrapreneurship) is even more challenging, given the lack of accessible measures.

Assessment has been an area of scholarly concentration since the introduction of entrepreneurship education in engineering curricula. Initially, assessment focused on the formative and summative assessment of learning objectives (Wise & Rzasa, 2004; Wang & Kleppe, 2001) or output metrics, such as enrolment, GPA, or retention (Gilmartin et al., 2016; Ohland et al., 2004). As entrepreneurship education became more widely adopted within engineering, efforts were made to capture how the educational experiences impacted learning (Shartrand et al., 2008) or familiarity with terms and concepts with respect to becoming an entrepreneur (e.g., finance and accounting, people and human resources, sales and marketing, and product ideation and development) (Shartrand et al., 2008; Besterfield-Sacre et al., 2012). Beyond knowledge and skills, engineering student attitudes, behaviors, self-efficacy, and perceptions of programs and faculty as they relate to entrepreneurship have also been explored (Duval-Couetil et al., 2012).

Questions surrounding research designs and valid measures linger (Zappe, 2018; Huang-Saad et al., 2018; Yi & Duval-Couetil, 2021). Programs are inconsistent in their definitions of entrepreneurship and desired outcomes (Gilmartin et al., 2016; Zappe et al., 2021). Without shared definitions and goals for entrepreneurship programs within engineering, assessment measures cannot be identified (Zappe, 2018), particularly when outcomes can vary from the intent to start a business to more psychosocial constructs associated with entrepreneurial mindset development (Huang-Saad et al., 2018). Engineering education scholars have reviewed the assessment literature, converging on similar findings and highlighting the inconsistencies in entrepreneurship assessment practice, including stated purposes, methods, and instruments (Da Silva et al., 2015; Purzer et al., 2016; Huang-Saad et al., 2018; Zappe et al., 2021). There have been efforts to develop rigorous instruments to assess the entrepreneurial mindset, such as with Brunhaver et al.'s (2018) entrepreneurial mindset assessment (ESEMA). However, the work is informed by existing frameworks rather than theory. This is not uncommon (Huang-Saad et al., 2018) and is similar to what is occurring in entrepreneurship education, more generally, where there are calls for better measures (Liñán & Chen, 2009; Maritz & Brown, 2013) informed by theory-driven research (Fayolle & Gailly, 2013; Fayolle et al., 2016) as well as rigorous experimental design (Yi & Duval-Couetil, 2021).

In light of these challenges, evaluating the impact of entrepreneurship education rests largely on self-reported, subjective measures that can be collected in the near term through surveys or interviews (e.g., intention to start a business, self-efficacy) rather than longer-term, more-objective ones (e.g., number of firms created and business performance; Nabi et al., 2017). A recent review of university programs found that entrepreneurship education outcomes fell into four categories: attitudes and motivations, knowledge and skills, behavioral and action-oriented outcomes, and actual start-up activities (Yi & Duval-Couetil, 2021). The authors found that the impacts of entrepreneurship education are most often gauged by increases in measures, such as entrepreneurial intention, motivation, and self-efficacy; however, results are mixed and include some negative effects (Dickson et al., 2008; Martin et al., 2013; Nabi et al., 2017), reinforcing the need for more rigorous research.

While subjective measures do demonstrate the value of entrepreneurship education to students, behaviors and actions that reflect actual entrepreneurial activity offer more accurate measures of impact because they reduce potential bias in respondents' self-reported ratings. However, in the review study cited earlier, only 4 out of 61 studies reviewed used objective measures, likely due to the challenges associated with conducting longitudinal research (e.g., low survey response rates, the time and resources required to track alumni, and long timelines for publishing). This means it is unclear how intending to become an entrepreneur relates to becoming or acting like one. To move the field forwards, scholars suggest examining outcomes in relation to context, pedagogical methods, and audiences; exploring emotion and mindset approaches (e.g., impacts on optimism and ambiguity tolerance); examining the relationship between intention and behavior through longitudinal research; and investigating competence-based pedagogical models (Nabi et al., 2017).

5 Integrating Leadership Into Engineering Education

5.1 Defining Leadership and Its Importance in Engineering

Companies employing engineers call on academe to prepare future engineers with the requisite knowledge, skills, and values to exhibit leadership in a dynamic workforce. Driven by human, technology, and societal interactions, the nature of engineering work is constantly evolving. Leadership is inextricably linked to successfully completing work which involves bringing out the best of people while navigating change, uncertainty in sociotechnical systems, and power dynamics. As technical innovations become more complex, practicing engineers need to navigate a dizzying array of sociotechnical complexities while contributing to more and larger projects with diverse stakeholders, costly-to-change deadlines, firm budgets, and defined quality standards. Engineers with leadership preparation that includes both technical and professional leadership-coupled competencies are better able to effectively handle human and technological challenges, resulting in positive outcomes for themselves, their companies, and society (Clegorne et al., 2021).

Motivated to develop leadership competencies in future engineers, engineering educators can draw upon theories of leadership and leadership development from the well-established field of leadership studies (Dinh et al., 2014). Given that conceptualizing leadership is context-dependent (Northouse, 2018), however, engineering educators first need an understanding of how leadership is defined in engineering. In their review of engineering research literature, Simmons et al. (2017) found no single definition of leadership and identified the dominant approach towards leadership as focused on the individual who leads a group of followers. This traditional, leader-centric approach contrasts with more contemporary approaches in leadership theory that emphasize a holistic lens and privilege group dynamics, changing roles, and integrated professional and technical competencies (Simmons et al., 2017, 2020). In explorations of leadership in engineering, as defined by industry leaders and professionals, researchers have noted an expanding view of leadership, beyond the traditional, individual-as-leader approach, suggesting that in practice, leadership in engineering may be in the process of being redefined (Clegorne et al., 2021; Garahan et al., 2020; Simmons et al., 2020).

To support the apparent shift to a more holistic view of leadership, the field of engineering has been encouraged to embrace leadership development over leader development (Simmons et al., 2017, 2021). Whereas leader development focuses on training an elite cadre of managers and supervisors, leadership development emphasizes leadership as a shared, context-dependent process with a focus on understanding and developing a culture of leadership (Heifetz et al., 2009; Jepson, 2009; Western, 2010). Specifically, the suggestion here is to de-emphasize developing traits of a leader and emphasize creating an environment in which a group of people share leadership. Simmons et al. (2017, p. 1) described the goal of this type of leadership education:

> Within the leadership paradigm, one seeks not to create the perfect leader, but rather to develop a culture of better teammates and role players who pass off leadership and followership as needed when context shifts, thus creating a team more resilient to adversity of contextual shifts.
>
> *(Robledo et al., 2012)*

5.2 Engineering Leadership Education

Engineering leadership education aims to teach students how to effectively handle human and technological challenges. When industry began calling for engineering graduates to not only have more professional competencies but also be provided with opportunities to gain these competencies through integration with technical content (ABET, n.d.), colleges of engineering responded in different ways. The characteristics of leadership programs vary across multiple measures, including the use of theory, access to courses, integration with technical content, and perspective on the nature of leadership on a continuum from leader-centric to shared. Founded in 1987, Tufts Gordon Institute is one of the oldest engineering leadership programs and offers courses leading to intensive and highly integrated minors and graduate degrees to undergraduate and graduate students (Gordon Institute, n.d.). For more than 25 years, Penn State's College of Engineering has offered the Engineering Leadership Development minor with courses theoretically informed by situational leadership (Schuhmann et al., 2015). Brigham Young's engineering leadership program (which began in 2011) has a required component for second-year engineering students using a shared leadership perspective. Conversely, Northwestern University's Field Study in Leadership course (which began in 1990) requires students to apply to be admitted to the course and frames content from a positional, leader-centered leadership perspective (Schuhmann et al., 2015). While the majority of engineering leadership education has been shown to be occurring in the United States (Graham et al., 2009; Khattak et al., 2012), a few programs with an explicit focus on leadership have been reported in Australia, Belgium, Germany, and Spain, with additional programs in Australia and Europe identified as having a non-explicit focus on leadership (Khattak et al., 2012).

Although many leadership development models exist, no significant models are rooted in engineering and reflect the important knowledge, skills, and attitudes that employers expect of engineering graduates. To fill this void and enable a culture of leadership, Simmons et al. (2020, 2021) found shared engineering leadership is enabled by the following competencies: adaptability, ambition/drive, assertiveness, big picture thinking, communication skills, computer skills, critical thinking/problem-solving, economic principles/trends, ethics/responsibility, humility, legal knowledge, management, people focus, professionalism, quality control, safety and risk management, self-awareness, teamwork, collaboration, networking, and time management. While taking on positional leadership experiences while in college can be helpful in gaining critical competencies, structured education is necessary to provide formal instruction and assessment.

5.3 Engineering Leadership Pedagogical Approaches

Despite industry placing increasing importance on leadership and growing recognition of the need for formal leadership instruction, engineering educators typically receive little to no formal training to teach leadership (Shulman, 2005). Engineering faculty, instead, point to work with advisory boards, interaction with professional contacts, their own experience in industry, leadership opportunities through their academic position, and personal experience as sources of leadership knowledge (Groen et al., 2018; Polmear et al., 2022). With a limited background in formal leadership training, engineering educators face the daunting task of preparing engineers for the complexities of a changing profession while also debating if professional competencies such as leadership can be taught, and if so, what approach (i.e., leader vs. leadership development) should be taken. In addition, educators face shrinking required credit hours for engineers to earn an undergraduate engineering degree, leaving little room for additional coursework. As such, integration of leadership across the curriculum holds promise to effectively achieve the desired learning outcomes around leadership. Leadership education is effective when taught within a context and an integrated teaching approach is aligned with many ABET student outcomes, most notably "an ability to function effectively on a team whose members together provide leadership, create a collaborative and inclusive environment, establish goals, plan tasks, and meet objectives" (ABET, n.d.).

An eco-leadership approach to leadership education in engineering can guide pedagogical approaches to support integration of leadership development across the existing curriculum (Garahan et al., 2020). Drawing from the theory of distributed leadership and the field of ecology, eco-leadership rejects a hierarchical view of leadership and emphasizes leveraging the talents and skills of all team members. Understanding leadership through an ecological approach is beneficial as it supports inclusion, strengthens the entire team and organization, and moves the field of engineering beyond traditional, leader-centric approaches. Table 20.1 provides a list of recommendations for teaching engineering leadership and displays suggested practices based on the contrasting approaches of traditional, leader-centric leadership and an eco-leadership approach.

5.4 How Is Engineering Leadership Education Assessed

Similar to what was discussed for entrepreneurship, there is no consistent assessment of engineering leadership development. As a result, engineering educators cannot know if students are evolving in their perspectives and development of leadership competencies throughout college without naming, integrating, and assessing these competencies. Leadership studies and leadership development in other fields can provide possible assessment models. For example, the leadership identity development (LID) theory and model (Komives et al., 2005, 2006, 2009) offers an understanding of how college students progress in their leadership identity formation across six stages as shaped by personal, group, and developmental influences.

In addition to developing and incorporating leadership assessment into engineering education, engineering educators can also help reframe ideas of leadership and leadership development among students, higher education administrators, industry professionals, and other engineering education stakeholders. The differentiation between leader development and leadership development is critical and can be supported through pedagogy, as described earlier. Additionally, another aspect of the engineering culture regarding leadership is related to norms around professional formation.

Table 20.1 Recommendations for Teaching Engineering Leadership

Recommendation	*Leader-Centric Approach*	*Eco-Leadership Approach*
Divorce the concepts of leadership and leader and highlight the distinction between the two.	Focus on the skills, attributes, and behaviors of individual leaders and how they are able to influence, persuade, or control their subordinates.	Focus on the skills, attributes, and behaviors of individual leaders and how they are able to influence, persuade, or control their subordinates.
Emphasize the interconnectedness between leadership and teamwork.	Highlight the behavior and role of team leaders.	Focus on the fluidity and social context of certain projects. Explore different team roles and team dynamics. Consider personal attributes that are associated with different types of team roles.
Highlight the eco-ethical approach inherent in the discussion of sustainability by connecting outcomes of sustainability and leadership.	Emphasize the skills and behaviors of leaders that enable them to engage others.	Explore the broader contexts (e.g., economic, social, environmental) of a project and its common vision. Consider how trust is developed and fostered within a team and its community partners. Examine how aspects of a project and roles of team members and leaders interrelate.

Source: Adapted from Garahan et al. (2020).

Engineering students are often not exposed to the idea of professional formation as a lifelong process that is largely in the hands of the individual. With a clearer vision of the practice of engineering, students can put undergraduate education in its proper place – an initial introduction to the field – and recognize it is not a full preparation for the vastness of areas of practice with its intricate specialties, unique and evolving context, wicked problems, and the needs of people.

6 Discussion

As the world has continued to become more complex, so have the demands on the educational system preparing today's students. Unfortunately, the ability of educational institutions to keep pace with these changes and demands for reform has lagged. To date, most reform in engineering education has been implemented in a piecemeal fashion, adding to an already-intensive, theory-based math- and science-based curriculum (Froyd et al., 2012). This is evident in how calls for more creativity, entrepreneurship, and leadership have been operationalized in engineering curriculum, often relying on design classes, electives, or co-curricular activities to fulfill these needs. While each domain has been lauded as an important part of engineering curriculum, each domain has its own history and implementation in an already-overburdened curriculum. At the same time, each domain is so broad that effective means of assessment continue to be debated in the literature. We use this as an opportunity to encourage the community to consider a more comprehensive approach to engineering education reform for the 21st century. In doing so, we present six main reflections.

6.1 Reflection No. 1: We Are Not Effectively Preparing Students for Their Future Careers in a Changing Global Environment

While it is commonly accepted engineers will play a critical role in economic development as future innovators, approaches as to how we should train students in the innovation process and to what outcomes are still uncertain and inconsistent. This largely stems from the fact that the constructs of creativity, entrepreneurship, and leadership are challenging to define, with debates existing about how to operationalize each construct (e.g., Pichot et al., 2020; Plucker et al., 2004; Simmons et al., 2017; Zappe, 2018). All are complex, often overlapping with each other and with other constructs. For example, the entrepreneurial mindset, a term advocated heavily by KEEN, is often operationalized by using different attributes that include creativity, leadership, and opportunity recognition, among others. Zappe argues that the entrepreneurial mindset is not a construct by itself but rather subsumes other constructs (Zappe, 2018). In a systematic review of research on the impact of STEM entrepreneurship programs, Zappe et al. (2021) found entrepreneurship programs focused on a variety of intended outcomes, including entrepreneurial mindset, teamwork, communication, creativity, failure, and risk. In a critical review of the literature to examine how leadership is taught in undergraduate engineering programs, Simmons et al. (2017) found there is no common theoretical approach informing leadership (instruction) in engineering. This inconsistent use of theory is also evident in engineering entrepreneurship education as well (Huang-Saad et al., 2018). Because of the difficulty with operationalizing the constructs, all are also difficult to measure and assess (e.g., Barbot et al., 2019; Said-Metwaly et al., 2017). Due to the challenges with measurement, indirect methods, such as self-assessments, are often used, which have inherent limitations. In addition to the challenges in measuring creativity, entrepreneurship, and leadership, the terms are often used in everyday contexts, resulting in significant misconceptions.

6.2 Reflection No. 2: There Are Many Reasons That Creativity, Leadership, and Entrepreneurship Are Not Integrated Into the Curriculum

Cropley (2017) theorizes engineering uses a reductionist tradition which focuses on understanding phenomena in terms of its parts rather than as a whole. As a result, Cropley argues that "[e]ngineering curricula continue to focus on traditional topics, taught in traditional ways, and these make little room for the creativity that almost everyone agrees is critical to engineering education" (p. 215). Another potential reason for the lack of emphasis on these domains in engineering may be related to the many myths associated with each. For example, some of the major myths associated with creativity include (1) creativity is innate and cannot be improved; (2) creativity is related to negative aspects of psychology, such as nonconformity, drug use, and mental illness; and (3) creativity is a "fuzzy, soft" construct related only to the arts or music (Plucker et al., 2004). Other hypotheses include that the organizational climate of engineering colleges, the classroom environment of many engineering courses, and the most often used instructional techniques are not conducive for eliciting creative behaviors or requiring leadership skills. As Cropley (2015) noted, most engineering programs emphasize "convergent, analytic work, and passive knowledge acquisition" (p. 163).

Each of the three domains explored here has its own challenges that may be barriers into further integration into engineering. One of these challenges is the difficulty arriving at operationalized definitions for these constructs, as all three are very complex and overlap with constructs from psychology or other fields. Within engineering education, the lack of consensus in defining each construct can lead to widely divergent intended outcomes and goals across programs as well as challenges relating to measurement and assessment. Another challenge is that each domain is saddled with common misconceptions which can lead to instructional and institutional barriers. Another challenge is that most engineering faculty are not knowledgeable about these or perhaps feel that they do not belong in the core engineering curriculum.

6.3 Reflection No. 3: Lack of Change in the Curriculum Can Have Negative Consequences

Regardless of the reasons for the lack of integration of these professional skills into the curriculum, the consequences can be damaging. Creativity drives innovative solutions. Engineers need to learn how to identify opportunities and advocate for their ideas. And it is only the technology developers who can lead the process. Without contextualizing engineering education in the innovation process, future innovators will struggle with the humanistic responsibility they have with the technological solutions that they will create. At the same time, without acknowledging the non-technical aspects of engineering, we are deliberately narrowing the pipeline for future engineers. As an example, students who are drawn to creativity may not enter engineering programs or fail to persist. In fact, this finding is supported by Atwood and Pretz (2016), who found a negative relationship between students' creative self-efficacy and persistence in engineering programs. If we desire to create an engineering student body and workforce that is diverse, inclusive to all, and prepared to tackle our world's major problems, perhaps it is time for engineering as a discipline and an education system to change.

6.4 Reflection No. 4: We Need More Research in How to Teach for and Assess Creativity, Leadership, and Entrepreneurship

A consistent theme in the literature across all domains is the lack of or limited use of educational theory in exploring the pedagogies used to promote these domains in the classroom. While engineering education has leveraged experiential learning, largely through design, outcomes and assessment of outcomes have been limited. Linkages to theoretical and practical work in other disciplines can be helpful to guide research in engineering education. Investigations on the development of these domains in students are also an area where research is needed.

Within each of the domains, the field has a need for systematic reviews of the SOTL literature on how faculty are teaching for creativity, entrepreneurship, and leadership in engineering. What are the general approaches being taken, and how effective are these? A systematic review of instructional approaches in the applied engineering education (SOTL) literature as well as in the design literature would reveal patterns of approaches being used to teach for creativity, leadership, and entrepreneurship. Additional research topics include better understanding of the relationship between each domain and diversity, equity, and inclusion. What is the relationship between pedagogical approaches used to teach creativity, leadership, and entrepreneurship and those intended to promote inclusion and belonging? If technical engineering classes are taught with these domains in mind, could students' sense of belonging or engineering identity increase? Could we draw in a more diverse student population if the field were structured to bring in more experiences intended to elicit creativity or leadership?

Additional theory-based research as well as investigations into the applied SOTL literature in engineering education would be helpful to better understand the current landscape as well as future directions.

Relatedly, as mentioned repeatedly earlier, assessment is a challenge across all three domains. Guidance on how to assess each construct as well as access to psychometrically sound, valid instruments are needed in the engineering education community.

6.5 Reflection No. 5: Engineering Instructors Need to Be Provided with Ideas and Approaches for Integrating Creativity, Entrepreneurship, and Leadership Into an Already-Full Curriculum

One of the major challenges of integrating these domains into the engineering curriculum is that it is already so full. It is difficult to squeeze in additional topics when students' schedules are so full and course objectives aim to include so much. We argue that we need to consider structuring our courses and curriculum in a new way. At the less-challenging end, we can integrate entrepreneurship and leadership into design courses. At the more challenging end, a curricular or programmatic rehauling to remove courses that no longer serve students and to create a new program of study that more fully prepares students for future careers. For example, we can consider removing courses that have historically served as gatekeepers to engineering but are no longer serving a purpose in preparing students for our modern society. Another approach could be weaving design into all engineering courses. Work in design courses can be integrated with other technical courses students are taking concurrently, which could give students the necessary agency to succeed in more challenging courses. Restructuring the engineering curriculum would allow space for topics such as creativity, leadership, and entrepreneurship that are critical for our engineering graduates. We also need to encourage administrators and faculty to partner with students to create the curriculum. Students need to have a voice in this path rather than having curricular decisions made solely by the university.

6.6 Reflection No. 6: We Have No Choice but to Reconsider Our Approach

Sorby et al. (2021) described the current state of engineering education as "stuck in 1955," which is not structured to meet the needs of a changing world. The authors argue for a "revolution" in engineering education that is more motivating and more inclusive to all students. As they note:

> Not only are engineering curricula often unattractive to women and students of color, but they also fail to prepare all students for their future careers. How many creative problem-solvers, who would have become excellent engineers, have been driven from our programs over the years? How many potential inventors and entrepreneurs have not been inspired to join our ranks? How many out-of-the-box thinkers have been lost from engineering due to the rigidity of the engineering curricula? The true loss of human talent from engineering disciplines is impossible to calculate.

After reviewing the current state of engineering education literature, we argue that creativity, entrepreneurship, and leadership need to be better integrated into the curriculum. Not only will this integration better prepare students for their future careers, but including these perspectives may also help broaden who is interested in and motivated to study engineering.

References

Abdulla Alabbassi, A. M., Tadik, H., Acar, S., & Runco, M. A. (2021). Birth order and divergent thinking: A meta-analysis. *Creative Research Journal, 33*(4), 331–346.

ABET. Accreditation Policy and Procedure Manual (APPM), 2021–2022. (n.d.). Retrieved December 19, 2021, from www.abet.org/accreditation/accreditation-criteria/criteria-for-accrediting-engineering-programs-2021-2022/.

Advances in Engineering Education. (2018). 7(1), entire issue.

Åstebro, T., Bazzazian, N., Braguinsky, S., Asstebro, T., Bazzazian, N., & Braguinsky, S. (2012). Startups by recent university graduates and their faculty: Implications for university entrepreneurship policy. *Research Policy, 41*(4), 663–677. http://dx.doi.org/10.1016/j.respol.2012.01.004

Atwood, S. A., & Pretz, J. E. (2016). Creativity as a factor in persistence and academic achievement of engineering undergraduates. *Journal of Engineering Education, 105*(4), 540–559.

Azoulay, P., Jones, B. F., Kim, J. D., & Miranda, J. (2020). Age and high-growth entrepreneurship. *American Economic Review: Insights, 2*(1), 65–82.

Bae, T. J., Qian, S., Miao, C., & Fiet, J. O. (2014). The relationship between entrepreneurship education and entrepreneurial intentions: A meta – analytic review. *Entrepreneurship Theory and Practice, 38*(2), 217–254.

Barbot, B., Hass, R. W., & Reiter-Palmon, R. (2019). Creativity assessment in psychological research: (Re)Setting the standards. *Psychology of Aesthetics, Creativity, and the Arts, 13*(2), 233–240. doi:10.1037/aca0000233

Barile, S., Saviano, M., & Simone, C. (2015). Service economy, knowledge, and the need for T-shaped innovators. *World Wide Web, 18*(4), 1177–1197.

Bauer, K. W. (2004). Conducting longitudinal studies. *Cross-Cultural Research*, 75–90.

Becker, M. (1995). 19th century foundations of creativity research. *Creativity Research Journal, 8*, 219–229.

Beghetto, R. A. (2010). Creativity in the classroom. In J. C. Kaufman & R. J. Sternberg (Eds.), *The Cambridge handbook of creativity* (pp. 447–463). Cambridge University Press.

Besterfield-Sacre, M. E., Robinson, A. M., Ozaltin, O., Shuman, L. J., Shartrand, A. M., & Weilerstein, P. (2012). Essential factors related to entrepreneurial knowledge in the engineering curriculum. In *Proceedings of the ASEE annual conference and exposition.*

Blessing, J., Mekemson, K., & Pistrui, D. (2008). Building an entrepreneurial engineering ecosystem for future generations: The Kern Entrepreneurship Education Network. In *Proceedings of the ASEE annual conference and exposition.*

Browder, R. E., Aldrich, H. E., & Bradley, S. W. (2017). Entrepreneurship research, makers, and the maker movement. *Academic Management Proceedings, 1*, 14361.

Brunhaver, S. R., Bekki, J. M., Carberry, A. R., London, J. S., & McKenna, A. F. (2018). Development of the engineering student entrepreneurial mindset assessment (ESEMA). *Advances in Engineering Education*, 7(1), 1–12.

Byers, T., Seelig, T., Sheppard, S., & Weilerstein, P. (2005). Entrepreneurship: Its role in engineering education. *The Bridge*, 35–40.

Charyton, C., & Merrill, J. A. (2009). Assessing general creativity and creative engineering design in first year engineering students. *Journal of Engineering Education, 98*(2), 145–156. https://doi.org/10.1002/j.2168-9830.2009.tb01013.x

Clegorne, N. A., Simmons, D. R., & McCall, C. J. (2021). A Delphi study of industry perspectives. *Journal of Leadership Education, 20*(1), 46–61.

Cropley, D. H. (2015). Promoting creativity and innovation in engineering education. *Psychology of Aesthetics, Creativity, and the Arts, 9*(2), 161–171.

Cropley, D. H. (2017). Nurturing creativity in the engineering classroom. In R. A. Beghetto & J. C. Kaufman (Eds.), *Nurturing creativity in the classroom.* Cambridge University Press.

Da Silva, G. B., Costa, H. G., & de Barros, M. D. (2015). Entrepreneurship in engineering education: A literature review. *International Journal of Engineering Education, 31*(6), 1701–1710.

Denson, C. D., Buelin, J. K., Lammi, M. D., & D'Amico, S. (2015). Developing instrumentation for assessing creativity in engineering design. *Journal of Technology Education, 27*(1), 23–40.

Dickson, P. H., Solomon, G. T., & Weaver, K. M. (2008). Entrepreneurial selection and success: Does education matter? *Journal of Small Business and Enterprise Development, 15*(2), 239–258. doi:10.1108/14626000810871655

Dinh, J. E., Lord, R. G., Gardner, W. L., Meuser, J. D., Liden, R. C., & Hu, J. (2014). Leadership theory and research in the new millennium: Current theoretical trends and changing perspectives. *The Leadership Quarterly, 25*(1), 36–62.

Duval-Couetil, N. (2013). Assessing the impact of entrepreneurship education programs: Challenges and approaches. *Journal of Small Business Management, 51*(3), 394–409. http://doi.wiley.com/10.1111/jsbm.12024

Duval-Couetil, N., Reed-Rhoads, T., & Haghighi, S. (2012). Engineering students and entrepreneurship education: Involvement, attitudes and outcomes. *International Journal of Engineering Education, 28*(2), 425–435.

Duval-Couetil, N., Shartrand, A., & Reed-Rhoads, T. (2016). The role of entrepreneurship program models and experiential activities on engineering student outcomes. *Advances in Engineering Education, 5*(1), 1–27.

Entrepreneurship Education and Pedagogy. (2020) *3*(1), entire issue.

Duval-Couetil, N., & Wheadon, J. (2013). The value of entrepreneurship to recent engineering graduates: A qualitative perspective. In *Proceedings of the frontiers in education conference*.

Duval-Couetil, N., Wheadon, J., Kisenwether, E., & Tranquillo, J. (2013). Entrepreneurship and ABET accreditation: How and where does it fit? In *Proceedings of the frontiers in education conference*.

Eesley, C. E., & Miller, W. F. (2018). Impact: Stanford University's economic impact via innovation and entrepreneurship. *Foundations and Trends in Entrepreneurship*, *14*(2), 130–278.

Epicenter. (2016). *National center for engineering pathways to innovation*. http://epicenter.stanford.edu/

Epstein, A., Huang-Saad, A., & Duval-Couetil, N. (2020). Faculty and trainee perceptions of NSF I-corps technology commercialization training. *IEEE Enginering Management Review*, *49*(1), 97–105.

Espinoza, C., Mardones, C., Sáez, K., & Catalán, P. (2019). Entrepreneurship and regional dynamics: The case of Chile. *Entrepreneurship & Regional Development*, *31*(9–10), 755–767. https://doi.org/10.1080/08985626.2019.1565421

Essig, L., & Guevara, J. (2016). *A landscape of arts entrepreneurship in US higher education*. Pave Program in Arts Entrepreneurship. https://herbergerinstitute.asu.edu/sites/default/files/a_landscape_of_arts_entrepreneurship_in_us_higher_education_0.pdf

European Commission. (2015). *Entrepreneurship education: A road to success*. Publications Office of The European Union.

Falkäng, J., & Alberti, F. (2000). The assessment of entrepreneurship education. *Industry and Higher Education*, *14*(2), 101–108.

Fayolle, A., & Gailly, B. B. (2013). The impact of entrepreneurship education on entrepreneurial attitudes and intention: Hysteresis and persistence. *Journal of Small Business Management*, *53*(1), 1–19.

Fayolle, A., Gailly, B. B., & Lassas-Clerc, N. (2006). Assessing the impact of entrepreneurship education programmes: A new methodology. *Journal of European Industrial Training*, *30*, 701–720.

Fayolle, A., Verzat, C., & Wapshott, R. (2016). In quest of legitimacy: The theoretical and methodological foundations of entrepreneurship education research. *International Small Business Journal: Researching Entrepreneurship*, *34*(7), 895–904.

Forest, C. R., Moore, R. A., Jariwala, A. S., Fasse, B. B., Linsey, J., Newstetter, W., et al. (2014). The invention studio: A university maker space and culture. *Advances in Engineering Education*, *4*(2), 1–32.

Froyd, J. E., Wankat, P. C., & Smith, K. A. (2012). Five major shifts in 100 years of engineering education. *Proceedings of the IEEE*, *100*, 1344–1360.

Garahan, K. L., Clegorne, N. A., & Simmons, D. R. (2020). Understanding leadership through an ecological lens: A rhetorical cluster analysis of the civil engineering body of knowledge. *International Journal of Engineering Education*, *36*(1A), 130–141.

Gibb, A. (2002). In pursuit of a new "enterprise" and "entrepreneurship" paradigm for learning: Creative destruction, new values, new ways of doing things and new combinations of knowledge. *International Journal of Management Reviews*, *4*(3), 233–269. http://doi.wiley.com/10.1111/1468-2370.00086

Gibb, A., & Price, A. A. (2014). Compendium of pedagogies for teaching entrepreneurship. *International Entrepreneurship Educators Programme*. National Centre for Entrepreneurship in Education.

Gilmartin, S. K., Chen, H. L., & Estrada, C. (2016). Investigating entrepreneurship program models in undergraduate engineering education. *International Journal of Engineering Education*, *32*(5), 2048–2065.

Gilmartin, S. K., Shartrand, A., Chen, H. L., Estrada, C., & Sheppard, S. (2014). *US-based entrepreneurship programs for undergraduate engineers: Scope, development, goals, and pedagogies*. Epicenter Technical Brief.

Glăveanu, V. P., & Beghetto, R. (2021). Creative experience: A non-standard definition of creativity. *Creativity Research Journal*, *33*(2), 75–80.

Gordon Institute. (n.d.). Retrieved December 19, 2021, from https://gordon.tufts.edu/

Graham, R., Crawley, E., & Mendelsohn, B. R. (2009). *Engineering leadership education: A snapshot review of international good practice*. White paper sponsored by the Bernard M. Gordon-MIT Engineering Leadership Program.

Groen, C. J., Simmons, D. R., & Clegorne, N. A. (2018). Faculty ways of knowing, valuing and assessing leadership in the undergraduate engineering curriculum. In *Proceedings of the ASEE annual conference and exposition*.

Heifetz, R. A., Grashow, A., & Linsky, M. (2009). *The practice of adaptive leadership: Tools and tactics for changing your organization and the world*. Harvard Business Press.

Howard, T. J., Culley, S. J., & Dekoninck, E. (2008). Describing the creative design process by the integration of engineering design and the cognitive psychology literature. *Design Studies*, *29*(2), 160–180.

Huang-Saad, A. (2009). Fostering the entrepreneurial mindset in the engineering classroom. In *Proceedings of the frontiers in education conference*.

Huang-Saad, A., Bodnar, C., & Carberry, A. R. (2020). Examining current practice in engineering entrepreneurship education. *Entrepreneurship Education and Pedagogy*, *3*(1), 5–14.

Huang-Saad, A., Morton, C. S., & Libarkin, J. C. (2018). Entrepreneurship in higher education: A research review for engineering education researchers. *Journal of Engineering Education*, *107*(2), 263–290.

Huffstickler, M., Zappe, S. E., Manning, K. B., & Slattery, M. J. (2017). Impact of a biomedical engineering undergraduate research program on student and faculty perceptions of creativity. In *Proceedings of the ASEE annual conference and exposition*.

Hytti, U., & O'Gorman, C. (2004). What is "enterprise education"? An analysis of the objectives and methods of enterprise education programmes in four European countries. *Education + Training*, *46*(1), 11–23.

Jaussi, K. S., Randel, A. E., & Dionne, S. D. (2007). I am, I think I can, and I do. The role of personal identity, self-efficacy, and cross-application of experiences in creativity at work. *Creativity Research Journal*, *19*(2–3), 247–258.

Jepson, D. (2009). Studying leadership at cross-country level: A critical analysis. *Leadership*, *5*(1), 61–80. https://doi.org/10.1177/1742715008098310

Kakouris, A., & Liargovas, P. (2021). On the about/for/through framework of entrepreneurship education: A critical analysis. *Entrepreneurship Education and Pedagogy*, *4*(3), 396–421.

Katz, J. A. (2003). The chronology and intellectual trajectory of American entrepreneurship education 1876–1999. *Journal of Business Venturing*, *18*(2), 283–300.

Kazerounian, K., & Foley, S. (2007). Barriers to creativity in engineering education: A study of instructors and students perceptions. *Journal of Mechanical Design*, *129*, 761–768.

Khattak, H., Ku, H., & Goh, S. (2012). Courses for teaching leadership capacity in professional engineering degrees in Australia and Europe. *European Journal of Engineering Education*, *37*(3), 279–296.

Komives, S. R., Longerbeam, S. D., Mainella, F., Osteen, L., Owen, J. E., & Wagner, W. (2009). Leadership identity development. *Journal of Leadership Education*, *8*(1), 11–47. https://doi.org/10.1353/csd.2006.0048

Komives, S. R., Longerbeam, S. D., Owen, J. E., Mainella, F. C., & Osteen, L. (2006). A leadership identity development model: Applications from a grounded theory. *Journal of College Student Development*, *47*(4), 401–418. https://doi.org/10.1353/csd.2006.0048

Komives, S. R., Owen, J. E., Longerbeam, S. D., Mainella, F. C., & Osteen, L. (2005). Developing a leadership identity: A grounded theory. *Journal of College Student Development*, *46*(6), 593–611. https://doi.org/10.1353/csd.2005.0061

Kryshtanovych, M., Bilyk, V., Hanushchyn, S., Sheremet, I., & Vasylenko, K. (2021). Modelling the ways to increase the creativity of psychology students as a basic factor in professional development. *Creativity Studies*, *14*(1), 34–50. https://doi.org/10.3846/cs.2021.12571

Kuratko, D. F. (2009). The entrepreneurial imperative of the 21st century. *Business Horizons*, *52*(5), 421.

Lackéus, M. (2015). *Entrepreneurship in Education: What, why, when, how*. Retrieved October 27, 2022, from www.oecd.org/cfe/leed/BGP_Entrepreneurship-in-Education.pdf

Li, J., Zhang, Y., & Matlay, H. (2003). Entrepreneurship education in China. *Education and Training*, *45*(8–9), 495–505.

Li, W., & Li, C. (2015). Entrepreneurship education in China. In J. C. Sanchez-Garcia (Ed.), *Entrepreneurship education and training* (pp. 27–37). IntechOpen.

Liñán, F., & Chen, Y. W. (2009). Development and cross-cultural application of a specific instrument to measure entrepreneurial intentions. *Entrepreneurship Theory and Practice*, *33*(3), 593–617.

Litzinger, T., Zappe, S., Hunter, S., & Mena, I. (2015). Increasing integration of the creative process across engineering curricula. *International Journal of Engineering Education*, *31*(1B), 335–342.

London, J. S., Bekki, J. M., Brunhaver, S. R., Carberry, A. R., & McKenna, A. F. (2018). A framework for entrepreneurial mindsets and behaviors in undergraduate engineering students: Operationalizing the Kern Family Foundation's "3Cs." *Advances in Engineering Education*, 7(1), 1–12.

Lucena, J., Downey, G., Jesiek, B., & Elber, S. (2008). Competencies beyond countries: The re-organization of engineering education in the United States, Europe and Latin America. *Journal of Engineering Education*, *97*(4), 433–447.

Maritz, A., & Brown, C. R. (2013). Illuminating the black box of entrepreneurship education programs. *Education + Training*, *55*(3), 234–252.

Martin, B. C., McNally, J. J., & Kay, M. J. (2013). Examining the formation of human capital in entrepreneurship: A meta-analysis of entrepreneurship education outcomes. *Journal of Business Venturing*, *28*(2), 211–224.

McKenna, A., Lichtenstein, G., Weilerstein, P., & Monroe-White, T. (2018). The entrepreneurial mindset: Using the questions of what, why, and how as an organizing framework. *Advances in Engineering Education*, 7(1), 1–8.

Mejia, C., D'Ippolito, B., & Kajikawa, Y. (2021). Major and recent trends in creativity research: An overview of the field with the aid of computational methods. *Creativity and Innovation Management*, *30*(3), 475–497. https://doi.org/10.1111/caim.12453

Mi, S., Bi, H., & Lu, S. (2020). Trends and foundations of creativity research in education: A method based on text mining. *Creativity Research Journal*, *32*(3), 215–227.

Miller, S. R., Hunter, S. T., Starkey, E., Ramachandran, S., Ahmed, F., & Fuge, M. (2021). How should we measure creativity in engineering design? A comparison between social science and engineering approaches. *Journal of Mechanical Design*, *143*(3). https://doi.org/10.1115/1.4049061

Morris, M., Kuratko, D., & Cornwall, J. (2013). *Entrepreneurship programs and the modern university*. Edward Elgar Pub.

Mumford, M. D., Medeiros, K. E., & Partlow, P. J. (2012). Creative thinking: Processes, strategies, and knowledge. *The Journal of Creative Behavior*, *46*(1), 30–47.

Mumford, M. D., Mobley, M. J., Reiter-Palmon, R., Uhlman, C. E., & Doares, L. M. (1991). Process analytic models of creative capacities. *Creativity Journal*, *4*(2), 91–122.

Nabi, G., Liñán, F., Fayolle, A., Krueger, N., & Walmsley, A. (2017). The impact of entrepreneurship education in higher education: A systematic review and research agenda. *Academy of Management Learning & Education*, *16*(2), 277–299.

National Academy of Engineering. (2005). *The engineer of 2020: Visions of engineering in the new century*. The National Academies Press. https://doi.org/10.17226/10999

National Academy of Sciences, National Academy of Engineering, & Institute of Medicine. (2007). *Rising above the gathering storm: Energizing and employing America for a brighter economic future*. The National Academies Press. https://doi.org/10.17226/11463

National Academy of Sciences, National Academy of Engineering, & Institute of Medicine. (2010). *Rising above the gathering storm: Revisited*. The National Academies Press.

Neck, H. M., Greene, P. G., Branson, R., & Ash, M. K. (2011). Entrepreneurship education: Known worlds. *Journal of Small Business Management*, *49*(1), 55–70.

Neumark, D., Wall, B., & Zhang, J. (2011). Do small businesses create more jobs? New evidence for the United States from the national establishment time series. *Review of Economics and Statistics*, *93*(1), 16–29.

Nnakwe, C., Cooch, N., & Huang-Saad, A. (2018). Investing in academic technology innovation and entrepreneurship: Moving beyond research funding through the NSF I-corps program. *Technology & Innovation*, *19*(4), 773–786.

Northouse, P. G. (2018). *Leadership: Theory and practice*. Sage Publishing.

Ohland, M. W., Frillman, S. A., Zhang, G., Brawner, C. E., & Miller, III T. K. (2004). The effect of an entrepreneurship program on GPA and retention. *Journal of Engineering Education*, *94*(4), 293–301.

Pichot, N., Bonetto, E., Pavani, J., Arciszewski, T., Bonnardel, N., & Weisberg, R. W. (2020). The construct validity of creativity: Empirical arguments in favor of novelty as the basis for creativity. *Creativity Research Journal*, *34*(1), 2–13. https://doi.org/10.1080/10400419.2021.1997176

Plucker, J. A., Behgetto, R. A., & Dow, G. T. (2004). Why isn't creativity more important to educational psychologists? Potentials, pitfalls, and future directions in creativity research. *Educational Psychologist*, *39*(2), 83–96.

Polmear, M., Volpe, E., Simmons, D. R., Clegorne, N., & Weisenfeld, D. (2022). Leveraging faculty knowledge, experience, and training for leadership education in engineering undergraduate curricula. *European Journal of Engineering Education*, 1–20.

Purzer, S., Fila, N. D., & Nataraja, K. (2016). Evaluation of current assessment methods in engineering entrepreneurship education. *Advances in Engineering Education*, *5*(1), 1–13.

Rasmussen, E., & Sørheim, R. (2006). Action-based entrepreneurship education. *Technovation*, *26*(2), 185–194.

Rideout, E. C., & Gray, D. O. (2013). Does entrepreneurship education really work? A review and methodological critique of the empirical literature on the effects of university-based entrepreneurship education. *Journal of Small Business Management, 51*(3), 329–351.

Roberts, E. B., Murray, F., & Kim, J. D. (2015). Entrepreneurship and innovation at MIT: Continuing global growth and impact – an updated report. *Foundations and Trends in Entrepreneurship, 15*(1), 1–55.

Robledo, I. C., Peterson, D. R., & Mumford, M. D. (2012). Leadership of scientists and engineers: A three-vector model. *Journal of Organizational Behavior, 33*(1), 140–147.

Royal Academy of Engineering. (2007). *Creating systems that work*. Royal Academy of Engineering.

Royal Academy of Engineering. (2013). *Educating engineers for the 21st century*. Retrieved October 27, 2022, from www.raeng.org.uk/publications/reports/educating-engineers-21st-century

Rover, D. T. (2005). New economy, new engineer. *Journal of Engineering Education, 94*(4), 427–428.

Runco, M. A., & Albert, R. S. (2010). Creativity research: A historical view. In J. C. Kaufman & R. J. Sternberg (Eds.), *Cambridge handbook of creativity*. Cambridge University Press.

Sá, C., & Holt, C. (2019). Profiles of entrepreneurship students: Implications for policy and practice. *Education + Training, 61*(2), 122–135.

Said-Metwaly, S., Van den Noortgate, W., & Kyndt, E. (2017). Methodological issues in measuring creativity: A systematic literature review. *Creativity: Theory-Research-Applications, 4*(2), 276–301.

Schuelke-Leech, B. A. (2021). Engineering entrepreneurship teaching and practice in the United States and Canada. *IEEE Transactions on Engineering Management, 68*(6), 1570–1589.

Schuhmann, R. J., Erdman, A. M., Matson, J. V., Lang, D. H., Soper, J. G., & Horner, D. H. (2015). Engineering leadership education-the path forward. In *Proceedings of the ASEE annual conference & exposition*.

Shartrand, A., Weilerstein, P., Besterfield-Sacre, M., & Golding, K. (2010). Technology entrepreneurship programs in U.S. engineering schools: Course and program characteristics at the undergraduate level. In *Proceedings of the ASEE annual conference & exposition*.

Shartrand, A., Weilerstein, P., Besterfield-Sacre, M., & Olds, B. M. (2008). Assessing student learning in technology entrepreneurship. In *Proceedings of the frontiers in education conference*.

Shulman, L. S. (2005). Signature pedagogies in the professions. *Daedalus, 134*(3), 52–59.

Simmons, D. R., Clegorne, N. A., Polmear, M., Scheidt, M., & Godwin, A. (2021). Connecting engineering students' perceptions of professional competencies and their leadership development. *Journal of Civil Engineering Education, 147*(2).

Simmons, D. R., Clegorne, N. A., & Woods-Wells, T. (2017). Leadership paradigms in construction: Critical review to inform research and practice. *Journal of Management in Engineering, 33*(4). https://doi.org/10.1061/(ASCE)ME.1943-5479.0000518

Simmons, D. R., McCall, C., & Clegorne, N. A. (2020). Leadership competencies for construction professionals as identified by construction industry executives. *Journal of Construction Engineering and Management, 146*(9). https://doi.org/10.1061/(ASCE)CO.1943-7862.0001903

Sorby, S., Fortenberry, N. L., & Bertoline, G. (2021). Stuck in 1955: Engineering education needs a revolution. *Issues in Science and Technology*. Retrieved December 20, 2021, from https://issues.org/engineering-education-change-sorby-fortenberry-bertoline/

Souitaris, V., Zerbinati, S., & Al-Laham, A. (2007). Do entrepreneurship programmes raise entrepreneurial intention of science and engineering students? The effect of learning, inspiration and resources. *Journal of Business Venturing, 22*(4), 566–591.

Starkey, E., Toh, C. A., & Miller, S. R. (2016). Abandoning creativity: The evolution of creative ideas in engineering design course projects. *Design Studies, 47*, 47–72. https://doi.org/10.1016/j.destud.2016.08.003

Sternberg, R. (2007). Creativity as a habit. In A. Tan (Ed.), *Creativity: A handbook for teachers*. World Scientific.

Tierney, P., & Farmer, S. M. (2002). Creative self-efficacy: Its potential antecedents and relationship to creative performance. *Academy of Management Journal, 45*(6), 1137–1148.

Törnkvist, S. (1998). Creativity: Can it be taught? The case of engineering education. *European Journal of Engineering Education, 23*(1), 5–12.

Torrance, W. E. F. (2013). *Entrepreneurial campuses: Action, impact, and lessons learned from the Kauffman campuses initiative*. Ewing Marion Kauffman Foundation Research Paper.

U.S. Small Business Administration. (2004). *Entrepreneurship in the 21st century*. U.S. Small Business Administration.

Vesper, H., & Gartner, B. (1997). Measuring progress in entrepreneurship education. *Journal of Business Venturing*, *12*, 403–421.

Waller, D. R., & Strong, D. S. (2017). An investigation of student definitions and value of creativity in engineering education. Proceedings of the 2016 Canadian engineering education association. Dalhousie University. https://doi.org/10.24908/pceea.v0i0.6543

Wang, E. L., & Kleppe, J. A. (2001). Teaching invention, innovation, and entrepreneurship in engineering. *Journal of Engineering Education*, *90*(4), 565–570.

Weilerstein, P., & Shartrand, A. (2008). A decade of technological innovation: A retrospective view of the first decade of the NCIIA. In *Proceedings of the ASEE annual conference & exposition*.

Western, S. (2010). Eco-leadership: towards the development of a new paradigm. In B. W. Redekop (Ed.), *Leadership for environmental sustainability* (pp. 36–54). Routledge.

Wheadon, J., & Duval-Couetil, N. (2017). Elements of entrepreneurially-minded learning. *Journal of Engineering Entrepreneurship*, 7(3), 17–25.

Wilson, D., & Purushothaman, R. (2006). Dreaming with BRICs: The path to 2050. In S. C. Jain (Ed.), *Emerging economies and the transformation of international business*. Edward Elgar Publishing.

Wilson, K. E. (2008). Entrepreneurship education in Europe. In *Entrepreneurship and higher education*. OECD.

Wise, J. C., & Rzasa, S. E. (2004). Institutionalizing the assessment of engineering entrepreneurship. In *Proceedings of the frontiers in education conference*.

Woodcock, C. S. E., Shekhar, P., & Huang-Saad, A. (2019). Examining project based entrepreneurship and engineering design course professional skills outcomes. *International Journal of Engineering Education*, *35*(2), 631–644.

Yi, S., & Duval-Couetil, N. (2021). Standards for evaluating impact in entrepreneurship education research: Using a descriptive validity framework to enhance methodological rigor and transparency. *Entrepreneurship Theory and Practice*, *46*(6), 1685–1716.

You, J. W. (2021). Enhancing creativity in team project-based learning amongst college students: The moderating role of psychological safety. *Innovations in Education and Teaching International*, *58*(2), 135–145.

Yurtseven, H. O. (2002). How does the image of engineering affect student recruitment and retention? A perspective from the USA. *Global Journal of Engineering Education*, *6*(1), 17–23.

Zappe, S. E. (2018). Avoiding construct confusion: An attribute-focused approach to assessing entrepreneurial mindset. *Advances in Engineering Education*, 7(1).

Zappe, S. E., Cutler, S. L., & Gase, L. (2021). A systematic review of the impacts of entrepreneurial support programs in science, technology, engineering, and math fields. *Entrepreneurship Education and Pedagogy*. doi:10.1177/25151274211040422

Zappe, S. E., Hochstedt, K. S., & Kisenwether, E. C. (2013a). Faculty beliefs of entrepreneurship and design education: An exploratory study comparing entrepreneurship and design faculty. *Journal of Engineering Entrepreneurship*, *4*(1), 55–78.

Zappe, S. E., Hochstedt, K. S., & Kisenwether, E. C. (2013b). Teaching to innovate: Beliefs and perceptions of instructors who teach entrepreneurship to engineering students. *International Journal of Engineering Education*, *4*(1), 55–79.

Zappe, S. E., Mena, I., & Litzinger, T. (2013). Creativity is not a purple dragon. In *Proceedings of the annual national collegiate inventors and innovators alliance conference*.

Zappe, S. E., Reeves, P. M., Mena, I. B., & Litzinger, T. A. (2015). A cross-sectional study of engineering students' creative self-concepts: An exploration of creative self-efficacy, personal identity, and expectations. In *Proceedings of the ASEE annual conference & exposition*.

Zappe, S. E., & Tise, J. (2019). Work in progress: Engineering students' changing conceptions of the value of creativity. *Proceedings of the eighth research in engineering education symposium*.

21

Educating the Whole Engineer by Integrating Engineering and the Liberal Arts

Sarah Jayne Hitt, Amy Banzaert, and Olga Pierrakos

1 Introduction

Ensuring that engineering education leads to engineering professionals worldwide who are technically skilled, globally oriented, and socially responsible – the *whole* engineer – requires professional and academic institutions alike to promote and develop deep integration between engineering and the liberal arts (e.g., ABET, 2020; Adams et al., 2011; Ambrose, 2013; Lattuca et al., 2006; National Academy of Engineering, 2004; Walther et al., 2017). The need for engineering students to understand their work in context via a connection with and understanding of the liberal arts has never been more urgent. Recognizing this, professional bodies such as the National Academy of Engineering and the Royal Academy of Engineering have strongly articulated the need for engineers to understand the social, ethical, cultural, political, and environmental contexts of their work (National Academy of Engineering, 2004; Engineering Council, 2017). Engineering accreditors have continued to refine guidelines to ensure engineering education engages with the broader societal and ethical discourse (ABET, 2020; Engineers Australia, 2008; ENAEE, 2021; Engineering Council, 2020).

It is not hard to argue that the United Nations (U.N.) Sustainable Development Goals (United Nations, 2020) and the Grand Challenges of Engineering (Olson, 2016) are impossible to meet without bridging the divide between engineering and the liberal arts (Herkert & Banks, 2012). Additionally, more closely aligning these fields of study could help restore widespread understanding of the criticality of the liberal arts, improve engineering education, and offer liberal arts students insight and access to engineering approaches and concepts.

The de-prioritization of liberal arts study in higher education through defunding and calls for "useful" degrees and skills agendas that prioritize technical competence ahead of the historical liberal arts (Geiger et al., 2015; Jones & Hearn, 2018) similarly present both an opportunity and challenge for this call to action. As we will show, such calls as well as responses have existed for decades, if not centuries, yet there is clear evidence that far more work is needed to achieve education for the whole engineer. We argue that better integration of educational research and practice is necessary in order to achieve this outcome and discuss opportunities therein. This chapter describes the various motivations for embracing such integration as well as the inherent challenges in doing so. We summarize a selected few of the large number of prior efforts, explore the existing educational research underpinning those undertakings, and posit that these efforts seem to be not fully recognized or deemed a success because empirical research has not been conducted as rigorously as in other areas of

DOI: 10.4324/9781003287483-25

engineering education. We outline summaries of inspirational practice and point to new directions for research that would have the potential to create impact.

2 Terminology: Liberal Arts and Integrative Learning

The concept of a liberal arts education has a multiplicity of definitions, with a wide range of understanding found both in the historical record and in the present day. As such, providing an unqualified definition of the liberal arts is far beyond the scope of this chapter. That being said, key elements that are reasonably consistent across many sources include a focus on both broad and deep learning for the sake of the individual learner as well as for the broader community or society; deeply reflective, thoughtful, inquiry-based practices; emphasis on disciplines within the arts and sciences that emphasize intellectual gains rather than professional skill development and preparation; and a learning community – often a small, residential, undergraduate-oriented college campus – that involves both faculty and students engaging with each other both in small classes (<20) and outside of class (e.g., Ferrall & Ferrall, 2011; Roche, 2010; Kimball, 2010; Princeton University, n.d.).

What has been termed "traditional" engineering education – focused on the development and practice of applying scientific and mathematical theory and knowledge to the design and development of structures, systems, and processes – is often framed as distinct from and perhaps even orthogonal to the liberal arts. As will be discussed in this chapter, engineering education today typically emphasizes professional preparation and skill development to ensure that students achieve a capacity for and orientation towards problem-solving largely constrained to technological solutions.

While these distinctions can seem clear-cut, many, including Riley (2015), argue that this bifurcation reinforces unnecessary and even harmful disciplinary and pedagogical silos and may limit creativity, interdisciplinarity, and exploration. We agree with this perspective and, given the widespread perception of the educational approaches as distinct, find it necessary to discuss the question somewhat within this binary in order to contemplate moving beyond it, which we propose is possible with the use of integrative learning approaches.

Integrative learning encompasses the rich connections that can be made across liberal arts, disciplinary boundaries, and professional domains. It is generally defined as the ability to recognize and make connections between disparate knowledge, concepts, or contexts in ways that strengthen depth and breadth of learning. Depending on the discipline, integrative learning can be defined as a synonym of interdisciplinary learning or as an overarching concept, of which interdisciplinary learning is a key example. Integrative learning is not new to undergraduate education or to engineering education (Blackshields et al., 2014; Huber et al., 2007; National Academies of Sciences, Engineering, and Medicine, 2018; Newell, 1999), though its level of priority has ebbed and flowed over time and place as the practice has evolved over millennia.

3 Historical Perspectives on the Integration of Liberal Arts and Engineering

Leonardo da Vinci has been pointed to as the historical figure who most embodies the integration of the liberal arts and engineering. This recognition is connected to the flourishing of the "renaissance engineer" rhetoric, which holds that looking backward to the idea of the "renaissance man" and his ability to integrate knowledge from all disciplines was essential to being able to develop engineering solutions going forward in the 21st century (Bomke, 2007). While there are problematic elements with the historical totem of the renaissance engineer (for instance, it alludes to mostly male and mostly Western contributions to the field), its motivation is useful in helping to reframe the liberal arts and engineering as united within approaches to engineering as well as within the interests, talents, and motivations of engineering students (Dabby, 2001).

During da Vinci's time, however, engineering was not in itself a formalized discipline. While the "liberal arts" as an educational framework arguably traces back to the Lyceums of Aristotle, or even further, "engineering" as a defined subject area only emerged through the French École Polytechnique in the late 18th century, though humans had of course been developing and teaching engineering practices for millennia through apprenticeship models (Jørgensen, 2007). The transition to engineering education as an academic field of study rather than simply a learned set of skills emerged at different rates and with different priorities in the United States, Japan, England, France, Germany, and Sweden (Meiskin & Smith, 1996). These transitions were informed by the particular cultures of countries and regions and by the different meanings of engineering and engineering work in those places (Lucena et al., 2008).

Jørgensen (2007) demonstrates that the divergence between apprenticeship-based and university-rooted learning has its roots in a debate over the value of theory versus practice in education, and these tensions of situating engineering within the academy alongside the liberal arts and natural sciences "were evident from the beginning" (p. 223). Koshland (2010) concurs that:

> [T]he separation between basic and applied science created, in the minds of many in the liberal arts, a separation between science and engineering. Academe viewed the latter as applied and vocational, and hence determined that such fields had no place in a liberal arts institution.
>
> *(p. 53)*

Russo (2007) characterizes the humanities offerings in Milan's Politecnico and Turin's Scuola d'Ingegneria as having a "merely decorative function" by the 1860s (p. 1); around the same time, "students majoring in engineering or science at Harvard and Yale did not have equal status with the more elite students in the arts" (Grayson, 1980, p. 377). We can see that a perceived divide between engineering and the liberal arts was in some senses culturally and institutionally baked into the modern educational system from the beginning.

In the United States, however, scholars (e.g., Akera, 2011; Harris et al., 1994) have noted that a review of the history of engineering education in that country shows that a constant and consistent call for integration and for foundational liberal arts education for engineering students has existed alongside these divisions from the beginning. In describing how some of the United States' first engineering schools, such as the United States Military Academy at West Point, Rensselaer Polytechnic Institute, and the Massachusetts Institute of Technology, derived their approach from the French Grande Ecole system of the 18th century, Angulo (2012) outlines the inclusion of foundational course subjects in the liberal arts, such as rhetoric, literature, philosophy, history, and foreign languages, during the mid-1800s. The perceived disciplinary divide also did not prevent liberal arts colleges from developing engineering majors in the mid-1800s, including Union College in 1845 and "Swarthmore graduating its first [engineering] major in 1874" (Koshland, 2010, p. 58). Additionally, Akera (2011) shows how the 1862 Morrill Act in the United States placed "engineering education on a four-year undergraduate model that combined technical training with liberal education" (p. 3).

This liberal arts grounding for mid-19th-century engineers in the United States appears to be distinct from the way that engineering education evolved in Europe at the same time (Jørgensen, 2007).[1] Yet in both places, the late-19th-century changes in society, technologies, and economies wrought structural changes to the educational system that resulted in a de-prioritization of the liberal arts in engineering. Sample (1988) notes that "changes in engineering education have been undergirded by, and perhaps even driven by, important changes in the larger society" (p. 55), and we can see evidence of this in the shift to prioritize technical curricular content in engineering programs during the era that coincided with immigration, industrialization, and urbanization in Europe and the United States. For instance, Grayson (1980) describes schools in the United States turning away from European traditions to focus more on vocational and professional preparation after 1880, and Jørgensen (2007) reports that in the late 19th century, "technical universities developed in Germany and the Scandinavian countries to meet local institutional traditions" in crafts and apprenticeships (p. 220). As Dewey noted in 1944, the "liberal arts were sharply contrasted with the useful arts [which were] acquired by means of sheer apprenticeship in fixed routines" (p. 391).

The creation of quality standards through accreditation of engineering education that began in the United States in the 1930s is one of the important structural changes that further emphasized this contrast. While accrediting organizations "expressed a desire to see that engineering students are exposed to a larger picture than only technical courses can provide," none of the original criteria for accrediting engineering programs made any "mention of non-technical matters" (Stephan, 2001, pp. 155–156). Crucially, Jørgensen (2007) points out that the US system of accreditation derived from the British one that "emphasized practical skills and engineering experience, and it also supported the idea that engineering competencies were of a different nature than the academic qualifications given by universities" (p. 221). And as engineering education transformed again in response to the demands of World Wars I and II, accrediting organizations, engineering societies, academic institutions, and governments were all concerned with the professionalization of engineering and its ability to meet defense and industrial needs. Research funds awarded through federal subsidies after World War II continued to require US engineering schools to have "faculty members with strong mathematical and scientific research capabilities," meaning, that funding became more closely tied to technical expertise rather than general education (Prados et al., 2005, p. 167). The emphasis on specialization also continued, so that when instructors were allowed "to focus only on the delivery of their specific subject," this contributed to "the separation of the humanistic-social stem from the scientific-technological stem" (Akera, 2011, p. 5). This focus doesn't mean that calls to maintain liberal arts within the engineering curriculum ended during this time, as we can see from Swarthmore's critique of its engineering college in 1967, which stated that engineering is "the profession that links the values of the humanists, the discoveries of the scientists and the analyses of the social scientists" (as cited in Koshland, 2010, p. 60). However, it does mean that in many cases, they weren't emphasized as the boundary between the "two cultures" developed and became more rigid (Snow, 1959). In Europe, too, the late-20th-century emphasis on scientific research and professional work experience for engineering students created an educational focus that prioritized technical subjects, and industry demands for specialization "created tension between generalized engineering knowledge" and specific "domains of engineering practice" such as rail engineering or biomechanical engineering (Jørgensen, 2007, p. 230). Thus, the structural and societal pressures that affected the engineering education system served to reinforce the two cultures' divide late into the 20th century.

In the 1980s, several publications began to renew calls to unite liberal arts and engineering education and to highlight the challenges that needed to be overcome in order to do so. An ABET report at the start of the decade declared that "the humanities and social science requirement for engineering programs is well intentioned on the part of ABET but at the majority of engineering schools the execution is disappointing, to understate the case" (Cunningham, 1980, p. 6). In 1988, the *Chronicle of Higher Education* (1988) published an article titled "Engineering Students Are Said to Get Incoherent Education in the Liberal Arts" as a call to action and cited a lack of cooperation among engineering and liberal arts faculty as well as inadequate student advising as major barriers to broadening the scope of engineering education. The book *Unfinished Design: The Humanities and Social Sciences in Undergraduate Engineering Education* (Johnston et al., 1988) served as a major motivating factor for the *Chronicle of Higher Education* article. In it, the authors pointed to the value of humanities and social sciences in engineering education, highlighted 13 liberal arts engineering programs that already existed, and offered a discussion of the numerous obstacles to broader adoption. While other similar perspectives were published in the late 1980s (e.g., Florman, 1987; Useem, 1989), influential interdisciplinary voices went quiet until after the turn of the century (van de Beemt et al., 2020). Stephan (2001) shows how during this time many US institutions relied on general education courses in order to meet non-technical ABET criteria and therefore sidestepped any meaningful integration of liberal arts with the technical curriculum.

4 Contemporary Motivation for Integrating the Liberal Arts and Engineering

By the end of the 20th century, voices advocating for action that could unite the liberal arts and engineering grew louder. Two engineers at the Colorado School of Mines presented a paper at the Conference on College Composition and Communication in 1993, stating, "[A]lthough we acknowledge that scientists, engineers, and humanists live and function in different discourse communities, we also believe that some connections among these communities must be made if we are to avoid a 21st century Tower of Babel" (Olds & Miller, 1993, p. 3). This harbinger of disaster avoidance is perhaps why it is during another time of great societal and technological change that a resurgence of calls to integrate and prioritize the liberal arts in engineering education occurred. In the *European Journal of Education*, Ruprecht (1997) aligned the value of openness that emerged after the fall of the Soviet Union with the practice of opening the engineering curriculum to integrate the humanities so that students are prepared to face the demands of a changing world with an attitude of humility and responsibility. The turn of the millennium coincided with the growth of the digital world, rapid globalization, post-industrial capitalism, as well as its resultant environmental degradation, and scholars began to define "gigaton problems that require gigaton solutions" (Xu et al., 2010, p. 4037). Solving these "wicked problems" would require more than just technical know-how. Indeed, by the 1990s,

> some critics demanded a humanistic input into the curriculum . . . based on the assumption that engineering students, through confrontation with alternate positions and opportunities to discuss social and ethical issues, would be better prepared to meet the challenges of technology.
> *(Jørgensen, 2007, p. 232)*

Yet most of the scholarship on the topic produced during this time could be characterized as descriptions of the value of the humanities or calls to action rather than formalized research.

In the early 2000s, more publications emerged that described practice and that provided potential research directions for the integrative interventions that scholars advocated. For instance, Turbak and Berg (2002) describe teaching robotic design in a liberal arts environment, and Wilson (2000) provides an account of the obstacles to a liberal engineering education. Other notable perspectives in the first decade of the 21st century include Ollis et al.'s (2004) *Liberal Engineering in the Twenty-First Century*, a targeted response to ABET 2000 criteria that – for the first time – integrated social criteria, and Gorbet et al.'s (2008) seminal research on best practices for enabling interdisciplinary learning in project groups.

This work was one of the first empirical perspectives on the liberal arts in engineering education. Two decades on, *The Chronicle of Higher Education* followed up their 1988 cautionary article with the much more optimistic "Engineering and the Liberal Arts: Strangers No Longer" (Christ et al., 2008). Movements such as STEAM (science, technology, engineering, arts, and mathematics) (Perignat & Katz-Buonincontro, 2019) and the Engineer of 2020 (National Academy of Engineering, 2004), developed at this time, too, articulating an approach of deep integrative learning to unite engineering and the liberal arts more closely.

Simultaneously, engineering accreditation boards globally refined their criteria to require engineering education to engage with the broader societal and ethical discourse. Table 21.1 shows a selection of criteria for engineering program accreditation worldwide, highlighting the focus on

Table 21.1 Engineering Accreditation Criteria Related to Integrative Learning

Board	*Criteria*	*Governing Countries*[a]
ABET[b]	1. An ability to apply engineering design to produce solutions that meet specified needs, with consideration of public health, safety, and welfare, as well as global, cultural, social, environmental, and economic factors. 2. An ability to communicate effectively with a range of audiences. 3. An ability to recognize ethical and professional responsibilities in engineering situations and make informed judgments, which must consider the impact of engineering solutions in global, economic, environmental, and societal contexts. 4. An ability to function effectively on a team whose members together provide leadership, create a collaborative and inclusive environment, establish goals, plan tasks, and meet objectives.	United States, Austria, Bahrain, Brazil, Brunei, Chile, China, Ecuador, Egypt, India, Indonesia, Jamaica, Jordan, Kazakhstan, Korea, Kuwait, Lebanon, Mexico, Morocco, Oman, Palestinian Territory, Peru, Philippines, Poland, Portugal, Qatar, Russia, Saudi Arabia, South Africa, Spain, Turkey, UAE, Vietnam
AEAC[c]	Engineering application work should be representative of the field of practice and include technical and non-technical considerations. A key objective should be to develop an appreciation of the interactions between technical systems and the social, cultural, ethical, legal, political, environmental, and economic context in which they operate.	Australia, Hong Kong SAR, China, Kuwait, Malaysia, Singapore, Vietnam
ENAEE[d]	Awareness of non-technical – societal, health and safety, environmental, economic, and industrial – implications of engineering practice to inform judgments that include reflection on relevant social and ethical issues.	Kyrgyz Republic, Switzerland, Chile, Russia, Spain, Romania, Germany, France, United Kingdom, Ireland, Finland, Peru, Jordan, Poland, Republic of Kazakhstan, Turkey, Portugal, Italy, Slovak Republic
AHEP[e]	Design solutions for broadly defined problems that meet a combination of societal, user, business, and customer needs as appropriate. This will involve consideration of applicable health and safety, diversity, inclusion, cultural, societal, environmental, and commercial matters, codes of practice, and industry standards. Engineering activity can have a significant societal impact, and engineers must operate in a responsible and ethical manner, recognize the importance of diversity, and help ensure that the benefits of innovation and progress are shared equitably and do not compromise the natural environment or deplete natural resources to the detriment of future generations.	United Kingdom, Australia, Brunei, China, France, Greece, Hong Kong SAR, China, Hungary, Indonesia, India, Ireland, Jordan, Macao SAR, China, Malaysia, Malta, Mauritius, Myanmar, Netherlands, New Zealand, Oman, Russia, Serbia, Singapore, Spain, Sri Lanka, Trinidad and Tobago, Ukraine, United Arab Emirates

[a] Accreditation.org (n.d.).
[b] Accreditation Board for Engineering and Technology Inc. (ABET, 2020).
[c] Australian Engineering Accreditation Centre (Engineers Australia, 2008).
[d] European Network for Accreditation of Engineering Education (ENAEE, 2021).
[e] Accreditation of Higher Education Programmes (Engineering Council, 2020).

outcomes related to integrative learning. In response, we see the conversation on liberal arts and engineering transitioning from a focus on the inherent benefit of the liberal arts in engineering education to a focus on methods of integrating them.

Work on the potential of teaching engineering from a liberal arts perspective began to gain significant momentum and specificity with publications such as King's (2011) treatise on restructuring engineering education, which summarizes perspectives from nearly two dozen proponents of change. Further, Bucciarelli and Drew's (2015) design plan for liberal studies in engineering called for a redefinition of engineering fundamentals to include the humanities, arts, and social sciences. The second decade of the 21st century also saw a burgeoning of accounts of experiences establishing programs that blend engineering and liberal arts traditions (e.g., Daly et al., 2019; Gillette et al., 2014). The creation of new institutions "established from a blank slate with a distinctive and integrated educational approach, such as SUTD [Singapore University of Technology and Design], Olin College of Engineering, Iron Range Engineering and Charles Sturt University," demonstrates the extent to which stakeholders and investors are willing to support these initiatives (Graham, 2018, p. 31).

These efforts have shifted the narrative: from simply providing engineering students with foundational liberal arts courses in order to develop well-rounded engineering graduates to educating engineers in a more holistic, collaborative, and interdisciplinary way so that we create more socially and globally responsive and responsible engineering practice. Given this progress, it is curious that calls to better integrate the liberal arts and engineering education persist. We will argue later in this chapter that more robust research is required to achieve full integration and recognition of these practices.

5 Notable Exemplars of the Integration of Liberal Arts and Engineering Education

Efforts to integrate the liberal arts into engineering are therefore prevalent and ongoing. Yet we must also recognize the parallel effort to integrate more engineering and quantitative concepts into liberal arts study, such as the Sloan Foundation's New Liberal Arts (NLA) Program (1980–1992), which invested significant resources to include quantitative reasoning and technology into many liberal arts curriculums (Goldberg, 1986). More recently, Bucciarelli and Drew's (2015) conceptual Liberal Studies in Engineering degree program also explored whether liberal arts courses might inject engineering content into them. As has been previously noted, many liberal arts–focused colleges and universities have long had engineering programs. These efforts in liberal arts contexts must be read alongside those that focus on technical contexts in order to gain a full understanding of the possibilities and potential for meaningful integration. In what follows, we identify four areas in which the literature demonstrates this occurring in many places over several decades and indicate where more research is needed to fully elicit the effectiveness and impact of these approaches.

5.1 Category 1: Single Courses or Modules, or One-Off Interventions and Experiences

This category includes integration that occurs via one course or through a single learning experience. The literature shows a trend towards embedding the liberal arts through broader topic areas relevant to engineering, such as design, ethics, and sustainability. For example, Nieusma (2008) describes an effort at Rensselaer Polytechnic Institute to bring aesthetic and sociotechnical perspectives into a product design studio as a way to embed liberal arts learning into engineering design. At Uppsala University, the Department of Civil and Industrial Engineering's ethics course uses drama (including role-play and performance) to elicit engineering awareness and judgment related to ethical situations (Birch & Lennerfors, 2020). The Sustainability Enrichment Week for students in Edinburgh Napier University's School of Engineering and the Built Environment demonstrates another integrative strategy using a co-curricular enrichment week of experiential learning to situate engineering explicitly in cultural, historical, political, or environmental contexts (Hitt et al., 2022).

This category also includes examples of team-taught elective courses for engineers that bring together social scientists or humanists with STEM faculty teaching on a particular theme, such as Olin's Sustainability: Science, Society, and Systems course (Olin College of Engineering, 2022). Finally, the emergence of the UN Sustainable Development Goals and the National Academy of Engineering's Grand Challenges spurred other types of integrative courses (Carlson & Wong, 2020; Wood et al., 2019). Indeed, there are many examples of liberal arts and engineering integration in the literature that fit into this category, and the journal *Engineering Studies* published a special issue focusing on this theme in 2015. However, most of the available scholarship tends toward more descriptions of or reports on practice rather than a study of the methods and effectiveness of the integration. A systematic review of these types of efforts in both liberal arts and engineering contexts would be valuable to elicit a synthesis of some conclusions that could help frame future research. These examples show that this is an avenue through which educators are able to develop their own integrative opportunities as well as influence further integration and can act as a spur for others to build upon.

5.2 Category 2: Programmatic Interventions

This category refers to degree programs that adopt systematic approaches to integrating the liberal arts into engineering. Such programs refer to majors and minors where intentional integration of engineering and liberal arts learning is an identified goal. In some cases, the integration is at the curricular structure level, where major or minor requirements dictate a minimum set of courses from engineering and a minimum set of courses from traditional liberal arts courses to be completed. In other cases, the integration is at the course level and across several courses in a curriculum where programmatic experiences expose students to engineering and traditional liberal arts learning outcomes with a course experience. There are also examples of thematic integrations across programs where key themes like ethics are intentionally integrated across a curriculum.

Programmatic examples that target *curricular structures* to achieve integration between engineering and liberal arts can be broken out into bachelor of arts (BA) and bachelor of science (BS) programs. While the BA programs do not target ABET accreditation, such programs tend to have more liberal arts course requirements than BS programs that have to meet ABET requirements and therefore contain a minimum number of engineering credit hours. There are many institutions that offer BA programs that intentionally bring together engineering courses and liberal arts courses, and some notable examples include Brown University, Johns Hopkins, Trinity College, University of Rochester, Rice University, Lehigh University, Dartmouth College, and Worcester Polytechnic Institute. Dual degree programs also bring BA liberal arts degrees and BS engineering degrees together, such as at Monash University, Union College, and Carnegie Mellon University. Liberal arts and engineering minors also demonstrate curricular structures that are integrated. Examples include the McBride Honors Program in Public Affairs at the Colorado School of Mines and the art, music, philosophy, and languages minors at Rose-Hulman Institute of Technology. The nascent BS engineering degree at Wake Forest University (WFU) has also built programmatic curricular links to liberal arts minors by leveraging the minimum required 45 credit hours (out of a 120 credits) ABET expects for a broad education. In the most recent graduating class (2022), 75% of WFU engineering majors pursued a minor in arts, humanities, or sciences.

Programmatic examples that target *learning experiences within new or existing courses* to achieve integration between engineering and liberal arts also exist. Union College has been motivated to create programmatic approaches to bring technical and non-technical majors to work together in tackling grand challenges that truly require multidisciplinary integration to lead to a solution (Traver & Klein, 2010). Union College has also been hosting the Engineering and Liberal Education Symposium, and thus a national conversation, for many years as an effort to bring scholars in engineering and the liberal arts together. Worcester Polytechnic Institute (WPI), as another example, invested institutionally in an integrative approach that requires interdisciplinary projects across the four-year undergraduate experience. In such experiences, humanities students work with engineering students to tackle real-world challenges. Another example is at University College London, which launched in 2012 a new liberal arts and sciences program where engineering thinking is a first-year option for students (Bell et al., 2019).

Programmatic examples that target *learning experiences within an engineering degree* to achieve integration between engineering and liberal arts are also emerging and serve as models for other institutions. With support from a National Science Foundation Revolutionizing Engineering and Computer Science Departments (IUSE/PFE: RED) award, the School of Engineering at the University of San Diego is integrating the liberal arts directly into engineering courses within a new general engineering curriculum. Such an approach enables a reframing of traditional engineering context to have broader societal context by bringing a humanistic approach to engineering practice, integrating liberal arts topics such as social responsibility and social justice into the curriculum (Chen & Hoople, 2017).

Wake Forest University's engineering degree also took a programmatic approach to integrating liberal arts across the engineering curriculum. Liberal arts topics contextualized and infused across the engineering curriculum include professional identity development (bringing psychology to engineering), a historical perspective of the beginnings of engineering and modern engineering (bringing history to engineering), virtue ethics and character education (bringing philosophy to engineering), social and environmental justice (bringing policy to engineering). One effective strategy to achieve such integration has been to invite guest speakers from other departments (e.g., anthropology, art, communication, English, environmental studies, history, policy, psychology, religious studies, sustainability, writing, etc.) in core engineering courses to support the learning of engineering students (Kenny et al., 2021; Pierrakos & Stottlemyer, 2019). According to senior exit survey results, such experiences are strongly valued by engineering students because it makes visible to them the important context and knowledge that can be contributed by experts outside of engineering. An effort funded by the Kern Family Foundation that has also led to fruitful integrations for WFU Engineering is infusing character education across the curriculum (Pierrakos et al., 2019; Koehler et al., 2020; Gross et al., 2021). External funding enabled the hiring of full-time scholars with background in religious studies and psychology to teach together with engineering faculty across the curriculum. Courses most fertile for cross-disciplinary co-teaching have been the project-based courses, such as capstone design, which inherently enable a broad and rich array of knowledge integrations and professional contexts to take place. It is also not uncommon to find students outside of engineering (e.g., art, biology, chemistry, economics, environmental studies, and Spanish) working side by side with engineering faculty and students on research and design projects. The WFU model of liberal arts education and engineering education has supported student agency because engineering students are encouraged to pursue liberal arts minors, get technical credit for research and study-abroad experiences, and customize their four-year experience within and beyond engineering. Such flexibility and agility have spurred a very diverse student body – over 40% of the student body are women, and over 20% are students of color; the program is also supported by a diverse faculty body, noted by ABET as a programmatic strength.

Like the first category of single classes or interventions, much of the scholarship that exists in this area is descriptive rather than empirically studied. Research that synthesizes these programmatic efforts would also be a valuable addition to the literature.

5.3 Category 3: Integration That Targets Non-Engineers

Just as engineering students benefit from holistic education and thinking, efforts have also increased to ensure students focused on traditional liberal arts subjects have ready access to engineering content through single courses (e.g., d'Entremont et al., 2017; Flath & Michelfelder, 2017; Mikic & Voss, 2006) and engineering minors. Smith College, as an example, has been offering a minor in engineering science for many years to complement academic studies across any liberal arts degree and to offer non-engineering students the ability to explore areas such as robotics, energy, water, health, etc. The Smith College engineering science minor requires five engineering courses, two of which are introductory courses. As another more recent example, an engineering minor launched in 2020 at Wake Forest University enables students not majoring in engineering to explore engineering knowledge, skills, and mindsets. In this minor, students are required to complete the two project-based first-year engineering courses (that do not have associated pre- or co-requisites), complete a calculus course, complete a physics course, and complete sophomore-level course offerings or work on interdisciplinary engineering project courses (like capstone design or independent study research project courses). To date, majors from business, chemistry, economics, environmental science, mathematics, mathematical economics, politics and international affairs, and studio art have pursued the engineering minor.

Taking a different approach, Wellesley College, an elite women's liberal arts college in the northeast United States, created a unique engineering faculty position and an associated set of introductory engineering courses to ensure that its students could expand their perspective to include engineering concepts and thinking – realms that many students felt were out of reach for a variety of reasons, including the alienation many girls and women experience in STEM subjects (Thom, 2001). Banzaert and Ducas (2016) describe Wellesley's engineering courses as compelling to a wider range of students through a number of pedagogical and engagement best practices, including a highly hands-on, project-based curriculum that often includes community-based projects. Although this program does support the modest number of students who pursue a degree in engineering through a dual degree program or wish to go on to graduate-level engineering programs, these courses are taken primarily by students who are not seeking engineering careers (Banzaert & Ducas, 2016). Many students report that these exploratory courses help them feel empowered and prepared to enter into discourse and work addressing the role of technology in society.

In order for more traditional liberal arts colleges like those at Wellesley, many of which do not have a Department of Engineering or engineering major, to invite engineering educators to provide engineering knowledge that could also enrich the traditional liberal arts experience, research on the effectiveness of this approach is required. While the concept of considering engineering as a liberal art, or as belonging within liberal arts, is not a new argument (see, for example, Corfield, 1993; Koshland, 2010), the authors are not aware of any study which compares the efforts of these programs with those that integrate the liberal arts into engineering, but this could be a promising area of research.

5.4 Category 4: Institutional Integration

There have also been larger-scale efforts at integration at the college or institutional level, which date back to the mid-20th century in the United States at both Harvey Mudd College and Smith College. Harvey Mudd's first course catalog of 1957 places the liberal arts as central to the education of an engineer, stating that "a special need exists for physical scientists and engineers with broad enough training in the social sciences and humanities to assume technical responsibility with an understanding of the relation of technology to the rest of society" (qtd. in Dym & Bright, 2004). At Smith, engineering is seen as a liberal art, and its structural integration within the rest of the college enables concomitant "educational benefits" to students both within and without engineering (Christ, 2010). More recently, Olin in the United States (opened in 2002) and NMITE in the UK (opened in 2021) saw the inclusion of liberal arts within engineering as a critical component of the innovative approaches these new institutions chose to take (Schwartz, 2007; Usher & Sheppard, 2017). While the curricular structures and pedagogical approaches privileging liberal arts integration at these institutions have been described (e.g., Somerville et al., 2005; Rogers et al., 2021), there seem to be no research studies focusing explicitly on their liberal arts emphasis or on the student response to this approach both during and after their education. Research of this type could illuminate what effects uniting the liberal arts and engineering on an institutional scale has yielded. Additionally, studies could be developed that compare them with institutions that do not deliberately and explicitly integrate these areas.

Finally, we note that within all these categories, examples are not limited to the United States and Europe. In the past 15 years, a conversation has begun around the world about the value of and potential approaches to integrating the liberal arts and engineering, such as in Japan, Russia, India, and China (Nohara et al., 2008; Rudskoy et al., 2021; Bakilapadavu & Shekhavat, 2013; Tang et al., 2016). Thus, there is quite a large body of work outlining visions for integration as well as describing examples of practice. However, these practices have not necessarily resulted in a substantive body of research that studies, analyzes, and evaluates that practice so that others can learn from and build upon it. Most critically, more research is needed on these exemplars through the lens of liberal arts teaching and learning; indeed, in all these categories, there is a need to clarify what differences there might be between liberal arts *content* (learning distributed across the humanities, arts, natural sciences, and social sciences) versus liberal arts *methods* (such as Socratic discussion, experiential and student-centered learning, and critical questioning) (Becker, 2014). It is unclear which content or methods have yielded the most success or, indeed, how best to define success in this context. However, there are now enough institutional exemplars explicitly and deliberately integrating the liberal arts in engineering that a robust research study could be developed to help address this lack of clarity.

6 Ongoing Challenges and Future Directions

Despite the long history of this conversation and the many available examples of approaches, the academy is still grappling with establishing research and best practices for integrative education in general and for integrating liberal arts and engineering education in particular. This seeming stagnation likely has several causes, including "the ideological separation of technical and social engineering competencies" (Cech, 2013) and the fact that despite recent reform of accreditation criteria, the changes that are intended to create flexibility that could support more inclusion of the liberal arts might actually "once again reproduce engineering's traditional instrumental educational emphasis" (Seron & Silbey, 2009). Additionally, Lyall (2019) highlights the structural challenges to interdisciplinarity for both educators and institutions alike. Indeed, van den Beemt and colleagues (2020) recently published a review of the literature on interdisciplinary engineering education with clear implications for and distinctions from integrative learning. They concluded that clear learning goals and assessment strategies have not yet been identified and note that while international engineering education scholars agree that some level of integration is required to count as interdisciplinary, specific processes are not well understood or documented. Winebrake (2015) also highlights significant structural and cultural barriers to integrating the liberal arts and engineering, including lack of resources, disciplinary ignorance or arrogance, and bureaucracy. Lyall (2019) notes that there is a "mismatch between interdisciplinary expectations and the prevailing norms of discipline-based scholarship." Higher education, after all, is still grappling with a siloed approach to undergraduate education that many argue does not reflect real-world professional practice that is increasingly interdisciplinary, and institutions and departments still struggle with the integration that some have described as an essential vehicle towards innovation (Selznick et al., 2021). Given these realities, it is understandable why resistance to integration persists. Nevertheless, we argue that the time has come to overcome those barriers.

The cross-institutional Integrative Learning Project (ILP), sponsored by AAC&U and the Carnegie Foundation for the Advancement of Teaching, offers six guiding principles to develop a context where interdisciplinary efforts like uniting engineering and the liberal arts can flourish: (1) make integrative learning a campus-wide concern; (2) design initiatives strategically; (3) support faculty creatively; (4) make a commitment to knowledge-building; (5) recognize that institutionalization is a long-term process; and (6) build networks beyond campus for collaboration and exchange (Huber et al., 2007). Given these principles, we see four key future directions for integrating the liberal arts and engineering. These, all of which will benefit from more research and support, are further detailed in this section.

6.1 Future Direction 1: Strengthening Efforts to Integrate the Liberal Arts via Ethics Education

Whether viewed from the perspectives of integrating engineering into liberal arts study or from integrating the liberal arts into engineering, both sustainability (encompassing ecology, politics, and economics) and ethics are essential subjects in which all students should develop competency. In the case of ethics, subgroups dedicated to its focus have existed as part of engineering education organizations on and off for decades, beginning as early as 1940 with the Committee on Principles of Engineering Ethics of the Engineers' Council for Professional Development (Stephan, 2001). However, as Herkert (2000) points out, by the late 1990s, "nearly 80% of engineering graduates attend[ed] schools that [did] not have an ethics-related course requirement for all students" (p. 303). While this situation has improved markedly since then, with "engineering ethics [accepted] as an essential component of the engineering curriculum" (Kim et al., 2021, p. 1), the actual practice of embedding ethics in engineering education can be seen as a microcosm that reveals many of the challenges to and opportunities for integration.

Engineering ethics itself is still often set aside from the technical curriculum and delivered in stand-alone courses or via guest speakers that do not meaningfully integrate it into engineering courses and real-world problems. This practice is ironic, given that the liberal arts have always been a fundamental component of engineering ethics. Weil (1984) describes the field as emerging "in the mid-1970s when scholars from engineering and philosophy joined" and notes that "engineering ethics is the offspring of these two disciplinary areas primarily, but the field draws from other disciplines as well: law, behavioral and management sciences, history, and religious studies." Kline (2001) also shows that engineering ethics "pedagogical methods come from moral philosophy, history, and sociology." Additionally, apart from the obvious connection between the discipline of moral philosophy and engineering, ethics education is also a doorway that can open towards many other liberal arts areas. For instance, liberal arts–centered educational approaches to engineering ethics, such as arts-based methods, value-sensitive design, biomimicry, stakeholder engagement, cultural and political considerations, and corporate social responsibility, can all be embedded in engineering education by incorporating ethics in engineering practice.

Yet while Finelli et al. (2012) show that students find ethics learning integrated into advanced engineering courses to be the most effective pedagogy, this practice raises the aforementioned tension of who has disciplinary expertise and authority to teach it. That is, does a professor in an upper-level technical course have sufficient knowledge of ethics as a discipline and experience with ethics pedagogy to appropriately teach the material? If philosophers and educators in the humanities are not engaged in developing and delivering this content, it is understandable they could feel threatened or devalued, and situations like this should raise questions of quality and rigor among course assessors and accreditors. Additionally, while integrating knowledge of ethical theories is critical to effective and responsible engineering practice, this approach is not common for most engineering programs (Pierrakos et al., 2019). To avoid these tensions, we posit that the most effective pedagogies and learning experiences in engineering ethics are those that draw from deep and meaningful engagement with experts from both disciplines who welcome the opportunity for integrative curricula.

Ethics is not and should not be the only way that liberal arts are integrated into engineering education, but if, as is increasingly true, accreditors and quality assurance organizations are requiring engineering students to have a grounding in ethics, then it is a powerful portal for the inclusion of other liberal arts fields, such as psychology, drama, history, and more, as well as for further research into strategies for doing so that are most effective in achieving learning outcomes.

6.2 Future Direction 2: Continuing and Supporting Professional Development for Faculty

The value of integrative curricular innovations, including those that support the integration of engineering and the liberal arts, depends on the pedagogies and assessment practices that enable them. Pedagogical professional development is therefore essential and critical to sustain such efforts. Critical facets of pedagogy not only include learning models but also co-teaching models to support the inherent complexity of integrative learning. There are various co-teaching models that could authentically support integrative learning; we must ensure institutions see co-teaching as a meaningful professional development activity and a valuable investment (Rytivaara & Kershner, 2012). Faculty require specialized professional development to learn strategies for integrative teaching in support of integrative learning: centers of teaching and learning are crucial to support these initiatives.

Experiential learning pedagogies that bring the external world to the classroom are conducive to supporting integrative learning; engineering education is well-positioned to continue and improve such learning pedagogies through existing cornerstone and capstone projects, service-learning experiences, and project-based learning. Such experiences lay a fertile ground for both engineering and the liberal arts to be infused (Rogers et al., 2021), but faculty need to be able to access research and training opportunities that can prepare them to do so. Similarly, co-curricular and extracurricular experiential experiences like study abroad, internships, and undergraduate research have the potential – with appropriate structure and reflection – to serve as additional, integrated experiences to facilitate integrative learning and ensure that connections are made between coursework and community, theory, and practice. However, these experiences require substantial faculty time and resources to develop and deliver.

Assessments must support the scaffolding that is essential to knowledge integration and learning and enable self-awareness, self-direction, and metacognition. Rubrics and portfolios can be powerful assessment tools. Self-assessment approaches offer a vehicle toward intentional learning, where the assessment tools themselves enhance the depth of reflection and offer a structured framework to what many students perceive as an ill-structured problem or experience (Huber & Hutchings, 2004). However, assessment tools that effectively support integrative learning can require more time to develop and validate, as they require significant collaboration among faculty from diverse disciplines. More research is needed to test and evaluate the essential integration so that instructors and educational leaders can best tackle curricular and programmatic change that more meaningfully integrates the liberal arts and engineering education. Funding and other incentives are essential to these initiatives so that faculty have the time, resources, motivation, and support to develop, refine, and evaluate them.

6.3 Future Direction 3: Improving Institutional Structures, Policies, and Processes

While individual faculty efforts can make an immense impact and are essential to building bridges between engineering and liberal arts, institutional efforts are required to sustain integrative learners over the duration of their curricular, co-curricular, and extracurricular higher education experiences. Institutional and organizational structures are critical in achieving desired outcomes. The current state of most higher education institutional structures continues to represent a fragmented curricular landscape of general education courses and major courses which are frequently disconnected from co-curricular experiences and extracurricular, real-world experiences beyond the campus. This fragmentation repeats itself in academic structures and policies which are linked to curricular processes, staffing, rewards, space, reporting structures, leadership, and more. In order to achieve sustained change, administrative leaders could reimagine and improve institutional structures, policies, resource allocations, and processes creatively with transparency and inclusive participation of faculty, staff, students, and external stakeholders. Institutions should leverage internal and external funding opportunities to support this institutional approach of integrative learning.

In this work, there is value in prioritizing the role that future and former students can have in promoting integrative learning. For example, both Olin College of Engineering and the New Model Institute for Technology and Engineering made use of student co-design years in institutional and curricular development (Kerns et al., 2005; Rogers et al., 2021). As integrative learning can be seen as fundamentally inclusive learning, students should have a central role in both curricular change and the institutional redesign required to support widespread, effective integrative learning.

6.4 Future Direction 4: Leveraging Existing Efforts Through Research and Building Connections

Despite the many areas where integration of the liberal arts and engineering is currently being employed, calls to do more persist in popular dialogue (Osgood, 2017; Moustafa, 2018; Wadhwa, 2018). Our review of research in this area finds this narrative to be almost identical to the conversation of 40, 75, and even 100 years ago (Deloughry, 1988; Hammond, 1950; Richardson, 1908), suggesting that the many and varied efforts at integrating these two strands of education that we have highlighted here are either not recognized as successful or not recognized at all. Interestingly, more recent think pieces relate calls for more liberal arts in engineering to the type of human development that higher education provides, or about the type of engineering worker it is seen as valuable to produce. This discussion not only relates to the expectations of the public (in terms of the way a holistically educated engineer serves society) but also how civic, industrial, and professional bodies can encourage and reward this type of integrated education most effectively.

While accreditation boards globally now require engineering education to engage with the broader societal and ethical discourse, as shown in Table 21.1 earlier in this chapter, it is clear from the repeated calls that such requirements have not created ubiquitous, well-recognized educational programs integrating the liberal arts and engineering. We see opportunities for empirical research initiatives to establish best practices in this work, including determining criteria for success and better understanding of the reasons that certain initiatives have not persisted.

Similarly, much of the research we were able to identify is centered on the United States, with a few exceptions. What could be learned from other countries and cultures about approaches to meeting these learning outcomes that invite meaningful reflection and opportunities? Support for integrating the liberal arts and engineering may also be found by working collaboratively with international organizations, governments, funding agencies, and employers of our students. Knowledge building and sharing with external stakeholders like these could ensure the sustained and intentional change as well as recognition required to achieve these outcomes.

7 A Path Forward

While engineering education is not new to integrative learning (Rover, 2007), engineering educators can still gain insight into integrative learning from other disciplines. The emergence of medical humanities is a case in point: Dolan's (2015) chapter "One Hundred Years of Medical Humanities" reveals uncanny correlations and similarities between the broad societal concerns and educational approaches of medical education and engineering education in the last century. He shows that the concentrated, global effort at integrating the liberal arts and medicine resulted in "institutional expansion and professionalization of medical humanities" as a discipline and brought about profound changes to curriculum and other educational practices that resulted in a more interdisciplinary, more holistic approach to medical education (p. 21). The concerns and tensions are similar to those found in discussions about integrating the liberal arts in engineering education; therefore, the opportunities for learning and collaboration promise to be just as fruitful.

Engineering educators can thus contribute immensely to this dynamic and growing field of interdisciplinary scholarship. For instance, might engineering faculty collaborate closely with colleagues from liberal arts disciplines to create opportunities that ensure engineering students receive education in the liberal arts from disciplinary experts? This effort could simultaneously help bolster enrollment in liberal arts courses by demonstrating the applications of liberal arts to STEM fields. Interdisciplinary learning provides benefits to both subject areas, and there is value in creating or expanding opportunities for liberal arts students to learn engineering. The professional nature of the engineering degree is truly a fertile ground for integrative learning, where technical knowledge, professional knowledge, and personal growth come together at the intersections of engineering and liberal arts. Perhaps, in time, engineering can be widely acknowledged as a modern liberal art.

At a time when humanity faces the unprecedented challenge of climate change coupled with destabilizing forces, including political unrest, widespread immigration, and the COVID-19 pandemic, we find new momentum for this call to integration and also new challenges. The role of engineering in providing benefits to humanity – for example, in minimizing COVID spread, providing sustainable energy solutions, and offering improved quality of life for people around the world – is often celebrated and recognized. However, the harms that can result from engineering practice going unchecked (e.g., unintended and unanticipated consequences) are becoming more and more apparent: for example, in fomenting misinformation through social media and creating the technologies that are exacerbating climate change. Educating engineers more broadly so as to better encompass both engineering and liberal arts habits of mind and perspectives – and carefully studying best practices and conducting research to connect them to learning outcomes – offers the possibility of ensuring that these risks can be minimized, the benefits maximized, and the practice of integration widely recognized as valuable and critical to the work.

Note

1 There is little research that discusses the history of engineering education through the lens of the liberal arts outside of the United States and Europe, likely because the "liberal arts" as a higher education construct comes from that cultural heritage. The authors have therefore had to limit the historical perspective to these regions.

References

ABET. (2020). *Criteria for accrediting engineering programs*. www.abet.org/accreditation/accreditation-criteria/criteria-for-accrediting-engineering-programs-2020-2021/

Accreditation.org. (n.d.). *Accrediting bodies*. https://accreditation.org/explore-accreditation/accrediting-bodies

Adams, R., Evangelou, D., English, L., Dias de Figueiredo, A., Mousoulides, N., Pawley, A. L., Schiefellite, C., Stevens, R., Svinicki, M., Martin-Trenor, J., & Wilson, D. M. (2011). Multiple perspectives on engaging future engineers. *Journal of Engineering Education, 100*(1), 48–88. https://doi.org/10.1002/j.2168-9830.2011.tb00004.x

Akera, A. (2011, June 26–29). Liberal learning revisited: A historical examination of the underlying reasons, frustrations, and continued prospects for engineering and liberal arts integration. In *ASEE annual conference and exposition*. https://doi.org/10.18260/1-2–18282

Ambrose, S. A. (2013). Undergraduate engineering curriculum: The ultimate design challenge. *The Bridge, 43*(2), 16–23. www.nae.edu/81221/Summer-issue-of-The-Bridge-on-Undergraduate-Engineering-Education

Angulo, A. J. (2012). The polytechnic comes to America: How French approaches to science instruction influenced mid-nineteenth century American higher education. *History of Science, 50*(3), 315–338. https://doi.org/10.1177/007327531205000304

Bakilapadavu, G., & Shekhavat, S. (2013). Integrating humanities and liberal arts in engineering curriculum: Need, experiences and new directions. *Pertanika Journal of Social Sciences & Humanities, 21*(1), 373–382. www.researchgate.net/publication/290747816_Integrating_humanities_and_liberal_arts_in_engineering_curriculum_Need_experiences_and_new_directions

Banzaert, A., & Ducas, T. (2016). *Technology and engineering at Wellesley College: A case study*. ASEE.

Becker, J. (2014). *What a liberal arts and sciences education is . . . and is not*. https://tools.bard.edu/wwwmedia/files/6366602/1/Jonathan%20Becker%20What%20a%20Liberal%20Arts%20Education%20is%20and%20is%20Not.pdf

Bell, S., Chilvers, A., Jones, L., & Badstuber, N. (2019). Evaluating engineering thinking in undergraduate engineering and liberal arts students. *European Journal of Engineering Education, 44*(3), 429–444. https://doi.org/10.1080/03043797.2018.1552663

Birch, P., & Lennerfors, T. (2020, October). Teaching engineering ethics with drama. In *IEEE frontiers in education conference*. https://doi.org/10.1109/FIE44824.2020.9274160

Blackshields, D., Cronin, J. G., Higgs, B., Kilcommins, S., McCarthy, M., & Ryan, A. (2014). *Integrative learning*. Taylor & Francis.

Bomke, W. (2007). Engineers of the renaissance and the concept of 'the new renaissance engineer.' In S. H. Christensen, M. Meganck, & B. Delahousse (Eds.), *Philosophy in engineering* (pp. 29–40). Academia. www.researchgate.net/publication/278016115_Philosophy_in_Engineering

Bucciarelli, L. L., & Drew, D. E. (2015). Liberal studies in engineering – a design plan. *Engineering Studies*, 7(2–3), 103–122. https://doi.org/10.1080/19378629.2015.1077253

Carlson, C., & Wong, C. (2020, June). If Engineers solve problems, why are there still so many problems to solve? Getting beyond technical "solutions" in the classroom. In *2020 ASEE virtual annual conference*. https://doi.org/10.18260/1-2–34747

Cech, E. A. (2013). Ideological wage inequalities? The technical/social dualism and the gender wage gap in engineering. *Social Forces, 91*(4), 1147–1182. www.jstor.org/stable/43287496

Chen, D. A., & Hoople, G. D. (2017, June). Contextualizing a new general engineering curriculum in the liberal arts. In *2017 ASEE annual conference & exposition*. https://doi.org/10.18260/1-2–28073

Christ, C. T. (2010). What is happening in liberal education? In D. Grasso & M. B. Burkins (Eds.), *Holistic engineering education*. Springer. https://doi.org/10.1007/978-1-4419-1393-7_6

Christ, C. T., Clark, R., & Ainlay, S. C. (2008). Engineering and the liberal arts: Strangers no longer. *The Chronicle of Higher Education, 55*(2), A76.

Corfield, K. (1993). Engineering as a liberal art. *Australasian Journal of Engineering Education, 4*(1), 61–67. https://search.informit.org/doi/10.3316/aeipt.114623

Cunningham, R. G. (1980). President's report to ECPD. *48th annual report for the year ending September 30, 1980*, ABET.

Dabby, D. (2001, June 24–27). Heroes for the renaissance engineer: Leonardo, Nabokov, Bach and Borodin. In *2001 ASEE annual conference and exposition*. https://doi.org/10.18260/1-2–9329

Daly, S. R., Mosyjowski, E. A., & Seifert, C. M. (2019). Teaching creative process across disciplines. *The Journal of Creative Behavior, 53*(1), 5–17.

Deloughry, T. J. (1988). Engineering students are said to get incoherent education in the liberal arts. *The Chronicle of Higher Education, 34*, 15.

D'Entremont, A., Kirchen, P., Ellis, N., Grecu, C., & Green, S. (2017). The art of the possible: Developing a course to introduce engineering to non-engineers. *Proceedings of the Canadian engineering education association (CEEA)*. https://doi.org/10.24908/pceea.v0i0.7356

Dewey, J. (1944). The function of the liberal arts college in a democratic society. *The American Scholar, 13*(4), 391–393. www.jstor.org/stable/i40054120

Dolan, B. (2015). One hundred years of medical humanities: A thematic overview. In B. Dolan (Ed.), *Humanitas: Readings in the development of the medical humanities* (pp. 1–30). University of California Medical Humanities Press. https://escholarship.org/uc/item/3gq0j8fm

Dym, C., & Bright, A. (2004). General engineering at Harvey Mudd: 1957–2003. In *ASEE 2004 annual conference*. https://doi.org/10.18260/1-2–13721

ENAEE. (2021). *EUR-ACE® framework standards and guidelines*. European Accreditation of Engineering Programs. www.enaee.eu/wp-content/uploads/2022/03/EAFSG-04112021-English-1-1.pdf

Engineering Council. (2017). *Statement of ethical principles*. www.engc.org.uk/standards-guidance/guidance/statement-of-ethical-principles/

Engineering Council. (2020). *The Accreditation of higher education programmes (AHEP)* (4th ed.). www.engc.org.uk/media/3464/ahep-fourth-edition.pdf

Engineers Australia. (2008). *Accreditation management system: Education programs at the level of professional engineer document no. S02: Accreditation criteria summary*. www.engineersaustralia.org.au/sites/default/files/content-files/2016-12/S02_Accreditation_Criteria_Summary.pdf

Ferrall, V. E. J., & Ferrall, V. E. (2011). *Liberal arts at the brink*. Harvard University Press.

Finelli, C. J., Holsapple, M. A., Ra, E., Bielby, R. M., Burt, B. A., Carpenter, D. D., Harding, T. S., & Sutkus, J. A. (2012). An assessment of engineering students' curricular and co-curricular experiences and their ethical development. *Journal of Engineering Education, 101*(3), 469–494. https://doi.org/10.1002/j.2168-9830.2012.tb00058.x

Flath, D., & Michelfelder, D. P. (2017, June 24–28). Thinking like an engineer: Responding to the engineering blind spot at liberal arts colleges. In *2017 ASEE annual conference & exposition*. https://doi.org/10.18260/1-2–29027

Florman, S. C. (1987). *The civilized engineer*. St. Martin's Press.

Geiger, R. L., Melin, C., Kleinman, D. L., Woodward, K., Moses, Y., & Rothblatt, S. (2015). *A new deal for the humanities: Liberal arts and the future of public higher education*. Rutgers University Press.

Gillette, D. D., Lowham, E., & Haungs, M. (2014). When the hurly-burly's done, of battles lost and won: How a hybrid program of study emerged from the toil and trouble of stirring liberal arts into an engineering cauldron at a public polytechnic. *Engineering Studies, 6*(2), 108–129. https://doi.org/10.1080/19378629.2014.944186

Goldberg, S. (1986). The Sloan foundation's new liberal arts program. *Change: The Magazine of Higher Learning, 18*(2), 14–15. https://doi.org/10.1080/00091383.1986.9937051

Gorbet, R. B., Schoner, V., & Taylor, B. (2008). Best practices for enabling effective cross-disciplinary learning in interdisciplinary project groups. *International Journal of Engineering Education, 24*(3), 495–508.

Graham, R. (2018). *The global state of the art in engineering education*. Massachusetts Institute of Technology. www.cti-commission.fr/wp-content/uploads/2017/10/Phase-1-engineering-education-benchmarking-study-2017.pdf

Grayson, L. P. (1980). A brief history of engineering education in the United States. *IEEE Transactions on Aerospace and Electronic Systems, 16*(3), 373–392.

Gross, M. D., Wiinikka-Lydon, J., Lamb, M., Pierrakos, O., & Yeaman, A. (2021, July). The Virtues of teamwork: A course module to cultivate the virtuous team worker. In *2021 ASEE virtual annual conference*. https://peer.asee.org/37905

Hammond, H. P. (1950). Contributions of the liberal arts to the training of engineers. *The Journal of Higher Education, 21*(4), 189–193. https://doi.org/10.2307/1976366

Harris, J. G., DeLoatch, E. M., Grogan, W. R., Peden, I. C., & Whinnery, J. R. (1994). Journal of engineering education round table: Reflections on the Grinter report. *Journal of Engineering Education, 83*(1), 69–94. https://doi.org/10.1002/j.2168-9830.1994.tb00120.x

Herkert, J. R. (2000). Engineering ethics education in the USA: Content, pedagogy and curriculum. *European Journal of Engineering Education, 25*(4), 303–313. https://doi.org/10.1080/03043790050200340

Herkert, J. R., & Banks, D. A. (2012). I have seen the future! Ethics, progress, and the grand challenges for engineering. *International Journal of Engineering, Social Justice, and Peace, 1*(2), 109–122. https://doi.org/10.24908/ijesjp.v1i2.4306

Hitt, S. J., Hairstans, R., Leitch, K., & Connell-Skinner, K. (2022, September 19–22). Developing a strategic partnership through a sustainability enrichment week. In *Proceedings of the 50th SEFI annual conference*.
Huber, M. T., & Hutchings, P. (2004). Integrative learning: Mapping the terrain. The academy in transition. *Association of American Colleges and Universities*. www.researchgate.net/publication/254325229_Integrative_Learning_Mapping_the_Terrain
Huber, M. T., Hutchings, P., Gale, R., Miller, R., & Breen, M. (2007). Leading initiatives for integrative learning. *Liberal Education, 93*(2), 46–51. https://doi.org/10.1080/23752696.2019.1686647
Johnston, J. S., Shaman, S., & Zemsky, R. M. (1988). *Unfinished design: The humanities and social sciences in undergraduate engineering education*. Association of American Colleges.
Jones, B. N., & Hearn, J. C. (2018). Under siege: The future of the liberal arts at State U. In J. Blanchard (Ed.), *Controversies on campus: Debating the issues confronting American universities in the 21st century* (pp. 162–175). ABC-CLIO.
Jørgensen, U. (2007). Historical accounts of engineering education. In E. Crawley, J. Malmqvist, S. Östlund, & D. Brodeur (Eds.), *Rethinking engineering education: The CDIO approach* (pp. 216–240). Springer.
Kenny, M. C., Pierrakos, O., & O'Connell, M. (2021, July). Infusing the liberal arts in first-year engineering: A module on history, professional identity, and courage. In *2021 ASEE virtual annual conference*. https://peer.asee.org/37336
Kerns, S. E., Miller, R. K., & Kerns, D. V. (2005). The development of the initial Olin college curriculum. In *Educating the engineer of 2020: Adapting engineering education to the new century* (pp. 98–113). National Academies Press.
Kim, D., Jesiek, B. K., & Loui, M. C. (2021, October 13–16). Moral narratives of practicing engineers across industry sectors for engineering ethics education. In *2021 IEEE frontiers in education conference (FIE)*. https://doi.org/10.1109/FIE49875.2021.9637275
Kimball, B. A. (2010). *The liberal arts tradition: A documentary history*. UPA.
King, C. J. (2011). Restructuring engineering education: Why, how and when? *Research & Occasional Paper Series: CSHE, 12*(11). https://cshe.berkeley.edu/sites/default/files/publications/rops.12.11.king.engineeringed.9.29.11.pdf
Kline, R. R. (2001). Using history and sociology to teach engineering ethics. *IEEE Technology and Society Magazine, 4*(20), 13–20. https://doi.org/10.1109/44.974503
Koehler, J., Pierrakos, O., Lamb, M., Demaske, A., Santos, C., Gross, M. D., & Brown, D. F. (2020, June). What can we learn from character education? A literature review of four prominent virtues in engineering education. In *2020 ASEE virtual annual conference*. https://doi.org/10.18260/1-2–35497
Koshland, C. P. (2010). Liberal arts and engineering. In D. Grasso & M. B. Burkins (Eds.), *Holistic engineering education* (pp. 53–67). Springer.
Lattuca, L. R., Terenzini, P. T., Volkwein, J. F., & Peterson, G. D. (2006). The changing face of engineering education. *The Bridge, 36*(2), 5–13. www.nae.edu/7631/TheValue-AddedApproachtoEngineeringEducationAnIndustryPerspective
Lucena, J., Downey, G., Jesiek, B., & Elber, S. (2008). Competencies beyond countries: The re-organization of engineering education in the United States, Europe, and Latin America. *Journal of Engineering Education, 97*(4), 433–447. https://doi.org/10.1002/j.2168-9830.2008.tb00991.x
Lyall, C. (2019). *Being an interdisciplinary academic : How institutions shape university careers* (1st ed.). Springer International Publishing. https://doi.org/10.1007/978-3-030-18659-3
Meiskin, P., & Smith, C. (1996). *Engineering labour: Technical workers in comparative perspective*. Verso.
Mikic, B., & Voss, S. E. (2006). Engineering for everyone: Charging students with the task of designing creative solutions to the problem of technology literacy. In *Engineering: Faculty publications*. Smith College. https://scholarworks.smith.edu/egr_facpubs/73 10.18260/1–2–323
Moustafa, A. (2018, March 5). *Do engineers really need a liberal arts education?* [Video]. YouTube. https://youtu.be/NAFc_ZAVB-E
National Academies of Sciences, Engineering, & Medicine. (2018). *The integration of the humanities and arts with sciences, engineering, and medicine in higher education: Branches from the same tree*. The National Academies Press. https://doi.org/10.17226/24988
National Academy of Engineering. (2004). *The engineer of 2020: Visions of engineering in the new century*. The National Academies Press. https://doi.org/10.17226/10999

Newell, W. H. (1999). The promise of integrative learning. *About Campus, 4*(2), 17–23. https://doi.org/10.1177/108648229900400205

Nieusma, D. (2008). Integrating technical, social, and aesthetic analysis in the product design studio: A Case study and model for a new liberal education for engineers. In *ASEE annual conference & exposition*. https://doi.org/10.18260/1-2-4086

Nohara, K., Kawamoto, S., & Kusakabe, O. (2008). An attempt of international liberal arts education for nurturing "global engineers" at Tokyo institute of technology. *Journal of Japanese Society for Engineering Education, 56*(4), 114–122.

Olds, B. M., & Miller, R. L. (1993, March 31–April 3). *A model for professional education in the 21st century: Integrating humanities and engineering through writing*. Paper presentation, conference on college composition and communication.

Olin College of Engineering. (2022, July 6). *It's all systems go with 4S*. www.olin.edu/articles/its-all-systems-go-4s

Ollis, D. F., Neeley, K. A., & Luegenbiehl, H. C. (Eds.). (2004). *Liberal education in twenty-first century engineering: Responses to ABET/EC 2000 criteria*. Peter Lang.

Olson, S. (2016). *Grand challenges for engineering: Imperatives, prospects, and priorities: Summary of a forum*. The National Academies Press. https://doi.org/10.17226/23440

Osgood, K. (21 May 2017). Engineers need the liberal arts too. *The Chronicle of Higher Education*. www.chronicle.com/article/engineers-need-the-liberal-arts-too/

Perignat, E., & Katz-Buonincontro, J. (2019). STEAM in practice and research: An integrative literature review. *Thinking Skills and Creativity, 31*, 31–43. https://doi.org/10.1016/j.tsc.2018.10.002

Pierrakos, O., Prentice, M., Silverglate, C., Lamb, M., Demaske, A., & Smout, R. (2019, October). Reimagining engineering ethics: From ethics education to character education. In *2019 IEEE frontiers in education conference*. https://doi.org/10.1109/FIE43999.2019.9028690

Pierrakos, O., & Stottlemyer, E. (2019, October). Liberal arts engineering education: Reflecting on the past and envisioning the future. In *2019 IEEE frontiers in education conference*. https://doi.org/10.1109/FIE43999.2019.9028604

Prados, J. W., Peterson, G. D., & Lattuca, L. R. (2005). Quality assurance of engineering education through accreditation: The impact of engineering criteria 2000 and its global influence. *Journal of Engineering Education, 94*(1), 165–184. https://doi.org/10.1002/j.2168-9830.2005.tb00836.x

Princeton University. (n.d.). *What does liberal arts mean?* Retrieved July 24, 2022, from https://admission.princeton.edu/academics/what-does-liberal-arts-mean

Richardson, C. (1908). A plea for the broader education of the chemical engineer. *Science, 27*(681), 81–89. https://doi.org/10.1126/science.27.681.81

Riley, D. (2015). Facepalms and cringes: Liberal education misapprehended. *Engineering Studies*, 7(2–3). https://doi.org/10.1080/19378629.2015.1062499

Roche, M. W. (2010). *Why choose the liberal arts?* University of Notre Dame Press.

Rogers, H. L., Hitt, S. J., & Allan, D. G. (2021). Problem-based learning in institutional and curricular design at the new model institute for technology and engineering (NMITE). *Journal of Problem-Based Learning in Higher Education, 9*(2). https://doi.org/10.5278/ojs.jpblhe.vi.6814

Rover, D. (2007). Integrative learning. *Journal of Engineering Education, 96*(3), 275–277. https://doi.org/10.1002/j.2168-9830.2007.tb00936.x

Rudskoy, A. I., Borovkov, A. I., & Romanov, P. I. (2021). Is the transfer of russian engineering education to the American liberal arts system relevant? *Vysshee Obrazovanie v Rossii = Higher Education in Russia, 30*(6), 47–59. https://doi.org/10.31992/0869-3617-2021-30-6-47-59

Ruprecht, R. (1997). Humanities in engineering education. *European Journal of Engineering Education, 22*(4), 363–375. https://doi.org/10.1080/03043799708923468

Russo, M. T. (2007). The newcomers: Humanities in engineering education. In *Proceedings of the international conference on engineering education, Coimbra, Portugal*. http://icee2007.dei.uc.pt/proceedings/papers/144.pdf

Rytivaara, A., & Kershner, R. (2012). Co-teaching as a context for teachers' professional learning and joint knowledge construction. *Teaching and Teacher Education, 28*(7), 999–1008. https://doi.org/10.1016/j.tate.2012.05.006

Sample, S. B. (1988). Engineering education and the liberal arts tradition. *IEEE Transactions on Education, 31*(2), 54–57.

Schwartz, J. (2007, September 30). Re-engineering engineering. *The New York Times Magazine, 94*(L).

Selznick, B. S., Mayhew, M. J., Zhang, L., & McChesney, E. T. (2021). Creating an organizational culture in support of innovation education: A Canadian case study. *Journal of College Student Development, 62*(2), 219–235. https://doi.org/10.1353/csd.2021.0018

Seron, C., & Silbey, S. S. (2009). The dialectic between expert knowledge and professional discretion: Accreditation, social control and the limits of instrumental logic. *Engineering Studies, 1*(2), 101–127. https://doi.org/10.1080/19378620902902351

Snow, C. P. (1959). *The two cultures and the scientific revolution*. Cambridge University Press.

Somerville, M., Anderson, D., Berbeco, H., Crisman, J., Dabby, D., Donis-Keller, H., Holt, S. S., Kerns, S., Kerns, D. V., Martello, R., Miller, R. K., Moody, M., Pratt, G., Pratt, J. C., Shea, C., Schiffinan, S., Spence, S., Stein, L. A., Stolk, J. D., Storey, B. D., Tilley, B., Vandiver, B., & Zastavker, Y. (2005). The Olin curriculum: Thinking toward the future. *IEEE Transactions on Education, 48*(1), 198–205. https://doi.org/10.1109/TE.2004.842905

Stephan, K. D. (2001). All this and engineering too: History of accreditation requirements for non-technical curriculum content in U.S. engineering education 1933–2000. *Proceedings International Symposium on Technology and Society*, 155–161. https://doi.org/10.1109/ISTAS.2001.937733

Tang, X., & Zhu, Q., & Pang, H. (2016, June). Toward a cross-cultural conversation: Liberal arts education for engineers in China and the U.S. In *2016 ASEE annual conference & exposition*. https://doi.org/10.18260/p.27055

Thom, M. (2001). *Balancing the equation: Where are women and girls in science, engineering and technology?* National Council for Research on Women. https://eric.ed.gov/?id=ED455348

Traver, C., & Klein, J. D. (2010, June). Integration of engineering and the liberal arts: A two way street. In *2010 ASEE annual conference & exposition*. https://doi.org/10.18260/1-2–16036

Turbak, F., & Berg, R. (2002). Robotic design studio: Exploring the big ideas of engineering in a liberal arts environment. *Journal of Science Education and Technology, 11*(3), 237–253.

United Nations. (2020). *The sustainable development goals report 2020*. https://unstats.un.org/sdgs/report/2020/The-Sustainable-Development-Goals-Report-2020.pdf

Useem, M. (1989). Liberal education and the corporation: The Hiring and advancement of college graduates. Aldine de Gruyter.

Usher, K., & Sheppard, D. (2017). AIMLED – A New approach to engineering higher education. In *New approaches to engineering in higher education: Proceedings of the conference* (pp. 65–70).

van den Beemt, A., MacLeod, M., van der Veen, J., Van de Ven, A., van Baalen, S., Klaassen, R., & Boon, M. (2020). Interdisciplinary engineering education: A review of vision, teaching, and support. *Journal of Engineering Education, 109*(3), 508–555. https://doi.org/10.1002/jee.20347

Wadhwa, V. (2018, June 12) Why liberal arts and the humanities are as important as engineering. *The Washington Post*. www.washingtonpost.com/news/innovations/wp/2018/06/12/why-liberal-arts-and-the-humanities-are-as-important-as-engineering/

Walther, J., Miller, S. E., & Sochacka, N. W. (2017). A model of empathy in engineering as a core skill, practice orientation, and professional way of being. *Journal of Engineering Education, 106*(1), 123–148. http://dx.doi.org/10.1002/jee.20159

Weil, V. (1984). The rise of engineering ethics. *Technology in Society, 6*, 341–345.

Wilson, M. C. (2000, June). Obstacles to a liberal engineering education. In *2000 ASEE annual conference*. https://doi.org/10.18260/1-2–8601

Winebrake, J. J. (2015). The integrative liberal arts and engineering – The 'grand challenge' of curricular implementation. *Engineering Studies*, 7(2–3), 193–195. https://doi.org/10.1080/19378629.2015.1062486

Wood, A., Arslan, S., Barrett, J., Brownell, S. A., Herbert, A. M., Marshall, M., Oates, K. K., Spanagel, D., Winebrake, J. J., & Zastavker, Y. V. (2019, June). Work in progress: Transformation through liberal arts-focused grand challenges scholars programs. In *2019 ASEE annual conference & exposition*. https://doi.org/10.18260/1-2–33657

Xu, M., Crittenden, J. C., Yongsheng, C., Thomas, V. M., Noonan, D. S., Desroches, R., Brown, M. A., & French, S. P. (2010). Gigaton problems need gigaton solutions. *Environmental Science & Technology, 44*(11), 4037–4041. https://dx.doi.org/10.1021/es903306e

Part 5

Engineering Education at the Intersection of Technology and Computing

22

Engineering Education and Online Learning

Andrea Gregg and Nada Dabbagh

1 Introduction

Historically, engineering education has had an ambivalent relationship to online education, generally being outpaced by other disciplines in terms of leveraging this approach to reach students (Bourne et al., 2005; Kinney, 2015; Seaman et al., 2021; Tabas et al., 2012). While some of this concern has been based on legitimate pedagogical challenges, such as the hands-on nature of lab courses and the importance of collaborative project work, it also stems from concerns about the changing role of faculty, a lack of positive experiences teaching online, and concerns about the quality of online learning overall (Bourne et al., 2005; Kinney, 2015; Seaman et al., 2021). Additionally, higher education's response to the COVID global pandemic brought online engineering education top of mind for engineering faculty and students, which unfortunately for many led to the problematic conflation of emergency remote teaching (ERT) with intentionally designed online courses (Hodges et al., 2020). Importantly, online distance learning has an established history of both practice and research, with well-established instructional design methods, foundational learning theories, and corresponding quality standards.

In the 2014 edition of the *Cambridge Handbook of Engineering Education Research*, it was suggested that:

> Designing learning environments without learning theory is comparable to designing a bridge without mechanical laws and principles. In both cases, the goal is unlikely to be accomplished; the learner fails to change in desired ways and the bridge collapses.
>
> *(Newstetter & Svinicki, 2014, p. 29)*

In this updated *Handbook of Engineering Education Research*, we extend Newstetter and Svinicki's (2014) analogy to suggest that offering online engineering education *without* consulting the corresponding online learning theories and frameworks as well as instructional design best practices contributes to problematic online teaching and learning experiences. These problematic experiences frustrate both faculty and students and perpetuate the off-stated but ill-supported belief that *online learning just doesn't work for engineering*. A primary goal of this chapter, therefore, is to invite a mindset shift when it comes to how engineering educators think about online learning. If engineering educators embrace the same innovative thinking that has led to so many other amazing advances, from

DOI: 10.4324/9781003287483-27

the gas turbine to the electric car, we can become a leader with regard to the potential of online education to expand the positive reach of engineering education globally. This chapter is organized in three major sections: (1) history, context, and theory of online learning; (2) online learning within engineering education; and (3) looking to the future.

In the first section – *history, context, and theory of online education* – we establish the pedagogical history and contemporary context for online learning, building on centuries of research and practice in distance education. We distinguish emergency remote teaching (ERT), which largely involves the often-rushed remote delivery of a course designed to be taught residentially, from intentionally designed online courses, which are learning experiences designed specifically to maximize the affordances and minimize the constraints of online learning contexts. We then highlight theoretical frameworks which guide the intentional design of engaged and meaningful online teaching and learning experiences. In the next section – *online learning within engineering education* – we focus specifically on engineering education and its place within this online learning landscape, looking at successful examples of online offerings as well as significant ways in which engineering education is arguably behind where it could be, missing opportunities to innovate and reach more students, given the gap in qualified engineers. We look at legitimate challenges posed by elements, such as hands-on labs and project-based learning as well as potential solutions to doing these things online. In the final section – *looking to the future* – we turn our attention to what is (and should be) next for engineering education, arguing the need for engineering education to embrace the innovation at the heart of the discipline to not only leverage but also make better online education.

2 History, Context, and Theory of Online Education

2.1 History

Distance education was established in the 1800s in Europe, when it took the form of "correspondence courses" or "extension courses," crossing the Atlantic in 1873 (Schlosser & Simonson, 2003; Simonson et al., 2009). Correspondence courses were delivered primarily through print media, with the content segmented into manageable units providing a lot of structure to ensure success (Dabbagh & Bannan-Ritland, 2005). The original target groups of distance education efforts were adults with occupational, social, and family commitments, and the focus was on "individuality of learning and flexibility in both time and place of study" (Hanson et al., 1997, p. 4). Guided readings, frequent tests, and free pacing of progress through the program by the student were key elements of classic distance education. This instructional approach seems to align with current practices of self-paced online courses and adaptive learning technologies, or what is referred to in some contexts as "personalized" or "individualized" learning (Domenech et al., 2016). These models of distance education can be found in online settings, like LinkedIn Learning, Khan academy, and many massive open online courses or MOOCs. The emphasis is typically not on interaction with others but instead on interaction between the learner and the learning materials, that is, learner–content interaction.

These correspondence courses benefited from the planning, guidance, and pedagogical practices of an educational organization without being under the continuous and immediate supervision of teachers present with their students in lecture rooms or on the premises. Moore (1994) refers to this type of correspondence study as non-autonomous or teacher-determined and gauges the degree of learner autonomy by determining how much guidance a learner needs in formulating objectives, identifying sources of information, and measuring objectives. Moore notes that in most conventional educational programs, resident or distance, the learner is very dependent on the teacher for guidance and the teacher is active while the student is passive.

These types of learning environments are known as *directed* learning environments (Hannafin et al., 1997), because they embody conventional instructional approaches typically found in face-to-face classroom learning and classic forms of computer-assisted instruction (CAI) (Dabbagh & Bannan-Ritland, 2005). Directed learning environments reflect an objectivist epistemology, where learning is driven by externally generated objectives (objectives generated independently of the learner) and via explicit activities and practice, with the goal of transmitting a discrete and well-defined body of knowledge. Content is therefore separated from the contexts in which it naturally occurs and is structured according to tasks, objectives, and prerequisites.

There is, however, another perspective on traditional distance learning in which the program is more responsive to students' needs and goals, and the student "accepts a high degree of responsibility for the conduct of the learning program" (Hanson et al., 1997, p. 9). Moore (1994) and Wedemeyer (1981) use the term "independent study" for such programs, acknowledging the very important characteristic of distance education: the independence of the student. This independence is traditionally characterized by self-study, pacing, and progress and, in some cases, extends beyond the self-study autonomy by allowing the student to select the learning objectives, resources, context of study, and deliverables. This "independence" is also known as self-directed learning, where the student is an active participant in the learning process, reflecting a more constructivist and personalized approach to distance learning.

Throughout the 1900s, radio broadcasting and, later, satellite television delivered content to physically distanced students. But the creation of the Internet in the late 1900s changed everything. With the Internet, *online learning* became possible. Online learning, in its simplest form, can be described as any learning that takes place using the Internet as a delivery system (Dabbagh et al., 2019). However, Lowenthal and Wilson (2010) argue that definitions of online learning are continuously emerging. Terminology such as e-Learning, microlearning, and blended or hybrid learning are often used interchangeably and inconsistently across practitioners, researchers, and policymakers (Moore et al., 2011; Mayadas et al., 2015). Therefore, it is best to think about online learning as a range of pedagogical practices or methods made accessible by the Internet that include learning experiences where students work primarily independently or autonomously, experiencing little or no interaction with an instructor or other learners, to learning experiences where students are highly engaged in interactive and collaborative learning with the instructor and peers. Just as there is no singular in-person course experience – consider the differences between hands-on labs, project-based courses, and lecture courses – there is no singular online course learning experience.

2.2 Contemporary Context for Online Learning

Learning technology, specifically information communication technology, or ICT, has played a significant role in realizing the modern meaning of distance education. Recent advances in ICT have redefined the boundaries and interactional pedagogies of a traditional distance learning environment by stretching its scope and deepening its interconnectedness. New learning interactions that were not perceived possible before can now be facilitated, such as the coupling of experts from all around the world with novices, the accessibility of global resources, the opportunity to publish to a world audience, the opportunity to take virtual field trips and participate in virtual labs, the opportunity to communicate with a wider range of people, and the ability to share and compare information, negotiate meaning, and co-construct knowledge. These activities emphasize learning as a function of interactions with others and with the shared tools of the community. In other words, learning can be viewed as a social process in which social interaction plays an integral part in the learning process and the emphasis is on acquiring useful knowledge through enculturation (understanding how knowledge is used by a group of practitioners or members of a community). Even MOOCs, which can be predominately or wholly self-study, rely on ICT to connect experts with learners, highlighting the human element of teaching through largely video-based instructional delivery, and are often accompanied by global learning communities.

2.2.1 Transactional Distance Theory

In educational settings, these educational interactions are manifested in learner–instructor interaction, learner–content interaction, and learner–learner interaction (Moore & Kearsley, 1995), also known as Moore's theory of transactional distance (1993, 2019). Moore's theory of transactional distance (1993, 2019) explains and quantifies the learning relationship between instructor and student in an online learning context where there is a substantial physical or temporal distance between the two. First formulated in 1993, transactional distance theory, or TDT, considered the many different forms of distance learning interactions – such as learner–instructor, learner–content, and learner–learner – perceived as necessary for enhancing social learning skills, such as communication or group-process skills. These learning interactions are also perceived as tools or activities that promote higher-order thinking and sustain motivation in online distance education (Navarro & Shoemaker, 2000). Research on types of interactions is one of the more robust bodies of research in online learning (Hodges et al., 2020). It shows that the presence of each of these types of interactions, when meaningfully integrated, increases achievement of the learning outcomes.

As Hodges et al. (2020) purport:

> Careful planning for online learning includes not just identifying the content to cover but also carefully tending to how you're going to support different types of interactions that are important to the learning process. This approach recognizes learning as both a social and a cognitive process, not merely a matter of information transmission.

Based on transactional distance theory, we conceptualize online learning as the "deliberate organization and coordination of distributed forms of interaction and learning activities to achieve a shared goal" (Dabbagh & Bannan-Ritland, 2005, p. 12). The following attributes apply to this definition:

1. Globalization and learning as a social process are inherent and enabled through telecommunications technology.
2. The concept of a learning group is fundamental in achieving and sustaining learning.
3. The concept of distance is relatively unimportant or blurred and does not necessarily imply the "long-distance" physical separation of the learner and the instructor.
4. Teaching and learning events (or course events) are distributed over time and place, occurring synchronously and/or asynchronously using different media.
5. Learners are engaged in multiple forms of interaction: learner–learner, learner–group, learner–content, and learner–instructor.
6. Internet, web-based, and mobile technologies and applications are used to support the teaching and learning process and to facilitate learning and knowledge building through meaningful action and interaction.

In order to better understand how to support these types of learning interactions online, we describe the types of delivery formats and modalities that constitute the core of online learning in higher education contexts.

2.2.2 Delivery Formats of Online Learning

A major consideration when defining online learning in higher education contexts is based on the amount of time spent in a physical classroom compared to the amount of time spent in online activities. This is known as the course delivery format. For example, a web-enhanced or web-supported course utilizes technology as a supplement to traditional classroom activities – usually no more than

20%. Blended or hybrid courses retain some face-to-face elements, but a large portion of instruction takes place online, typically 50–80%. The degree of face-to-face versus online instruction is largely determined by individual institutions who may use blended courses for a variety of reasons, including freeing classroom space to offer more course sections or to allow some flexibility for students (Dabbagh et al., 2019). In contrast, a fully online course consists of no face-to-face contact: all learning takes place via the Internet. There is also the HyFlex course delivery format that combines face-to-face and online learning. A HyFlex course allows interaction between students meeting face-to-face with an instructor in a physical classroom as well as with students at a remote location using synchronous and asynchronous technologies. Students can attend live in-person, live online (synchronously), or they can participate asynchronously (Milman et al., 2020).

2.2.3 Communication Synchrony of Online Learning

Communication synchrony, or modality, of online learning refers to the type of communication that happens online. There are two types of online communication: synchronous and asynchronous. Asynchronous communication allows learners to complete course activities at their own pace, time, and choosing. There may be time constraints on when assignments must be completed (e.g., one-week learning units, project, or discussion posts due dates), but within those constraints, the learner can work on their own schedule to complete coursework. Asynchronous communication methods include audio/video recorded lectures, discussion forums, interactive video lessons, collaborative wikis, social media learning activities, and games and simulations. In contrast, synchronous communication requires learners to be online at designated scheduled times. Synchronous communication methods include instant messaging, live chats, videoconferencing, and live broadcasting. Zoom, Microsoft Teams, and Skype are among the technologies that can be used to facilitate synchronous online learning. While there are advantages to this modality of online learning, such as engaging learners in live discussions, there are some disadvantages that could impede participation. A synchronous learning environment may not be an issue when learners have compatible schedules and are in geographical proximity; however, the lack of flexibility can be limiting. For example, working professionals with family and other obligations may find it difficult to be online at a prescribed time. If synchronous scheduling is planned, offering choices to accommodate learners is suggested (Dabbagh et al., 2019). An exception to this would be fully online programs in which students entered the program aware of the synchronous requirements and are able to meet them.

Overall, synchronous online learning can be more impactful, engaging, and community-centered, fostering high social presence, while asynchronous online learning offers more flexibility, time on task, more time to think deeply about the content and reflect on one's learning and is more self-directed and self-paced. Hence, there are instructional benefits (and drawbacks) to offering synchronous or asynchronous online learning. That is why it is critical to intentionally design an online course. And by *intentionally* we mean applying the systematic process of instructional design that considers several factors and moderating variables when designing an online course (Hodges et al., 2020). These factors include but are not limited to online communication synchrony (asynchronous only, synchronous only, blend of both), learning outcomes, the target audience, subject matter, delivery modality (online, blended, HyFlex), pacing, student–instructor ratio, pedagogy (instructional approach), instructor role online, student role online, role of online assessments, sources of feedback, and last but not the least, the learning technology.

2.2.4 Online Learning Defined

As mentioned earlier, *online learning*, in its simplest form, can be described as any learning that takes place using the Internet as a delivery system. However, the more contemporary and research-based definition describes online learning as an "open and distributed learning environment that uses pedagogical and technological tools, to facilitate learning and knowledge building through meaningful action and interaction" (Dabbagh et al., 2019). More specifically, online learning is conceptualized as the deliberate or intentional organization and coordination of distributed forms of interaction and learning activities to achieve a shared goal. Based on this understanding, it becomes extremely critical to differentiate between emergency remote teaching (ERT) and online learning. As Hodges et al. (2020), postulate:

> Well-planned online learning experiences are meaningfully different from courses offered online in response to a crisis or disaster. Colleges and universities working to maintain instruction during the COVID-19 pandemic should understand those differences when evaluating this emergency remote teaching.

Forced to go totally online during the COVID-19 pandemic, professors and teachers in higher education and K–12 contexts rushed to transition their courses and lesson plans to a fully online format without having the luxury of time, knowledge, or support to engage in effective and principled online learning design. The result was suboptimal learning experiences adding to the stigma that online learning is subpar or lower quality than face-to-face learning, a stigma that online learning already carries, despite research showing otherwise (Hodges et al., 2020).

Even prior to ERT due to the COVID-19 pandemic, many instructors in higher education contexts made and continue to make the mistake of thinking that an online course can be easily created by uploading lecture notes, creating online tests, and including some PowerPoint files and web links (Dabbagh et al., 2019). While it is true that a course can be easily developed that way, it certainly will not result in an effective and engaging learning experience. Students would likely be disengaged from each other, forfeiting opportunities for learning with and from one another. Such courses, developed without sufficient planning and with no instructional design guidance, model passive learning, where students are receiving information to remember and restate without any real thinking or application. The instructor decides what is to be learned, and students have no incentive to engage deeply with the concepts. This contrasts with active learning described as "involving students in doing things and thinking about the things they are doing" (Bonwell & Eison, 1991, p. 2), which is highly important in engineering education (Prince, 2004). Thus, online learning must be deliberately and intentionally designed to be meaningful and engaging, and by *meaningful* we mean learning that has value, purpose, and significance (Meaningful, n.d.) and learning that is grounded in learning theory.

2.3 Online Learning Theories, Models, and Frameworks

There are several learning theories and frameworks that are foundational to online learning. In this section, we highlight the most current and consequential in supporting *meaningful* online learning. This is consistent with how we defined online learning earlier in this chapter and with what we believe online engineering education should aspire to be. Meaningful learning contains elements of several constructivist and social constructivist learning theories which grew in part from the work of Dewey and Piaget. One such learning theory is situated cognition, which is based on Bandura's social learning theory (1977). Situated cognition emphasizes the importance of context and posits that knowledge and skills must be connected to the context in which they will be used rather than being taught in isolation (Lave & Wenger, 1991). Instructional models that support the principles of situated cognition include:

- Communities of practice (CoP), where novice learners are interacting with experts, learning from each other through observation, imitation, and modeling;
- Cognitive apprenticeships (CA), where experts coach and scaffold novice learners in a real-world context (Brown et al., 1989);

- Situated learning (SL), which enables the exploration of authentic scenarios, cases, or problems;
- Problem-based learning (PBL), which emphasizes collaborative problem-solving of complex problems; and
- Goal-based scenarios (GBS), simulations, and game-based learning (GBL) environments that emphasize learning by doing and immersive learning.

These constructivist-based instructional models are particularly relevant for engineering education, where students need **authentic and experiential** learning environments to demonstrate their expertise and learn how to think and act like engineers. For example, PBL is being increasingly adopted in engineering education to provide opportunities for engineering students to learn real-world problem-solving skills. As Newstetter and Svinicki (2014) posit, "too often in engineering classrooms, the instructional activities required of the students are not aligned with the kind of knowledge those activities are intended to foster" (p. 43). Constructivist-based instructional models bring the situative, cognitivist, and social together in instructional design. Additionally, what characterizes these learning environments is active engagement, interaction between instructors and students, and high-quality learning experiences that must be considered when designing online learning. We briefly describe two foundational instructional design models that support the design and development of meaningful online learning experiences: the Community of Inquiry (COI) framework and the Meaningful Online Learning Design (MOLD) framework. Both COI and MOLD can be used to design authentic and experiential online engineering education.

2.3.1 Community of Inquiry (COI) Framework

The Community of Inquiry (COI) framework is a social constructivist model of learning processes in online and blended environments (see Figure 22.1). First developed by Garrison et al. (2000) and further conceptualized by Garrison and Anderson (2003), and later by Garrison (2017) to reflect its implications on e-Learning, the model describes how learning takes place for a group of individual learners through the educational experience that occurs at the intersection of social, cognitive, and teaching presence. COI helps online instructors create online communities using three essential elements:

- *Teaching presence.* Defined as the design, facilitation, and direction of cognitive and social processes for the realization of meaningful learning. This involves the (1) instructional design and organization of the course and activities, (2) facilitation of the course and activities, and (3) direct instruction.
- *Social presence.* Refers to the ability to perceive others in an online environment as "real" and the projection of oneself as a real person. Social presence involves open communication, affective expression, and group cohesion.
- *Cognitive presence.* Refers to the extent to which learners are able to construct and confirm meaning through sustained reflection and discourse.

The ultimate goal of the COI framework is to build a solid foundation of social presence and teaching presence to stimulate cognitive presence in a course (Huang et al., 2020).

2.3.2 Meaningful Online Learning Design (MOLD) Framework

The Meaningful Online Learning Design (MOLD) framework supports the design of constructivist learning environments, such as the ones described earlier (e.g., COP, PBL, CA, GBS, GBL), through the integration of active, constructive, intentional, authentic, and cooperative learning experiences

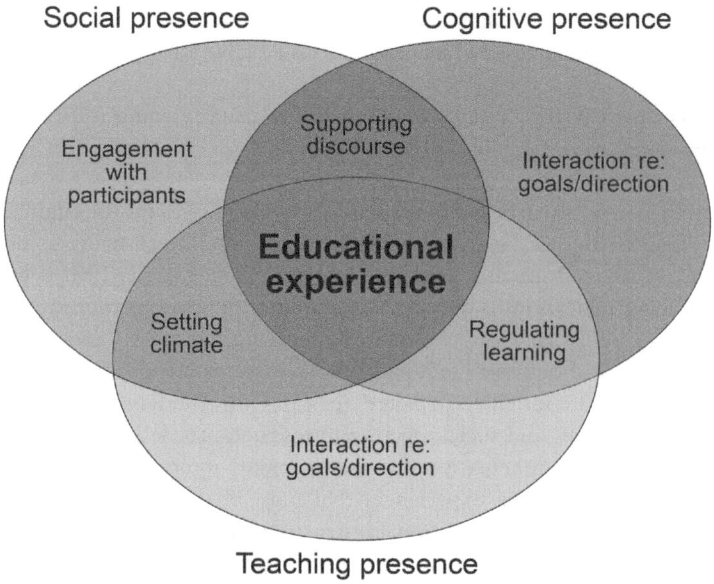

Figure 22.1 Community of Inquiry model. CC BY-SA 3.0.
Source: https://creativecommons.org/licenses/by-sa/3.0, via Wikimedia Commons.

using a variety of instructional strategies and learning technologies (Dabbagh et al., 2019). Creating meaningful learning opportunities in engineering education is not only possible but also essential to fostering authentic and experiential learning.

So how do we create meaningful learning experiences in online settings? We ensure that:

- Learning is active (manipulative/observant), which means learners are intimately engaged with the environment, taking on the role of an informal scientist and observing the consequences and results of their actions. We ensure that the learning activities we design engage the learner in manipulative and observant behavior.
- Learning is constructive (articulate/reflective), which means learners should be given the opportunity to articulate their understanding of the subject matter and reflect on their learning. We ensure that the learning activities we design engage learners in constructing their own simple mental models that explain what they observe, and with experience, support, and more reflection, their mental models become increasingly complex.
- Learning is intentional (goal directed/regulatory), which means when learners are actively and deliberately working toward a cognitive goal, they think and learn more, because they are fulfilling a personal intention. Self-directed learning requires the learner to engage in metacognitive self-regulatory strategies that include organizing, time management, and self-discipline.
- Learning is authentic (complex/contextual), which means we create learning tasks that are situated in some meaningful real-world task or simulated in some case-based or problem-based learning environment so that learners not only better understand the material but are also able to transfer this understanding to new situations.
- Learning is cooperative (collaborative/conversational), which means we create learning activities that engage learners in social activity, working together in communities and taking advantage of each other's skills and knowledge to support their goals and actions.

The Meaningful Online Learning Design (MOLD) framework (see Figure 22.2) posits that to create active, engaging, and authentic online learning environments, designers must consider the meaningful learning characteristics of active, constructive, intentional, authentic, and cooperative learning when selecting instructional strategies, learning technologies, and learning activities, the three components represented in the triangle in Figure 22.2.

What differentiates MOLD from other instructional design models is that it allows the instructional designer or online instructor the flexibility to *start* the design of online learning interactions and experiences with any of the three components depicted in the triangle. Once a starting point is selected, instructional designers or online course instructors can proceed to integrate the other two components based on the pedagogical affordances these components instantiate or support. The decision regarding which component to consider first is largely based on the particular instructional context and the expertise of the instructional designer or online instructor. For example, if a learning technology for online learning has already been selected by the institution or organization, such as the learning management system (LMS), the instructional designer or instructor can begin exploring the pedagogical potential of the learning technology and proceed to select appropriate instructional strategies and learning activities to ensure overall instructional effectiveness and compatibility of the learning design. Alternatively, a college professor who may be more experienced in pedagogical approaches related to the subject matter of instruction can choose to start with a familiar instructional strategy and proceed to explore corresponding learning activities and learning technologies to create an integrated learning design. Another unique feature of MOLD is its emphasis on learning technologies as a key component in the overall design process. Rather than treating technology as a delivery vehicle or a transmissive educational technology (Jonassen, 2000), technology is placed on an equal footing with the other two components to ensure that the pedagogical affordances that technology brings forth to a learning situation are given appropriate consideration.

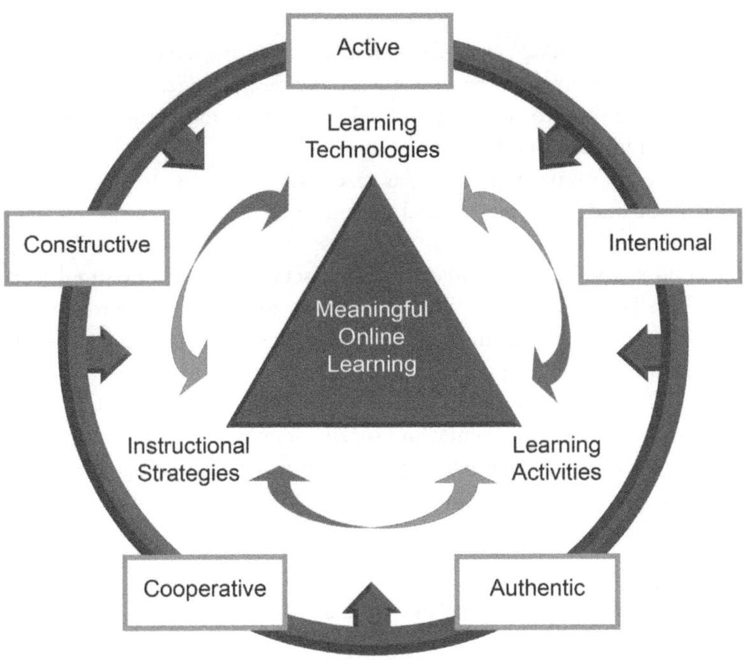

Figure 22.2 Meaningful Online Learning Design (MOLD) framework.

The following scenario provides a context for implementing the exploratory instructional strategies of *problem-solving, hypotheses testing, exploration and creation, and role-playing* in an online setting to support meaningful online learning (Dabbagh et al., 2019, pp. 44–45).

A sculptor's goal is to creatively express his or her vision in three dimensions. However, the ultimate success or failure of the work also depends on the sculptor's ability to envision the many practical variables that influence the final outcome. Fine arts students in the studios at L'Ecole usually discover this the hard way. For instance, a student often finds out only after the final casting or sculpture that his or her selection of materials does not support the design model. Works of art are not exempt from the laws of physics, installation parameters, budgets, material availability, or applicability. Currently, students have no reliable methods for analyzing the feasibility of their constructions. How can they predict whether a 3D object of significant scale can stand erect, hang, or fly, or whether a 100-foot-by-100-foot-by-1-inch curvilinear shape can be constructed from Styrofoam with no visible means of support? Furthermore, fine arts students usually fail to calculate whether an installation can be accomplished within the given budget. To eliminate the trial-and-error approach students often use in the design process, L'Ecole is investing in a technology-based environment to assist students with creative and practical planning issues. The system will help students visualize their concept in three dimensions and make decisions that determine "if" and "how" the piece can exist in the real world. Students can manipulate design aspects within the system and see the effects of their decisions on the end product. By understanding the many facets of a sculptor's work, the students can ensure that their creative visions will take shape.

Potential learning outcomes for this scenario include enabling students to:

- Predict real-world structural integrity of prototypes.
- Analyze the physical properties and integrity of materials used in their sculptures.
- Judge the applicability of various materials to a particular project.
- Develop abstract mental models of the relationship between art concepts and materials or media.
- Organize and balance a multiplicity of artistic considerations, such as scale, texture, and surface.
- Follow a sound decision-making process during the design stage.
- Use problem-solving skills to address potential production and design problems.
- Analyze the spatial impacts and limitations of installation.
- Select the best tools for sculpting particular materials.
- Determine realistic budgets for individual pieces.

Considering the context and learning outcomes of this scenario, an online simulation or 3D immersive virtual world can be developed with the following pedagogical affordances: engage learners in exploratory and experiential learning through learning by doing and role-playing, provide a controlled environment in which hypotheses testing and problem-solving can occur and learning from trial-and-error is supported, provide a wide variety of versatile tools that learners may use to interact with and manipulate environmental and social parameters, provide simple ideas and methods that are grounded in an authentic context, support personalized learning experiences, and support just-in-time feedback. These types of online simulations and 3D immersive virtual environments are also fully applicable to engineering education.

3 Online Learning Within Engineering Education

Having established the history, context, and foundational learning theories for online education, we now look specifically at online learning within engineering education considering the relative pace of adoption, concerns of some engineering educators regarding online learning, a possible model for online engineering education from another applied discipline, and successful examples of online learning in engineering contexts.

3.1 Pace of Adoption

In terms of a big-picture understanding of the online higher education landscape, national survey reports combined with the Integrated Postsecondary Education Data System (IPEDS) data show that student enrollments in online undergraduate and graduate courses and programs have consistently increased since 2002.[1] And since 2012, growth in online enrollments has been outpacing growth in residential enrollments (Seaman et al., 2018). Prior to the COVID pandemic, the number of students studying in-person on campus had declined significantly between 2012 and 2019 by nearly 2 million people, going from 18.3 million to 16.4 million. At the same time, students taking at least one online course increased by roughly 2 million during that same time from 5.4 million to 7.4 million (Seaman & Seaman, 2019).

Considered in the context of higher education overall, engineering has been proportionally slower to leverage online teaching and learning (Allen & Seaman, 2008; Allen & Seaman, 2011; Seaman et al, 2021; Bourne et al., 2005; Kocdar et al., 2020; Kinney, 2015). As of this writing, there are currently only 3, out of a total of 684, institutions offering ABET-accredited engineering programs available to fully online students (ABET, n.d.a). While there are likely many more institutions with individual engineering hybrid and online courses, less than 0.5% of ABET-accredited engineering programs offer online engineering degrees. This <0.5% is especially notable when considering that as of 2019, 14.4% of all higher education students were enrolled in fully online programs (Seaman & Seaman, 2019).

Perhaps because of the relatively low numbers of online engineering offerings, there also seems to be a limited emphasis on online learning and education within engineering education research. The *Cambridge Handbook for Engineering Education Research* (CHEER) described the domain of engineering education research as focused on "virtually all aspects of formal and informal learning systems," including elements such as key stakeholders (students and faculty), system constraints (economic, social, and political), assessment, curriculum, and teaching and learning (Fortenberry, 2014, p. xiv). CHEER was also described as "the critical reference source for the growing field of engineering education research, featuring the work of world experts writing to define and inform this emerging field" (Johri & Olds, 2014, front material). Of the 35 CHEER chapters, there were only two chapters that referenced either of the terms "online learning" or "online education." Notably, the reference to "online learning" was solely to clarify that online learning would not be addressed in the chapter. By way of comparison, there were six chapters that referenced "active learning," 19 chapters that referenced the term "lab" or "laboratory," and 19 chapters that referenced the term "lecture."

Like *CHEER*, the *Journal of Engineering Education* (JEE), the flagship journal in the field of engineering education, also identifies a broad research agenda with an emphasis on areas such as learning, systems, diversity, and assessment. Using the Wiley Online Library interface to search all abstracts across all dates within the JEE database for the term "online learning" or "online education" yielded 5 publications, 2 of which were specific to COVID. Using the same search parameters to search JEE with the term "active learning" yielded 31 publications, "lab" yielded 24 publications, and "lecture" yielded 77 papers. The *CHEER* and *JEE* cases provided here clearly do not represent a systematic literature review but simply provide a glimpse into the engineering education scholarly discussion of online learning relative to other topics.

3.2 Primary Concerns

By way of potential explanation for some of the gap between national online higher education trends and engineering education, there are key elements central to engineering that have and can pose challenges in online contexts. As the editor of the *Journal for Online Engineering Education* wrote in its inaugural issue: "I often find that when I talk to colleagues about online engineering education the first response is about what cannot be taught online" (Reynolds, 2010). In their heavily cited JEE publication, Bourne et al. (2005) asked, "[W]hy has undergraduate engineering education lagged behind some other fields in adopting online methodologies?" (p. 132) and suggested that the answer lies, at least in part, in the hands-on labs and the heavy use of quantitative reasoning central to engineering. Regarding engineering courses heavy in diagraming and math notations, certainly, prior to the invention and proliferation of smartphones, tablet devices, and specialized quantitative keyboards, these were especially challenging (ElSheikh & Najdi, 2013; Smith et al., 2004). Now, however, students have more options, including the use of traditional "pen-and-paper" combined with digital scans/images that are uploaded, working directly on tablet devices, and/or faxing or scanning paper-based solutions. Highly collaborative project-based courses have also been raised as problematic in online contexts when team members are separated by both time and geography (Beneroso & Robinson, 2022; Scholes, 2021). Additionally, hands-on labs have long been a central element of engineering education. Options for approaching both project- and lab-based courses online are further discussed in the following.

These pedagogical concerns about online education are well-captured in the *International Journal of Engineering Education*'s (IJEE) call for its special issue, "Engineering Education Everywhere: Good Practices for Emergency Situations and Remote Regions," which begins by emphasizing the importance of in-person engineering education.

> Higher education in general greatly benefits from face-to-face interactions between students and professors, for making the educational experience more effective, efficient and human. In the particular case of engineering education, the fundamental relevance of hands-on activities, of practical tasks in workshops and laboratories and of employing software resources in collaborative working environments, among others, support the acquisition of important professional skills.
>
> *(Lantada & Nuñez, 2020)*

Regarding online labs, a systematic literature review employing text-mining on 120 peer-reviewed publications with a focus on online learning and engineering education identified "virtual and remote labs as a learning environment" as a key theme throughout the literature (Kocdar et al., 2020). Hands-on lab experiences where undergraduate engineering students demonstrate that they have "an ability to develop and conduct appropriate experimentation, analyze and interpret data, and use engineering judgment to draw conclusions is a hallmark of engineering education" (ABET, n.d.b). We direct your attention to Chapter 24, Online Laboratories in Engineering Education Research and Practice, in this Handbook, for in-depth coverage of this topic.

While some of the gap between national online higher education trends and engineering education specifically results, at least in part, from legitimate pedagogical challenges, there also persists misinformation and a general lack of understanding about the history and reality of online learning (Bourne et al., 2005; Kinney, 2015). This is supported by the findings of a mixed-methods study investigating engineering faculty perspectives on online learning which found that the faculty members who had taught at least one course online prior to completing a survey related to the potential of online learning were more likely to believe that the same learning outcomes as a face-to-face course could be achieved in an online course. In contrast, those who had never taught online did not think equal outcomes were possible (Kinney, 2015).

It has been suggested by some within engineering education that online education is an "unproven pedagogical approach" (Tabas et al., 2012). In reality, online learning is no more "unproven" than its in-person counterpart. As discussed at length in the first part of this chapter, online education has a long history, building on centuries of distance education, in which learners and instructors have been geographically and often temporally separated and also leveraging ICT, as well as corresponding theoretical frameworks and models such as transactional distance, community of inquiry, and meaningful online learning.

It has also been asserted that "[t]he level of which online education is successful compared with that of the traditional face-to-face model lacks a true measure" (Tabas et al., 2012). This statement is belied by the reality that during the past decades of growth in online higher education, online education has undergone intense and persistent scrutiny in terms of quality. For decades, scholars have conducted comparative research between online and in-person courses considering outcomes such as student performance, retention, and satisfaction. There are active research centers dedicated to the study of online learning which catalog these efficacy studies. The Distance Education and Technology Advancements (DETA) center manages the *No Significant Difference* database (No Significant Difference, n.d.), and Oregon State's Ecampus Research Unit facilitates the *Online Learning Efficacy Research* database (Oregon State University eCampus Research Unit, n.d.).

Additionally, meta-analyses have found, when looking across these comparative studies, that no significant differences exist between the two environments (Bernard et al., 2004; Russell, 1999). In addressing the variability of the individual studies in terms of their comparative findings, Bernard et al. (2004) explain that:

> [M]any applications of DE [distance education] outperform their classroom counterparts and that many [applications of DE] perform more poorly.

In other words, factors other than the course setting alone (i.e., online, blended, face-to-face) impact quality and efficacy measures. For instance, the experiences of instructors and students of ERT during the COVID pandemic should not be taken to reflect the overall quality of online learning. It should not be surprising that courses designed to be taught in face-to-face settings that were rapidly transitioned for delivery online in a time of unprecedented global stress with little to no instructional design were experienced by many as less-than-ideal educational experiences.

Toward this end, there are quality models like the Quality Matters rubrics and the Online Learning Consortium quality framework for evaluating online courses against research-supported standards specific to online learning rather than comparing those courses to in-person practices. Some have suggested that the time has come to stop comparing online learning to the classroom as if the classroom model represented *the* ideal for teaching and learning. In this vein, Abrami et al. (2011) ask:

> How far would our understanding of automotive technology have progressed, for instance, if cars (i.e., "horseless carriages") were still designed as alternatives to horses?
>
> *(p. 98)*

This is an especially apt analogy for engineering educators. And as engineering education continues to better leverage opportunities offered by online education, it can be useful to look at successful models from another applied field.

3.3 Lessons from Nursing Education

While the differences are more easily recognized – nursing is female-dominated, while engineering is the opposite; nursing is classified as a "soft" discipline, while engineering is "hard" – nursing education and engineering education have a lot of important similarities, especially when it comes to considering successful models for online teaching and learning. Nursing, like engineering, involves significant elements of hands-on work and the need to be able to function successfully in real-world settings. Much like engineering, nursing is an applied field which "requires discipline-specific psychomotor, cognitive, and affective skills" (Jones et al., 2020). Additionally, both disciplines struggle to recruit and retain enough students to meet societal demands.

Online nursing education, especially for those with a registered nursing (RN) degree but without a bachelor's of science (BS) degree, continues to grow (Smith et al., 2009), potentially in part in response to the current shortage of qualified nurses (Haddad et al., 2022). Some challenges for nursing education present in online contexts include the necessity of establishing an active and engaged community of students and instructors in asynchronous settings (Jones et al., 2020; Smith et al., 2009), concerns about the isolation of nursing students, and the need to provide hands-on, real-world experiences essential for both accreditation purposes and for graduating nurses with the requisite skills and abilities (Mitchell & Delgado, 2014).

Online nursing education has approached the requirement of a practicum experience where students work in a clinical setting under the guidance of a preceptor through a very structured process. This process includes program support of students to identify local clinical settings and qualified preceptors. The online students can continue studying at a distance while still gaining the requisite local clinical experience (Mitchell & Delgado, 2014). There are, of course, challenges, such as the difficulty of locating clinical settings for all students and finding qualified preceptors, and this is where the program support is essential. Another model used in online nursing education across the BS degree involves the separation of non-clinical work, which is completed through online learning, and that of clinical work, which requires in-person applications and evaluations (Ota et al., 2018). This model does restrict students who may not be able to travel for the clinical requirements but still provides a predominately online learning experience without sacrificing the psychomotor and applied elements of the degree. Nursing education, like engineering education, requires an active learning community where students do not simply passively receive content but also actively construct it, which requires strategies for teaching and social presence, such as instructor communication and the modeling of constructive interactions (Jones et al., 2020).

3.4 Successful Examples of Online Engineering Education

Engineering education does not have to rely only on other disciplines for good examples, as there are also important cases of successful online learning already taking place within engineering. One such example is Stanford University's Architecture/Engineering/Construction (AEC) online course, which is part of Stanford's AEC-Global Teamwork Master's Program (Araújo, 2019). The AEC course is a two-quarter interdisciplinary course that engages yearly about 30 architecture, structural engineering, building systems MEP (mechanical, electrical, and plumbing) engineering, life cycle financial management, and construction management majors from many countries worldwide (Fruchter, 2014). The course uses project-based learning as an active teaching strategy for online learning. Project-based learning is a variant of problem-based learning (PBL), mentioned earlier in this chapter, that focuses on real-world projects, preceded by theoretical lectures related to the problem at hand, and engages students in the "know how" and "know why" approach for training professional functions and methodological skills of problem analysis and application. The course uses an immersive (3D) virtual learning platform called Terf. Terf provides the creation of virtual and immersive 3D collaborative spaces for sharing content and generating an environment for the co-creation of information and the reuse of data and knowledge gathered by the students in a work group. The immersive platform allows team members to "plunge" into a virtual 3D environment in which they share and work collaboratively in the organization of their problem-solving projects while interacting through tools such as voice, videos, text editors, whiteboards, SmartBoards, and slideshows (Araujo & Arantes, 2014). For more information on these type of immersive technologies, please see Chapter 23, The Use of Extended Reality (XR), Wearable, and Haptic Technologies for Learning Across Engineering Disciplines, in this Handbook.

While the Stanford example earlier highlights technology allowing students entrance into virtual 3D worlds, this level of immersion is not a requirement for quality online engineering education. High-quality online learning environments are those in which learners are active participants in the learning process and experience connections to and interactions with instructors, the curriculum, and other students. For example, Penn State's well-established online mechanical engineering master of science program is learner-centered and designed to scaffold graduate education for working engineers. While it relies on sophisticated audio and video capture and rendering technology that allows distance learners full access to high-quality pedagogical materials, from the students' perspectives, the course technology is relatively simple. They generally use the Canvas learning management system (LMS) tools to access course content, much of which is high-quality video; interact asynchronously with their peers and instructors; and complete learning activities and assessments. Synchronous interactions typically take place in Zoom. Key program decisions ranging from curriculum development, including research requirements, to online course instructional design specifics and multimedia standard are grounded in theoretical orientations and pedagogical models appropriate for graduate engineering education that support authentic and experiential learning: academic literacies theory, cognitive apprenticeship, and community of practice theory (Gregg et al., 2021). The online program is also marked by continuous improvement. Data collected from annual student surveys and exit interviews with graduating students, as well as faculty input, inform program improvement decisions to ensure high-quality teaching and learning experiences in this online program.

Additional successful examples of online engineering education include programs that have employed the in-person residency as a key element, much like the nursing practicum and residency examples discussed previously. These further cultivate learning community and connections between learners, instructors, and their institution and also give students opportunities to apply skills in context (Descoteaux et al., 2009; Nepal & Lawrence, 2011). Norwich University offers a master of civil engineering program fully online except for a residency week required at the end of the program. This residency program has two outcomes: to develop and demonstrate public presentation of technical material and to further establish the academic community among and between students and their instructors (Descoteaux et al., 2009). The master in industrial distribution engineering technology program offered at the Texas A&M University similarly requires a one-week residency for its distance students to enable them to complete an intensive project working on a company-sponsored project (Nepal & Lawrence, 2011). One caution when considering this residency approach is to avoid "overrequiring" them, as there are distinct advantages to online courses made up of working engineering professionals, and requiring excessive in-person experiences can become prohibitive to those with full-time employment, which continue to be most online students.

While the earlier examples focus on formal and traditional education (i.e., online degree programs and courses), there is also an entire landscape of online engineering education in the realm of informal and "nontraditional" learning. These "disruptive" education approaches include massive open online courses (MOOCs), Khan Academy, YouTube tutorials, and open educational resources (OERs). MOOCs, such as those offered through EdX and Coursera, can usually be taken for free or at a low cost and have a significant number of offerings focused on engineering topics (Iqbal et al., 2015), many offered by schools with strong engineering programs. While MOOCs tend to be offered by universities, Khan Academy continues to provide free educational videos. Much of the focus is on STEM, and there are a wide number of offerings in engineering areas.

4 Looking to the Future: Embracing Innovative Teaching and Learning Approaches

The increasingly complex future we all face requires equally innovative and adaptive solutions. In this final part of the chapter, we suggest drawing on the innovation already at the heart of engineering to not only leverage but also **make better online education**. Online engineering education is not an "end all" or singular solution but can be another necessary tool among many. Additionally, online learning can have implications for expanding the reach of engineering education and is also an area in need of further research exploration.

4.1 Broadening Participation

There is a renewed focus on the importance of broadening participation in engineering for multiple reasons, including a gap between engineering needs and the number of qualified people; historical systemic exclusions that continue today; and limited access to resources among many potential engineers. While online learning will certainly not solve these issues, it is potentially one option to make engineering education more available to those with limited resources. Consider students in rural high schools without calculus, which can often then restrict them from enrolling in engineering majors. As the ASEE president, who came from a high school without calculus, has argued:

> [W]e tend to accept students that have had a lot of opportunity, so they come well-prepared and we don't do a very good job of attracting and retaining educating students that haven't had opportunities. So, we filter for opportunity not abilities.
>
> *(ASEE, 2022)*

Instead of the traditional "weed out" model, Carpenter argues that we need to move to a "weed in" focus (ASEE, 2022).

Relying solely on the traditional in-person model of engineering education within higher education may not be sufficient to meet these needs. Distance education has its roots in serving the underserved, those "learning at the backdoor" (Wedemeyer, 1981), and there are ways in which online learning can continue this tradition by continuing to offer free and low-cost options, such as MOOCs, Khan Academy tutorials, YouTube educational videos, and other open educational resources. These resources are certainly not unproblematic – currently tending to largely serve those who already have means (Pollack Ichou, 2018) – and ensuring that they do expand the reach for engineering education requires significant work, but much like the library democratized books, some of these resources have the potential to expand the reach of engineering education.

4.2 Learning Engineering

Advances in ICT and digital media have pushed online learning to new frontiers, enabling the design of highly interactive and engaging learning experiences. Coupled with these advances is the rise of online learning as a major form of education for adults (Dede et al., 2019). The COVID-19 pandemic has accelerated the delivery of online education, but with little attention to the factors that impact the instructional design of quality online learning experiences for students. The field of instructional design, or what is now also referred to as learning design and technology, is continuously evolving to keep up with technology trends and expand online learning design such that it takes into account factors that impact online learning design, such as modality, pacing, pedagogy, student instructor ratio, role of online assessments, role of the online instructor, role of the online student, online communication synchrony, feedback strategies, and personalization of learning, to name a few. Learning engineering has come to the forefront as a powerful method to guide the design of online learning experiences and optimize its effectiveness and efficiency (Dede et al., 2019). Learning engineering applies evidence-based engineering methodologies to develop learning technologies and infrastructures to better support learners and learning (Wagner et al., 2018). It uses big data to iteratively improve the design of learning experiences and leverages knowledge of the learning sciences, learning analytics, educational data mining, artificial intelligence, machine learning, design-based research, and theories of human development to design, construct, and deploy new learning technologies and architectures (Wagner et al., 2018). In other words, learning engineering recognizes that the development of new tools and architectures to help advance learning can benefit from engineering expertise. This, in turn, will benefit online engineering education.

Engineers are already using learning engineering practices, such as design-based research (DBR) methodology, to solve real-world problems by designing and enacting new and innovative interventions, extending theories, and refining design principles to meet the needs of the future. Although engineers may not refer to how they innovate as DBR, they are engaged in research that is "(a) pragmatic; (b) grounded; (c) interactive, iterative and flexible; (d) integrative; and (e) contextual" (p. 7), as Wang and Hannafin (2005) would put it. So why not apply this research to online engineering education? *DBR* is defined as:

> A systematic but flexible methodology aimed to improve educational practices through iterative analysis, design, development, and implementation, based on collaboration among researchers and practitioners in real-world settings, and leading to contextually-sensitive design principles and theories.
>
> *(Wang & Hannafin, p. 6)*

As technology has continued to advance, engineering is an increasingly essential requirement for learning and development initiatives that methodologically depend upon data science, computer science, and learning science. Engineering educators and curriculum developers can adopt learning engineering practices and DBR to develop learning technologies and digital infrastructures to better support the teaching and learning of engineering disciplines in in-person and online settings.

4.3 Future Research

In terms of research areas for engineering educators to explore when it comes to teaching and learning in online contexts, potential topics include virtual labs (see Chapter 24, in this Handbook, for more information on virtual labs), active learning, metacognition, and online quality standards. While both active learning and metacognition have received significant attention in the engineering education literature, their coverage in online engineering contexts remains less thorough (Brent et al., 2021; Gregg & O'Connor, 2022; Prince et al., 2020).

Another area which would benefit from further research and development is that of quality standards specific to online engineering. As discussed previously, the issue of *quality* has long been central to online education. This includes faculty concerns about quality (Tabas et al., 2012), studies to compare quality between online and in-person education settings (Abrami et al., 2011; Bernard et al., 2004; Russell, 1999), and models specific to quality in online learning, such as the Quality Matters rubric and Online Learning Consortium's (OLC) quality scorecard.

Exploring quality in online engineering education is especially relevant precisely because

> [t]hese [quality] standards often blur the line between course design, informed by instructional design theory and practice, and the academic province of faculty who are responsible for curriculum delivery and who also typically do not have specific training in pedagogical method or course design.
> *(McCurry & Lampe, 2019)*

This is another area in which engineering education can look to online nursing, as they have already developed a discipline-specific model for quality standards (Authement & Dormire, 2020).

5 Conclusion

There is no uniform agreement on precisely what engineering *is* and what an engineer *does* (National Academy of Engineering, 2019; Pauley, 2009), yet keywords and concepts frequently associated with engineering include *innovation*, *technology*, *tools*, and *problem-solving* (National Academy of Engineering, 2018). Engineering is an enterprise grounded in innovative problem-solving as well as the creation and use of new tools and technology to transform society. We believe online learning offers engineering educators a crucial context in which to innovate to improve teaching and learning as well as expand access beyond its traditional demographics, and towards that end, we close with a still-relevant call from the National Academy of Engineering (2004).

> If the United States is to maintain its economic leadership and be able to sustain its share of high-technology jobs, it must prepare for a new wave of change. While there is no consensus at this stage, it is agreed that innovation is the key and engineering is essential to this task; but engineering will only contribute to success if it is able to continue to adapt to new trends and educate the next generation of students so as to arm them with the tools needed for the world as it will be, not as it is today.

Note

1 One challenge in fully understanding the landscape of online higher education is that there is no singular database or repository that can be relied on to accurately identify all formal degrees or certificates, let alone individual courses, offered online. The United States Department of Education (USDE) did not start collecting data about distance education until 2012, through the *Integrated Postsecondary Education Data System* (IPEDS). IPEDS collects data from all higher education institutions that receive federal funding and tracks metrics such as institutional characteristics, costs, enrollment, and student success metrics like retention and graduation rates. When it comes to online distance education specifically, there are well-documented methodological challenges, due in part to how colleges and universities define and report distance learning (Becker, 2016; Miller et al., 2017; Poulin & Straut, 2016). Additionally, the ways in which IPEDS traditionally reported graduation rates for distance education serving institutions have been problematic as it has relied on a traditional four-year-to-graduation model which does not align with the part-time working adult learner's academic trajectory. This has suggested, inaccurately, exceedingly poor retention and graduation rates for online students (Poulin & Straut, 2018), further contributing to societal misunderstandings about the efficacy of online learning. To counter this, organizations like WICHE Cooperative for Educational Technologies (WCET) and Online Learning Consortium (OLC) have worked actively with the USDE to ensure that there is a more nuanced and accurate understanding of the realities of online distance education.

References

ABET. (n.d.a). *Accredited programs*. https://amspub.abet.org/aps/category-search

ABET. (n.d.b). *Criteria for accrediting engineering programs, 2022 – 2023*. https://www.abet.org/accreditation/accreditation-criteria/criteria-for-accrediting-engineering-programs-2022-2023/

Abrami, P. C., Bernard, R. M., Bures, E. M., Borokhovski, E., & Tamim, R. M. (2011). Interaction in distance education and online learning: Using evidence and theory to improve practice. *Journal of Computing in Higher Education, 23*(2), 82–103. https://doi.org/10.1007/s12528-011-9043-x

Allen, I. E., & Seaman, J. (2008). *Staying the course – online education in the United States, 2008*. The Sloan Consortium. http://sloanconsortium.org/publications/survey/staying_course

Allen, I. E., & Seaman, J. (2011). *Going the distance: Online education in the United States, 2011*. The Sloan Consortium. http://sloanconsortium.org/publications/survey/going_distance_2011

Araújo, U. F. (2019). 2019.3D immersive platforms and problem-based learning projects: A search for quality in education. In M. Moallem, W. Hung, & N. Dabbagh (Eds.), *Wiley handbook of problem-based learning* (pp. 575–592). Wiley. https://doi.org/10.1002/9781119173243.ch25

Araujo, U. F., & Arantes, V. A. (2014). Re-inventing school to develop active citizens. In A. S. Castro (Ed.), *Positive psychology in Latin America, cross cultural advancements in positive psychology* (1st ed., Vol. 10, pp. 241–254). Springer. https://doi.org/10.1007/978-94-017-9035-2_14

ASEE. (2022). *Jenna carpenter on "weed out" culture in engineering* [video]. YouTube. www.youtube.com/watch?v=89LiWKHfLr0

Authement, R. S., & Dormire, S. L. (2020). Introduction to the online nursing education best practices guide. *SAGE Open Nursing, 6*, https://doi.org/10.1177/2377960820937290

Bandura, A. (1977). *Social Learning theory*. Prentice Hall.

Becker, J. (2016). *The fuzzy math of online learning enrollments*. www.jonbecker.net/the-fuzzy-math-of-online-learning-enrollments/

Beneroso, D., & Robinson, J. (2022). Online project-based learning in engineering design: Supporting the acquisition of design skills. *Education for Chemical Engineers, 38*, 38–47. https://doi.org/10.1016/j.ece.2021.09.002

Bernard, R. M., Abrami, P. C., Lou, Y., Borokhovski, E., Wade, A., Wozney, L., Wallet, P. A., Fiset, M., & Huang, B. (2004). How does distance education compare with classroom instruction? A meta-analysis of the empirical literature. *Review of Educational Research, 74*(3), 379–439. https://doi.org/10.3102/00346543074003379

Bonwell, C., & Eison, J. (1991). Active learning: Creating excitement in the classroom. *ASHEERIC higher education report no. 1*. The George Washington University, School of Education and Higher Education.

Bourne, J., Harris, D., & Mayadas, F. (2005). Online engineering education: Learning anywhere, anytime. *Journal of Engineering Education, 94*(1), 131–146. https://doi.org/10.1002/j.2168-9830.2005.tb00834.x

Brent, R., Prince, M., & Felder, R. (2021). Promoting and managing student-student interactions in online STEM classes. *International Journal of Engineering Education, 37*(3), 797–813.

Brown, J. S., Collins, A., & Duguid, P. (1989). Situated cognition and the culture of learning. *Educational Researcher, 18*(1), 32–42. https://doi.org/10.3102/0013189X018001032

Dabbagh, N., & Bannan-Ritland, B. (2005). *Online learning: Concepts, strategies, and application*. Prentice Hall, Pearson Education.

Dabbagh, N., Howland, J., & Marra, R. (2019). *Meaningful online learning: Integrating strategies, activities, and learning technologies for effective designs*. Routledge. https://doi.org/10.4324/9781315528458

Dede, C., Richards, J., & Saxberg, B. (Eds.). (2019). *Learning engineering for online education: Theoretical contexts and design-based examples*. Taylor and Francis. https://doi.org/10.4324/9781351186193

Descoteaux, T., Muckerman, D., & Sabol, S. (2009, June). The importance of an on campus residency experience in distance education programs. In *ASEE annual conference & exposition proceedings*. https://doi.org/10.18260/1-2–4527

Domenech, D., Sherman, M., & Brown, J. (2016). *Personalizing 21st century education: A framework for student success*. Jossey-Bass.

ElSheikh, R. M., & Najdi, S. D. (2013). Math keyboard symbols and its effect in improving communication in math virtual classes. *International Journal of Information and Education Technology, 3*(6), 638–642. https://doi.org/10.7763/IJIET.2013.V3.352

Fortenberry, N. L. (2014). Foreword. In A. Johri & B. M. Olds (Eds.), *Cambridge handbook of engineering education research* (pp. xii–xiv). Cambridge University Press. https://doi.org/10.1017/CBO9781139013451.001

Fruchter, R. (2014). Transformative 3D immersive collaboration environment. In support of AEC global teamwork. *Computing in Civil and Building Engineering*, 1425–1432. https://doi.org/10.1061/9780784413616.177

Garrison, D. R. (2017). *E-Learning in the 21st century: A community of inquiry framework for research and practice* (3rd ed.). Routledge/Taylor and Francis. http://dx.doi.org/10.4324/9781315667263

Garrison, D. R., & Anderson, T. (2003). *E-Learning in the 21st century: A framework for research and practice.* Routledge. https://doi.org/10.4324/9780203166093

Garrison, D. R., Anderson, T., & Archer, W. (2000). Critical inquiry in a text-based environment: Computer conferencing in higher education. *The Internet and Higher Education, 2*(2–3), 87–105. https://doi.org/10.1016/S1096-7516(00)00016-6

Gregg, A., Berdanier, C. G., & Thole, K. A. (2021, July), *Theoretical and applied perspectives on online graduate engineering education: Learning-centered vision, administration, and course design.* Paper presented at 2021 ASEE virtual annual conference content access, virtual conference. https://peer.asee.org/37907

Gregg, A., & O'Connor, J. (2022). Reflecting while doing: Integrating active learning and metacognitive activities in a fully online thermodynamics course. In *2022 ASEE annual conference & exposition.*

Haddad, L. M., Annamaraju, P., & Toney-Butler, T. J. (2022). *Nursing shortage.* StatPearls Publishing. PMID: 29630227.

Hannafin, M. J., Hill, J. R., & Land, S. M. (1997). Student-centered learning and interactive multimedia: Status, issues, and implication. *Contemporary Education, 68*(2), 94–99.

Hodges, C., Moore, S., Lockee, B., Trust, T., & Bond, A. (2020, March 27). The difference between emergency remote teaching and online learning. *Educause Review.* https://er.educause.edu/articles/2020/3/the-difference-between-emergency-remoteteaching-and-online-learning

Huang, W., Hurt, A., Richardson, J. C., Swan, K., & Caskurlu, S. (2020). *Community of inquiry framework.* Purdue Repository for Online Teaching and Learning: Purdue University.

Iqbal, S., Zang, X., Zhu, Y., Hussain, D., Zhao, J., Gulzar, M. M., & Rasheed, S. (2015). towards MOOCs and their role in engineering education. In *2015 7th international conference on information technology in medicine and education (ITME)* (pp. 705–709). https://doi.org/10.1109/ITME.2015.89

Johri, A., & Olds, B. M. (2014). *Cambridge handbook of engineering education research.* Cambridge University Press. https://doi.org/10.1017/CBO9781139013451

Jonassen, D. H. (2000). Toward a design theory of problem solving. *Educational Technology Research & Development, 48*(4), 63–85. doi:10.1007/BF02300500

Jones, K., Raynor, P., & Polyakova-Norwood, V. (2020). Faculty caring behaviors in online nursing education: An integrative review. *Distance Education, 41*(4), 559–581. https://doi.org/10.1080/01587919.2020.1821601

Kinney, L. S. (2015). *Faculty perceptions of online learning in engineering education* [Doctoral Dissertation, University of Texas]. University of Texas Scholar Works https://repositories.lib.utexas.edu/handle/2152/30923

Kocdar, S., Bozkurt, A., & Dogan, K. G. (2020). Engineering through distance education in the time of the fourth industrial revolution: Reflections from three decades of peer reviewed studies. *Computer Applications in Engineering Education, 29*(4), 931–949. https://doi.org/10.1002/cae.22367

Lantada, A. D., & Nuñez, M. (Ed.). (2020). Engineering education everywhere: Good practices for emergency situations and remote regions. [Special Issue Call for Papers]. *International Journal of Engineering Education.* www.ijee.ie/callfp/CfP_Emergencies%20and%20Remote.pdf

Lave, J., & Wenger, E. (1991). *Situated learning: Legitimate peripheral participation.* Cambridge University Press. https://doi.org/10.1017/CBO9780511815355

Lowenthal, P., & Wilson, B. G. (2010). Labels do matter! A critique of AECT's redefinition of the field. *TechTrends, 54,* 38–46. http://dx.doi.org/10.1007/s11528-009-0362-y

Mayadas, F., Miller, G., & Sener, J. (2015). *Definitions of E-learning courses and programs version 2.0.* http://onlinelearningconsortium.org/updated-e-learning-definitions-2/

Meaningful. (n.d.). *Dictionary.com Unabridged.* Dictionary.com website www.dictionary.com/browse/meaningful

McCurry, D. S., & Lampe, M. (2019). Bringing faculty to online course quality standards. In *Distance learning administration ANNUAL conference* (pp. 127–132).

Miller, A., Topper, A. M., & Richardson, S. (2017). *Suggestions for improving IPEDS distance education data collection. U.S. department of education.* National Postsecondary Education Cooperative. Retrieved December 18, 2021, from http://nces.ed.gov/pubsearch

Milman, N., Irvine, V., Kelly, K., Miller, J., & Saichaie, K. (2020). *7 things you should know about the HyFlex course model.* EDUCAUSE Learning Initiative (ELI). https://library.educause.edu/resources/2020/7/7-things-you-should-know-about-the-hyflex-course-model

Mitchell, M. M., & Delgado, C. (2014). Online learning: Advantages and challenges in implementing an effective practicum experience. *Open Journal of Nursing, 4*(2014), 379–384. https://doi.org/10.4236/ojn.2014.46044

Moore, J., Dickson-Deane, C., & Galyen, K. (2011). e-Learning, online learning, and distance learning environments: Are they the same? *Internet and Higher Education, 14*, 129–135. https://doi.org/10.1016/j.iheduc.2010.10.001

Moore, M. G. (1993). Theory of transactional distance. In D. Keegan (Ed.), *Theoretical principles of distance education*. Routledge.

Moore, M. G. (1994). Autonomy and interdependence. *The American Journal of Distance Education, 8*(2), 1–5.

Moore, M. G. (2019). The theory of transactional distance. In M. Moore & W. Diehl (Eds.), *Handbook of distance education*. Routledge.

Moore, M. G., & Kearsley, G. (1995). *Distance education: A systems view*. Wadsworth Publishing.

National Academy of Engineering. (2004). *The engineer of 2020: Visions of engineering in the new century*. National Academies Press.

National Academy of Engineering. (2018). *Understanding the educational and career pathways of engineers*. The National Academies Press. https://doi.org/10.17226/25284

National Academy of Engineering. (2019). *President's perspective: What is engineering?* www.nae.edu/221278/Presidents-Perspective-What-Is-Engineering

Navarro, P., & Shoemaker, J. (2000). Performance and perceptions of distance learners in cyberspace. In M. G. Moore & G. T. Cozine (Eds.), *Web-based communications, the internet and distance education*. The American Center for the Study of Distance Education: The Pennsylvania State University.

Nepal, B., & Lawrence, F. (2011, June). Project-based residency course for online graduate program. In *ASEE annual conference & exposition proceedings*. https://doi.org/10.18260/1-2–18724

Newstetter, W. C., & Svinicki, M. D. (2014). Learning theories for engineering education practice. In A. Johri & B. Olds (Eds.), *Cambridge handbook of engineering education research* (pp. 29–46). Cambridge University Press. https://doi.org/10.1017/CBO9781139013451.005

No Significant Difference Database. (n.d.). *Distance education and technological advancements (DETA) research center*. https://detaresearch.org/research-support/no-significant-difference/

Oregon State University eCampus Research Unit. (n.d.). *Online learning efficacy research database*. https://ecampus.oregonstate.edu/research/projects/online-learning-efficacy-research/

Ota, M., Peck, B., & Porter, J. (2018). Evaluating a blended online learning model among undergraduate nursing students. *CIN: Computers, Informatics, Nursing, 36*(10), 507–512. https://doi.org/10.1097/CIN.0000000000000456

Pawley, A. L. (2009). Universalized narratives: Patterns in how faculty members define "engineering". *Journal of Engineering Education, 98*(4), 309–319. https://doi.org/10.1002/j.2168-9830.2009.tb01029.x

Pollack Ichou, R. (2018). Can MOOCs reduce global inequality in education? *Australasian Marketing Journal, 26*(2), 116–120. https://doi.org/10.1016/j.ausmj.2018.05.007

Poulin, R., & Straut, T. (2016). *WCET distance education enrollment report 2016*. WICHE Cooperative for Educational Technologies website. http://wcet.wiche.edu/initiatives/research/WCET-Distance-Education-Enrollment-Report-2016

Poulin, R., & Straut, T. (2018). *Count all students! New outcome measures now include non-traditional students*. https://wcetfrontiers.org/2018/04/10/count-all-students/

Prince, M. (2004). Does active learning work? A review of the research. *Journal of Engineering Education, 93*(3), 223–231. https://doi.org/10.1002/j.2168-9830.2004.tb00809.x

Prince, M., Felder, R., & Brent, R. (2020). Active student engagement in online STEM classes: Approaches and recommendations. *Advances in Engineering Education, 8*(4), 1–25.

Reynolds, M. (2010). The future of engineering education. *Journal for Online Engineering Education, 1*(1).

Russell, T. L. (1999). *The no significant difference phenomenon*. Office of Instructional Telecommunications, University of North Carolina.

Schlosser, L. A., & Simonson, M. (2003). *Distance education:* towards *a definition and glossary of terms*. Sponsored by the Definitions and Terminology Committee Association for Educational Communications and Technology.

Scholes, C. A. (2021). Chemical engineering design project undertaken through remote learning. *Education for Chemical Engineers, 36*, 65–72. https://doi.org/10.1016/j.ece.2021.03.003

Seaman, J. E., Allen, I. E., & Ralph, N. (2021). Teaching online: STEM education in the time of COVID. *Bay View Analytics*. www.bayviewanalytics.com/reports/stem_education_in_the_time_of_covid.pdf

Seaman, J. E., Allen, I. E., & Seaman, J. (2018). *Grade increase: Tracking distance education in the United States*. Babson Survey Research Group. https://files.eric.ed.gov/fulltext/ED580852.pd

Seaman, J. E., & Seaman, J. (2019). Distance education state almanac 2019. *Bay View Analytics*. www.bayviewanalytics.com/reports/almanac/national_almanac2019.pdf

Simonson, M., Smaldino, S., Albright, M., & Zvacek, S. M. (2009). *Teaching and learning at a distance: Foundations of distance education* (4th ed.). Allyn & Bacon.

Smith, G. G., Ferguson, D., & Gupta, S. (2004). Diagrams and math notation in elearning: Growing pains of a new generation. *International Journal of Mathematical Education in Science and Technology*, *35*(1), 681–695. https://doi.org/10.1080/0020739042000232583

Smith, G. G., Passmore, D., & Faught, T. (2009). The challenges of online nursing education. *The Internet and Higher Education*, *12*(2), 98–103. https://doi.org/10.1016/j.iheduc.2009.06.007

Tabas, J. M., LeMay, C. M., & Freije, E. (2012, June). *Online education: The end of learning*. Paper presented at 2012 ASEE annual conference & exposition. https://doi.org/10.18260/1-2–21761

Wagner, E., Barr, A., Blake-Plock, S., & Robson, R. (2018). *7 things you should know about learning engineering*. EDUCAUSE Learning Initiative (ELI). https://library.educause.edu/resources/2018/9/7-things-you-should-know-about-learning-engineering

Wang, F., & Hannafin, M. J. (2005). Design-based research and technology-enhanced learning environments. *ETR&D*, *53*, 5–23. https://doi.org/10.1007/BF02504682

Wedemeyer, C. A. (1981). *Learning at the back door: Reflections on non-traditional learning in the lifespan*. University of Wisconsin Press.

23

The Use of Extended Reality (XR), Wearable, and Haptic Technologies for Learning Across Engineering Disciplines

Diana Bairaktarova, Andrew Valentine, and Rami Ghannam

1 Introduction

Industries are obliged to invest in continuous innovation in an increasingly competitive global market. For example, technology-enhanced environments have been implemented in industries, including, but not limited to, military, medical, and entertainment – all with different objectives but similar affordances. Most important here, technologies have been widely integrated into many aspects of teaching and learning in higher education, often helping enhance student performance when used in ways that have been found to be empirically effective. Therefore, it is important for tertiary-level educators to know and understand what technologies may help with learning and teaching in specific disciplines, such as engineering, as well as the appropriate learning activities.

In engineering education, it is important for educators to understand the skills and knowledge industry is seeking in graduates and how technologies may be used to effectively help students learn these skills and obtain this knowledge. In addition, it is also important for educators to know if such technologies are already being used for training purposes in industry and, if so, how they can be adopted for education, with the intention of allowing students to experience relevant industry-based practices when possible. The results will be twofold: engineering education will more closely resemble engineering practices, and its graduates will be more familiar with industry expectations.

Research has found that the engineering industry perceives that graduates need a foundation in design to be effective practitioners. Technology can be a powerful tool as its use during the design process can allow for unconstrained creativity not found in a physical environment, expanding the possible designs one can formulate and test. With these possibilities, merging digital and physical information can recast how engineers design and, more importantly, how engineering students learn the intricacies of the design process. Emerging technologies have enabled revolutionary changes in the way engineers interact with objects by allowing them to engage in immersive 3D experiences, enhancing their perceptual experience.

Virtual reality (VR) is a technology of total immersion in which the user interacts in a world that is completely computer-generated (Nilsson et al., 2018; Stuerzlinger et al., 2006). In contrast, *augmented reality (AR)* provides a perception of the world created by overlaying the physical world with a digital lens, allowing users to remain in the physical world but providing them with computer-generated objects to interact with perceptually. *Wearable technologies* and *neurophysiological (haptic)*

DOI: 10.4324/9781003287483-28

technologies are worn by users, enabling them to seamlessly interface and interact with machines and computers.

Review of the current literature suggests that there is a gap in understanding how emerging technologies can be used to bridge theory to practice in engineering education. Collective adaptation to novel user interfaces and technologies is vital for creating effective learning environments and promoting active participation through interactive and collaborative pedagogy methods (Srinivasan & Bairaktarova, 2018). Engineering educators are interested in applying innovative pedagogies, including relying on technology, to achieve effective ways of teaching and learning. In such collective adaptation, instructors and students are creators of new knowledge – this unique symbiosis paves the way for both the learning of the subject matter and the building of a special relationship among educators, students, technology, and the learning environment (Pakala & Bairaktarova, 2020).

In this chapter, we present an overview of how four technologies (VR, AR, wearable technologies, haptic technologies) have been used in engineering education and engineering practice for teaching, learning, and training purposes. Our intention is to provide the reader with an introduction to these technologies and to highlight how these technologies may be empirically applied in engineering education. In some sections of the chapter, we reference and explicitly refer the reader to helpful additional resources, such as comprehensive literature reviews, for in-depth explanations on certain topics which may not be extensively discussed here due to space limitations. This chapter has been designed to act as an introduction and a starting point for researchers and educators interested in learning about these four technologies in engineering education.

We begin by first discussing selected learning theories, specifically situated cognition, communities of practice, and distributed cognition, and their application in the context of technologies being used for enhancing learning and teaching. We then discuss each in detail in separate sections. It is important to note that although these technologies are presented in separate sections, all four are not always separate categories; thus, we offer examples where these technologies interact and overlap, including describing the disadvantages and limitations of interchangeable use. Finally, the chapter concludes with a summary of the lessons learned, their implications, and suggestions for future work within the scope of this chapter.

2 Learning Theories

Research has shown that a learning environment enriched with stimuli and artifacts, concrete or augmented, that closely map the engineering practitioner's workplace shows promise in enhancing engineering students' performance on design-related tasks (Bairaktarova et al., 2017; Bairaktarova & Johri, 2016). Scholars further suggest that the situated perspective on learning is applicable in bridging engineering education research and the learning sciences (Johri & Olds, 2011; Johri et al., 2014).

An abundance of work in psychology (Leont'ev, 1978; Luria, 1978; Newman et al., 1989; Norman, 1991; Salomon, 1993; Scribner, 1984; Vygotsky & Cole, 1978), anthropology (Chaiklin & Lave, 1993; Flor & Hutchins, 1991; Gantt & Nardi, 1992; Hutchins, 1995b; Lave, 1988; B. Nardi & Miller, 1990; Nardi & Miller, 1991; Suchman, 1987), and computer science (Clement, 1990; Mackay, 1990; MacLean et al., 1990) has found that any attempts at understanding how people learn and work is incomplete when the unit of analysis is the "unaided individual with no access to other people or artifacts for accomplishing the task at hand" (Nardi, 1996, p. 35). Theories of situated and distributed cognition have purposefully been used to understand learning, knowing, and doing as context-specific social processes by characterizing cognition as being socially shared (Clartcey, 2008; Lave & Wenger, 1991a). These concepts are rooted in both Vygotskian's understandings of higher mental processes as internalized social relationships (Palincsar, 1998) and Dewey's early objections to stimulus–response theory. Because both situated and distributed learning exhibit many similarities, scholars have synergistically used both as frameworks to guide their research design and analysis (Ogunseiju et al., 2020a; Ogunseiju et al., 2022).

2.1 Situated Cognition and Communities of Practice

Proponents of a situated perspective posit that knowledge is "distributed among people and their environments, including objects, artifacts, tools, books, and the communities of which they are a part" (Greeno et al., 1996, p. 17). In this paradigm, knowledge emerges dynamically through individuals' interactions with one another and their surrounding environment, meaning, that knowledge is co-constructed and reinterpreted within a given societal context (Clartcey, 2008). The situated perspective maintains that knowing and learning are simultaneously an individual and communal quest (Newstetter & Svinicki, 2014). These groups, or more accurately, communities of practice (CoP), as Lave and Wegner (1991) called them, are working towards a shared goal that is achievable by leveraging the group's collective knowledge and through the use of tools. The goal of these CoPs is often characterized by what is valued in the community, blending the boundaries of the material, social, and cultural values (Johri & Olds, 2011). Per this perspective, *knowing* means that one is able to participate meaningfully in the community, while *learning* means that one can move from peripheral forms of participation to full participation (Newstetter & Svinicki, 2014). Full participation, which is enabled through mentorship from a more experienced member of the community, is achieved as the novice is afforded opportunities to observe and practice a particular activity.

Lave (1991) described the various views on cognition held by situative theorists, which he referred to as cognition plus, interpretive, and situated social practice. In the cognition plus view, researchers focus on unchallenged ideas of individual cognition but expand to consider how the individual is influenced by social factors, while based on the interpretive view, situatedness is located in language and social interaction and cognition is located in the quest for social enterprise rather than dependent upon a physical dimension. Lastly, the situated social practice view, which shares several tenets with the interpretive view, emphasizes the "relational interdependency of agent and world, activity, meaning, cognition, learning and knowing" (Lave & Wenger, 1991b, p. 67), where cognition is located in the historical development of an ongoing activity (Moore & Rocklin, 1998).

In addition to Lave's (1991) characterization, Robbins and Aydede (2009) categorized the different views of situated cognition found in the *Cambridge Handbook of Situated Cognition* into three main ideas used by researchers: (1) the embodiment thesis, where cognition depends not only on the brain but also on the body; (2) the embedding thesis, where cognitive activity utilizes the natural and social structures in the environment; and (3) the extension thesis, where cognition extends beyond individual agents to encompass the physical and social environment.

2.2 Distributed Cognition

While situated cognition allows us to observe ways in which engineering work is situated within the virtual or augmented realities context rich with its own material, social, and cultural dimensions, the distributed cognition framework allows us to operationalize collaborations between cognitive subsystems. At its essence, distributed cognition posits that all cognitive processes within a system are distributed. Hutchins (1995a) pioneered this framework in his book *Cognition in the Wild*, detailing naval navigation by demonstrating how it is the result of a coordinated process between sailors, instruments, and their interactions under social organization (Cheon, 2014). Per this view, a cognitive system comprises "agents, traditions of practice, material artifacts, devices, and instruments, the characteristics of which in multiple ways support the dissemination of information across the system" (Osbeck & Nersessian, 2014, p. 83). Embracing the same tenets of situated cognition, distributed cognition extends it to include the radical notion that not only are tools and artifacts helpful in enhancing individuals' cognitive process but that they can also become part of a cognitive system by integrating into its cognitive architecture. In this way, it posits that what is accomplished by the system cannot be accomplished by any of its parts alone. Distributed cognition has been found to be an appropriate framework for analyzing scientific practices, such as the use of the Hubble telescope as a distributed system (Giere, 2006), high-energy physics research (Knorr-Cetina, 1999), the manipulation of chemical formula representations (Giere & Moffatt, 2003), bioengineering and biomedical research (Magnus, 2007; Nersessian, 2005, 2006), and particularly relevant for this chapter, human–computer interactions (Hollan et al., 2000). Hollan et al.'s (2000) approach purposefully includes characteristics of situated cognition in their framework of distributed cognition. In their framework specific to human–computer interaction, they posit that cognition is socially distributed, embodied, and culturally embedded. Additionally, Hollan et al. (2000) have identified the following three processes as types of distributions within the sociotechnical systems they have analyzed: (1) cognitive processes may be distributed across members of a social group, (2) cognitive processes may involve coordination between internal and external (material or environmental) structures, and (3) cognitive processes may be distributed through time with a compounding effect, in which the results of earlier events transform the nature of later events.

3 Virtual Reality

3.1 What Is Virtual Reality, and How Does It Work?

Virtual reality (VR) can be classified as a three-dimensional computer-generated simulation (or world) that allows a user to be immersed in an artificial environment (Pantelidis, 1997). This simulated world may either be based on a real-world location or be newly constructed. Typically, the user is able to move around and interact with this virtual world through some type of input controls. A VR computer simulation is designed with the intention that the user is convinced that they are located within the computer-generated world (Bell & Fogler, 1997), and VR environments that have highly realistic graphics and interactivity are better at achieving this goal. The different types of VR, for example, CAVE (cave automatic virtual environment), HMD (head-mounted display), and desktop, are suitable for different purposes; more about these different types of VR are discussed in a following subsection. Each type of VR requires both hardware and software, including the simulation, to run, with each varying based on the type of VR. Specific to our purpose here, virtual reality can provide a way to engage students in learning about concepts which may be difficult to teach effectively using traditional methods (Radianti et al., 2020).

3.2 Growth of Virtual Reality in Engineering Education

Virtual reality has been widely used in engineering education for more than two decades. For example, in construction engineering education, the number of papers about VR published annually increased from zero to three between 1997 and 2010 to consistently more than four between 2011 and 2016 (with an increasing trend) (Wang et al., 2018). Head-mounted display VR technology has seen a rapid increase in adoption in engineering education since 2015, with the review conducted by Huang and Roscoe (2021) identifying only five papers before 2015 and only one paper per year in 1995, 2005, 2006, 2013, 2014. However, between 2015 and 2020, this number increased to between three and eight papers published each year. Historically, the field of engineering has used virtual reality in education to a greater extent than many other fields (Radianti et al., 2020).

The recent shift to online learning due to COVID has also seen increased use of technologies to support student learning. As a recent study found, virtual reality was preferred by civil engineering students for learning compared to videos, but instructor-aided learning was still preferred over virtual reality (Try et al., 2021).

3.3 Understanding Virtual Reality in Education More Broadly

Over the last five years, numerous literature reviews have investigated and summarized the use of virtual reality in education, many focusing on specific contexts, including K–12 education (Di Natale et al., 2020; Maas & Hughes, 2020; Tilhou et al., 2020), higher education (Di Natale et al., 2020; Radianti et al., 2020), post-secondary education (Concannon et al., 2019), computer science education (Agbo et al., 2021; Pirker et al., 2020), engineering education (di Lanzo et al., 2020; Huang & Roscoe, 2021; Soliman et al., 2021), and education more generally (Freina & Ott, 2015; Hamilton et al., 2021; Jensen & Konradsen, 2018; Kavanagh et al., 2017). Meta-analyses have also investigated the efficacy of virtual reality on students' learning performance in K–12 and higher education (Merchant et al., 2014; B. Wu et al., 2020) and training more generally (Kaplan et al., 2021). These review and meta-analysis studies are highlighted here to demonstrate the extensive work that has been conducted over the past couple of decades in the area of virtual reality in the context of education generally and, specifically of interest here, in the context of engineering education. Educators can refer to these existing literature reviews to understand the wide diversity of ways that virtual technology can be adapted for enhancing teaching and learning.

One of the major challenges with VR is that while there have been rapid developments and use of VR in educational contexts, research, especially in the latter, has lagged behind (Bower & Jong, 2020), meaning, that "educators and researchers do not have the evidence base they need to determine the 'when,' 'why' and 'how'" they can most effectively make use of VR in educational contexts (Bower & Jong, 2020, p. 1981). Further, some of this research is based on anecdotal evidence, making it challenging to understand whether virtual reality should be adopted in a specific educational context. When considering the possibility of introducing virtual reality technology into an engineering course, there is, however, existing information and experiences that can be drawn on for insight and inspiration.

3.4 What Are Learning Situations Where VR Can Help? Examples of the Impact of VR on Learning

In considering whether to use VR in engineering education, we first need to understand whether it typically helps students learn more effectively. Although virtual reality has been widely adopted in education, this does not mean that it is universally effective. To evaluate the efficacy of the technology, we can refer to the relevant meta-analysis studies that have been published. Merchant et al. (2014) concluded that different types of virtual reality (games, simulations, virtual worlds) were all effective at improving learning outcomes, although games were more effective than simulations and virtual worlds. On the other hand, Kaplan et al. (2021) concluded that training in extended reality (including virtual reality) did not lead to different outcomes compared to traditional (non-simulated) methods of training. However, their results do not necessarily mean that virtual reality training is not effective; they clearly show that virtual reality is at least as effective as traditional training, but it is the context of when to use virtual reality that may need to be more selective. Specific to head-mounted virtual reality, Wu et al. (2020) concluded that simulations of this type of VR increased students' learning performance compared to non-immersive approaches. The review by Di Natale et al. (2020) demonstrated that virtual reality can also promote students' motivation and engagement, both of which are linked to enhanced learning performance. Based on these findings, we can reasonably conclude that virtual reality can have a positive influence on student learning, but the decision of when to use it may not be as straightforward.

3.5 Virtual Reality Is Used in Many Different Engineering Disciplines, but Learning Activities Are Often Discipline-Specific

Based on previous literature reviews, virtual reality has been used in a wide range of engineering subdisciplines in engineering education, including, but not limited to, civil, electrical, mechanical, chemical, industrial, and construction (Concannon et al., 2019; di Lanzo et al., 2020; Huang & Roscoe, 2021; Pellas et al., 2020). However, while many engineering education disciplines use virtual reality, it is more prevalent in some than in others. The data provided here from relevant literature reviews give a general overview of the disciplines that have used virtual reality the most: the review conducted by di Lanzo et al. (2020) included 17 studies covering civil (4), electrical (2), industrial (2), mechanical (4), pneumatic (1), software (1), and unspecified (6), while Concannon et al.'s (2019) review included 14 studies from civil (3), computer (2), electrical (1) general (4), mechanical (2), numerical control (1), and pneumatic (1). As these findings suggest, virtual reality is most commonly used in civil, mechanical, and electrical engineering.

While virtual reality is used in a wide range of disciplines, individual learning activities often have a narrow focus designed to teach students a specific concept, which may be relevant only to a certain engineering discipline. For example, chemical engineers may learn about processes in a chemical processing plant, while civil engineers may learn about road or building design. As these examples suggest, the learning activity is typically discipline-specific and may not be suitable for students outside of it. This limitation is important for educators to consider as virtual reality learning activities can be expensive to develop, and it may be difficult to justify the resource cost when the activity may be applicable to only a restricted number of engineering students.

In contrast are situations where an educator is teaching a general engineering class taken by students from many disciplines or a general education class which may even include students from outside the engineering field. In this case, the virtual reality learning activity may focus on the development of generic skills (also known as soft or professional skills) that are applicable to all engineering disciplines, such as communication skills or ethical decision-making. We refer the reader to the informative literature reviews by Concannon et al. (2019), di Lanzo et al. (2020), Huang and Roscoe (2021), and Pellas et al. (2020) for additional examples of discipline-specific and non-discipline-specific virtual reality learning activities.

3.6 Different Types of Virtual Reality Used in Engineering Education

Various types of VR have been used in engineering education including cave automatic virtual environment (CAVE), desktop computers, mobile, and head-mounted device (HMD) (di Lanzo et al., 2020). Overall, HMDs are the most prevalent type of VR used. They are more immersive than other types as they completely block out the user's view of the real world and typically lead to enhanced benefits for skill and knowledge development in educational contexts compared to less-immersive VR (B. Wu et al., 2020), such as desktop- and mobile-based VR. There was an increase in HMD VR uptake in engineering education during the mid-2010s due to the availability of new cost-effective commercial devices, including the HTC VIVE, produced by the High-Tech Computer Corporation, and Oculus Rift, produced by the Oculus Corporation (Huang & Roscoe, 2021). However, the need for specialized hardware and extensive space means it can be quite resource-intensive to develop and facilitate learning activities using this type of VR, concerns that mean HMD also has difficulties in terms of scalability and can be very difficult to adopt in classes with a large enrollment.

In a CAVE environment, a user stands in the middle of a space or room, often in the shape of a cube, surrounded by large screens or walls with projections on them, typically on all sides. With users standing in the middle of the screens, their view of the real world is replaced with that of the virtual environment projected onto the screens, meaning, the user feels that they are within the virtual environment. The images on the screens may be static, meaning, that the user cannot move in the virtual environment, or the images may change as the user interacts with the virtual environment (which could be controlled by a wide range of devices, such as a joystick or even 3D motion tracking, which tracks how the user walks around the space). This type of VR can be challenging to adopt as it requires both a large space with a specific shape and specialized equipment, including the screens or projectors. These requirements also make it difficult to scale for use in large classes.

In a desktop computer VR environment, a user sits in front of a standard desktop computer screen. This type of environment is less immersive than CAVE- or HMD-based systems because the user can still see the real world in their peripheral vision, reducing their level of immersion in the virtual world. In this environment, users are often able to interact with the virtual world using the computer keyboard and mouse. One of the key advantages of this type of VR is that the barrier to entry is low, as desktop computers are widely accessible. Virtual reality learning activities developed for a desktop computer can be easily distributed and used by many students, including remotely in their own homes. The cost of development is also often less expensive than for CAVE and HMD because there is no need for additional specialized hardware. Testing the VR environment is also easier as it can be conducted using any standard desktop computer. Therefore, we suggest that desktop-based VR environments may be a suitable entry point for educators looking to use VR learning activities in their courses because they are easier to design, develop, distribute, and use in classes of any size.

3.7 Reasons Engineering Educators Use Virtual Reality in Engineering Education

While past research highlights a diversity of reasons for engineering educators to use virtual reality, teaching course content knowledge was most common (Huang & Roscoe, 2021). More specifically, in their review of virtual reality systems in engineering education, Huang and Roscoe (2021) concluded that the primary reasons engineering educators use virtual reality were instruction and training of course content knowledge, cognitive training, skill training, and motivating students to explore STEM careers, while di Lanzo et al. (2020) concluded that engineering educators may use virtual reality because of expected educational benefits, application to distance or remote learning, the immersive nature of the simulation learning environment, and recent advances in virtual reality technology, and Radianti et al. (2020) found that virtual reality was used in engineering for teaching analytical and problem-solving skills and soft skills, including communication and collaboration, as well as practical and declarative knowledge.

One of the predominant reasons for the adoption and evaluation of virtual reality in education is experiential and situational learning (Di Natale et al., 2020). Virtual reality facilitates students being able to practice procedures and learn new skills, providing a means for experiential learning in situations where it may otherwise be unfeasible or too dangerous for in-person training (Freina & Ott, 2015; Kaplan et al., 2021; Slater & Sanchez-Vives, 2016). Similarly, virtual reality also provides an opportunity in engineering for training users to both experience and practice managing unexpected and hazardous scenarios (Kumar et al., 2021). One example in mechanical engineering involves students taking on the role of a vehicle-loading crane operator in the context of learning about safety in design. In this case, the learning activity highlighted the problematic design of the loading crane controls, resulting in a dangerous situation for the operator (Valentine et al., 2021). In addition, chemical engineering widely makes use of virtual laboratories to train new staff and students in a consistent and safe manner (Domingues et al., 2010). One of the benefits of using virtual means for this purpose is the ability to train a large number of users fairly easily (Potkonjak et al., 2016).

Virtual reality can also provide a cost-effective way for educators to provide students with access to situations that would otherwise be impractical due to cost constraints (Kaplan et al., 2021; Slater &

Sanchez-Vives, 2016), such as actual work sites or locations; for example, Ijaz et al.'s (2017) engineering students at an Australian university experienced village settings in rural Niger in west Africa as they made observations with a specific engineering focus (water, energy, transport, or building construction).

However, the use of virtual reality is not limited to teaching technical skills. For example, students can also use simulations to improve communication skills (McGovern et al., 2020; Shorey et al., 2020), learn about building empathy, often through experiencing another person's viewpoint (Barbot & Kaufman, 2020; Bujic et al., 2020), and build intercultural understanding (Hickman & Akdere, 2018). As these example show, virtual reality can be very useful in building the important non-technical skills critical to the work of a professional engineer but often given limited attention in the engineering curricula (Trevelyan, 2019).

3.8 Challenges With Using Virtual Reality Technology

Although virtual reality can have many benefits for education, it also involves important logistical and practical challenges that must be considered. One of the primary issues which users encounter is cybersickness (McGill et al., 2017; Munafo et al., 2017), caused by the brain becoming confused by the disconnect between what is seen in the simulation and the sensory experience in the real world. Another pressing issue is the notable increase in class sizes throughout many parts of the world, challenging because of the logistical issues involved in ensuring all students are able to experience the simulation. There are issues with procuring and setting up enough sets of equipment for all students and training teaching staff how to teach these classes and troubleshoot potential issues. Educators have highlighted concerns about the level of support they receive, challenges with administering the equipment, and a lack of time (Bower et al., 2020) as reasons they may be averse to using virtual reality.

Typically, such activities also require a dedicated area to house the virtual reality equipment, and many institutions may not have the space as the equipment is cumbersome and expensive (Taxen & Naeve, 2002). Ijaz et al. (2017) describe a dedicated learning space with 26 sets of Oculus Rift equipment in one classroom, an environment which can serve an entire class of students, each with their own computer, using a simulation at the same time. These researchers used this environment in several courses over a year to engage students in a range of activities, including 360-degree videos, 3D models, VR applications, and AR applications. While this works, the range of simulations is typically limited to those which can be completed by sitting. Other simulations, those designed for the HTC VIVE, for example, require a space of about four meters square set of equipment. A possible solution to the issue of large classes may be to have one student complete the simulation while others watch the interaction on a nearby screen. Valentine et al. (2021) investigated this possibility, finding that students' learning performance was similar whether they used the equipment themselves or watched another student.

4 Augmented Reality

4.1 What Is Augmented Reality, and How Does It Work?

Augmented reality (AR) is a technology that adds additional 3D computer-generated objects to the real world seen by a user. In general, the user observes the real world through a device such as specially made glasses or a tablet computer or mobile phone with a camera. Each AR device has a screen or something similar, fitted so that it covers part or all of the user's view of the real world. The AR device then adds computer-generated 3D objects (e.g., door, pencil, car, dog) onto a specific part of the screen. When the user looks through the screen of the AR device, they see a view of the real world, including the new computer-generated 3D objects placed at specified locations within it. These computer-generated objects can be designed so that the user can interact with them, meaning, that although the users remain in the physical world, the AR provides computer-generated objects they interact with perceptually. Compared to a VR, the AR environment offers a more realistic feeling for the users because it includes aspects of the real world rather than just computer graphics, providing a safer and more comfortable environment that can be beneficial in learning.

Although AR technology is newer than VR, it has already played a role in industry across several fields, specifically the automotive industry, interior design, architecture, engineering, and construction (Ghannam et al., 2020; Palmarini et al., 2018; Siltanen, 2017). Because the interfaces using AR are still new to supporting creation and modification, there is a gap in understanding how AR can be used to bridge theory to practice in engineering. The interactions among workers and between them and their environments can be mapped into these AR systems to provide a more intuitive and spatial feel regarding an assembly's design.

4.2 Uses of Augmented Reality in Engineering Education

In construction, civil, and mechanical engineering, students have several key challenges with abstract engineering concepts. First, students have difficulty forming a mental image of engineering tasks and operations, and the interaction between resources, the prerequisite knowledge for identifying project performance, such as cost, quality, schedule, and safety risks (Bairaktarova, 2018). This challenge results because classroom illustrations of engineering operations and associated risks are often fragmented, and the subsequent exercises tend to be ill-structured. Even when supplemented with videos, it is not feasible for students to isolate each of the engineering tasks to perform critical analyses or simulations of the influence of the on-site risks. Secondly, students have limited access to engineering sites to collaboratively try or examine the role of sensing systems for addressing the project risks. Such experiences are usually difficult to provide because of the hazardous nature of, for example, construction sites and constraints, such as limited access, scheduling, and weather.

However, extant studies have shown the need for supporting and enhancing engineering education with field experiences (Mihelcic et al., 2006) that provide students with opportunities to develop increased understanding of how engineering principles and theories are put into practice. This is particularly important in construction and civil engineering management education, where students learn how labor, materials, and equipment interact to facilitate the assembly of building and civil infrastructure systems. More importantly, field experiences are also necessitated by the increasing need to improve the productivity and safety of construction projects using data sensing technologies (Tang et al., 2012). These and similar challenges have prompted increasing interest in the exploration of augmented environments for classroom instruction (Dib & Adamo-Villani, 2014) and interactive simulations (Alexander et al., 2005; Goulding et al., 2012; Li et al., 2012).

One such example of augmented reality (AR), holographic scenes, hs been found to enhance cognitive learning in engineering education (Balamuralithara & Woods, 2009; Ieronutti & Chittaro, 2007; Nikolic et al., 2011; Sampaio et al., 2010). Using augmented three-dimensional (3D) objects in the form of holograms, students can visualize and interact with digital representations of construction sites and sensing technologies in the physical classroom environment. Thus, students can feel present on construction site environments as they can move naturally and explore in three dimensions the spatial distribution, boundaries, dependencies, and interaction between tasks. In addition, students can perform selective analysis of construction tasks, operations, and resources. Using the HoloLens, a head-mounted display designed by Microsoft, the AR-produced holograms appear as 3D objects existing in the physical space and respond to gaze, gestures, and voice commands. With the HoloLens, construction sites can be projected in front of a student or group of students, who can then touch and tag resources (e.g., equipment), position sensors (e.g., laser scanners, drones, and real-time location sensors), and observe ergonomic exposures of construction workers. Furthermore, students can collaboratively navigate a construction site and buildings, observing both indoor and outdoor activities, with each student having different points of view (Ogunseiju et al., 2020b).

Ogunseiju and colleagues (2020b, 2022) designed and evaluated a learning environment with holographic scenes utilizing the theory of cognitive apprenticeship, which explains learning as a situative cognitive process. This study exemplifies the application of the situative cognition framework by explaining that learning can be enhanced through modeling, scaffolding, and reflection on the learner's problem-solving skills (Ogunseiju et al., 2020b; Ogunseiju et al., 2022).

A recent systematic literature review of augmented reality in engineering education (Vasquez-Carbonell, 2022) focuses on the advantages of AR as well as some of its disadvantages, primarily (1) the high-cost devices used in AR, (2) the high cost and difficulty in developing AR apps, (3) the limited information about actual student learning achievement using AR, and (4) the difficulties in using AR. Based on the literature review, the author provides answers to some of the basic questions revolving around AR technology, the most relevant information being that Unity 3D and Blender are among the most frequently used software for AR app development, while smartphones, tablets, and Microsoft HoloLens are the most frequently used hardware to run AR. However, AR-specific devices like the MS HoloLens are still quite expensive, and because of their shortcomings, mobile devices are the best options for using AR.

More specifically focusing on the pedagogical use of AR, researchers at Sohar University and Memorial University propose it for delivering instruction on the maintenance and repair of mechanical and computer systems (Yousif, 2022). In addition to pointing to several critical research gaps regarding AR in education, they also compare the advantages and disadvantages of using VR technology, noting that this technology is not yet fully integrated into student learning approaches. They conclude by discussing their development of a virtual student training environment that is comparatively inexpensive and received positive feedback from students who used it (Yousif, 2022).

In an earlier study, scholars at Asian Technological University created 3D AR images of PC motherboard components through a system development method by capturing the actual hardware components using their 3D conversion in the open-source Selva3D app to create the Assembly app along with markers for AR activation through smartphone cameras (Enzai et al., 2021). Their development process exemplifies a simple, systematic way to build an AR environment that can visualize parts such as hard drives, computer chips, and graphics card in 3D. A survey investigating the effectiveness of this method of instruction found positive feedback from students as well as increased student interest in learning; as a result, they plan to include the AR learning environment in all microprocessor courses in electrical engineering at their institution. Their resources are available to engineering educators for in-class demonstrations.

By providing a learning environment that allows students to apply theoretical knowledge to real-world problems, educators can cultivate communication problem-solving, creativity, and collaborative skills in their students (Elliott & Bruckman, 2002). Although VR and AR technologies have been used in education only recently, some of the literature suggests that immersion environments create a more stimulating learning environment as students become active participants in this process (Behzadan & Kamat, 2012). However, other studies reveal that many people are still uncomfortable navigating and interacting with a fully virtual world (H.-K. Wu et al., 2013). One of the advantages of AR is that it does not eliminate the real world from a user's experience, and hence, users have a more realistic sense of presence in the visualization experience as AR incorporates several important aspects of visualization, including an effective alignment of real and virtual worlds and real-time interaction and feedback. In addition, AR provides a convenient interface for constructivism and discovery-based learning, spatial understanding, and social interaction, while at the same time allowing users to learn through mistakes without having to worry about real-world consequences (Dunleavy et al., 2009).

As a result, several studies have validated the technological effectiveness of AR in the learning process (Kaur et al., 2020; Nesterov et al., 2017), a significant and important finding, as past research has also found that the implementation of AR technology in engineering education is imperative, given the transition of the current industrial model to the "Industry 4.0" (Nesterov et al., 2017). Acknowledging the widespread use of smartphones among students and workers, researchers anticipate a future where it will be possible to interact with machines through AR and mixed reality (MR), a blend of physical and digital worlds, interventions. In addition, students will be able to get industry specialized training through updated product models, although this will require collaboration between academia and industry. According to these researchers, there is no need for radical changes in teaching methods; rather, they propose developing AR laboratory complexes as an additional academic resource to effectively harness the AR/MR technologies by utilizing their space for AR markers and QR codes. The AR technology also has potential to promote students' self-learning primarily due to the readily available updated 3D resources from the industry.

Finally, it comes as no surprise that recent research on AR technology has focused on its use during COVID. For example, a study completed at the Universidad de Las Americas and Universidad de´ Alicante investigated the design and use of a NetAR app for engineering students to use to study remotely (Kaur et al., 2020). They captured a set of images to form a scene, then converted it to a 3D AR environment using Unity3D, and finally loaded the models into the NetAR app using a menu. This app focused on teaching the complex concepts of networking using both a basic and an advanced level, depending on the priority of the student using it. Feedback from the students indicated that they found the study of networking systems, particularly the routing and the flow of information through gateways, more effective in 3D using AR techniques (Kaur et al., 2020).

5 Wearable and Haptic

5.1 Wearable Technologies for Learning

Advances in photonics, microtechnology, and machine learning have led to the development of wearable devices that enable users to participate in various tasks by seamlessly interfacing with desktop computers, smartphones, a touch- or gesture-based system, or more advanced types of technologies, such as virtual reality (VR) and augmented reality (AR) (Kaminska et al., 2019). Moreover, these technologies can be worn on a range of body locations, as shown in Figure 23.1, and can play a significant role in learning and education (Khosravi et al., 2021).

In fact, the literature suggests that student learning improves as a result of using wearable devices in the classroom, with benefits being demonstrated for a wide variety of subjects and age limits, from K–12 to tertiary-level higher education. Moreover, Usha Goswami suggested that it may be possible to detect certain neurological biomarkers in the brain, which can help consolidate cognitive methods used to "measure" learning (Goswami, 2009).

In this chapter, we discuss and present the range of wearable and sensor technologies that have been used to enhance the teaching and learning of various engineering disciplines. Based on their location, we have divided these wearable technologies into three broad categories, head-worn, wrist-worn, and chest-worn, discussing the building blocks and architectures of these technologies as well as their impact on teaching and learning.

5.1.1 Building Blocks of Wearable Technologies

Before discussing how wearable devices have been used and their benefits to student learning, it is important to define and identify the main components or building blocks of a wearable computing device. Figure 23.2 is a concept diagram showing that wearable devices are being used to capture

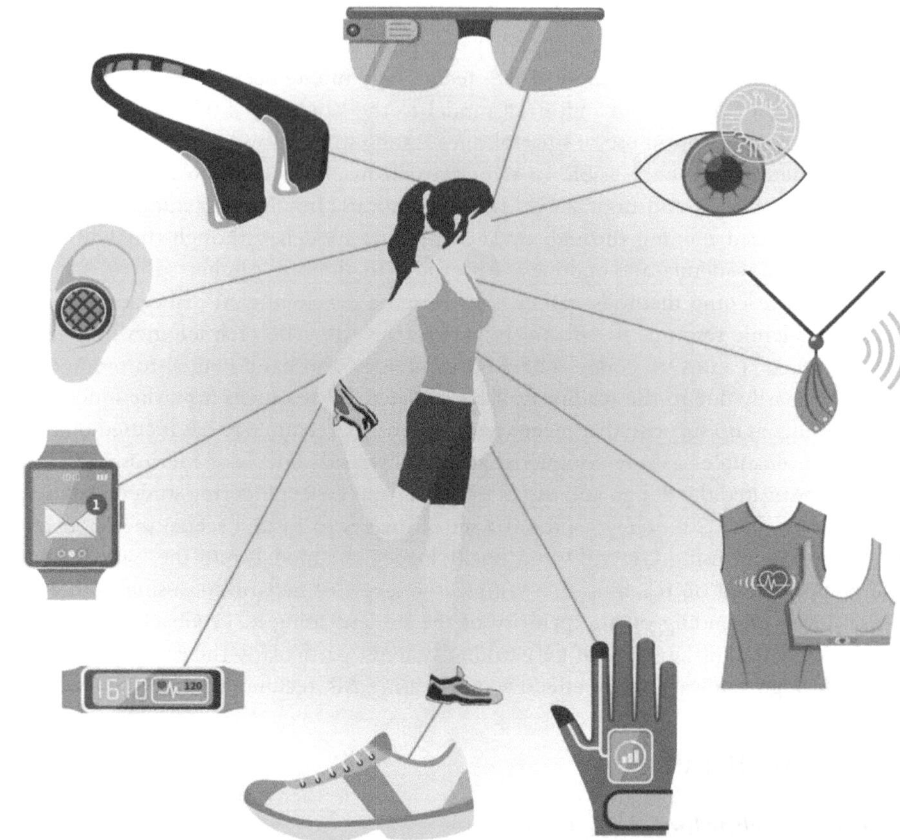

Figure 23.1 Range of wearable devices and their location on the human body.
Source: Image adopted from Ghannam et al. (2020).

information from the brain as well as from the human senses for various tasks. These building blocks have remained basically the same since the early HMDs from the 1990s (Martin et al., 2021). Most recently, wearable devices have been used to detect hand and eye gestures using low-power processors and high–energy density lithium-ion batteries (Liang et al., 2018; Tanwear et al., 2020).

The building blocks of a wearable computing device consist of seven main modules, the interfacing, sensing, communications, energy harvesting, power management, battery storage, and signal processing modules (Ghannam et al., 2020; Khosravi & Ghannam, 2021b). The interfacing module is responsible for collecting information via sensors or other input devices as well as providing output information to the wearer. Similarly, the communications module is responsible for transferring this information between the wearable and its wearer, while the signal processing module is responsible for processing and managing the data collected from the interfacing and sensing modules, and the energy-harvesting and battery storage modules are responsible for harvesting and storing the wearable's energy supply. Finally, the power management module efficiently connects and manages these modules.

Due to advancements in packaging and nanofabrication, it is now possible to embed these modules into a small area and at a relatively low cost (Heidari et al., 2021). Moreover, the emergence of the fifth-generation (5G) wireless network will enable these wearable computing systems to be efficiently connected to us and perhaps to experienced educators. Furthermore, as sensors and

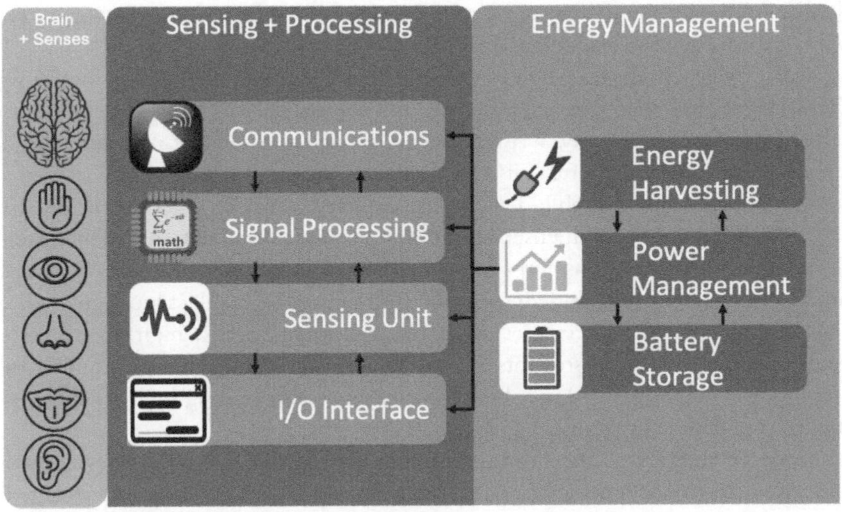

Figure 23.2 Essential building blocks of a wearable computing device. Sensors are used to collect vital information from the human brain and senses, which is then processed and communicated via the signal processing and communications blocks. All the electronic blocks are powered via the energy-harvesting and power-conditioning blocks. Finally, the input/output (I/O) interface is used to communicate these findings back to the user.

processing power have improved in computers, so has the development of wearable computing devices that allow humans to interact with the environment.

Based on these building blocks, Ghannam et al. (2020) and Ghannam et al. (2020) surveyed a range of undergraduate and postgraduate engineering programs to determine if students were acquiring the essential technical skills for designing and developing wearable technologies that could address neuroengineering problems. According to their review, European institutions such as the Technical University of Munich (TUM), Ecole Polytechnique Federale de Lausanne (EPFL), and Politecnico di Milanoengineering offered comprehensive training programs focused on teaching the skills needed for developing wearable devices for neurological diseases, such as Parkinson's, dementia, and Alzheimer.

5.1.2 Wearables in Engineering Education

All three categories of wearable devices, head-, chest-, and wrist-worn, have been used for teaching and learning, with most research based on laboratory experiments and only a small number based on actual user studies, as highlighted by Billinghurst et al. in their ten-year systematic review on AR (Dey et al., 2018). Examples of wearable devices worn on each of these body parts during teaching and learning in engineering programs are provided.

- Chest-worn wearables are embedded with different sensors that can collect and store data such as heart and respiration rates. For example, sociometric sensors can be used to collect important social interaction data from users. Moreover, combining these sensors with infrared sensors as well as an accelerometer and microphones enables us to capture learner information, including speech and conversation dynamics, body movement, posture, and social proximity (Zhou et al., 2020). This combination of sensors has been used to collect real-time social interaction data for predicting collaboration quality and creative fluency in a graduate-level mechanical engineering design course at the University of California (Zhou et al., 2020).

- Wrist-worn. Wearable devices such as wristbands, which can collect physiological signals from the human body, have been proven beneficial for educational purposes. Due to their popularity and their location on the body, wristbands can be used with a large number of students for a group assessment. Furthermore, wristbands offer users the added flexibility of free movement in comparison to other wearable devices that are head-worn or chest-worn (de Arriba Perez et al., 2018). They can sense the user's hand motions (Liang et al., 2018, 2019) as well as distinguish different types of actions and objects by their sounds (Starner, 2016; Ward et al., 2006). Examples in the literature include wristbands with sensors that collect bio-signals for estimating stress in students (de Arriba Perez et al., 2018), important because high stress levels caused by burnout, a common trait in students at universities, can result in a high number of dropouts (Maslach & Jackson, 1981) and may be a contributing factor leading to fewer students enrolling in engineering degree programs (Khosravi & Ghannam, 2021a).
- Head-worn. In addition to the previously mentioned HMDs, wearables devices worn on the head include electroencephalogram (EEG) sensors, which are small sensors attached to a person's scalp that measure electrical signals in the brain. In terms of their use in higher education, the University of South Australia found evidence to support the impact of real-time information overlay on learning using HMDs (Marner et al., 2013). EEG sensors have been used in educational design programs to assess brain activity through a wearable plug-and-play headset, combined with Oculus Rifts VR to conduct spatial assessments (Van Goethem et al., 2019), while Emotiv EPOC® EEG head-mounted gaming systems have been used in cognitive and brain science to measure brain activity in a study conducted at Macquire University (Alvarez et al., 2016), and Akbulut et al. (2018) investigated the use of VR on the performance of computer engineering bachelor of science students.

While all three types of wearable devices have been the subject of past research, most studies on learning applications have perhaps involved HMDs due to the nature of the human visual system and its importance in processing information. In addition, HMDs have several advantages over wrist-worn or chest-worn devices since they are entirely hands-free. In comparison, a wrist-worn device requires at least one arm to check the display and often another to manipulate the user interface (UI). Moreover, HMDs are mounted closer to the wearer's primary senses of sight and hearing, providing a unique first-person view of the world, matching the user's perspective (Starner, 2016), as will be discussed in the next section.

5.1.3 Human Visual System and Current State-of-the-Art

Humans are visual in their nature (Aparicio & Costa, 2015), and our eyes recognize visuals first, followed by printed text (Dyrud & Worley, 2006). In fact, people perceive 80% of their information visually (Sokolov et al., 2020), and more than half of the human brain is involved in processing visual information (Diamant, 2008). Furthermore, in the brain itself, hundreds of millions of neurons are devoted to visual processing, comprising approximately 30% of the cortex compared to 8% for touch and just 3% for hearing (Pillars, 2015; Sherwood et al., 2012). It comes, therefore, as no surprise that the majority of wearable technologies which have been used for enhanced learning are head-mounted, in particular, head-mounted displays for VR and AR applications.

Moreover, to create effective technology that provides a strong sense of reality in VR and AR requires an understanding of how the brain processes information from its senses. Table 23.1 lists the current technology capabilities and what we need to achieve a truly immersive experience (Cuervo et al., 2018). According to the literature, we are likely to achieve these technical specifications by 2030.

Table 23.1 Comparison between Typical HMD Specifications and Our Eye Capabilities

Property	*Our Eyes*	*HMD Specs*
Field of view	210° × 135°	110° × 110°
Angular pixel density	>600 pixels per degree	10–13 pixels per degree
Frame rate	1,800 Hz	90 Hz
Dynamic range	96 bits	24 bits
Stereo overlap	120°	110°

5.2 Haptic Devices in Engineering Education

Multimodal learning is a constructivist perspective that considers learners as sensemakers who work to select, organize, and assimilate new information with existing knowledge by interacting with learning environments that integrate verbal with non-verbal information (Moreno & Mayer, 2007). Based on this premise, Moreno and Mayer (2007) developed a cognitive-affective theory of learning with media (CATLM) that considers learners' interaction and learning with multimodal environments through four tenets: (a) learners process different external modes through separate processing modalities; (b) each modality has a limited processing capacity within the working memory; (c) meaningful learning occurs when newly processed information is appropriately selected, organized, and integrated with already-existing knowledge; and (d) learners' cognitive engagement with the multimodal environment is mediated by motivational factors. Haptic technologies are emerging technologies that can enhance students' educational experiences by supporting these four tenets.

Haptic refers to the sense of touch, which enables us to perform a wide variety of tasks in the real world:

> In virtual worlds, the sense of touch must be artificially recreated by stimulating the human body (typically the hands) in a manner that produces the salient features of touch needed to enhance realism and human performance.
>
> *(Culbertson et al., 2018, p. 385)*

There are three categories of haptic systems: graspable (kinesthetic: force-feedback), wearable (tactile: cutaneous), and touchable (encountered displays). These three categories describe the interaction modalities for kinesthetic and cutaneous stimulation in interactive haptic devices. Figure 23.3 presents the three types of haptic systems.

Haptic technologies play an important role in enabling humans to touch and interact with the contents of virtual and augmented environments. In engineering practice, engineers perform a variety of manual tasks in physical environments through the coordination of touch sensation, perception, and movement. In VR and AR environments, the performance of these skilled tasks is possible only if they together comprise a haptic modality. An emerging body of research investigates the efficacy of haptic technologies for VR and AR in education settings and looks at how to create haptic interfaces that allow learners to touch and feel realistic virtual objects.

In engineering education, haptic technology has been used in teaching concepts related to "invisible forces," such as magnetism and buoyancy (Park et al., 2012; Sanchez et al., 2013; Young et al., 2011), and mechanical properties such as resonance in a dynamic system (Okamura et al., 2002). For example, simulations of van der Waals forces at the nanoscale have been developed to give learners a hands-on experience of phenomena such as "snap to contact," thereby providing a more visceral understanding of the forces at play during scanning (Park et al., 2012). Although students are more engaged in learning activities with haptic feedback and report that the haptic information helps them

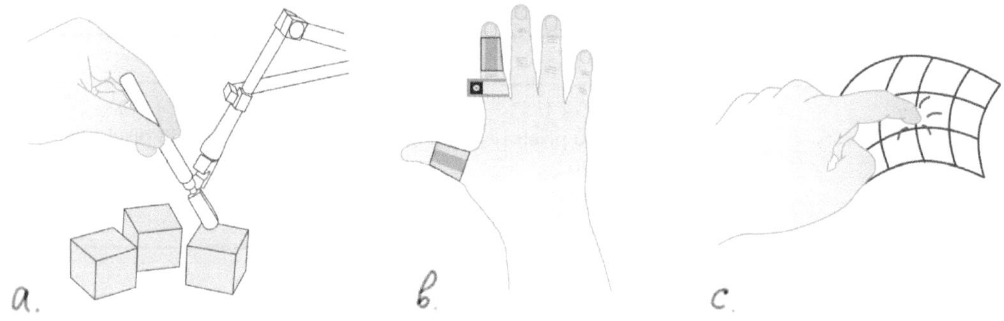

Figure 23.3 Examples of (a) graspable, (b) wearable, and (c) touchable haptic systems.
Source: Image adapted from Culbertson et al. (2018).

visualize abstract concepts better, results on the efficacy of haptic information is somewhat mixed in the sense that not all pre- and post-tests show a significant difference (Jones et al., 2016). This calls for additional studies on what types of haptic information brings additional learning benefits and a more nuanced understanding of when to engage haptic feedback in education.

The synergy between situated and distributed cognition in the application of haptic technology can be helpful to arrive at a blended framework characterized by the following description: cognition is embodied, embedded, and extended within a sociotechnical system, and it is distributed among its cognitive agents, cognitive artifacts, and through time. Advances in cognitive science posit that incorporating body activity during the learning process may foster deeper learning (Lindgren & Johnson-Glenberg, 2013). As Wilson (2002) suggests, "embodied cognition holds that cognitive processes are deeply rooted in the body's interactions with the world" (Wilson, 2002, p. 626). This theory emphasizes that sensory and motor functions are relevant for successful interaction with an environment.

Embodied cognition describes learning as occurring through bodily experiences as sensorimotor systems interact with the environment (Barsalou, 2003; Fischer & Zwaan, 2008; Lakoff & Johnson, 1999; Wilson, 2002), with touch being seen as an active discovery sense and more meaningful than passively seeing and hearing (Taylor et al., 1973). Further, research has shown that coordination and transformation across multiple modes, such as visual, linguistic, and hands-on or sensorimotor experiences, can play an important role in developing understanding (Jewitt et al., 2001), with sensory experiences increasing active engagement, motivation (Bergin et al., 1996), and attention to learning.

Haptic technology makes it possible to extend students' interactions with digital visualizations in a more immersive environment than conventional learning contexts (Brooks Jr et al., 1990; Jones et al., 2003). This evidence, along with assumptions of embodied cognition, such as (a) cognition is situated, (b) cognition is off-loaded onto the environment, and (c) cognition is for action, supports the use of haptic-based interaction to foster design and spatial thinking.

6 Concluding Thoughts

In this chapter, we have reviewed a number of VR, AR, and multimodal applications using wearable and haptic devices, which have been found to have great potential in engineering education. Empirical research supports that higher degree of modalities provided additional channels for information presentation and delivery, facilitating the sensemaking process in learning and teaching. The integration of VR, AR, and wearable and haptic devices into learning environments creates immersive user experiences, thus enhancing user engagement. We envision they have great potential in increasing learning-based applicability and in teaching and learning engineering.

As emerging technologies become more prevalent, instructional designers and educators will continue to leverage devices to deliver instruction. Dede and colleagues have outlined several emerging practices, documenting the idiosyncratic set of definitions, conceptual frameworks, and methods (Dede, 2011). In the case of AR technology, for example, they claim, "AR is an instructional approach looking for contexts where it will be the most effective tool." While the challenge of facilitating collaborative and experiential learning may be the instructional problem addressed by AR, Dede and colleagues challenge researchers to continue exploring how this approach might impact other persistent educational problems, such as the limitations within the expanding ecology of pedagogies.

There are several areas which we did not discuss due to space limitations, but these areas also call for future research. These include, for example, the advantages and disadvantages of each of the engines used for creating virtual environments (e.g., unity, usability, etc.) and the process of technology-enhanced learning environments, that is, design, implementation, and evaluation. Considering that eye tracking and haptic feedback both impact on user's learning efficiency and cognitive process, there are several issues that need to be noted, especially in their design and implementation, and even though studies describing how to set up the learning environments exist, more research is needed. More studies investigating factors that have it challenging to engineer haptic technologies for augmented and virtual realities considering the extraordinary spatial and temporal tactile acuity and the complex interplay between continuum mechanics, haptic perception, and interaction should be made.

It should be noted that discussing these technologies as learning environments is challenging as they are not always explicit categories and there are areas where they can interact and overlap. For example, AR can include haptic technologies, and wearables can be part of a VR experience. We have briefly noted that there are hybrid systems; however, it was something we did not address in much detail in this chapter.

This chapter suggests several directions for research-based instructional practices and adds to the situated perspective on engineering learning as well as provides a basis for future studies of engineering teaching and learning in immersive environments and haptic and wearable technologies across the engineering curriculum and disciplines. The applications of technology in education are constantly increasing to support and impact engineering processes across a range of industries. However, state-of-the-art research on immersive environments and haptic and wearable technologies applied to engineering learning focuses on the technology from the point of view of its functionality, usability, and aesthetic factors. Another direction for research could be to consider the perspective on how the challenges mentioned earlier could be addressed through convergent research on AR, VR, wearables, and haptic perception, mechanics, electronics, and material technologies for ultimate benefits in teaching and learning engineering.

Furthermore, one of the recurring issues in research on learning supported by VR and AR is the way in which these technologies improve learners' abilities. Therefore, research on how technology-supported pedagogies can support engineering ability, including spatial and creative abilities, must be a priority for contemporary engineering programs. Methodically studying the ways technology can be used to support learning will enable educators to prepare engineering graduates for the future by improving the connection between engineering education and real-world engineering practice.

References

Agbo, F., Sanusi, I., Oyelere, S., & Suhonen, J. (2021). Application of virtual reality in computer science education: A systemic review based on bibliometric and content analysis methods. *Education Sciences*, *11*(3), 1–23. doi:10.3390/educsci11030142

Akbulut, A., Catal, C., & Yıldız, B. (2018). On the effectiveness of virtual reality in the education of software engineering. *Computer Applications in Engineering Education*, *26*(4), 918–927.

Alexander, A. L., Brunye, T., Sidman, J., & Weil, S. A. (2005). From gaming to training: A review of studies on fidelity, immersion, presence, and buy-in and their effects on transfer in pc-based simulations and games. *DARWARS Training Impact Group*, *5*, 1–14.

Alvarez, V., Bower, M., de Freitas, S., Gregory, S., & De Wit, B. (2016). The use of wearable technologies in Australian universities: Examples from environmental science, cognitive and brain sciences and teacher training. *Mobile Learning Futures – Sustaining Quality Research and Practice in Mobile Learning*, 25–32.

Aparicio, M., & Costa, C. J. (2015). Data visualization. *Communication Design Quarterly Review*, *3*(1), 7–11.

Bairaktarova, D. (2018). Collaborative research: vObjects-understanding their utility to enhance learning of abstract and complex engineering concepts. In 2018 *ASEE annual conference & exposition proceedings.*

Bairaktarova, D., Graziano, W., & Cox, M. (2017). Enhancing engineering students' performance on design task: The box of parts. *Journal of Mechanical Design*, *139*(5).

Bairaktarova, D., & Johri, A. (2016). The role of virtual objects in performing engineering related task. In *2016 IEEE frontiers in education conference (fie)* (pp. 1–5).

Balamuralithara, B., & Woods, P. C. (2009). Virtual laboratories in engineering education: The simulation lab and remote lab. *Computer Applications in Engineering Education*, *17*(1), 108–118.

Barbot, B., & Kaufman, J. C. (2020, October). What makes immersive virtual reality the ultimate empathy machine? Discerning the underlying mechanisms of change. *Computers in Human Behavior*, *111*, 106431. doi:10.1016/j.chb.2020.106431

Barsalou, L. (2003). Situated simulation in the human conceptual system. *Language and Cognitive Processes*, *18*(5–6), 513–562.

Behzadan, A. H., & Kamat, V. R. (2012). A framework for utilizing context-aware augmented reality visualization in engineering education. In *12th international conference on construction application of virtual reality* (pp. 5–8). Citeseer.

Bell, J. T., & Fogler, H. S. (1997). *Ten steps to developing virtual reality applications for engineering education.* Paper presented at 1997 Annual Conference, Milwaukee, Wisconsin.

Bergin, J., Patino-Martinez, M., Brodie, K., McNally, M., Goldweber, M., Naps, T., Jimenez-Peris, R., Rodger, S., Khuri, S., & Wilson, J. (1996). An overview of visualization: Its use and design: Report of the working group in visualization. *ACM SIGCSE Bulletin*, *28*(SI), 192–200.

Bower, M., DeWitt, D., & Lai, J. W. (2020). Reasons associated with preservice teachers' intention to use immersive virtual reality in education. *British Journal of Educational Technology*, *51*(6), 2215–2233. doi:10.1111/bjet.13009

Bower, M., & Jong, M. S.-Y. (2020). Immersive virtual reality in education. *British Journal of Educational Technology*, *51*(6), 1981–1990. doi:10.1111/bjet.13038

Brooks Jr, F. P., Ouh-Young, M., Batter, J. J., & Jerome Kilpatrick, P. (1990). Project gropehaptic displays for scientific visualization. *ACM SIGGraph Computer Graphics*, *24*(4), 177–185.

Bujic, M., Salminen, M., Macey, J., & Hamari, J. (2020, January). "Empathy machine": How virtual reality affects human rights attitudes. *Internet Research*, *30*(5), 1407–1425. (Publisher: Emerald Publishing Limited) doi:10.1108/INTR-07-2019-0306.

Chaiklin, S., & Lave, J. (1993). *Understanding practice: Perspectives on activity and context.* Cambridge University Press.

Cheon, H. (2014). Distributed cognition in scientific contexts. *Journal for General Philosophy of Science*, *45*(1), 23–33.

Clartcey, W. J. (2008). Scientific antecedents of situated cognition. *The Cambridge Handbook of Situated Cognition*, 11.

Clement, A. (1990). Cooperative support for computer work: A social perspective on the empowering of end users. In *Proceedings of the 1990 ACM conference on computer supported cooperative work* (pp. 223–236).

Concannon, B. J., Esmail, S., & Roduta Roberts, M. (2019). Head-mounted display virtual reality in post-secondary education and skill training. In *Frontiers in education* (Vol. 4, p. 80). Frontiers Media SA.

Cuervo, E., Chintalapudi, K., & Kotaru, M. (2018). Creating the perfect illusion: What will it take to create life-like virtual reality headsets? In *Proceedings of the 19th international workshop on mobile computing systems & applications* (pp. 7–12). https://cuervo.dev/wp-content/uploads/2020/05/perfect-illusion.pdf

Culbertson, H., Schorr, S. B., & Okamura, A. M. (2018). Haptics: The present and future of artificial touch sensation. *Annual Review of Control, Robotics, and Autonomous Systems*, *1*(1), 385–409.

de Arriba Perez, F., Santos-Gago, J. M., Caeiro-Rodr´ıguez, M., & Iglesias, M. J. F. (2018). Evaluation of commercial-off-the-shelf wrist wearables to estimate stress on students. *JoVE (Journal of Visualized Experiments)* (136), e57590. doi:10.3791/57590

Dede, C. (2011). Developing a research agenda for educational games and simulations. *Computer Games and Instruction, 3*(2), 233–250.

Dey, A., Billinghurst, M., Lindeman, R. W., & Swan, J. E. (2018, April). A systematic review of 10 years of augmented reality usability studies: 2005 to 2014. *Frontiers in Robotics and AI, 5*. doi:10.3389/frobt.2018.00037

Diamant, E. (2008). Unveiling the mystery of visual information processing in human brain. *Brain Research, 1225*, 171–178. doi:10.1016/j.brainres.2008.05.017

Dib, H., & Adamo-Villani, N. (2014). Serious sustainability challenge game to promote teaching and learning of building sustainability. *Journal of Computing in Civil Engineering, 28*(5), A4014007–1–A4014007–11.

di Lanzo, J., Valentine, A., Sohel, F., Yapp, A., Muparadzi, K., & Abdelmalek, M. (2020). A review of the uses of virtual reality in engineering education. *Computer Applications in Engineering Education, 28*(3), 748–763. doi:10.1002/cae.22243

Di Natale, A. F., Repetto, C., Riva, G., & Villani, D. (2020). Immersive virtual reality in k12 and higher education: A 10-year systematic review of empirical research. *British Journal of Educational Technology, 51*(6), 2006–2033.

Domingues, L., Rocha, I., Dourado, F., Alves, M., & Ferreira, E. C. (2010, May). Virtual laboratories in (bio)chemical engineering education. *Education for Chemical Engineers, 5*(2), e22–e27. doi:10.1016/j.ece.2010.02.001

Dunleavy, M., Dede, C., & Mitchell, R. (2009). Affordances and limitations of immersive participatory augmented reality simulations for teaching and learning. *Journal of Science Education and Technology, 18*(1), 7–22.

Dyrud, M. A., & Worley, R. B. (2006). Visual communication. *Business Communication Quarterly, 69*(4), 397–399.

Elliott, J., & Bruckman, A. (2002). Design of a 3D interactive math learning environment. In *Proceedings of the 4th conference on Designing interactive systems: Processes, practices, methods, and techniques* (pp. 64–74).

Enzai, N. I. M., Ahmad, N., Ghani, M. A. H. A., Rais, S. S., & Mohamed, S. (2021). Development of augmented reality (ar) for innovative teaching and learning in engineering education. *Asian Journal of University Education, 16*(4), 99–108.

Fischer, M. H., & Zwaan, R. A. (2008). Embodied language: A review of the role of the motor system in language comprehension. *Quarterly Journal of Experimental Psychology, 61*(6), 825–850.

Flor, N. V., & Hutchins, E. L. (1991). Analyzing distributed cognition in software teams: A case study of team programming during. In *Proceedings of the fourth annual workshop on empirical studies of programmers* (pp. 36–59). Ablex Publishing.

Freina, L., & Ott, M. (2015). A literature review on immersive virtual reality in education: state of the art and perspectives. In *The international scientific conference e-learning and software for education* (Vol. 1, pp. 10–1007).

Gantt, M., & Nardi, B. A. (1992). Gardeners and gurus: Patterns of cooperation among cad users. In *Proceedings of the SIGCHI conference on human factors in computing systems* (pp. 107–117).

Ghannam, R., Curia, G., Brante, G., Fan, H., & Heidari, H. (2020). Wearable electronics for neurological applications: A review of undergraduate engineering programmes. In *2020 transnational engineering education using technology (treet)*, pp. 1–4. doi:10.1109/TREET50959.2020.9189753

Ghannam, R., Curia, G., Brante, G., Khosravi, S., & Fan, H. (2020). Implantable and wearable neuroengineering education: A review of postgraduate programmes. *IEEE Access, 8*, 212396–212408. doi:10.1109/ACCESS.2020.3040064

Giere, R. N. (2006). *Cognition in the wild*. The University of Chicago Press.

Giere, R. N., & Moffatt, B. (2003). Distributed cognition: Where the cognitive and the social merge. *Social Studies of Science, 33*(2), 301–310.

Goswami, U. (2009). Mind, brain, and literacy: Biomarkers as usable knowledge for education. *Mind, Brain, and Education, 3*(3), 176–184.

Goulding, J., Nadim, W., Petridis, P., & Alshawi, M. (2012). Construction industry offsite production: A virtual reality interactive training environment prototype. *Advanced Engineering Informatics, 26*(1), 103–116.

Greeno, J. G., Collins, A. M., & Resnick, L. B. (1996). Cognition and learning. In D. C. Berliner & R. C. Calfee (Eds.), *Handbook of educational psychology* (pp. 15–46). MacMillian.

Hamilton, D., McKechnie, J., Edgerton, E., & Wilson, C. (2021, March). Immersive virtual reality as a pedagogical tool in education: A systematic literature review of quantitative learning outcomes and experimental design. *Journal of Computers in Education*, *8*(1), 1–32. doi:10.1007/s40692-020-00169-2

Heidari, H., Ozt¨ urk, M., Ghannam, R., Law, M.-K., Khanbareh, H., & Miah, A. H. (2021). IEEE access special section editorial: Energy harvesting technologies for wearable and implantable devices. *IEEE Access*, *9*, 91324–91327. doi:10.1109/ACCESS.2021.3088622

Hickman, L., & Akdere, M. (2018, February). Developing intercultural competencies through virtual reality: Internet of things applications in education and learning. In *2018 15th learning and technology conference (LT)* (pp. 24–28). doi:10.1109/LT.2018.8368506

Hollan, J., Hutchins, E., & Kirsh, D. (2000). Distributed cognition: Toward a new foundation for human-computer interaction research. *ACM Transactions on Computer-Human Interaction (TOCHI)*, 7(2), 174–196.

Huang, W., & Roscoe, R. (2021). Head-mounted display-based virtual reality systems in engineering education: A review of recent research. *Computer Applications in Engineering Education*, *29*(5), 1420–1435. doi:10.1002/cae.22393

Hutchins, E. (1995a). *Cognition in the wild*. MIT Press.

Hutchins, E. (1995b). How a cockpit remembers its speeds. *Cognitive Science*, *19*(3), 265–288.

Ieronutti, L., & Chittaro, L. (2007). Employing virtual humans for education and training in x3d/vrml worlds. *Computers & Education*, *49*(1), 93–109.

Ijaz, K., Marks, B., Hartley, T., Gibbens, P., & Thomas, J. (2017). The immersive learning laboratory: Employing virtual reality technology in teaching. In *28th annual conference of the Australasian association for engineering education (AAEE 2017)* (pp. 974–981).

Jensen, L., & Konradsen, F. (2018). A review of the use of virtual reality head-mounted displays in education and training. *Education and Information Technologies*, *23*(4), 1515–1529. doi:10.1007/s10639-017-9676-0

Jewitt, C., Kress, G., Ogborn, J., & Tsatsarelis, C. (2001). Exploring learning through visual, actional and linguistic communication: The multimodal environment of a science classroom. *Educational Review*, *53*(1), 5–18.

Johri, A., & Olds, B. M. (2011). Situated engineering learning: Bridging engineering education research and the learning sciences. *Journal of Engineering Education*, *100*(1), 151–185.

Johri, A., Olds, B. M., & O'Connor, K. (2014). Situative frameworks for engineering learning research. *Cambridge Handbook of Engineering Education Research*, 47–66.

Jones, M. G., Andre, T., Superfine, R., & Taylor, R. (2003). Learning at the nanoscale: The impact of students' use of remote microscopy on concepts of viruses, scale, and microscopy. *Journal of Research in Science Teaching: The Official Journal of the National Association for Research in Science Teaching*, *40*(3), 303–322.

Jones, M. G., Childers, G., Emig, B., Chevrier, J., Stevens, V., & Tan, H. (2016). The efficacy of visuohaptic simulations in teaching concepts of thermal energy, pressure, and random motion. In *Insights from research in science teaching and learning* (pp. 73–86). Springer.

Kaminska, D., Sapi´ nski, T., Wiak, S., Tikk, T., Haamer, R. E., Avots, E., . . . Anbarjafari, G. (2019). Virtual reality and its applications in education: Survey. *Information*, *10*(10), 318.

Kaplan, A. D., Cruit, J., Endsley, M., Beers, S. M., Sawyer, B. D., & Hancock, P. A. (2021, June). The effects of virtual reality, augmented reality, and mixed reality as training enhancement methods: A meta-analysis. *Human Factors*, *63*(4), 706–726. doi:10.1177/0018720820904229

Kaur, D. P., Mantri, A., & Horan, B. (2020). Enhancing student motivation with use of augmented reality for interactive learning in engineering education. *Procedia Computer Science*, *172*, 881–885.

Kavanagh, S., Luxton-Reilly, A., Wuensche, B., & Plimmer, B. (2017). A systematic review of virtual reality in education. *Themes in Science & Technology Education*, *10*(2), 85–119. https://eric.ed.gov/?id=EJ1165633

Khosravi, S., Bailey, S. G., Parvizi, H., & Ghannam, R. (2021). Learning enhancement in higher education with wearable technology. arXiv preprint arXiv:2111.07365.

Khosravi, S., & Ghannam, R. (2021a). Calling all engineers: We need you! *IEEE Potentials*, *40*(5), 14–17. doi:10.1109/MPOT.2021.3073885

Khosravi, S., & Ghannam, R. (2021b). Enhancing STEM education using wearable biofeedback. In *Third IEEE UK&I YP postgrad STEM research symposium*. https://ieeeukiyp.org/3rdstem/schedule/

Knorr-Cetina, K. (1999). *Epistemic cultures: How the sciences make knowledge*. Harvard University Press.

Kumar, V. V., Carberry, D., Beenfeldt, C., Andersson, M. P., Mansouri, S. S., & Gallucci, F. (2021, July). Virtual reality in chemical and biochemical engineering education and training. *Education for Chemical Engineers, 36*, 143–153. doi:10.1016/j.ece.2021.05.002

Lakoff, G., & Johnson, M. (1999). *Philosophy in the flesh: The embodied mind and its challenge to western thought* (Vol. 640). Basic Books.

Lave, J. (1988). *Cognition in practice*. Cambridge University Press.

Lave, J. (1991). Situating learning in communities of practice. In L. B. Resnick, J. M. Levine, & S. D. Teasley (Eds.), *Perspectives on socially shared cognition* (pp. 63–82). American Psychological Association.

Lave, J., & Wenger, E. (1991). *Situated learning: Legitimate peripheral participation*. Cambridge University Press.

Leont'ev, A. N. (1978). *Activity, consciousness, and personality*. Prentice-Hall.

Li, H., Chan, G., & Skitmore, M. (2012). Visualizing safety assessment by integrating the use of game technology. *Automation in Construction, 22*, 498–505.

Liang, X., Ghannam, R., & Heidari, H. (2018). Wrist-worn gesture sensing with wearable intelligence. *IEEE Sensors Journal, 19*(3), 1082–1090.

Liang, X., Li, H., Wang, W., Liu, Y., Ghannam, R., Fioranelli, F., & Heidari, H. (2019). Fusion of wearable and contactless sensors for intelligent gesture recognition. *Advanced Intelligent Systems, 1*(7), 1900088.

Lindgren, R., & Johnson-Glenberg, M. (2013). Emboldened by embodiment: Six precepts for research on embodied learning and mixed reality. *Educational Researcher, 42*(8), 445–452.

Luria, A. R. (1978). *The making of mind: A personal account of soviet psychology*. Harvard University Press.

Maas, M., & Hughes, J. (2020). Virtual, augmented and mixed reality in K–12 education: A review of the literature. *Technology, Pedagogy and Education, 29*(2), 231–249. doi:10.1080/1475939X.2020.1737210

Mackay, W. E. (1990). Patterns of sharing customizable software. In *Proceedings of the 1990 ACM conference on computer-supported cooperative work* (pp. 209–221).

MacLean, A., Carter, K., Lovstrand, L., & Moran, T. (1990). User-tailorable systems: press-¨ ing the issues with buttons. In *Proceedings of the SIGCHI conference on human factors in computing systems* (pp. 175–182).

Magnus, P. D. (2007). Distributed cognition and the task of science. *Social Studies of Science, 37*(2), 297–310.

Marner, M. R., Irlitti, A., & Thomas, B. H. (2013). Improving procedural task performance with augmented reality annotations. In *2013 IEEE international symposium on mixed and augmented reality (ISMAR)* (pp. 39–48).

Martin, T., Starner, T., Siewiorek, D., Kunze, K., & Laerhoven, K. (2021, July). 25 years of ISWC: Time flies when you're having fun. *IEEE Pervasive Computing, 20*(3), 72–78. doi:10.1109/MPRV.2021.3094898

Maslach, C., & Jackson, S. E. (1981). The measurement of experienced burnout. *Journal of Organizational Behavior, 2*(2), 99–113.

McGill, M., Ng, A., & Brewster, S. (2017, May). I am the passenger: How visual motion cues can influence sickness for in-car VR. In *Proceedings of the 2017 CHI conference on human factors in computing systems* (pp. 5655–5668). Association for Computing Machinery. doi:10.1145/3025453.3026046

McGovern, E., Moreira, G., & Luna-Nevarez, C. (2020, October). An application of virtual reality in education: Can this technology enhance the quality of students' learning experience? *Journal of Education for Business, 95*(7), 490–496. doi:10.1080/08832323.2019.1703096

Merchant, Z., Goetz, E., Cifuentes, L., Keeney-Kennicutt, W., & Davis, T. (2014). Effectiveness of virtual reality-based instruction on students' learning outcomes in K-12 and higher education: A meta-analysis. *Computers and Education, 70*, 29–40. doi:10.1016/j.compedu.2013.07.033

Mihelcic, J. R., Phillips, L. D., & Watkins Jr, D. W. (2006). Integrating a global perspective into education and research: Engineering international sustainable development. *Environmental Engineering Science, 23*(3), 426–438.

Moore, J. L., & Rocklin, T. R. (1998). The distribution of distributed cognition: Multiple interpretations and uses. *Educational Psychology Review, 10*(1), 97–113.

Moreno, R., & Mayer, R. (2007). Interactive multimodal learning environments. *Educational Psychology Review, 19*(3), 309–326.

Munafo, J., Diedrick, M., & Stoffregen, T. A. (2017, March). The virtual reality headmounted display Oculus Rift induces motion sickness and is sexist in its effects. *Experimental Brain Research, 235*(3), 889–901. doi:10.1007/s00221-016-4846-7

Nardi, B. A. (1996). Studying context: A comparison of activity theory, situated action models, and distributed cognition. *Context and Consciousness: Activity Theory and Human-Computer Interaction, 69102*, 35–52.

Nardi, B. A., & Miller, J. (1990). The spreadsheet interface: A basis for end user programming. In *Proceedings of interact' 90* (pp. 977–983). Cambridge University Press.

Nardi, B. A., & Miller, J. R. (1991). Twinkling lights and nested loops: Distributed problem solving and spreadsheet development. *International Journal of Man-Machine Studies*, *34*(2), 161–184.

Nersessian, N. J. (2005). Interpreting scientific and engineering practices: Integrating the cognitive, social, and cultural dimensions. In M. E. Gorman, R. D. Tweney, D. C. Gooding, & A. P. Kincannon (Eds.), *Scientific and technological thinking* (pp. 17–56). Erlbaum.

Nersessian, N. J. (2006). Model-based reasoning in distributed cognitive systems. *Philosophy of Science*, *73*(5), 699–709.

Nesterov, A., Kholodilin, I., Shishkov, A., & Vanin, P. (2017). Augmented reality in engineering education: Opportunities and advantages. *Communications-Scientific Letters of the University of Zilina*, *19*(4), 117–120.

Newman, D., Griffin, P., & Cole, M. (1989). *The construction zone: Working for cognitive change in school.* Cambridge University Press.

Newstetter, W. C., & Svinicki, M. D. (2014). Learning theories for engineering education practice. *Cambridge Handbook of Engineering Education Research*, 29–46.

Nikolic, D., Jaruhar, S., & Messner, J. I. (2011). Educational simulation in construction: Virtual construction simulator. *Journal of Computing in Civil Engineering*, *25*(6), 421–429.

Nilsson, N. C., Serafin, S., Steinicke, F., & Nordahl, R. (2018). Natural walking in virtual reality: A review. *Computers in Entertainment (CIE)*, *16*(2), 1–22.

Norman, D. (1991). *Designing interaction: Psychology at the human computer interface.* (chap. Cognitive artifacts). Cambridge University Press.

Ogunseiju, O., Akanmu, A., & Bairaktarova, D. (2020a). Holographic learning environment for bridging the technical skill gap of the future smart construction engineering students. In *Enabling the development and implementation of digital twins: Proceedings of the 20th international conference on construction applications of virtual reality.*

Ogunseiju, O., Akanmu, A., & Bairaktarova, D. (2020b). Holographic learning environment for bridging the technical skill gap of the future smart construction engineering students. *Enabling the development and implementation of digital twins: Proceedings of the 20th international conference on construction applications of virtual reality.*

Ogunseiju, O. R., Gonsalves, N., Akanmu, A. A., Bairaktarova, D., Bowman, D. A., & Jazizadeh, F. (2022). Mixed reality environment for learning sensing technology applications in construction: A usability study. *Advanced Engineering Informatics*, *53*, 101637.

Okamura, A. M., Richard, C., & Cutkosky, M. R. (2002). Feeling is believing: Using a forcefeedback joystick to teach dynamic systems. *Journal of Engineering Education*, *91*(3), 345–349.

Osbeck, L. M., & Nersessian, N. J. (2014). Situating distributed cognition. *Philosophical Psychology*, *27*(1), 82–97.

Pakala, K., & Bairaktarova, D. (2020). Leveraging technology to elevate pedagogy in mechanical engineering teaching and learning. In C. E. Baukal, Jr. (Ed.), *Mechanical engineering education handbook.* Nova Science Publishers, Inc.

Palincsar, A. S. (1998). Social constructivist perspectives on teaching and learning. *Annual Review of Psychology*, *49*(1), 345–375 (Publisher: Annual Reviews 4139 El Camino Way, PO Box 10139, Palo Alto, CA 94303–0139, USA).

Palmarini, R., Erkoyuncu, J. A., Roy, R., & Torabmostaedi, H. (2018). A systematic review of augmented reality applications in maintenance. *Robotics and Computer-Integrated Manufacturing*, *49*, 215–228 (Publisher: Elsevier).

Pantelidis, V. S. (1997). Virtual reality and engineering education. *Computer Applications in Engineering Education*, *5*(1), 3–12. doi:10.1002/(SICI)1099-0542(1997)5:1⟨3::AID-CAE1⟩3.0.CO;2-H.

Park, J., Doxon, A. J., Provancher, W. R., Johnson, D. E., & Tan, H. Z. (2012). Haptic edge sharpness perception with a contact location display. *IEEE Transactions on Haptics*, *5*(4), 323–331.

Pellas, N., Dengel, A., & Christopoulos, A. (2020). A scoping review of immersive virtual reality in stem education. *IEEE Transactions on Learning Technologies*, *13*(4), 748–761.

Pillars, W. (2015). *Visual note-taking for educators: A teacher's guide to student creativity.* WW Norton & Company.

Pirker, J., Dengel, A., Holly, M., & Safikhani, S. (2020). Virtual reality in computer science education: A systematic review. In *Proceedings VRST' 20: 26th ACM symposium on virtual reality software and technology* (pp. 1–8). doi:10.1145/3385956.3418947

Potkonjak, V., Gardner, M., Callaghan, V., Mattila, P., Guetl, C., Petrovic, V. M., & Jovanovic, K. (2016, April). Virtual laboratories for education in science, technology, and engineering: A review. *Computers & Education, 95*, 309–327. doi:10.1016/j.compedu.2016.02.002

Radianti, J., Majchrzak, T., Fromm, J., & Wohlgenannt, I. (2020). A systematic review of immersive virtual reality applications for higher education: Design elements, lessons learned, and research agenda. *Computers and Education, 147*, 103778. doi:10.1016/j.compedu.2019.103778

Robbins, P., & Aydede, M. (2009). *The Cambridge handbook of situated cognition* (pp. 3–10). Cambridge University Press.

Salomon, G. (1993). *Distributed cognitions: Psychological and educational considerations*. Cambridge University Press.

Sampaio, A. Z., Henriques, P. G., & Martins, O. P. (2010). Virtual reality technology used in civil engineering education. *The Open Virtual Reality Journal, 2*(1).

Sanchez, K. L., Magana, A. J., Sederberg, D., Richards, G. P., Jones, M. G., & Tan, H. Z. (2013). Investigating the impact of visuohaptic simulations for conceptual understanding in electricity and magnetism. In *2013 ASEE annual conference & exposition* (pp. 23–831).

Scribner, S. (1984). *Everyday cognition: Its development in social context (in chap. studying working intelligence)*. Harvard University Press.

Sherwood, L., Klandorf, H., & Yancey, P. (2012). *Animal physiology: From genes to organisms*. Cengage Learning.

Shorey, S., Ang, E., Ng, E. D., Yap, J., Lau, L. S. T., & Chui, C. K. (2020, November). Communication skills training using virtual reality: A descriptive qualitative study. *Nurse Education Today, 94*(104592), 1–10. doi:10.1016/j.nedt.2020.104592

Siltanen, S. (2017). Diminished reality for augmented reality interior design. *The Visual Computer, 33*(2), 193–208.

Slater, M., & Sanchez-Vives, M. (2016). Enhancing our lives with immersive virtual reality. *Frontiers Robotics AI, 3*(74), 1–47. doi:10.3389/frobt.2016.00074

Sokolov, A., Ostromukhov, R., Vezhenkova, I., Kovalevskaya, A., Kustov, T., JimenezCastaneda, R., . . . Al-Zoubi, A. (2020). Virtual ecological laboratory to develop a˜ pv module recycling workshop. In *2020 IEEE global engineering education conference (educon)* (pp. 434–441).

Soliman, M., Pesyridis, A., Dalaymani-Zad, D., Gronfula, M., & Kourmpetis, M. (2021, January). The application of virtual reality in engineering education. *Applied Sciences, 11*(6), 2879. Retrieved December 9, 2021, from www.mdpi.com/2076-3417/11/6/2879 (Number: 6 Publisher: Multidisciplinary Digital Publishing Institute) doi:10.3390/app11062879

Srinivasan, A., & Bairaktarova, D. (2018). Learning thermodynamics with iPad: Adaptation and learning opportunities for both instructor and students. *The International Journal of Engineering Education, 34*(5), 1615–1625.

Starner, T. (2016). Wearable computing: Meeting the challenge. In W. Barfield (Ed.), *Fundamentals of wearable computing* (pp. 13–29). CRC Press.

Stuerzlinger, W., Zaman, L., Pavlovych, A., & Oh, J.-Y. (2006). The design and realization of covid: A system for collaborative virtual 3D design. *Virtual Reality, 10*(2), 135–147.

Suchman, L. (1987). *Plans and situated actions*. Cambridge University Press.

Tang, P., Aktan, H. M., & Polasek, J. S. (2012). Integrating sensing technology and building information modeling into a construction engineering curriculum. In *2012 ASEE annual conference & exposition* (pp. 25–806).

Tanwear, A., Liang, X., Liu, Y., Vuckovic, A., Ghannam, R., Bohnert, T., . . . Heidari, H.¨ (2020). Spintronic sensors based on magnetic tunnel junctions for wireless eye movement gesture control. *IEEE Transactions on Biomedical Circuits and Systems, 14*(6), 1299–1310.

Taxen, G., & Naeve, A. (2002, August). A system for exploring open issues in VR-based´ education. *Computers & Graphics, 26*(4), 593–598. doi:10.1016/S0097-8493(02)00112-7

Taylor, M., Lederman, S., & Gibson, R. (1973). Tactual perception of texture. *Handbook of Perception, 3*, 251–272.

Tilhou, R., Taylor, V., & Crompton, H. (2020). 3D virtual reality in K-12 education: A thematic systematic review. In S. Yu, M. Ally, & A. Tsinakos (Eds.), *Emerging technologies and pedagogies in the curriculum* (pp. 169–184). Springer. doi:10.1007/978-981-15-0618-5 10

Trevelyan, J. (2019). *The making of an expert engineer*. CRC Press.

Try, S., Panuwatwanich, K., Tanapornraweekit, G., & Kaewmoracharoen, M. (2021). Virtual reality application to aid civil engineering laboratory course: A multicriteria comparative study. *Computer Applications in Engineering Education, 29*(6), 1771–1792. doi:10.1002/cae.22422

Valentine, A., der Veen, T., Kenworthy, P., Hassan, G., Guzzomi, A., Khan, R., & Male, S. (2021). Using head mounted display virtual reality simulations in large engineering classes: Operating vs observing. *Australasian Journal of Educational Technology, 37*(3), 119–136. doi:10.14742/ajet.5487

Van Goethem, S., Adema, K., van Bergen, B., Viaene, E., Wenborn, E., & Verwulgen, S. (2019). A test setting to compare spatial awareness on paper and in virtual reality using EEG signals. In *International conference on applied human factors and ergonomics* (pp. 199–208).

Vasquez-Carbonell, M. (2022). A systematic literature review of augmented reality in engineering education: Hardware, software, student motivation & development recommendations. *Digital Education Review* (41), 249–267.

Vygotsky, L. S., & Cole, M. (1978). *Mind in society: Development of higher psychological processes.* Harvard University Press.

Wang, P., Wu, P., Wang, J., Chi, H.-L., & Wang, X. (2018). A critical review of the use of virtual reality in construction engineering education and training. *International Journal of Environmental Research and Public Health, 15*(6), 1204.

Ward, J. A., Lukowicz, P., Troster, G., & Starner, T. E. (2006). Activity recognition of assembly tasks using body-worn microphones and accelerometers. *IEEE Transactions on Pattern Analysis and Machine Intelligence, 28*(10), 1553–1567.

Wilson, M. (2002). Six views of embodied cognition. *Psychonomic Bulletin & Review, 9*(4), 625–636.

Wu, B., Yu, X., & Gu, X. (2020). Effectiveness of immersive virtual reality using headmounted displays on learning performance: A meta-analysis. *British Journal of Educational Technology, 51*(6), 1991–2005. doi:10.1111/bjet.13023

Wu, H.-K., Lee, S. W.-Y., Chang, H.-Y., & Liang, J.-C. (2013). Current status, opportunities and challenges of augmented reality in education. *Computers & Education, 62*, 41–49.

Young, J. J., Stolfi, C., Tan, H. Z., Chevrier, J., Dick, B., & Bertoline, G. (2011). Learning force concepts using visual trajectory and haptic force information at the elementary school level. In *2011 IEEE world haptics conference* (pp. 391–396).

Yousif, M. (2022). Vr/ar environment for training students on engineering applications and concepts. *Artificial Intelligence & Robotics Development Journal*, 173–186.

Zhou, N., Kisselburgh, L., Chandrasegaran, S., Badam, S. K., Elmqvist, N., & Ramani, K. (2020). Using social interaction trace data and context to predict collaboration quality and creative fluency in collaborative design learning environments. *International Journal of Human-Computer Studies, 136*, 102378.

24

Online Laboratories in Engineering Education Research and Practice

Dominik May, Gustavo R. Alves, Alexander A. Kist, and Susan M. Zvacek

1 Introduction

Over the last decades, technology development and the Internet have shaped how we play and work. Among other innovations, these advances have made their way into education and have led to the development of online laboratories. The advent of those new forms of technology-enhanced instruction in the area of laboratory-based teaching and learning was briefly discussed in the first edition of this very handbook. Johri and Olds's (2014) one chapter discussed recent developments on the use of information technology in engineering education at that time (Madhavan & Lindsay, 2014). However, only one part of that chapter formed the discussion of remotely accessible experimentation technology, which was seen primarily to overcome the barrier of co-location during the learning process (Madhavan & Lindsay, 2014, p. 643).

The chapter in this new volume specifically builds on this prior discussion and examines online laboratories in more detail, considering broader educational research. On the one hand, online laboratories can be seen as just one other instructional tool out of the many options information technology offers for educational settings. In this chapter, on the other hand, we want to display and discuss the affordances and challenges online laboratories bring to the table for modern, innovative engineering education (Shor et al., 2011). We, therefore, dedicate this chapter to an in-depth reflection on the increased relevance of online laboratories for the engineering education landscape. This reflection also includes a discussion of the historical context of online laboratories, a reflection of the wider pedagogical considerations, and thoughts concerning the future trajectory of the field.

Writing this chapter following the COVID-19 disruption clearly offers the opportunity to go about online laboratories in instructional settings in two ways. Firstly, one could describe the specific impact COVID-19 had on the online learning community and, with that, on the perceived importance of instructional online laboratory solutions. Secondly, one could take a broader perspective and discuss online laboratories from a more holistic standpoint. We decided to mostly follow the second approach and add considerations in context with the pandemic disruption where applicable. By doing so, we hope to make this chapter helpful and interesting for a more diverse audience, including online laboratory experts and complete newcomers to this field. We also believe that many of the innovations and developments around online laboratories in the context of COVID-19 have not yet been sufficiently assessed outside of that unique situation due to the lack of time. Hence, the insights might not yet be of great value for the times after the pandemic. Nevertheless, we do

DOI: 10.4324/9781003287483-29

foresee further research results coming out soon that show very practical and also more theoretical research outcomes based on scholarly activities during the last two years. Finally, we also refrain from going too much into detail in terms of the technical development of online laboratories. This would require a completely different approach and mostly address laboratory developers instead of the general engineering education community.

So far, we have used the term *online laboratories* very naturally and without going into greater detail. However, as is true for many other terminologies, it is essential to define what the authors intend when using specific terms. Hence, we want to provide a working definition of that very term before we move on with more detailed considerations.

1.1 Online Laboratories Defined

> [Instructional] Laboratories allow the application and testing of theoretical knowledge in practical learning situations. Active working with experiments and problem solving does help learners to acquire applicable knowledge that can be used in practical situations. That is why courses in the sciences and engineering incorporate laboratory experimentation as an essential part of educating students.
>
> (Auer & Pester, 2007, p. 285)

Building on this broad understanding of the instructional laboratory itself, online laboratories are instructional laboratories in which students and equipment are not co-located in the same physical location or space. The opposite to that are traditional, hands-on laboratories, in which students use the equipment by manually operating it while being physically situated in front of or in close proximity to it. This broad understanding of online laboratories includes remote, virtually represented, fully simulated, and otherwise-emulated laboratories (Nickerson et al., 2007; Kennepohl & Moore, 2016; Auer et al., 2018; May, 2020). As for the instructional laboratory itself, the terminologies *online*, *virtual*, or *remote laboratory* can take on different meanings. These differences can be seen even when comparing international research communities or groups. In that context, several terms and expressions appeared over time, such as *cyberlab*, *web-based lab*, *weblab*, *web-accessible lab*, *online lab*, *virtual lab*, *iLab*, *remote-controlled laboratory* (RCL), and *remote access laboratory* (RAL), among others (Alves et al., 2007). Typically, differences in the use of the term *online laboratories* can be attributed to the specific technical setup in use, including gear and control interface.

The possibility to perform experiments remotely, in laboratories shared by different higher education institutions, was first proposed by Aburdene et al. (1991), the same year the first-ever web server was installed. Quoting Aburdene et al. (1991, p. 589), "sharing laboratories among universities is one possible solution . . . laboratory experiments can be operated remotely." The expression "remote laboratory" as an important subcategory of online laboratories was, however, later coined by Aktan et al. (1996) in an IEEE Transactions on Education article. On the same note, Froyd et al. (2012) wrote later:

> Remote laboratories, a method that can at least partially replace live experimentation, was first developed by Aktan et al. In a remote laboratory, students use a computer to control an actual experiment that is in a different physical space. . . . Remote laboratories allow institutions to share expensive equipment, and equipment downtime is reduced.
>
> *(Froyd et al., 2012, p. 1354)*

Research and development efforts for remote laboratories specifically have been the main driver for the international online laboratory community for a long time. However, the use of fully virtual laboratories and simulations also gained attention and led to a diversification of the field (Balamuralithara & Woods, 2009; de Jong et al., 2013; Potkonjak et al., 2016; Auer et al., 2018). Nowadays, there are many different types and subtypes of online laboratories: remote laboratories with live usage of real equipment (Reid et al., 2022), remote laboratories using pre-recorded experiment videos (KC et al., 2021), virtual desktop-based laboratories using simulated data (Makransky & Petersen, 2019), and even fully immersive virtual laboratories based on virtual reality technology (Franzluebbers et al., 2020; Kumar et al., 2021), to name just a few.

However, in many publications, the term "online laboratories" still refers to setups where physically existing equipment is controlled remotely via a web interface. Sometimes, such remote laboratories are even administered live and like traditional, hands-on laboratories with an instructor or laboratory assistant in the lab, students located off campus, and the equipment controlled remotely by students. This has been the case especially when remote access to laboratory equipment for students needed to be set up quickly without the necessary time to develop a fully functional remote laboratory (e.g., during the COVID-19 pandemic and its imposed social contact restrictions). Another permutation of this setup is the hybrid laboratory, where some students are in the laboratory and others are joining in via the Internet. However, HyFlex learning environments (Beatty, 2014) and pedagogical approaches mixing face-to-face experiences and online experimentation lie outside the scope of this chapter.

Nevertheless, discussing only remotely controlled laboratories in this chapter would not be sufficient to cover the current landscape of online laboratory technology and educational research. Fully virtual or partially simulated laboratories have gained attention as well. Even though these laboratories are technically very different from "classical" remote laboratories, we consider them conceptually close enough to be included without losing focus. Thus, for the sake of this chapter, *online laboratories* refer to both fully virtual or simulated laboratory equipment or experiences and to remotely accessible experimentation equipment for the purpose of laboratory-based instruction. This definition excludes technical solutions, like take-home lab kits that can be connected to a web server and augmented reality laboratory solutions. This exclusion is not intended to devalue those other solutions by any means. However, lab kits and augmented reality solutions typically co-locate the experimenter and the experiment. The inclusion of those laboratories in the discussion here would simply blur the focus of this chapter.

Similarly, to the terminology, various definitions and classifications for online laboratories have been proposed over the past two decades, for example, by Dormido-Bencomo (2004), Maiti et al. (2017). Specifically, Dormido-Bencomo (2004) proposed a simplified classification of laboratory environments based on two criteria: type of access to the laboratory resource (local, remote) and nature of the accessed resource (real, simulated). Table 24.1 is drawn based on the combination of these criteria and has been adopted by many authors. Although this represents a simplified and widely accepted classification, there are also examples of online laboratories that lie on the border between two environments, for example, laboratories that combine remote access to real equipment with the existence of simulated parts (Bruns & Erbe, 2004) and remote laboratories that return data recorded from real experiments, allowing simultaneous access by multiple users (Columbia-CTL, 2021; GOLC, 2021).

Another more recently published typology offered by May (2020) and Terkowsky et al. (2019) uses a framework categorizing online laboratory solutions along the three dimensions of "pedagogical approach," "degree of virtualization," and "laboratory distribution" (see Figure 24.1, based on Zutin et al. (2010) and Zutin (2018)). In this framework, the authors also include augmented reality laboratories as an intermediate stage of virtualization between real hands-on laboratories and fully remote laboratories, still measured along the degree of experiment virtualization. In other words, one can also distinguish instructional online laboratories along the continuum of physical reality (hands-on laboratories), augmented reality (augmented reality laboratories), mediated reality

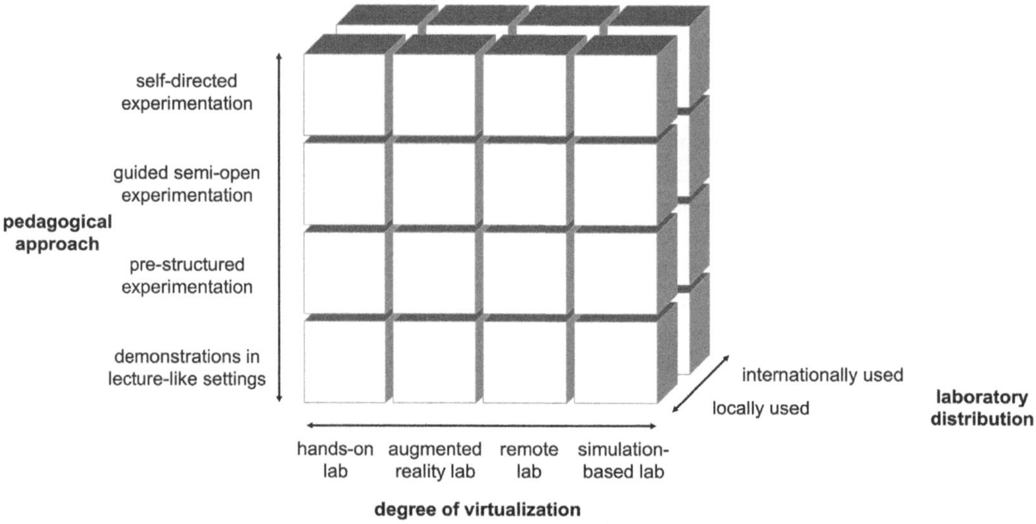

Figure 24.1 Three-dimensional framework for online laboratories.
Source: May (2020).

Table 24.1 Typology of Different (Online) Laboratory Environments

		Nature of the Resource	
		Real	*Simulated*
Access Type	***Local***	Traditional hands-on laboratory	Virtual laboratory with local user access
	Remote	Remote laboratory	Virtual laboratory with distributed user access

Source: Adapted from Dormido-Bencomo (2004).

(remote laboratories), and simulated reality (simulation-based laboratories). The third dimension in this framework stems from the specific context of the author's work at that time, which examined laboratory usage distributed locally and internationally.

Most of the online laboratory classifications like the ones shown earlier include various dimensions, such as the location of the experiment versus the location of the user, multi-user versus single-user, hands-on versus mediated, face-to-face versus online, and simulated versus real, to just name a few. In everyday instructional practice, there are even combined forms of the aforementioned online laboratory types possible, which makes categorizing the different forms of online laboratories and their classroom application even more blurry at times. This is also the reason that we won't use any of the category systems later in this chapter to systemize existing online labs. We still believe that the brief discussion earlier will help the reader frame a conceptual understanding of different online laboratory typologies.

In summary, online laboratories as a concept can be described as a conceptual, instructional space in which students undertake a laboratory-like learning activity without requiring direct but somehow mediated access to real or virtual equipment (Kist et al., 2012). In the following, this chapter focuses on online laboratories as a general concept in the context of instructional laboratories in which students are either (1) not co-located with the experimental equipment during the laboratory experience but use online technology to control experimentation equipment remotely or (2) make use of fully virtual, simulated instruments for their laboratory experiences.

1.2 Chapter Overview

As mentioned earlier, early motivations for deploying online laboratories in general and remote laboratories particularly included providing more frequent access to experimental equipment and laboratory experiences, utilizing expensive equipment more efficiently, sharing equipment among institutions, broadening access to equipment, and allowing students who are not on campus access to laboratory-based instruction (Gustavsson, 2001; Madhavan & Lindsay, 2014). While remotely controlled experiments have been an active field of research for well over 20 years, more recent events, like the restrictions in context with the COVID-19 pandemic, have brought the affordances of online laboratory teaching to the forefront. Lockdowns and access restrictions have clearly triggered a leap in innovation and, in some cases, creative solutions.

As a starting point for our discussion, Section 2 in this chapter partly zooms out from the perspective on online laboratories only and unpacks why laboratories in general are used in engineering education and discusses the learning rationale for using such practical activities. However, offering learning activities in laboratory spaces is a resource-intensive endeavor for any educational institution. Naturally, laboratory spaces and specific equipment are required, and depending on student numbers, several copies of the same equipment are needed to offer enough seats in a laboratory course. Furthermore, both laboratory spaces and the equipment need to be maintained and supported by laboratory staff, faculty, supervisors, and tutors. It was no wonder that sharing laboratory equipment to provide affordable, flexible access was another early driver of online laboratory developments (Aburdene et al., 1991). Finally, it is also the goal of that section to display both the affordances and also the challenges for online laboratory–based instruction.

With this chapter, it is our intent to take a broader perspective on the overall online laboratory research landscape instead of focusing too much on very recent developments. Our main reason for this approach is the observation that online laboratories gained significant attention during the pandemic years, but much of it has been communicated as emergency remote teaching approaches instead of seeing the general affordances online laboratories can bring to the table (e.g., Fox et al., 2020; Sandi-Urena, 2020; Kruger et al., 2022). However, it is our strong belief that while displaying online laboratories as a fallback option in cases of emergency, the community would do a disservice to the great potential online laboratories have for the instructional landscape. Thus, Section 3 provides a broader perspective and a summary of educational research into online laboratories in the context of different types of instructional online laboratories. That section explores how the research has evolved and summarizes open research questions in the field.

Even today, many publications around online laboratory research focus on technical development and implementation details instead of instructional design, successful learning, and pedagogy, for example, in the form of educational research studies (Lindsay, 2005; Nickerson et al., 2007; Heradio et al., 2016; Post et al., 2019). However, instructional design considerations are at least as crucial as technical considerations for the successful design and delivery of online laboratories and are, therefore, discussed in Section 4 more broadly for laboratory-based instruction and for the specific context of online laboratories.

Given the close relationship between the scientific fields and the technical challenges in delivering online experiments, it is not surprising that many of the first online laboratories were proposed in the field of electrical engineering, information technology, and robotics, for example (Aktan et al., 1996). More recently, a wider range of examples has emerged in disciplines such as chemical engineering and bioengineering (Hossain et al., 2015; Faulconer & Gruss, 2018; Jones et al., 2021). Section 5 provides a timeline of the development of online laboratories and shares international examples of online laboratories and their respective working groups. Not all the laboratory examples are still active. That fact already illustrates a major challenge for the field: keeping online laboratories active and running independently from funded projects or even individual persons at the institutions.

We will come back to that discussion later in this chapter. Nevertheless, it is worth having a look at those examples as they display the diversity in online laboratory solutions and broaden the perspective to international projects. Section 6, finally, discusses future considerations for technology development and educational research around online laboratories based on the previous chapters and draws respective conclusions.

2 Rationale Behind Online Laboratories in Engineering Education

There are, of course, many reasons to require laboratory activities in engineering degree programs. Laboratory work can provide the opportunity for students to demonstrate achievement of such learning goals as engaging in experimentation, gathering and analyzing data, solving problems, and identifying the relationships between theoretical and applied knowledge (Feisel & Rosa, 2005). In addition to discipline-specific outcomes, laboratories can facilitate the so-called professional skills that transcend procedural skills and technical proficiency. These cross-disciplinary abilities, including written and verbal communication skills, teamwork, and creativity, for example, are particularly critical in the work environment. Professionals with both in-depth knowledge of their field and relevant professional skills are often considered "T-shaped" (Guest, 1991). This metaphor proposes that disciplinary knowledge forms the vertical stroke of the T, while cross-disciplinary skills form the horizontal bar at the top. This configuration enables engineers to collaborate with non-engineers, explain their thinking to others, and generate creative solutions to ill-defined challenges (Tranquillo, 2017).

The value of laboratory work can additionally be confirmed with a review of accreditation guidelines for engineering education. Both the Accreditation Board for Engineering and Technology (ABET) (ABET, 2020) and the European Accredited Engineer (EUR-ACE) label (ENAEE, 2021) criteria for engineering degree programs include outcomes that can be addressed with laboratory-based instruction. The question is what laboratories and, specifically, online laboratories bring to the table to develop this broad set of skills. Feisel and Rosa (2005, p. 121) noted that "[w]hile there seems to be general agreement that laboratories are necessary, little has been said about what they are expected to accomplish." Feisel and Rosa's list of intended learning objectives in the laboratory is, hence, particularly helpful beyond the accreditation criteria to gain an understanding of both the relevance of the instructional setting "laboratory" for the engineering domain as a whole and the high diversity of learning outcomes that can ideally be achieved through laboratory-based instruction (Feisel & Rosa, 2005, p. 127)

Having a closer look at Feisel and Rosa's list of learning objectives for the laboratory reveals that not all of them can be sufficiently represented in an online laboratory setting. Without discussing each of those outcomes in detail here, it is obvious that outcome 1, "instrumentation," can hardly be addressed in an online laboratory setting as the selection of applied sensors and instruments is mostly predefined in online laboratories. In contrast to that, outcome 2, "models," surely can be achieved in an online setting. A similar distinction can even be made in the context of only one learning objective. Outcome 3, "experiment," for example, includes aspects of the design of an experimental procedure (specify appropriate equipment and procedures) and the analysis of gathered data (interpret the resulting data). Whereas the first part is difficult to implement in online laboratories (only if designing the experimental setup is part of a simulation for example), data analysis and interpretation can be done in online laboratory settings too. Online laboratories may even be better suited in that sense, as they typically offer the opportunity to do more experiments and, with that, gather more data by mitigating practical constraints traditional, hands-on instructional laboratories typically entail.

At this point, it needs to be noted that back in 2005, the authors already pointed out suggestions for further inquiry and future research concerning online laboratories (Feisel & Rosa, 2005, p. 128).

They clearly stated that there is a need for assessing the effectiveness of remote laboratories, comparing the effectiveness of simulations vs. remote access of real equipment, and developing of laboratory simulations that include "noise" in terms of non-ideal parameters and data.

So far, this section has looked at the rationale for laboratory-based learning and the importance of laboratories in engineering education in general. The following sections unpack drivers for offering practical laboratory-based learning activities online. Advantages can be classified as educational drivers relating to students and operational drivers that link to institutional requirements. Following the discussion of advantages is a description of the challenges facing the incorporation of these nontraditional approaches.

2.1 Advantages of Online Laboratories for Student Learning

For student learning, online laboratories offer the potential for flexibility, more individual time on task, tightly coupled theoretical and practical learning activities, learning analytics, and access to remote resources. Online laboratories give students more flexibility to complete the exercises when and where they like, and in many cases, this freedom extends to allowing students to self-pace their learning. In an intriguing example of flexibility, Craifaleanu and Craifaleanu (2022) described their instructional co-creation, with students, of virtual laboratory activities, offering an innovative option they felt was optimal for the online environment. In-person laboratory access is often time-limited due to access and supervisory constraints. Online laboratory activities remove this constraint and allow students 24/7 access, although care must be taken to ensure that timely support is available for students when hitting roadblocks.

Online laboratories are generally accessible to students individually or in collaborative groups and are often available 24/7, whereas in-person laboratories are time-tabled and frequently must be completed in groups, due to resource constraints. This means that not all group members get time to interact with and to control the apparatuses. This limitation does not apply in the case of on-demand online laboratories, providing greater instructional design flexibility by enabling both individual and group activities. Rubim et al. (2019), in a review of 99 articles from 59 journals, summarized their findings by noting that "[t]he direction of research points to the use of remote laboratories as a means of inclusion, as an alternative for those whose access to experimentation is restricted" (p. 827), recognizing the value of remote or virtual experimentation beyond the traditional environment. Traditionally, laboratories are scheduled to loosely match the timing of the delivery of the theoretical curriculum. However, students may view theoretical background work and practical activities as only marginally related. Online laboratories allow embedding laboratory learning within the curriculum, tightly coupled with the delivery of theoretical content and practical skill acquisition. Achuthan et al. (2021), in their exploration of the effectiveness of a remote laboratory for mechanical engineering, found that students spent significantly more time interacting with equipment and made more experimental attempts than in the physical laboratory. Interestingly, this occurred even though the average time to complete the activity was notably shorter for students using the remote laboratories. The researchers also recorded more frequent interactions between students and instructors on topics related to theory.

This tighter integration and the fact that access to experiments is mediated by technology allows detailed, individual data collection, which in turn can enable and support learning analytics. Analytics can be done at an aggregated level to analyze how well an experiment operates and how users interact with the experiment. This can also be implemented at an individual level to understand better how individuals engage with the experiments, assess their limitations, and trigger support if they are stuck. Raman et al. (2021) found that within the online laboratory system, pages with theoretical content were viewed for a longer time than other pages, on average, and corresponding simulation pages often were viewed during the same session. This may indicate that students were familiarizing themselves with the conceptual material as a type of foundational preparation before attempting the activities.

Finally, some experiments are inherently risky and cannot be performed safely by students, and others are not possible because of where the activity is located. For example, using lasers may endanger students, and live observations of the southern night sky cannot be done from the Northern Hemisphere. In these cases, a virtual laboratory can allow students to use resources they should not or could not otherwise access, which in turn also offers operational advantages.

2.2 Advantages of Online Laboratories for Operational Considerations

Operational drivers to use online laboratories include cost benefits through reduced staffing and space requirements, higher equipment utilizations, and safety considerations. Laboratory spaces with student access are typically staffed by technical and learning support staff, but this is not required in most remotely controlled environments. While support staff is still required to maintain the equipment, it is not coupled with when students use the experiments. In a discussion of factors influencing a move to instructional online laboratories in higher education, Radhamani et al. (2021) described cost-effectiveness as a situational (i.e., "mooring") effect driving the adoption of remote laboratories. Although online laboratory activities might be considered more passive than in-person experimentation, these researchers listed the potential for increased interaction within the laboratory activities as one of the factors attracting programs to these online systems.

Engineering and specialized equipment can be expensive to purchase, have specific space requirements, and might not be readily available at all institutions. At the same time, laboratory spaces and equipment are often underutilized. Online laboratories provide an opportunity to decrease space requirements and significantly increase equipment utilization. This can lead to significant operational cost savings. In addition to that, the equipment may have specific safe operating procedures that do not allow students to control the experiment in person. For safety reasons or to avoid damage to the rig, some equipment is controlled by an instructor. Operational limits can be implemented in an online setting, and boundaries can be enforced. This allows the design of inherently dangerous experiments to be operated by students within safe limits. Equipment with online access operated as part of an online laboratory management system can be shared between institutions across international borders.

In the case of distance education coursework, students are not located on campus and do not have ready access to laboratory spaces. With the increasing availability of online degree programs, this will apply across a more significant proportion of the sector. While not all learning outcomes can be addressed through online laboratories, they certainly will reduce the time distance students must be on campus. In many industries, the practice has shifted toward remote control and online operation. Using remote laboratories offers an opportunity to upskill students and prepare them for their future workplace.

2.3 Challenges for Online Laboratory-Based Instruction

While online laboratories have significant benefits, they also present challenges. As discussed previously, online laboratories can address similar, but not always identical, learning outcomes to those of face-to-face laboratories. They require a more careful and purposeful pedagogical design than traditional laboratory activities in which a group of students completes most practical activities (Zacharia & De Jong, 2018). Besides addressing the intended learning outcomes regarding collaboration and teamwork, this practice also has the advantage that students can support each other when completing the practice activities. This is not the case for most online laboratories. It is therefore essential to provide on-demand support when individual students struggle and recognize these limitations when the learning activities are designed. As two example limitations, in their SWOT analysis of the VISIR remote lab, Alves et al. (2022) identified the need for specialized support as a weakness and teachers' resistance as a threat. Likewise, professional development for staff is necessary to use the affordances of remote laboratories effectively. Traditional laboratory manuals may have gaps readily filled in by the instructor during face-to-face sessions. However, in an online environment, missing steps can throw students off and lead to significant frustration.

Not all skills can be embedded in an online laboratory experiment, such as hands-on tactile experiences, as noted by Deniz et al. (2022), Rubim et al. (2019), and pedagogical conversations about why these skills are essential may offer alternative activities that students can undertake to acquire similar skills. There is also some expectation management required with faculty and students. Learners may perceive online laboratories as inauthentic and low stakes. If the activities are perceived as games, students may not apply the same effort and rigor that they typically would in a physical space where actual equipment is at stake and academic and peer pressure is applied.

While it's easy to assume that online learning and other technology-facilitated options are widely accepted, many instructors still consider remote or simulated lab options a poor substitute (at best). As recently as 2021, Keller argued that "[s]cience, one of the most difficult courses to teach remotely, has spawned a plethora of fake labs, also known as virtual lab simulations" (Keller, 2021). His op-ed was, in fact, a solid argument for well-designed science simulations, whether he intended it to be or not. Unfortunately, not all readers will see beyond the headline claiming that "[v]irtual lab simulations don't teach science." The challenge to adopt nontraditional labs across a department or beyond an occasional assignment can be exacerbated significantly as a result.

Initial capital investment in designing and building online laboratories is often significant, and there are also substantial costs involved in taking research projects to the operational state. In the past, many online laboratories evolved from personal interest and from research projects funded for a limited time (see examples later in this chapter). However, supporting laboratories in an ongoing fashion cannot be done at zero cost. This can be particularly difficult when academics are not rewarded for continuing support and maintenance but for innovation (Alves et al., 2022). To share laboratory resources beyond research projects requires service agreements and payment plans between organizations. Supporting cutting-edge web technologies over the long term can be difficult when the environment is in constant flux. Changes in technology can make existing implementations obsolete, as seen in the case of Adobe Flash Player–based solutions.

Another challenge can be the integration with existing university systems for authentication and learning management system support, for example. ICT systems and remote laboratories are inherently complex. Supporting these within a corporate network with tight security and regular updates can be challenging when production systems have significant ongoing service and support needs.

3 Educational Research on Online Laboratories

Over the last three decades, technical research papers and pilot studies, educational case studies, and educational research work covering online laboratories have been published across a variety of scholarly outlets. The referenced publications and research in the following subsections yet give an overview on specific online lab developments (see, for example, Nickerson et al., 2007; Brinson, 2015; Heradio et al., 2016; Potkonjak et al., 2016; Nikolic et al., 2021). As discussed earlier in this chapter, the early years of research were much dominated by the technical perspective and the attempt to effectively bring instructional hands-on laboratory work online and make it accessible remotely (De Jong & Van Joolingen, 1998). These technical research efforts and discussions are still present in the worldwide online laboratory research community. However, the focus has widened significantly in recent years. Starting around the beginning of this century, many researchers shifted their attention from technical considerations of online experimentation to education and instructional design.

In a bibliometric analysis paper, Heradio et al. (2016) examined and summarized the literature on virtual and remote laboratories from its beginnings to the year 2015, identifying the most influential publications, the most researched topics, and how the interest in those topics had evolved along the way. To do so, bibliographical data was gathered from ISI Web of Science, Scopus, and GRC2014 (Zappatore et al., 2015). Based on their work, the authors identified five main areas of research into which the most influential work could be organized: general overview work on online laboratories (i.e., remote and virtual laboratories); approaches to build, manage, and share online laboratories; descriptions of particular online laboratories; collaborative learning with online laboratories; and assessing the educational effectiveness of online laboratories. The latter two categories focus on the educational research perspective and serve as the nexus for this section.

3.1 20 Years of Online Laboratory Research

The results from much of the research currently available serve to reinforce the long-standing argument, initiated by Clark (1983, 1994), that technologies do not directly influence learning achievement. Rather, they provide instructional designers an array of affordances that enable teaching and learning strategies that do make a difference in educational outcomes. These affordances can be thought of as the possibilities offered by a device or application, such as the ability to use remote equipment for more long-term experimentation than might be possible otherwise. Decades of research has shown that simply substituting a technological option for a traditional approach, without any change in the learning strategy, produces a "no significant difference" result in learning achievement (NRCDETA, 2019). A classic example is that students using paper flashcards for memorization remember just as much as students using an online flash card app in the same way (e.g., Sage et al., 2020). In other words, it's not the technology (online laboratories) that makes a difference; it's what students are doing as learners that can and does. Nevertheless, studies comparing the learning gains attributed to a technological solution versus a non-technological one are still being conducted and will be discussed in the following because they represent the majority of scholarly discussion in this field.

Ma and Nickerson (2006) compiled one of the first literature review papers in the context of online laboratories and discussed research on hands-on, simulated, and remote laboratories. The authors drew several conclusions with regard to the research state-of-the-art at that time. For example, the authors recognized that hands-on laboratory advocates emphasized design skills, while remote laboratory advocates focused on conceptual thinking and understanding. Ma and Nickerson made clear that students learn not only from the interaction with equipment but also from interaction with peers and teachers. Recognizing that technology would be implemented in laboratories even more in the following years, they made clear that it was of focal importance that teamwork and peer interaction needed to remain part of the instructional experience in online laboratory settings, as in hands-on laboratories. In their conclusion, the authors reflected on how students don't need only conceptual understanding but also cognitive immersion to maximize the learning potential in the laboratory environment and that the psychology of presence during an experiment may be as important as the technology itself. The authors furthermore summarized that the boundaries among the three laboratory types (remote, virtual, and hands-on) started to blur in the sense that most laboratories were already mediated by information technology and that combinations of hands-on and remote or virtual experiences in one and the same course setting were tested.

In a large-scale, multi-year, randomized study, Corter et al. (2011) compared learning activities and outcomes for hands-on, remotely operated, and simulation-based educational laboratories in an undergraduate engineering course in which the students typically worked in teams. Study data in this work showed that in the hands-on laboratory format, higher learning outcomes were achieved when the students collected experiment data as a group instead of individually. In contrast, remote laboratories seemed to work better in terms of learning outcome achievement when the students worked individually (although learner collaboration is easily facilitated with online laboratory activities). The pattern of time spent with the laboratory activity also suggested that working with real instead of simulated data (e.g., when comparing a remote laboratory with an entirely simulated lab) may induce higher levels of student motivation. Subsequently, the specific way new technologies in lab-based education were used for instructional purposes – in terms of instructional context, course requirements, and student interaction levels – largely determined their effectiveness, according to the authors.

Two further examples of such comparison studies include Brinson (2015) and Faulconer and Gruss (2018). The former synthesized a large number of empirical studies that focused on directly comparing learning outcome achievement when using traditional (in-person, hand-on) laboratories and nontraditional (remote or virtual) laboratories. This review summarized post-2005 research results in terms of student learning outcome achievement, learning outcome assessment, and respective assessment tools to evaluate student learning outcome achievement. Overall, findings suggested that student learning outcome achievement was at least equal or higher in nontraditional versus traditional laboratories across all learning outcome areas. However, outcomes and assessment tools were not consistent across all studies, and the majority of studies focused on learning outcomes related to content knowledge instead of conceptual understanding by using quizzes and tests as the most common assessment instrument.

Faulconer and Gruss (2018) also compared the effectiveness of traditional hands-on, face-to-face laboratories versus nontraditional, online, remote, or distance laboratories. Their article laid out the existing benefits and drawbacks of the different instructional laboratory modes using existing literature. Their review supported Brinson's work and found that a well-designed, nontraditional laboratory can be just as effective as a traditional, face-to-face laboratory experience when measuring either content knowledge acquisition or student opinions as the metric for equivalence. This is very much in line with works discussed by authors beyond Brinson (2017). Furthermore, these authors noted that there is little to no evidence to suggest that traditional laboratories are better at developing practical skills in comparison to nontraditional laboratories. However, nontraditional laboratories have the advantage in cost, accessibility, and safety, but traditional laboratories have the advantage in future safety concerns and group work. In other words, studies indicate that nontraditional laboratories can provide as many benefits as traditional laboratories.

3.2 Knowing What We Do Not Yet Know

Even though the previously discussed reviews seem to conclude very much in favor of online experimentation versus traditional laboratories, there is a constant critique in the research community of a significant lack of empirically comparable and scalable research results. In a recently published review by Nikolic et al. (2021), the authors examined assessment implementations to measure student achievement or learning by comparing published work on remote, simulation, and traditional teaching laboratories, with a particular focus on engineering. It was observed by the authors that empirical evidence around online laboratories so far is built primarily around students' subjective perceptions of their learning (e.g., Corter et al., 2011) or experiences collected via superficial post-intervention surveys, which only in some cases are complemented with more-advanced and validated quantitative assessment instruments. With the laboratory as a multifaceted, multi-domain learning environment also covering the psychomotor and affective domains, such observations suggest that the empirical data being collected and published so far is providing only an incomplete analysis. Based on their review, Nikolic et al. (2021) argued that in many studies, the research assessment was focused on the cognitive domain from the students' potentially subjective perspective, underselling the learning actually being achieved by the learners.

Nikolic et al. (2021), in some sense, also provided a superb example for the take-away message many other review papers offer in their conclusions (see also for example, Potkonjak et al., 2016; Post et al., 2019): the empirical bases of knowledge and in-depth educational research results are still somewhat weak and lacking general and replicable research results. Results are, in some cases, even contradictory, though the general direction of research shows that online laboratories offer great potential for engineering instruction. However, a general and broad assessment about online laboratories simply to compare them with hands-on laboratories is not helpful in the long run because of the diversity of online laboratory solutions, application settings, and possible learning outcomes.

At this point, before we zoom in on certain online laboratory examples in Section 5, we first want to zoom out even a bit more and shed a light on instructional design considerations for traditional, hands-on, and online laboratories. This seems to be important, as the instructional design, following Clark (1983) again, remains to be the cornerstone for a successful design and introduction for online laboratories and, hence, is needed to discuss the whole picture.

4 Instructional Design of Online Laboratories

The typical scenario for engineering instructors is to complete their own academic work, possibly up to a terminal degree, engage in nonacademic professional activity (maybe), then enter a teaching role. It is unlikely that these new faculty members have studied teaching and learning, coming into their new position with only their own student experiences to guide them. Therefore, it is unsurprising that laboratory work, online or in-person, has an inconsistent record of success when it comes to engaging students in higher-order thinking or linking theory and practice. Duderstadt (2008, p. 33) argued that highly structured laboratory courses did little to teach "the most important technical skills of engineering: the integration of knowledge, synthesis, design, and innovation." An intriguing result was found for example by Jones (2018) and Hamadani et al. (2022), however. Students who used Labster (a commercial provider for virtual laboratories in science education) during his biochemistry course scored highly on test questions that required higher-order thinking and the application of learned ideas but poorly on their recall of facts and definitions. Clearly, more research is needed to delve into this phenomenon, although it begs the question of the value of simple recall to begin with. While consistent use of validated instructional design models is recommended for all laboratory work, it is especially critical for the online environment, where students may have limited access to an instructor, TA, or course peers who could provide motivation, address their questions, or clarify instructions.

For this section, a framework based loosely on Gagne (1977) and his classic "nine events of instruction" model is proposed. His work expanded on instructional design models that focused on determining desired outcomes, planning instructional "interventions" to enable students to achieve the outcomes, and creating assessment instruments to measure student progress toward or mastery of the outcomes. The specificity of Gagne's model has been broadened and subsumed in the four categories of the MOST framework: *motivation, objectives, strategies, and tools and resources* (Zvacek, 2021). As noted by Clark (1983, 1994), the specific technologies used for the implementation of instruction are less significant than the instructional components. This model, therefore, can be applied to any type of laboratory-based instruction, with its relevance to remote and virtual activities addressed in each section.

4.1 Motivation

The constant refrain, "My students aren't motivated," echoes through the halls of academia the world over. This condition can be alleviated, however, by relying on what psychological research has to say about what motivation consists of and how to facilitate it. One of the key reasons humans persist in a difficult activity is that they recognize its relevance to their long-term goals (e.g., "Learning how to calculate angle of repose will help me learn how to design bridges") (Albrecht & Karabenick, 2018). Hand in hand with relevance is confidence, or self-efficacy (Keller, 2016). The student may see the relevance of the assignment but have little confidence they can successfully complete it. Moderately challenging tasks encourage confidence ("I can do this if I work hard"), but assignments that are too easy ("This is just busywork") or too difficult ("I won't get this no matter how hard I try") erode the student's willingness to invest effort in the task (Ryan & Deci, 2020). Two additional variables, context and transfer, position motivation within a larger perspective, beyond the assignment (context) and beyond the course (transfer) (Blume et al., 2010). A laboratory assignment needs to fit into a meaningful sequence of activities that facilitates linking new learning to already-acquired knowledge and skills while simultaneously setting the stage for more advanced content and tasks to come. Laboratory work that occurs in isolation from other learning reduces motivation and inhibits the development of robust cognitive networks. Similarly, a critical element of motivation for laboratories is the expectation that one can transfer the new skills to challenges that may be encountered in the professional workplace, thus increasing task relevance as well. Each of these components (relevance, confidence, context, and transfer) contributes to motivation and can improve the motivational capacity of laboratory work.

How are these variables related to remote or virtual laboratories? First, as noted by Peck et al. (2018), motivation remains one of the most challenging aspects of online learning for many students. They may feel isolated from their peers, especially if laboratory work that was traditionally completed in groups now is done individually. Many students lack time management skills, which becomes obvious when online coursework requires a high degree of self-regulation, reducing their confidence and, consequently, their motivation. In addition, many students are surprised when they discover that online coursework is not easier than in-person learning. This demotivating realization can inhibit effort and even lead to students dropping the course. Addressing motivational variables up front is an essential part of designing online laboratories.

4.2 Objectives

There is little argument concerning the value of identifying desired outcomes for an instructional lesson or unit. Objectives help students know what is expected and help teachers keep instruction focused on the course's most important concepts. Accreditation criteria or lists of intended learning outcomes for the laboratory (Feisel & Rosa, 2005), as noted earlier, can provide broad guidance for those outcomes, such as experimentation, design, and problem-solving. The difficulty comes when there is a mismatch between our own big-picture goals and the more specific breakdown of tasks that lead to that outcome. If you're looking for higher-order thinking in your students, do your objectives reflect that? More importantly, do your assessment activities require students to exhibit those skills or simply respond to easy-to-measure basic knowledge questions? Additionally, are your objectives explained directly in assignments, or hidden away in the syllabus, which may or may not be read by students?

This is important for any type of instruction, but especially pertinent for online teaching (Simonson et al., 2019). Students who don't have the luxury of catching the instructor in the hall or after class to get clarification on course expectations may find themselves guessing what constitutes a successful demonstration of knowledge and skills for a particular assignment or activity. Along these lines, the PhET Interactive Simulations Project was designed intentionally to "[optimize] understanding by giving students a lightly guided system to explore" (Perkins, K. in Jones, 2018) and was noted by Borish et al. (2022), who found that students overwhelmingly rated clear expectations and guidance from instructors as critical to their success with virtual laboratory work.

4.3 Strategies

There are a variety of educational approaches and frameworks that can be used to structure learning activities for laboratory assignments (Zvacek, 2015). For example, a problem-based approach presents a challenge that students might address with data gathering and analysis, design, or collaboration. A cognitive apprenticeship framework emphasizes the development of student autonomy by scaffolding the learning activities from high levels of guidance to independent decision-making (Clark & Mahboobin, 2018; Collins et al., 1987; Dennen & Burner, 2007; Frank et al., 2017; Pinto & Zvacek, 2022). It is important to remember that strategies are what students are doing to learn the content and skills that enable them to achieve the objectives, not what instructors are doing to teach.

The types of strategies most effective with online laboratories are those that involve practice and feedback. Many times, the strategy incorporates an instructional wrapper around the use of the remote or simulated equipment. For example, students may be required to do a pre-laboratory activity where they predict the results of their experimentation. Feedback would occur when comparing their prediction to the actual results that were obtained, followed by an opportunity to revise their work based on the feedback. Feedback may be as simple as activities in which students calculate how randomly assigned variables will influence performance and then confirm (or not) their calculations based on the resulting data. These types of before-and-after wrappings can position the manipulation of equipment or materials as part of a broader context from which students draw conclusions.

Some strategies may rely on students working with others to solve problems or apply specific design principles. Working with peers can develop skills of consensus building, communication, and negotiation. Peers can also provide feedback on one another's work as part of a learning strategy that benefits both participants. Borish et al. (2022), for example, noted that students who worked in a group as part of their virtual laboratory activities reported a greater sense of community within the course. An especially effective strategy for laboratory work is to require students to explain their decision-making process or problem solution to a peer who then shares their work, followed by a discussion of how and why they agreed or disagreed. Such explanations could be written, spoken, or expressed as images or concept maps (Zvacek et al., 2013). The practice of explaining their thinking requires that students know the content well enough to articulate it clearly to someone else, while acting as a potential peer teaching activity. A bonus is the strengthening of communication skills that are necessary for collaborating with others.

Formative assessment and instructor feedback strategies were noted by Van den Beemt et al. (2022) as a crucial element for students in a systems and control engineering course. Students reported that their follow-up progress meetings with instructors after completing remote laboratory activities on their own or in groups were a valuable part of their course success. Another assessment strategy, screen-captured videos, takes advantage of online tools to measure progress toward or mastery of laboratory objectives. Such videos, in which students conduct experiments and gather data as part of the online activity, can be uploaded to the learning management system (or other repository) and viewed by an instructor at a later time. For synchronous assessments, the ubiquity of videoconferencing systems can make presentations, demonstrations, or real-time data analysis by students readily accessible and convenient, whether for individuals or collaborative groups (Simonson et al., 2019). In general, online laboratories provide most of the same assessment opportunities as their face-to-face counterparts while addressing the challenges of space, time, equipment access, and cost.

4.4 Tools and Resources

The final component of the MOST framework asks, "What must be available to facilitate motivation, help students achieve the desired learning outcomes, and implement the strategies?" Four types of the necessary elements include content objects, materials and equipment, expertise/time, and tech support.

Content objects are the media that provide declarative, conceptual, and procedural knowledge within an organizational structure that facilitates comprehension and application of that knowledge. The most typical forms of content are books, articles, videos, and images, as well as students' notes taken during lectures or demonstrations. As with learning strategies, it is crucial to choose content resources that enable students to complete the laboratory tasks successfully and that are accessible to all learners (see, for example, Costa et al., 2015; Mourão & Netto, 2019). Resources that appear irrelevant or lack connection to the assignments may end up ignored or (maybe worse) encourage students to disregard other course resources as well. It may fall to the instructor to ensure that content objects are presented with guidance on what to do with them or why they're important.

It may be necessary to provide raw materials, if any, that will be used for the laboratory tasks. For virtual or remote laboratories, however, there may not be any materials required, or those materials may be accessed remotely, along with the equipment. For the purposes of this discussion, *equipment* includes the devices that students manipulate during the activity and the means of accessing the devices, such as a robust and reliable Internet connection. Bernhard (2018) argued that instructional strategies and purposes must be considered when choosing laboratory equipment. The affordances offered by specific experimental technologies "may shape students' experience of focal phenomena . . . and this mediating role is often neglected" (p. 819). In addition, it is crucial that instructional designers or instructors recognize that while remote activities can facilitate learning for students who find traveling to campus a challenge, not all students have easy access from their home to remote equipment, and that provisions for such barriers be addressed ahead of time.

A type of resource that is easily taken for granted is expertise, especially with a traditional in-person, hands-on laboratory configuration. When using remote or virtual laboratories, however, instructional designers or instructors might not have the technical skills necessary to establish the required connections, program a simulation, or create a virtual environment (Khan & Abid, 2021). Even if they do, it may take a significant block of time for which they should be compensated. Neither instructors nor instructional designers are expected to write course textbooks without additional remuneration, and the labor-intensive task of creating remote or virtual laboratories is no different. Instructors must also determine how student support for the learning activity will be provided. Although the equipment may be available 24/7, virtual office hours represent another consumer of faculty time to consider as a necessary resource.

Finally, tech support must be considered a student and instructor resource. Although the upfront design of a remote or virtual laboratory may involve specialized expertise, provisions for ongoing technical assistance are also required. Issues related to equipment access, operation, and troubleshooting must be considered, with a support plan in place before implementing the laboratory activities. Determining whose responsibility it is to ensure ongoing availability and operability of equipment, as well as how (or if) they will be compensated, may ultimately need to be addressed by upper administration.

5 International Examples of Online Laboratories

This section will provide an overview of the historical genesis and growth of online laboratories over time and share different strands of developments and use cases across the globe. In that light, we will display international examples of multi-institutional projects and respective research groups in the field of online laboratories. It is necessary mentioning at this point that those exemplary use cases are not on the level of individual online laboratory solutions but represent collaborative efforts (in some cases, even internationally) to either collect and curate many online laboratory solutions for an overarching portal, or combine different online laboratories at one institution to a wider set of experiments, or connect multiple institutions through a specific shared online laboratory that can be used across those institutions.

Currently, 26 years after the seminal article by Aktan et al. (1996), there are examples of remote and virtual laboratories in practically all disciplines (or sub-areas) of science, technology, engineering, and mathematics (STEM). In an early work, Zutin et al. (2010), described a repository (called "Lab2Go") of aggregated, searchable information about remote and virtual laboratory resources available in open or restricted format or pending prior request. This work does not list specific laboratories but marks an early endeavor to develop a space, through which multiple online laboratories can be shared. In another work, Endean and Braithwaite (2012) listed about 160 remote and virtual experiments offered by a single institution, in areas ranging from chemical engineering to materials science to electrical circuits. A later, more extensive work (Gröber et al., 2013) documented at least 335 remote laboratories, the majority in the field of engineering (n = 64%), with the remainder in the field of physics (n = 36%). Brinson (2017) classified the nontraditional laboratories presented in Ma and Nickerson (2006) and in Brinson (2015) according to the area of use (distinguishing engineering and natural sciences). According to Brinson (2017), there was an evolution from 2006 reporting a total of 60 nontraditional laboratories (NTLs) with a majority in the engineering area (n = 39, 65%) and a minority in the area of natural sciences (n = 13, 22%), to the 56 NTLs described in Brinson (2015) in the area of natural sciences (n = 46, 82 %) and a minority in the field of engineering or computer science (n = 9, 16%). Although Brinson (2017) noted that most remote and virtual laboratories are not accessible in an open or commercially available format, the trend has been toward open access, due not only to public funding for the development of online educational resources but also to the recent COVID-19 pandemic. This trend was also reported in Esposito et al. (2021), which presented a review of 40 NTLs and Lab Network Initiatives (or "federated laboratories"). In a very recent book on the use of remote laboratories in STEM education, García-Zubía (2021) described 15 remote laboratories in areas ranging from control and automation to mechanics.

Being virtually impossible to present all the previously referred remote and virtual laboratories in detail, at this point we would like to refer to the previously referenced overview and review articles (also see Section 3.1) and point out that those manuscripts provide excellent lists of the examined labs. In the following, we will display another set of exemplary online laboratories. The selection was made based on a recent publication by Raman et al. (2022) and is based on the overall relevance and impact by the online laboratories themselves, the connected research group, or the underlying research project for the international community.

5.1 Online Laboratory Research Groups and Solutions Around the Globe

The following subsections present an initial set of four exemplary use cases across the globe that have received public funding, represent multi-institutional working groups (Go-Lab, Next-Lab, LabShare, and Virtual Labs), or have been able to spread internationally beyond their region of origin. Building on that, an additional set of six specific online laboratories (GOLDi, NCSLab, RexLab, UNILabs, VISIR, and WebLab-Deusto) extends the global coverage. Raman's (2022) work provides a historical and bibliometric analysis of the past three decades of online laboratories development. The publication includes an overview with the top 18 contributing institutions, based on the total publications. The overview also displays the top 15 authors, based on total publications, total citations, and total publications with attention.

The two sets of exemplary use cases presented here cover half of the top contributing institutions and all major authors assessed by total publications. The exemplary use cases also provide a global overview in terms of the geographical and timely distribution of online laboratory developments and research activities. Furthermore, the majority (6 out of 10) of the presented examples are also part of the 40 Lab Network Initiatives (LTI) presented in Esposito et al. (2021). For each example of the first set of major use cases, a brief description and the URL are provided. The six examples forming the second set are briefly described and summarized in a graphic (Figure 24.2). This graphic also maps the respective home institutions and relevant authors (Raman et al., 2022).

5.1.1 Multi-Institutional Working Groups

5.1.1.1 EUROPE (GO-LAB, NEXT-LAB, AND GO-GA)

The Go-Lab portal (www.golabz.eu) was developed in the context of the Go-Lab project (2012–2016) and continued through the Next-Lab (2017–2019) and GO-GA projects (2018–2020), all funded by the European Commission and offering over 1,000 remote and virtual experiments. The entry page of this repository includes a menu, through which it is possible to check the number of experiments by subject domains, type (remote, virtual, dataset), target age group (<7, 7–8, 9–10, 11–12, 13–14, 15–16, >16 years), and language of presentation. Using the list of experiments by subject domains, it was possible at the time of this writing to verify that the vast majority (n = 700, 64%) were in physics, in the topics of electricity and magnetism (n = 131, 12%) and forces and motion (n = 323, 29%).

In Go-Lab, the online laboratories are part of Inquiry Learning Spaces (ILS), where students learn about a STEM concept from an investigative perspective divided into five phases: contextualization, conceptualization, experimentation, conclusion, and discussion. The ILS concept is grounded in work developed by Ton de Jong, who coordinated both projects (De Jong & Van Joolingen, 1998; Pedaste et al., 2015). Those European projects clearly mark one of the largest and longest-lasting endeavors to collect a high number of online laborites of any type and collectively offer them for usage for instruction. However, some of those online laboratories are not working anymore, which clearly shows a not-yet-solved challenge in the development of online laboratories: securing long-term support and further development of online laboratories independently from individuals and project-based funding. We will touch on this aspect in our final section, but letting online laboratories mature from their initial support is clearly one of the major, so far unsolved, tasks in the community.

5.1.1.2 AUSTRALIA (LABSHARE)

LabShare was an Australian government–funded project (2008–2011) that aimed to create a national network of shared remotely accessible laboratories. It was led by David Lowe, then affiliated with the University of Technology Sydney, who published several papers about the project (Lowe et al., 2009a, 2009b). The project website (www.labshare.edu.au) is no longer active, although still visible through the Internet Archive Wayback Machine. Although not pertaining to the original project consortium, the University of Sydney has also installed the Remote Laboratory Management System (RLMS) developed in the context of the LabShare project at https://labshare.sydney.edu.au.

This project marks an example of a project-based online laboratory development which did not survive after both project support and funding ended, even though the project was set up as a nationwide endeavor. One would think that the inclusion of several institutions mitigates the risk of project results, such as developed laboratory setups being lost after the project period, but this specific case proves that, if online laboratory infrastructure is not made part of the universities' general laboratory infrastructure, it is difficult to maintain long-term support for it. However, it is still worth mentioning LabShare in this context here because this project was one of the drivers for remote laboratory developments back in those days.

5.1.1.3 INDIA (VIRTUAL LABS)

Virtual Labs is an Indian nationwide initiative which is supported by the Ministry of Human Resources in India. The Virtual Labs initiative (2008–2011) brought together 11 engineering education institutions, with a predominance of Indian Institutes of Technology (IIT). Under Virtual Labs, over 100 online laboratories consisting of approximately 700+ web-enabled experiments were designed for remote operation and viewing. As per the homepage information itself, the Virtual Labs initiative has registered more than four million experiments as of 2021.

Virtual Labs displays the potential relevance and power online laboratories can play for an education sector in a specific country or region. Connecting several institutions over long distances by sharing infrastructure can be of mutual benefit to all participating partners. As of today, it still remains to be proven if this initiative solves funding and support challenges and stays active in the long run, though.

5.1.1.4 UNITED STATES (ILAB MIT)

The iLab project (2001–2019) at the Massachusetts Institute of Technology (MIT) offered several online laboratories for instruction in electrical engineering and computer science, civil engineering, and chemical engineering. According to its coordinators, Del Alamo et al. (2002), some of these laboratories were shared with students from universities in North America, Europe, Asia, and Africa.

Although centered on a single institution, MIT, this network is probably one of the best-known cases of a federation of online laboratories, with installations reported in institutions like the Obafemi Awolowo University in Nigeria, the Makerere University in Uganda, the Carinthia University of Applied Sciences in Austria, or the University of Brasov in Romania (García-Zubía & Alves, 2011). One of the reasons for its success may come from the fact it was well supported by MIT and received generous funding from several sponsors, including Microsoft. Regarding the iLab (MIT) and the previously named LabShare projects, García-Zubía (2021) noted:

> iLAB and LabShare were excellent examples of remote laboratories directed by Judson Harward and David Lowe at MIT (USA) and UTS (Australia), respectively, and in their day were a world reference due to both the sophistication of their experiments and the quality of their RLMS, which permitted scalability, universality and federation from their core. However, both are more or less inactive, and neither can be used in class in a secure fashion.
>
> *(p. 74)*

5.1.2 Specific Online Laboratories

The following six more specific examples represent online laboratories which have been developed and introduced for the first time at specific institutions (see Figure 24.2). However, some of them, like the VISIR lab, for example, have outgrown their local applications and are now used at several institutes across the globe. In that sense, VISIR is an outstanding example of one lab that is shared across institutions in the sense of shared infrastructure and even started international research collaborations and development efforts.

5.1.2.1 VISIR

The Virtual Instrument Systems in Reality (VISIR) project started in 1999 at the Blekinge Institute of Technology (BTH), Sweden, under the leadership of Gustavsson (2001). VISIR is also an acronym for the associated remote laboratory that allows users to perform remote experiments with electrical and electronic circuits in less than a second, thus supporting the concurrent access

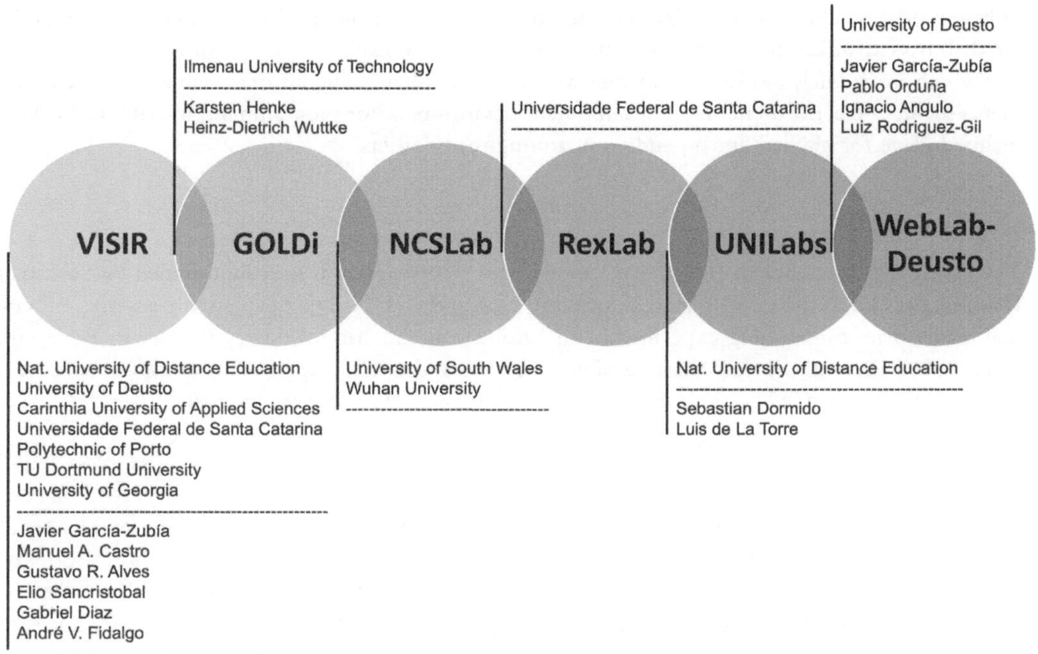

Figure 24.2 Specific exemplary online laboratory use cases (GOLDi, NCSLab, RexLab, UNILabs, VISIR, and WebLab-Deusto).

of several users (http://openlabs.bth.se). Presently, the VISIR remote laboratory is installed in all five continents, specifically in Argentina, Austria, Australia, Brazil, Costa Rica, Georgia, Germany, India, Morocco, Portugal, Spain, Sweden, and the United States. García-Zubía (2021, p. 82) noted that "VISIR is perhaps the most powerful, spectacular and frequently used remote experiment in the world, and it has won various awards."

5.1.2.2 A-ZUBGOLDI

The Grid of Online Laboratory Devices Ilmenau (GOLDi) was developed by Karsten Henke and Heinz-Dietrich Wuttke, based on work that can be traced back to 2003 (Henke et al., 2003). It uses a grid concept to implement a remote laboratory infrastructure based on the iLab architecture of MIT. As of 2022, GOLDi is still in operation (www.goldi-labs.net).

5.1.2.3 NCSLAB

The Networked Control System Laboratory (NCSLab) is a remote laboratory that integrates various test rigs and experimental facilities of control systems around the world which was established in the University of South Wales, UK, in 2006. It is presently based at the University of Wuhan, China (www.powersim.whu.edu.cn/ncslab/), and is still quite active, with new developments regarding its user interface (Lei et al., 2021).

5.1.2.4 REXLAB

The Remote Experimentation Laboratory (RexLab) was initially founded by João Bosco da Mota Alves in 1997 at the Federal University of Santa Catarina (UFSC), Brazil, as a result of the MSc thesis of Juarez Bento da Silva, who developed a remote debugger for the 8051 microcontroller. Presently, RexLab has expanded the number of remote experiments available through its site (http://relle.ufsc.br), which are publicly available to anyone wishing to use them. The list of available remote experiments includes electrical circuits, a development environment for programming in ARDUINO, an inclined plane for physics, and a pendulum, among many others.

5.1.2.5 UNILABS

The University Network of Interactive Laboratories (UNILabs) was initially founded by Sebastian Dormido at the National Distance Education University (UNED), Spain, with a series of virtual laboratories for running experiments on automation and control theory (Dormido-Bencomo et al., 2000). Presently, it supports a number of remote and virtual experiments on those same areas (https://unilabs.dia.uned.es), being one of the most successful exemplary use cases.

5.1.2.6 WEBLAB-DEUSTO

The foundations of WebLab-Deusto can be traced back to 2004, in a publication authored by Javier García-Zubía (García-Zubía, 2004). Presently, WebLab-Deusto offers a series of remote laboratories, mainly for supporting digital design, robot and ARDUINO programming, and remote experiments with electrical and electronic circuits (using its own VISIR node), among others (https://weblab.deusto.es/). LabsLand, a company that offers services based on remote laboratories, is a spin-off of WebLab-Deusto.

At this point, we want to halt displaying specific examples and again refer to the articles referenced at the beginning of this subsection and in Section 3. It needs to be stated that the research and development community working on online laboratories is still highly volatile. In addition to the online laboratories named up to this point, there are many more initiatives covering everything from very course-specific solutions to broader efforts across engineering curricula, and we personally expect even more case studies and online laboratory examples to be published based on work that happened during the COVID-19 interruption. It clearly may be too early to draw a conclusion on how the years 2020 and 2021 impacted the international online laboratory community. There was a sharp spike in online laboratory efforts detectable (see for example, Abumalloh et al., 2021; Mohammed et al., 2020; Vasiliadou, 2020; Vergara et al., 2022), but it is not yet clear to what extent this spike will lead to a fundamental change in the application and wider use of online laboratories in the broader engineering education landscape. In the following, we want to wrap up this chapter by pointing out future possible developments and challenges with regard to online laboratories.

6 Future Perspectives on Online Laboratories

Gravier et al. (2008) presented a review of the state-of-the-art of remote laboratories, covering an initial 10-year period (1997–2007) of developments, to then identify possible evolutions for the next generation of remote laboratories. Authors identified reusability, interoperability, opportunity to collaborate, and convergence with LMSs as four major issues for the leverage of remote laboratories. Many of these aspects were later addressed in the IEEE 1876–2019 Standard for Networked Smart Learning Objects for Online Laboratories (IEEE, 2019).

An article by Martins-Ferreira and Graven (2014), "Rise and Fall of Remote Labs: Or Perhaps Not?" presented a framework to delineate a plan of action for repositioning remote laboratories as technology-enhanced educational tools able to add value to teaching and learning processes. The plan of action specified four criteria defined by the authors: (1) "institutional networking" should be considered a priority for every institution active in this field; (2) "pedagogical value" represents an area where failure to improve may dictate the fall of remote laboratories; and (3) "availability" and (4) "accessibility" represent areas where successful research and development projects should generate relevant results to convert remote laboratories into a mainstream educational technology. These four criteria are met in projects like Go-Lab, Next-Lab, Virtual Labs, or VISIR, as already described in Section 5. In general terms, meeting these criteria has been a path to the successful evolution of online laboratories. In addition to that, long-term success for online laboratories needs to be seen in disconnecting the laboratories from both individual developers or researchers and extra-mural funding. It has been proven many times that no online laboratory can survive in the long run if it is not introduced into the institution's general IT infrastructure and if financial as well as technical support is not coming from inside the institution. Otherwise, online laboratories remain "only" a temporarily finite project that dies after external funding ends or faculty move on to the next project. Actually, the switch from setting up online laboratories as part of a funded project to making them a long-term part of the curriculum is absolutely critical for success and has been a stumbling block for many, now-defunct laboratories.

Correia et al. (2021) proposed a graphical evolution model for remote and virtual laboratories that was validated against several existing and extinct online laboratories. A major aspect contributing to the endurance and evolution of online laboratories was the existence of several positive feedback loops, including a start-up. This was the case of WebLab-Deusto, which led to a start-up named LabsLand (2021). Some interesting aspects of LabsLand are that it uses the prosumer concept, where educational institutions may provide their own remote laboratories and/or use remote laboratories provided by other institutions ("institutional networking"); it provides additional didactical/pedagogical support, including integration with an LMS (Gravier et al., 2008) and "pedagogical value"; and it guarantees the "availability" and "accessibility" of all provided remote laboratories, complying with the third condition proposed by Martins-Ferreira and Graven (2014). These examples support the idea that, in some cases, it is possible to pinpoint specific aspects that can contribute to the positive evolution of online laboratories.

6.1 A(n) (Un)Certain Future?

The COVID-19 pandemic was a boost to many online educational solutions, including online laboratories, as part of several "emergency responses" described in recent literature. In the words of Pablo Orduña, co-founder and CEO of LabsLand:

> The usage of LabsLand remote laboratories has increased substantially since the beginning of the pandemic. In 2020, both the number of sessions and users was 7 times higher, and it is keeping the growing trend in 2021.
>
> *(Personal communication, November 5, 2021)*

This exponential growth was triggered by an unforeseen and exceptional reason; nevertheless, it supports the idea that if the conditions are favorable to the strengths of online laboratories (24/7 availability, online access, existence of supporting pedagogical materials, etc.), then its use will shift from being an option to being the option.

Any answer to "How will online laboratories impact the future of both face-to-face and online engineering education and how will they shape lab-based instruction as a whole?" faces the prime challenge associated with any prediction, that is, getting it right. In any case, recent emergency responses to the COVID-19 pandemic, in the engineering education sector, showed an unprecedented interest in and use of online laboratories, in parallel with alternative solutions, like visualized experiments or take-home laboratories, also called pocket laboratories. This justifies why the question is not "if online laboratories will impact the future of both face-to-face and online engineering education" but rather "how they will impact it." In a normal (i.e., non-emergency) situation, it's likely that the experience gained during the pandemic will not be lost, and many institutions/teachers/students will consider online laboratories a viable technology-enhanced tool able to support the acquisition of experimental skills and practical knowledge. Other emergency-proven solutions like pocket laboratories and visualized (or ultraconcurrent) laboratories are likely to be part of a sort of "laboratory palette," where each laboratory type may be used according to a set of conditions defined by the institution and/or the teachers. In any case, it will always be important to think first of what the intended learning outcomes are (see Feisel and Rosas's list), to then consider the overall instructional design (see the MOST framework), then investigate the available options and the characteristics (see the referenced SWOT analysis by Alves et al., 2022) of each option and proceed with a reasonable instructional approach. In other words, one must be aware of the ten commandments of remote experimentation proposed by García-Zubía (2021) that include, for example, "Think about the curriculum and you will succeed" and "The (remote) experiment should help, it should not in itself be a challenge." Recommendations from engineering education experts, such as Douglas (2020), are also relevant: "So, my recommendation is, the very first thing to think about is what were the learning objectives associated with that laboratory? What were the learning goals?" (0:23).

Exactly this mismatch between a more technology-driven development of online laboratories and the lack of in-depth pedagogical considerations may be one of the major reasons for the fact that online laboratories are still somewhat of a niche in the engineering education research community. So far, and this may change with the long-term impact of COVID-19, online laboratories have not yet gained a level of widespread attention in the instructional community, specifically in higher engineering education. It seems like there is still a dire need for further knowledge development that goes beyond the sheer technical development of individual labs and their somewhat-superficial, student perception–based evaluation. Further in-depth educational research is still needed to develop results that scholarly underpin and guide both the development and application of online laboratories.

6.2 Concluding Remarks

In summary, many research findings suggest that online laboratories can serve as an effective instructional tool for engineering education. Drawing on the advantages offered by online laboratories may help solve existing shortcomings of traditional curricula and hands-on laboratories, such as safety or capacity issues. However, review studies also underscore that comparative evaluations of different online laboratory technologies are difficult and may be of little use unless educationally relevant variables, such as student ability, time on task, and cooperative work patterns, are measured and controlled. Unfortunately, studies focused on the use of specific learning strategies with online laboratories and research examining metacognitive effects, time on task, teamwork skills, universal design, and learner self-efficacy (to name just a few examples) are not well represented in the literature. Variables related to the use of online laboratories also could include return on investment, efficiency, or instructor perceptions of usability. On that note, it is interesting that students typically rate remotely operated laboratories as less effective than simulated laboratories, even when learning achievement favors the former. Nevertheless, online laboratories have advantages in availability, cost–benefit, and sometimes learner inclusivity, which explains the underlying satisfaction ratings in many studies.

This chapter explored the value and challenges of online laboratories for engineering education. While some of the challenges may be mitigated with advances in learning technologies, the need for technologically savvy instructors and collaboration-minded institutions will remain. However, the benefits associated with learning and operational considerations are likely to outweigh, in the long term, inherent limitations that may dissuade potential adopters. In addition, online laboratories present an opportunity for institutions that have not initiated online degree programs because addressing the need for laboratories was deemed an insuperable hurdle. This alone has significant implications for sweeping change in academia. The future of engineering education will require flexibility, access, rigor, and creativity. Online laboratories will accommodate and complement those goals.

References

Ab Abumalloh, R. A., Asadi, S., Nilashi, M., Minaei-Bidgoli, B., Nayer, F. K., Samad, S., Mohd, S., & Ibrahim, O. (2021). The impact of coronavirus pandemic (COVID-19) on education: The role of virtual and remote laboratories in education. *Technology in Society, 67*, 101728.

ABET. (2020). *Criteria for accrediting engineering programs: 2021–2022. ABET engineering accreditation commission*. www.abet.org/accreditation/accreditation-criteria/criteria-for-accrediting-engineering-programs-2021-2022/

Aburdene, M., Mastascusa, E., & Massengale, R. (1991). *A proposal for a remotely shared control systems laboratory*. Paper presented at the proceedings frontiers in education twenty-first annual conference, engineering education in a new world order.

Achuthan, K., Raghaven, D., Shankar, B., Francis, S. P., & Kani Kolil, V. (2021). Impact of remote experimentation, interactivity, and platform effectiveness on laboratory learning outcomes. *International Journal of Educational Technology in Higher Education, 18*(1), 38.

Aktan, B., Bohus, C. A., Crowl, L. A., & Shor, M. H. (1996). Distance learning applied to control engineering laboratories. *IEEE Transactions on Education, 39*(3), 320–326.

Albrecht, J. R., & Karabenick, S. A. (2018). Relevance for learning and motivation in education. *Journal of Experimental Education, 86*(1), 1–10.

Alves, G. R., Gericota, M. G., Silva, J. B., & Alves, J. B. (2007). Large and small scale networks of remote labs: A survey. *Advances on Remote Laboratories and e-Learning Experiences, 15*.

Alves, G. R., Marques, M. A., Fidalgo, A. V., Garcia-Zubia, J., Castro, M., Hernandez-Jayo, U., Garcia-Loro, F., & Kreiter, C. (2022). A roadmap for VISIR remote lab. *European Journal of Engineering Education*. https://doi.org/10.1080/03043797.2022.2054312

Auer, M. E., Azad, A. K., Edwards, A., & De Jong, T. (Eds.). (2018). *Cyber-physical laboratories in engineering and science education*. Springer.

Auer, M. E., & Pester, A. (2007). Toolkit for distributed online-lab grids. In J. García-Zubía & L. Gomes (Eds.), *Advances on remote laboratories and e-learning experiences*. Universidad Deusto.

Balamuralithara, B., & Woods, P. C. (2009). Virtual laboratories in engineering education: The simulation lab and remote lab. *Computer Applications in Engineering Education, 17*(1), 108–118.

Beatty, B. (2014). Hybrid courses with flexible participation: The hyflex course design. In L. Kyei-Blankson & E. Ntuli (Eds.), *Practical applications and experiences in K-20 blended learning environments* (pp. 153–177). IGI Global.

Bernhard, J. (2018). What matters for students' learning in the laboratory? Don't neglect the role of experimental equipment! *Instructional Science, 46*, 819–846.

Blume, B. D., Ford, J. K., Baldwin, T. T., & Huang, J. L. (2010). Transfer of training: A meta-analytic review. *Journal of Management, 36*(4), 1065–1105.

Borish, V., Werth, A., Sulaiman, N., Fox, M., Hoehn, J., & Lewandowski, H. (2022). Undergraduate student experiences in remote lab courses during the COVID-19 pandemic. *Physical Review Physics Education Research, 18*(2), 1–26.

Brinson, J. R. (2015). Learning outcome achievement in non-traditional (virtual and remote) versus traditional (hands-on) laboratories: A review of the empirical research. *Computers & Education, 87*, 218–237.

Brinson, J. R. (2017). A further characterization of empirical research related to learning outcome achievement in remote and virtual science labs. *Journal of Science Education and Technology, 26*(5), 546–560.

Bruns, F. W., & Erbe, H.-H. (2004). Mixed reality with hyper-bonds: A means for remote labs. *IFAC Proceedings Volumes, 37*(4), 551–556.

Clark, R. E. (1983). Reconsidering research on learning from media. *Review of Educational Research, 53*(4), 445–459.

Clark, R. E. (1994). Media will never influence learning. *Educational Technology Research and Development, 42*(2), 21–29.

Clark, R. E., & Mahboobin, A. (2018). Scaffolding to support problem-solving performance in a bioengineering lab: A case study. *IEEE Transactions in Education, 61*, 1–10.

Collins, A., Brown, J. S., & Newman, S. E. (1987). *Cognitive apprenticeship: Teaching the craft of reading, writing and mathematics (technical report no. 403)*. BBN Laboratories.

Columbia-CTL. (2021). *Planning for online lab sessions*. Columbia University in the City of New York: Office of the Provost (Center for Teaching and Learning Columbia University). https://ctl.columbia.edu/resources-and-technology/teaching-with-technology/teaching-online/online-labs/

Correia, R. C., Alves, G. R., & Fonseca, J. M. (2021). *An evolution model for remote and virtual labs*. Paper presented at the 2021 4th international conference of the Portuguese society for engineering education (CISPEE).

Corter, J. E., Esche, S. K., Chassapis, C., Ma, J., & Nickerson, J. V. (2011). Process and learning outcomes from remotely-operated, simulated, and hands-on student laboratories. *Computers & Education*, *57*(3), 2054–2067.

Costa, L. C. P., Correa, A. G. D., Salmon, D. L., Zuffo, M. K., & Lopes, R. D. (2015). Accessible education digital book on tablets for people with visual impairment. *IEEE Transactions on Consumer Electronics*, *61*(3), 271–278.

Craifaleanu, A., & Craifaleanu, I. (2022). A co-creation experiment for virtual laboratories of mechanics in engineering education. *Computer Applications in Engineering Education*. Wiley Online Library. https://doi.org/10.1002/cae.22498

De Jong, T., Linn, M. C., & Zacharia, Z. C. (2013). Physical and virtual laboratories in science and engineering education. *Science*, *340*(6130), 305–308.

De Jong, T., & Van Joolingen, W. R. (1998). Scientific discovery learning with computer simulations of conceptual domains. *Review of Educational Research*, *68*(2), 179–201.

Del Alamo, J., Brooks, L., McLean, C., Hardison, J., Mishuris, G., Chang, V., & Hui, L. (2002). *The MIT microelectronics WebLab: A web-enabled remote laboratory for microelectronic device characterization*. Paper presented at the World Congress on Networked Learning in a Global Environment.

Deniz, S., Müller, U. C., Steiner, I., & Sergi, T. (2022). Online (remote) teaching for laboratory based courses using "digital twins" of the experiments. *Journal of Engineering for Gas Turbines and Power*, *144*(5), 051016.

Dennen, V. P., & Burner, K. J. (2007). The cognitive apprenticeship model in educational practice. In J. M. Spector, M. D. Merrill, J. van Merriënboer, et al. (Eds.), *Handbook of research on educational communications and technology* (pp. 425–439). Routledge.

Dormido-Bencomo, S. (2004). Control learning: Present and future. *Annual Reviews in Control*, *28*(1), 115–136.

Dormido-Bencomo, S., Sánchez-Sánchez, J., & Morilla-García, F. (2000). *Laboratorios virtuales remotos para la práctica a distancia de la automática*. XXI Jornadas de Automática, Conferencia plenaria.

Douglas, K. (Producer). (2020). Moving classroom lab instruction online [Video]. www.youtube.com/watch?v=ZgDegfWqkj0

Duderstadt, J. J. (2008). *Engineering for a changing world: A roadmap to the future of engineering practice, research, and education*. Flexner.

ENAEE. (2021). EUR-ACE framework standards and guidelines. *European Network for Accreditation of Engineering Education*. www.enaee.eu/eur-ace-system/standards-and-guidelines/

Endean, M., & Braithwaite, N. (2012). *Online practical work for science and engineering students – A collaborative scoping activity between the UK Open University and East China University of Science and Technology*. eSTEeM final report. www.open.ac.uk/scholarship-and-innovation/esteem/projects/themes/teaching-laboratory-practice-online/online-practical-work-science-and-engineering

Esposito, G., Mezzogori, D., Reverberi, D., Romagnoli, G., Ustenko, M., & Zammori, F. (2021). Non-traditional labs and lab network initiatives: A review. *International Journal of Online & Biomedical Engineering*, *17*(5), 4–32.

Faulconer, E. K., & Gruss, A. B. (2018). A review to weigh the pros and cons of online, remote, and distance science laboratory experiences. *International Review of Research in Open and Distributed Learning*, *19*(2), 156–168.

Feisel, L. D., & Rosa, A. J. (2005). The role of the laboratory in undergraduate engineering education. *Journal of Engineering Education*, *94*(1), 121–130.

Fox, M. F., Werth, A., Hoehn, J. R., & Lewandowski, H. J. (2020). Teaching labs during a pandemic: Lessons from Spring 2020 and an outlook for the future. arXiv preprint arXiv:2007.01271.

Frank, B., Simper, N., & Kaupp, J. (2017). Formative feedback and scaffolding for developing complex problem solving and modelling outcomes. *European Journal of Engineering Education*, *43*, 552–568.

Franzluebbers, A., Tuttle, A. J., Johnsen, K., Durham, S., & Baffour, R. (2020, February). Collaborative virtual reality training experience for engineering land surveying. In *International conference on remote engineering and virtual instrumentation* (pp. 411–426). Springer.

Froyd, J. E., Wankat, P. C., & Smith, K. A. (2012). Five major shifts in 100 years of engineering education. *Proceedings of the IEEE, 100*(Special Centennial Issue), 1344–1360.

Gagne, R. M. (1977). *The conditions of learning*. Holt, Rinehart, & Winston Inc.

García-Zubía, J. (2004). Programmable logic and weblabs. *Microelectronics Education*, 277–281.

García-Zubía, J. (2021). *Remote laboratories: Empowering STEM education with technology*. World Scientific.

García-Zubía, J., & Alves, G. R. (Eds.). (2011). *Using remote labs in education: Two little ducks in remote experimentation*. Universidad Deusto.

GOLC. (2021). *GOLC: Online lab award*. http://online-engineering.org/GOLC_online-lab-award.php

Gravier, C., Fayolle, J., Bayard, B., Ates, M., & Lardon, J. (2008). State of the art about remote laboratories paradigms: Foundations of ongoing mutations. *International Journal of Online Engineering, 4*(1).

Gröber, S., Eckert, B., & Jodl, H.-J. (2013). A new medium for physics teaching: Results of a worldwide study of remotely controlled laboratories (RCLs). *European Journal of Physics, 35*(1), 018001.

Guest, D. (1991). The hunt is on for the renaissance man of computing. *The Independent*, 17(09).

Gustavsson, I. (2001). *Laboratory experiments in distance learning*. Paper presented at the international conference on engineering education, ICEE.

Hamadani, K. M., Jiang, Y., Ahmadinia, A., Hadaegh, A., Moraleja-Garcia, J., Mendez, A., . . . & Sheperd, M. (2022, May). Framework for scalable content development in hands-on virtual and mixed reality science labs. In *2022 8th international conference of the immersive learning research network (iLRN)* (pp. 1–6). IEEE.

Henke, K., Wuttke, H. D., & Hellbach, S. (2003). Laboratory via Internet—New ways in education and research. *International Journal of Computers and Applications, 25*(3), 157–163.

Heradio, R., De La Torre, L., Galan, D., Cabrerizo, F. J., Herrera-Viedma, E., & Dormido, S. (2016). Virtual and remote labs in education: A bibliometric analysis. *Computers & Education, 98*, 14–38.

Hossain, Z., Jin, X., Bumbacher, E. W., Chung, A. M., Koo, S., Shapiro, J. D., Truong, C. Y., Choi, S., Orloff, N. D., Blikstein, P., & Riedel-Kruse, I. H. (2015). *Interactive cloud experimentation for biology: An online education case study*. Paper presented at the proceedings of the 33rd annual ACM conference on human factors in computing systems.

IEEE. (2019). IEEE standard for networked smart learning objects for online laboratories (IEEE Standards Committee). In *IEEE std 1876–2019* (Vol. IEEE std 1876–2019).

Johri, A., & Olds, B. M. (2014). *Cambridge handbook of engineering education research*. Cambridge University Press.

Jones, E. V., Shepler, C. G., & Evans, M. J. (2021). Synchronous online-delivery: A novel approach to online lab instruction. *Journal of Chemical Education, 98*(3), 850–857.

Jones, N. (2018, October 3). Simulated labs are booming. *Nature, 562*. S5–S7. https://doi.org/10.1038/d41586-018-06831-1

Narasimhamurthy, K. C. Orduna, P., Rodríguez-Gil, L., GC, B., Susheen Srivatsa, C. N., & Mulamuttal, K. (2021). Analog electronic experiments in ultra-concurrent laboratory. In *Cross reality and data science in engineering: Proceedings of the 17th international conference on remote engineering and virtual instrumentation 17* (pp. 37–45). Springer International Publishing.

Keller, H. (2021). *Virtual lab simulations don't teach science*. eLearning Inside. https://news.elearninginside.com/virtual-lab-simulations-dont-teach-science/

Keller, J. M. (2016). Motivation, learning, and technology: Applying the ARCS-V motivation model. *Participatory Educational Research, 3*(2), 1–15.

Kennepohl, D., & Moore, M. G. (2016). *Teaching science online: Practical guidance for effective instruction and lab work*. Stylus Publishing.

Khan, Z. H., & Abid, M. I. (2021). Distance learning in engineering education: Challenges and opportunities during COVID-19 pandemic crisis in Pakistan. *The International Journal of Electrical Engineering & Education*, 0020720920988493.

Kist, A. A., Maxwell, A. D., & Gibbings, P. D. (2012). *Expanding the concept of remote access laboratories*. Paper presented at the 2012 ASEE annual conference & exposition.

Kruger, K., Wolff, K., & Cairncross, K. (2022). Real, virtual, or simulated: Approaches to emergency remote learning in engineering. *Computer Applications in Engineering Education, 30*(1), 93–105.

Kumar, V. V., Carberry, D., Beenfeldt, C., Andersson, M. P., Mansouri, S. S., & Gallucci, F. (2021). Virtual reality in chemical and biochemical engineering education and training. *Education for Chemical Engineers, 36*, 143–153.

LabsLand. (2021). *Real laboratories, on the internet.* https://labsland.com/en

Lei, Z., Zhou, H., Hu, W., Liu, G.-P., Deng, Q., Zhou, D., Liu, Z.-W., & Gao, X. (2021). Unified 3-D interactive human-centered system for online experimentation: Current deployment and future perspectives. *IEEE Transactions on Industrial Informatics, 17*(7), 4777–4787.

Lindsay, E. D. (2005). *The impact of remote and virtual access to hardware upon the learning outcomes of undergraduate engineering laboratory classes.* University of Melbourne, Department of Mechanical and Manufacturing Engineering,

Lowe, D., Murray, S., Lindsay, E., & Liu, D. (2009a). Evolving remote laboratory architectures to leverage emerging internet technologies. *IEEE Transactions on Learning Technologies, 2*(4), 289–294.

Lowe, D., Murray, S., Weber, L., de la Villefromoy, M., Johnston, A., Lindsay, E., Nageswaran, W., & Nafalski, A. (2009b). *LabShare:* towards *a national approach to laboratory sharing.* Paper presented at the proceedings of the 20th annual conference for the ustralasian association for engineering education.

Ma, J., & Nickerson, J. V. (2006). Hands-on, simulated, and remote laboratories: A comparative literature review. *ACM Computing Surveys (CSUR), 38*(3), 7-es.

Madhavan, K., & Lindsay, E. D. (2014). Use of information technology in engineering education. In *Cambridge handbook of engineering education research* (pp. 633–654). Cambridge University Press.

Maiti, A., Zutin, D. G., Wuttke, H.-D., Henke, K., Maxwell, A. D., & Kist, A. A. (2017). A framework for analyzing and evaluating architectures and control strategies in distributed remote laboratories. *IEEE Transactions on Learning Technologies, 11*(4), 441–455.

Makransky, G., & Petersen, G. B. (2019). Investigating the process of learning with desktop virtual reality: A structural equation modeling approach. *Computers & Education, 134*, 15–30.

Martins-Ferreira, J. M., & Graven, O. H. (2014). *Rise and fall of remote labs: Or perhaps not?* Paper presented at the 2014 11th international conference on remote engineering and virtual instrumentation (REV).

May, D. (2020). Cross reality spaces in engineering education: Online laboratories for supporting international student collaboration in merging realities. *International Journal of Online & Biomedical Engineering, 16*(3), 4–26.

Mohammed, A. K., El Zoghby, H. M., & Elmesalawy, M. M. (2020, October). Remote controlled laboratory experiments for engineering education in the post-COVID-19 era: Concept and example. In *2020 2nd novel intelligent and leading emerging sciences conference (NILES)* (pp. 629–634). IEEE.

Mourão, A. B., & Netto, J. F. M. (2019). SIMROAA multi-agent recommendation system for recommending accessible learning objects. In *2019 IEEE frontiers in education conference (FIE)* (pp. 1–9). Doi:10.1109/FIE43999.2019.9028504

Nickerson, J. V., Corter, J. E., Esche, S. K., & Chassapis, C. (2007). A model for evaluating the effectiveness of remote engineering laboratories and simulations in education. *Computers & Education, 49*(3), 708–725.

Nikolic, S., Ros, M., Jovanovic, K., & Stanisavljevic, Z. (2021). Remote, simulation or traditional engineering teaching laboratory: A systematic literature review of assessment implementations to measure student achievement or learning. *European Journal of Engineering Education*, 1–22.

NRCDETA. (2019). No significant difference. *National Research Center for Distance Education and Technological Advancements.* https://detaresearch.org/research-support/no-significant-difference/

Peck, L., Stefaniak, J. E., & Shah, S. J. (2018). The correlation of self-regulation and motivation with retention and attrition in distance education. *Quarterly Review of Distance Education, 19*(3), 1–80.

Pedaste, M., Mäeots, M., Siiman, L. A., De Jong, T., Van Riesen, S. A., Kamp, E. T., Manoli, C. C., Zacharia, Z. C., & Tsourlidaki, E. (2015). Phases of inquiry-based learning: Definitions and the inquiry cycle. *Educational Research Review, 14*, 47–61.

Pinto, S., & Zvacek, S. M. (2022). Cognitive apprenticeship and T-shaped instructional design in computational fluid mechanics: Student perspectives on learning. *International Journal of Mechanical Engineering Education, 50*(1), 51–77.

Post, L. S., Guo, P., Saab, N., & Admiraal, W. (2019). Effects of remote labs on cognitive, behavioral, and affective learning outcomes in higher education. *Computers & Education, 140*, 103596.

Potkonjak, V., Gardner, M., Callaghan, V., Mattila, P., Guetl, C., Petrović, V. M., & Jovanović, K. (2016). Virtual laboratories for education in science, technology, and engineering: A review. *Computers & Education, 95*, 309–327.

Radhamani, R., Kumar, D., Nizar, N., Achuthan, K., Nair, B., & Diwakar, S. (2021). What virtual laboratory usage tells us about laboratory skill education pre- and post-COVID-19: Focus on usage, behavior, intention, and adoption. *Education and Information Technologies, 26*, 7477–7495.

Raman, R., Achuthan, K., Nair, V. K., & Nedungadi, P. (2022). Virtual laboratories – A historical review and bibliometric analysis of the past three decades. *Education and Information Technologies*, 27. https://doi.org/10.1007/s10639-022-11058-9

Raman, R., Vinuesa, R., & Nedungadi, P. (2021). Acquisition and user behavior in online science laboratories before and during the COVID-19 pandemic. *Multimodal Technologies and Interaction, 5*(46). https://doi.org/10.3390/mti5080046

Reid, D. P., Burridge, J., Lowe, D. B., & Drysdale, T. D. (2022). Open-source remote laboratory experiments for controls engineering education. *International Journal of Mechanical Engineering Education*, 03064190221081451.

Rubim, J. P., Mota, V. P., Garcia, L. G., de Brito, G. L. R., & dos Santos, G. F. (2019). The role of remote experimentation as a teaching tool: A literature review. *International Journal of Information and Education Technology, 9*(11), 826–830.

Ryan, R. M., & Deci, E. L. (2020). Intrinsic and extrinsic motivation from a self-determination theory perspective: Definitions, theory, practices, and future directions. *Contemporary Educational Psychology, 61*, 101860.

Sandi-Urena, S. (2020). Experimentation skills away from the chemistry laboratory: Emergency remote teaching of multimodal laboratories. *Journal of Chemical Education, 97*(9), 3011–3017.

Sage, K., Piazzini, M., Downey IV, J. C., & Ewing, S. (2020). Flip it or click it: Equivalent learning of vocabulary from paper, laptop, and smartphone flashcards. *Journal of Educational Technology Systems, 49*(2), 145–169.

Shor, M. H., Bohus, C., & Atkan, B. (2011). Second best to being there: An historical perspective. In J. García-Zubía & G. R. Alves (Eds.), *Using remote labs in education: Two little ducks in remote experimentation* (pp. 27–52). Universidad Deusto.

Simonson, M., Zvacek, S. M., & Smaldino, S. (2019). *Teaching and learning at a distance: Foundations of distance education* (7th ed.). Information Age Publishing.

Terkowsky, C., Frye, S., & May, D. (2019). Online engineering education for manufacturing technology: Is a remote experiment a suitable tool to teach competences for "Working 4.0"? *European Journal of Education, 54*(4), 577–590.

Tranquillo, J. (2017). The T-shaped engineer. *Journal of Engineering Education Transformations, 30*(4), 12–24.

Van den Beemt, A., Groothuijsen, S., Ozkan, L., & Hendrix, W. (2022, August 12). Remote labs in higher engineering education: Engaging students with active learning pedagogy. *Journal of Computing in Higher Education*, 1–21. Doi:10.1007/s12528-022-09331-4

Vasiliadou, R. (2020). Virtual laboratories during coronavirus (COVID-19) pandemic. *Biochemistry and Molecular Biology Education, 48*(5), 482–483.

Vergara, D., Fernández-Arias, P., Extremera, J., Dávila, L. P., & Rubio, M. P. (2022). Educational trends post COVID-19 in engineering: Virtual laboratories. *Materials Today: Proceedings, 49*, 155–160.

Wuttke, H.-D., & Henke, K. (2002). *Living pictures – tool-oriented learning modules and laboratory for teaching digital via Internet.* Paper presented at the proceedings of the ICEE-2002 international conference on engineering education (UMIST).

Zacharia, Z. C., & de Jong, T. (2018). Pedagogy of cyber-physical experimentation. In M. E. Auer, A. K. Azad, A. Edwards, & T. De Jong (Eds.), *Cyber-physical laboratories in engineering and science education*. Springer.

Zappatore, M., Longo, A., & Bochicchio, M. A. (2015). *The bibliographic reference collection GRC2014 for the Online Laboratory Research community*. Paper presented at the proceedings of 2015 12th international conference on remote engineering and virtual instrumentation (REV).

Zutin, D. G. (2018). Online laboratory architectures and technical considerations. In M. E. Auer, A. K. M. Azad, A. Edwards, & T. de Jong (Eds.), *Cyber-physical laboratories in engineering and science education* (pp. 5–16). Springer.

Zutin, D. G., Auer, M. E., Maier, C., & Niederstätter, M. (2010). *Lab2go – A repository to locate educational online laboratories*. Paper presented at the IEEE Educon 2010 conference.

Zvacek, S. M. (2015). *From know-how to know-why: Lab-created learning*. Paper presented at the 2015 3rd experiment international conference (exp. At'15).
Zvacek, S. M. (2021, November 22–24). *Lab-created learning: Designing assignments for meaningful learning [Conference Session]*. Exp.atWS'21, Online conference.
Zvacek, S. M., Restivo, M. T., & Chouzal, M. F. (2013). Concept mapping for higher order thinking. *International Journal of Engineering Pedagogy*, *3*, 6–10.

25

Towards a Techno-Social Realist Approach in Primary and Secondary Computing Education

Aman Yadav and Michael Lachney

1 Introduction

The demarcation between computing education and engineering education has always been unstable in primary and secondary education around the world. This is especially so in the case of programmable hardware, which has been a foundational part of computing education at least since Seymour Papert's computer-controlled Logo Turtle robots in the 1960s. Today, the legacy of Papert's turtles is seen in popular educational technologies, such as LEGO Mindstorms, MakeyMakey, LilyPad Arduino, and many other tools that are used in primary and secondary engineering teaching and learning. Indeed, at the American Society for Engineering Education, LEGO Mindstorms has been reported on, discussed, or cited in hundreds of conference proceedings since the turn of the century. To a lesser extent, the same is true of programmable software, such as MIT's App Inventor and Scratch. At the same time, one of the flagship conferences for computer science education, the Association for Computing Machinery's (ACM) Special Interest Group on Computer Science Education (SIGCSE), also reports on these technologies, with educational robotics being a recurrent theme. The point of identifying the fluidity of these educational technologies between engineering and computing is not to point out a problem but instead to show the importance of computing education to engineering educators.

In this chapter, we provide an introductory overview of a largely Western history of primary and secondary computing education, where it is today, and where current research is going, with specific attention to the current focus on issues of diversity, equity, and inclusivity in the field (issues also becoming increasingly important to engineering education (e.g., Slaton, 2015). We acknowledge, along with other scholars, that for many years the academic discipline of computer science was co-constructed with electrical engineering (Ceruzzi, 1988), but that it has also developed its own unique identity and problem spaces, with branching and looping arrays of research programs (Tedre et al., 2018). Computing education represents one iterative branch of the discipline, and primary and secondary computing education is just one subbranch of that. But unlike other branches, the fact that children are the recipients means that it has received much attention from decision makers in the private and public sectors, as well as the general public. Subsuming computing into mainstream primary and secondary educational discourses and policies has shaped its meanings, endowing it with the hopes, dreams, fears, and anxieties that circulate around education generally, with or without computing. In this chapter we attend to the multiple meanings that have competed over computing

DOI: 10.4324/9781003287483-30

within the primary and secondary education landscape. The goal of the chapter is to introduce general-level discursive sensibilities that frame computing in education. We focus largely on Western contexts but also attend to how these sensibilities get exported around the globe.

Generally, we show how computing researchers and educators have a long history of centering technologies as agents for educational, if not social and economic, change. This has led to a strong sense of techno-optimism in computing education theory, practice, and research (e.g., Negroponte, 1995). Like so many other technological projects, computing education gets wrapped up in *progress* narratives that take for granted that computers will inevitably reform or revolutionize teaching and learning (Selwyn, 2016; Lachney et al., 2021). However, as long as there has been optimism, there has been pessimism. Indeed, as "disruptors" and "innovators" flood markets with new systems, devices, and applications, traditional assumptions and norms around privacy, ethics, and surveillance are challenged (e.g., Zuboff, 2019). The issue of electronic waste produced by schools alone is enough to call into question whether computing education really is progressive (Lachney et al., 2018). Such issues have led some scholars to argue for the need to "distrust" educational computing and technology generally (Selwyn, 2014).

We highlight the general discourses of techno-optimism and techno-pessimism that have shaped the meanings of computing education in the West and around the world, while also acknowledging that the material realities are much more nuanced than such simplistic demarcations. Indeed, we currently see movement beyond the dichotomic optimism/pessimism discourses within computing education research and theory. More so today, computing education researchers and scholars are constructing techno-social perspectives that remind us how the technical is always social and the social is always technical. These perspectives may be better characterized as *realist* because they do not fall into simplistic demarcations. They accept the social limitations and social impact of computing while also not giving up on the role of computing in teaching and learning.

We turn to five instances where we see techno-social realism in computing education today: (1) equity-oriented professional development for computer science teachers, (2) culturally responsive computing, (3) universal design for learning, (4) issues of computing and environmental (un)sustainability, and (5) translanguaging. We end with some ideas for future trajectories for techno-social realist research on educational technologists and curriculum designers, understudied actors in the field of computing education.

2 20th-Century Primary and Secondary Computing Education

This section will cover historical perspectives on Western computing education at primary and secondary levels, largely between the 1960s and 1990s. Specifically, we will discuss how the current discussion around computational thinking and computer science in primary and secondary schools is grounded in earlier work of computer scientists like Peter Naur, Donald Knuth, and Seymour Papert. Before narrowing in on these levels of education, it is important to briefly mention general computing education history at the university level.

In the 1940s and 1950s, computing in higher education was largely about training specialists to make machines run consistently and reliably (Tedre et al., 2018). By the 1960s, professional organizations and university programs started to appear. This decade saw these organizations and disciplines competing over the shape, identity, and role of computing in the academy (Tedre et al., 2018). Mathematicians, engineers, interdisciplinary scholars, and others worked to shape the goals and images of computer science (e.g., would programming be emphasized over theory, or were they co-constitutive?). This diversity was reproduced in educational priorities, which ranged from systems specialists and researchers to programming and beyond (Keenan, 1964). The ACM SIGCSE was founded in 1968 at a time that also saw a proliferation of educational CS materials at the university level, including those that sought to shift computing away from being math-oriented to being more discipline-diverse. As Tedre et al. (2018) explained, the "theory-vs.-practice" debate divided educators and researchers: "[m]any theoretically-oriented people saw computing as little more than a numerical extension to mathematics and logic . . . but many practicing programmers felt that the connection between programming competence and mathematical skills was weak" and that theory was not necessary to be an expert programmer (p. 170). The 1980s marked a shift in computer science's methodological and epistemological orientations, with the turn toward the computational sciences, especially simulating natural processes and modeling agent-based behaviors (Tedre & Denning, 2017). From the 1980s through the 2000s, computing became focused on computational problem-solving, with curricula that were social and design-oriented.

We now turn away from computer science education at the university level and toward some pivotal moments at the primary and secondary levels. While there has been a recent push for computer science education in formal primary and secondary schooling, it has a long history dating back to the 1960s, when computing was still in its infancy as a discipline. Caeli and Yadav (2020) provided a short history of computing education starting in the 1960s with the work of Alan Perlis and Donald Knuth, two computer science professors in the United States, and Peter Naur, a Danish computer scientist. All three of these early pioneers argued for the need to teach computer science in primary and secondary schools because of its "educational side-effects" and use it as a tool for problem-solving and deeper understanding of disciplinary ideas (Caeli & Yadav, 2020).

For example, Naur introduced the idea of *datalogy* instead of computer science as it centered human thinking rather than computers and something that should be incorporated into formal schooling (Tannert et al., 2021). In his book *Computing as a Human Activity*, Naur (1992) discussed his writings from the 1960s, where he defined *datalogy* as the discipline of "data, their nature and use," and *datamatics* as the "processing of data by automatic means" (p. 176). He further argued that datalogy and datamatics are analogous to linguistics and mathematics when placed in education. Naur suggested that it was important for school-age students to learn notions of data representation and formal data processes, which would serve as a foundation for university education. From Naur's perspective, computer science as a field needed to be designated as *computology* or *datalogy*, with the goal of students experiencing "how significant aspects of the world can be modelled by processing of data in computers" (p. 61). Thus, computing as a tool is only as good as people using it to solve problems they've identified (Caeli & Yadav, 2020).

Harel and Papert (1991) paralleled many of these ideas to propose *constructionism* as a learning theory that uses the power of computational tools to learn by making and doing. The desire for immersive learning based on physical manipulatives was a direct challenge to the "skill and drill" educational technologies of the time. Unlike previous theories of computers in education, where computers act as teachers, Papert's educational programming language, Logo – which he developed with Wally Feurzeig and Cynthia Solomon in the late 1960s – was designed for the student to become a tutor and the computer as the tutee (Taylor, 1980). These notions of using computing as a tool to explore mathematics, science, and language arts through design was a key aspect of constructionism that flattens the Piagetian socio-cultural hierarchy of concrete and formal thinking and prioritizes learning through building, design, and making with computational and non-computational artifacts. Indeed, Papert was on one of Jean Piaget's teams in the late 1950s and early 1960s, where his knowledge of child development deepened and was eventually applied to his work with Logo at MIT in the 1960s. But unlike Piaget, who focused on children's linear stages of development (moving from the concrete operational stage to the formal operative stage), Papert (1980) was much more interested in the diversity of *ways* children learn, especially ways in which concrete and formal thinking blend and blur together.

He argued that computational tools like Logo can be used to engage students in learning, as it allows them to *make* rather than *use* educational software (i.e., turning students from consumers into producers). For example, Papert (1980) discussed how a fifth grader created screen graphics using LogoWriter by working at the intersection of mathematics, arts, and computing. The computational tool in this case broke down the disciplinary silos and, as the student said, "put art and math together" (Papert, 1980, p. 5). Similarly, Resnick and Ocko (1991) argued that LEGO/Logo design activities were most beneficial when children are put in control and given freedom to design objects that are mathematically significant but also meaningful to them. These multiple paths of learning that allow students to use their own creativity and perspectives in the design process are much more powerful than a canned or instructionist "recipe-based" approach (Resnick & Ocko, 1991). Thus, computers should be used as *convivial tools* that give "each person who uses them the greatest opportunity to enrich the environment with the fruits of his or her vision" (Falbel, 1991, p. 32).

As a foundation for constructionism, Turkle and Papert (1991) argued for the need to challenge the "hegemony of the abstract, formal, and logical as the privileged canon in scientific thought." (p. 161). This argument was grounded in emerging literature at that time from feminists (e.g., Keller, 1984) and ethnographers (e.g., Traweek, 1988) who were studying the historical and social practices of scientific knowledge production. For Turkle and Papert, this literature revealed diverse ways of doing science and knowledge production, including ways that went beyond hierarchies between concrete thinking and abstract thinking (Lachney & Foster, 2020). If the hierarchy between concrete and abstract is not always suited for professional science, then why maintain it in education? In studying and celebrating the affordances of computers for supporting diverse modes of learning, Turkle and Papert (in their own work and together (Papert, 1980; Turkle, 1984; Turkle & Papert, 1991)) argued that it was important to accept education that is epistemologically pluralistic – that is, multiple ways of knowing and thinking – something formal computer science education sometimes lacks. Within the formal study of computer science, learners are often forced to approach the computer in "one right way" that emphasizes control through structure and planning. Rarely does the study of computer science allow students to use computers as an "expressive medium that different people can make their own in their own way" (Turkle & Papert, 1991, p. 165).

Not long after Papert's book *Mindstorms* was published, scholars and researchers started to question the legitimacy of Papert's ideas. There were arguments that the Logo programming environment was perceptually impoverished for children: while it simulates plant growth or bird flight, these simulations are decontextualized and reinforced through instrumental reasoning (Davy, 1984). In addition, some theorists argued that Papert's conflation of Piagetian operative stages was proposing "violence" to crucial stages of "natural" cognitive development (Zajonc, 1984). Others had concerns that Papert's push to move the use of computers beyond drill-and-practice activities confused how children learn. For instance, Dreyfus and Dreyfus (1984) suggest that with knowledge building and skill acquisition, it might be the opposite: "from abstract rules to particular cases" (p. 44).

Empirical studies of Logo showed both positive (Clements, 1990) and negative (Littlefield et al., 1989) results, with more detailed recommendations on what it would take to teach programming (Pea & Kurland, 1984). A review of the research on Logo from the 1980s suggested that skill transfer was most probable in highly structured environments, where time to acquire technical skills could be learned and practiced (Salomon & Perkins, 1987). This was antagonistic to much of Papert's open-ended design philosophy of constructionism. Despite clear controversy around Logo, by the 1990s it had a global presence, including in Brazil, Russia, Costa Rica, and other countries. But Logo failed to live up to its promise to revolutionize schooling, given that its purposes and goals lacked uniformity. The question of whether the "spirit" of Logo can be preserved when used as part of "regular" math and science curricula was an ongoing topic of debate within the constructionist education and research communities (Hoyles & Noss, 1987; Lemerise, 1990).

3 21st-Century Primary and Secondary Computing Education

This section will cover the current state of computing education at primary and secondary level, including the diffusion of CS over the last decade through its integration in other disciplines as well as stand-alone computing curricula. It then looks at the rise of CS education in the 21st century through the trends of curriculum standardization and spread of constructionist pedagogies. Since the early 2000s, computer science education has grown considerably across the globe with movements like CSforAll in the United States, Computing at School in the United Kingdom, and inclusive learning by conceptualizing computing concepts in native languages in India. The push for CS in primary and secondary education started with the need to have more students pursue CS at the undergraduate level to have enough people for unfilled computing jobs. Even today, the data from Code.org suggests that there are 609,151 computing-related jobs open in the United States, while only 47% of high schools offer computer science and 71,226 CS students graduated into the workforce (Code.org, n.d.).

As noted earlier, the United States is not the only country seeking to increase its computing workforce through initiatives at primary and secondary levels in the 21st century. In their comparative study of 14 case studies of primary and secondary CS education from across the globe, Hubwieser et al. (2015) noted that approaches and priorities vary greatly across national contexts. They found that in France, CS education is often related to information communication technology education, whereas in Sweden it was focused more on programming and web technologies. They also found that teacher education for CS varies. In Israel, for example, a Bachelor of Science is required to teach CS, but not in Bavaria, New Zealand, and many other countries that they surveyed. Recently, Yadav and colleagues (2022) reported similar findings for teacher certification across several European countries, States in the U.S, and New Zealand as well as challenges of recruiting pre-service teachers in CS teacher preparation programs.

To increase participation and awareness of CS at primary and secondary levels around the globe, CS has shifted from just being framed as coding or designing with technology to something that includes a "range of mental tools" (Wing, 2006, p. 33). Wing argued for computational thinking (CT) as a way to introduce CS concepts and practices in primary and secondary classrooms, so students understand what it means to think computationally. While there is no consensus on the definition of CT, certain practices have been widely suggested as being key to engaging in CT. The practices include abstraction, decomposition, algorithms, debugging, and pattern recognition (Krauss et al., 2017). The argument has been made that as students engage in these CT practices and as CT gets added to their toolkit in ways similar to literacy and mathematics, they will be prepared to live and work in a world driven more and more by computing (Wing, 2006; Yadav, Hong, et al., 2016).

As a result of this push to broaden conceptions of computer science beyond programming in the United States, the College Board, a non-profit organization that develops and assesses curricula for college readiness, launched a new secondary CS course, Computer Science Principles (CSP), in 2016. CSP is designed to be an approachable CS course that focuses on seven big ideas of computing and CT practices, which include creativity, abstraction, data and information, algorithms, programming, the Internet, and global impact (College Board, 2020). The latest data suggests that the number of students taking CSP has significantly increased since the course was launched, with 114,188 students taking the corresponding exam, a 62% increase since 2017. At the same time, the other high school introductory CS course (CSA) that uses the Java programming language has only seen a 13% increase over the same period. While CS has grown over the last few years, the latest report on the state of CS education found that girls make up only 31% of high school students enrolled in foundational computer science, 44% of the middle school students, and 49% of the elementary students enrolled in computer science (Code.org, CSTA, & ECEP Alliance, 2021). While the number of students taking CSP has increased considerably over the past five years, there remain large gaps in who has access and who is successful in the course. Data from 2020 suggest that CSP is one of the worst courses that is skewed towards males, and pass rates for Black and Hispanic students are significantly lower than for White and Asian students (Ericson, 2020).

Just as there has been an increase in more traditional CS course content, there persists a focus on creativity and personal expression that can be traced back to constructionism. In order for students to use computing to express creativity (Yadav & Cooper, 2017), there has been a consistent focus on design and sociality. In the 2000s, this has been most prominently represented by researchers, technologists, and educators with direct or indirect connections to the MIT Media Lab, where the goals and ideas of Papert and constructionism are still strongly propagated (Resnick, 2018). Kafai and Burke (2014) provide an important overview of 21st-century computing education that is largely grounded in this constructionist tradition. Reporting on the design and implementation of programmable hardware and software, they highlight techno-social desires and means for young people to shift from being consumers of media and technology to being producers. For example, the globally used visual programming environment Scratch and its online community draws on the history of hip-hop and theater arts to encourage students to create original content (e.g., animations, games, music, visual designs, etc.) and remix content from their peers (Resnick et al., 2009). Or consider the growing popularity of programming hardware for making electronic textiles that young people can use to refashion their clothes and other youth culture accessories (Fields et al., 2018). These examples highlight how 21st-century computing education blends the technical with the social and cultural.

Towards this shift, Kafai (2016) builds on Wing's (2006) idea of computational thinking by proposing a framework for *computational participation*: moving primary and secondary computing education beyond just a focus on individuals working with tools and code to people working in communities and contexts in computationally social and creative ways. This has implications for issues of equity. As Kafai and Burke (2014) explain, "When computation is thought of in terms of participation and not just thinking, it becomes clear that there is a tremendous discrepancy in who gets to participate" (p. 9–10). Indeed, computational participation is part of a larger move within technology and media literacy education to shift from a problem space around the digital divide to the "participation gap" – a gap between those who have the knowledge and skills to participate in digital communities and the information economy and those who do not (Jenkins et al., 2009). Yet computational participation has its own limitations when it comes to equity. For example, even though online and offline communities for computational participation have existed for over a decade, many Black, Brown, and Indigenous communities continue to be underrepresented in computing disciplines and fields (Lachney et al., 2019a).

3.1 Techno-Optimism vs. Techno-Pessimism

As noted earlier, computational thinking, constructionism, datalogy, CSP, etc. frame computing and computers in optimistic terms, sometimes reformist and other times revolutionary (Agalianos et al., 2006), but always as a force to change teaching and learning for the better and improving education more generally, including CS education. This techno-optimism and its support come from all directions. In government, politicians such as the US president Bill Clinton claimed computers were the "great equalizers" for education (quoted in Selwyn, 2016, p. 26); private sector actors like the Danish LEGO Group design programmable hardware to sell as part of computing education around the world (Lachney, 2014); tech companies focus on increasing access to computational devices to diversify their markets and seek out future customers (Selwyn, 2012); and non-governmental organizations, such as One Laptop Per Child, flood countries in the Global South with computers in hopes of spurring children's upward mobility (Ames, 2019).

Consider, for example, one of the most prominent and successful global computing and engineering education interventions across the world, the FIRST Lego League (FLL) competition for 9- to 16-year-old children, depending on the country. FLL is part of the larger nonprofit organizational network For Inspiration and Recognition of Science and Technology (FIRST). FIRST hosts robotic design challenges for children of all ages, with specific educational technologies used for each, from VEX Robotics to LEGO WeDO. FLL uses LEGO MINDSTORMS, a programmable robotics kit that is produced and distributed by the LEGO Group. FLL has a global presence, as evidenced by the fact that 216,480 children from 69 countries participated in the program during 2021 (FIRST, 2021). FIRST has reason to be celebratory and optimistic in their global engineering and computer science education efforts. For example, female alumni from FIRST competitions are more likely to declare engineering and computer science majors when compared to young women who did not participate in FIRST (FIRST firstlegoleague.org., 2021).

At the same time, techno-optimism is often unquestioning in assuming that computers and access to them are inherently beneficial, as well as inevitable, due to the continued growth of computing power (Lachney et al., 2021). In addition, this optimism is often rooted in market-oriented reforms (i.e., neoliberal reforms), which are driven by corporate actors like the LEGO Group instead of local interests or community goals (Lachney, 2014). While this optimism often appears warranted on the surface, once technology fixes for education are explored in any depth, more nuanced pictures emerge.

For example, consider the One Laptop Per Child (OLPC) nonprofit initiative. While OLPC and the flagship device the OX laptop are said to be inspired by Papert's constructionism, the story of OLPC sometimes begins with Nicholas Negroponte – MIT professor and one of the founding members of the MediaLab – and his trip to the 2005 World Economic Forum in Davos (Ames, 2019). Within a global crowd who was firmly grounded in the celebratory atmosphere of neoliberal solutionism, Negroponte pitched the idea of a hundred-dollar laptop to support technology access to children around the world. By the end of the following year, OLPC had support from the United Nations Development Program, Google, New Corporation, eBay, and many other international actors (Ames, 2019, pp. 3–4). As OX laptops – anthropomorphic green-and-white machines that were designed to be durable and self-contained – found their way into the hands of people in Africa, North America, South America, Asia, and Europe, the program was often celebrated as an "intuitive and common-sensical idea that transcends any further debate," and Papert's largely Western-inspired theory of constructionism was exported around the world (Selwyn, 2013, p. 128).

The OX laptops were designed to be highly versatile in terms of physical durability, power sources, and software capabilities. The laptops often came with a variety of programming languages, including Python and JavaScript. However, the techno-optimism was so fundamental to the project that Negroponte scoffed at the idea of even evaluating the OX laptop's local impact. When speaking about this, he explained the idea away by treating the machines as akin to a utility like electricity, which requires no controlled studies to understand its impact and benefit (Selwyn, 2013, p. 140). This type of techno-optimism is exactly what leads people to overlook the sociotechnical limitations of any computing device or system. This initiative is what Falbel (1991) would call *exposure* to computing that puts computers *within reach* for children; however, this *exposure* does not allow them to enter the professional computer culture, as they do not also have access to people who have knowledge about computers.

While the benefits of OLPC are certainly worth mentioning – for example, highlighting the need for low-cost computing technologies and the innovations to meet that need – the unquestioned assumption that the OX laptops would be accepted, used, and maintained unproblematically appears naive in retrospect. Selwyn (2013) noted that far from being neutral, the program is grounded in a Western ideology of romantic individualism, which emerged in its focus on the child, as opposed to the community or system, as an avenue for change. At a material level, Warschauer and Ames (2010) explained how they found disconnections between the universalist approach of OLPC and the particular conditions of local communities, with some communities struggling to meet the costs of running the OX laptops and maintaining the infrastructure for use. Exploring the racial politics of OLPC, Fouché (2012) explained how the program's racial and cultural agnosticism is part of a long history of framing Western technologists in savior terms when they transfer devices into non-White and non-European contexts as solutions for local problems. These ideas of technology as the savior for non-White folks are deeply rooted in education, as evidenced by Skinner's programmable teaching machines, which he wanted to test on poor Black children (Watters, 2021). Far from being a radical break from the past, OLPC has reproduced and become entrenched in existing power, political, and ideological structures at local, national, and international levels (Ames, 2019).

The OLPC examples illustrate how, as techno-optimistic perspectives of computing education proliferate and more stakeholders in education and research are introduced to curricula and technologies that are designed for primary- and secondary-age students, challenges concerning inequity and injustice have made the idea of computers as positive change agents questionable. This has led to an increasing sense of techno-pessimism, where the push for computing education is seen as being driven by economic and nationalistic competition that limits the focus on using computing power, computational thinking, and computer skills as generative sources for empowering students and their communities (Lachney et al., 2021). Of course, such pessimism about technology and education is nothing new. As Cuban (1986) and Noble (2001) point out across all levels of education, as long as there have been people claiming that technologies are going to revolutionize education, there has been repeated disappointment, not to mention the fact that computers are oversold to schools only to be underused (Cuban, 2001).

What is more, techno-pessimists often point out that a focus on technology misses the mark of what people value about education in the first place, not interactions with devices, but interactions with people. Philosopher of technology Winner (2009) makes this point, explaining that when people recall their positive experiences in school, they rarely say, "The experiences that inspired me most deeply were the wonderful hours I spent pushing the Logo turtle around the screen," but instead are more likely to talk about "that special teacher who changed a person's life" (p. 590). But this is not to suggest that teachers and technologies are always easily separable parts of students' experiences in the classroom; the expertise and framing that teachers bring to technologies and, conversely, how technologies shape teaching can be co-constitutive.

Consider, for example, the now-foundational study *Stuck in the Shallow End: Education, Race, & Computing* by Margolis et al. (2008, 2017) that puts these relationships into perspective. Indeed, while the study is over a decade old, it remains pertinent for explaining why techno-optimism increasingly appears naive when it comes to CS education, at least in a US context. The study shows how larger social issues in the United States, such as racialization and class stratification, are structured into and reproduced through education systems and computing pedagogy. Key findings from their study of computing education across three schools in the Los Angeles Unified School District highlight how having access to technology is not enough and that tech fixes in education policy often fail to create meaningful change at local levels. Their conclusions echo Anyon's (2014) research, which suggests that unless policymakers and activists address poverty and wealth inequities, educational reforms will continue to feel Sisyphean.

What is more, Margolis et al. (2008, 2017) found that even in a school with access to computing materials and expertise, the small numbers of African American and Latinx students in a computer science course were not just the result of histories of the classist racialization in policies at federal, state, and district levels but also the result of teachers', counselors', administrators', and even students' own beliefs about who belonged in computing based on ideas about "natural" abilities. Stories from Margolis et al. (2008, 2017) detail heartbreaking situations that point to the reproduction of cultural and biological essentialism: for example, teachers singling out African American or Latinx students as not belonging in CS courses, stating they have low expectations of them, and assuming that they lack natural ability or the cultural background to succeed in CS. These attitudes did not go unnoticed by students who sometimes reproduced harmful societal and cultural stereotypes about White and Asian students being good at math and computing, and Black and Latinx students being deficient.

Another perspective within techno-pessimist framing of technology are the disproportionately harmful effects of technology on Black and Brown communities (Benjamin, 2019; Noble, 2018). From the facial recognition technologies that misidentify darker-skinned people (Buolamwini, 2017) to recidivism software in the criminal justice system that recommends higher prison sentences for Black defendants, bias in the design and deployment of technologies is ever present. Technologies are mostly developed for nationalistic or capitalistic reasons. Take the example of Simon Ramo, the vice president of an aeronautical manufacturer, who stated in 1957, "[W]e are in rapid transition today to a new world which threatens to be dominated by technological advance" (as cited in Watters, 2021, p. 149). Ramo went on to design the first ballistic missile. Given this, the question arises on what the goals of computing education need to be and what kinds of technologies are needed for a just world.

The techno-pessimistic and techno-optimistic perspectives both give meaning and shape to computing education for primary- and secondary-age children around the world. But just as a strict optimistic line is limiting, so, too, is a strict pessimistic one. A realistic perspective would embrace qualities of both – for example, the amazing ways that a computer can support children's creativity while understanding that the same computer does not exist in isolation of larger sociopolitical structures, cultural contexts, economic conditions, and ecologies of expertise.

3.2 Techno-Social Realism in Computing Education

On their own, neither techno-pessimism nor techno-optimism are adequate for understanding the sociotechnical reality of computers in education for primary- and secondary-age students. While the techno-optimists place too much weight on the role of individuals and machines for creating change, they do accurately recognize that human–computer interactions can foster meaningful learning experiences. While the techno-pessimists sometimes overlook the roles of computers in positive grassroots efforts, they accurately recognize that no technological device or system is neutral or value-free.

In this section we move beyond this dichotomy and look at those who take what we call a *techno-social realist* approach to computing education. We argue that the techno-social realists accept the contradictory conditions of computing education. They do not put too much weight on the technology of computing education itself, situating the agency of computer devices and systems as just some parts of quality computing and computing education. The techno-social realist perspective does not make technology central to computer education but, instead, treats computer devices and systems as actors within larger networks of humans and nonhumans who collectively constitute learning environments.

The coinage of the term *techno-social realist* draws from the increasing number of computer education researchers who are recognizing that the social and technological are always co-constituted, with neither the social solely determining the technological nor the technological solely determining the social (Magenheim & Schulte, 2006; Mertala, 2021). Discussions of the sociotechnical or the techno-social highlight this recognition and provide new ways to conceptualize human–machine relationships at multiple scales, most readily in ways that shift away from techno-solutionism and towards more ecological perspectives that are needed to confront 21st-century global challenges (Easterbrook, 2014).

Indeed, when this framing is applied, the current techno-social situation quickly becomes rather bleak as the macro-issues that enter schools are often beyond any individual's control, sometimes placing teachers in a double bind (Lachney et al., 2018). For example, the same devices that support rich learning also bring up ethical issues, like how computing is contributing to environmental degradation, upstream and downstream (Mayhew & Patitsas, 2021); how technology companies are extracting data from online behaviors to aggregate and sell for making predictions about future behaviors in ways that might shape market demands and elections (Zuboff, 2019); how the goal of the *CSforAll* movement is rather obligatory for many, but that computing disciplines and fields continue to reproduce anti-Black racism (Rankin et al., 2021). It has also long been recognized that educational computing technologies have intimate connections to US militarism (Noble, 1991).

Taking these conditions and others into account while also seeking to design quality computing education for young people has been foundational for research on equity-oriented professional development for computing teachers and research on culturally responsive computing. Both areas make issues of justice, ethics, and social responsibility central to computing education by situating computing and computer technologies as only some of the nodes within larger techno-social ecologies. In addition, computing education research that explores the topics of universal design for learning, the environmental impact of computing, and translanguaging also take techno-social perspectives. We explain these five areas of research and then how they exemplify the techno-social realist position before discussing a glaring gap in the literature pertaining to curriculum developers and technologists.

3.2.1 Equity-Oriented Professional Development

Building on their prior findings about teachers' beliefs and recognizing that computing education reproduces injustice, including racial injustice, Margolis and colleagues (2008, 2017) have been leaders in reporting on computer teacher professional development for their high school Exploring Computer Science (ECS) program, which includes, among other things, an explicit focus on equity (e.g., Ryoo et al., 2021; Fields et al., 2018; Margolis et al., 2015). While recognizing that "it takes a village" (Ryoo et al., 2015) of not just teachers and students but also administrators, counselors, policymakers, and others to support broadening participation efforts in computing, their focus on teachers is meant to help challenge deficit thinking – that is, thinking that frames racially marginalized students and their communities as deficient and problematic in academic pursuits – in schools and stereotypes about who belongs in computing. Their goal has been to move beyond access to a focus on pedagogy and epistemology (Margolis et al., 2012). For example, Goode and Ryoo (2019) outline four dimensions to support teachers' knowledge of and practices for designing inclusive learning environments for computing education: (1) having specialized knowledge of computing, (2) developing pedagogical practices that impact student learning, (3) drawing on students' cultural background and knowledge in the computing classroom, and (4) understanding how education systems and policies shape educational experiences and outcomes. The ECS professional development programs have been successful in expanding CS education in the United States, including the Los Angeles Unified School District (Margolis et al., 2008, 2017), and supporting teachers' confidence in delivering engaging CS content (Goode et al., 2019).

However, their research also suggests that supporting CS teachers' – especially White teachers – racial literacies and understanding of racial equity's relationship to computing and the CS classroom has its own unique challenges. In research on teachers in ECS professional development programs and classroom implementations, Goode and colleagues (2020) reported on how even when the topics of race and racism were infused into the ECS curriculum and training, White teachers tended to avoid the topics or reproduced deficit orientations about students and their families and communities. In contrast, teachers of color and Black teachers engaged with the topics directly, not only naming race and racism as part of computing and computer education, but also identifying how their own roles as educators might come to bear on these issues in the classroom and beyond. Given that the majority of US teachers are White (US Department of Education, 2020), addressing individualistic, deflective, and avoidant discourses about race within computing education is crucial for constructing equitable futures. Goode and colleagues (2021) explain that long-term professional development is key for helping White teachers move beyond these reductive discourses and into equity-centered CS education.

3.2.2 Culturally Responsive Computing

Like equity-oriented research on professional development for computing teachers, research in culturally responsive computing (CRC) does not center technology for technology's sake or dismiss the role that computers can play in transforming educational experiences. Instead, it builds on culturally relevant pedagogy (Ladson-Billings, 2014), culturally responsive teaching (Gay, 2018), and culturally sustaining pedagogies (Paris, 2012). CRC seeks to challenge deficit modes of education – that is, education that frames the families, communities, heritages, interests, and cultures of students, especially Black, Brown, and Indigenous students, as antithetical to academics – within computing and technology education (Lachney, 2017).

Scott et al. (2014) explained four goals that CRC should try and meet. First, CRC should motivate and improve the STEM and CS learning experiences of racially marginalized children. Second, this should be accomplished through deep explorations of heritage knowledge and vernacular culture, as well as space for critiquing power and celebrating cultural diversity. Third, the first two goals are meant to come together to break down the barriers between what is traditionally considered scientific and technical and what is traditionally considered cultural. Fourth, this work should be done not only to represent community and cultural knowledge and identities but also school and curricular demands.

To further clarify, consider how Eglash et al. (2013) outlined four (often overlapping) areas that CRC research and practice can engage with: (1) *Indigenous knowledge* – developing epistemological connections between Indigenous design practices and computing activities to demonstrate the mathematical and computational sophistication in designs and epistemologies to help resist primitivist stereotypes and myths of genetic determinism (Moreno Sandoval, 2013; López-Quiñones et al., 2023); (2) *vernacular culture* – the computational modeling of vernacular linguistic, design, and/or artistic practices with computational thinking, computer programming, or educational technologies to explore young people's existing interests, whether rap, graffiti, jewelry making, or another aspect of popular youth culture, in computational contexts (Pinkard, 1999; Eglash et al., 2006; Gilbert et al., 2008; Bennett, 2016); (3) *civic and political cultures* – the use of computational tools and computational thinking to support participation in community life, grassroots advocacy, social justice, political processes, and/or consciousness raising to take part in justice projects, such as raising awareness about and eliminating the gender pay gap (Scott et al., 2014; Scott & Garcia, 2016; Cooke et al., 2019); (4) *hacking cultures* – culturally and socially situated repurposing of existing technologies or engagement in DIY practices through programming, engineering, media production, and/or computational thinking that build on the "hacker ethic" (Levy, 1984) or, by extension, "maker mindset" (Dougherty, 2016) values of freedom and openness, decentralization, and anti-authoritarianism, (Eglash et al., 2004; Santo, 2011).

3.2.3 Universal Design for Learning

Universal design for learning (UDL) is an educational framework that is often used to address issues of inclusivity and access for students with and without disabilities, largely to make learning environments usable to the broadest array of children (Israel et al., 2022). But its basic principles build on the universal design (UD) movement that emerged from disability justice and social movement activism in the 1960s (Hamraie, 2017). The UD movement focused on making the built environment more accessible to people with disabilities by standardizing ramps and automatic doors, but this also had the result of making the environment more accessible for people with strollers, those who use crutches, and many others. Hence, designing for those with the least access has a benefit to others. UDL extends this form of UD into education with three principles:

> (1) multiple means of engaging and motivating learners; (2) multiple means of representing information so students can perceive and comprehend that information; and (3) multiple means for providing students opportunities to express what they know and navigate the environment in which they learn.
>
> *(Israel et al., 2022, p. 3)*

Within primary and secondary schools, UDL has been applied to make computing education more flexible for students with and without disabilities (Bouck & Yadav, 2021). For teachers to develop and employ UDL practices in their classrooms, they required sustained and substantial professional development; otherwise, implementation may appear shallow (Ray et al., 2018). Israel et al. (2022) found that when primary teachers used UDL to support computing lessons, it was mostly by trying to represent content through multiple media and formats. These types of studies situate UDL within productive tension between the desires for universality and the need to be responsive to local contexts. This constant negotiation is what makes it fit into a techno-social realist perspective, as it does not get caught up in prioritizing either the global or the local.

3.2.4 Computing and Environmental (Un)Sustainability

The materiality of computer hardware, the energy requirements for computational maintenance and innovation, and the mining of minerals that are in all our computing devices have all brought into focus how computing education is implicated in the environmental crisis that humanity faces. While the green computing movement has sought to confront the harmful environmental results for computing technologies that have been around since the turn of the century (Smith, 2013), there has been little research into its application in primary or secondary computing education. Indeed, when it comes to addressing issues such as electronic waste, educational researchers have only started to reckon with how they are implicated in its production and exportation to largely poor nations (Lachney et al., 2018). There has been little research on how schools teach about computing's impact on the environment and the role of computing education in making a more sustainable world.

Eglash et al. (2013) is one exception, with the "e-waste to makerspace" project, where children learn about the environmental harms of e-waste while repurposing e-waste as part of maker-type projects (e.g., automated watering cans). There has been much more research on computing and environmental (un)sustainability in higher education that is worth mentioning. Easterbrook (2014), for example, argued that computational thinking is too limiting for addressing the need to make computing more sustainable and suggested a shift to systems thinking in undergraduate curriculum. Mayhew and Patitsas (2021) argue that bringing sustainability and environmental issues into undergraduate computing education is complicated by the fact that hegemonic computing education lacks a focus on materiality. They argued against an "add sustainability and stir" approach to greening the curriculum and instead imply that larger-scale educational change is needed to truly address these issues. Primary and secondary computing education can learn from these researchers and others on what to do and not to do when bringing issues of sustainability into their classrooms, but research on this topic is sorely lacking.

3.2.5 Translanguaging

Finally, we want to point to an exciting area of research that attends to how computing education can support making multilingual students' rich linguistic skills assets for learning. Vogel (2021) uses the theory of translanguaging as a foundation for doing this in the computing classroom. Translanguaging describes how children use diverse linguistic and semiotic resources in fluid and flexible ways, despite traditional education systems' demand to often silo and isolate them (García et al., 2014). It may be that in everyday speech, for example, multilingual children use Spanish and English together in ways that blend or break down traditional demarcations. Vogel (2021) revealed the ways that computational learning environments that support personalization can support multilingual students' sense of agency in ways that challenge the hegemonic linguistic standards of schooling. Given that most programming languages used around the world are English-based, this research has the potential to not only describe how programmers are already translanguaging but also how normative applications of translanguaging in learning environments can prepare children for working in such contexts.

3.3 Future Research for the Techno-Social Realist Agenda

All these areas of research exemplify techno-realism by not falling into either strict pessimism or strict optimism. They recognize the affordances and limitations of the role of computing in improving not only schooling but also the world more generally without centering technology as the only agent of change. But for Lachney and Yadav (2020), this is only part of the story. Building on the idea that "it takes a village" to construct equity-oriented CS education (Ryoo et al., 2015; Lachney, Bennett et al., 2021; Lachney, Eglash et al., 2021), they visualize the key actors needed to create deep forms of CRC: educators, cultural experts, and technologists (see Figure 25.1). However, while there has been much research on educators' understandings and beliefs about supporting equity in CS (e.g., Yadav, Gretter, et al., 2016) and there is a growing literature on supporting cultural experts in collaborating in broadening participation efforts (e.g., Lachney, Yadav et al., 2021), there has been much less descriptive or normative work on technologists. In addition, Lachney and Yadav (2020) do not mention another key actor: curriculum designers.

As a future research trajectory, computing education researchers must start to empirically interrogate the beliefs and practices of technologists who create programmable hardware and software, as well as the curriculum designers who connect these technologies to academic content and contexts. A growing body of literature on computer science and artificial intelligence is reinforcing the idea that "artifacts have politics" (Winner, 1986/2020). Scholars in the United States have pointed to how, for example, the "master-slave" command from computer science and computer engineering is racialized (Eglash, 2007). More generally, Benjamin (2019) shows how the designs of algorithms and devices can reproduce structural racism (e.g., predictive algorithms that use already racially biased police data) and interpersonal racism (e.g., the unconscious biases of designers whose facial recognition software does not recognize Black faces). Yet there has been very little research that critically studies how educational technologists and curriculum designers think about racism, sexism, classism, ableism, etc. in technology design, development, and implementation. As part of this work, computing education researchers must interrogate how technologists' and designers' assumptions get baked into the technologies and curricula that many educators and researchers take for granted in their work with computing education.

This work should speak to how it is important to acknowledge the oppressive and harmful impact of technologies on Black and Brown communities. Within a techno-realist perspective, we need to help students understand the role of computer science in the design and implementation of technologies that maintain and amplify existing social hierarchies. The technologies are not going

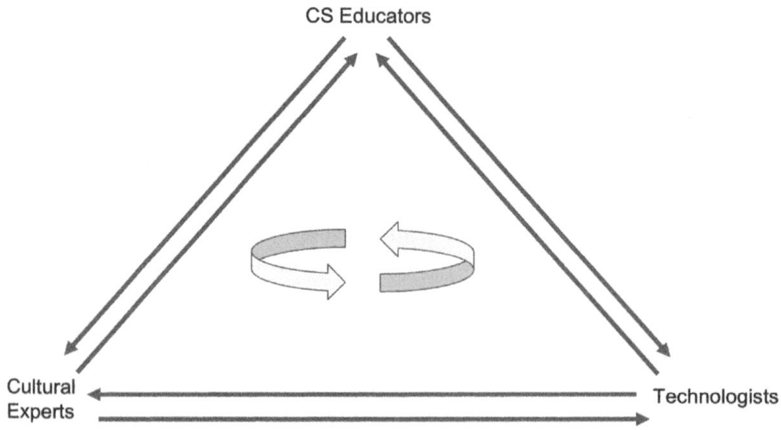

Figure 25.1 A diagram about fostering deep culturally responsive computing collaborations.
Source: Lachney and Yadav (2020).

away, so the question is, How should we prepare students to think critically about the design of computing technologies and question what goals are being served with such technologies? In order to accomplish this curriculum, teacher preparation needs to be revised in ways that move away from top-down approaches for computer science education that prioritizes capitalistic goals of the tech industry (Yadav & Heath, 2022). Instead, we need to highlight how computer science can be used as a tool for expressing personal agency, creativity, and addressing issues important to students and their communities.

Conclusion

Computing education and engineering education at the primary and secondary levels have long been co-mingled. This is apparent not only in the overlap in the technologies that are employed but also by the organizations that support them (e.g., FLL). Still, we have sought to provide an overview of computing education as having its own unique identity that is still growing. While computer programming has been part of primary and secondary education since the late 1960s, the 21st century has seen an explosion of computing technologies, curricula, and private–public partnerships to prioritize computer science education. This work often finds rhetorical justification in nationalistic claims that include the need to prepare students to work in an information economy founded on computing power, computational thinking, and computer skills. At the same time, a growing number of K–12 computer science education researchers are attending to the inequities and injustices that computer science itself contributes to, especially those issues that disproportionately impact low-income Black, Brown, and Indigenous communities, as well as people with disabilities and who live in environmentally compromised areas. In this chapter, we have framed the need for computer science education to move away from the tensions of techno-optimism and techno-pessimism and instead use a techno-social realist perspective that sees computer science education in a broader context.

References

Agalianos, A., Whitty, G., & Noss, R. (2006). The social shaping of Logo. *Social Studies of Science*, *36*(2), 241–267.

Ames, M. G. (2019). *The charisma machine: The life, death, and legacy of one laptop per child.* MIT Press.

Anyon, J. (2014). *Radical possibilities: Public policy, urban education, and a new social movement*. Routledge.

Babbitt, W., Lachney, M., Bulley, E., & Eglash, R. (2015). Adinkra mathematics: A study of ethnocomputing in Ghana. *Multidisciplinary Journal of Educational Research*, *5*(2), 110–135.

Benjamin, R. (2019). *Race after technology: Abolitionist tools for the new Jim code*. Polity Press.

Bennett, A. G. (2016). Ethnocomputational creativity in STEAM education: A cultural framework for generative justice. *Teknokultura*, *13*(2), 587–612. doi:10.5209/rev_TEKN.2016.v13.n2.52843

Bennett, A. G., Eglash, R., Lachney, M., & Babbitt, W. (2016). Design agency: Diversifying computer science at the intersections of creativity and culture. In *Revolutionizing Education through web-based instruction* (pp. 35–56). IGI Global.

Bouck, E. C., & Yadav, A. (2021). Providing access and opportunity for computational thinking and computer science to support mathematics for students with disabilities. *Journal of Special Education Technology*. doi:10.1177/0162643420978564

Buolamwini, J. A. (2017). *Gender shades: Intersectional phenotypic and demographic evaluation of face datasets and gender classifiers* (PhD Dissertation). Massachusetts Institute of Technology.

Caeli, E. N., & Yadav, A. (2020). Unplugged approaches to computational thinking: A historical perspective. *TechTrends*, *64*, 29–36. doi:10.1007/s11528-019-00410-5

Ceruzzi, P. (1988). Electronics technology and computer science, 1940–1975: A coevolution. *Annals of the History of Computing*, *10*(4), 257–275.

Clements, D. H. (1990). Metacomponential development in a logo programming environment. *Journal of Educational Psychology*, *82*(1), 141–149.

Code.org. (n.d.). Code.org. Retrieved February 23, 2023, https://code.org/

Code.org, CSTA, & ECEP Alliance. (2021). *2021 State of computer science education: Accelerating action through advocacy*. https://advocacy.code.org/stateofcs

College Board. (2020). *AP computer science principles*. https://apcentral.collegeboard.org/media/pdf/ap-computer-science-principles-course-and-exam-description.pdf

Cooke, L., Vogel, S., Lachney, M., & Santo, R. (2019, February). Culturally responsive computing: Supporting diverse justice projects in/as computer science education. In *2019 research on equity and sustained participation in engineering, computing, and technology (RESPECT)* (pp. 1–2). IEEE.

Cuban, L. (1986). *Teachers and machines: The classroom use of technology since 1920*. Teachers College Press.

Cuban, L. (2001). *Oversold and underused*. Harvard University Press.

Cuban, L. (2011). *Oversold and underused*. Harvard University Press.

Davy, J. (1984). Mindstorms in the lamplight. *The Teachers College Record*, *85*(4), 549–558.

Dougherty, D. (2016). *Free to make: How the maker movement is changing our schools, our jobs, and our minds*. North Atlantic Books.

Dreyfus, D., & Stuart, D. (1984). Putting computers in their proper place: Analysis versus intuition in the classroom. *The Teachers College Record*, *85*(4), 578–601.

Easterbrook, S. (2014*). From computational thinking to systems thinking: A conceptual toolkit for sustainability computing*. ICT for Sustainability (ICT4S-14). doi:10.2991/ict4s-14.2014.28.

Ericson, B. (2020). *A deeper look at women of color* (Black, Hispanic/Latinx, and Native American) and AP CS in 2020. https://cs4all.home.blog

Eglash, R. (2007). Broken metaphor: The master-slave analogy in technical literature. *Technology and Culture*, *48*(2), 360–369.

Eglash, R., Bennett, A., Cooke, L., Babbitt, W., & Lachney, M. (2021). Counter-hegemonic computing: Computer science education for value emancipation. *ACM Transactions on Computing Education*, *21*(2), 1–28.

Eglash, R., Bennett, A., O'Donnell, C., Jennings, S., & Cintorino, M. (2006). Culturally situated design tools: Ethnocomputing from field site to classroom. *American Anthropologist*, *108*(2), 347–362. doi:10.1525/aa.2006.108.2.347

Eglash, R., Croissant, J. L., Di Chiro, G., & Fouch, R. (2004). *Appropriating Technology: Vernacular Science and Social Power*. University of Minnesota Press.

Eglash, R., Gilbert, J. E., & Foster, E. (2013). Toward culturally responsive computing education. *Communications of the ACM*, *56*(7), 33–36.

Eglash, R., Lachney, M., Babbitt, W., Bennett, A., Reinhardt, M., & Davis, J. (2020). Decolonizing education with Anishinaabe arcs: Generative STEM as a path to indigenous futurity. *Educational Technology Research and Development*, *68*(3), 1569–1593.

Falbel, A. (1991). The computer as a convivial tool. *Constructionism*, 29–37.

Fields, D. A., Kafai, Y., Nakajima, T., Goode, J., & Margolis, J. (2018). Putting making into high school computer science classrooms: Promoting equity in teaching and learning with electronic textiles in exploring computer science. *Equity & Excellence in Education, 51*(1), 21–35.

FIRST firstlegoleague.org. (2021). *IRST 2021 Annual impact report.* Retrieved from February 23, 2023, https://www.firstinspires.org/sites/default/files/uploads/annual-report/fy2021-annual-impact-report.pdf

Fouché, R. (2012). Black inventors to OLPC. *Race After the Internet*, 61–85.

Galison, P. (1997). *Image and logic: A material culture of microphysics.* University of Chicago Press.

García, O., Wei, L., García, O., & Wei, L. (2014). Translanguaging in education: Principles, implications and challenges. In *Translanguaging: Language, bilingualism and education*, 119–135. Springer.

Gay, G. (2018). *Culturally responsive teaching: Theory, research, and practice.* Teachers College Press.

Gilbert, J. E., Arbuthnot, K., Hood, S., Grant, M. M., West, M. L., McMillian, Y., . . . Eugene, W. (2008). Teaching algebra using culturally relevant virtual instructors. *International Journal of Virtual Reality*, 7(1), 21–30.

Goode, J., Ivey, A., Johnson, S. R., Ryoo, J. J., & Ong, C. (2021). Rac (e) ing to computer science for all: How teachers talk and learn about equity in professional development. *Computer Science Education, 31*(3), 374–399.

Goode, J., Johnson, S. R., & Sundstrom, K. (2020). Disrupting colorblind teacher education in computer science. *Professional Development in Education, 46*(2), 354–367.

Goode, J., Peterson, K., Malyn-Smith, J., & Chapman, G. (2019, February). Online professional development for high school teachers: Building the capacity of a national cadre of teachers to broaden participation in computing. In *2019 research on equity and sustained participation in engineering, computing, and technology (RESPECT)* (pp. 1–5). IEEE.

Goode, J., & Ryoo, J. J. (2019). *24 teacher knowledge for inclusive computing learning.*

Hamraie, A. (2017). *Building access: Universal design and the politics of disability.* University of Minnesota Press.

Harel, I. E., & Papert, S. E. (1991). *Constructionism.* Ablex Publishing.

Hoyles, C., & Noss, R. (1987). Synthesizing mathematical conceptions and their formalization through the construction of a Logo-based school mathematics curriculum. *International Journal of Mathematical Education in Science and Technology, 18*(4), 581–595.

Hubwieser, P. (2013, February). The Darmstadt model: A first step towards a research framework for computer science education in schools. In *International conference on informatics in schools: Situation, evolution, and perspectives* (pp. 1–14). Springer.

Hubwieser, P., Giannakos, M. N., Berges, M., Brinda, T., Diethelm, I., Magenheim, J., . . . & Jasute, E. (2015). A global snapshot of computer science education in K-12 schools. In *Proceedings of the 2015 ITiCSE on working group reports* (pp. 65–83). ACM Press.

Israel, M., Kester, B., Williams, J. J., & Ray, M. J. (2022). Equity and inclusion through UDL in K-6 computer science education: Perspectives of teachers and instructional coaches. *ACM Transactions on Computing Education, 22*(3), 1–22.

Jenkins, H., Purushotma, R., Weigel, M., Clinton, K., & Robison, A. J. (2009). Confronting the challenges of participatory culture: Media education for the 21st century. In *The John D. and Catherine T. MacArthur foundation reports on digital media and learning.* MIT Press.

Jeremijenko, N. (2002-Present). *Feral robotic dogs.* www.nyu.edu/projects/xdesign/feralrobots

Kafai, Y. B. (2016). From computational thinking to computational participation in K–12 education. *Communications of the ACM, 59*(8), 26–27.

Kafai, Y. B., & Burke, Q. (2014). *Connected code: Why children need to learn programming.* MIT Press.

Kafai, Y. B., Lee, E., Searle, K., Fields, D., Kaplan, E., & Lui, D. (2014). A crafts-oriented approach to computing in high school: Introducing computational concepts, practices, and perspectives with electronic textiles. *ACM Transactions on Computing Education (TOCE), 14*(1), 1–20.

Kafai, Y. B., Searle, K., Martinez, C., & Brayboy, B. (2014, March). Ethnocomputing with electronic textiles: Culturally responsive open design to broaden participation in computing in American Indian youth and communities. In *Proceedings of the 45th ACM technical symposium on computer science education* (pp. 241–246). ACM. doi:10.1145/2538862.2538903

Keenan, T. A. (1964). Computers and education. *Communications of the ACM*, 7(4), 205–209.

Keller, E. F. (1984). *A feeling for the organism, 10th anniversary edition: The life and work of Barbara McClintock*. Macmillan.

Knuth, D. E. (1974). Computer programming as an art. *Communications of the ACM*. https://doi.org/10.1145/361604.361612

Krauss, J., Prottsman, K., & Yongpradit, P. (2017). *Computational thinking and coding for every student: The teacher's getting-started guide*. Corwin.

Lachney, M. (2014). Building the Lego classroom. In M. J. P. Wolf (Ed.), *Lego studies: Examining the building blocks of a transmedial phenomenon* (pp. 166–186). Routledge.

Lachney, M. (2017). Culturally responsive computing as brokerage: Toward asset building with education-based social movements. *Learning, Media and Technology*, *42*(4), 420–439.

Lachney, M., Babbitt, W., Bennett, A., & Eglash, R. (2019a). Generative computing: African-American cosmetology as a link between computing education and community wealth. *Interactive Learning Environments*, 1–21.

Lachney, M., Babbitt, W., Bennett, A., & Eglash, R. (2019b). "A voice to talk about it": Cosmetologists as STEM experts in educational technology design and implementation. *EURODL*, *22*(2), 41–55.

Lachney, M., Bennett, A. G., Eglash, R., Yadav, A., & Moudgalya, S. (2021). Teaching in an open village: A case study on culturally responsive computing in compulsory education. *Computer Science Education*, 1–27.

Lachney, M., Boltz, L. O., Dillman, B., Robertson, C., & Yadav, A. (2018). Local classrooms, global technologies: Toward the integration of sociotechnical macroethical issues into teacher education. *Bulletin of Science, Technology & Society*, *38*(1–2), 13–22.

Lachney, M., Eglash, R., Bennett, A., Babbitt, W., Foy, L., Drazin, M., & Rich, K. M. (2021). pH empowered: Community participation in culturally responsive computing education. *Learning, Media and Technology*, 1–22.

Lachney, M., & Foster, E. K. (2020). Historicizing making and doing: Seymour Papert, Sherry Turkle, and epistemological foundations of the maker movement. *History and Technology*, *36*(1), 54–82.

Lachney, M., Ryoo, J., & Santo, R. (2021). Introduction to the special section on justice-centered computing education, part 1. *ACM Transactions on Computing Education (TOCE)*, *21*(2), 1–15.

Lachney, M., & Yadav, A. (2020). Computing and community in formal education. *Communications of the ACM*, *63*(3), 18–21.

Lachney, M., Yadav, A., Drazin, M., Allen, M. C., & Babbitt, W. (2021). Culturally responsive debugging: A method to support cultural experts' early engagement with code. *TechTrends*, 1–14.

Ladson-Billings, G. (2014). Culturally relevant pedagogy 2.0: Aka the remix. *Harvard Educational Review*, *84*(1), 74–84.

Lemerise, T. (1990). Can we integrate logo into the regular mathematics curriculum and sill persevere the logo spirit? *Learning of Mathematics*, *10*(3), 17–19.

Levy, S. (1984). *Hackers: Heroes of the computer revolution*. Anchor Press.

Littlefield, J., Delclos, V. R., Bransford, J. D., Clayton, K. N., & Franks, J. J. (1989. Some prerequisites for teaching thinking: Methodological issues in the study of LOGO programming. *Cognition and Instruction*, *6*(4), 331–366.

López-Quiñones, A., Martinez-Lopez, M., Moreno Sandoval, C. D., Carroll-Miranda, J., Lindala, A. E., Chatman, M. C., . . . & Flores-Reyes, E. (2023). Ancestral computing for sustainability: Centering indigenous epistemologies in researching computer science education. *TechTrends*, 1–11.

Lyles, D., Lachney, M., Foster, E., & Zatz, Z. (2016). Generative contexts: Generating value between community and educational settings. *Teknokultura*, *13*(2), 613–637.

Magenheim, J., & Schulte, C. (2006). Social, ethical and technical issues in informatics – An integrated approach. *Education and Information Technologies*, *11*(3), 319–339. doi:10.1007/s10639-006-9012-6

Magerko, B., Freeman, J., Mcklin, T., Reilly, M., Livingston, E., Mccoid, S., & Crews-Brown, A. (2016). Earsketch: A steam-based approach for underrepresented populations in high school computer science education. *ACM Transactions on Computing Education (TOCE)*, *16*(4), 1–25.

Margolis, J., Estrella, R., Goode, J., Holme, J., & Nao, K. (2008). *Stuck in the shallow end: Race, education, and computing*. MIT Press.

Margolis, J., Estrella, R., Goode, J., Holme, J. J., & Nao, K. (2017). *Stuck in the shallow end*. MIT Press.

Margolis, J., Goode, J., & Chapman, G. (2015). An equity lens for scaling: A critical juncture for exploring computer science. *ACM Inroads*, *6*(3), 58–66.

Margolis, J., Ryoo, J. J., Sandoval, C. D., Lee, C., Goode, J., & Chapman, G. (2012). Beyond access: Broadening participation in high school computer science. *ACM Inroads*, *3*(4), 72–78.

Mayhew, E. J., & Patitsas, E. (2021). *Materiality matters in computing education: A duoethnography of two digital logic educator.*

Mertala, P. (2021). The pedagogy of multiliteracies as a code breaker: A suggestion for a transversal approach to computing education in basic education. *British Journal of Educational Technology*, *52*(6), 2227–2241. doi:10.1111/bjet.13125

Moreno Sandoval, C. D. (2013). Critical ancestral computing: A culturally relevant computer science education. *PsychNology Journal*, *11*(1), 91–112.

Naur, P. (1992). *Computing: A human activity*. ACM.

Negroponte, N. (1995). *Being digital*. Vintage Press.

Noble, D. D. (1991). *The classroom arsenal: Military research, information technology and public education*. Routledge.

Noble, D. F. (2001). *Digital diploma mills: The automation of higher education*. Aakar Books.

Noble, S. U. (2018). *Algorithms of oppression*. New York University Press.

Papert, S. (1980). *Mindstorms: Children, computers, and powerful ideas*. Basic Books.

Papert, S. (1993). *The children's machine: Rethinking school in the age of the computer*. Basic Books.

Paris, D. (2012). Culturally sustaining pedagogy: A needed change in stance, terminology, and practice. *Educational Researcher*, *41*(3), 93–97.

Pea, R. D., & Kurland, D. M. (1984). On the cognitive effects of learning computer programming. *New Ideas in Psychology*, *2*(2), 137–168.

Pinkard, N. D. (1999). Lyric reader: An architecture for creating intrinsically motivating and culturally responsive reading environments. *Interactive Learning Environments*, 7(1), 1–30. doi:10.1076/ilee.7.1.1.3599

Pinkard, N. D., Erete, S., Martin, C. K., & McKinney de Royston, M. (2017). Digital youth divas: Exploring narrative-driven curriculum to spark middle school girls' interest in computational activities. *Journal of the Learning Sciences*, *26*(3), 477–516.

Rankin, Y. A., Thomas, J. O., & Erete, S. (2021). Black women speak: Examining power, privilege, and identity in CS education. *ACM Transactions on Computing Education (TOCE)*, *21*(4), 1–31.

Ray, M., Israel, M., Lee, C., & Do, V. (2018). A cross-case analysis of instructional strategies to support participation of K-8 Students with disabilities in CS for all. In *Proceedings of the 49th ACM technical symposium on computer science education*.

Resnick, M., Maloney, J., Monroy-Hernández, A., Rusk, N., Eastmond, E., Brennan, K., . . . Kafai, Y. (2009). Scratch: Programming for all. *Communications of the ACM*, *52*(11), 60–67.

Resnick, M. (2017). *Lifelong kindergarten: Cultivating creativity through projects, passion, peers, and play*. MIT Press.

Resnick, M., & Ocko, S. (1991). LEGO/Logo: Learning through and about design. In I. Harel & S. Papert (Eds.), *Constructionism* (pp. 141–150). Ablex Publishing.

Ryoo, J., Goode, J., & Margolis, J. (2015). It takes a village: Supporting inquiry-and equity-oriented computer science pedagogy through a professional learning community. *Computer Science Education*, *25*(4), 351–370.

Ryoo, J., Morris, A., & Margolis, J. (2021). What happens to the Raspado man in a cash-free society?" Teaching and learning socially responsible computing. *ACM Transactions on Computing Education (TOCE)*, *21*(2), 1–28.

Salomon, G., & Perkins, D. N. (1987). Transfer of cognitive skills from programming: When and how? *Journal of Educational Computing Research*, *3*(2), 149–169.

Santo, R. (2011). Hacker literacies: Synthesizing critical and participatory media literacy frameworks. *International Journal of Learning and Media*, *3*(3).

Schulte, C., & Budde, L. (2018). A Framework for Computing Education: Hybrid Interaction System: The need for a bigger picture in computing education. In *18th Koli calling international conference on computing education research (Koli Calling '18)*.

Scott, K. A., & Garcia, P. (2016). Techno-social change agents: Fostering activist dispositions among girls of color. *Meridians*, *15*(1), 65–85.

Scott, K. A., Sheridan, K. M., & Clark, K. (2014). Culturally responsive computing: A theory revisited. *Learning, Media and Technology*, *40*(4), 412–436. doi:10.1080/17439884.2014.924966

Scott, K. A., & White, M. A. (2013). COMPUGIRLS' standpoint culturally responsive computing and its effect on girls of color. *Urban Education, 48*(5), 657–681.

Searle, K. A., & Kafai, Y. B. (2015). Boys' needlework: Understanding gendered and indigenous perspectives on computing and crafting with electronic textiles. In *ICER* (pp. 31–39). doi:10.1145/2787622.2787724

Selwyn, N. (2012). *Education in a digital world: Global perspectives on technology and education*. Routledge.

Selwyn, N. (2013). *Distrusting educational technology: Critical questions for changing times*. Routledge.

Selwyn, N. (2014). *Distrusting educational technology: Critical questions for changing times*. Routledge.

Selwyn, N. (2016). *Is technology good for education?* John Wiley & Sons.

Slaton, A. E. (2015). Meritocracy, technocracy, democracy: Understandings of racial and gender equity in American engineering education. In *International perspectives on engineering education* (pp. 171–189). Springer,

Smith, B. E. (2013). *Green computing: Tools and techniques for saving energy, money, and resources*. CRC Press.

Tannert, M., Lorentzen, R. F., & Berthelsen, U. D. (2021). Computational Thinking as Subject Matter: As an Independent Subject or Integrated across Subjects? In A. Yadav & U.D. Berthelsen (Eds.), *Computational thinking in education: A pedagogical perspective* (pp. 73–89). Routledge.

Taylor, R. (1980). *The computer in the school: Tutor, tool, tutee*. Teachers College Press.

Tedre, M., & Denning, P. J. (2017). Shifting identities in computing: From a useful tool to a new method and theory of science. In *Informatics in the Future* (pp. 1–16). Springer.

Tedre, M., Malmi, S., & Malmi, L. (2018). Changing aims of computing education: A historical survey. *Computer Science Education, 28*(2), 158–186.

Traweek, S. (1988). *Beamtimes and lifetimes*. Harvard University Press.

Turkle, S. (1984). *The second self: Computers and the human spirit*. Mit Press.

Turkle, S., & Papert, S. (1990). Epistemological pluralism: Styles and voices within the computer culture. *Signs, 16*(1), 128–157. doi:10.1086/494648

Turkle, S., & Papert, S. (1991). Epistemological pluralism and the revaluation of the concrete. In I. Harel & S. Papert (Eds.), *Constructionism, research, reports, and essays, 1985–1990* (pp. 161–191). Ablex Publishing Company.

Turkle, S., & Papert, S. (1992). Epistemological pluralism and the revaluation of the concrete. *Journal of Mathematical Behavior, 11*(1), 3–33.

U.S. Department of Education. (2020). *Race and ethnicity of public school teachers and their students*. Retrieved November 23, 2021, from https://nces.ed.gov/pubs2020/2020103.pdf

Vogel, S. (2021). "Los Programadores Debieron Pensarse Como Dos Veces": Exploring the intersections of language, power, and technology with bi/multilingual students. *ACM Transactions on Computing Education (TOCE), 21*(4), 1–25.

Warschauer, M., & Ames, M. (2010). Can one laptop per child save the world's poor? *Journal of International Affairs*, 33–51.

Watters, A. (2021). *Teaching machines: The history of personalized learning*. MIT Press.

Wing, J. M. (2006). Computational thinking. *Communications of the ACM, 49*(3), 33–35.

Winner, L. (2009). Information technology and educational amnesia. *Policy Futures in Education*, 7(6), 587–591.

Winner, L. (1986/2020). *The whale and the reactor: A search for limits in an age of high technology* (2nd ed.). University of Chicago Press.

Yadav, A., Connolly, C., Berges, M., Chytas, C., Franklin, C., Hijón-Neira, R., . . . & Warner, J. R. (2022). A review of international models of computer science teacher education. In *Proceedings of the 2022 working group reports on innovation and technology in computer science education*, 65–93. ACM Press.

Yadav, A., & Cooper, S. (2017). Fostering creativity through computing. *Communications of the ACM, 60*(2), 31–33. doi:10.1145/3029595

Yadav, A., Gretter, S., Hambrusch, S., & Sands, P. (2016). Expanding computer science education in schools: Understanding teacher experiences and challenges. *Computer Science Education, 26*(4), 235–254.

Yadav, A., & Heath, M. (2022). Breaking the code: Confronting racism in computer science through community, criticality, and citizenship. *TechTrends*. doi:10.1007/s11528-022-00734-9

Yadav, A., Hong, H., & Stephenson, C. (2016). Computational thinking for all: Pedagogical approaches to embedding a 21st century problem solving in K-12 classrooms. *TechTrends*, *60*, 565–568. doi: 10.1007/s11528-016-0087-7

Zajonc, A. (1984). Computer pedagogy: Questions concerning the new educational technology. *The Teachers College Record*, *85*(4), 569–577.

Zuboff, S. (2019). *The age of surveillance capitalism: The fight for the future at the new frontier of power*. Public Affairs Books.

26

A Selective Review of Computing Education Research

Lauri Malmi and Aditya Johri

1 Introduction

Whether the goal is to simulate, design, or analyze, computational data and algorithms are core to engineering. Consequently, it is common for all engineering students to take at least one or more courses related to computing during their degree program. For example, programming is required for most engineering majors, and so are courses on the use of computational tools such as Matlab™ for engineering tasks. Given this dependence between engineering and computing, it is important that education scholars in both fields develop a better understanding of the other field to help prepare a stronger future engineering and computing workforce and to be able to conduct new and innovative research. Computing Education Research (CER) is a broad field and has a significant research alignment with EER, including a focus on (engineering) epistemologies, learning mechanisms, learning systems, diversity and inclusiveness, and assessment (Adams et al., 2006). Since we cannot do justice to the entire range of work in CER, this chapter is a selective overview of CER for engineering education scholars. A similar review of EER for computing education researchers with some comparison of research traditions and culture of the fields can be found in Loui and Borrego (2019).

The chapter begins with a brief history of computing education, followed by a discussion of the scope of CER, and the main publication venues and practices in CER. The subsequent section briefly discusses the role and use of theoretical frameworks in CER. Thereafter, we present the two subareas of CER that we selected as particularly relevant from an EER perspective: research on programming education and research on software tools to support computing education. A brief conclusion ends the chapter. Our review focuses on research in tertiary or higher education contexts (refer to Chapter 25, this volume, by Yadav and Lachney (2023) for more information on computing education in K–12 level), mainly in the Western cultural context (United States, Europe, Australasia).

2 Brief History and Overview of Computing Education Research (CER)

Computer science (CS) is a young discipline compared to natural sciences and many fields of engineering. Consequently, computing education is a younger field than engineering education. Computing education emerged in the mid-1950s as computers started being used in industry. This

DOI: 10.4324/9781003287483-31

created a need for specialists, resulting in company-level training programs. In the late 1950s, universities started building educational programs for training computing professionals towards specific jobs but without a shared vision of the profession or of education goals, course requirements, or learning resources needed. In the United States, the Association for Computing Machinery (ACM) set up a curriculum committee in 1968 that was explicitly aimed at advancing academic computing programs in universities (Atchison et al., 1968). The initial recommendation emphasized mathematical approaches, but a decade later, new guidelines put an emphasis on a more practice-based, hands-on approach, including programming and applications (Austing et al., 1979). The purview was further broadened when a new curriculum recommendation, in collaboration with IEEE Computer Society, acknowledged the significance of the social and professional context of computing (Tucker & Barnes, 1990). The increasing breadth of the field resulted in independent recommendations for different subareas of computing after 2000. The Computing Curricula (2005) defined five different subareas: computer engineering, computer science, software engineering, information technology, and information systems. Computing Curricula (2020) further added two more areas, cybersecurity and data science. For simplicity, we use the term "computing" to denote any of these fields. However, it must be noted that the focal areas covered in this chapter belong to computer science (CS), and "computing" in common parlance is used as synonym of it.[1]

Denning et al. (1989) characterized computing as a discipline that combines three tightly intertwined aspects, including theory, abstraction (modelling), and design. Tedre and Sutinen (2008, p. 153) built on this when they discussed the characteristics of computing: "Those aspects rely on three different intellectual traditions (the task force called them paradigms): the mathematical (or analytical, theoretical, or formalist) tradition, the scientific (or empirical) tradition, and the engineering (or technological) tradition." This characterization well demonstrates the analogy between computing and engineering as disciplines, which is naturally reflected in CER and EER too. However, despite these similar approaches, computing should not be considered a subdiscipline of engineering. Computing and computing education have been developed both in engineering schools as well as in non-technical higher education institutes. While EER venues have published numerous CER papers, much of CER takes place and is published in venues and communities which are distinct from EER.

The development of CER as a Discipline-Based Education Research (DBER) field can be traced to several related computing education initiatives (Guzdial & du Boulay, 2019; Tedre et al., 2018). First, formal computing education saw scholarly development in academic institutions with research focused on future computing professionals. Second, for decades, there has been research exploring professional computing, as well as research on how children learn computing in K–12 education. Finally, research in the field of human–computer interaction and related areas such as computer-supported cooperative work has given much focus on computing education. The most significant community focusing solely on CER is ACM Special Interest Group in Computing Education (SIGCSE). However, CER papers are also published in many other venues, such as those focusing on software engineering education, human–computer interaction, educational technology, and e-learning, as well as in numerous educational research journals and, as noted earlier, in many EER venues.

Research on professionals in the field has been ongoing at least since the early 1970s. Weinberg's *Psychology of Programming* (Weinberg, 1971) was published in 1971, and in the 1980s, Soloway carried out seminal research on experts' programming plans (e.g., Soloway & Ehrlich, 1984). Important early venues for presenting such research were the Empirical Studies of Programmers (ESP) conferences and Psychology of Programming Interest Group (PPIG) workshops, which both started in 1986. The ESP conferences ceased in the 1990s, but PPIG is still active.

In the last ten years, computing education in schools has boomed internationally, when many countries have included more computing courses and content in K–12 curricula. Research on children's learning of computing also has a long history. For example, the Logo language was designed in the late 1960s, and Papert's (2020) classic book, *Mindstorms*, which addressed children's learning of programming, turtle graphics, and Logo, was first published in 1980. More elaborative discussions of the history of computing and computing education are available in Denning and Tedre (2019), Guzdial and du Boulay (2019), and Tedre et al. (2018).

2.1 Structure of the Field

Currently, the most comprehensive source for work in CER is *The Cambridge Handbook of Computing Education Research* (Fincher & Robins, 2019), which divides research in the field into the four following topical areas.

Systemic issues are topics which persist in the field, including research on novice or introductory programming, more advanced programming, assessment and plagiarism, various pedagogical approaches, as well as questions on equity and diversity.

New milieux address more recent issues that have arisen with computing's spread beyond the "traditional" university settings (formal classrooms and departments of computing). Such work covers research on computational thinking and computing in schools (K–12), as well as computing for other disciplines and new programming paradigms.[2]

Systems software and technology is an area focusing on how software and hardware tools can support learning, including tangible computing and integrated learning environments.

The final section on *teacher and student knowledge* investigates issues that concern the production and acquisition of computing knowledge, such as teacher knowledge, teacher training and professional development, learning outside classrooms, student knowledge and misconceptions, students' motivation, attitudes and dispositions, as well as students as teachers and communicators.

The Cambridge Handbook of CER (Fincher & Robins, 2019) does not give a canonical definition of CER, nor does it prescribe what should be included in the field. This would be difficult, as the field continuously evolves following the rapid development of computing itself, as well as its ever-growing penetration in society. We could, however, give the following characterization as a draft which widely covers the research topics in the field: *CER investigates complex phenomena related to the teaching and learning of computing, including actors (students, teachers, organizations), curricular content, learning resources and technologies, as well as recruitment and retention in both formal and informal educational settings. CER also investigates research in the field itself by analyzing ongoing or published research and develops domain-specific theories and methods to support the field.*

CER is an interdisciplinary field which draws on methods, theories, and content from several disciplines. It leverages research methods and theoretical frameworks from social sciences, especially from educational sciences and psychology, but also from sociology, anthropology, philosophy, and ethics. It also addresses learning content from all areas of computing and widely applies methods and technologies from computing to develop, apply, and analyze software-based learning resources, tools, and data collected from their use.

2.2 Publication Venues

While CER papers are published in numerous conferences and journals, there are a few venues which focus explicitly on computing education. ACM Special Interest Group of Computer Science Education (SIGCSE) was established after publishing the first ACM computing curriculum in 1968, and it launched its first annual conference, SIGCSE Technical Symposium in 1970. Thereafter, SIGCSE has launched several new conferences, which all have their own profile. In addition, there are a number of other conferences organized by other organizations which have started as regional conferences and grown to international venues later. Table 26.1 lists these venues, followed by a few other significant conferences which also publish CER papers among research from other areas.

Table 26.1 Computing Education–Related Conferences and Journals

Acronym	*Conference Title*	*Since*	*Audience*	*Organized at Locations*
ACM SIGCSE Conferences				
SIGCSE	SIGCSE Technical Symposium	1970	Teachers and researchers	US/Canada
ITiCSE	Innovation and Technology in Computer Science Education	1996	Teachers and researchers	Europe
ICER	International Computing Education Research Conference	2005	Researchers	US/Canada, Europe, Australasia
Non-ACM Conferences with primary focus on computing education (*denotes in-cooperation conferences with proceedings in ACM DL)				
ACE*	Australasian Computing Education Conference	1996	Teachers and researchers	Australasia
Koli Calling*	Koli Calling – International Computing Education Research Conference	2001	Researchers	Finland
WIPCSE*	Workshop in Primary and Secondary Computing Education	2006	Researchers	Europe
ISSEP	International Conference on Informatics in Secondary Schools	2006	Teachers and researchers	Europe
Other conferences that also publish computing education–related research				
ICSE	International Conference on Software Engineering			
CHI	ACM CHI Conference on Human Factors in Computing Systems			
SEFI	European Engineering Education Conference			
ASEE	American Society for Engineering Education Annual Conference			
IEEE-associated conferences that also publish computing education–related research				
FIE	Frontiers in Education			
EDUCON	Global Engineering Education Conference			
TALE	Teaching, Assessment, and Learning in Engineering			
EDUNINE	World Engineering Education Conference			
Journals publishing computing education–related research				
ACM Transactions on Computing Education (TOCE)				
Computer Science Education				
IEEE Transactions on Education				
IEEE Transactions on Learning Technologies				
Computers & Education				
Journal of Educational Computing Research				
Journal of Information Technology Education				

Two journals, *ACM Transactions on Computing Education* and *Computer Science Education*, focus solely on publishing work in CER, while there are many others which publish CER papers in addition to other education-related papers. We list some of the more commonly known ones in Table 26.1, but there are many others.

It should be noted that CER-focused conferences follow the publication tradition of computing sciences, where conference papers are considered almost equally valuable scientific contributions as journal papers. Contributions are submitted as full papers and undergo a rigorous review process (for an overview, see Petre et al. (2020)). In the past few years, publishing CER papers has become more competitive, resulting in lower acceptance rates.

3 Theoretical Frameworks in Computing Education Research

As CER has grown as a field, the use of theories from social sciences in research has received considerable attention (Malmi et al., 2014; Lishinski et al., 2016; Szabo et al., 2019; Szabo & Sheard, 2023; Malmi et al., 2020), along with domain-specific theoretical developments within CER itself (Malmi et al., 2019, 2023). The interest reflects the maturation of CER as a DBER field as methodological rigor and use of theoretical contributions come to be valued more (Tenenberg & Malmi, 2022). Also, see Chapter 7, this volume, by Goncher et al. (2023) for more on the use of theory and theoretical frameworks.

Currently, there are no dominant theoretical frameworks within the field; rather, there is a richness of theories that have been adopted. A recent survey by Szabo et al. (2019) extensively reviewed CER papers in the entire ACM digital library, looking for references to a predefined list of 75 learning theories. The top 10 most cited theories were Csikszentmihalyi's flow, learning styles, mental models, self-efficacy theory, progression of early computational thinking, constructivism, problem-based learning, metacognition, mindsets, and communities of practice. These covered roughly two-thirds of the over 15,000 citations found in the library to some learning theory in their list; each of these theories were cited in 4–13% of the total pool of citations. The other 65 theories covered the remaining one-third of citations.

Malmi et al. (2023) explored the development of new domain-specific theoretical constructs that address learning processes, studying, or learning performance in computing. They identified 85 constructs first published in three major CER venues during 2005–2020. Further investigation into how these constructs had been used in papers citing the original paper, however, revealed that only a small fraction of the citing papers actually used the construct to guide further research or developed the construct further. It seems that CER is still in an early phase of developing its own domain-specific theory base. Another common finding in this work was that the terms *theory*, *model*, and *theoretical framework* are often used quite loosely in CER literature (Malmi et al., 2014). Tedre and Pajunen (2023) discuss this rather in depth in their paper. It is also worth noting that in computer science, the concept *theory* is often associated with mathematical theorems and logical proofs that are heavily used in theoretical computer science, algorithms research, and machine learning research, a very different interpretation from theories in social sciences.

In what follows, we give a few examples of how some theoretical frameworks have been applied within CER to address challenges in learning computing. We encourage the reader to look at the original references for more details.

Research of human cognition states that humans' *working memory* capacity is highly limited and it is impossible to address more than a few items at the same time. *Schema theory*, which we discuss more later, gives one explanation for differences between novice and expert programmers; experts, when compared to novices, employ a wide variety of schemas that they can recognize and apply in different abstraction levels. *Cognitive load theory*, a commonly used theory in CER (Duran et al., 2022), can be used to build on this to explain why learning programming is difficult. For novices, programming tasks have many tightly connected topics (e.g., syntax, language constructs and how they are combined to build an algorithm to implement some goal, underlying notional machine, programming tools that must be used) which together cause high *intrinsic load* for the student that is difficult to reduce. On the other hand, *extrinsic load* (concerning presentation of the task and environment) can be reduced. Mayer's *cognitive theory for multimedia learning* (Mayer & Mayer, 2005) provides many guidelines for this when designing learning resources. These guidelines have frequently been applied when developing learning resources for computing.

Motivation theories have received substantial interest in CER, especially in programming education research. For instance, *self-efficacy* has been widely used as a factor associated with programming success and retention (Lishinski & Yadav, 2019). Kinnunen and Simon (2011) carried out a semester-long study in an introductory programming course finding how self-efficacy and emotional reactions varied during the course. Task difficulty and *goal orientation* may moderate these changes. Specific instruments have been developed to measure self-efficacy in computing contexts (Ramalingan & Wiedenbeck, 1998; Danielsiek et al., 2017). *Metacognition* and *self-regulation theories* have been used to explore students' metacognitive strategies (Falkner et al., 2014) for identifying positive associations with programming ability (Bergin & Reilly, 2005). See Loksa et al. (2022) for many other examples.

Bloom and *Solo taxonomies*, which describe and classify knowledge and skills, have been used in research seeking to evaluate programming task complexity. The difficulty of tasks depends on participants' previous knowledge as well as the programming language used. Thus, more elaborate models of knowledge representation have been developed (e.g., Duran et al., 2018). Lister and Teague applied the *Neo-Piagetian framework* to analyze students' progress in programming, thus giving explanations for observations they had made in a longitudinal think-aloud study of novice programmers (Teague & Lister, 2014).

For interested readers, we recommend exploring more examples and references for applying theoretical frameworks in the chapters in *Cambridge Handbook* that discuss theories from learning sciences (Margulieux et al., 2019), cognitive sciences (Robins et al., 2019), as well as motivation, attitudes, and dispositions (Lishinski & Yadav, 2019).

4 Review of Selected Subareas in CER

As discussed in Section 3.1, CER covers a broad selection of topics. Most of these areas have also been addressed in EER. The goal of this chapter is not to contrast the results in CER and EER. Rather, we wish to highlight two characteristic areas of CER which have a long trajectory within it while they are less discussed in EER contexts, though they are relevant there as well. First, *research on programming* knowledge and skills is at the core of CER, and programming, at least in the introductory level, is essential in engineering programs. Second, we will discuss *tools research*, which gives CER a specific profile among other DBER fields, as many CER practitioners develop their own tools (naturally domain-specific educational software, such as simulation tools, are developed and used in other fields as well).

4.1 Research on Programming in CER

CER has, from its inception, had a strong interest in programming education, especially in the introductory programming courses. More advanced aspects of software development education are addressed in research as well, but not to the same extent. The same applies to other topics in computing, such as data structures and algorithms, databases, computing security, human–computing interaction, computing graphics, data mining, machine learning, or artificial intelligence. An obvious reason for this biased emphasis is that novice programming courses are often large in the number of enrolled students, and many students struggle to pass them or drop out. Another reason is that understanding programming, software development, and acquiring good programming skills are core competencies needed for further studies in the field as well as professional work regardless of whether one is directly involved in software projects.

The role of programming for computing studies could be comparable to the role of calculus in engineering studies. While, in professional life, engineers generally use tools, such as Matlab, to carry out computing, understanding the underlying mathematics is still needed.

First-year introductory programming courses are often labeled as Computer Science 1 or 101 and 2 or 201 (CS1/CS2), or some variation thereof. However, there is no common agreement on what these courses should cover. As the field has developed, "advances in the field have led to an even more diverse set of approaches in introductory courses than the models set out in Computing Curriculum, 2001" (ACM/IEEE-CS, 2013). The body of research work on introductory programming is extensive; therefore, we narrowed our scope to rely on two recent review papers. Robins (2019) presents an extensive summary of various focal areas in research on introductory programming and looks at the history of the research from the 1970s until the current day; Luxton-Reilly et al. (2018) present a systematic literature survey in introductory programming education, identifying 1,666 papers addressing this area published in numerous journals and conferences between 2003 and 2017. In the following, we discuss some of the main challenges identified in research. The list is not comprehensive, and we recommend readers to search for more results from the aforementioned survey papers and the cited original research papers.

4.1.1 Variation in Learning Outcomes

For decades, it has been widely agreed that learning programming is difficult, which is demonstrated by high failure and drop-out rates in introductory courses. In a survey by Bennedsen and Caspersen (2007), data was collected from 63 institutions internationally, including numbers of enrolled students who withdraw from the course, skip the final exam, or sit and fail. The results showed an average failure rate of roughly 33% and often up to 50% or higher. Subsequently, Watson and Li (2014) surveyed literature relating to 51 institutions and reported very similar results. In 2019, a new study was carried out, which compared pass rates in introductory courses in STEM disciplines, including computing (Simon et al., 2019). In their findings, the pass rates in computing were, on average, 75%, which was on the low end of this comparison, but not alarmingly low. The results of these studies must be interpreted with some caution due to issues with data and large institutional differences; however, the low pass rate of students taking programming is a consistent finding.

While low pass rates are a serious concern, there are other equally relevant challenges. There is much evidence that learning among students who pass basic courses is weak, that is, it is relatively short-term or conceptually deficient. This was established in the 1980s (Soloway et al., 1983; Kurland et al., 1989) and reaffirmed by McCracken et al. (2001) when their large multinational study demonstrated serious deficiencies in students' ability to solve common problems that students in any type of CS program should hypothetically be able to solve. The study was repeated ten years later, giving better support for students (i.e., a test harness they could use to check their code correctness), finding both poorly and well-performing student groups (Utting et al., 2013). In another large multinational study, Lister et al. (2004) investigated students' tracing skills, that is, how well students could read code and trace its execution, and they found much evidence of weak and fragile learning.

A common finding is that in many programming courses, the results are not normally distributed. In addition to low-performing students, typical CS1 courses have high rates of very well-performing students, and the grade distribution can have two clearly separate peaks, often called "bimodal" distribution (Robins, 2019). This complicates the picture and challenges the simple conclusion that programming is just difficult to learn (Kölling, 2009). Some explanations given to this phenomenon include the cumulative nature of the learning content in CS1. If one fails to learn enough in the early weeks of the course, one faces more and more troubles later because the whole course is building on the previously covered content in the course (Robins, 2010). Moreover, Höök and Eckerdal (2015) presented a finding that students who failed an exam had studied considerably less time at the computer than did well performing students, and despite their interest in the topic, they failed to learn programming by focusing on lectures and reading the course textbook. Pedagogical approaches that build on weekly compulsory exercises with adequate feedback and support can address this challenge.

On the other hand, the whole finding of bimodality has also been questioned by carrying out a rigorous analysis of grade distributions and noting that the claim may reflect instructors' confirmation bias and beliefs of their students (Patitsas et al., 2019).

4.1.2 Difficulty Due to Complex Learning Goals

Du Boulay (1986) summarized the main challenges to learning programming as follows. First, there is *orientation*, what programming is for and what kind of problems can be addressed with this skill. Second, programs are abstract entities; to understand program execution, one must understand the underlying *notional machine*, that is, an abstract description of program execution (Sorva, 2013). Third, programs are written in formal languages with strict syntax and semantics, which regulate what kind of programs are syntactically correct and what operations are acceptable. Fourth, programming proficiency requires developing knowledge of and applying a large number of programming *schemas*, that is, standard methods for implementing common goals. Finally, programming tasks must be carried out with tools (compilers, interpreters, integrated development environments, debuggers, etc.). Though introductory courses largely focus on the used programming language, the other topics listed previously cannot be ignored or fully isolated, which increased the complexity of learning for students. Cognitive load theory (Sweller et al., 1998; Sweller, 2010) provides an explanation for these challenges as *high element interactivity*, the extent to which the task involves interacting elements that must be held in working memory simultaneously. The situation thus causes high intrinsic load for students, and it is difficult to reduce this load.

Robins et al. (2003) reviewed programming education literature and summarized the following learning goals and challenges in programming. Students need to acquire *knowledge* of programming language and tools, they need to learn *strategies* on how to apply this knowledge appropriately, and they need to construct and compare *mental models* of program state, that is, what happens "under the hood" when a program is executed. Moreover, all these aspects are relevant in three phases, *designing* the program, *implementing* it, and finally, *evaluating* the result, which covers testing and debugging the program. Achieving these goals is not a minor task for students.

Whereas most papers have historically considered the challenge of teaching programming from a teacher's point of view, researchers have also investigated students' perspectives and what they have perceived as difficulties. Compared with communicating with a natural language, which is flexible for presenting things, programs are written in formal languages, which have a strict syntax. Indeed, syntax errors are frequently reported as challenges for students (e.g., Robins et al., 2006). Moreover, it is challenging for students to write programs that are without error or ambiguity, and students often have difficulties understanding the task, designing the program structure, and using some language constructs, such as loops and arrays (Robins et al., 2006). In a large-scale study involving six European universities, students reported that the most difficult aspects of programming were understanding how to design a program to solve a certain task, how to divide functionality into procedures, and how to find bugs in their own programs (Lahtinen et al., 2005).

4.1.3 Novice and Expert Knowledge Differences

Soloway and Spohrer (1989) explored the differences between novice and expert programmers. Their findings indicated that novices had deficiencies in understanding some key language constructs, such as loops, arrays, and recursion. Winslow (1996) found that novices were often limited to surface and superficially organized knowledge. They lacked adequate programming schemas and mental models, and they considered programs line by line instead of focusing on larger meaningful structures. On the other hand, programming experts have a large selection of schemas (also called chunks, plans, or scripts) which organize knowledge. For example, there are schemas that store data into an array, schemas that browse array content, and schemas that find the largest item in the array. When reading, tracing, and writing code, experts can operate with these schemas, which can be very complex, while novices struggle with low-level language constructs and how to assemble or put things together. Thus, learning programming could be described as a process of creating, applying, modifying, combining, and evaluating schemas (Rist, 2004). Furthermore, expert programmers can operate with schemas in very different abstraction levels; for example, when managing bit operations in low-level programming, managing multidimensional arrays, implementing graph algorithms, finding and using various available library functions, designing meaningful class structures, or selecting appropriate software architecture models. For novices, it takes years of learning and practice to build such a selection of multilevel schemas which they could use efficiently. An additional challenge is that schemas are often language-dependent because different programming languages provide different constructs for implementing similar goals. Thus, knowledge of schemas learned with one language does not necessarily transfer when learning a new language (Kao et al., 2022).

Closely related to schemas are *mental models*, a concept adopted from cognitive science and widely used in CER. Mental models are personal internal models of how something works. Greca and Moreira (2000, p. 5) contrast them with teachers' *conceptual models*, as follows:

> [C]onceptual models are precise and complete representations that are coherent with scientifically accepted knowledge. That is, whereas mental models are internal, personal, idiosyncratic, incomplete, unstable and essentially functional, conceptual models are external representations that are shared by a given community, and have their coherence with the scientific knowledge of that community. These external representations can materialize as mathematical formulations, analogies, or as material artifacts.

Viable mental models can be useful, as they provide means for explaining and predicting interaction of subjects with the world. However, mental models are implicit, incomplete, imprecise, and sometimes inconsistent with conceptual models. Not all mental models are viable, which can challenge novice students trying to comprehend program execution. Ma et al. (2007) explored the viability of students' mental models in a CS1 course in Java and found that one-third of students had non-viable mental models of value assignment and only one-sixth had a viable mental model of reference assignment. Considering how central these constructs are in programming, it is understandable how novices struggle in tracing program execution.

Ben-Ari (2001, pp. 56, 60), in his critique of constructivism in computer science education, argued that "a model of a computer . . . must be explicitly taught and discussed, not left to haphazard construction and not glossed over with facile analogies," because "novice computer science students have no effective model of a computer," and "the computer forms an accessible ontological reality." Program execution can be taught with the help of a *notional machine* that is an abstract conception of how software and hardware are working during program execution. Fincher et al. (2020, p. 22) define *notional machine* (NM) as follows:

> An NM has a pedagogical purpose, its generic function is to draw attention to, or make salient, some hidden aspect of programs or computing. It will have a specific focus within programs or computing, and will adopt a particular representation that highlights specific aspects of the focus.

Notional machines can be presented at different abstraction levels, and they are language-dependent (Sorva, 2013). In teaching, they are often presented with visualizations, which abstract away details and allow students to grasp the dynamic process and how data is presented and manipulated in the computer memory.

4.1.4 Challenges in Development of Programming Skills

A highly important aspect of programming is that it is a dynamic process. Basic knowledge of language constructs and schemas is not sufficient. Conceptual knowledge ("knowing what") goes hand in hand with practical knowledge, that is, strategies and skills to apply it ("knowing how"). Practices include, for example, utilizing problem-solving strategies, using patterns or analogies in design, evaluating the impact of the structure of the program, implementing or designing algorithms, understanding the pros and cons of different data structures and algorithms, selecting programming tools, and applying testing and debugging strategies. Experts can apply a wide selection of strategies, while novices use only a small set of rudimentary strategies (Robins et al., 2003). Eckerdal (2009) summarizes that concepts and practices are equally important parts of learning goals, and they are equally difficult for students to learn.

Perkins et al. (1989) studied novices' programming process and found three types of behavior groups: stoppers, movers, and tinkerers. Stoppers simply stopped and gave up when facing difficulties or lacking clear directions on how to proceed. Movers, on the other hand, kept trying, experimenting by modifying their code to try to find ways forward. Tinkerers also modified code frequently, but without understanding it, working more or less randomly. These results reflect students' insufficient pool of strategies for addressing challenges they face in programming.

It is important to understand that reading and writing code are separate, though related, skills. Reading and tracing skills (on code execution) are prerequisite for code writing (Xie et al., 2019). If they are not mastered well enough, students' coding process may end up in endless tinkering. Students write code, but when they do not fully understand how it works, they may try random changes in the code when facing difficulties in getting it to work correctly. Moreover, code-reading skills are essential in later phases of study and in professional work in the context of code reviews and code maintenance.

How students learn code reading and tracing is not yet well understood. Luxton-Reilly et al. (2018, p. 60) suggest that "[g]iven the evidence that has been found for the value of code-reading skills in novice programmers, there is an ongoing need to explore further ways of encouraging the development of these skills."

4.1.5 Factors Influencing Students' Learning

Researchers have sought to find out which student-related factors influence students' learning. For decades, there have been observations that some professional programmers are much more productive than others. This has led to assumptions that programming skill is an innate characteristic for some people, a claim for which later work has found little support. There has been a lot of research which has tried to identify factors predicting academic success, including success in learning programming (Hellas et al., 2018). However, the results concerning programming are inconclusive. Robins (2019, p. 349) summarizes, "In short, no factor or combination of factors which clearly predict success in learning a first programming language has been found."

Despite this, there is considerable work investigating the role of various psychological factors associated with success in learning programming (Lishinski & Yadav, 2019; Malmi et al., 2020), in line with similar research that examines learning in other domains. Much work has addressed students' *self-efficacy* in programming contexts, finding out that higher levels of self-efficacy are associated with greater student performance in computing, which is analogous to findings in other STEM fields (Lishinski & Yadav, 2019). Self-efficacy varies during the first programming course due to emotional reactions to work done (Kinnunen & Simon, 2011). Another relevant factor is Dweck's notion of *mindset*, one's belief whether one can grow and develop. Some students risk developing a fixed mindset, thus failing to believe that they can learn programming well (Murphy & Thomas, 2008). *Engagement* and *self-regulated learning* have also received considerable attention in research literature, and there are numerous attempts to explore the impact of various pedagogical interventions on student engagement (Luxton-Reilly et al., 2018, p. 63; Loksa et al., 2022).

While the aforementioned factors are theory-based, substantial research has focused on analyzing students' programming processes, which can be carried out by analyzing log data from programming environments and automatic assessment systems, even at the keystroke level. Much of this work falls under *learning analytics* and *educational data mining* (Grover & Korhonen, 2017). There are some large-scale data repositories (e.g., Brown et al., 2018) which have supported a generation of statistical models on programming behavior. Much of this analysis has focused on seeking to identify students at risk of dropping out, and some results indicate that behavior-based analysis is better in this sense than test-based analysis (Watson et al., 2014).

4.1.6 Teaching Approaches in Programming

Finally, another area that should be briefly mentioned here is teaching methods and approaches. For the interested reader, we recommend the reviews by Falkner and Sheard (2019) and Luxton-Reilly et al. (2018, Section 6). Two unique and relevant CER-related approaches are discussed here. *Pair programming* is a technique frequently used in professional programming, where two people work together on the same program. One person (driver) writes and edits the code, while the other one (navigator or observer) reviews the written code. Roles are switched regularly. Empirical research has reported that pair programming provides students with better support to produce a higher quality of work and improves pass rates among students with low academic performance. It also improves student enjoyment, though conflicts may still appear. For more-experienced students, there is less evidence of benefits. Forming the pairs with the right balance of skills is important (Luxton-Reilly et al., 2018; Umapathy & Ritzhaupt, 2017). *Peer reviewing* of code, when carried out in collaborative settings, also shows clear benefits in identifying errors and discussing higher-level design and implementation issues (Hundhausen et al., 2009).

Media computation (Guzdial, 2003, 2016) is an approach targeted initially to non-computer science majors to set up a motivating application context where computation skills also increase creativity. Media, including text, images, sound, and videos, is now digital data and thus provides ample possibilities for meaningful programming tasks. Empirical evaluation studies have found evidence that media computation increases retention and increases the sense of relevance of programming studies (Guzdial, 2013).

4.2 Research and Development of Tools and Software-Based Learning Resources in CER

Since the inception of the CER field, many computing educators have developed their own software tools to support their own teaching and their students' learning. This line of work continues to this day, and it is a fairly distinctive characteristic of computing education and CER, given the sheer volume of such software.[3] Valentine (2004) analyzed 20 years of papers (1984–2003) addressing CS1/CS2 courses in SIGCSE Technical symposium and found that, out of the 444 papers in this pool, 22% focused on various self-developed tools to support education. A broader review by Luxton-Reilly et al. (2018) identified over 250 papers addressing tools for introductory programming education with a growing trend towards tool development. Overall, research on tools is a mix of educational development, software development, and empirical research. Even though many papers in this area lack rigorous empirical analysis, it is important to examine tools research as this work sheds light on the CER field more broadly. Furthermore, several of the tool categories presented in what follows address analogical goals which are present in engineering education too. We discuss this more at the end of the section.

The use of tools in computing education can be explained by two forces: (1) adequate software development skills among computing researchers and faculty to develop educational software and (2) easy access to skilled expertise in the form of students who support development through capstone projects, summer internships, or thesis work. While educational software is certainly developed in other fields as well, teachers in those fields are likely to have less options for developing their own tailored software. Although the development and use of tools is common, most tools are used only in the context where they were originally developed, and very few have gained a wide international dissemination, when we exclude commercial tools.[4] Some notable exceptions[5] include BlueJ for learning object-oriented programming in Java, Python Tutor for interactive visualization of program execution for multiple programming languages, JFLAP for learning theoretical computer science topics, and Web-CAT for automatic assessment of programming exercises (Edwards & Perez-Quinones, 2008).

Tools can be categorized broadly based on their technical underpinnings or their functional use. From a technical standpoint, there are tools that are *software applications*, which can be downloaded and installed, such as BlueJ and Web-CAT. Then, there are a number of tools that provide some *service at a specific website*, such as Python tutor or PeerWise[6] (Denny et al., 2008), where students can generate multiple-choice questions for other students and respond to and rate the available questions. In addition to these popular services, there are hundreds of small interactive applications that demonstrate the workings of individual concepts, such as a particular data structure or sorting algorithm, which can be found online. A third category of tools is *software frameworks*, which support building smart, interactive learning content (Brusilovsky et al., 2014). An example is the Jsvee framework (Sirkiä, 2016) that enables building program visualizations of the execution of Python programs. Another one is jsParsons (Helminen et al., 2012), a tool for building Parsons problems, also called *programming puzzles*. In this learning activity, textual program code is split into sentence-level visual blocks which are given to the student in random order, and the student should drag and drop the blocks through a direct manipulation interface into the right order to generate a working program. Finally, even *programming languages* can be considered tools; Logo (Solomon et al., 2020) and Pascal (Wirth, 1971), for instance, were initially designed to be simple programming languages targeted at children or novice programmers. Modern block-based languages, such as Scratch™ and App Inventor™, have a similar goal.

4.2.1 Functional Categories of Tools

From a functional perspective, *educational programming environments* include tools specifically tailored for programming education, which aim for simplification by excluding most of the complexities in professional tools. Specific environments have been developed for many programming languages, mostly for C, C++, Java, and Python (see Luxton-Reilly et al., 2018, p. 74). A comparable group of tools includes various *libraries* and *application programming interfaces (APIs)*, which have been developed to simplify complexities of professional languages. For instance, implementing graphical user interfaces with Java using basic libraries, such as Swing, is a fairly complex process; therefore, many simplified libraries focusing on basic operations have been developed for programming courses (e.g., DoodlePad, Squint for Java, and cs1graphics for Python).[7] Finally, a natural part of programming education is learning to use professional development tools; typical examples of these are integrated development environments, such as Eclipse, IntelliJ IDEA, and Visual Studio, as well as basic command-level tools, such as gcc or javac.[8] However, there are no conclusive results on whether students learn better in a professional or educational environment (Luxton-Reilly et al., 2018).

The second category of tools based on functionality includes *visualization tools* that demystify the software and help the user understand the structure of complex software and the execution process, which are inherently abstract and invisible for the user. They are tools that can be used to teach and learn notional machines for a specific programming language.

There are two kinds of visualization tools: *program visualization tools* and *algorithm visualization tools*. Program visualization tools, for example, Python Tutor, visualize how code execution proceeds step by step or in larger steps and how the memory content, that is, states of variables, program's runtime stack, and heap change during the execution (see Sorva et al., 2013). Algorithm visualization tools visualize execution on a more abstract level and typically show dynamic visualizations of data structures that a specific algorithm is manipulating; for example, demonstrating how a sorting algorithm switches contents of an array or how a search algorithm proceeds in a binary search tree (see Shaffer et al., 2010). An important aspect of both program and algorithm visualization is how students interact with the learning content. In a meta-analysis of empirical research on algorithm visualization, Hundhausen et al. (2002) found that students who worked actively with visualization – to edit, change, adapt it – learned better compared with those who just viewed it. Overall, visualization tools respond to the need for simplifying and making the learning of programming more accessible for students developing this skill.

A wide category of tools focuses on *automatic assessment or feedback* on students' academic work. Such tools have been extensively used to address challenges in programming education. For reviews, see Ala-Mutka (2005), Ihantola et al. (2010), Keuning et al. (2018), Lajis et al. (2018), and Paiva et al. (2022). Introductory programming courses (not only MOOCs[9]) are very large in many institutes, ranging from hundreds to even thousands of enrolled students. Automatic assessment tools can be used to address a very large share of the work needed for assessment and giving formative or summative feedback on students' solutions to programming exercises. This enables teachers to give more guidance to students when they solve the exercises or do not understand or disagree with the automatic feedback.

Automatic assessment tools are most often used to check program correctness, that is, whether a student's program passes test cases defined by the teacher. However, some tools have also been developed to give feedback on programming style, program structure, use of specific programming language constructs, program runtime efficiency, or how well the program has been tested. Some tools work on a more abstract level, giving feedback on algorithmic tasks (Malmi et al., 2004). Automatic and human assessment can also be combined for better support of learning (Ala-Mutka et al., 2004; Novak et al., 2019); for example, the system can check that a student's program passes given requirements before it is forwarded to the teacher for closer evaluation.

The previous examples emphasized the teacher's point of view of using tools. From students' point of view, this group of tools provides several benefits: (1) the systems allow students to revise their solution based on the feedback and resubmit their work several times; (2) the feedback from the tool is available immediately anywhere they are working with an Internet connection; and (3) the feedback is available anytime 24/7, unless the teachers wish to limit the time. Some tools provide more specific advantages, such as enhanced error messages. A well-known problem for novice programmers is struggling with programming language syntax, and a part of the problem is that compiler error messages are not always easy to comprehend. Thus, some tools provide students tailored *enhanced compiler error messages*. Despite the seemingly obvious advantages, the results for improved student performance are still inconclusive (Pettit et al., 2017). *Drill-and-practice systems* allow students to train their skills with a systematically ordered set of exercises or lessons. CodeWorkout and CodeWrite are examples of such systems; the latter one includes an additional feature that students can themselves create new exercises with test cases, and the new exercise can be rated by other students (Edwards & Murali, 2017; Denny et al., 2011). *Intelligent tutoring systems (ITS)* provide additional context and adaptive feedback for students during the programming process. They can, for example, integrate additional learning resources, for example, videos and exercises, in a structured order that matches learning objectives and support problem-solving (Gross & Pinkwart, 2015), or they can track student progress and adapt their feedback accordingly (Pullan et al., 2013). Modern trends include integrating AI techniques into the programming system to provide improved hints for students (e.g., Rivers & Koedinger, 2017). Recent advances in AI technologies can even solve typical CS1 course programming exercises, which creates new challenges for programming pedagogy (Finnie-Ansley et al., 2022).

One specific aspect of tools research concerns *e-books*, that is, interactive online books, which integrate a mix of smart learning content with static content, such as text, images, and videos. In this context, the interactive components could be automatically assessed exercises, code visualization elements, interactive code execution demonstrations where students can modify the code, algorithm simulation exercises, etc. Examples of such learning resources are the CS principles book[10] for learning programming (Ericson et al., 2016) and OpenDSA[11] (Fouh et al., 2014) for learning data structures and algorithms. While these resources are highly valuable in education, building and maintaining such resources is more complex than authoring traditional printed textbooks or static online books.

For more information of tools for the interested reader, we recommend Luxton-Reilly et al. (2018, Section 6.4) and Malmi et al. (2019).

4.2.2 Challenges in Tools Research and Implementation

While tools can greatly support both students' and teachers' work, there are also significant challenges involved in developing and using them. Developing new learning resources which heavily employ smart learning content requires resource-intensive software engineering practices (Haaranen et al., 2020). In particular, sustained use of any software requires maintenance and persistent updates to the technology environment, which include installing new hardware, new versions of operating systems or other system software, bug fixes, or improving data and communication security, as well as keeping the systems compatible with other in-house or external systems. A second challenge is reliability of services. For example, if the automatic assessment system breaks for any reason during the weekend or just before submission deadlines, it is likely difficult to organize support at a short notice. The alternative is to use commercial providers, which includes costs and reduced opportunities for tailored system development. It is also worth recognizing that, if the teacher has invested significant effort into developing software-based learning resources based on products from a specific vendor, there exists a risk of becoming too dependent on the service. Vendor lock-in implies that work cannot be transferred to another system without significant new effort if the service shuts down at a short notice or the vendor's cost policy changes, making costs become too high.

4.2.3 Perspectives for Tools Research in Engineering Education

The previous presentation of tools research included several categories where analogical tools exist or can be developed for engineering education. Automatic assessment tools can be used naturally in many cases where engineering students submit program code. On the other hand, when exercises deal with mathematical expressions, automatic assessment tools developed for mathematics education, for example, Stack[12] or Numbas,[13] may apply. Regardless of tools, there are similar types of benefits and challenges for teachers, and engineering education may benefit from pedagogical results of these kind of tools in CER literature.

Program and algorithm visualization tools are used for visualizing dynamic processes, and there has been substantial work to investigate their pedagogical use cases. Dynamic processes and simulations are present in many areas of engineering. Despite the different visual presentations, the pedagogical findings may apply across both fields, and they may learn from each other. Intelligent tutoring systems can naturally be developed for engineering education contexts, and the technologies on which they are built are applicable and can be adapted to present and analyze engineering knowledge. Finally, within computing education research, many technological solutions have been built for interactive e-books, such that can be applied for building e-books in engineering domains.

5 Conclusion

In this chapter, we provided a selective overview of CER, highlighting a few unique and important aspects of the work in the field, including programming, tools, and environments. Through this review, we want to present a window into the growing field of CER and hope that this chapter will appeal to newcomers in both CER and EER, as well as allow those who are familiar with either (or both) of the fields to find common ground for future research and scholarship. We acknowledge that in a single chapter we can cover only a small share of work in CER, and therefore, we have given many references to review papers and other relevant research as examples of the work carried out in CER. We encourage readers to familiarize themselves with the examples in the original reviews and publications.

Acknowledgments

This work was partly supported by a Nokia-Fulbright Distinguished Chair award and by US National Science Foundation Awards No. EEC-1941186, 1937950, 1939105. Any opinions, findings, and conclusions or recommendations expressed in this material are those of the authors and do not necessarily reflect the views of the funding agencies.

Notes

1 *Computing* and *computer science* are widely used terms in the English-speaking world. Other terms are also used elsewhere, such as *informatics or information technology*. For simplicity, we use the previous ones systematically.
2 Programming paradigms basically characterize the principles of how programs are structured. Commonly identified paradigms include procedural programming, object-oriented programming, functional programming, and declarative programming. As modern programming languages support different ways of building programs, the whole concept of paradigm has been questioned (Krishnamurthi & Fisler, 2019).
3 Computing teachers naturally use many other educational technologies, too, including generic learning management systems (LMS) and tools for specific learning activities, for example, discussion forums or data visualization, but we are not discussing them here.
4 Some commercial tools have originally been developed in universities, before they were commercialized, such as Blackboard (originally WebCT), or CodeGrade (www.codegrade.com/).
5 www.bluej.org/; https://pythontutor.com/; www.jflap.org/; https://web-cat.org/.
6 https://peerwise.cs.auckland.ac.nz.
7 https://doodlepad.org/; http://dept.cs.williams.edu/~tom/weavingCS/s07/doc/squintDoc/; www.cs1graphics.org/.
8 https://www.eclipse.org; www.jetbrains.com/idea/; https://visualstudio.microsoft.com/.
9 There is a lot of research on MOOCs, including computing-specific MOOCs, most of which is disseminated in aligned fields, such as educational data mining and learning analytics.
10 https://runestone.academy/runestone/books/published/StudentCSP/index.html. Runestone Academy (https://runestone.academy/ns/books/index) has a wide set of free e-books for computer science.
11 https://opendsa-server.cs.vt.edu/.
12 https://stack-assessment.org/.
13 www.numbas.org.uk/.

References

ACM/IEEE – CS Joint Task Force on Computing Curricula. (2013). *Computer science curricula 2013*. ACM Press and IEEE Computer Society Press.

Adams, R., Aldridge, D., Atman, C., Barker, L., Besterfield-Sacre, M., Bjorklund, S., & Young, M. (2006). The research agenda for the new discipline of engineering education. *Journal of Engineering Education, 95*(4), 259–261.

Ala-Mutka, K. M. (2005). A survey of automated assessment approaches for programming assignments. *Computer Science Education, 15*(2), 83–102.

Ala-Mutka, K., & Jarvinen, H. M. (2004). Assessment process for programming assignments. In *IEEE international conference on advanced learning technologies, 2004. Proceedings*. (pp. 181–185). IEEE.

Atchison, W. F., Conte, S. D., Hamblen, J. W., Hull, T. E., Keenan, T. A., Kehl, W. B., . . . Young Jr, D. M. (1968). Curriculum 68: Recommendations for academic programs in computer science: A report of the ACM curriculum committee on computer science. *Communications of the ACM, 11*(3), 151–197.

Austing, R. H., Barnes, B. H., Bonnette, D. T., Engel, G. L., & Stokes, G. (1979). Curriculum'78: Recommendations for the undergraduate program in computer science – a report of the ACM curriculum committee on computer science. *Communications of the ACM, 22*(3), 147–166.

Ben-Ari, M. (2001). Constructivism in computer science education. *Journal of Computers in Mathematics and Science Teaching, 20*(1), 45–73.

Bennedsen, J., & Caspersen, M. E. (2007). Failure rates in introductory programming. *ACM SIGCSE Bulletin, 39*(2), 32–36.

Bergin, S., & Reilly, R. (2005). Programming: Factors that influence success. *ACM SIGCSE Bulletin, 37*(1), 411–415.

Brown, N. C., Altadmri, A., Sentance, S., & Kölling, M. (2018, August). Blackbox, five years on: An evaluation of a large-scale programming data collection project. In *Proceedings of the 2018 ACM conference on international computing education research* (pp. 196–204).

Brusilovsky, P., Edwards, S., Kumar, A., Malmi, L., Benotti, L., Buck, D., Ihantola, P., Prince, R., Sirkiä, T., Sosnovsky, S., & Urquiza, J. (2014). Increasing adoption of smart learning content for computer science education. In *Proceedings of the working group reports of the 2014 on innovation & technology in computer science education conference* (pp. 31–57).

Computing Curricula. (2005). *ACM*. www.acm.org/education/curricula-recommendations

Computing Curricula. (2020). *ACM*. www.acm.org/education/curricula-recommendations

Danielsiek, H., Toma, L., & Vahrenhold, J. (2017). An instrument to assess self-efficacy in introductory algorithms courses. In *Proceedings of the 2017 ACM conference on international computing education research (ICER '17)* (pp. 217–225). ACM.

Denning, P. J. (Chairman), Comer, D. E., Gries, D., Mulder, M. C., Tucker, A., Turner, A. J., & Young, P. R. (1989). Computing as a discipline. *Communications of the ACM, 32*(1), 9–23.

Denning, P. J., & Tedre, M. (2019). *Computational thinking*. MIT Press.

Denny, P., Hamer, J., Luxton-Reilly, A., & Purchase, H. (2008). PeerWise: Students sharing their multiple choice questions. In *Proceedings of the fourth international workshop on computing education research* (pp. 51–58).

Denny, P., Luxton-Reilly, A., Tempero, E., & Hendrickx, J. (2011). Codewrite: Supporting student-driven practice of java. In *Proceedings of the 42nd ACM technical symposium on Computer science education* (pp. 471–476).

Du Boulay, B. (1986). Some difficulties of learning to program. *Journal of Educational Computing Research, 2*(1), 57–73.

Duran, R., Sorva, J., & Leite, S. (2018). towards an analysis of program complexity from a cognitive perspective. In *Proceedings of the 2018 ACM conference on international computing education research* (pp. 21–30).

Duran, R., Zavgorodniaia, A., & Sorva, J. (2022). Cognitive load theory in computing education research: A review. *ACM Transactions on Computing Education (TOCE), 22*(4).

Eckerdal, A. (2009). *Novice programming students' learning of concepts and practice*. (Doctoral dissertation). Acta Universitatis Upsaliensis.

Edwards, S. H., & Murali, K. P. (2017). CodeWorkout: Short programming exercises with built-in data collection. In *Proceedings of the 2017 ACM conference on innovation and technology in computer science education* (pp. 188–193).

Edwards, S. H., & Perez-Quinones, M. A. (2008). Web-CAT: Automatically grading programming assignments. In *Proceedings of the 13th annual conference on innovation and technology in computer science education* (pp. 328–328).

Ericson, B. J., Rogers, K., Parker, M., Morrison, B., & Guzdial, M. (2016). Identifying design principles for CS teacher Ebooks through design-based research. In *Proceedings of the 2016 ACM conference on international computing education research* (pp. 191–200).

Falkner, K., & Sheard, J. (2019). Pedagogic approaches. In S. A. Fincher & A. V. Robins (Eds.). *The Cambridge handbook of computing education research*. Cambridge University Press.

Falkner, K., Vivian, R., & Falkner, N. (2014). Identifying computer science self-regulated learning strategies. In *Proceedings of the 2014 conference on innovation & technology in computer science Education (ITiCSE '14)* (pp. 291–296). ACM.

Fincher, S., Johan Jeuring, J., Miller, C., Donaldson, P., Du Boulay, B. Hauswirth, M., Hellas, A., et al. (2020). Notional machines in computing education: The education of attention. In *Proceedings of the working group reports on innovation and technology in computer science education* (pp. 21–50).

Fincher, S. A., & Robins, A. V. (Eds.). (2019). *The Cambridge handbook of computing education research*. Cambridge University Press.

Finnie-Ansley, J., Denny, P., Becker, B. A., Luxton-Reilly, A., & Prather, J. (2022). The robots are coming: Exploring the implications of OpenAI codex on introductory programming. In *Australasian computing education conference* (pp. 10–19).

Fouh, E., Karavirta, V., Breakiron, D. A., Hamouda, S., Hall, S., Naps, T. L., & Shaffer, C. A. (2014). Design and architecture of an interactive eTextbook – The OpenDSA system. *Science of Computer Programming, 88*, 22–40.

Goncher, A., Hingle, A., Johri, A., & Case, J. (2023). The role and use of theory in engineering education research. In A. Johri (Ed.), *International handbook of engineering education research* (pp. 137–155). Routledge.

Greca, I. M., & Moreira, M. A. (2000). Mental models, conceptual models, and modelling. *International Journal of Science Education, 22*(1), 1–11.

Gross, S., & Pinkwart, N. (2015, July). towards an integrative learning environment for Java programming. In *2015 IEEE 15th international conference on advanced learning technologies* (pp. 24–28). IEEE.

Grover, S., & Korhonen, A. (2017). Unlocking the potential of learning analytics in computing education. *ACM Transactions on Computing Education, 17*(3), (pp. 1–4).

Guzdial, M. (2003, June). A media computation course for non-majors. In *Proceedings of the 8th annual conference on innovation and technology in computer science education* (pp. 104–108).

Guzdial, M. (2013, August). Exploring hypotheses about media computation. In *Proceedings of the ninth annual international ACM conference on international computing education research* (pp. 19–26).

Guzdial, M. (2016). Learner-centered design of computing education: Research on computing for everyone. *Synthesis Lectures on Human-Centered Informatics, 8*(6), 1–165.

Guzdial, M., & du Boulay, B. (2019). The history of computing education research. *The Cambridge Handbook of Computing Education Research (2019)*, pp. 11–39.

Haaranen, L., Mariani, G., Sormunen, P., & Lehtinen, T. (2020). Complex online material development in CS courses. In *Koli calling'20: Proceedings of the 20th Koli calling international conference on computing education research* (pp. 1–5).

Hellas, A., Ihantola, P., Petersen, A., Ajanovski, V. V., Gutica, M., Hynninen, T., Knutas, A., Leinonen, J., Messom, C., & Liao, S. N. (2018, July). Predicting academic performance: a systematic literature review. In *Proceedings companion of the 23rd annual ACM conference on innovation and technology in computer science education* (pp. 175–199).

Helminen, J., Ihantola, P., Karavirta, V., & Malmi, L. (2012). How do students solve parsons programming problems? An analysis of interaction traces. In *Proceedings of the ninth annual international conference on international computing education research* (pp. 119–126).

Hundhausen, C., Agrawal, A., Fairbrother, D., & Trevisan, M. (2009). Integrating pedagogical code reviews into a CS 1 course: An empirical study. *ACM SIGCSE Bulletin, 41*(1), 291–295.

Hundhausen, C. D., Douglas, S. A., & Stasko, J. T. (2002). A meta-study of algorithm visualization effectiveness. *Journal of Visual Languages & Computing, 13*(3), 259–290.

Höök, L. J., & Eckerdal, A. (2015). On the bimodality in an introductory programming course: An analysis of student performance factors. In *Learning and teaching in computing and engineering (LaTiCE 2015)* (pp. 79–86). IEEE.

Ihantola, P., Ahoniemi, T., Karavirta, V., & Seppälä, O. (2010, October). Review of recent systems for automatic assessment of programming assignments. In *Proceedings of the 10th Koli calling international conference on computing education research* (pp. 86–93).

Kao, Y., Matlen, B., & Weintrop, D. (2022). From one language to the next: Applications of analogical transfer for programming education. *ACM Transactions on Computing Education (TOCE), 22*(4).

Keuning, H., Jeuring, J., & Heeren, B. (2018). A systematic literature review of automated feedback generation for programming exercises. *ACM Transactions on Computing Education (TOCE), 19*(1), 1–43.

Kinnunen, P., & Simon, B. (2011). CS majors' self-efficacy perceptions in CS1: Results in light of social cognitive theory. In *Proceedings of the seventh international workshop on computing education research (ICER '11)* (pp. 19–26). ACM.

Krishnamurthi, S., & Fisler, K. (2019). Programming paradigms and beyond. *The Cambridge handbook of computing education research*, 377–413.

Kurland, D. M., Pea, R. D., Clement, C., & Mawby, R. (1989). A study of the development of programming ability and thinking skills in high school students. In E. Soloway & J. C. Spohrer (Eds.), *Studying the novice programmer* (pp. 83–112). Lawrence Erlbaum.

Kölling, M. (2009). *Quality – oriented teaching of programming*. Retrieved November 11, 2022, from https://blogs.kcl.ac.uk/proged/2009/09/04/quality-oriented-teaching-of-programming/

Lahtinen, E., Ala – Mutka, K., & Järvinen, H. M. (2005). A study of the difficulties of novice programmers. *ACM SIGCSE Bulletin, 37*(3), 14–18.

Lajis, A., Baharudin, S. A., Ab Kadir, D., Ralim, N. M., Nasir, H. M., & Aziz, N. A. (2018). A review of techniques in automatic programming assessment for practical skill test. *Journal of Telecommunication, Electronic and Computer Engineering (JTEC), 10*(2–5), 109–113.

Lishinski, A., Good, J., Sands, P., & Yadav, A. (2016). Methodological rigor and theoretical foundations of CS education research. In *Proceedings of the 2016 ACM conference on international computing education research* (pp. 161–169).

Lishinski, A., & Yadav, A. (2019). Motivation, attitudes, and dispositions. In S. A. Fincher & A. V. Robins (Eds.), *The Cambridge handbook of computing education research* (pp. 801–826). Cambridge University Press.

Lister, R., Adams, E. S., Fitzgerald, S., Fone, W., Hamer, J., Lindholm, M., McCartney, R., Moström, J. E., Sanders, K., Seppälä, O., & Simon, B. (2004). A multi – national study of reading and tracing skills in novice programmers. *ACM SIGCSE Bulletin, 36*(4), 119–150.

Loksa, D., Margulieux, L., Becker, B. A., Craig, M., Denny, P., Pettit, R., & Prather, J. (2022). Metacognition and self-regulation in programming education: Theories and exemplars of use. *ACM Transactions on Computing Education (TOCE), 22*(4).

Loui, M. C., & Borrego, M. (2019). Engineering education research. In S. A. Fincher & A. V. Robins (Eds.), *The Cambridge handbook of computing education research* (pp. 292–322). Cambridge University Press.

Luxton-Reilly, A., Albluwi, I., Becker, B. A., Giannakos, M., Kumar, A. N., Ott, L., Paterson, J., Scott, M. J., Sheard, J., & Szabo, C. (2018). Introductory programming: A systematic literature review. In *Proceedings companion of the 23rd annual ACM conference on innovation and technology in computer science education* (pp. 55–106).

Ma, L., Ferguson, J., Roper, M., & Wood, M. (2007). Investigating the viability of mental models held by novice programmers. *ACM SIGCSE Bulletin, 39*(1), 499–503.

Malmi, L., Karavirta, V., Korhonen, A., Nikander, J., Seppälä, O., & Silvasti, P. (2004). Visual algorithm simulation exercise system with automatic assessment: TRAKLA2. *Informatics in Education, 3*(2), 267–288.

Malmi, L., Sheard, J., Bednarik, R., Helminen, J., Kinnunen, P., Korhonen, A., Myller, N., Sorva, J., & Taherkhani, A. (2014). Theoretical underpinnings of computing education research: What is the evidence? In *Proceedings of the tenth annual conference on international computing education research* (pp. 27–34).

Malmi, L., Sheard, J., Kinnunen, P., Simon, & Sinclair, J. (2019). Computing education theories: What are they and how are they used? In *Proceedings of the 2019 ACM conference on international computing education research* (pp. 187–197).

Malmi, L., Sheard, J., Kinnunen, P., Simon, & Sinclair, J. (2020). Theories and models of emotions, attitudes, and self-efficacy in the context of programming education. In *Proceedings of the 2020 ACM conference on international computing education research* (pp. 36–47).

Malmi, L., Sheard, J., Kinnunen, P., & Sinclair, J. (2023). Development and use of domain-specific learning theories, models and instruments in computing education. *ACM Transactions on Computing Education (TOCE), 23*(1).

Margulieux, L. E., Dorn, B., & Searle, K. A. (2019). *Learning sciences for computing education*. In S. A. Fincher & A. V. Robins (Eds.), *The Cambridge handbook of computing education research* (pp. 208–230). Cambridge University Press.

Mayer, R., & Mayer, R. E. (Eds.). (2005). *The Cambridge handbook of multimedia learning*. Cambridge University Press.

McCracken, M., Almstrum, V., Diaz, D., Guzdial, M., Hagan, D., Kolikant, Y. B., Laxer, C., Thomas, L., Utting, I., & Wilusz, T. (2001). A multi-national, multi-institutional study of assessment of programming skills of first-year CS students. *ACM SIGCSE Bulletin, 33*, 125–180.

Murphy, L., & Thomas, L. (2008, June). Dangers of a fixed mindset: Implications of self-theories research for computer science education. In *Proceedings of the 13th annual conference on innovation and technology in computer science education* (pp. 271–275).

Novak, M., Joy, M., & Kermek, D. (2019). Source-code similarity detection and detection tools used in academia: A systematic review. *ACM Transactions on Computing Education (TOCE), 19*(3), 1–37.

Paiva, J. C., Leal, J. P., & Figueira, Á. (2022). Automated assessment in computer science education: A state-of-the-art review. *ACM Transactions on Computing Education (TOCE), 22*(3), 1–40.

Papert, S. A. (2020). *Mindstorms: Children, computers, and powerful ideas*. Basic Books.

Patitsas, E., Berlin, J., Craig, M., & Easterbrook, S. (2019). Evidence that computer science grades are not bimodal. *Communications of the ACM, 63*(1), 91–98.

Perkins, D. N., Hancock, C., Hobbs, R., Martin, F., & Simmons, R. (1989). Conditions of learning in novice programmers. In E. Soloway & J. C. Spohrer (Eds.), *Studying the novice programmer* (pp. 261–279). Lawrence Erlbaum.

Petre, M., Sanders, K., McCartney, R., Ahmadzadeh, M., Connolly, C., Hamouda, S., Harrington, B., Lumbroso, J., Maguire, J., Malmi, L., & McGill, M. M. (2020). Mapping the landscape of peer review in computing education research. In *Proceedings of the working group reports on innovation and technology in computer science education* (pp. 173–209).

Pettit, R. S., Homer, J., & Gee, R. (2017, March). Do enhanced compiler error messages help students? Results inconclusive. In *Proceedings of the 2017 ACM SIGCSE technical symposium on computer science education* (pp. 465–470).

Pullan, W., Drew, S., & Tucker, S. (2013, September). An integrated approach to teaching introductory programming. In *2013 second international conference on E-learning and E-technologies in education (ICEEE)* (pp. 81–86). IEEE.

Ramalingam, V., & Wiedenbeck, S. (1998). Development and validation of scores on a computer programming self-efficacy scale and group analyses of novice programmer self-efficacy. *Journal of Educational Computing Research, 19*(4), 367–381.

Rist, R. S. (2004). Learning to program: Schema creation, application, and evaluation. In S. Fincher & M. Petre (Eds.), *Computer science education research* (pp. 175–195). Taylor & Francis.

Rivers, K., & Koedinger, K. R. (2017). Data-driven hint generation in vast solution spaces: A self-improving python programming tutor. *International Journal of Artificial Intelligence in Education, 27*(1), 37–64.

Robins, A. V. (2010). Learning edge momentum: A new account of outcomes in CS1. *Computer Science Education, 20*(1), 37–71.

Robins, A. V. (2019). Novice programmers and introductory programming. In *The Cambridge handbook of computing education research* (pp. 327–377). Cambridge University Press.

Robins, A. V., Haden, P., & Garner, S. (2006). Problem distributions in a CS1 course. In *Proceedings of the eighth Australasian computing education conference (ACE 2006)*, CRPIT, 52 (pp. 165–173). Australian Computer Society.

Robins, A. V., Margulieux, L. E., & Morrison, B. B. (2019). Cognitive sciences for computing education. In S. A. Fincher & A. V. Robins (Eds.), *The Cambridge handbook of computing education research* (pp. 231–275). Cambridge University Press.

Robins, A. V., Rountree, J., & Rountree, N. (2003). Learning and teaching programming: A review and discussion. *Computer Science Education, 13*(2), 137–172.

Shaffer, C. A., Cooper, M. L., Alon, A. J. D., Akbar, M., Stewart, M., Ponce, S., & Edwards, S. H. (2010). Algorithm visualization: The state of the field. *ACM Transactions on Computing Education (TOCE), 10*(3), 1–22.

Simon, Luxton-Reilly, A., Ajanovski, V. V., Fouh, E., Gonsalvez, C., Leinonen, J., Parkinson, J., Poole, M., & Thota, N. (2019). Pass rates in introductory programming and in other STEM disciplines. In *Proceedings of the working group reports on innovation and technology in computer science education* (pp. 53–71).

Sirkiä, T. (2016). October. Jsvee & Kelmu: Creating and tailoring program animations for computing education. In *2016 IEEE working conference on software visualization (VISSOFT)* (pp. 36–45). IEEE.

Solomon, C., Harvey, B., Kahn, K., Lieberman, H., Miller, M. L., Minsky, M., Papert, A., & Silverman, B. (2020). History of Logo. *Proceedings of the ACM on Programming Languages, 4*(HOPL), 1–66.

Soloway, E., & Ehrlich, K. (1984). Empirical studies of programming knowledge. *Transactions on Software Engineering* (5), 595–609.

Soloway, E., Ehrlich, K., Bonar, J., & Greenspan, J. (1983). What do novices know about programming? In B. Shneiderman & A. Badre (Eds.), *Directions in human – computer interactions* (pp. 27–54). Ablex.

Soloway, E., & Spohrer, J. C. (Eds.). (1989). *Studying the novice programmer.* Lawrence Erlbaum.

Sorva, J. (2013). Notional machines and introductory programming education. *ACM Transactions on Computing Education, 13*(2), 1–31.

Sorva, J., Karavirta, V., & Malmi, L. (2013). A review of generic program visualization systems for introductory programming education. *ACM Transactions on Computing Education (TOCE), 13*(4), 1–64.

Sweller, J. (2010). Element interactivity and intrinsic, extraneous, and germane cognitive load. *Educational Psychology Review, 22*(2), 123–138.

Sweller, J., Van Merrienboer, J. J., & Paas, F. G. (1998). Cognitive architecture and instructional design. *Educational Psychology Review, 10*(3), 251–296.

Szabo, C., Falkner, N., Petersen, A., Bort, H., Cunningham, K., Donaldson, P., Hellas, A., Robinson, J., & Sheard, J. (2019). Review and use of learning theories within computer science education research: Primer for researchers and practitioners. In *Proceedings of the working group reports on innovation and technology in computer science education* (pp. 89–109).

Szabo, C., & Sheard, J. (2023). Learning theories use and relationships in computing education research. *ACM Transactions on Computing Education, 23*(1), 1–34.

Teague, D., & Lister, R. (2014). Longitudinal think aloud study of a novice programmer. In *Proceedings of the sixteenth Australasian computing education conference (ACE2014), conferences in research and practice in information technology series* (Vol. 148).

Tedre, M., & Pajunen, J. (2023). Grand theories or design guidelines? Perspectives on the role of theory in computing education research. *ACM Transactions on Computing Education (TOCE), 23*(1).

Tedre, M., Simon, & Malmi, L. (2018). Changing aims of computing education: A historical survey. *Computer Science Education, 28*(2), 158–186.

Tedre, M., & Sutinen, E. (2008). Three traditions of computing: What educators should know. *Computer Science Education, 18*(3), 153–170.

Tenenberg, J., & Malmi, L. (2022). Editorial: Conceptualizing and use of theory in computing education research. *ACM Transactions on Computer Science Education, 22*(4).

Tucker, A. B., & Barnes, B. H. (1990). *Computing curricula 1991–Report of the ACM/IEEE-CS joint curriculum task force.* Jointly published by ACM Press, IEEE Computer Society Press.

Umapathy, K., & Ritzhaupt, A. D. (2017). A meta-analysis of pair-programming in computer programming courses: Implications for educational practice. *ACM Transactions on Computing Education (TOCE), 17*(4), 1–13.

Utting, I., Tew, A. E., McCracken, M., Thomas, L., Bouvier, D., Frye, R., Paterson, J., Caspersen, M., Kolikant, Y. B. D., Sorva, J., & Wilusz, T. (2013). A fresh look at novice programmers' performance and their teachers' expectations. In *Proceedings of the ITiCSE 2013* (pp. 15–32).

Valentine, D. W. (2004). CS educational research: A meta-analysis of SIGCSE technical symposium proceedings. *ACM SIGCSE Bulletin, 36*(1), 255–259.

Watson, C., & Li, F. W. (2014). Failure rates in introductory programming revisited. In *Proceedings of the 2014 conference on innovation & technology in computer science education* (pp. 39–44). ACM.

Watson, C., Li, F. W., & Godwin, J. L. (2014). No tests required: Comparing traditional and dynamic predictors of programming success. In *Proceedings of the 45th ACM technical symposium on Computer science education* (pp. 469–474).

Weinberg, G. M. (1971). *The psychology of computer programming*. Van Nostrand Reinhold.

Winslow, L. E. (1996). Programming pedagogy – A psychological overview. *ACM SIGCSE Bulletin*, *28*(3), 17–22.

Wirth, N. (1971). The programming language Pascal. *Acta Informatica*, *1*(1), 35–63.

Xie, B., Loksa, D., Nelson, G. L., Davidson, M. J., Dong, D., Kwik, H., Tan, A. H., Hwa, L., Li, M., & Ko, A. J. (2019). A theory of instruction for introductory programming skills. *Computer Science Education*, *29*(2–3), 205–253.

Yadav, A., & Lachney, M. (2023). Towards a techno-social realist approach in primary and secondary computing education. In A. Johri (Ed.), *International handbook of engineering education research* (pp. 553–572). Routledge.

Part 6

Engineering Education Research Methods and Assessment

27

Delineating Anti-Blackness in Engineering Education Research Methodology

James Holly Jr., Kelly J. Cross, and Walter C. Lee

1 Introduction

Racism, anti-Black racism in particular, has noticeably become a central part of the lexicon within engineering education in the United States following the racially traumatizing events of 2020. Despite this increase in discussing racism, there remains a lack of engagement in our discipline with the meanings and implications of this social phenomenon. As Black engineering education scholars, we are both encouraged by the increased dialogue and dismayed by the shortsighted discourse taking place. While many may agree that racism is a non-biological social construction, few have reckoned with the implications of that truth for a field whose methodological practices preserve Whiteness and sustain harmful narratives about Black people (Holly, Jr. & Thomas Quigley, 2022). Accordingly, the first purpose of this chapter is to illuminate how current standard research methods and methodologies in engineering education ignore the racialized history under which these knowledge production procedures were developed (Zuberi, 2001; Zuberi & Bonilla-Silva, 2008). We were inspired by the words of Du Bois (quoted earlier), who asserted that racial ignorance, or the intentional evasion and distortion of social reality (Mueller, 2020), has perpetuated myths and misrepresentations of Black people. These lies inhibit our ability to accurately portray and analyze social reality in engineering education, especially as it relates to enacting accountability for racial injustice. The second purpose for this chapter is to describe how engineering education research is a domain of socialization and indoctrination into a White supremacy culture of knowledge production. We will problematize research practices that may appear as ordinary and natural so the extent to which they are *social constructions* rooted in anti-Blackness, and are thus changeable, is more apparent.

Though the practices of individual researchers may vary, dominant research practices (i.e., approaches to knowledge production) in engineering education reveal themselves in places such as academic journals, funding solicitations, annual meetings, and doctoral programs. These approaches are not always explicitly named, written, or taught. However, multiple scholars in the United States have offered critiques that begin to reveal some of these prominent features as it relates to social identities and systems of oppression. For example, Pawley (2017) critiqued the need for authors to justify a focus on diversity in the introductions of their papers, emphasizing how current expectations about the need to justify why diversity is valuable enable Whiteness and maleness to remain the "default" position in engineering education. More recently, Haverkamp et al. (2021) critiqued dominant discourse and research approaches regarding gender, stating that current practices obscure the existence

DOI: 10.4324/9781003287483-33

and experiences of transgender and gender nonconforming people. Common across such critiques is the desire to see engineering education research (EER) live up to many of its espoused values related to diversity, equity, inclusion, and social justice.

We continue this conversation with a specific focus on anti-Black racism. We build on prior critiques of US engineering education research offered by Holly Jr. (2020) about the extent to which the field is entangled with anti-Blackness, defined as Blackness/people being conceived as subhuman and antithetical to Whiteness/people. In a guest editorial for the *Journal of Engineering Education*, Holly Jr. critiques the existence of policies and practices that tear down Black people in engineering and engineering education research. He argues that the normalcy of anti-Blackness has led to silent complicity and calls for direct counteraction against anti-Black research. Holly Jr. offers a list of research practices as a starting point towards freeing research in engineering education from entanglement with anti-Blackness. Herein, we similarly argue that anti-Black research practices (and knowledge production practices more broadly) are embedded within the current educational research training received by many burgeoning engineering education researchers. We understand anti-Blackness is a global phenomenon, often referred to as Afrophobia; still, we locate this argument within the United States to retain specificity in the exposition of the ways White supremacy was constructed and is maintained in this nation (Roberts, 2011).

More specifically, we offer a counternarrative to the dominant discourse about scientific inquiry in US engineering education. Because of the ways White supremacy has shaped research procedures considered to be the most trustworthy and reliable, racial prejudice is largely hidden due to either silence or ignorance. We intend to make an explicit critique of these norms. We took this direct approach to disrupt the unreflective use of standard research methods because the Black experience in education, and engineering education more specifically, continues to be misunderstood, misrepresented, and misinterpreted. The framing of "underrepresentation" misrepresents the unjust and violent circumstances that have always limited, and for many years totally excluded, Black people's participation in higher education (Mustaffa, 2017). Historically White colleges/universities (HWCUs) initially denied engineering education to Black people (Slaton, 2010), but even when access was granted, Black engineers "had to overcome professional and personal barriers to success in order to make the contributions that constitute the legacies they have left for the generations that have followed" (Slaughter, 2015, p. 1). These institutional barriers are part of an oppressive educational system (Au, 2015; McGee, 2020a; McGee et al., 2022) constructed to maintain Whiteness and homogenization; thus, the attrition of Black engineering students is misinterpreted as inability, and the resilience of those who remain is misunderstood as contentment. The inattention of engineering education researchers to Black suffering in education and society more broadly, and in engineering educational contexts specifically, legitimizes rationales that equate the lower academic performance of Black engineering students with cognitive inferiority. Such is the essence of anti-Blackness in engineering education research.

We present a narrative of knowledge production in engineering education from our vantage point as a pathway to broadening research practices towards diverse approaches to what the engineering education community accepts as knowledge and truth. Drawing on Black intellectuals from various traditions (Roberts, 1999; Toldson, 2019; Yancy, 2016), we delineate the ways race and racism affect decision-making along the entire life cycle of research projects. We present this counternarrative in four parts. In Part 1, we discuss the importance of understanding the historical context in which a researcher is embedded. In Part 2, we argue that knowledge is contextually constructed and demonstrate the importance of interjecting one's humanity into education research methods and methodology. In Part 3, we present knowledge production as a socialization process and critique how engineering education training socializes scholars to ask and answer questions, make meaning of our answers, and communicate these results. And in Part 4, we call for alternative practices that center the humanity, cultural knowledge, and lived experience of Black people as a method of transgressing and challenging epistemological norms. Across these parts, we bring humanness to the forefront of research rather than dehumanizing or ignoring the human actor as the researcher conducting the study or guiding the inquiry. In doing so, we speak directly to the engineering education research community that has yet to address the false notion of colorblindness and unbiased research methodologies. We also speak in solidarity with the many Black scholars from various disciplines trying to be heard and have their contributions recognized. Our intended audience are Black engineering education researchers committed to producing scholarship grounded in the ways of knowing cultivated through the struggle, resilience, and ingenuity of Black people, and non-Black engineering education researchers committed to accomplishing racial justice in engineering.

In sharing these insights, we have four primary intentions. Firstly, we intend to assist scholars that want to affirm the intellectual richness of Blackness in engineering education. We recognize shortcomings in our own training and see value in offering insights to present and future scholars. Secondly, we seek to uplift the experiences of all Black scholars that have experienced marginalization in engineering education. We situate these ideas in our shared historical foundation of inequality in the United States while recognizing that Black people are not a monolith. Thirdly, we intend to colorize the ontological and epistemological foundations of engineering education research. We recognize work done by those who come before us and introduce previously excluded concepts and ideas into the engineering education research community lexicon. Lastly, we intend to inspire those who are firmly entrenched in the engineering education culture to release some of the unproductive practices. We recognize the difficulty associated with decolonizing the mind (i.e., freeing oneself from thoughts, preferences, and values that uphold imperialism and colonialism) while maintaining the need to resist and have hope.

2 Part 1 – Deconstructing White Supremacy

The importance of discussing racism is not simply acknowledging its presence but instead also developing understanding of how it works in order to abolish its existence. Many books have been written explaining the origins of Whiteness and its mutations over time (Fields & Fields, 2014; Kendi, 2016; Roberts, 2011), but these resources have been treated like cultural artifacts rather than heuristics for making sense of our social reality. In the following section, we provide some basic points about White supremacy in broader society to assist the reader in understanding manifestations of White supremacy in engineering education knowledge production. Whereas racism is endemic to the social structure of the United States, we will situate these characteristics within the epistemological norms of engineering education research. Neely Fuller Jr. made a profound statement in declaring misconceptions of White supremacy as a fundamental source of confusion, and his words seem prophetic when examining the abundance of unsuccessful initiatives to broaden the participation of Black people in engineering (London et al., 2020, 2022). Therefore, we offer insight into the mechanisms of White supremacy to strengthen the ability of engineering education researchers to better assess the plight of Black people in engineering.

As Cross (2020) argues, racism is a manifestation of White supremacy. Because the concept of anti-Blackness is interconnected with White supremacy ideology, we position anti-Blackness as the manifestation of racist research practices based on White supremacy. The characteristics of White supremacy culture ensure that anything framed by this cognitive paradigm is constrained and destructive. Yet this paradigm – alongside other systems of oppression – is exactly what has generated a research economy governed by white logic and white methods (Zuberi & Bonilla-Silva, 2008) and dictates the standards of rigor in engineering education research (Riley, 2017). Consequently, we assert knowledge production as a socialization process currently dictated by White supremacy.

2.1 What Is White Supremacy?

White supremacy is a term used to capture the all-encompassing dimensions of White privilege, dominance, and assumed superiority in society. It is a social phenomenon that has ideological, institutional, social, cultural, historical, political, economic, interpersonal, and educational dimensions (Alexander, 2010; Anderson, 2016; Bonilla-Silva, 2006; Rasmussen et al., 2001). It refers to the ruling class or power elite using the pseudo-scientific concept of race to create Whiteness and a hierarchy of racialized value in the colonies of what has since become the United States (Fields & Fields, 2014; Roberts, 2011). White supremacy disconnects and divides while establishing Whiteness as property tied to Anglo-Saxon exceptionalism (Harris, 1993). At its core, White supremacy is about power and sustaining control by declaring Whiteness as the standard to judge the fabricated racial hierarchy. And Black people who openly challenge the roots of White supremacy are currently and historically denied access, ostracized, overpoliced, brutalized, or otherwise punished (Fraser & Griffin, 2020; Hesse, 2014; Smith & Garrett-Scott, 2021).

2.2 What Are Some Characteristics of White Supremacy Culture?

Culture reflects institutionalized beliefs, values, norms, and standards. Therefore, White supremacy culture is the widespread ideology of systemic racism teaching us both overtly and covertly that Whiteness is distinct and holds the highest value. The White supremacist culture is covered well by other authors (Du Bois, 1903a; Kendi, 2016; Marable, 1983; Muhammad, 2019; Zinn, 1980), so we will not do a comprehensive description of the culture here. Instead, we define four key concepts that are embedded in the culture and are relevant to our research standards: (1) *White power,* (2) *White insecurity*, (3) *White immunity*, and (4) *complicity and collusion*. These summaries help explain the functionality of these concepts in how we engage in meaning-making and legitimizing knowledge production practices.

White power, which is more commonly referred to as either White privilege or Whiteness as property, is a set of social advantages accrued throughout this nation's history. These advantages are unattainable by those not racialized as White, and they give White people the capacity to exercise control over others (Harris, 1993). The marginalization of non-White people is a by-product of placing Whiteness as the default to which all non-White people are compared. White power can exist even when folks claim to be working against it, because its normalization removes the necessity of intent and helps maintain the racial hierarchy in this country. The anti-Black manifestation of White power is placing the concerns and perspectives of Black engineers as peripheral, suggesting Black people are not worthy of even the basic resources, and not supporting Black intellectual contributions (Holly, Jr. & Thomas Quigley, 2022; McGee, 2020b).

White insecurity, which is more commonly referred to as either White fragility or White guilt, is a state in which even a minimum amount of racial reckoning becomes intolerable for White people due to a perceived right to be emotionally comfortable. When responsibility for racial harm is sought, this can trigger a range of defensive moves as White people exhibit a fear of open conflict (Okun, 2021). These moves include the outward display of emotions (such as anger, fear, and guilt) and behaviors (such as argumentation, silence, and leaving). These behaviors, in turn, function to reinstate White racial equilibrium, where Whites can confine themselves to White ideology by living, learning, and working in predominantly White spaces or by refusing to engage with the realities of race. The anti-Black manifestation of White insecurity is claiming ignorance or good intentions when Black engineers pronounce the harmful effects of White people's actions; other instances include crying or being overly apologetic without rectifying the damage caused.

White immunity refers to White people being able to avoid accountability for their harmful actions. The concept of White immunity was developed in response to the ways notions of White privilege impaired society's understanding of racism despite amplifying its discourse (Cabrera, 2017). It suggests that non-White people are not granted the same rights and equal treatment under law as White people. While this construct is defined individually, the political infrastructure that secures its operation is rooted in the historic practices of the legislature. White people do not only engage in mistreatment of non-White people; they are also legally protected in doing so. Some anti-Black circumstances where White immunity has manifested is after explicitly questioning Black engineering students' intelligence (McGee & Martin, 2011), neglecting or belittling Black student advisees (Burt et al., 2018), and avoidance of critique Black researchers endure for solely focusing on Black students (Griffin et al., 2013).

Complicity and collusion refer to the participation in (and, therefore, advancement of) systems of discrimination and oppression. These concepts are related yet distinct. When people are complicit in institutionalized oppression, they understand how they benefit from improper actions, but they may act alone. When people participate in collusion, they engage in secret agreements with other people to carry out improper actions. The anti-Black manifestation of White complicity shows up through silence regarding racist social norms and traditions in promotion review processes (Settles et al., 2021) or demonstrating partiality when Black students seek help and are deterred from being in the program (Newman, 2011). The anti-Blackness manifestation of collusion shows up in the collaborative critique of Black scholars without adequate expertise through the denial of Black scholarship and publications and devaluing the published work of Black scholars, particularly those that are counter to the accepted knowledge of a White scholar (Delgado Bernal & Villalpando, 2002).

2.3 What Are the Effects of White Supremacy Culture?

White supremacy culture is reflected in the disproportionate harm and violence directed towards Black people and communities in all aspects of engineering education and practice. This can easily be identified by reviewing scholarship on Black people in engineering, which is mostly about our suffering and the coping mechanisms we employ to persist (Fletcher et al., 2021; Freeman & Huggans, 2009; McGee et al., 2019; Moore et al., 2003). In the United States, Whiteness governs the criteria used to determine whose ideas, actions, or experiences have merit (Ahmed, 2012; Dotson, 2014). White supremacy culture concurrently teaches us that Blackness is valueless, dangerous, and threatening (Ladson-Billings & Donnor, 2005; Sharpe, 2016). Thus, intellectual contributions by Black people in engineering receive more scrutiny and receive less acclaim even after overcoming intensified scrutiny. Black researchers that examine racialized experiences of students are pigeonholed as diversity scholars – which is considered inferior – or their work is allegedly tainted by bias, even when the essence of their work centers the technological knowledge deemed critical for professional formation (Hendrix, 2002). Alternatively, White faculty are esteemed for doing things considered diversity scholarship, in spite of blatant deficiencies in their expertise and harm to Black people being perpetuated by their work. This infringement on the epistemic agency of Black scholars is detrimental to everyone; beyond the obvious implications for promotion and tenure, it damages "the state of social knowledge and shared epistemic resources" (Dotson, 2012, p. 24) among our scholarly community.

Perhaps our richest epistemic resource is ourselves; our lived experiences shape how we make sense of the world around us. Unfortunately, many of us are unaware of how White supremacy culture influences our perspectives and sensemaking. Some of us are aware but have succumbed to false notions of objectivity and view inserting ourselves into our research practice as a contaminating act. In the next section, we discuss research paradigms and the role of positionality in our scholarship. We must understand the role we play in knowledge discovery and dissemination as an essential step to reclaiming our epistemic agency.

3 Part 2 – Research Paradigms and Researcher Positionality

What does it mean to do research? And what does it mean to *be* a researcher? We move to these questions after providing a brief overview of White supremacy because people come to engineering education research with various answers, perspectives, experiences, and preconceived notions about social science research. In short, they arrive with different worldviews and positionalities. We provide a brief discussion of each in the following sections.

3.1 What Is a Research Paradigm?

In academe, we often refer to one's worldview and its relation to their work as a *research paradigm*. According to Patton (2014), a *paradigm* is a way of thinking about and making sense of the world. Paradigms inform what research we consider pursuing, whether consciously or unconsciously. Overall, a researcher's paradigm determines what they see and how they make sense of what they see. A researcher paradigm is the filter through which we see the field of engineering education and thus impacts how we define the research problem (e.g., who and what we study), how we develop and approach to answer research questions (e.g., what theory and methodology we choose), and how we interpret and communicate the results or information learned over the course of the study (e.g., where we publish). They are normative and deeply embedded within how we learn to conduct research. For example, because engineering favors paradigms that focus on causal and quantitative relationships, people who were trained in engineering often have trouble transitioning to qualitative and interpretive approaches when they transition to engineering education research.

Consequently, one's paradigm will likely shift as they become a member of the engineering education community and learn how to meet the agreed-upon discipline standards, including what is important, what is legitimate, and what is reasonable. In our case, we span the paradigms of being constructivist (Cross), critical (Holly, Jr.), and pragmatic (Lee). While none of us fit neatly into one of these categories, we tend to express one more readily. This also means thst through working collaboratively, we influence each other's paradigms, for example, Author Lee mentioned while working on this project he noticed himself developing more of a critical disposition.

3.2 What Is Our Researcher Positionality?

The epistemological position of objectivity and the false notion of meritocracy in US engineering might compel one to divorce themselves from the work and seek "truth" above all else. Similar to other engineering education scholars who have promoted the idea of disclosing positionality (e.g., Hampton et al., 2021; Secules et al., 2021), we encourage others to interject their humanity into their research methods and methodology as a manner of intellectual honesty. We advocate re-politicizing methodological practices; more specifically, we're calling for work that not only pursues truth but also considers what such truth means in relation to improving the social conditions of Black people.

For example, we come to this work as two Black men and a Black woman who earned their doctorate in engineering education at the two oldest US engineering education programs. During our graduate education, we were each advised and primarily trained by White researchers. When we joined the engineering education community, we each already had racial pride and confidence from our personal upbringings. Holly Jr. and Cross's graduate research was directly focused on race, whereas Lee's research was tangentially focused on it. Each of us has published engineering education research. We also bring editorial and review experiences with journals in the field. As we write this chapter, we each currently work as faculty members and are evaluated based on engineering education research. It is through these experiences that we have come to intimately know engineering education research methodology and become rooted firmly in the US engineering education culture. As such, we had to interrogate our own training and look outside ourselves to identify what about the training we received could have been different in the past and should be different going forward. This chapter is our attempt to share these insights with others.

James Holly Jr.: My expertise is conceived from two elements: personal experience and Black studies. I unite my experiential knowledge with the sociological understanding provided by Black thinkers who have conceptualized the powerful ways Whiteness has historically shaped social systems within this country and the ways it continues to do so, particularly in the educational system. Taken together, these perspectives formulate the lens through which I make visible, and counteract, the ways disciplinary knowledge and forms of racialized power have been co-constructed to minoritize and exclude Black people.

Kelly J. Cross: I have conducted workshops on managing personal bias in STEM and promoting inclusion in higher education. I also bring the perspective of intersectionality to this work, as I have multiple marginalized identities that impact my daily experiences. Overall, my research agenda is to broaden participation in engineering, and collaboration is a key component of all my professional activities. My previous research investigated the experiences of various marginalized groups, including women of color and members of the LGBTQ+ spectrum. I am a culturally responsive practitioner, researcher, and educational leader who will continue to create space for the voiceless in engineering.

Walter C. Lee: My admiration and respect for both political- and scholar-activists have grown, alongside my understanding of the world. While I can say I do many things differently than my non-Black colleagues, I cannot say I am as intentional about my teaching and use of research methodology as I could be. In many ways, I have been complicit. It is in this spirit that I joined this project, to spend time learning alongside Holly Jr. and Cross as part of an ongoing effort to improve my understanding of concepts relevant to my sociopolitical interest and research agenda.

4 Part 3 – Anti-Blackness in Engineering Education Research

The discipline of engineering education research is relatively young, and already the educational research practices that predominate the discipline reflect anti-Black patterns in broader social science scholarship (Zuberi, 2000). These trends can be seen in scenarios like situating White engineering students as the standard for other racial groups to be measured by, considering information gained from studies with only Black people or small numbers as non-generalizable, and diminishing citations of non-White scholars outside the field. Although the so-called underrepresentation of Black engineering students has long been an issue, the scholarship of Black educational theorists and researchers is not adequately utilized to study the educational experiences of Black engineering students. We consider this to be expected, given the societal context of a White supremacy culture in which EER exists and was formed by; however, this socialization process by which even we were trained need not be permanent. We characterize research training in EER as a socialization process to accentuate the consistency of the social structure and the agency we have to disrupt these norms.

We situate our understanding of our own experiences in this process using Harro's (2000) cycle of socialization. In the engineering education context, the beginning of socialization where we began to embrace the values and behaviors and acquire requisite skills represents our entry into the research community. This entry point is where institutional and cultural socialization formalized our initiation into engineering ways of thinking, communicating, and problem-solving. While our indoctrination did not require intentionality or visibility, its results influenced our participation within this academic community. Once fully in the cycle, we conducted our work based on the research norms that we encountered, and these norms provided a means to reinforce our socialization. Lastly, our interpretations of our interactions and research results continued to support the dominant narrative.

We write this chapter in an effort to break this cycle. We are kept in this cycle through ignorance, insecurity, confusion, obliviousness, and fear (Harro, 2000). When we are unaware of what beliefs, values, and ideologies regulate our research practice, we lack agency in combating the dominant tendencies shaping knowledge production. Instead, we advocate for active disruption as a way to make us question our beliefs, critically look at the process, and see the flaws within the accepted process of socialization. To facilitate such disruption, we critique several aspects of the socialization process in engineering education research to elucidate the structures that regulate EER by teaching us how to ask questions, answer questions, make meaning, and communicate.

4.1 How Do Engineering Education Researchers Learn to Ask Questions?

How we were taught to ask questions is important because it determined which people's interests we prioritized. The exhibition or suppression of our power as researchers is in operation from the outset of our research design. As Ruha Benjamin suggests, "It's at that [stage] of forming the question and posing the problem that the power dynamics are already being laid that set us off in one trajectory or another" (Vox, 2021). In this section, we briefly discuss two ways in which we were socialized to ask questions: our introduction to problem statements and funding opportunities. Combined, both of these elements of the knowledge production process dictate what issues are deemed worthy of study and whose interests influence how these issues are examined.

4.1.1 Problem Statements

Which questions are worth asking? In addition to being trained how to answer questions, we were also taught how to construct a purpose statement and frame research questions. The training we received impacted which problems we could see and considered worthy of scientific inquiry. It influenced our likelihood of "studying up" (Nader, 1972) or expanding our field of inquiry to include those whose position in the power structure would enable them to make engineering more equitable. More often than not, we were taught to pose questions about engineering students as opposed to faculty or administrators. These lessons were not explicitly taught; rather, it was implied through the absence of such studies within coursework, syllabi, journals, and conference proceedings. For example, Author Cross has an article looking at engineering student stress. Although the study relates engineering student stress to specific student outcomes and characteristics of department culture, the role of faculty and the administration in creating the engineering stress culture was removed from the article during the review process. Despite previous studies (Harper & Hurtado, 2007; Mwangi et al., 2018) suggesting engineering is a hostile climate for racially minoritized students, little research has considered the role of the leadership that is responsible for creating and sustaining that hostile environment. We noticed the tendency to study racially minoritized students through a deficit lens (Harper, 2010) and an unwillingness to explicitly disrupt power imbalances. Similar to researchers in fields such as science, medical, and educational research, we were encouraged to discuss disparities but not similarly encouraged to explore the source or cause of the disparities.

4.1.2 Funding Opportunities

What questions are fundable or solicited? Because external funding occupies such a prominent role in the US research enterprise, resource allocation practices had a substantial impact on how we were trained and our ability to do research focused on Black people's uplift. We learned to frame questions by reviewing the requests or solicitations issued by funding agencies. Though not explicitly written anywhere, we learned that it can be difficult to obtain resources for projects that challenge the political, economic, or intellectual status quo, projects that can be characterized as unrelated to core engineering knowledge or experiences. Consequently, we learned to frame questions about Black people in regard to how reducing their struggles would also be beneficial to the White majority population, and we were introduced to the business case for diversity (Herring, 2009; Robinson & Dechant, 1997) or even interest convergence as methods to be heard. We were not exposed to explicit requests for anti-racist projects and were led to believe such issues were not of priority to the discipline.

4.2 How Do Engineering Education Researchers Learn to Answer Questions?

How we were taught to answer questions is important because it determined how we viewed prior work and conceived of future work. The training we received impacted which methodologies we considered legitimate scientific inquiry. Audre Lorde posed an insightful question that we used to reflect on our training: "What does it mean when the tools of a racist patriarchy are used to examine the fruits of that same patriarchy?" (Lorde, 1983, p. 25). We must similarly grapple with what it means to use the tools of White supremacy to address matters associated with racism or Black people. Doing so is vital because the research approaches we select have underlying propositions about how we determine what is true and how we can deepen our understanding of various phenomena. In this section, we briefly discuss three ways in which we were socialized to answer questions in engineering education: how we were trained to use data collection methods, view our role as the researcher, and engage with research participants. Combined, each of these elements of the knowledge production process dictates what information should be considered and how.

4.2.1 The Role of the Researcher

What role should a researcher have in the research process? The training we received suggested that it was more important to consider the "researcher-as-instrument" in qualitative research than quantitative research, despite the role of the researcher needing to be considered regardless of methodology. As discussed earlier, who is doing the research influences the research design itself and should be acknowledged through positionality statements that disclose the researcher paradigm applied in all engineering education research. This training influenced the extent to which we disregarded the personhood of non-Black researchers and positioned them as objective outsiders, or embraced the personhood of Black researchers and positioned them as insiders who share the characteristic, role, or experience being examined. We encountered questions about who should be studying what and what expertise qualifies a person to pursue or advance a given topic. But we were also taught to racialize some researchers and ignore the race of others. For example, White scholars are permitted to take a race-neutral approach to research where race has significant analytical power, but research done by Black scholars is discounted when they elevate how their race influenced their decision to investigate racial inequity in engineering. We were never instructed to explicitly name Whiteness. We received inconsistent interpretations of the role ascribed to positionality in the research process.

4.2.2 The Role of the Participant

What role should a participant have in the research process? The training we received influenced how we interacted with the people and communities we chose to study. We were taught to relegate participants as research objects. While a more humane approach is not only referring to people involved in our study as participants rather than subjects but also positioning them as partners in the knowledge production process. While we were introduced to some of these discussions, we were not encouraged to position participants as experts on their lived experiences (Toliver, 2021; Toole,

2022). They may not be familiar with a particular research methodology, but they are able to provide detail and nuance about their own lived experience that we, as a research community, could not understand otherwise.

4.2.3 Methods

Which methods and methodologies are credible? The training we received impacted how we valued (or devalued) conceptual and empirical methods, how we approached sampling, and our desire for generalizability. These ideas were reinforced through graduate courses and dissertation milestones because one hallmark of a research field are the methods and data sources it leverages as well as the variables it measures. Researchers in engineering tend to favor a positivist epistemology defined as a set of beliefs and community practices (e.g., worldview of the nature of research) that define engineering research (i.e., scientific method) as based on the statistical analysis of experimental results, and those results are typically in the form of continuous data that reveal truths about and relationships between social phenomena (Creswell & Creswell, 2018). A positivist epistemology leads to an overreliance on experimentation and statistics to bolster arguments of objectivity. This default training limits our ability to capture the complexity of reality because it assumes the researcher and the context are divorced from the results. A positivist epistemology also creates a hierarchy between quantitative and qualitative research approaches. In addition to debates over the value of qualitative, quantitative, and mixed methods, we were briefly introduced to certain methodologies within each of these traditions, primarily borrowing from fields such as psychology, sociology, learning sciences, and higher education. Just as there are culturally sensitive instructional approaches, there are culturally sensitive research approaches (Tillman, 2002) that need greater recognition for mitigating misrepresentation of social reality across the nuances of racism.

4.3 How Do Engineering Education Researchers Learn to Make Meaning?

How we were taught to make meaning is important because it guides the significance of our intellectual contributions, specifically, who or what we tried to impact and in what way. Said differently, the lens we learned to use provided frames for conceiving the relevance of what we learned. What we were taught to leave unnamed was just as consequential as what we learned to name. Kwame Ture made this clear when he proclaimed:

> Anytime you make an analysis of an oppressed people, in any aspect of their life, and you leave out the enemy, you will never come to a correct analysis. On the contrary, you will blame the oppressed for all of their problems.
>
> *[Kwame Ture in Branch, 2013]*

When learning theoretical frameworks, we were not encouraged to directly analyze the influence of the oppressor or view White supremacy as our enemy. In this section, we briefly discuss two ways in which we were socialized to make meaning in engineering education: our exposure to theoretical frameworks and research implications. Combined, both of these elements of the knowledge production process influence our ability to make sense of experiences and identify paths forward.

4.3.1 Theoretical Frameworks

How should theory be used in our research? The training we received influenced which constructs, concepts, and phenomena we considered worth exploring and were capable of seeing. The theories at our disposal ultimately determined the observations and inquiries we were able to make. Though consensus does not exist regarding what constitutes strong versus weak theory in social science (Sutton & Staw, 1995), we encounter a variety of theoretical perspectives through being introduced to theories, theoretical frameworks, conceptual models, etc. We were not introduced to a plethora of theoretical frameworks that had the ability to consider race, much less were we introduced to a theory born out of a Black epistemology (e.g., endarkened epistemology, Black feminist epistemology); fortunately, Black scholars are shifting this trend in EER (Nicole, 2022; Thomas et al., 2016). Yet scholars beware the perpetrators of Black epistemicide – systematic destruction or devaluing of rival forms of knowledge – seeking relevance amid the rise of anti-racist rhetoric. As in other disciplines (Motsa, 2017), well-known White scholars in EER hijack the epistemic of burgeoning Black scholars (i.e., epistemological theft) without the theoretical grounding to accurately contextualize the issues under study. Again, the harm extends beyond the individuals involved; we encourage White scholars who want to assist this work to offer their financial and personnel resources to support the epistemological resources Black students and faculty possess.

4.3.2 Research Implications

What do research results mean for research, practice, and policy? The training we received – or did not receive – influenced how we make sense of our data and what we thought should be done next. In addition to collecting and analyzing data, we were tasked with interpreting our results and assigning meaning to them. We received explicit reminders to consider what our findings meant for future researchers, students, and pedagogical practices. But we were not similarly taught how to consider or put forth implications related to the redistribution of power. The positivist research perspective and White supremacist foundation limit the opportunity for implications to be put forth that engage the nuance and complexity of the Black experience in standard research communication. The interests of Black participants or general scholarship of Black authors is discounted and heavily misunderstood, so interpretations that fall outside of stereotypical conceptions of Black people are deemed unrealistic.

4.4 How Do Engineering Education Researchers Learn How to Communicate?

How we were taught to communicate our scholarship is important because it influenced the manner in which we stimulated discourse with fellow scholars. While sharing our work was a point of aspiration, we sometimes found ourselves compelled to anticipate and appease the White gaze (i.e., the reactions of White readers or observers). Regarding the White gaze, Toni Morrison offers an important question worth considering: "What happens to the writerly imagination of a black author who is at some level *always* conscious of representing one's race to, or in spite of, a race of readers that understands itself to be 'universal' or race-free?" (emphasis original, 1992, p. xii). We similarly ask ourselves how Black researchers and the ability to do engineering education research that affirms Black conceptions of social reality are affected by the prominence of White people governing every publishing outlet within the discipline and dominating the review process. In this section, we briefly discuss two ways in which we were socialized to communicate in engineering education: our experiences producing academic writing and citing previous work. Combined, both of these elements of the knowledge production process influence our ability to enhance the literature of our discipline.

4.4.1 "Good" Academic Writing

The standard uses of the English language and current accepted forms of technical reporting might compel one to reject scholarship that does not readily conform to these communication practices.

For example, herein, language presented a limit on writing this chapter because there are lived experiences within Black American culture that we did not have the language to explain (Somé, 1995). There are concepts unfamiliar to our Eurocentric-dominated speech that we cannot express in terms of our thoughts and feelings but might have been able to through more artistic forms of expression or representation that are often rejected as being "scientific." As such, we encourage you to recognize other forms of expression and communication styles as valid and legitimate.

Whose perspectives are centered in academic writing practices? The research training we receive stressed the importance of being able to communicate in linguistic practices familiar to engineering education and aligned with the master narratives (e.g., Stanley, 2007) of the academic discipline. We were taught to produce conference papers and journal articles. As such, socialization occurred through the processes of interacting with editors and reviewers as well as the advisers and instructors who help you refine your work. In many ways, we were at the mercy of this collection of people, who would ultimately decide whether our work was "good" or credible work (Coley et al., 2021). We were subjected to what scholars describe as White linguistic supremacy (Baker-Bell, 2020). This challenge was exacerbated by the criteria for what is acceptable varying widely across venues, and there were seldom mechanisms in place to equip scholars with the skills needed to construct quality reviews of other scholars' work. While the theoretical and methodological expertise of the reviewers was often taken into account, we could not be sure that their understanding of race or racism was used as a screening mechanism.

4.4.2 Citation Practices

How should your work be situated in existing literature? The training we received influenced the extent to which we defaulted to the White literary canon that occupied much of our "literature review" space. We were taught practices that affirmed certain journals and embedded certain voices in our work. Despite deciding which literature to reference being such a central part of doing research (Penders, 2018), we were seldom taught practices for equitable and responsible referencing. This shortcoming is particularly applicable as it relates to research affirming Black intellectualism because the neglect of citing Black scholarship permits irresponsibility when scholars present their work as seminal or repeat flawed interventions due to expediency. We reinforce citation politics through the process of developing manuscripts or preparing proposals, particularly as it relates to the literature review sections (Kim, 2020).

5 Part 4 – Humanizing Practices and Recommendations

Those committed to using humanizing research practices must rebel, push back, disrupt, and refuse to comply with the socialization process described in Part 3 of this chapter. The institutional and systemic nature of anti-Blackness also requires systematic opposition. To assist in these efforts, we offer recommendations for members of the engineering education research community who wish to resist White supremacy culture. Our recommendations are rooted in two perspectives: (1) resisting White supremacy culture and (2) critical transformation of engineering education. These perspectives put forth specific self-sustaining characteristics and patterns of manifestation (e.g., ways of thinking and showing up in a culture). They help us define both our roles and the prescribed response.

5.1 We Need Research Questions That Affirm Black People

How we are socialized to ask questions is vital to our ability to examine the effects of racism because its causes and how it manifests remain a point of contention in how scholars determine its role in social phenomena. Asking questions about Black people without centering Black perspectives reinforces anti-Blackness. Zuberi and Bonilla-Silva (2008) explain the crux of what it means to study race:

> It is not a question of how a person's race causes disadvantage and discrimination. The real issue is the way the society responds to an individual's racial identification. The question has more to do with society itself, not the innate makeup of individuals. Racial identity is about shared social status, not shared individual characteristics. Race is not about an individual's skin color. Race is about an individual's relationship to other people within the society. While racial identification may be internalized and appear to be the result of self-designation, it is, in fact, a result of the merging of self-imposed choice within an extensor imposed context.
>
> *(p. 7)*

Being Black is not a problem. The problem is how society views Blackness, and our questions must indicate such. Instead of examining characteristics of Black students, it would be more sensible to examine the social arrangements within engineering that give low value to Black students' ways of existing within engineering learning environments. Noticing the disparities that disenfranchise Black students requires intentionality in how inquiries about the causes and effects of these disparities influence educational outcomes.

5.2 We Need of Research Methodologies That Affirm Black People

How we are socialized to answer questions is vital to our ability to study Black people in humane ways. Answering questions about Black people without centering Black perspectives and agency reinforces anti-Blackness. Black scholars, such as W. E. B. Du Bois, have long proposed approaches to conducting research that would make the study and analysis of Black people's existence in the United States more humane (Monteiro, 2000, 2008). Scholarship that dignifies Black social reality requires different ways of inquiry than what has traditionally been taught in engineering education doctoral programs. The field needs methodologies that directly confront the epistemic injustices that suggest Black people's testimony is unsuitable for determining what is true (Fricker, 2007). Researchers must go beyond active listening to reckoning with their own fallibility while simultaneously empowering participants to have more control of the research process. An example of this approach to research is author Holly Jr.'s (2021) work, where he centered his own voice and experience in recounting how he used his agency to teach engineering to Black boys. Outside of EER, we can learn much from Dillard's (2000) endarkened feminist epistemology, Reviere's (2001) conceptualization of Afrocentrism as a research methodology.

5.3 We Need Data Interpretations That Affirm Black People

How people are socialized to make meaning is arguably the most important aspect of scholarship. Making meaning about the experiences of Black people without a multilayered understanding of racism, Blackness, and liberation will reinforce anti-Blackness. Consider the following assertion regarding the intricacy of the Black experience:

> Understanding this distinction between a theory of racism and a theory of blackness (in an anti-Black world) is key: whereas the former may invoke Black examples, and even rely on Black experience of racism in the formation of its tenets, only critical theorization of blackness confronts the specificity of antiblackness, as a social construction, as an embodied lived experience of social suffering and resistance, and perhaps most importantly, as an antagonism, in which the

> Black is a despised thing-in-itself (but not person for herself or himself) in opposition to all that is pure, human(e), and White.
>
> *(Dumas & Ross, 2016, p. 416)*

Scholars like Dumas and Ross (2016) have asserted a more specific analysis of Blackness to explicate the sociopolitical context Black people endure. Doing engineering education research on Black people with integrity will require this specificity. It must place the current experiences of Black people within broader national and international history, allowing nuance where traditional characterizations are monolithic.

5.4 We Need Communication Practices That Affirm Black People

How we communicate our ideas is important because it restricts our ability to authentically represent our thoughts. Disregarding the devaluation of Black literature and linguistics reinforces anti-Blackness. For example, given the symbolic capital afforded by citations and metrics such as the H-Index, researchers are at the mercy of a predominantly White readership as it relates to citations. Without explicit mechanisms to counteract White linguistic supremacy, Black authors, particularly those doing work rooted in Blackness, may see no option other than adhering to the White gaze. The burden of navigating the White gaze while also trying to contribute to the liberation of Black people is similar to the notion of enduring double consciousness. As explained by Du Bois (1903): "One ever feels his twoness, – an American, a Negro; two souls, two thoughts, two unreconciled strivings; two warring ideals in one dark body, whose dogged strength alone keeps it from being torn asunder" (p. 38). Publishing in engineering education and conducting research through the lens of Blackness could be seen as two unreconciled strivings. Though an uncomfortable truth, the ideas of White researchers are deemed inherently better and will likely go underscrutinized. Storytelling is an ancient tradition across the African diaspora used for various purposes; the dearth of storytelling across the educational spectrum is a particular inhibition for Black students and scholars. Toliver (2021) asserts that "the embeddedness of storytelling in Black people's lives means researching responsibly would require storytelling in research, from our data collection to our data representation, especially when working with Black populations" (p. xvii). Our aspiration is not just to inspire creative approaches to the participants' work but to instigate new ways of thinking about engineering teaching and educational research rooted in Black cultural ways of thinking, knowing, interpreting, and representing our work.

5.5 We Need to Transform Engineering Education Research Practices

Harro's (2000) cycle of liberation describes liberation concepts that we operationalize as resistance to anti-Blackness and White supremacy socialization. For the scope of this chapter, we describe five liberation concepts: (1) critical incidents, (2) building community, (3) plan change, (4) creating change, and (5) maintaining change. Similar to White supremacy socialization, we divide the five concepts into behaviors and mindsets, where the first three concepts of liberation are related to a mindset or ways of thinking and the last two are related to observable behaviors.

The first concept we articulate is the need to experience a critical incident that challenges our current beliefs or values. For example, some researchers rejected arguments about the influence of anti-Black racism until they witnessed Derek Chauvin murder George Floyd. The reluctance to interrogate social inequity in a productive way will limit your ability to learn from others and requires us to experience certain things to appreciate the complexity of aspects of culture, such as racism.

The next concept of liberation is the building of community. Community building is key not only for the emotional and psychological support of the members of a group but also in direct opposition to White supremacist culture that seeks to divide based on race. The engineering education community has generally adopted the community of practice as a core tenet; however, stratification still occurs within the desired community. The history of Black activism also displays there is power in numbers when seeking transformative change.

The final mindset concept of the cycle of liberation is the process of developing a plan to enact the change desired (Harro, 2000). The planning change step of the liberation cycle requires organizing, fundraising, educating to define appropriate actions and assessment metrics, identifying key leadership roles to redefine policies, and shifting to practices to foster healing. The three concepts previously mentioned are approaches to allow a shift in the mindset of those who wish to engage in the cycle of liberation.

The first practical action is creating change. Creating change is related to planning change, as it is the actionable steps taken in response to the systemic change plan or approach followed. Specifically, creating the change requires the power and skill to guide the change to establish new social and intellectual practices. Creating change includes creating space for groups to dialogue in a meaningful way that leads to healing and sharing both power and accountability. Are there any venues for refuting or critiquing work published about Black people that are considered racist or otherwise problematic? Does an author-to-editor appeal process exist for scholars who encounter racist or otherwise-uninformed critiques of their work? Are there processes in place to enable post-publication reviews, responses, rebuttals, retractions of anti-Black work? Are there venues that enable scholars to condemn, categorically reject, or use their scholarly ability to show where published anti-Black scholarship went wrong? At the time of writing these words, we would largely say no.

Finally, maintaining change requires the integration of Blackness into all engineering scholarship where we model authenticity, integrity, and wholeness of a person with the appropriate response and rewards. Additionally, we must minimize the risk of scholars engaging in a Black-centered perspective in their scholarship and acknowledge the value that is brought when we amplify marginalized identities. That is to say, we implement the policies and procedures that allow the natural evolution of academic thought rather than forcing a hegemonic agenda based on White supremacy culture and exclusionary practices. Therefore, the two related actionable concepts of creating and maintaining change are non-prescriptive but guiding thoughts to engage researchers' scholarship and professional activities differently.

5.6 We Need Accomplices Who Will Embrace the Risk of Promoting Blackness

According to Powell and Kelly (2017), risk separates allies from accomplices, where the latter "seeks to locate ourselves in the movement not as benevolent supporters, but as risk-takers who aim to destabilize white supremacy in ourselves, families, schools, communities, and within the judicial system" (p. 43). In this piece, we call for both the privileged and the oppressed to engage in activities to become an accomplice for appreciating Blackness in engineering. Specific to our White colleagues, our recommendations for accomplices aim to address five concepts that keep us in the cycle of socialization: ignorance, insecurity, confusion, obliviousness, and fear.

Despite living in the information age, too many in the engineering education community remain ignorant or have a lack of understanding of the complexity of race, racism, and racial issues within the context of the US culture that is echoed, replicated, and reinforced within the engineering culture. As a result, we recommend you take intentional engagement and actively seek understanding, not just information, to initiate a continual process of learning how race is enacted within the engineering culture and its impact. A specific action could include starting a discussion group to discuss topics of disrupting inequity or critical topics introduced by podcast.

Insecurity prevents many from intervening when they witness impropriety; whether in a meeting or in the classroom, when anti-Blackness is expressed, we need engagement and dialogue. We recommend you consider positionality and power in all your work. You must be reflective and take time to clarify your intentions with specific and accurate descriptions. You may have racial power or positional power; enact your agency such that your capacity will benefit more than yourself and support Black people.

Confusion results from the myths and misinformation we have been fed, though knowledge alone is inadequate. Black people often know what afflicts us and what will be truly supportive, but others remain confused because they are unwilling to adhere to our guidance. You must develop connections with the people you wish to assist, and be willing to work in solidarity. You must be open to an alternative point of view that is not only different but also contrary to your perception of reality. An action would include reflecting on your discomfort with acknowledging systemic racism and the negative impact it has on scholars of color. Actively seek to understand if those around you share your perception of an equitable culture in your research group, department, or academic unit. Invite a critical scholar to present their research to your research group or graduate program.

Obliviousness refers to the intentional disengagement or acknowledgment of racism within the engineering culture. Obliviousness is often paired with a psychological paralysis that prevents actions, behavior adjustments, or identifying opportunities to engage. We implore you to commit to affirming Blackness with asset-driven language that centers race in the context of power in all your scholarship and concurrently continue efforts to disrupt White dominance. Specifically, we encourage you to prioritize the vulnerable and the oppressed with the focus of amplifying their voices.

Fear of losing control has always been essential to maintaining racism and division among the artificial categories of race and ethnicity. However, the existential danger that racism presents in the form of violence is very real and tangible. Thus, we stress developing strategies for your participation in transformative action and realize that your comfort with dialoguing about anti-Black racism does not make it go away; only your active confronting of the issue will make it less painful.

6 Conclusion

The canon of EER has traditionally accepted rationale offered by White scholars as to why Black students' performance tends to be lower than their White peers', while rationale offered by Black scholars is less cited by the research community. The devaluing of Black scholarship is not simply offensive; it permits the reproduction of erroneous framings of the problems we claim to address. Historically, Black scholarship was intentionally excluded; hence, Carter G. Woodson started the first Black academic journal to publish Black scholarship. Presently, publishing procedures mute the voice and logic offered by Black scholars and do not position them as the expert on a topic despite our demonstrated expertise. Whereas White scholars are quickly accepted as experts on a topic with no, or minimal, demonstrated expertise. With this chapter, we sought to portray the mechanisms that normalize this disparity and profess its consequences for the intellectual economy of research production. We encourage aspiring EER scholars, particularly those who are Black, to resist this contaminated socialization process. The problem extends much deeper than citing Black scholars, and we hope this chapter can serve as a resource for reconstructing new perspectives on the process and product of research in engineering education.

Acknowledgments

The authors are grateful for the feedback from Drs. Ruby Mendenhall and Monique S. Ross. Author Cross would like to shout-out her Privilege and Power Team for their assistance.

References

Ahmed, S. (2012). *On being included: Racism and diversity in institutional life*. Duke University Press.

Alexander, M. (2010). The new Jim crow: Mass incarceration in the age of colorblindness. The New Press.

Anderson, C. (2016). *White rage: The unspoken truth of our racial divide*. Bloomsbury Publishing USA.

Au, W. (2015). High-stakes testing: A tool for white supremacy for over 100 years. In B. Picower & E. Mayorga (Eds.), *What's race got to do with it: How current school reform policy maintains racial and economic inequality* (pp. 21–43). Peter Lang.

Baker-Bell, A. (2020). *Linguistic justice: Black language, literacy, identity, and pedagogy*. Routledge.

Bonilla-Silva, E. (2006). *Racism without racists: Color-blind racism and the persistence of racial inequality in the United States*. Rowman & Littlefield Publishers.

Branch, P. S. (2013, November 5). *Dr. Kwame Ture: No revolution without organization* [Video]. YouTube. https://youtu.be/ROew63_7sg4

Burt, B. A., Williams, K. L., & Smith, W. A. (2018). Into the storm: Ecological and sociological impediments to Black males' persistence in engineering graduate programs. *American Educational Research Journal, 55*(5), 965–1006.

Cabrera, N. L. (2017). White immunity: Working through the pedagogical pitfalls of privilege. *Journal Committed to Social Change on Race and Ethnicity, 3*(1), 74–86.

the margins: Leaning in to an antiracist review process. *Journal of Engineering Education, 110*(1), 8–14.

Creswell, J. W., & Creswell, J. D. (2018). *Research design: Qualitative, quantitative, and mixed methods approaches* (5th ed.). Sage Publishing.

Cross, K. J. (2020). Racism is the manifestation of White supremacy and antiracism is the answer. *Journal of Engineering Education, 109*(4), 625–628.

Delgado Bernal, D., & Villalpando, O. (2002). An apartheid of knowledge in academia: The struggle over the "legitimate" knowledge of faculty of color. *Equity & Excellence in Education, 35*(2), 169–180. doi:10.1080/713845282

Dillard, C. B. (2000). The substance of things hoped for, the evidence of things not seen: Examining an endarkened feminist epistemology in educational research and leadership. *International journal of qualitative studies in education, 13*(6), 661–681.

Dotson, K. (2012). A cautionary tale: On limiting epistemic oppression. *Frontiers: A Journal of Women Studies, 33*(1), 24–47.

Dotson, K. (2014). Conceptualizing epistemic oppression. *Social Epistemology, 28*(2), 115–138.

Du Bois, W. E. B. (1903). *The souls of Black folk*. A. C. McClurg & Co.

Dumas, M. J., & Ross, K. M. (2016). "Be real Black for me" imagining BlackCrit in education. *Urban Education, 51*(4), 415–442.

Fields, K. E., & Fields, B. J. (2014). *Racecraft: The soul of inequality in American life*. Verso Trade.

Fletcher, T. L., Jefferson, J. P., Boyd, B. N., & Cross, K. J. (2021). Missed opportunity for diversity in engineering: Black women and undergraduate engineering degree attainment. *Journal of College Student Retention: Research, Theory & Practice*, 1–28.

Fraser, R. J., & Griffin, M. (2020). "Why sit ye here and die"? Counterhegemonic histories of the Black female intellectual in Nineteenth-century America. *Journal of American Studies, 54*(5), 1005–1031.

Freeman, T. L., & Huggans, M. A. (2009). Persistence of African-American male community college students in engineering. In H. T. Frierson, W. Pearson, Jr., & J. H. Wyche (Eds.), *Black American males in higher education: Diminishing proportions* (pp. 229–252). Emerald Group.

Fricker, M. (2007). *Epistemic injustice: Power and the ethics of knowing*. Oxford University Press.

Griffin, K. A., Bennett, J. C., & Harris, J. (2013). Marginalizing merit? Gender differences in Black faculty D/discourses on tenure, advancement, and professional success. *The Review of Higher Education, 36*(4), 489–512.

Hampton, C., Reeping, D., & Ozkan, D. S. (2021). Positionality statements in engineering education research: A look at the hand that guides the methodological tools. *Studies in Engineering Education, 1*(2).

Harper, S. R. (2010). An anti-deficit achievement framework for research on students of color in STEM. *New Directions for Institutional Research, 2010*(148), 63–74.

Harper, S. R., & Hurtado, S. (2007). Nine themes in campus racial climates and implications for institutional transformation. *New Directions for Student Services, 2007*(120), 7–24.

Harris, C. I. (1993). Whiteness as property. *Harvard Law Review, 106*(8), 1707–1791.

Harro, B. (2000). The cycle of socialization. In M. Adams, W. J. Blumenfeld, R. Castañeda, H. W. Hackman, M. L. Peters, & X. Zuñiga (Eds.), *Readings for diversity and social justice: An anthology on racism, antisemitism, sexism, heterosexism, ableism, and classism* (pp. 15–20). Routledge.

Haverkamp, A., Bothwell, M., Montfort, D., & Driskill, Q. L. (2021). Calling for a paradigm shift in the study of gender in engineering education. *Studies in Engineering Education, 1*(2), 55–70.

Hendrix, K. G. (2002). "Did being Black introduce bias into your study?" Attempting to mute the race-related research of black scholars. *Howard Journal of Communications, 13*(2), 153–171.

Herring, C. (2009). Does diversity pay? Race, gender, and the business case for diversity. *American Sociological Review, 74*(2), 208–224.

Hesse, B. (2014). Escaping liberty: Western hegemony, black fugitivity. *Political Theory, 42*(3), 288–313.

Holly Jr, J. (2020). Disentangling engineering education research's anti-Blackness. *Journal of Engineering Education, 109*(4), 629–635.

Holly Jr., J. (2021). Equitable pre-college engineering education: Teaching with racism in mind. *Journal of Pre-College Engineering Education Research, 11*(1), 155–171.

Holly Jr, J., & Thomas Quigley, T. (2022). Reckoning with the harm of anti-Blackness in engineering education: A reparatory justice research approach. *Journal of Women and Minorities in Science and Engineering.* http://doi.org/10.1615/JWomenMinorScienEng.2022036667

Kendi, I. X. (2016). *Stamped from the beginning: The definitive history of racist ideas in America.* Nation Books.

Kim, A. L. (2020). The politics of citation. *Diacritics, 48*(3), 4–9.

Ladson-Billings, G., & Donnor, J. (2005). Waiting for the call: The moral activist role of critical race theory. In N. Denzin & Y. Lincoln (Eds.), *Handbook of qualitative research* (3rd ed., pp. 279–301). Sage.

London, J. S., Lee, W. C., Phillips, C., Van Epps, A. S., & Watford, B. A. (2020). A systematic mapping of scholarship on broadening participation of African Americans in engineering and computer science. *Journal of Women and Minorities in Science and Engineering, 26*(3), 199–243.

London, J. S., Lee, W. C., Watford, B. A., Ash, C. H., Holloman, T., Pee, C. M., & Hampton, C. (2022). Climbing uphill: Toward a common agenda for the advancement of Black Americans in engineering. *Journal of Women and Minorities in Science and Engineering, 28*(3), 101–118.

Lorde, A. (1983). The master's tools will never dismantle the master's house. In R. Lewis & S. Mills (Eds.), *Feminist postcolonial theory: A reader* (pp. 25–28). Routledge.

Marable, M. (1983). *How capitalism underdeveloped Black America.* South End Press.

McGee, E. O. (2020a). *Black, brown, bruised: How racialized STEM education stifles innovation.* Harvard Education Press.

McGee, E. O. (2020b). Interrogating structural racism in STEM higher education. *Educational Researcher, 49*(9), 633–644.

McGee, E. O., Botchway, P. K., Naphan-Kingery, D. E., Brockman, A. J., Houston, S., & White, D. T. (2022). Racism camouflaged as impostorism and the impact on black STEM doctoral students. *Race Ethnicity and Education, 25*(4), 487–507.

McGee, E. O., Griffith, D. M., & Houston, S. L. (2019). "I know I have to work twice as hard and hope that makes me good enough": Exploring the stress and strain of Black doctoral students in engineering and computing. *Teachers College Record, 121*(4), 1–38.

McGee, E. O., & Martin, D. B. (2011). "You would not believe what I have to go through to prove my intellectual value!" Stereotype management among academically successful Black mathematics and engineering students. *American Educational Research Journal, 48*(6), 1347–1389.

Monteiro, A. (2000). Being an African in the world: The Du Boisian epistemology. *The Annals of the American Academy of Political and Social Science, 568*(1), 220–234.

Monteiro, A. (2008). WEB Du Bois and the study of black humanity: A rediscovery. *Journal of Black Studies, 38*(4), 600–621.

Moore III, J. L., Madison-Colmore, O., & Smith, D. M. (2003). The prove-them-wrong syndrome: Voices from unheard African-American males in engineering disciplines. *The Journal of Men's Studies, 12*(1), 61–73.

Morrison, T. (1992). *Playing in the dark: Whiteness and the literary imagination*. Harvard University Press.

Motsa, Z. (2017). When the lion tells the story: A response from South Africa. *Higher Education Research & Development, 36*(1), 28–35.

Mueller, J. C. (2020). Racial ideology or racial ignorance? An alternative theory of racial cognition. *Sociological Theory, 38*(2), 142–169.

Muhammad, K. G. (2019). *The condemnation of Blackness: Race, crime, and the making of modern urban America*. Harvard University Press.

Mustaffa, J. B. (2017). Mapping violence, naming life: A history of anti-Black oppression in the higher education system. *International Journal of Qualitative Studies in Education, 30*(8), 711–727.

Mwangi, C. A. G., Thelamour, B., Ezeofor, I., & Carpenter, A. (2018). "Black elephant in the room": Black students contextualizing campus racial climate within US racial climate. *Journal of College Student Development, 59*(4), 456–474.

Nader, L. (1972). *Up the anthropologist: Perspectives gained from studying up*. Political Science. Corpus ID: 141077709.

Newman, C. B. (2011). Engineering success: The role of faculty relationships with African American undergraduates. *Journal of Women and Minorities in Science and Engineering, 17*(3), 193–207.

Nicole, F. (2022, August 27). *We deserve education without trauma: The occurrence of spirit-murdering on Black womxn in engineering doctoral programs* [Conference Paper]. In *2022 ASEE annual conference & exposition*.

Okun, T. (2021). *White supremacy culture characteristics*. White Supremacy Culture. www.whitesupremacyculture.info/characteristics.html

Patton, M. Q. (2014). *Qualitative research & evaluation methods: Integrating theory and practice*. Sage publications.

Pawley, A. L. (2017). Shifting the "default": The case for making diversity the expected condition for engineering education and making whiteness and maleness visible. *Journal of Engineering Education, 106*(4), 531–533.

Penders, B. (2018). Ten simple rules for responsible referencing. *PLOS Computational Biology, 14*(4), e1006036. www.ncbi.nlm.nih.gov/pmc/articles/PMC5896885/

Powell, J., & Kelly, A. (2017). Accomplices in the academy in the age of Black Lives Matter. *Journal of Critical Thought and Praxis, 6*(2), 42–65.

Rasmussen, B. B., Klinenberg, E., Nexica, I. J., & Wray, M. (Eds.). (2001). *The making and unmaking of whiteness*. Duke University Press.

Reviere, R. (2001). Toward an Afrocentric research methodology. *Journal of Black Studies, 31*(6), 709–728.

Riley, D. (2017). Rigor/Us: Building boundaries and disciplining diversity with standards of merit. *Engineering Studies, 9*(3), 249–265.

Roberts, D. (1999). *Killing the black body: Race, reproduction, and the meaning of liberty*. Vintage.

Roberts, D. (2011). *Fatal invention: How science, politics, and big business re-create race in the twenty-first century*. New Press.

Robinson, G., & Dechant, K. (1997). Building a business case for diversity. *Academy of Management Perspectives, 11*(3), 21–31.

Secules, S., McCall, C., Mejia, J. A., Beebe, C., Masters, A. S., Sánchez-Peña, M. L., & Svyantek, M. (2021). Positionality practices and dimensions of impact on equity research: A collaborative inquiry and call to the community. *Journal of Engineering Education, 110*(1), 19–43.

Settles, I. H., Jones, M. K., Buchanan, N. T., & Dotson, K. (2021). Epistemic exclusion: Scholar(ly) devaluation that marginalizes faculty of color. *Journal of Diversity in Higher Education, 14*(4), 493–507.

Sharpe, C. (2016). *In the wake: On Blackness and being*. Duke University Press.

Slaton, A. E. (2010). *Race, rigor, and selectivity in US engineering: The history of an occupational color line*. Harvard University Press.

Slaughter, J. B. (2015). Introduction. In J. B. Slaughter, Y. Tao, & W. Pearson Jr (Eds.), *Changing the face of engineering: The African American experience* (pp. 1–10). Johns Hopkins University Press.

Smith, C. A., & Garrett-Scott, D. (2021). "We are not named": Black women and the politics of citation in anthropology. *Feminist Anthropology, 2*(1), 18–37.

Somé, M. P. (1995). *Of water and the spirit: Ritual, magic, and initiation in the life of an African shaman*. Penguin.

Stanley, C. A. (2007). When counter narratives meet master narratives in the journal editorial-review process. *Educational Researcher, 36*(1), 14–24.

Sutton, R. I., & Staw, B. M. (1995). What theory is not. *Administrative Science Quarterly*, 371–384.

Thomas, L. D., & Watt, D. L., & Cross, K. J., & Magruder, J. A., & Easley, C. R., & Monereau, Y. J., & Phillips, M. R., & Benjamin, A. M. (2016, June 25). *As purple is to lavender: Exploring womanism as a theoretical framework in engineering education* [Conference Paper]. In *2016 ASEE annual conference & exposition*.

Tillman, L. C. (2002). Culturally sensitive research approaches: An African-American perspective. *Educational Researcher, 31*(9), 3–12.

Toldson, I. A. (2019). *No BS (Bad Stats): Black people need people who believe in Black people enough not to believe every bad thing they hear about Black people*. Brill Sense.

Toliver, S. R. (2021). *Recovering Black storytelling in qualitative research: Endarkened storywork*. Routledge.

Toole, B. (2022). Demarginalizing standpoint epistemology. *Episteme, 19*(1), 47–65.

Vox. (2021, March 31). *Are we automating racism* [Video]. Youtube. https://youtu.be/Ok5sKLXqynQ

Yancy, G. (2016). *Black bodies, white gazes: The continuing significance of race in America*. Rowman & Littlefield.

Zinn, H. (1980). *A people's history of the United States*. HarperCollins.

Zuberi, T. (2000). Deracializing social statistics: Problems in the quantification of race. *The Annals of the American Academy of Political and Social Science, 568*(1), 172–185.

Zuberi, T. (2001). *Thicker than blood: How racial statistics lie*. University of Minnesota Press.

Zuberi, T., & Bonilla-Silva, E. (Eds.). (2008). *White logic, white methods: Racism and methodology*. Rowman & Littlefield Publishers.

28

Broadening Dissemination Genres to Share Hidden Insight via Design Cases in Engineering Education Research

Vanessa Svihla, Lizandra C. Godwin, Ardeshir Raihanian Mashhadi, and Jennifer R. Brown

1 Introduction: Hidden Insight

Typical publications offer little insight into the educational designs – such as curricula, learning activities, outreach events, and courses – we create, be it as instructors teaching our classes or directing engineering outreach activities or as engineering education researchers developing interventions to study how certain kinds of experiences result in more inclusive or effective outcomes. Journals offer limited space for reporting such work, resulting in concise summaries of settings, materials, and procedures that offer little to go on. For instance, in their study framing preschool children playing with blocks as engineering design, Gold et al. (2021) describe their intervention as offering "two small (2-inch) figurines from the TV program, 'Bob the Builder,' and ask[ing] children to briefly discuss and plan something they wanted to build for their figurines" (p. 809). While certainly sufficient for their study purpose, this unsatisfyingly brief account leaves many questions unanswered: What happens before this moment? How much have the children played with these blocks before? Are they all familiar with "Bob the Builder"? Does it matter? Were other toys or shows considered and ruled out? Why not use a situation instead of a show? None of these questions need to be answered for the particular study reported, but they matter for the reader who wants to try this in their own context, whether as a new study or as an outreach program. In light of the so-called "replication crisis" in psychology, in which replication studies fail to find the same results (Wiggins & Christopherson, 2019), one consideration is the adequacy of the broader design that is disseminated in these research studies.

While typical engineering education research (EER) publications provide understanding about learning, participation, and development processes and theories, a great deal of insight gained in the process of developing the learning experiences remains covert. Specifically, although it is becoming more common to publish datasets and more complete methods as supplemental files, the learning experiences and environments that we design, and especially the decisions we make in the process, are seldom reported. Even related disciplines, such as decision-based system design, focus on rationalizing decisions by connecting the engineering attributes to consumer-oriented product attributes and do not report the paths designers take in a vicarious way (Chen et al., 2012). Yet empirical

DOI: 10.4324/9781003287483-34

insights into how and what we design are a valuable source of information that can supplement and enrich more common dissemination practices. Further, as many of us have had abundant experiences as learners with "traditional" instruction, design cases can also serve as a source of inspiration for instructors to counter this preponderance of exposure to traditional learning designs.

Design cases are a dissemination genre that shares vivid and vicarious accounts of *what* and *how* we design (Boling, 2010). In this chapter, we primarily focus on the design of learning experiences (e.g., curricula, learning environments, instructional tools, assessments, etc.), building on the practice in fields like architecture, fashion, and graphic design of disseminating cases that illuminate specific design processes and products (Svihla & Boling, 2020).

While typical EER publications report on the *impact* of our learning/instructional designs, design cases share *how* we designed the learning experiences – usually related as an engaging story about the decisions and challenges, and with figures depicting drafts and final versions. Design cases are "a form of knowledge-sharing" that is "common to – and critical in – fields where design is a central practice" (Svihla & Boling, 2020, p. 630). Engineering faculty may be familiar with two design fields: engineering design and learning/instructional design. In this chapter, we focus on the latter.

While many engineering faculty have only experienced formal design education in their own undergraduate capstone course, others draw upon industry or start-up experience or teach design in some fashion (e.g., first-year design, design challenges threaded in core courses, mentoring capstone or intramural teams, facilitating K–12 outreach design experiences). Even a distant capstone class may provide a framework for thinking about design and approaching the development of learning and outreach experiences *as designers*. Engineering educators commonly share insights at conferences about their innovations in teaching and outreach, yet we seldom learn about the decisions and challenges encountered along the way. By publishing design cases, the field stands to articulate and amass critical knowledge that can inform both researchers and educators – knowledge that can reap gains in the support of a productive learning environment and a sustainable engineering curriculum. In our experiences reading design cases and supporting students and colleagues to do so, we have noticed that we come away with new insights for our own ways of designing learning experiences and new inspiration for the kinds of learning experiences we might create. While repositories of teaching cases and other sources of teaching materials (e.g., TeachEngineering, KEEN Cards, LearnChemE, etc.) offer materials to download and use, compared to design cases, these resources offer little insight into the decisions made in the process of developing such materials. Yet sharing that process and the insights gained along the way can help others that seek to adopt these resources.

In this chapter, we describe what design cases are and offer practical advice for scholars interested in writing a design case. Throughout this chapter, we make reference to the design case template (Appendix) to further elaborate on the specific sections of a design case. The template was originally developed by the first author, who teaches graduate students to write design cases. We adapted this template as we embarked on writing this chapter, as co-authors annotated the original version with questions and wonderings. In particular, the design case template may be useful for engineering faculty whose training is primarily technical, as there is a steep learning curve when trying to make sense of education research–related terminology. In our experience supporting those new to the genre, we have noticed that investing some time to learn about this style of writing and why many find it valuable is helpful.

We also illustrate the value of design cases by examining exemplars from engineering education and contrast design cases with other similar-sounding genres, such as design-based research and case study.

2 What Are Design Cases?

Design cases are a genre of scholarly, empirical writing in which the process of designing some specific solution is shared. While some fields may publish design cases that report on technical designs (e.g., the design of a surgical device, a novel sensor, a system for monitoring wastewater), our primary focus in this chapter is on designs that support learning and, to a lesser degree, other human-centered design fields. In the former, the designers might be instructional designers or faculty who design courses or outreach programs, and the stakeholders are usually learners and sometimes other instructors.

Typically offering a vivid depiction of the context and designed solution, design cases share a vicarious narrative – usually from the point of view of the designer (Boling, 2010). Many canonical design fields like architecture, product design, and graphic design share and curate their designs in ways that highlight the process and product (Ferris et al., 2004; Gardner & Fishel, 2009; Kim et al., 2004). Venues like the *International Journal of Designs for Learning* (*IJDL*) offer insights into how people design learning tools and experiences for diverse settings, like museums, camps, university courses, and workplaces – and notably offer few design cases in engineering. Design cases might also feature justice, equity, diversity, and inclusivity (JEDI) innovations (Brown et al., 2018; Suarez-Grant & Haras, 2022). In addition to the design cases published in IJDL, we find design cases in the fields of human–computer interaction (HCI) and computer-supported collaborative work (CSCW).

Here, we share a few examples from engineering education research. In the section that follows, we discuss them in more detail to highlight their utility and to draw some distinctions about variants within this genre. We selected these papers by asking each author to use their own interests in engineering education as a guide for choosing a design case. We encourage readers to access the cases and review them in tandem with our analysis.

Bull et al. (2018) describe the design of Make-to-Learn Electric Motor Invention Kits – henceforth referred to as the "motor kits design case" – that strive to teach students about engineering concepts as well as the process of invention through reconstruction of historic working models. This design case details the development of engineering activities that can feasibly be done in makerspaces, also sharing what did not work. Readers get direct insight about the decision-making process as well as the challenges faced throughout the project, including discovering that building a functioning motor was more challenging than expected. The careful and detailed documentation makes the case relatable to engineers, and such comprehensive information is not easily communicated through other forms of scholarship.

Wilson and Bruni-Bossio (2020) share how they use visual schematics and graphics to give overviews of courses, making connections between the course objectives, tasks, and outcomes, thereby providing an alternative way to help students understand the purpose of a given course. We henceforth refer to this as the "course schematics design case." The process of creating the graphics also gave the instructor an opportunity to reflect on the course holistically and make sure the course was logically constructed. As engineering faculty, we are accustomed to using figures, graphics, and schematics in our research papers and grant proposals as a means to organize our own thoughts (during the writing process) to tell the story or build our case. This design case demonstrates that the same approach would be useful in our teaching, both for our own process in course design and for student learning.

Ahn et al. (2016) describe designing a "struggle-oriented" learning experience – henceforth referred to as the "struggle STEM design case" – to motivate diverse students to pursue STEM degrees in college, a relatable and enduring issue. The instruction was developed to challenge the myth that only smart people create scientific knowledge and belong in STEM disciplines. Ahn and colleagues treated research on achievement motivation by McClelland (1978) as part of their precedent and organized a set of principles based on published research. For instance, they emphasized humanizing science, making struggle explicit, and connecting struggle to beneficial outcomes. As the learning experience is very different from much of the content-focused curricula faculty encounter, such guidelines can clarify the values the design team held and the ways others might apply them in similar contexts.

Wan et al. (2016) details the prevalence of dementia, citing research that characterizes stages of dementia, with attention to the roles of technology for extending autonomy and tracking those who wander, and reasons such technology has faced skepticism or limited uptake. Following an extensive literature review, Wan et al. (2016) offer justification and a detailed account of their data collection and analysis methods for conducting an exploratory study into the phenomenon they sought to design for – namely, developing a GPS-based technology and website solution for patient wandering, which we henceforth refer to as the "dementia design case." They detailed challenges faced in reporting their results to the industry partners, who expressed concern about its representativeness. Yet the authors shared the results of their exploratory study, detailing diversity across three contexts. They identified differences in how stakeholders perceived wandering-prevention measures and technological solutions. For instance, some objected to hiding doors behind curtains, and some described circular never-ending paths as enhancing patient confusion. These study results became a set of five design implications for their original idea. In addition, their observations that staff have little time to sit and monitor a website on a computer were developed into three design implications for a mobile app. They described their prototype concisely and then detailed a five-month pilot test across three sites. In this design case, we get a comprehensive account of challenges encountered in deploying a prototype and collecting data on its use.

3 What Makes Design Cases a Valuable Form of Scholarship?

Given the demands many of us face, why should we make space to read or write design cases, especially as they clearly fail to meet the criteria Streveler and Smith (2006) set for "rigorous" research? Scholars have critiqued the word "rigor" both for its inequitable impacts and the narrowed scope of what constitutes educational scholarship (Riley, 2017; Walther et al., 2017). Just as quality standards differ between studies that employ a randomized controlled trial, a case study, or a systematic literature review, the standards that make design cases scholarly and useful exist and differ from other genres.

To illustrate these, we draw attention to some differences across four design cases and explore features common to design cases that make them useful. In doing so, we share guidelines from foundational and recent publications *about* design cases (Gray, 2020; Howard, 2011; Reeve & Svihla, 2016; Svihla & Boling, 2020; Svihla & Reeve, 2016b; Wulf et al., 2011). Although the three articles published in IJDL focus on learning-related designs and the fourth example is an HCI-related design, we focus our comparisons on what makes them distinctive as design cases.

First, design cases vary in how much they reference academic literature and the ways in which they use it (Howard, 2011). In the motor kits design case, there are 12 citations; in the course schematics design case, there are 25 citations; while there are more than 80 citations in the struggle STEM design case and the dementia design case. However, as some design cases may have fewer and even no citations, we draw attention to the varied ways these authors treat the work they cite (Appendix Sections 11 and 16), especially in contrast to typical empirical educational research. In typical works, the aim in citing published works may be to establish what is already known, to situate the problem under study as significant to the scholarly community, to build an academic argument, to explain or develop theory, to identify gaps in understanding, and/or to discuss results in light of extant scholarship (Table 28.1). Such aims understandably foreground scholarship, but sometimes to the detriment of worldly practice and needs (Svihla & Reeve, 2016a). This is the root of the difference in how design cases typically use citations – abductively[1] rather than inductively or deductively (Dorst, 2011). Some design cases do not even mention learning theory or research on learning, though they do offer their reasoning, sometimes sharing experiential accounts for decisions (Cole Harmon et al., 2021).

Table 28.1 Contrasting Approaches to Using Citations

Aim	*In Typical Empirical Educational Research*	*In Design Cases*
Problem definition	Authors use scholarship to establish what is known about a problem. Citations are sufficient for establishing a scholarly problem, and uncited information is untrustworthy. Citations may situate the problem under study as significant to the scholarly community. Review of published work may identify gaps in understanding.	Authors may choose to cite scholarship and published data as part of an argument about an existing problem and its impacts, but such external information is not sufficient. Authors also describe gathering of local, contextual, or timely information, such as through needs assessment. (Appendix Sections 11 and 16.)
Solution development	Authors use published scholarship to build an academic argument, explain or develop theory about mechanisms for how people learn and/or develop. They may use this to account for a phenomenon or justify an intervention deductively.	Authors may elect to cite research and/or theory, as well as past designs, as *precedent* that shapes a designed solution. In doing so, outcomes of scholarship need not serve as justification, but as inspiration, used abductively. Here, how the research inspired a design idea in context is key. (Appendix Section 11.)
Evaluation	Authors discuss their results in light of extant research, situating their contribution in generalizing or transferability terms.	Authors seldom discuss their design process, solution, or the degree to which their solution functioned in light of extant research, though they may choose to suggest complications to or draw comparisons between unresolved or emergent issues in their design case and published information. (Appendix Sections 11, 13.)

Second, design cases vividly depict the designed solution (Boling, 2010). In typical empirical research, such information is given short shrift. Design cases sometimes embed interactive multimedia, link to a full design, or illustrate a single, repeated cycle in detail (Appendix Section 12). This commitment allows scholars to take inspiration from design ideas offered in a design case. In our examples, we note that in the motor kits design case, an instructional sequence, figures, and part lists for kits in the sequence is depicted (Bull et al., 2018). In the course schematics design case, figures offer examples of what faculty produced in their workshops, how faculty used these in their teaching, and a synthesis about how they treat the workshops as a formalized process. In the struggle STEM design case, figures depict their most recent iteration, as well as their formalization of their design as a set of five guidelines. Wan et al. (2016) share figures showing the prototypes and illustrate their use in three settings. Notably, in all four examples, the authors first depict their design process, generally illustrating an initial design, then summarize finalized designs.

Third, design cases offer a storied account of design processes and practices (Boling, 2010). This vicarious narrative approach engages readers in ways that allow them to bring these experiences into their set of precedent and use them in their own designing (Howard et al., 2012) – be it for learning, teaching, or research purposes (Appendix Sections 9–11). Importantly, these narratives provide insight into designers' framing agency – the decisions they make to frame and solve the problem (Svihla et al., 2019, 2021). This is the core of what makes design cases so valuable. While typical empirical research may offer a research-based justification for certain design decisions, design is not a logical, rational, or deterministic process (Jonassen, 2000). At the heart of design is abductive reasoning, in which designers use their judgment and preferences to fill in unknowns (Dorst, 2003). This means that only warranting a design using published, peer-reviewed research is likely a partial or post hoc accounting and tidying-up of what is commonly a co-evolutionary, dynamic, and responsive process (Dorst, 2019; Svihla & Reeve, 2016a).

The forthrightness of these accounts, in which authors share failures, challenges, and rejected ideas, is surprisingly compelling and useful information that is seldom shared in traditional publishing (Svihla & Reeve, 2016a). As such, design cases are useful even though they are not arguments for whether a particular designed learning experience supported learning (Appendix Sections 11, 13, and 14). For instance, in the motor kits design case, the authors offer a forthright account of the team's assumptions that historic motors would be straightforward to assemble, an assumption that proved false and led to major redesign (Bull et al., 2018). In the course schematics design case, Wilson and Bruni-Bossio (2020) describe the challenges of supporting faculty to take up student points of view in developing visuals that would be useful to students, as well as contending with their own limited capacity to design visuals. In the struggle STEM design case, the authors share the difficulties they faced integrating the stories they developed with typical course content (Ahn et al., 2016). In the dementia design case, Wan et al. (2016) share challenges encountered during pilot testing, such as needing to create a low-battery alarm, as well as challenges they faced in partnering.

Boling (2010) argues that the value of design cases comes from sharing knowledge about our design processes and decision-making. We agree that this genre offers inspiration about how and what to design. As such, we also consider some educational uses of design cases. Our purpose here is not to detail whether case-based learning works – as others have explicated this (Kolodner, 2002; Kolodner et al., 2005; Watson & Marir, 1994), including in engineering education (Fleddermann, 2000; Lambert et al., 2016; Nespoli & Lambert, 2010; Yadav et al., 2010), and published compilations to support case-based learning (Ertmer et al., 2019; Lambert et al., 2016). However, while instructional cases may be based in authentic narratives, they are usually edited to make the moral of the story clear, and they may even be fictionalized accounts. Such sanitized accounts, while they focus on a particular lesson, mask the complexity of practice, especially of design. Design cases can shed light on ways designers respond to emergent challenges and, therefore, can be a useful instructional tool for students and professionals alike. For instance, for those who typically design *for* others, reading a design case that describes co-design or participatory practices can offer insight into challenges designers encounter in navigating team dynamics (e.g., Friesen & Jacobsen, 2021; Vahey et al., 2018).

Fourth, design cases do not conclude with typical conclusions, results, and implications that traditional empirical studies offer (Appendix Sections 13–15). Instead, authors may reflect on what they considered to be distinctive in their case. They may draw attention to idiosyncratic challenges or features encountered or reflect on their own process and offer the advice they wish they had begun their design process with. Design cases typically avoid efforts to generalize, though they may offer *transferable* practices, methods, or frameworks. For instance, in the motor kits design case, Bull et al. (2018) first reflect on their design process and lessons learned, including that they needed broader knowledge and more time than they initially expected, and then offer a concise summary of the impact of their effort in terms of teachers and students using what they developed. In the course schematics design case, Wilson and Bruni-Bossio (2020) likewise reflect on what they learned in the process, including that they enjoyed sharing ideas across disciplines and having opportunities for feedback. They then offer a conclusion that summarizes their design case concisely, foregrounding the most salient aspects without drawing conclusions or implications beyond their context. The struggle STEM design case offers lessons learned and a formalization of their approach as a set of five guidelines. Finally, the dementia design case likewise offers lessons learned about partnering and managing a project, as well as for their specific context of managing dementia using technology solutions. They also provide a concise summary of their design case, much as in the course schematics design case.

Because of the detail about context, aims, and decisions, others can take ideas from design cases and transfer them to their own situations. This aligns with longstanding attention in the field of design research on transferability (Chow & Ruecker, 2006). Thus, design cases offer readers a chance to peer into the complex decision-making process of design, expanding the set of precedents that they can then access in their own design work. Design cases offer inspiration about both design processes and designed solutions. It is this distinctive value that can help bridge the gap between engineering education research and practice.

4 Planning and Writing a Design Case

A design case may complement other publication efforts on a project. While studies published and presented in other venues report on the impacts and insights gained through a research study, the description of the learning intervention and its design can be conveyed usefully in a design case. Yet writing a design case presents challenges, in part because most of our work presents a sanitized account (Svihla & Reeve, 2016a). Academics typically struggle to write in a forthright, narrative style and to share their judgments and preferences rather than aiming at objectivity (Howard, 2014; Smith, 2010). We offer practical advice about planning a design case. Our first suggestion is to read several design cases and reflect on their style and what that style offers.

We provide guidance on norms and variability within norms for writing design cases, drawn from published guidelines (Howard, 2011; Li et al., 2021; Wulf et al., 2011, 2015), and experience writing, editing, and reviewing design cases. Design cases typically provide detail about the context and problem, the particular design process as it unfolded, a vivid depiction of the design itself, and a concluding reflection (Howard, 2014). They might also provide insight on a collaborative design that has been refined through multiple iterations illuminating how feedback was incorporated into their design process (Friesen & Jacobsen, 2021; Matuk et al., 2016).

4.1 Who Are the Authors? Who Was Involved in the Design Case?

In the introduction, design cases typically position the authors (Appendix Section 3). Some cases are written by the designers themselves, or a subset of a design team, yet others are told by someone external to the team, such as when writing historic design cases (Howard & Gray, 2014). Often, authors provide too little detail about their roles and identities (Howard, 2014), an issue that has long been resolved in qualitative research via positionality statements. Positionality statements can increasingly be found in EER, even in quantitative research (Hampton et al., 2021; Secules et al., 2021); these statements help readers understand how social and professional roles might have shaped the study, including how the authors gained access to the study site, the trustworthiness of data, and the experiences brought to bear on the interpretation process.

We encourage design case authors to include detail in positioning themselves, helping the reader understand how their professional and social roles and identities may have afforded them greater access to particular information, and how these roles and identities may have shaped their design decisions or perceptions of decisions (Appendix Section 3). And in describing the process of designing, authors should offer insight into their decision-making. Even those decisions warranted by findings from research are picked subjectively – there are *many* research findings that could be incorporated, after all. The design decisions made based on research findings are also brought into particular contexts in ways shaped by personal experiences and preferences. If all four authors of this chapter decided to design an intervention based on the results of a study showing that contextualizing student learning of slope did not result in learning gains but did change students' beliefs that learning about slope is important (Bowen & Peterson, 2019), we would produce different designs.

4.2 What If I Cannot Back Design Decisions Rationally?

Another challenge for authors relates to tensions between subjectivity and objectivity, especially as many engineering faculty hold positivist/objectivist epistemologies (Beddoes, 2014; Claris & Riley, 2012). Faculty accustomed to using passive voice and being able to measure or model almost all the variables that might impact outcomes of their technical research may find the first-person narratives that admit to subjectivity challenging to craft (Appendix Section 4). Compared to the well-understood laboratory setting, where accounting for contextual effects might be limited to environmental conditions (e.g., ambient temperature, humidity, altitude) and perhaps apparatus issues (e.g., calibration data, actual versus expected functionality of equipment), there are many more potential factors at play when considering context involved in learning. For instance, some students may have had too little sleep, some may already understand the topic and be bored, others too little and be confused; some students may be distracted by external events, and others may draw inspiration to focus from their life experiences. And while there is research that may offer guidance on many of these factors, even if we knew all the contextual factors at play, we would not have hard-and-fast rules for whether, how, and how much they impact each learner. Yet designers commonly consider these factors, often leveraging their own experiences as learners and as instructors as they make design decisions. They may even discount research on learning if they cannot envision how it could function in the context they know well. Authors of design cases embrace the idea that design decisions often come from their experiences. They show ownership of their decisions and reflect on the information that shaped their decisions (Appendix Sections 4 and 11).

In contrast, in typical EER, learning experiences are presented concisely as an intervention or treatment based in the literature review, with few or no details about design process. For example, in *J-PEER*, authors explained the match between their literature review and treatment as "structured around the concept of authenticity for content delivery; that is, as a means to attach relevance to the concept of slope and demonstrate how it might be used to design engineering structures and solve engineering-related problems," and then detailed the treatment in three paragraphs (Bowen & Peterson, 2019, pp. 3–4), enough information for a reader to get the gist, but not enough information to support replication. Even in more elaborated examples that share a design justification, we seldom detect even a trace of subjectivity, and when pilot testing and revision are described, the outcomes and nature of revisions are left to the readers' imaginations. Our purpose here is not to critique specific authors for not sharing forthright accounts but rather to acknowledge that this practice is pervasive and linked to notions of objectivity.

Given the subjectivity expressed in design cases, readers may wonder whether design cases can be considered "rigorous." Smith (2010), in articulating what makes design cases rigorous, relates design cases to qualitative research. She suggests trustworthiness stems from prolonged engagement with the design, the inclusion of rejected ideas and attempts, employing thick description, and triangulation across sources of information, such as emails, sketches, photographs, and drafts or prototypes, as well as designers' accounts of the process. For this reason, we advise developing strategies for documenting design process, such as developing and using a file-naming protocol to track versions, making design memos to trace decision-making, audio-recording design conversations when permissible (and using these recordings in accordance with ethical standards for research on humans and based in guidance from the ethics or institutional review board). For instance, in a design case written by one of the authors, we often recorded and transcribed conversations with teachers as we were designing projects (Svihla et al., 2016). These recordings helped us recreate the sequence of decisions, including the many on-the-fly decisions that happened responsively within teaching.

4.3 What Should I Include?

Authors often encounter challenges in deciding what to include and exclude, from describing the context and learners to detailing an interesting and understandable story (Appendix Section 5). Smith (2010) argues that the trustworthiness of a design case depends in part on "thick description," a notion drawn from ethnographic research (Geertz, 2002). Given this, our advice is to include more detail than typically feels comfortable and then let a peer reviewer advise on what seems unnecessary. Alternatively, the author may develop a practice of asking, "Have I provided enough information that the reader can really picture the context?" and "Are there details that I have left out that, had they been quite different, would have had an important impact on my design process or designed solution?" For instance, as an author, I might not think to describe details about my learners because they seem so normal to me.

Design cases also include detailed insight about what prompts revision (Appendix Sections 13 and 14), compared to typical EER publications. In an example of the latter, Langman et al. (2019) explained revisions concisely:

> Revisions based on this pilot testing led to the next draft, which was then piloted with high school students from an urban parochial high school in an honors-level biology class. The module was revised again and piloted with a college-prep biology class at the same parochial high school.
>
> *(p. 6)*

In contrast, design cases offer detailed and honest accounts, including of failures.

> This plan was founded on a significant misconception; the design team believed that the members of the advisory board would be able to successfully construct a motor. [. . .] Despite this belief, none of the members of the advisory board were able to construct a working model of a rotary motor.
>
> *(Bull et al., 2018, p. 4)*

4.4 Should I Include Data about Student Engagement and Learning?

Those following the norms for design cases in HCI and CSCW should also detail "appropriation," meaning, what happens to the design in the hands of stakeholders in the context of use (Appendix Section 14). Wulf et al. (2011) argue that studying the implementation provides critical design knowledge and, additionally, that planning to document implementation can orient the designers to better anticipate how stakeholders might (mis)use or adapt designs, enabling the designers to be more proactive (de Carvalho, 2021; Wulf et al., 2015). We offer advice and concerns about this practice.

The practice of examining how designs function in real-world situations, whether as a pilot or full-scale implementation, can provide critical feedback that is impossible to get in other ways. However, when stakeholders adapt designs in unexpected ways – especially in ways that go against our vision – our impulse may be to redesign to force fidelity and reduce stakeholder agency (Beever & Brightman, 2016). These trends are present often in educational settings and tend to reduce the professionalism of instructors and treat learners as homogenous. Instead, we favor the idea of developing designs that are intended to be malleable in stakeholders' hands (Cabitza & Simone, 2017) and educative, helping them understand how specific aspects of the design function (Davis et al., 2014), fostering buy-in.

A second concern about studying implementation relates to the complexity and ethics of research on humans. For those who are new to doing research on teaching and learning, it may be surprising to realize that such studies should be reviewed by their institutional review board (IRB). As this is often an intimidating process, one response is to simplify data collection strategies, relying on anonymous, voluntary surveys. While such information is often safer because it normally cannot be connected to a particular student, it typically provides thin data. IRBs vary in how they interpret relevant laws and in how risk-averse they act. The first author, who has worked on research across the United States and Canada and dealt with IRBs across eight institutions, received highly varied guidance about consent and studying their own course, from one university treating it as exempt to another viewing it as deception. Following your own design, even into your own course, can therefore pose complications. We recommend partnering with more experienced education researchers if you want to share evaluation and implementation data that include learner data or restricting what you share to your own perceptions. We also encourage designers to co-author with others involved, such as teaching assistants, and if the learners are adults, including some of them as authors can be a means to share the design case in a polyvocal format, offering multiple points of view on the experience. However, be attentive to power imbalances when involving students, such as by providing a means for them to share any concerns about the process with a third party.

5 What about Case Study and Design-Based Research?

While the terms may sound similar, and while design cases, case studies, and design-based research all result in empirical scholarly publications, they result from distinct activities. Keep in mind that a design case is not a research method as we distinguish these forms of scholarship.

5.1 Design-Based Research

Design-based research (DBR) originated in the learning sciences field but has been utilized in EER to study learning (Johri & Olds, 2011; Koretsky et al., 2021; Lyon & Magana, 2021). DBR aims to build theory by "instantiating" theory into a learning design and repeatedly testing it under real-world conditions (Brown, 1992; Svihla, 2014; The Design-Based Research Collective, 2003). DBR, rather than focusing on "what works," explores the conditions under which a design and its related theory might and might not support development and learning by taking into account context. DBR prioritizes data that treat learning as a *process*, including artifacts such as student work and recordings of interactions.

DBR is iterative, with further refinement of the theory at each step, often in the form of a set of conjectures about how certain kinds of participation can be encouraged, and in turn, how such participation supports learning (Sandoval, 2014). DBR can be considered to be a response to the issues of well-controlled laboratory studies that failed to function "in the blooming, buzzing confusion" of actual classrooms (Brown, 1992, p. 141). Because of the open stance DBR researchers take, emergent and unexpected events can reshape the design and theory. Thus, DBR is a shift away from established education research methods where the variables are known *a priori* (Collins, 1992).

While these characteristics should set DBR apart from design cases, there are some ambiguities we address. First, learning scientists have used multiple terms to describe DBR, such as "design experiments," "educational design research," and "design research." We find the last term particularly confusing, as there is a field named "design research" that publishes the journal *Design Studies*. We prefer the more distinctive term, *design-based research*, in part because the primary aim of this methodology is theory work, and few DBR publications offer insight into design processes (Svihla & Reeve, 2016b). However, as DBR has become increasingly popular, it is also common to find studies labeled as DBR that neither develop nor even make explicit how theory relates to a designed intervention but simply document a single use of a design in real-world settings. Most such papers are also not design cases, as they do not make clear how and why they designed the particular intervention.

A second issue, and this shows up in design cases as well, is the creation of so-called "design principles." Classically, and across multiple fields, design principles are based in human visual perception and treated as universal (Lin, 2013). While those in fields like design research have developed more expansive views of design principles, they are still consistently about how people design, and are "fundamental, primary, or general laws or truths" that guide the kinds of design methods or practices designers might use (Blizzard & Klotz, 2012, p. 468), rather than about cognition and learning. A review of the kinds of information often labeled in DBR studies and design cases as design principles, however, shows this information to be about learning, not designing (Kali, 2008). For instance, "humanize content knowledge by providing the stories" and "make the learning process vivid with explicit actions and strategies" are guidance about the learning experience, not guidance about how to design (Ahn et al., 2016, p. 78). Furthermore, few of these principles are "fundamental, primary, or general laws or truths" (Blizzard & Klotz, 2012, p. 468). Rather than aiming at such generalization, one alternative would be to adapt the notion of "design patterns" which are recurrent problems paired with general solutions that can be used "a million times over, without ever doing it the same way twice" (Alexander et al., 1977, p. x). Originally proposed in architecture and planning and taken up in fields like HCI, the notion of design patterns as reusable and adaptable heuristics fits the nature of design work described in both DBR and design cases. Sharing design patterns and the conditions and contexts under which they might transfer would be an appropriate focus for a section of a design case detailing lessons learned.

5.2 Case Study

Case study is a notoriously varied approach, a situation enhanced by the generality of the term *case* (Ragin & Becker, 1992). In fields like business, it may refer to something like a design case or an instructional case, and in medicine, cases may document an unusual illness. We focus on case study as used in education research, where it is typically defined as a systematic investigation of a bounded situation (Stake, 1995, 2013; Yin, 2003), commonly using a single in-depth account or multiple focused accounts to explore, compare, or synthesize experiences of or within that situation. Case study, unlike DBR, typically does not focus on an intervention designed by the authors. Research questions in case study begin with *how* or *why* (Yin, 2003) and shed light on how context, culture, and communities shape learning environments and experiences.

The small sample size may lead some engineering faculty to feel skeptical about the methodology. However, this small number of participants provides an opportunity to understand a phenomenon in depth and to surface new knowledge about the role of contextual factors. The specific, context-dependent knowledge is where a case study draws its methodological strength (Case & Light, 2011). Like most qualitative methodologies, the primary purpose of case study is not to make generalizations about populations but rather to illustrate and understand how processes like learning, participation, and development unfold over time and are perceived by participants (Creswell, 2013; Denzin & Lincoln, 2008). As such, case study has become increasingly common in EER, providing valuable insight into complex processes like professional engineering identity formation in response to course norms (Danielak et al., 2014).

Case studies are conducted in their natural settings rather than in a closely controlled laboratory, where the variables are limited. A key attribute of case studies is the triangulation of multiple sources of evidence (Stake, 1995, 2013). However, the omnivorous nature of case study, in which almost anything can count as data (Stake, 1995, 2013), contributes to confusion about this method. Fortunately, scholars have offered guidelines to address common confusions and critiques (Flyvbjerg, 2006; Ragin & Becker, 1992).

5.3 How Do Design Cases Differ from DBR and Case Study?

There are a number of similarities between design cases, case study, and DBR. All three report on empirical scholarship *related* to learning as it happens *in situ*, rather than in well-controlled conditions. Authors may take up varied epistemologies, though an interpretivist stance is common across all three. Publications of design cases, DBR, and case studies may make use of any form of data, from surveys and assessments to video recordings and written reflections. We summarize ways they differ in purpose and form in Table 28.2 to illustrate that, fundamentally, design cases offer insights

Table 28.2 Common Purposes and Procedures across Design Cases, DBR, and Case Study in the Broad Context of Teaching and Learning and as They Can Be Used in EER

	Design Cases	*Design-Based Research*	*Case Study*
Purposes/role of inquiry	Illuminate instructional/learning design process and solutions; build knowledge in community of design practices	Build learning theory and instructional/learning designs that instantiate that theory	Understand the how and why of a phenomenon from participant point of view
Role of authors	Designers (or historians)	Designer-scholars	Scholars
Role of context	A focus of investigation in order to frame and solve an instructional/learning problem	Aspects may be foregrounded and theorized to "humble" an existing theory; theory may change responsively to context	A focus of investigation to understand participants and their perceptions; may help bound the case
Role of designing	Design process is the primary focus and shared as vicarious, first-person narrative	Study design (which sites, which data to collect, etc.) may reflect design process (about the designed intervention, how many iterations, how and why changes were made responsively)	Restricted to study design, such as how many cases, which data to collect, and how to analyze data
Role of designed solution/intervention	Depicted vividly to serve as precedent for others	To make theory testable in real-world conditions	None, though pre-existing designs may provide a context for or suggest bounding for a case study
Relation to justice, equity, diversity, inclusivity (JEDI)	Can illustrate JEDI design practices and equitable/inclusive learning designs (e.g., Brown et al., 2018)	Increasingly common to hold JEDI aims, involving participatory design practices (e.g., Gutiérrez & Jurow, 2016)	Uncommon, but from a critical stance, cases could illustrate (in)equitable phenomena (e.g., Allen, 2008)
Use in EER	Rare	Increasingly common	Increasingly common

into how people design by sharing accounts of design practice with depictions of what was designed. Compared to other forms of scholarship, design cases offer inspiration and precedent, providing insight into the complex practices designers use. This can be particularly potent when aiming to expand one's repertoire of practices to meet goals related to justice, equity, diversity, and inclusivity (JEDI), offering insight into participatory and co-design practices that mitigate the impacts of power imbalances and offering inspiration about what JEDI learning designs might be like.

Conclusions

In this chapter, we explored what design cases are: they offer vivid depiction of designed learning experiences paired with vicarious narratives of the particular design process that led to a particular learning/instructional design. We articulated that design cases are valuable for their capacity to illustrate design practices and for their potential to offer inspiration – both for how and what we design. We shared solutions to common challenges for those new to writing design cases, including using active voice; detailing the authors, learners, and context; engaging with literature not to justify decisions but as potential sources of inspiration; and displaying framing agency – the capacity to make decisions that are consequential to framing the problem (Svihla et al., 2019, 2021) – by owning and accounting for the subjective decisions made. We contrasted design cases with other scholarship, including design-based research and case study, to illustrate what makes design cases distinctive, including that design cases are empirical, but are not a form of research methodology. We close with suggestions for authors, journals, and conference organizers that might build capacity in this form of writing.

We encourage authors of design cases to seek out venues specific to engineering education and to query whether they would consider a design case, such as the *Journal of Engineering Education, European Journal of Engineering Education, and Australasian Journal of Engineering Education*. In our own experience, it is feasible to publish a design case in conferences hosted by the *American Society for Engineering Education* (Kellam et al., 2021). In fact, a design case represents a comparatively small change for authors who publish reports of teaching innovations without a systematic study of their impact. Sharing the decisions and design process would enhance such papers by supplementing innovations with insight into the design inspiration and paths not taken.

Likewise, certain types of grants lend themselves to the development of design cases. For instance, projects that place teachers in laboratories and then support teachers to develop activities for their own classrooms could be published as design cases. Grants that develop new courses or outreach programs could offer useful information for future grant-seekers by sharing the ins and outs of their process.

Authors can consider ways to complement other publications with a design case. For instance, Gold et al. (2021) recently reported concisely on an intervention to uncover engineering-related aspects in young children's play. They identified six related behaviors, including explaining how things work. They offer brief justification for their use of wooden blocks, figures, and a framing for the task, appropriate to the paper type. Yet such an intervention could also be detailed into a design case. The authors might share personal anecdotes that influenced their ideas for the study intervention or, as they continue the line of work, share the various forms explored but that did not lend themselves to supporting engineering behaviors in play. Likewise, Marra et al. (2021) studied the impact of an innovative undergraduate engineering program using think-aloud protocols and interviews. Although they share more detailed information about the program than many studies do, a design case detailing the program's development would offer greater insight about the nuances of the program and, importantly, how and why specific choices were made.

Second, we encourage journals and conferences that publish teaching innovations – like ASEE, technical conferences that offer education sessions, discipline-based education journals like *Chemical*

Engineering Education, and *Advances in Engineering Education* – to explore and adapt the guidelines set by IJDL (https://scholarworks.iu.edu/journals/index.php/ijdl/about/submissions) for design cases, including developing clear rubrics that shape and focus feedback and that discourage inadvertent IRB violations. However, it is important to note that authors and reviewers alike may benefit from support (Howard, 2011). This could include workshops at conferences that help authors plan each section of their design case, with opportunities to then participate in a community of practice as they craft their manuscripts. Likewise, pairing more- and less-experienced reviewers and allowing them access to each other's reviews and the editor's comments can help them develop expertise in reviewing design cases. In editing design cases for a book – including from authors who had not written one previously – we recruited experienced design case authors to serve as mentors and co-authors. As design cases commonly include, and depend heavily on, illustration and even multimedia, one challenge some journals may face are limitations posed by print media. However, this could be addressed by publishing such materials as a supplement, and perhaps with QR codes in the print versions.

Ultimately, it is our hope that engineering educators and engineering education researchers discover value in the scholarly genre of design cases, contributing to an expanded set of inspirational precedent and more detailed knowledge of design practices. By sharing these aspects of our work in vivid and vicarious ways, we stand to enhance the capacity of the community to design more innovative and equitable learning opportunities.

Acknowledgments

This material is based in part upon work supported by the National Science Foundation under Grant EEC 1751369. Any opinions, findings, and conclusions or recommendations expressed in this material are those of the authors and do not necessarily reflect the views of the National Science Foundation. The work by Ardeshir Raihanian Mashhadi was done prior to joining Amazon.

Note

1 Abductive reasoning is used when there is incomplete information about the problem and how to go about solving it. Rather than deducing or inducing solutions, tentative conjectures are tested, providing additional information about both problem and solution. *Abductive reasoning* is synonymous with "design thinking" in the field of design research (Kolko, J. (2010). Abductive thinking and sensemaking: The drivers of design synthesis [Article]. *Design Issues*, *26*(1), 15–28. https://doi.org/10.1162/desi.2010.26.1.15).

References

Ahn, J., Luna-Lucero, M., Lamnina, M., Nightingale, M., Novak, D., & Lin-Siegler, X. (2016). Motivating students' STEM learning using biographical information. *International Journal of Designs for Learning*, 7(1). https://doi.org/10.14434/ijdl.v7i1.19409

Alexander, C., Ishikawa, S., Silverstein, M., Jacobson, M., Fiksdahl-King, I., & Angel, S. (1977). *A pattern language: Towns, buildings, constructions*. Oxford University Press.

Allen, A. (2008). Learning to teach, learning to play the game: A case study of a minority teacher candidate teaching for equity, diversity and social justice. *Comparative and International Education*, *37*(2). https://doi.org/10.5206/cie-eci.v37i2.9120

Beddoes, K. (2014). Methodology discourses as boundary work in the construction of engineering education. *Social Studies of Science*, *44*(2), 293–312. https://doi.org/10.1177/0306312713510431

Beever, J., & Brightman, A. O. (2016). Reflexive principlism as an effective approach for developing ethical reasoning in engineering. *Science and Engineering Ethics*, *22*(1), 275–291. https://doi.org/10.1007/s11948-015-9633-5

Blizzard, J. L., & Klotz, L. E. (2012). A framework for sustainable whole systems design. *Design Studies, 33*(5), 456–479. https://doi.org/10.1016/j.destud.2012.03.001

Boling, E. (2010). The need for design cases: Disseminating design knowledge. *International Journal of Designs for Learning, 1*(1), 1–8. https://doi.org/10.14434/ijdl.v1i1.919

Bowen, B., & Peterson, B. (2019). Exploring authenticity through an engineering-based context in a project-based learning mathematics activity. *Journal of Pre-College Engineering Education Research (J-PEER), 9*(1), 1. https://doi.org/10.7771/2157-9288.1073

Brown, A. L. (1992). Design experiments: Theoretical and methodological challenges in creating complex interventions in classroom settings. *The Journal of the Learning Sciences, 2*(2), 141–178. https://doi.org/10.1207/s15327809jls0202_2

Brown, J., Ring-Whalen, E., Roehrig, G., & Ellis, J. (2018). Advancing culturally responsive science education in secondary classrooms through an induction course. *International Journal of Designs for Learning, 9*(1), 14–33. https://doi.org/10.14434/ijdl.v9i1.23297

Bull, G., Garofalo, J., Littman, M., & Hoffman, M. (2018). The make to learn electric motor design sequence. *International Journal of Designs for Learning, 9*(1), 1–13. https://doi.org/10.14434/ijdl.v9i1.22790

Cabitza, F., & Simone, C. (2017). Malleability in the hands of end-users. In F. Paternò & V. Wulf (Eds.), *New perspectives in end-user development* (pp. 137–163). Springer. https://doi.org/10.1007/978-3-319-60291-2_7

Case, J. M., & Light, G. (2011). Emerging research methodologies in engineering education research. *Journal of Engineering Education, 100*(1), 186–210. https://doi.org/10.1002/j.2168-9830.2011.tb00008.x

Chen, W., Hoyle, C., & Wassenaar, H. J. (2012). *Decision-based design: Integrating consumer preferences into engineering design*. Springer Science & Business Media. https://doi.org/10.1007/978-1-4471-4036-8

Chow, R., & Ruecker, S. (2006). Transferability – A wonder on the ground of design research. In *Proceedings of the design research society international conference*. https://dl.designresearchsociety.org/drs-conference-papers/drs2006/researchpapers/97

Claris, L., & Riley, D. (2012). Situation critical: Critical theory and critical thinking in engineering education. *Engineering Studies, 4*(2), 101–120. https://doi.org/10.1080/19378629.2011.649920

Cole Harmon, R., Hospelhorn, M., Gutierrez, E., Velarde, C., Fetrow, M., & Svihla, V. (2021). Mission to Mars amidst a pandemic. *International Journal of Designs for Learning, 12*(1), 158–170. https://doi.org/10.14434/ijdl.v12i1.31295

Collins, A. (1992). Toward a design science of education. In E. Scanlon & T. O'Shea (Eds.), *New directions in educational technology* (pp. 15–22). Springer-Verlag.

Creswell, J. W. (2013). *Qualitative inquiry and research design: Choosing among five approaches*. Sage.

Danielak, B. A., Gupta, A., & Elby, A. (2014). Marginalized identities of sense-makers: Reframing engineering student retention. *Journal of Engineering Education, 103*(1), 8–44. https://doi.org/10.1002/jee.20035

Davis, E. A., Palincsar, A. S., Arias, A. M., Bismack, A. S., Marulis, L., & Iwashyna, S. (2014). Designing educative curriculum materials: A theoretically and empirically driven process. *Harvard Educational Review, 84*(1), 24–52. https://doi.org/10.17763/haer.84.1.g48488u230616264

de Carvalho, A. F. P. (2021). Mastering design case studies for grounded design. *Proceedings of the European conference on computer-supported cooperative work* (pp. 1–8). https://doi.org/10.18420/ecscw2021-wsmc07

Denzin, N. K., & Lincoln, Y. S. (2008). *Strategies of qualitative inquiry* (Vol. 2). Sage.

The Design-Based Research Collective. (2003). Design-based research: An emerging paradigm for educational inquiry. *Educational Researcher, 32*(1), 5–8. https://doi.org/10.3102/0013189X032001005

Dorst, K. (2003). The problem of design problems. *Design Thinking Research Symposium Proceedings, 17*(19.11). www.creativityandcognition.com/cc_conferences/cc03Design/papers/23DorstDTRS6.pdf

Dorst, K. (2011). The core of 'design thinking' and its application. *Design Studies, 32*(6), 521–532. https://doi.org/10.1016/j.destud.2011.07.006

Dorst, K. (2019). Co-evolution and emergence in design. *Design Studies, 65*, 60–77. https://doi.org/10.1016/j.destud.2019.10.005

Ertmer, P. A., Quinn, J., & Glazewski, K. D. (2019). *The ID casebook: Case studies in instructional design*. Routledge. https://doi.org/10.4324/9781315148083

Ferris, K., Bannon, L., Ciolfi, L., Gallagher, P., Hall, T., & Lennon, M. (2004). Shaping experiences in the Hunt museum: A design case study. In *Proceedings of the conference on designing interactive systems: Processes, practices, methods, and techniques* (pp. 205–214). https://doi.org/10.1145/1013115.1013144

Fleddermann, C. B. (2000). Engineering ethics cases for electrical and computer engineering students. *IEEE Transactions on Education, 43*(3), 284–287. https://doi.org/10.1109/13.865202
Flyvbjerg, B. (2006). Five misunderstandings about case-study research. *Qualitative Inquiry, 12*(2), 219–245. https://doi.org/10.1177/1077800405284363
Friesen, S., & Jacobsen, M. (2021). Collaborative design of professional graduate programs in education. *International Journal of Designs for Learning, 12*(1), 64–76.
Gardner, B., & Fishel, C. (2009). *LogoLounge 5: 2,000 international identities by leading designers.* Rockport Publishers.
Geertz, C. (2002). Thick description: Toward an interpretive theory of culture. In C. Jenks (Ed.), *Culture: Critical concepts in sociology* (pp. 173–196). Routledge.
Gold, Z. S., Elicker, J., Evich, C. D., Mishra, A. A., Howe, N., & Weil, A. E. (2021). Engineering play with blocks as an informal learning context for executive function and planning. *Journal of Engineering Education, 110*(4), 803–818. https://doi.org/10.1002/jee.20421
Gray, C. M. (2020). Markers of quality in design precedent. *International Journal of Designs for Learning, 11*(3), 1–12. https://doi.org/10.14434/ijdl.v11i3.31193
Gutiérrez, K. D., & Jurow, A. S. (2016). Social design experiments: Toward equity by design. *Journal of the Learning Sciences, 25*(4), 565–598. https://doi.org/10.1080/10508406.2016.1204548
Hampton, C., Reeping, D., & Ozkan, D. S. (2021). Positionality statements in engineering education research: A look at the hand that guides the methodological tools. *Studies in Engineering Education, 1*(2). https://doi.org/10.21061/see.13/
Howard, C. D. (2011). Writing and rewriting the instructional design case: A view from two sides. *International Journal of Designs for Learning, 2*(1). https://doi.org/10.14434/ijdl.v2i1.1104
Howard, C. D. (2014). The rhetoric of instructional design cases: Knowledge building via examples of process and product. In B. Hokanson & A. Gibbons (Eds.), *Design in educational technology design thinking, design process, and the design studio* (pp. 107–124). Springer. https://doi.org/10.1007/978-3-319-00927-8_7
Howard, C. D., Boling, E., Rowland, G., & Smith, K. M. (2012). Instructional design cases and why we need them. *Educational Technology, 52*(3), 34–38. www.jstor.org/stable/44430039
Howard, C. D., & Gray, C. M. (2014). Introduction to the special issue on historic design cases. *International Journal of Designs for Learning, 5*(2). https://doi.org/10.14434/ijdl.v5i2.13257
Johri, A., & Olds, B. M. (2011). Situated engineering learning: Bridging engineering education research and the learning sciences. *Journal of Engineering Education, 100*(1), 151–185. https://doi.org/10.1002/j.2168-9830.2011.tb00007.x
Jonassen, D. H. (2000). Toward a design theory of problem solving. *Educational Technology Research and Development, 48*(4), 63–85. https://doi.org/10.1007/BF02300500
Kali, Y. (2008). The design principles database as means for promoting design-based research. In A. E. Kelly, R. Lesh, & J. Y. Baek (Eds.), *Handbook of design research methods in education: Innovations in science, technology, engineering, and mathematics learning and teaching* (pp. 423–438). https://doi.org/10.4324/9781315759593.ch21
Kellam, N., Svihla, V., Davis, S. C., Susan, S., & Desiderio, J. (2021). Using power, privilege, and intersectionality to understand, disrupt, and dismantle oppressive structures within academia: A design case. *Proceedings of CoNECD: The Collaborative Network for Engineering and Computing Diversity*, 1–25. https://cms.jee.org/36136
Kim, S. W., Kim, M. C., Park, S. H., Jin, Y. K., & Choi, W. S. (2004). Gate reminder: A design case of a smart reminder. In *Proceedings of the 5th conference on designing interactive systems: Processes, practices, methods, and techniques* (pp. 81–90). https://doi.org/10.1145/1013115.1013128
Kolko, J. (2010). Abductive thinking and sensemaking: The drivers of design synthesis [Article]. *Design Issues, 26*(1), 15–28. https://doi.org/10.1162/desi.2010.26.1.15
Kolodner, J. L. (2002). Analogical and case-based reasoning: Their implications for education. *Journal of the Learning Sciences, 11*(1), 123–126. https://doi.org/10.1207/S15327809JLS1101_5
Kolodner, J. L., Cox, M. T., & González-Calero, P. A. (2005). Case-based reasoning-inspired approaches to education. *The Knowledge Engineering Review, 20*(3), 299–303. https://doi.org/10.1017/S0269888906000634
Koretsky, M. D., McColley, C. J., Gugel, J. L., & Ekstedt, T. W. (2021). Aligning classroom assessment with engineering practice: A design-based research study of a two-stage exam with authentic assessment. *Journal of Engineering Education.* https://doi.org/10.1002/jee.20436

Lambert, S., Newton, C., & Effa, D. (2016). Ten years of enhancing engineering education with case studies: Insights, lessons and results from a design chair. *Proceedings of the Canadian Engineering Education Association*, 1–6. https://doi.org/10.24908/pceea.v0i0.6508

Langman, C., Zawojewski, J., McNicholas, P., Cinar, A., Brey, E., Bilgic, M., & Mehdizadeh, H. (2019). Disciplinary learning from an authentic engineering context. *Journal of Pre-College Engineering Education Research (J-PEER)*, *9*(1), 5. https://doi.org/10.7771/2157-9288.1178

Li, Q., Tolmie, P., Weibert, A., Schorch, M., Müller, C., & Wulf, V. (2021). E-portfolio: Value tensions encountered in documenting design case studies. *Ethics and Information Technology*, 1–5. https://doi.org/10.1007/s10676-020-09533-3

Lin, J. (2013). Development of scales for the measurement of principles of design. *International Journal of Human-Computer Studies*, *71*(12), 1112–1123. https://doi.org/10.1016/j.ijhcs.2013.08.003

Lyon, J. A., & Magana, A. J. (2021). The use of engineering model-building activities to elicit computational thinking: A design-based research study. *Journal of Engineering Education*, *110*(1), 184–206. https://doi.org/10.1002/jee.20372

Marra, R. M., Hacker, D. J., & Plumb, C. (2021). Metacognition and the development of self-directed learning in a problem-based engineering curriculum. *Journal of Engineering Education*. https://doi.org/10.1002/jee.20437

Matuk, C., McElhaney, K., Chen, J., Lim-Breitbart, J., Kirkpatrick, D., & Linn, M. (2016). Iteratively refining a science explanation tool through classroom implementation and stakeholder partnerships. *International Journal of Designs for Learning*, *7*(2).

McClelland, D. C. (1978). Managing motivation to expand human freedom. *American Psychologist*, *33*(3), 201.

Nespoli, O., & Lambert, S. (2010). Engineering design case studies: Effective and sustainable development methods. *Proceedings of the American Society for Engineering Education Annual Conference & Exposition*, *15*(480), 481–415. https://doi.org/10.18260/1-2-16420

Ragin, C., & Becker, H. (1992). *What is a case? Exploring the foundations of social inquiry*. Cambridge University Press.

Reeve, R., & Svihla, V. (2016). Designing: The unseen dimension of our scholarship. In V. Svihla & R. Reeve (Eds.), *Design as scholarship: Case studies from the learning sciences* (pp. 140–149). Routledge.

Riley, D. (2017). Rigor/Us: Building boundaries and disciplining diversity with standards of merit. *Engineering Studies*, *9*(3), 249–265. https://doi.org/10.1080/19378629.2017.1408631

Sandoval, W. (2014). Conjecture mapping: An approach to systematic educational design research. *Journal of the Learning Sciences*, *23*(1), 18–36. https://doi.org/10.1080/10508406.2013.778204

Secules, S., McCall, C., Mejia, J. A., Beebe, C., Masters, A. S., Sánchez-Peña, M. L., & Svyantek, M. (2021). Positionality practices and dimensions of impact on equity research: A collaborative inquiry and call to the community. *Journal of Engineering Education*, *110*(1), 19–43. https://doi.org/10.1002/jee.20377

Smith, K. M. (2010). Producing the rigorous design case. *International Journal of Designs for Learning*, *1*(1). https://doi.org/10.14434/ijdl.v1i1.917

Stake, R. E. (1995). *The art of case study research*. Sage Publications.

Stake, R. E. (2013). *Multiple case study analysis*. Guilford Press.

Streveler, R. A., & Smith, K. A. (2006). Conducting rigorous research in engineering education. *Journal of Engineering Education*, *95*(2), 103–105. https://doi.org/10.1002/j.2168-9830.2006.tb00882.x

Suarez-Grant, J., & Haras, C. (2022). The inclusive teaching program: A seven-module online course for faculty of higher education. *International Journal of Designs for Learning*, *13*(1), 139–152.

Svihla, V. (2014). Advances in design-based research in the learning sciences. *Frontline Learning Research*, *2*(4), 35–45. https://doi.org/10.14786/flr.v2i4.114

Svihla, V., & Boling, E. (2020). Introduction to design case chapters. In *Handbook of research in educational communications and technology* (pp. 629–645). Springer. https://doi.org/10.1007/978-3-030-36119-8_28

Svihla, V., Gomez, J. R., Watkins, M. A., & Peele-Eady, T. B. (2019). Characterizing framing agency in design team discourse. In *Proceedings of the American society for engineering education annual conference & exposition* (pp. 1–19). https://doi.org/10.18260/1-2–32505

Svihla, V., Peele-Eady, T. B., & Gallup, A. (2021). Exploring agency in capstone design problem framing. *Studies in Engineering Education*, *2*(2), 96–119. https://doi.org/10.21061/see.69

Svihla, V., & Reeve, R. (Eds.). (2016a). *Design as scholarship: Case studies from the learning sciences*. Routledge. www.routledge.com/Design-as-Scholarship-Case-Studies-from-the-Learning-Sciences/Svihla-REEVE/p/book/9781138891661

Svihla, V., & Reeve, R. (2016b). Untold stories. In V. Svihla & R. Reeve (Eds.), *Design as scholarship: Case studies from the learning sciences* (pp. 1–10). Routledge.

Svihla, V., Reeve, R., Field, J., Lane, W., Collins, J., & Stiles, A. (2016). Framing, reframing and teaching: Design decisions before, during and within a project-based unit. *International Journal of Designs for Learning*, 7(1), 19–36. https://doi.org/10.14434/ijdl.v7i1.19427

Vahey, P., Reider, D., Orr, J., Presser, A. L., & Dominguez, X. (2018). The evidence based curriculum design framework: Leveraging diverse perspectives in the design process. *International Journal of Designs for Learning*, *9*(1), 135–148. https://doi.org/10.14434/ijdl.v9i1.23080

Walther, J., Sochacka, N. W., Benson, L. C., Bumbaco, A. E., Kellam, N., Pawley, A. L., & Phillips, C. M. L. (2017). Qualitative research quality: A collaborative inquiry across multiple methodological perspectives. *Journal of Engineering Education*, *106*(3), 398–430. https://doi.org/10.1002/jee.20170

Wan, L., Müller, C., Randall, D., & Wulf, V. (2016). Design of A GPS monitoring system for dementia care and its challenges in academia-industry project. *ACM Transactions on Computer-Human Interaction*, *23*(5), 1–36. https://doi.org/10.1145/2963095

Watson, I., & Marir, F. (1994). Case-based reasoning: A review. *The Knowledge Engineering Review*, *9*(4), 327–354. https://doi.org/10.1017/S0269888900007098

Wiggins, B. J., & Christopherson, C. D. (2019). The replication crisis in psychology: An overview for theoretical and philosophical psychology. *Journal of Theoretical and Philosophical Psychology*, *39*(4), 202.

Wilson, J., & Bruni-Bossio, V. (2020). Using visual mapping to communicate connections between learning outcomes and student tasks. *International Journal of Designs for Learning*, *11*(3), 13–22. https://doi.org/10.14434/ijdl.v11i3.26101

Wulf, V., Müller, C., Pipek, V., Randall, D., Rohde, M., & Stevens, G. (2015). Practice-based computing: Empirically grounded conceptualizations derived from design case studies. In *Designing socially embedded technologies in the real-world* (pp. 111–150). Springer.

Wulf, V., Rohde, M., Pipek, V., & Stevens, G. (2011). Engaging with practices: Design case studies as a research framework in CSCW. In *Proceedings of the ACM conference on computer supported cooperative work* (pp. 505–512). https://doi.org/10.1145/1958824.1958902

Yadav, A., Shaver, G. M., & Meckl, P. (2010). Lessons learned: Implementing the case teaching method in a mechanical engineering course. *Journal of Engineering Education*, *99*(1), 55–69. https://doi.org/10.1002/j.2168-9830.2010.tb01042.x

Yin, R. K. (2003). *Case study research: Design and methods*. Sage Publications Inc.

1 Appendix: Design Case Template

For clarity, we include section numbers below that are referenced in the chapter.

1 **Abstract**
The abstract should provide a brief overview of the design case. It should answer questions such as: What context did you design for? Who are the learners? What specific learning needs do your learners have? What did you design? What inspired your design ideas? How did the designed solution address the problem and meet needs?

2 **Introduction**
Orient the reader to the flow of your design case in sequence, with one sentence each about the context, the designed product, and the particular design process. Foreshadow something interesting or distinctive that your case will share.

3 Position authors clearly. Were you the sole designer, part of a team, some other stakeholder, or are you writing as a historian? If the case is written by a team of designers and non-designers, offer conventions to explain this. Position yourself in terms of access you had to the design process and discuss the social and role identities that may have shaped your perceptions and actions related to the design process. If the process was participatory, involving stakeholders as co-designers, make this clear.

4 Use active, attributional language ("I designed," "We designed," "They designed") throughout. Cases written from a first-person point of view (I, we) are more compelling.

5 **Context**
Every design exists in a specific context. Describe the context you designed for. Discuss relevant characteristics to provide a vivid and thick description of the context that enables the reader to picture it and compare it to their own experiences.

6 Describe the learners in ways that enable readers to consider how your learners differ from those they design for. Consider the following questions:
- What are the learner demographics (age/educational level, gender, race/ethnicity, cultural/language/nationality, etc.)?
- What prior knowledge, experience, and skills do the learners have? Do they have everyday and cultural experiences that the designer considered?

- What attitudes, perceptions, and dispositions do the learners have that are salient to the design problem? For instance, are they known to be apprehensive about technology and the design is a technological design?
- Do the learners have any disabilities that were taken into account or that impacted their engagement?

7 You may foreshadow how characteristics of the context and learners shaped your instructional design decisions or discuss particular challenges that were idiosyncratic, notorious, persistent, or commonplace.

8 Describe designs that existed prior to your design and what about them was not satisfactory.

9 **Design Process**

Present a narrative account of the particular design process and design decisions. This is typically organized sequentially with headings that orient the reader to phases. If you used a specific design model, the steps of the model might be useful headings. However, if the step involves many substeps, as is the case with "analysis" in ADDIE (analysis, design, development, implementation, and evaluation, a well-known instructional design model), consider using substeps as headings (e.g., learner analysis, task analysis). If the design process was iterative, or if you engaged in design practices or methods but did not follow a particular model, just use descriptive headings.

10 Include figures to depict the design process. This might include sketches or prototypes, if such exist, evidencing steps towards the final design. Include a timeline showing stages in the design process and any critical moments.

11 The narrative typically answers questions like:

- How and why did you decide to embark on the design process?
- How did you come to understand the needs?
- What other information did you gather or consider?
- What constraints or requirements shaped your process?
- What key decisions did you make and based on what information?
- Did you base decisions on learning theory or research on learning? If so, be specific about how it served as inspiration, not as justification. Note that design cases are not required to include a theoretical framework or research backing. Rather than using citations to warrant decisions, design case authors need to explain what about the research inspired them and how they decided to make use of the research or theory.
- Did you base decisions on past experience, your own intuition, or others' experiences? Share this.
- How did you move toward solution?
- How did you narrow or rule ideas out?
- Did you prototype and test your design?
- Did you evaluate your design and make changes based on this evaluation?
- What challenges did you encounter?

12 **Final Designed Solution**

Concisely depict the final designed solution vividly. If there is not a "final" design, depict the most recent iteration/version. Normally, the full design cannot easily be depicted; consider the following strategies:

- Include a table that shows the overall sequence, noting what learners might encounter/do as well as what instructors do. Provide information about duration of activities listed in the sequence.
- Provide a use-case that narrates the experience from the point of view of a single learner or group of learners.

- Illustrate the learning model or cycle if there is one. If the same sequence is repeated multiple times, this can be illustrated in a cycle or timeline, with text that explains a single example.
- Help the reader understand what the learning activities and assessments are.
- Include figures showing examples, such as screenshots of a module or a worksheet/activity.

13 **Challenges/Lessons Learned**

Design cases include a section detailing specific challenges encountered or insights gained during the design process (and during implementation, if applicable). While these should be apparent in the narrative of the design process, this section draws out salient aspects in a reflective manner. Consider the following types of questions here:

- What made the design process go particularly smoothly? What could have made it more challenging? Or what made the design process particularly challenging? How did you navigate the challenges? What advice would you offer yourself next time?
- How did collaboration or power dynamics shape your design process?
- How did constraints, like time or budget, impinge upon or support your design process?
- What role did your own and others' understanding of design as an ill-structured process impact your efforts?
- Sometimes, authors initially place blame on a particular person for challenges but then find they cannot publish the design case without potentially causing harm to that person. Instead, take a structural approach: What structures and practices shaped that behavior? Aim critique at institutional policies and norms that prevent effective design teamwork.
- Did something unexpected happen that altered progress?
- Were there failures along the way?
- Resist the urge to issue "design principles." Consider, instead, discussing the design methods or practices that were particularly productive, or describe "design patterns" (Alexander et al., 1977) – that is, reusable and adaptable heuristics.

14 Some design cases share details about implementation or pilot testing, commonly from the designer or instructor point of view, because sharing information from learners should only be done if the university's ethics or institutional review board has provided approval; human data includes photos of students, interviews with instructors or students, instructors' accounts, student outcomes, student work, and even summaries of student work. Keeping this in mind, consider questions that direct attention to the design process or designed solution, such as:

- Was the design implemented as intended? What does the implementation suggest, in terms of changes that should or should not be made to the design?
- If this was a pilot test, did the context or learners differ in ways that make using information from piloting challenging?
- What kinds of adaptations were made? What do these modifications suggest?
- Were any challenges or uncertainties resolved?

15 **Concluding Thoughts**

Conclude with a concise summary of the entire case. Provide a characterization of the particular design process, such as its agility, emergence, iteration, oscillation, uncertainty, constrainedness, etc. Summarize the author's reflection about the particular design process and solution, lessons learned, and implications for the author's design practice. Be humble in concluding, sharing failures and challenges.

16 **References**

Design cases seldom have long lists of references. You do not need to back design decisions with citations, as this is not where the rigor of design cases comes from. Instead, authors typically explain what inspired them, even if that inspiration was from a journal article.

29

Considerations for Engineering Education Research Using Quantitative Methods

Margret A. Hjalmarson, Geoffrey L. Herman, and Kerrie A. Douglas

> But it seems important to wrestle with these challenges to ensure that we develop research designs that capture what is important rather than only what is easily measurable.
>
> (Coburn, 2003, p. 9)

While education research has a long history of different types of evidence gathering, our focus here is on quantitative data and the challenges and opportunities such data presents for engineering education research. What Coburn's quote addresses is the need to not only measure but also consider the meaning and impact of our research methodologies. Rather than overviewing statistical methods in depth because of the wealth of other sources specific to quantitative methods in education and introductory statistics (e.g., Coe et al., 2021; Creswell, 2018; Krathwohl, 2004; Lomax & Hahs-Vaughn, 2020), our goal is to explore what kinds of questions quantitative evidence is useful for answering and the process of designing and carrying out a study using quantitative data. We focus on data as a source of evidence (Becker, 2017; Gough, 2021) for building an argument about the nature of a phenomenon. Our stance is that quantitative data is one tool in the toolbox of possible evidence for understanding educational phenomena. Whether quantitative methods are appropriate depends on the research questions and the data that might be available (Ercikan & Roth, 2006). In this chapter, we will explore what quantitative research explains and what the limitations of quantitative studies might be as we explore the process of research design. We explore the design choices and questions researchers need to grapple with to design and carry out studies.

We begin the chapter by situating quantitative methods as an option for gathering evidence about educational phenomenon. Then, we explore the phases of quantitative research by attempting to explain aspects unique to quantitative studies, but also exploring how quantitative research needs to be explained in transparent ways that are clear about the series of decisions that ultimately take the project from research questions through conclusions. Then, we describe the phases of literature review and identify theory, data collection, and data analysis in more detail.

1 Situating Quantitative Methods

The purpose of engineering education research is to advance knowledge and inform the field about the state of engineering education and possibilities for enhancing engineering education. So what

DOI: 10.4324/9781003287483-35

is the role of quantitative methods in engineering education? We have sometimes observed a bias towards quantitative information as "more objective" or "better" evidence. We have also observed quantitative methods that are carried out without attending to the complexity of quantitative measurement, analysis, and interpretation. As a result, here we explore how to make quantitative data a meaningful evidence source that contributes to the advancement of engineering education research and practice. As a general goal, the purpose of quantitative methods is to represent a phenomenon numerically to analyze it with statistical models or summarized quantitatively. Often, the purpose is to use a sample to make inferences about the larger population or group.

The process of quantitative research includes explaining how and why the data collection and analysis have resulted in the claims and findings. So did the researcher only find what they were hoping to find? Did they organize the data collection to set up certain findings? Does the article present an accounting not just of the expected findings but also of unexpected findings? The description of the research methods should help readers understand how and why the conclusions, claims, and findings are being made (Ming & Goldenberg, 2021; Allen-Platt et al., 2021). We also need to avoid only using new information to reinforce our existing beliefs, a phenomenon known as confirmation bias (Klayman, 1995). A purpose of research methods, qualitative or quantitative, is to help researchers support a transparent decision-making process while generating knowledge and reporting their findings. Within the decision-making process, the researcher has perceptions, values, and beliefs that influence the investigation (Ercikan & Roth, 2006; Gough, 2021). Research methods should help us challenge ourselves or check counterproductive impulses as an appropriate acknowledgment of our own limitations as researchers and our participation in an ongoing research conversation rather than isolating engineering education research. Quantitative methods formalize the process of challenging/checking our biases and perceptions via statistical procedures for analysis, such as significance tests and effect sizes.

Quantitatively interrogating our beliefs can be particularly challenging because educational phenomena are often not directly visible or easily countable – knowledge, motivation, self-efficacy, learning, etc. (Ko & Fincher, 2019). The measurement of knowledge attempts to document what an individual knows, but the researcher has to make choices about how to document that knowledge. The quantitative research process is collecting data that represents the phenomenon and then modeling that data in ways that are meaningful. The goal is to develop an argument that the models and representations can be used to understand and interpret the phenomenon (Sloane & Gorard, 2003; Sloane & Wilkins, 2017). As a model of a phenomenon, every representation has flaws and limitations – things it represents well and things it misses. One question for quantitative research methods is whether the process of research (modeling and representation) is leading to findings that are meaningful and credible while acknowledging the limitations. Discussing limitations is not a sign of weakness but, rather, evidence of quality research, demonstrating the thoughtfulness and skepticism of the researcher to the reviewers and readers.

1.1 Quantitative Research, Large Datasets, and Measuring Change

One common set of inquiries where quantitative research methods are often necessary and helpful is for large datasets. We will leave the questions of "what is large?" aside momentarily to consider questions of scale and how to describe patterns within data. The empirical questions that are of educational interest can be things like "Does it work?" or "Will it work in lots of different places?" or "How can we affect change among large groups of people?" Engineering education is regularly interested in change in local systems and organizations, but also in describing larger patterns or phenomena. For example, questions of diversity, equity, and inclusion in the system of engineering education might be examined via looking at patterns that emerge across many institutions (e.g., Flores et al., 2021; Main et al., 2020; Jewett & Chen, 2022; Navarro et al., 2019; Skvoretz et al.,

2020) or patterns about students' pathways through engineering at one institution (Main et al., 2022) or in one state (Knight et al., 2020). Large datasets across multiple institutions can illuminate patterns that may or may not be unique to a particular type of institution. Alternatively, we may also want to identify outlier institutions or organizations that vary from the common pattern. For example, a recent Mathematical Association of America study of calculus instruction first conducted a large-scale national survey to understand students' progression through calculus from calculus I to calculus II. Then, they conducted in-depth case studies of institutions that were markedly successful at supporting students through the calculus trajectory to try to understand what they might be doing organizationally (Bressoud et al., 2015)[1]. So scale can mean looking for larger trends and seeking places that are not following the trend.

Studying education at scale also means investigating change. So what might change about a phenomenon and why? Quantitative research can be used to describe patterns of change in response to either interventions introduced by researchers or naturally occurring changes. For instance, if a university changes a policy for when students need to declare an engineering major, researchers might be interested in understanding how that impacts recruitment and retention in different engineering disciplines. The research questions that accompany questions of change include understanding what changes occurred (intended and unintended) and understanding the mechanisms of change.

A risk of questions at scale is that there are many things to count that might be easy to count but are not informative or meaningful. We should pursue not only those questions that are "easily measurable" but also explore mechanisms for measurement that will be meaningful. For example, gathering grades can be a metric of "success" in engineering. However, large-scale studies have examined how grades are not necessarily predictors of future engagement in STEM (Bressoud et al., 2015; Seymour & Hunter, 2019). So, while grades might be relatively easy to gather, they are not necessarily the most important measure of complex phenomenon, like recruitment and retention in engineering, where multiple factors at the societal, institutional, and individual levels might be at play. Theory, existing research, and knowledge of the context should inform what measures might be meaningful. Godwin et al.'s chapter in this volume will examine emergent quantitative methods that are particularly relevant for large datasets.

We provide one critical warning about a common misuse of quantitative methods: the pull towards simplistic binary claims. When encountering complexity, we have a cognitive bias toward trying to fit those complexities into simple narratives that help us wrap our minds around the phenomenon (Gilbert, 1998; Krauss, 2017; Slotta & Chi, 2006). Engineering education questions sometimes appear simple: "Do the students understand more about the topic using one teaching method vs. another teaching method?" It is tempting to want simple answers, like, "Yes, teaching method A is better than teaching method B," except the endeavor of teaching and learning is influenced by the contexts, experiences, and motivations of instructors and learners (among other parties) so the answers are rarely so simple. There is a pull towards simplistic interpretations of statistics and quantitative information. For instance, a p-value from an experimental study can too quickly become a binary that claims a hypothesis has statistically significant evidence supporting it (indicating that the hypothesis is true) or not. However, it is rare (if ever truly possible) that simplistic, powerful narratives exist for understanding the complexity of education.

Throughout this chapter, we will encourage the reader to use quantitative methods as tools that encourage critical thinking about how likely it is that patterns exist or how strong those patterns really are. We are ultimately not trying to determine a simple binary of what is true or false but rather trying to build consensus within a community about what explanation of a phenomenon is most compelling or the most useful (Krathwohl, 2004). In addition, being clear about the research methods used helps a study contribute to the larger understanding of a phenomenon. Rarely (if ever) does one study present a definitive conclusion on a topic. Rather, an individual study is one part of a larger, field-level conversation that attempts to move forward in understanding a phenomenon.

2 Phases of Quantitative Research

The next sections describe the phases of quantitative research. While they appear linear, researchers may move between them as the project proceeds, and we see them as a recursive process (Figure 29.1). For instance, unexpected findings might require returning to a literature review to understand inferences about different groups of students. As the study proceeds, returning to the theoretical framing at each phase should examine how the findings are situated within the theory and how they contribute to understanding the theory. We include theory and literature as separate influences to distinguish between other research studies (i.e., the research literature) and the theories or conceptual frameworks that guide a study (i.e., theory). After piloting a measure, the researchers might need to revise the measurement approach in later rounds of data collection to improve reliability. In short, as a knowledge-generating activity, researchers should be learning throughout the project about the phenomenon being investigated. We suggest a multi-step process for quantitative educational research: (1) articulate research objectives; (2) determine desired claims; (3) identify evidence to warrant claim; (4) determine population, units of analysis, sample, and sampling frame; (5) determine measurement methods; (6) plan data collection procedures; (7) plan analysis and (8) interpretation.

2.1 Foundations of a Quantitative Research Project

At the beginning of any research endeavor, we should be able to identify what it is we are trying to investigate or understand to guide design choices. For this reason, studies often begin with questions, hypotheses, conjectures, and/or propositions that then shape the investigation. We make choices about a topic to investigate that then inform choices about the evidence to gather. The first question to ask when considering quantitative data is whether a phenomenon can be described with quantitative evidence and how to gather appropriate quantitative evidence for the phenomenon. This begins with the notion that how the phenomenon is measured is important for considering how the phenomenon is understood (Daly et al., 2012). Quantitative analyses are not neutral – they are representations of context, patterns, and phenomenon. So, understanding what the questions are about the context or the participants is important to organizing the hypotheses or conjectures one might have about that phenomenon.

Table 29.1 Phases of Quantitative Research

Foundations of a quantitative project: research objectives and purpose statements	What is puzzling about a situation? What needs to be understood? How are these conjectures grounded in existing work? What variables are relevant? Are hypotheses testable (if/then)?
Theory and Literature • Determine claims to investigate • Identify evidence to warrant claims • Define constructs and relationship	How should the phenomenon be framed with theory? How are constructs defined to support measurement? What is already known about this topic in prior literature? How are different aspects of the phenomenon related?
Data collection • Determine population, units of analysis, sampling • Measurement methods	Is it valid? Is it feasible? Is it ethical? What are limitations?
Data analysis	Is it valid? What is the unit of analysis? What are appropriate models/techniques for analysis? What analyses are meaningful?
Interpreting findings	What does the analysis mean in light of the theory? Are the findings credible? Are the findings meaningful? How should findings be reported?

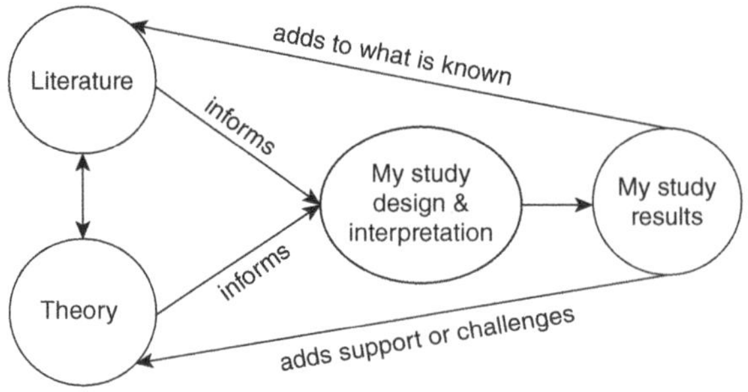

Figure 29.1 Recursive study design.

For instance, to ask whether students have "learned" something, we need to know what they know already in order to measure any change that might have occurred because of an intervention. If research claims students learned something, what evidence is needed to support that claim? Similarly, hypotheses about the phenomenon inform questions that might be asked. For example, how long would we expect students to take to learn the topic or engineering practice? Does the knowledge have subcomponents or prerequisite knowledge? We should also be able to articulate what characteristics of a learning environment might help students learn the topic. Examining the learning environment generates questions about how to design tasks, how to assess students' understanding, or how to structure the learning environment (e.g., organizing collaborative teams, providing resources at different points, presenting new content, designing projects or formative assessments).

Drawing on existing research and knowledge of the context, researchers might have conjectures or propositions about what might need to be changed (Sandoval, 2014). Educational studies are often grounded in attempts to design something new that will change students' learning experience (e.g., Newstetter, 2005). If we are to learn as a community from studies, we need to not only know that a change happened but also explore how and why the learning experience was designed in a particular way to support change. Reporting on the hypotheses and conjectures the researcher had about the learning environment allows the community to understand how those conjectures might be explored in other settings. Our emphasis here is that in order for knowledge from research to be shareable, researchers need to explain and critically examine the conjectures and propositions that underlie the curricular, pedagogical, or policy changes they are investigating.

2.2 Articulate Research Objectives

Educational research begins with a purpose statement based on understanding of relevant literature and theory. When we begin to conduct research in an area that is new to us, it is especially important to consider what is currently known and what questions are still left unanswered and are important to know. There are very few topics in engineering education so nascent that a body of literature either in engineering education or another research community has not considered them. For engineering education researchers to truly advance what is known about topics of importance to our community and related research communities, it is essential that we do due diligence in locating relevant literature (Gough, 2021). Thus, researchers new to an area of study add value by situating their work within the context of other studies, akin to joining a conversation that has been happening by first listening and understanding before offering comments. Quantitatively, this research conversation includes the selection of variables, the tools for measurement and how they are used to gather and interpret data. Quantitative research might be needed to identify areas for further investigation qualitatively or might build upon qualitative research that is pointing toward phenomenon to investigate at a larger scale. Thus, empirical engineering education research should both be informed by empirical studies and theory as well as contribute to what is known in the literature and theory development. It is a recursive relationship, where our research is based on theoretical relationships and what others have found, but then we join in the conversation by clearly articulating how our findings add to what is known and whether our findings are supportive or challenge current theoretical understandings. Figure 29.1 shows the recursive relationship between existing literature on a topic, the theoretical framework, and an individual study.

Booth and colleagues (2016) provide a template for writing a purpose statement: (1) name your topic, (2) add an indirect question to explain why, and (3) state the importance of the problem (i.e., the "so what"). Locke and colleagues (2014) provide a template diagram of logic to aid researchers in writing their purpose statement and the rationale. The approach starts with identification of what is known from literature about the factors and relationships in your topic that support why you think the particular relationship between variables needs to be tested. There is no one right way to go about developing a statement of purpose; however, whether the study is being proposed for funding or you have results written up for conference proceedings or journal publication, readers will want to know that you designed your quantitative study based on (1) a problem of significance from others' point of view (e.g., existing research, National Academies, UNESCO, other educational institutes, etc.), (2) how others have studied it, (3) that you have a theoretical framework for understanding the relationships of under study, (4) who your audience is, and (5) what you hope will be done with the information found.

2.3 Identify Claims to Investigate and Type of Study

Along with your purpose statement, consider what type of conclusions you are expecting or what types of claims you will be investigating. In other words, when the study is over, what do you hope to be able to say? Perhaps a researcher would like to claim that a particular curricular intervention led to sophomore-level materials engineering students' increase in conceptual understanding of plastic deformation at the atomic level. This type of claim is an example of studying an intervention, where an *experimental* or *quasi-experimental* research design would provide the strongest evidence that the intervention led to desired results. In engineering education research, true experimental designs are somewhat uncommon, and for good reason – it is very difficult to randomly assign students to a "treatment" group. Additionally, ethical concerns arise if a researcher has literature and theory to support the curriculum is actually a better learning opportunity than the "control" group's curriculum. Quasi-experimental research methods are implemented when some causal inference is desired but the researcher is not able to randomly assign participants to specific conditions. Leppink (2019) discusses different types of experimental designs and common statistical methods in each context. When designing an experimental study in education, it is very important to consider what variables can be controlled, and what variables are uncontrollable. In the case of a quasi-experimental study on curriculum designed to increase materials engineering students' conceptual understanding of plastic deformation at the atomic level, the researcher could implement the curriculum in their own section, but maybe the other instructors of different sections are less likely to implement the curriculum as intended. So the researcher would not control which students received or did not receive the intervention. There may be differences in students' understanding across the sections or differences in how the material was presented. A pre/post-test design would help account for differences that exist in the students' understanding prior to the curriculum intervention. The researcher could then compare the gains that students made between the courses.

Other claims are more descriptive in nature. For example, a researcher might want to claim engineering doctoral students have few opportunities to prepare them for professional (nonacademic) work settings. Another example would be a researcher who wants to understand the experiences of women tenure-track faculty in different higher education settings in engineering from an intersectional perspective (Aldridge et al., 2019). These examples are of descriptive studies where survey research methods would be appropriate. Blair et al. (2013) describe different types of surveys and how to determine your sampling strategies.

3 Role of Theory

The first question many doctoral students encounter in becoming researchers is, "What's a theory?" Multiple definitions and framings for theory exist in education research and social sciences more generally (Godwin et al., 2021; Magana, 2022). Here, we will focus on how theory both guides the development of quantitative studies and how theory is developed by studies. One framing of theory is that it situates a particular project within the larger space of research. So a study about students' learning is situated in how the researchers are defining learning in the context. For example, a common learning experience for engineering students are design projects. Researchers have defined aspects of the design process within theories about learning to design in an engineering setting (e.g., Crismond & Adams, 2012). The theory of the engineering design process then guides the creation of a learning experience. The research should test theories about design by gathering evidence about the students' learning experience. For instance, existing measures might be used or adapted to understand how the design projects influenced students' perceptions and interest in engineering.

Theory guides the methodological choices and the interpretation of the findings. In developing an experimental or quasi-experimental study, the researcher must define what the experiment is intended to accomplish. So what variables are changing? What are hypotheses about what might influence students' learning in different ways? In comparing outcomes across different conditions, what makes those conditions comparable? Where might underlying patterns influence results? For instance, students are not evenly distributed to different course sections (Marbouti et al., 2018), and there are underlying phenomena that would influence comparison across course sections or instructors. The enacted curriculum might vary from the intended curriculum (Clements, 2007; Coburn, 2003), so the study needs to document what was taught in different sections to understand if they are truly comparable. As another example, critical theories in education (e.g., critical race theory, critical feminist theory) are often connected to qualitative methods; however, critical quantitative methods engage quantitative research design using a critical theory framework (e.g., Frank et al., 2021; Sablan, 2019). So what are the ways in which data is gathered, analyzed, and represented that change how interpretations are made (Godwin et al., 2021)? How do researchers define the population that guides sample selection? What are the conceptual metaphors that are used to describe constructs like "broadening participation" in different ways (Lee, 2019)? This leads us to the data collection process which we frame as evidence gathering.

4 Data Collection Design

There are a number of decisions to be made while planning data collection in a quantitative study. While running statistical analyses can be done relatively quickly, it takes considerable time and intentionality to design a strong quantitative study. Quantitative researchers tend to spend the bulk of their research time designing and planning for the data collection. Considerations include questions shown in Table 29.2.

4.1 Identify Evidence to Warrant Potential Claims and Possible Rebuttals

When you clearly define the research purpose, the research questions, and the types of claims the study should develop, then you can ask, *What evidence is needed to support the potential claim? How might*

Table 29.2 Considerations for Designing Quantitative Educational Research

- What is the overall research objective?
- At the end of the study, what do I hope to know? How will I know it?
- What type of study is it (e.g., experimental, descriptive, correlational)?
- What variables am I interested in?
- What relationships between variables do I expect?
- What type of measures will I use to collect variables of interest?
- What population (whether people or objects) do I want to represent?
- What will be the unit(s) of analysis?
- How large of a sample size do I need to collect?
- How will I collect my data?
- Should I offer participant incentives?

my claims be rebutted? What are the limitations? In the example question mentioned earlier – that a particular curricular intervention might lead to an increase in sophomore-level materials engineering students' conceptual understanding of plastic deformation at the atomic level – evidence to support the claim might be pre/post-test changes in an assessment of conceptual understanding of plastic deformation at the atomic level. If the pre/post-tests were only administered to the classes with the new curriculum, if there were significant increases, one rebuttal might be that it is unknown if the change in understanding was the same, more, or less than the former curriculum. Another rebuttal might be that it is unknown whether the students would have improved in post-test scores simply because they saw the questions previously (known as instrumentation effects). The researcher might then consider a comparison of students' pre/post-test gain or allow the pre-test to be a covariate in a comparison between new curriculum and former curriculum sections to determine if students in the new curriculum sections increased significantly more than students in the old curriculum. One potential rebuttal to evidence might be that instructor's implementation and classroom instruction were also responsible for students' learning more from the new curriculum. The researcher would then need to examine if the sample was large enough to consider instructor effects and analyze students' data from a nested perspective.

All studies have limitations. It is impossible to design a perfect study in education that accounts for all uncontrolled variables (Glass, 1976). However, the researcher needs to do due diligence to consider what is the strongest evidence to support the possible claims and how the research design limits the interpretation and inferences made from the study. When drawing on existing literature, researchers need to understand when a new study might resolve a limitation of prior work (e.g., by collecting data in a different way or using a new measure), so reporting on limitations is critical to quantitative projects.

4.2 Measurement

A critical component of quantitative data collection is selecting the tools for quantifying the information. Engineering education researchers often find themselves asking questions about what students' know, can do, perceive, believe, have attitudes about, are motivated by, identify with, and many other variables that are not directly observable. Whether the researcher is studying students' ability to solve ill-structured engineering problems or students' engineering identity development – these variables are defined by the researcher, and how evidence is collected to represent the variable must be clear. The variables and the tools used to measure them should be consistent with the purpose of the study and the underlying theory about the phenomenon. Furthermore, there is no such thing as a perfect measure, because we cannot ever directly observe the variable to confirm our approach is fully perfect. This is a concern in engineering education research because the common statistical methods that engineers are familiar with have underlying assumptions that there is no measurement error. Therefore, whenever it is feasibly possible, statistical approaches that allow for latent variable analysis are desired. Latent variable analyses, such as factor analysis and structural equation modeling, take into account the measurement error in the full model. While full considerations of educational measurement in quantitative research are beyond the scope of this chapter, we suggest reading foundational works of Kane (1992, 2001, 2016), Douglas and Purzer (2015), National Research Council (2014) and the assessment chapter on validity (this volume). The chapter on assessment discusses foundational terms of validity, reliability, and fairness in educational assessment as well as approaches to measurement models.

4.3 Population, Sampling Frame, and Sample

In quantitative education research, the purpose is to use statistical methods to make inferences about specific populations. Thus, it is very important to clearly articulate what population is being studied and how well results can be generalized for the whole population. While there are probabilistic and non-probabilistic sampling strategies (described in detail by Blair et al., 2013), inferential statistics are based on the assumption of random sampling. Therefore, while convenience sampling (e.g., posting to a listserv) might be a quicker method for obtaining survey responses, it is unknown how well that sample of participants represents the whole population being studied. The sampling frame is the identified pool of potential participants to your study. In the example of a researcher who wanted to study engineering doctoral students' opportunities to develop professional skills, the desired population to represent is engineering doctoral students. Engineering doctoral students attend research universities with engineering doctoral programs. So the researcher might decide the sampling frame would be the list of ABET-accredited engineering PhD programs. The researcher then needs to determine the sampling strategy based on what subgroups existed within the sampling frame that may be of interest for comparison, the sample size needed to reach adequate representation of the population, and the practical costs of the study. Additionally, as the researcher is narrowing in on the statistical methods appropriate to the data and that would support appropriate claims, conducting a power analysis helps determine the sample size necessary (Cohen, 1992; Chow et al, 2007). There are several software packages that can support determining the appropriate sample size based on expected variance in the dependent variable and type of experimental design. For example, optimal design is a tool to aid in determining the sample size needed at multiple levels of a nested study in education (Spybrook et al., 2011).

4.4 Data Collection Procedures

Data collection involves a series of decisions that impact the inferences and interpretations that can be made about the data. Some of these decisions may seem minor or obvious; however, there is usually research about methodological approaches to guide the design of data collection. For instance, the decision of whether or not to offer participant incentives and what type of incentive is an important early decision an educational researcher makes. If completion rates are low for a survey, it is difficult to justify whether the results actually represent the population the study was designed to represent or if there are differences between respondents and non-respondents that could impact the conclusions. In particular, students and faculty are asked to take a lot of surveys. For example, within one academic year, the same engineering student might be asked to take surveys regarding attitudes about physics, math anxiety, information literacy, mental health awareness, participation in mentoring programs, attitudes about alcohol, preferences for in-person or online learning. So survey fatigue might impact response rates. While calculating your target sample size, it is important to consider what your likely response rate will be. Without an incentive, response rates can be quite low (e.g., Kost & da Rosa, 2018; Pit et al., 2014). Some researchers try to encourage responses by using lottery-type incentives that offer students the chance to win a cash prize or gift. While this is fairly common practice, research indicates lottery-based incentives do little to increase participation. Trespalacios and Perkins (2016) note the literature supports that the following are effective at significantly increasing response rates: multiple contacts in recruitment (Dillman et al., 2014; Klofstad et al., 2008), incentives (Church, 1993; Heerwegh, 2006; Porter & Whitcomb, 2003), personalized invitation (Heerwegh et al., 2007; Joinson et al., 2007), and how the email recruitment is written regarding the trustworthiness of the sender and appealing to participants' sense of altruism (Trouteaud, 2004). Typically, a combination of approaches is most effective, rather than relying on only one approach.

More effective approaches include appealing to participants' sense of contribution – that they are helping an important effort – and offering an incentive at the time the survey is taken. The amount of compensation should be based on research and length of the survey. Kost and da Rosa (2018) found that compensating participants $20 for completing a long survey resulted in a 50% response rate, and the diversity of participants also increased. Our point is that recruitment is a study design choice, and researchers need to consider and carefully document how they have recruited participants to complete instruments in light of the goals of the research, the desired participants, and the limitations of each option.

4.5 Preparing Data for Analysis

Between data collection and data analysis is a critical stage that includes important decisions. In quantitative data collection, even the most well-designed survey or tool will require data cleaning. For instance, some respondents will start a survey but not finish the survey. Then, the researcher needs to decide what to do with missing or incomplete data. Is it enough that there are responses to some sets of questions but not to others? Responses to demographic questions might be optional, but if the data about how respondents identified their race or gender are missing, how will the responses be included or represented in the analysis? There are procedures for addressing missing data that are beyond the scope of this paper, but the important thing to understand is that data cleaning is an important step between the data collection and the analysis. The methods section for the publications should include information about the process of preparing the data for analysis or "data cleaning." These are design questions that should be informed by other research, guidance from research methods in similar studies, and the research goals and purposes.

Because of the variety of design decisions, we encourage engineering educators to connect to quantitative methods experts and research literature about methods. Just as the research questions are guided by existing research, the research methods should be guided by research, and the methods sections should reference the sources that were used to made research design choices. Researchers will also need to attend to emerging options for gathering quantitative data, some of which are discussed in the chapter on advanced quantitative research methods in this handbook.

5 Analysis and Interpretation of Quantitative Data

While data analysis and interpretation are an end point of a research study, the interpretation planning should accompany the data collection and analysis planning (Creswell, 2018; Krathwohl, 2004). The research questions and data collection methods impact the observations and interpretations that are possible. The types of claims we want to make must inform the types of analyses we want to perform, which in turn inform how to design our study. There are three major considerations for analyses and interpretation: the structure of the data, the selection of appropriate statistical analysis procedures, and the reporting of results. Because there are numerous high-quality and freely available discussions of quantitative research methods and constructs such as "what is a p-value?" (e.g., Jupp, 2006; Sirkin, 2006), we encourage readers to find referenced articles about most methods and constructs discussed. We will instead focus on discussing important considerations, common mistakes, controversies, and epistemological stances towards these methods and constructs and how to navigate the vagaries and challenges of developing and publishing quantitative findings.

5.1 Checking Assumptions

Every quantitative method requires making some simplifying assumptions about our data that ultimately both empower and limit the study (Hathaway, 1995). By making simplifying assumptions, we gain a structured framework for reasoning about our data and enable comparisons with other studies that provide further insights into our topic. However, these assumptions also strip some of the nuance from the data, potentially squashing important observations about small subgroups or caveats to generalized findings. Before reporting results from a statistical test, we must first understand and interrogate the assumptions that statistical tests make and how those assumptions inform how interpretations are made.

We will use a running example of a common statistical test, a two-sample Student's t-test (Sirkin, 2006; Student, 1908), to explore the implications of the various assumptions that must be made to apply the test and when we might relax those assumptions. For simplicity, suppose we select the t-test to analyze the effects of a randomized, controlled trial study where a control group receives no educational intervention and a treatment group receives an educational intervention. Participants in each group complete a pre-test and a post-test to measure learning, and we plan to use t-tests to measure how much students learned over the semester and to compare the means of each population. To use a two-sample Student's t-test, we must verify the assumptions that the samples in our data are independent, the measurement scale is parametric, and samples are normally distributed about a mean and have equal variance (Sirkin, 2006). What do each of these assumptions tell us about our data and how we interpret the results from a t-test?

5.1.1 Independence

The assumption of independence means that any two observations in our dataset do not affect each other or are in any way related other than their membership in our study populations (i.e., control vs. treatment group). Verifying the assumption of independence is essential for making a strong claim that a difference in two populations is a result of their membership and not because of some unobserved/undisclosed confounding variable. This assumption is often difficult, if not impossible, to hold in educational settings.

For example, if you have ever left an exam feeling like you gained a deeper understanding of the material by taking the exam, then you have experienced the testing effect (Roediger III et al., 2011). The testing effect is where the act of being tested solidifies existing learning and accelerates future learning. The act of measuring students' learning affects students' learning. In our t-test example, the pre- and post-tests are likely not truly independent observations – the tests themselves are a confounding variable. Similar problems exist for "non-cognitive" measurements, where the position of demographic questions may affect response rates (Green et al., 2000; Teclaw et al., 2012) or other survey-taking behaviors. This lack of independence in studying humans has been elsewhere codified through ideas such as Campbell's law and the Hawthorne effect (Campbell, 1979; Parsons, 1974).

It's likely that no study will ever have truly independent samples, but as education researchers, we may choose to use statistical tests that require an assumption of independence. When doing so, we need to acknowledge the limitations of those decisions. We may need to acknowledge that relaxing the condition of independence a little may accentuate or attenuate the effect. We need to carefully consider what potentially confounding variables continue to exist, and discuss alternative interpretations of our findings in light of those confounders. Let's say there is not a statistically significant difference between the groups in our running example. You now have three potential interpretations: (1) you do not have evidence that the intervention had an effect, (2) the intervention might have had an effect but it was overshadowed by the testing effect, or (3) there was some other confounding variable that was not measured or included in the model. This is where the researcher may need to return to data collection, the literature, or the theory to interpret the outcomes of the analysis.

5.1.2 Types of Measurements

The measurement scale (e.g., nominal, ordinal, interval, ratio) we use for our data determines what mathematical operations or analytic techniques we can perform on our data and, in turn, what types of language we can use to articulate our interpretations (Krathwohl, 2004). The measurement scale affects the degree to which the research questions align with the type of data we have collected. We must consider the meaning of numbers in the scale and the relationship between the numbers. For example, for an ordinal scale (when a numerical number implies only order and not relative magnitude), operations like addition or multiplication do not have meaning, which in turn means we cannot make claims such as "Class A was 1.5 times more motivated than Class B." However, these restrictions are often not strictly held in education research. These relationships also inform assumptions about the distribution of the population data. If we can defensibly describe a population using the parameters of a distribution (e.g., mean and standard deviation), then we can use parametric statistics. Otherwise, when an observed distribution violates the assumptions of statistical distributions, then we must use non-parametric statistics.

When trying to study phenomena that cannot be directly observed, we often rely on surveys that use Likert scale questions (Likert, 1932). These questions often take the form: "How often do you go to class? (1 – Never; 2 – Rarely; 3 – Sometimes; 4 – Often; 5 – Always)." Researchers commonly debate about how to statistically interpret these questions (e.g., Carifio & Perla, 2008). According to a strict interpretation of the assumptions of the Student's t-test, we should not use the t-test to analyze students' responses to Likert scale questions, but instead, we should use an equivalent non-parametric test, such as the Mann Whitney-U. However, ample papers argue that using the t-test is appropriate, even preferred, despite the violation of assumptions (Carifio & Perla, 2008; Norman, 2010; Sullivan & Artino, 2013). These papers argue that most parametric tests are robust enough and that Likert scale questions are equidistant enough that the statistical test generally gives the "right" answer. As with any data, the mean and median present different information. For example, a survey item completed at two time points by participants might have a pre-test mean of 2.8 and a median of 3. The post-test mean might be 3.4, but the median is still 3. The question is which statistics is informative in the context of the study.

So what is the right thing to do? Consider the data skeptically and be upfront about whatever limitations might arise from the decision. Some ordinal scale questions can be reasonably understood as being approximately equidistant (e.g., strongly disagree, disagree, neither agree nor disagree, agree, strongly agree), so a parametric test is more defensible. If the scale could be more readily interpreted as not equidistant (e.g., never, rarely, sometimes, often, always), then using non-parametric tests is probably better. Ultimately, the decision needs to be defended and explained. In order to report the means of Likert scale questions and use parametric tests, we recommend the following: cite some articles that support the decision, then explain in the methods section that results are reported using only one significant digit past the decimal point (i.e., 3.2 rather than 3.16) as an acknowledement that the scale may not be precise enough to warrant greater precision. In the discussion, try to avoid statements that imply that the Likert scale question is a ratio scale (e.g., Class A was twice as motivated as Class B), because these statements further stretch the limitations of the types of claims that are possible.

5.1.3 Distribution Types

We can quickly summarize quantitative data using descriptive statistics, such as a measure of central tendency (i.e., mode, median, or mean), the shape of a distribution (e.g., normal, skewed, uniform), and the spread of the distribution (e.g., variance or inner-quartile range). While central tendency is often the focus of statistical tests like the Student's t-test, paying attention to the shape and spread of distributions is critical for fully understanding whether two populations are truly similar or different.

Suppose the Student's t-test is used to study student learning at multiple institutions. If a t-test indicates that the students' performance on a measure of learning at the two institutions are is significantly different, does that mean the two institutions are basically the same? Not necessarily! If assessment scores at one institution have a normal distribution while the other institution has a bimodal distribution of scores, then these differences in shapes would indicate that these institutions might be admitting different populations of students into their programs. In this situation, it may have been inappropriate to use the t-test, as it would inappropriately obscure a potentially important difference. For example, is it possible that an intervention was helping some populations of students while hurting others at the second institution? Likewise, if the variances of the two distributions are unequal, then the Student's t-test may inappropriately obscure the difference in homogeneity of the two institutions.

5.2 Hypothesis Testing, Effect Sizes, and Practical Significance

The hypotheses we test with quantitative methods never exist within a vacuum but within our preexisting theories and value systems that led us to believe that those hypotheses were worth testing (or which hypotheses we are afraid to test). These value systems both create a bias or expectation to find that our expected hypotheses are true and provide a framework for what we perceive as the cost of making the wrong decision about the veracity of our hypotheses. We could correctly reject or accept our hypothesis, or we could make one of two types of errors: type 1 error or false positive (associated with probability α), where we incorrectly accept a false hypothesis, or type 2 error or false negative (associated with probability β), where we incorrectly reject a true hypothesis. The evidence we need to accept or reject a hypothesis depends on a myriad of factors beyond the simple statistical facts of our data.

When deciding how to interpret and analyze the results of a statistical test, we must likewise consider the ramifications and societal effects of drawing the wrong conclusion about the veracity of our hypothesis. This is another point at which researchers should attend to the limitations or the boundaries of the study. Claims should be carefully situated in the context of the study and be cautious about overgeneralizing.

5.2.1 The Controversial p-Value

While it is common practice to describe any p-value that is less than α as statistically significant, this practice of significance testing is coming under increased scrutiny because of the way that it is often misused and misinterpreted (Goodman, 1999; Lehmann, 1993; Trafimow & Marks, 2015; Wainer & Robinson, 2003; Wasserstein & Lazar, 2016). To begin:

> The p value is the probability of obtaining an effect equal to or more extreme than the one observed considering the null hypothesis is true. This effect can be a difference in a measurement between two groups or any measure of association between two variables.
>
> *(Biau et al., 2010, p. 886)*

Be careful to not conflate the p-value with the size of an effect and avoid practices that can encourage this type of misinterpretation (Ellis, 2010). A small p-value indicates that we can reject the null hypothesis with confidence, but it does not indicate that the effect of that difference is large or meaningful. Avoid misleading phrases, like "more significant" or "marginally significant," which can imply an effect size. Likewise, report only one α value per statistical test. Reporting multiple statistical significance levels with varying numbers of stars (e.g., *$p<0.05$, **$p<0.01$) can be misinterpreted as implying a result is more "true." The *ASA [American Statistical Association] Statement on p-Values: Context, Process, and Purpose* explores the reporting of p-values in more detail and emphasizes the importance of considering the context and design of the study as well as other possible statistics (Wasserstein & Lazar, 2016).

Report actual p-values to two or three decimal places (e.g., $p = 0.002$ or $p = 0.10$) unless smaller than 0.001 rather than the size of p relative to α (e.g., $p > 0.05$ or $p < 0.01$) (American Psychological Association, 2020). This level of detail empowers the reader and other researchers to make their own determinations about whether to reject or accept the null hypothesis. Alternatively, consider reporting confidence intervals in addition to, or in place of, *p*-values (American Psychological Association, 2020; Gardner & Altman, 1986; Wasserstein & Lazar, 2016). Retaining information about the variance of the sample improves the transparency about the range of possible interpretations about the practical significance of the findings. The more detail a publication has about the study, the better the contribution the study can make to the larger conversation about the research topic.

5.2.2 Multiple Statistical Significance Tests

When performing multiple statistical significance tests, the likelihood of type 1 error increases. For example, in an experimental study, we might want to compare three populations (e.g., online vs. in-person vs. hybrid). If using only pairwise comparisons, we would need to run a family of Students' t-tests comparing every possible pair of populations (online vs. in-person, online vs. hybrid, in-person vs. hybrid). Each additional test in the family is another chance that we find a statistically significant difference by random chance rather than because of a true difference: three tests would increase our type 1 error rate to $1 - (1 - 0.05)^3 = 0.143$ from our desired 0.05 error rate. This phenomenon is known as the family-wise error rate. While there is general consensus that researchers need to take the increased likelihood of type 1 error into consideration when interpreting their statistical findings, there is less consensus about the best way to do this.

If appropriate, use tests that make multiple comparisons at once, such as an analysis of variance (ANOVA) test rather than a t-test, or Wilcoxon rank-sum test rather than a Mann Whitney U. These tests enable comparisons of all populations as a group, mitigating the issue of family-wise error rates. If a multiple-comparison test is not possible, some methodologists argue that α should be "adjusted" or "corrected" to be smaller. There are myriad mathematical techniques, such as the Holm–Bonferroni procedure, the Hochberg procedure, or the Sidak procedure. Others argue that any form of correction is unnecessary, arguing instead for careful interpretation rather than formulas. For example, if a study with 20 comparisons finds 2 comparisons result in $p < 0.05$, the researchers would simply acknowledge that there is not strong evidence for rejecting the null hypothesis because of the high likelihood of type 1 error. Our goal is not to settle this debate or argue for the "best" approach but to raise awareness of the issue. When performing multiple comparisons, we recommend disclosing whether a correction was made and providing citations and arguments to support the decision.

5.3 Effect Sizes, Correlation Coefficients, and Practical Significance

To richly interpret our studies, we need to move beyond only statistical significance and deeply explore effect sizes and the potential societal impact of our findings (e.g., Ellis, 2010; Ming & Goldenberg, 2021; Reese, 2004). While a *p*-value helps determine *whether* two populations are different or two phenomena are related, effect sizes provide measures that help us understand *how much* two phenomena are related or two different populations are different or *how likely* one outcome is relative to another.

5.3.1 Correlation Coefficients

A common effect size measure is a correlation coefficient, which describes how strongly two phenomena are related (Cohen, 1988). In statistics textbooks or tutorials, there are often guidelines that describe the relative strength of correlations using words like *weak*, *moderate*, or *strong*, though these ranges and the words used vary from text to text (Kozak, 2009). However, the interpretation of the strength of a correlation will vary considerably across studies, and interpreting effect sizes in education depends on multiple characteristics of the study and its context (Jacob et al., 2019). In highly controllable contexts, like physics, with precise measurements, correlations need to be much higher to be meaningful than in social science and education settings, where there are always numerous confounders and often only indirect or imprecise measurements. Consider contextualizing effect sizes by referencing other studies rather than using vague, contextless words like "weak" or "strong." By comparing the correlation coefficients with similar studies in the domain, you can help yourself and readers understand why the particular findings are important within your context.

5.3.2 Absolute Effect Sizes

When comparing means or medians between two populations, we can simply describe the difference in absolute terms. These types of effect sizes are helpful when trying to convey an intuitive sense of the magnitude of a finding, potentially even without specific knowledge of a study. Absolute effect sizes can be particularly powerful when the measure being explored has social weight and meaning, such as a grade point average, but they lose their power without that context. For example, if you say that students' motivation improved by 0.3 on a 5-point Likert scale, the reader is unlikely to have any sense of whether that is a big improvement or a small improvement, whereas saying that student attrition from engineering was halved, dropping from 20% to 10%, readily conveys an enormous and impactful improvement in student outcomes.

5.3.3 Standardized Effect Sizes

In addition to limitations of context, absolute effect sizes provide limited support for comparing effect sizes between different measures of the same phenomenon (e.g., grade letter in a course versus score on a 25-point validated assessment) (Ellis, 2010). They also do not convey information about how big the effect size is relative to the inherent spread or noisiness of the samples being studied. For example, a difference in means of 0.3 points on a 5-point Likert scale might be trivial if the standard deviation is 1.5, or it might be enormous if the standard deviation is 0.1. Standardized effect sizes complement the absolute effect size by scaling it by the variance of the samples using various mathematical formulas. Some common effect size measures include Cohen's *d*, Hedge's *g*, Glass's delta, Omega squared (ω^2), and Eta squared (η^2). Reporting these standardized effect sizes is particularly helpful for other researchers who are trying to understand how robust a finding is across a wide range of contexts and studies. Like with statistical significance tests, reviewing the assumptions required by the measure can help you determine which standardized effect size to use. Reporting standardized effect sizes is generally a good practice and is especially helpful for comparing findings across studies or in meta-analyses.

Standardized effect sizes are particularly important because they offer a helpful corrective for a common misinterpretation of p-values for high-powered studies. With a large-enough sample, even trivially small differences can be statistically significant. For p-values, a smaller p-value does not necessarily mean that one study is "more significant" than another study with a larger p-value. Standardized effect sizes are not sensitive to sample size and are thus easier to interpret: a larger effect size in one study indicates a larger effect than another study.

Like correlation coefficients, standardized effect sizes often have some defined range of possible values that determine how you interpret them. For example, effect sizes like Cohen's *d* and Hedge's *g*, commonly used in meta-analyses (Ellis, 2010), can be interpreted as follows: 0 implies no difference in means, 1 implies that the means of the two samples are one standard deviation apart, and 2 implies that they are two standard deviations apart, and so forth. Some texts or researchers attach qualitative words, such as "small" or "large," to different ranges or values (Cohen, 1988; Ellis, 2009). These guidelines may be helpful for getting general bearings for interpretation, but as with correlation coefficients, it may be more important to compare effects sizes from our studies with other similar studies in our discipline to determine the practical significance and importance of our findings (Cohen, 1988; Sawilowsky, 2009).

5.3.4 Statistical Power

Ideally, we conducted a power analysis prior to conducting our research study to determine an appropriate sample size for our study, given our tolerance for type 1 and type 2 zerror and desired/expected effect size. Unfortunately, research studies are often not as straightforward as planned. Perhaps our study had unexpectedly high attrition of participants or our sample had much higher variance than reported in the literature, decreasing the statistical power of our study. This lowered power may lead to unclear findings (e.g., sufficiently large effect sizes for practical significance that are not statistically significant). Re-running or referring back to your original power analysis can help you decide whether your study was underpowered and more sampling is needed for strong inferences from the data or whether there was simply insufficient evidence to support your hypothesis.

5.3.5 Practical Significance

Reporting a p-value and an effect size is still insufficient to fully interpret our data. Determining the practical significance of our findings requires that we return to discussing and considering our value systems as researchers, the theories that inform our research, and the potential societal impact our findings may suggest (Ming & Goldenberg, 2021). Arguing that a finding is important, contributes to foundational knowledge, or warrants further action is not fundamentally a quantitative argument but about our beliefs and values as researchers, the connection between our study and other research, and understanding the context of the study.

6 Findings, Results, and Representations

Each statistical test will produce results that you need to interpret. Typically, the findings are complex in education research. The researcher needs to make decisions about what the findings mean in light of the measurement, the purpose of the research, and the underlying theory. The representations of the data and the analysis will influence how other researchers interpret the findings and understand the meaning of the study. A recent *Journal of Engineering Education* editorial points to considerations for visualizations in terms of diversity and accessibility (Schimpf & Beddoes, 2021). With any representation, the researcher should ask whether it clarifies, illuminates, or helps explain the phenomenon under investigation in a meaningful way. While researchers should be cautious that representations are not misleading or inaccurate, every representation highlights some features of the data and obscures other features. Each representation should help the field understand and interpret the data in order to make sense of the investigation. There are multiple resources about reporting statistical data and creating representations (American Psychological Association, 2020; Cooper, 2020).

In interpreting the findings, you need to return to the theory and other research. So if the statistical test indicates that the students who experienced the curricular intervention did increase in conceptual understanding of plastic deformation at the atomic level, this finding likely is not the end of the story. The researcher should examine the features of the curricular intervention as connected to other similar interventions to discuss the finding. The researcher should also consider pedagogical implications that can be drawn about instruction. As important as what we can claim, there also needs to be a discussion of what cannot be claimed or the limitations of the study. What is still unknown? What might be other mitigating factors? What might be confounding factors? What are the boundaries of the study? For example, was a similar curricular intervention happening in another course the students were taking at the same time that might be further enhancing their learning? Were there underlying phenomena that were not measured? A common outcome is that students' knowledge of one aspect of the concept might increase more than other aspects. The researcher should discuss why that might have occurred. Is their knowledge of plastic deformation at the atomic level connected to a mathematics course that some might have taken and others did not?

Discussion and Conclusion

In this chapter, we have attempted to guide engineering education researchers towards the kinds of decisions, design choices, and procedures that need to be considered in quantitative studies. We also emphasize the need for the study to be part of the conversation with the field of engineering education and related disciplines. In large part, this is because each study exists in relation to other work and transparent reporting supports analysis of findings across multiple studies (Gough, 2021). Given the complexity of educational phenomena, we also present quantitative methods as one tool among many for understanding them. There are numerous sources for how quantitative data can be used with qualitative data (i.e., in "mixed methods" studies) and ongoing advances in quantitative data analysis (some of these are described in the chapter on advanced quantitative methods in this volume). Engineering education research should continually strive to improve the methods we are using in our research and consider how different sources of evidence contribute to understanding engineering education.

We also reiterate that, when using quantitative methods, the data exists, draws from, and influences the contexts from which it is taken. Researchers need to be aware of their own biases, the biases of the field, and be conscious of how work is reported to the field in ways that are ethical and help advance engineering education research in productive ways. The numbers do not speak for themselves but rather are representations of underlying patterns in phenomenon. The researchers' responsibility is to be transparent about how the numbers were generated and how they are used to explore and explain educational phenomenon. We challenge the field to continue to learn, to develop, and to adapt new research methods in order to better understand engineering education, engineering students at all levels, the engineering profession, and engineering as part of the larger society.

Disclaimer

This chapter was written while Margret A. Hjalmarson served as a program director at the National Science Foundation. Any opinions, findings, and conclusions or recommendations expressed in this material are those of the authors and do not necessarily reflect the views of the National Science Foundation.

Note

1 www.maa.org/programs/faculty-and-departments/curriculum-development-resources/national-studies-college-calculus.

References

Aldridge, J., Yoon, S. Y., Cox, M. F., Main, J. B., & McGee, E. O. (2019). Development of the persistence of engineers in the academy survey (PEAS). In *2019 ASEE annual conference & exposition*. https://doi.org/10.18260/1-2--32169

Allen-Platt, C., Gerstner, C.-C., Boruch, R., & Ruby, A. (2021). Toward a science of failure analysis: A narrative review. *Review of Research in Education, 45*(1), 223–252. https://doi.org/10.3102/0091732X20985074

American Psychological Association. (2020). *Publication manual of the American Psychological Association* (7th ed.). APA.

Becker, H. S. (2017). *Evidence*. University of Chicago Press.

Biau, D. J., Jolles, B. M., & Porcher, R. (2010). P value and the theory of hypothesis testing: An explanation for new researchers. *Clinical orthopedics and related research, 468*(3), 885–892. https://doi.org/10.1007/s11999-009-1164-4

Blair, J., Czaja, R. F., & Blair, E. A. (2013). *Designing surveys: A guide to decisions and procedures* (3rd ed.). SAGE Publications Ltd. https://us.sagepub.com/en-us/nam/designing-surveys/book235701

Booth, A., Sutton, A., & Papaioannou, D. (2016). *Systematic approaches to a successful literature review*. Sage.

Bressoud, D. M., Mesa, V., & Rasmussen, C. (Eds.). (2015). *Insights and recommendations from the MAA national study of college calculus*. MAA Press. www.maa.org/sites/default/files/pdf/cspcc/InsightsandRecommendations.pdf

Campbell, D. T. (1979). Assessing the impact of planned social change. *Evaluation and Program Planning, 2*(1), 67–90. https://doi.org/10.1016/0149-7189(79)90048-X

Carifio, J., & Perla, R. (2008). Resolving the 50-year debate around using and misusing Likert scales. *Medical Education, 42*(12), 1150–1152. https://doi.org/10.1111/j.1365-2923.2008.03172.x

Chow, S.-C., Wang, H., & Shao, J. (2007). *Sample size calculations in clinical research* (2nd ed.). Chapman and Hall/CRC.

Church, A. H. (1993). Estimating the effect of incentives on mail survey response rates: A meta-analysis. *Public Opinion Quarterly, 57*(1), 62–79. https://doi.org/10.1086/269355

Clements, D. H. (2007). Curriculum research: Toward a framework for "research-based curricula." *Journal for Research in Mathematics Education, 38*(1), 35–70. www.jstor.org/stable/30034927

Coburn, C. E. (2003). Rethinking scale: Moving beyond numbers to deep and lasting change. *Educational Researcher, 32*(6), 3–12. https://doi.org/10.3102/0013189X032006003

Coe, R., Waring, M., Hedges, L. V., & Day Ashley, L. (Eds.). (2021). *Research methods and methodologies in education* (3rd ed.). https://us.sagepub.com/en-us/nam/research-methods-and-methodologies-in-education/book271175

Cohen, J. (1988). *Statistical power analysis for the behavioral sciences*. Routledge. https://doi.org/10.4324/9780203771587

Cohen, J. (1992). A power primer. *Psychological Bulletin*, *112*(1), 155–159. https://doi.org/10.1037/0033-2909.112.1.155

Cooper, H. (2020). *Reporting quantitative research in psychology: How to meet APA style journal article reporting standards* (2nd ed.). APA.

Creswell, J. W. (2018). *Educational research planning, conducting, and evaluating quantitative and qualitative research*. Pearson.

Crismond, D. P., & Adams, R. S. (2012). The informed design teaching and learning matrix. *Journal of Engineering Education*, *101*(4), 738–797. https://doi.org/10.1002/j.2168-9830.2012.tb01127.x

Daly, S. R., Adams, R. S., & Bodner, G. M. (2012). What does it mean to design? A qualitative investigation of design professionals' experiences. *Journal of Engineering Education*, *101*(2), 187–219. https://doi.org/10.1002/j.2168-9830.2012.tb00048.x

Dillman, D. A., Smyth, J. D., & Christian, L. M. (2014). *Internet, phone, mail, and mixed-mode surveys: The tailored design method* (4th ed.). Wiley.

Douglas, K. A., & Purzer, Ş. (2015). Validity: Meaning and relevancy in assessment for engineering education research: Assessment validity for engineering education research. *Journal of Engineering Education*, *104*(2), 108–118. https://doi.org/10.1002/jee.20070

Ellis, P. D. (2009). *Thresholds for interpreting effect sizes*. www.polyu.edu.hk/mm/effectsizefaqs/thresholds_for_interpreting_effect_sizes2.html

Ellis, P. D. (2010). *The essential guide to effect sizes: Statistical power, meta-analysis, and the interpretation of research results*. Cambridge University Press. https://doi.org/10.1017/CBO9780511761676

Ercikan, K., & Roth, W.-M. (2006). What good is polarizing research into qualitative and quantitative? *Educational Researcher*, *35*(5), 14–23. https://doi.org/10.3102/0013189X035005014

Flores, L. Y., Navarro, R. L., Lee, B. H., Hu, X., Diaz, D., & Martinez, L. (2021). Social cognitive predictors of Latinx and White engineering students' academic satisfaction and persistence intentions: Exploring interactions among social identities and institutional context. *Journal of Vocational Behavior*, *127*, 103580. https://doi.org/10.1016/j.jvb.2021.103580

Frank, T. J., Powell, M. G., View, J. L., Lee, C., Bradley, J. A., & Williams, A. (2021). Exploring racialized factors to understand why black mathematics teachers consider leaving the profession. *Educational Researcher*, *50*(6), 381–391. https://doi.org/10.3102/0013189X21994498

Gardner, M. J., & Altman, D. G. (1986). Confidence intervals rather than P values: Estimation rather than hypothesis testing. *British Medical Journal (Clinical Research Ed.)*, *292*(6522), 746–750. https://doi.org/10.1136/bmj.292.6522.746

Gilbert, P. (1998). The evolved basis and adaptive functions of cognitive distortions. *British Journal of Medical Psychology*, *71*(4), 447–463. https://doi.org/10.1111/j.2044-8341.1998.tb01002.x

Glass, G. V. (1976). Primary, secondary, and meta-analysis of research. *Educational Researcher*, *5*(10), 3–8. https://doi.org/10.3102/0013189X005010003

Godwin, A., Benedict, B., Rohde, J., Thielmeyer, A., Perkins, H., Major, J., Clements, H., & Chen, Z. (2021). New epistemological perspectives on quantitative methods: An example using topological data analysis. *Studies in Engineering Education*, *2*(1), 16–34. https://doi.org/10.21061/see.18

Goodman, S. N. (1999). Toward evidence-based medical statistics. 1: The P value fallacy. *Annals of Internal Medicine*, *130*(12), 995–1004. https://doi.org/10.7326/0003-4819-130-12-199906150-00008

Gough, D. (2021). Appraising evidence claims. *Review of Research in Education*, *45*(1), 1–26. https://doi.org/10.3102/0091732X20985072

Green, R. G., Murphy, K. D., & Snyder, S. M. (2000). Should demographics be placed at the end or at the beginning of mailed questionnaires? An empirical answer to a persistent methodological question. *Social Work Research*, *24*(4), 237. https://doi.org/10.1093/swr/24.4.237

Hathaway, R. S. (1995). Assumptions underlying quantitative and qualitative research: Implications for institutional research. *Research in Higher Education*, *36*(5), 535–562. https://doi.org/10.1007/BF02208830

Heerwegh, D. (2006). An investigation of the effect of lotteries on web survey response rates. *Field Methods*, *18*(2), 205–220. https://doi.org/10.1177/1525822X05285781

Heerwegh, D., Abts, K., & Loosveldt, G. (2007). Minimizing survey refusal and noncontact rates: Do our efforts pay off? *Survey Research Methods*, *1*(1), 3–10. https://doi.org/10.18148/srm/2007.v1i1.46

Jacob, R. T., Doolittle, F., Kemple, J., & Somers, M.-A. (2019). A framework for learning from null results. *Educational Researcher*, *48*(9), 580–589. https://doi.org/10.3102/0013189X19891955

Jewett, E. C., & Chen, R. (2022). Examining the relationship between AP STEM course-taking and college major selection: Gender and racial differences. *Journal of Engineering Education*, *111*(3), 512–530. https://doi.org/10.1002/jee.20464

Joinson, A. N., Woodley, A., & Reips, U.-D. (2007). Personalization, authentication and self-disclosure in self-administered Internet surveys. *Computers in Human Behavior*, *23*(1), 275–285. https://doi.org/10.1016/j.chb.2004.10.012

Jupp, V. (2006). *The SAGE dictionary of social research methods*. SAGE Publications, Ltd. https://doi.org/10.4135/9780857020116

Kane, M. T. (1992). An argument-based approach to validity. *Psychological Bulletin*, *112*(3), 527–535. https://doi.org/10.1037/0033-2909.112.3.527

Kane, M. T. (2001). Current concerns in validity theory. *Journal of Educational Measurement*, *38*(4), 319–342. https://doi.org/10.1111/j.1745-3984.2001.tb01130.x

Kane, M. T. (2016). Explicating validity. *Assessment in Education: Principles, Policy & Practice*, *23*(2), 198–211. https://doi.org/10.1080/0969594X.2015.1060192

Klayman, J. (1995). Varieties of confirmation bias. In J. Busemeyer, R. Hastie, & D. L. Medin (Eds.), *Psychology of learning and motivation* (Vol. 32, pp. 385–418). Academic Press. https://doi.org/10.1016/S0079-7421(08)60315-1.

Kline, R. B. (2004). *Beyond significance testing: Reforming data analysis methods in behavioral research*. American Psychological Association. https://doi.org/10.1037/10693-000

Klofstad, C. A., Boulianne, S., & Basson, D. (2008). Matching the message to the medium: Results from an experiment on internet survey email contacts. *Social Science Computer Review*, *26*(4), 498–509. https://doi.org/10.1177/0894439308314145

Knight, D. B., Grohs, J. R., Bradburn, I. S., Kinoshita, T. J., Vaziri, S. M., Matusovich, H., & Carrico, C. (2020). Illuminating inequality in access: Variation in enrollment in undergraduate engineering programs across Virginia's high schools. *Journal of Engineering Education*, *109*(4), 665–684. https://doi.org/10.1002/jee.20352

Ko, A. J., & Fincher, S. A. (2019). A study design process. In A. V. Robins & S. A. Fincher (Eds.), *The Cambridge handbook of computing education research* (pp. 81–101). Cambridge University Press. https://doi.org/10.1017/9781108654555.005

Kost, R. G., & da Rosa, J. C. (2018). Impact of survey length and compensation on validity, reliability, and sample characteristics for ultrashort-, short-, and long-research participant perception surveys. *Journal of Clinical and Translational Science*, *2*(1), 31–37. https://doi.org/10.1017/cts.2018.18

Kozak, M. (2009). What is strong correlation? *Teaching Statistics*, *31*(3), 85–86. https://doi.org/10.1111/j.1467-9639.2009.00387.x

Krathwohl, D. R. (2004). *Methods of educational and social science research: An integrated approach* (2nd ed.). Waveland Press Inc.

Krauss, A. (2017). Understanding child labour beyond the standard economic assumption of monetary poverty. *Cambridge Journal of Economics*, *41*(2), 545–574. https://doi.org/10.1093/cje/bew019

Lee, W. C. (2019). Pipelines, pathways, and ecosystems: An argument for participation paradigms. *Journal of Engineering Education*, *108*(1), 8–12. https://doi.org/10.1002/jee.20241

Lehmann, E. L. (1993). The Fisher, Neyman-Pearson theories of testing hypotheses: One theory or two? *Journal of the American Statistical Association*, *88*(424), 1242–1249. https://doi.org/10.2307/2291263

Leppink, J. (2019). *Statistical methods for experimental research in education and psychology*. Springer International Publishing. https://doi.org/10.1007/978-3-030-21241-4_1

Likert, R. (1932). A technique for the measurement of attitudes. *Archives of Psychology*, *22*(140), 55–55.

Locke, L. F., Spirduso, W. W., & Silverman, S. J. (2014). *Proposals that work: A guide for planning dissertations and grant proposals* (6th edn.). Sage.

Lomax, R. G., & Hahs-Vaughn, D. L. (2020). *An introduction to statistical concepts* (4th ed.). Routledge.

Magana, A. J. (2022). The role of frameworks in engineering education research. *Journal of Engineering Education, 111*(1), 9–13. https://doi.org/10.1002/jee.20443

Main, J. B., Griffith, A. L., Xu, X., & Dukes, A. M. (2022). Choosing an engineering major: A conceptual model of student pathways into engineering. *Journal of Engineering Education, 111*(1), 40–64. https://doi.org/10.1002/jee.20429

Main, J. B., Tan, L., Cox, M. F., McGee, E. O., & Katz, A. (2020). The correlation between undergraduate student diversity and the representation of women of color faculty in engineering. *Journal of Engineering Education, 109*(4), 843–864. https://doi.org/10.1002/jee.20361

Marbouti, F., Shafaat, A., Ulas, J., & Diefes-Dux, H. A. (2018). Relationship between time of class and student grades in an active learning course. *Journal of Engineering Education, 107*(3), 468–490. https://doi.org/10.1002/jee.20221

Ming, N. C., & Goldenberg, L. B. (2021). Research worth using: (Re)framing research evidence quality for educational policymaking and practice. *Review of Research in Education, 45*(1), 129–169. https://doi.org/10.3102/0091732X21990620

National Research Council. (2014). *Developing assessments for the next generation science standards*. The National Academies Press. https://doi.org/10.17226/18409.

Navarro, R. L., Flores, L. Y., Legerski, J.-P., Brionez, J., May, S. F., Suh, H. N., Slivensky, D. R., Tapio, F., Lee, H.-S., Garriott, P. O., Hunt, H. K., Desjarlais, C. D., Lee, B.-H., Diaz, D., Zhu, J., & Jung, A.-K. (2019). Social cognitive predictors of engineering students' academic persistence intentions, satisfaction, and engagement. *Journal of Counseling Psychology, 66*(2), 170–183. https://doi.org/10.1037/cou0000319

Newstetter, W. C. (2005). Designing cognitive apprenticeships for biomedical engineering. *Journal of Engineering Education, 94*(2), 207–213. https://doi.org/10.1002/j.2168-9830.2005.tb00841.x

Norman, G. (2010). Likert scales, levels of measurement and the "laws" of statistics. *Advances in Health Sciences Education, 15*(5), 625–632. https://doi.org/10.1007/s10459-010-9222-y

Parsons, H. M. (1974). What happened at Hawthorne? *Science, 183*(4128), 922–932. https://doi.org/10.1126/science.183.4128.922

Pit, S. W., Vo, T., & Pyakurel, S. (2014). The effectiveness of recruitment strategies on general practitioner's survey response rates – a systematic review. *BMC Medical Research Methodology, 14*(1), 1–14. https://doi.org/10.1186/1471-2288-14-76

Porter, S. R., & Whitcomb, M. E. (2003). The impact of contact type on web survey response rates. *Public Opinion Quarterly, 67*(4), 579–588. https://doi.org/10.1086/378964

Reese, R. A. (2004). Does significance matter? *Significance, 1*(1), 39–40. https://doi.org/10.1111/j.1740-9713.2004.00009.x

Roediger III, H. L., Putnam, A. L., & Smith, M. A. (2011). Chapter one – ten benefits of testing and their applications to educational practice. In J. P. Mestre & B. H. Ross (Eds.), *Psychology of learning and motivation* (Vol. 55, pp. 1–36). Academic Press. https://doi.org/10.1016/B978-0-12-387691-1.00001-6

Sablan, J. R. (2019). Can you really measure that? Combining critical race theory and quantitative methods. *American Educational Research Journal, 56*(1), 178–203. https://doi.org/10.3102/0002831218798325

Sandoval, W. A. (2014). Conjecture mapping: An approach to systematic educational design research. *Journal of the Learning Sciences, 23*(1), 18–36. https://doi.org/10.1080/10508406.2013.778204

Sawilowsky, S. (2009). New effect size rules of thumb. *Journal of Modern Applied Statistical Methods, 8*(2). https://doi.org/10.22237/jmasm/1257035100

Schimpf, C., & Beddoes, K. (2021). Designing equitable and inclusive visualizations: An underexplored facet of best practices for research and publishing. *Journal of Engineering Education, 110*(2), 289–293. https://doi.org/10.1002/jee.20388

Seymour, E., & Hunter, A.-B. (Eds.). (2019). *Talking about leaving revisited: Persistence, relocation, and loss in undergraduate STEM education*. Springer International Publishing. https://doi.org/10.1007/978-3-030-25304-2_1

Skvoretz, J., Kersaint, G., Campbell-Montalvo, R., Ware, J. D., Smith, C. A. S., Puccia, E., Martin, J. P., Lee, R., MacDonald, G., & Wao, H. (2020). Pursuing an engineering major: Social capital of women and underrepresented minorities. *Studies in Higher Education, 45*(3), 592–607. https://doi.org/10.1080/03075079.2019.1609923

Sirkin, M. R. (2006). *Statistics for the social sciences*. SAGE Publications, Inc. https://doi.org/10.4135/9781412985987

Sloane, F. C., & Gorard, S. (2003). Exploring modeling aspects of design experiments. *Educational Researcher, 32*(1), 29–31. https://doi.org/10.3102/0013189X032001029

Sloane, F. C., & Wilkins, J. L. M. (2017). Aligning statistical modeling with theories of learning in mathematics education research. In J. Cai (Ed.), *Compendium for research in mathematics education* (pp. 183–207). National Council of Teachers of Mathematics.

Slotta, J. D., & Chi, M. T. H. (2006). Helping students understand challenging topics in science through ontology training. *Cognition and Instruction, 24*(2), 261–289. https://doi.org/10.1207/s1532690xci2402_3

Spybrook, J., Bloom, H., Congdon, R., Hill, C., Martinez, A., & Raudenbush, S. (2011). *Optimal design plus empirical evidence: Documentation for the "optimal design" software*. William T. Grant Foundation. http://wtgrantfoundation.org/resource/optimal-design-with-empirical-information-od

Student. (1908). The probable error of a mean. *Biometrika, 6*(1), 1–25. https://doi.org/10.2307/2331554

Sullivan, G. M., & Artino, A. R., Jr. (2013). Analyzing and interpreting data from Likert-type scales. *Journal of Graduate Medical Education, 5*(4), 541–542. https://doi.org/10.4300/JGME-5-4-18

Teclaw, R., Price, M. C., & Osatuke, K. (2012). Demographic question placement: Effect on item response rates and means of a veterans health administration survey. *Journal of Business and Psychology, 27*(3), 281–290. https://doi.org/10.1007/s10869-011-9249-y

Trafimow, D., & Marks, M. (2015). Editorial. *Basic and Applied Social Psychology, 37*, 1–2.

Trespalacios, J. H., & Perkins, R. A. (2016). Effects of personalization and invitation email length on web-based survey response rates. *TechTrends, 60*(4), 330–335. https://doi.org/10.1007/s11528-016-0058-z

Trouteaud, A. R. (2004). How you ask counts: A test of internet-related components of response rates to a web-based survey. *Social Science Computer Review, 22*(3), 385–392. https://doi.org/10.1177/0894439304265650

Wainer, H., & Robinson, D. H. (2003). Shaping up the practice of null hypothesis significance testing. *Educational Researcher, 32*(7), 22–30. https://doi.org/10.3102/0013189X032007022

Wasserstein, R. L., & Lazar, N. A. (2016). The ASA's statement on p-values: Context, process, and purpose. *The American Statistician, 70*(2), 129–133. https://doi.org/10.1080/00031305.2016.1154108

30

Advanced Considerations in Quantitative Methods for New Directions in Engineering Education Research

Andrew Katz, Allison Godwin, and Cory Brozina

1 Introduction

Engineering education often trains graduates in statistics associated with experimental and controlled systems (Montgomery & Runger, 2010). Because of this training, many engineering education researchers may tend to default to modeling approaches that align with classical approaches to null hypothesis significance testing dating back over 100 years. Often, these approaches include pairwise testing across groups (e.g., *t*-tests and chi square tests) or simple models (e.g., multivariate logistic or linear regression models, which fall under the broad category of generalized linear models). However, social science research engages in studies of complex phenomena and ecosystems (Harvey & Reed, 1997). Many times, meaningful research questions can only be partially answered with these traditional approaches. Additionally, these quantitative approaches can perpetuate overgeneralizations, particularly for systemically marginalized populations (i.e., Black, Latino/a/x, Indigeneous, women, first-generation students, LGBTQIA+ people, and numerous other groups).

More recent conversations in engineering education have considered how quantitative research might leverage emerging methods to add depth and nuance to the types of claims that can be made (Gero & Milanovic, 2020; Godwin et al., 2021; Hu & Shealy, 2019; Villanueva et al., 2019). In addition, researchers have also emphasized the reality that a researcher has a significant influence on multiple aspects of study design, data collection, data analysis, and interpretation, and their role in both making and handling data should be explicitly considered in this research (Godwin, 2020; Hampton et al., 2021). This chapter will discuss important considerations for quantitative research in study design, data collection and preparation, data analysis, and data equity that move beyond the generalized linear model. We envision that this chapter will be most useful to individuals who are familiar with quantitative methods and who are considering ways to push research forward with advanced quantitative methods. This chapter is a survey of selected advanced quantitative considerations and methods that we believe offer the highest potential for new directions for research in engineering education. This chapter is not exhaustive or systematic in the choice of the areas selected. In part, the methods discussed are chosen because the authors have used them in their own work and found them useful for gleaning new insights. Moreover, we do not delve far into the specific quantitative details or technical methods of each consideration; instead, we introduce areas for further attention,

DOI: 10.4324/9781003287483-36

discuss advantages and limitations of each consideration or approach, emphasize resources that can be used to explore these topics, and present an example of recent studies using these methods in engineering education and social science research.

2 Study Design

Study design includes broad considerations for quality, causal inference, and other complex relationships and patterns that should be considered at the outset of a quantitative research study. These considerations must be revisited throughout the planning and execution of a study. Beyond these two, there are of course many more considerations: the National Research Council (NRC, 2002) provides a framework for considering quality in quantitative research that we find useful. The report emphasizes properly contextualized research questions, research grounded in relevant theory, research methods aligned with the research questions, a well-reasoned interpretation of findings, and the generalizability and application of findings linked to the bounds and limitations of the data. Herein we discuss considerations of quality, particularly in measurement and causal inference as two common pitfalls in study design.

2.1 Quality

Many of the NRC considerations are not new to researchers; however, when working with human subjects, there are an infinite number of factors that could influence data collection quality and interpretation – particularly measurement. These considerations include question wording and subsequent interpretation as well as participant's mood, prior experiences, and other factors. Some of this variation may be random, but other parts of the variation in observed data may also be systematic. The issue of measurement is also complicated by the latent or underlying nature of many of the phenomena engineering education researchers choose to study, which are called constructs (e.g., learning, motivation, metacognition, etc.). Most often, constructs do not exist in and of themselves but are operationalized via a measurement process. In measuring constructs, with standard tools like surveys, observations, or assessments, inference and validation are a constant facet of quantitative research to which the researcher must attend.

Inference is the leap from an observed, measured value to an estimate of underlying standing on a construct (Walters, 1970). For example, one might make an inference about a students' level of motivation based on a score on a survey. *Validity* is defined in the *Standards for Educational and Psychological Testing* (APA et al., 2014) as

> the degree to which evidence and theory support the interpretations of test scores entailed by proposed uses of tests. Validity is, therefore, the most fundamental consideration in developing and evaluating tests. The process of validation involves accumulating evidence to provide a sound scientific basis for the proposed score interpretations. It is the interpretations of test scores for proposed uses that are evaluated, not the test itself.
>
> *(p. 11)*

Validity is not a static set of checkboxes or statistics (Cronbach & Meehl, 1955). Instead, validation is the continual and ongoing process of evaluating the evidence available for the use of a measure in the given context (Douglas & Purzer, 2015; Kane, 1992). Validation considers multiple sources of evidence (e.g., content validity, construct validity, concurrent validity, predictive validity, reliability, and the fair use of the measure) but is an overall and single argument for the use of a measure. Despite the updated definitions from major research societies, validation continues to be a key concern in the making of data: "For a concept that is the foundation of virtually all aspects of our measurement work, it seems that the term validity continues to be one of the most misunderstood or widely misused of all" (Frisbie, 2005, p. 21).

We would argue that the norm of taking "validated" scales and using them without question does not consider the full range of quality considerations in measurement. Often, these scales are developed and tested on general engineering populations, which overrepresents White cismen (Pawley, 2017). In psychology research, this is part of a broader problem of studying one specific subset of a population and is known as "WEIRD psychology" because of studying samples in Western, educated, industrialized, rich, and democratic countries (Henrich et al., 2010). To wit, measures used in one context (i.e., with a psychology pool of first-year engineers) may not translate to another (e.g., upper-division mechanical engineering students). First and foremost, considerations about validation must be incorporated into the types of data collected utilized in quantitative modeling. Validation is also part of the process of ensuring research quality throughout the research design process. While we emphasize measurement, the process of validation is essential to the entire research process. This topic has excellent resources that provide a longer discussion of the considerations and practices; refer to *Standards for Educational and Psychological Testing* (APA et al., 2014) and *Scale Development* (DeVellis & Thorpe, 2021) for measurement, specifically.

2.2 Causal Inference

Causal inference is another complex phenomenon of interest to inform study designs in engineering education research (and social science research more generally). Prototypical causal questions in engineering education can include asking if a teaching method improves a student learning outcome or whether a change in state financial aid policy will improve access to engineering education for students from underserved backgrounds. The implicit assumption is that teaching a certain way causes differential learning outcomes or instituting a particular policy will cause changes in access to engineering degrees, respectively.

Many researchers will know the adage that correlation does not imply causation. At the same time, the causal impact of various teaching strategies on student learning in science (Schroeder et al., 2007), a makerspace on student experiences (Galaledin et al., 2016), a new technology program on student achievement (Gulek & Demirtas, 2005), a pre-collegiate summer program on retention (Maggio et al., 2005), and a new policy on systemic behaviors (Prados et al., 2005) are examples of causal questions that engineering education researchers may want to answer. While there are some methods for addressing these from a qualitative perspective (Maxwell, 2004, 2012), there is an array of options available to quantitative researchers as well. Causal inference can be considered an advanced topic in quantitative research because common models used in basic quantitative research are typically effective for establishing statistical associations between variables but are not necessarily designed for identifying causal association between variables. To establish causal associations, one needs to engage particular study designs or data analysis techniques. While experimental designs that utilize randomization address this need and are commonly taught in introductory quantitative courses, their use is not so straightforward, as introductory coursework may initially present them, particularly with human subjects. In those courses, one may also encounter the idea of conditioning on certain variables by adding them to a regression model; however, adding more variables can further exacerbate the situation, as discussed next.

Biases from conditioning on too many variables (or omitting variables) are notoriously difficult to detect (and we discuss the implications of this bias in model error later on in the chapter). Diagrammatic representations of the hypothesized relationships between variables, such as causal directed acyclic graphs (DAGs; Pearl, 1995; Spirtes, 2010), can help researchers identify potential issues otherwise created by either omitting variables or unobserved confounders (Steiner & Kim, 2016). In a DAG, variables are represented as nodes in a graph, and causal relationships between variables are represented as arrows, with the arrow originating from the cause and terminating at the effect. Figures 30.2 and 30.3 in the following sections show two simple DAGs for general settings of instrumental variables and the potential causes of scholarship receipt, respectively. Drawing these kinds of causal DAGs can help the researcher document their hypotheses about which variable(s) are causally associated with other variables. In the veterans and scholarships example that follows, without drawing a DAG to depict the relationships between these three variables, it can be difficult to identify the induced collider bias from conditioning on scholarship receipt with such a model. For a helpful guide demonstrating how DAGs can illustrate threats to validity (discussed earlier), see Matthay and Glymour (2020).

In engineering education, causal questions might fit under one of the three types of questions outlined in the National Research Council's (2002) report on education research. Those three types were descriptive questions, causal questions, and mechanistic questions. To answer causal questions, researchers can use a variety of study designs, such as randomized control trials (RCTs), difference-in-difference, regression discontinuity, instrumental variables, and propensity score matching (Arellano & Bover, 1995; Angrist et al., 1996; Benedetto et al, 2018; Listl et al., 2016; Rosenbaum & Rubin, 1983). Table 30.1 lists some of their benefits and drawbacks, and the following paragraphs elaborate on their main ideas.

Many individuals will be familiar with the basics of RCTs, as already mentioned, which have historically been considered the quintessential study design for establishing causal associations between variables (Cartwright, 2007). In a simple version, such a design would entail random assignment of some participants to a "treatment" group and others to a "control" group. Theoretically, random assignment helps ensure any observed differences between the averages of the two groups on some outcome variable of interest (e.g., grades, enrollment rate) are because of the treatment and not some other observed (or unobserved) confounding variables. While easy to implement in theory,

Table 30.1 Trade-Offs for Common Causal Inference Techniques

Causal Inference Technique	*Advantages*	*Limitations*
Randomized control trial (RCT)	Balances potential confounders to enable simple estimation of causal effect of factor of interest	Concerns about equitable participant treatment; often not practical to implement in education settings
Difference-in-difference (DiD)	Simple method to estimate causal effect over time when matched control exists	Can be difficult to identify adequate match with equivalent trends
Regression discontinuity (RD)	Estimate causal effect when there is a threshold that determines access to intervention	Requires threshold cutoff to determine exposure to intervention
Instrumental variable (IV)	Can handle unobserved confounding well	Strong instrumental variables can be difficult to identify in practice; if assumptions not satisfied, can lead to more biased estimates than simple ordinary least squares
Propensity score method – propensity score matching (PSM) and inverse probability weighting (IPW)	Simple method when potential confounders are observed; easily implemented in statistical programming languages (e.g., R)	Identifying which covariates to include in propensity score model is challenging; matching methods can be inefficient and exclude observations

in practice, these kinds of study designs can be infeasible for engineering education researchers, as others have highlighted (Thomas, 2016). There are issues around equity (e.g., random assignment when one intervention is believed to be more effective), logistics (e.g., selection bias and quasi-experiments – those where there is no random assignment), and confounder imbalance (Deaton & Cartwright, 2018) that raise three obstacles to this study design.

Recognizing RCTs have their own limitations, especially in education settings (Connolly et al., 2018), there are other options available to researchers interested in answering causal questions, as mentioned earlier. Many of them offer ways to address the challenge of not being able to randomize assignment to treatment and control groups, meaning, that experimental designs are infeasible. Instead, these designs tend to work with observational data or data from natural experiments – settings where one group randomly receives some treatment but not because the researcher assigned it that way (Craig et al., 2017; Leatherdale, 2019). An example of a study design that leverages this arrangement is a difference-in-difference (DiD) design. The idea here is to match participants or units of observation based on similar characteristics and then track them over time as one subset (e.g., the intervention group) receives some treatment or exposure while the other group (i.e., the control group) does not (Donald & Lang, 2007; Abadie, 2005). The researcher then observes some outcome(s) of interest to measure if there is a difference in how those groups change over time (Wing et al., 2018). The assumption is that the treatment group would have followed the same trend in the change in their outcome of interest as the control group if they had not been exposed to the intervention of interest. Therefore, any deviation from that hypothetical trend is due to the intervention. For example, Deschacht and Goeman (2015) wanted to know the effects of a blended learning environment on adult learner persistence and course performance. In their DiD study design, students in 2009 and 2012 (the control group) did not have the blended learning format, while students in 2010 and 2011 (the intervention group) did have the blended learning format. The assumption is that the students in the control group and intervention group were similar except in their exposure to the blended learning environment, so any changes in the intervention group's course persistence or performance that deviate from the trend one might expect (as observed in the control group) were due to that changed course format. For example, hypothetically there may have been a 10% attrition rate in the program in the control group. If the blended learning environment group then showed a 15% attrition rate, then one might conclude the blended learning environment caused a 5% increase in the attrition rate as calculated from the difference in the differences (15–10%). A typical plot one might expect to see that illustrates the basic idea of difference-in-difference designs is shown in Figure 30.1. This plot shows trends in a treatment group and control group over time. If the treatment group did not, in fact, receive the treatment, then one might have expected them to

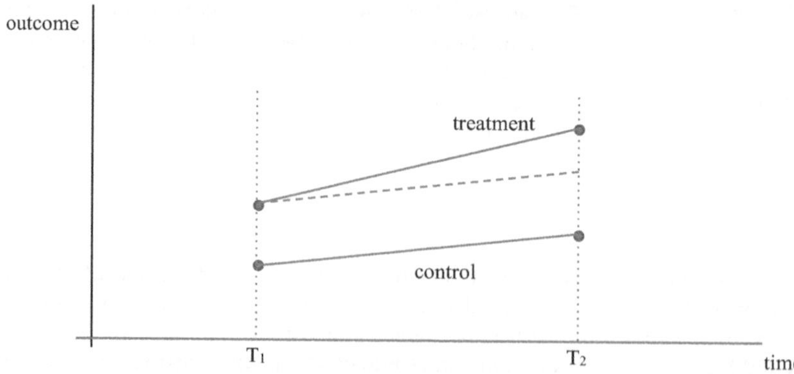

Figure 30.1 Stylized plot showing how differences in differences can appear in DiD study designs.

follow the same trend as the control group – the parallel trends assumption. Creating these kinds of plots with the outcome plotted against time helps illustrate the contrast between the treatment and control group.

A second alternative to RCTs for estimating causal effects is a regression discontinuity design. Regression discontinuity involves forming control and treatment groups based on participants' relative closeness in some threshold continuous variable, for example, a test score (Imbens & Lemieux, 2008). Participants in one group lie to one side of the cutoff, and participants in the other group lie close to the other side of the cutoff. The classic example for this design was the Thistlewaite and Campbell study from the 1960 trying to estimate the impacts of public recognition on student motivation (Thistlewaite & Campbell, 2017). Other common examples arise from researchers wanting to estimate the effect of participating in an academic program, and participation is determined based on a GPA test score (Jacob & Lefgren, 2004; Matsudaira, 2008). In these kinds of studies, students above the threshold cutoff are admitted into the program, while students below the cutoff do not partake in the program. In a regression discontinuity design, a researcher would take students just above the cutoff as members of the treatment group, and students just below the cutoff as members of the control group. The idea is that students close to the cutoff on either side should be similar to each other in other potential confounding factors, and therefore, any differences in future outcomes would be due to the program (or intervention of interest). For a review of regression discontinuity designs in education, see (Valentine et al., 2017; McCall & Bielby, 2012).

A third alternative for answering causal questions is with instrumental variables (IVs). An IV design addresses the issue of confounding variables by identifying a variable (called the IV) that is associated with the explanatory variable of interest (i.e., the treatment or intervention), but not with the outcome variable (Angrist & Imbens, 1995; Angrist et al., 1996; Greenland, 2000; Martens et al., 2006). Assumptions around IV designs focus on this isolation of the IV from other potential confounders and the outcome such that any statistical association the IV has with the outcome is only mediated through the treatment variable. A simple version of this is shown in Figure 30.2. The figure illustrates how the instrumental variable is associated with the treatment/intervention but not with any potential confounders. Additionally, any effect from the instrumental variable on the outcome is entirely mediated through the treatment, which is why there is no arrow directly from the instrument to the outcome.

For an example of this design, Block et al. (2013) wanted to estimate the effect of education on choices to pursue entrepreneurship. There are several potential confounding variables that would make working with observational data to generate this causal estimate biased. To address that challenge, the authors used family demographic background as their instrument in the study. Specifically, they used parents' social class as their IV. Under the assumptions of IV designs, which they demonstrated in their paper (as is always suggested), parent social class is associated with a child's educational attainment (i.e., the explanatory variable of interest as the treatment or intervention

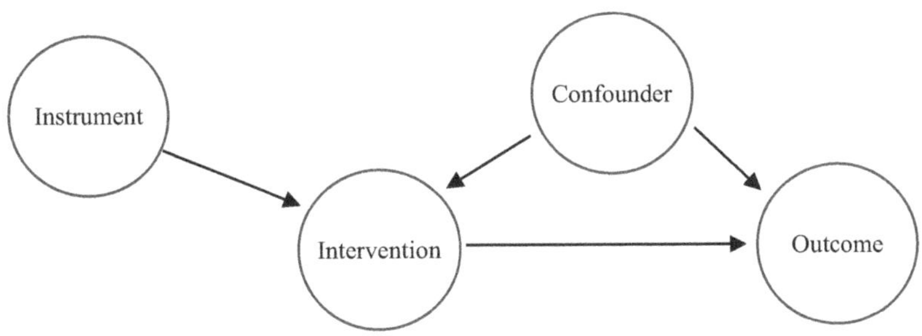

Figure 30.2 Instrumental variable DAG.

variable), but not with other control variables of the outcome variable of interest (i.e., a binary variable of entrepreneurial intent). For more on the use of IVs in education research, see (Orosz et al., 2020; Salas-Velasco, 2006).

Finally, a fourth avenue towards estimating causal effects without randomization is through propensity score, either propensity score matching or inverse probability weighting. Propensity scores estimate the probability of treatment based on other variables in the dataset (Rosenbaum & Rubin, 1983). The idea is to use these propensity scores as a way to match each participant in the intervention group with one or more participants in the control group based on their propensity score. For example, if someone in the intervention group has a propensity score of 0.6 (remember, because these are probabilities, they should range between 0 and 1), then the researcher might try to match that participant with a participant in the control group who also has a propensity score of 0.6. That would be an example of nearest neighbor matching. There are other methods for matching based on kernel functions, intervals, or radii, as discussed in Caliendo and Kopeinig (2008). If the propensity score function has met a couple of technical assumptions related to covariate balance and treatment assignment, then these matched participants should be similar to each other except that one is exposed to the intervention and the other is not (Abadie & Imbens, 2016). The researcher can then examine the differences in the outcome variable between matched participants to estimate the average causal effect of the intervention. An example of propensity score matching in engineering education is Olitsky's (2014) estimation of the effects on STEM versus non-STEM major choice (the intervention) on student wage earnings (the outcome of interest). In that study, the author found that major choice accounted for 5 to 28% of a change in after-college earnings, with estimates varying across student gender and academic achievement. For practical guidance on implementing PSM and more examples in education, see (Caliendo & Kopeinig, 2008; Powell et al., 2020).

Rather than using any of these alternative study designs, some researchers may be tempted simply to add as many control variables to their models as possible to isolate the marginal effect of a particular explanatory variable on the outcome variable. Their assumption would be that the marginal effect produced from their regression model is what they have isolated to be the causal effect. Unfortunately, including as many variables as possible can generate misleading statistical associations by creating collider bias (Cole et al., 2010; Elwert & Winship, 2014). Collider bias – also known as endogenous selection bias – is a phenomenon in which a spurious statistical association is created between two independent random variables because they share a common effect, which has been conditioned upon. For example, consider the scenario of modeling student scholarship receipt as a function of their veteran status and their grades. Students may receive a scholarship through their veteran status or through their grades, depicted in Figure 30.3. However, there may be little a priori statistical association between veteran status and grades. Yet when conditioning on the common effect (i.e., receiving a scholarship), there can appear to be a spurious relationship between those two.

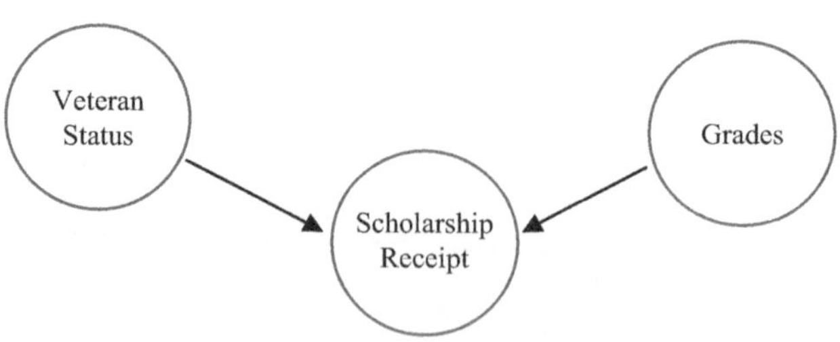

Figure 30.3 DAG of scholarship receipt.

When answering questions of causal inference, researchers should also consider whether the explanatory variables of interest are manipulable. This kind of consideration would not show up in a DAG. Such graphical representations are agnostic about additional information about the variables, such as the type of variables being represented by the nodes, and require a level of background knowledge from the researcher. On the topic of manipulability, some make the claim that there can be "no causation without manipulation" (Holland, 1986). This phrase is especially notable with demographic variables such as race (Holland, 2003). More often, these will be treated as mediating or moderating variables. On the other hand, there are instances such as smoking where there is a consensus around causality that was established without random assignment of participants to a smoking and non-smoking group. Here, the general advice to researchers is to proceed with caution when attempting to make causal claims about these non-manipulable variables.

3 Data Collection and Preparation

Often, quantitative research relies on existing data (i.e., academic transcripts or clickstream data) or collected data through surveys. There are several resources that describe considerations for survey data collection and validation in more detail that we will not discuss in this chapter (indeed, this discussion could be an entire chapter in and of itself; refer to DeVellis & Thorpe, 2021; Douglas et al., 2023). In the following, we present some newer methods for data collection as well as how to handle common but vexing problems with educational data, including missingness and error.

3.1 Multimodal Data

Multimodal quantitative methods are new data collection techniques. Multimodal techniques rely on multiple forms of quantitative data that combine direct measures (e.g., biological measurements, clickstream data, eye tracking) and indirect measures more typical of educational methods (e.g., survey data, interviews, focus groups, reflections, etc.). Villanueva et al. (2019) conducted a study with multimodal approaches to data collection, including interviews and electrodermal activity sensor data, from 12 womxn students to study psychophysiological responses to academic mentoring. Gero and Milanovic (2020) proposed measuring design thinking using a combination of brain imaging, electrodermal activity, eye movements, protocol analysis, surveys, and interviews to understand both cognition and emotion simultaneously. Two of the biggest challenges of multimodal approaches are the effort (i.e., time, cost, etc.) associated with data collection and synthesis of heterogeneous data. However, we note these methods as advanced ways to consider complex phenomena and the combination of these data sources as underutilized in engineering education research. Engineering education researchers may consider multimodal methods in data collection to better capture multiple examples of and synthesize the student experience.

3.2 Missing Data Methods

In data collection, whether at a single time point or over multiple, missing data are a common problem. Often, the default in simple analyses and in software is pair or listwise deletion. However, these approaches can lead to biased estimates and lack of precision in statistical analyses (Newman, 2014). It is essential to properly deal with missing data. Typically, if fewer than 5% of the observations are missing, then the risk of biased estimates is low; otherwise, the missingness needs to be addressed.

There are three types of missingness in data: missing completely at random (MCAR), missing at random (MAR), and missing not at random (MNAR). MCAR data occurs when the missing cases are no different than non-missing cases in terms of the analysis being performed. A potential observation is missing completely at random if, say, an individual decides whether to answer survey questions on the basis of a completely random event (e.g., coin flips). This type of missingness is rare. For example, if women are less likely to answer a question about gender bias in an engineering context, then the data are not missing completely at random. In the unlikely event that the process is missing completely at random, then inferences based on listwise deletion are unbiased, but inefficient because some cases are lost. MAR depends on known values that are described by variables observed in the dataset. Following the earlier example about women's non-responses on a gender bias question, the data are missing at random so long as women generally respond to the question identifying their gender. If data are missing at random, then inferences based on listwise deletion will be biased and inefficient (Bell et al., 2009). The final type of missing data is missing not at random (MNAR; also known as not missing at random, NMAR). This type of data is particularly problematic in analysis. The missing data depends on events or items which the researcher has not measured. This type of missing data would occur if women in the earlier example are not only less likely to refuse to answer a survey question about gender bias but also their gender (i.e., there are no other variables that can predict which respondents are women). This type of missingness can be damaging to conclusions drawn from incorrect analysis. The missingness can be accounted for, but an explicit external model from the collected data must be used rather than an imputation (McKnight et al., 2007).

When data are MAR, imputation techniques can be used to recover power and avoid biased estimates. Among the imputation options available to researchers, we caution against mean or regression substitution. Mean substitution uses the mean of all data for missing data. This approach adds no new information, and it decreases error and distribution added to potential bias (Li et al., 2015). Regression substitution treats each missing value as a linear variable and substitutes the trend value for the missing value. This approach can show trends that are not true of the raw data and can lead to erroneous conclusions (Cole, 2008).

More robust methods include full information maximum likelihood (FIML) and multiple imputation techniques (Allison, 2012). FIML does not impute any data; instead, it uses each case's available data to compute maximum likelihood estimates. The maximum likelihood estimate of a parameter is the most likely value to have resulted in the observed data if a question was answered. The likelihood is calculated separately for cases with complete data on some variables and cases with complete data on all variables. These two likelihoods are then maximized together to find the estimates. This method addresses concerns about biased parameter estimates and standard errors. One advantage of this approach is that it does not require the careful selection of variables used to impute values, because it uses the full set of data available. It also results in the same answer each time the imputation is conducted. It is, however, limited to linear models.

Multiple imputation by chained equations fills in estimates for missing data by estimating values multiple times with bootstrapped error (Buuren, 2012; Azur et al., 2011). The result of these techniques provides multiple datasets with identical values for the non-missing data and slightly different imputed values in each dataset. Then, the statistical analyses are performed on each imputed dataset separately, and the resulting estimates, standard error, etc. are recombined. Because of the variation in imputed data, there will be differences in the statistical test parameter estimates, which results in appropriate estimates of standard errors and p-values in testing (Allison, 2000).

Most statistical software programs have these techniques available, and the book *Missing data: A gentle introduction* (McKnight et al., 2007) provides a useful discussion and starting point about the assumptions, trade-offs, and steps of missing data techniques.

3.3 Understanding Error

It is important to understand that all data collected, and thus all models created, will have errors associated with them in the form of measurement error and model error. The goal is to decrease error as much as possible so that estimates from statistical testing reflect the "true" value. We put *truth* in scare quotes to indicate that the goal of error reduction is an accurate estimate based on the data and practical implications of that value but also that any study cannot fully capture the entire essence of a phenomenon. There are many ways to reduce the error in both data collection and analysis so that the models which are created are more useful and the resulting interpretations are accurate and fair. These approaches depend on the type of model used and the data.

Measurement error can be in the form of random error and systematic error. Random error happens more often when not enough data is collected, while systematic error occurs when the instrument used for data collection is misaligned with the end goal. One of the ways to detect error in data is to first visually inspect the data for outliers. Visualization can be accomplished by plotting a histogram or box plot to quickly inspect the data for any data that are different from the rest. There are various ways to deal with outliers, such as deleting the data (depending on the reason for it, such as a data entry error) or using more robust methods, which take into consideration outliers. For example, Huber weights reduce the contribution of outliers in calculating regression estimates in regression calculations (Lambert-lacroix & Zwald, 2011). Additionally, for latent measurements, particular statistical techniques that disaggregate measurement error from model error can be used (e.g., structural equation modeling discussed later in the chapter). Any decision to delete outliers should be carefully weighed with the implications of this approach. In research that considers policies or interventions, outliers may reflect unique experiences of individuals that will not be best served by the particular effort under study. If those individuals are also systemically marginalized, results can perpetuate inequity and even cause the opposite outcome from the intention.

Model error can come from several sources. In most studies, measurement error and model error are combined in a single error term. Some techniques described later on in the chapter are able to disaggregate these sources and provide ways to better understand where error originates in the research. One main source of model error is misspecification, sometimes called "type 3 error," or variable omission bias (Mosteller, 2006). This error can arise, for example, in regression models when the functional form (the algebraic form of the relationship between a dependent variable and regressors) does not adequately reflect the underlying data relationships (Dennis et al., 2019). This error can result from omitting relevant variables, for example, and applies across different model types. We acknowledge that all possible independent variables are unlikely to be a part of a study; however, the DAG process described earlier may provide ways to consider the range of important variables from literature and theory. This technique is useful even in accounting for relationships that are not causal. Another source of error is overfitting the model to the dataset. Overfitting is when a model too closely matches the limited data on which it was trained and it is unable to be used accurately with additional data (in other words, it lacks predictive validity evidence for its use). This type of model error occurs commonly with machine learning approaches. A common approach to reduce this is the use of cross-validation and information criteria (Shao, 1993). Cross-validation methods entail using a portion of the data for estimating the model, and the remaining portion for testing the model. One can repeat this cycle of estimation and testing to eventually identify an optimal set of model parameters that fits the data well but also fits unseen (in cross-validation, the data reserved for testing) data well. This approach allows one to generate reasonable model estimates that can be generalized to similar data and increases utility of models and reduces error. Another source of model error for regression can be collinearity, which is when two or more predictors in a model are so highly related that the design matrix for the regression model becomes singular (in other words, the independent variables are so highly related that the estimates are biased and cannot independently predict the value of the dependent variable). This issue can be addressed by using typical assessments of collinearity like tolerance and a variance inflation factor (VIF; Alin, 2010) to identify collinear variables and removing one of the redundant predictor variables from the model. A small tolerance value indicates that the variable under consideration is almost a perfect linear combination of the independent variables already in the equation and that it should not be added to the regression equation. The VIF measures the impact of collinearity among the variables in a regression model. The variance inflation factor (VIF) is 1/Tolerance; it is always greater than or equal to 1. There is no formal VIF value for determining presence of multicollinearity, but a value less than 2.5, 5, or 10 is often used (Gareth et al., 2013). The actual impact depends on the model, including sample size, variable omission bias, number of predictors, etc. (Lavery et al., 2019). More generally, checking for multicollinearity is one of several assumption checks that one should conduct and report. Together, these areas provide some of the common sources of model error that may be able to be reduced in study design and measurement considerations.

4 Data Analysis

We now move from general considerations in study design and data quality to novel methods in quantitative data analysis. In moving beyond the typical methods historically studied and used in engineering education research (i.e., simple hypothesis testing or regression), researchers have many options that open avenues for different types of research questions to be studied. In what follows, we outline some of the new(er) options available to encourage the community to consider them in their own work. We have selected these because of their diversity in applications and kinds of data they enable researchers to analyze, as elaborated in Table 30.2. As mentioned in the introduction, for each of these methods we provide an introduction, advantages, limitations, and example and point to further resources.

4.1 LCA/Clustering

4.1.1 Introduction

Latent class analysis (LCA, or, for non-categorical outcomes, latent profile analysis, LPA) is a technique that identifies individuals that share common characteristics or response patterns and establishes mutually exclusive subgroups or classes within data (Hagenaars & McCutcheon, 2002). This

Table 30.2 Advanced Methods Reviewed in This Chapter

Method	*Application*
Latent class analysis (LCA)	Reveals underlying groups (classes) in data based on individuals' response patterns. Estimates measurement error separately from model error.
Structural equation modeling (SEM)	Set of techniques to simultaneously examine both direct and indirect relationships between measured variables. Estimates measurement error separately from model error.
Social network analysis (SNA)	Captures interpersonal dynamics and relationships in networks. Maps connections between entities (e.g., people) to reveal underlying information flows and hierarchies that can help explain social phenomena.
Topological data analysis (TDA)	Novel person-centered approach that reveals underlying data structure and connections (i.e., the shape of the data) even in complex and noisy data.
Monotonic effects models	More realistic modeling assumptions about explanatory variables that capture monotonicity of effects without assuming the same magnitude of effect. Moving from 1 to 2 on the predictor variable does not have the same predicted effect on the outcome as moving from 2 to 3.
Bayesian data analysis	Express prior knowledge about data generating processes. In addition to incorporating prior information, one can also directly model uncertainty in estimates through capturing entire probability distributions rather than point estimates.

technique assumes that membership in one of these classes causes or explains particular patterns across multiple variables (i.e., survey questions, scales, assessment items, etc.; Muthén & Muthén, 2000). In other words, individuals' scores on a set of variables originate from their class membership, and this membership is identified through LCA.

LCA is similar to another person-centered analysis, cluster analysis. However, these two approaches make different assumptions about the data. In cluster analysis, the assumption is that cases with the closest (as a distance, correlation, or density metric) patterns across variables belong in the same cluster, that is, the results are driven by the data rather than the underlying class membership (Hennig et al., 2015). Latent class analysis also assigns individuals to classes based on a probability score, while clustering may be either hard (i.e., a point belongs to a cluster or not) or soft (i.e., a probability of cluster membership is calculated, like LCA).

4.1.2 Advantages

LCA/LPA account for measurement error in the estimation of classes. Additionally, this technique may be better for particular kinds of research questions because there is an underlying statistical model. The choice of which aspects on which to group individuals is determined by this model over clustering. In addition, because the technique is based on this underlying model, there are several statistical tests to assess model fit and determine if the resulting classes fit the data. Finally, there are formal criteria to determine the number of classes used in this analysis (Karnowski, 2017).

In comparison, clustering does not require an underlying model and, instead, is solely based on observed similarities. This approach may be useful in exploratory data analyses or when multiple group memberships are desired.

4.1.3 Limitations

LCA assigns participants to classes based on probability determined from patterns on the measured variables (B.O. Muthén & Muthén, 2000). As such, accurate class assignment is not guaranteed. Additionally, because the assignment is conducted based on probabilities, the exact number of class members cannot be determined (unless an arbitrary probability threshold is decided upon by the researcher; Weller et al., 2020). Also, researchers often name the resulting classes, but due to the complexity of underlying patterns that determine classes, they may misname or oversimplify the general characteristics of class membership.

Clustering has a different set of limitations (Everitt et al., 2011). First, the different methods of clustering, for example, hierarchical, k-means, agglomerative, often give very different results due to the linking process. As such, it is particularly important to consider methodological trade-offs for the purpose of the study. Additionally, except for single-linkage clustering, the ordering of variables used in the model will affect the results. These models also depend on the data used, so dropping or adding cases will change the results. Some clustering algorithms such as k-means also have an element of stochasticity to them, so cluster assignments may not be stable unless the researcher sets the same seed for their computer's random number generator. Finally, determining the best model for clustering can be challenging. Cluster validation is often best when a known set of clusters can be used for comparison. In engineering education research, there are often not known classifications/clusters. Clustering often needs more robust support for decision-making and claims made.

4.1.4 Example

A study of 629 adolescents' (average age of 16) perceptions of the prevalence and severity of sexism in STEM used LCA and found three distinct classes that varied by perceived prevalence and severity of sexism in STEM (Robnett & John, 2020). Notably, one of the classes considered sexism as relatively common and a serious issue; participants in this class were more likely to be girls and had a higher value for STEM study. Follow-on qualitative analyses indicated that these participants were more likely to describe the specific ways in which sexism can harm girls and women and to view sexism as having deep, systemic implications.

4.1.5 Further Resources

There are several considerations in LCA covered in this comprehensive discussion from Weller et al. (2020). To learn more about clustering, refer to Everitt et al. (2011) for a strong reference text. For a practical guide of implementing cluster analysis in R, Kassambara (2017) is an excellent resource.

4.2 SEM

4.2.1 Introduction

Structural equation modeling (SEM) allows researchers to combine both path models and measurement models to test a system of regression equations simultaneously. The structural model showcases the relationship between latent variables, and the measurement model showcases the relationship between the latent variables and their corresponding observed indicators, usually in the form of survey questions. The basis of any structural model must derive from a theoretical perspective that SEM is used to test the validity of that theory. There are five distinct steps to utilizing SEM: (1) model specification, (2) model identification, (3) model estimation, (4) model testing, and (5) model modification (Schumacker & Lomax, 2010). The overarching goals of these steps are to build a model that supports theory building and at the same time is parsimonious – a simpler model is usually better. The results of SEM produce fit indices that indicate whether the model is supported by the data. Like many quantitative research endeavors, if the data are not properly collected, the results will not be valid. There is an array of software that will allow users to input raw data to compute SEM analyses, such as R, LISREL, and AMOS, among others. SEM is often associated with causal inference; however, just like with any quantitative method, it is the study design and not the statistical method that permits causal hypotheses to be adequately tested (Bullock et al., 1994). Overall, if you have multiple latent and observed variables, using SEM to validate a theoretical model is a major contribution to the field.

4.2.2 Advantages

The biggest advantage of using SEM is that it accounts for measurement error in the latent constructs, whereas using just a simple path model to determine relationships between variables does not account for measurement error. The ability to deal with latent variables, such as identity or motivation constructs (Jones et al., 2014), is critical to engineering education researchers involved in the social sciences. These types of models also account for both direct and indirect effects helping researchers identify how different variables interrelate.

4.2.3 Limitations

As with any method, there are limitations that exist for SEM as well. Those limitations fall within four main categories, as outlined by Kline (2015), and include specification, data preparation, analysis and respecification, and interpretation. As mentioned earlier, model specification is a critical first step in conducting SEM. Just having latent constructs of interest does not necessitate the use of SEM.

The model design should be built on underlying theory. As with any quantitative method, understanding your data and the data generating process is just as paramount to SEM as any other method. Whether that may be ensuring the accuracy of the data or ignoring the pattern of missing data, it is important to have data that accurately represent your sample. With SEM, there are also many ways to misstep when conducting the analyses. It will be key to understand which fit indices to use, use a large-enough sample size, and report all the information necessary. Lastly, it is also important to not overstate your findings from the model. SEM is one step in building a case for a more robust theory.

4.2.4 Example

In Brozina (2015), SEM was used to analyze the model of domain identification (Osborne & Jones, 2011). The author used the relationships between pre-college variables of SAT math, SAT verbal, high school GPA, and identity and motivation constructs along with course engagement depicted as an observed variable from mean usage of a learning management system. The model also included final grade as an observed variable. The analysis had a sample size of 714, with results indicating that SAT math and motivation had a significant direct effect on final grade in this first-year engineering course.

4.2.5 Further Resources

For further reading on SEM, see Schumacker and Lomax (2010) and Kline (2016).

4.3 SNA

4.3.1 Introduction

Social network analysis (SNA) provides another person-centered approach to data analysis. In this approach, SNA uses networks and graph theory to understand social structures. This method is based on the idea that individuals are embedded in layers of social networks and interactions that shape and explain key outcomes and behaviors. Network analysis has been used in a wide range of applications in social science, particularly since the early 2000s (Borgatti et al., 2009). The method helps examine the structure of interactions or relationships captured in data. In social science research, the participant is often the unit of analysis, and their position within a network is considered. The network consists of nodes that represent the actors (again, typically individual participants, but conceivably also organizations or other entities), and the edges represent the relationships between those actors. Edges can be directed (i.e., has a direction, like help given in a class to one person by another person) or undirected (i.e., does not have a direction, like students who work together on course assignments). Once the graph is created, various properties can be analyzed, including the distribution of individuals within the network (i.e., nodes), through centrality measures, graph density, strength of connections, path analysis, and identification of hubs (Wasserman & Faust, 1994); the relationships between nodes (i.e., edges) through various tie metrics, including similarities to others, social relations, interactions, and flows (Borgatti et al., 2009); and the shape of the network through clustering for community detection. These metrics can also be used to predict particular outcomes (Scott, 2011). Network analyses can also be used with other units of analyses to understand patterns within the literature of citation networks and patterns, co-authorship, or shared keywords (Mejia et al., 2021).

4.3.2 Advantages

An SNA approach allows researchers to focus on dynamics in groups and the potential effects of networks on outcomes of interest (Ushakov & Kukso, 2015). Given the large number of interpersonal interactions in education, this can help highlight the effects of different network structures, relationship types, and information flows on student, faculty, and administrator experiences. More advanced approaches to SNA also expand the kind of networks that one may consider to capture dynamics between different kinds of actors (e.g., students and teachers; participants and learning resources) rather than unipartite networks that only support having one kind of node (actor).

4.3.3 Limitations

Limitations to SNA include difficulty in data collection, availability of data, potential privacy concerns, and capturing changes in the nature of relationships or network structure over time. The first three concerns arise from how one plans to collect data on network structures. One can survey participants, but that kind of recall data depends on the quality of participants to report all the actors in their networks, which could be an extensive process. Alternative methods include looking at social media sites (Gruzd et al., 2016) or institutional trace data (Goggins et al., 2010), but those methods raise privacy concerns if participants did not consent to participate in a study (if the data are not publicly available).

4.3.4 Example

A study of 209 senior capstone mechanical engineering students used social networks to characterize team leadership distribution (Novoselich & Knight, 2018). The authors found that leadership was more shared among teammates and that more teams used effective teaming skills. The authors concluded that engineering educators should consider leadership models in training and forming teams. This approach provided novel ways to consider complex team dynamics.

4.3.5 Further Resources

For further reading on SNA in education, see (Carolan, 2013; Grunspan et al., 2014).

4.4 Topological Data Analysis

4.4.1 Introduction

Topological data analysis is a recently developed technique that uses topology and geometry to infer information about the structure of multidimensional data (Chazal & Michel, 2021). The results of this method are descriptive results of data progressions that indicate the persistent structure of the data. Similar to clustering methods, the analysis technique relies on similarity measures.

4.4.2 Advantages

This approach provides ways to preserve the whole participant response in the data, which offers opportunities to re-evaluate the epistemic norms of quantitative research that often draw conclusions within some quantifiable certainty, consider multiple measures simultaneously, and avoid treating individuals as reducible to responses on a single variable or as outliers in data (Godwin et al., 2021). It also handles complex, noisy data well.

4.4.3 Limitations

Some limitations of this approach are that the results reveal relationships rather than definitive groupings or clusters. Data are most often visual relationships within data that can be analyzed for features of those relationships for additional analysis. The decisions of how the relationships are analyzed and subsequently used can have significant impact on the interpretation of data. It is important to define these structures in ways that provide relevant information about the data (Chazal & Michel, 2021).

4.4.4 Example

As this statistical technique is one of the newest reviewed in this chapter, there are few examples of its use in engineering education. One project used survey data from 2,916 first-year engineering students at four US institutions to understand how students with non-normative engineering identities navigate engineering (Benson et al., 2017). The authors used a range of attitudinal measures to characterize particular patterns in the data. Results revealed a large dense grouping of students, which the authors named the "normative" group, and seven other distinct groups (the "non-normative" groups). Results of group comparisons and follow-up qualitative research emphasize unique experiences and patterns of pathways through engineering.

4.4.5 Further Resources

For an introduction to the method, refer to Lum et al. (2013). For a discussion of decisions and steps, refer to Chazal and Michel (2021).

4.5 Monotonic Effect Models

4.5.1 Introduction

Many quantitative methods make assumptions about the underlying data. For instance, consider the case of an ordinal variable – a variable that has a natural ordering to its categories (Stevens, 1946). These are often found on Likert-style survey items (Likert, 1932). An example of an ordinal variable might be frequency of times a topic was covered in a course, with options including "never," "rarely," "sometimes," "frequently," and "always," or the extent to which a participant agrees with options ranging from "strongly disagree" to "strongly agree." There is a clear ordering, but the gap between one option and the next might not have the same meaning for each jump. Despite this possibility, some researchers will use models that assume a linear relationship between the ordinal explanatory variable and an outcome variable of interest (e.g., level of engagements, as operationalized by a scale score) (Helwig, 2017). Using such models will assume a one-unit change in the predictor variable (e.g., frequency of some pedagogical technique's use in class) leads to the same X unit change in the outcome variable (e.g., student engagement), no matter where someone is moving from/to on that predictor scale (NB: ordinal variables as outcome variables can also create modeling challenges, and ordinal logistic regression models may be more appropriate than linear models in those settings (Liu & Koirala, 2012)). This can be problematic for several reasons, one of which is the observation that moving from one point on the ordinal scale to another may not always have the same effect. In other words, moving from "never" to "rarely" may not be the same as moving from "frequently" to "always." Classic approaches that assume constant effect, however, do not account for that nuance. To address the issues associated with that assumption, one could use monotonic effect models (Bürkner et al., 2021). These models make a more relaxed assumption: moving from one level of the explanatory variable (e.g., rarely) to another level (e.g., sometimes) will have a *monotonic* statistical effect on the outcome variable (in this case, level of engagement) but not necessarily the same magnitude of the effect as moving from "sometimes" to "frequently." In other words, both moves would have the same sign (positive or negative), but not the same value.

4.5.2 Advantages

Monotonic effects models allow researchers to more accurately model ordinal data and drop the assumption of constant effects from moving along values on an ordinal scale. Using a more realistic model can lead to less-biased parameter estimates, reduction in loss of statistical power, and a reduction in false positive errors (type 1 error rate) (Liddell & Kruschke, 2018).

4.5.3 Limitations

Although these models are flexible for handling ordinal data, they can sometimes produce wide uncertainty intervals due to the large number of observations needed to estimate marginal effects. Another limitation is their relative novelty, which means the software infrastructure to implement them is still in nascent stages of development and not commonly available across statistical computing software packages. The good news is that free sources such as the brms package in R do allow them at this time (Bürkner et al., 2022).

4.5.4 Example

An example of their use in engineering education was from Milovanovic et al. (2021), in which they modeled associations of career interests in sustainability with various instructional activities and student self-perceived proficiency in design skills. The self-perceived design proficiency and pedagogical techniques used in their design courses were measured using ordinal variables on a 5-point scale ranging from "strongly disagree" to "strongly agree" for various statements. These design skills and pedagogical strategies were the predictor variables, which is why the authors used the monotonic effects model. Doing so, they found positive associations between students' self-perceived design proficiency and interest in sustainable design careers.

4.5.5 Further Resources

For further reading on the theoretical properties of these models and a case study demonstrating implementation, see Bürkner and Charpentier (2020).

4.6 Bayesian Data Analysis

4.6.1 Introduction

Bayesian data analysis is another approach to complicating the traditional frequentist statistics taught and used in many social science disciplines (Efron, 2005). Problems with frequentist statistical testing revolve around interpretation of results and common constructs in that paradigm. For example, *p* values, despite their controversy (Wasserstein & Lazar, 2016), are still often reported in results sections. Technically, a *p* value is the probability that the value of a test statistic of the data would be as large or larger under some assumed statistical model (i.e., that which is implied by a null hypothesis). When the *p* value is below some predetermined threshold, then that is considered sufficient evidence to reject the null hypothesis in that study. This is not an intuitive concept, and even individuals with statistical training tend to misinterpret *p* values (Goodman, 2008; Gagnier & Morgenstern, 2017; Haller & Krauss, 2002; Lyu et al., 2018). The same misapprehension has been suggested of interpretations of confidence intervals (Hoekstra et al., 2014; Morey et al., 2016) and significance tests (Falk & Greenbaum, 1995). In each example, difficulties have tended to arise with the common concepts in frequentist analysis. Results under the Bayesian paradigm have the benefit of being more consistent with how people actually interpret *p* values and confidence intervals (Vandekerckhove et al., 2018). A Bayesian approach does this by giving direct estimates of the uncertainty around parameter values in a statistical model.

So what is this alternative all about? At a high level, the approach is about using conditional probabilities via Bayes's rule to characterize the probability distribution of the parameter(s) of interest given observed data and prior information about those parameters. Researchers fit models to capture the uncertainty around the distribution of model parameters by incorporating prior knowledge about the specific domain, with a likelihood function they believe represents the underlying data generating process. This procedure ultimately produces an entire probability distribution over those unknown parameters (rather than single point estimates, as often is done under frequentist approaches). The step of creating a probability distribution also points to an underlying philosophical commitment about the nature of the parameters. In a frequentist paradigm, the unknown parameter (e.g., difference between means of two groups, regression coefficient) is treated as a fixed (i.e., non-random) value that we are trying to estimate. Our knowledge of that parameter may be imperfect, but that is a function of the data we have collected.

For an example of where prior knowledge would be relevant, consider the scenario where a researcher wants to model student grades using explanatory variables, including previous test scores and number of hours the student studied for an assessment. If one were modeling the outcome variable (i.e., student's grade) as a Z-score, then the researcher would be relatively confident that the parameter estimate for the effect of the "number of hours studied" on the grade Z-score should have a relatively small standard deviation, since it might be unusual for number of hours studied to move a student's grade by multiple standard deviations on the Z-score. This information about the small effect of "number of hours studied" is incorporated into the analysis in the form of a strong prior over the value of the parameter that is to be estimated via Bayes's rule to generate the posterior probability density. The point here is that the researcher is not approaching their analysis with a blank slate, but frequentist analysis using maximum likelihood estimation actually does implicitly make that assumption (in the form of uninformative priors).

Bayesian data analysis can be used in most settings where classical frequentist models (e.g., models using maximum likelihood estimation) are used. Additionally, given their flexibility, Bayesian models fit settings where there is limited information in the data (i.e., small sample sizes (Winter & Depaoli, 2020; McNeish, 2016; Smid et al., 2020), complex nesting structures (i.e., mixed effects, especially with a small number of clusters; McNeish & Stapleton, 2016a, 2016b), and where the traditional maximum likelihood estimation approach does not work (e.g., structural equation models with small samples; Lee & Song, 2004).

4.6.2 Advantages

Beyond allowing researchers to incorporate prior information about the problem into their analysis, there are also simulation studies which suggest that Bayesian approaches might be more powerful (in a statistical sense) than classical frequentist tests when used in intervention studies (Chen & Fraser, 2017) and more robust when using smaller sample sizes in multilevel modeling (Stegmueller, 2013). This means that a Bayesian approach could help reduce the incidence of false negatives (e.g., detecting true effects of an intervention on an outcome of interest in the classroom) and biased parameter estimation. There is also evidence to suggest that Bayesian methods can even outperform widely used options, such as *t*-tests (Kruschke, 2013).

4.6.3 Limitations

There are philosophical and practical limitations to a Bayesian approach. Philosophically, one might object to the interpretation of probability as degree of belief or an epistemic expression rather than an empirical expression of frequency, as is the interpretation in the frequentist probability tradition (Gelman, 2011; Etz & Vandekerckhove, 2018). Practically, there are also limitations in that implementing Bayesian models can have a steep learning curve. Those steep learning curves are both conceptually for understanding some of the math behind Bayesian data analysis if the researcher is not familiar with fundamental building blocks such as probability theory and Monte Carlo methods. There are additional practical limitations if the researcher is not familiar with statistical computing environments (e.g., R, Python) in which many Bayesian methods are implemented. There are some alternative methods that require less programming (e.g., JASP; van Doorn et al., 2021), but the models implemented in those options tend to be limited.

4.6.4 Example

In engineering education, Bayesian analysis has been used to study topics ranging from a meta-analysis of the effects of computer-based scaffolding on problem-based learning (Kim et al., 2018), factor analysis in studies of engineers' capacity to innovate (Ferguson et al., 2018), and design education (Milovanovic et al., 2021). These approaches are not to be confused with studies using Bayesian networks, which are probabilistic graphical models (i.e., DAGs) that can be used in causal inference settings, as mentioned earlier. Vaziri et al. (2022) used a Bayesian regression approach to study student motivation in a business analytics course. They framed their study using Jones's MUSIC model of student motivation and found that components of that motivation model were positively associated with student effort, final grade, and course rating.

4.6.5 Further Resources

For an accessible introduction to Bayesian data analysis, see McElreath (2016). For a more classic text on the topic, see Gelman et al. (2014). See Levy (2016) for a discussion focused on applications in education, and Gelman et al. (2020) for a discussion on practical implementation and workflow considerations.

5 Data Equity

A pressing consideration in the minds of these authors is how researchers can do justice to the variation in lived experiences of participants through quantitative measures. The history of quantitative methods is steeped in White supremacy and eugenics (Zuberi, 2001; Zuberi & Bonilla-Silva, 2008). Often, goals of generalizing to an engineering population or finding generalizable trends continue to reify a White male representation of results and perpetuate inequity (Godwin, 2020; Holly Jr., 2020; Pawley, 2017). This topic could be a whole chapter in and of itself. In what follows, we raise considerations for the process of research and reference this ongoing conversation in social science and education research.

Quantitative research is typically framed as if it is certain or knowable within some margin of error (or that it is "objective"; Bryman, 2008). However, the design decisions at each step of the research process is influenced by the researcher (i.e., the research questions, research design, theoretical framing, population of study, sample, measurement, analysis, interpretations, acknowledgment of limitations, etc.), which influences the numeric results and changes whose lived experiences are prioritized. For example, a study of differences in academic performance in a course by race/ethnicity in the United States can have several points at which value-laden decisions about persons and methods are made that influence the claims from research. First is the way in which the difference is described. It could be a deficit-based statement that Black students perform worse than White students or an "academic gap." However, this statement problematizes students rather than the systems of disadvantage that create disparate student outcomes (Ladson-Billings, 2007). Additionally, identifying who is Black and who is White is often based on categories common in education, which are not devoid of historical issues of racism (Horton & Sykes, 2008). Additionally, a tension in quantitative research is having power to detect differences. Often, there are tensions of how disaggregation can be accomplished, particularly in discussions of race/ethnicity, due to how small the sample of particular groups is within engineering systems. The analysis itself could use comparisons that focus on a single variable and mean comparisons that obscure extreme cases important to understanding the phenomenon. Additionally, important aspects of students' lived experiences may not be included in the model or treated as separable variables (e.g., gender; Ro & Loya, 2015). Additionally, measures of deeply personal psychological factors (e.g., motivation, interest, identities, etc.) are limited; a person's internal states are not fully knowable through qualitative or quantitative approaches. Finally, the use of the results could lead to policy changes to provide support (positive) or decisions to reduce funding for "underperforming" schools (punitive). These are just a few examples of how the researcher's decisions shape research findings. These examples are just a few of the key decision points in research design.

In each of these decisions, the researcher is present and influential, and the confluence of all these decisions can either address data equity or not. From the preceding example, there might be concerns that data equity cannot be achieved in quantitative research. We argue that in any research, there are tensions and limitations. We encourage the community to consider lessons from qualitative research methods, which acknowledge and explicitly include descriptions of how the researcher was engaged in the decision-making and interpretation process of research (Secules et al., 2021). Additionally, there are opportunities to critically interrogate and revise methods to consider how systemic inequity is present in and replicated by quantitative work. For example, several researchers have made calls for better ways to measure demographics (Fernandez et al., 2016; Haverkamp et al., 2021; Viano & Baker, 2020). Some newer quantitative methods begin to disrupt traditional statistical norms with possible data equity application through analytical approaches (e.g., person-centered analyses), while others reshape epistemic and methodological norms (e.g., FemQuant and QuantCrit). As more robust ways of making data and using methods to address the complexity of social data continue to push the types of questions and studies that can be conducted, an ethical consideration of how participants are represented in and discussed in quantitative research is paramount. Often, this conundrum is framed as a quality issue in quantitative research: if all confounding variables could be conceptualized and measured, a best-fitting model could be developed. However, the purpose of traditional quantitative research methods to describe or make inference, explain a particular phenomenon, or generalize inference across groups creates a tension between general trends and specificity (Babbie, 2020). This tension may arise in part because most quantitative research studies center variables rather than people as the unit of analysis (Wang et al., 2013). Variable-centered research focuses on relating variables to one another, for example, in regression analyses. Participants in these analyses are treated as non-unique data carriers that could be replaced by other randomly sampled individuals. The results of these analyses are aggregate statements that are not often applicable to an individual (Eye & Widermann, 2015). Some emerging methods and methodologies begin to disrupt these norms and offer ways to consider the whole person or response patterns in data. These approaches also provide opportunities to address key ethical and inclusion issues in quantitative research.

Person-centered methodologies assume that the population under study is heterogeneous. It is important to emphasize that if variation or distinct groups is expected in data, aggregation of raw data, as in traditional variable-centered quantitative analyses, can result in inaccurate conclusions, such as those from the ecological fallacy. We have already discussed some of these methods earlier, including LCA, clustering, and topological data analysis. The results of such studies focus on preserving the variation in individual's responses, resulting in authentic groupings of individuals, as opposed to imposing superficial characterizations of groups (Laursen & Hoff, 2006; Morin et al., 2018). In these analyses, parameters are first estimated for each individual, and if generalization is the goal, parameters are aggregated (rather than raw) data (Eye & Widermann, 2015). The results of these types of analyses are significantly different from variable-centered ones. These approaches are relatively new in social science research with increased use with the availability of computing resources (Laursen & Hoff, 2006). For a more comprehensive review of person-centered approaches and opportunities for novel research approaches, refer to Godwin et al., 2021.

Finally, FemQuant and QuantCrit offer both theoretical and methodological approaches to addressing issues in quantitative research. These approaches are more comprehensive than the person-centered methods in that they have a particular epistemic standpoint, a set of guiding tenets, and commitments to using methods to frame anti-deficit, anti-racist, and feminist research questions that speak to the overarching structure of inequity (Sablan, 2019). Both of these approaches are critical quantitative methodologies that draw on principles and histories of feminist and critical race theories (i.e., FemQuant and QuantCrit, respectively). FemQuant and QuantCrit share common framings but were developed separately, with FemQuant developing earlier than QuantCrit. There are numerous books and excellent studies that give a more thorough discussion of these approaches (refer to McCall, 2002; Oakley, 1998; Sprague, 2005; Sprague & Zimmerman, 1989; and a special issue edited by Gillborn, 2018).

Both approaches are based in several tenets that frame research (summarized in Godwin et al., 2021) and adapted from prior work (Bowleg, 2008; Gillborn et al., 2018; Hesse-Biber & Piatelli, 2012; Sigle-Rushton, 2014; Sprague & Zimmerman, 1989). The first is that domination is a central component of society that is not natural but rather is socially constructed and supported through multiple dimensions of difference or categories that quantitative research employs. These framings also are based on the premise that quantitative research is not neutral and that numbers are representations of domination based on local or global meanings relating to differences in human bodies. As such, neutrality often parallels naturality in that what is deemed natural is often connected to political ideology (Oakley, 1998). Another tenet is the importance of intersectionality, which is that inequality exists beyond an individual's social position. In addition, inequality is multiplicative rather than additive for persons experiencing multiple inequalities, and that multiplicative effect cannot be represented by simple variables or identities (Covarrubias, 2011; López et al., 2018). The fourth tenet is the humanity participants. Data cannot "speak for itself" or act anthropomorphically in any other way. Rather, data is interpreted by researchers through their scientific understandings and global enculturation. This concept is related to how social processes like inequality may be reflected in statistical differences in data by race/ethnicity and/or gender, but that these measures are not natural, neutral, or causal. (Holland, 2008; Gillborn et al., 2018). Next, quantification that unduly supports assumptions that there is an average, or dominant, group from which systematically marginalized individuals simply differ is not acceptable. Quantitative measurement must also be complemented with counter-stories (quantitative or qualitative), which challenge the assumptions of difference. Person-centered analyses may provide quantitative methods able to identify narratives that are counter to what may be extracted from traditional variable-oriented engineering education work. Similarly, qualitative data may also identify quantitative measures unaccounted or wrongly accounted or models misspecified (Sigle-Rushton, 2014). Finally, the last tenet of reflexivity requires actively examining the role of the researcher. Consistent with the prior tenets, research by the nature of it being conducted by people within social systems has values within it. A transparent discussion of research decision-making and the role of the researcher in this process, as well as how the data perpetuate or disrupt political and social discourse, must be included.

These tenets not only guide FemQuant and QuantCrit research but also necessitate a critical consideration of how methods are used and interpretations of results are made. Currently, there are numerous areas of quantitative methods that have not been used in this scholarship (Sablan, 2019). Most often, descriptive statistics or demographic statistics are used. These approaches emphasize important differences between groups, but they may do little to establish underlying causes or motivations that can guide policy change or the implementation of interventions. Many of the advanced quantitative approaches for examining complex social phenomena and person-centered analyses discussed in this chapter can provide additional ways of expanding these critical approaches to quantitative methods. Additionally, measurement theory can be used to question and push forward critical perspectives (refer to Douglas et al., 2023). FemQuant and QuantCrit are not new, but they have only recently begun to be taken up in STEM education research, and there are significant opportunities for more work in engineering education research.

Conclusions

The concepts and ideas covered in this chapter are some of the many issues that quantitative researchers should consider as they move from basic quantitative studies to more complex designs. Issues of data quality, causal inference, graphical representations, analysis methods, and data equity in data decision-making are all important for shaping how the engineering education research community learns about the lived experiences of participants in these ecosystems. While basic introductions to quantitative data analysis can help a researcher get started, there are additional complications and considerations. Around data collection, these can include issues of data quality and modeling quality. Around causal inference, this involves recognizing alternatives to RCTs for estimating causal effects, such as regression discontinuity, instrumental variables, and propensity score–based methods like propensity score matching and inverse probability weighting. Around analysis, this translates to considering ways that other methods may provide new ways to ask and answer research questions. We provide a starting list of promising methods, but we also acknowledge that there are many other approaches not described in this chapter. Finally, there is the important issue of the researcher's reflexivity and role throughout the entire process. In total, we hope the concepts and resources here can extend the researchers' toolbox as they embark on their quantitative investigations. We encourage the engineering education research community to use and develop new approaches in quantitative methods. In presenting results from these newer approaches, there may be an additional burden of transparency in the research process and descriptions of steps taken and arguments for why these steps meet the key assumptions and align with the research questions, theory, methods, and interpretation for a study as the field continues to mature.

References

Abadie, A. (2005). Semiparametric difference-in-differences estimators. *The Review of Economic Studies*, *72*(1), 1–19.

Abadie, A., & Imbens, G. W. (2016). Matching on the estimated propensity score. *Econometrica*, *84*(2), 781–807. https://doi.org/10.3982/ECTA11293

Alin, A. (2010). Multicollinearity. *Wiley Interdisciplinary Reviews: Computational Statistics*, *2*(3), 370–374. https://doi.org/10.1002/wics.84

Allison, P. D. (2000). Multiple imputation for missing data: A cautionary tale. *Sociological Methods & Research*, *28*(3), 301–309. https://doi.org/10.1177/0049124100028003003

Allison, P. D. (2012, April). Handling missing data by maximum likelihood. In *SAS global forum* (Vol. 2012, No. 312, pp. 1038–1021). https://statisticalhorizons.com/wp-content/uploads/2022/01/MissingData-ByML.pdf

Angrist, J. D., & Imbens, G. W. (1995). *Identification and estimation of local average treatment effects* (Working Paper No. 118). National Bureau of Economic Research. https://doi.org/10.3386/t0118

Angrist, J. D., Imbens, G. W., & Rubin, D. B. (1996). Identification of causal effects using instrumental variables. *Journal of the American Statistical Association*, *91*(434), 444–455. https://doi.org/10.1080/01621459.1996.10476902

Arellano, M., & Bover, O. (1995). Another look at the instrumental variable estimation of error-components models. *Journal of Econometrics, 68*(1), 29–51. https://doi.org/10.1016/0304-4076(94)01642-D

Azur, M. J., Stuart, E. A., Frangakis, C., & Leaf, P. J. (2011). Multiple imputation by chained equations: What is it and how does it work? *International Journal of Methods in Psychiatric Research, 20*(1), 40–49.

Babbie, E. R. (2020). *The practice of social research*. Cengage learning.

Bell, B., Kromrey, J., & Ferron, J. (2009). *Missing data and complex samples: The impact of listwise deletion vs. Subpopulation analysis on statistical bias and hypothesis test results when data are MCAR and MAR*. Joint Statistical Meeting.

Benedetto, U., Head, S. J., Angelini, G. D., & Blackstone, E. H. (2018). Statistical primer: Propensity score matching and its alternatives†. *European Journal of Cardio-Thoracic Surgery, 53*(6), 1112–1117. https://doi.org/10.1093/ejcts/ezy167

Benson, L., Potvin, G., Kirn, A., Godwin, A., Doyle, J., Rohde, J., Verdin, D., & Boone, H. (2017, June 1). *Board # 9: Characterizing student identities in engineering: Attitudinal profiles of engineering majors*. https://doi.org/10.18260/1-2-27950

Block, J. H., Hoogerheide, L., & Thurik, R. (2013). Education and entrepreneurial choice: An instrumental variables analysis. *International Small Business Journal, 31*(1), 23–33. https://doi.org/10.1177/0266242611400470

Borgatti, S. P., Mehra, A., Brass, D. J., & Labianca, G. (2009). Network analysis in the social sciences. *Science, 323*(5916), 892–895.

Bowleg, L. (2008). When black + lesbian + woman ≠ black lesbian woman: The methodological challenges of qualitative and quantitative intersectionality research. *Sex Roles, 59*(5–6), 312–325. https://doi.org/10.1007/s11199-008-9400-z

Brozina, C. (2015). *Learning analytics: Understanding first-year engineering students through connected student-centered data* (Doctoral Dissertation). Virginia Tech.

Bryman, A. (2008). Of methods and methodology. *Qualitative Research in Organizations and Management: An International Journal, 3*(2), 159–168.

Bullock, H. E., Harlow, L. L., & Mulaik, S. A. (1994). Causation issues in structural equation modeling research. *Structural Equation Modeling: A Multidisciplinary Journal, 1*(3), 253–267. https://doi.org/10.1080/10705519409539977

Bürkner, P.-C., & Charpentier, E. (2020). Modelling monotonic effects of ordinal predictors in Bayesian regression models. *British Journal of Mathematical and Statistical Psychology, 73*(3), 420–451. https://doi.org/10.1111/bmsp.12195

Bürkner, P.-C., Gabry, J., Weber, S., Johnson, A., Modrak, M., Badr, H. S., Weber, F., Ben-Shachar, M. S., & Rabel, H. (2022). *Brms: Bayesian regression models using "stan"* (2.17.0). https://CRAN.R-project.org/package=brms

Buuren, S. van. (2012). *Flexible imputation of missing data*. CRC Press.

Caliendo, M., & Kopeinig, S. (2008). Some practical guidance for the implementation of propensity score matching. *Journal of Economic Surveys, 22*(1), 31–72. https://doi.org/10.1111/j.1467-6419.2007.00527.x

Carolan, B. V. (2013). *Social network analysis and education: Theory, methods & applications*. SAGE Publications.

Cartwright, N. (2007). Are RCTs the gold standard? *BioSocieties, 2*(1), 11–20. https://doi.org/10.1017/S1745855207005029

Chazal, F., & Michel, B. (2021). An introduction to topological data analysis: Fundamental and practical aspects for data scientists. *Frontiers in Artificial Intelligence, 4*. www.frontiersin.org/articles/10.3389/frai.2021.667963

Chen, D.-G., & Fraser, M. W. (2017). A Bayesian perspective on intervention research: Using prior information in the development of social and health programs. *Journal of the Society for Social Work and Research, 8*(3), 441–456. https://doi.org/10.1086/693432

Cole, J. C. (2008). How to deal with missing data: Conceptual overview and details for implementing two modern methods. In J. W. Osborne (Ed.), *Best practices in quantitative methods* (pp. 214–238). Sage Publications.

Cole, S. R., Platt, R. W., Schisterman, E. F., Chu, H., Westreich, D., Richardson, D., & Poole, C. (2010). Illustrating bias due to conditioning on a collider. *International Journal of Epidemiology, 39*(2), 417–420. https://doi.org/10.1093/ije/dyp334

Connolly, P., Keenan, C., & Urbanska, K. (2018). The trials of evidence-based practice in education: A systematic review of randomised controlled trials in education research 1980–2016. *Educational Research, 60*(3), 276–291. https://doi.org/10.1080/00131881.2018.1493353

Covarrubias, A. (2011). Quantitative intersectionality: A critical race analysis of the Chicana/o educational pipeline. *Journal of Latinos and Education, 10*(2), 86–105. https://doi.org/10.1080/15348431.2011.556519

Craig, P., Katikireddi, S. V., Leyland, A., & Popham, F. (2017). Natural experiments: An overview of methods, approaches, and contributions to public health intervention research. *Annual Review of Public Health, 38*, 39–56. https://doi.org/10.1146/annurev-publhealth-031816-044327

Cronbach, L. J., & Meehl, P. E. (1955). Construct validity in psychological tests. *Psychological Bulletin, 52*(4), 281–302. https://doi.org/10.1037/h0040957

Deaton, A., & Cartwright, N. (2018). Understanding and misunderstanding randomized controlled trials. *Social Science & Medicine, 210*, 2–21. https://doi.org/10.1016/j.socscimed.2017.12.005

Dennis, B., Ponciano, J. M., Taper, M. L., & Lele, S. R. (2019). Errors in statistical inference under model misspecification: Evidence, hypothesis testing, and AIC. *Frontiers in Ecology and Evolution*, 7, 372.

Deschacht, N., & Goeman, K. (2015). The effect of blended learning on course persistence and performance of adult learners: A difference-in-differences analysis. *Computers & Education, 87*, 83–89. https://doi.org/10.1016/j.compedu.2015.03.020

DeVellis, R. F., & Thorpe, C. T. (2021). *Scale development: Theory and applications*. Sage Publications.

Donald, S. G., & Lang, K. (2007). Inference with difference-in-differences and other panel data. *The Review of Economics and Statistics, 89*(2), 221–233. https://doi.org/10.1162/rest.89.2.221

Douglas, K. A., Neumann, K., & Oliveri, M. E. (2023). Contemporary approaches to assessment of engineering competencies for diverse learners. In A. Johri (Ed.), *International handbook of engineering education research* (pp. XX–XX). Publisher.

Douglas, K. A., & Purzer, Ş. (2015). Validity: Meaning and relevancy in assessment for engineering education research. *Journal of Engineering Education, 104*(2), 108–118. https://doi.org/10.1002/jee.20070

Efron, B. (2005). Bayesians, frequentists, and scientists. *Journal of the American Statistical Association, 100*(469), 1–5. https://doi.org/10.1198/016214505000000033

Elwert, F., & Winship, C. (2014). Endogenous selection bias: The problem of conditioning on a collider variable. *Annual Review of Sociology, 40*(1), 31–53. https://doi.org/10.1146/annurev-soc-071913-043455

Etz, A., & Vandekerckhove, J. (2018). Introduction to Bayesian inference for psychology. *Psychonomic Bulletin & Review, 25*(1), 5–34. https://doi.org/10.3758/s13423-017-1262-3

Everitt, B., Landau, S., Leese, M., & Stahl, D. (Eds.). (2011). *Cluster analysis* (5th ed.). Wiley.

Eye, A., & Wiedermann, W. (2015). Person-centered analysis. In *Emerging trends in the social and behavioral sciences* (pp. 1–18). John Wiley & Sons, Inc. doi: https://doi.org/10.1002/9781118900772.etrds0251

Falk, R., & Greenbaum, C. W. (1995). Significance tests die hard: The amazing persistence of a probabilistic misconception. *Theory & Psychology, 5*(1), 75–98. https://doi.org/10.1177/0959354395051004

Ferguson, D. M., Navin, D. D., Thompson, M. L., Phillips, A. D., Ohland, M. W., & Jablokow, K. W. (2018). Using Bayesian analysis to refine the measurement of the innovative capacities of engineers. In *2018 IEEE frontiers in education conference (FIE)* (pp. 1–4). https://doi.org/10.1109/FIE.2018.8658698

Fernandez, T., Godwin, A., Doyle, J., Verdin, D., Boone, H., Kirn, A., Benson, L., & Potvin, G. (2016). More comprehensive and inclusive approaches to demographic data collection. In *ASEE annual conference & exposition proceedings*. https://doi.org/10.18260/p.25751.

Frisbie, D. A. (2005). Measurement 101: Some fundamentals revisited. *Educational Measurement: Issues and Practice, 24*(3), 21–28.

Gagnier, J. J., & Morgenstern, H. (2017). Misconceptions, misuses, and misinterpretations of P values and significance testing. *JBJS, 99*(18), 1598–1603. https://doi.org/10.2106/JBJS.16.01314

Galaleldin, M., Bouchard, F., Anis, H., & Lague, C. (2016). The impact of makerspaces on engineering education. *Proceedings of the Canadian Engineering Education Association (CEEA)*. https://doi.org/10.24908/pceea.v0i0.6481

Gareth, J., Daniela, W., Trevor, H., & Robert, T. (2013). *An introduction to statistical learning: With applications in R*. Springer.

Gelman, A. (2011). Induction and deduction in bayesian data analysis. *Markets, and Morals: RMM, 2*, 67–78. https://doi.org/10.22029/jlupub-374

Gelman, A. (2014). *Bayesian data analysis* (3rd ed.). CRC Press.

Gelman, A., Vehtari, A., Simpson, D., Margossian, C. C., Carpenter, B., Yao, Y., Kennedy, L., Gabry, J., Bürkner, P.-C., & Modrák, M. (2020). Bayesian workflow (arXiv:2011.01808). arXiv. http://arxiv.org/abs/2011.01808

Gero, J. S., & Milovanovic, J. (2020). A framework for studying design thinking through measuring designers' minds, bodies and brains. *Design Science*, *6*, e19.

Gillborn, D. (2018). QuantCrit: Rectifying quantitative methods through Critical Race Theory [Special Issue]. *Race Ethnicity and Education*, *21*(2), 149–273. https://doi.org/10.1080/13613324.2017.1377675

Gillborn, D., Warmington, P., & Demack, S. (2018). QuantCrit: education, policy, 'Big Data' and principles for a critical race theory of statistics. *Race Ethnicity and Education*, *21*(2), 158–179. https://doi.org/10.1080/13613324.2017.1377417

Godwin, A. (2020). Sitting in the tensions: Challenging whiteness in quantitative research. *Studies in Engineering Education*, *1*(1), 78–82. https://doi.org/10.21061/see.64

Godwin, A., Benedict, B., Rohde, J., Thielmeyer, A., Perkins, H., Major, J., Clements, H., & Chen, Z. (2021). New epistemological perspectives on quantitative methods: An example using topological data analysis. *Studies in Engineering Education*, *2*(1), 16. https://doi.org/10.21061/see.18

Goggins, S. P., Galyen, K., & Laffey, J. (2010). Network analysis of trace data for the support of group work: Activity patterns in a completely online course. In *Proceedings of the 16th ACM international conference on supporting group work* (pp. 107–116). https://doi.org/10.1145/1880071.1880089

Goodman, S. (2008). A dirty dozen: Twelve P-value misconceptions. *Seminars in Hematology*, *45*(3), 135–140. https://doi.org/10.1053/j.seminhematol.2008.04.003

Greenland, S. (2000). An introduction to instrumental variables for epidemiologists. *International Journal of Epidemiology*, *29*(4), 722–729. https://doi.org/10.1093/ije/29.4.722

Grunspan, D. Z., Wiggins, B. L., & Goodreau, S. M. (2014). Understanding classrooms through social network analysis: A primer for social network analysis in education research. *CBE – Life Sciences Education*, *13*(2), 167–178. https://doi.org/10.1187/cbe.13-08-0162

Gruzd, A., Paulin, D., & Haythornthwaite, C. (2016). Analyzing social media and learning through content and social network analysis: A faceted methodological approach. *Journal of Learning Analytics*, *3*(3), 46–71. https://doi.org/10.18608/jla.2016.33.4

Gulek, J. C., & Demirtas, H. (2005). Learning with technology: The impact of laptop use on student achievement. *The Journal of Technology, Learning and Assessment*, *3*(2), Article 2. https://ejournals.bc.edu/index.php/jtla/article/view/1655

Hagenaars, J. A., & McCutcheon, A. L. (Eds.). (2002). *Applied latent class analysis*. Cambridge University Press.

Haller, H., & Krauss, S. (2002). Misinterpretations of significance: A problem students share with their teachers? *Methods of Psychological Research Online*, 7, 1–20.

Hampton, C., Reeping, D., & Ozkan, D. S. (2021). Positionality statements in engineering education research: A look at the hand that guides the methodological tools. *Studies in Engineering Education*, *1*(2), 126–141. https://doi.org/10.21061/see.13

Harvey, D. L., & Reed, M. (1997). *Social science as the study of complex systems* (pp. 295–323). The University of Michigan Press.

Haverkamp, A., Bothwell, M., Montfort, D., & Driskill, Q.-L. (2021). Calling for a paradigm shift in the study of gender in engineering education. *Studies in Engineering Education*, *1*(2), 55–70. http://doi.org/10.21061/see.34

Helwig, N. E. (2017). Regression with ordered predictors via ordinal smoothing splines. *Frontiers in Applied Mathematics and Statistics*, *3*. www.frontiersin.org/articles/10.3389/fams.2017.00015

Hennig, C., Meila, M., Murtagh, F., & Rocci, R. (Eds.). (2015). *Handbook of cluster analysis*. CRC Press.

Henrich, J., Heine, S. J., & Norenzayan, A. (2010). Most people are not WEIRD. *Nature*, *466*(7302), 29–29. https://doi.org/10.1038/466029a

Hesse-Bibber, S. N., & Piatelli, D. (2012). The feminist practice of holistic reflexivity. In S. Hesse-Biber, *Handbook of feminist research: Theory and praxis* (pp. 557–582). Sage Publications, Inc. https://doi.org/10.4135/9781483384740.n27

Hoekstra, R., Morey, R. D., Rouder, J. N., & Wagenmakers, E.-J. (2014). Robust misinterpretation of confidence intervals. *Psychonomic Bulletin & Review*, *21*(5), 1157–1164. https://doi.org/10.3758/s13423-013-0572-3

Holland, P. W. (1986). Statistics and causal inference. *Journal of the American Statistical Association, 81*(396), 945–960. https://doi.org/10.2307/2289064

Holland, P. W. (2003). *Causation and race causation and race*. Educational Testing Service.

Holly Jr, J. (2020). Disentangling engineering education research's anti-Blackness. *Journal of Engineering Education, 109*(4), 629–635.

Horton, H. D., & Sykes, L. L. (2008). Critical demography and the measurement of racism. In *White logic, White methods: Racism and methodology* (pp. 239–250). Rowman & Littlefield Publishers.

Hu, M., & Shealy, T. (2019). Application of functional near-infrared spectroscopy to measure engineering decision-making and design cognition: Literature review and synthesis of methods. *Journal of Computing in Civil Engineering, 33*(6), 04019034.

Imbens, G. W., & Lemieux, T. (2008). Regression discontinuity designs: A guide to practice. *Journal of Econometrics, 142*(2), 615–635. https://doi.org/10.1016/j.jeconom.2007.05.001

Jacob, B. A., & Lefgren, L. (2004). Remedial education and student achievement: A regression-discontinuity analysis. *The Review of Economics and Statistics, 86*(1), 226–244. https://doi.org/10.1162/003465304323023778

Jones, B. D., Osborne, J., Paretti, M., & Matusovich, H. (2014). Relationships among students' perceptions of a first-year engineering design course and their engineering identification, motivational beliefs, course effort, and academic outcomes. *International Journal of Engineering Education, 30*(6), 1340–1356.

Kane, M. T. (1992). An argument-based approach to validity. *Psychological Bulletin, 112*(3), 527–535. https://doi.org/10.1037/0033-2909.112.3.527

Karnowski, V. (2017). Latent class analysis. In *The international encyclopedia of communication research methods* (pp. 1–10). John Wiley & Sons, Ltd. https://doi.org/10.1002/9781118901731.iecrm0130

Kassambara, A. (2017). *Practical guide to cluster analysis in R: Unsupervised machine learning* (1st ed.). STHDA.

Kim, N. J., Belland, B. R., & Walker, A. E. (2018). Effectiveness of computer-based scaffolding in the context of problem-based learning for stem education: Bayesian meta-analysis. *Educational Psychology Review, 30*(2), 397–429. https://doi.org/10.1007/s10648-017-9419-1

Kline, R. B. (2016). *Principles and practice of structural equation modeling* (4th ed.). The Guilford Press.

Kruschke, J. K. (2013). Bayesian estimation supersedes the t test. *Journal of Experimental Psychology: General, 142*(2), 573–603. https://doi.org/10.1037/a0029146

Ladson-Billings, G. (2007). Pushing past the achievement gap: An essay on the language of deficit. *The Journal of Negro Education, 76*(3), 316–323. www.jstor.org/stable/40034574

Lambert-Lacroix, S., & Zwald, L. (2011). Robust regression through the Huber's criterion and adaptive lasso penalty. *Electronic Journal of Statistics, 5*(none), 1015–1053. https://doi.org/10.1214/11-EJS635

Laursen, B., & Hoff, E. (2006). Person-centered and variable-centered approaches to longitudinal data. *Merrill-Palmer Quarterly, 52*(3), 377–389. https://www.jstor.org/stable/23096200

Lavery, M. R., Acharya, P., Sivo, S. A., & Xu, L. (2019). Number of predictors and multicollinearity: What are their effects on error and bias in regression? *Communications in Statistics-Simulation and Computation, 48*(1), 27–38. https://doi.org/10.1080/03610918.2017.1371750

Leatherdale, S. T. (2019). Natural experiment methodology for research: A review of how different methods can support real-world research. *International Journal of Social Research Methodology, 22*(1), 19–35. https://doi.org/10.1080/13645579.2018.1488449

Lee, S.-Y., & Song, X.-Y. (2004). Evaluation of the Bayesian and maximum likelihood approaches in analyzing structural equation models with small sample sizes. *Multivariate Behavioral Research, 39*(4), 653–686. https://doi.org/10.1207/s15327906mbr3904_4

Levy, R. (2016). Advances in Bayesian modeling in educational research. *Educational Psychologist, 51*(3–4), 368–380. https://doi.org/10.1080/00461520.2016.1207540

Li, P., Stuart, E. A., & Allison, D. B. (2015). Multiple imputation: A flexible tool for handling missing data. *JAMA, 314*(18), 1966. https://doi.org/10.1001/jama.2015.15281

Liddell, T. M., & Kruschke, J. K. (2018). Analyzing ordinal data with metric models: What could possibly go wrong? *Journal of Experimental Social Psychology, 79*, 328–348. https://doi.org/10.1016/j.jesp.2018.08.009

Likert, R. (1932). A technique for the measurement of attitudes. *Archives of Psychology, 22*(140), 55–55.

Listl, S., Jürges, H., & Watt, R. G. (2016). Causal inference from observational data. *Community Dentistry and Oral Epidemiology, 44*(5), 409–415. https://doi.org/10.1111/cdoe.12231

Liu, X., & Koirala, H. (2012). Ordinal regression analysis: Using generalized ordinal logistic regression models to estimate educational data. *Journal of Modern Applied Statistical Methods, 11*(1). https://doi.org/10.22237/jmasm/1335846000

López, N., Erwin, C., Binder, M., & Chavez, M. J. (2018). Making the invisible visible: Advancing quantitative methods in higher education using critical race theory and intersectionality. *Race Ethnicity and Education, 21*(2), 180–207. https://doi.org/10.1080/13613324.2017.1375185

Lum, P. Y., Singh, G., Lehman, A., Ishkanov, T., Vejdemo-Johansson, M., Alagappan, M., Carlsson, J., & Carlsson, G. (2013). Extracting insights from the shape of complex data using topology. *Scientific Reports, 3*(1), 1236. https://doi.org/10.1038/srep01236

Lyu, Z., Peng, K., & Hu, C.-P. (2018). P-value, confidence intervals, and statistical inference: A new dataset of misinterpretation. *Frontiers in Psychology, 9*, 868. https://doi.org/10.3389/fpsyg.2018.00868

Maggio, J., White, W. G., Molstad, S., & Kher, N. (2005). Prefreshman summer programs' impact on student achievement and retention. *Undefined.* www.semanticscholar.org/paper/Prefreshman-Summer-Programs%27-Impact-on-Student-and-Maggio-White/bf526c4b9143dfd7c0ee35c51b392bfec732c597

Martens, E. P., Pestman, W. R., de Boer, A., Belitser, S. V., & Klungel, O. H. (2006). Instrumental variables: Application and limitations. *Epidemiology, 17*(3), 260–267.

Matsudaira, J. D. (2008). Mandatory summer school and student achievement. *Journal of Econometrics, 142*(2), 829–850. https://doi.org/10.1016/j.jeconom.2007.05.015

Matthay, E. C., & Glymour, M. M. (2020). A graphical catalog of threats to validity: Linking social science with epidemiology. *Epidemiology (Cambridge, MA), 31*(3), 376–384. https://doi.org/10.1097/EDE.0000000000001161

Maxwell, J. A. (2004). Causal explanation, qualitative research, and scientific inquiry in education. *Educational Researcher, 33*(2), 3–11. https://doi.org/10.3102/0013189X033002003

Maxwell, J. A. (2012). The importance of qualitative research for causal explanation in education. *Qualitative Inquiry, 18*(8), 655–661. https://doi.org/10.1177/1077800412452856

McCall, B. P., & Bielby, R. M. (2012). Regression discontinuity design: Recent developments and a guide to practice for researchers in higher education. In J. C. Smart & M. B. Paulsen (Eds.), *Higher education: Handbook of theory and research: Volume 27* (pp. 249–290). Springer Netherlands. https://doi.org/10.1007/978-94-007-2950-6_5

McCall, L. (2002). *Complex inequality: Gender, class, and race in the new economy*. Routledge. https://doi.org/10.4324/9780203902455

McElreath, R. (2016). *Statistical rethinking: A Bayesian course with examples in R and Stan*. CRC Press/Taylor & Francis Group.

McKnight, P. E., McKnight, K., Sidani, S., & Figueredo, A. J. (Eds.). (2007). *Missing data: A gentle introduction*. Guilford Press.

McNeish, D. M. (2016). On using Bayesian methods to address small sample problems. *Structural Equation Modeling: A Multidisciplinary Journal, 23*(5), 750–773. https://doi.org/10.1080/10705511.2016.1186549

McNeish, D. M., & Stapleton, L. M. (2016a). The effect of small sample size on two-level model estimates: A review and illustration. *Educational Psychology Review, 28*(2), 295–314. https://doi.org/10.1007/s10648-014-9287-x

McNeish, D. M., & Stapleton, L. M. (2016b). Modeling clustered data with very few clusters. *Multivariate Behavioral Research, 51*(4), 495–518. https://doi.org/10.1080/00273171.2016.1167008

Mejia, C., Wu, M., Zhang, Y., & Kajikawa, Y. (2021). Exploring topics in bibliometric research through citation networks and semantic analysis. *Frontiers in Research Metrics and Analytics, 6*, 742311. https://doi.org/10.3389/frma.2021.742311

Milovanovic, J., Shealy, T., & Katz, A. (2021). Higher perceived design thinking traits and active learning in design courses motivate engineering students to tackle energy sustainability in their careers. *Sustainability, 13*(22), 12570. https://doi.org/10.3390/su132212570

Montgomery, D. C., & Runger, G. C. (2010). *Applied statistics and probability for engineers*. John Wiley & Sons.

Morey, R. D., Hoekstra, R., Rouder, J. N., Lee, M. D., & Wagenmakers, E.-J. (2016). The fallacy of placing confidence in confidence intervals. *Psychonomic Bulletin & Review, 23*(1), 103–123. https://doi.org/10.3758/s13423-015-0947-8

Morin, A. J., Bujacz, A., & Gagné, M. (2018). Person-centered methodologies in the organizational sciences: Introduction to the feature topic. *Organizational Research Methods, 21*(4), 803–813.

Mosteller, F. (2006). A k-sample slippage test for an extreme population. In *Selected papers of Frederick Mosteller* (pp. 101–109). Springer.

Muthén, B., & Muthén, L. K. (2000). Integrating person-centered and variable-centered analyses: Growth mixture modeling with latent trajectory classes. *Alcoholism: Clinical and Experimental Research, 24*(6), 882–891. https://doi.org/10.1111/j.1530-0277.2000.tb02070.x

National Research Council (U.S.), Shavelson, R. J., & Towne, L. (Eds.). (2002). *Scientific research in education.* National Academy Press.

Newman, D. A. (2014). Missing data: Five practical guidelines. *Organizational Research Methods, 17*(4), 372–411. https://doi.org/10.1177/1094428114548590

Novoselich, B. J., & Knight, D. B. (2018). Shared leadership in capstone design teams: Social network analysis. *Journal of Professional Issues in Engineering Education and Practice, 144*(4), 04018006. https://doi.org/10.1061/(ASCE)EI.1943-5541.0000376

Oakley, A. (1998). Gender, methodology and people's ways of knowing: Some problems with feminism and the paradigm debate in social science. *Sociology, 32*(4), 707–731. https://doi.org/10.1177/0038038598032004005

Olitsky, N. H. (2014). How do academic achievement and gender affect the earnings of STEM majors? A propensity score matching approach. *Research in Higher Education, 55*(3), 245–271. https://doi.org/10.1007/s11162-013-9310-y

Orosz, K., Proteasa, V., & Crăciun, D. (2020). The use of instrumental variables in higher education research. In J. Huisman & M. Tight (Eds.), *Theory and method in higher education research* (Vol. 6, pp. 61–80). Emerald Publishing Limited. https://doi.org/10.1108/S2056-375220200000006005

Osborne, J. W., & Jones, B. D. (2011). Identification with academics and motivation to achieve in school: How the structure of the self influences academic outcomes. *Educational Psychology Review, 23*(1), 131–158. https://doi.org/10.1007/s10648-011-9151-1

Pawley, A. L. (2017). Shifting the "default": The case for making diversity the expected condition for engineering education and making whiteness and maleness visible. *Journal of Engineering Education, 106*(4), 531–533. https://doi.org/10.1002/jee.20181

Pearl, J. (1995). Causal diagrams for empirical research. *Biometrika, 82*(4), 669–688. https://doi.org/10.1093/biomet/82.4.669

Powell, M. G., Hull, D. M., & Beaujean, A. A. (2020). Propensity Score Matching for Education Data: Worked Examples. *The Journal of Experimental Education, 88*(1), 145–164. https://doi.org/10.1080/00220973.2018.1541850

Prados, J. W., Peterson, G. D., & Lattuca, L. R. (2005). Quality assurance of engineering education through accreditation: The impact of engineering criteria 2000 and its global influence. *Journal of Engineering Education, 94*(1), 165–184. https://doi.org/10.1002/j.2168-9830.2005.tb00836.x

Ro, H. K., & Loya, K. I. (2015). The effect of gender and race intersectionality on student learning outcomes in engineering. *The Review of Higher Education, 38*(3), 359–396. https://doi.org/10.1353/rhe.2015.0014

Robnett, R. D., & John, J. E. (2020). "It's wrong to exclude girls from something they love." Adolescents' attitudes about sexism in science, technology, engineering, and math. *Child Development, 91*(1), e231–e248.

Rosenbaum, P. R., & Rubin, D. B. (1983). The central role of the propensity score in observational studies for causal effects. *Biometrika, 70*(1), 41–55. https://doi.org/10.1093/biomet/70.1.41

Sablan, J. R. (2019). Can you really measure that? Combining critical race theory and quantitative methods. *American Educational Research Journal, 56*(1), 178–203.

Salas-Velasco, M. (2006). Private returns to an university education: An instrumental variables approach. *Higher Education, 51*(3), 411–438. https://doi.org/10.1007/s10734-004-6413-6

Schroeder, C. M., Scott, T. P., Tolson, H., Huang, T.-Y., & Lee, Y.-H. (2007). A meta-analysis of national research: Effects of teaching strategies on student achievement in science in the United States. *Journal of Research in Science Teaching, 44*(10), 1436–1460. https://doi.org/10.1002/tea.20212

Schumacker, R. E., & Lomax, R. G. (2010). *A beginner's guide to structural equation modeling* (3rd ed.). Routledge.

Scott, J. (2011). Social network analysis: Developments, advances, and prospects. *Social Network Analysis and Mining, 1*, 21–26.

Secules, S., McCall, C., Mejia, J. A., Beebe, C., Masters, A. S., Sánchez-Peña, M. L., & Svyantek, M. (2021). Positionality practices and dimensions of impact on equity research: A collaborative inquiry and call to the community. *Journal of Engineering Education, 110*(1), 19–43.

Shao, J. (1993). Linear model selection by cross-validation. *Journal of the American Statistical Association, 88*(422), 486–494.

Sigle-Rushton, W. (2014). Essentially quantified? Towards a more feminist modeling strategy. In M. Evans, C. Hemmings, M. Henry, H. Johnstone, S. Madhok, A. Plomien, & S. Wearing (Eds.), *The SAGE handbook of feminist theory* (pp. 431–445). Sage Publications Ltd. https://doi.org/10.4135/9781473909502.n29

Smid, S. C., McNeish, D., Miočević, M., & van de Schoot, R. (2020). Bayesian versus frequentist estimation for structural equation models in small sample contexts: A systematic review. *Structural Equation Modeling: A Multidisciplinary Journal, 27*(1), 131–161. https://doi.org/10.1080/10705511.2019.1577140

Spirtes, P. (2010). Introduction to causal inference. *Journal of Machine Learning Research, 11*(54), 1643–1662.

Sprague, J. (2005). How feminists count: Critical strategies for quantitative methods. In J. Sprague (Ed.), *Feminist methodology for critical researchers: Bridging differences* (1st ed., pp. 81–117). Rowman & Littlefield.

Sprague, J., & Zimmerman, M. K. (1989). Quality and quantity: Reconstructing feminist methodology. *The American Sociologist, 20*(1), 71–86.

Stegmueller, D. (2013). How many countries for multilevel modeling? A comparison of frequentist and Bayesian approaches. *American Journal of Political Science, 57*(3), 748–761. https://doi.org/10.1111/ajps.12001

Steiner, P. M., & Kim, Y. (2016). The mechanics of omitted variable bias: Bias amplification and cancellation of offsetting biases. *Journal of Causal Inference, 4*(2), 20160009. https://doi.org/10.1515/jci-2016-0009

Stevens, S. S. (1946). On the theory of scales of measurement. *Science, 103*(2684), 677–680. https://doi.org/10.1126/science.103.2684.677

Thistlewaite, D. L., & Campbell, D. T. (2017). Regression-discontinuity analysis: An alternative to the ex-post facto experiment. *Observational Studies, 3*(2), 119–128. https://doi.org/10.1353/obs.2017.0000

Thomas, G. (2016). After the gold rush: Questioning the "gold standard" and reappraising the status of experiment and randomized controlled trials in education. *Harvard Educational Review, 86*(3), 390–411. https://doi.org/10.17763/1943-5045-86.3.390

Ushakov, K. M., & Kukso, K. N. (2015). Advantages of social network analysis in educational research. *Russian Education & Society, 57*(10), 871–888. https://doi.org/10.1080/10609393.2015.1148967

Valentine, J. C., Konstantopoulos, S., & Goldrick-Rab, S. (2017). What happens to students placed into developmental education? A meta-analysis of regression discontinuity studies. *Review of Educational Research, 87*(4), 806–833. https://doi.org/10.3102/0034654317709237

van Doorn, J., van den Bergh, D., Böhm, U., Dablander, F., Derks, K., Draws, T., Etz, A., Evans, N. J., Gronau, Q. F., Haaf, J. M., Hinne, M., Kucharský, Š., Ly, A., Marsman, M., Matzke, D., Gupta, A. R. K. N., Sarafoglou, A., Stefan, A., Voelkel, J. G., & Wagenmakers, E.-J. (2021). The JASP guidelines for conducting and reporting a Bayesian analysis. *Psychonomic Bulletin & Review, 28*(3), 813–826. https://doi.org/10.3758/s13423-020-01798-5

Vandekerckhove, J., Rouder, J. N., & Kruschke, J. K. (2018). Editorial: Bayesian methods for advancing psychological science. *Psychonomic Bulletin & Review, 25*(1), 1–4. https://doi.org/10.3758/s13423-018-1443-8

Vaziri, S., Vaziri, B., Novoa, L. J., & Torabi, E. (2022). Academic motivation in introductory business analytics courses: A Bayesian approach. *INFORMS Transactions on Education, 22*(2), 121–129. https://doi.org/10.1287/ited.2021.0247

Viano, S., & Baker, D. (2020). How administrative data collection and analysis can better reflect racial and ethnic identities. *Review of Educational Research, 44*(1), 301–331. https://doi.org/10.3102%2F0091732X20903321

Villanueva, I., Di Stefano, M., Gelles, L., Osoria, P. V., & Benson, S. (2019). A race re-imaged, intersectional approach to academic mentoring: Exploring the perspectives and responses of womxn in science and engineering research. *Contemporary Educational Psychology, 59*, 101786.

Walters, A. A. (1970). Statistical inference. In A. A. Walters (Ed.), *An introduction to econometrics* (pp. 32–78). Palgrave Macmillan UK. https://doi.org/10.1007/978-1-349-15277-3_3

Wang, Y. C., Burke, M., & Kraut, R. E. (2013, April). Gender, topic, and audience response: An analysis of user-generated content on Facebook. In *Proceedings of the SIGCHI conference on human factors in computing systems* (pp. 31–34).

Wasserman, S., & Faust, K. (1994). *Social network analysis: Methods and applications*. Cambridge University Press.

Wasserstein, R. L., & Lazar, N. A. (2016). The ASA statement on p-values: Context, process, and purpose. *The American Statistician*, *70*(2), 129–133. https://doi.org/10.1080/00031305.2016.1154108

Weller, B. E., Bowen, N. K., & Faubert, S. J. (2020). Latent class analysis: A guide to best practice. *Journal of Black Psychology*, *46*(4), 287–311. https://doi.org/10.1177/0095798420930932

Wing, C., Simon, K., & Bello-Gomez, R. A. (2018). Designing difference in difference studies: Best practices for public health policy research. *Annual Review of Public Health*, *39*(1), 453–469. https://doi.org/10.1146/annurev-publhealth-040617-013507

Winter, S. D., & Depaoli, S. (2020). An illustration of Bayesian approximate measurement invariance with longitudinal data and a small sample size. *International Journal of Behavioral Development*, *44*(4), 371–382. https://doi.org/10.1177/0165025419880610

Wludyka, P. (2012). Study designs and their outcomes. In J. P. McDonough & K. Macha (Eds.), *Epidemiology for advanced nursing practice* (pp. 81–114). Jones & Bartlett Learning.

Zeger, S. L., Irizarry, R., & Peng, R. D. (2006). On time series analysis of public health and biomedical data. *Annual Review of Public Health*, *27*(1), 57–79. https://doi.org/10.1146/annurev.publhealth.26.021304.144517

Zuberi, T. (2001). *Thicker than blood: How racial statistics lie*. University of Minnesota Press.

Zuberi, T., & Bonilla-Silva, E. (2008). Causation and race. In *White logic, White methods: Racism and methodology*. Rowman & Littlefield Publishers.

31

Contemporary Approaches to Assessment of Engineering Competencies for Diverse Learners

Kerrie A. Douglas, Knut Neumann, and Maria Elena Oliveri

1 Introduction

This chapter describes a sociocultural approach to develop assessments of multidimensional engineering competencies. We built the approach informed by four seminal works: (1) Pellegrino and colleagues' (2014) introduction of evidence-centered design (ECD) to the engineering education community (ECD) was originally proposed by (2) Mislevy et al. (2003), (3) Harris and colleagues' (2019) framework for development of knowledge-in-use assessment tasks which incorporate disciplinary ideas, cross-cutting concepts, and science practice, and (4) Oliveri and colleagues' (2019) sociocultural-informed ECD (SCI-ECD) to support the development of culturally and linguistically responsive assessments. Each of these assessment frameworks is complimentary and contains crucial aspects for increasing the validity, reliability, and fairness of assessments and their associated interpretations. By integrating the state-of-the-art of assessment development techniques, we propose an approach that tackles two challenges for high-quality engineering assessments: enabling diverse groups of learners to express what they know and can do in multiple ways, and becoming responsive to global demands for engineers with deeper levels of understanding. Thus, the approach proposed in this chapter is particularly for those engineering education researchers interested in assessing diverse groups of engineering students' multidimensional learning (i.e., capable of integrating knowledge and skills needed from more than one domain), as required by today's global economy.

Today's globalized work requires not only engineers prepared with technical training but also those who can utilize their technological and scientific knowledge and respective skills to engage in engineering practices (e.g., design decision-making, modeling, problem-solving, etc.) while working with diverse multicultural teams. Engineering educators worldwide recognize the need to prepare learners to integrate multiple perspectives associated with differing stakeholders' considerations, including environmental, legal, and physical aspects of work, and conduct projects in multicultural contexts (e.g., Auer & Rüütmann, 2021; Kim & Care, 2020; National Research Council, 2012; UNESCO, 2016). Consequently, educational policies have shifted away from expectations of rote memorization of knowledge and basic application of skills to expecting students to use their knowledge in conjunction with multiple skills to complete complex tasks (Kim & Care, 2020; Pellegrino, 2012).

Educators and researchers increasingly recognize that students who cannot complete complex tasks are unprepared for the upcoming challenges they will face in their work. For instance, the 23rd

DOI: 10.4324/9781003287483-37

International Conference on Interactive Collaborative Learning in Estonia in 2020 focused on *Educating Engineers for Future Industrial Revolutions* (Auer & Rüütmann, 2021). Conference proceedings elucidated the multidimensional nature of the competencies required to solve ill-defined problems, collaborate in multinational teams, and keep up with increasing automation and rapidly changing technological advancements. Having knowledge even of a broad range of topics is insufficient to meet the real challenges engineers must solve – they must possess core disciplinary understanding, alongside professional skills (e.g., teaming, communication, etc.) and engineering practices (e.g., problem scoping, modeling, etc.). Educators increasingly seek to support the development of a combination of professional skills such as creativity and collaboration (Geisinger, 2016) and engineering practices such as design or problem-solving, in conjunction with deep disciplinary knowledge (ABET Criterion 3 Outcomes 1–7, www.abet.org). Additionally, preparing for the engineering workforce means students are prepared to work in unfamiliar contexts – beyond the problems they might find in a textbook (Jorion et al., 2015).

To realize such dramatic shifts in engineering education requires assessment approaches aligned with the competencies educators aim to impart and with recognition that students can demonstrate competency in different ways. In light of global shifts in engineering education, we define engineering competencies as multidimensional constructs necessitating the integration of knowledge and skills from different domains. Rather than emphasizing a broad range of knowledge or skills from the engineering sciences, engineering competencies require integration of knowledge about the core concepts with a broad range of professional skills from different domains to make sense of a given problem and then engage in further learning to develop effective solutions (Neumann & Nordine, 2022). Furthermore, as engineers are expected to work in diverse, multinational, and multidisciplinary teams and to value the different perspectives team members create, engineering education must enable students to demonstrate their multidimensional engineering competencies in diverse ways.

2 Foundational Concepts: Validity, Reliability, and Fairness

The *Standards for Educational and Psychological Testing* (AERA et al., 2014) state clearly that validity, reliability, and fairness are equally important and foundational to assessment. More recently, the educational assessment community has been engaged in conversations regarding equitable assessment (e.g., Randall, 2021). Attending to each of these processes requires continual iterative refinement, which implies that developing an effective assessment that leads to valid score inferences for diverse populations is a multiyear process.

2.1 Validity

Validity is the overarching evaluation regarding the extent to which an assessment score(s) is measuring the construct the developers had intended to measure (Douglas & Purzer, 2015). The evaluation of validity is built on multiple sources of evidence which speak to the plausibility that the resulting score represents test-takers' targeted knowledge, skills, and abilities (Kane, 1992, 2015). Through a chain of reasoning, assessment developers can follow scientific argumentation that if the assessment instrument truly measures (specific construct) in (the desired population), then it will measure evidence of (validity source). The determination of evidence is designed to create a strong argument for validity long before the first actual assessment item is written.

Validity begins with how the developers define the construct to be assessed and continues through to the consequences of the assessment being used (Messick, 1989). The *Standards for Educational and Psychological Testing* (AERA et al., 2014) clearly outline that the sources of validity sought, and evidence required, are dependent on the degree to which personal consequences are made from the test scores. Validity in educational assessment is not a checklist of psychometric procedures (Songer & Ruiz-Primo, 2012) or simply studied with psychometrics post-data collection using the assessment instrument. This more modern view of validity is holistic and requires intentionality to enable the identification of the most salient construct elements required for assessment.

2.2 Reliability

Building from the *Standards* (AERA et al., 2014), we understand that for an assessment to be considered valid for a particular purpose, the assessment must also demonstrate evidence of reliability. The term *reliability* is understood conceptually as consistency and precision. Construct irrelevance introduces variance in scores which decrease the assessment's consistency and precision of measurement.

2.3 Fairness

Fairness has many social connotations and field-specific meanings that require explanation, yet there is not one definition of fairness in educational assessment. Fairness is fundamentally a validity issue because construct irrelevant variance (CIV) can occur in numerous ways that prohibit students from truly being assessed based on the construct of interest. CIV, also known as construct irrelevance, "refers to the degree to which test scores are affected by processes that are extraneous to the test's intended purpose" (AERA et al., 2014, p. 12). In other words, CIV is the variance in test scores not related to the construct purportedly being assessed. CIV and issues of fairness are not clearly delineated; however, the lens of fairness highlights how minoritize subgroups of students are more likely to be harmed from the CIV when test scores are used. Furthermore, *fairness* as a term invokes considerations of equity, inclusiveness, and just assessment (Douglas et al., 2022). The *Standards* identifies four dimensions of fairness: (1) equal treatment during the testing process, (2) lack of measurement bias, (3) test-takers' access to construct measured, and (4) validity of individual test score interpretations for intended uses. The *Standards* list four main threats to fairness: (1) test content, (2) test context, (3) confusion about the type of response expected for the question, and (4) opportunity to learn.

While much of the evidence for or against fairness in educational assessment has been obtained through psychometric modules examining differential item functioning (DIF; long after the assessment has been administered), our goal is to develop assessments that are purposefully inclusive. Researchers have found that cultural and linguistic diversity, neurological and physical differences, poor representation in psychometric models, stereotype threat, linguistic diversity, and socialization differences all contribute to whether an assessment score is fair for the population that is assessed (Arbuthnot, 2009, 2017; Cobb & Russell, 2015; Freedle, 2010; Helms, 2006; Nguyen & Ryan, 2008). Thus, discussion of approaches to find bias in assessment tasks is beyond the scope of our chapter. However, we point readers to Zumbo et al.'s (2015) discussion in the next-generation work on fairness and the third generation of DIF analyses and ecological modeling (Bronfenbrenner & Morris, 2006) that not only is flagging DIF important when developing assessments but also is identifying its potential sources and understanding why (and how) items function differentially across groups. It is also important to oversample minority groups in piloting phases to ensure the psychometric models used to evaluate item functioning represent all learner responses, not only the dominant culture/racial/linguistic group of test-takers.

Understanding how people learn and represent their learning in diverse ways is foundational to developing fair assessments. Thus, we approach fairness from a multidisciplinary perspective. Void of such understanding, the assessment developer runs the risk of only assessing students in the way the developer would demonstrate understanding, not in ways that the students would demonstrate understanding.

2.3.1 A Multidisciplinary Perspective on Fairness

Social theories of learning describe human development and learning as mediated by cultural and social contexts (Vygotsky, 1978). Learning is supported through the use of participatory tools, such as the use of language and physical objects situated within the learners' identity and culture. Learning is constructed through various interactions that cannot be separated from the historical, social, and cultural contexts in which learners participate (Hooks, 1992; Mignolo, 2011). Thus, our individual unique realities take shape through activity patterns and reflection with the world around us. Assessment models that decontextualize learners from their lived experiences will inevitably fail at producing fair assessment results. Instead, the principle of situatedness (i.e., acknowledging the situated nature of learning) guides assessment designers to design tasks that are situated within students' experiences, thus promoting fairer assessment of diverse learner subgroups.

Cognitive scientists define situatedness as the foundational premise that the mind is embodied in the body, that cognitive activity is embedded in the natural and social environment, and that individual cognitive boundaries extend beyond the individual (Robbins & Aydede, 2008). The mind is therefore embodied, embedded, and extended. Assembling empirical research associated with this premise in updating its earlier report on learning processes, *How People Learn: Brain, Mind, Experience, and School*, the National Academies of Sciences, Engineering, and Medicine (2018) acknowledged the complex influence of culture. Authors of the report demonstrate that culture – the socially transmitted, learned behavior of a group carried across generations – is central to understanding variation in instructional practices and individual differences among students to be able to meaningfully assess diverse populations' understanding of key concepts taught in engineering education.

3 Challenges with Traditional Assessments

Traditional assessments in engineering education are subject to two major challenges resulting from the most long-standing threats to assessment validity: construct underrepresentation and CIV (Messick, 1995). These two threats are associated with adequate representation of what is intended to be measured (construct representation) and variance in answers based on factors other than knowledge of the construct (CIV). CIV is an often-overlooked aspect in engineering assessments, especially in terms of gender, race, and ethnicity (Douglas et al., 2016), which has implications for fairness of assessments' use. While other concerns about traditional assessments exist, these two are particularly thorny because of the difficulty in precisely scoping what will be assessed and how to ensure all engineering learners are fairly assessed.

3.1 Construct Representation Issues

Traditional assessment models (e.g., assessments that rely primarily on multiple choice, calculation, etc.) rely on narrow construct definitions centered on either the recall of content knowledge or recipe-style (low-level) skill application (Harris et al., 2019). The traditional approach to designing assessments begins with clear learning objectives which are then aligned with Bloom's taxonomy or the revised Bloom's (Anderson & Krathwohl, 2001; Bloom et al., 1956) to decide what type of item is to be written. The approach leads to a very narrow construct definition of what is to be assessed, and each aspect of competence (i.e., knowledge, skills) is typically assessed independently.

While the ability to align closely with narrowly and well-written learning objectives can be a strength of traditional assessments, they tend to fall short of the multidimensional skills that engineers use on the job (AAAS, 2017). Even assessments well aligned to curriculum learning objectives can place undue balance on recall or low-level application rather than in-depth understanding or how the learner would be expected to use this knowledge in future work. For example, concept inventories (CIs) are designed to assess whether students have a conceptual understanding of a foundational scientific/engineering concept (Klymkowsky & Garvin-Doxas, 2020) but not necessarily how the student would use that conceptual understanding in their engineering practice.

3.2 Construct Irrelevance in Assessing Multicultural Populations

Assessments that are not purposefully designed for multicultural groups of learners do not allow for valid inferences regarding students' knowledge because of CIV (Oliveri et al., 2019). CIV can occur due to several reasons, with consequences of disadvantaging diverse student groups. One problem with overemphasizing traditional item types is that the item formats do not account for how various cultures represent knowledge in different ways (various ways of knowing and using funds of knowledge), which leads to items being inherently differentially easier or harder for certain student groups (Freedle & Kostin, 1990). Moreover, traditional assessment models do not explicitly consider the diversity of learners, resulting in score discrepancies between racial, ethnic, and socioeconomic groups that, in part, could be explained by CIV. CIV's presence may differentially affect test-takers, such as disadvantaging test-takers that are from cultures that are more distant from the test developers' culture. For example, if a UK-based fundamentals of electronics course assessment uses language difficulty beyond the English proficiency required at admissions, learners' English proficiency becomes CIV, explaining a portion of non-native-English-speaker assessment scores (e.g., Douglas et al., 2018). Another example related to opportunity to learn is presenting a problem context to engineering students that was not covered in the course activities and only students from some cultures would have familiarity with.

Closed-response formats, not only including multiple-choice items, but also tasks that are designed to be solved in one specific way, while popular in the assessment of specific engineering science and skills, are less suitable to assess true competence, where students are pulling from knowledge and practices in multiple areas to solve a problem. The questions are so specific that students are barred from drawing on their depth of knowledge and experiences in multiple engineering settings to solving the item. Such assessment questions effectively disadvantage students that may be missing the specific pattern of resources the item is designed to assess. For example, students may lack specific linguistic (i.e., particular language skills), cultural (i.e., experience with particular item formats), or substantive resources (i.e., a gap in knowledge). This issue is particularly problematic if the user of the assessment knows little about the instruction students have received or students' cultural background (Mislevy, 2016).

Language used on the test items and tasks can create unnecessary challenges for culturally and linguistically diverse test-takers and then becomes misconstrued as a lack of knowledge of the content assessed (Oliveri, 2019). Without consideration of the linguistic diversity of test-takers, an assessment's passages and questions are less accessible to some test-taker populations, potentially leading to score misinterpretation, disengagement with the task, and increased cognitive load. Thus, linguistic complexity can become a source of CIV that goes unnoticed. Addressing these issues is important, and its investigation falls under larger foundational concepts, as explained in the next section.

4 Using Socioculturally Informed ECD to Assess Multidimensional Engineering Competencies

ECD (Mislevy et al., 2017) is a principled approach to guide the development of engineering competency assessments. Mislevy and colleagues designed ECD to describe the central processes embedded in standardized testing to achieve the necessary validity evidence – from conceptualization to delivery architecture in computerized testing to probability-based reasoning in accumulating

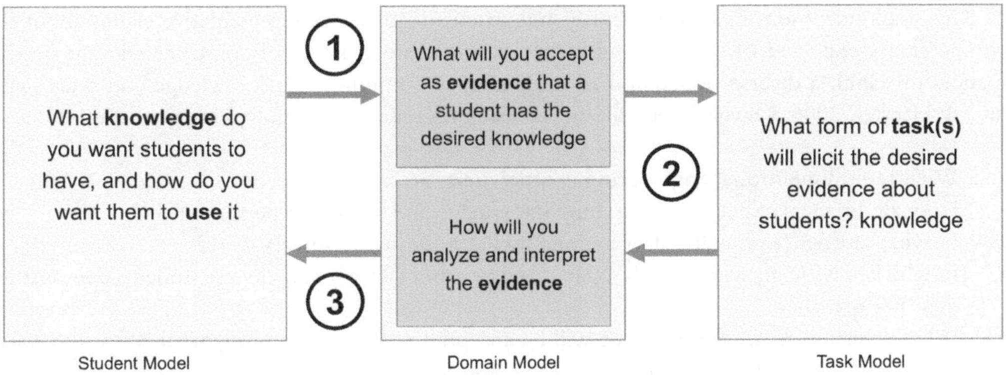

Figure 31.1 Evidence-centered design.
Source: Adapted from Pellegrino and colleagues (2014).

evidence of student learning across diverse tasks. ECD prompts critical reflection and articulation of whom we assess, for what purposes, and under what constraints (Oliveri et al., 2019).

Pellegrino and colleagues (2014) presented a version of ECD for the engineering education research community which entails a *student model, domain model*, and *task model* (Figure 31.1). They described the process of assessment development as three steps involving the delineation of (1) the desired score-based assessment claims, including what students are expected to know and turning them in evidentiary statements; (2) identifying the ways in which to obtain such evidence and how to evaluate it; and (3) describing how to interpret the evidence in light of the claims made. We build from that work to explicitly address sociocultural considerations.

4.1 *Identifying and Delineating Multidimensional Engineering Competencies*

The first step involves identifying the multidimensional engineering competencies and breaking them down into smaller claims (i.e., subclaims) through a process referred to as domain analysis. In domain analysis, claims about what will be assessed are made as well as the bounds to which it will assessed. Competencies are typically identified from standards or other high-level documents (e.g., ABET, O*NET) as well as observations and interviews of the focal jobs needing those competencies (see www.uniformguidelines.com/). Domain analysis also requires considering how the competencies have been predefined and what they mean in the current assessment context, which is particularly important for construct representation. The outcomes of the domain analysis will reveal the combination of knowledge (e.g., thermodynamics, statics, etc.) as well as engineering (e.g., designing, mathematical modeling, etc.) and professional skills (e.g., communication, leadership, teaming, etc.), representing the competencies to be assessed. That is, domain analysis will not only delineate the knowledge and skills entailed in competencies but also describe how diverse students are expected to combine the knowledge and skills to demonstrate their competency. To ensure the assessment speaks to a diverse population of learners, it is important to not only describe one (idealized) way in which students are expected to demonstrate their competency but also allow for multiple ways in which students may demonstrate the competency and various ways to collect evidence of and interpret differential levels of competency attainment.

This step's purpose is to clearly identify and articulate construct-representative claims about the competencies expected of students and clearly define a broad range of construct-relevant performances in which a diverse population of students may combine their knowledge and skills (Mislevy & Haertel, 2006; Oliveri et al., 2019). This step is guided by the following questions:

- What multidimensional engineering competencies are to be assessed?
- How are learners expected to combine knowledge and skills into performances?
- In what settings (e.g., cultural or national) will the future engineers work?
- In which ways (e.g., with regards to language or other factors) may diverse students demonstrate their performance?
- What diverse ways of performance will be accepted as evidence to support the claim that students have mastered the competencies to be assessed?

Answering the posed questions then would lead to (1) explicit claims about students' desired competence, including the different ways students may combine knowledge and skills, and (2) statements about how diverse subgroups of students would provide evidence of mastering the competencies. These claims (and evidence) must accurately capture diverse groups' ways of thinking and doing.

If students were expected to work as engineers in locations across a large geographical region, the assessment would need to reflect the different local conditions. For instance, engineers might need specific knowledge of the soil characteristics in the area where they are working, whether they are designing a building that will be placed in rural/urban settings, and the type of gases emanating from soils within the area, which differ considerably based on geographic region. Such considerations contribute to construct relevance. Thus, an assessment should include questions related to the presence of radon levels, a lagoon versus a septic tank, and the soil permeability. The assessment may also prompt students to make design decisions based on the soil characteristics and local building codes.

To cater to increasing diversity in student populations, assessments must become more linguistically accessible; this will help minimize CIV. In this step of the development process, it can help to include experts on linguistically diverse learners to ensure that the claims and evidence statements sufficiently capture the different ways in which different language skills will play out in the competencies and different ways in which students can perform to provide evidence about their competencies. If language skills are not relevant to a respective competency, this has substantial influence on what can or should be considered evidence and how that evidence is gathered (see next section).

4.2 Developing and Scoring Tasks Assessing Multidimensional Engineering Competencies

This step has two purposes: (1) develop assessment tasks and associated item formats (e.g., open-ended problems, multiple-choice questions, essays, etc.) that can capture multiple ways in which diverse student groups can perform, as defined by the evidence statements, and (2) define procedures of evaluating students' performance to draw conclusions about students' competency levels and how that information will be fed back to students. In this step, the following questions are asked:

- What type of tasks will be required to elicit evidence about the different ways in which students are expected to perform?
- How will the different ways in which students can perform on these tasks be evaluated to draw conclusions about claims on students' competencies?
- What type of feedback will be provided to the learners to enhance opportunity to learn for populations with different skill levels?

The focus in developing assessment tasks is to connect the knowledge and skills with the potential ways of collecting evidence of the competencies (Mislevy & Haertel, 2006; Oliveri et al., 2019). Assessment tasks must be chosen based on type, scope, and complexity of the assessed competencies. Since traditional task formats such as multiple choice are more amenable to assessing the recall of facts or the ability to perform low-level procedures, additional task types need to be considered for a valid, reliable, and fair assessment of multidimensional engineering competencies.

The choice of assessment tasks matters because of the implications on what type of information is construct-relevant and what is construct-irrelevant, as discussed in the section on CIV. Particularly poignant is the long-standing research demonstrating that multiple-choice questions favor White students over other racial and ethnic groups (Klein et al., 1997; Oliveri et al., 2018). However, researchers have found that contextualized tasks and use of performance-based assessments can decrease disparities in assessment scores between racial and ethnic groups (Guignard et al., 2016; Sternberg, 2010). Thus, consideration of task types and formats is important when assessments are administered to multicultural engineering populations.

4.2.1 Open-Ended Problems

Open-ended problems resemble tasks that engineers typically encounter in their everyday work (Diefes-Dux et al., 2010). These tasks cannot be reduced to testable facts but instead should resemble real-world problems that require careful thought to solve and can be approached in diverse ways. Gutiérrez Ortiz et al. (2021) describe two ways such tasks differ from traditional tasks: (1) they lack relevant information, so students need to identify what information is needed and how to find it, or to determine what assumptions can be made in place of exact information; (2) information surplus, effectively creating an excess of data so that students have to screen the information, identify the information needed, and make a decision about what information to use. This way, tasks can be clearly defined (enhancing construct representation) and yet allow for different ways of solving them (reducing construct irrelevance for diverse groups of learners).

One alternative to having students solve an open-ended problem is to have them describe how they would solve it (Bailey & Szabo, 2007). Kalyuga and Sweller (2004), for example, asked students to articulate the first step they would take to solve the problem. While novices typically employ a trial-and-error approach or low-level fine-grained step-by-step procedures, experts exhibit higher-level overarching solution schemes (Kalyuga, 2006). The respective research found these approaches reliably and validly differentiated between different levels of competence in students.

Evaluating student responses to open-ended problems is not without challenges (Diefes-Dux et al., 2010). A central challenge is to identify (all) the diverse ways in which students might solve the task and to design scoring rubrics that can be utilized to evaluate student responses across these various ways (Davis et al., 2002). One approach to solve this issue is to score the complexity of the ways in which students have approached the solution of the task in relation to the possible ways of solving the task. Fortus et al. (2019), for example, used a scoring rubric to assess students' three-dimensional learning about energy (based on the framework for K–12 science education). The rubric delineated different levels of integration of scientific and engineering practices, disciplinary core ideas, and cross-cutting concepts. The lowest level marked no integration (i.e., application of the practice with no connection to energy ideas or presentation of energy ideas with no engagement in the practice), and the highest level marked full integration (i.e., engagement in the practice using energy ideas).

4.2.2 Portfolios, Written Memos, and Technical Reports

Whereas open-ended problems are purposefully ambiguous to allow for multiple solutions, portfolios, written memos, and technical reports all begin by clearly formulating the task, describing what students are supposed to do and communicating expectations about students' performance. The problem types require greater effort to score, limiting reliability between graders, yet have some advantage over open-ended problems. The advantages are that students can be provided with unambiguous information about the ways in which they are expected to solve the task through the rubrics, which are then used to score their solutions. In addition, students can be provided with templates or detailed instructions as to what information to include in what section.

Portfolios are collections of students' (course) work. Hence, portfolios present evidence about students' competencies over a longer period of time than other assessment methods. Newstetter and Khan (1997), for example, draw on portfolios to assess students' competency in problem-solving through design, asking students to include all design iterations. The resulting portfolios, in addition to the final product, describe students' solution processes. The evaluation rubric used six criteria: (1) portfolio design as an artifact of communication and persuasion as well as evidence of (2) problem structuring, (3) problem decomposition, (4) incremental development of the design solution, (5) constraint setting, and (6) different abstraction levels and various types of representation of the problem space. These criteria were the diverse ways in which students as designers were expected to engage in the task.

Written memos are a similar technique often found at engineers' workplaces (Amare & Brammer, 2005). Wertz et al. (2013), for example, tasked students with examining the buildings in which they lived to identify areas where energy efficiency could be improved and write a memo to recommend ways to increase the building's sustainability. The memo's quality was assessed in a multistage process, including the classification of citations, assignment of audience and purpose to each source, assessment of documentation of each information source, use of sources to create argument, and overall memo quality – again acknowledging the diverse ways in which students can respond to these tasks.

Technical reports are essentially a combination of portfolios and memos. They describe the process, progress, or results of technical work or state of a technical problem and also may include recommendations or conclusions. Hence, depending on the exact task, technical reports can also be evaluated using scoring rubrics.

4.2.3 Computer-Based Tasks

Computer-based tasks (CBTs) include a variety of assessment tasks that are delivered through a computer-mediated environment. Recent advancements in technology used for CBTs have the potential to address the challenges associated with allowing students to use or come up with a broad range of different solutions. Early implementation of CBTs were mostly on-screen versions of traditional paper-and-pencil tests (Bennett, 2008). Recently, researchers have used CBTs to assess more complex constructs using task types, such as simulations (Pellegrino & Quellmalz, 2010), or scenario-based tasks applied to the assessment of workplace competencies (Oliveri et al., 2021). Research shows these CBTs are able to interactively assess the integration of knowledge structures and respective inquiry skills (Quellmalz et al., 2012; Oliveri et al., 2021).

The advantage of simulations is that although they represent a complex system, they also limit the possible actions, effectively limiting the patterns of resources students may demonstrate. Simulations typically offer students the possibility to interact with a (simulated) environment. These simulated environments can come at different levels of abstraction, from simply presenting students with opportunities to define selected parameters of a particular system (e.g., a building's heating, ventilation, and air-conditioning system; Rampazzo & Beghi, 2018), to a realistic depiction of a real-world environment (e.g., a virtual lab in which students interact in ways indistinguishable from a real lab) or virtual workplace (in which students can learn to effectively communicate in diverse workplace environments with virtual peers in design environments; Slomp et al., 2021). In fact, in engineering education, simulations are an actual tool that students work with (e.g., simulating the energy efficiency of buildings). Hence, CBTs using simulations not only allow for a more valid assessment of learning outcomes (compared to paper-and-pencil tasks) but also are sometimes the only way to assess select learning outcomes.

As with any performance assessment, the evaluation might be product- or process-based. The product-based evaluation focuses on the final solution submitted by the student; the process-based evaluation focuses on the student's actions during the solution process. The process-based evaluation allows for more valid conclusions (e.g., Baxter et al., 1996) but is more resource-intensive in terms of data collection and analysis. Recent research suggests that students' performance can be automatically scored not only by using knowledge engineering (i.e., a priori defined rules encapsulating specific behaviors, for example, Buckley et al., 2006, 2010) but also by using machine learning (i.e., having a machine learn features of specific behaviors from human labeling). Sao Pedro et al. (2013), for example, developed machine learning–based detectors of inquiry strategies (e.g., control-of-variables strategy) by having humans label replayed episodes from log-files of students working on a simulation and using a decision tree algorithm to identify sets of features (e.g., total number of trials, incomplete number of trials, and total number of actions) predicting the occurrence of the respective strategies.

4.2.4 Engineering Projects

Projects require students to work on a particular task over a longer period of time than other assignments. These are often complex, real-world tasks resembling those students will meet in their later work lives. Hence, students need to draw on a broad range of knowledge and skills. Projects requiring knowledge and skills developed across multiple courses or the whole engineering program are referred to as capstone projects or, since such projects are often part of the program's final year, final-year projects. To that end, capstone projects are also suitable as a checkpoint for assessing program learning outcomes (Abu Salem et al., 2020).

In engineering, projects often involve a design component as part of the solution (Lutz & Paretti, 2017). The exact nature of the tasks in a project is subject to the respective education program. One feature, however, is common to all projects: students must deliver a product. This product can, for example, be a device or specific piece of technology, a design or specification, or a particular procedure or method. Beyond the actual product, the assignment often requires a report documenting the process and progress of developing the product or detailing the final product and how it meets the requirements laid out in the task. This allows for assessing not only the final product but also how students approached the development of the product. Among other advantages, this approach provides insights into the complex patterns of resources students marshaled in design processes.

The evaluation of students' projects is subject to a range of challenges (Oehlers, 2006). For one, projects often, in addition to a broad range of existing knowledge and skills, expect students to acquire new knowledge and skills. Any evaluation strategy must, hence, adhere to this variety of knowledge and skills that students may have activated in creating the project, including those that students have acquired throughout the project. One strategy in assessing projects is to evaluate the product. Sobek and Jain (2004), for example, rate products with respect to cost (i.e., person-hours devoted to the project), time (i.e., weeks the project took to complete), and quality (i.e., customer satisfaction and product quality). Another strategy is to focus on the process by collecting artifacts created throughout. Such artifacts can include proposals, plans, literature reviews, final reports, and oral presentations (Dugan, 2011; Lohman, 2006; Rees et al., 2019). Assessing different aspects through smaller artifacts can provide information about the process without jeopardizing the outcomes (Tubino et al., 2020). Yousafzai et al. (2015) and Damaj and Yousafzai (2019) present a unified framework for evaluating process and product quality of project work and develop a set of rubrics to score the quality of process and product together with a formula to generate a single sum score (see Abu Salem, 2020).

Since projects are commonly assigned to groups of students, one concern is the assessment of individual students (McKenzie et al., 2004; Rees Lewis et al., 2019). In response to these concerns, Tubino et al. (2020) present a rubric that enables assessing contributions by individual students while still encouraging teamwork. The rubric is divided into three categories: (1) criteria students need to comply with to demonstrate engagement in the project, (2) project practices that need to be enacted, and (3) adaptability, or approaches that allow teams to respond to novel or changing situations. For each criterion in these categories, the authors define three levels beyond failure: pass (P), credit (C), and distinction (D). Students must provide evidence for their contribution. For example, to achieve a P for a category 2 criterion (communication), students need to demonstrate they responded to queries in a timely manner in accordance with team agreements; to achieve a C, they need to demonstrate evidence of discussing project-relevant information with project stakeholders; and for a D, they must present evidence of successfully communicating key project information to external audiences (see Tubino et al., 2020 for details). From an initial evaluation, the authors conclude that the rubrics were suitable to assess individual students' work, effectively addressing the concerns that projects do not allow for individual grading of students.

4.3 Interpreting Student Performance and Drawing Conclusions About Their Competencies

Effectively assessing students' engineering competencies requires not only careful task authoring but also appropriately selecting strategies for analyzing students' performance. Although inferring students' competence from their responses appears straightforward, scoring their responses and, more importantly, drawing conclusions about their level of task mastery are not easy. When assessing multidimensional engineering competencies, we are not solely interested in comparing students' performance; we are also interested in how students differ in their mastery, so we are able to provide students with feedback. Feedback may include understanding what is missing from a subpar performance, whether students lack an important piece of knowledge and whether they lack any of the skills they need to use their knowledge.

The purpose of this step is to transform the assumption about how students' competencies can play out in different ways in which different students can perform on the tasks and to a specific model that links students' scores on individual tasks to a measure of their competency based on these assumptions. For this step, these questions should be considered:

- What competencies (i.e., knowledge and skills) are to be assessed at what mastery levels?
- How is the mastery of competencies reflected in individual task scores?
- What is the measurement model that best suits the assessed competencies, and how are they reflected in task scores?

It is challenging to discern from student responses how different students differ in their competency, or how different students with the same level of competency differ in their performance. We typically face a large amount of data from complex and not very well-defined tasks. Sometimes, these data are conflicting; students can exhibit mastery of one task but not another assessing the same competency (Hadenfeldt et al., 2014). To help sort these issues out, we use measurement models. Measurement models help us decide to which extent students' responses (i.e., the data) exhibit evidence for the different ways in which students can demonstrate mastery of a competency (or not).

The choice of the measurement model depends on the analytic strategy. The analytic strategy requires reflecting our assumptions about students' competencies based on how students' performance was scored. Different measurement models reflect different assumptions about the construct and the reality of the data we have (Mislevy & Huang, 2007). These ways of thinking can be broadly categorized into three different data analytic: the performance-centered, the resources-focused, and the process-based strategy.

4.3.1 Performance-Centered Strategy

The performance-centered strategy builds on classical test theory and essentially corresponds to calculating the sum or mean score across items as a composite measure of student competence, sometimes individually for different groups of students (i.e., diverse learners) or tasks (assessing different aspects of students' competence). The sum or mean score, then, is a function of students' true competence plus an error term. One assumption of classical test theory is that errors are uncorrelated across tasks, that is, in summarizing or averaging across items, the error terms cancel each other out. However, this is often not the case. For example, some tasks are more text-heavy, or students less-proficient in the test language may struggle to answer those tasks, leading to correlated errors. Other examples that introduce correlations into error terms are specific (numeric or otherwise) representations used that may be unfamiliar to some students, or typical misconceptions that apply to select tasks across different contexts. In some situations, the unfamiliarity or the familiarity with the task may influence performance, but other times, the same type of tasks can reflect different concepts relevant to the domain that may be more or less difficult for students to master (for example, mastering energy forms appears to be substantially easier than mastering energy conservation, Neumann et al., 2013). Hence, classical test theory rests on assumptions that are generally not met in reality which introduce a potential bias into analysis.

4.3.2 Resources-Focused Strategy

The resources-focused strategy acknowledges the shortcomings of the performance-centered strategy by building on latent variable models. These models assume that one or more (latent) variables underlie students' performance across different tasks. In the following, we will discuss the three families of latent variable models that have received substantial attention in educational assessment. While there are differing approaches, of most importance is that the measurement model chosen aligns to the construct definition. As we advance that engineering competence is a multidimensional construct, we present measurement models that can be used to increase validity, reliability, and fairness of engineering assessments measuring multidimensional constructs.

4.3.2.1 MULTIDIMENSIONAL ITEM RESPONSE (MIRT) MODELS

MIRT models allow for modeling the role of different dimensions in students' competencies. These models provide one ability estimate for each dimension, plus an estimation of the correlations between the dimensions in addition to item difficulty parameters for each item relative to each dimension. However, multidimensionality has two fundamentally different types. The first type, which is also known as between-item dimensionality, refers to assessments in which we have different task sets measuring different constructs. For example, one task set might measure design skills, while another measure understanding of statics. Sometimes, however, individual tasks measure different aspects of a single construct. For example, students may be required to integrate their understanding of statics with design skills to solve an engineering problem. This is also known as within-item dimensionality. Students do not necessarily have to be competent with respect to each dimension in order to perform successfully in such cases, as their ability in one dimension compensates for their lack of ability in another. Such multidimensionality and the respective models are referred to as compensatory models.

4.3.3 Structural Equation Models

Structural equation models (SEMs) are models aimed at identifying the (correlational) structure underlying a range of observable variables (Kaplan, 2008). SEMs differentiate between measurement models, which measure students' performance on a series of observable variables (also known as indicators) as a function of an underlying latent factor, and structural models, which build on two or more measurement models and model the relationships between the latent factors. These relationships can be anything from simple correlational relationships to complex chains of causal relations, potentially including mediating or moderating effects.

Multi-group SEMs allow for investigating how relationships between latent variables vary across different groups of students (e.g., students from different backgrounds or with different native languages). SEMs even allow us to constraint the relationships between certain factors to specific values. This can be useful, for example, to investigate specific hypotheses about the relationships, for example, to determine measurement invariance across groups (van de Schoot et al., 2012; for an example, see Opitz et al., 2019). Alternatively, constraining relationships can be used to demonstrate how a set of tasks measures a construct (e.g., engineering competencies) beyond another construct (e.g., black racial identity). For example, Ford and Helms (2012) demonstrated that Black racial identity was predictive of SAT scores, more so than high school GPA. The higher the scores of students identifying as Black, the lower their SAT scores. Ford and Helms argued that was evidence of unfairness as the test was disadvantaging students with healthy levels of racial identity. To test whether Black racial identity is also prejudiced in engineering exams, through a SEM approach, we would need two task sets, for example, one measuring engineering competency, the other measuring Black racial identity development. Now, if we assume an underlying factor, Black racial identity, drives performance on the tasks measuring Black racial identity development and students' performance on the tasks measuring competence in engineering, and engineering competence drives students' task performance measuring competence, we can test, using the data, if engineering competence explains variance in the items. To this end, we constrain the correlation between both factors to zero to ensure that the latter factor is indeed representing the factor underlying solely a part of students' performance on each task not explained by students' racial identity. This type of model is known as two- or bi-factor models (Gunnell & Gaudreau, 2015).

4.3.4 Diagnostic Classification Models

In some assessments, the dimensions underlying students' competencies of interest do not represent a latent continuum; instead, they mark knowledge or skills that students do or do not have. In a CI, for example, we are often interested in whether students have mastered a particular concept like the energy concept. This assumption is reflected in the so-called diagnostic classification models (DCMs; Rupp & Templin, 2008). In DCMs, students' competence is delineated into a set of dichotomous latent variables representing specific resources (i.e., KSAs), the existence (or non-existence) of which determines their performance.

One use of DCMs is in the analysis of data on student competence collected through CIs (e.g., Jorion et al., 2015). Jorion and colleagues (2015), for example, highlight the potential of DCMs to examine students' conceptual profiles, that is, which of the concepts the CI assesses students have mastered and which ones they have not. DCMs could also be utilized to detect if students hold a particular misconception or not (e.g., Bradshaw & Templin, 2014). Most DCMs can only detect if students have mastered or not mastered a particular concept, or if students hold a particular misconception, but not both. However, an incorrect response to a task may be due to a student lacking mastery of a concept, possessing some misconceptions, or both. To address this issue, Kuo et al. (2018) recently proposed a DCM that can be used to simultaneously assess mastery of a concept (or any kind of skill, for that matter) and the existence of a misconception.

4.3.5 Process-Based Analytics Strategies

Assessments in engineering education and beyond are often based on products. Evaluation only of products created by students, however, falls short of capturing all the available information on students' competencies (Zlatkin-Troitschanskaia et al., 2015). In the case of a design task, for example, information on how students proceeded in designing the solution would allow for valuable insights about students' competencies. Some students may take a trial-and-error approach, while others may pursue a more systematic approach, indicating different levels of competency. CBTs that allow students to design a solution to a problem and test its functionality can provide such information (Dabbagh & Beattie, 2010; see also Gobert et al., 2013).

However, such information is often too large in volume and too heterogenous to be analyzed by humans. Hence, researchers recently started to use data science techniques to make sense of them. There are two fundamentally different approaches to using data science techniques in the analysis of students' achievement. The first approach seeks to replace the human rater. A growing body of research shows that, if adequately trained, computers can achieve a similar agreement with humans as human rater with each other (for an overview, see Zhai et al., 2020; for an example, Zhai et al., 2021). This development facilitates the efficient scoring of student responses in the large scale. However, this approach does not unlock the actual potential of data science techniques in assessment, as assessments that can be actually considered too large-scale to efficiently score responses using human raters are rare and the use of data science techniques in analyzing student responses poses some challenges (e.g., Cheuk, 2021). The second approach appears more promising when it comes to the assessment of student competence. This second approach does not try to replace the human rater (necessarily) but rather to amend the rater. Data science techniques are a promising amendment in situations, in which the data we are looking at are too large in volume, heterogeneous, or complex. This applies in many cases where we are looking at students' interaction with CBTs, especially those that draw on simulations or microworlds. As students engage with complex tasks, the possible interactions become manifold and so become the data – up to the point where a meaningful analysis of the data and linking it to the product outcomes becomes impossible for a human rater. Here, data science techniques can play out their full potential.

One widespread example of the use of data science techniques in assessment is to analyze any kind of traditional-style input that would usually require coding and, more importantly, is impossible to code in terms of how the input unfolds. This may include, for example, students writing an essay, writing source code (in software engineering), or drawing a model. A standard technique for analyzing any kind of textual data (e.g., essays or source code) is natural language processing (NLP; Bird et al., 2009). NLP refers to the process of automated, computerized analysis of natural language data. NLP may involve different methods or tools pending the analytic focus. Rahimi et al. (2017), for example, use NLP to investigate students' use of evidence and their organization of ideas and evidence in support of their claim as part of the response-to-text assessment (RTA, Correnti et al., 2013).

In the RTA, students write an essay in response to a text read aloud and discussed with them. The researchers score the essays using two rubrics, one for the use of evidence and one rubric for the organization which is built from a series of features obtained through NLP. These features include superficial text features, such as the number of paragraphs and sentence lengths. However, the features also include more complex features, such as the length of topic chains (a list of occurrences of a given topic in the essay), indicating how well a topic is elaborated in the text or text coherence determined as the latent semantic analysis similarity of subsequent sentences; more similar sentences indicate a more coherent organization of the text.

Simulations, such as those found in www.nanoHUB.org, in sketchtivity (Castro et al., 2021), and in microworlds, present the opportunity to analyze how students interact with the tools. The most common approach to analyzing student process data in simulation-based tasks is supervised machine learning (see Zhai et al., 2020). In supervised machine learning, a machine is trained to score any kind of task using human-scored examples. That is, human raters score students' performance, for example, as indicative of the application of a particular problem-solving strategy (e.g., trial-and-error). The computer is then fed an array of data with respect to select features of students' interaction, including, for example, the number of trials, the sequence of actions taken, or the time per action. Using respective algorithms, the computer identifies typical patterns of features predictive of the human raters' score. Gobert et al. (2015), for example, use a J48 decision trees algorithm to detect whether students have mastered the skill of designing controlled experiments using a broad array of features, such as all actions count, complete trials count, or simulation pause count, providing a solution for assessing this (ill-defined) skill in the context of complex assessment tasks. Interestingly, the authors found the same set of features can be used to assess this skill in different domains (Sao Pedro et al., 2012, 2013), indicating that there are generalizable features determining how students engage in the design of controlled experiments (and possibly other inquiry or higher-order thinking skills).

5 Conclusion

Assessments are a major tool in engineering education but, just like any other tool, can be used in ways that benefit or harm. While there are many calls for increasing the diversity and inclusion in engineering education, there has been very little discussion of how to practically design assessments that demonstrate properties of validity, reliability, and fairness for multicultural groups of students. These considerations begin when an assessment is conceptualized and continue through to the use and consequences from use of an assessment score.

In this chapter, we have provided an approach to designing assessments from a sociocognitive perspective to capture students' deeper levels of understanding content and how they use their understanding in engineering practices and skills. Acknowledgment of the situated nature of learning allows assessment designers to bridge technical and public concerns regarding fairness by acknowledging the need for greater attention to not only the technical aspects of measurement (e.g., psychometric modeling) but also clear guidance on how the test scores are interpreted in defined contexts. Multiple-choice and calculation-based assessment questions are readily scored, yet achieving complex engineering competence requires opportunities for feedback afforded through open-ended problems, space to justify or explain reasoning, and using content knowledge alongside the practices and skills of engineering. Additionally, we suggest that an increase in contextualized tasks, use of performance-based assessments, and inclusion of engineering practices might help reduce differences in opportunity to learn and create opportunities for higher representation in engineering education by diverse populations.

Acknowledgments

The authors would like to thank Drs. Jim Pellegrino and Bob Mislevy for feedback on earlier versions of this chapter. This work was supported in part by the US NSF, EEC No. 2047420. Any opinions, findings, and conclusions or recommendations expressed in this material are those of the authors and do not necessarily reflect the views of the National Science Foundation.

References

Abu Salem, F., Damaj, I., Hamandi, L., & Zantout, R. (2020). Effective assessment of computer science capstone projects and student outcomes. *International Journal of Engineering Pedagogy (iJEP), 10*(2), 72. https://doi.org/10.3991/ijep.v10i2.11855

Amare, N., & Brammer, C. (2005). Perceptions of memo quality: A case study of engineering practitioners, professors, and students. *Journal of Technical Writing and Communication, 35*(2), 179–190. https://doi.org/10.2190/ML5N-EYG1-T3F7-RER6

American Association for the Advancement of Science (AAAS). (2017). *Annual report: The force for science*. www.aaas.org/resources/aaas-annual-report-2017

American Educational Research Association, American Psychological Association, & National Council on Measurement in Education. (2014). *Standards for educational and psychological testing*. American Educational Research Association.

Anderson, L. W., & Krathwohl, D. R. (2001). *A taxonomy for learning, teaching, and assessing: A revision of Bloom's taxonomy of educational objectives*. Allyn & Bacon.

Arbuthnot, K. (2009). The effects of stereotype threat on standardized mathematics test performance and cognitive processing. *Harvard Educational Review, 79*(3), 448.

Arbuthnot, K. (2017). *Global perspectives on educational testing: Examining fairness, high-stakes and policy reform*. Emerald Group Publishing.

Auer, M. E., & Rüütmann, T. (Eds.). (2021). *Educating engineers for future industrial revolutions* (Vol. 1). Springer International Publishing. www.vitalsource.com/products/educating-engineers-for-future-industrial-v9783030681982

Bailey, R., & Szabo, Z. (2007). Assessing engineering design process knowledge. *International Journal of Engineering Education, 22*(3), 508.

Baxter, G. P., Elder, A. D., & Glaser, R. (1996). Knowledge-based cognition and performance assessment in the science classroom. *Educational Psychologist, 31*(2), 133–140.

Bennett, R. E. (2008). *Technology for large-scale assessment*. ETS Report No. RM-08–10. Princeton, NJ: Educational Testing Service (ETS).

Bird, S., Klein, E., & Loper, E. (2009). *Natural language processing with Python*. O'Reilly.

Bloom, B. S., Engelhart, M. D., Furst, E. J., Hill, W. H., & Krathwohl, D. R. (1956). *Taxonomy of educational objectives: The classification of educational goals. Handbook 1: Cognitive Domain*. David McKay.

Bradshaw, L., & Templin, J. (2014). Combining item response theory and diagnostic classification models: A psychometric model for scaling ability and diagnosing misconceptions. *Psychometrika, 79*(3), 403–425. https://doi.org/10.1007/s11336-013-9350-4

Bronfenbrenner, U., & Morris, P. A. (2006). The bioecological model of human development. In R. M. Lerner & W. Damon (Eds.), *Handbook of child psychology, vol. 1: Theoretical models of human development* (6th ed., pp. 793–828). Wiley.

Buckley, B. C., Gobert, J. D., & Horwitz, P. (2006). Using log files to track students' model-based inquiry. In *Proceedings of the 7th international conference on learning sciences, ICLS 2006* (pp. 57–63). Erlbaum.

Buckley, B. C., Gobert, J. D., Horwitz, P., & O'Dwyer, L. (2010). Looking inside the black box: Assessments and decision-making in biologica. *International Journal of Learning Technology, 5*(2), 166–190.

Castro, L. M. C., Ray, S., Merzdorf, H. E., Douglas, K. A., & Hammond, T. (2021, October). A metalearning approach to personalized automatic assessment of rectilinear sketches. In *2021 IEEE frontiers in education conference (FIE)* (pp. 1–6). doi:10.1109/FIE49875.2021.9637255

Cheuk, T. (2021). Can AI be racist? Color-evasiveness in the application of machine learning to science assessments. *Science Education, 105*(5), 825–836.

Cobb, F., & Russell, N. M. (2015). Meritocracy or complexity: Problematizing racial disparities in mathematics assessment within the context of curricular structures, practices, and discourse. *Journal of Education Policy, 30*(5), 631–649.

Correnti, R., Matsumura, L. C., Hamilton, L., & Wang, E. (2013). Assessing students' skills at writing analytically in response to texts. *The Elementary School Journal*, *114*(2), 142–177. https://doi.org/10.1086/671936

Dabbagh, N., & Beattie, M. (2010). Student and instructor perceptions of the usefulness of computer-based microworlds in supporting the teaching and assessment of computer networking skills: An exploratory study. *Advances in Engineering Education*, *2*(1).

Damaj, I., & Yousafzai, J. (2019). Effective assessment of student outcomes in computer engineering programs using a minimalistic framework. *The International Journal of Engineering Education*, *35*(1), 59–75.

Davis, D. C., Gentili, K. L., Trevisan, M. S., & Calkins, D. E. (2002). Engineering design assessment processes and scoring scales for program improvement and accountability. *Journal of Engineering Education*, *91*(2), 211–221. https://doi.org/10.1002/j.2168-9830.2002.tb00694.x

Diefes-Dux, H. A., Zawojewski, J. S., & Hjalmarson, M. A. (2010). Using educational research in the design of evaluation tools for open-ended problems. *International Journal of Engineering Education*, *26*(4), 807.

Douglas, K. A., Cross, K., & Purzer, Ş. (2022, October). The assessment dilemma: Examining our practices through equity and fairness. In *2022 IEEE frontiers in education conference*.

Douglas, K. A., Fernandez, T., Purzer, Ş., Fosmire, M., & Van Epps, A. S. (2018). The critical-thinking engineering information literacy test (CELT): A validation study for fair use among diverse students. *International Journal of Engineering Education*, *34*(4), 1347–1362.

Douglas, K. A., & Purzer, Ş. (2015). Validity: Meaning and relevancy in assessment for engineering education research. *Journal of Engineering Education*, *104*(2), 108–118. https://doi.org/10.1002/jee.20070

Douglas, K. A., Purzer, Ş., Rynearson, A. G, & Strobel, J. (2016). Reliability, validity, and fairness: A content analysis of assessment instrument publications in major engineering education journals. *International Journal of Engineering Education*, *32*(5), 1960–1971.

Ford, D. Y., & Helms, J. E. (2012). Overview and introduction: Testing and assessing African Americans: "Unbiased" tests are still unfair. *The Journal of Negro Education*, *81*(3), 186–189. https://doi-org.ezproxy.lib.purdue.edu/10.7709/jnegroeducation.81.3.0186

Fortus, D., Kubsch, M., Bielik, T., Krajcik, J., Lehavi, Y., Neumann, K., Nordine, J., Opitz, S., & Touitou, I. (2019). Systems, transfer, and fields: Evaluating a new approach to energy instruction. *Journal of Research in Science Teaching*, *56*(10), 1341–1361. https://doi.org/10.1002/tea.21556

Freedle, R. O. (2010). On replicating ethnic test bias effects: The Santelices and Wilson study. *Harvard Educational Review*, *80*(3), 394–403, 435. Research Library. www.proquest.com/scholarly-journals/on-replicating-ethnic-test-bias-effects/docview/756747242/se-2?accountid=13360

Freedle, R. O., & Kostin, I. (1990). Item difficulty of four verbal item types and an index of differential item functioning for black and white examinees. *Journal of Educational Measurement*, *27*(4), 329–343. https://doi.org/10.1111/j.1745-3984.1990.tb00752.x

Geisinger, K. F. (2016). 21st century skills: What are they and how do we assess them? *Applied Measurement in Education*, *29*(4), 245–249. https://doi.org/10.1080/08957347.2016.1209207

Gobert, J. D., Baker, R. S., & Wixon, M. B. (2015). Operationalizing and detecting disengagement within online science microworlds. *Educational Psychologist*, *50*(1), 43–57. https://doi.org/10.1080/00461520.2014.999919

Gobert, J. D., Sao Pedro, M., Raziuddin, J., & Baker, R. S. (2013). From log files to assessment metrics: Measuring students' science inquiry skills using educational data mining. *Journal of the Learning Sciences*, *22*(4), 521–563. https://doi.org/10.1080/10508406.2013.837391

Guignard, J.-H., Kermarrec, S., & Tordjman, S. (2016). Relationships between intelligence and creativity in gifted and non-gifted children. *Learning and Individual Differences*, *52*, 209–215. https://doi.org/10.1016/j.lindif.2015.07.006

Gunnell, K. E., & Gaudreau, P. (2015). Testing a bi-factor model to disentangle general and specific factors of motivation in self-determination theory. *Personality and Individual Differences*, *81*, 35–40. https://doi.org/10.1016/j.paid.2014.12.059

Gutiérrez Ortiz, F. J., Fitzpatrick, J. J., & Byrne, E. P. (2021). Development of contemporary engineering graduate attributes through open-ended problems and activities. *European Journal of Engineering Education*, *46*(3), 441–456. https://doi.org/10.1080/03043797.2020.1803216

Hadenfeldt, J. C., Repenning, B., & Neumann, K. (2014). Die kognitive Validität von Ordered Multiple Choice Aufgaben zur Erfassung des Verständnisses von Materie. *Zeitschrift für Didaktik der Naturwissenschaften*, *20*(1), 57–68.

Harris, C. J., Krajcik, J. S., Pellegrino, J. W., & DeBarger, A. H. (2019). Designing knowledge-in-use assessments to promote deeper learning. *Educational Measurement: Issues and Practice, 38*(2), 53–67. https://doi.org/10.1111/emip.12253

Helms, J. E. (2006). Fairness is not validity or cultural bias in racial-group assessment: A quantitative perspective. *American Psychologist, 61*(8), 845–859. pdh. https://doi.org/10.1037/0003-066X.61.8.845

Hooks, B. (1992). *Black looks: Race and representation.* South End Press.

Jorion, N., Gane, B. D., James, K., Schroeder, L., DiBello, L. V., & Pellegrino, J. W. (2015). An analytic framework for evaluating the validity of concept inventory claims. *Journal of Engineering Education, 104*(4), 454–496. https://doi.org/10.1002/jee.20104

Kalyuga, S. (2006). Rapid assessment of learners' proficiency: A cognitive load approach. *Educational Psychology, 26*(6), 735–749. https://doi.org/10.1080/01443410500342674

Kalyuga, S., & Sweller, J. (2004). Measuring knowledge to optimize cognitive load factors during instruction. *Journal of Educational Psychology, 96*(3), 558–568. https://doi.org/10.1037/0022-0663.96.3.558

Kane, M. T. (1992). An argument-based approach to validity. *Psychological Bulletin, 112*(3), 527–535. https://doi.org/10.1037/0033-2909.112.3.527

Kane, M. T. (2015). Explicating validity. *Assessment in Education: Principles, Policy & Practice, 23*(2), 198–211. https://doi.org/10.1080/0969594X.2015.1060192

Kaplan, D. (2008). *Structural equation modeling: Foundations and extensions.* Sage Publications.

Kim, H., & Care, E. (2020). *Capturing 21st century skills: Analysis of assessment in selected sub-Saharan African countries* (p. 55). UNESCO Office Dakar and Regional Bureau for Education in Africa and Brookings Institution. www.brookings.edu/research/capturing-21st-century-skills/

Klein, S. P., Jovanovic, J., Stecher, B. M., McCaffrey, D., Shavelson, R. J., Haertel, E., Solano-Flores, G., & Comfort, K. (1997). Gender and racial/ethnic differences on performance assessments in science. *Educational Evaluation and Policy Analysis, 19*(2), 83–97. https://doi.org/10.3102/01623737019002083

Klymkowsky, M. W., & Garvin-Doxas, K. (2020). Concept inventories: Design, application, uses, limitations, and next steps. In J. J. Mintzes & E. M. Walter (Eds.), *Active learning in college science: The case for evidence-based practice* (pp. 775–790). Springer International Publishing. https://doi.org/10.1007/978-3-030-33600-4_48

Kuo, B.-C., Chen, C.-H., & de la Torre, J. (2018). A cognitive diagnosis model for identifying coexisting skills and misconceptions. *Applied Psychological Measurement, 42*(3), 179–191. https://doi.org/10.1177/0146621617722791

Lohman, M. C. (2006). Factors influencing teachers' engagement in informal learning activities. *Journal of Workplace Learning, 18*(3), 141–156. https://doi.org/10.1108/13665620610654577

Lutz, B., & Paretti, M. C. (2017). Exploring student perceptions of capstone design outcomes. *The International Journal of Engineering Education, 33*(5), 1521–1533.

McKenzie, L. J., Trevisan, M. S., Davis, D. C., & Beyerlein, S. W. (2004, June). Capstone design courses and assessment: A national study. In *Proceedings of the 2004 American society of engineering education annual conference & exposition* (pp. 1–14).

Messick, S. (1989). Validity. In R. L. Linn (Ed.), *Educational measurement* (3rd ed.). Macmillan Publishing Co. www.scirp.org/reference/ReferencesPapers.aspx?ReferenceID=2394847

Messick, S. (1995). Standards of validity and the validity of standards in performance asessment. *Educational Measurement: Issues and Practice, 14*(4), 5–8. https://doi-org.ezproxy.lib.purdue.edu/10.1111/j.1745-3992.1995.tb00881.x

Mignolo, W. D. (2011). *The darker side of western modernity.* De Gruyter.

Mislevy, R. J. (2016). How developments in psychology and technology challenge validity argumentation: How psychology and technology challenge validity. *Journal of Educational Measurement, 53*(3), 265–292. https://doi.org/10.1111/jedm.12117

Mislevy, R. J., Almond, R. G., & Lukas, J. F. (2003). A brief introduction to evidence-centered design. *ETS Research Report Series, 2003*(1), i–29. https://doi.org/10.1002/j.2333-8504.2003.tb01908.x

Mislevy, R. J., & Haertel, G. D. (2006). Implications of evidence-centered design for educational testing. *Educational Measurement: Issues and Practice, 25*(4), 6–20. https://doi.org/10.1111/j.1745-3992.2006.00075.x

Mislevy, R. J., Haertel, G. D., Riconscente, M., Rutstein, D. W., & Ziker, C. (2017). *Assessing model-based reasoning using evidence-centered design: A suite of research-based design patterns.* Springer.

Mislevy, R., & Huang, C.-W. (2007). Measurement models as narrative structures. In M. von Davier & C. H. Carstensen (Eds.), *Multivariate and mixture distribution Rasch models: Extensions and applications* (pp. 15–35). Springer Verlag.

National Academies of Sciences, Engineering, and Medicine. (2018). *How people learn II: Learners, contexts, and cultures*. National Academies Press. https://doi.org/10.17226/24783

National Research Council. (2012). *A framework for K-12 science education: Practices, crosscutting concepts, and core ideas*. The National Academies Press. https://doi.org/10.17226/13165

Neumann, K., & Nordine, J. (2022). Instructional coherence and the development of student competence in physics. In *Physics education* (pp. 151–173). Springer.

Neumann, K., Viering, T., Boone, W. J., & Fischer, H. E. (2013). towards a learning progression of energy. *Journal of Research in Science Teaching, 50*(2), 162–188. https://doi.org/10.1002/tea.21061

Newstetter, W. C., & Khan, S. (1997). A developmental approach to assessing design skills and knowledge. *Proceedings Frontiers in Education 1997 27th Annual Conference. Teaching and Learning in an Era of Change, 2*, 676–680. https://doi.org/10.1109/FIE.1997.635892

Nguyen, H.-H. D., & Ryan, A. M. (2008). Does stereotype threat affect test performance of minorities and women? A meta-analysis of experimental evidence. *Journal of Applied Psychology, 93*(6), 1314.

Oehlers, D. J. (2006). Sequential assessment of engineering design projects at university level. *European Journal of Engineering Education, 31*(4), 487–495. https://doi.org/10.1080/03043790600676760

Oliveri, M. E. (2019). Evaluating text complexity to improve the design of formative educational scenario-based assessments for culturally and linguistically diverse populations. In M. Asil, K. Ercikan, & J. Gorin (Eds.), *In cultural contexts and priorities in assessment special issue*. Frontiers in Education.

Oliveri, M. E., Lawless, R. R., & Mislevy, R. J. (2019). Using evidence-centered design to support the development of culturally and linguistically sensitive collaborative problem-solving assessments. *International Journal of Testing*. doi:10.1080/15305058.2018.1543308. www.tandfonline.com/eprint/kczjtG2hTJDDvKKpIfIz/full?target=10.1080/15305058.2018.1543308

Oliveri, M. E., Lawless, R. R., Robin, F., & Bridgeman, B. (2018). An exploratory analysis of differential item functioning and its' possible sources in a higher education admissions context. *Applied Measurement in Education, 31*, 1–16. doi:10.1080/08957347.2017.1391258

Oliveri, M. E., Slomp, D. H., Rupp, A. A., & Mislevy, R. J. (2021). Principled development of workplace English communication Part 3: An integrated design and appraisal framework. *Journal of Writing Analytics, 5*, 109–141. https://doi.org/10.37514/JWA-J.2021.5.1.04

Opitz, S. T., Neumann, K., Bernholt, S., & Harms, U. (2019). Students' energy understanding across biology, chemistry, and physics contexts. *Research in Science Education, 49*, 521–541. https://doi.org/10.1007/s11165-017-9632-4

Pellegrino, J. W. (2012). Assessment of science learning: Living in interesting times. *Journal of Research in Science Teaching, 49*(6), 831–841. https://doi.org/10.1002/tea.21032

Pellegrino, J. W., DiBello, L. V., & Brophy, S. P. (2014). The science and design of assessment in engineering education. In A. Johri & B. M. E. Olds (Eds.), *Cambridge handbook of engineering education research* (pp. 571–598). Cambridge University Press. https://doi.org/10.1017/CBO9781139013451.036

Pellegrino, J. W., & Quellmalz, E. S. (2010). Perspectives on the integration of technology and assessment. *Journal of Research on Technology in Education, 43*(2), 119–134. doi:10.1080/15391523.2010.10782565

Quellmalz, E. S., Timms, M. J., Silberglitt, M. D., & Buckley, B. C. (2012). Science assessments for all: Integrating science simulations into balanced state science assessment systems. *Journal of Research in Science Teaching, 49*(3), 363–393.

Rahimi, Z., Litman, D., Correnti, R., Wang, E., & Matsumura, L. C. (2017). Assessing students' use of evidence and organization in response-to-text writing: Using natural language processing for rubric-based automated scoring. *International Journal of Artificial Intelligence in Education, 27*(4), 694–728. https://doi.org/10.1007/s40593-017-0143-2

Rampazzo, M., & Beghi, A. (2018). Designing and teaching of an effective engineering continuing education course: Modeling and simulation of HVAC systems. *Computer Applications in Engineering Education, 26*(4), 739–748. https://doi.org/10.1002/cae.21916

Randall, J. (2021). "Color-neutral" is not a thing: Redefining construct definition and representation through a justice-oriented critical antiracist lens. *Educational Measurement: Issues and Practice, 40*(4), 82–90. https://doi.org/10.1111/emip.12429

Rees Lewis, D. G., Gerber, E. M., Carlson, S. E., & Easterday, M. W. (2019). Opportunities for educational innovations in authentic project-based learning: Understanding instructor perceived challenges to design for adoption. *Educational Technology Research and Development*, *67*(4), 953–982. https://doi.org/10.1007/s11423-019-09673-4

Robbins, P., & Aydede, M. (2008). A short primer on situated cognition. In P. Robbins & M. Ayede (Eds.), *The Cambridge handbook of situated cognition* (pp. 3–10). Cambridge University Press.

Rupp, A. A., & Templin, J. L. (2008). Unique characteristics of diagnostic classification models: A comprehensive review of the current state-of-the-art. *Measurement: Interdisciplinary Research & Perspective*, *6*(4), 219–262. https://doi.org/10.1080/15366360802490866

Sao Pedro, M. A., de Baker, R. S. J., Gobert, J. D., Montalvo, O., & Nakama, A. (2013). Leveraging machine-learned detectors of systematic inquiry behavior to estimate and predict transfer of inquiry skill. *User Modeling and User-Adapted Interaction*, *23*(1), 1–39. https://doi.org/10.1007/s11257-011-9101-0

Sao Pedro, M. A., Gobert, J., & Baker, R. (2012). Assessing the learning and transfer of data collection inquiry skills using educational data mining on students' log files. In *Annual meeting of the American educational research association*.

Slomp, D. H., Oliveri, M. E., & Elliot, N. (2021). Meeting the challenges of workplace English communication in the 21st century. *The Journal of Writing Analytics*, *5*, 342–370. https://doi.org/10.37514/JWA-J.2021.5.1.11

Sobek, D., & Jain, V. (2004, June). Two instruments for assessing design outcomes of capstone projects. In *2004 annual conference* (pp. 9–1327).

Songer, N. B., & Ruiz-Primo, M. A. (2012). Assessment and science education: Our essential new priority? *Journal of Research in Science Teaching*, *49*(6), 683–690. https://doi.org/10.1002/tea.21033

Sternberg, R. J. (2010). Assessment of gifted students for identification purposes: New techniques for a new millennium. *Learning and Individual Differences*, *20*(4), 327–336.

Tubino, L., Cain, A., Schneider, J.-G., Thiruvady, D., & Fernando, N. (2020). Authentic individual assessment for team-based software engineering projects. In *Proceedings of the ACM/IEEE 42nd international conference on software engineering: Software engineering education and training*, 71–81. https://doi.org/10.1145/3377814.3381702

UNESCO. (2016). *Education 2030: Incheon declaration and framework for action for the implementation of sustainable development goal 4: Ensure inclusive and equitable quality education and promote lifelong learning opportunities for all.* UneSCO. https://unesdoc.unesco.org/ark:/48223/pf0000245656

van de Schoot, R., Lugtig, P., & Hox, J. (2012). A checklist for testing measurement invariance. *European Journal of Developmental Psychology*, *9*(4), 486–492. https://doi.org/10.1080/17405629.2012.686740

Vygotsky, L. S. (1978). *Mind in society: The development of higher psychological processes*. Harvard University Press.

Wertz, R. E., Purzer, Ş., Fosmire, M. J., & Cardella, M. E. (2013). Assessing information literacy skills demonstrated in an engineering design task. *Journal of Engineering Education*, *102*(4), 577–602. https://doi-org.ezproxy.lib.purdue.edu/10.1002/jee.20024

Yousafzai, J., Damaj, I., & El Abd, M. (2015). A unified approach for assessing capstone design projects and student outcomes in computer engineering programs. In *2015 IEEE global engineering education conference (EDUCON)* (pp. 333–339). https://doi.org/10.1109/EDUCON.2015.7095993

Zhai, X., Shi, L., & Nehm, R. H. (2021). A meta-analysis of machine learning-based science assessments: Factors impacting machine-human score agreements. *Journal of Science Education and Technology*, *30*(3), 361–379.

Zhai, X., Yin, Y., Pellegrino, J. W., Haudek, K. C., & Shi, L. (2020). Applying machine learning in science assessment: A systematic review. *Studies in Science Education*, *56*(1), 111–151.

Zlatkin-Troitschanskaia, O., Shavelson, R. J., & Kuhn, C. (2015). The international state of research on measurement of competency in higher education. *Studies in Higher Education*, *40*(3), 393–411. https://doi.org/10.1080/03075079.2015.1004241

Zumbo, B. D., Liu, Y., Wu, A. D., Shear, B. R., Olvera Astivia, O. L., & Ark, T. K. (2015). A methodology for Zumbo's third generation DIF analyses and the ecology of item responding. *Language Assessment Quarterly*, *12*(1), 136–151. doi:10.1080/15434303.2014.972559

32

The Future of Engineering Education Research

Jeffrey Buckley, Patric Wallin, Esther Matemba, Jason Power, Atasi Mohanty, and Gunter Bombaerts

1 Introduction

For many engineering education research (EER) scholars, their journey began as an engineering faculty member who then subsequently developed an interest in educational scholarship. Others started as scholars in neighboring fields, such as physics or computer science education, and then shifted their research interests towards engineering education or continued with their research interests but were able to shift their identity to that of an "engineering education" researcher because of blurry disciplinary boundaries. In relative terms, and at least in terms of scale, it is only in recent years that people are beginning their academic journeys as "engineering education researchers."

It can also be said that over these last two decades, the field of EER has undergone considerable change on a global level. In some parts of the world, EER could still be considered as an emerging area of scholarship. In other areas, such as the United States (US) and parts of Europe where there is a longer engineering education tradition, the last two decades in particular have seen a shift towards increased scientific rigor (Adams et al., 2006; Borrego, 2007) and scholarliness (Edström et al., 2018). Given this, it is very pertinent to have international handbooks in the field, such as this volume and the previously published collection edited by Johri and Olds (2014). These handbooks have several functions, such as acting as "lines in the sand" to guide continued investigation and innovation and as being consultable texts for researchers to gain access to comprehensive summaries of critical topics across the field which have been curated by experts (Johri et al., 2022).

This handbook specifically reflects the current state of thinking and evidence through its 30 topical chapters. Having reviewed the included chapters, we find this handbook to be a very useful contribution to the engineering education literature base. But even an extended handbook such as this has its limitations and cannot address everything of disciplinary importance. We therefore want to use the opportunity of this concluding chapter to highlight other elements we believe to be important. We further offer a synthesizing narrative to some of the concepts treated in isolation, not as a critique of this handbook, but to frame for the reader that engineering education is and can (always) be much more. In doing so, we offer our concluding thoughts through the predominant framing of a few implications for future EER and practice.

Specifically, in this chapter we discuss the multiple levels of engineering education in an effort to synthesize some of the discourse offered through the more explicit topical chapters; we comment on the concept of a "global engineer" as a multidimensional aim of engineering education which

DOI: 10.4324/9781003287483-38

encompasses the concepts elaborated on throughout this handbook; and lastly, we discuss the need for engineering education researchers to look further afield in terms of geographical, conceptual, philosophical, and methodological knowledge in order to guide the next decade engineering education evolution.

2 Considering Engineering Education as a Multilevel System

A classical focus of EER and of higher education research more generally is on pedagogical methods and embedding such methods in enacted practice (Evans et al., 2021). However, engineering work is very complex (cf. Buckley, Trevelyan et al., 2022). Engineering has many subfields, there are many types of engineer, and there is significant variability in what constitutes the activity of engineering even within engineering subfields and types of engineer (Buckley, Gumaelius et al., 2019). Engineering is an amorphous discipline. As a result, both engineering education and EER must be able to reflect and control for this complexity. This, in turn, often necessitates both engineering education and EER themselves becoming complex endeavors. This handbook and the pertinent literature present many different dimensions, perspectives, and levels which can and need to be adopted by engineering education practitioners and researchers. Approaching engineering education through the lens of a multilevel system approach can both illustrate and aid in organizing this complexity (Banathy, 1992; Baraldi & Corsi, 2016; Vanderstraeten, 2000). More specifically, such a multilevel system approach of engineering education shows at least four levels:

- The student level (individual students, student groups, classes, year groups, and non-curricular student organizations)
- The institutional personnel level (educators at various academic grades, researchers, institution support staff, management)
- The institutional culture and industry level (broader university ecosystems, national educational approaches, industry partners)
- The societal level (communities, governments, national university networks, regional and international cultures and networks)

Beyond these systems, there is a clear influence of disciplines and disciplinary dynamics (Becher, 1994). Engineering education is often treated narrowly and singularly, but there are similar disciplines, such as "computing education" (see Malmi and Johri, chapter 26, and Yadav and Lachney, chapter 25) or "industrial design," which are distinct but closely related. Hitt, Banzaert, and Pierrakos (chapter 21) also introduce the integration of liberal arts into engineering education, which is even more different but just as relevant. Such disciplinary boundary spanning has implications at all levels, such as attracting very different students, staff, and projects, influencing departmental cultures through either silos or collaboration, and implicating how societies view engineering education in terms of purpose, function, and accessibility.

2.1 The Student Level

Much engineering education discourse has focused on disciplinary and transversal knowledge and skills, and indeed, this handbook also has chapters which address these topics. But recent years have seen a growth in interest in a broader array of conative factors. It is beyond dispute that concepts such as emotion (Lönngren, Direito, Tormey, and Huff, Chapter 8), identity (Huff and Ross, Chapter 9), motivation (Bombaerts & Spahn, 2021), and creativity (Zappe, Huang-Saad, Duval-Couetil, and Simmons, Chapter 20) are of key importance to engineering education. However, it is very difficult to interpret these concepts in a precise and evidence-informed way, particularly in terms of their operationalization through assessment. This, we believe, will be an important agenda for engineering education researchers across the next decade as efforts are continually invested in understanding students on an individual level. Intuitively, it is clear that there are large differences in emotions, motivations, and creativity among engineering students, and while authors have stressed the importance of individual differences across these and related concepts (e.g., Felder & Brent, 2005), it is remarkable that little research exists in engineering education that looks at how engineering education should respond to associated diversity in individual differences across these concepts. Students have different basic psychological needs and motivations (Ryan & Deci, 2020), different attitudes about teaching and learning (Bombaerts & Vaessen, 2022), and respond differently to specific classroom environments and instructional practices. While these differences may not lead to different learning styles – as this theory has been reliably disproven through replication (Kirschner, 2017) – this interaction can lead to different learning preferences (e.g., individual or collaborative), different approaches to learning (e.g., surface, deep, or strategic), and different intellectual developmental levels (e.g., mindset and capacity to acquire and evaluate knowledge). A developed understanding of student individual differences means we no longer find ourselves dealing with an individual–group dichotomy when considering pedagogy. Engineering education needs to cater for individual needs while acknowledging group dynamics.

In trying to achieve this balance, individual differences can be seen as an opportunity. Students can learn from each other as peers in collaborative or cooperative learning and can support each other in engaging with increasingly complex material (Kirschner et al., 2018). The results of strategically designed collaborative learning approaches can be increased deep learning, more developed collaborative skills, and persistence in STEM fields (Mercier, Goldstein, Baligar, and Rajarathinam, Chapter 19). This is certainly an important reason to use group work in higher education, next to, of course, the mechanism for the upscaling or massification of courses. However, as a note of caution, it is often too easily assumed that group learning is an easy "win-win" sum of individual learnings, and that by placing students in a group, they will automatically learn from each other. It is possible that instead of positive attitudes, negative views can be reinforced, and that misconceptions and falsehoods can be perpetuated (Bombaerts & Nickel, 2017).

In-group differences, and related opportunities and challenges, can and often should be deliberately increased. Students of different levels and types of expertise, or from differing engineering programs, can be brought together, or engineering students can collaborate with students from non-engineering programs, such as the humanities, philosophy, or liberal arts students (Hitt, Banzaert, and Pierrakos, Chapter 21). These compositions can be strategically designed, or they can emerge naturally through, for example, non-curricular or non-degree activities (Chen, Kusimo, and Cardenas-Navia, Chapter 16). A broader composition can also deliberately be designed, for example, in courses in which postgraduate students work together with and give feedback to undergraduate students. Additionally, the time dimension of education is an important aspect in EER, and it is not surprising that there is a lot of attention to this in this handbook. The career of students is quite nicely mapped across the included chapters: primary and secondary school when Yadav and Lachney (Chapter 25) discuss computing education, professional learning for pre-college engineering teachers (Carberry, Klein-Gardner, Lottero-Perdue, and Shirey, Chapter 12), and graduate education in the United States (Fleming, Borrego, and Knight, Chapter 13). In this timeline, however, a few important phases are missing – bachelor education with its challenges of the transition from secondary school to university, the crucial step from university to the first job and the complex learning young engineers go through, and then lifelong learning. Gaining an understanding of these different stages has become more of a priority for engineering education researchers in recent years (Asplund & Flening, 2022; Beddoes, 2022; Johri, 2022). Within the career timeline, phases of education as students progress through a course also merit consideration. In particular, there is a difference among the introductory phase in which students get acquainted to many aspects of the particular course, the middle phase in which the necessary building blocks should be processed, and the ending phase in which a general overview and both broader and more nuanced insights need to emerge (Buckley, Trevelyan et al., 2022).

2.2 The Institutional Personnel Level

A second group of agents in the engineering education system are those who professionally support education and student learning. These are the people who engineering students encounter during their studies. These can be paid (student) teaching assistants, postgraduate research students (master and doctoral), postdoctoral researchers, junior or senior faculty (teachers, coaches, etc.), administrators, educational support personnel, or management. Cutler and Strong (Chapter 14) analyze how job responsibilities, appointment types, and other factors impact the actual education and the career pathways of those involved.

The diversity of roles presented here creates its own system dynamics. Instructors differ in many aspects, for example, in their roles and hierarchy, their personality types, interests, and views of what "good" engineering education is, and their areas of expertise. Typical pedagogical questions arise from this, both at the applied and theoretical levels. For example, should a philosophy of mathematics course in electrical engineering be taught by someone from the electrical engineering discipline or by a non-disciplinary expert in the philosophy of mathematics? The presence of instructors and management gives rise to tensions between optimizing pedagogical strength and financial feasibility. Beyond this, it is important to note that engineering education has been criticized for not being inclusive (enough) in its acceptance and accommodation of diverse learners. It can be argued that this pertains to the institutional culture and societal levels – and it does – but it is worth mentioning here to highlight the capacity that institutional personnel have in being agents of chance. This handbook presents many ideas in this regard, such as leveraging the hidden curriculum to develop social capital in minority groups (Alarcón, Sellers, Paul, and Smith, Chapter 18), adopting more informal learning approaches to encourage engagement and the development of competencies (Polmear, Chance, Hadgraft, and Shaw, Chapter 15), and considering contemporary approaches to assessment of engineering learning that are inclusive to diverse learners (Douglas, Neumann, and Oliveri, Chapter 31). There is a clear need for more research in engineering education to focus on interactions that emerge through institutional personnel, and it appears that comparative research (Tang, Case and Pineda, Chapter 3) and critical cultural analysis (Secules, Perez, Johri, and Pea, Chapter 10) may be auspicious methodologies for such investigation.

2.3 The Institutional Culture and Industry Level

Much of engineering education is not under the control of individuals, be they students, instructors, or administrators. Institutional characteristics strongly determine the boundary conditions of which modes of education are preferable and which are not (Pantzos et al., 2022). The continuous digitalization of engineering education is a clear example of how institutions adopt different approaches. For example, Bairaktarova, Valentine, and Ghannam (Chapter 23) illustrate the use of virtual reality, augmented reality, wearables, and haptic technologies in engineering education, and May, Alves, Kist, and Zvacek (Chapter 24) focus on online laboratories. These and other developments underpin the plea made by Gregg and Dabbagh (Chapter 22) for engineering education to become the innovative leader in the digitalization of higher education. This also impacts specific elements of the curriculum, as shown by Douglas, Neumann, and Oliveri (Chapter 31), who provide concrete examples of how to assess engineering competencies for diverse learners.

Many other aspects of engineering education are also dependent on institutional culture, such as curricula in the first instance, detailed aspects of different pedagogies which are implemented (Bekkers & Bombaerts, 2017), and assessment. For example, consider that there are CDIO (conceive, design, implement, and operate) institutions, faculties, and departments. Important choices have to be made here. Lindsay, Hadgraft, Ulseth, and Boyle (Chapter 6) focus on the necessity of flexibility in terms of engineering curricula to answer the current disruptive nature of the Industry 4.0. This, in turn, implicates the need to consider the collaborative relationship between academia and industry. For example, the types of links that engineering education providers have with industry often determine their financial abilities, and there is certainly a difference in the capacity to organize desired and envisioned engineering education provision in more- or less-affluent institutions, as labs, materials, and immersive learning can be costly. As links with industry are considered important for the financial situation of institutions, the societal role of a technical university or engineering department is often a delicate balance. How (far) should such an institution go in taking up a societal role? For example, should a university go against "big tech" to try to change the attention economy (high-tech companies making users addicted), should they engage in ethical hacking, or should they react against energy companies with whom they work whose practices are having a negative environmental impact? Importantly, this is not only a role of management. Engineering personnel can also play an active role as pedagogical approaches which stress the role of external stakeholders, such as project-based (MacLeod & van der Veen, 2020; Servant-Miklos & Kolmos, 2022), case-based (Martin et al., 2019), or challenge-based (Bombaerts et al., 2021; Doulougeri, Bombaerts et al., 2022; Doulougeri, Vermunt et al., 2022) learning, can create system dynamics that have impact beyond the institution.

2.4 The Societal Level

Finally, engineering education institutions play an important societal role. Many chapters in this handbook testify to this, such as those pertaining to social justice (Chen, Hoople, Leydens, and Rottmann, Chapter 17), anti-Blackness (Holly Jr., Cross, and Lee, Chapter 27), decolonization (Cicek, Masta, Goldfinch, and Kloot, Chapter 4), and diversity and equity (Mejia and Martin, Chapter 11). Each of these chapters – we believe – brings important societal issues to the fore of EER discourse. However, we would also advocate the need for engineering education researchers to consider the generalizability and/or transferability of these issues to, for example, different geographical regions and cultural groups. Parts of EER have been noted as being US-centric (cf. Williams et al., 2018), and for topics with such implication, it is critical that global perspectives are acquired. In this regard, higher education institutions have societal responsibility – whether they want it or not. For engineering education providers, it seems often difficult to realize that and act upon this responsibility. The more rational engineering approach, reducing engineering work to finding micro-solutions and considering knowledge and facts as strictly separated from society and ethics (Latour, 1993; Silvast et al., 2020), makes it difficult at a very fundamental (epistemological and ontological) level. Of course, the hidden curriculum (Alarcón, Sellers, Paul, and Smith, Chapter 18) and the informal learning (Polmear, Chance, Hadgraft, and Shaw, Chapter 15) that emerge in the way the university (including their internal organizations, such as makerspaces, innovation hubs, etc.) interacts with its ecosystem will inform students of what is and what is not an acceptable way of being a critical global engineer.

Engineering education certainly has its influence on the institutional and the university's ecosystem level, but certain habits and cultures reach further. National and international university alliances emerge worldwide, but there are also national and regional levels to be determined in engineering education. National and international university alliances answer a trend of internationalization and upscaling (Fuchs et al., 2022; Vukasovic & Stensaker, 2018), and the impact on engineering education certainly warrants analysis. This handbook has a strong US focus. Nevertheless, Martin, Gwynne-Evans, Kazakova, and Zhu (Chapter 5) address the challenge for engineering ethics education, showing regional authorship patterns and discussing recent pedagogical and institutional practices to broaden engineering ethics education towards global institutions. The authors iterate from the individual to the global, where regional levels are important in emancipatory processes. We, for example, see a strong merit in exchanges within (sub)regions to develop their own identity at the regional level, like Central Asia, Eastern Europe, or Central Africa. This overarching view on the need to consider the remit (local to global) of an engineer underpins the importance of the concept of a "global engineer." This concept, in many ways, ties together several of the concepts which are individually presented throughout this handbook into two overlapping aims: (1) for engineering, to develop "global" engineers, and (2) for EER, to investigate what it means to be a "global" engineer.

3 Developing Global Engineers as an Aim of Engineering Education

Engineering education needs to be responsive to global issues of relevance as they emerge. It also needs to be proactive and support the development of engineers who not only view the world as it is but also view it as it could be. Today's engineers need to see themselves as global engineers and global citizens; they need to take responsibility for their work and its potential impact on the wider society and its multiple stakeholders. Critically, engineers today need to consider not only their impact on the present but also how what they do could impact the lives of future generations. There is therefore a need for engineers to adopt multiple perspectives of the world, one of which being that of a social activist or changemaker (cf. Martin et al., Chapter 5).

Engineers need to embrace global challenges which are complex and evolving and understand and try to resolve these in systematic and holistic ways. In order to meet these challenges, the next generation of engineers will need to utilize both cutting-edge knowledge and technology but also develop new competencies that will allow them to understand the unique contexts of these challenges so that they can achieve sustainable solutions. The development of engineers that are capable of collaboration, problem-solving, and ethical decision-making has become a priority for engineering education providers. Through this handbook, it is clear that this is a call coming from and supported by academics, but it is echoed by professional bodies. Indeed, both the UK Engineering Council and the US National Academy of Engineering have expressed the need for engineers to appreciate and understand the political, social, cultural, and economic impact and contexts of their work (Engineering Council, 2022; National Academy of Engineering, 2004). With this in mind, we need to consider what a global engineer is and, importantly, how engineering education providers can design pedagogies and curricula to support them.

A *global engineer* has been defined by Giovannelli and Sandekian (2017, p. 1) as someone who practices engineering:

- With forethought of its far-reaching consequences, both physical and social;
- With an appreciation of international colleagues and/or in international offices; and
- With cultural sensitivity, so that personal interactions are both pleasant and effective.

Several of the chapters in this handbook are dedicated to individual skills and competencies which, if developed, would support engineers practicing in these ways. Throughout the book, they are treated in isolation, as each chapter has a specific purpose; however, taken as a collective, they present a connectivity which illustrates the attributes of a contemporary global engineer. For example, Zappe, Huang-Saad, Duval-Couetil, and Simmons (Chapter 20) discuss pedagogical approaches to developing creativity, entrepreneurship, and leadership in engineering education. Further to this, current engineers need to be sensitive to diversity, practice inclusivity, and understand social justice, as outlined by Hitt, Banzaert, Pierrakos (Chapter 21). This has resulted in the introduction of new concepts into EER in recent years, such as emotions (Lönngren, Direito, Tormey, and Huff, Chapter 8) and the need to consider learner diversity in practice, such as through assessment (Douglas, Neumann, and Oliveri, Chapter 31). One aspect which is thematic across these chapters is that global engineering requires engineers to develop a high level of proactive agency, and that in being proactive, it is necessary that global engineers have a nuanced appreciation for ethics. While there are many competencies and characteristics noted throughout this handbook, Hitt, Banzaert, and Pierrakos (Chapter 21) offer the overarching idea that engineering education for global engineers requires educating the *whole* engineer. In light of this, it could be beneficial for engineering education researchers to establish a consensus framework of the competencies that define a global whole engineer, inclusive of cognitive, conative, physical, and physiological individual differences where relevant (Manichander, 2016). Such a framework would then permit the co-locating of pertinent pedagogical research which aims to examine the development including knowledge, skills, and attitudes.

Regarding relevant pedagogical approaches for developing global engineers described in this handbook, Martin, Gwynne-Evans, Kazakova, and Zhu (Chapter 5) discuss the teaching of engineering ethics education as part of the development of engineering identity. Specifically, they note that engineering practices requiring engagement, agency, and tenacity are at the core of the development of a global engineer identity, and that this can often occur outside the formal learning environment. One insight we noted of this handbook is that across several chapters, the traditional model of engineering education is witnessing significant augmentation in response to societal needs. One only needs to glance at the table of contents to see the inclusion of chapters on informal learning (Polmear, Chance, Hadgraft, and Shaw, Chapter 15), integrating the liberal arts into engineering education (Hitt, Banzaert, Pierrakos, Chapter 21), and transforming engineering education through social capital and the hidden curriculum (Alarcón, Sellers, Paul, and Smith, Chapter 18), to name just a few. Moreso, from a more focused pedagogical perspective, Gregg and Dabbagh (Chapter 22) describe how engineering educators first adopted, and now in many instances push, the cutting edge of online learning pedagogies. The related adoption of cutting-edge haptic and augmented reality technology adoption is outlined then by Bairaktarova, Valentine, and Ghannam (Chapter 23). One thing is clear – engineering education practice is evolving and methodologically diversifying in terms of pedagogy, and this evolution in practice is aligned with efforts to support the development of engineers who can operate effectively in a changing society. This evolution could also be considered as a result of efforts to disrupt prevailing practices in engineering education. This particular lens is elaborated on by Lindsay, Hadgraft, Ulseth, and Boyle (Chapter 6), who, based on the Doblin innovation framework (Keeley et al., 2013), present "five themes that are emergent, unifying ideas . . . giving shape to the changes we see in contemporary engineering programs":

- Development of the "student engineer" identity from early in the program.
- Understanding perceptions and expectations from students, employers, academics, and others.
- Adopting a flexible curriculum.
- Modifying the learning environment.
- Starting with a greenfield site – entirely new programs.

In considering the implications of the work described through these chapters, it is worth considering the role and responsibility we (stakeholders of engineering education research and practice) have in these educational transactions. There has been a development in our understanding of what constitutes relevant engineering knowledge and skills, and also in how modern engineering students engage with a knowledge base that is increasing in volume and complexity at an exponential rate. Gone are the days when a lone engineer could design a product, or even a component. Due to the increasing complexity of the field, we see increasing specialization within the profession. This has an inevitable impact on industry, where larger teams are required to solve problems, which in turn has resulted in our current priority of increasing collaborative capacities and supporting the rounded development of our graduates. In practice, this is reflected in our pedagogies. We see a consistent increase in the use of problem and project-based learning methodologies that seek to replicate the complex demands of modern engineering. These approaches, championed by organizations such as the CDIO initiative, tend to focus on skill development and knowledge retrieval capacities rather than the more traditional procedural or rote approaches. In many ways, these approaches signify a fundamental shift in how we see ourselves as educators. Barger et al. (2016) outline how student epistemologies – views on the nature and origin of knowledge – can be changed by factors within the control of the educator. Decisions based on the aspects of engineering education described earlier, such as incorporation of collaborative learning strategies or the inclusion of problem-based learning, can fundamentally shift how our students perceive the world. This is a profound privilege and responsibility. It is important to recognize the level of agency that we are afforded within these large systems and even more important that we make informed decisions that are cognizant of the global challenges that are quickly becoming existential threats.

4 Engineering Education Research as a Mirror of Engineering Education Practice: Philosophical, Theoretical, and Methodological Implications

Much like how engineering education practice is amorphous and can be considered to have blurry disciplinary boundaries, EER is also a complex melding of disciplines. Klassen and Case (2022) consider the position of the discipline relative to the strength of its boundaries to fields from which methods and theory are regularly drawn. They conclude that the argument for EER as a distinct field has served a purpose in establishing a clear identity, but it has not led to a unique knowledge base that would fully legitimize this distinct status and suggest that this could be overcome by looking to parent disciplines for methodological and theoretical direction. This outward-looking approach suggests a greater consideration of relevant research from other disciplines. There is some evidence to suggest that EER has begun to consider relevant evidence from neighboring disciplines (Williams et al., 2018); however, this is not universal and appears to be more focused on relevant theories and less so on the adoption of methods. In reading the topic chapters, one thought we consistently arrived at related to the outward perspective of engineering education researchers – or at times the lack thereof. Is this a problem? Not necessarily, but it does present natural limitations and challenges from which can be borne considerable opportunity. We think it is important to consider this from both a theoretical and methodological perspective.

4.1 Translating Foundational Theories to the Applied Context of Engineering Education Research

One interesting tension that emerges across the different chapters in this handbook is how research is theoretically framed and approached. Where some chapters provided very nuanced and balanced accounts of concepts, theories, and phenomena, others focus more on providing solutions, concrete approaches, and recipes. However, even in chapters that focus more on concepts, theories, and phenomena, there often appears to be a goal to solve some kind of problem, overcome a perceived barrier, or at least contribute to making engineering education "better."

In some ways, this is not surprising, as engineering practice also tends to be more on the applied rather than on the basic or foundational side. EER, being rooted in engineering practice, takes with it ideas and philosophies from the discipline. Thus, engineering practice does not only define the context for EER but also influences research methodologies, perspectives, and approaches. EER is intrinsically context-bound and defined by the field of engineering. To some extent, we argue, it is grounded in educational practices within engineering rather than phenomena-driven research that is predominantly occupied with specific facts, concepts, or theories. In this respect, EER differs from higher education research and research on university pedagogy, which is oftentimes defined by the phenomenon that is studied or the disciplinary perspective that informs research. For example, there is an entire field that focuses on critical perspectives in higher education, another that is concerned with social interactions, or power dynamics. EER, like other discipline-based educational research (DBER) disciplines, oftentimes asks similar questions but is contextually bound to engineering education. With this in mind, we want to advise researchers to look beyond the work published in EER journals and engage more broadly with the research literature in higher education.

It is important, we believe, to provide some underpinning for this advice. There have been several bibliometric analyses of EER prior to now and they tend to conclude a similar cautionary note that EER is broadening but is perhaps not as outward-looking as it could be – yet. Williams et al. (2018), for example, conducted a co-citation analysis of EER and suggested that while European engineering education research has spread beyond its regional origins in terms of international collaborations and citations, US engineering education remains relatively siloed. Wankat et al. (2014) outlined the evolution of the *European Journal of Engineering Education* and the *Journal of Engineering Education* and noted an increase in citations of psychology and education sources increased over time. However, they also warned that a silo effect was evident where research on education relevant to specific engineering disciplines was not typically influencing general EER work. To add to this and to provide a slightly different perspective, Apiola et al. (2021) analyzed the 50 years of publications from the Frontiers in Education (FIE) conference. While they didn't examine citation sources in general, they did identify that there is a low tendency for FIE authors to cite previous FIE publications and interestingly question whether a stronger link would even be desirable. We add this as a precursor to the brief and rapid analysis we present next. We do think engineering education researchers need to look more outwards in general but encourage readers to consider the relevance of this suggestion with regards to specific disciplinary research agendas.

We searched the Scopus database for all publications over the previous five years (January 1, 2018, to October 20, 2022) that were published in the following sources:

- *International Journal of Engineering Education*
- *European Journal of Engineering Education*
- *Journal of Engineering Education*
- *Global Journal of Engineering Education*
- *Australasian Journal of Engineering Education*

This search returned a total of 1,407 articles. These articles referred to 45,309 sources, which were what we considered to be of interest rather than the original articles themselves. The Scopus database only had bibliometric details of 13,693 of these sources.

Figure 32.1 identifies the top 25 sources (of the 13,693 available in Scopus) based on the frequency of which they were referenced by the 1,407 articles returned in the search. It is immediately evident that a very large portion of references are made to EER outlets. References to the *American Society of Engineering Education (ASEE) Annual Conference and Exposition*, the *International Journal of Engineering Education*, the *Journal of Engineering Education*, the *European Journal of Engineering Education*, *Frontiers in Education*, and *IEEE Transactions on Education* accounted for 33% of all citations. Further, of the top 25 cited sources, 79% of citations are to DBER sources. This analysis is reminiscent of the analysis conducted by Bruce et al. (2017). They presented a co-citation analysis of research relating to spatial ability in psychology, neuroscience, and mathematics education. What became

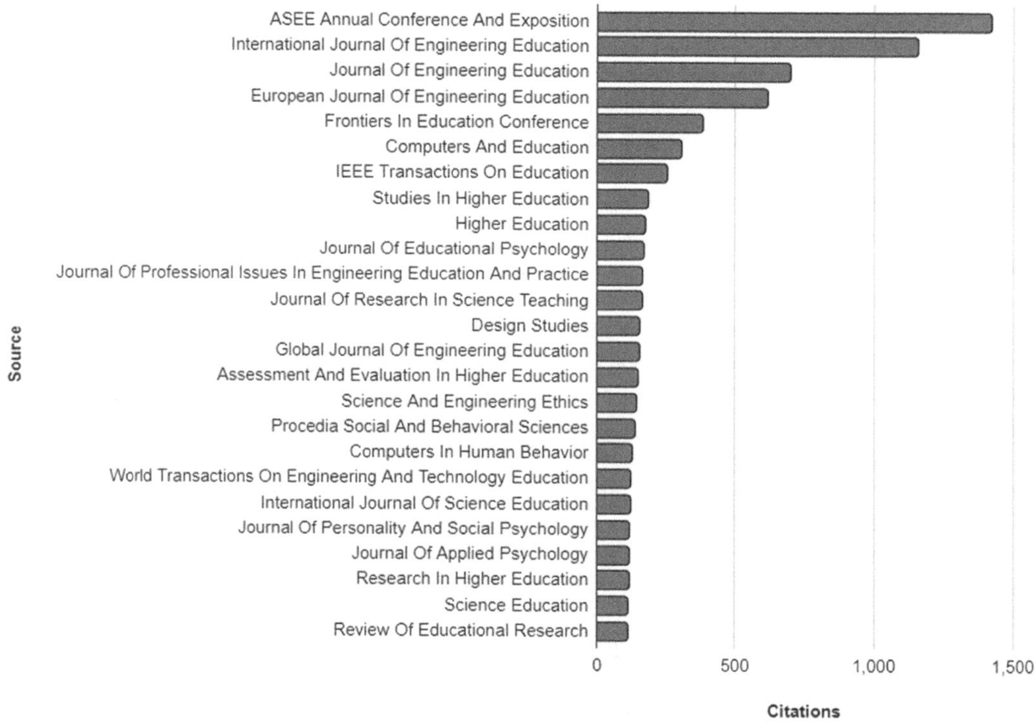

Figure 32.1 The top 25 most cited sources from the *International Journal of Engineering Education*, the *European Journal of Engineering Education*, the *Journal of Engineering Education*, the *Global Journal of Engineering Education*, and the *Australasian Journal of Engineering Education* for the last five years (January 1, 2018, to October 20, 2022).

Source: Data retrieved from Scopus.

immediately apparent when they plotted their analysis was that mathematics education researchers who were conducting research into spatial ability were not connected with the psychology or neuroscience literature, and the psychology and neuroscience researchers were not connected with the mathematics education research. This, of course, would present a set of limitations for any discipline, but for the mathematics educators, it illustrated a disconnect between their applied research and the foundational work pertaining to the phenomenon at the core of their inquiry. Viewing engineering education as akin to mathematics education in this vein, we want to emphasize the importance for engineering education scholars to not only engage with applied literature which often presents adaptations of foundational constructs but also engage with the original material as well.

One risk that emerges from the desire to come up with research-based solutions is that it can potentially overshadow the ability to spend time on understanding and exploring a problem from multiple perspectives. This stands in contrast to research that mainly focuses on understanding a phenomenon, situation, or context without the clear desire to solve any perceived problems or overcome barriers. For example, Lönngren, Direito, Tormey, and Huff, in their chapter on emotions (Chapter 8), provide a nuanced account about the importance of emotions in engineering and the fallacy of seeing engineering as purely rational practice. It highlights how engineering is embedded in complex social structures and draws on the broader literature to describe and discuss different perspectives, approaches, and research methods in relation to emotions. At the same time, the chapter only briefly touches on how the concept of emotion is covered in the humanistic and philosophical literature and does not critically discuss the phenomenon itself and how it is used. Rather, there appears to be a clear desire to improve engineering education through research on emotions that subsequently informs teaching practices. This is already highlighted in the title, "with great potential to assist engineering education reform for the 21st century." The risk, then, is that emotions are reduced to a tool, a means to an end, rather than being seen as intrinsically interesting and important to study and understand. This might not necessarily be what the authors intended, but it illustrates the importance for both readers and authors to critically reflect.

In contrast, Huff and Ross, in their chapter on identity (Chapter 9), provide a very balanced account of the concept of identity. They provide ample links to the literature on identity from a variety of other fields, as well as critical perspectives on the concept itself and how it is used. At the same time, the authors make the concept of identity relevant for EER without a direct link to how a focus on identity might improve engineering education. By critically discussing implications of choosing between different frameworks and conceptualizations of identity, the chapter opens the concept for other researchers as a phenomenon rather than a tool. It actively supports the reader in reflecting on the concept.

While these two chapters are chosen as examples, the trend to highlight the *use-value* and impact of the EER is apparent across the different chapters. In other words, research is oftentimes thought of as applied rather than fundamental. Even the chapters that cover more fundamental aspects and describe and discuss phenomena from multiple perspectives have a tendency to make a clear connection as to why a focus on this phenomenon is important for developing engineering education. In this way, it appears that the use-value of the research is put to the forefront and there is a clear link established between research and development. Overall, it appears across the chapters that there remains a strong utilitarian and instrumental dimension in EER, where there is a clear goal to solve perceived problems. Research becomes a tool to advance engineering education, and the use-value of research easily gets reduced and being judged from how applicable it is. In other words, research is not necessarily seen as being of value on its own, but its relevance is determined by the impact it has on practice (Davies, 1999).

One potential challenge that arises from this is that questions might get framed too narrowly by focusing on problems to solve. Rather than seeing what emerges from the research and how it might or might not be of interest for engineering education, the anticipated findings might influence the research design process. In addition, more nuanced accounts might be reduced to pseudo-causal relations that are, in fact, strongly context-dependent. Applied and fundamental research should, however, not be seen as in opposition to each other or as exclusive binaries. They are deeply interdependent of each other and span a vast research landscape that covers endless nuances and approaches. What is important though, we argue, is that researchers in the field of engineering education are aware of these risks and reflect on the broader research landscape and how it influences their own studies and approaches. Furthermore, this awareness is necessary to navigate the intertwined mesh of research, development, practice, and policy (Gornitzka, 2013).

The applied nature of much of EER often leads to a tighter integration of research, development, and practice and can inform policy processes directly. On the one hand, this closeness means that research potentially is more readily used to inform practice and that the way from practitioner to research is shorter than in other cases. On the other hand, however, this closeness potentially blurs the boundary between research and evaluation. In contrast to research, the focus in evaluations explicitly lies clearly on improving the quality of education and directly guides the revision of practices. Evaluation thus has a different intention to research, which is conceptually and practically important (Sandars et al., 2017). The blurriness between evaluation and research in EER demands that researchers reflect carefully on their research design, direction, and motivation and are aware how different stakeholders influence and shape research processes in explicit or implicit ways. There is certainly a risk that rather than contributing to research-informed policy making, policies are shaping research processes and favor certain types of questions and designs. Eventually, this can lead to research losing its critical edge, no longer questioning central underlying premises in engineering education, and taking boundary conditions for granted. To counteract this, we argue that engineering education researchers need to ask themselves what we are researching and why, who is framing the boundary conditions, and how engineering education researchers can challenge those boundaries.

With this in mind, we want to urge engineering education researchers to engage in critical reflections around their own positionality and reflexivity, and how they inform and shape their research questions, framings, methodologies, and analysis, among other elements of the research process (Holmes, 2020). *Positionality* refers to a researcher's worldview and their position with respect to the research activities' social and political contexts. Worldviews, here, include ontological assumptions (beliefs about the nature of reality), epistemological assumptions (beliefs about the nature of knowledge), methodological assumptions (beliefs about how knowledge can be generated), and axiological assumptions (beliefs about value of knowledge and action). Positionality is informed by *reflexivity*, which captures the process through which one reflects on one's own worldviews and position, as well as how positionality shapes the research process. In this way, reflexivity is concerned with questioning assumptions and becoming aware of how researchers themselves influence the research. In other words, reflexivity is about what researchers do with and how they reflect on their own positionality (Finlay, 1998). Together, both positionality and reflexivity influence how research is conducted, its outcomes, results, and dissemination (Rowe, 2014).

While the importance of positionality and reflexivity is often discussed in relation to qualitative research, we argue that this is equally important in quantitative research and should be part of the general foundation in all EER. Research in engineering education and higher education, in more general terms, is seldomly value-free. Research and education are always political (Freire, 1970). As outlined before, EER is embedded in a complex social, cultural, and political mesh, as are researchers themselves. Rather than trying to avoid that their own worldviews and dispositions influence their research, which is impossible, we wish to encourage researchers to acknowledge their own positionality and approach it with critical reflexivity. Furthermore, we argue that it is important to go beyond the individual dimension of positionality and reflexivity and move towards critical reflection on EER and its boundary conditions. This is a crucial step in approaching the questions we posed before: what we are researching, who is framing the boundary conditions, and how engineering education researchers can challenge those boundaries. Following Brookfield's (2017) work on critical reflection, we want to emphasize that reflection becomes critical when it questions assumptions and common practices, and when it focuses on power relations. There is a risk that EER contributes to someone else's agenda if researchers do not critically reflect on their research approaches and question their own approaches.

4.2 Translating Methodologies Developed in Foundational Disciplines to the Applied Context of Engineering Education Research

Like our view that broadening our theoretical outlook is important for engineering education researchers, the adoption of methodologies from other disciplines also needs to be viewed through a critical lens. EER is described by Goncher et al. (Chapter 7) as an applied social science, for which they then adopt Neuman's (2011, p. 1) definition of applied social science as "research that attempts to solve a concrete problem or address a specific policy question and that has a direct, practice application." Two fundamental questions arise when EER is considered in this way: (1) Is EER a "science," and (2) what challenges and opportunities arise through the contextual application of foundational research?

The first question is rather intentionally provocative and is perhaps primarily related to nomenclature, but its intention is to highlight an important omission of this handbook – that of a general chapter on inductive qualitative research focused on theory generation. The term "social science" was formerly used to refer to sociology, the science of societies and individual relationships within societies. The origins of sociology can be traced back to the French philosopher of science Auguste Comte and French sociologist Émile Durkheim, and it was notably rooted in positivism and quantitative methods. However, as society modernized, scholars – notably Karl Marx – began to reject positivism, and sociology broadened paradigmatically, theoretically, and methodologically. Today, the term *social science* is used in a broader sense to describe several branches of research, including anthropology, economics, education, political science, psychology, and sociology, to name a few. Social scientists now operate under various paradigms, such as postpositivism, interpretivism, and pragmatism, engage with a myriad of theoretical perspectives, such as critical, feminist, and social cognitive theories, and adopt methodological approaches which are quantitative, qualitative, and mixed methods. Where the positivist inception of the social sciences lends to them being considered a science, today there is a position which can be taken to move away from the term social science and to instead consider it as "social research" in response to the now-broader nature of scholarship the term represents. This handbook, however, which contains two very useful chapters dedicated to quantitative research in general (Hjalmarson, Herman, and Douglas, Chapter 29; Katz, Godwin, and Brozina, Chapter 30), is noticeably missing a chapter on qualitative methodology in general (although we believe there was an intention to have such a chapter, but for various reasons, it never made it into the final version). There, of course, are qualitative methods examined throughout the handbook – such as in Tang et al.'s insightful chapter (Chapter 3) on comparative research – but there are none which take the form of, for example, "Considerations for Engineering Education Research Using Qualitative Methods" (to play on the title of Hjalmarson et al.'s chapter [Chapter 29]). We note this as an important point of discussion as we conclude this handbook, as there is a wealth of credible qualitative and "mixed methods" research available for review through dedicated EER academic journals and conference proceedings. We would also note that in a previous handbook (Johri & Olds, 2014), there were chapters on qualitative methods in EER in general (J. Case & Light, 2014) and on ethnographic methods more specifically (Johri, 2014) which readers could consult.

The second question possibly raises considerably more interest – that of the challenges and opportunities which emerge through the adoption of methodologies which were developed in more foundational fields (e.g., psychology, anthropology, or sociology) and which have been translated for use in EER. Before reflecting on methodologies discussed in this handbook, I (Jeff) am reminded of a comment made to me by an engineering education PhD candidate at a doctoral research symposium. In a special interest group (SIG) discussion on research methods, this person commented that if you want to learn how to conduct research in engineering education, the last place you should look for guidance is published EER. As the conversation progressed, this sentiment was elaborated on and highlighted (1) the shortcomings of published works in explaining methodological nuances, that is, there was a lack of transparency in how methods were reported (cf. Buckley, Adams et al., 2022), and (2) that important methodological dimensions often get lost when a methodology is applied in a field it did not originate from. Hearing this from a person who was actively investing efforts in learning about different methodologies to inform their own impending work was a stark indicator of the potential to take important details for granted.

At least the first of these issues – improving methodological transparency – is straightforward to address through developing and implementing methods to ensure published works contain all essential information, and Svihla et al., in their chapter (Chapter 28), begin to address this issue. Specifically, Svihla et al. note that often engineering education researchers give relatively rich descriptions of methodological procedures but relatively little information on educational settings or interventions that are integral to their investigation. In response, they provide a template for the publication of "learning design cases" and speak about the dissemination of EER scholarship through this lens.

The second issue is more complex – what is methodologically lost when methods from other disciplines are applied in EER, and what is gained? Using the research regarding spatial ability as an example to keep consistency with the previous citing of the work of Bruce et al. (2017), an illustrative example can be provided. Spatial ability is a psychological construct described as "the ability to make use of simulated mental imagery to solve problems – perceiving, discriminating, manipulating, and recalling nonlinguistic images in the 'mind's eye'" (Schneider & McGrew, 2018, p. 125). It is not directly measurable, and therefore, psychometric tests have been developed as indicators of different sub-dimensions of the construct known as spatial factors. One of these, the visualization factor, has been the subject of extensive EER over the past three decades (Atit et al., 2020; Buckley et al., 2018; Buckley, Seery et al., 2019; Hyland et al., 2021; Munoz-Rubke et al., 2021; Ramey & Uttal, 2017; Sorby et al., 2018). Outside of engineering education, there is much more research on other spatial factors; however, visualization is the factor that has been identified as the most useful for engineering students and engineers and therefore receives the most attention. More importantly, to this point, as psychometric tests serve primarily as indicators of the construct they purport to measure, in psychological research, there is a view that multiple instruments should be used and a composite score of these used in data analysis (Buckley, 2022; Moreau & Wiebels, 2021). This is not the usual practice in EER, where, instead, generally, only one measure is used, or where there are multiple tests used, they are treated individually. Acknowledging that time limitations can exist in terms of access to participants, which can translate to methodological limitations, in this example and for many other constructs examined throughout EER, we believe there is a lot of value to be gained from consulting more regularly with literature from parental disciplines where methods and methodologies are conceived and developed without an applied context.

It is likely that the growth of theoretical links to neighboring disciplines will ultimately lead to increased use of related methods. Case and Light (2011) provide an overview of a diverse range of methods and examples of their use within EER. A related introduction for newcomers is presented by Goncher, Hingle, Johri, and Case (Chapter 7). It is encouraging to note that many of the examples of methodological use within EER include mixed and qualitative methods, as earlier trends suggested a dominance of descriptive quantitative approaches (Malmi et al., 2018). Many of the examples presented by Case and Light (2011), including ethnography, action research, phenomenography, discourse analysis, and narrative analysis, utilized systems and approaches that have been developed in neighboring disciplines over decades. A broader argument for the need to consider more diverse methods is presented by Borrego et al. (2009), who outline the many potential benefits. However, as we begin to adopt these approaches, it is important to consider the need for professional development support to ensure methodological fidelity and broader best practice. Malmi et al. (2018) suggests that the execution of adopted methods requires further improvement. This, in many ways, is to be expected, and the capacity of a discipline to provide critical self-evaluations should be considered a strength.

Importantly, the call to be more considerate of the methodologies we use as they are implemented in their originating disciplines also has the potential to support EER in its own definition. Borrego (2007) argued that clear classification of EER is essential if we are to continue advancement. In subsequent work, Borrego et al. (2014, p. 46) identified key areas that must be further developed in order to create a solid foundation:

- Progression: evidence that researchers are informed by previous studies and build upon or deepen understanding.
- Model publications: publications that other researchers hold up as models of conduct and presentation of research studies in the field.

- Seminal publications: publications recognized as important or definitive because they marked new directions or provided new insights.

These prerequisites are, in many ways, dependent on further adoptions of methods and theory from parental disciplines. Borrego et al. (2014) argued that this could result in a higher consensus within the discipline and support the pursuit of more complex questions. Jesiek et al. (2009) suggested that the ambiguity associated with engineering education researchers has led many to recognize challenges both in terms of wider academic recognition and in the scaling of future research efforts. Examples of methods and systems adopted from other disciplines have shown potential in both addressing this perceived ambiguity and in supporting consensus. One now widely used example is the rapid adoption of systematic reviews and associated protocols originally developed for use within medical research (Power, 2021b). In addition, open science practices and supporting infrastructure such as the Open Science Framework have increased in popularity within EER (Power, 2021a). These examples demonstrate a capacity for engineering education to be informed by developments in neighboring disciplines, but also to adapt to our specific contexts and, in some instances, lead their further development at the highest levels. In these and other examples, we can benefit from the work of colleagues in neighboring disciplines. There is perhaps no better example than the evolution of science education as a research discipline (Fensham, 2004).

Science education researchers, being from an older discipline, have encountered and, in many instances, overcome similar issues that now face engineering education researchers. This is in part due to an outward shift in perspective and the adoption of many methods that would have originated in social science disciplines. This disruption to engineering education is examined by Lindsay, Hadgraft, Ulseth, and Boyle (Chapter 6), where the impacts of innovation are considered alongside the broader shifting purpose of engineering education. While there are lessons to be learned from the historical experiences of science education as an emerging discipline, it should not be considered as a sequence to be followed. The process was necessarily messy, and the value of debate then and now is a primary factor in its development. In this manner, the emergence of science education research offers insight, not instruction. Instead, the emergence of EER must be considered in the context of its own current position and about any influencing factors of the current international education environment.

5 Concluding Thoughts

In concluding this conclusion, we are posed with the question of how we evaluate the handbook as a whole.

As noted earlier, we find it a very useful core text within the engineering education literature, in which the editors have curated a selection of accessible and informative topical chapters at the frontier of engineering education discourse. This book opens discussions and provides direction for researchers and educators to help them navigate and contribute to contemporary research agendas and practice. The included chapters have built on developing topics in EER and can potentially contribute to transforming engineering education towards developing more globally competent and socially responsible engineers. Across the chapters, there are calls for more empirical and rigorous research in underdeveloped areas of EER, and these are balanced with comprehensive summaries of the work which has been conducted here to date.

Throughout this chapter, we have noted some limitations of this handbook, and this was to be an inevitable situation. Some limitations are quite apparent (such as there not being a dedicated chapter on general qualitative or mixed-methods research in EER), and others were more subtle (such as there being a dominant US focus across some chapters with a global voice missing). But these limitations in themselves provoked us to reflect on the nature and current state of engineering education research and practice. So from that perspective, even in its limitations, this book has been successful in achieving one of its likely primary intentions of acting as a platform from which to inform the future of the field.

We therefore consider it fitting to conclude with a summary of our thoughts and our main take-home message(s) after reading and reflecting on the included chapters. There are different ways of thinking and working in engineering education. What is central is that as a scholar of EER, or as an engineering educator, we are aware of the choices we make, how we position ourselves, and who influences us, so that we can be critically reflexive. We need to ask ourselves at a macro-level, what are engineering education research and practice, what could they become, and what do we want them to be (cf. Ramnath, Bix, Winberg, Pevkur & Conlon, Chapter 2)? Prevailing practices have changed before and are emerging and solidifying again now. There is a strong movement towards evidence-based work – with rigor and scholarliness becoming keywords – and much progress has been made. However, we would contend that there is still much that can be improved upon and there is a need to ensure that the evidence base we are constructing is not mono-dimensional. Ultimately, going forward, there is a need to balance the standing of the shoulders of (disciplinary) giants with looking back at our origins and being aware of simultaneously occurring foundational advances.

Reference List

Adams, R., Aldridge, D., Atman, C., Barker, L. J., Besterfield-Sacre, M., Bjorklund, S., Bransford, J., Breslow, L., Brophy, S., Burrows, V., Busch-Vishniac, I., Carlson, M., Carlson, L., Chen, H., Dahleh, M., Diefes-Dux, H., Duffy, T., Evangelou, D., Felder, R., . . . Watford, B. (2006). The research agenda for the new discipline of engineering education. *Journal of Engineering Education, 95*(1), 259–261. https://doi.org/10.1002/j.2168-9830.2006.tb00900.x

Apiola, M., Tedre, M., López-Pernas, S., Saqr, M., Daniels, M., & Pears, A. (2021). A scientometric journey through the FIE bookshelf: 1982–2020. In *2021 IEEE frontiers in education conference (FIE)* (pp. 1–9). https://doi.org/10.1109/FIE49875.2021.9637209

Asplund, F., & Flening, E. (2022). Boundary spanning at work placements: Challenges to overcome, and ways to learn in preparation for early career engineering. *European Journal of Engineering Education, 47*(1), 50–69. https://doi.org/10.1080/03043797.2021.1889467

Atit, K., Uttal, D. H., & Stieff, M. (2020). Situating space: Using a discipline-focused lens to examine spatial thinking skills. *Cognitive Research: Principles and Implications, 5*. https://doi.org/10.1186/s41235-020-00210-z

Banathy, B. H. (1992). *A systems view of education: Concepts and principles for effective practice*. Educational Technology Publications.

Baraldi, C., & Corsi, G. (2016). *Niklas Luhmann: Education as a social system* (1st ed.). Springer.

Barger, M. M., Wormington, S. V., Huettel, L. G., & Linnenbrink-Garcia, L. (2016). *Learning and Individual Differences, 48*, 1–8. https://doi.org/10.1016/j.lindif.2016.04.002

Becher, T. (1994). The significance of disciplinary differences. *Studies in Higher Education, 19*(2), 151–161. https://doi.org/10.1080/03075079412331382007

Beddoes, K. (2022). Gender as structure in the organisational socialisation of newcomer civil engineers. *European Journal of Engineering Education, 47*(1), 102–116. https://doi.org/10.1080/03043797.2021.1915251

Bekkers, R., & Bombaerts, G. (2017). Introducing broad skills in higher engineering education: The patents and standards courses at Eindhoven university of technology. *Technology & Innovation, 19*(2), 493–507. https://doi.org/10.21300/19.2.2017.493

Bombaerts, G., Doulougeri, K., Tsui, S., Laes, E., Spahn, A., & Martin, D. A. (2021). Engineering students as co-creators in an ethics of technology course. *Science and Engineering Ethics, 27*(4), 48. https://doi.org/10.1007/s11948-021-00326-5

Bombaerts, G., & Nickel, P. J. (2017). Feedback for relatedness and competence: Can feedback in blended learning contribute to optimal rigor, basic needs, and motivation? In *2017 IEEE global engineering education conference (EDUCON)* (pp. 1089–1092). https://doi.org/10.1109/EDUCON.2017.7942984

Bombaerts, G., & Spahn, A. (2021). Simplify! Using self-determination theory to prioritise the redesign of an ethics and history of technology course. *European Journal of Engineering Education, 46*(2), 210–226. https://doi.org/10.1080/03043797.2019.1702924

Bombaerts, G., & Vaessen, B. (2022). Motivational dynamics in basic needs profiles: Toward a person-centered motivation approach in engineering education. *Journal of Engineering Education, 111*(2), 357–375. https://doi.org/10.1002/jee.20448

Borrego, M. (2007). Development of engineering education as a rigorous discipline: A study of the publication patterns of four coalitions. *Journal of Engineering Education, 96*(1), 5–18. https://doi.org/10.1002/j.2168-9830.2007.tb00911.x

Borrego, M., Douglas, E. P., & Amelink, C. T. (2009). Quantitative, qualitative, and mixed research methods in engineering education. *Journal of Engineering Education, 98*(1), 53–66. https://doi.org/10.1002/j.2168-9830.2009.tb01005.x

Borrego, M., Foster, M. J., & Froyd, J. E. (2014). Systematic literature reviews in engineering education and other developing interdisciplinary fields. *Journal of Engineering Education, 103*(1), 45–76. https://doi.org/10.1002/jee.20038

Brookfield, S. D. (2017). *Becoming a critically reflective teacher*. John Wiley & Sons.

Bruce, C. D., Davis, B., Sinclair, N., McGarvey, L., Hallowell, D., Drefs, M., Francis, K., Hawes, Z., Moss, J., Mulligan, J., Okamoto, Y., Whiteley, W., & Woolcott, G. (2017). Understanding gaps in research networks: Using "spatial reasoning" as a window into the importance of networked educational research. *Educational Studies in Mathematics, 95*(2), 143–161. https://doi.org/10.1007/s10649-016-9743-2

Buckley, J. (2022). The need to consider the predictive capacity of intelligence and its malleability within design and technology education research. *International Journal of Technology and Design Education, 32*(1), 1–15. https://doi.org/10.1007/s10798-020-09588-9

Buckley, J., Adams, L., Aribilola, I., Arshad, I., Azeem, M., Bracken, L., Breheny, C., Buckley, C., Chimello, I., Fagan, A., Fitzpatrick, D. P., Garza Herrera, D., Gomes, G. D., Grassick, S., Halligan, E., Hirway, A., Hyland, T., Imtiaz, M. B., Khan, M. B., . . . Zhang, L. (2022). An assessment of the transparency of contemporary technology education research employing interview-based methodologies. *International Journal of Technology and Design Education, 32*(4), 1963–1982. https://doi.org/10.1007/s10798-021-09695-1

Buckley, J., Gumaelius, L., Hyland, T., Seery, N., & Pears, A. (2019). A comparison of Swedish and Irish secondary students' conceptions of engineers and engineering using the Draw-an-Engineer Test. In *126th ASEE annual conference & exposition* (pp. 1–23). https://peer.asee.org/a-comparison-of-swedish-and-irish-secondary-students-conceptions-of-engineers-and-engineering-using-the-draw-an-engineer-test

Buckley, J., Seery, N., & Canty, D. (2019). Spatial cognition in engineering education: Developing a spatial ability framework to support the translation of theory into practice. *European Journal of Engineering Education, 44*(1–2), 164–178. https://doi.org/10.1080/03043797.2017.1327944

Buckley, J., Seery, N., Canty, D., & Gumaelius, L. (2018). Visualization, inductive reasoning, and memory span as components of fluid intelligence: Implications for technology education. *International Journal of Educational Research, 90*(1), 64–77. https://doi.org/10.1016/j.ijer.2018.05.007

Buckley, J., Trevelyan, J., & Winberg, C. (2022). Perspectives on engineering education from the world of practice. *European Journal of Engineering Education, 47*(1), 1–7. https://doi.org/10.1080/03043797.2021.2000694

Case, J. M., & Light, G. (2011). Emerging research methodologies in engineering education research. *Journal of Engineering Education, 100*(1), 186–210. https://doi.org/10.1002/j.2168-9830.2011.tb00008.x

Case, J. M., & Light, G. (2014). Framing qualitative methods in engineering education research: Established and emerging methodologies. In A. Johri & B. Olds (Eds.), *Cambridge Handbook of Engineering Education Research* (pp. 535–550). Cambridge University Press.

Davies, P. (1999). What is evidence-based education? *British Journal of Educational Studies, 47*(2), 108–121. https://doi.org/10.1111/1467-8527.00106

Doulougeri, K., Bombaerts, G., Martin, D., Watkins, A., Bots, M., & Vermunt, J. D. (2022). Exploring the factors influencing students' experience with challenge-based learning: A case study. In *2022 IEEE global engineering education conference (EDUCON)* (pp. 981–988). https://doi.org/10.1109/EDUCON52537.2022.9766574

Doulougeri, K., Vermunt, J. D., Bombaerts, G., & Bots, M. (2022, September). Analyzing student-teacher interactions in challenge-based Learning. In *SEFI 50th annual conference*. https://doi.org/10.1109/EDUCON52537.2022.9766574

Edström, K., Bernhard, J., & van den Bogaard, M. (2018). Updating the aims and scope. *European Journal of Engineering Education, 43*(5), 651–651. https://doi.org/10.1080/03043797.2018.1501840

Engineering Council. (2022, February 11). *Statement of ethical principles.* www.engc.org.uk/standards-guidance/guidance/statement-of-ethical-principles/

Evans, C., Kandiko Howson, C., Forsythe, A., & Edwards, C. (2021). What constitutes high quality higher education pedagogical research? *Assessment & Evaluation in Higher Education, 46*(4), 525–546. https://doi.org/10.1080/02602938.2020.1790500

Felder, R. M., & Brent, R. (2005). Understanding student differences. *Journal of Engineering Education, 94*(1), 57–72. https://doi.org/10.1002/j.2168-9830.2005.tb00829.x

Fensham, P. J. (2004). *Defining an identity: The evolution of science education as a field of research* (Vol. 20). Springer Netherlands. https://doi.org/10.1007/978-94-010-0175-5

Finlay, L. (1998). Reflexivity: An essential component for all research? *British Journal of Occupational Therapy, 61*(10), 453–456. https://doi.org/10.1177/030802269806101005

Freire, P. (1970). *Pedagogy of the oppressed: 30th anniversary edition translated by Myra Bergman Ramos.* Continuum.

Fuchs, L., Cuevas-Garcia, C., Mottl, P., & Bombaerts, G. (2022). Alliance formation and moral reflexivity: towards responsible European engineering universities? In *Frontiers in education conference.*

Giovannelli, L., & Sandekian, R. (2017). Global engineering: What do we mean by it and how are we preparing our students for It? In *2017 ASEE annual conference & exposition proceedings* (p. 28410). https://doi.org/10.18260/1-2--28410

Gornitzka, Å. (2013). The interface between research and policy – a note with potential relevance for higher education. *European Journal of Higher Education, 3*(3), 255–264. https://doi.org/10.1080/21568235.2013.816469

Holmes, A. G. D. (2020). Researcher positionality – A consideration of its influence and place in qualitative research – A new researcher guide. *Shanlax International Journal of Education, 8*(4), 1–10.

Hyland, T., Dunbar, R., Murray, N., & Buckley, J. (2021). *Spatial skills development for industrial training.* First Polymer Training Skillnet. www.skillnetireland.ie/wp-content/uploads/2021/12/Spatial-Skills-Development-for-Industrial-Training-First-Polymer-Training-Skillnet.pdf

Jesiek, B. K., Newswander, L. K., & Borrego, M. (2009). Engineering education research: Discipline, community, or field? *Journal of Engineering Education, 98*(1), 39–52. https://doi.org/10.1002/j.2168-9830.2009.tb01004.x

Johri, A. (2014). Conducting interpretive research in engineering education using qualitative and ethnographic methods. In A. Johri (Ed.), *Cambridge handbook of engineering education research* (pp. 551–570). Cambridge University Press.

Johri, A. (2022). Lifelong and lifewide learning for the perpetual development of expertise in engineering. *European Journal of Engineering Education, 47*(1), 70–84. https://doi.org/10.1080/03043797.2021.1944064

Johri, A., Edström, K., Du, X., Mitchell, J., & May, D. (2022). Knowledge construction in engineering education research – assessing the role of journals, books, conferences, and other products of research. In *Proceedings of FIE 2022.*

Johri, A., & Olds, B. (2014). *Cambridge handbook of engineering education research.* Cambridge University Press.

Keeley, L., Walters, H., Pikkel, R., & Quinn, B. (2013). *Ten types of innovation: The discipline of building breakthroughs.* John Wiley & Sons.

Kirschner, P. (2017). Stop propagating the learning styles myth. *Computers and Education, 106*(1), 166–171.

Kirschner, P., Sweller, J., Kirschner, F., & Zambrano, R. J. (2018). From cognitive load theory to collaborative cognitive load theory. *International Journal of Computer-Supported Collaborative Learning.* https://doi.org/10.1007/s11412-018-9277-y

Klassen, M., & Case, J. M. (2022). Productive tensions? Analyzing the arguments made about the field of engineering education research. *Journal of Engineering Education, 111*(1), 214–231. https://doi.org/10.1002/jee.20440

Latour, B. (1993). *We have never been modern.* Harvard University Press.

MacLeod, M., & van der Veen, J. T. (2020). Scaffolding interdisciplinary project-based learning: A case study. *European Journal of Engineering Education, 45*(3), 363–377. https://doi.org/10.1080/03043797.2019.1646210

Malmi, L., Adawi, T., Curmi, R., de Graaff, E., Duffy, G., Kautz, C., Kinnunen, P., & Williams, B. (2018). How authors did it – A methodological analysis of recent engineering education research papers in the European journal of engineering education. *European Journal of Engineering Education, 43*(2), 171–189. https://doi.org/10.1080/03043797.2016.1202905

Manichander, T. (2016). *Psychology of the learner.* Lulu Press Inc.

Martin, D. A., Conlon, E., & Bowe, B. (2019). The role of role-play in student awareness of the social dimension of the engineering profession. *European Journal of Engineering Education, 44*(6), 882–905. https://doi.org/10.1080/03043797.2019.1624691

Moreau, D., & Wiebels, K. (2021). Assessing change in intervention research: The benefits of composite outcomes. *Advances in Methods and Practices in Psychological Science, 4*(1). https://doi.org/10.1177/2515245920931930

Munoz-Rubke, F., Will, R., Hawes, Z., & James, K. H. (2021). Enhancing spatial skills through mechanical problem solving. *Learning and Instruction, 75*, 101496. https://doi.org/10.1016/j.learninstruc.2021.101496

National Academy of Engineering. (2004). *The engineer of 2020: Visions of engineering in the new century.* The National Academies Press. https://doi.org/10.17226/10999

Neuman, W. L. (2011). *Basics of social research: Qualitative and quantitative approaches* (3rd ed.). Pearson.

Pantzos, P., Gumaelius, L., Buckley, J., & Pears, A. (2022). Engineering students' perceptions of the role of work industry-related activities on their motivation for studying and learning in higher education. *European Journal of Engineering Education*, 1–19. https://doi.org/10.1080/03043797.2022.2093167

Power, J. (2021a). Enhancing engineering education through the integration of open science principles: A strategic approach to systematic reviews. *Journal of Engineering Education, 110*(3), 509–514. https://doi.org/10.1002/jee.20413

Power, J. (2021b). Systematic reviews in engineering education: A catalyst for change. *European Journal of Engineering Education, 46*(6), 1163–1174. https://doi.org/10.1080/03043797.2021.1980770

Ramey, K., & Uttal, D. (2017). Making sense of space: Distributed spatial sensemaking in a middle school summer engineering camp. *Journal of the Learning Sciences, 26*(2), 277–319.

Rowe, W. (2014). Positionality. In D. Coghlan & M. Brydon-Miller (Eds.), *The SAGE encyclopedia of action research*. SAGE Publications Ltd. https://doi.org/10.4135/9781446294406.n277

Ryan, R. M., & Deci, E. L. (2020). Intrinsic and extrinsic motivation from a self-determination theory perspective: Definitions, theory, practices, and future directions. *Contemporary Educational Psychology, 61*, 101860. https://doi.org/10.1016/j.cedpsych.2020.101860

Sandars, J., Brown, J., & Walsh, K. (2017). Research or evaluation – does the difference matter? *Education for Primary Care, 28*(3), 134–136. https://doi.org/10.1080/14739879.2017.1305873

Schneider, J., & McGrew, K. (2018). The Cattell-Horn-Carroll theory of cognitive abilities. In D. Flanagan & E. McDonough (Eds.), *Contemporary intellectual assessment: Theories, tests, and issues* (pp. 73–163). The Guilford Press.

Servant-Miklos, V. F. C., & Kolmos, A. (2022). Student conceptions of problem and project based learning in engineering education: A phenomenographic investigation. *Journal of Engineering Education, 111*(4), 792–812. https://doi.org/10.1002/jee.20478

Silvast, A., Laes, E., Abram, S., & Bombaerts, G. (2020). What do energy modellers know? An ethnography of epistemic values and knowledge models. *Energy Research & Social Science, 66*, 101495. https://doi.org/10.1016/j.erss.2020.101495

Sorby, S., Veurink, N., & Streiner, S. (2018). Does spatial skills instruction improve STEM outcomes? The answer is 'yes'. *Learning and Individual Differences, 67*(1), 209–222. https://doi.org/10.1016/j.lindif.2018.09.001

Vanderstraeten, R. (2000). Luhmann on socialization and education. *Educational Theory, 50*(1), 1–23. https://doi.org/10.1111/j.1741-5446.2000.00001.x

Vukasovic, M., & Stensaker, B. (2018). University alliances in the Europe of knowledge: Positions, agendas and practices in policy processes. *European Educational Research Journal, 17*(3), 349–364. https://doi.org/10.1177/1474904117724572

Wankat, P. C., Williams, B., & Neto, P. (2014). Engineering education research in European journal of engineering education and journal of engineering education: Citation and reference discipline analysis. *European Journal of Engineering Education, 39*(1), 7–17. https://doi.org/10.1080/03043797.2013.867316

Williams, B., Wankat, P. C., & Neto, P. (2018). Not so global: A bibliometric look at engineering education research. *European Journal of Engineering Education, 43*(2), 190–200. https://doi.org/10.1080/03043797.2016.1153043

Index

Note: Page numbers in *italic* indicate a figure and page numbers in **bold** indicate a table on the corresponding page.

For Product Safety Concerns and Information please contact our EU representative GPSR@taylorandfrancis.com Taylor & Francis Verlag GmbH, Kaufingerstraße 24, 80331 München, Germany